D1261529

SEP - 9 2009

WATER-RESOURCES ENGINEERING

Second Edition

David A. Chin

Upper Saddle River, New Jersey 07458

Library of Congress Cataloging-in-Publication Data

Chin, David A.
   Water resources engineering / David A. Chin.–2nd ed.
     p. cm.
   Includes bibliographical references and index.
   ISBN 0-13-148192-4
    1. Hydraulics. 2. Hydraulic engineering. 3. Hydrology. I. Title.

TC160.C52 2006
627–dc22

                                      2006041989

Vice President and Editorial Director, ECS: *Marcia Horton*
Senior Editor: *Holly Stark*
Editorial Assistant: *Nicole Kunzmann*
Executive Managing Editor: *Vince O'Brien*
Managing Editor: *David A. George*
Production Editor: *Daniel Sandin*
Director of Creative Services: *Paul Belfanti*
Art Director: *Jayne Conte*
Cover Designer: *Bruce Kenselaar*
Art Editor: *Greg Dulles*
Manufacturing Manager: *Alexis Heydt-Long*
Manufacturing Buyer: *Lisa McDowell*

© 2006 Pearson Education, Inc.
Pearson Prentice Hall
Pearson Education, Inc.
Upper Saddle River, New Jersey 07458

All rights reserved. No part of this book may be reproduced in any form or by any means, without permission in writing from the publisher.

Pearson Prentice Hall™ is a trademark of Pearson Education, Inc.

The author and publisher of this book have used their best efforts in preparing this book. These efforts include the development, research, and testing of the theories and programs to determine their effectiveness. The author and publisher make no warranty of any kind, expressed or implied, with regard to these programs or the documentation contained in this book. The author and publisher shall not be liable in any event for incidental or consequential damages in connection with, or arising out of, the furnishing, performance, or use of these programs.

Printed in the United States of America

10 9 8 7 6 5 4 3

ISBN: 0-13-148192-4

Pearson Education Ltd., *London*
Pearson Education Australia Pty. Ltd., *Sydney*
Pearson Education Singapore, Pte. Ltd.
Pearson Education North Asia Ltd., *Hong Kong*
Pearson Education Canada, Inc., *Toronto*
Pearson Educación de Mexico, S.A. de C.V.
Pearson Education—Japan, *Tokyo*
Pearson Education Malaysia, Pte. Ltd.
Pearson Education, Inc., *Upper Saddle River, New Jersey*

To my wife, Linda Sue, for her love and support

Contents

Preface

The field of water-resources engineering covers the design of systems to control the quantity, quality, timing, and distribution of water to support both human habitation and the needs of the environment. Water-resources engineers are typically trained in civil or environmental engineering programs, and practice in a variety of subspecialties.

The technical and scientific bases for most water-resources specializations are found in the areas of hydraulics and hydrology, and this text covers these areas with depth and rigor. The fundamentals of closed-conduit flow, open-channel flow, surface-water hydrology, ground-water hydrology, and water-resources planning and management are all covered in detail. Applications include the design of water-distribution systems, hydraulic structures, sanitary sewer systems, stormwater-management systems, and water-supply wellfields. All the design protocols presented in this book are consistent with the relevant ASCE, WEF, and AWWA Manuals of Practice. The topics covered constitute much of the technical background expected of water-resources engineers. This text is appropriate for undergraduate and first-year graduate courses in hydraulics and hydrology, and practitioners will also find the material to be a useful reference. Users of this book should ideally have taken courses in calculus through differential equations and a first course in fluid mechanics.

The book begins with an introduction to the field of water-resources engineering (Chapter 1), and orients the reader to the depth and breadth of water-resources engineering. Chapter 2 covers the fundamentals of flow in closed conduits, including a detailed exposition on the selection of pumps and the design of water-supply systems. Chapter 3 covers flow in open channels from basic principles, including the computation of water-surface profiles, and the performance of hydraulic structures. Applications of this material to the design of lined and unlined drainage channels are presented along with the design of sanitary-sewer systems. Many of the analytical methods used by water-resources engineers are based on the theory of probability and statistics, and Chapter 4 presents elements of probability and statistics that are relevant to the practice of water-resources engineering. Useful probability distributions, hydrologic data analysis, and frequency analysis are all covered, and the applications of these techniques to risk analysis in engineering design are illustrated by examples. Chapter 5 covers surface-water hydrology and focuses mostly on urban design applications. The ASCE Manuals of Practice on the design of surface-water management systems (ASCE, 1992) and urban runoff quality management (ASCE, 1998) were used as bases for much of the material presented. Coverage includes the specification of design rainfall, runoff models, routing models, and water-quality models. Applications of this material to the design of both minor and major components of stormwater-management systems are presented. Chapter 6 covers ground-water hydrology, including the basic equations of ground-water flow, analytic solutions describing flow in aquifers, saltwater intrusion, and ground-water flow in the unsaturated zone. Applications to the design of municipal wellfields and individual water-supply wells, the delineation of wellhead protection areas, the design of aquifer pumping tests, and the design of exfiltration trenches are presented. Chapter 7 covers water-resources planning and management, with emphasis on water-supply planning, floodplain management, and drought management.

This book is a reflection of the author's belief that water-resources engineers must gain a firm understanding of the depth and breadth of the technical areas that are fundamental to their discipline, and by so doing will be more innovative, view water-resource systems holistically, and be technically prepared for a lifetime of learning. On the basis of this vision, the material contained in this book is presented mostly from first principles, is rigorous, is relevant to the practice of water-resources engineering, and is reinforced by detailed presentations of design applications.

DAVID A. CHIN
University of Miami

CHAPTER 1

Introduction

1.1 Water-Resources Engineering

Water-resources engineering is concerned with the analysis and design of systems to control the quantity, quality, timing, and distribution of water to meet the needs of human habitation and the environment. Aside from the engineering and environmental aspects of water-resource systems, their feasibility from legal, economic, financial, political, and social viewpoints must also be considered in the development process. In fact, the successful operation of an engineered system usually depends as much on nonengineering analyses (e.g., economic and social analyses) as on sound engineering design (Delleur, 2003); this is particularly true in developing countries (Brookshire and Whittington, 1993). Examples of water-resource systems include domestic and industrial water supply, wastewater treatment, irrigation, drainage, flood control, salinity control, sediment control, pollution abatement, and hydropower-generation systems.

The waters of the earth are found on land, in the oceans, and in the atmosphere. The core science of water-resources engineering is *hydrology*, which deals with the occurrence, distribution, movement, and properties of water on earth. Engineering hydrologists are primarily concerned with water on land and in the atmosphere, from its deposition as atmospheric precipitation, such as rainfall and snowfall, to its inflow into the oceans and its vaporization into the atmosphere. The technical areas that are fundamental to water-resources engineering can be grouped into the following five categories:

1. Subsurface hydrology
2. Surface water and climate
3. Hydrogeochemistry and water chemistry
4. Erosion, sedimentation, and geomorphology
5. Water policy, economics, and systems analysis

Subsurface hydrology is concerned with the occurrence and movement of water below the surface of the earth; surface water and climate studies are concerned with the occurrence and movement of water above the surface of the earth; hydrogeochemistry is concerned with the chemical changes in water that is in contact with earth materials; erosion, sedimentation, and geomorphology deal with the effects of sediment transport on landforms; and water policy, economics, and systems analyses are concerned with the political, economic, and environmental constraints in the design and operation of water-resource systems. The quantity and quality of water are inseparable issues in design, and the modern practice of water-resources engineering demands that practitioners be technically competent in understanding the physical processes that govern the movement of water, the chemical and biological processes that affect the quality of water, the economic and social considerations that must be taken into account, and the environmental impacts associated with the construction and operation of water-resource projects.

The design of water-resource systems usually involves interaction with government agencies. Collection of hydrologic and geologic data, granting of development permits, specification of design criteria, and use of government-developed computer models of water-resource systems are some of the many areas in which water-resource engineers interact with government agencies. The following are some of the key water-resources agencies in the United States:

- **U.S. Geological Survey (USGS).** The Water Resources Division of the USGS has primary federal responsibility for collection and dissemination of measurements of stream discharge and stage, reservoir and lake stage and storage, ground-water levels, well and spring discharge, and the quality of surface and ground water in the United States. USGS maintains a network of thousands of stream gages and ground-water monitoring wells. Most of this data can be accessed using the National Water Information System II at the Web site waterdata.usgs.gov/nwis/. USGS constructs and distributes $7\frac{1}{2}$-minute quadrangle topographic maps, which are useful in hydrologic studies; the agency also produces geological maps of subsurface formations.

- **National Climatic Data Center (NCDC).** NCDC, the world's largest active source of weather data, produces numerous climate publications and responds to data requests from all over the world. Most of the data available from NCDC is collected and analyzed by the National Weather Service (NWS), Military Services, Coast Guard, Federal Aviation Administration, and cooperative observers. Data collected by NWS include rainfall and evaporation measurements at over 10,000 locations in the United States.

- **U.S. Bureau of Reclamation (USBR).** USBR is responsible for planning, construction, operation, and maintenance of a variety of water-resource facilities whose objectives include irrigation, power generation, recreation, fish and wildlife preservation, and municipal water supply. Most activities are confined to the 17 states west of the Mississippi River. Besides being the largest wholesale supplier of water in the United States, USBR is the sixth largest hydroelectric supplier in the United States.

- **U.S. Environmental Protection Agency (USEPA).** USEPA is responsible for the implementation and enforcement of federal environmental laws. The agency's mission is to protect public health and to safeguard and improve the natural environment—air, water, and land—upon which human life depends.

- **U.S. Natural Resources Conservation Service (NRCS).** NRCS, formerly the Soil Conservation Service, works with landowners on private lands to conserve natural resources. NRCS provides technical and financial assistance to farmers and ranchers for flood protection, recreation, and water-supply development in small watersheds (less than 100,000 ha). The NRCS publishes general soil maps for each state and detailed soil maps for each county in the United States.

- **U.S. Army Corps of Engineers (USACE).** USACE is responsible for the planning, construction, operation, and maintenance of a variety of water-resource facilities whose objectives include navigation, flood control, water supply, recreation, hydroelectric power generation, water-quality control, and other purposes.

TABLE 1.1: Selected Internet Sites Relevant to Water-Resources Engineering in the United States

Organization	Web address
National Climatic Data Center (NCDC)	www.ncdc.noaa.gov
U.S. Army Corps of Engineers (USACE)	www.usace.army.mil
U.S. Bureau of Reclamation (USBR)	www.usbr.gov
U.S. Environmental Protection Agency (EPA)	www.epa.gov
U.S. Geological Survey (USGS)	www.usgs.gov
U.S. Natural Resources Conservation Service (NRCS)	www.nrcs.usda.gov

These government agencies provide a wealth of information on water resources, relevant government regulations, and useful computer software that can be found on the internet. Several of the more useful Web sites currently in use (and likely to be around for the foreseeable future) are listed in Table 1.1.

1.2   The Hydrologic Cycle

The *hydrologic cycle* is defined as the pathway of water as it moves in its various phases through the atmosphere, to the earth, over and through the land, to the ocean, and back to the atmosphere (National Research Council, 1991). The movement of water in the hydrologic cycle is illustrated in Figure 1.1, where the relative magnitudes of various hydrologic processes are given in units relative to a value of 100 for the rate of precipitation on land. The relative magnitudes are based on global annual averages (Chow et al., 1988).

A description of the hydrologic cycle can start with the evaporation of water from the oceans driven by energy from the sun. The evaporated water, in the form of water vapor, rises by convection, condenses in the atmosphere to form clouds, and precipitates onto land and ocean surfaces, predominantly as rain or snow. Precipitation on land surfaces is partially intercepted by surface vegetation, partially stored in surface

FIGURE 1.1: Hydrologic cycle.
Source: Chow et al. (1988).

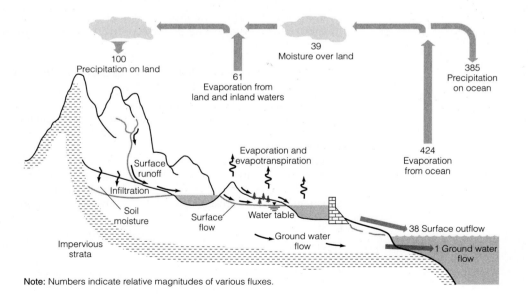

Note: Numbers indicate relative magnitudes of various fluxes.

depressions, partially infiltrated into the ground, and partially flows over land into drainage channels and rivers that ultimately lead back to the ocean. Precipitation that is intercepted by surface vegetation is eventually evaporated into the atmosphere; water held in depression storage either evaporates or infiltrates into the ground; and water that infiltrates into the ground contributes to the recharge of ground water, which either is utilized by plants or becomes subsurface flow that ultimately emerges as recharge to streams or directly to the ocean. *Ground water* is defined as the water below the land surface; water above the land surface (in liquid form) is called *surface water*. In urban areas, the ground surface is typically much less pervious than in rural areas, and surface runoff is mostly controlled by constructed drainage systems. Surface waters and ground waters in urban areas also tend to be significantly influenced by the water-supply and wastewater removal systems that are an integral part of urban developments. Since humanmade systems are part of the hydrologic cycle, it is the responsibility of the water-resources engineer to ensure that systems constructed for water use and control are in harmony with the needs of the natural environment.

The quality of water varies considerably as it moves through the hydrologic cycle, with contamination resulting from several sources. Classes of contaminants commonly found in water, along with some examples, are listed in Table 1.2. The effects of the quantity and quality of water on the health of terrestrial ecosystems and the value of these ecosystems in the hydrologic cycle are often overlooked. For example, the modification of free-flowing rivers for energy or water supply, and the drainage of wetlands, can have a variety of adverse effects on aquatic ecosystems, including losses in species diversity, floodplain fertility, and biofiltration capability (Gleick, 1993).

On a global scale, the distribution of the water resources of the earth is given in Table 1.3, where it is clear that the vast majority of the earth's water resources is contained in the oceans, with most of the fresh water being contained in ground water and polar ice. The amount of water stored in the atmosphere is relatively small, although the flux of water into and out of the atmosphere dominates the hydrologic cycle. Wood (1991) estimated that the typical residence time for atmospheric water is on the order of a week, the typical residence time for soil moisture is on the order of weeks to months, and the typical residence time in the oceans is on the order of tens of thousands of years. The estimated fluxes of precipitation, evaporation, and runoff

TABLE 1.2: Classes of Water Contaminants

Contaminant class	Example
Oxygen-demanding wastes	Plant and animal material
Infectious agents	Bacteria and viruses
Plant nutrients	Fertilizers, such as nitrates and phosphates
Organic chemicals	Pesticides, such as DDT, detergent molecules
Inorganic chemicals	Acids from coal mine drainage, inorganic chemicals such as iron from steel plants
Sediment from land erosion	Clay silt on streambed, which may reduce or even destroy life forms living at the solid-liquid interface
Radioactive substances	Waste products from mining and processing of radioactive material, radioactive isotopes after use
Heat from industry	Cooling water used in steam generation of electricity

TABLE 1.3: Estimated World Water Quantities

Item	Volume ($\times 10^3$ km^3)	Percent total water (%)	Percent fresh water (%)
Oceans	1,338,000.	96.5	
Ground water			
Fresh	10,530.	0.76	30.1
Saline	12,870.	0.93	
Soil moisture	16.5	0.0012	0.05
Polar ice	24,023.5	1.7	68.6
Other ice and snow	340.6	0.025	1.0
Lakes			
Fresh	91.	0.007	0.26
Saline	85.4	0.006	
Marshes	11.47	0.0008	0.03
Rivers	2.12	0.0002	0.006
Biological water	1.12	0.0001	0.003
Atmospheric water	12.9	0.001	0.04
Total water	1,385,984.61	100.	
Fresh water	35,029.21	2.5	100.

Source: USSR National Committee for the International Hydrological Decade (1978).

TABLE 1.4: Fluxes in Global Hydrologic Cycle

Component	Oceanic flux (mm/year)	Terrestrial flux (mm/year)
Precipitation	1270	800
Evaporation	1400	484
Runoff to ocean (rivers plus ground water)	—	316

Source: USSR National Committee for the International Hydrological Decade (1978).

within the global hydrologic cycle are given in Table 1.4. These data indicate that the global average precipitation over land is on the order of 800 mm/yr (31 in./yr), of which 484 mm/yr (19 in./yr) is returned to the atmosphere as evaporation and 316 mm/yr (12 in./yr) is returned to the ocean via surface runoff. On a global scale, large variations from these average values are observed. In the United States, for example, the highest annual rainfall is found at Mount Wai'ale'ale on the Hawaiian island of Kauai with an annual rainfall of 1168 cm (460 in.), while Greenland Ranch in Death Valley, California, has the lowest annual average rainfall—4.5 cm (1.78 in.)—in the United States. Two of the most widely used climatic measures are the mean annual rainfall and the mean annual potential evapotranspiration. A *climatic spectrum* appropriate for subtropical and midlatitudinal regions is given in Table 1.5, and water-resource systems tend to differ substantially between climates. For example, forecasting and planning for drought conditions is particularly important in semiarid climates, whereas droughts are barely noticeable in very humid areas.

TABLE 1.5: Climate Spectrum

Climate	Mean annual precipitation (mm)	Mean annual evapotranspiration (mm)	Length of rainy season (months)
Superarid	<100	<3,000	<1
Hyperarid	100–200	2400–3600	1–2
Arid	200–400	2000–2400	2–3
Semiarid	400–800	1600–2000	3–4
Subhumid	800–1600	1200–1600	4–6
Humid	1600–3200	1200	6–9
Hyperhumid	3200–6400	1200	9–12
Superhumid	≥6400	1200	12

Source: Ponce et al. (2000).

On regional scales, water resources are managed within topographically defined areas called *watersheds* or *basins*. These areas are typically enclosed by topographic high points in the land surface, and within these bounded areas the path of the surface runoff can usually be controlled with a reasonable degree of coordination.

1.3 Design of Water-Resource Systems

The uncertainty and natural variability of hydrologic processes require that most water-resource systems be designed with some degree of *risk*. Approaches to designing such systems are classified as either *frequency-based design*, *risk-based design*, or *critical-event design*. In frequency-based design, the exceedance probability of the design event is selected a priori and the water-resource system is designed to accommodate all lesser events up to and including an event with the selected exceedance probability. The water-resource system will then be expected to fail with a probability equal to the exceedance probability of the design event. The frequency-based design approach is commonly used in designing the minor structures of urban drainage systems. For example, urban storm-drainage systems are typically designed for precipitation events with return periods of 10 years or less, where the *return period* of an event is defined as the reciprocal of the (annual) exceedance probability of the event. In risk-based design, systems are designed such that the sum of the capital cost and the cost of failure is minimized. Capital costs tend to increase and the cost of failure tends to decrease with increasing system capacity. Because any threats to human life are generally assigned extremely high failure costs, structures such as large dams are usually designed for rare hydrologic events with long return periods and commensurate small failure risks. In some extreme cases, where the consequences of failure are truly catastrophic, water-resource systems are designed for the largest possible magnitude of a hydrologic event. This approach is called critical-event design, and the value of the design (hydrologic) variable in this case is referred to as the *estimated limiting value* (ELV).

Water-resource systems can be broadly categorized as *water-control systems* or *water-use systems*, as shown in Table 1.6, but it should be noted that these systems are not mutually exclusive. A third category of *environmental restoration systems* has been suggested by Mays (1996). The following sections present a brief overview of the design objectives in water-control and water-use systems.

TABLE 1.6: Water-Resource Systems

Water-control systems	Water-use systems
Drainage	Domestic and industrial water supply
Flood control	Wastewater treatment
Salinity control	Irrigation
Sediment control	Hydropower generation
Pollution abatement	

1.3.1 Water-Control Systems

Water-control systems are primarily designed to control the spatial and temporal distribution of surface runoff resulting from rainfall events. Flood-control structures and storage impoundments reduce the peak flows in streams, rivers, and drainage channels, thereby reducing the occurrence of floods. A *flood* is defined as a high flow that exceeds the capacity of a stream or drainage channel, and the elevation at which the flood overflows the embankments is called the *flood stage*. A *floodplain* is the normally dry land adjoining rivers, streams, lakes, bays, or oceans that is inundated during flood events. Typically, flows with return periods from 1.5 to 3 years represent bankfull conditions, with larger flows causing inundation of the floodplain (McCuen, 1989). The 100-year flood has been adopted by the U.S. Federal Emergency Management Agency (FEMA) as the base flood for delineating floodplains, and the area inundated by the 500-year flood is sometimes delineated to indicate areas of additional risk. Encroachment onto floodplains reduces the capacity of the watercourse and increases the extent of the floodplain. Approximately 7%–10% of the land in the United States is in a floodplain, and in the 1970s flood-related deaths were 200 per year, with another 80,000 per year forced from their homes (Wanielista and Yousef, 1993).

The largest floodplain areas in the United States are in the South; the most populated floodplains are along the north Atlantic coast, the Great Lakes region, and in California (Viessman and Lewis, 1996).

In urban settings, water-control systems include storm-sewer systems for collecting and transporting surface runoff, and storage reservoirs that attenuate peak runoff rates and reduce pollutant loads in drainage channels. Urban stormwater control systems are typically designed to prevent flooding from runoff events with return periods of 10 years or less. For larger runoff events, the capacity of these systems is exceeded and surface (street) flooding usually results.

1.3.2 Water-Use Systems

Water-use systems are designed to support human habitation and include water-treatment systems, water-distribution systems, wastewater-collection systems, and wastewater-treatment systems. The design capacity of these systems is generally dictated by the population of the service area, commercial and industrial requirements, and the economic design life of the system. Water-use systems are designed to provide specified levels of service: Water-treatment systems, for example, must produce water of sufficient quality to meet drinking water standards, water-distribution systems must deliver peak demands while sustaining adequate water pressures, wastewater-collection systems must have sufficient capacity to transport wastes without overflowing into the streets, and wastewater-treatment systems must provide a sufficient level of treatment that effluent discharges will not degrade the aquatic environment. In agricultural areas, the water requirements of plants are met by a combination of rainfall and irrigation. The design of irrigation systems requires the estimation of crop evapotranspiration rates and leaching requirements in agricultural areas, with the portion of these requirements that are not met by rainfall being met by irrigation systems. In rivers where there is sufficient available energy, such as behind large dams or in rapidly flowing rivers, hydroelectric power generation may be economically feasible.

1.4 About this Book

The fundamental technical aspects of water-resources engineering derive mostly from hydraulics, hydrology, probability and statistics, and economics. A good understanding of these subject areas is the foundation on which engineers build sound designs and operational protocols for water-resource systems. This book addresses each of these fundamental areas in detail.

Chapter 2 presents the hydraulics of flow in closed conduits along with detailed coverage of the application of closed-conduit flow principles to the design of water-distribution systems. The estimation of water demand and design flows, pipeline selection, pump selection, storage-reservoir design, and estimation of service pressures are all covered. The hydraulics of flow in open channels is rigorously presented in Chapter 3, from basic principles to the computation of water-surface profiles, with design applications including hydraulic structures, channels, and sanitary-sewer systems. Chapter 4 reviews probability-distribution functions, the estimation of probability distributions from measured hydrologic data, and applications of probability and statistics to risk analysis in water-resource systems. Surface-water hydrology is presented in Chapter 5 from an urban perspective through the specification of design rainfall and abstraction models and the computation of the quantity and quality of surface runoff resulting from a specified rainfall. The design of urban stormwater-management systems is covered in detail, and the estimation of evapotranspiration, which is fundamental to the design of agricultural irrigation systems, is also covered. Chapter 6 covers the basic principles of flow in porous media, with particular emphasis on the flow fields induced by pumping wells. The fundamentals of saltwater intrusion and flow in the unsaturated zone are also covered. Design applications of ground-water hydrology include the design of municipal wellfields, the design and construction of water-supply wells, the delineation of wellhead-protection areas, the design of aquifer pumping tests, and the design of exfiltration trenches for ground-water injection. Chapter 7 describes the conventional water-resources planning process, legal and regulatory issues that must be addressed, and techniques for assessing economic feasibility. The application of water-resources planning and management to water-supply projects, floodplain management, drought management, irrigation systems, dams and reservoirs, hydropower projects, and navigation are all covered.

In summary, this book covers the subject areas that are fundamental to the practice of water-resources engineering. A firm grasp of the material covered in this book along with complementary practical experience are the foundations on which water-resources engineering is practiced.

Problems

1.1. Search the World Wide Web to determine the mean annual rainfall and evapotranspiration of Boston, Massachusetts, and Santa Fe, New Mexico. Classify the climate in these cities.

CHAPTER 2

Flow in Closed Conduits

2.1 Introduction

Flow in closed conduits includes all cases where the flowing fluid completely fills the conduit. The cross sections of closed conduits can be of any shape or size and the conduits can be made of a variety of materials. Engineering applications of the principles of flow in closed conduits include the design of municipal water-supply systems and transmission lines. The basic equations governing the flow of fluids in closed conduits are the continuity, momentum, and energy equations, and the most useful forms of these equations for application to pipe flow problems are derived in this chapter. The governing equations are presented in forms that are applicable to any fluid flowing in a closed conduit, but particular attention is given to the flow of water.

The computation of flows in pipe networks is a natural extension of the flows in single pipelines, and methods of calculating flows and pressure distributions in pipeline systems are also described here. These methods are particularly applicable to the analysis and design of municipal water-distribution systems, where the engineer is frequently interested in assessing the effects of various modifications to the system. Because transmission of water in closed conduits is typically accomplished using pumps, the fundamentals of pump operation and performance are also presented in this chapter. A sound understanding of pumps is important in selecting the appropriate pump to achieve the desired operational characteristics in water-transmission systems. The design protocol for municipal water-distribution systems is presented as an example of the application of the principles of flow in closed conduits. Methods for estimating water demand, design of the functional components of distribution systems, network analysis, and the operational criteria for municipal water-distribution systems are all covered.

2.2 Single Pipelines

The governing equations for flows in pipelines are derived from the conservation laws of mass, momentum, and energy, and the forms of these equations that are most useful for application to closed-conduit flow are derived in the following sections.

2.2.1 Steady-State Continuity Equation

Consider the application of the continuity equation to the control volume illustrated in Figure 2.1. Fluid enters and leaves the control volume normal to the control surfaces, with the inflow velocity denoted by $v_1(\mathbf{r})$ and the outflow velocity by $v_2(\mathbf{r})$, where $\mathbf{r}$ is the radial position vector originating at the centerline of the conduit. Both the inflow and outflow velocities vary across the control surface. The steady-state continuity

FIGURE 2.1: Flow through closed conduit

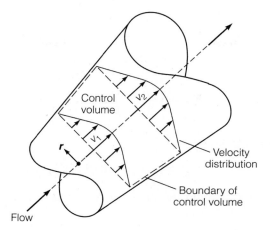

equation for an incompressible fluid can be written as

$$\int_{A_1} v_1 \, dA = \int_{A_2} v_2 \, dA \qquad (2.1)$$

Defining V_1 and V_2 as the average velocities across A_1 and A_2, respectively, where

$$V_1 = \frac{1}{A_1} \int_{A_1} v_1 \, dA \qquad (2.2)$$

and

$$V_2 = \frac{1}{A_2} \int_{A_2} v_2 \, dA \qquad (2.3)$$

the steady-state continuity equation becomes

$$\boxed{V_1 A_1 = V_2 A_2 \ (= Q)} \qquad (2.4)$$

The terms on each side of Equation 2.4 are equal to the volumetric flowrate, Q. The steady-state continuity equation simply states that the volumetric flowrate across any surface normal to the flow is a constant.

EXAMPLE 2.1

Water enters a pump through a 150-mm diameter intake pipe and leaves through a 200-mm diameter discharge pipe. If the average velocity in the intake pipeline is 1 m/s, calculate the average velocity in the discharge pipeline. What is the flowrate through the pump?

Solution   In the intake pipeline, $V_1 = 1$ m/s, $D_1 = 0.15$ m and

$$A_1 = \frac{\pi}{4} D_1^2 = \frac{\pi}{4}(0.15)^2 = 0.0177 \ \text{m}^2$$

In the discharge pipeline, $D_2 = 0.20$ m and

$$A_2 = \frac{\pi}{4} D_2^2 = \frac{\pi}{4}(0.20)^2 = 0.0314 \ \text{m}^2$$

According to the continuity equation,

$$V_1 A_1 = V_2 A_2$$

Therefore,

$$V_2 = V_1 \left(\frac{A_1}{A_2} \right) = (1) \left(\frac{0.0177}{0.0314} \right) = 0.56 \text{ m/s}$$

The flowrate, Q, is given by

$$Q = A_1 V_1 = (0.0177)(1) = 0.0177 \text{ m}^3\text{/s}$$

The average velocity in the discharge pipeline is 0.56 m/s, and the flowrate through the pump is 0.0177 m^3/s.

2.2.2   Steady-State Momentum Equation

Consider the application of the momentum equation to the control volume illustrated in Figure 2.1. Under steady-state conditions, the component of the momentum equation in the direction of flow (x-direction) can be written as

$$\sum F_x = \int_A \rho v_x \mathbf{v} \cdot \mathbf{n} \, dA \tag{2.5}$$

where $\sum F_x$ is the sum of the x-components of the forces acting on the fluid in the control volume, ρ is the density of the fluid, v_x is the flow velocity in the x-direction, and $\mathbf{v} \cdot \mathbf{n}$ is the component of the flow velocity normal to the control surface. Since the unit normal vector, $\mathbf{n}$, in Equation 2.5 is directed outward from the control volume, the momentum equation for an incompressible fluid (ρ = constant) can be written as

$$\sum F_x = \rho \int_{A_2} v_2^2 \, dA - \rho \int_{A_1} v_1^2 \, dA \tag{2.6}$$

where the integral terms depend on the velocity distributions across the inflow and outflow control surfaces. The velocity distribution across each control surface is generally accounted for by the *momentum correction coefficient*, β, defined by the relation

$$\boxed{\beta = \frac{1}{AV^2} \int_A v^2 \, dA} \tag{2.7}$$

where A is the area of the control surface and V is the average velocity over the control surface. The momentum coefficients for the inflow and outflow control surfaces, A_1 and A_2, are given by β_1 and β_2, where

$$\beta_1 = \frac{1}{A_1 V_1^2} \int_{A_1} v_1^2 \, dA \tag{2.8}$$

$$\beta_2 = \frac{1}{A_2 V_2^2} \int_{A_2} v_2^2 \, dA \tag{2.9}$$

Substituting Equations 2.8 and 2.9 into Equation 2.6 leads to the following form of the momentum equation

$$\sum F_x = \rho\beta_2 V_2^2 A_2 - \rho\beta_1 V_1^2 A_1 \tag{2.10}$$

Recalling that the continuity equation states that the volumetric flowrate, Q, is the same across both the inflow and outflow control surfaces, where

$$Q = V_1 A_1 = V_2 A_2 \tag{2.11}$$

then combining Equations 2.10 and 2.11 leads to the following form of the steady-state momentum equation

$$\sum F_x = \rho\beta_2 Q V_2 - \rho\beta_1 Q V_1 \tag{2.12}$$

or

$$\sum F_x = \rho Q(\beta_2 V_2 - \beta_1 V_1) \tag{2.13}$$

In many cases of practical interest, the velocity distribution across the cross section of the closed conduit is approximately uniform, in which case the momentum coefficients, β_1 and β_2, are approximately equal to unity and the steady-state momentum equation becomes

$$\sum F_x = \rho Q(V_2 - V_1) \tag{2.14}$$

Consider the common case of flow in a straight pipe with a uniform circular cross section illustrated in Figure 2.2, where the average velocity remains constant at each cross section,

$$V_1 = V_2 = V \tag{2.15}$$

then the steady-state momentum equation becomes

$$\sum F_x = 0 \tag{2.16}$$

The forces that act on the fluid in a control volume of uniform cross section are illustrated in Figure 2.2. At Section 1, the average pressure over the control surface is equal to p_1 and the elevation of the midpoint of the section relative to a defined datum is equal to z_1, at Section 2, located a distance L downstream from Section 1, the pressure is p_2, and the elevation of the midpoint of the section is z_2. The average shear stress exerted on the fluid by the pipe surface is equal to τ_0, and the total shear force opposing flow is $\tau_0 P L$, where P is the perimeter of the pipe. The fluid weight acts vertically downward and is equal to $\gamma A L$, where γ is the specific weight of the fluid and A is the cross-sectional area of the pipe. The forces acting on the fluid system that have components in the direction of flow are the shear force, $\tau_0 P L$; the weight of the fluid in the control volume, $\gamma A L$; and the pressure forces on the upstream and downstream faces, $p_1 A$ and $p_2 A$, respectively. Substituting the expressions for the forces into the momentum equation, Equation 2.16, yields

$$p_1 A - p_2 A - \tau_0 P L - \gamma A L \sin\theta = 0 \tag{2.17}$$

where θ is the angle that the pipe makes with the horizontal and is given by the relation

$$\sin\theta = \frac{z_2 - z_1}{L} \tag{2.18}$$

FIGURE 2.2: Forces on flow in closed conduit

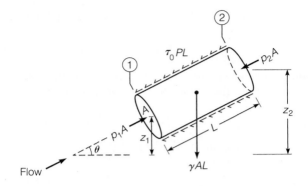

Combining Equations 2.17 and 2.18 yields

$$\frac{p_1}{\gamma} - \frac{p_2}{\gamma} - z_2 + z_1 = \frac{\tau_0 PL}{\gamma A} \tag{2.19}$$

Defining the *total head*, or energy per unit weight, at Sections 1 and 2 as h_1 and h_2, where

$$h_1 = \frac{p_1}{\gamma} + \frac{V^2}{2g} + z_1 \tag{2.20}$$

and

$$h_2 = \frac{p_2}{\gamma} + \frac{V^2}{2g} + z_2 \tag{2.21}$$

then the *head loss* between Sections 1 and 2, Δh, is given by

$$\Delta h = h_1 - h_2 = \left(\frac{p_1}{\gamma} + z_1\right) - \left(\frac{p_2}{\gamma} + z_2\right) \tag{2.22}$$

Combining Equations 2.19 and 2.22 leads to the following expression for head loss:

$$\Delta h = \frac{\tau_0 PL}{\gamma A} \tag{2.23}$$

In this case, the head loss, Δh, is entirely due to pipe friction and is commonly denoted by h_f. In the case of pipes with circular cross sections, Equation 2.23 can be written as

$$h_f = \frac{\tau_0 (\pi D) L}{\gamma (\pi D^2/4)} = \frac{4\tau_0 L}{\gamma D} \tag{2.24}$$

where D is the diameter of the pipe. The ratio of the cross-sectional area, A, to the perimeter, P, is defined as the *hydraulic radius*, R, where

$$R = \frac{A}{P} \tag{2.25}$$

and the head loss can be written in terms of the hydraulic radius as

$$h_f = \frac{\tau_0 L}{\gamma R} \tag{2.26}$$

The form of the momentum equation given by Equation 2.26 is of limited utility in that the head loss, h_f, is expressed in terms of the boundary shear stress, τ_0, which is not an easily measurable quantity. However, the boundary shear stress, τ_0, can be expressed in terms of measurable flow variables using dimensional analysis, where τ_0 can be taken as a function of the mean flow velocity, V; density of the fluid, ρ; dynamic viscosity of the fluid, μ; diameter of the pipe, D; characteristic size of roughness projections, ε; characteristic spacing of the roughness projections, ε'; and a (dimensionless) form factor, m, that depends on the shape of the roughness elements on the surface of the conduit. This functional relationship can be expressed as

$$\tau_0 = f_1(V, \rho, \mu, D, \varepsilon, \varepsilon', m) \tag{2.27}$$

According to the Buckingham pi theorem, this relationship between eight variables in three fundamental dimensions can also be expressed as a relationship between five nondimensional groups. The following relation is proposed:

$$\frac{\tau_0}{\rho V^2} = f_2\left(\text{Re}, \frac{\varepsilon}{D}, \frac{\varepsilon'}{D}, m\right) \tag{2.28}$$

where Re is the Reynolds number defined by

$$\text{Re} = \frac{\rho V D}{\mu} \tag{2.29}$$

The relationship given by Equation 2.28 is as far as dimensional analysis goes, and experiments are necessary to determine an empirical relationship between the nondimensional groups. Nikuradse (1932; 1933) conducted a series of experiments in pipes in which the inner surfaces were roughened with sand grains of uniform diameter, ε. In these experiments, the spacing, ε', and shape, m, of the roughness elements (sand grains) were constant and Nikuradse's experimental data fitted to the following functional relation:

$$\frac{\tau_0}{\rho V^2} = f_3\left(\text{Re}, \frac{\varepsilon}{D}\right) \tag{2.30}$$

It is convenient for subsequent analysis to introduce a factor of 8 into this relationship, which can then be written as

$$\frac{\tau_0}{\rho V^2} = \frac{1}{8} f\left(\text{Re}, \frac{\varepsilon}{D}\right) \tag{2.31}$$

or simply

$$\frac{\tau_0}{\rho V^2} = \frac{f}{8} \tag{2.32}$$

where the dependence of the *friction factor*, f, on the Reynolds number, Re, and relative roughness, ε/D, is understood. Combining Equations 2.32 and 2.24 leads to the following form of the momentum equation for flows in circular pipes:

$$\boxed{h_f = \frac{fL}{D}\frac{V^2}{2g}} \tag{2.33}$$

This equation, called the *Darcy–Weisbach equation*,* expresses the frictional head loss, h_f, of the fluid over a length L of pipe in terms of measurable parameters, including the pipe diameter (D), average flow velocity (V), and the friction factor (f) that characterizes the shear stress of the fluid on the pipe. Some references name Equation 2.33 simply as the Darcy equation; however, this is inappropriate, since it was Julius Weisbach who first proposed the exact form of Equation 2.33 in 1845, with Darcy's contribution on the functional dependence of f on V and D in 1857 (Brown, 2002; Rouse and Ince, 1957). The differences between laminar and turbulent flow were later quantified by Osbourne Reynolds[†] in 1883 (Reynolds, 1883).

Based on Nikuradse's (1932, 1933) experiments on sand-roughened pipes, Prandtl and von Kármán established the following empirical formulae for estimating the friction factor in turbulent pipe flows:

$$
\begin{aligned}
&\text{Smooth pipe}\left(\frac{k}{D} \approx 0\right): \quad \frac{1}{\sqrt{f}} = -2\log\left(\frac{2.51}{\mathrm{Re}\sqrt{f}}\right) \\[2mm]
&\text{Rough pipe}\left(\frac{k}{D} \gg 0\right): \quad \frac{1}{\sqrt{f}} = -2\log\left(\frac{k/D}{3.7}\right)
\end{aligned}
\tag{2.34}
$$

where k is the roughness height of the sand grains on the surface of the pipe. The variables k and ε are used equivalently to represent the roughness height, although k is more used in the context of an equivalent roughness height and ε as an actual roughness height. Turbulent flow in pipes is generally present when $\mathrm{Re} > 4000$; transition to turbulent flow begins at about $\mathrm{Re} = 2300$. The pipe behaves like a *smooth pipe* when the friction factor does not depend on the height of the roughness projections on the wall of the pipe and therefore depends only on the Reynolds number. In *rough pipes*, the friction factor is determined by the relative roughness, k/D, and becomes independent of the Reynolds number. The smooth-pipe case generally occurs at lower Reynolds numbers, when the roughness projections are submerged within the viscous boundary layer. At higher values of the Reynolds number, the thickness of the viscous boundary layer decreases and eventually the roughness projections protrude sufficiently far outside the viscous boundary layer that the shear stress of the pipe boundary is dominated by the hydrodynamic drag associated with the roughness projections into the main body of the flow. Under these circumstances, the flow in the pipe becomes *fully turbulent*, the friction factor is independent of the Reynolds number, and the pipe is considered to be (hydraulically) rough. The flow is actually turbulent under both smooth-pipe and rough-pipe conditions, but the flow is termed *fully turbulent* when the friction factor is independent of the Reynolds number. Between the smooth- and rough-pipe conditions, there is a transition region in which the friction factor depends on both the Reynolds number and the relative roughness. Colebrook (1939) developed the following relationship that asymptotes to the Prandtl

---

*Henry Darcy (1803–1858) was a nineteenth-century French engineer; Julius Weisbach (1806–1871) was a German engineer of the same era. Weisbach proposed the use of a dimensionless resistance coefficient, and Darcy carried out the tests on water pipes.

[†]Osbourne Reynolds (1842–1912).

and von Kármán relations:

$$\frac{1}{\sqrt{f}} = -2\log\left(\frac{k/D}{3.7} + \frac{2.51}{\text{Re}\sqrt{f}}\right) \tag{2.35}$$

This equation is commonly referred to as the *Colebrook equation* or *Colebrook–White* equation. Equation 2.35 can be applied in the transition region between smooth-pipe and rough-pipe conditions, and values of friction factor, f, predicted by the Colebrook equation are generally accurate to within 10–15% of experimental data (Finnemore and Franzini, 2002; Alexandrou, 2001). The accuracy of the Colebrook equation deteriorates significantly for small pipe diameters, and it is recommended that this equation not be used for pipes with diameters smaller than 2.5 mm (Yoo and Singh, 2005).

Commercial pipes differ from Nikuradse's experimental pipes in that the heights of the roughness projections are not uniform and are not uniformly distributed. In commercial pipes, an *equivalent sand roughness*, k_s, is defined as the diameter of Nikuradse's sand grains that would cause the same head loss as in the commercial pipe. The equivalent sand roughness, k_s, of several commercial pipe materials is given in Table 2.1. These values of k_s apply to clean new pipe only; pipe that has been in service for a long time usually experiences corrosion or scale buildup that results in values of k_s orders of magnitude larger than the values given in Table 2.1 (Echávez, 1997; Gerhart et al., 1992). The rate of increase of k_s with time depends primarily on the quality of the water being transported, and the roughness coefficients for older water mains are usually determined through field testing (AWWA, 1992). The expression for the friction factor derived by Colebrook (Equation 2.35) was plotted by Moody (1944) in what is commonly referred to as the *Moody diagram*,[*] reproduced in Figure 2.3. The Moody diagram indicates that for Re $\leq$ 2000, the flow is laminar and the friction factor is given by

$$f = \frac{64}{\text{Re}} \tag{2.36}$$

which can be derived theoretically based on the assumption of laminar flow of a Newtonian fluid (Daily and Harleman, 1966). For $2000 < \text{Re} \leq 4000$ there is no fixed relationship between the friction factor and the Reynolds number or relative roughness, and flow conditions are generally uncertain (Wilkes, 1999). Beyond a Reynolds number of 4000, the flow is turbulent and the friction factor is controlled by the thickness of the laminar boundary layer relative to the height of the roughness projections on the surface of the pipe. The dashed line in Figure 2.3 indicates the boundary between the fully turbulent flow regime, where f is independent of Re, and the transition regime, where f depends on both Re and the relative roughness, k_s/D. The equation of this dashed line is given by (Mott, 1994)

$$\frac{1}{\sqrt{f}} = \frac{\text{Re}}{200(D/k_s)} \tag{2.37}$$

---

[*]This type of diagram was originally suggested by Blasius in 1913 and Stanton in 1914 (Stanton and Pannell, 1914). The Moody diagram is sometimes called the *Stanton diagram* (Finnemore and Franzini, 2002).

TABLE 2.1: Typical Equivalent Sand Roughness for Various New Materials

Material	Equivalent sand roughness, k_s (mm)
Asbestos cement:	
Coated	0.038
Uncoated	0.076
Brass	0.0015–0.003
Brick	0.6
Concrete:	
General	0.3–3.0
Steel forms	0.18
Wooden forms	0.6
Centrifugally spun	0.13–0.36
Copper	0.0015–0.003
Corrugated metal	45
Glass	0.0015–0.003
Iron:	
Cast iron	0.19–0.26
Ductile iron	0.26
Lined with bitumen	0.12
Lined with spun concrete	0.030–0.038
Galvanized iron	0.15
Wrought iron	0.046 –0.06
Lead	0.0015
Plastic (PVC)	0.0015–0.03
Steel	
Coal-tar enamel	0.0048
New unlined	0.045–0.076
Riveted	0.9–9.0
Wood stave	0.18

Sources: Haestad Methods, Inc. (2002), Moody (1944), Sanks (1998).

The line in the Moody diagram corresponding to a relative roughness of zero describes the friction factor for pipes that are hydraulically smooth.

Although the Colebrook equation (Equation 2.35) can be used to calculate the friction factor in lieu of the Moody diagram, this equation has the drawback that it is an *implicit equation* for the friction factor and must be solved iteratively. This minor inconvenience was circumvented by Jain (1976), who suggested the following explicit equation for the friction factor:

$$\frac{1}{\sqrt{f}} = -2\log\left(\frac{k_s/D}{3.7} + \frac{5.74}{\text{Re}^{0.9}}\right), \quad 10^{-6} \le \frac{k_s}{D} \le 10^{-2}, \; 5000 \le \text{Re} \le 10^8 \quad (2.38)$$

where, according to Jain (1976), Equation 2.38 deviates by less than 1% from the Colebrook equation within the entire turbulent-flow regime, provided that the restrictions on k_s/D and Re are honored. The Jain equation (Equation 2.38) can be more

FIGURE 2.3: Moody diagram

Source: Moody (1944).

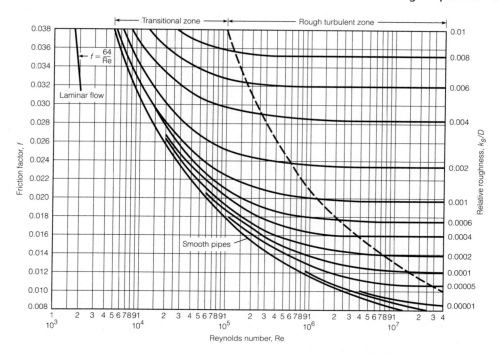

conveniently written as

$$f = \frac{0.25}{\left[\log\left(\frac{k_s}{3.7D} + \frac{5.74}{\mathrm{Re}^{0.9}}\right)\right]^2}, \qquad 10^{-6} \le \frac{k_s}{D} \le 10^{-2}, \, 5000 \le \mathrm{Re} \le 10^8 \qquad (2.39)$$

According to Franzini and Finnemore (1997) and Granger (1985), values of the friction factor calculated using the Colebrook equation are generally accurate to within 10% to 15% of experimental data. Uncertainties in relative roughness and in the data used to produce the Colebrook equation make the use of several-place accuracy in pipe flow problems unjustified. As a rule of thumb, an accuracy of 10% in calculating friction losses in pipes is to be expected (Munson et al., 1994; Gerhart et al., 1992).

EXAMPLE 2.2

Water from a treatment plant is pumped into a distribution system at a rate of 4.38 m³/s, a pressure of 480 kPa, and a temperature of 20°C. The pipe has a diameter of 750 mm and is made of ductile iron. Estimate the pressure 200 m downstream of the treatment plant if the pipeline remains horizontal. Compare the friction factor estimated using the Colebrook equation to the friction factor estimated using the Jain equation. After 20 years in operation, scale buildup is expected to cause the equivalent sand roughness of the pipe to increase by a factor of 10. Determine the effect on the water pressure 200 m downstream of the treatment plant.

Solution According to the Darcy–Weisbach equation, the difference in total head, Δh, between the upstream section (at exit from treatment plant) and the downstream section (200 m downstream from the upstream section) is given by

$$\Delta h = \frac{fL}{D}\frac{V^2}{2g}$$

where f is the friction factor, L is the pipe length between the upstream and downstream sections (= 200 m), D is the pipe diameter (= 750 mm), and V is the velocity in the pipe. The velocity, V, is given by

$$V = \frac{Q}{A}$$

where Q is the flowrate in the pipe (= 4.38 m³/s) and A is the area of the pipe cross section given by

$$A = \frac{\pi}{4}D^2 = \frac{\pi}{4}(0.75)^2 = 0.442 \text{ m}^2$$

The pipeline velocity is therefore

$$V = \frac{Q}{A} = \frac{4.38}{0.442} = 9.91 \text{ m/s}$$

The friction factor, f, in the Darcy–Weisbach equation is calculated using the Colebrook equation:

$$\frac{1}{\sqrt{f}} = -2\log\left[\frac{k_s}{3.7D} + \frac{2.51}{\text{Re}\sqrt{f}}\right]$$

Here Re is the Reynolds number and k_s is the equivalent sand roughness of ductile iron (= 0.26 mm). The Reynolds number is given by

$$\text{Re} = \frac{VD}{\nu}$$

where ν is the kinematic viscosity of water at 20°C, which is equal to 1.00×10^{-6} m²/s. Therefore

$$\text{Re} = \frac{VD}{\nu} = \frac{(9.91)(0.75)}{1.00 \times 10^{-6}} = 7.43 \times 10^6$$

Substituting into the Colebrook equation leads to

$$\frac{1}{\sqrt{f}} = -2\log\left[\frac{0.26}{(3.7)(750)} + \frac{2.51}{7.43 \times 10^6\sqrt{f}}\right]$$

or

$$\frac{1}{\sqrt{f}} = -2\log\left[9.37 \times 10^{-5} + \frac{3.38 \times 10^{-7}}{\sqrt{f}}\right]$$

This is an implicit equation for f, and the solution is

$$f = 0.016$$

The head loss, Δh, between the upstream and downstream sections can now be calculated using the Darcy–Weisbach equation as

$$\Delta h = \frac{fL}{D}\frac{V^2}{2g} = \frac{(0.016)(200)}{0.75}\frac{(9.91)^2}{(2)(9.81)} = 21.4 \text{ m}$$

Using the definition of head loss, Δh,

$$\Delta h = \left(\frac{p_1}{\gamma} + z_1\right) - \left(\frac{p_2}{\gamma} + z_2\right)$$

where p_1 and p_2 are the upstream and downstream pressures, γ is the specific weight of water, and z_1 and z_2 are the upstream and downstream pipe elevations. Since the pipe is horizontal, $z_1 = z_2$ and Δh can be written in terms of the pressures at the upstream and downstream sections as

$$\Delta h = \frac{p_1}{\gamma} - \frac{p_2}{\gamma}$$

In this case, $p_1 = 480 \text{ kPa}$, $\gamma = 9.79 \text{ kN/m}^3$, and therefore

$$21.4 = \frac{480}{9.79} - \frac{p_2}{9.79}$$

which yields

$$p_2 = 270 \text{ kPa}$$

Therefore, the pressure 200 m downstream of the treatment plant is 270 kPa. The Colebrook equation required that f be determined from an implicit equation, but the explicit Jain approximation for f is given by

$$\frac{1}{\sqrt{f}} = -2\log\left[\frac{k_s}{3.7D} + \frac{5.74}{\text{Re}^{0.9}}\right]$$

Substituting for k_s, D, and Re gives

$$\frac{1}{\sqrt{f}} = -2\log\left[\frac{0.26}{(3.7)(750)} + \frac{5.74}{(7.43 \times 10^6)^{0.9}}\right]$$

which leads to

$$f = 0.016$$

This is the same friction factor obtained using the Colebrook equation within an accuracy of two significant digits.

After 20 years, the equivalent sand roughness, k_s, of the pipe is 2.6 mm, the (previously calculated) Reynolds number is 7.43×10^6, and the Colebrook equation gives

$$\frac{1}{\sqrt{f}} = -2\log\left[\frac{2.6}{(3.7)(750)} + \frac{2.51}{7.43 \times 10^6\sqrt{f}}\right]$$

or

$$\frac{1}{\sqrt{f}} = -2\log\left[9.37 \times 10^{-4} + \frac{3.38 \times 10^{-7}}{\sqrt{f}}\right]$$

which yields

$$f = 0.027$$

The head loss, Δh, between the upstream and downstream sections is given by the Darcy–Weisbach equation as

$$\Delta h = \frac{fL}{D}\frac{V^2}{2g} = \frac{(0.027)(200)}{0.75}\frac{(9.91)^2}{(2)(9.81)} = 36.0 \text{ m}$$

Hence the pressure, p_2, 200 m downstream of the treatment plant is given by the relation

$$\Delta h = \frac{p_1}{\gamma} - \frac{p_2}{\gamma}$$

where $p_1 = 480$ kPa, $\gamma = 9.79$ kN/m^3, and therefore

$$36.0 = \frac{480}{9.79} - \frac{p_2}{9.79}$$

which yields

$$p_2 = 128 \text{ kPa}$$

Therefore, pipe aging over 20 years will cause the pressure 200 m downstream of the treatment plant to decrease from 270 kPa to 128 kPa. This is quite a significant drop and shows why velocities of 9.91 m/s are not used in these pipelines, even for short lengths of pipe.

The problem in Example 2.2 illustrates the case where the flowrate through a pipe is known and the objective is to calculate the head loss and pressure drop over a given length of pipe. The approach is summarized as follows: (1) calculate the Reynolds number, Re, and the relative roughness, k_s/D, from the given data; (2) use the Colebrook equation (Equation 2.35) or Jain equation (Equation 2.38) to calculate f; and (3) use the calculated value of f to calculate the head loss from the Darcy–Weisbach equation (Equation 2.33), and the corresponding pressure drop from Equation 2.22.

Flowrate for a given head loss.   In many cases, the flowrate through a pipe is not controlled but attains a level that matches the pressure drop available. For example, the flowrate through faucets in home plumbing is determined by the gage pressure in the water main, which is relatively insensitive to the flow through the faucet. A useful approach to this problem that uses the Colebrook equation has been suggested by Fay (1994), where the first step is to calculate $\text{Re}\sqrt{f}$ using the rearranged Darcy–Weisbach equation

$$\text{Re}\sqrt{f} = \left(\frac{2gh_fD^3}{v^2L}\right)^{\frac{1}{2}} \tag{2.40}$$

Using this value of $\text{Re}\sqrt{f}$, solve for Re using the rearranged Colebrook equation

$$\text{Re} = -2.0(\text{Re}\sqrt{f})\log\left(\frac{k_s/D}{3.7} + \frac{2.51}{\text{Re}\sqrt{f}}\right) \tag{2.41}$$

Using this value of Re, the flowrate, Q, can then be calculated by

$$Q = \frac{1}{4}\pi D^2 V = \frac{1}{4}\pi D\nu\text{Re} \tag{2.42}$$

This approach must necessarily be validated by verifying that Re > 2300, which is required for application of the Colebrook equation. Swamee and Jain (1976) combine Equations 2.40 to 2.42 to yield

$$\boxed{Q = -0.965D^2\sqrt{\frac{gDh_f}{L}}\ln\left(\frac{k_s/D}{3.7} + \frac{1.784\nu}{D\sqrt{gDh_f/L}}\right)} \tag{2.43}$$

EXAMPLE 2.3

A 50-mm diameter galvanized iron service pipe is connected to a water main in which the pressure is 450 kPa gage. If the length of the service pipe to a faucet is 40 m and the faucet is 1.2 m above the main, estimate the flowrate when the faucet is fully open.

Solution   The head loss, h_f, in the pipe is estimated by

$$h_f = \left(\frac{p_{\text{main}}}{\gamma} + z_{\text{main}}\right) - \left(\frac{p_{\text{outlet}}}{\gamma} + z_{\text{outlet}}\right)$$

where $p_{\text{main}} = 450$ kPa, $z_{\text{main}} = 0$ m, $p_{\text{outlet}} = 0$ kPa, and $z_{\text{outlet}} = 1.2$ m. Therefore, taking $\gamma = 9.79$ kN/m^3 (at 20°C) gives

$$h_f = \left(\frac{450}{9.79} + 0\right) - (0 + 1.2) = 44.8 \text{ m}$$

Also, since $D = 50$ mm, $L = 40$ m, $k_s = 0.15$ mm (from Table 2.1), and $\nu = 1.00 \times 10^{-6}$ m^2/s (at 20°C), the Swamee–Jain equation (Equation 2.43) yields

$$Q = -0.965D^2\sqrt{\frac{gDh_f}{L}}\ln\left(\frac{k_s/D}{3.7} + \frac{1.784\nu}{D\sqrt{gDh_f/L}}\right)$$

$$= -0.965(0.05)^2\sqrt{\frac{(9.81)(0.05)(44.8)}{40}}\ln\left[\frac{0.15/50}{3.7} + \frac{1.784(1.00 \times 10^{-6})}{(0.05)\sqrt{(9.81)(0.05)(44.8)/40}}\right]$$

$$= 0.0126 \text{ m}^3/\text{s} = 12.6 \text{ L/s}$$

The faucet can therefore be expected to deliver 12.6 L/s when fully open.

Diameter for a given flowrate and head loss.   In many cases, an engineer must select a size of pipe to provide a given level of service. For example, the maximum flowrate and maximum allowable pressure drop may be specified for a water delivery pipe, and the engineer is required to calculate the minimum diameter pipe that will satisfy these design constraints. Solution of this problem necessarily requires an iterative procedure. The following steps are suggested (Streeter and Wylie, 1985)

1. Assume a value of f.
2. Calculate D from the rearranged Darcy–Weisbach equation,

$$D = \sqrt[5]{\left(\frac{8LQ^2}{h_f g\pi^2}f\right)} \tag{2.44}$$

where the term in parentheses can be calculated from given data.

3. Calculate Re from

$$\text{Re} = \frac{VD}{\nu} = \left(\frac{4Q}{\pi\nu}\right)\frac{1}{D} \tag{2.45}$$

where the term in parentheses can be calculated from given data.

4. Calculate k_s/D.
5. Use Re and k_s/D to calculate f from the Colebrook equation.
6. Using the new f, repeat the procedure until the new f agrees with the old f to the first two significant digits.

EXAMPLE 2.4

A galvanized iron service pipe from a water main is required to deliver 200 L/s during a fire. If the length of the service pipe is 35 m and the head loss in the pipe is not to exceed 50 m, calculate the minimum pipe diameter that can be used.

Solution

Step 1.   Assume $f = 0.03$

Step 2.   Since $Q = 0.2$ m³/s, $L = 35$ m, and $h_f = 50$ m, then

$$D = \sqrt[5]{\left[\frac{8LQ^2}{h_f g\pi^2}\right]f} = \sqrt[5]{\left[\frac{8(35)(0.2)^2}{(50)(9.81)\pi^2}\right](0.03)} = 0.147 \text{ m}$$

Step 3.   Since $\nu = 1.00 \times 10^{-6}$ m²/s (at 20°C), then

$$\text{Re} = \left[\frac{4Q}{\pi\nu}\right]\frac{1}{D} = \left[\frac{4(0.2)}{\pi(1.00 \times 10^{-6})}\right]\frac{1}{0.147} = 1.73 \times 10^6$$

Step 4.   Since $k_s = 0.15$ mm (from Table 2.1, for new pipe), then

$$\frac{k_s}{D} = \frac{1.5 \times 10^{-4}}{0.147} = 0.00102$$

Step 5. Using the Colebrook equation (Equation 2.35) gives

$$\frac{1}{\sqrt{f}} = -2\log\left(\frac{k_s/D}{3.7} + \frac{2.51}{\text{Re}\sqrt{f}}\right)$$

$$= -2\log\left(\frac{0.00102}{3.7} + \frac{2.51}{1.73 \times 10^6\sqrt{f}}\right)$$

which leads to

$$f = 0.020$$

Step 6. $f = 0.020$ differs from the assumed f ($= 0.03$), so repeat the procedure with $f = 0.020$.

Step 2. For $f = 0.020$, $D = 0.136$ m

Step 3. For $D = 0.136$ m, Re $= 1.87 \times 10^6$

Step 4. For $D = 0.136$ m, $k_s/D = 0.00110$

Step 5. $f = 0.020$

Step 6. The calculated f ($= 0.020$) is equal to the assumed f. The required pipe diameter is therefore equal to 0.136 m or 136 mm. A commercially available pipe with the closest diameter larger than 136 mm should be used.

The iterative procedure demonstrated in the previous example converges fairly quickly, and does not pose any computational difficulty. Swamee and Jain (1976) have suggested the following explicit formula for calculating the pipe diameter, D:

$$D = 0.66\left[k_s^{1.25}\left(\frac{LQ^2}{gh_f}\right)^{4.75} + \nu Q^{9.4}\left(\frac{L}{gh_f}\right)^{5.2}\right]^{0.04} \tag{2.46}$$

$$3000 \leq \text{Re} \leq 3 \times 10^8, \ 10^{-6} \leq \frac{k_s}{D} \leq 2 \times 10^{-2}$$

Equation 2.46 will yield a D within 5% of the value obtained by the method using the Colebrook equation. This method is illustrated by repeating the previous example.

EXAMPLE 2.5

A galvanized iron service pipe from a water main is required to deliver 200 L/s during a fire. If the length of the service pipe is 35 m, and the head loss in the pipe is not to exceed 50 m, use the Swamee–Jain equation to calculate the minimum pipe diameter that can be used.

Solution   Since $k_s = 0.15$ mm, $L = 35$ m, $Q = 0.2$ m^3/s, $h_f = 50$ m, $\nu = 1.00 \times 10^{-6}$ m^2/s, the Swamee–Jain equation gives

$$D = 0.66 \left[k_s^{1.25} \left(\frac{LQ^2}{gh_f} \right)^{4.75} + \nu Q^{9.4} \left(\frac{L}{gh_f} \right)^{5.2} \right]^{0.04}$$

$$= 0.66 \left\{ (0.00015)^{1.25} \left[\frac{(35)(0.2)^2}{(9.81)(50)} \right]^{4.75} + (1.00 \times 10^{-6})(0.2)^{9.4} \left[\frac{35}{(9.81)(50)} \right]^{5.2} \right\}^{0.04}$$

$$= 0.140 \text{ m}$$

The calculated pipe diameter (140 mm) is about 3% higher than calculated by the Colebrook equation (136 mm).

2.2.3   Steady-State Energy Equation

The steady-state energy equation for the control volume illustrated in Figure 2.4 is given by

$$\frac{dQ_h}{dt} - \frac{dW}{dt} = \int_A \rho e \, \mathbf{v} \cdot \mathbf{n} \, dA \tag{2.47}$$

where Q_h is the heat added to the fluid in the control volume, W is the work done by the fluid in the control volume, A is the surface area of the control volume, ρ is the density of the fluid in the control volume, and e is the internal energy per unit mass of fluid in the control volume given by

$$e = gz + \frac{v^2}{2} + u \tag{2.48}$$

where z is the elevation of the fluid mass having a velocity v and internal energy u. By convention, the heat added to a system and the work done by a system are positive quantities. The normal stresses on the inflow and outflow boundaries of the control volume are equal to the pressure, p, with shear stresses tangential to the boundaries of the control volume. As the fluid moves across the control surface with velocity $\mathbf{v}$, the power (= rate of doing work) expended by the fluid against the external pressure

FIGURE 2.4: Energy balance in closed conduit

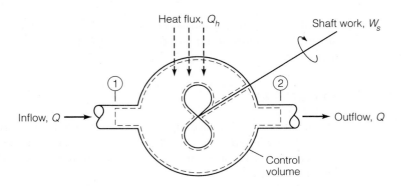

forces is given by

$$\frac{dW_p}{dt} = \int_A p\mathbf{v} \cdot \mathbf{n}\, dA \tag{2.49}$$

where W_p is the work done against external pressure forces. The work done by a fluid in the control volume is typically separated into work done against external pressure forces, W_p, plus work done against rotating surfaces, W_s, commonly referred to as the *shaft work*. The rotating element is called a *rotor* in a gas or steam turbine, an *impeller* in a pump, and a *runner* in a hydraulic turbine. The rate at which work is done by a fluid system, dW/dt, can therefore be written as

$$\frac{dW}{dt} = \frac{dW_p}{dt} + \frac{dW_s}{dt} = \int_A p\mathbf{v} \cdot \mathbf{n}\, dA + \frac{dW_s}{dt} \tag{2.50}$$

Combining Equation 2.50 with the steady-state energy equation (Equation 2.47) leads to

$$\frac{dQ_h}{dt} - \frac{dW_s}{dt} = \int_A \rho\left(\frac{p}{\rho} + e\right)\mathbf{v} \cdot \mathbf{n}\, dA \tag{2.51}$$

Substituting the definition of the internal energy, e, given by Equation 2.48 into Equation 2.51 yields

$$\frac{dQ_h}{dt} - \frac{dW_s}{dt} = \int_A \rho\left(h + gz + \frac{v^2}{2}\right)\mathbf{v} \cdot \mathbf{n}\, dA \tag{2.52}$$

where h is the enthalpy of the fluid defined by

$$h = \frac{p}{\rho} + u \tag{2.53}$$

Denoting the rate at which heat is being added to the fluid system by $\dot{Q}$, and the rate at which work is being done against moving impervious boundaries (shaft work) by $\dot{W}_s$, the energy equation can be written in the form

$$\dot{Q} - \dot{W}_s = \int_A \rho\left(h + gz + \frac{v^2}{2}\right)\mathbf{v} \cdot \mathbf{n}\, dA \tag{2.54}$$

Considering the terms $h + gz$, where

$$h + gz = \frac{p}{\rho} + u + gz = g\left(\frac{p}{\gamma} + z\right) + u \tag{2.55}$$

and γ is the specific weight of the fluid, Equation 2.55 indicates that $h + gz$ can be assumed to be constant across the inflow and outflow openings illustrated in Figure 2.4, since a hydrostatic pressure distribution across the inflow/outflow boundaries guarantees that $p/\gamma + z$ is constant across the inflow/outflow boundaries normal to the flow direction, and the internal energy, u, depends only on the temperature, which can be assumed constant across each boundary. Since $\mathbf{v} \cdot \mathbf{n}$ is equal to zero over the impervious boundaries in contact with the fluid system, Equation 2.54 can be integrated to yield

$$\dot{Q} - \dot{W}_s = (h_1 + gz_1)\int_{A_1} \rho\mathbf{v}\cdot\mathbf{n}\,dA + \int_{A_1} \rho\frac{v^2}{2}\mathbf{v}\cdot\mathbf{n}\,dA + (h_2 + gz_2)\int_{A_2} \rho\mathbf{v}\cdot\mathbf{n}\,dA$$

$$+ \int_{A_2} \rho\frac{v^2}{2}\mathbf{v}\cdot\mathbf{n}\,dA$$

$$= -(h_1 + gz_1)\int_{A_1} \rho v_1\,dA - \int_{A_1} \rho\frac{v_1^3}{2}\,dA + (h_2 + gz_2)\int_{A_2} \rho v_2\,dA$$

$$+ \int_{A_2} \rho\frac{v_2^3}{2}\,dA \tag{2.56}$$

where the subscripts 1 and 2 refer to the inflow and outflow boundaries, respectively, and the negative signs result from the fact that the unit normal points out of the control volume, causing $\mathbf{v}\cdot\mathbf{n}$ to be negative on the inflow boundary and positive on the outflow boundary.

Equation 2.56 can be simplified by noting that the assumption of steady flow requires that rate of mass inflow to the control volume is equal to the mass outflow rate and, denoting the mass flow rate by $\dot{m}$, the continuity equation requires that

$$\dot{m} = \int_{A_1} \rho v_1\,dA = \int_{A_2} \rho v_2\,dA \tag{2.57}$$

Furthermore, the constants α_1 and α_2 can be defined by the equations

$$\int_{A_1} \rho\frac{v^3}{2}\,dA = \alpha_1\rho\frac{V_1^3}{2}A_1 \tag{2.58}$$

$$\int_{A_2} \rho\frac{v^3}{2}\,dA = \alpha_2\rho\frac{V_2^3}{2}A_2 \tag{2.59}$$

where A_1 and A_2 are the areas of the inflow and outflow boundaries, respectively, and V_1 and V_2 are the corresponding mean velocities across these boundaries. The constants α_1 and α_2 are determined by the velocity profile across the flow boundaries, and these constants are called *kinetic energy correction factors*. If the velocity is constant across a flow boundary, then it is clear from Equation 2.58 that the kinetic energy correction factor for that boundary is equal to unity; for any other velocity distribution, the kinetic energy factor is greater than unity. Combining Equations 2.56 to 2.59 leads to

$$\dot{Q} - \dot{W}_s = -(h_1 + gz_1)\dot{m} - \alpha_1\rho\frac{V_1^3}{2}A_1 + (h_2 + gz_2)\dot{m} + \alpha_2\rho\frac{V_2^3}{2}A_2 \tag{2.60}$$

Invoking the continuity equation requires that

$$\rho V_1 A_1 = \rho V_2 A_2 = \dot{m} \tag{2.61}$$

and combining Equations 2.60 and 2.61 leads to

$$\dot{Q} - \dot{W}_s = \dot{m}\left[\left(h_2 + gz_2 + \alpha_2\frac{V_2^2}{2}\right) - \left(h_1 + gz_1 + \alpha_1\frac{V_1^2}{2}\right)\right] \qquad (2.62)$$

which can be put in the form

$$\frac{\dot{Q}}{\dot{m}g} - \frac{\dot{W}_s}{\dot{m}g} = \left(\frac{p_2}{\gamma} + \frac{u_2}{g} + z_2 + \alpha_2\frac{V_2^2}{2g}\right) - \left(\frac{p_1}{\gamma} + \frac{u_1}{g} + z_1 + \alpha_1\frac{V_1^2}{2g}\right) \quad (2.63)$$

and can be further rearranged into the useful form

$$\left(\frac{p_1}{\gamma} + \alpha_1\frac{V_1^2}{2g} + z_1\right) = \left(\frac{p_2}{\gamma} + \alpha_2\frac{V_2^2}{2g} + z_2\right) + \left[\frac{1}{g}(u_2 - u_1) - \frac{\dot{Q}}{\dot{m}g}\right] + \left[\frac{\dot{W}_s}{\dot{m}g}\right]$$
$$(2.64)$$

Two key terms can be identified in Equation 2.64: the (shaft) work done by the fluid per unit weight, h_s, defined by the relation

$$h_s = \frac{\dot{W}_s}{\dot{m}g} \qquad (2.65)$$

and the energy loss per unit weight, commonly called the head loss, h_L, defined by the relation

$$h_L = \frac{1}{g}(u_2 - u_1) - \frac{\dot{Q}}{\dot{m}g} \qquad (2.66)$$

Combining Equations 2.64 to 2.66 leads to the most common form of the steady-state *energy equation*

$$\boxed{\left(\frac{p_1}{\gamma} + \alpha_1\frac{V_1^2}{2g} + z_1\right) = \left(\frac{p_2}{\gamma} + \alpha_2\frac{V_2^2}{2g} + z_2\right) + h_L + h_s} \qquad (2.67)$$

where a positive head loss indicates an increase in internal energy (manifested by an increase in temperature) and/or a loss of heat, and a positive value of h_s is associated with work being done by the fluid, such as in moving a turbine runner. Many practitioners incorrectly refer to Equation 2.67 as the *Bernoulli equation*, which bears some resemblance to Equation 2.67 but is different in several important respects. Fundamental differences between the energy equation and the Bernoulli equation are that the Bernoulli equation is derived from the momentum equation, which is independent of the energy equation, and the Bernoulli equation does not account for fluid friction.

Energy and hydraulic grade lines.   The *total head*, h, of a fluid at any cross section of a pipe is defined by

$$\boxed{h = \frac{p}{\gamma} + \alpha\frac{V^2}{2g} + z} \qquad (2.68)$$

where p is the pressure in the fluid at the centroid of the cross section, γ is the specific weight of the fluid, α is the kinetic energy correction factor, V is the average velocity across the pipe cross section, and z is the elevation of the centroid of the pipe cross section. The total head measures the average energy per unit weight of the fluid flowing across a pipe cross section. The energy equation, Equation 2.67, states that changes in the total head along the pipe are described by

$$h(x + \Delta x) = h(x) - (h_L + h_s) \tag{2.69}$$

where x is the coordinate measured along the pipe centerline, Δx is the distance between two cross sections in the pipe, h_L is the head loss, and h_s is the shaft work done by the fluid over the distance Δx. The practical application of Equation 2.69 is illustrated in Figure 2.5, where the head loss, h_L, between two sections a distance Δx apart is indicated. At each cross section, the total energy, h, is plotted relative to a defined datum, and the locus of these points is called the *energy grade line*. The energy grade line at each pipe cross section is located a distance $p/\gamma + \alpha V^2/2g$ vertically above the centroid of the cross section, and between any two cross sections the elevation of the energy grade line falls by a vertical distance equal to the head loss caused by pipe friction, h_L, plus the shaft work, h_s, done by the fluid. The *hydraulic grade line* measures the hydraulic head $p/\gamma + z$ at each pipe cross section. It is located a distance p/γ above the pipe centerline and indicates the elevation to which the fluid would rise in an open tube connected to the wall of the pipe section. The hydraulic grade line is therefore located a distance $\alpha V^2/2g$ below the energy grade line. In most water-supply applications the velocity heads are negligible and the hydraulic grade line closely approximates the energy grade line.

FIGURE 2.5: Head loss along pipe

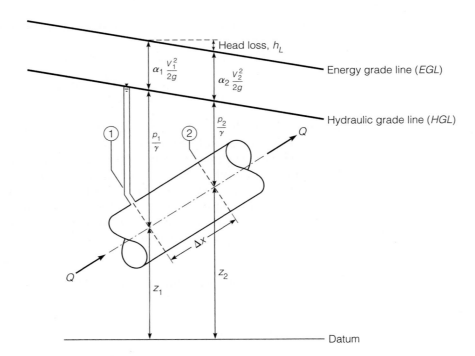

Both the hydraulic grade line and the energy grade line are useful in visualizing the state of the fluid as it flows along the pipe and are frequently used in assessing the performance of fluid-delivery systems. Most fluid-delivery systems, for example, require that the fluid pressure remain positive, in which case the hydraulic grade line must remain above the pipe. In circumstances where additional energy is required to maintain acceptable pressures in pipelines, a pump is installed along the pipeline to elevate the energy grade line by an amount h_s, which also elevates the hydraulic grade line by the same amount. This condition is illustrated in Figure 2.6. In cases where the pipeline upstream and downstream of the pump are of the same diameter, the velocity heads $\alpha V^2/2g$ both upstream and downstream of the pump are the same, and the head added by the pump, h_s, goes entirely to increase the pressure head, p/γ, of the fluid. It should also be clear from Figure 2.5 that the pressure head in a pipeline can be increased by simply increasing the pipeline diameter, which reduces the velocity head, $\alpha V^2/2g$, and thereby increases the pressure head, p/γ, to maintain the same approximately total energy at the pipe section. Expansion losses will cause a reduction in total energy.

Velocity profile.   The momentum and energy correction factors, α and β, depend on the cross-sectional velocity distribution. The velocity profile in both smooth and rough pipes of circular cross section can be estimated by the semi-empirical equation

$$v(r) = \left[(1 \ + \ 1.326\sqrt{f}) \ - \ 2.04\sqrt{f} \log \left(\frac{R}{R \ - \ r} \right) \right] V \qquad (2.70)$$

FIGURE 2.6: Pump effect on flow in pipeline

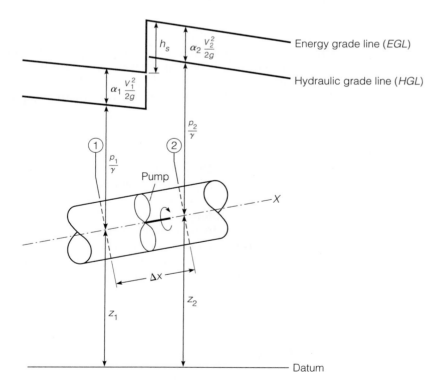

where $v(r)$ is the velocity at a radial distance r from the centerline of the pipe, R is the radius of the pipe, f is the friction factor, and V is the average velocity across the pipe.

The velocity distribution given by Equation 2.70 agrees well with velocity measurements in both smooth and rough pipes. This equation, however, is not applicable within the small region close to the centerline of the pipe and is also not applicable in the small region close to the pipe boundary. This is apparent since at the axis of the pipe dv/dr must be equal to zero, but Equation 2.70 does not have a zero slope at $r = 0$. At the pipe boundary v must also be equal to zero, but Equation 2.70 gives a velocity of zero at a small distance from the wall, with a velocity of $-\infty$ at $r = R$. The energy and momentum correction factors, α and β, derived from the velocity profile are (Moody, 1950)

$$\alpha = 1 + 2.7f \tag{2.71}$$

$$\beta = 1 + 0.98f \tag{2.72}$$

Another commonly used equation to describe the velocity distribution in turbulent pipe flow is the empirical *power law* equation given by

$$\boxed{v(r) = V_0 \left(1 - \frac{r}{R}\right)^{\frac{1}{n}}} \tag{2.73}$$

where V_0 is the centerline velocity and n is a function of the Reynolds number, Re. Values of n typically range between 6 and 10 and can be approximated by (Fox and McDonald, 1992; Schlichting, 1979)

$$n = 1.83 \log \mathrm{Re} - 1.86 \tag{2.74}$$

The power law is not applicable within $0.04R$ of the wall, since the power law gives an infinite velocity gradient at the wall. Although the profile fits the data close to the centerline of the pipe, it does not give zero slope at the centerline. The kinetic energy coefficient, α, derived from the power law equation is given by

$$\alpha = \frac{(1 + n)^3 (1 + 2n)^3}{4n^4 (3 + n)(3 + 2n)} \tag{2.75}$$

For n between 6 and 10, α varies from 1.08 to 1.03. In most engineering applications, α and β are taken as unity (see Problem 2.14).

Head losses in transitions and fittings.   The head losses in straight pipes of constant diameter are caused by friction between the moving fluid and the pipe boundary and are estimated using the Darcy–Weisbach equation. Flow through pipe fittings, around bends, and through changes in pipeline geometry causes additional head losses, h_0, that are quantified by an equation of the form

$$\boxed{h_0 = \sum K \frac{V^2}{2g}} \tag{2.76}$$

where K is a loss coefficient that is specific to each fitting and transition, and V is the average velocity at a defined location within the transition or fitting. The loss

FIGURE 2.7: Loss coefficients for transitions and fittings

Source: Roberson and Crowe (1997).

Description	Sketch	Additional Data		K	
Pipe entrance	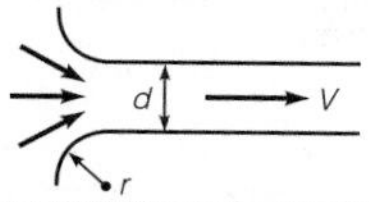	r/d		K	
		0.0		0.50	
		0.1		0.12	
		>0.2		0.03	
Contraction	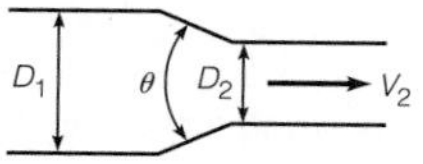	D_2/D_1		K $\theta = 60°$	K $\theta = 180°$
		0.0		0.08	0.50
		0.20		0.08	0.49
		0.40		0.07	0.42
		0.60		0.06	0.27
		0.80		0.06	0.20
		0.90		0.06	0.10
Expansion	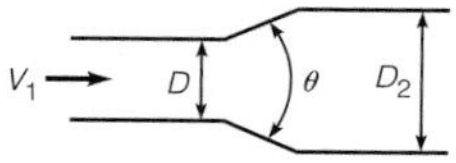	D_1/D_2		K $\theta = 20°$	K $\theta = 180°$
		0.0			1.00
		0.20		0.30	0.87
		0.40		0.25	0.70
		0.60		0.15	0.41
		0.80		0.10	0.15
90° miter bend	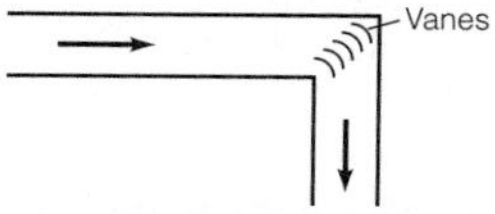	Without vanes		K = 1.1	
		With vanes		K = 0.2	
90° smooth bend	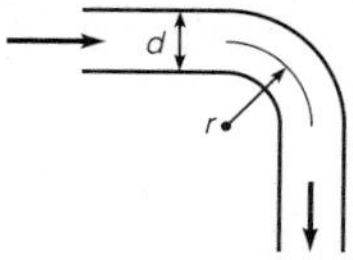	r/d		K $\theta = 90°$	
		1		0.35	
		2		0.19	
		4		0.16	
		6		0.21	
Threaded pipe fittings	Globe valve — wide open			K	
				10.0	
	Angle valve — wide open			5.0	
	Gate valve — wide open			0.2	
	Gate valve — half open			5.6	
	Return bend			2.2	
	Tee				
	straight-through flow			0.4	
	side-outlet flow			1.8	
	90° elbow			0.9	
	45° elbow			0.4	

coefficients for several fittings and transitions are shown in Figure 2.7. Head losses in transitions and fittings are also called *local head losses* or *minor head losses*. The latter term should be avoided, however, since in some cases these head losses are a significant portion of the total head loss in a pipe. Detailed descriptions of local head losses in various valve geometries can be found in Mott (1994), and additional data on local head losses in pipeline systems can be found in Brater and colleagues (1996).

EXAMPLE 2.6

A pump is to be selected that will pump water from a well into a storage reservoir. In order to fill the reservoir in a timely manner, the pump is required to deliver 5 L/s when the water level in the reservoir is 5 m above the water level in the well. Find the head that must be added by the pump. The pipeline system is shown in Figure 2.8. Assume that the local loss coefficient for each of the bends is equal to 0.25 and that the temperature of the water is 20°C.

FIGURE 2.8: Pipeline system

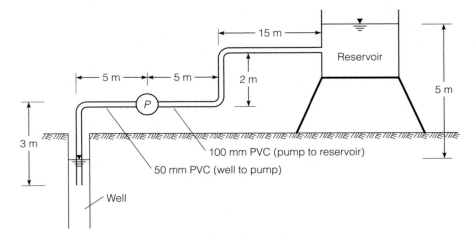

Solution Taking the elevation of the water surface in the well to be equal to 0 m, and proceeding from the well to the storage reservoir (where the head is equal to 5 m), the energy equation (Equation 2.67) can be written as

$$0 - \frac{V_1^2}{2g} - \frac{f_1 L_1}{D_1} \frac{V_1^2}{2g} - K_1 \frac{V_1^2}{2g} + h_p - \frac{f_2 L_2}{D_2} \frac{V_2^2}{2g} - (K_2 + K_3) \frac{V_2^2}{2g} - \frac{V_2^2}{2g} = 5$$

where V_1 and V_2 are the velocities in the 50-mm ($=D_1$) and 100-mm ($=D_2$) pipes, respectively; L_1 and L_2 are the corresponding pipe lengths; f_1 and f_2 are the corresponding friction factors; K_1, K_2, and K_3 are the loss coefficients for each of the three bends; and h_p is the head added by the pump. The cross-sectional areas of each of the pipes, A_1 and A_2, are given by

$$A_1 = \frac{\pi}{4} D_1^2 = \frac{\pi}{4}(0.05)^2 = 0.001963 \text{ m}^2$$

$$A_2 = \frac{\pi}{4} D_2^2 = \frac{\pi}{4}(0.10)^2 = 0.007854 \text{ m}^2$$

When the flowrate, Q, is 5 L/s, the velocities V_1 and V_2 are given by

$$V_1 = \frac{Q}{A_1} = \frac{0.005}{0.001963} = 2.54 \text{ m/s}$$

$$V_2 = \frac{Q}{A_2} = \frac{0.005}{0.007854} = 0.637 \text{ m/s}$$

PVC pipe is considered smooth ($k_s \approx 0$) and therefore the friction factor, f, can be estimated using the Jain equation

$$f = \frac{0.25}{\left[\log_{10} \dfrac{5.74}{\text{Re}^{0.9}} \right]^2}$$

where Re is the Reynolds number. At $20°C$, the kinematic viscosity, ν, is equal to 1.00×10^{-6} m^2/s and for the 50-mm pipe

$$\text{Re}_1 = \frac{V_1 D_1}{\nu} = \frac{(2.54)(0.05)}{1.00 \times 10^{-6}} = 1.27 \times 10^5$$

which leads to

$$f_1 = \frac{0.25}{\left[\log_{10} \dfrac{5.74}{(1.27 \times 10^5)^{0.9}}\right]^2} = 0.0170$$

and for the 100-mm pipe

$$\text{Re}_2 = \frac{V_2 D_2}{\nu} = \frac{(0.637)(0.10)}{1.00 \times 10^{-6}} = 6.37 \times 10^4$$

which leads to

$$f_2 = \frac{0.25}{\left[\log_{10} \dfrac{5.74}{(6.37 \times 10^4)^{0.9}}\right]^2} = 0.0197$$

Substituting the values of these parameters into the energy equation yields

$$0 - \left[1 + \frac{(0.0170)(8)}{0.05} + 0.25\right]\frac{2.54^2}{(2)(9.81)} + h_p$$

$$- \left[\frac{(0.0197)(22)}{0.10} + 0.25 + 0.25 + 1\right]\frac{0.637^2}{(2)(9.81)} = 5$$

which leads to

$$h_p = 6.43 \text{ m}$$

Therefore the head to be added by the pump is 6.43 m.

Local losses are frequently neglected in the analysis of pipeline systems. As a general rule, neglecting local losses is justified when, on average, there is a length of 1000 diameters between each local loss (Streeter et al., 1998).

Head losses in noncircular conduits. Most pipelines are of circular cross section, but flow of water in noncircular conduits is commonly encountered. The hydraulic radius, R, of a conduit of any shape is defined by the relation

$$R = \frac{A}{P} \tag{2.77}$$

where A is the cross-sectional area of the conduit and P is the wetted perimeter. For circular conduits of diameter D, the hydraulic radius is given by

$$R = \frac{\pi D^2/4}{\pi D} = \frac{D}{4} \tag{2.78}$$

or

$$D = 4R \tag{2.79}$$

Using the hydraulic radius, R, as the length scale of a closed conduit instead of D, the frictional head losses, h_f, in noncircular conduits can be estimated using the Darcy–Weisbach equation for circular conduits by simply replacing D by $4R$, which yields

$$h_f = \frac{fL}{4R}\frac{V^2}{2g} \tag{2.80}$$

where the friction factor, f, is calculated using a Reynolds number, Re, defined by

$$\text{Re} = \frac{\rho V(4R)}{\mu} \tag{2.81}$$

and a relative roughness defined by $k_s/4R$.

Characterizing a noncircular conduit by the hydraulic radius, R, is necessarily approximate, since conduits of arbitrary cross section cannot be described with a single parameter. Secondary currents that are generated across a noncircular conduit cross section to redistribute the shears are another reason why noncircular conduits cannot be completely characterized by the hydraulic radius (Liggett, 1994). However, using the hydraulic radius as a basis for calculating frictional head losses in noncircular conduits is usually accurate to within 15% for turbulent flow (Munson et al.., 1994; White, 1994). This approximation is much less accurate for laminar flows, where the accuracy is on the order of ±40% (White, 1994). Characterization of noncircular conduits by the hydraulic radius can be used for rectangular conduits where the ratio of sides, called the *aspect ratio*, does not exceed about 8:1 (Olson and Wright, 1990), although some references state that aspect ratios must be less than 4:1 (Potter and Wiggert, 2001).

EXAMPLE 2.7

Water flows through a rectangular concrete culvert of width 2 m and depth 1 m. If the length of the culvert is 10 m and the flowrate is 6 m³/s, estimate the head loss through the culvert. Assume that the culvert flows full.

Solution   The head loss can be calculated using Equation 2.80. The hydraulic radius, R, is given by

$$R = \frac{A}{P} = \frac{(2)(1)}{2(2 + 1)} = 0.333 \text{ m}$$

and the mean velocity, V, is given by

$$V = \frac{Q}{A} = \frac{6}{(2)(1)} = 3 \text{ m/s}$$

At $20°C$, $\nu = 1.00 \times 10^{-6}$ m²/s, and therefore the Reynolds number, Re, is given by

$$\text{Re} = \frac{V(4R)}{\nu} = \frac{(3)(4 \times 0.333)}{1.00 \times 10^{-6}} = 4.00 \times 10^6$$

A median equivalent sand roughness for concrete can be taken as $k_s = 1.6$ mm (Table 2.1), and therefore the relative roughness, $k_s/4R$, is given by

$$\frac{k_s}{4R} = \frac{1.6 \times 10^{-3}}{4(0.333)} = 0.00120$$

Substituting Re and $k_s/4R$ into the Jain equation (Equation 2.39) for the friction factor gives

$$f = \frac{0.25}{\left[\log \left(\frac{ks/4R}{3.7} + \frac{5.74}{\text{Re}^{0.9}} \right) \right]^2}$$

$$= \frac{0.25}{\left[\log \left(\frac{0.00120}{3.7} + \frac{5.74}{(4.00 \times 10^6)^{0.9}} \right) \right]^2}$$

which yields

$$f = 0.0206$$

The frictional head loss in the culvert, h_f, is therefore given by the Darcy–Weisbach equation as

$$h_f = \frac{fL}{4R} \frac{V^2}{2g} = \frac{(0.0206)(10)}{(4 \times 0.333)} \frac{3^2}{2(9.81)} = 0.0709 \text{ m}$$

The head loss in the culvert can therefore be estimated as 7.1 cm.

Empirical friction-loss formulae.   Friction losses in pipelines should generally be calculated using the Darcy–Weisbach equation. However, a minor inconvenience in using this equation to relate the friction loss to the flow velocity results from the dependence of the friction factor on the flow velocity; therefore, the Darcy–Weisbach equation must be solved simultaneously with the Colebrook equation. In modern engineering practice, computer hardware and software make this a very minor inconvenience. In earlier years, however, this was considered a real problem, and various empirical head-loss formulae were developed to relate the head loss directly to the flow velocity. Those most commonly used are the *Hazen–Williams formula* and the *Manning formula*.

The *Hazen–Williams formula* (Williams and Hazen, 1920) is applicable only to the flow of water in pipes and is given by

$$\boxed{V = 0.849 C_H R^{0.63} S_f^{0.54}} \tag{2.82}$$

where V is the flow velocity (in m/s), C_H is the Hazen–Williams roughness coefficient, R is the hydraulic radius (in m), and S_f is the slope of the energy grade line, defined by

$$S_f = \frac{h_f}{L} \tag{2.83}$$

where h_f is the head loss due to friction over a length L of pipe. Values of C_H for a variety of commonly used pipe materials are given in Table 2.2. Solving Equations 2.82 and 2.83 yields the following expression for the frictional head loss:

$$h_f = 6.82 \frac{L}{D^{1.17}} \left(\frac{V}{C_H} \right)^{1.85} \tag{2.84}$$

TABLE 2.2: Pipe Roughness Coefficients

Pipe material	C_H		n	
	Range	Typical	Range	Typical
Ductile and cast iron:				
New, unlined	120–140	130	—	0.013
Old, unlined	40–100	80	—	0.025
Cement lined and seal coated	100–140	120	0.011–0.015	0.013
Steel:				
Welded and seamless	80–150	120	—	0.012
Riveted	—	110	0.012–0.018	0.015
Mortar lining	120–145	130	—	—
Asbestos cement		140	—	0.011
Concrete	100–140	120	0.011–0.015	0.012
Vitrified clay pipe (VCP)	—	110	0.012–0.014	—
Polyvinyl chloride (PVC)	135–150	140	0.007–0.011	0.009
Corrugated metal pipe (CMP)	—	—	—	0.025

Sources: Velon and Johnson (1993); Wurbs and James (2002).

where D is the diameter of the pipe. The Hazen–Williams equation is applicable to the flow of water at $16°C$ in pipes with diameters between 50 mm and 1850 mm, and flow velocities less than 3 m/s (Mott, 1994). Outside of these conditions, use of the Hazen–Williams equation is strongly discouraged. To further support these quantitative limitations, Street and colleagues (1996) and Liou (1998) have shown that the Hazen–Williams coefficient has a strong Reynolds number dependence, and is mostly applicable where the pipe is relatively smooth and in the early part of its transition to rough flow. Furthermore, Jain and colleagues (1978) have shown that an error of up to 39% can be expected in the evaluation of the velocity by the Hazen–Williams formula over a wide range of diameters and slopes. In spite of these cautionary notes, the Hazen–Williams formula is frequently used in the United States for the design of large water-supply pipes without regard to its limited range of applicability, a practice that can have very detrimental effects on pipe design and could potentially lead to litigation (Bombardelli and García, 2003). In some cases, engineers have calculated correction factors for the Hazen–Williams roughness coefficient to account for these errors (Valiantzas, 2005).

A second empirical formula that is sometimes used to describe flow in pipes is the Manning formula, which is given by

$$V = \frac{1}{n}R^{\frac{2}{3}}S_f^{\frac{1}{2}} \qquad (2.85)$$

where V, R, and S_f have the same meaning and units as in the Hazen–Williams formula, and n is the Manning roughness coefficient. Values of n for a variety of commonly used pipe materials are given in Table 2.2. Solving Equations 2.85 and 2.83 yields the following expression for the frictional head loss:

$$h_f = 6.35 \frac{n^2 L V^2}{D^{\frac{4}{3}}} \qquad (2.86)$$

The Manning formula applies only to rough turbulent flows, where the frictional head losses are controlled by the relative roughness. Such conditions are delineated by Equation 2.37.

EXAMPLE 2.8

Water flows at a velocity of 1 m/s in a 150-mm diameter new ductile iron pipe. Estimate the head loss over 500 m using: (a) the Hazen–Williams formula, (b) the Manning formula, and (c) the Darcy–Weisbach equation. Compare your results and assess the validity of each head-loss equation.

Solution

(a) The Hazen–Williams roughness coefficient, C_H, can be taken as 130 (Table 2.2), $L = 500$ m, $D = 0.150$ m, $V = 1$ m/s, and therefore the head loss, h_f, is given by Equation 2.84 as

$$h_f = 6.82 \frac{L}{D^{1.17}} \left(\frac{V}{C_H} \right)^{1.85} = 6.82 \frac{500}{(0.15)^{1.17}} \left(\frac{1}{130} \right)^{1.85} = 3.85 \text{ m}$$

(b) The Manning roughness coefficient, n, can be taken as 0.013 (approximation from Table 2.2), and therefore the head loss, h_f, is given by Equation 2.86 as

$$h_f = 6.35 \frac{n^2 L V^2}{D^{\frac{4}{3}}} = 6.35 \frac{(0.013)^2 (500)(1)^2}{(0.15)^{\frac{4}{3}}} = 6.73 \text{ m}$$

(c) The equivalent sand roughness, k_s, can be taken as 0.26 mm (Table 2.1), and the Reynolds number, Re, is given by

$$\text{Re} = \frac{VD}{\nu} = \frac{(1)(0.15)}{1.00 \times 10^{-6}} = 1.5 \times 10^5$$

where $\nu = 1.00 \times 10^{-6}$ m^2/s at 20°C. Substituting k_s, D, and Re into the Colebrook equation yields the friction factor, f, where

$$\frac{1}{\sqrt{f}} = -2 \log \left[\frac{k_s}{3.7D} + \frac{2.51}{\text{Re}\sqrt{f}} \right] = -2 \log \left[\frac{0.26}{3.7(150)} + \frac{2.51}{1.5 \times 10^5 \sqrt{f}} \right]$$

which yields

$$f = 0.0238$$

The head loss, h_f, is therefore given by the Darcy–Weisbach equation as

$$h_f = f \frac{L}{D} \frac{V^2}{2g} = 0.0238 \frac{500}{0.15} \frac{1^2}{2(9.81)} = 4.04 \text{ m}$$

It is reasonable to assume that the Darcy–Weisbach equation yields the most accurate estimate of the head loss. In this case, the Hazen–Williams formula gives a head loss 5% less than the Darcy–Weisbach equation, and the Manning formula yields a head loss 67% higher than the Darcy–Weisbach equation.

From the given data, Re = 1.5×10^5, D/k_s = 150/0.26 = 577, and Equation 2.37 gives the limit of rough turbulent flow as

$$\frac{1}{\sqrt{f}} = \frac{\text{Re}}{200(D/k_s)} = \frac{1.5 \times 10^5}{200(577)} \quad \rightarrow \quad f = 0.591$$

Since the actual friction factor (=0.0238) is much less than the minimum friction factor for rough turbulent flow (=0.591), the flow is not in the rough turbulent regime and the Manning equation is not valid. Since the pipe diameter (=150 mm) is between 50 mm and 1850 mm, and the velocity (=1 m/s) is less than 3 m/s, the Hazen–Williams formula is valid. The Darcy–Weisbach equation is unconditionally valid. Given these results, it is not surprising that the Darcy–Weisbach and Hazen–Williams formulae are in close agreement, with the Manning equation giving a significantly different result. These results indicate why application of the Manning equation to closed-conduit flows is strongly discouraged.

2.3 Pipe Networks

Pipe networks are commonly encountered in the context of water-distribution systems. The performance criteria of these systems are typically specified in terms of minimum flow rates and pressure heads that must be maintained at the specified points in the network. Analyses of pipe networks are usually within the context of: (1) designing a new network, (2) designing a modification to an existing network, and/or (3) evaluating the reliability of an existing or proposed network. The procedure for analyzing a pipe network usually aims at finding the flow distribution within the network, with the pressure distribution being derived from the flow distribution using the energy equation. A typical pipe network is illustrated in Figure 2.9, where the boundary conditions consist of inflows, outflows, and constant-head boundaries such as storage reservoirs. Inflows are typically from water-treatment facilities, and outflows from consumer withdrawals or fires. All outflows are assumed to occur at network junctions.

The basic equations to be satisfied in pipe networks are the continuity and energy equations. The continuity equation requires that, at each node in the network, the sum of the outflows is equal to the sum of the inflows. This requirement is expressed

FIGURE 2.9: Typical pipe network

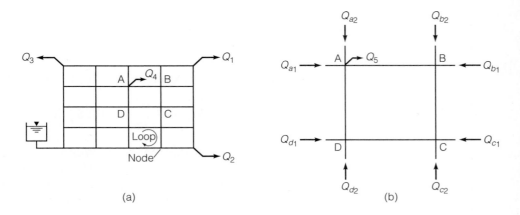

(a)   (b)

by the relation

$$\sum_{i=1}^{NP(j)} Q_{ij} - F_j = 0, \qquad j = 1, \ldots, NJ \tag{2.87}$$

where $NP(j)$ is the number of pipes meeting at junction j; Q_{ij} is the flowrate in pipe i at junction j (inflows positive); F_j is the external flow rate (outflows positive) at junction j; and NJ is the total number of junctions in the network. The energy equation requires that the heads at each of the nodes in the pipe network be consistent with the head losses in the pipelines connecting the nodes. There are two principal methods of calculating the flows in pipe networks: the nodal method and the loop method. In the nodal method, the energy equation is expressed in terms of the heads at the network nodes, while in the loop method the energy equation is expressed in terms of the flows in closed loops within the pipe network.

2.3.1 Nodal Method

In the nodal method, the energy equation is written for each pipeline in the network as

$$h_2 = h_1 - \left(\frac{fL}{D} + \sum K_m\right)\frac{Q|Q|}{2gA^2} + \frac{Q}{|Q|}h_p \tag{2.88}$$

where h_2 and h_1 are the heads at the upstream and downstream ends of a pipe; the terms in parentheses measure the friction loss and local losses, respectively; and h_p is the head added by pumps in the pipeline. The energy equation given by Equation 2.88 has been modified to account for the fact that the flow direction is in many cases unknown, in which case a positive flow direction in each pipeline must be assumed, and a consistent set of energy equations stated for the entire network. Equation 2.88 assumes that the positive flow direction is from node 1 to node 2. Application of the nodal method in practice is usually limited to relatively simple networks.

EXAMPLE 2.9

The high-pressure ductile-iron pipeline shown in Figure 2.10 becomes divided at point B and rejoins at point C. The pipeline characteristics are given in the following tables.

Pipe	Diameter (mm)	Length (m)
1	750	500
2	400	600
3	500	650
4	700	400

Location	Elevation (m)
A	5.0
B	4.5
C	4.0
D	3.5

FIGURE 2.10: Pipe network

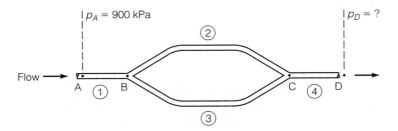

If the flowrate in Pipe 1 is 2 m³/s and the pressure at point A is 900 kPa, calculate the pressure at point D. Assume that the flows are fully turbulent in all pipes.

Solution   The equivalent sand roughness, k_s, of ductile-iron pipe is 0.26 mm, and the pipe and flow characteristics are as follows:

Pipe	Area (m²)	Velocity (m/s)	k_s/D	f
1	0.442	4.53	0.000347	0.0154
2	0.126	—	0.000650	0.0177
3	0.196	—	0.000520	0.0168
4	0.385	5.20	0.000371	0.0156

where it has been assumed that the flows are fully turbulent. Taking $\gamma = 9.79$ kN/m³, the head at location A, h_A, is given by

$$h_A = \frac{p_A}{\gamma} + \frac{V_1^2}{2g} + z_A = \frac{900}{9.79} + \frac{4.53^2}{(2)(9.81)} + 5 = 98.0 \text{ m}$$

and the energy equations for each pipe are as follows

Pipe 1: $\quad h_B = h_A - \dfrac{f_1 L_1}{D_1} \dfrac{V_1^2}{2g} = 98.0 - \dfrac{(0.0154)(500)}{0.75} \dfrac{4.53^2}{(2)(9.81)}$

$\qquad\qquad = 87.3 \text{ m}$ $\hfill (2.89)$

Pipe 2: $\quad h_C = h_B - \dfrac{f_2 L_2}{D_2} \dfrac{Q_2^2}{2g A_2^2} = 87.3 - \dfrac{(0.0177)(600)}{0.40} \dfrac{Q_2^2}{(2)(9.81)(0.126)^2}$

$\qquad\qquad = 87.3 - 85.2 Q_2^2$ $\hfill (2.90)$

Pipe 3: $\quad h_C = h_B - \dfrac{f_3 L_3}{D_3} \dfrac{Q_3^2}{2g A_3^2} = 87.3 - \dfrac{(0.0168)(650)}{0.50} \dfrac{Q_3^2}{(2)(9.81)(0.196)^2}$

$\qquad\qquad = 87.3 - 29.0 Q_3^2$ $\hfill (2.91)$

Pipe 4: $\quad h_D = h_C - \dfrac{f_4 L_4}{D_4} \dfrac{Q_4^2}{2g A_4^2} = h_C - \dfrac{(0.0156)(400)}{0.70} \dfrac{Q_4^2}{(2)(9.81)(0.385)^2}$

$\qquad\qquad = h_C - 3.07 Q_4^2$ $\hfill (2.92)$

and the continuity equations at the two pipe junctions are

$$\text{Junction B:} \quad Q_2 + Q_3 = 2 \text{ m}^3/\text{s} \tag{2.93}$$

$$\text{Junction C:} \quad Q_2 + Q_3 = Q_4 \tag{2.94}$$

Equations 2.90 to 2.94 are five equations in five unknowns: h_C, h_D, Q_2, Q_3, and Q_4. Equations 2.93 and 2.94 indicate that

$$Q_4 = 2 \text{ m}^3/\text{s}$$

Combining Equations 2.90 and 2.91 leads to

$$87.3 - 85.2Q_2^2 = 87.3 - 29.0Q_3^2$$

and therefore

$$Q_2 = 0.583Q_3 \tag{2.95}$$

Substituting Equation 2.95 into Equation 2.93 yields

$$2 = (0.583 + 1)Q_3$$

or

$$Q_3 = 1.26 \text{ m}^3/\text{s}$$

and from Equation 2.95

$$Q_2 = 0.74 \text{ m}^3/\text{s}$$

According to Equation 2.91

$$h_C = 87.3 - 29.0Q_3^2 = 87.3 - 29.0(1.26)^2 = 41.3 \text{ m}$$

and Equation 2.92 gives

$$h_D = h_C - 3.07Q_4^2 = 41.3 - 3.07(2)^2 = 29.0 \text{ m}$$

Therefore, since the total head at D, h_D, is equal to 29.0 m, then

$$29.0 = \frac{p_D}{\gamma} + \frac{V_4^2}{2g} + z_D = \frac{p_D}{9.79} + \frac{5.20^2}{(2)(9.81)} + 3.5$$

which yields

$$p_D = 236 \text{ kPa}$$

Therefore, the pressure at location D is 236 kPa.

   This problem has been solved by assuming that the flows in all pipes are fully turbulent. This is generally not known a priori, and therefore a complete solution would require repeating the calculations until the assumed friction factors are consistent with the calculated flowrates.

2.3.2 Loop Method

In the loop method, the energy equation is written for each loop of the network, in which case the algebraic sum of the head losses within each loop is equal to zero. This requirement is expressed by the relation

$$\sum_{j=1}^{NP(i)} (h_{L,ij} - h_{p,ij}) = 0, \qquad i = 1, \ldots, NL \tag{2.96}$$

where $NP(i)$ is the number of pipes in loop i, $h_{L,ij}$ is the head loss in pipe j of loop i, $h_{p,ij}$ is the head added by any pumps that may exist in line ij, and NL is the number of loops in the network. Combining Equations 2.87 and 2.96 with an expression for calculating the head losses in pipes, such as the Darcy–Weisbach equation, and the pump characteristic curves, which relate the head added by the pump to the flowrate through the pump, yields a complete mathematical description of the flow problem. Solution of this system of flow equations is complicated by the fact that the equations are nonlinear, and numerical methods must be used to solve for the flow distribution in the pipe network.

Hardy Cross method. The Hardy Cross method (Cross, 1936) is a simple technique for hand solution of the loop system of equations governing flow in pipe networks. This iterative method was developed before the advent of computers, and much more efficient algorithms are now used for numerical computations. In spite of this, the Hardy Cross method is presented here to illustrate the iterative solution of the loop equations in pipe networks. The Hardy Cross method assumes that the head loss, h_L, in each pipe is proportional to the discharge, Q, raised to some power n, in which case

$$h_L = rQ^n \tag{2.97}$$

where typical values of n range from 1 to 2, where $n = 1$ corresponds to viscous flow and $n = 2$ to fully turbulent flow. The proportionality constant, r, depends on which head-loss equation is used and the types of losses in the pipe. Clearly, if all head losses are due to friction and the Darcy–Weisbach equation is used to calculate the head losses, then r is given by

$$r = \frac{fL}{2gA^2D} \tag{2.98}$$

and $n = 2$. If the flow in each pipe is approximated as $\hat{Q}$, and ΔQ is the error in this estimate, then the actual flowrate, Q, is related to $\hat{Q}$ and ΔQ by

$$Q = \hat{Q} + \Delta Q \tag{2.99}$$

and the head loss in each pipe is given by

$$
\begin{aligned}
h_L &= rQ^n \\
&= r(\hat{Q} + \Delta Q)^n \\
&= r\left[\hat{Q}^n + n\hat{Q}^{n-1}\Delta Q + \frac{n(n-1)}{2}\hat{Q}^{n-2}(\Delta Q)^2 + \cdots + (\Delta Q)^n\right]
\end{aligned} \tag{2.100}
$$

If the error in the flow estimate, ΔQ, is small, then the higher-order terms in ΔQ can be neglected and the head loss in each pipe can be approximated by

$$h_L \approx r\hat{Q}^n + rn\hat{Q}^{n-1}\Delta Q \tag{2.101}$$

This relation approximates the head loss in the flow direction. However, in working with pipe networks, it is required that the algebraic sum of the head losses in any loop of the network (see Figure 2.9) must be equal to zero. We must therefore define a positive flow direction (such as clockwise), and count head losses as positive in pipes when the flow is in the positive direction and negative when the flow is opposite to the selected positive direction. Under these circumstances, the sign of the head loss must be the same as the sign of the flow direction. Further, when the flow is in the positive direction, positive values of ΔQ require a positive correction to the head loss; when the flow is in the negative direction, positive values in ΔQ also require a positive correction to the calculated head loss. To preserve the algebraic relation among head loss, flow direction, and flow error (ΔQ), Equation 2.101 for each pipe can be written as

$$h_L = r\hat{Q}|\hat{Q}|^{n-1} + rn|\hat{Q}|^{n-1}\Delta Q \tag{2.102}$$

where the approximation has been replaced by an equal sign. On the basis of Equation 2.102, the requirement that the algebraic sum of the head losses around each loop be equal to zero can be written as

$$\sum_{j=1}^{NP(i)} r_{ij}Q_j|Q_j|^{n-1} + \Delta Q_i \sum_{j=1}^{NP(i)} r_{ij}n|Q_j|^{n-1} = 0, \qquad i = 1,\ldots,NL \tag{2.103}$$

where $NP(i)$ is the number of pipes in loop i, r_{ij} is the head-loss coefficient in pipe j (in loop i), Q_j is the estimated flow in pipe j, ΔQ_i is the flow correction for the pipes in loop i, and NL is the number of loops in the entire network. The approximation given by Equation 2.103 assumes that there are no pumps in the loop, and that the flow correction, ΔQ_i, is the same for each pipe in each loop. Solving Equation 2.103 for ΔQ_i leads to

$$\Delta Q_i = -\frac{\sum_{j=1}^{NP(i)} r_{ij}Q_j|Q_j|^{n-1}}{\sum_{j=1}^{NP(i)} nr_{ij}|Q_j|^{n-1}} \tag{2.104}$$

This equation forms the basis of the Hardy Cross method.

The steps to be followed in using the Hardy Cross method to calculate the flow distribution in pipe networks are:

1. Assume a reasonable distribution of flows in the pipe network. This assumed flow distribution must satisfy continuity.

2. For each loop, i, in the network, calculate the quantities $r_{ij}Q_j|Q_j|^{n-1}$ and $nr_{ij}|Q_j|^{n-1}$ for each pipe in the loop. Calculate the flow correction, ΔQ_i, using Equation 2.104. Add the correction algebraically to the estimated flow in each pipe. [*Note*: Values of r_{ij} occur in both the numerator and denominator of Equation 2.104; therefore, values proportional to the actual r_{ij} may be used to calculate ΔQ_i.]

3. Repeat step 2 until the corrections (ΔQ_i) are acceptably small.

The application of the Hardy Cross method is best demonstrated by an example.

EXAMPLE 2.10

Compute the distribution of flows in the pipe network shown in Figure 2.11(a), where the head loss in each pipe is given by

$$h_L = rQ^2$$

and the relative values of r are shown in Figure 2.11(a). The flows are taken as dimensionless for the sake of illustration.

Solution   The first step is to assume a distribution of flows in the pipe network that satisfies continuity. The assumed distribution of flows is shown in Figure 2.11(b), along with the positive-flow directions in each of the two loops. The flow correction for each loop is calculated using Equation 2.104. Since $n = 2$ in this case, the flow correction formula becomes

$$\Delta Q_i = -\frac{\sum_{j=1}^{NP(i)} r_{ij}Q_j|Q_j|}{\sum_{j=1}^{NP(i)} 2r_{ij}|Q_j|}$$

The calculation of the numerator and denominator of this flow correction formula for loop I is tabulated as follows:

Loop	Pipe	Q	$rQ|Q|$	$2r|Q|$
I	4–1	70	29,400	840
	1–3	35	3675	210
	3–4	−30	−4500	300
			28,575	1350

The flow correction for loop I, ΔQ_I, is therefore given by

$$\Delta Q_I = -\frac{28,575}{1350} = -21.2$$

and the corrected flows are

Loop	Pipe	Q
I	4–1	48.8
	1–3	13.8
	3–4	−51.2

Moving to loop II, the calculation of the numerator and denominator of the flow correction formula for loop II is given by

FIGURE 2.11: Flows in pipe network

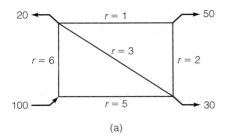

(a)

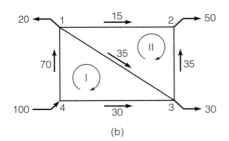

(b)

Loop	Pipe	Q	$rQ|Q|$	$2r|Q|$
II	1–2	15	225	30
	2–3	−35	−2450	140
	3–1	−13.8	−574	83
			−2799	253

The flow correction for loop II, ΔQ_{II}, is therefore given by

$$\Delta Q_{II} = -\frac{-2799}{253} = 11.1$$

and the corrected flows are

Loop	Pipe	Q
II	1–2	26.1
	2–3	−23.9
	3–1	−2.7

This procedure is repeated in the following table until the calculated flow corrections do not affect the calculated flows, to the level of significant digits retained in the calculations.

Iteration	Loop	Pipe	Q	$rQ|Q|$	$2r|Q|$	ΔQ	Corrected Q
2	I	4–1	48.8	14,289	586		47.7
		1–3	2.7	22	16		1.6
		3–4	−51.2	−13,107	512		−52.3
				1204	1114	−1.1	
	II	1–2	26.1	681	52		29.1
		2–3	−23.9	−1142	96		−20.9
		3–1	−1.6	−8	10		1.4
				−469	157	3.0	
3	I	4–1	47.7	13,663	573		47.7
		1–3	1.4	6	8		1.4
		3–4	−52.3	−13,666	523		−52.3
				3	1104	0.0	
	II	1–2	29.1	847	58		29.2
		2–3	−20.9	−874	84		−20.8
		3–1	1.4	6	8		1.5
				−21	150	0.1	
4	I	4–1	47.7	13,662	573		47.7
		1–3	1.5	7	9		1.5
		3–4	−52.3	−13,668	523		−52.3
				1	1104	0.0	
	II	1–2	29.2	853	58		29.2
		2–3	−20.8	−865	83		−20.8
		3–1	1.5	7	9		1.5
				−5	150	0.0	

The final flow distribution, after four iterations, is given by

Pipe	Q
1–2	29.2
2–3	−20.8
3–4	−52.3
4–1	47.7
1–3	−1.5

It is clear that the final results are fairly close to the flow estimates after only one iteration.

As the above example illustrates, complex pipe networks can generally be treated as a combination of simple loops, with each balanced in turn until compatible flow conditions exist in all loops. Typically, after the flows have been computed for all pipes in a network, the elevation of the hydraulic grade line and the pressure are computed for each junction node. These pressures are then assessed relative to acceptable operating pressures.

2.3.3 Practical Considerations

In practice, analyses of complex pipe networks are usually done using computer programs that solve the system of continuity and energy equations that govern the flows in the network pipelines. These computer programs, such as EPANET (Rossman, 2000), generally use algorithms that are computationally more efficient than the Hardy Cross method, such as the linear theory method, the Newton–Raphson method, and the gradient algorithm (Lansey and Mays, 1999).

The methods described in this text for computing steady-state flows and pressures in water distribution systems are useful for assessing the performance of systems under normal operating conditions. Sudden changes in flow conditions, such as pump shutdown/startup and valve opening/closing, cause hydraulic transients that can produce significant increases in water pressure—a phenomenon called *water hammer*. The analysis of transient conditions requires a computer program to perform a numerical solution of the one-dimensional continuity and momentum equation for flow in pipelines, and is an essential component of water-distribution system design (Wood, 2005d). Transient conditions will be most severe at pump stations and control valves, high-elevation areas, locations with low static pressures, and locations that are far from elevated storage reservoirs (Friedman, 2003). Appurtenances used to mitigate the effects of water hammer include valves that prevent rapid closure, pressure-relief valves, surge tanks, and air chambers. Detailed procedures for transient analysis in pipeline systems can be found in Martin (2000).

2.4 Pumps

Pumps are hydraulic machines that convert mechanical energy to fluid energy. They can be classified into two main categories: (1) positive displacement pumps, and (2) rotodynamic or kinetic pumps. Positive displacement pumps deliver a fixed quantity of fluid with each revolution of the pump rotor, such as with a piston or cylinder, while rotodynamic pumps add energy to the fluid by accelerating it through the action of a

FIGURE 2.12: Types of pumps

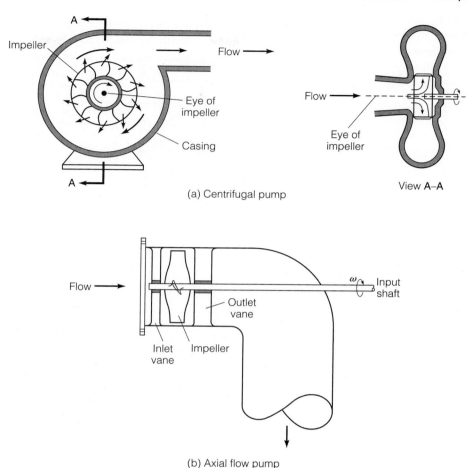

(a) Centrifugal pump

View **A–A**

(b) Axial flow pump

rotating impeller. Rotodynamic pumps are far more common in engineering practice and will be the focus of this section.

Three types of rotodynamic pumps commonly encountered are centrifugal pumps, axial-flow pumps, and mixed-flow pumps. In centrifugal pumps, the flow enters the pump chamber along the axis of the impeller and is discharged radially by centrifugal action, as illustrated in Figure 2.12(a). In axial-flow pumps, the flow enters and leaves the pump chamber along the axis of the impeller, as shown in Figure 2.12(b). In mixed-flow pumps, outflows have both radial and axial components. Typical centrifugal and axial-flow pump installations are illustrated in Figure 2.13. Key components of the centrifugal pump are a foot valve installed in the suction pipe to prevent water from leaving the pump when it is stopped and a check valve in the discharge pipe to prevent backflow if there is a power failure. If the suction line is empty prior to starting the pump, then the suction line must be *primed* (filled) prior to startup. Unless the water is known to be very clean, a strainer should be installed at the inlet to the suction piping. The pipe size of the suction line should never be smaller than the inlet connection on the pump; if a reducer is required, it should be of the eccentric type, since concentric reducers place part of the supply pipe above the

FIGURE 2.13: Centrifugal and axial-flow pump installations

Source: Finnemore and Franzini (2002).

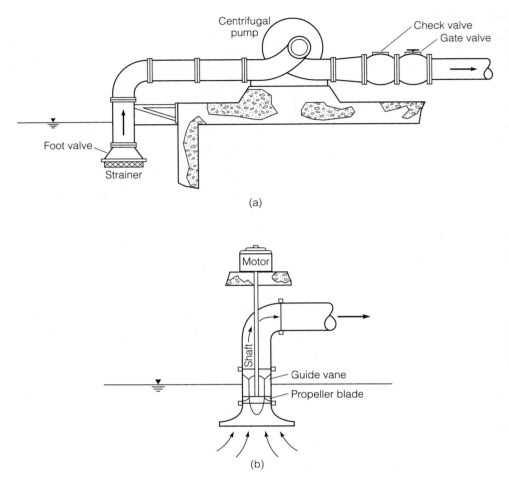

pump inlet where an air pocket could form. The discharge line from the pump should contain a valve close to the pump to allow service or pump replacement.

The pumps illustrated in Figure 2.13 are both *single-stage* pumps, which means that they have only one impeller. In *multistage* pumps, two or more impellers are arranged in series in such a way that the discharge from one impeller enters the eye of the next impeller. If a pump has three impellers in series, it is called a three-stage pump. Multistage pumps are typically used when large pumping heads are required, and are commonly used when extracting water from deep underground sources.

The performance of a pump is measured by the head added by the pump and the pump efficiency. The head added by the pump, h_p, is equal to the difference between the total head on the discharge side of the pump and the total head on the suction side of the pump, and is sometimes referred to as the *total dynamic head*. The efficiency of the pump, η, is defined by

$$\eta = \frac{\text{power delivered to the fluid}}{\text{power supplied to the shaft}} \tag{2.105}$$

Pumps are inefficient for a variety of reasons, such as frictional losses as the fluid moves over the solid surfaces, separation losses, leakage of fluid between the impeller and the casing, and mechanical losses in the bearings and sealing glands of the pump. The pump-performance parameters, h_p and η, can be expressed in terms of the fluid properties and the physical characteristics of the pump by the functional relation

$$gh_p \quad \text{or} \quad \eta = f_1(\rho, \mu, D, \omega, Q) \tag{2.106}$$

where the energy added per unit mass of fluid, gh_p, is used instead of h_p (to remove the effect of gravity), f_1 is an unknown function, ρ and μ are the density and dynamic viscosity of the fluid, respectively, Q is the flowrate through the pump, D is a characteristic dimension of the pump (usually the inlet or outlet diameter), and ω is the angular speed of the pump impeller. Equation 2.106 is a functional relationship between six variables in three dimensions. According to the Buckingham pi theorem, this relationship can be expressed as a relation between three dimensionless groups as follows:

$$\frac{gh_p}{\omega^2 D^2} \quad \text{or} \quad \eta = f_2\left(\frac{Q}{\omega D^3}, \frac{\rho \omega D^2}{\mu}\right) \tag{2.107}$$

where $gh_p/\omega^2 D^2$ is called the *head coefficient*, and $Q/\omega D^3$ is called the *flow coefficient* (Douglas et al., 1995). In most cases, the flow through the pump is fully turbulent and viscous forces are negligible relative to the inertial forces. Under these circumstances, the viscosity of the fluid is neglected and Equation 2.107 becomes

$$\frac{gh_p}{\omega^2 D^2} \quad \text{or} \quad \eta = f_3\left(\frac{Q}{\omega D^3}\right) \tag{2.108}$$

This relationship describes the performance of all (geometrically similar) pumps in which viscous effects are negligible, but the exact form of the function in Equation 2.108 depends on the geometry of the pump. A series of pumps having the same shape (but different sizes) are expected to have the same functional relationships between $gh_p/(\omega^2 D^2)$ and $Q/(\omega D^3)$ as well as η and $Q/(\omega D^3)$. A class of pumps that have the same shape (i.e., are geometrically similar) is called a *homologous series*, and the performance characteristics of a homologous series of pumps are described by curves such as those in Figure 2.14. Pumps are selected to meet specific design conditions and, since the efficiency of a pump varies with the operating condition, it is usually desirable to select a pump that operates at or near the point of maximum efficiency, indicated by the point P in Figure 2.14.

FIGURE 2.14: Performance curves of a homologous series of pumps.

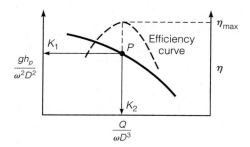

The point of maximum efficiency of a pump is commonly called the *best-efficiency point* (BEP), and sometimes the *nameplate* or *design point*. Maintaining operation near the BEP will allow a pump to function for years with little maintenance, and as the operating point moves away from the BEP, pump thrust and radial loads increase, which increases the wear on the pump bearings and shaft. For these reasons, it is generally recommended that pump operation should be maintained between 70% and 130% of the BEP flowrate (Lansey and El-Shorbagy, 2001). At the BEP in Figure 2.14,

$$\frac{gh_p}{\omega^2 D^2} = K_1 \quad \text{and} \quad \frac{Q}{\omega D^3} = K_2 \tag{2.109}$$

Eliminating D from these equations yields

$$\frac{\omega Q^{\frac{1}{2}}}{(gh_p)^{\frac{3}{4}}} = \sqrt{\frac{K_2}{K_1^{\frac{3}{2}}}} \tag{2.110}$$

The term on the righthand side of this equation is a constant for a homologous series of pumps and is denoted by the *specific speed*, n_s, defined by

$$n_s = \frac{\omega Q^{\frac{1}{2}}}{(gh_p)^{\frac{3}{4}}} \tag{2.111}$$

where any consistent set of units can be used. The specific speed, n_s, is also called the *shape number* (Hwang and Houghtalen, 1996; Wurbs and James, 2002) or the *type number* (Douglas et al., 2001). In SI units, ω is in rad/s, Q in m³/s, g in m/s², and h_p in meters. The most efficient operating point for a homologous series of pumps is therefore specified by the specific speed. This nomenclature is somewhat unfortunate, since the specific speed is dimensionless and hence does not have units of speed. It is common practice in the United States to define the specific speed by N_s, as

$$\boxed{N_s = \frac{\omega Q^{\frac{1}{2}}}{h_p^{\frac{3}{4}}}} \tag{2.112}$$

where N_s is not dimensionless, ω is in revolutions per minute (rpm), Q is in gallons per minute (gpm), and h_p is in feet (ft). Although N_s has dimensions, the units are seldom stated in practice. The required pump operating point gives the flowrate, Q, and head, h_p, required from the pump; the rotational speed, ω, is determined by the synchronous speeds of available motors; and the specific speed calculated from the required operating point is the basis for selecting the appropriate pump. Since the specific speed is independent of the size of a pump, and all homologous pumps (of varying sizes) have the same specific speed, then the calculated specific speed at the desired operating point indicates the type of pump that must be selected to ensure optimal efficiency.

The types of pump that give the maximum efficiency for given specific speeds, n_s, are listed in Table 2.3 along with typical flowrates delivered by the pumps. Table 2.3 indicates that centrifugal pumps have low specific speeds, $n_s < 1.5$; mixed-flow pumps have medium specific speeds, $1.5 < n_s < 3.7$; and axial-flow pumps have high specific

TABLE 2.3: Pump Selection Guidelines

Type of pump	Range of specific speeds, n_s*	Typical flowrates (L/s)	Typical efficiencies (%)
Centrifugal	0.15–1.5 (400–4000)	<60	70–94
Mixed flow	1.5–3.7 (4000–10,000)	60–300	90–94
Axial flow	3.7–5.5 (10,000–15,000)	>300	84–90

*The specific speeds in parentheses correspond to N_s given by Equation 2.112, with ω in rpm, Q in gpm, and h_p in ft.

speeds, $n_s > 3.7$. This indicates that centrifugal pumps are most efficient at delivering low flows at high heads, while axial flow pumps are most efficient at delivering high flows at low heads. The efficiencies of radial-flow (centrifugal) pumps increase with increasing specific speed, while the efficiencies of mixed-flow and axial-flow pumps decrease with increasing specific speed. Pumps with specific speeds less than 0.3 tend to be inefficient (Finnemore and Franzini, 2002). Since axial-flow pumps are most efficient at delivering high flows at low heads, this type of pump is commonly used to move large volumes of water through major canals, and an example of this application is shown in Figure 2.15, where there are three axial-flow pumps operating in parallel, and these pumps are driven by motors housed in the pump station.

Most pumps are driven by standard electric motors. The standard speed of AC synchronous induction motors at 60 cycles and 220 to 440 volts is given by

$$\text{Synchronous speed (rpm)} = \frac{3600}{\text{no. of pairs of poles}} \qquad (2.113)$$

A common problem is that, for the motor speed chosen, the calculated specific speed does not exactly equal the specific speed of available pumps. In these cases, it is recommended to choose a pump with a specific speed that is close to and greater than the required specific speed. In rare cases, a new pump may be designed to meet the design conditions exactly; however, this is usually very costly and only justified for very large pumps.

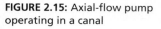

FIGURE 2.15: Axial-flow pump operating in a canal

2.4.1 Affinity Laws

The performance curves for a homologous series of pumps are illustrated in Figure 2.14. Any two pumps in the homologous series are expected to operate at the same efficiency when

$$\left(\frac{Q}{\omega D^3}\right)_1 = \left(\frac{Q}{\omega D^3}\right)_2 \quad \text{and} \quad \left(\frac{h_p}{\omega^2 D^2}\right)_1 = \left(\frac{h_p}{\omega^2 D^2}\right)_2 \qquad (2.114)$$

These relationships are sometimes called the *affinity laws for homologous pumps*. An affinity law for the power delivered to the fluid, P, can be derived from the affinity relations given in Equation 2.114, since P is defined by

$$P = \gamma Q h_p \qquad (2.115)$$

which leads to the following derived affinity relation:

$$\left(\frac{P}{\omega^3 D^5}\right)_1 = \left(\frac{P}{\omega^3 D^5}\right)_2 \qquad (2.116)$$

In accordance with the dimensional analysis of pump performance, Equation 2.107, the affinity laws for scaling pump performance are valid as long as viscous effects are negligible. The effect of viscosity is measured by the Reynolds number, Re, defined by

$$\mathrm{Re} = \frac{\rho \omega D^2}{\mu} \qquad (2.117)$$

and scale effects are negligible when $\mathrm{Re} > 3 \times 10^5$ (Gerhart et al., 1992). In lieu of stating a Reynolds number criterion for scale effects to be negligible, it is sometimes stated that larger pumps are more efficient than smaller pumps and that the scale effect on efficiency is given by (Moody and Zowski, 1969; Stepanoff, 1957)

$$\frac{1 - \eta_2}{1 - \eta_1} = \left(\frac{D_1}{D_2}\right)^{\frac{1}{4}} \qquad (2.118)$$

where η_1 and η_2 are the efficiencies of homologous pumps of diameters D_1 and D_2, respectively. The effect of changes in flowrate on efficiency can be estimated using the relation

$$\frac{0.94 - \eta_2}{0.94 - \eta_1} = \left(\frac{Q_1}{Q_2}\right)^{0.32} \qquad (2.119)$$

where Q_1 and Q_2 are corresponding homologous flowrates.

EXAMPLE 2.11

A pump with a 1200-rpm motor has a performance curve of

$$h_p = 12 - 0.1Q^2$$

where h_p is in meters and Q is in cubic meters per minute. If the speed of the motor is changed to 2400 rpm, estimate the new performance curve.

Solution   The performance characteristics of a homologous series of pumps is given by

$$\frac{gh_p}{\omega^2 D^2} = f\left(\frac{Q}{\omega D^3}\right)$$

For a fixed pump size, D, for two different motor speeds, ω_1 and ω_2, the general performance curve can be written as

$$\frac{h_1}{\omega_1^2 D^2} = f\left(\frac{Q_1}{\omega_1 D^3}\right)$$

and

$$\frac{h_2}{\omega_2^2 D^2} = f\left(\frac{Q_2}{\omega_2 D^3}\right)$$

where h_1 and Q_1 are the head added by the pump and the flowrate, respectively, when the speed is ω_1; and h_2 and Q_2 are the head and flowrate when the speed is ω_2. Since the pumps are part of a homologous series, then, neglecting scale effects, the function f is fixed, and therefore when (for a fixed D)

$$\frac{Q_1}{\omega_1} = \frac{Q_2}{\omega_2}$$

then

$$\frac{h_1}{\omega_1^2} = \frac{h_2}{\omega_2^2}$$

These relations are simply statements of the affinity laws for a fixed pump size, D. In the present case, $\omega_1 = 1200$ rpm and $\omega_2 = 2400$ rpm and the affinity laws give that

$$Q_1 = \frac{\omega_1}{\omega_2} Q_2 = \frac{1200}{2400} Q_2 = 0.5 Q_2$$

$$h_1 = \frac{\omega_1^2}{\omega_2^2} h_2 = \frac{1200^2}{2400^2} h_2 = 0.25 h_2$$

Since the performance curve of the pump at speed ω_1 is given by

$$h_1 = 12 - 0.1 Q_1^2$$

then the performance curve at speed ω_2 is given by

$$0.25 h_2 = 12 - 0.1(0.5 Q_2)^2$$

which leads to

$$h_2 = 48 - 0.1 Q_2^2$$

The performance curve of the pump with a 2400-rpm motor is therefore given by

$$h_p = 48 - 0.1 Q^2$$

2.4.2 Pump Selection

In selecting a pump for any application, consideration must be given to the required pumping rate, commercially available pumps, characteristics of the system in which the pump operates, and the physical limitations associated with pumping water.

Commercially available pumps. Specification of a pump generally requires selection of a manufacturer, model (homologous) series, impeller size, D, and rotational speed, ω. For each model series and rotational speed, pump manufacturers provide a *performance curve* or *characteristic curve* that shows the relationship between the head, h_p, added by the pump and the flowrate, Q, through the pump. A typical example of a set of pump characteristic curves (h_p versus Q) provided by a manufacturer for a homologous series of pumps is shown in Figure 2.16. In this case, the homologous series of pumps (Model 3409) has impeller diameters ranging from 12.1 in. (307 mm) to 17.5 in. (445 mm) with a rotational speed of 885 revolutions per minute. Superimposed on the characteristic curves are constant-efficiency lines for efficiencies ranging from 55% to 86%, and (dashed) isolines of *required net positive suction head*, which is the minimum allowable difference between the pressure head on the suction side of the pump and the pressure head at which water vaporizes (i.e., saturation vapor pressure). In Figure 2.16, the required net positive suction head ranges from 16 ft (4.9 m) for higher flowrates down to approximately zero, which is indicated by a bold line that meets the 55% efficiency contour. Also shown in Figure 2.16, below the characteristic

FIGURE 2.16: Pump performance curve

Source: Goulds Pumps (www.gouldspumps.com).

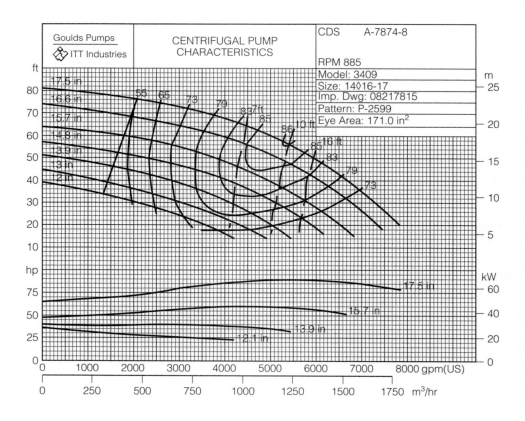

curves, is the power delivered to the pump (in kW) for various flowrates and impeller diameters. This power input to the pump shaft is called the *brake horsepower*.

System characteristics.   The goal in pump selection is to select a pump that operates at a point of maximum efficiency and with a net positive suction head that exceeds the minimum allowable value. Pumps are placed in pipeline systems such as that illustrated in Figure 2.17, in which case the energy equation for the pipeline system requires that the head, h_p, added by the pump is given by

$$h_p = \Delta z + Q^2 \left[\sum \frac{fL}{2gA^2D} + \sum \frac{K_m}{2gA^2} \right] \tag{2.120}$$

where Δz is the difference in elevation between the water surfaces of the source and destination reservoirs, the first term in the square brackets is the sum of the head losses due to friction, and the second term is the sum of the local head losses. Equation 2.120 gives the required relationship between h_p and Q for the pipeline system, and this relationship is commonly called the *system curve*. Because the flowrate and head added by the pump must satisfy both the system curve and the pump characteristic curve, Q and h_p are determined by simultaneous solution of Equation 2.120 and the pump characteristic curve. The resulting (solution) values of Q and h_p identify the *operating point* of the pump. The location of the operating point on the performance curve is illustrated in Figure 2.18.

FIGURE 2.17: Pipeline system

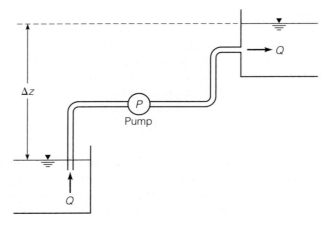

FIGURE 2.18: Operating point in pipeline system

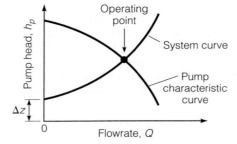

Limits on pump location.   If the absolute pressure on the suction side of a pump falls below the saturation vapor pressure of the fluid, the water will begin to vaporize, and this process of vaporization is called *cavitation*. Cavitation is usually a transient phenomenon that occurs as water enters the low-pressure suction side of a pump and experiences the even lower pressures adjacent to the rotating pump impeller. As the water containing vapor cavities moves toward the high-pressure environment of the discharge side of the pump, the vapor cavities are compressed and ultimately implode, creating small localized high-velocity jets that can cause considerable damage to the pump machinery. Collapsing vapor cavities have been associated with jet velocities on the order of 110 m/s, and pressures of up to 800 MPa when the jets strike a solid wall (Finnemore and Franzini, 2002; Knapp et al., 1970). The damage caused by collapsing vapor cavities usually manifests itself as *pitting* of the metal casing and impeller, reduced pump efficiency, and excessive vibration of the pump. The noise generated by imploding vapor cavities resembles the sound of gravel going through a centrifugal pump. Since the saturation vapor pressure increases with temperature, a system that operates satisfactorily without cavitation during the winter may have problems with cavitation during the summer.

The potential for cavitation is measured by the net positive suction head, NPSH, defined as the difference between the head on the suction side of the pump (at the inlet to the pump) and the head when cavitation begins, hence

$$\text{NPSH} = \left(\frac{p_s}{\gamma} + \frac{V_s^2}{2g} + z_s \right) - \left(\frac{p_v}{\gamma} + z_s \right) = \frac{p_s}{\gamma} + \frac{V_s^2}{2g} - \frac{p_v}{\gamma} \qquad (2.121)$$

where p_s, V_s, and z_s are the pressure, velocity, and elevation of the fluid at the suction side of the pump, and p_v is the saturation vapor pressure of water at the temperature of the fluid. In cases where water is being pumped from a reservoir, the NPSH can be calculated by applying the energy equation between the reservoir and the suction side of the pump; in this case the calculated NPSH is called the *available net positive suction head*, NPSH$_A$, and is given by

$$\text{NPSH}_A = \frac{p_0}{\gamma} - \Delta z_s - h_L - \frac{p_v}{\gamma} \qquad (2.122)$$

where p_0 is the pressure at the surface of the reservoir (usually atmospheric), Δz_s is the difference in elevation between the suction side of the pump and the fluid surface in the source reservoir (called the *suction lift* or *static suction head* or *static head*), h_L is the head loss in the pipeline between the source reservoir and suction side of the pump (including local losses), and p_v is the saturation vapor pressure. In applying either Equation 2.121 or 2.122 to calculate NPSH, care must be taken to use a consistent measure of the pressures, using either gage pressures or absolute pressures. Absolute pressures are usually more convenient, since the vapor pressure is typically given as an absolute pressure. A pump requires a minimum NPSH to prevent the onset of cavitation within the pump, and this minimum NPSH is called the *required net positive suction head*, NPSH$_R$, which is generally supplied by the pump manufacturer.

EXAMPLE 2.12

Water at $20°$C is being pumped from a lower to an upper reservoir through a 200-mm pipe in the system shown in Figure 2.17. The water-surface elevations in the source and destination reservoirs differ by 5.2 m, and the length of the steel pipe ($k_s = 0.046$ mm) connecting the reservoirs is 21.3 m. The pump is to be located 1.5 m above the water surface in the source reservoir, and the length of the pipeline between the source reservoir and the suction side of the pump is 3.5 m. The performance curves of the 885-rpm homologous series of pumps being considered for this system are given in Figure 2.16. If the desired flowrate in the system is 0.315 m³/s (= 5000 gpm), what size and specific-speed pump should be selected? Assess the adequacy of the pump location based on a consideration of the available net positive suction head. The pipe intake loss coefficient can be taken as 0.1.

Solution   For the system pipeline: $L = 21.3$ m, $D = 200$ mm $= 0.2$ m, $k_s = 0.046$ mm, and (neglecting local losses) the energy equation for the system is given by

$$h_p = 5.2 + \frac{fL}{2gA^2D}Q^2 \tag{2.123}$$

where h_p is the head added by the pump (in meters), f is the friction factor, Q is the flowrate through the system (in m³/s), and A is the cross-sectional area of the pipe (in m²) given by

$$A = \frac{\pi}{4}D^2 = \frac{\pi}{4}(0.2)^2 = 0.03142 \text{ m}^2 \tag{2.124}$$

The friction factor, f, can be calculated using the Jain formula (Equation 2.39),

$$f = \frac{0.25}{\left[\log\left(\frac{k_s}{3.7D} + \frac{5.74}{Re^{0.9}}\right)\right]^2} \tag{2.125}$$

where Re is the Reynolds number given by

$$Re = \frac{VD}{\nu} = \frac{QD}{A\nu} \tag{2.126}$$

and $\nu = 1.00 \times 10^{-6}$ m²/s at 20°C. Combining Equations 2.125 and 2.126 with the given data yields

$$f = \frac{0.25}{\left[\log\left(\frac{4.6 \times 10^{-5}}{3.7(0.2)} + \frac{5.74}{\left(\frac{Q(0.2)}{(0.03142)(1.00 \times 10^{-6})}\right)^{0.9}}\right)\right]^2} \tag{2.127}$$

which simplifies to

$$f = \frac{1}{4[\log(6.216 \times 10^{-5} + 4.32 \times 10^{-6}Q^{-0.9})]^2} \tag{2.128}$$

Combining Equations 2.123 and 2.128 gives the following relation:

$$h_p = 5.2 + \frac{(21.3)}{2(9.81)(0.03142)^2(0.2)(4)[\log(6.216 \times 10^{-5} + 4.32 \times 10^{-6}Q^{-0.9})]^2}Q^2$$

$$= 5.2 + \frac{1375Q^2}{[\log(6.216 \times 10^{-5} + 4.32 \times 10^{-6}Q^{-0.9})]^2} \tag{2.129}$$

This relation is applicable for h_p in meters and Q in m^3/s. Since 1 m = 3.281 ft and 1 m^3/s = 15,850 gpm, Equation 2.129 can be put in the form

$$\frac{h_p}{3.281} = 5.2 + \frac{1375\left(\frac{Q}{15,850}\right)^2}{\left[\log\left(6.216 \times 10^{-5} + 4.32 \times 10^{-6}\left(\frac{Q}{15,850}\right)^{-0.9}\right)\right]^2}$$

which simplifies to

$$h_p = 17.1 + \frac{1.79 \times 10^{-5}Q^2}{[\log(6.216 \times 10^{-5} + 7.96 \times 10^{-3}Q^{-0.9})]^2} \tag{2.130}$$

where h_p is in ft and Q in gpm. Equation 2.130 is the "system curve" which relates the head added by the pump to the flowrate through the system, as required by the conservation-of-energy equation. Since the pump characteristic curve must also be satisfied, the operating point of the pump is at the intersection of the system curve (Equation 2.130) and the pump characteristic curve given in Figure 2.16. The system curve and the pump curves are both plotted in Figure 2.19, and the operating points for the various pump sizes are listed in the following table:

Pump Size (in.)	Operating Point (gpm)
12.1	2900
13	3400
13.9	3900
14.8	4400
15.7	4850
16.6	5450
17.5	5850

Since the desired flowrate in the system is 5000 gpm (0.315 m^3/s), the 16.6-in. (42.2-cm) pump should be selected. This 16.6-in. pump will deliver 5450 gpm (0.344 m^3/s) when all system valves are open, and can be throttled down to 5000 gpm as required. If a closer match between the desired flowrate and the operating point is desired for the given system, then an alternative series of homologous pumps should be considered. For the selected 16.6-in. pump, the maximum efficiency point is at Q = 5000 gpm, h_p = 52 ft (15.8 m), and ω = 885 rpm; hence the specific speed, N_s, of the selected pump (in U.S. Customary units) is given by

$$N_s = \frac{\omega Q^{\frac{1}{2}}}{h_p^{\frac{3}{4}}} = \frac{(885)(5000)^{\frac{1}{2}}}{(52)^{\frac{3}{4}}} = 3232$$

Comparing this result with the pump-selection guidelines in Table 2.3 confirms that the pump being considered must be a centrifugal pump.

FIGURE 2.19: Pump operating points

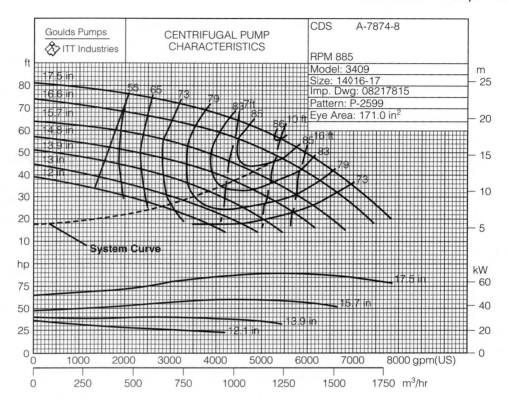

The available net positive suction head, NPSH_A is defined by Equation 2.122 as

$$\text{NPSH}_A = \frac{p_0}{\gamma} - \Delta z_s - h_L - \frac{p_v}{\gamma} \tag{2.131}$$

Atmospheric pressure, p_0, can be taken as 101 kPa; the specific weight of water, γ, is 9.79 kN/m^3; the suction lift, Δz_s, is 1.5 m; and at 20°C, the saturated vapor pressure of water, p_v, is 2.34 kPa. The head loss, h_L, is estimated as

$$h_L = \left(0.1 + \frac{fL}{D}\right)\frac{V^2}{2g} \tag{2.132}$$

where the entrance loss at the pump intake is $0.1\ V^2/2g$. For a flowrate, Q, equal to 5450 gpm (0.344 m^3/s), Equation 2.128 gives the friction factor, f, as

$$f = \frac{1}{4[\log(6.216 \times 10^{-5} + 4.32 \times 10^{-6}Q^{-0.9})]^2}$$

$$= \frac{1}{4[\log(6.216 \times 10^{-5} + 4.32 \times 10^{-6}(0.344)^{-0.9})]^2}$$

$$= 0.00366$$

and the average velocity of flow in the pipe, V, is given by

$$V = \frac{Q}{A} = \frac{0.344}{0.03142} = 10.9 \text{ m/s}$$

Substituting $f = 0.00366$, $L = 3.5$ m, $D = 0.2$ m, and $V = 10.9$ m/s into Equation 2.132 yields

$$h_L = \left(0.1 + \frac{(0.00366)(3.5)}{0.2}\right)\frac{10.9^2}{2(9.81)} = 3.44 \text{ m}$$

and hence the available net positive suction head, NPSH_A, (Equation 2.131) is

$$\text{NPSH}_A = \frac{101}{9.79} - 1.5 - 3.44 - \frac{2.34}{9.79} = 5.13 \text{ m}$$

According to the pump properties given in Figure 2.19, the required net positive suction head, NPSH_R, for the 16.6-in. pump at the operating point is 12 ft (= 3.66 m). Since the available net positive suction head (5.13 m) is greater than the required net positive suction head (3.66 m), the pump location relative to the intake reservoir is adequate and cavitation problems are not expected.

2.4.3 Multiple-Pump Systems

In cases where a single pump is inadequate to achieve a desired operating condition, multiple pumps can be used. Combinations of pumps are referred to as *pump systems*, and the pumps within these systems are typically arranged either in series or in parallel. The characteristic curve of a pump system is determined by the arrangement of pumps. Consider the case of two identical pumps in series, illustrated in Figure 2.20(a). The flow through each pump is equal to Q, and the head added by each pump is h_p. For the two-pump system, the flow through the system is equal to Q and the head added by the system is $2h_p$. Consequently, the characteristic curve of the two-pump (in series) system is related to the characteristic curve of each pump in that for any flow Q the head added by the system is twice the head added by a single pump, and the relationship between the single-pump characteristic curve and the two-pump characteristic curve is illustrated in Figure 2.20(b). This analysis can be extended to cases where the pump system contains n identical pumps in series, in which case the n-pump characteristic curve is derived from the single-pump characteristic curve by

FIGURE 2.20: Pumps in series

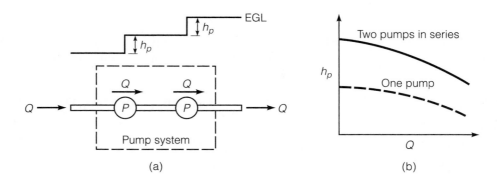

(a)

(b)

FIGURE 2.21: Pumps in parallel

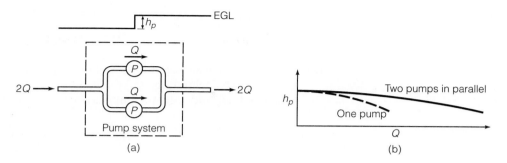

(a)   (b)

multiplying the ordinate of the single-pump characteristic curve (h_p) by n. Pumps in series are used in applications involving unusually high heads.

The case of two identical pumps arranged in parallel is illustrated in Figure 2.21. In this case, the flow through each pump is Q and the head added is h_p; therefore, the flow through the two-pump system is equal to $2Q$, while the head added is h_p. Consequently, the characteristic curve of the two-pump system is derived from the characteristic curve of the individual pumps by multiplying the abscissa (Q) by two. This is illustrated in Figure 2.21(b). In a similar manner, the characteristic curves of systems containing n identical pumps in parallel can be derived from the single-pump characteristic curve by multiplying the abscissa (Q) by n. Pumps in parallel are used in cases where the desired flowrate is beyond the range of a single pump and also to provide flexibility in pump operations, since some pumps in the system can be shut down during low-demand conditions or for service. This arrangement is common in sewage pump stations and water-distribution systems, where flowrates vary significantly during the course of a day.

When pumps are placed either in series or in parallel, it is usually desirable that these pumps be identical; otherwise, the pumps will be loaded unequally and the efficiency of the pump system will be less than optimal. In cases where nonidentical pumps are placed in series, the characteristic curve of the pump system is obtained by summing the heads added by the individual pumps for a given flowrate. In cases where nonidentical pumps are placed in parallel, the characteristic curve of the pump system is obtained by summing the flowrates through the individual pumps for a given head.

EXAMPLE 2.13

If a pump has a performance curve described by the relation

$$h_p = 12 - 0.1Q^2$$

then what is the performance curve for: (a) a system having three of these pumps in series; and (b) a system having three of these pumps in parallel?

Solution

(a) For a system with three pumps in series, the same flow, Q, goes through each pump, and each pump adds one-third of the head, H_p, added by the pump system. Therefore,

$$\frac{H_p}{3} = 12 - 0.1Q^2$$

and the characteristic curve of the pump system is

$$H_p = 36 - 0.3Q^2$$

(b) For a system consisting of three pumps in parallel, one-third of the total flow, Q, goes through each pump, and the head added by each pump is the same as the total head, H_p added by the pump system. Therefore

$$H_p = 12 - 0.1\left(\frac{Q}{3}\right)^2$$

and the characteristic curve of the pump system is

$$H_p = 12 - 0.011Q^2$$

2.5 Design of Water-Distribution Systems

Water-distribution systems move water from treatment plants to homes, offices, industries, and other consumers. The major components of a water-distribution system are pipelines, pumps, storage facilities, valves, and meters. The primary requirements of a distribution system are to supply each customer with a sufficient volume of water at adequate pressure, to deliver safe water that satisfies the quality expectations of customers, and to have sufficient capacity and reserve storage for fire protection and emergency conditions (AWWA, 2003c).

2.5.1 Water Demand

Major considerations in designing water-supply systems are the water demands of the population being served, the fire flows needed to protect life and property, and the proximity of the service area to sources of water. There are usually several categories of water demand within any populated area, and these sources of demand can be broadly grouped into residential, commercial, industrial, and public. Residential water use is associated with houses and apartments where people live; commercial water use is associated with retail businesses, offices, hotels, and restaurants; industrial water use is associated with manufacturing and processing operations; and public water use includes governmental facilities that use water. Large industrial requirements are typically satisfied by sources other than the public water supply.

A typical distribution of per-capita water use for an average city in the United States is given in Table 2.4. These rates vary from city to city as a result of differences in local conditions that are unrelated to the efficiency of water use. Water consumption is frequently stated in terms of the average amount of water delivered per day (to all categories of water use) divided by the population served, which is called the *average per-capita demand*. The distribution of average per-capita rates among 392 water-supply systems serving approximately 95 million people in the United States is shown in Table 2.5. The average per-capita water usage in this sample was 660 L/d, with a standard deviation of 270 L/d. Generally, high per-capita rates are found in water-supply systems servicing large industrial or commercial sectors, affluent communities, arid and semi-arid areas, and communities without water meters (Dziegielewski et al., 1996).

TABLE 2.4: Typical Distribution of Water Demand

Category	Average use (liters/day)/person	Percent of total
Residential	380	56
Commercial	115	17
Industrial	85	12
Public	65	9
Loss	40	6
Total	685	100

Source: Solley (1998).

TABLE 2.5: Distribution of Per-Capita Water Demand

Range (liters/day)/person	Number of systems	Percent of systems
190–370	30	7.7
380–560	132	33.7
570–750	133	33.9
760–940	51	13.0
950–1130	19	4.8
>1140	27	6.9

Source: Derived from data in AWWA (1986).

In the planning of municipal water-supply projects, the water demand at the end of the design life of the project is usually the basis for design. For existing water-supply systems, the American Water Works Association (AWWA, 1992) recommends that every 5 or 10 years, as a minimum, water-distribution systems be thoroughly reevaluated for requirements that would be placed on it by development and reconstruction over a 20-year period into the future. The estimation of the design flowrates for components of the water-supply system typically requires prediction of the population of the service area at the end of the design life, which is then multiplied by the per-capita water demand to yield the design flowrate. Whereas the per-capita water demand can usually be assumed to be fairly constant, the estimation of the future population typically involves a nonlinear extrapolation of past population trends.

A variety of methods are used in population forecasting. The simplest models treat the population as a whole, fit empirical growth functions to historical population data, and forecast future populations based on past trends. The most complex models disaggregate the population into various groups and forecast the growth of each group separately. A popular approach that segregates the population by age and gender is *cohort analysis* (Sykes, 1995). High levels of disaggregation have the advantage of making forecast assumptions very explicit, but these models tend to be complex and require more data than the empirical models that treat the population as a whole. Over relatively short time horizons, on the order of 10 years, detailed disaggregation models may not be any more accurate than using empirical extrapolation models of the population as a whole.

FIGURE 2.22: Growth phases in populated areas

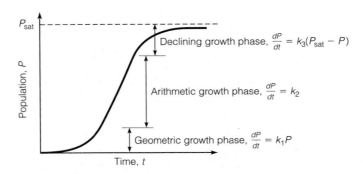

Populated areas tend to grow at varying rates, as illustrated in Figure 2.22. In the early stages of growth, there are wide-open spaces. Population, P, tends to grow geometrically according to the relation

$$\frac{dP}{dt} = k_1 P \tag{2.133}$$

where k_1 is a growth constant. Integrating Equation 2.133 gives the following expression for the population as a function of time:

$$\boxed{P(t) = P_0 e^{k_1 t}} \tag{2.134}$$

where P_0 is the population at some initial time designated as $t = 0$. Beyond the initial geometric growth phase, the rate of growth begins to level off and the following arithmetic growth relation may be more appropriate:

$$\frac{dP}{dt} = k_2 \tag{2.135}$$

where k_2 is an arithmetic growth constant. Integrating Equation 2.135 gives the following expression for the population as a function of time:

$$\boxed{P(t) = P_0 + k_2 t} \tag{2.136}$$

where P_0 is the population at $t = 0$. Ultimately, the growth of population centers becomes limited by the resources available to support the population, and further growth is influenced by the saturation population of the area, P_{sat}, and the population growth is described by a relation such as

$$\frac{dP}{dt} = k_3(P_{sat} - P) \tag{2.137}$$

where k_3 is a constant. This phase of growth is called the *declining-growth* phase. Almost all communities have zoning regulations that control the use of both developed and undeveloped areas within their jurisdiction (sometimes called a master plan), and a review of these regulations will yield an estimate of the saturation population of the undeveloped areas. Integrating Equation 2.137 gives the following expression for the population as a function of time:

$$\boxed{P(t) = P_{sat} - (P_{sat} - P_0)e^{-k_3 t}} \tag{2.138}$$

where P_0 is the population at $t = 0$.

The time scale associated with each growth phase is typically on the order of 10 years, although the actual duration of each phase can deviate significantly from this number. The duration of each phase is important in that population extrapolation using a single-phase equation can only be justified for the duration of that growth phase. Consequently, single-phase extrapolations are typically limited to 10 years or less, and these population predictions are termed *short-term projections* (Viessman and Welty, 1985). Extrapolation beyond 10 years, called *long-term projections*, involve fitting an S-shaped curve to the historical population trends and then extrapolating using the fitted equation. The most commonly fitted S-curve is the so-called *logistic curve*, which is described by the equation

$$P(t) = \frac{P_{sat}}{1 + ae^{bt}} \tag{2.139}$$

where a and b are constants. The conventional methodology to fit the population equations to historical data is to plot the historical data, observe the trend in the data, and fit the curve that best matches the population trend. Using extrapolation methods, errors less than 10% can be expected for planning periods shorter than 10 years, and errors greater than 50% can be expected for planning periods longer than 20 years (Sykes, 1995).

EXAMPLE 2.14

You are in the process of designing a water-supply system for a town, and the design life of your system is to end in the year 2020. The population in the town has been measured every 10 years since 1920 by the U.S. Census Bureau, and the reported populations are tabulated here. Estimate the population in the town using (a) graphical extension, (b) arithmetic growth projection, (c) geometric growth projection, (d) declining growth projection (assuming a saturation concentration of 600,000 people), and (e) logistic curve projection.

Year	Population
1920	125,000
1930	150,000
1940	150,000
1950	185,000
1960	185,000
1970	210,000
1980	280,000
1990	320,000

Solution   The population trend is plotted in Figure 2.23, where a geometric growth rate approaching an arithmetic growth rate is indicated.

(a) A growth curve matching the trend in the measured populations is indicated in Figure 2.23. Graphical extension to the year 2020 leads to a population estimate of 530,000 people.

(b) Arithmetic growth is described by Equation 2.136 as

$$P(t) = P_0 + k_2 t \tag{2.140}$$

FIGURE 2.23: Population trend

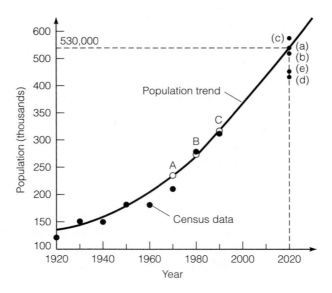

where P_0 and k_2 are constants. Consider the arithmetic projection of a line passing through points B and C on the approximate growth curve shown in Figure 2.23. At point B, $t = 0$ (year 1980) and $P = 270,000$; at point C, $t = 10$ (year 1990) and $P = 330,000$. Applying these conditions to Equation 2.140 yields

$$P = 270,000 + 6000\,t \qquad (2.141)$$

In the year 2020, $t = 40$ years and the population estimate given by Equation 2.141 is

$$P = 510,000 \text{ people}$$

(c) Geometric growth is described by Equation 2.134 as

$$P = P_0 e^{k_1 t} \qquad (2.142)$$

where k_1 and P_0 are constants. Using points A and C in Figure 2.23 as a basis for projection, then at $t = 0$ (year 1970), $P = 225,000$, and at $t = 20$ (year 1990), $P = 330,000$. Applying these conditions to Equation 2.142 yields

$$P = 225,000 e^{0.0195 t} \qquad (2.143)$$

In the year 2020, $t = 50$ years and the population estimate given by Equation 2.143 is

$$P = 597,000 \text{ people}$$

(d) Declining growth is described by Equation 2.138 as

$$P(t) = P_{\text{sat}} - (P_{\text{sat}} - P_0)e^{-k_3 t} \qquad (2.144)$$

where P_0 and k_3 are constants. Using points A and C in Figure 2.23, then at $t = 0$ (year 1970), $P = 225,000$, at $t = 20$ (year 1990), $P = 330,000$, and $P_{\text{sat}} = 600,000$.

Applying these conditions to Equation 2.144 yields

$$P = 600{,}000 - 375{,}000e^{-0.0164\,t} \qquad (2.145)$$

In the year 2020, $t = 50$ years and the population given by Equation 2.145 is given by

$$P = 434{,}800 \text{ people}$$

(e) The logistic curve is described by Equation 2.139. Using points A and C in Figure 2.23 to evaluate the constants in Equation 2.139 ($t = 0$ in 1970) yields

$$P = \frac{600{,}000}{1 + 1.67e^{-0.0357\,t}} \qquad (2.146)$$

In 2020, $t = 50$ years and the population given by Equation 2.146 is

$$P = 469{,}000 \text{ people}$$

These results indicate that the population projection in 2020 is quite uncertain, with estimates ranging from 597,000 for geometric growth to 434,800 for declining growth. The projected results are compared graphically in Figure 2.23. Closer inspection of the predictions indicates that the declining and logistic growth models are limited by the specified saturation population of 600,000, while the geometric growth model is not limited by saturation conditions and produces the highest projected population.

The multiplication of the population projection by the per-capita water demand is used to estimate the *average daily demand* for a municipal water-supply system.

Variations in demand.   Water demand generally fluctuates on a seasonal and a daily basis, being below the average daily demand in the early-morning hours and above the average daily demand during the midday hours. Typical daily cycles in water demand are shown in Figure 2.24. On a typical day in most communities, water use is lowest

FIGURE 2.24: Typical daily cycles in water demand

Source: Linsley et al. (1992).

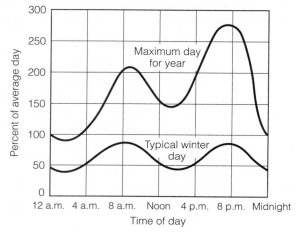

at night (11 p.m. to 5 a.m.) when most people are asleep. Water use rises rapidly in the morning (5 a.m. to 11 a.m.) followed by moderate usage through midday (11 a.m. to 6 p.m.). Use then increases in the evening (6 p.m. to 10 p.m.) and drops rather quickly around 10 p.m. Overall, water-use patterns within a typical 24-hour period are characterized by demands that are 25% to 40% of the average daily demand during the hours between midnight and 6:00 a.m. and 150% to 175% of the average daily demand during the morning or evening peak periods (Velon and Johnson, 1993).

The range of demand conditions that are to be expected in water-distribution systems is specified by *demand factors* or *peaking factors* that express the ratio of the demand under certain conditions to the average daily demand. Typical demand factors for various conditions are given in Table 2.6, where the *maximum daily demand* is defined as the demand on the day of the year that uses the most volume of water, and the *maximum hourly demand* is defined as the demand during the hour that uses the most volume of water. The demand factors in Table 2.6 should serve only as guidelines, with the actual demand factors in any distribution system being best estimated from local measurements. In small water systems, demand factors may be significantly higher than those shown in Table 2.6.

Fire demand.   Besides the fluctuations in demand that occur under normal operating conditions, water-distribution systems are usually designed to accommodate the large (short-term) water demands associated with fighting fires. Although there is no legal requirement that a governing body must size its water-distribution system to provide fire protection, the governing bodies of most communities provide water for fire protection for reasons that include protection of the tax base from destruction by fire, preservation of jobs, preservation of human life, and reduction of human suffering. Flowrates required to fight fires can significantly exceed the maximum flowrates in the absence of fires, particularly in small water systems. In fact, for communities with populations less than 50,000, the need for fire protection is typically the determining factor in sizing water mains, storage facilities, and pumping facilities (AWWA, 2003c). In contrast to urban water systems, many rural water systems are designed to serve only domestic water needs, and fire-flow requirements are not considered in the design of these systems (AWWA, 2003c).

Numerous methods have been proposed for estimating fire flows, the most popular of which was proposed by the Insurance Services Office, Inc. (ISO, 1980), which is an organization representing the fire-insurance underwriters. The required

TABLE 2.6: Typical Demand Factors

Condition	Range of demand factors	Typical value
Daily average in maximum month	1.10–1.50	1.20
Daily average in maximum week	1.20–1.60	1.40
Maximum daily demand	1.50–3.00	1.80
Maximum hourly demand	2.00–4.00	3.25
Minimum hourly demand	0.20–0.60	0.30

Source: Velon and Johnson (1993). Reprinted by permission of The McGraw-Hill Companies.

fire flow for individual buildings can be estimated using the formula (ISO, 1980)

$$\boxed{\text{NFF}_i = C_i O_i (X + P)_i} \tag{2.147}$$

where NFF_i is the *needed fire flow* at location i, C_i is the *construction factor* based on the size of the building and its construction, O_i is the *occupancy factor* reflecting the kinds of materials stored in the building (values range from 0.75 to 1.25), and $(X + P)_i$ is the sum of the *exposure factor* (X_i) and *communication factor* (P_i) that reflect the proximity and exposure of other buildings (values range from 1.0 to 1.75). The construction factor, C_i, is the portion of the NFF attributed to the size of the building and its construction and is given by

$$C_i = 220F\sqrt{A_i} \tag{2.148}$$

where C_i is in L/min; A_i (m^2) is the effective floor area, typically equal to the area of the largest floor in the building plus 50% of the area of all other floors; and F is a coefficient based on the class of construction, given in Table 2.7. The maximum value of C_i calculated using Equation 2.148 is limited by the following: 30,000 L/min for construction classes 1 and 2; 23,000 L/min for construction classes 3, 4, 5, and 6; and 23,000 L/min for a one-story building of any class of construction. The minimum value of C_i is 2000 L/min, and the calculated value of C_i should be rounded to the nearest 1000 L/min. The occupancy factors, O_i, for various classes of buildings are given in Table 2.8. Detailed tables for estimating the exposure and communication factors, $(X + P)_i$, can be found in AWWA (1992), and values of $(X + P)_i$ are typically on the order of 1.4. The NFF calculated using Equation 2.147 should not exceed 45,000 L/min, nor be less than 2000 L/min. According to AWWA (1992), 2000 L/min is the minimum amount of water with which any fire can be controlled and suppressed

TABLE 2.7: Construction Coefficient, F

Class of construction	Description	F
1	Frame	1.5
2	Joisted masonry	1.0
3	Noncombustible	0.8
4	Masonry, noncombustible	0.8
5	Modified fire resistive	0.6
6	Fire resistive	0.6

Source: AWWA (1992).

TABLE 2.8: Occupancy Factors, O_i

Combustibility class	Examples	O_i
C-1 Noncombustible	Steel or concrete products storage	0.75
C-2 Limited combustible	Apartments, churches, offices	0.85
C-3 Combustible	Department stores, supermarkets	1.00
C-4 Free burning	Auditoriums, warehouses	1.15
C-5 Rapid burning	Paint shops, upholstering shops	1.25

Source: AWWA (1992).

TABLE 2.9: Needed Fire Flow for One- and Two-Family Dwellings

Distance between buildings (m)	Needed fire flow (L/min)
>30	2000
9.5–30	3000
3.5–9.5	4000
<3.5	6000

Source: AWWA (1992).

TABLE 2.10: Required Fire-Flow Durations

Required fire flow (L/min)	Duration (h)
<9000	2
11,000–13,000	3
15,000–17,000	4
19,000–21,000	5
23,000–26,000	6
26,000–30,000	7
30,000–34,000	8
34,000–38,000	9
38,000–45,000	10

Source: AWWA (1992).

safely and effectively. The NFF should be rounded to the nearest 1000 L/min if less than 9000 L/min, and to the nearest 2000 L/min if greater than 9000 L/min. For one- and two-family dwellings not exceeding two stories in height, the NFF listed in Table 2.9 should be used. For other habitable buildings not listed in Table 2.9, the NFF should not exceed 13,000 L/min maximum.

Usually the local water utility will have a policy on the upper limit of fire protection that it will provide to individual buildings. Those wanting higher fire flows need to either provide their own system or reduce fire-flow requirements by installing sprinkler systems, fire walls, or fire-retardant materials (Walski, 1996; AWWA, 1992). Estimates of the needed fire flow calculated using Equation 2.147 are used to determine the fire-flow requirements of the water-supply system, where the needed fire flow is calculated at several representative locations in the service area, and it is assumed that only one building is on fire at any time (Sykes, 1995). The design duration of the fire should follow the guidelines in Table 2.10. If these durations cannot be maintained, insurance rates are typically increased accordingly. A more detailed discussion of the requirements for fire protection has been published by the American Water Works Association (AWWA, 1992).

EXAMPLE 2.15

Estimate the flowrate and volume of water required to provide adequate fire protection to a 10-story noncombustible building with an effective floor area of 8000 m^2.

Solution   The NFF can be estimated by Equation 2.147 as

$$\text{NFF}_i = C_i O_i (X + P)_i$$

where the construction factor, C_i, is given by

$$C_i = 220 \, F \sqrt{A_i}$$

For the 10-story building, $F = 0.8$ (Table 2.7, noncombustible, Class 3 construction), and $A_i = 8000 \text{ m}^2$, hence

$$C_i = 220(0.8) \sqrt{8000} = 16,000 \text{ L/min}$$

where C_i has been rounded to the nearest 1000 L/min. The occupancy factor, O_i, is given by Table 2.8 as 0.75 (C-1 noncombustible); $(X + P)_i$ can be estimated by the median value of 1.4; and hence the needed fire flow, NFF, is given by

$$\text{NFF}_i = (16,000)(0.75)(1.4) = 17,000 \text{ L/min}$$

This flow must be maintained for a duration of four hours (Table 2.10). Hence the required volume, V, of water is given by

$$V = 17,000 \times 4 \times 60 = 4.08 \times 10^6 \text{ L} = 4080 \text{ m}^3$$

Fire hydrants are placed throughout the service area to provide either direct hose connections for firefighting or connections to pumper trucks (also known as fire engines). A single-hose stream is generally taken as 950 L/min, and hydrants are typically located at street intersections or spaced 60–150 m apart (McGhee, 1991). In high-value districts, additional hydrants may be necessary in the middle of long blocks to supply the required fire flows. Fire hydrants may also be used to release air at high points in a water-distribution system and blow off sediments at low points in the system.

Design flows.   The design capacities of various components of the water-supply system are given in Table 2.11, where *low-lift pumps* refer to low-head, high-rate units that convey the raw-water supply to the treatment facility and *high-lift pumps* deliver finished water from the treatment facility into the distribution network at suitable pressures. The required capacities shown in Table 2.11 consist of various combinations of the maximum daily demand, maximum hourly demand, and the fire demand. Typically, the delivery pipelines from the water source to the treatment plant, as well as the treatment plant itself, are designed with a capacity equal to the maximum daily demand. However, because of the high cost of providing treatment, some utilities have trended toward using peak flows averaged over a longer period than one day to design water-treatment plants (such as 2 to 5 days), and relying on system storage to meet peak demands above treatment capacity. This approach has serious water-quality implications and should be avoided if possible (AWWA, 2004). The flowrates and pressures in the distribution system are analyzed under both maximum daily plus fire demand and the maximum hourly demand, and the larger flowrate governs the design. Pumps are sized for a variety of conditions from maximum daily to maximum hourly demand, depending on their function in the distribution system.

TABLE 2.11: Design Periods and Capacities in Water-Supply Systems

Component	Design period (years)	Design capacity
1. Source of supply:		
River	Indefinite	Maximum daily demand
Wellfield	10–25	Maximum daily demand
Reservoir	25–50	Average annual demand
2. Conveyance:		
Intake conduit	25–50	Maximum daily demand
Conduit to treatment plant	25–50	Maximum daily demand
3. Pumps:		
Low-lift	10	Maximum daily demand, one reserve unit
High-lift	10	Maximum hourly demand, one reserve unit
4. Treatment plant	10–15	Maximum daily demand
5. Service reservoir	20–25	Working storage plus fire demand plus emergency storage
6. Distribution system:		
Supply pipe or conduit	25–50	Greater of (1) maximum daily demand plus fire demand, or (2) maximum hourly demand
Distribution grid	Full development	Same as for supply pipes

Source: Gupta (2001).

Additional reserve capacity is usually installed in water-supply systems to allow for redundancy and maintenance requirements.

EXAMPLE 2.16

A metropolitan area has a population of 130,000 people with an average daily demand of 600 L/d/person. If the needed fire flow is 20,000 L/min, estimate: (a) the design capacities for the wellfield and the water-treatment plant; (b) the duration that the fire flow must be sustained and the volume of water that must be kept in the service reservoir in case of a fire; and (c) the design capacity of the main supply pipeline to the distribution system.

Solution

(a) The design capacity of the wellfield should be equal to the maximum daily demand (Table 2.11). With a demand factor of 1.8 (Table 2.6), the per-capita demand on the maximum day is equal to $1.8 \times 600 = 1080$ L/day/person. Since the population served is 130,000 people, the design capacity of the wellfield, Q_{well}, is given by

$$Q_{\text{well}} = 1080 \times 130,000 = 1.4 \times 10^8 \text{ L/d} = 1.62 \text{ m}^3/\text{s}$$

The design capacity of the water-treatment plant is also equal to the maximum daily demand, and therefore should also be taken as 1.62 m³/s.

(b) The needed fire flow, Q_{fire}, is 20,000 L/min = 0.33 m³/s. According to Table 2.10, the fire flow must be sustained for 5 hours. The volume, V_{fire}, required for the fire flow will be stored in the service reservoir and is given by

$$V_{fire} = 0.33 \times 5 \times 3600 = 5940 \text{ m}^3$$

(c) The required flowrate in the main supply pipeline is equal to the maximum daily demand plus fire demand or the maximum hourly demand, whichever is greater.

$$\text{Maximum daily demand + fire demand} = 1.62 + 0.33 = 1.95 \text{ m}^3/\text{s}$$

$$\text{Maximum hourly demand} = \frac{3.25}{1.80} \times 1.62 = 2.92 \text{ m}^3/\text{s}$$

where a demand factor of 3.25 has been assumed for the maximum hourly demand. The main supply pipe to the distribution system should therefore be designed with a capacity of 2.92 m³/s. The water pressure within the distribution system must be above acceptable levels when the system demand is 2.92 m³/s.

2.5.2 Pipelines

Water-distribution systems typically consist of connected pipe loops throughout the service area. Pipelines in water-distribution systems include *transmission lines*, *arterial mains*, and *distribution mains*. Transmission lines carry flow from the water-treatment plant to the service area, typically have diameters greater than 600 mm, and are usually on the order of 3 km apart. Arterial mains are connected to transmission mains and are laid out in interlocking loops with the pipelines not more than 1 km apart and diameters in the range of 400–500 mm. Distribution mains form a grid over the entire service area, with diameters in the range of 150–300 mm, and supply water to every user. Pipelines in distribution systems are collectively called *water mains*, and a pipe that carries water from a main to a building or property is called a *service line*. Water mains are normally installed within the rights-of-way of streets. Dead ends in water-distribution systems should be avoided whenever possible, since the lack of flow in such lines may contribute to water-quality problems.

Pipelines in water-distribution systems are typically designed with constraints relating to the minimum pipe size, maximum allowable velocity, and commercially available materials that will perform adequately under operating conditions.

Minimum Size. The size of a water main determines its carrying capacity. Main sizes must be selected to provide the capacity to meet peak domestic, commercial, and industrial demands in the area to be served, and must also provide for fire flow at the necessary pressure. For fire protection, insurance underwriters typically require a minimum main size of 150 mm for residential areas and 200 mm for high-value districts (such as sports stadiums, shopping centers, and libraries) if cross-connecting mains are not more than 180 m apart. On principal streets, and for all long lines not connected at frequent intervals, 300-mm and larger mains are required. Fire hydrants require a minimum main size of 150 mm.

Service Lines. Service lines are pipes, including accessories, that carry water from the main to the point of service, which is normally a meter setting or curb stop located at the property line. Service lines can be any size, depending on how much water is required to serve a particular customer. Single-family residences are most commonly served with 20-mm (3/4-in.) diameter service lines, while larger residences and buildings located far from the main connection should have a 25-mm (1-in.) or larger service line. To properly size service lines it is essential to know the peak demands than any service tap will be called on to serve. A common method to estimate service flows is to sum the *fixture units* associated with the number and type of fixtures served by the service line and then use the *Hunter curve* to relate the peak flowrate to total fixture units. This relationship is included in most local plumbing codes and is contained in the Uniform Plumbing Code (UPC). Recent research has indicated that the peak flows estimated from fixture units and the Hunter curve provide conservative estimates of peak flows (AWWA, 2004). Irrigation demands that occur simultaneously with peak domestic demands must be added to the estimated peak domestic demands. Service lines are sized to provide an adequate service pressure downstream of the water meter when the service line is delivering the peak flow. This requires that the pressure and elevation at the tap, length of service pipe, head loss at the meter, elevation at the water meter, valve losses, and desired pressure downstream of the meter be known. Using the energy equation, the minimum service-line diameter is calculated using this information. It is usually better to overdesign than underdesign a service line because of the cost of replacing the line if service pressures turn out to be inadequate. To prevent water hammer, velocities greater than 3 m/s should be avoided. Materials used for service-line pipe and tubing are typically either copper (tubing) or plastic, which includes polyvinyl chloride (PVC) and polyethylene (PE). Type K copper is the most commonly used material for copper service lines. Older service lines used lead and galvanized iron, which are no longer recommended. The valve used to connect a small-diameter service line to a water main is called a *corporation stop*, sometimes loosely referred to as the corporation cock, corporation tap, corp stop, corporation, or simply corp or stop (AWWA, 2003c). Tapping a water main and inserting a corporation stop directly into the pipe wall requires a tapping machine, and taps are typically installed at the 10 or 2 o'clock position on the pipe. Guidelines for designing water service lines and meters are given in AWWA Manual M22 (AWWA, 2004). Good construction practices must be used when installing service lines to avoid costly repairs in the future. This includes burying the pipe below frost lines, maintaining proper ditch conditions, proper backfill, trench compaction, and protection from underground structures that may cause damage to the pipe.

Allowable Velocities. Maximum allowable velocities in pipeline systems are imposed to control friction losses and hydraulic transients. Maximum allowable velocities of 0.9 to 1.8 m/s are common in water-distribution pipes, and the American Water Works Association recommends a limit of about 1.5 m/s under normal operating conditions, but velocities may exceed this guideline under fire-flow conditions (AWWA, 2003c). The importance of controlling the maximum velocities in water distribution systems is supported by the fact that a sudden change in velocity of 0.3 m/s in water transmission and distribution systems can increase the pressure in a pipe by approximately 345 kPa, while the standard design for ductile iron pipe includes only a 690-kPa allowance (AWWA, 2003d).

Material. Pipeline materials should be selected based on a consideration of service conditions, availability, properties of the pipe, and economics. In selecting pipe materials the following considerations should be taken into account:

- Most water-distribution mains in older cities in the United States are made of (gray) cast iron pipe (CIP), with many cities having CIP over 100 years old and still providing satisfactory service (Mays, 2000; AWWA, 2003d). CIP is no longer manufactured in the United States.

- For new distribution mains, ductile iron pipe (DIP) is most widely used for pipe diameters up to 760 mm (30 in.). DIP has all the good qualities of CIP plus additional strength and ductility. DIP is manufactured in diameters from 76 to 1625 mm (3–64 in.). For diameters from 100 to 500 mm (4–20 in.), standard commercial sizes are available in 50-mm (2-in.) increments, while for diameters from 600 to 1200 mm (24–48 in.), the size increments are 150 mm (6 in.). The standard lengths of DIP are 5.5 m (18 ft) and 6.1 m (20 ft). DIP is usually coated (outside and inside) with a bituminous coating to minimize corrosion. An internal cement-mortar lining 1.5–3 mm thick is common, and external polyethylene wraps are used to reduce corrosion in corrosive soil environments. DIP used in water systems in the United States is provided with a cement-mortar lining unless otherwise specified by the purchaser. The design of DIP and fittings is covered in AWWA Manual M41 (AWWA, 2003d) and guidance for DIP lining is covered in AWWA Standard C105 (latest edition). Tests conducted by the Ductile Iron Pipe Research Association (DIPRA) suggest that a Hazen–Williams C-value of 140 is appropriate for the design of cement-mortar lined DIP. A variety of joints are available for use with DIP, which includes push-on (the most common), mechanical, flanged, ball-and-socket, and numerous joint designs. A stack of DIP is shown in Figure 2.25, where the bell and spigot pipe ends that facilitate push-on connection are apparent. A rubber gasket, to ensure a tight fit, is contained in the bell side of the pipe.

- Steel pipe usually compares favorably with DIP for diameters larger than 400 mm (16 in.). As a consequence, steel pipe is primarily used for transmission lines in water-distribution systems. Steel pipe is available in diameters from 100 to 3600 mm (4–144 in.). The standard length of steel pipe is 12.2 m (40 ft). The interior of steel pipe is usually protected with either cement mortar or epoxy, and the exterior is protected by a variety of plastic coatings, bituminous materials, and polyethylene tapes, depending on the degree of protection required. Guidance for the design of steel pipe is covered in AWWA Manual M11 (latest edition), and linings for steel pipe are covered under AWWA Standard C205 (latest edition) and AWWA Standard C210 (latest edition).

- Plastic materials used for fabricating water-main pipe include polyvinyl chloride (PVC), polyethylene (PE), and polybutylene (PB). PVC pipe is by far the most widely used type of plastic pipe material for small-diameter

FIGURE 2.25: Ductile iron pipe

water mains. The American Water Works Association standard (C900) for PVC pipe in sizes from 100 to 300 mm and laying lengths of 6.1 m (20 ft) is based on the same outside diameter as for DIP. In this way, standard DIP fittings can be used with PVC pipe. PVC pipe is commonly available in diameters from 100 to 914 mm. Extruded PE and PB pipe is primarily used for water service pipe in small sizes; however, the use of PB has decreased significantly because of structural difficulties caused by premature pipe failures. Research has documented that pipe materials such as PVC, PE, and PB may be subject to permeation by lower-molecular-weight organic solvents or petroleum products (AWWA, 2002b). If a water pipe must pass through an area subject to contamination, caution should be used in selecting PVC, PE, and PB pipes. In the hydraulic design of PVC pipes, a roughness height of 0.0015 mm or a Hazen–Williams C-value of 150 is appropriate for design (AWWA, 2002b). Details of large-diameter PE pipe are found in AWWA Standard C906 (latest edition) and information on PVC water-main pipe is available in AWWA Manual M23 (2002b).

- Asbestos-cement (A-C) pipe has been widely installed in water-distribution systems, especially in areas where metallic pipe is subject to corrosion, such as in coastal areas. It has also been installed in remote areas, where its light weight makes it much easier to install than DIP. Common diameters are in the range of 100 to 890 mm. The U.S. Environmental Protection Agency banned most uses of asbestos in 1989 and, due to the manufacturing ban, new A-C pipe is no longer being installed in the United States.

- Fiberglass pipe is available for potable water in sizes from 25 to 3600 mm. Advantages of fiberglass pipe include corrosion resistance, light weight, low installation cost, ease of repair, and hydraulic smoothness. Disadvantages include susceptibility to mechanical damage, low modulus of elasticity, and lack of standard joining system. Fiberglass pipe is covered in AWWA Standard C950 (latest edition).

- The use of concrete pressure pipe has grown rapidly since 1950. Concrete pressure pipe provides a combination of the high tensile strength of steel and the high compressive strength and corrosion resistance of concrete.

The pipe is available in diameters ranging from 250 to 6400 mm and in standard lengths from 3.7 to 12.2 m. The design of concrete pressure pipe is covered in AWWA Manual M9 (latest edition). Concrete pipe is available with various types of liners and reinforcement; the four types in common use in the United States and Canada are: prestressed concrete cylinder pipe, bar-wrapped concrete cylinder pipe, reinforced concrete cylinder pipe, and reinforced concrete noncylinder pipe. The manufacture of prestressed concrete cylinder pipe is covered under AWWA Standard C301 (latest edition) and AWWA Standard C304 (latest edition), bar-wrapped concrete cylinder pipe is covered under AWWA Standard C303 (latest edition), reinforced concrete cylinder pipe is covered under AWWA Standard C300 (latest edition), and reinforced concrete noncylinder pipe is covered under AWWA Standard C302 (latest edition).

Pipelines in water-distribution systems should be buried to a depth below the frost line in northern climates and at a depth sufficient to cushion the pipe against traffic loads in warmer climates (Clark, 1990). In warmer climates, a cover of 1.2 m to 1.5 m is used for large mains and 0.75 m to 1.0 m for smaller mains. In areas where frost penetration is a significant factor, mains can have as much as 2.5 m of cover. Trenches for water mains should be as narrow as possible while still being wide enough to allow for proper joining and compaction around the pipe. The suggested trench width is the nominal pipe diameter plus 0.6 m; in deep trenches, sloping may be necessary to keep the trench wall from caving in. Trench bottoms should be undercut 15 to 25 cm, and sand, clean fill, or crushed stone installed to provide a cushion against the bottom of the excavation, which is usually rock (Clark, 1990). Standards for pipe construction, installation, and performance are published by the American Water Works Association in its C-series standards, which are continuously being updated.

2.5.3 Pumps

Service pressures are typically maintained by pumps, with head losses and increases in pipeline elevations acting to reduce pressures, and decreases in pipeline elevations acting to increase pressures. When portions of the distribution system are separated by long distances or significant changes in elevation, *booster pumps* are sometimes used to maintain acceptable service pressures. In some cases, *fire-service pumps* are used to provide additional capacity for emergency fire protection. Pumps operate at the intersection of the pump performance curve and the system curve. Since the system curve is significantly affected by variations in water demand, there is a significant variation in pump operating conditions. In most cases, the range of operating conditions is too wide to be met by a single pump, and multiple-pump installations or variable-speed pumps are required (Velon and Johnson, 1993).

2.5.4 Valves

Valves in water-distribution systems are designed to perform several different functions. Their primary functions are to start and stop flow, isolate piping, regulate pressure and throttle flow, prevent backflow, and relieve pressure. *Shutoff valves* or *gate valves* are typically provided at 350-m intervals so that areas within the system can be isolated for repair or maintenance; *air-relief valves* or *air-and-vacuum relief valves* are required at high points to release trapped air; *blowoff valves* or *drain valves* may be required at low points; and *backflow-prevention devices* are required by applicable regulations to prevent contamination from backflows of nonpotable water into the distribution system from system outlets. To maintain the performance of water-distribution systems it is recommended that each valve be operated through a full cycle and then returned to its normal position on a regular schedule. The time interval between operations should be determined by the manufacturer's recommendations, size of the valve, severity of the operating conditions, and the importance of the installation (AWWA, 2003d).

2.5.5 Meters

The water meter is a changeable component of a customer's water system. Unlike the service line and water tap, which when incorrectly sized will generally require expensive excavation and retapping, water meters can usually be changed less expensively. Selection of the type and size of a meter should be based primarily on the range of flow, and the pressure loss through the meter should also be a consideration. For many single-family residences, a 20-mm service line with a 15-mm meter is typical, while in areas where irrigation is prevalent, 20-mm or 25-mm meters may be more prevalent. Undersizing the meter can cause pressure-related problems, and oversizing can result in reduced revenue and inaccurate meter recordings (since the flows do not register). Some customers, such as hospitals, schools, and factories with processes requiring uninterrupted water service, should have bypasses installed around the meter so that meter test and repair activities can be performed at scheduled intervals without inconvenience to either the customer or the utility. The bypass should be locked and valved appropriately.

2.5.6 Fire Hydrants

Fire hydrants are one of the few parts of a water distribution system that are visible to the public, so keeping them well maintained can help a water utility project a good public image. Fire hydrants are direct connections to the water mains and, in addition to providing an outlet for fire protection, they are used for flushing water mains, flushing sewers, filling tank trucks for street washing, tree spraying, and providing a temporary water source for construction jobs. A typical fire hydrant is shown in Figure 2.26, along with the pipe connection between the water main and the fire hydrant. The vertical pipe connecting the water main to the fire hydrant is commonly called a *riser*. The water utility is usually responsible for keeping hydrants in working order, although fire departments sometime assume this responsibility. Standard practice is to install hydrants only on mains 150 mm or larger, however, larger mains are often necessary

FIGURE 2.26: Fire hydrant and connection to water main

to ensure that the residual pressure during fire flow remains greater than 140 kPa. Guidelines for the placement of hydrants are as follows (AWWA, 2003c):

- Not too close to buildings, since fire fighters will not position their fire (pumper) trucks where a building wall could fall on them.
- Preferably located near street intersections, where the hose can be strung to fight a fire in any of several directions.
- Far enough from a roadway to minimize the danger of their being struck by vehicles.
- Close enough to the pavement to ensure a secure connection with the pumper and hydrant without the risk of the truck getting stuck in mud or snow.
- In areas of heavy snow, hydrants must be located where they are least liable to be covered by plowed snow or struck by snow-removal equipment.
- Hydrants should be high enough off the ground that valve caps can be removed with a standard wrench, without the wrench hitting the ground.

Fire hydrants should be inspected and operated through a full cycle on a regular schedule, and the hydrant should be flushed to prevent sediment buildup in the hydrant or connecting piping.

2.5.7 Water-Storage Reservoirs

Water usually enters the system at a fairly constant rate from the treatment plant. To accommodate fluctuations in demand, a storage reservoir is typically located at the head of the system to store the excess water during periods of low demand and provide water during periods of high demand. In addition to the operational storage required to accommodate diurnal (24-hour cycle) variations in water demand, storage facilities are also used to provide storage to fight fires, to provide storage for emergency conditions, and to equalize pressures in water-distribution systems.

Storage facilities are classified as either elevated storage, ground storage, or standpipes. The function and relative advantages of these types of systems are as follows:

Elevated Storage Tanks. *Elevated storage tanks* are constructed above ground such that the height of the water in the tank is sufficient to deliver water to the distribution system at the required pressure. The storage tank is generally supported by a steel or concrete tower, the tank is directly connected to the distribution system through a vertical pipe called a *riser*, the water level in the tank is equal to the elevation of the hydraulic grade line in the distribution system (at the outlet of the storage tank), and the elevated storage tank is said to *float* on the system. Elevated storage is useful in the case of fires and emergency conditions, since pumping of water from elevated tanks is not necessary, although the water must generally be pumped into the tanks. Occasionally, system pressure could become so high that the tank would overflow; to prevent this, altitude valves must be installed on the tank fill line. Elevated tanks are usually made of steel.

Ground Storage Reservoirs. *Ground storage reservoirs* are constructed at or below ground level and usually discharge water to the distribution system through pumps. These systems, which are sometimes referred to simply as *distribution-system reservoirs* or *ground-level tanks*, are usually used where very large quantities of water must be stored or when an elevated tank is objectionable to the public. When a ground-level or buried reservoir is located at a low elevation on the distribution system, water is admitted through a remotely operated valve, and a pump station is provided to transfer water into the distribution system. Completely buried reservoirs are often used where an above-ground structure is objectionable, such as in a residential neighborhood. In some cases, the land over a buried reservoir can be used for recreational facilities such as a ball field or tennis court. Ground storage reservoirs are typically constructed of steel or concrete.

Standpipes. A tank that rests on the ground with a height that is greater than its diameter is generally referred to as a *standpipe*. In most installations, only water in the upper portion of the tank will furnish usable system pressure, so most larger standpipes are equipped with an adjacent pumping system that can be used in an emergency to pump water to the system from the lower portion of the tank. Standpipes combine the advantages of elevated storage with the ability to store large quantities of water, and they are usually constructed of steel. Standpipes taller than 15 m are usually uneconomical, since it tends to be more economical to build an elevated tank than to accommodate the dead storage that must be pumped into the system.

Storage facilities in a distribution system are required to have sufficient volume to meet the following criteria: (1) adequate volume to supply peak demands in excess of the maximum daily demand using no more than 50% of the available storage capacity; (2) adequate volume to supply the critical fire demand in addition to the volume required for meeting the maximum daily demand fluctuations; and (3) adequate volume to supply the average daily demand of the system for the estimated duration of a possible emergency. Conventional design practice is to rely on pumping to meet the daily operational demands up to the maximum daily demand; where detailed demand data is not available, the storage available to supply the peak demands should equal 20% to 25% of the maximum daily demand volume. Sizing the storage volume for fire protection is based on the product of the critical fire flow and duration for the service area. In extremely large systems, where fire demands may be only a small fraction of the maximum daily demand, fire storage may not be necessary (Walski, 2000). Emergency storage is generally necessary to provide water during power outages, breaks in water mains, problems at treatment plants, unexpected shutdowns of water-supply facilities, and other sporadic events. Emergency volumes for most municipal water-supply systems vary from one to two days of supply capacity at the average daily demand. The recommended standards for water works developed by the Great Lakes Upper Mississippi River Board of State Public Health and Environmental Managers suggest a minimum emergency storage capacity equal to the average daily system demand.

In cases where elevated storage tanks are used, the minimum acceptable height of water in a tank is determined by computing the minimum acceptable piezometric head in the service area and then adding to that figure an estimate of the head losses between the critical service location and the location of the elevated service tank, under the condition of average daily demand. The maximum height of water in the elevated tank is then determined by adding the minimum acceptable piezometric head to the head loss between the tank location and the critical service location under the condition of maximum hourly demand. The difference between the calculated minimum and maximum heights of water in the elevated storage tank is then specified as the normal operating range within the tank. The normal operating range for water in elevated tanks is usually limited to 4.5 to 6 meters, so that fluctuations in pressure are limited to 35 to 70 kPa. In most cases, the operating range is located in the upper half of the storage tank, with storage in the lower half of the tank reserved for firefighting and emergency storage. Any water stored in elevated tanks less than 14 m (46 ft) above ground is referred to as *ineffective storage* (Walski et al., 1990), since the pressure in connected distribution pipes will be less than the usual minimum acceptable pressure during emergency conditions of 140 kPa (20 psi). Operational storage in elevated tanks is normally at elevations of more than 25 m (81 ft) above the ground, since under these conditions the pressure in connected distribution pipes will exceed the usual minimum acceptable pressure during normal operations of 240 kPa (35 psi). A typical elevated storage tank is illustrated in Figure 2.27. These types of storage facilities generally have only a single pipe connection to the distribution system, and this single pipe handles both inflows and outflows from the storage tank. This piping arrangement is in contrast to ground storage reservoirs, which have separate inflow and outflow piping. The inflow piping delivers the outflow from the water-treatment facility to the reservoir, while the outflow piping delivers the water from the reservoir to the pumps that input water into the distribution system. The largest elevated storage tank in the United States has a volume of 15,520 m^3 (ASCE, 2000). Elevated storage tanks are best placed on the downstream side of the largest demand from the source, with the advantages that: (1) if a pipe breaks near the source, the break will not result in disconnecting all the storage from the customers; and (2) if flow reaches the center of demand from more than one direction, the flow carried by any individual pipe will be lower and pipe sizes will generally be smaller, with associated cost savings (Walski, 2000). If there are multiple storage tanks in the distribution system, the tanks should be placed roughly the same distance from the source or sources, and all tanks should have approximately the same overflow elevation (otherwise, it may be impossible to fill the highest tank without overflowing or shutting off the lower tanks).

FIGURE 2.27: Elevated storage tank

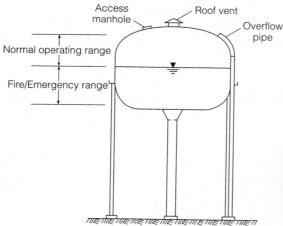

EXAMPLE 2.17

A service reservoir is to be designed for a water-supply system serving 250,000 people with an average demand of 600 L/d/capita, and a needed fire flow of 37,000 L/min. Estimate the required volume of service storage.

Solution The required storage is the sum of three components: (1) volume to supply the demand in excess of the maximum daily demand, (2) fire storage, and (3) emergency storage.

The volume to supply the peak demand can be taken as 25% of the maximum daily demand volume. Taking the maximum daily demand factor as 1.8 (Table 2.6), then the maximum daily flowrate, Q_m, is given by

$$Q_m = (1.8)(600)(250,000) = 2.7 \times 10^8 \text{ L/d} = 2.7 \times 10^5 \text{ m}^3/\text{d}$$

The storage volume to supply the peak demand, V_{peak}, is therefore given by

$$V_{peak} = (0.25)(2.7 \times 10^5) = 67,500 \text{ m}^3$$

According to Table 2.10, the 37,000 L/min (= 0.62 m^3/s) fire flow must be maintained for at least 9 hours. The volume to supply the fire demand, V_{fire}, is therefore given by

$$V_{fire} = 0.62 \times 9 \times 3600 = 20,100 \text{ m}^3$$

The emergency storage, V_{emer}, can be taken as the average daily demand, in which case

$$V_{emer} = 250,000 \times 600 = 150 \times 10^6 \text{ L} = 150,000 \text{ m}^3$$

The required volume, V, of the service reservoir is therefore given by

$$V = V_{peak} + V_{fire} + V_{emer}$$
$$= 67,500 + 20,100 + 150,000$$
$$= 237,600 \text{ m}^3$$

The service reservoir should be designed to store 238,000 m^3 of water. This large volume will require a ground storage tank (recall that the largest elevated-tank volume in the United States is 15,520 m^3), and it is interesting to note that most of the storage in the service reservoir is reserved for emergencies.

EXAMPLE 2.18

A water-supply system is to be designed in an area where the minimum allowable pressure in the distribution system is 300 kPa. A hydraulic analysis of the distribution network under average-daily-demand conditions indicates that the head loss between the low-pressure service location, which has a pipeline elevation of 5.40 m, and the location of the elevated storage tank is 10 m. Under maximum-hourly-demand conditions, the head loss between the low-pressure service location and the elevated storage tank is 12 m. Determine the normal operating range for the water stored in the elevated tank.

Solution Under average demand conditions, the elevation z_0 of the hydraulic grade line (HGL) at the reservoir location is given by

$$z_0 = \frac{p_{min}}{\gamma} + z_{min} + h_L$$

where $p_{min} = 300$ kPa, $\gamma = 9.79$ kN/m^3, $z_{min} = 5.4$ m, and $h_L = 10$ m, which yields

$$z_0 = \frac{300}{9.79} + 5.4 + 10 = 46.0 \text{ m}$$

Under maximum-hourly-demand conditions, the elevation z_1 of the HGL at the service reservoir is given by

$$z_1 = \frac{300}{9.79} + 5.4 + 12 = 48.0 \text{ m}$$

Therefore, the operating range in the storage tank should be between elevations 46.0 m and 48.0 m.

It is important to keep in mind that the best hydraulic location and most economical design are not always the deciding factors in the location of an elevated tank. In some cases, the only acceptable location will be in an industrial area or public park. In cases where public opinion is very strong, a water utility may have to construct ground-level storage, which is more aesthetically acceptable.

2.5.8 Performance Criteria for Water-Distribution Systems

The primary functions of water-distribution systems are to: (1) meet the water demands of users while maintaining acceptable pressures in the system; (2) supply water for fire protection at specific locations within the system, while maintaining acceptable pressures for normal service throughout the remainder of the system; and (3) provide a sufficient level of redundancy to support a minimum level of reliable service during emergency conditions, such as an extended loss of power or a major water-main failure (Zipparro and Hasen, 1993). Real-time operation of water distribution systems is typically based on remote measurements of pressures and storage-tank water levels within the distribution system. The pressure and water-level data are typically transmitted to a central control facility via telemetry, and adjustments to the distribution system are made from the central facility by remote control of pumps and valves within the distribution system. These electronic control systems are generally called *supervisory control and data acquisition* (SCADA) systems (Chase, 2000). Operating criteria for service pressures and storage facilities are described below.

Maintaining adequate pressures in the distribution system while supplying the service demands requires that the system be analyzed on the basis of allowable pressures. Minimum acceptable pressures are necessary to prevent contamination of the water supply from cross connections. Criteria for minimum acceptable service pressures recommended by the Great Lakes Upper Mississippi River Board of State Public Health and Environmental Managers (GLUMB, 1987) and endorsed by the American Water Works Association (AWWA, 2003c) are typical of most water-distribution systems, and they are listed in Table 2.12. During main breaks, when the pressure in water-supply pipelines can drop below 140 kPa, it is not uncommon for

TABLE 2.12: Minimum Acceptable Pressures in Distribution Systems

Demand condition	Minimum acceptable pressure (kPa)
Average daily demand	240–410
Maximum daily demand	240–410
Maximum hourly demand	240–410
Fire situation	>140
Emergency conditions	>140

Source: GLUMB (1987).

a water utility to issue a "boil water" advisory because of the possibility of system contamination from cross connections (Chase, 2000). There are several considerations in assessing the adequacy of service pressures, including: (1) the pressure required at street level for excellent flow to a 3-story building is about 290 kPa; (2) flow is adequate for residential areas if the pressure is not reduced below 240 kPa; (3) the pressure required for adequate flow to a 20-story building is about 830 kPa, which is not desirable because of the associated leakage and waste; (4) very tall buildings are usually served with their own pumping equipment; and (5) it is usually desirable to maintain normal pressures of 410–520 kPa, since these pressures are adequate for the following purposes:

- To supply ordinary consumption for buildings up to 10 stories.
- To provide adequate sprinkler service in buildings of 4 to 5 stories.
- To provide direct hydrant service for quick response.
- To allow larger margin for fluctuations in pressure caused by clogged pipes and excessive length of service pipes.

Pressures higher than 650 kPa should be avoided if possible because of excessive leakage and water use, and the added burden of installing and maintaining pressure-reducing valves and other specialized equipment (Clark, 1990). Customers do not generally like high pressure because water comes out of a quickly opened faucet with too much force (AWWA, 2003c). In addition, excessive pressures decrease the life of water heaters and other plumbing fixtures.

2.5.9   Water Quality

The quality of water delivered to consumers can be significantly influenced by various components of a water-distribution system. The principal factors affecting water quality in distribution systems are the quality of the treated water fed into the system; the material and condition of the pipes, valves, and storage facilities that make up the system; and the amount of time that the water is kept in the system (Grayman et al., 2000; AWWA, 2003c). Key processes that affect water quality within the distribution system usually include the loss of disinfection residual, with resulting microbial regrowth, and the formation of disinfection byproducts such as trihalomethanes. Water-quality deterioration is often proportional to the time the water is resident in the distribution system. The longer the water is in contact with the pipe walls and is held in storage facilities, the greater the opportunity for water-quality changes. Generally, a hydraulic detention time of less than 7 days in the distribution system is recommended (AWWA, 2003c).

The velocity of flow in many water mains is very low, particularly in dead-end mains or in areas of low water consumption. As a result, corrosion products and other solids tend to settle on the pipe bottom. These deposits can be a source of color, odor, and taste in the water when they are stirred up by an increase in flow velocity or a reversal of flow in the distribution system. To prevent these sediments from accumulating and causing water-quality problems, pipe flushing is a typical maintenance routine. Flushing involves opening a hydrant located near the problem area, and keeping it open as long as needed to flush out the sediment, which typically requires the removal of up to three pipe volumes (AWWA, 2003c). Experience will teach an operator how often or how long certain areas should be flushed. Some systems find that dead-end mains must be flushed as often as weekly to avoid customer complaints of rusty water. The flow required for effective flushing is in the range of 0.75–1.1 m/s, with velocities limited to less than 3.1–3.7 m/s to avoid excessive scouring (AWWA, 2003c). If flushing proves to be inadequate for cleaning mains, air purging or cleaning devices, such as swabs or pigs, may need to be used.

In recognition of the influence of the water-distribution system on water quality, water-quality regulations in the United States require that water to be sampled at the entry point to the distribution system, at various points within the distribution system, and at consumers' taps (Kirmeyer et al., 1999). The computer program EPANET (Rossman, 2000) is widely used to simulate the water quality in distribution networks.

2.5.10 Network Analysis

Methodologies for analyzing pipe networks were discussed in Section 2.3, and these methods can be applied to any given pipe network to calculate the pressure and flow distribution under a variety of demand conditions. In complex pipe networks, the application of computer programs to implement these methodologies is standard practice (Haestad Methods Inc., 1997a; 1997b; 2002). Computer programs allow engineers to easily calculate the hydraulic performance of complex networks, the age of water delivered to consumers, and the origin of the delivered water. Water age, measured from the time the water enters the system, gives an indication of the overall quality of the delivered water. Steady-state analyses are usually adequate for assessing the performance of various components of the distribution system, including the pipelines, storage tanks, and pumping systems, while time-dependent (transient) simulations are useful in assessing the response of the system over short time periods (days or less), evaluating the operation of pumping stations and variable-level storage tanks, performing energy consumption and cost studies, and water-quality modeling (Velon and Johnson, 1993; Haestad Methods, Inc., 2002). Modelers frequently refer to time-dependent simulations as *extended-period simulations*, and several examples can be found in Larock et al. (2000).

An important part of analyzing large water-distribution systems is the *skeletonizing* of the system, which consists of representing the full water-distribution system by a subset of the system that includes only the most important elements. For example, consider the case of a water supply to the subdivision shown in Figure 2.28(a), where the system shown includes the service connections to the houses. A slight degree of skeletonization could be achieved by omitting the household service pipes (and their associated head losses) from consideration and accounting for the water demands at

FIGURE 2.28: Skeletonizing a water-distribution system

Source: Haestad Methods, 1997, *Practical Guide: Hydraulics and Hydrology*, pp. 61–62. Copyright © 1997 by Haestad Methods, Inc. Reprinted by permission.

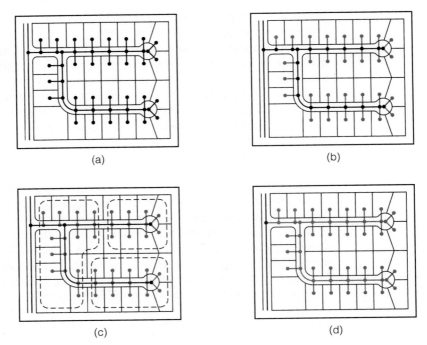

the tie-ins, as shown in Figure 2.28(b). This reduces the number of junctions from 48 to 19. Further skeletonization can be achieved by modeling just 4 junctions, consisting of the ends of the main piping and the major intersections, as shown in Figure 2.28(c). In this case, the water demands are associated with the nearest junctions to each of the service connections, and the dashed lines in Figure 2.28(c) indicate the service areas for each junction. A further level of skeletonization is shown in Figure 2.28(d), where the water supply to the entire subdivision is represented by a single node, at which the water demand of the subdivision is attributed.

Clearly, further levels of skeletonization could be possible in large water-distribution systems. As a general guideline, larger systems permit more degrees of skeletonization without introducing significant error in the flow conditions of main distribution pipes.

The results of a pipe-network analysis should generally include pressures and/or hydraulic grade line elevations at all nodes, flow, velocity, and head loss through all pipes, as well as rates of flow into and out of all storage facilities. These results are used to assess the hydraulic performance and reliability of the network, and they are to be compared with the guidelines and specifications required for acceptable performance.

2.6 Computer Models

Several good computer models are available for simulating flow in closed conduits, the majority of them developed primarily for computing flows and pressure distributions in water-supply networks. In engineering practice, the use of computer models to apply the fundamental principles covered in this chapter is usually essential. In choosing a model for a particular application, there are usually a variety of models to choose from; however, in doing work that is to be reviewed by regulatory bodies, models developed and maintained by agencies of the United States government have the greatest credibility and, perhaps more importantly, are almost universally acceptable in supporting permit applications and defending design protocols. A secondary guideline in choosing a model is that the simplest model that will accomplish the design objectives should be given the most serious consideration.

EPANET. EPANET is a water-distribution-system modeling package developed by the U.S. Environmental Protection Agency's Water Supply and Water Resources Division. It performs extended-period simulation of hydraulic and water-quality behavior within pressurized pipe networks. A more detailed description of EPANET can be found in Rossman (2000), and the program can be downloaded from the World Wide Web at: `www.epa.gov/ORD/NRMRL/wswrd/epanet.html`.

Summary

The hydraulics of flow in closed conduits is the basis for designing water-supply systems and other systems that involve the transport of water under pressure. The fundamental relationships governing flow in closed conduits are the conservation laws of mass, momentum, and energy; the forms of these equations that are most useful in engineering applications are derived from first principles. Of particular note is the momentum equation, the most useful form of which is the Darcy–Weisbach equation. Techniques for analyzing flows in both single and multiple pipelines, using the nodal and loop methods, are presented. Flows in closed conduits are usually driven by pumps, and the fundamentals of pump performance using dimensional analysis and similitude are presented. Considerations in selecting a pump include the specific speed under design conditions, the application of affinity laws in adjusting pump performance curves, the computation of operating points in pump-pipeline systems, practical limits on pump location based on the net positive suction head, and the performance of pump systems containing multiple units.

Water-supply systems are designed to meet service-area demands during the design life of the system. Projection of water demand involves the estimation of per-capita demands and population projections. Over short time periods, populations can be expected to follow either geometric, arithmetic, or declining-growth models, while over longer time periods a logistic growth curve or disaggregation methods might be more appropriate. Components of water-supply systems must be designed to accommodate daily fluctuations in water demand plus potential fire and emergency flows. The design periods and capacities of various components of water-supply systems are listed in Table 2.11. Key considerations in designing water-distribution systems include required service pressures (Table 2.12), pipeline selection and installation, and provision of adequate storage capacity to meet fire demands and emergency conditions.

Problems

2.1. Water at 20°C is flowing in a 100-mm diameter pipe at an average velocity of 2 m/s. If the diameter of the pipe is suddenly expanded to 150 mm, what is the new velocity in the pipe? What are the volumetric and mass flowrates in the pipe?

2.2. A 200-mm diameter pipe divides into two smaller pipes, each of diameter 100 mm. If the flow divides equally between the two smaller pipes and the velocity in the

200-mm pipe is 1 m/s, calculate the velocity and flowrate in each of the smaller pipes.

2.3. The velocity distribution in a pipe is given by the equation

$$v(r) = V_0 \left[1 - \left(\frac{r}{R} \right)^2 \right] \tag{2.149}$$

where $v(r)$ is the velocity at a distance r from the centerline of the pipe, V_0 is the centerline velocity, and R is the radius of the pipe. Calculate the average velocity and flowrate in the pipe in terms of V_0.

2.4. Calculate the momentum correction coefficient, β, for the velocity distribution given in Equation 2.149.

2.5. Water is flowing in a horizontal 200-mm diameter pipe at a rate of 0.06 m³/s, and the pressures at sections 100 m apart are equal to 500 kPa at the upstream section and 400 kPa at the downstream section. Estimate the average shear stress on the pipe and the friction factor, f.

2.6. Water at 20° C flows at a velocity of 2 m/s in a 250-mm diameter horizontal ductile iron pipe. Estimate the friction factor in the pipe, and state whether the pipe is hydraulically smooth or rough. Compare the friction factors derived from the Moody diagram, the Colebrook equation, and the Jain equation. Estimate the change in pressure over 100 m of pipeline. How would the friction factor and pressure change be affected if the pipe were not horizontal but 1 m lower at the downstream section?

2.7. Show that the Colebrook equation can be written in the (slightly) more convenient form:

$$f = \frac{0.25}{\{\log[(k_s/D)/3.7 + 2.51/(\mathrm{Re}\sqrt{f})]\}^2}$$

Why is this equation termed "(slightly) more convenient"?

2.8. If you had your choice of estimating the friction factor either from the Moody diagram or from the Colebrook equation, which one would you pick? Explain your reasons.

2.9. Water leaves a treatment plant in a 500-mm diameter ductile iron pipeline at a pressure of 600 kPa and at a flowrate of 0.50 m³/s. If the elevation of the pipeline at the treatment plant is 120 m, estimate the pressure in the pipeline 1 km downstream where the elevation is 100 m. Assess whether the pressure in the pipeline would be sufficient to serve the top floor of a 10-story building (approximately 30 m high).

2.10. A 25-mm diameter galvanized iron service pipe is connected to a water main in which the pressure is 400 kPa. If the length of the service pipe to a faucet is 20 m and the faucet is 2.0 m above the main, estimate the flowrate when the faucet is fully open.

2.11. A galvanized iron service pipe from a water main is required to deliver 300 L/s during a fire. If the length of the service pipe is 40 m and the head loss in the pipe is not to exceed 45 m, calculate the minimum pipe diameter that can be used. Use the Colebrook equation in your calculations.

2.12. Repeat Problem 2.11 using the Swamee–Jain equation.

2.13. Use the velocity distribution given in Problem 2.3 to estimate the kinetic energy correction factor, α, for turbulent pipe flow.

2.14. The velocity profile, $v(r)$, for turbulent flow in smooth pipes is sometimes estimated by the seventh-root law, originally proposed by Blasius (1911):

$$v(r) = V_0 \left(1 - \frac{r}{R} \right)^{\frac{1}{7}}$$

where V_0 is the maximum (centerline) velocity and R is the radius of the pipe. Estimate the energy and momentum correction factors corresponding to the seventh-root law.

2.15. Show that the kinetic energy correction factor, α, corresponding to the power-law velocity profile is given by Equation 2.75. Use this result to confirm your answer to Problem 2.14.

2.16. Water enters and leaves a pump in pipelines of the same diameter and approximately the same elevation. If the pressure on the inlet side of the pump is 30 kPa and a pressure of 500 kPa is desired for the water leaving the pump, what is the head that must be added by the pump, and what is the power delivered to the fluid?

2.17. Water leaves a reservoir at 0.06 m³/s through a 200-mm riveted steel pipeline that protrudes into the reservoir and then immediately turns a 90° bend with a local (minor) loss coefficient equal to 0.3. Estimate the length of pipeline required for the friction losses to account for 90% of the total losses, which includes both friction losses and so-called "minor losses." Would it be fair to say that for pipe lengths shorter than the length calculated in this problem, the word "minor" should not be used?

2.18. The top floor of an office building is 40 m above street level and is to be supplied with water from a municipal pipeline buried 1.5 m below street level. The water pressure in the municipal pipeline is 450 kPa, the sum of the local loss coefficients in the building pipes is 10.0, and the flow is to be delivered to the top floor at 20 L/s through a 150-mm diameter PVC pipe. The length of the pipeline in the building is 60 m, the water temperature is 20°C, and the water pressure on the top floor must be at least 150 kPa. Will a booster pump be required for the building? If so, what power must be supplied by the pump?

2.19. Water is pumped from a supply reservoir to a ductile iron water-transmission line, as shown in Figure 2.29. The high point of the transmission line is at point A, 1 km downstream of the supply reservoir, and the low point of the transmission line is at point B, 1 km downstream of A. If the flowrate through the pipeline is 1 m³/s, the diameter of the pipe is 750 mm, and the pressure at A is to be 350 kPa, then: (a) estimate the head that must be added by the pump; (b) estimate the power supplied by the pump; and (c) calculate the water pressure at B.

2.20. A pipeline is to be run from a water-treatment plant to a major suburban development 3 km away. The average daily demand for water at the development is 0.0175 m³/s, and the peak demand is 0.578 m³/s. Determine the required diameter of ductile iron pipe such that the flow velocity during peak demand is 2.5 m/s. Round the pipe diameter upward to the nearest 25 mm (i.e., 25 mm, 50 mm, 75 mm, ...). The water pressure at the development is to be at least 340 kPa during average demand conditions, and 140 kPa during peak demand. If the water at the treatment plant is stored in a ground-level reservoir where the level of the water is 10.00 m NGVD and the ground elevation at the suburban development is 8.80 m NGVD,

FIGURE 2.29: Problem 2.19

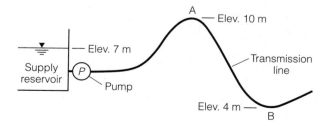

determine the pump power (in kilowatts) that must be available to meet both the average daily and peak demands.

2.21. Water flows at 5 m³/s in a 1 m × 2 m rectangular concrete pipe. Calculate the head loss over a length of 100 m.

2.22. Water flows at 10 m³/s in a 2 m × 2 m square reinforced-concrete pipe. If the pipe is laid on a (downward) slope of 0.002, what is the change in pressure in the pipe over a distance of 500 m?

2.23. Derive the Hazen–Williams head-loss relation, Equation 2.84, starting from Equation 2.82.

2.24. Compare the Hazen–Williams formula for head loss (Equation 2.84) with the Darcy–Weisbach equation for head loss (Equation 2.33) to determine the expression for the friction factor that is assumed in the Hazen–Williams formula. Based on your result, identify the type of flow condition incorporated in the Hazen–Williams formula (rough, smooth, or transition).

2.25. Derive the Manning head-loss relation, Equation 2.86.

2.26. Compare the Manning formula for head loss (Equation 2.86) with the Darcy–Weisbach equation for head loss (Equation 2.33) to determine the expression for the friction factor that is assumed in the Manning formula. Based on your result, identify the type of flow condition incorporated in the Manning formula (rough, smooth, or transition).

2.27. Determine the relationship between the Hazen–Williams roughness coefficient and the Manning roughness coefficient.

2.28. Given a choice between using the Darcy–Weisbach, Hazen–Williams, or Manning equations to estimate the friction losses in a pipeline, which equation would you choose? Why?

2.29. Water flows at a velocity of 2 m/s in a 300-mm new ductile iron pipe. Estimate the head loss over 500 m using: (a) the Hazen–Williams formula; (b) the Manning formula; and (c) the Darcy–Weisbach equation. Compare your results. Calculate the Hazen–Williams roughness coefficient and the Manning coefficient that should be used to obtain the same head loss as the Darcy–Weisbach equation.

2.30. Reservoirs A, B, and C are connected as shown in Figure 2.30. The water elevations in reservoirs A, B, and C are 100 m, 80 m, and 60 m, respectively. The three pipes connecting the reservoirs meet at the junction J, with pipe AJ being 900 m long, BJ 800 m long, CJ 700 m long, and the diameter of all pipes equal to 850 mm. If all pipes are made of ductile iron and the water temperature is 20°C, find the flow into or out of each reservoir.

2.31. The water-supply network shown in Figure 2.31 has constant-head elevated storage tanks at A and B, with inflows and withdrawals at C and D. The network is on flat terrain, and the pipeline characteristics are as follows:

FIGURE 2.30: Problem 2.30

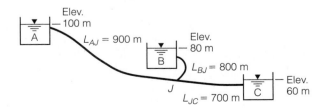

FIGURE 2.31: Problem 2.31

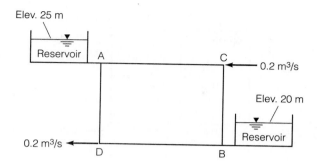

Pipe	L (km)	D (mm)
AD	1.0	400
BC	0.8	300
BD	1.2	350
AC	0.7	250

If all pipes are made of ductile iron, calculate the inflows/outflows from the storage tanks. Assume that the flows in all pipes are fully turbulent.

2.32. Consider the pipe network shown in Figure 2.32. The Hardy Cross method can be used to calculate the pressure distribution in the system, where the friction loss, h_f, is estimated using the equation

$$h_f = rQ^n$$

and all pipes are made of ductile iron. What value of r and n would you use for each pipe in the system? The pipeline characteristics are as follows:

Pipe	L (m)	D (mm)
AB	1000	300
BC	750	325
CD	800	200
DE	700	250
EF	900	300
FA	900	250
BE	950	350

You can assume that the flow in each pipe is hydraulically rough.

FIGURE 2.32: Problem 2.32

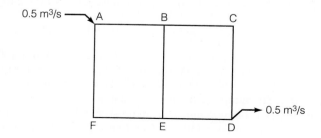

FIGURE 2.33: Problem 2.33

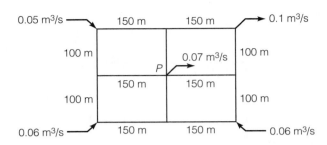

2.33. A portion of a municipal water-distribution network is shown in Figure 2.33, where all pipes are made of ductile iron and have diameters of 300 mm. Use the Hardy Cross method to find the flowrate in each pipe. If the pressure at point P is 500 kPa and the distribution network is on flat terrain, determine the water pressures at each pipe intersection.

2.34. What is the constant that can be used to convert the specific speed in SI units (Equation 2.111) to the specific speed in U.S. Customary units (Equation 2.112)?

2.35. What is the highest synchronous speed for a motor driving a pump?

2.36. Derive the affinity relationship for the power delivered to a fluid by two homologous pumps. [*Note*: This affinity relation is given by Equation 2.116.]

2.37. A pump is required to deliver 150 L/s (±10%) through a 300-mm diameter PVC pipe from a well to a reservoir. The water level in the well is 1.5 m below the ground surface and the water surface in the reservoir is 2 m above the ground surface. The delivery pipe is 300 m long, and local losses can be neglected. A pump manufacturer suggests using a pump with a performance curve given by

$$h_p = 6 - 6.67 \times 10^{-5}Q^2$$

where h_p is in meters and Q in L/s. Is this pump adequate?

2.38. A pump is to be selected to deliver water from a well to a treatment plant through a 300-m long pipeline. The temperature of the water is $20°$C, the average elevation of the water surface in the well is 5 m below the ground surface, the pump is 50 cm above the ground surface, and the water surface in the receiving reservoir at the water-treatment plant is 4 m above the ground surface. The delivery pipe is made of ductile iron (k_s = 0.26 mm) with a diameter of 800 mm. If the selected pump has a performance curve of $h_p = 12 - 0.1Q^2$, where Q is in m³/s and h_p is in m, then what is the flowrate through the system? Calculate the specific speed of the required pump (in U.S. Customary units), and state what type of pump will be required when the speed of the pump motor is 1200 rpm. Neglect local losses.

2.39. A pump lifts water through a 100-mm diameter ductile iron pipe from a lower to an upper reservoir (Figure 2.34). If the difference in elevation between the reservoir surfaces is 10 m, and the performance curve of the 2400-rpm pump is given by

$$h_p = 15 - 0.1Q^2$$

where h_p is in meters and Q in L/s, then estimate the flowrate through the system. If the pump manufacturer gives the required net positive suction head under these operating conditions as 1.5 m, what is the maximum height above the lower reservoir that the pump can be placed and maintain the same operating conditions?

2.40. Water is being pumped from reservoir A to reservoir F through a 30-m long PVC pipe of diameter 150 mm (see Figure 2.35). There is an open gate valve located at

FIGURE 2.34: Problem 2.39

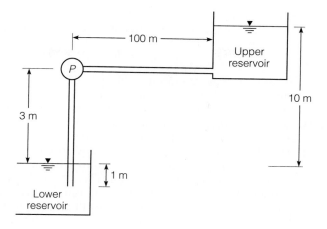

FIGURE 2.35: Problem 2.40

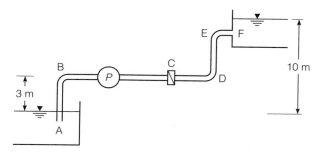

C; 90° bends (threaded) located at B, D, and E; and the pump performance curve is given by

$$h_p = 20 - 4713Q^2$$

where h_p is the head added by the pump in meters and Q is the flowrate in m³/s. The specific speed of the pump (in U.S. Customary units) is 3000. Assuming that the flow is turbulent (in the smooth, rough, or transition range) and the temperature of the water is 20°C, then (a) write the energy equation between the upper and lower reservoirs, accounting for entrance, exit, and local losses between A and F; (b) calculate the flowrate and velocity in the pipe; (c) if the required net positive suction head at the pump operating point is 3.0 m, assess the potential for cavitation in the pump (for this analysis you may assume that the head loss in the pipe is negligible between the intake and the pump); and (d) use the affinity laws to estimate the pump performance curve when the motor on the pump is changed from 800 rpm to 1600 rpm.

2.41. If the performance curve of a certain pump model is given by

$$h_p = 30 - 0.05Q^2$$

where h_p is in meters and Q is in L/s, what is the performance curve of a pump system containing n of these pumps in series? What is the performance curve of a pump system containing n of these pumps in parallel?

2.42. A pump is placed in a pipe system in which the energy equation (system curve) is given by

$$h_p = 15 + 0.03Q^2$$

where h_p is the head added by the pump in meters and Q is the flowrate through the system in L/s. The performance curve of the pump is

$$h_p = 20 - 0.08Q^2$$

What is the flowrate through the system? If the pump is replaced by two identical pumps in parallel, what would be the flowrate in the system? If the pump is replaced by two identical pumps in series, what would be the flowrate in the system?

2.43. Derive an expression for the population, P, versus time, t, where the growth rate is: (a) geometric, (b) arithmetic, and (c) declining.

2.44. The design life of a planned water-distribution system is to end in the year 2030, and the population in the town has been measured every 10 years since 1920 by the U.S. Census Bureau. The reported populations are tabulated below. Estimate the population in the town using: (a) graphical extension, (b) arithmetic growth projection, (c) geometric growth projection, (d) declining growth projection (assuming a saturation concentration of 100,000 people), and (e) logistic curve projection.

Year	Population
1920	25,521
1930	30,208
1940	30,721
1950	37,253
1960	38,302
1970	41,983
1980	56,451
1990	64,109

2.45. A city founded in 1950 had a population of 13,000 in 1960; 125,000 in 1975; and 300,000 in 1990. Assuming that the population growth follows a logistic curve, estimate the saturation population of the city.

2.46. The average demand of a population served by a water-distribution system is 580 L/d/capita, and the population at the end of the design life is estimated to be 100,000 people. Estimate the maximum daily demand and maximum hourly demand at the end of the design life.

2.47. Estimate the flowrate and volume of water required to provide adequate fire protection to a five-story office building constructed of joisted masonry. The effective floor area of the building is 5000 m^2.

2.48. What is the maximum fire flow and corresponding duration that can be estimated for any building?

2.49. A water-supply system is being designed to serve a population of 200,000 people, with an average per-capita demand of 600 L/d/person and a needed fire flow of 28,000 L/min. If the water supply is to be drawn from a river, what should be the design capacity of the supply pumps and water-treatment plant? For what duration must the fire flow be sustained, and what volume of water must be kept in the service reservoir to accommodate a fire? What should be the design capacity of the distribution pipes?

2.50. What is the minimum acceptable water pressure in a distribution system under average-daily-demand conditions?

2.51. Calculate the volume of storage required for the elevated storage reservoir in the water-supply system described in Problem 2.49.

CHAPTER 3

Flow in Open Channels

3.1 Introduction

In open-channel flows the water surface is exposed to the atmosphere. This type of flow is typically found in sanitary sewers, drainage conduits, canals, and rivers. Open-channel flow, sometimes referred to as *free-surface* flow, is more complicated than closed-conduit flow, since the location of the free surface is not constrained, and the depth of flow depends on such factors as the discharge and the shape and slope of the channel. Flows in conduits with closed sections, such as pipes, may be classified as either open-channel flow or closed-conduit flow, depending on whether the conduit is flowing full. A closed pipe flowing partially full is an open-channel flow, since the water surface is exposed to the atmosphere. Open-channel flow is said to be *steady* if the depth of flow at any specified location does not change with time; if the depth of flow varies with time, the flow is called *unsteady*. Most open-channel flows are analyzed under steady-flow conditions. The flow is said to be *uniform* if the depth of flow is the same at every cross section of the channel; if the depth of flow varies, the flow is *nonuniform* or *varied*. Uniform flow can be either steady or unsteady, depending on whether the flow depth changes with time; however, uniform flows are practically nonexistent in nature. More commonly, open-channel flows are either steady nonuniform flows or unsteady nonuniform flows. Open channels are classified as either *prismatic* or *nonprismatic*. Prismatic channels are characterized by an unvarying shape of the cross section, constant bottom slope, and relatively straight alignment. In nonprismatic channels, the cross section, alignment, and/or bottom slope changes along the channel. Constructed drainage channels such as pipes and canals tend to be prismatic, while natural channels such as rivers and creeks tend to be nonprismatic.

This chapter covers the basic principles of open-channel flow and derives the most useful forms of the continuity, momentum, and energy equations. These equations are applied to the computation of water-surface profiles, predicting the performance of hydraulic structures, and designing both lined and unlined open channels.

3.2 Basic Principles

The governing equations of flow in open channels are the continuity, momentum, and energy equations. Any flow in an open channel must satisfy all three of these equations. Analysis of flow can usually be accomplished with the control-volume form of the governing equations, and the most useful forms of these equations for steady open-channel flows are derived in the following sections.

FIGURE 3.1: Flow in an open channel

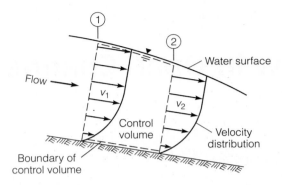

3.2.1 Steady-State Continuity Equation

Consider the case of steady nonuniform flow in the open channel illustrated in Figure 3.1. The flow enters and leaves the control volume normal to the control surfaces, with the inflow velocity distribution denoted by v_1 and the outflow velocity distribution by v_2; both the inflow and outflow velocities vary across the control surfaces. The steady-state continuity equation can be written as

$$\int_{A_1} \rho v_1 \, dA = \int_{A_2} \rho v_2 \, dA \tag{3.1}$$

where ρ is the density of the fluid, which can be taken as constant for most applications involving water. Defining V_1 and V_2 as the average velocities across A_1 and A_2, respectively, where

$$V_1 = \frac{1}{A_1} \int_{A_1} v_1 \, dA \tag{3.2}$$

and

$$V_2 = \frac{1}{A_2} \int_{A_2} v_2 \, dA \tag{3.3}$$

then, for an incompressible fluid (ρ = constant) such as water, the steady-state continuity equation (Equation 3.1) can be written as

$$\boxed{V_1 A_1 = V_2 A_2} \tag{3.4}$$

which is the same expression that was derived for steady flow of an incompressible fluid in closed conduits.

3.2.2 Steady-State Momentum Equation

Consider the case of steady nonuniform flow in the open channel illustrated in Figure 3.2. The steady-state momentum equation for the control volume shown in Figure 3.2 is given by

$$\sum F_x = \int_A \rho v_x \mathbf{v} \cdot \mathbf{n} \, dA \tag{3.5}$$

where F_x represents the forces in the flow direction, x; A is the surface area of the control volume; v_x is the flow velocity in the x-direction, and $\mathbf{n}$ is a unit normal directed

FIGURE 3.2: Steady nonuniform flow in an open channel

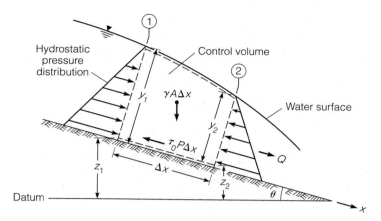

outward from the control volume. Since the velocities normal to the control surface are nonzero only for the inflow and outflow surfaces, Equation 3.5 can be written as

$$\sum F_x = \int_{A_2} \rho v_x^2 \, dA - \int_{A_1} \rho v_x^2 \, dA \tag{3.6}$$

where A_1 and A_2 are the upstream and downstream areas of the control volume, respectively. If the velocity is uniformly distributed (i.e., constant) across the control surface, then Equation 3.6 becomes

$$\sum F_x = \rho v_2^2 A_2 - \rho v_1^2 A_1 \tag{3.7}$$

where v_1 and v_2 are the velocities on the upstream and downstream faces of the control volume, respectively. In reality, velocity distributions in open channels are never uniform, and so it is convenient to define a *momentum correction coefficient*, β, by the relation

$$\beta = \frac{\int_A v^2 \, dA}{V^2 A} \tag{3.8}$$

where V is the mean velocity across the channel section of area A. The momentum correction coefficient, β, is sometimes called the *Boussinesq coefficient*, or simply the *momentum coefficient*. Applying the definition of the momentum correction coefficient to Equation 3.6 leads to the following form of the momentum equation:

$$\sum F_x = \rho \beta_2 V_2^2 A_2 - \rho \beta_1 V_1^2 A_1 \tag{3.9}$$

where β_1 and β_2 are the momentum correction coefficients at the upstream and downstream faces of the control volume, respectively. Values of β can be expected to be in the range 1.03–1.07 for regular channels, flumes, and spillways, and in the range 1.05–1.17 for natural streams (Chow, 1959).

Since the continuity equation requires that the discharge, Q, is the same at each cross section, then

$$Q = A_1 V_1 = A_2 V_2 \tag{3.10}$$

and the momentum equation (Equation 3.9) can be written as

$$\sum F_x = \rho \beta_2 Q V_2 - \rho \beta_1 Q V_1 \tag{3.11}$$

By definition, values of β must be greater than or equal to unity. In practice, however, deviations of β from unity are second-order corrections that are small relative to the uncertainties in the other terms in the momentum equation. By assuming

$$\beta_1 \approx \beta_2 = 1$$

the momentum equation can be written as

$$\sum F_x = \rho Q V_2 - \rho Q V_1 = \rho Q (V_2 - V_1) \tag{3.12}$$

Considering the forces acting on the control volume shown in Figure 3.2, then Equation 3.12 can be written as

$$\gamma A \Delta x \sin\theta - \tau_0 P \Delta x + \gamma A (y_1 - y_2) = \rho Q (V_2 - V_1) \tag{3.13}$$

where γ is the specific weight of the fluid; A is the average cross-sectional area of the control volume; Δx is the length of the control volume; θ is the inclination of the channel; τ_0 is the average shear stress on the control surface; P is the average (wetted) perimeter of the cross section of the control volume; and y_1 and y_2 are the upstream and downstream depths, respectively, at the control volume. The three force terms on the lefthand side of Equation 3.13 are the component of the weight of the fluid in the direction of flow, the shear force exerted by the channel boundary on the moving fluid, and the net hydrostatic force. If z_1 and z_2 are the elevations of the bottom of the channel at the upstream and downstream faces of the control volume, then

$$\sin\theta = \frac{z_1 - z_2}{\Delta x} \tag{3.14}$$

Combining Equations 3.13 and 3.14 and rearranging leads to

$$\tau_0 = -\gamma \frac{A}{P} \frac{\Delta z}{\Delta x} - \gamma \frac{A}{P} \frac{\Delta y}{\Delta x} - \gamma \frac{A}{P} \frac{V}{g} \frac{\Delta V}{\Delta x} \tag{3.15}$$

where Δz, Δy, and ΔV are defined by

$$\Delta z = z_2 - z_1, \quad \Delta y = y_2 - y_1, \quad \Delta V = V_2 - V_1 \tag{3.16}$$

and $V(= Q/A)$ is the average velocity in the control volume. The ratio A/P is commonly called the *hydraulic radius*, R, where

$$R = \frac{A}{P} \tag{3.17}$$

Combining Equations 3.15 and 3.17 and taking the limit as $\Delta x \to 0$ yields

$$\begin{aligned}
\tau_0 &= -\gamma R \left[\lim_{\Delta x \to 0} \frac{\Delta z}{\Delta x} + \lim_{\Delta x \to 0} \frac{\Delta y}{\Delta x} + \frac{V}{g} \lim_{\Delta x \to 0} \frac{\Delta V}{\Delta x} \right] \\
&= -\gamma R \left[\frac{dz}{dx} + \frac{dy}{dx} + \frac{V}{g} \frac{dV}{dx} \right] \\
&= -\gamma R \frac{d}{dx} \left[y + z + \frac{V^2}{2g} \right]
\end{aligned} \tag{3.18}$$

The term in brackets is the energy per unit weight of the fluid, E, defined as

$$E = y + z + \frac{V^2}{2g} \qquad (3.19)$$

It should be noted that the energy per unit weight of a fluid element is usually defined as $p/\gamma + z' + V^2/2g$, where z' is elevation of the fluid element relative to a defined datum. If the pressure is hydrostatic across the cross section, then $p/\gamma + z' = \text{constant} = y + z$, where y is the water depth and z is the elevation of the bottom of the channel. The energy per unit weight, E, can therefore be written as $y + z + V^2/2g$. A plot of E versus the distance along the channel is called the *energy grade line*. The momentum equation, Equation 3.18, can now be written as

$$\tau_0 = -\gamma R \frac{dE}{dx} \qquad (3.20)$$

or

$$\boxed{\tau_0 = \gamma R S_f} \qquad (3.21)$$

where S_f is equal to the slope of the energy grade line, which is taken as positive when it slopes downward in the direction of flow.

3.2.2.1   Darcy–Weisbach equation

In practical applications, it is useful to express the average shear stress, τ_0, on the boundary of the channel in terms of the flow and surface-roughness characteristics. A functional expression for the average shear stress, τ_0, can be expressed in the following form

$$\tau_0 = f_0(V, R, \rho, \mu, \varepsilon, \varepsilon', m, s) \qquad (3.22)$$

where V is the mean velocity in the channel, R is the hydraulic radius, ρ is the fluid density, μ is the dynamic viscosity of the fluid, ε is the characteristic size of the roughness projections on the channel boundary, ε' is the characteristic spacing of the roughness projections, m is a dimensionless form factor that describes the shape of the roughness elements, and s is a channel shape factor that describes the shape of the channel cross section. In accordance with the Buckingham pi theorem, the functional relationship given by Equation 3.22 between nine variables in three dimensions can also be expressed as a relation between six nondimensional groups as follows:

$$\frac{\tau_0}{\rho V^2} = f_1\left(\text{Re}, \frac{\varepsilon}{4R}, \frac{\varepsilon'}{4R}, m, s\right) \qquad (3.23)$$

where Re is the Reynolds number defined by the relation

$$\text{Re} = \frac{\rho V(4R)}{\mu} \qquad (3.24)$$

and the variable $4R$ is used instead of R for convenience in subsequent analyses.

The relationship given by Equation 3.23 is as far as dimensional analysis goes, and experimental data are necessary to determine an empirical relationship between the nondimensional groups. The problem of determining an empirical expression for the boundary shear stress in open-channel flow is similar to the problem faced in determining an empirical expression for the boundary shear stress in pipe flow, where $4R$ for circular conduits is equal to the pipe diameter. If the influences of the shape of the cross section and the arrangement of roughness elements on the boundary shear stress, τ_0, are small relative to the influence of the size of the roughness elements, then the shear stress can be expressed in the following functional form:

$$\boxed{\frac{\tau_0}{\rho V^2} = \frac{1}{8} f\left(\text{Re}, \frac{\varepsilon}{4R}\right)} \tag{3.25}$$

where the function f can be expected to closely approximate to the Darcy friction factor in pipes.

In reality, the friction factor, f, in Equation 3.25 has been observed to be a function of channel shape, decreasing roughly in the order of rectangular, triangular, trapezoidal, and circular channels (Chow, 1959). As channels become very wide or otherwise depart radically from the circle or semicircle, the friction factors derived from pipe experiments become less applicable to open channels (Daily and Harleman, 1966). Myers (1991) has shown that friction factors in wide rectangular open channels are as much as 45% greater than in narrow sections with the same Re and $\varepsilon/4R$. The question of how to account for the shape of an open channel in estimating the friction factor remains open (Pillai, 1997), and the assumption that the friction factor is independent of the arrangement and shape of the roughness projections has been shown to be invalid in gravel-bed streams with high boulder concentrations (Ferro, 1999). The transition from laminar to turbulent flow in open channels occurs at a Reynolds number of about 600, but laminar free-surface flows are seldom encountered in nature.

It is convenient to define three types of turbulent flow: *smooth*, *transition*, and *rough*. The flow is classified as "smooth" when the roughness projections on the channel boundary are submerged within a laminar sublayer, in which case the friction factor in open channels can be estimated by (Henderson, 1966)

$$f = \frac{0.316}{\text{Re}^{\frac{1}{4}}}, \quad \text{Re} < 10^5 \tag{3.26}$$

$$\frac{1}{\sqrt{f}} = -2.0 \log_{10}\left(\frac{2.51}{\text{Re}\sqrt{f}}\right), \quad \text{Re} > 10^5 \tag{3.27}$$

These relations are the same as the Blasius and Prandtl–von Kármán equations for flow in pipes. However, owing to the free surface and the interdependence of the hydraulic radius, discharge, and slope, the relationship between f and Re in open channel flow is not identical to that for pipe flow (Chow, 1959). The flow is classified as "rough" when the roughness projections on the channel boundary extend out of the laminar sublayer, creating sufficient turbulence that the friction factor depends only on the relative roughness. Under these conditions, the friction factor can be estimated by (ASCE, 1963)

$$\frac{1}{\sqrt{f}} = -2 \log_{10}\left(\frac{k_s}{12R}\right) \tag{3.28}$$

where k_s is the equivalent sand roughness of the open channel. Equation 3.28 is derived from the integration of the Nikuradse velocity distribution for fully rough turbulent flow over a trapezoidal open channel cross section (Keulegan, 1938), and gives a higher friction factor than the Prandtl–von Kármán equation that is used in pipe flow. In the "transition" region between (hydraulically) smooth and rough flow, the friction factor depends on both the Reynolds number and the relative roughness and can be approximated by (ASCE, 1963)

$$\frac{1}{\sqrt{f}} = -2\log_{10}\left(\frac{k_s}{12R} + \frac{2.5}{\mathrm{Re}\sqrt{f}} \right) \tag{3.29}$$

This relation differs slightly from the Colebrook equation for transition flow in closed conduits and can be applied in both smooth and rough flow. Equation 3.29 was originally suggested by Henderson (1966) for wide open channels (width/depth $\geq$ 10), and others have suggested similar formulations with different constants (Yen, 1991).

For rock-bedded channels, such as those in some natural streams or unlined canals, the larger rocks produce most of the resistance to flow. In those cases, Limerinos (1970) has shown that the friction factor, f, can be estimated from the size of the rock in the stream bed using the relation

$$\frac{1}{\sqrt{f}} = 1.2 + 2.03\log_{10}\left(\frac{R}{d_{84}} \right) \tag{3.30}$$

where d_{84} is the 84-percentile size of the rocks on the stream bed. A current review of open-channel flow resistance can be found in Yen (2002).

The three types of flow (smooth, transition, rough) can be identified using the *shear velocity Reynolds number*, $k_s u_*/\nu$, where u_* is the *shear velocity* defined by

$$u_* = \sqrt{\frac{\tau_0}{\rho}} = \sqrt{gRS_f} \tag{3.31}$$

and ν is the kinematic viscosity of the fluid. The transition flow region can be defined by (Henderson, 1966)

$$4 \leq \frac{u_* k_s}{\nu} \leq 100 \tag{3.32}$$

where the lower limit defines the end of smooth flow ($u_* k_s/\nu < 4$) and the upper limit defines the beginning of rough flow ($u_* k_s/\nu > 100$). There is still some debate in the defining the transition-flow region by the limits in Equation 3.32; for example, Yang (1996) defines the transition region as $5 \leq u_* k_s/\nu \leq 70$ and Rubin and Atkinson (2001) define it as $5 \leq u_* k_s/\nu \leq 80$.

Combining Equations 3.21 and 3.25 leads to the following form of the momentum equation, which is most commonly used in practice:

$$V = \sqrt{\frac{8g}{f}}\sqrt{RS_f} \tag{3.33}$$

where f is a function of the relative roughness and Reynolds number of the flow given by Equation 3.29. In cases where the flow is uniform, the slope of the energy grade line, S_f, is equal to the slope of the channel, S_0, since under these conditions

$$S_f = -\frac{d}{dx}\left(y + z + \frac{V^2}{2g}\right) = -\frac{dz}{dx} = S_0$$

where the depth, y, and average velocity, V, are constant and independent of x under uniform flow conditions. The depth of flow under steady uniform-flow conditions is called the *normal* depth of flow.

In many practical cases, determination of the flow conditions in open-channel flow requires simultaneous solution of Equations 3.29 and 3.33. In these cases, it is convenient to express the momentum equation, Equation 3.33, in the form

$$V\sqrt{f} = \sqrt{8gRS_0} \tag{3.34}$$

and the friction factor, f, given by Equation 3.29, can be expressed in the form

$$\frac{1}{\sqrt{f}} = -2\log_{10}\left(\frac{k_s}{12R} + \frac{2.5}{\frac{\rho V(4R)}{\nu}\sqrt{f}}\right) = -2\log_{10}\left(\frac{k_s}{12R} + \frac{0.625\mu}{\rho RV\sqrt{f}}\right) \tag{3.35}$$

Combining Equations 3.34 and 3.35 yields

$$Q = VA = -2A\sqrt{8gRS_0}\log_{10}\left(\frac{k_s}{12R} + \frac{0.625\mu}{\rho R^{3/2}\sqrt{8gS_0}}\right) \tag{3.36}$$

This derived relationship is particularly useful in relating the flowrate, Q, to the flow area, A, and hydraulic radius, R, for a given channel slope, S_0, and roughness height, k_s.

EXAMPLE 3.1

Water flows at a depth of 1.83 m in a trapezoidal, concrete-lined section ($k_s = 1.5$ mm) with a bottom width of 3 m and side slopes of 2:1 (H:V). The slope of the channel is 0.0005 and the water temperature is 20°C. Assuming uniform flow conditions, estimate the average velocity and flowrate in the channel.

Solution   The flow in the channel is illustrated in Figure 3.3. From the given data: $S_0 = 0.0005$, $A = 12.2$ m², $P = 11.2$ m, and

$$R = \frac{A}{P} = \frac{12.2}{11.2} = 1.09 \text{ m}$$

For concrete, $k_s = 1.5$ mm $= 0.0015$ m, and at 20°C, the density, ρ, and dynamic viscosity, μ, of water are given by $\rho = 998.2$ kg/m³, and $\mu = 0.00100$ N·s/m².

FIGURE 3.3: Flow in a trapezoidal channel

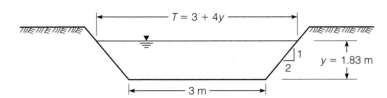

Substituting these data into Equation 3.36 gives the flowrate, Q, as

$$Q = -2A\sqrt{8gRS_0} \log_{10}\left(\frac{k_s}{12R} + \frac{0.625\mu}{\rho R^{3/2}\sqrt{8gS_0}} \right)$$

$$= -2(12.2)\sqrt{8(9.81)(1.09)(0.0005)} \log_{10}$$

$$\times \left(\frac{0.0015}{12(1.09)} + \frac{0.625(0.00100)}{(998.2)(1.09)^{3/2}\sqrt{8(9.81)(0.0005)}} \right) = 19.8 \text{ m}^3/\text{s}$$

and the average velocity, V, is given by

$$V = \frac{Q}{A} = \frac{19.8}{12.2} = 1.62 \text{ m/s}$$

Therefore, for the given flow depth in the channel, the flowrate is 19.8 m³/s and the average velocity is 1.62 m/s.

3.2.2.2 Manning equation

To fully appreciate the advantage of using the Darcy–Weisbach equation (Equation 3.33) compared with other flow equations used in practice, some historical perspective is needed. The Darcy–Weisbach equation is based primarily on the pipe experiments of Nikuradse and Colebrook, which were conducted between 1930 and 1940; however, observations on rivers and other large open channels began much earlier. In 1775, Chézy* proposed the following expression for the mean velocity in an open channel

$$V = C\sqrt{RS_f} \tag{3.37}$$

where C was referred to as the *Chézy coefficient*. Equation 3.37 has exactly the same form as Equation 3.33 and was derived in the same way, except that the functional dependence of the Chézy coefficient on the Reynolds number and the relative roughness was not considered. Comparing Equations 3.33 and 3.37, the Chézy coefficient is related to the friction factor by

$$C = \sqrt{\frac{8g}{f}} \tag{3.38}$$

In 1869, Ganguillet and Kutter (1869) published an elaborate formula for C that became widely used. In 1890, Manning (1890) demonstrated that the data used by Ganguillet and Kutter were fitted just as well by a simpler formula in which C varies as the sixth root of R, where

$$C = \frac{R^{\frac{1}{6}}}{n} \tag{3.39}$$

and n is a coefficient that is characteristic of the surface roughness alone. Since C is not a dimensionless quantity, values of n were specified to be consistent with length

---

*Antoine de Chézy (1718–1798) was a French engineer.

units measured in meters and time in seconds. If the length units are measured in feet, then Equation 3.39 becomes

$$C = 1.486 \frac{R^{\frac{1}{6}}}{n} \tag{3.40}$$

where 1.486 is the cube root of 3.281, the number of feet in a meter. When either Equation 3.39 or 3.40 is combined with the Chézy equation, the resulting expression is called the *Manning equation* or *Strickler equation* (in Europe) and is given by

$$\boxed{V = \frac{1}{n} R^{\frac{2}{3}} S^{\frac{1}{2}} \qquad \text{(SI units)}} \tag{3.41}$$

or

$$\boxed{V = \frac{1.486}{n} R^{\frac{2}{3}} S^{\frac{1}{2}} \qquad \text{(U.S. Customary units)}} \tag{3.42}$$

where $S = S_f = S_0$ under uniform flow conditions, and $1/n$ is called the *Strickler coefficient* when Equation 3.41 or 3.42 is called the Strickler equation (Douglas et al., 2001). The coefficient 1.486 in Equation 3.42 is much too precise considering the accuracy with which n is known and should not be written any more precisely than 1.49 or even 1.5 (Henderson, 1966). Typical values of the roughness coefficient, n, used in engineering practice are given in Table 3.1, where lower values of n are for surfaces in good condition, and higher values are for surfaces in poor condition. Based on a review of the literature, Johnson (1996) indicated that Manning n values estimated from field measurements typically have errors in the range of 5% to 35%.

Williamson (1951) investigated the consistency between the Manning equation and the Darcy–Weisbach equation. After making some minor adjustments to Niku-radse's data, Williamson found that for rough flow, the functional relation between the friction factor and the relative roughness, k_s/R, could be approximated by the relation

$$f = 0.113 \left(\frac{k_s}{R}\right)^{\frac{1}{3}} \tag{3.43}$$

Since the Chézy coefficient, C, is given by Equation 3.38 in terms of the friction factor, combining Equations 3.38 and 3.43 leads to (in SI units)

$$C = \frac{R^{\frac{1}{6}}}{0.038 d^{\frac{1}{6}}} \tag{3.44}$$

where the equivalent sand roughness, k_s, has been replaced by d, which represents the characteristic stone or gravel size on the bed of the open channel. Equation 3.44 is identical to the Manning expression for C if the roughness coefficient, n, is given by

$$\boxed{n = 0.038 d^{\frac{1}{6}}} \tag{3.45}$$

where d is measured in meters. Equation 3.45 was also proposed by Meyer-Peter and Muller (1948), where d is the 90-percentile size of bed material. White (1994) has shown that Equation 3.45 is valid in the range $2.5 < R/d < 250$ and that it is very

TABLE 3.1: Manning Coefficient for Open Channels

Channel type	Manning n	Range
Lined channels:		
Brick, glazed	0.013	0.011–0.015
Brick	—	0.012–0.018
Concrete, float finish	0.015	0.011–0.020
Asphalt	—	0.013–0.02
Rubble or riprap	—	0.020–0.035
Concrete, concrete bottom	0.030	0.020–0.035
Gravel bottom with riprap	0.033	0.023–0.036
Vegetal	—	0.030–0.40
Excavated or dredged channels:		
Earth, straight and uniform	0.027	0.022–0.033
Earth, winding, fairly uniform	0.035	0.030–0.040
Rock	0.040	0.035–0.050
Dense vegetation	—	0.05 –0.12
Unmaintained	0.080	0.050–0.12
Natural channels:		
Clean, straight	0.030	0.025–0.033
Clean, irregular	0.040	0.033–0.045
Weedy, irregular	0.070	0.050–0.080
Brush, irregular	—	0.07 –0.16
Floodplains:		
Pasture, no brush	0.035	0.030–0.050
Brush, scattered	0.050	0.035–0.070
Brush, dense	0.100	0.070–0.160
Timber and brush	—	0.10 –0.20

Source: ASCE (1982); Wurbs and James (2002); Bedient and Huber (2002).

close to the empirical relation proposed by Strickler (1923), where

$$n = 0.0417 d^{\frac{1}{6}}$$
(3.46)

which was derived from studies in gravel-bed streams in Switzerland, where d is the 50-percentile (median) size of the bed material in meters.

Any expression that relates the Manning roughness coefficient, n, to a roughness height, d, is based on the assumption that the friction factor, f, depends only on the relative roughness, d/R, in which case the flow condition is hydraulically rough and, according to Equation 3.32, this requires that

$$\frac{u_* d}{\nu} \geq 100 \qquad \text{or} \qquad \frac{(\sqrt{gRS_f})d}{\nu} \geq 100$$
(3.47)

This restriction can be combined with Equation 3.45 and the kinematic viscosity, ν, of water at 20°C to be put in the form (French, 1985)

$$\boxed{n^6 \sqrt{RS_f} \geq 9.6 \times 10^{-14}}$$
(3.48)

If this inequality is satisfied, then hydraulically rough conditions exist and the Manning equation can be applied.

EXAMPLE 3.2

Water flows at a depth of 1.83 m in a trapezoidal, concrete-lined section with a bottom width of 3 m and side slopes of 2:1 (H:V). The slope of the channel is 0.0005 and the water temperature is 20°C. Use the Manning equation to estimate the average velocity and flowrate in the channel, and assess the validity of using the Manning equation in this case.

Solution   The flow in the channel is illustrated in Figure 3.3, and the Manning equation gives the average velocity, V, as

$$V = \frac{1}{n} R^{\frac{2}{3}} S_0^{\frac{1}{2}}$$

From the given data, the channel slope, S_0, is 0.0005; the flow area, A, is 12.2 m²; the wetted perimeter, P, is 11.2 m; the hydraulic radius, R, is 1.09 m; and Table 3.1 indicates that a midrange roughness coefficient for concrete is $n = 0.015$. Using these data, the average velocity, V, given by the Manning equation is

$$V = \frac{1}{0.015}(1.09)^{\frac{2}{3}}(0.0005)^{\frac{1}{2}} = 1.58 \text{ m/s}$$

and the corresponding flowrate, Q, is

$$Q = AV = (12.2)(1.58) = 19.3 \text{ m}^3/\text{s}$$

According to Equation 3.48, the flow is hydraulically rough when

$$n^6\sqrt{RS_0} \geq 9.6 \times 10^{-14}$$

and in this case,

$$n^6\sqrt{RS_0} = (0.015)^6\sqrt{(1.09)(0.0005)} = 2.66 \times 10^{-13}$$

Since $n^6\sqrt{RS_0} \geq 9.6 \times 10^{-14}$, the flow is hydraulically rough and this condition for the validity of the Manning equation is met. As a second condition, the Manning equation assumes that $n = 0.038d^{1/6}$, which is valid in the range $2.5 < R/d < 250$ (White, 1994). In this case, d can be estimated by

$$d = \left(\frac{n}{0.038}\right)^6 = \left(\frac{0.015}{0.038}\right)^6 = 0.0038 \text{ m}$$

and therefore

$$\frac{R}{d} = \frac{1.09}{0.0038} = 290$$

Since $R/d > 250$, the linear relation between n and $d^{1/6}$ is not valid according to White (1994), and therefore the Manning equation is not strictly applicable in this case.

Sturm (2001) has presented an alternate analysis that improves the often-cited derivation of the relationship between the Manning roughness coefficient and the roughness height. Noting that the roughness coefficient, n, and friction factor, f, are related by

$$\frac{R^{1/6}}{n} = \sqrt{\frac{8g}{f}} \qquad (3.49)$$

and substituting the ASCE-recommended expression for the roughness coefficient (Equation 3.28) into Equation 3.49 and rearranging yields

$$\frac{n}{k_s^{1/6}} = \frac{\frac{1}{\sqrt{8g}}\left(\frac{R}{k_s}\right)^{\frac{1}{6}}}{2.0\log\left(12\frac{R}{k_s}\right)} \qquad (3.50)$$

Taking $g = 9.81$ m/s^2, Equation 3.50 is plotted in Figure 3.4, which illustrates that value of $n/k_s^{1/6}$ is effectively constant over a wide range of values of R/k_s, which is an essential assumption of the Manning equation. Equation 3.50 (with $g = 9.81$ m/s^2) gives a minimum value of

$$\boxed{\frac{n}{k_s^{1/6}} = 0.039} \qquad (3.51)$$

and $n/k_s^{1/6}$ is within $\pm 5\%$ of a constant value over a range of R/k_s given by $4 < R/k_s < 500$ as shown by Yen (1991). This range of R/k_s for the validity of the Manning equation differs somewhat from the range given by Hager (1999) as $3.6 < R/k_s < 360$. It is important to note that the R/k_s criterion for the validity of the Manning equation relies on the assumption that the flow is fully turbulent. Considering the fully turbulent criteria proposed by Henderson (1966), Yang (1996), and Rubin and Atkinson (2001), a liberal estimate of the turbulence criterion is that $u_* k_s/\nu > 70$. For water at $20°$C,

FIGURE 3.4: Variation of $n/k_s^{\frac{1}{6}}$ in fully turbulent flow.

Source: Sturm (2001).

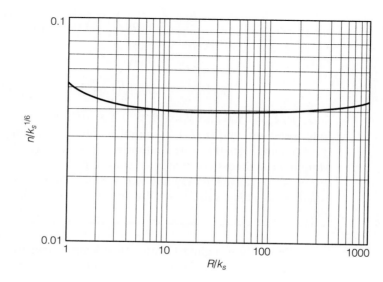

this turbulence requirement can be put in the form

$$\boxed{k_s\sqrt{RS_f} > 2.2 \times 10^{-5}} \tag{3.52}$$

For sand-bed channels, a variety of sediment sizes have been suggested by various investigators for estimating the value of k_s: Meyer-Peter and Muller (1948) suggested $k_s = d_{90}$, Einstein (1950) suggested $k_s = d_{65}$, and Simons and Richardson (1966) suggested $k_s = d_{85}$. In many natural river channels, the values of k_s are in the range of 3 cm–90 cm and are much larger than the actual diameters of the bed materials to account for boundary irregularities and bed forms (U.S. Army Corps of Engineers, 2002).

In summary, the validity of the Manning equation relies on two assumptions: (1) $n/k_s^{\frac{1}{6}}$ is constant; and (2) the flow is fully turbulent. These assumptions require that $4 < R/k_s < 500$ and that Equation 3.52 be satisfied. If the conditions for the validity of the Manning equation are met, the estimated velocities using the friction factor and Manning equation are consistent and interchangeable. It should be noted that the sixth-root relationship between the roughness height, k_s, and the roughness coefficient, n, means that large relative errors in estimating k_s result in a much smaller relative errors in estimating n. For example, a thousandfold change in the roughness height results in about a threefold change in n.

EXAMPLE 3.3

Water flows at a depth of 3.00 m in a trapezoidal, concrete-lined section with a bottom width of 3 m and side slopes of 2 : 1 (H : V). The slope of the channel is 0.0005 and the water temperature is 20°C. Assess the validity of using the Manning equation using $n = 0.014$.

Solution   From the given data: $n = 0.014, S_0 = 0.0005, A = 27$ m^2, $P = 16.42$ m, and $R = A/P = 1.64$ m. Assuming that the Manning equation is valid, then Equation 3.51 gives

$$k_s = \left(\frac{n}{0.039}\right)^6 = \left(\frac{0.014}{0.039}\right)^6 = 0.0021 \text{ m}$$

$$\frac{R}{k_s} = \frac{1.64}{0.0021} = 781$$

$$k_s\sqrt{RS_0} = 0.0021\sqrt{(1.64)(0.0005)} = 6.01 \times 10^{-5}$$

Since $R/k_s > 500$ and $k_s\sqrt{RS} > 2.2 \times 10^{-5}$ these results indicate that the flow is fully turbulent, and $(n/k_s)^{1/6}$ cannot be taken as a constant. Based on these results, the Manning equation formulation is not strictly valid in this case. Specifically, the Manning roughness coefficient, n, cannot be taken as a constant and independent of the depth of flow. Unfortunately, it is still common practice to accept the Manning equation as being valid based on the turbulence criterion ($k_s\sqrt{RS} > 2.2 \times 10^{-5}$) alone.

Where the roughness varies significantly over the perimeter of a channel, the *composite roughness* or *equivalent roughness*, n_e, of the channel is usually computed

TABLE 3.2: Composite Roughness Formulae

Formula	Assumption	Reference
$n_e = \left(\dfrac{\sum_{i=1}^{N} P_i n_i^{3/2}}{P} \right)^{2/3}$	Total cross-sectional mean velocity equal to subarea mean velocity	Horton (1933a), Einstein (1934)
$n_e = \left(\dfrac{\sum_{i=1}^{N} P_i n_i^{2}}{P} \right)^{1/2}$	Total resistance force equal to the sum of subarea resistance forces	Einstein and Banks (1951), Muhlhofer (1933), Pavlovskii (1931)
$n_e = \dfrac{PR^{5/3}}{\sum_{i=1}^{n} \frac{P_i R_i^{5/3}}{n_i}}$	Total discharge is sum of subarea discharges	Lotter (1933)[†]
$\ln n_e = \dfrac{\sum_{i=1}^{N} P_i y_i^{3/2} \ln n_i}{\sum_{i=1}^{N} P_i y_i^{3/2}}$	Logarithmic velocity distribution over depth y for wide channel	Krishnamurthy and Christensen (1972)[‡]

[†] P and R are the perimeter and hydraulic radius of the entire cross-section, respectively.
[‡] y_i is the average flow depth in Section i.

by first subdividing the channel section into N smaller sections, where the ith section has a roughness n_i, wetted perimeter, P_i, and hydraulic radius, R_i. Such channels are commonly called *compound channels* or *composite channels*, and several commonly used formulae for calculating n_e are listed in Table 3.2. The equations in Table 3.2 are based on differing assumptions on the distribution of flow in the compound channel. For example, the equation proposed by Horton (1933a) and Einstein (1934) assumes that each subdivided area has the same mean velocity, and the equation proposed by Lotter (1933) assumes that the total flow is equal to the sum of the flows in the subdivided areas. Several additional formulae for estimating the composite roughness in compound channels can be found in Yen (2002). It is interesting to note that the equations for estimating composite roughness require both the hydraulic radius and wetted perimeter, and hence depend on the way the subsections are divided and the relative amount of wetted perimeter of the subsections. However, according to Yen (2002), the differences between the composite roughness equations far exceed the differences due to subarea division methods. Motayed and Krishnamurthy (1980) used data from 36 natural-channel cross sections in Maryland, Georgia, Pennsylvania, and Oregon to assess the relative performance of Einstein and Banks (1951), Lotter (1933), and Krishnamurthy and Christensen (1972) formulae (shown in Table 3.2) and concluded that the formula proposed by Lotter (1933) performed best. In a broader context, the very limited data available in the literature show considerable scattering, and they are insufficient to identify which composite-roughness equations are most promising (Yen, 2002). Under these circumstances, it is prudent to determine the composite roughness using all of the proposed models, and select a representative composite roughness based on these estimates.

A composite roughness is usually necessary in analyzing flood flows in which a river bank is overtopped and the flow extends into the adjacent *floodplain*. The extent and geometry of the floodplain can be estimated using digital terrain models (Norman

TABLE 3.3: Manning Roughness Coefficient for Floodplains

Surface	Minimum	Normal	Maximum
Pasture, no brush:			
Short grass	0.025	0.030	0.035
High grass	0.030	0.035	0.050
Cultivated areas:			
No crop	0.020	0.030	0.040
Mature row crops	0.025	0.035	0.045
Mature field crops	0.030	0.040	0.050
Brush:			
Scattered brush, heavy weeds	0.035	0.050	0.070
Light brush and trees, in winter	0.035	0.050	0.060
Light brush and trees, in summer	0.040	0.060	0.080
Medium to dense brush, in winter	0.045	0.070	0.110
Medium to dense brush, in summer	0.070	0.100	0.160
Trees:			
Dense willows, summer, straight	0.110	0.150	0.200
Cleared land with tree stumps, no sprouts	0.030	0.040	0.050
Cleared land with tree stumps, heavy growth of sprouts	0.030	0.040	0.050
Heavy stand of timber, a few down trees, little undergrowth, flood stage below branches	0.080	0.100	0.120
Heavy stand of timber, a few down trees, little undergrowth, flood stage reaching branches	0.100	0.120	0.160

Source: Dodson (1999).

et al., 2003), and the roughness elements in the floodplain, typically consisting of shrubs, trees, and possibly houses, are usually much larger the roughness elements in the river channel. Typical values of the Manning roughness coefficient in floodplains are shown in Table 3.3.

EXAMPLE 3.4

The floodplain shown in Figure 3.5 can be divided into sections with approximately uniform roughness characteristics. The Manning n values for each section are as follows:

Section	n
1	0.040
2	0.030
3	0.015
4	0.013
5	0.017
6	0.035
7	0.060

Use the formulae in Table 3.2 to estimate the composite roughness.

FIGURE 3.5: Flow in a floodplain

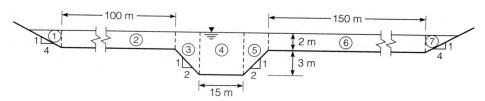

Solution   From the given shape of the floodplain (Figure 3.5), the following geometric characteristics are derived:

Section, i	P_i (m)	A_i (m²)	R_i (m)	n_i	y_i (m)
1	8.25	8.00	0.97	0.040	1.00
2	100	200	2.00	0.030	2.00
3	6.71	21	3.13	0.015	3.50
4	15.0	75	5.00	0.013	5.00
5	6.71	21	3.13	0.017	3.50
6	150	300	2.00	0.035	2.00
7	8.25	8.00	0.97	0.060	1.00
	295	633			

It should be noted that the internal water lines dividing the subsections are not considered a part of the wetted perimeter in computing the subsection hydraulic radius, R_i. This is equivalent to assuming that the internal shear stresses at the dividing water lines are negligible compared with the bottom shear stresses. For the given shape of the floodplain, the total perimeter, P, of the (compound) channel is 295 m, the total area, A, is 633 m², and hence the hydraulic radius, R, of the compound section is given by

$$R = \frac{A}{P} = \frac{633}{295} = 2.15 \text{ m}$$

Substituting these data into the formulae listed in Table 3.2 yields the following results:

Formula	n_e
Horton/Einstein	0.033
Einstein and Banks/Muhlhofer	0.033
Lotter	0.022
Krishnamurthy and Christensen	0.026
Average	0.029

It is apparent from this example that estimates of n_e can vary significantly. A conservative (high) estimate of the composite roughness is 0.033, and the average composite roughness predicted by the models is 0.029.

It should be noted that the composite-roughness approach to calculating the flowrates in compound channels is not entirely satisfactory, since the large-scale turbulence generated by the velocity shear between the main channel and overbank portion of the channel is not accounted for in the Manning equation (Bousmar and Zech, 1999).

The Manning equation is commonly used to estimate flow in rivers based on the cross-sectional geometry, slope, and a calibrated or estimated roughness coefficient. This approach may not be feasible in rivers at inaccessible locations, where estimating flows using remotely sensed hydraulic information might be the only practical alternative. Such estimates based on aerial or satellite observations of water-surface width and maximum channel width, and channel slope data obtained from topographic maps, have been shown to provide discharge estimates within a factor of 1.5–2 (Bjerklie et al., 2005).

3.2.2.3 Velocity distribution in open channels

In *wide channels*, lateral boundaries have negligible effects on the velocity distribution in the central portion of the channel. A wide channel is typically defined as one having a width that exceeds 10 times the flow depth (Franzini and Finnemore, 1997). The velocity distribution, $v(y)$, in wide open channels can be approximated by the relation (Vanoni, 1941),

$$v(y) = V + \frac{1}{\kappa}\sqrt{gdS_0}\left(1 + 2.3\log\frac{y}{d}\right) \qquad (3.53)$$

where V is the depth-averaged velocity, κ is the von Kármán constant (≈ 0.4), d is the depth of flow, S_0 is the slope of the channel, and y is the distance from the bottom of the channel. In channels that are not wide, the geometry of the lateral boundaries must be considered in estimating the velocity distribution, leading to much more complex expressions (e.g., Wilkerson and McGahan, 2005). Equation 3.53 indicates that the average velocity, V, in wide open channels occurs at $y/d = 0.368$, or at a distance of $0.368d$ above the bottom of the channel. This result is commonly approximated by the relation

$$V = v(0.4d) \qquad (3.54)$$

The average velocity, V, can also be related to the velocities at two depths using Equation 3.53, which yields

$$V = \frac{v(0.2d) + v(0.8d)}{2} \qquad (3.55)$$

It is standard practice of the U.S. Geological Survey to use measurements at $0.2d$ and $0.8d$ with Equation 3.55 to estimate the average velocity in channel sections with depth greater than 0.75 m (2.5 ft) and to use measurements at $0.4d$ with Equation 3.54 to estimate the average velocity in sections with depth less than 0.75 m (2.5 ft).

The velocity distribution given by Equation 3.53 indicates that the maximum velocity occurs at the water surface. However, the maximum velocity in open channels often occurs below the water surface, presumably as a result of air drag (Chiu et al., 2002; Finnemore and Franzini, 2002). In the Mississippi River, the maximum velocity has been observed to occur as much as one-third the water depth below the water surface (Gordon, 1992).

3.2.3 Steady-State Energy Equation

The steady-state energy equation for the control volume shown in Figure 3.2 is

$$\frac{dQ_h}{dt} - \frac{dW}{dt} = \int_A \rho e \, \mathbf{v} \cdot \mathbf{n} \, dA \qquad (3.56)$$

where Q_h is the heat added to the fluid in the control volume, W is the work done by the fluid in the control volume, A is the surface area of the control volume, ρ is the density of the fluid in the control volume, and e is the internal energy per unit mass of fluid in the control volume given by

$$e = gz + \frac{v^2}{2} + u \tag{3.57}$$

where z is the elevation of a fluid mass having a velocity v and internal energy u. The normal stresses on the inflow and outflow boundaries of the control volume are equal to the pressure, p, with shear stresses tangential to the control-volume boundaries. As the fluid moves with velocity $\mathbf{v}$, the power expended by the fluid is given by

$$\frac{dW}{dt} = \int_A p\mathbf{v} \cdot \mathbf{n}\, dA \tag{3.58}$$

No work is done by the shear forces, since the velocity is equal to zero on the channel boundary and the flow direction is normal to the direction of the shear forces on the inflow and outflow boundaries. Combining Equation 3.58 with the steady-state energy equation (Equation 3.56) leads to

$$\frac{dQ_h}{dt} = \int_A \rho\left(\frac{p}{\rho} + e\right)\mathbf{v} \cdot \mathbf{n}\, dA \tag{3.59}$$

Substituting the definition of the internal energy, e, (Equation 3.57) into Equation 3.59 gives the following form of the energy equation:

$$\frac{dQ_h}{dt} = \int_A \rho\left(h + gz + \frac{v^2}{2}\right)\mathbf{v} \cdot \mathbf{n}\, dA \tag{3.60}$$

where h is the enthalpy of the fluid defined by

$$h = \frac{p}{\rho} + u \tag{3.61}$$

Denoting the rate at which heat is being added to the fluid system by $\dot{Q}$, then the energy equation becomes

$$\dot{Q} = \int_A \rho\left(h + gz + \frac{v^2}{2}\right)\mathbf{v} \cdot \mathbf{n}\, dA \tag{3.62}$$

Considering the term $h + gz$, then

$$h + gz = \frac{p}{\rho} + u + gz = g\left(\frac{p}{\gamma} + z\right) + u \tag{3.63}$$

where γ is the specific weight of the fluid. Equation 3.63 indicates that $h + gz$ can be assumed to be constant across the inflow and outflow control surfaces, since the hydrostatic pressure distribution across the inflow/outflow boundaries guarantees that $p/\gamma + z$ is constant across the boundaries and since the internal energy, u, depends only on the temperature, which can be assumed constant across each boundary. Since

$\mathbf{v} \cdot \mathbf{n}$ is equal to zero over the impervious boundaries in contact with the fluid system, Equation 3.62 simplifies to

$$\dot{Q} = (h_1 + gz_1) \int_{A_1} \rho \mathbf{v} \cdot \mathbf{n} \, dA + \int_{A_1} \rho \frac{v^2}{2} \mathbf{v} \cdot \mathbf{n} \, dA$$

$$+ (h_2 + gz_2) \int_{A_2} \rho \mathbf{v} \cdot \mathbf{n} \, dA + \int_{A_2} \rho \frac{v^2}{2} \mathbf{v} \cdot \mathbf{n} \, dA \tag{3.64}$$

where the subscripts 1 and 2 refer to the inflow and outflow boundaries, respectively. Equation 3.64 can be further simplified by noting that the assumption of steady state requires that rate of mass inflow, $\dot{m}$, to the control volume is equal to the mass outflow rate, where

$$\dot{m} = \int_{A_2} \rho \mathbf{v} \cdot \mathbf{n} \, dA = - \int_{A_1} \rho \mathbf{v} \cdot \mathbf{n} \, dA \tag{3.65}$$

where the negative sign comes from the fact that the unit normal points out of the control volume. Also, the kinetic energy correction factors, α_1 and α_2, can be defined by the equations

$$\int_{A_1} \rho \frac{v^3}{2} \, dA = \alpha_1 \rho \frac{V_1^3}{2} A_1 \tag{3.66}$$

$$\int_{A_2} \rho \frac{v^3}{2} \, dA = \alpha_2 \rho \frac{V_2^3}{2} A_2 \tag{3.67}$$

where A_1 and A_2 are the areas of the inflow and outflow boundaries, respectively, and V_1 and V_2 are the corresponding mean velocities across these boundaries. The kinetic energy correction factor, α, is sometimes called the *Coriolis coefficient* or the *energy coefficient*. The kinetic energy correction factors, α_1 and α_2, are determined by the velocity profile across the flow boundaries. In regular channels, flumes, and spillways, α is typically in the range 1.1–1.2, while in natural channels α is typically in the range 1.1–2.0 (Wurbs and James, 2002). Combining Equations 3.64 to 3.67 leads to

$$\dot{Q} = -(h_1 + gz_1)\dot{m} - \alpha_1 \rho \frac{V_1^3}{2} A_1 + (h_2 + gz_2)\dot{m} + \alpha_2 \rho \frac{V_2^3}{2} A_2 \tag{3.68}$$

where the negative signs come from the fact that the unit normal points out of the inflow boundary, making $\mathbf{v} \cdot \mathbf{n}$ negative for the inflow boundary in Equation 3.64. Invoking the steady-state continuity equation

$$\rho V_1 A_1 = \rho V_2 A_2 = \dot{m} \tag{3.69}$$

and combining Equations 3.68 and 3.69 leads to

$$\dot{Q} = \dot{m} \left[\left(h_2 + gz_2 + \alpha_2 \frac{V_2^2}{2} \right) - \left(h_1 + gz_1 + \alpha_1 \frac{V_1^2}{2} \right) \right] \tag{3.70}$$

which can be put in the form

$$\frac{\dot{Q}}{\dot{m}g} = \left(\frac{p_2}{\gamma} + \frac{u_2}{g} + z_2 + \alpha_2\frac{V_2^2}{2g}\right) - \left(\frac{p_1}{\gamma} + \frac{u_1}{g} + z_1 + \alpha_1\frac{V_1^2}{2g}\right) \tag{3.71}$$

where p_1 is the pressure at elevation z_1 on the inflow boundary and p_2 is the pressure at elevation z_2 on the outflow boundary. Equation 3.71 can be further rearranged into the form

$$\left(\frac{p_1}{\gamma} + \alpha_1\frac{V_1^2}{2g} + z_1\right) = \left(\frac{p_2}{\gamma} + \alpha_2\frac{V_2^2}{2g} + z_2\right) + \left[\frac{1}{g}(u_2 - u_1) - \frac{\dot{Q}}{\dot{m}g}\right] \tag{3.72}$$

The energy loss per unit weight or *head loss*, h_L, is defined by the relation

$$h_L = \frac{1}{g}(u_2 - u_1) - \frac{\dot{Q}}{\dot{m}g} \tag{3.73}$$

Combining Equations 3.72 and 3.73 leads to a useful form of the energy equation:

$$\left(\frac{p_1}{\gamma} + \alpha_1\frac{V_1^2}{2g} + z_1\right) = \left(\frac{p_2}{\gamma} + \alpha_2\frac{V_2^2}{2g} + z_2\right) + h_L \tag{3.74}$$

The *head*, h, of the fluid at any cross section is defined by the relation

$$h = \frac{p}{\gamma} + \alpha\frac{V^2}{2g} + z \tag{3.75}$$

where p is the pressure at elevation z, γ is the specific weight of the fluid, α is the kinetic energy correction factor, and V is the average velocity across the channel. The head, h, measures the average energy per unit weight of the fluid flowing across a channel cross section, where the piezometric head, $p/\gamma + z$, is taken to be constant across the section, assuming a hydrostatic pressure distribution normal to the direction of the flow. In this case, the piezometric head can be written in terms of the invert (i.e., bottom) elevation of the cross section, z_0, and the depth of flow, y, as

$$\frac{p}{\gamma} + z = y\cos\theta + z_0 \tag{3.76}$$

where θ is the angle that the channel makes with the horizontal. Combining Equations 3.75 and 3.76 leads to the following expression for the head, h, at a flow boundary of a control volume:

$$h = y\cos\theta + \alpha\frac{V^2}{2g} + z_0 \tag{3.77}$$

and therefore the energy equation (Equation 3.74) can be written as

$$\left(y_1\cos\theta + \alpha_1\frac{V_1^2}{2g} + z_1\right) = \left(y_2\cos\theta + \alpha_2\frac{V_2^2}{2g} + z_2\right) + h_L \tag{3.78}$$

where y_1 and y_2 are the flow depths at the upstream and downstream sections of the control volume respectively, and z_1 and z_2 are the corresponding invert elevations.

Equation 3.78 is the most widely used form of the energy equation in practice and can be written in the summary form

$$h_1 = h_2 + h_L \qquad (3.79)$$

where h_1 and h_2 are the heads at the inflow and outflow boundaries of the control volume, respectively. A rearrangement of the energy equation (Equation 3.78) gives

$$y_2 \cos \theta + \alpha_2 \frac{V_2^2}{2g} = y_1 \cos \theta + \alpha_1 \frac{V_1^2}{2g} + (z_1 - z_2) - h_L \qquad (3.80)$$

which can also be written in the more compact form

$$\boxed{\left[y \cos \theta + \alpha \frac{V^2}{2g} \right]_2^1 = (S - S_0 \cos \theta)L} \qquad (3.81)$$

where L is the distance between the inflow and outflow sections of the control volume, S is the head loss per unit length ($=$ slope of the energy grade line) given by

$$S = \frac{h_L}{L} \qquad (3.82)$$

and S_0 is the slope of the channel defined as

$$S_0 = \frac{z_1 - z_2}{L \cos \theta} \qquad (3.83)$$

In contrast to our usual definition of slopes, downward slopes are generally taken as positive in open-channel hydraulics. The relationship between S_0 and $\cos \theta$ is shown in Table 3.4, where it is clear that for open-channel slopes less than 0.1 (10%), the error in assuming that $\cos \theta = 1$ is less than 0.5%. Since this error is usually less than the uncertainty in other terms in the energy equation, the energy equation (Equation 3.81) is frequently written as

$$\left[y + \alpha \frac{V^2}{2g} \right]_2^1 = (S - S_0)L, \qquad S_0 < 0.1 \qquad (3.84)$$

and the range of slopes corresponding to this approximation is frequently omitted. The slopes of rivers and canals in plain areas are usually on the order of 0.01% to 1%, while the slopes of mountain streams are typically on the order of 5% to 10% (Montes, 1998; Wohl, 2000). Channels with slopes in excess of 1% are commonly regarded as

TABLE 3.4: S_0 Versus $\cos \theta$

S_0	$\cos \theta$
0.001	0.9999995
0.01	0.99995
0.1	0.995
1	0.707

steep (Hunt, 1999). In cases where the head loss is entirely due to frictional resistance, the energy equation is written as

$$\left[y \, + \, \alpha \frac{V^2}{2g} \right]_2^1 = (S_f \, - \, S_0)L, \qquad S_0 < 0.1 \tag{3.85}$$

where S_f is the frictional head loss per unit length. Equation 3.85 is sometimes written in the expanded form

$$\left(y_1 \, + \, \alpha_1 \frac{V_1^2}{2g} \, + \, z_1 \right) = \left(y_2 \, + \, \alpha_2 \frac{V_2^2}{2g} \, + \, z_2 \right) + h_f, \qquad S_0 < 0.1 \tag{3.86}$$

where h_f is the head loss due to friction. Equation 3.86 is superficially similar to the Bernoulli equation, but the two equations are fundamentally different, because the Bernoulli equation is derived from the momentum equation (not the energy equation) and does not contain a head-loss term.

3.2.3.1   Energy grade line

The head at each cross section of an open channel, h, is given by

$$h = y \, + \, \alpha \frac{V^2}{2g} \, + \, z_0, \qquad S_0 < 0.1 \tag{3.87}$$

where y is the depth of flow, V is the average velocity over the cross section, z_0 is the elevation of the bottom of the channel, and S_0 is the slope of the channel. As stated previously, in most cases the slope restriction ($S_0 < 0.1$) is met, and this restriction is not explicitly stated in the definition of the head. When the head, h, at each section is plotted versus the distance along the channel, this curve is called the *energy grade line*. The point on the energy grade line corresponding to each cross section is located a distance $\alpha V^2/2g$ vertically above the water surface; between any two cross sections the elevation of the energy grade drops by a distance equal to the head loss, h_L, between the two sections. The energy grade line is useful in visualizing the state of a fluid as it flows along an open channel and especially useful in visualizing the performance of hydraulic structures in open-channel systems.

3.2.3.2   Specific energy

The *specific energy*, E, of a fluid is defined as the energy per unit weight of the fluid measured relative to the bottom of the channel and is given by

$$E = y \, + \, \alpha \frac{V^2}{2g} \tag{3.88}$$

or

$$E = y \, + \, \alpha \frac{Q^2}{2gA^2} \tag{3.89}$$

where Q is the volumetric flowrate and A is the cross-sectional area. The specific energy, E, was originally introduced by Bakhmeteff (1932), and appears explicitly in the energy equation, Equation 3.85, which can be written in the form

$$\boxed{E_1 - E_2 = h_L + \Delta z}$$ (3.90)

where E_1 and E_2 are the specific energies at the upstream and downstream sections, respectively, h_L is the head loss between sections, and Δz is the change in elevation between the upstream and downstream sections. In many cases of practical interest, E_1, h_L, and Δz can be calculated from given upstream flow conditions and channel geometry, E_2 can be calculated using Equation 3.90, and the downstream depth of flow, y, calculated from E_2 using Equation 3.88. For a given shape of the channel cross section and flowrate, Q, the specific energy, E, depends only on the depth of flow, y. The typical relationship between E and y given by Equation 3.89, for a constant value of Q, is shown in Figure 3.6. The salient features of Figure 3.6 are: (1) there is more than one possible flow depth for a given specific energy; and (2) the specific energy curve is asymptotic to the line

$$y = E$$ (3.91)

In accordance with the energy equation (Equation 3.90), the specific energy at any cross section can be expressed in terms of the specific energy of the fluid at an upstream section, change in the elevation of the bottom of the channel, and energy loss due to friction. The fact that there can be more than one possible depth for a given specific energy leads to the question of which depth will exist. The specific energy diagram, Figure 3.6, indicates that there exists a depth, y_c, at which the specific energy is a minimum. At this point, Equation 3.89 indicates that

$$\frac{dE}{dy} = 0 = 1 - \frac{Q^2}{gA_c^3}\frac{dA}{dy}$$ (3.92)

where A_c is the flow area corresponding to $y = y_c$, and the kinetic energy correction factor, α, has been taken equal to unity.

Referring to the general open-channel cross-section shown in Figure 3.7, it is clear that for small changes in the flow depth, y,

$$dA = T\,dy$$ (3.93)

FIGURE 3.6: Typical specific energy diagram

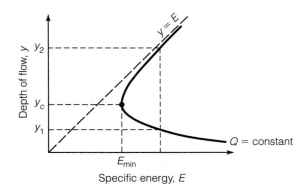

FIGURE 3.7: Typical channel cross section

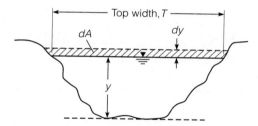

where dA is the increase in flow area resulting from a change in depth dy, and T is the top width of the channel when the flow depth is y. Equation 3.93 can be written as

$$\frac{dA}{dy} = T \tag{3.94}$$

which can be combined with Equation 3.92 to yield the following relationship under *critical flow* conditions:

$$\boxed{\frac{Q^2}{g} = \frac{A_c^3}{T_c}} \tag{3.95}$$

where T_c is the top width of the channel corresponding to $y = y_c$. Equation 3.95 forms the basis for calculating the *critical flow depth*, y_c, in a channel for a given flowrate, Q, since the lefthand side of the equation is known and the righthand side is a function of y_c for a channel of given shape. In most cases, an iterative solution for y_c is required (Patil et al., 2005). The specific energy under critical flow conditions, E_c, is then given by

$$E_c = y_c + \frac{A_c}{2T_c} \tag{3.96}$$

Defining the *hydraulic depth*, D, by the relation

$$D = \frac{A}{T} \tag{3.97}$$

then a Froude number, Fr, can be defined by

$$\mathrm{Fr}^2 = \frac{V^2}{gD} = \frac{Q^2 T}{gA^3} \tag{3.98}$$

Combining this definition of the Froude number with the critical flow condition given by Equation 3.95 leads to the relation

$$\boxed{\mathrm{Fr}_c = 1} \tag{3.99}$$

where Fr_c is the Froude number under critical flow conditions. When $y > y_c$, Equation 3.98 indicates that Fr < 1, and when $y < y_c$, Equation 3.98 indicates that Fr > 1. Flows where $y < y_c$ are called *supercritical*; where $y > y_c$, flows are called *subcritical*. It is apparent from the specific energy diagram, Figure 3.6, that when the flow conditions are close to critical, a relatively large change of depth occurs with small variations in specific energy. Flow under these conditions is unstable, and excessive wave action or undulations of the water surface may occur. Experiments in rectangular channels have shown that these instabilities can be avoided if Fr < 0.86 or Fr > 1.13 (U.S. Army Corps of Engineers, 1995).

In channels that have very wide and flat overbanks, and in cases where the energy coefficient depends on the depth of flow, multiple critical depths are possible (Jain, 2001; U.S. Army Corps of Engineers, 2002).

EXAMPLE 3.5

Determine the critical depth for water flowing at 10 m³/s in a trapezoidal channel with bottom width 3 m and side slopes of 2 : 1 (H : V).

Solution   The channel cross section is illustrated in Figure 3.8, where the depth of flow is y, and the top width, T, and flow area, A, are given by

$$T = 3 + 4y$$

$$A = 3y + 2y^2$$

Under critical flow conditions

$$\frac{Q^2}{g} = \frac{A_c^3}{T_c}$$

and since $Q = 10$ m³/s and $g = 9.81$ m/s², under critical flow conditions

$$\frac{10^2}{9.81} = \frac{(3y_c + 2y_c^2)^3}{(3 + 4y_c)}$$

Solving for y_c yields

$$y_c = 0.855 \text{ m}$$

Therefore, the critical depth of flow in the channel is 0.855 m. Flow under this condition is unstable and is generally avoided in design applications.

The critical flow condition described by Equation 3.95 can be simplified considerably in the case of flow in rectangular channels, where it is convenient to deal with the flow per unit width, q, given by

$$q = \frac{Q}{b} \tag{3.100}$$

where b is the width of the channel. The flow area, A, and top width, T, are given by

$$A = by \tag{3.101}$$

$$T = b \tag{3.102}$$

FIGURE 3.8: Trapezoidal cross section

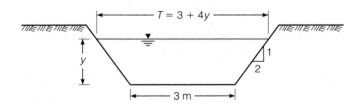

The critical flow condition given by Equation 3.95 then becomes

$$\frac{(qb)^2}{g} = \frac{(by_c)^3}{b}$$

(3.103)

which can be solved to yield the critical flow depth

$$y_c = \left(\frac{q^2}{g}\right)^{\frac{1}{3}}$$

(3.104)

and corresponding critical energy

$$E_c = y_c + \frac{q^2}{2gy_c^2}$$

(3.105)

Combining Equations 3.104 and 3.105 leads to the following simplified form of the minimum specific energy in rectangular channels:

$$E_c = \frac{3}{2} y_c$$

(3.106)

The specific energy diagram illustrated in Figure 3.9 for a rectangular channel is similar to the nonrectangular case shown in Figure 3.6, with the exception that the specific energy curve for a rectangular channel corresponds to a fixed value of q rather than Q in a nonrectangular channel. The specific energy diagram in Figure 3.9 indicates that the specific energy curve shifts upward and to the right for increasing values of q. This form of the specific energy diagram is particularly informative in understanding what happens to the flow in a rectangular channel when there is a constriction, such as when the channel width is narrowed to accommodate a bridge. Suppose that the flow per unit width upstream of the constriction is q_1 and the depth of flow at this location is y_1. If the channel is constricted so that the flow per unit width becomes q_2, then, provided energy losses in the constriction are minimal, Figure 3.9 indicates that there are two possible flow depths in the constricted section, y_2 and y_2'. Neglecting energy losses in the constriction is reasonable if the constriction is smooth and takes place over a relatively short distance. Figure 3.9 indicates that it is physically

FIGURE 3.9: Specific energy diagram for rectangular channel

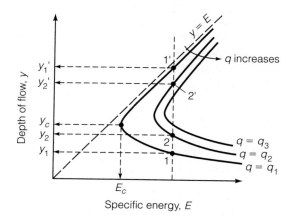

impossible for the flow depth in the constriction to be y_2', since this would require the flow per unit width, q, to increase and then decrease to reach y_2'. Since the flow per unit width decreases monotonically in a constriction, the flow depth can only go from y_1 to y_2. A similar case arises when the flow depth upstream of the constriction is y_1' and the possible flow depths at the constriction are y_2' and y_2. In this case, only y_2' is possible, since an increase and then a decrease in q would be required to achieve a flow depth of y_2.

Based on this analysis, it is clear that if the flow upstream of the constriction is subcritical ($y > y_c$ or Fr < 1), then the flow in the constriction must be either subcritical or critical, and if the flow upstream of the constriction is supercritical ($y < y_c$ or Fr > 1), then the flow in the constriction must be either supercritical or critical. If the flow upstream of the constriction is critical, then flow through the constriction is not possible under the existing flow conditions and the upstream flow conditions must necessarily change upon installation of the constriction in the channel.

Regardless of whether the flow upstream of the constriction is subcritical or supercritical, Figure 3.9 indicates that a maximum constriction will cause critical flow to occur at the constriction. A larger constriction (smaller opening) than that which causes critical flow to occur at the constriction is not possible based on the available specific energy upstream of the constriction. Any further constriction will result in changes in the upstream flow conditions to maintain critical flow at the constriction. Under these circumstances, the flow is said to be *choked*. Choked flows can result either from narrowing channels or from raising the invert elevation of the channel. Narrowing the channel increases the flowrate per unit width, q, while maintaining the available specific energy, E, whereas raising the channel invert maintains q and reduces E.

EXAMPLE 3.6

A rectangular channel 1.30 m wide carries 1.10 m³/s of water at a depth of 0.85 m. (a) If a 30-cm wide pier is placed in the middle of the channel, find the elevation of the water surface at the constriction. (b) What is the minimum width of the constriction that will not cause a rise in the upstream water surface?

Solution

(a) The cross sections of the channel upstream of the constriction and at the constriction are shown in Figure 3.10. Neglecting the energy losses between the constriction and the upstream section, the energy equation requires that the specific energy at the constriction be equal to the specific energy at the upstream section. Therefore,

$$y_1 + \frac{V_1^2}{2g} = y_2 + \frac{V_2^2}{2g}$$

where Section 1 refers to the upstream section and Section 2 refers to the constricted section. In this case, $y_1 = 0.85$ m, and

$$V_1 = \frac{Q}{A_1} = \frac{1.10}{(0.85)(1.30)} = 1.00 \text{ m/s}$$

FIGURE 3.10: Constriction in rectangular channel

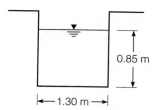

Upstream of constriction

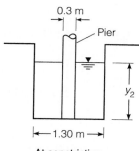

At constriction

The specific energy at Section 1, E_1, is

$$E_1 = y_1 + \frac{V_1^2}{2g} = 0.85 + \frac{1.00^2}{2(9.81)} = 0.901 \text{ m}$$

Equating the specific energies at Sections 1 and 2 yields

$$0.901 = y_2 + \frac{Q^2}{2gA^2}$$

$$= y_2 + \frac{1.10^2}{2(9.81)[(1.30 - 0.30)y_2]^2}$$

which simplifies to

$$y_2 + \frac{0.0617}{y_2^2} = 0.901$$

There are three solutions to this cubic equation: $y_2 = 0.33$ m, 0.80 m, and -0.23 m. Of the two positive depths, we must select the depth corresponding to the same flow condition as upstream. At the upstream section, $\text{Fr}_1 = V_1/\sqrt{gy_1} = 0.35$; therefore, the upstream flow is subcritical and the flow at the constriction must also be subcritical. The flow depth must therefore be

$$y_2 = 0.80 \text{ m}$$

where $\text{Fr}_2 = V_2/\sqrt{gy_2} = 0.49$. The other depth ($y_2 = 0.33$ m, $\text{Fr}_2 = 1.9$) is supercritical and cannot be achieved.

(b) The minimum width of constriction that does not cause the upstream depth to change is associated with critical flow conditions at the constriction. Under these conditions (for a rectangular channel)

$$E_1 = E_2 = E_c = \frac{3}{2} y_c = \frac{3}{2} \left(\frac{q^2}{g} \right)^{\frac{1}{3}}$$

If b is the width of the constriction that causes critical flow, then

$$E_1 = \frac{3}{2} \left[\frac{(Q/b)^2}{g} \right]^{\frac{1}{3}}$$

From the given data: $E_1 = 0.901$ m, and $Q = 1.10$ m³/s. Therefore

$$0.901 = \frac{3}{2} \left[\frac{1.10^2}{b^2(9.81)} \right]^{\frac{1}{3}}$$

which gives

$$b = 0.75 \text{ m}$$

If the constricted channel width is any less than 0.75 m, the flow will be choked and the upstream flow depth will increase.

The specific energy analyses covered in this section assume negligible energy losses and that the kinetic energy correction factor, α, is approximately equal to unity. Although these approximations are valid in many cases, in diverging transitions with angles exceeding 8°, flows tend to separate from the side of the channel, causing large energy losses and substantial increases in α (Montes, 1998). Under these conditions, the assumption of a constant specific energy and a value of α approximately equal to unity are not justified. To minimize energy losses in transitions, the U.S. Department of Agriculture (USDA, 1977) recommends that channel sides not converge at an angle greater than 14° or diverge at an angle greater than 12.5°.

For the general case of contractions and expansions in open channels, the head loss, h_e, is usually expressed in the form (U.S. Army Corps of Engineers, 2002)

$$h_e = C \left| \alpha_2 \frac{V_2^2}{2g} - \alpha_1 \frac{V_1^2}{2g} \right| \tag{3.107}$$

where C is either an expansion coefficient, C_e, or a contraction coefficient, C_c, α_1 and α_2 are the energy coefficients at the upstream and downstream sections, respectively, and V_1 and V_2 are the average velocities at the upstream and downstream sections, respectively. When the change in cross section is small, the coefficients C_c and C_e are typically on the order of 0.3 and 0.1, respectively; when the change in cross section is abrupt, such as at bridges, C_e and C_c may be as high as 0.8 and 0.6, respectively (U.S. Army Corps of Engineers, 2002).

3.3 Water-Surface Profiles

3.3.1 Profile Equation

The equation describing the shape of the water surface in an open channel can be derived from the energy equation, Equation 3.85, which is of the form

$$S_0 - S_f = \frac{\Delta\left(y + \alpha\frac{V^2}{2g}\right)}{\Delta x} \tag{3.108}$$

where S_0 is the slope of the channel, S_f is the slope of the energy grade line, y is the depth of flow, α is the kinetic energy correction factor, V is the average velocity, x is a coordinate measured along the channel (the flow direction defined as positive), and Δx is the distance between the upstream and downstream sections. Equation 3.108 can be further rearranged into

$$S_0 - S_f = \frac{\Delta y}{\Delta x} + \frac{\Delta\left(\alpha\frac{V^2}{2g}\right)}{\Delta x} \tag{3.109}$$

Taking the limit of Equation 3.109 as $\Delta x \to 0$, and invoking the definition of the derivative, yields

$$
\begin{aligned}
S_0 - S_f &= \lim_{\Delta x \to 0}\frac{\Delta y}{\Delta x} + \lim_{\Delta x \to 0}\frac{\Delta\left(\alpha\frac{V^2}{2g}\right)}{\Delta x} \\[2mm]
&= \frac{dy}{dx} + \frac{d}{dx}\left(\alpha\frac{V^2}{2g}\right) \\[2mm]
&= \frac{dy}{dx} + \frac{d}{dy}\left(\alpha\frac{V^2}{2g}\right)\frac{dy}{dx} \\[2mm]
&= \frac{dy}{dx}\left[1 + \frac{d}{dy}\left(\alpha\frac{Q^2}{2gA^2}\right)\right] \\[2mm]
&= \frac{dy}{dx}\left[1 - \alpha\frac{Q^2}{gA^3}\frac{dA}{dy}\right]
\end{aligned}
\tag{3.110}
$$

where Q is the (constant) flowrate and A is the (variable) cross-sectional flow area in the channel. Recalling that

$$\frac{dA}{dy} = T \tag{3.111}$$

where T is the top width of the channel and the hydraulic depth, D, of the channel is defined as

$$D = \frac{A}{T} \tag{3.112}$$

then the Froude number, Fr, of the flow can be written as

$$\text{Fr} = \frac{V}{\sqrt{gD}} = \frac{Q\sqrt{T}}{A\sqrt{gA}} = \frac{Q}{\sqrt{gA^3}}\sqrt{\frac{dA}{dy}} \tag{3.113}$$

or

$$\text{Fr}^2 = \frac{Q^2}{gA^3}\frac{dA}{dy} \tag{3.114}$$

Combining Equations 3.110 and 3.114 and rearranging yields

$$\boxed{\frac{dy}{dx} = \frac{S_0 - S_f}{1 - \alpha\text{Fr}^2}} \tag{3.115}$$

This differential equation describes the water-surface profile in open channels. To appreciate the utility of Equation 3.115, consider the relative magnitudes of the channel slope, S_0, and the friction slope, S_f. According to the Manning equation, for any given flowrate,

$$\frac{S_f}{S_0} = \left(\frac{A_n R_n^{\frac{2}{3}}}{AR^{\frac{2}{3}}}\right)^2 \tag{3.116}$$

where A_n and R_n are the cross-sectional area and hydraulic radius under normal flow conditions ($S_f = S_0$), and A and R are the actual cross-sectional area and hydraulic radius of the flow. Since $AR^{\frac{2}{3}} > A_n R_n^{\frac{2}{3}}$ when $y > y_n$, Equation 3.116 indicates that

$$S_f > S_0 \quad \text{when} \quad y < y_n, \quad \text{and} \quad S_f < S_0 \quad \text{when} \quad y > y_n \tag{3.117}$$

or

$$S_0 - S_f < 0 \quad \text{when} \quad y < y_n, \quad \text{and} \quad S_0 - S_f > 0 \quad \text{when} \quad y > y_n \tag{3.118}$$

It has already been shown that

$$\text{Fr} > 1 \quad \text{when} \quad y < y_c, \quad \text{and} \quad \text{Fr} < 1 \quad \text{when} \quad y > y_c \tag{3.119}$$

or

$$1 - \text{Fr}^2 < 0 \quad \text{when} \quad y < y_c, \quad \text{and} \quad 1 - \text{Fr}^2 > 0 \quad \text{when} \quad y > y_c \tag{3.120}$$

Based on Equations 3.118 and 3.120, the sign of the numerator in Equation 3.115 is determined by the magnitude of the flow depth, y, relative to the normal depth, y_n, and the sign of the denominator is determined by the magnitude of the flow depth, y, relative to the critical depth, y_c (assuming $\alpha \approx 1$). Therefore, the slope of the water surface, dy/dx, is determined by the relative magnitudes of y, y_n, and y_c.

3.3.2 Classification of Water-Surface Profiles

In hydraulic engineering, channel slopes are classified based on the relative magnitudes of the normal depth, y_n, and the critical depth, y_c. The hydraulic classification of slopes is shown in Table 3.5 and is illustrated in Figure 3.11. The range of flow depths for

FIGURE 3.11: Slope classifications

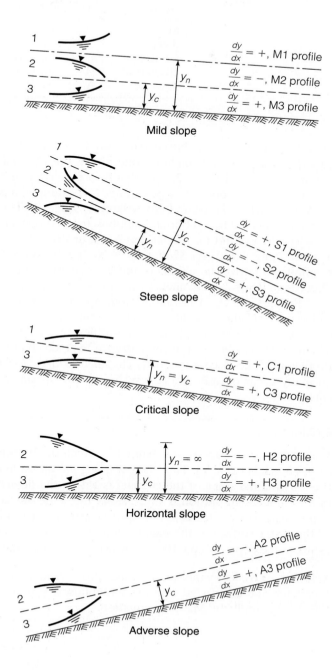

$\dfrac{dy}{dx} = +$, M1 profile

$\dfrac{dy}{dx} = -$, M2 profile

$\dfrac{dy}{dx} = +$, M3 profile

Mild slope

$\dfrac{dy}{dx} = +$, S1 profile

$\dfrac{dy}{dx} = -$, S2 profile

$\dfrac{dy}{dx} = +$, S3 profile

Steep slope

$\dfrac{dy}{dx} = +$, C1 profile

$\dfrac{dy}{dx} = +$, C3 profile

Critical slope

$\dfrac{dy}{dx} = -$, H2 profile

$\dfrac{dy}{dx} = +$, H3 profile

Horizontal slope

$\dfrac{dy}{dx} = -$, A2 profile

$\dfrac{dy}{dx} = +$, A3 profile

Adverse slope

TABLE 3.5: Hydraulic Classification of Slopes

Name	Type	Condition
Mild	M	$y_n > y_c$
Steep	S	$y_n < y_c$
Critical	C	$y_n = y_c$
Horizontal	H	$y_n = \infty$
Adverse	A	$S_0 < 0$

each slope can be divided into two or more zones, delimited by the normal and critical flow depths, where the highest zone above the channel bed is called Zone 1, the intermediate zone is Zone 2, and the lowest zone is Zone 3. Water-surface profiles are classified based on both the type of slope (for example, type M) and the zone in which the water surface is located (for example, Zone 2). Therefore, each water-surface profile is classified by a letter and number: for example, an M2 profile indicates a mild slope ($y_n > y_c$) and the actual depth y is in Zone 2 ($y_n < y < y_c$). In the case of nonsustaining slopes (Horizontal, Adverse), there is no Zone 1, as the normal depth is infinite in horizontal channels and is nonexistent in channels with adverse slope. There is no Zone 2 in channels with critical slope, as $y_n = y_c$.

Profiles in Zone 1 normally occur upstream of control structures such as dams and weirs, and these profiles are sometimes classified as *backwater curves*. Profiles in Zone 2, with the exception of S2 profiles, occur upstream of free overfalls, while S2 curves generally occur at the entrance to steep channels leading from a reservoir. Profiles in Zone 2 are sometimes classified as *drawdown curves*. Profiles in Zone 3 normally occur on mild and steep slopes downstream of control structures such as gates.

EXAMPLE 3.7

Water flows in a trapezoidal channel in which the bottom width is 5 m and the side slopes are 1.5:1 (H:V). The channel lining has an estimated Manning n of 0.04, and the longitudinal slope of the channel is 1%. If the flowrate is 60 m³/s and the depth of flow at a gaging station is 4 m, classify the water-surface profile, state whether the depth increases or decreases in the downstream direction, and calculate the slope of the water surface at the gaging station. On the basis of this water-surface slope, estimate the depth of flow 100 m downstream of the gaging station.

Solution To classify the water-surface profile, the normal and critical depths must be calculated and contrasted with the actual flow depth of 4 m. To calculate the normal flow depth, apply the Manning equation

$$Q = \frac{1}{n}A_n R_n^{2/3} S_0^{1/2} = \frac{1}{n}\frac{A_n^{5/3}}{P_n^{2/3}}S_0^{1/2}$$

where Q is the flowrate ($=60$ m³/s), A_n and P_n are the areas and wetted perimeters under normal flow conditions, and S_0 is the longitudinal slope of the channel ($=0.01$).

The Manning equation can be written in the more useful form

$$\frac{A_n^5}{P_n^2} = \left[\frac{Qn}{\sqrt{S_0}}\right]^3$$

where the lefthand side is a function of the normal flow depth, y_n and the righthand side is in terms of given data. Substituting the given parameters leads to

$$\frac{A_n^5}{P_n^2} = \left[\frac{(60)(0.04)}{\sqrt{0.01}}\right]^3 = 13,824$$

Since

$$A_n = (b + my_n)y_n = (5 + 1.5y_n)y_n$$

and

$$P_n = b + 2\sqrt{1 + m^2}\, y_n = 5 + 2\sqrt{1 + 1.5^2}\, y_n = 5 + 3.61y_n$$

then the Manning equation can be written as

$$\frac{(5 + 1.5y_n)^5 y_n^5}{(5 + 3.61y_n)^2} = 13,824$$

which yields

$$y_n = 2.25 \text{ m}$$

When the flow conditions are critical, then

$$\frac{A_c^3}{T_c} = \frac{Q^2}{g}$$

where A_c and T_c are the area and top width, respectively. The lefthand side of this equation is a function of the critical flow depth, y_c, and the righthand side is in terms of given data. Thus

$$\frac{A_c^3}{T_c} = \frac{60^2}{9.81} = 367$$

Since

$$A_c = (b + my_c)y_c = (5 + 1.5y_c)y_c$$

and

$$T_c = b + 2my_c = 5 + 2(1.5)y_c = 5 + 3y_c$$

then the critical flow equation can be written as

$$\frac{(5 + 1.5y_c)^3 y_c^3}{5 + 3y_c} = 367$$

which yields

$$y_c = 1.99 \text{ m}$$

Since y_n (=2.25 m) $> y_c$ (=1.99 m), then the slope is mild. Also, since y (=4 m) $> y_n > y_c$, the water surface is in Zone 1 and therefore the water-surface profile is an

M1 profile. This classification requires that the flow depth increases in the downstream direction.

The slope of the water surface is given by (assuming $\alpha = 1$)

$$\frac{dy}{dx} = \frac{S_0 - S_f}{1 - \text{Fr}^2}$$

where S_f is the slope of the energy grade line and Fr is the Froude number. According to the Manning equation, S_f can be estimated by

$$S_f = \left[\frac{nQ}{AR^{\frac{2}{3}}} \right]^2$$

and when the depth of flow, y, is 4 m, then

$$A = (b + my)y = (5 + 1.5 \times 4)(4) = 44 \text{ m}^2$$

$$P = b + 2\sqrt{1 + m^2}y = 5 + 2\sqrt{1 + 1.5^2}(4) = 19.4 \text{ m}$$

$$R = \frac{A}{P} = \frac{44}{19.4} = 2.27 \text{ m}$$

and therefore S_f is estimated to be

$$S_f = \left[\frac{(0.04)(60)}{(44)(2.27)^{\frac{2}{3}}} \right]^2 = 0.000997$$

The Froude number, Fr, is given by

$$\text{Fr}^2 = \frac{V^2}{gD} = \frac{(Q/A)^2}{g(A/T)} = \frac{Q^2 T}{gA^3}$$

where the top width, T, is given by

$$T = b + 2my = 5 + 2(1.5)(4) = 17 \text{ m}$$

and therefore the Froude number is given by

$$\text{Fr}^2 = \frac{(60)^2(17)}{(9.81)(44)^3} = 0.0732$$

Substituting the values for S_0 ($=0.01$), S_f ($=0.000997$), and Fr^2 ($=0.0732$) into the profile equation yields

$$\frac{dy}{dx} = \frac{0.01 - 0.000997}{1 - 0.0732} = 0.00971$$

The depth of flow, y, at a location 100 m downstream from where the flow depth is 4 m can be estimated by

$$y = 4 + \frac{dy}{dx}(100) = 4 + (0.00971)(100) = 4.97 \text{ m}$$

The estimated flow depth 100 m downstream could be refined by recalculating dy/dx for a flow depth of 4.97 m and then using an averaged value of dy/dx to estimate the flow depth 100 m downstream.

3.3.3 Hydraulic Jump

In some cases, supercritical flow must necessarily transition into subcritical flow, even though such a transition does not appear to be possible within the context of the water-surface profiles discussed in the previous section. An example is a case where water is discharged as supercritical flow from under a vertical gate into a body of water flowing under subcritical conditions. If the slope of the channel is mild, then the water emerging from under the gate must necessarily follow an M3 profile, where the depth increases with distance downstream. However, if the depth downstream of the gate is subcritical, then it is apparently impossible for this flow condition to be reached, since this would require a continued increase in the water depth through the M2 zone.

In reality, this transition is accomplished by an abrupt localized change in water depth called a *hydraulic jump*, which is illustrated in Figure 3.12(a), with the corresponding transition in the specific energy diagram shown in Figure 3.12(b). The supercritical (upstream) flow depth is y_1, the subcritical (downstream) flow depth is y_2, and the energy loss between the upstream and downstream sections is ΔE. If the energy loss were equal to zero, then the downstream depth, y_2, could be calculated by equating the upstream and downstream specific energies. However, the transition between supercritical and subcritical flow is generally a turbulent process with a significant energy loss that cannot be neglected. Applying the momentum equation to the control volume between Sections 1 and 2 leads to

$$P_1 - P_2 = \rho Q(V_2 - V_1) \tag{3.121}$$

where P_1 and P_2 are the hydrostatic pressure forces at Sections 1 and 2, respectively; Q is the flowrate; and V_1 and V_2 are the average velocities at Sections 1 and 2, respectively. Equation 3.121 neglects the friction forces on the channel boundary within the control volume, which is justified by the assumption that over a short distance the friction force will be small compared with the difference in upstream and downstream hydrostatic forces. The momentum equation, Equation 3.121, can be written as

$$\gamma \bar{y}_1 A_1 - \gamma \bar{y}_2 A_2 = \rho Q\left(\frac{Q}{A_2} - \frac{Q}{A_1}\right) \tag{3.122}$$

FIGURE 3.12: Hydraulic jump

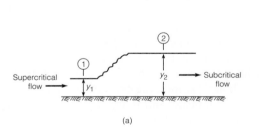

(a)

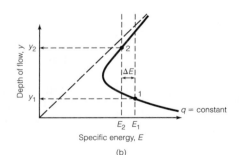

(b)

where $\bar{y}_1$ and $\bar{y}_2$ are the distances from the water surface to the centroids of Sections 1 and 2 and A_1 and A_2 are the cross-sectional areas at Sections 1 and 2, respectively. Equation 3.122 can be rearranged as

$$\frac{Q^2}{gA_1} + A_1\bar{y}_1 = \frac{Q^2}{gA_2} + A_2\bar{y}_2 \tag{3.123}$$

or

$$\boxed{\frac{Q^2}{gA} + A\bar{y} = \text{constant}} \tag{3.124}$$

The term on the lefthand side is called the *specific momentum*, and Equation 3.124 states that the specific momentum remains constant across the hydraulic jump.

In the case of a trapezoidal channel, the momentum equation, Equation 3.124, can be put in the form

$$\boxed{\frac{by_1^2}{2} + \frac{my_1^3}{3} + \frac{Q^2}{gy_1(b + my_1)} = \frac{by_2^2}{2} + \frac{my_2^3}{3} + \frac{Q^2}{gy_2(b + my_2)}} \tag{3.125}$$

where b is the bottom width of the channel, and m is the side slope. In the case of a rectangular channel, Equation 3.124 can be put in the form

$$\frac{q^2}{gy_1} + \frac{y_1^2}{2} = \frac{q^2}{gy_2} + \frac{y_2^2}{2} \tag{3.126}$$

where q is the flow per unit width. Equation 3.126 can be solved for y_2 to yield

$$y_2 = \frac{y_1}{2}\left(-1 + \sqrt{1 + \frac{8q^2}{gy_1^3}}\right) \tag{3.127}$$

which can also be written in the following nondimensional form:

$$\boxed{\frac{y_2}{y_1} = \frac{1}{2}\left(-1 + \sqrt{1 + 8\text{Fr}_1^2}\right)} \tag{3.128}$$

where Fr_1 is the upstream Froude number defined by

$$\text{Fr}_1 = \frac{V_1}{\sqrt{gy_1}} = \frac{q}{y_1\sqrt{gy_1}} \tag{3.129}$$

The depths upstream and downstream of a hydraulic jump, y_1 and y_2, are called the *conjugate depths* of the hydraulic jump, and experimental measurements have shown that Equation 3.128 yields values of y_2 to within 1% of observed values (Streeter et al., 1998). The theoretical relationship between conjugate depths of a hydraulic jump in a horizontal rectangular channel, Equation 3.128, can also be used for hydraulic jumps on sloping channels, provided the channel slope is less than about 5%. For larger channel slopes, the component of the weight of the fluid in the direction of flow becomes significant and must be incorporated into the momentum equation from which the hydraulic jump equation is derived. The lengths of hydraulic jumps are

TABLE 3.6: Characteristics of Hydraulic Jumps

Name	Fr_1	Energy dissipation	Characteristics
Undular jump	1.0–1.7	<5%	Standing waves
Weak jump	1.7–2.5	5–15%	Smooth rise
Oscillating jump	2.5–4.5	15–45%	Unstable; avoid
Steady jump	4.5–9.0	45–70%	Best design range
Strong jump	>9.0	70–85%	Choppy, intermittent

Source: USBR (1955).

around $6y_2$ for $4.5 < Fr_1 < 13$, and somewhat smaller outside this range. The length, L, of a hydraulic jump can also be estimated by (Hager, 1991)

$$\frac{L}{y_1} = 220 \tanh \frac{Fr_1 - 1}{22} \tag{3.130}$$

and the length, L_t, of the transition region between the end of the hydraulic jump and fully developed open-channel flow can be estimated by (Wu and Rajaratnam, 1996)

$$L_t = 10y_2 \tag{3.131}$$

Physical characteristics of hydraulic jumps in relation to the upstream Froude number, Fr_1, are listed in Table 3.6, and it is noteworthy that air entrainment commences when $Fr_1 > 1.7$ (Novak, 1994). A steady, well-established jump, with $4.5 < Fr_1 < 9.0$, is often used as an energy dissipator downstream of a dam or spillway and can also be used to mix chemicals, or act as an aeration mechanism (Potter and Wiggert, 1991). The energy loss in the hydraulic jump, ΔE, is given by

$$\Delta E = \left(y_1 + \frac{V_1^2}{2g} \right) - \left(y_2 + \frac{V_2^2}{2g} \right) \tag{3.132}$$

Combining Equation 3.132 with the equation relating the conjugate depths of the hydraulic jump, Equation 3.128, leads to the following expression for the head loss in a hydraulic jump in a rectangular channel:

$$\boxed{\Delta E = \frac{(y_2 - y_1)^3}{4y_1y_2}} \tag{3.133}$$

EXAMPLE 3.8

Water flows down a spillway at the rate of 12 m³/s per meter of width into a horizontal channel, where the velocity at the channel entrance is 20 m/s. Determine the (downstream) depth of flow in the channel that will cause a hydraulic jump to occur in the channel, and determine the power loss in the jump per meter of width.

Solution   In this case, $q = 12$ m²/s, and $V_1 = 20$ m/s. Therefore the initial depth of flow, y_1, is given by

$$y_1 = \frac{q}{V_1} = \frac{12}{20} = 0.60 \text{ m}$$

and the corresponding Froude number, Fr_1, is given by

$$Fr_1 = \frac{q}{y_1\sqrt{gy_1}} = \frac{12}{0.60\sqrt{(9.81)(0.60)}} = 8.24$$

which confirms that the flow is supercritical. The conjugate depth is given by Equation 3.128, where

$$\frac{y_2}{y_1} = \frac{1}{2}\left(-1 + \sqrt{1 + 8Fr_1^2}\right)$$
$$= \frac{1}{2}\left(-1 + \sqrt{1 + 8(8.24)^2}\right)$$
$$= 11.2$$

Therefore the conjugate downstream depth, y_2, is

$$y_2 = 11.2y_1 = 11.2(0.60) = 6.70 \text{ m}$$

The energy loss in the hydraulic jump, ΔE, is given by Equation 3.133 as

$$\Delta E = \frac{(y_2 - y_1)^3}{4y_1y_2} = \frac{(6.70 - 0.60)^3}{4(0.60)(6.70)} = 14.1 \text{ m}$$

and therefore the power loss, P, per unit width in the jump is given by

$$P = \gamma q \, \Delta E = 9790(12)(14.1) = 1.66 \times 10^6 \text{ W} = 1.7 \text{ MW}$$

The location of a hydraulic jump is important in determining channel wall heights as well as the flow conditions (subcritical or supercritical) in the channel. The mean location of the hydraulic jump is usually estimated by computing the upstream and downstream water-surface profiles; the jump is located where the upstream and downstream water depths are equal to the conjugate depths of the hydraulic jump. In many cases, the location of the hydraulic jump is controlled by installing baffle blocks, sills, drops, or rises in the bottom of the channel to create sufficient energy loss that the hydraulic jump forms at the location of these structures. Hydraulic structures that are specifically designed to induce the formation of hydraulic jumps are called *stilling basins*.

3.3.4 Computation of Water-Surface Profiles

The differential equation describing the water-surface profile in an open channel is given by Equation 3.115, which can be written in the form

$$\boxed{\frac{dy}{dx} = F(x,y)} \tag{3.134}$$

where y is the depth of flow in the channel, x is the distance along the channel, and $F(x, y)$ is a function defined by the relation

$$F(x, y) = \frac{S_0 - S_f}{1 - \alpha \text{Fr}^2} \tag{3.135}$$

where S_0 is the channel slope, S_f is the slope of the energy grade line, Fr is the Froude number of the flow, and α is the kinetic energy correction factor. The function $F(x, y)$ can be calculated using given values of y, Q, S_0, α, and channel geometry (which can be a function of x), using Equation 3.113 to estimate the Froude number and the Manning or Darcy–Weisbach head-loss formula to estimate the slope of the energy grade line, S_f.

A basic assumption is that the head loss between upstream and downstream sections can be estimated using either the Manning or Darcy–Weisbach head-loss formula, without regard to trends in depth. This approximation requires that flow conditions change gradually. Such flow conditions are called *gradually varied flow* (GVF). Conversely, flows that are not gradually varied are called *rapidly varied flow* (RVF). GVF can be further contrasted with RVF in that GVF is usually analyzed over longer distances, where friction losses due to boundary shear are significant, whereas RVF is analyzed over shorter distances, where boundary shear is less significant. Also, the pressure distribution in GVF is usually hydrostatic, whereas in RVF there is usually significant acceleration normal to the streamlines, causing a nonhydrostatic pressure distribution. An example of RVF is the hydraulic jump. Under GVF conditions, the Manning approximation (in SI units) to the friction slope, S_f, is given by

$$S_f = \left(\frac{nQ}{AR^{\frac{2}{3}}} \right)^2 \tag{3.136}$$

where n is the Manning roughness coefficient, Q is the flowrate, and A and R are the area and hydraulic radius of the cross section, respectively. It is sometimes convenient to define the *conveyance*, K, by the relation

$$K = \frac{1}{n} AR^{2/3} \tag{3.137}$$

in which case the friction slope, S_f, can be put in the form

$$S_f = \left(\frac{Q}{K} \right)^2 \tag{3.138}$$

The solution of the water-surface profile equation (Equation 3.134) is usually obtained using numerical integration, since an analytic solution is not possible and field information for calculating the function $F(x, y)$ is typically available only at discrete intervals along the channel. As an alternative to direct integration of Equation 3.134, water-surface profiles can also be calculated directly from the energy equation. Rearranging the form of the energy equation given by Equation 3.85 leads to

$$\boxed{\Delta L = \frac{\left[y + \alpha \frac{V^2}{2g} \right]_2^1}{\overline{S_f} - S_0}} \tag{3.139}$$

where ΔL is the distance between the upstream section (Section 1) and the downstream section (Section 2) and $\overline{S}_f$ is the mean slope of the energy grade line between Sections 1 and 2. The following alternative methods have been used to estimate $\overline{S}_f$:

1. Average conveyance

$$\overline{S}_f = \frac{Q^2}{\left[\frac{K_1 + K_2}{2}\right]^2} \tag{3.140}$$

2. Average friction slope

$$\overline{S}_f = \frac{S_{f1} + S_{f2}}{2} \tag{3.141}$$

3. Geometric mean friction slope

$$\overline{S}_f = \sqrt{S_{f1}S_{f2}} \tag{3.142}$$

4. Harmonic mean friction slope

$$\overline{S}_f = \frac{2S_{f1}S_{f2}}{S_{f1} + S_{f2}} \tag{3.143}$$

where K_1 and K_2 are the conveyances at the upstream and downstream sections, and S_{f1} and S_{f2} are the friction slopes at the upstream and downstream sections respectively. Method 1 is used as the default in the HEC-RAS computer code (U.S. Army Corps of Engineers, 2002), while Method 3 is the default used by the WSPRO computer code (U.S. Federal Highway Administration, 1998). Method 2 has been found to be most accurate for M1 profiles, while Method 4 is best for M2 profiles (U.S. Army Corps of Engineers, 1998). Differences between methods become smaller as the spacing between cross sections is reduced, and differences are typically minimal for cross-section spacings less than 150 m (U.S. Army Corps of Engineers, 1986; French, 2001).

In applying Equation 3.139 to calculate the water-surface profile, it is assumed that the contraction or expansion loss is negligible in comparison to the friction loss. Contraction and expansion losses can be accounted for by Equation 3.107, and the friction loss accounted for by $\overline{S}_f$. In the general case where friction and expansion/contraction losses are both accounted for, the appropriate energy equation is given by

$$\Delta L = \frac{\left[y + \alpha\frac{V^2}{2g}\right]_2^1}{\overline{S}_f + \frac{C}{\Delta L}\left|\alpha_2\frac{V_2^2}{2g} - \alpha_1\frac{V_1^2}{2g}\right| - S_0} \tag{3.144}$$

where C is the expansion or contraction coefficient between adjacent channel sections. The energy equation, given by Equation 3.144, is used for water-surface profile computations in the HEC-RAS computer code (U.S. Army Corps of Engineers, 2002). When the channel cross sections do not vary significantly, contraction and expansion losses between sections are much less than friction losses and, given the uncertainty in estimating friction losses, Equation 3.139 is an adequate representation of the energy equation in these cases.

Whenever the water surface passes through critical depth, the energy equation is not applicable. It is applicable only to gradually varied flows, and the transition from subcritical to supercritical or supercritical to subcritical is a rapidly varying flow. Such rapidly varying flows occur at significant changes in channel slope, some bridge constrictions, drop-structures and weirs, and some stream junctions. In many of these cases, empirical equations can be used (e.g., weirs), while at others it is necessary to apply the momentum equation to determine the changes in water-surface elevation.

Methods commonly used to determine the water-surface profile are the *direct-integration method*, *direct-step method*, and *standard-step method*. These methods are all based on the energy equation and yield essentially the same results; their differences are related to the ease and efficiency of the computations.

3.3.4.1 Direct-integration method

In applying the direct-integration method, the water-surface profile described by Equation 3.134 is expressed in the finite difference form

$$\frac{y_2 - y_1}{x_2 - x_1} = F(\bar{x}, \bar{y}) \tag{3.145}$$

where

$$\bar{x} = \frac{x_1 + x_2}{2} \quad \text{and} \quad \bar{y} = \frac{y_1 + y_2}{2} \tag{3.146}$$

and the subscripts refer to (adjacent) cross sections of the channel. A more convenient form of Equation 3.145 is

$$\boxed{y_2 = y_1 + F(\bar{x}, \bar{y})(x_2 - x_1)} \tag{3.147}$$

This equation is appropriate for computing the water-surface profile in the downstream direction. In most cases, however, water-surface profiles are computed in the upstream direction. A more appropriate form of Equation 3.147 is

$$\boxed{y_1 = y_2 - F(\bar{x}, \bar{y})(x_2 - x_1)} \tag{3.148}$$

In subcritical flow, calculations generally proceed in an upstream direction, while in supercritical flow calculations proceed in a downstream direction. In applying Equation 3.148, the following procedure is suggested:

1. Starting with known flow conditions at Section 2, assume a depth, y_1, at location x_1, and then calculate y_1 using Equation 3.148. On the first calculation, it is reasonable to assume that $y_1 = y_2$.
2. Repeat step 1 until the calculated value of y_1 is equal to the assumed value of y_1. This is then the depth of flow at x_1.

A similar procedure is used to apply Equation 3.147, with iterations on y_2 rather than y_1.

EXAMPLE 3.9

Water flows at $10 \text{ m}^3/\text{s}$ in a rectangular concrete channel of width 5 m and longitudinal slope 0.001. The Manning roughness coefficient, n, of the channel lining is 0.015, and the water depth is measured as 0.80 m at a gaging station. Use the direct-integration method to estimate the flow depth 100 m upstream of the gaging station.

Solution   From the given data: $Q = 10 \text{ m}^3/\text{s}$, $b = 5 \text{ m}$, $S_0 = 0.001$, $n = 0.015$, $y_2 = 0.80 \text{ m}$, $x_1 = 0 \text{ m}$, and $x_2 = 100 \text{ m}$. Assuming $y_1 = y_2 = 0.80 \text{ m}$, the hydraulic parameters of the flow are:

$$\bar{x} = \frac{x_1 + x_2}{2} = \frac{0 + 100}{2} = 50 \text{ m}$$

$$\bar{y} = \frac{y_1 + y_2}{2} = \frac{0.80 + 0.80}{2} = 0.80 \text{ m}$$

$$\bar{A} = b\bar{y} = (5)(0.80) = 4.0 \text{ m}^2$$

$$\bar{P} = b + 2\bar{y} = 5 + 2(0.80) = 6.60 \text{ m}$$

$$\bar{R} = \frac{\bar{A}}{\bar{P}} = \frac{4.0}{6.60} = 0.606 \text{ m}$$

$$\bar{S}_f = \left[\frac{nQ}{\bar{A}\,\bar{R}^{\frac{2}{3}}}\right]^2 = \left[\frac{(0.015)(10)}{(4.0)(0.606)^{\frac{2}{3}}}\right]^2 = 0.00274$$

$$\bar{D} = \frac{\bar{A}}{T} = \frac{b\bar{y}}{b} = \bar{y} = 0.80 \text{ m}$$

$$\bar{V} = \frac{Q}{\bar{A}} = \frac{10}{4} = 2.5 \text{ m/s}$$

$$\bar{\text{Fr}}^2 = \frac{\bar{V}^2}{g\bar{D}} = \frac{(2.5)^2}{(9.81)(0.80)} = 0.80$$

Using these results, and assuming $\alpha = 1$, Equation 3.135 gives

$$F(\bar{x}, \bar{y}) = \frac{S_0 - \bar{S}_f}{1 - \bar{\text{Fr}}^2} = \frac{0.001 - 0.00274}{1 - 0.80} = -0.0087$$

and the estimated depth 100 m upstream of the gaging station is given by Equation 3.148 as

$$y_1 = y_2 - F(\bar{x}, \bar{y})(x_2 - x_1) = 0.80 - (-0.0087)(100 - 0) = 1.67 \text{ m}$$

Since this calculated flow depth at Section 1 is significantly different from the assumed flow depth ($=0.80$ m), the calculations must be repeated, starting with the assumption that $y_1 = 1.67$ m. These calculations are summarized in the following table, where the assumed values of y_1 are given in Column 1 and the calculated values of y_1 (using Equation 3.148) in Column 4:

(1) y_1 (m)	(2) $\overline{S_f}$	(3) $\overline{Fr}^2$	(4) y_1 (m)
1.67	0.000761	0.216	0.77
1.5	0.000936	0.268	0.79
1.3	0.00122	0.352	0.83
1.1	0.00164	0.476	0.92
.	.	.	.
.	.	.	.
.	.	.	.
1.01	0.00193	0.559	1.01

These results indicate that the depth 100 m upstream from the gaging station is approximately equal to 1.01 m.

3.3.4.2   Direct-step method

In applying the direct-step method, the flow conditions at one section are known, the flow conditions at a second section are specified, and the objective is to find the distance between these two sections. With the flow conditions at two channel sections known, the terms on the righthand side of Equation 3.139 are evaluated to determine the distance ΔL between these sections. Computations then proceed to find the distance to another section with specified flow conditions, using the previously specified conditions as given and newly specified conditions to calculate the new interval ΔL.

The main drawbacks of the direct-step method are: (1) the water-surface profile is not computed at predetermined locations, and (2) the method is suitable only for prismatic channels, where the shape of the channel cross section is independent of the interval, ΔL. In cases where the flow conditions at specific locations in a prismatic or nonprismatic channel are required, the standard-step method should be used.

EXAMPLE 3.10

Water flows at 12 m³/s in a trapezoidal concrete channel ($n = 0.015$) of bottom width 4 m, side slopes 2:1 (H:V), and longitudinal slope 0.0009. If depth of flow at a gaging station is measured as 0.80 m, use the direct-step method to find the location where the depth is 1.00 m.

Solution   At the location where the depth is 1.00 m:

$$y_1 = 1.00 \text{ m}$$

$$A_1 = [4 + 2y_1]y_1 = [4 + 2(1.00)](1.00) = 6.00 \text{ m}^2$$

$$P_1 = 4 + 2\sqrt{5}y_1 = 4 + 2\sqrt{5}(1.00) = 8.47 \text{ m}$$

$$R_1 = \frac{A_1}{P_1} = \frac{6.00}{8.47} = 0.708 \text{ m}$$

$$V_1 = \frac{Q}{A_1} = \frac{12}{6.00} = 2.00 \text{ m/s}$$

$$S_{f1} = \left[\frac{nQ}{A_1 R_1^{\frac{2}{3}}}\right]^2 = \left[\frac{(0.015)(12)}{(6.00)(0.708)^{\frac{2}{3}}}\right]^2 = 0.00143$$

and where the depth is 0.80 m:

$$y_2 = 0.80 \text{ m}$$

$$A_2 = [4 + 2y_2]y_2 = [4 + 2(0.80)](0.80) = 4.48 \text{ m}^2$$

$$P_2 = 4 + 2\sqrt{5}\,y_2 = 4 + 2\sqrt{5}\,(0.80) = 7.58 \text{ m}$$

$$R_2 = \frac{A_2}{P_2} = \frac{4.48}{7.58} = 0.591 \text{ m}$$

$$V_2 = \frac{Q}{A_2} = \frac{12}{4.48} = 2.679 \text{ m/s}$$

$$S_{f2} = \left[\frac{nQ}{A_2 R_2^{\frac{2}{3}}}\right]^2 = \left[\frac{(0.015)(12)}{(4.48)(0.591)^{\frac{2}{3}}}\right]^2 = 0.00325$$

Substituting the hydraulic parameters at Sections 1 and 2 into Equation 3.139, and using Equation 3.141 to estimate the average friction slope, $\overline{S}_f$, gives

$$\Delta L = \frac{\left[y + \alpha \dfrac{V^2}{2g}\right]_2^1}{\overline{S}_f - S_0}$$

$$= \frac{\left(1.00 + \frac{2.000^2}{2\times9.81}\right) - \left(0.80 + \frac{2.679^2}{2\times9.81}\right)}{\left(\frac{0.00143 + 0.00325}{2}\right) - 0.0009}$$

$$= 26.4 \text{ m}$$

Hence the depth in the channel increases to 1.00 m at a location that is approximately 26.4 m upstream of the section where the depth is 0.80 m.

3.3.4.3   Standard-step method

In applying the standard-step method, the flow depth is known at one section and the objective is to find the flow depth at a second section a given distance away. The standard-step method is similar to the direct-step method in being based on the solution of Equation 3.139; in the standard-step method, however, ΔL is given and the flow conditions at the second section are unknown. The standard-step method is most applicable in natural channels, where the details of the cross sections are typically measured at locations that are easily accessible.

EXAMPLE 3.11

Water flows in an open channel whose slope is 0.04%. The Manning roughness coefficient of the channel lining is estimated to be 0.035, and the flowrate is 200 m³/s. At a given section of the channel, the cross section is trapezoidal, with a bottom width of 10 m, side slopes of 2:1 (H:V) and a depth of flow of 7 m. Use the standard-step method to calculate the depth of flow 100 m upstream from this section, where the cross section is trapezoidal, with a bottom width of 15 m and side slopes of 3:1 (H:V).

FIGURE 3.13: Calculation of
water-surface profile

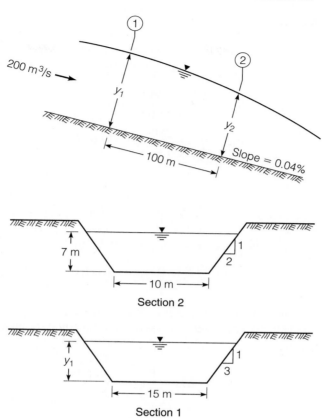

Section 2

Section 1

Solution   The channel along with the upstream and downstream channel sections is illustrated in Figure 3.13. In this case, $Q = 200 \text{ m}^3/\text{s}$, $n = 0.035$, and the flow conditions at Section 2 (downstream section) are given as $y_2 = 7 \text{ m}$, $b_2 = 10 \text{ m}$, $m_2 = 2$ (side slope is m_2:1). Therefore

$$A_2 = [b_2 + m_2 y_2]y_2 = [10 + (2)(7)](7) = 168 \text{ m}^2$$

$$P_2 = b_2 + 2y_2\sqrt{1 + m_2^2} = 10 + 2(7)\sqrt{1 + 2^2} = 41.3 \text{ m}$$

$$R_2 = \frac{A_2}{P_2} = \frac{168}{41.3} = 4.07 \text{ m}$$

$$V_2 = \frac{Q}{A_2} = \frac{200}{168} = 1.19 \text{ m/s}$$

$$S_{f2} = \left[\frac{nQ}{A_2 R_2^{\frac{2}{3}}}\right]^2 = \left[\frac{(0.035)(200)}{(168)(4.07)^{\frac{2}{3}}}\right]^2 = 0.000267$$

Assume $y_1 = 6.90$ m, then

$$A_1 = [b_1 + m_1 y_1] y_1 = [(15) + (3)(6.90)](6.90) = 246 \text{ m}^2$$

$$P_1 = b_1 + 2y_1 \sqrt{1 + m_1^2} = 15 + 2(6.90)\sqrt{1 + 3^2} = 58.6 \text{ m}$$

$$R_1 = \frac{A_1}{P_1} = \frac{246}{58.6} = 4.20 \text{ m}$$

$$V_1 = \frac{Q}{A_1} = \frac{200}{246} = 0.812 \text{ m/s}$$

$$S_{f1} = \left[\frac{nQ}{A_1 R_1^{\frac{2}{3}}} \right]^2 = \left[\frac{(0.035)(200)}{(246)(4.20)^{\frac{2}{3}}} \right]^2 = 0.000119$$

The average slope between Sections 1 and 2, $\overline{S}_f$, is therefore given by

$$\overline{S}_f = \frac{S_{f1} + S_{f2}}{2} = \frac{0.000119 + 0.000267}{2} = 0.000193$$

Substituting the flow parameters into the energy equation, Equation 3.139, and taking the velocity coefficients, α_1 and α_2, to be equal to unity yields

$$\Delta L = \frac{\left[y + \alpha \frac{V^2}{2g} \right]_2^1}{\overline{S}_f - S_0}$$

$$= \frac{\left[6.90 + \frac{0.812^2}{2(9.81)} \right] - \left[7.00 + \frac{1.19^2}{2(9.81)} \right]}{0.000193 - 0.0004}$$

$$= 670 \text{ m}$$

Since we are interested in finding the depth 100 m upstream of Section 2, we want to find the value of y_1 that yields $\Delta L = 100$ m, and repeated trials with the energy equation are necessary. These trials are easily implemented and lead to

$$y_1 = 7.02 \text{ m}$$

Therefore, the depth of the flow 100 m upstream is 7.02 m.

In computing water-surface profiles, significant errors can result if the selected cross sections are too far apart and the hydraulic properties of the flow change too radically from one cross section to the next. A common guideline is that the slope of the energy grade line should not decrease by more than 50% or increase by more than 100% between sections (Dodson, 1999). Selecting cross sections too far apart can result in there being no solution to the energy equation. In general, cross sections should be located at changes in channel geometry and slope, above and below major tributaries, and at structures such as bridges, submerged roads, and transitions. The computation of a water-surface profile generally begins at a section where the depth of flow is known. In most cases, the depth of flow is known at a section where there is a unique relationship between the depth and the flowrate. Such sections are called *control sections*. A typical control section requires that critical flow conditions occur at that section, in which case the depth of flow and the flowrate are related by Equation 3.95. Examples of control sections include free overfalls on mild channels, where the critical depth of flow occurs at a distance of three to four times the critical depth upstream of the brink of the overfall (Henderson, 1966), and channel constrictions that choke the flow to create critical conditions at the control section.

The computation of a water-surface profile is typically done to determine the water-surface elevations expected along a channel during a specified flood event, such as the 100-year flood event that is used to delineate a floodplain. Although the flow in a channel during any flood event is unsteady, it is typically assumed that the peak discharge rate occurs at the same time for the entire length of the channel and that the discharge rate changes along the channel only at major tributaries. In the case of 100-year flood events, computed water-surface profiles are typically used to establish floor elevations for buildings located in the floodplain of the channel.

The fundamental theories and methods of computing water-surface profiles are mature and have not changed in decades; however, advances in computer methods and Geographic Information Systems (GIS) to derive the fundamental data and present the results of analyses are continuing (Lovell and Atkinson, 2004).

3.4 Hydraulic Structures

Hydraulic structures are used to regulate, measure, and/or transport water in open channels. In cases where there is a fixed relationship between the flowrate through the structure and water-surface elevation upstream or downstream of the structure, these structures are called *control structures*. Hydraulic structures can generally be grouped into three categories: (1) flow-measuring structures, such as weirs; (2) regulation structures, such as gates and stilling basins; and (3) discharge structures, such as culverts. The performance of typical structures in each of these categories is discussed in the following sections.

3.4.1 Weirs

Weirs are elevated structures in open channels that are used to measure flow and control outflow from basins and drainage channels. In these structures, water flows through an opening of regular shape (typically rectangular, triangular, or trapezoidal) in which the relationship between flowrate and depth is fixed and known. The bottom edge of the opening is called the *crest*, and there are two types of weirs in common use: (1) sharp-crested weirs, and (2) broad-crested weirs.

3.4.1.1 Sharp-crested weirs

Sharp-crested, or *thin-plate*, weirs consist of a plastic or metal plate that is set vertically and across the width of the channel, and the main types of sharp-crested weirs are rectangular and V-notch weirs. For a weir to be considered sharp crested, the thickness of the crest and side plates should be between 1 and 2 mm (Martinez et al., 2005). If the plates are thicker than specified, the plate edges need to be beveled to an angle of at least $45°$; $60°$ is highly recommended for a V-notch weir (Bos, 1988). In *suppressed* (uncontracted) rectangular weirs, the rectangular opening spans the entire width of

FIGURE 3.14: Schematic diagrams of rectangular sharp-crested weirs

Source: Chadwick and Morfett (1993).

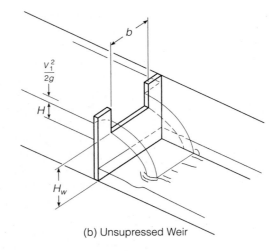

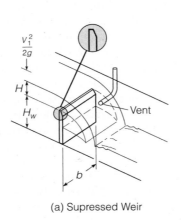

(a) Supressed Weir

(b) Unsupressed Weir

FIGURE 3.15: Operating rectangular sharp-crested rectangular weirs

the channel; in *unsuppressed* (contracted) weirs, the rectangular opening spans only a portion of the channel. Both suppressed and unsuppressed rectangular weirs are illustrated in Figure 3.14, and pictures of operating rectangular sharp-crested weirs are shown in Figure 3.15.

An elevation view of the flow over a sharp-crested weir is illustrated in Figure 3.16, where at Section 1, just upstream of the weir, the flow is approximately horizontal, the pressure distribution is approximately hydrostatic, and the head (or

FIGURE 3.16: Flow over a sharp-crested weir

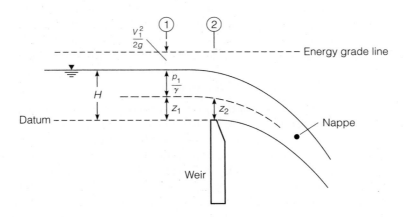

energy per unit weight) of the fluid, E_1, is given by

$$E_1 = H + \frac{V_1^2}{2g} \tag{3.149}$$

where H is the elevation of the water surface and V_1 is the average velocity at Section 1. At Section 2, over the crest of the weir, the flow can be approximated as horizontal, with a head, E_2, at any elevation, z_2, given by

$$E_2 = \frac{p_2}{\gamma} + \frac{v_2^2}{2g} + z_2 \tag{3.150}$$

where p_2/γ and v_2 are the pressure head and velocity, respectively, at elevation z_2. The jet of water that flows over the weir is commonly referred to as the *nappe*. The fluid pressure in Section 2 is equal to atmospheric pressure both at the top and bottom water surfaces, increasing above atmospheric pressure between the two water surfaces. The bottom water surface is where the water springs clear of the crest at Section 2. If the flow depth at Section 2 is small, then the pressure may be assumed equal to atmospheric pressure throughout the depth, and E_2 becomes

$$E_2 = \frac{v_2^2}{2g} + z_2 \tag{3.151}$$

Assuming that the head loss is negligible along a streamline crossing Section 1, at elevation z_1, and leaving Section 2 at elevation z_2, then

$$E_1 = E_2 \tag{3.152}$$

Combining Equations 3.149, 3.151, and 3.152 leads to

$$v_2 = \left[2g\left(H - z_2 + \frac{V_1^2}{2g}\right)\right]^{\frac{1}{2}} \tag{3.153}$$

The estimated flowrate, $\hat{Q}$, across Section 2 can be calculated by integrating the flowrates across elements of area $b\,dz_2$, where b is the width of the rectangular weir. Therefore

$$\hat{Q} = \int_0^H v_2 b\, dz_2$$

$$= \int_0^H \left[2g\left(H - z_2 + \frac{V_1^2}{2g}\right)\right]^{\frac{1}{2}} b\, dz_2$$

$$= \frac{2}{3}b\sqrt{2g}\left[\left(H + \frac{V_1^2}{2g}\right)^{\frac{3}{2}} - \left(\frac{V_1^2}{2g}\right)^{\frac{3}{2}}\right] \tag{3.154}$$

This equation assumes that the elevation of the water surface at Section 2 is equal to the elevation of the water surface at Section 1. This condition is physically impossible but does not necessarily lead to a significant error in the estimated flowrate at Section 2. If $V_1^2/2g$ is negligible compared with H, then Equation 3.154 reduces to

$$\hat{Q} = \frac{2}{3}\sqrt{2g}bH^{\frac{3}{2}} \tag{3.155}$$

This expression gives the flowrate, $\hat{Q}$, over the weir in terms of measurable quantities, b and H, and illustrates why the weir is a hydraulic structure that is used for measuring flowrates in open channels based on the upstream water stage, H.

The weir equation given by Equation 3.154, and approximated by Equation 3.155, was derived with the following theoretical discrepancies: (1) the pressure distribution in the water over the crest of the weir is not uniformly atmospheric; (2) the water surface does not remain horizontal as the water approaches the weir; and (3) viscous effects that cause a nonuniform velocity and a loss of energy between Sections 1 and 2 have been neglected. The error in the flowrate resulting from these theoretical discrepancies is handled by a *discharge coefficient*, C_d, defined by the relation

$$C_d = \frac{Q}{\hat{Q}} \tag{3.156}$$

where Q is the actual flowrate over the weir. Combining Equations 3.155 and 3.156 leads to the following expression for the flowrate over a weir in terms of the discharge coefficient:

$$\boxed{Q = \frac{2}{3}C_d\sqrt{2g}bH^{\frac{3}{2}}} \tag{3.157}$$

It can be shown by dimensional analysis that (Franzini and Finnemore, 2002)

$$C_d = f\left(\text{Re}, \text{We}, \frac{H}{H_w}\right) \tag{3.158}$$

where Re is a Reynolds number, We is a Weber number, and H_w is the height of the weir crest above the bottom of the channel, commonly referred to as the *crest height*. The Weber number is a nondimensional measure of the inertial force relative to the surface tension force, and is commonly denoted by

$$\text{We} = \frac{LV^2\rho}{g\sigma} \tag{3.159}$$

where L is the characteristic length scale, V is the velocity, ρ is the density, and σ is the surface tension. Experiments have shown that H/H_w is the most important variable affecting C_d, with We important only at low heads; Re is usually sufficiently high that viscous effects can be neglected. An empirical formula for C_d is (Rouse, 1946; Blevins, 1984)

$$\boxed{C_d = 0.611 + 0.075\frac{H}{H_w}} \tag{3.160}$$

which is valid for $H/H_w < 5$, and is approximate up to $H/H_w = 10$. For $H/H_w > 15$, the weir is called a *sill*, the discharge can be computed from the critical flow equation by assuming $y_c = H + H_w$ (Chaudhry, 1993), and the discharge coefficient is given by (Jain, 2001)

$$C_d = 1.06 \left(1 + \frac{H_w}{H} \right)^{3/2}$$ (3.161)

A general expression for C_d that is valid for any value of H/H_w has been proposed by Swamee (1988) and is given by

$$C_d = 1.06 \left[\left(\frac{14.14}{8.15 + H/H_w} \right)^{10} + \left(\frac{H/H_w}{H/H_w + 1} \right)^{15} \right]^{-0.01}$$ (3.162)

Equations 3.160 to 3.162 are valid only if the pressure under the nappe is atmospheric (Jain, 2001).

It is convenient to express the discharge formula, Equation 3.157, as

$$Q = C_w b H^{\frac{3}{2}}$$ (3.163)

where C_w is called the *weir coefficient* and is related to the discharge coefficient by

$$C_w = \frac{2}{3} C_d \sqrt{2g}$$ (3.164)

Taking $C_d = 0.62$ in Equation 3.164 yields $C_w = 1.83$, and Equation 3.163 becomes

$$Q = 1.83 b H^{\frac{3}{2}}$$ (3.165)

which gives good results if $H/H_w < 0.4$, which is within the usual operating range of most weirs (Franzini and Finnemore, 2002). Equation 3.165 is applicable in SI units, where Q is in m³/s, and b and H are in meters.

The accuracy of a weir-discharge formula depends significantly on the location of the gaging station for measuring the upstream head, H, and it is recommended that measurements of H be taken between $4H$ and $5H$ upstream of the weir (Ackers et al., 1978). The behavior of uncontracted weirs is complicated by the fact that air is trapped beneath the nappe, which tends to be entrained into the jet, thereby reducing the air pressure beneath the nappe and drawing the nappe toward the face of the weir. To avoid this effect, a vent is sometimes placed beneath the weir to maintain atmospheric pressure (see Figure 3.14a). In the case of unsuppressed (contracted) weirs, the air beneath the nappe is in contact with the atmosphere and venting is not necessary.

Experiments have shown that the effect of side contractions is to reduce the effective width of the nappe by $0.1H$, and that flowrate over the weir, Q, can be estimated by

$$Q = C_w (b - 0.1nH) H^{\frac{3}{2}}$$ (3.166)

where C_w is the weir coefficient calculated using Equation 3.164, b is the width of the contracted weir, and n is the number of sides of the weir that are contracted, usually

FIGURE 3.17: Cipolletti weir

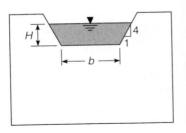

equal to 2. Equation 3.166 gives acceptable results as long as $b > 3H$. A type of contracted weir that is related to the rectangular sharp-crested weir is the *Cipolletti weir*, which has a trapezoidal cross section with side slopes 1 : 4 (H : V) and is illustrated in Figure 3.17. The advantage of using a Cipolletti weir is that corrections for end contractions are not necessary. The discharge formula can be written simply as

$$Q = C_w b H^{\frac{3}{2}} \qquad (3.167)$$

where b is the bottom-width of the Cipolletti weir. The minimum head on standard rectangular and Cipolletti weirs is 6 mm (0.2 ft), and at heads less than 6 mm (0.2 ft) the nappe does not spring free of the crest (Aisenbrey et al., 1974).

The sharp-crested weir is a *control structure*, since the flowrate is determined by the stage just upstream of the weir. This control relationship assumes that the water downstream of the weir, called the *tailwater*, does not interfere with the operation of the weir. If the tailwater elevation rises above the crest of the weir, then the flowrate becomes influenced by the downstream flow conditions, and the weir is *submerged*. The discharge over a submerged weir, Q_s, can be estimated in terms of the upstream and downstream heads on the weir using *Villemonte's formula* (Villemonte, 1947),

$$\frac{Q_s}{Q} = \left[1 - \left(\frac{y_d}{H} \right)^{\frac{3}{2}} \right]^{0.385} \qquad (3.168)$$

where Q is the calculated flowrate assuming the weir is not submerged, y_d is the head downstream of the weir, and H is the head upstream of the weir. The head downstream of the weir, y_d, is approximately equal to the difference between the downstream water-surface elevation and the crest of the weir. Consistent with these definitions, when $y_d = 0$, Equation 3.168 gives $Q_s = Q$. In using Equation 3.168 it is recommended that H be measured at least $2.5H$ upstream of the weir and that y_d be measured beyond the turbulence caused by the nappe (Brater et al., 1996). A more recent formula for calculating the discharge over a submerged weir is given by

Abu-Seida and Quraishi (1976) as

$$\boxed{\frac{Q_s}{Q} = \left(1 + \frac{y_d}{2H}\right)\sqrt{1 - \frac{y_d}{H}}}$$

(3.169)

Weirs should be designed to discharge freely rather than submerged because of greater measurement accuracy.

EXAMPLE 3.12

A weir is to be installed to measure flows in the range of 0.5–1.0 m³/s. If the maximum (total) depth of water that can be accommodated at the weir is 1 m and the width of the channel is 4 m, determine the crest height of a suppressed weir that should be used to measure the flowrate.

Solution   The flow over the weir is illustrated in Figure 3.18, where the crest height of the weir is H_w and the flowrate is Q. The height of the water over the crest of the weir, H, is given by

$$H = 1 - H_w$$

Assuming that $H/H_w < 0.4$, then Q is related to H by Equation 3.165, where

$$Q = 1.83\, bH^{\frac{3}{2}}$$

Taking $b = 4$ m, and $Q = 1$ m³/s (the maximum flowrate will give the maximum head, H), then

$$H = \left[\frac{Q}{1.83b}\right]^{\frac{2}{3}} = \left[\frac{1}{1.83(4)}\right]^{\frac{2}{3}} = 0.265 \text{ m}$$

The crest height, H_w, is therefore given by

$$H_w = 1 - 0.265 = 0.735 \text{ m}$$

and

$$\frac{H}{H_w} = \frac{0.265}{0.735} = 0.36$$

The initial assumption that $H/H_w < 0.4$ is therefore validated, and the height of the weir should be 0.735 m.

FIGURE 3.18: Weir flow

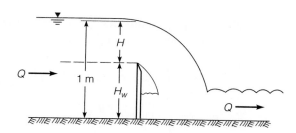

V-notch weirs.   A *V-notch weir* is a sharp-crested weir that has a V-shaped instead of a rectangular-shaped opening. These weirs, also called *triangular weirs*, are typically used instead of rectangular weirs under low-flow conditions, where rectangular weirs tend to be less accurate. V-notch weirs are usually limited to flows of 0.28 m³/s (10 cfs) or less, and are frequently found in small irrigation canals. An operating V-notch weir is shown in Figure 3.19. The basic theory of V-notch weirs is the same as for rectangular weirs, where the theoretical flowrate over the weir, $\hat{Q}$, is given by

$$\hat{Q} = \int_0^H v_2 b \, dz_2$$

$$= \int_0^H \left[2g\left(H - z_2 + \frac{V_1^2}{2g} \right) \right]^{\frac{1}{2}} b \, dz_2 \tag{3.170}$$

where b is the width of the V-notch weir at elevation z_2 and is given by

$$b = 2z_2 \tan\left(\frac{\theta}{2} \right) \tag{3.171}$$

where θ is the angle at the apex of the V-notch, as illustrated in Figure 3.20. Combining Equations 3.170 and 3.171 leads to

$$\hat{Q} = \int_0^H \left[2g\left(H - z_2 + \frac{V_1^2}{2g} \right) \right]^{\frac{1}{2}} 2z_2 \tan\left(\frac{\theta}{2} \right) dz_2 \tag{3.172}$$

The approach velocity, V_1, is usually negligible for the low velocities that are typically handled by V-notch weirs, and therefore Equation 3.172 can be approximated by

$$\hat{Q} = \int_0^H \left[2g(H - z_2) \right]^{\frac{1}{2}} 2z_2 \tan\left(\frac{\theta}{2} \right) dz_2$$

$$= \frac{8}{15}\sqrt{2g} \tan\left(\frac{\theta}{2} \right) H^{\frac{5}{2}} \tag{3.173}$$

FIGURE 3.20: Schematic diagram of V-notch weir

Source: Chadwick and Morfett (1993).

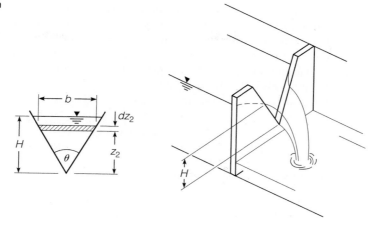

As in the case of a rectangular weir, the theoretical discharge given by Equation 3.173 is corrected by a discharge coefficient, C_d, to account for discrepancies in the assumptions leading to Equation 3.173. The actual flowrate, Q, over a V-notch weir is therefore given by

$$Q = \frac{8}{15} C_d \sqrt{2g} \tan\left(\frac{\theta}{2}\right) H^{\frac{5}{2}}$$

(3.174)

where C_d generally depends on Re, We, θ, and H.

The vertex angles used in V-notch weirs are usually between $10°$ and $90°$. Values of C_d for a variety of notch angles, θ, and heads, H, are plotted in Figure 3.21. The

FIGURE 3.21: Discharge coefficient in V-notch weirs

Source: Finnemore and Franzini (2002).

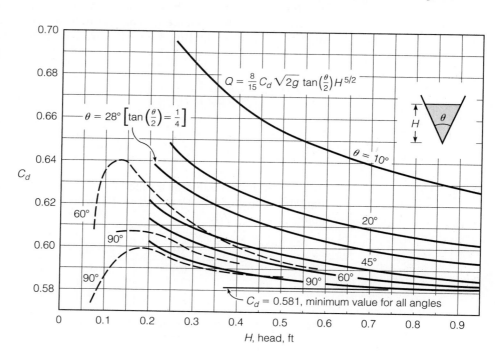

minimum discharge coefficient corresponds to a notch angle of 90°, the minimum value of C_d for all angles is 0.581, the rise in C_d at heads less than 0.5 ft (150 mm) is due to incomplete contraction, and at lower heads the frictional effects reduce the coefficient. Using $C_d = 0.58$ for engineering calculations is usually acceptable, provided that $20° < \theta < 100°$ and $H > 50$ mm (0.17 ft) (Potter and Wiggert, 1991; White, 1994). For $H < 0.17$ ft (50 mm), both viscous and surface-tension effects may be important and a recommended value of C_d is given by (White, 1994)

$$C_d = 0.583 + \frac{1.19}{(\text{ReWe})^{\frac{1}{6}}} \tag{3.175}$$

where Re is the Reynolds number defined by

$$\text{Re} = \frac{g^{\frac{1}{2}} H^{\frac{1}{2}}}{\nu} \tag{3.176}$$

and We is the Weber number defined by

$$\text{We} = \frac{\rho g H^2}{\sigma} \tag{3.177}$$

where ν is the kinematic viscosity and σ is the surface tension of water.

EXAMPLE 3.13

A V-notch weir is to be used to measure channel flows in the range 0.1 to 0.2 m³/s. What is the maximum head of water on the weir for a vertex angle of 45°?

Solution The maximum head of water results from the maximum flow, so $Q = 0.2$ m³/s will be used to calculate the maximum head. The relationship between the head and flowrate is given by Equation 3.174, which can be put in the form

$$H = \left[\frac{15Q}{8C_d\sqrt{2g}\,\tan(\theta/2)}\right]^{\frac{2}{5}} = \left[\frac{15(0.2)}{8C_d\sqrt{2(9.81)}\,\tan(45°/2)}\right]^{\frac{2}{5}} = \frac{0.530}{C_d^{2/5}}\text{ m}$$

or

$$H = \frac{1.74}{C_d^{2/5}}\text{ ft}$$

The discharge coefficient as a function of H for $\theta = 45°$ is given in Figure 3.21, and some iteration is necessary to find H. These iterations are summarized in the following table:

Assumed H (ft)	C_d (Fig. 3.21)	Calculated H (ft)
0.40	0.6	2.13
2.13	0.581	2.16
2.16	0.581	2.16

Therefore, the maximum depth expected at the V-notch weir is 2.16 ft = 0.66 m.

An alternative flow equation for V-notch weirs has been proposed by Kindsvater and Shen (USBR, 1997), where

$$Q = \frac{8}{15} C_d \sqrt{2g} \tan\left(\frac{\theta}{2}\right)(H + k)^{\frac{5}{2}} \qquad (3.178)$$

where C_d and k are both functions of the notch angle, θ. LMNO Engineering and Research Software (LMNO, 1999) developed the following analytic functions for C_d and k to match the graphical functions recommended by Kindsvater and Shen (USBR, 1997):

$$C_d = 0.6072 - 0.000874\theta + 6.1 \times 10^{-6}\theta^2 \qquad (3.179)$$

$$k = 4.42 - 0.1035\theta + 1.005 \times 10^{-3}\theta^2 - 3.24 \times 10^{-6}\theta^3 \qquad (3.180)$$

where θ is in degrees and k is in millimeters. The Kindsvater and Shen equation, Equation 3.178, has been recommended by several reputable organizations and researchers (International Organization for Standardization, 1980; American Society for Testing and Materials, 1993; U.S. Bureau of Reclamation, 1997; Martinez et al., 2005).

EXAMPLE 3.14

Repeat the previous example using the Kindsvater–Shen equation. Compare the results.

Solution   The maximum head of water results from the maximum flow, so $Q = 0.2$ m³/s will be used to calculate the maximum head. Using $\theta = 45°$, Equations 3.179 and 3.180 yield

$$C_d = 0.6072 - 0.000874\theta + 6.1 \times 10^{-6}\theta^2$$
$$= 0.6072 - 0.000874(45°) + 6.1 \times 10^{-6}(45°)^2$$
$$= 0.580$$
$$k = 4.42 - 0.1035\theta + 1.005 \times 10^{-3}\theta^2 - 3.24 \times 10^{-6}\theta^3$$
$$= 4.42 - 0.1035(45°) + 1.005 \times 10^{-3}(45°)^2 - 3.24 \times 10^{-6}(45°)^3$$
$$= 1.50 \text{ mm} = 0.0015 \text{ m}$$

The relationship between the head and flowrate is given by Equation 3.178, which can be put in the form

$$H = \left[\frac{15Q}{8C_d\sqrt{2g}\tan(\theta/2)}\right]^{\frac{2}{5}} - k$$

$$= \left[\frac{15(0.2)}{8(0.580)\sqrt{2(9.81)}\tan(45°/2)}\right]^{\frac{2}{5}} - 0.0015 = 0.657 \text{ m}$$

Therefore, the maximum depth expected at the V-notch weir is approximately 0.66 m. This is exactly the same result that was obtained using the graphical method. Since the Kindsvater–Shen equation does not require iteration, it is clearly preferable in this case.

The following guidelines have been suggested by the U.S. Bureau of Reclamation (USBR, 1997) for using V-notch weirs to measure flows in open channels:

- Head (H) should be measured at a distance of at least $4H$ upstream of the weir.
- The weir should be between 0.8 and 2 mm thick at the crest. If the bulk of the weir is thicker than 2 mm, the downstream edge of the crest can be chamfered at an angle greater than 45° (60° is recommended) to achieve the desired thickness of the edges. This should avoid having water cling to the downstream face of the weir.
- The water surface downstream of the weir should be at least 6 cm below the bottom of the crest to allow a free-flowing waterfall.
- Measured head (H) should be greater than 6 cm due to potential measurement error at such small heads and the fact that the nappe may cling to the weir.

In stormwater-management applications, V-notch weirs with invert angles less than 20° and heights less than 50 mm are usually undesirable, since such weirs are difficult to build and easy to clog (South Florida Water Management District, 1994).

In cases where the tailwater elevation rises above the crest of the weir, the flowrate is influenced by downstream flow conditions. Under this submerged condition, the discharge over the weir, Q_s, can be estimated in terms of the upstream and downstream heads on the weir using Villemonte's formula (Equation 3.168), with the exceptions that the exponent of y_d/H is taken as $\frac{5}{2}$ instead of $\frac{3}{2}$, and the unsubmerged discharge, Q, is calculated using Equation 3.174 or Equation 3.178 (Brater et al., 1996).

Compound weirs.   A *compound weir* is a combination of different types of weirs in a single structure. The most common compound weir consists of a rectangular weir with a V-notch in the middle as shown in Figure 3.22. These weirs are typically found in stormwater-management systems, where the V-notch is used to regulate the release of an initial volume of runoff stored in a detention area, while the rectangular component of the weir provides for the release of stormwater runoff from larger events. In these cases, the V-notch weir is frequently referred to as the *bleeder*. When the water level behind the compound weir is within the V-notch, the compound weir performs like a V-notch weir, with the discharge relation given by Equation 3.174 or 3.178. When the water level is above the V-notch, the weir discharge is taken as the sum of the discharge through the V-notch and the rectangular weir, with the V-notch discharge, Q_v, given by the orifice flow equation

$$Q_v = C_{vd}A\sqrt{2gH} \tag{3.181}$$

where C_{vd} is a discharge coefficient that accounts for head losses and flow contraction through the V-notch, A is the area of the V-notch, and H is the height of the water surface above the centroid of the V-notch. Combining the discharge over the rectangular weir, given by Equation 3.157, with the discharge through the V-notch,

FIGURE 3.22: Compound weir

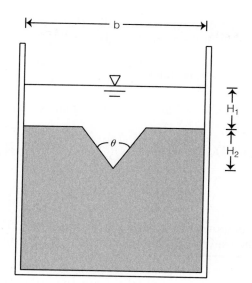

given by Equation 3.181, yields the following discharge equation for flow over a compound weir:

$$Q = \frac{2}{3} C_d \sqrt{2g} b H_1^{3/2} + C_{vd} A \sqrt{2g \left(H_1 + \frac{H_2}{3} \right)} \qquad (3.182)$$

where C_d is the discharge coefficient for the rectangular weir, given by Equation 3.160, H_1 and H_2 are the heights shown in Figure 3.22, and C_{vd} is typically taken as 0.6 (McCuen, 1998).

EXAMPLE 3.15

The outlet structure from a stormwater detention area is a compound weir consisting of a rectangular weir with a V-notch in the middle. The notch angle is 90°, the height of the notch is 20 cm, and the apex of the notch is placed at the bottom of the detention area. Determine the required crest length of the rectangular weir such that the discharge through the compound weir is 3.0 m³/s when the water depth in the detention area is 1.00 m.

Solution   From the given data: $\theta = 90°$, $H_2 = 20$ cm $= 0.20$ m, and $H_1 = 1.00 - H_2 = 1.00 - 0.20 = 0.80$ m. The discharge, Q, over the weir is given by Equation 3.182 as

$$Q = \frac{2}{3} C_d \sqrt{2g} b H_1^{3/2} + C_{vd} A \sqrt{2g \left(H_1 + \frac{H_2}{3} \right)} \qquad (3.183)$$

The discharge coefficient, C_d, is given by Equation 3.160 as

$$C_d = 0.611 + 0.075 \frac{H_1}{H_2} \qquad (3.184)$$

which yields

$$C_d = 0.611 + 0.075 \frac{0.80}{0.20} = 0.911$$

Since $H_1/H_2 = 4$, Equation 3.184 is valid, and C_d can be taken as 0.911 in the weir design. The area, A, of the V-notch is given by

$$A = H_2^2 \tan\left(\frac{\theta}{2}\right) = 0.20^2 \tan\left(\frac{90°}{2}\right) = 0.04 \text{ m}^2$$

Taking $C_{vd} = 0.6$ and $Q = 3$ m^3/s, substituting into Equation 3.183 gives

$$3 = \frac{2}{3}(0.911)\sqrt{2(9.81)}b(0.80)^{3/2} + 0.60(0.04)\sqrt{2(9.81)\left(0.80 + \frac{0.20}{3}\right)}$$

which yields

$$b = 1.51 \text{ m}$$

Therefore, a compound weir with a crest length of 1.51 m will yield a discharge of 3 m^3/s when the water depth in the detention area is 1 m.

A minor drawback in using the compound weir shown in Figure 3.22 is that the flow-head relations given by Equations 3.181 and 3.182 are discontinuous at $H = H_1$. Other less widely used compound-weir geometries have been proposed that do not have this discontinuity, such as a triangular weir with a V notch in the middle (Martinez et al., 2005).

3.4.1.2 Broad-crested weirs

Broad-crested weirs, also called *long-based weirs*, have significantly longer crest lengths than sharp-crested weirs. These weirs are usually constructed of concrete, have rounded edges, and are capable of handling much larger discharges than sharp-crested weirs. There are several different designs of broad-crested weirs, of which the rectangular (broad-crested) weir can be considered representative.

Rectangular (broad-crested) weirs. A typical rectangular weir is illustrated in Figure 3.23. These weirs operate on the theory that the elevation of the weir above the channel bottom is sufficient to create critical flow conditions over the weir. Under these circumstances, the estimated flowrate over the weir, $\hat{Q}$, is given by

$$\hat{Q} = y_c b V_c \tag{3.185}$$

FIGURE 3.23: Flow over a rectangular broad-crested weir

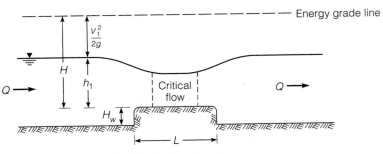

Longitudinal section view

where y_c is the critical depth of flow over the weir, b is the width of the weir, and V_c is the velocity at critical flow. If E_1 is the specific energy of the flow at Section 1 just upstream of the weir, and energy losses between this upstream section and the critical flow section over the weir are negligible, then the energy equation requires that

$$E_1 = H_w + y_c + \frac{V_c^2}{2g} \tag{3.186}$$

where H_w is the height of the weir crest above the upstream channel. Under critical flow conditions, the Froude number is equal to 1, hence

$$\frac{V_c}{\sqrt{gy_c}} = 1 \tag{3.187}$$

Combining Equations 3.186 and 3.187 yields the following expression for y_c:

$$y_c = \frac{2}{3}(E_1 - H_w) = \frac{2}{3}H \tag{3.188}$$

where H is the energy of the upstream flow measured relative to the weir-crest elevation. The upstream energy, H, can be written as

$$H = h_1 + \frac{V_1^2}{2g} \tag{3.189}$$

where h_1 is the elevation of the upstream water surface above the weir crest, and V_1 is the average velocity of flow upstream of the weir. The depth h_1 should be measured at least $2.5h_1$ upstream of the weir (Brater et al., 1996). Combining Equations 3.185, 3.187, and 3.188 leads to the following estimate of the flowrate over a rectangular weir:

$$\hat{Q} = \sqrt{g}b\left(\frac{2}{3}H\right)^{\frac{3}{2}} \tag{3.190}$$

In reality, energy losses over the weir are not negligible and the estimated flowrate, $\hat{Q}$, must be corrected to account for them. The correction factor is the discharge coefficient, C_d, and the actual flowrate, Q, over the weir is given by

$$\boxed{Q = C_d \sqrt{g}b\left(\frac{2}{3}H\right)^{\frac{3}{2}}} \tag{3.191}$$

where values of C_d can be estimated using the relation (Chow, 1959)

$$\boxed{C_d = \frac{0.65}{(1 + H/H_w)^{\frac{1}{2}}}} \tag{3.192}$$

Alternatively, Swamee (1988) proposed the relation

$$C_d = 0.5 + 0.1\left[\frac{\left(\frac{h_1}{L}\right)^5 + 1500\left(\frac{h_1}{L}\right)^{13}}{1 + 1000\left(\frac{h_1}{L}\right)^3}\right]^{0.1} \tag{3.193}$$

TABLE 3.7: Flow Conditions Over Broad-Crested Weirs

h_1/L	Weir classification	Flow condition
$h_1/L < 0.08$	Long-crested	The critical flow section is near the downstream end of the weir, the flow over the weir crest is subcritical, and the value of C_d depends on the resistance of the weir surface. This type of weir is of limited use for flow measurements.
$0.08 < h_1/L < 0.33$	Broad-crested	The region of parallel flow occurs near the middle section of the crest. The variation of C_d with h_1/L is small.
$0.33 < h_1/L < 1.5$	Narrow-crested	The streamlines are curved over the entire crest; there is no region of parallel flow over the crest.
$h_1/L > 1.5$	Sharp-crested	The flow separates at the upstream end and does not reattach to the crest. The flow is unstable.

To ensure proper operation of a broad-crested weir, flow conditions must be restricted to the operating range of $0.08 < H/L < 0.50$ (Bos, 1988) or $0.08 < H/L < 0.33$ (French, 1985) or $0.08 < h_1/L < 0.33$ (Sturm, 2001; Jain, 2001). These requirements are very similar, particularly in the usual circumstance where the velocity head upstream of the weir is negligible, in which case $h_1 \approx H$. Overall, it is recommended that broad-crested weirs be operated in the range $0.08 < h_1/L < 0.33$. Flow conditions for various values of h_1/L are listed in Table 3.7. It should be noted that using Equation 3.193 in the range $0.08 < h_1/L < 0.33$ yields $0.527 < C_d < 0.540$. This is the typical range C_d for properly operated broad-crested weirs.

An advantage of a broad-crested weir is that the tailwater can be above the crest of the weir without affecting the head-discharge relationship as long as the control section is unaffected. The limit of the head, y_d, downstream of the weir so that the discharge does not decrease by more than 1% is called the *modular limit*, which is usually expressed in terms of y_d/H. For rectangular broad-crested weirs, $y_d/H = 0.66$ can be taken as the modular limit (Bos, 1988).

EXAMPLE 3.16

A 20-cm high broad-crested weir is placed in a 2-m wide channel. Estimate the flowrate in the channel if the depth of water upstream of the weir is 50 cm.

Solution   Upstream of the weir, $h_1 = 0.5 \text{ m} - 0.2 \text{ m} = 0.30 \text{ m}$, and

$$H = h_1 + \frac{V_1^2}{2g} = h_1 + \frac{Q^2}{2gA_1^2} = 0.30 + \frac{Q^2}{2(9.81)(0.5 \times 2)^2} = 0.30 + 0.0510Q^2$$

The discharge coefficient, C_d, is given by

$$C_d = \frac{0.65}{(1 + H/H_w)^{\frac{1}{2}}} = \frac{0.65}{[1 + (0.30 + 0.0510Q^2)/0.2]^{\frac{1}{2}}} = \frac{0.65}{[2.5 + 0.255Q^2]^{\frac{1}{2}}}$$

where H_w has been taken as 0.2 m. The discharge over the weir is therefore given by

$$Q = C_d \sqrt{g}b \left(\frac{2}{3}H\right)^{\frac{3}{2}} = \frac{0.65}{[2.5 + 0.255Q^2]^{\frac{1}{2}}} \sqrt{9.81}(2) \left[\frac{2}{3}(0.30 + 0.0510Q^2)\right]^{\frac{3}{2}}$$

$$= 2.22 \left[\frac{(0.30 + 0.0510Q^2)^3}{2.5 + 0.255Q^2}\right]^{\frac{1}{2}}$$

which yields

$$Q = 0.23 \text{ m}^3/\text{s}$$

This solution assumes that the length of the weir is such that $0.08 < h_1/L < 0.33$.

3.4.2 Parshall Flumes

Although weirs are the simplest structures for measuring flowrates in open channels, the high head losses caused by weirs and the tendency for suspended particles to accumulate behind weirs may be important limitations. The Parshall flume, named after Ralph L. Parshall (1881–1960), provides a convenient alternative to weirs for measuring flowrates in open channels where high head losses and sediment accumulation are of concern. Such cases include flow measurement in wastewater treatment plants and irrigation channels. The Parshall flume, illustrated in Figure 3.24, consists of a converging section that causes critical flow conditions, followed by a steep throat section that provides for a transition to supercritical flow. The unique relationship between the depth of flow and the flowrate under critical flow conditions is the basic principle on which the Parshall flume operates. The transition from supercritical flow to subcritical flow at the exit of the flume usually occurs via a hydraulic jump, but under high-tailwater conditions the jump is sometimes submerged. Examples of Parshall flumes operating in the field are shown in Figure 3.25, where the flume on the left is made of cast-in-place concrete, while the flume on the right is a prefabricated unit.

Within the flume structure, water depths are measured at two locations, one in the converging section, H_a, and the other in the throat section, H_b. The flow depth in the throat section is measured relative to the bottom of the converging section, as illustrated in Figure 3.26. If the hydraulic jump at the exit of the Parshall flume is not submerged, then the discharge through the flume is related to the measured flow depth in the converging section, H_a, by the empirical discharge relations given in Table 3.8, where Q is the discharge in m^3/s, W is the width of the throat in m, and H_a is measured in meters.

Submergence of the hydraulic jump is determined by the ratio of the flow depth in the throat, H_b, to the flow depth in the converging section, H_a, and critical values for the ratio H_b/H_a are given in Table 3.9. Whenever H_b/H_a exceeds these critical values, the hydraulic jump is submerged and the discharge is reduced from the values given by the equations in Table 3.8. Corrections to the theoretical flowrates as a function of H_a and the percentage of submergence, H_b/H_a, are given in Figure 3.27 for a throat width of 0.30 m and in Figure 3.28 for a throat width of 3.0 m. Flow corrections for the 0.30-m flume are applied to larger flumes by multiplying the correction for the

FIGURE 3.24: Schematic diagram of Parshall flume

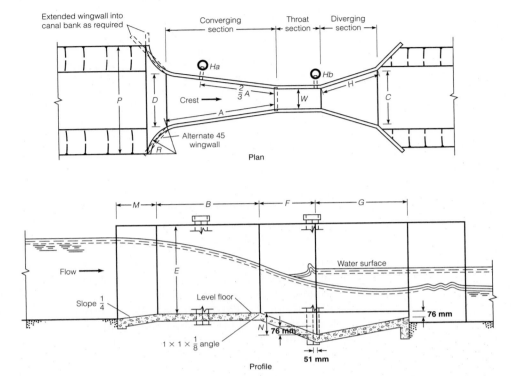

FIGURE 3.25: Operational Parshall flumes

0.30-m flume by a factor corresponding to the flume size given in Table 3.10. Similarly, flow corrections for the 3.0-m flume are applied to larger flumes by multiplying the correction for the 3.0-m flume by a factor corresponding to the flume size given in Table 3.11. According to Aisenbrey et al. (1974), correct zeroing and reading of the gages is necessary for accurate results, which are usually within 2% for free flows

FIGURE 3.26: Measured water depths in Parshall flume

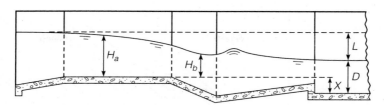

TABLE 3.8: Parshall Flume Discharge Equations

Throat width	Equation	Free-flow capacity (m^3/s)
8 cm	$Q = 0.177H_a^{1.547}$	0.0008–0.054
15 cm	$Q = 0.381H_a^{1.58}$	0.0014–0.11
23 cm	$Q = 0.535H_a^{1.53}$	0.0025–0.25
30 cm to 2.4 m	$Q = 0.372W(3.281H_a)^{1.570W^{0.026}}$	up to 4.0
3.0 m to 15.2 m	$Q = (2.29W + 0.474)H_a^{1.6}$	up to 56

TABLE 3.9: Submergence Criteria in Parshall Flumes

Throat Width	$(H_b/H_a)_{crit}$
2.5 cm, 5.1 cm, 7.6 cm	0.5
15.2 cm, 22.9 cm	0.6
30 cm to 2.4 m	0.7
2.4 m to 15.2 m	0.8

FIGURE 3.27: Parshall flume correction for submerged flow (throat width = 0.30 m)

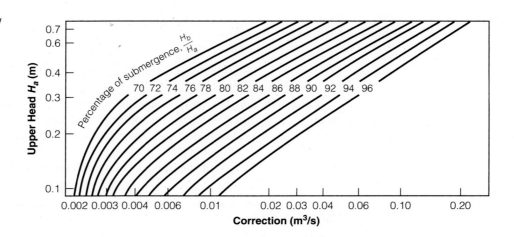

FIGURE 3.28: Parshall flume correction for submerged flow (throat width = 3.0 m)

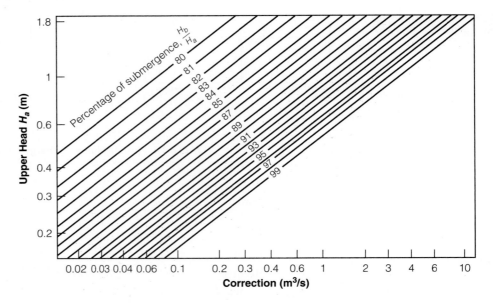

TABLE 3.10: Correction Factors for 0.30-m Parshall Flumes

Throat width (m)	Correction factor
0.30	1.0
0.46	1.4
0.61	1.8
0.91	2.4
1.22	3.1
1.83	4.3
2.44	5.4

and 5% for submerged flows. Parshall flumes do not reliably measure flowrates when the submergence ratio, H_b/H_a, exceeds 0.95.

Parshall flumes can be constructed of concrete, wood, fiberglass, galvanized metal, or any other construction material that can be built to specified dimensions in the field or prefabricated in a shop. As a general rule, the width of the Parshall flume should be about one-third to one-half of the top width of the water surface in the upstream channel at the design discharge and at normal depth (Aisenbrey et al., 1974).

EXAMPLE 3.17

Flow is being measured by a Parshall flume that has a throat width of 0.61 m. Determine the flowrate through the flume when the water depth in the converging section is 61.0 cm and the depth in the throat section is 51.8 cm.

TABLE 3.11: Correction Factors for
3.0-m Parshall Flumes

Throat width (m)	Correction factor
3.0	1.0
3.7	1.2
4.6	1.5
6.1	2.0
7.6	2.5
9.1	3.0
12.2	4.0
15.2	5.0

Solution From the given data: $W = 0.61$ m, $H_a = 0.610$ m, and $H_b = 0.518$ m. According to Table 3.8, the flowrate, Q, is given by

$$Q = 0.372W(3.281H_a)^{1.570W^{0.026}} = 0.372(0.61)(3.281 \times 0.610)^{1.570(0.61)^{0.026}} = 0.665 \text{ m}^3/\text{s}$$

In this case,

$$\frac{H_b}{H_a} = \frac{0.518}{0.610} = 0.85$$

Therefore, according to Table 3.9, the flow is submerged. Figure 3.27 gives the flowrate correction for a 0.30-m flume as 0.057 m³/s, and Table 3.10 gives the correction factor for a 0.61-m flume as 1.8. The flowrate correction, ΔQ, for a 0.61-m flume is therefore given by

$$\Delta Q = 0.057 \text{ m}^3/\text{s} \times 1.8 = 0.102 \text{ m}^3/\text{s}$$

and the flowrate through the Parshall flume is $Q - \Delta Q$, where

$$Q - \Delta Q = 0.665 \text{ m}^3/\text{s} - 0.102 \text{ m}^3/\text{s} = 0.563 \text{ m}^3/\text{s}$$

The flowrate is 0.563 m³/s.

3.4.3 Spillways

Spillways are used to discharge water that cannot be passed through a diversion system or stored in the reservoir behind a dam. Therefore, spillways typically function infrequently and only at times of flood, and their design adequacy is critical to the safety of the dam structure itself. Determination of the design flood for establishing the spillway capacity is an important design specification, and sometimes requires using the probable maximum flood (PMF).[*] Spillways are categorized as controlled or uncontrolled, depending on whether or not they are equipped with gates.

Criteria for spillway designs in the United States have been developed by the U.S. Army Corps of Engineers (1986) and the U.S. Bureau of Reclamation (1977). Commonly used types are overflow spillways, chute spillways, shaft spillways, side-channel spillways, and limited-service spillways. An *overflow spillway* is a section of

---

[*]See Chapter 5 for more detailed coverage of the probable maximum flood.

dam designed to permit water to pass over its crest. Overflow spillways are widely used on gravity, arch, and buttress dams,[†] and are generally an integral part of the dam structure. An example of an overflow spillway is shown in Figure 3.29. *Chute spillways*, also called *trough spillways*, are normally used with earth- or rock-filled embankment dams, and are normally designed to minimize excavation by setting the invert profile to approximate the profile of the natural ground. *Shaft spillways* include various configurations of crest designs, with or without gates, all of which transition into a closed conduit (tunnel) system immediately downstream from the crest. *Side-channel spillways* consist of an overflow weir discharging into a narrow channel in which the direction of flow is approximately parallel to the weir crest. *Limited-service spillways* are designed to operate very infrequently, and with the knowledge that some degree of damage or erosion will occur during operation. The most common type of spillway is the overflow spillway, and details of its design are given below. Design methodologies for other types of spillways can be found in Prakash (2004).

The shape of an overflow spillway is usually made to conform to the profile of a fully ventilated nappe of water flowing over a sharp-crested weir that is coincident with the upstream face of the spillway. This shape depends on the head over the crest, the inclination of the upstream face, and the velocity of approach. Crest shapes of ogee[‡] spillways have been studied extensively by the U.S. Bureau of Reclamation (USBR, 1977).

The head over the crest of a spillway is measured above the apex, which is point O in Figure 3.30. *High spillways* are defined as those in which the ratio of crest height, P, to design head, H_d, is greater than 1 (i.e., $P/H_d > 1$), and *low spillways* are those in which $P/H_d < 1$. When the actual head on the spillway is less than the design head, the trajectory of the nappe falls below the crest profile, creating positive pressures on the crest, and reducing the discharge coefficient. Conversely, when the actual head is higher than the design head, the nappe trajectory is higher than the crest, creating negative pressures on the crest and increasing the discharge. When negative pressures are generated on the crest, cavitation conditions are possible and

---

[†]See Chapter 7 for more detailed coverage on dams.

[‡]An *ogee* refers to an S-shaped profile.

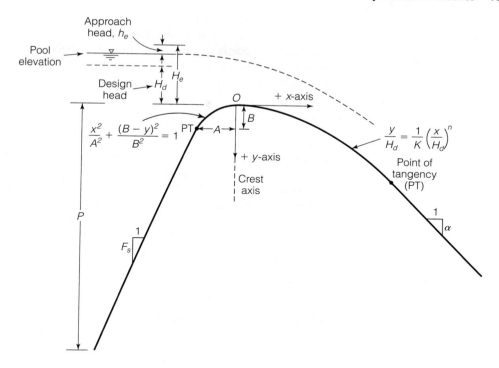

FIGURE 3.30: Ogee-crest spillway

should generally be avoided. Experiments have shown that cavitation on the spillway crest occurs whenever the pressure on the spillway is less than -7.6 m of water, and it is recommended that spillways be designed such that the maximum expected head will result in a pressure on the spillway crest of not less than -4.6 m of water (U.S. Department of the Army, 1986). It is generally desirable to design the crest shape of a high-overflow spillway for a design head, H_d, that is less than the head on the crest corresponding to the maximum reservoir level, H_e, and, to avoid cavitation problems, the U.S. Bureau of Reclamation (1987) recommends that H_e/H_d not exceed 1.33. Experiments have shown that, in cases where the pressure on the spillway is equal to -4.6 m, the relationship between the design head, H_d, and the maximum reservoir level, H_e, is given by (Reese and Maynord, 1987):

For $H_e \geq 9.1$ m,

$$H_d = 0.43 H_e^{1.22} \quad \text{(without piers)} \tag{3.194}$$

$$H_d = 0.41 H_e^{1.26} \quad \text{(with piers)} \tag{3.195}$$

and for $H_e < 9.1$ m,

$$H_d = 0.7 H_e \quad \text{(without piers)} \tag{3.196}$$

$$H_d = 0.74 H_e \quad \text{(with piers)} \tag{3.197}$$

where H_d and H_e are both in meters. When a crest with piers is designed for negative pressures, the piers must be extended downstream beyond the negative-pressure zone in order to prevent aeration of the nappe, nappe separation or undulation, and loss of the underdesign efficiency advantage (U.S. Army Corps of Engineers, 1986).

An important part of spillway design is specification of the shape of the spillway. The U.S. Bureau of Reclamation (1977) separates the shape of the spillway into two quadrants, one upstream and one downstream of the crest axis (apex) shown in Figure 3.30. The equation describing the surface of the spillway in the downstream quadrant is given by

$$\frac{y}{H_d} = \frac{1}{K}\left(\frac{x}{H_d}\right)^n \tag{3.198}$$

where (x, y) are the coordinates of the crest profile relative to the apex (point O) as shown in Figure 3.30, and K and n are constants that depend on the upstream-face inclination and velocity of approach. According to Murphy (1973), n can be taken as 1.85 in all cases, and K varies with the ratio of the approach depth, P, (shown in Figure 3.30) to the design head, H_d, as shown in Figure 3.31(a). Values of K range from 2.0 for a deep approach to 2.2 for a very shallow approach. The crest profile merges with the straight downstream section of slope α, at a location given by

$$\frac{X_{DT}}{H_d} = 0.485(K\alpha)^{1.176} \tag{3.199}$$

where X_{DT} is the horizontal distance from the apex to the downstream tangent point. An (elliptical) equation describing the surface of the spillway in the upstream quadrant is given by

$$\frac{x^2}{A^2} + \frac{(B - y)^2}{B^2} = 1 \tag{3.200}$$

FIGURE 3.31: Coordinate coefficients for spillway crests

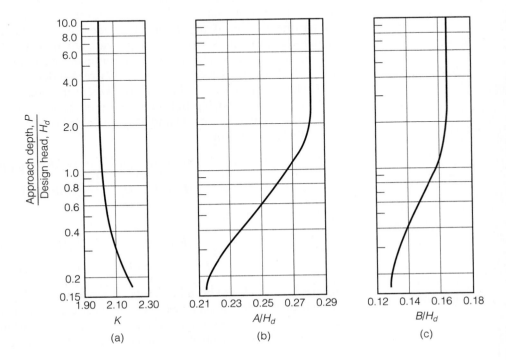

where (x, y) are the coordinates of the crest profile relative to the apex (point O) as shown in Figure 3.30, and A and B are equation parameters that can be estimated using Figures 3.31(b) and 3.31(c) for various values of the approach depth and design head. For an inclined upstream face with a slope of F_s, the point of tangency with the elliptical shape can be determined by the following equation:

$$X_{UT} = \frac{A^2 F_s}{(A^2 F_s^2 + B^2)^{1/2}} \qquad (3.201)$$

where X_{UT} is the horizontal distance from the apex to the upstream tangent point. The *chute* is that portion of the spillway that connects the crest curve to the terminal structure, and the flow in the spillway chute is generally supercritical. When the spillway is an integral part of a concrete gravity monolith, the chute is usually very steep, and slopes in the range from 0.6 : 1 to 0.8 : 1 (H : V) are common (Prakash, 2004). The terminal structure of a spillway is designed to dissipate the energy associated high flow velocities at the base of the spillway. The three most common types of energy-dissipation terminal structures are the stilling basin, roller bucket, and flip bucket. The stilling basin employs the hydraulic jump for energy dissipation (see Section 3.4.4), the roller bucket achieves energy dissipation in surface rollers (waves) over the bucket and ground rollers (waves) downstream of the bucket, and the flip bucket deflects the flow downstream, thereby transferring the energy to a position where impact, turbulence, and resulting erosion will not jeopardize the safety of the dam. The stilling basin is the most effective method for dissipating energy in flow over spillways.

Since the crest of a spillway is approximately conformed to the profile of the lower nappe surface from a (ventilated) sharp-crested weir, the discharge, Q, over a weir corresponding to a head H_e on the spillway crest is given by the relation

$$\boxed{Q = CL_e H_e^{3/2}} \qquad (3.202)$$

where C is the coefficient of discharge, and L_e is the effective length of the spillway crest. In cases where crest piers and abutments cause side contractions of the overflow, the effective length, L_e, is less than the actual crest length, L, according to the relation

$$\boxed{L_e = L - wN - 2(NK_p + K_a)H_e} \qquad (3.203)$$

where N is the number of piers, w is the width of each pier, K_p is the pier contraction coefficient, and K_a is the abutment contraction coefficient. Recommended values of K_p are: 0.02 for square-nose piers, 0.01 for round-nosed piers, and 0 for pointed-nose piers; and recommended values for K_a are: 0.2 for square abutments with headwalls at 90° to the flow direction, 0.1 for rounded abutments of radius of curvature, r, in the range $0.15H_d \leq r \leq 0.5H_d$, and 0 for well-rounded abutments with $r > 0.5H_d$ (U.S. Department of the Army, 1986; U.S. Bureau of Reclamation, 1987). The discharge coefficient, C, of an overflow spillway is influenced by a number of factors, including: (1) the crest height-to-head ratio or the velocity of approach; (2) the actual head being different from the design head; (3) the upstream face slope; and (4) the downstream submergence. The discharge coefficients for spillways can be estimated using relationships developed by the U.S. Bureau of Reclamation (1987) and shown in Figure 3.32. The curve shown in Figure 3.32(a) gives the basic discharge coefficient

FIGURE 3.32: Discharge coefficient for overflow spillways

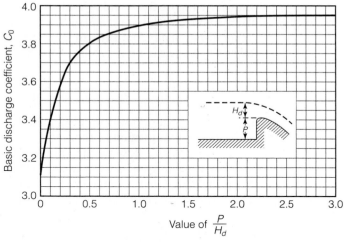

(a) Basic Discharge Coefficient

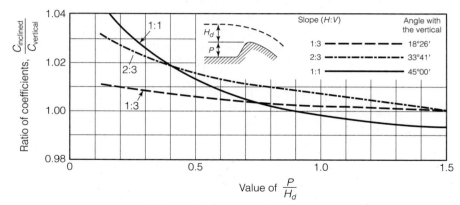

(b) Correction Factor for Sloping Upstream Face

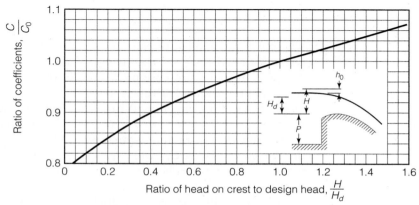

(c) Correction Factor for Other than Design Head

for a vertical-faced spillway with atmospheric pressure on the crest, the curve in Figure 3.32(b) gives the correction factor for a sloping upstream face, and the curve in Figure 3.32(c) gives the correction factor for a head other than the design head. The overall discharge coefficient for the spillway is equal to the product of the discharge coefficient in Figure 3.32(a) with the correction factors in Figures 3.32(b) and 3.32(c), all given in U.S. customary units. To convert this coefficient to SI units, multiply by 0.552. The increase in discharge coefficient, C, for heads greater than the design head is the basic reason for underdesigning spillway crests to obtain greater efficiency. The basic discharge coefficient given in Figure 3.32(a) is the same as for a vertical sharp-crested weir forming the upstream face of the spillway, with an adjustment to account for the fact that the head on a sharp-crested weir is measured relative to top of the vertical face of the weir, while the head on a spillway is measured relative to the crest of the spillway. For very high spillways ($P/H_d \gg 1$) the basic discharge coefficient is independent of P/H_d. A sloping upstream face can be used to prevent separation eddies that might occur on the vertical face of low spillways, and a sloping upstream face sometimes increases the discharge coefficient. For example, for a $1:1$ slope in the upstream face, Figure 3.32(b) indicates an increased discharge coefficient for $P/H_d < 0.9$. However, underdesigning does not result in increased discharge coefficients when $P/H_d < 0.5$ (U.S. Army Corps of Engineers, 1986).

EXAMPLE 3.18

Design a 60-m long overflow spillway that will discharge a design flood of 1500 m³/s at a maximum allowable pool elevation of 400 m. The bottom elevation behind the spillway is 350 m, the upstream face of the spillway is vertical, and the spillway chute is to have a slope of $1:2$ (H:V).

Solution   Assuming that the spillway is sufficiently high that the ratio of the spillway height, P, to the design head, H_d, is greater than 3, Figure 3.32(a) gives the basic discharge coefficient as $C_0 = 3.95$ (in U.S. Customary units) $= 0.552(3.95) = 2.18$ (in SI units). Assuming that the effective head, H_e, is less than 9.1 m, then for the limiting case where the water pressure on the spillway is equal to -4.6 m, Equation 3.196 requires that

$$\frac{H_e}{H_d} = \frac{1}{0.7} = 1.43$$

and Figure 3.32(c) gives $C/C_0 = 1.05$. The discharge coefficient, C, under maximum headwater conditions can now be calculated using the relation

$$C = \left(\frac{C}{C_0}\right)C_0 = (1.05)(2.18) = 2.29$$

From the given data, $Q = 1500$ m³/s, $L_e = 60$ m, and therefore the effective head, H_e, on the spillway is given by Equation 3.202 as

$$H_e = \left[\frac{Q}{CL_e}\right]^{2/3} = \left[\frac{1500}{(2.29)(60)}\right]^{2/3} = 4.92 \text{ m}$$

Since this calculated value of H_e is less than 9.1 m, the initial assumption that $H_e < 9.1$ m used in estimating the discharge coefficient is validated. The design head,

H_d, corresponding to $H_e = 4.92$ m is given by Equation 3.196 as

$$H_d = 0.7H_e = 0.7(4.92) = 3.44 \text{ m}$$

The depth of water, d, upstream of the spillway is given by

$$d = 400 \text{ m} - 350 \text{ m} = 50 \text{ m}$$

and the approach velocity, v_0, can be estimated by

$$v_0 = \frac{Q}{L_e d} = \frac{1500}{(60)(50)} = 0.5 \text{ m/s}$$

with a velocity head, h_0, given by

$$h_0 = \frac{v_0^2}{2g} = \frac{0.5^2}{2(9.81)} = 0.01 \text{ m}$$

Therefore, at the maximum pool elevation, the height of water above the spillway is $H_e - h_0 = 4.92 \text{ m} - 0.01 \text{ m} = 4.91 \text{ m}$, and the required crest elevation of the spillway is 400 m − 4.91 m = 395.09 m.

Having determined the required crest elevation of the spillway ($=395.09$ m) to pass the design flood ($=1500 \text{ m}^3/\text{s}$), we next determine the required shape of the spillway in the vicinity of the crest. Since the maximum pool elevation is 400 m and the bottom elevation behind the spillway is 350 m, the height of the spillway crest, P, is given by

$$P = 400 \text{ m} - 350 \text{ m} - 4.91 \text{ m} = 45.09 \text{ m}$$

and

$$\frac{P}{H_d} = \frac{45.09}{3.44} = 13.1$$

which indicates a "high spillway" with negligible approach velocity. Figure 3.31(a) gives $K = 2.0$ for the spillway coordinate coefficient, and the profile of the downstream quadrant of the spillway, taking $n = 1.85$, is given by

$$\frac{y}{H_d} = \frac{1}{K}\left(\frac{x}{H_d}\right)^n$$

$$\frac{y}{3.44} = \frac{1}{2.0}\left(\frac{x}{3.44}\right)^{1.85}$$

which simplifies to

$$y = 0.175x^{1.85}$$

Since the downstream slope of the spillway is 1:2 (H:V), then $\alpha = 2$ and the horizontal distance from the apex to the downstream tangent point, X_{DT}, is given by

Equation 3.199 as

$$\frac{X_{DT}}{H_d} = 0.485(K\alpha)^{1.176}$$

$$\frac{X_{DT}}{3.44} = 0.485(2.0 \times 2)^{1.176}$$

which gives

$$X_{DT} = 8.52 \text{ m}$$

Therefore, at a distance of 8.52 m downstream of the crest, the curved spillway profile merges into the linear profile of the spillway chute, which has a slope of 1 : 2 (H : V). The corresponding y value is $y = 0.175X_{DT}^{1.85} = 0.175(8.52)^{1.85} = 9.21$ m. Taking $P/H_d = 13.1$ in Figure 3.31(b) and (c) gives $A/H_d = 0.28$ and $B/H_d = 0.165$, which yields $A = 0.28(3.44) = 0.963$ m, $B = 0.165(3.44) = 0.568$ m, and the profile of the upstream quadrant of the spillway is given by Equation 3.200 as

$$\frac{x^2}{A^2} + \frac{(B-y)^2}{B^2} = 1$$

$$\frac{x^2}{0.963^2} + \frac{(0.568-y)^2}{0.568^2} = 1$$

which simplifies to

$$x^2 + 2.87(0.568 - y)^2 = 0.927$$

The upstream crest profile given by this equation intersects the vertical (upstream) spillway wall at $x = -A = -0.963$ m and $y = B = 0.568$ m.

Uncontrolled spillways provide an extremely reliable operation and require very low maintenance. The inclusion of gates on the crest of the spillway allows the crest to be placed significantly below the maximum operating reservoir level, thereby permitting the entire reservoir to be used for normal operating purposes, and resulting in a much narrower spillway facility. The decision on whether to use gates on a spillway must consider the time scale on which the pool elevation responds to rainfall events. In cases where the reservoir pool elevation responds significantly to rainfall events within 12 hours, gates should generally be avoided, and it is recommended that gates be considered only when pool response times to rainfall events exceed 24 hours. A gated spillway must include, as a minimum, two or preferably three spillway gates in order to satisfy safety concerns (U.S. Army Corps of Engineers, 1986), and both vertical-lift gates and Tainter gates are commonly used on spillway crests (see Section 3.4.5).

3.4.4 Stilling Basins

A *stilling basin* is typically used to force the occurrence of a hydraulic jump within the basin, stabilize the jump, and reduce the length required for the jump to occur. A stilling basin is typically located at the downstream end of a spillway or spillway chute, and is usually constructed of concrete. The low-velocity subcritical flow exiting a stilling basin prevents erosion and undermining of dam and spillway structures. The

U.S. Bureau of Reclamation has developed several standard stilling basin designs (U.S. Bureau of Reclamation, 1987), and a few of the more common types will be discussed here. For incoming Froude numbers in the range 1.7–2.5, the hydraulic jump is weak and no special appurtenances are necessary. This is called a *Type I* stilling basin. For Froude numbers in the range 2.5–4.5, a transition jump forms with considerable wave action. A *Type IV* stilling basin, shown in Figure 3.33, is recommended for this jump. This basin includes *chute blocks* and a solid end sill, which tends to hold the jump in the stilling basin. The recommended tailwater depth is 10% greater than the conjugate depth to help prevent sweepout of the jump. Because considerable wave action can remain downstream of the basin, this jump and basin are sometimes avoided altogether by widening the basin to increase the Froude number. For Froude numbers greater than 4.5, either *Type III* or *Type II* basins are recommended. The Type III basin, shown in Figure 3.34, includes baffle blocks, and so is limited to applications where the incoming velocity does not exceed 18 m/s.

For inflow velocities exceeding 18 m/s, the Type II basin, shown in Figure 3.35, is suggested. The Type II basin is slightly longer than the Type III basin, and the tailwater is recommended to be 5% greater than the conjugate depth (of the hydraulic jump) to help prevent sweepout.

Matching the tailwater and conjugate depth curves over a range of operating discharges is one of the most important aspects of stilling basin design. If the tailwater is lower than the conjugate depth of the jump, the jump may be swept out of the basin. On the other hand, a tailwater depth that is higher than the conjugate depth causes the jump to back up against the spillway chute and be drowned, no longer dissipating as much energy. The ideal situation is one in which the conjugate depth of the hydraulic jump perfectly matches the tailwater over the full range of operating discharges, but this is unlikely to occur. Instead, the basin floor elevation is normally set to match the conjugate depth and tailwater at the maximum design discharge, assuming that the conjugate depth is less than the tailwater depth when the discharge is less than the maximum discharge. In some cases, the tailwater depth equals the conjugate depth at a discharge less than the maximum discharge, and in these cases the stilling basin should be designed for the lower discharge. In no case should the tailwater depth be less than the conjugate depth within the operating range of the stilling basin.

Setting the floor elevation of the stilling basin and selecting the type of basin requires predicting the flow and velocity at the toe of the spillway, and hence the energy loss over the spillway. According to the U.S. Bureau of Reclamation (USBR, 1987), if the stilling basin is located immediately downstream of the crest of an overflow spillway, or if the spillway chute is no longer than the hydraulic head, energy losses can be neglected. In this case, the hydraulic head is defined as the difference in elevation between the reservoir water surface and the downstream water surface at the entrance to the stilling basin. If the spillway chute length is between one and five times the hydraulic head, an energy loss of 10% of the hydraulic head is recommended. For spillway chute lengths in excess of five times the hydraulic head, an energy loss of 20% of the hydraulic head is recommended. For more accurate estimates of head loss, the gradually varied flow equation can be solved along the spillway chute; however, it should be recognized that this equation is not valid in the vicinity of the spillway crest, where the flow is not gradually varied and the boundary layer is not fully developed.

FIGURE 3.33: Type IV stilling basin characteristics, 2.5 < Fr < 4.5.

Source: U.S. Bureau of Reclamation (1987).

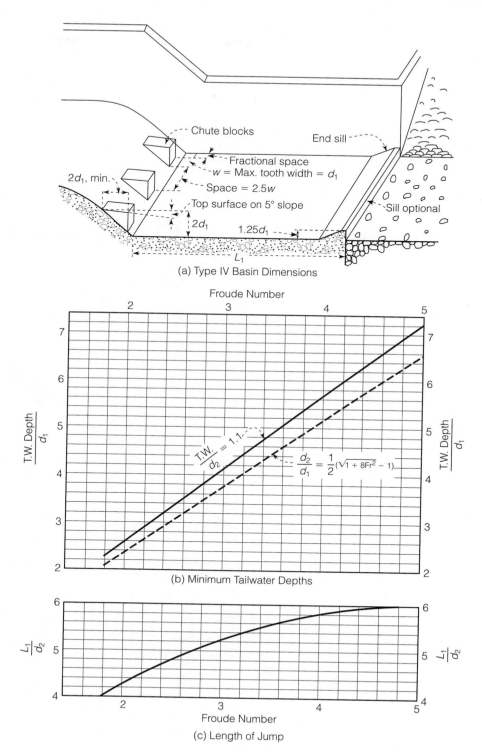

(a) Type IV Basin Dimensions

(b) Minimum Tailwater Depths

(c) Length of Jump

FIGURE 3.34: Type III stilling basin characteristics, Fr > 4.5 and $V_1 \leq 18$ m/s.

Source: U.S. Bureau of Reclamation (1987).

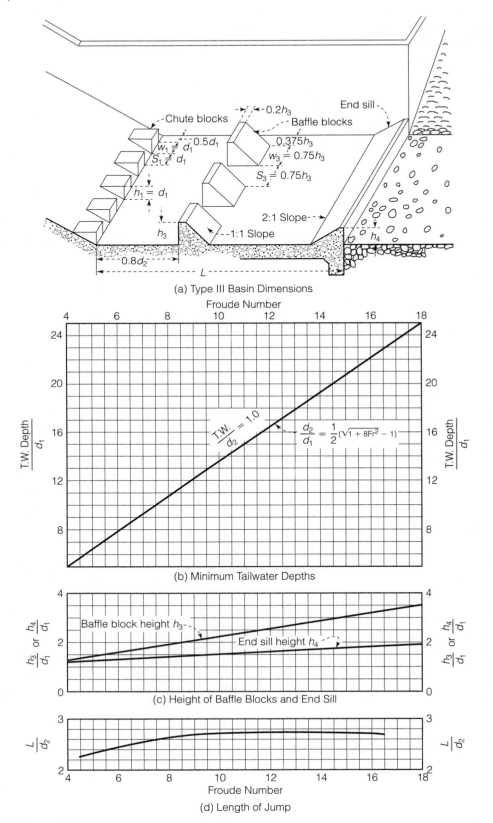

(a) Type III Basin Dimensions

(b) Minimum Tailwater Depths

$$\frac{\text{T.W.}}{d_2} = 1.0 \qquad \frac{d_2}{d_1} = \frac{1}{2}(\sqrt{1 + 8Fr^2} - 1)$$

(c) Height of Baffle Blocks and End Sill

(d) Length of Jump

FIGURE 3.35: Type II stilling basin characteristics, Fr > 4.5

Source: U.S. Bureau of Reclamation (1987).

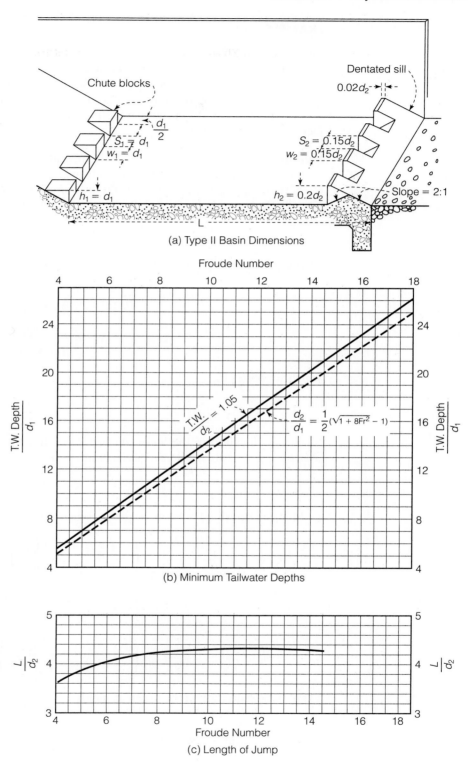

(a) Type II Basin Dimensions

(b) Minimum Tailwater Depths

$$\frac{d_2}{d_1} = \frac{1}{2}(\sqrt{1 + 8Fr^2} - 1)$$

(c) Length of Jump

EXAMPLE 3.19

The maximum design discharge over a spillway 12 m wide is 280 m³/s into a stilling basin of the same width. The reservoir behind the spillway has a surface elevation of 60.00 m, and the river water-surface elevation downstream of the stilling basin is 30.00 m. Assuming a 10% energy loss in the flow down the spillway, find the invert elevation of the floor of the stilling basin so that the hydraulic jump forms in the basin. Design the stilling basin.

Solution From the given data: $b = 12$ m, and $Q = 280$ m³/s. Since the water loses 10% of its energy flowing down the spillway, and the elevation behind the spillway is 60 m, then, taking the elevation of the bottom of the stilling basin as X, the specific energy, E_1, at entrance to the stilling basin (i.e., base of the spillway) is given by

$$E_1 = 0.90(60.00) - X = 54 - X$$

Taking y_1 as the depth of flow at the entrance to the stilling basin,

$$E_1 = y_1 + \frac{Q^2}{2g(by_1)^2}$$

$$54 - X = y_1 + \frac{280^2}{2(9.81)(12y_1)^2}$$

which simplifies to

$$X = 54 - y_1 - \frac{27.75}{y_1^2} \tag{3.204}$$

Since the elevation of the water surface downstream of the stilling basin is 30.00 m, the conjugate depth of the hydraulic jump, y_2, is given by

$$y_2 = 30 - X \tag{3.205}$$

and the relationship between y_1 and y_2 is given by the hydraulic jump equation

$$y_2 = \frac{y_1}{2}\left(-1 + \sqrt{1 + \frac{8q^2}{gy_1^3}}\right)$$

$$y_2 = \frac{y_1}{2}\left(-1 + \sqrt{1 + \frac{8(280/12)^2}{9.81y_1^3}}\right)$$

$$y_2 = \frac{y_1}{2}\left(-1 + \sqrt{1 + \frac{444.0}{y_1^3}}\right) \tag{3.206}$$

Combining Equations 3.204 to 3.206 yields

$$30 - 54 + y_1 + \frac{27.75}{y_1^2} = \frac{y_1}{2}\left(-1 + \sqrt{1 + \frac{444.0}{y_1^3}}\right)$$

which simplifies to

$$1.5y_1 + \frac{27.75}{y_1^2} - \frac{y_1}{2}\sqrt{1 + \frac{444.0}{y_1^3}} - 24 = 0$$

The solutions of this equation are $y_1 = 24.1$ m and $y_1 = 0.91$ m. These depths correspond to subcritical and supercritical flows, respectively (Fr = 0.1 and Fr = 8.6). Since the upstream depth of flow must be supercritical, take $y_1 = 0.91$ m, and Equation (3.204) gives

$$X = 54 - 0.91 - \frac{27.75}{0.91^2} = 19.6 \text{ m}$$

Therefore the stilling basin must have an elevation of 19.6 m for the hydraulic jump to occur within the stilling basin.

Since the Froude number of the flow entering the stilling basin is 8.6 and the entrance velocity is $280/(0.91 \times 12) = 26$ m/s, then a Type II basin is required. Using this type of stilling basin, it is recommended that the tailwater be 5% greater than the sequent depth to prevent sweepout. Since the elevation of the stilling basin was calculated assuming that the sequent depth and tailwater elevations were the same, then the invert elevation of the stilling basin should be recalculated taking the tailwater depth to be 5% greater than the sequent depth. In this case, Equation 3.205 becomes

$$1.05y_2 = 30 - X$$

and the required elevation of the stilling basin becomes 18.8 m, with an entering flow depth of 0.90 m, Froude number of 8.7, and velocity of 25.9 m/s. This indicates a Type II basin, for which the appropriate invert has now been calculated. The dimensions of the chute blocks and dentated sill in the Type II stilling basin are given in Figure 3.35 in terms of the entering depth (y_1) of 0.90 m and sequent depth (y_2) given by Equation 3.206 as 10.7 m. The variables y_1 and y_2 used here correspond to the variables d_1 and d_2 used in Figure 3.35, hence $d_1 = 0.90$ m and $d_2 = 10.7$ m. The length, L, of the stilling basin is derived from Figure 3.35(c), where Fr = 8.7, $L/d_2 = 4.3$, and hence $L = 4.3(10.7) = 46$ m.

3.4.5 Gates

Gates are used to regulate the flow in open channels. They are designed for either overflow or underflow operation, with underflow operation appropriate for channels in which there is a significant amount of floating debris. Two common types of gates are vertical gates and radial gates (also called *Tainter*[§] gates), which are illustrated in Figure 3.36. Vertical gates are supported by vertical guides with roller wheels, and large hydrostatic forces usually induce significant frictional resistance to raising and lowering the gates. An example of a vertical gate structure containing two gates and a close-up view of a gate lift are shown in Figure 3.37. Vertical gates are sometimes referred to as *vertical lift gates*, *sluice gates* or *vertical sluice gates*, where a *sluice* is an artificial channel for conducting water, with a gate to regulate the flow. The conventional radial (Tainter) gate consists of an arc-shaped face plate supported

---

[§]Tainter gates were patented in the United States by Jeremiah B. Tainter in 1886.

FIGURE 3.36: Types of gates

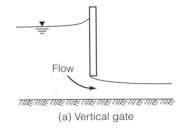

(a) Vertical gate   (b) Radial (Tainter) gate

FIGURE 3.37: Vertical gate structure (with two gates) and close-up view of gate lift

by radial struts that are attached to a central horizontal shaft called a *trunnion*, which transmits the hydrostatic force to the supporting structure. Since the vector of the resultant hydrostatic force passes through the axis of the horizontal shaft, only the weight of the gate needs to be lifted to open the gate. Radial gates are economical to install and are widely used in both underflow and overflow applications. Structural design guidelines for several types of gates can be found in Sehgal (1996).

Applying the energy equation to both vertical and radial gates, Figure 3.38, yields

$$y_1 + \frac{V_1^2}{2g} = y_2 + \frac{V_2^2}{2g} \tag{3.207}$$

where Sections 1 and 2 are upstream and downstream of the gate, respectively, and energy losses are neglected. Writing Equation 3.207 in terms of the flowrate, Q, leads to

$$y_1 + \frac{Q^2}{2gb^2y_1^2} = y_2 + \frac{Q^2}{2gb^2y_2^2}$$

FIGURE 3.38: Flow through gates.

Source: Chadwick and Morfett (1993).

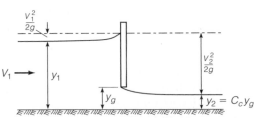
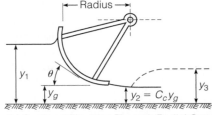

(a) Sectional Elevation Through Vertical Sluice Gate   (b) Sectional Elevation Through Radial Gate

and solving for Q gives

$$Q = by_1 y_2 \sqrt{\frac{2g}{y_1 + y_2}} \tag{3.208}$$

The depth of flow downstream of the gate, y_2, is less than the gate opening, y_g, since the streamlines of the flow contract as they move past the gate (see Figure 3.38). The downstream location where the depth of flow is most contracted is called the *vena contracta*, and y_2 denotes the depth at the vena contracta. Denoting the ratio of the downstream depth, y_2, to the gate opening, y_g, by the *coefficient of contraction*, C_c, where

$$C_c = \frac{y_2}{y_g} \tag{3.209}$$

then Equations 3.208 and 3.209 can be combined to yield the following expression for the discharge through a gate:

$$\boxed{Q = C_d b y_g \sqrt{2g y_1}} \tag{3.210}$$

where C_d is the *discharge coefficient* or *sluice coefficient* given by

$$\boxed{C_d = \frac{C_c}{\sqrt{1 + C_c \dfrac{y_g}{y_1}}}} \tag{3.211}$$

The form of the discharge equation given by Equation 3.210 expresses the discharge in terms of an "orifice-flow" velocity, $\sqrt{2g y_1}$, times the flow area through the gate, $b y_g$, times a discharge coefficient, C_d, to account for deviations from the orifice-flow assumption. On the basis of Equation 3.211, the discharge coefficient depends on the amount of flow contraction as measured by C_c and y_g/y_1. In the case of a vertical gate with a sharp edge, it has been reported that (Chadwick and Morfett, 1993)

$$C_c = 0.61 \tag{3.212}$$

whenever $0 < y_g/E_1 < 0.5$, where E_1 is the specific energy of the flow upstream of the gate, defined by $E_1 = y_1 + (V_1^2/2g)$. Recent experimental results by Lin et al. (2002) found values of C_c for sharp-edged vertical gates in range of 0.59 to 0.61, and these results are consistent with the reported range of 0.58–0.63 by Rajaratnam and Subramanya (1967), the suggested value of 0.60 by Henry (1960), and the suggested value of 0.61 by Henderson (1966) and Rajaratnam and Subramanya (1967). For vertical gates with rounded edges, values of C_c in the range 0.65 to 0.75 have been reported (Lin et al., 2002). In the case of radial gates, the contraction coefficient, C_c, is generally greater than 0.61 and is commonly expressed as a function of the angle θ (shown in Figure 3.38) as

$$C_c = 1 - 0.75\left(\frac{\theta}{90}\right) + 0.36\left(\frac{\theta}{90}\right)^2 \tag{3.213}$$

where θ is measured in degrees. Equation 3.213 gives results that are accurate to within $\pm 5\%$ provided that $\theta < 90°$ (Henderson, 1966).

Flows through gates may be free or submerged depending on the tailwater depth. If the tailwater depth is less than the conjugate depth of the vena contracta, the flow will be free, and if the tailwater depth is greater than the conjugate depth of the vena contracta, the flow through the gate will be submerged. The flow condition in which the tailwater depth is equal to the conjugate depth of the vena contracta is called the *distinguishing condition*. Under this condition, the flow through the vertical gate is given by Equation 3.210, and the tailwater depth, y_3, is related to the depth at the vena contracta, y_2, by the hydraulic jump equation (Equation 3.127) given by

$$y_3 = \frac{y_2}{2}\left(-1 + \sqrt{1 + \frac{8Q^2}{gb^2y_2^3}}\right) \tag{3.214}$$

Combining Equations 3.210, 3.211, and 3.214 yields the following expression for the tailwater depth, y_3, under the distinguishing condition,

$$y_3 = \frac{C_c y_g}{2}\left[\sqrt{1 + \frac{16}{\eta(1 + \eta)}} - 1\right] \tag{3.215}$$

where

$$\eta = C_c \frac{y_g}{y_1} \tag{3.216}$$

For a known gate opening, y_g, contraction coefficient, C_c, and upstream water depth, y_1, the maximum allowable downstream depth for free flow can be determined from Equation 3.215, which represents the distinguishing condition between free flow and submerged flow through the gate. For a given upstream depth and gate opening, the maximum allowable downstream tailwater depth under the free-flow condition increases with increasing contraction coefficient, implying that a gate with a large contraction coefficient, such as a radial gate, is more suitable for flow control under the free-flow condition (Lin et al., 2002).

EXAMPLE 3.20

A vertical sluice gate is opened 10 cm and the depth of water behind the gate is 1 m. If the width of the rectangular sluice is 1.5 m, estimate the free-flow discharge through the gate and the distinguishing condition for the tailwater depth.

Solution   From the given data: $y_g = 0.10$ m, $y_1 = 1$ m, and $b = 1.5$ m. Assuming $C_c = 0.61$, Equation 3.216 gives

$$\eta = C_c \frac{y_g}{y_1} = 0.61\frac{0.1}{1} = 0.061$$

and, for free-flow conditions, Equation 3.211 gives the discharge coefficient, C_d, as

$$C_d = \frac{C_c}{\sqrt{1 + \eta}} = \frac{0.61}{\sqrt{1 + 0.061}} = 0.592$$

The free-flow discharge through the gate is given by Equation 3.210 as

$$Q = C_d b y_g \sqrt{2gy_1} = 0.592(1.5)(0.1)\sqrt{2(9.81)(1)} = 0.393 \ \mathrm{m^3/s}$$

The distinguishing condition for the tailwater depth is given by Equation 3.215 as

$$y_3 = \frac{C_c y_g}{2}\left[\sqrt{1 + \frac{16}{\eta(1+\eta)}} - 1\right]$$

$$= \frac{0.61(0.10)}{2}\left[\sqrt{1 + \frac{16}{0.061(1+0.061)}} - 1\right] = 0.45 \ \mathrm{m}$$

Therefore, the flow depth through the gate will be equal to 0.393 m³/s as long as the tailwater depth is less than or equal to 0.45 m. If the tailwater depth exceeds 0.45 m, then the flow through the gate will be submerged and the flow will be less than 0.393 m³/s.

In cases where the discharge through the gate opening is supercritical and the depth of flow downstream of the gate exceeds the conjugate depth of the vena contracta (i.e., exceeds the distinguishing condition), the outflow will be submerged and the discharge equation given by Equation 3.210 will not be applicable. This condition is illustrated in Figure 3.39. An approximate analysis of the submerged-flow condition assumes that all head losses occur in the flow downstream of the gate, between Sections 2 and 3, in which case the energy equation can be written as

$$y_1 + \frac{Q^2}{2gb^2 y_1^2} = y + \frac{Q^2}{2gb^2 y_2^2} \tag{3.217}$$

where y is the depth of flow immediately downstream of the gate. Between Sections 2 and 3, the momentum equation (Equation 3.124) can be written as

$$\frac{y^2}{2} + \frac{Q^2}{gb^2 y_2} = \frac{y_3^2}{2} + \frac{Q^2}{gb^2 y_3} \tag{3.218}$$

and flowrate, Q, can be estimated by simultaneous solution of Equations 3.217 and 3.218, where y_1 and y_3 are usually known and y_2 is estimated by $C_c y_g$. Lin et al. (2002) provided the following analytic solution to Equations 3.217 and 3.218:

FIGURE 3.39: Submerged flow through gates.

Source: Chadwick and Morfett (1993).

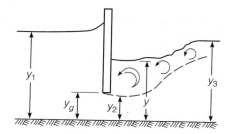

$$Q = C_c \frac{\left[\xi - \sqrt{\xi^2 - \left(\frac{1}{\eta^2} - 1\right)^2 \left(1 - \frac{1}{\lambda^2}\right)}\right]^{1/2}}{\frac{1}{\eta} - \eta} by_g\sqrt{2gy_1} \tag{3.219}$$

where

$$\lambda = \frac{y_1}{y_3} \tag{3.220}$$

$$\xi = \left(\frac{1}{\eta} - 1\right)^2 + 2(\lambda - 1) \tag{3.221}$$

and η is given by Equation 3.216. To estimate the flow through a gate under submerged-flow conditions, Equation 3.219 should be used.

EXAMPLE 3.21

Water is ponded behind a vertical gate to a height of 4 m in a rectangular channel of width 7 m. If the tailwater depth is 3.5 m and the gate opening is 1.15 m, what is the discharge through the gate?

Solution   From the given data: y_1 = 4 m, b = 7 m, y_g = 1.15 m, and y_3 = 3.5 m. It must first be determined whether free-flow or submerged-flow conditions exist. The distinguishing condition is given by Equation 3.215, where the contraction coefficient, C_c, can be assumed equal to 0.61, and η is given by Equation 3.216 as

$$\eta = C_c\frac{y_g}{y_1} = 0.61\frac{1.15}{4} = 0.175$$

Substituting into Equation 3.215 gives the distinguishing condition as

$$y_3 = \frac{C_c y_g}{2}\left[\sqrt{1 + \frac{16}{\eta(1 + \eta)}} - 1\right]$$

$$= \frac{0.61(1.15)}{2}\left[\sqrt{1 + \frac{16}{0.175(1 + 0.175)}} - 1\right] = 2.76 \text{ m}$$

Since the tailwater elevation (=3.5 m) exceeds 2.76 m, then the flow is submerged and Equation 3.219 must be used to calculate the flow through the gate. From the given data, Equations 3.220 and 3.221 give

$$\lambda = \frac{y_1}{y_3} = \frac{4}{3.5} = 1.14$$

$$\xi = \left(\frac{1}{\eta} - 1\right)^2 + 2(\lambda - 1) = \left(\frac{1}{0.175} - 1\right)^2 + 2(1.14 - 1) = 22.5$$

Substituting into Equation 3.219 gives

$$Q = C_c \frac{\left[\xi - \sqrt{\xi^2 - \left(\frac{1}{\eta^2} - 1\right)^2 \left(1 - \frac{1}{\lambda^2}\right)} \right]^{1/2}}{\frac{1}{\eta} - \eta} b y_g \sqrt{2gy_1}$$

$$= 0.61 \frac{\left[22.5 - \sqrt{22.5^2 - \left(\frac{1}{0.175^2} - 1\right)^2 \left(1 - \frac{1}{1.14^2}\right)} \right]^{1/2}}{\frac{1}{0.175} - 0.175} (7)(1.15)\sqrt{2(9.81)(4)}$$

$$= 19.1 \ \mathrm{m^3/s}$$

Therefore, under the given headwater, tailwater, and gate-opening conditions, the (submerged) discharge through the gate is 19.1 m³/s.

3.4.6 Culverts

Culverts are short conduits that are designed to pass peak flood discharges under roadways or other embankments. Because of the function they perform, culverts are commonly included in a class called *cross-drainage structures*, which also includes bridges. Culverts perform a similar function to that of bridges but, unlike bridges, they have spans less than 6 m (20 ft) and can be designed to have a submerged inlet. Typical cross sections of culverts include circular, arched, rectangular, and oval shapes. An arched corrugated metal culvert passing under a roadway is shown in Figure 3.40. Culverts with more than one barrel are frequently necessary to pass wide shallow streams under roadways.

Culvert design typically requires the selection of a barrel cross section that passes a given flowrate when the water is ponded to a given height at the culvert entrance. The hydraulic analysis of culverts is complicated by the fact that there are several possible flow regimes, with the governing flow equation being determined by the

FIGURE 3.40: Arched culvert under roadway

FIGURE 3.41: Flow through culvert with submerged entrance

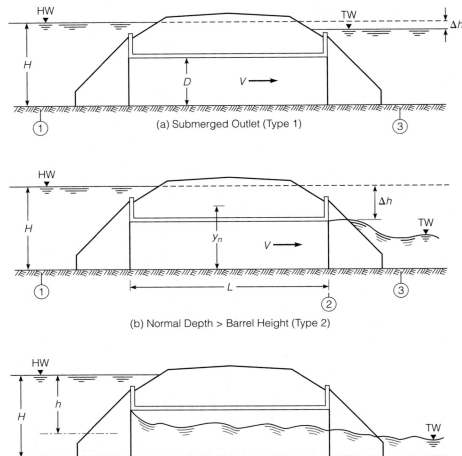

(a) Submerged Outlet (Type 1)

(b) Normal Depth > Barrel Height (Type 2)

(c) Entrance Control, Normal Depth < Barrel Height (Type 3)

flow regime. The U.S. Geological Survey (Bodhaine, 1976) classifies the flow regimes into six types, depending primarily on the headwater and tailwater elevations, and whether the slope is mild or steep. These six flow regimes can be grouped into either *submerged-entrance conditions* or *free-entrance conditions*, illustrated in Figures 3.41 and 3.42, respectively. The entrance to a culvert is regarded as submerged when the depth, H, of water upstream of the culvert exceeds $1.2D$, where D is the diameter or height of the culvert. Some engineers take this limit as $1.5D$ (e.g., French, 1985; Sturm, 2001), however, the water surface will impinge on the headwall when the headwater depth is about $1.2D$ if the critical depth, y_c, in the culvert occurs at the culvert entrance and $y_c \geq 0.8D$ (Finnemore and Franzini, 2002). Fundamentally, the minimum headwater depth to cause inlet submergence depends on the shape of the culvert entrance (Jain, 2001). The depth of water at the culvert entrance is called the *headwater depth*,[¶] and the depth of water at the culvert exit is the *tailwater depth*

[¶]Some authors define the headwater depth as the depth of water at the culvert entrance plus the velocity head (Tuncock and Mays, 1999).

FIGURE 3.42: Flow through culvert with free entrance

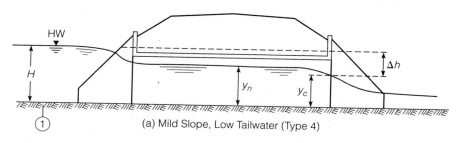

(a) Mild Slope, Low Tailwater (Type 4)

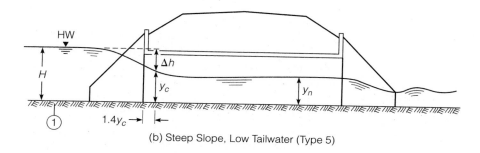

(b) Steep Slope, Low Tailwater (Type 5)

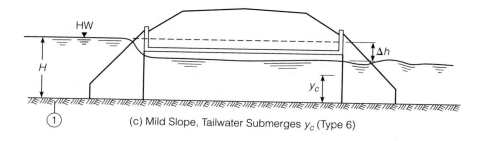

(c) Mild Slope, Tailwater Submerges y_c (Type 6)

(USFHWA, 1985). In the case of a submerged entrance, Figure 3.41 shows three possible flow regimes: (a) The outlet is submerged (Type 1 flow); (b) the outlet is not submerged and the normal depth of flow in the culvert is larger than the culvert height, D (Type 2 flow); and (c) the outlet is not submerged and the normal depth of flow in the culvert is less than the culvert height (Type 3 flow).

Types 1 and 2 flow.   In Type 1 flow, applying the energy equation between Sections 1 (headwater) and 3 (tailwater) leads to

$$\Delta h = h_i + h_f + h_o \tag{3.222}$$

where Δh is the difference between the headwater and tailwater elevations, h_i is the entrance loss, h_f is the head loss due to friction in the culvert, and h_o is the exit loss. Equation 3.222 neglects the velocity heads at Sections 1 and 3, which are usually small compared with the other terms, and also neglects the friction losses between Section 1 and the culvert entrance, and the friction loss between the culvert exit and Section 3.

Using the Manning equation to calculate h_f within the culvert, then

$$h_f = \frac{n^2 V^2 L}{R^{\frac{4}{3}}} \tag{3.223}$$

where n is the roughness coefficient, V is the velocity of flow, L is the length, and R is the hydraulic radius of the culvert. Head loss due to friction within the culvert is usually minor, except in long rough barrels on flat slopes (Bodhaine, 1976). The entrance loss, h_i, is given by

$$h_i = k_e \frac{V^2}{2g} \tag{3.224}$$

where k_e is the entrance loss coefficient. Entrance losses are caused by the sudden contraction and subsequent expansion of the stream within the culvert barrel, and the entrance geometry has an important influence on this loss. The exit loss, h_o, is given by (USFHWA, 1985)

$$h_o = k_o \left(\alpha \frac{V^2}{2g} - \alpha_d \frac{V_d^2}{2g} \right) \tag{3.225}$$

where k_o is the exit loss coefficient, α is the energy coefficient inside the culvert at the exit, α_d is the energy coefficient at the cross section just downstream of the culvert, and V_d is the average velocity at the cross section just downstream of the culvert. For a sudden expansion of flow, such as in a typical culvert, the exit loss coefficient, k_o, is normally set to 1.0, while in general exit loss coefficients can vary between 0.3 and 1.0. The exit loss coefficient should be reduced as the transition becomes less abrupt. It is commonplace for culverts to exit into large open bodies of water, in which case it is assumed that $k_o = 1.0$, $\alpha = \alpha_d = 1.0$, and $V_d = 0$, which yields the commonly used relation

$$h_o = \frac{V^2}{2g} \tag{3.226}$$

Combining Equations 3.222 to 3.224 with Equation 3.226 yields the following form of the energy equation between Sections 1 and 3:

$$\Delta h = \frac{n^2 V^2 L}{R^{\frac{4}{3}}} + k_e \frac{V^2}{2g} + \frac{V^2}{2g} \tag{3.227}$$

This equation can also be applied between Section 1 (headwater) and Section 2 (culvert exit) in Type 2 flow, illustrated in Figure 3.41(b), where the velocity head at the exit, $V^2/2g$, is equal to the exit loss in Type 1 flow. Equation 3.227 reduces to the following relationship between the difference in the water-surface elevations on both sides of the culvert, Δh, and the discharge through the culvert, Q:

$$Q = A \sqrt{\frac{2g\Delta h}{\frac{2gn^2 L}{R^{\frac{4}{3}}} + k_e + 1}} \tag{3.228}$$

where A is the cross-sectional area of the culvert. It should be noted that Δh is equal to the difference between the headwater and tailwater elevations for a submerged outlet (Type 1 flow), and Δh is equal to the difference between the headwater and the crown of the culvert exit when the normal depth of flow in the culvert exceeds the height of the culvert (Type 2 flow). If the Darcy–Weisbach equation is used to calculate the head loss in the culvert, then Equation 3.228 takes the form

$$Q = A \sqrt{\frac{2g\Delta h}{\frac{fL}{4R} + k_e + 1}} \qquad (3.229)$$

where f is the Darcy–Weisbach friction factor. Under both Type 1 and Type 2 conditions, the flow is said to be under *outlet control*, since the water depth at the outlet influences the discharge through the culvert. Equation 3.228 is the basis for the culvert-design nomographs developed by the U.S. Federal Highway Administration (1985).

Type 3 flow.   In Type 3 flow, the inlet is submerged and the culvert entrance will not admit water fast enough to fill the culvert. In this case, the culvert inlet behaves like an orifice and the discharge through the culvert, Q, is related to the head on the center of the orifice, h, by the relation

$$Q = C_d A \sqrt{2gh} \qquad (3.230)$$

where C_d is the coefficient of discharge and h is equal to the vertical distance from the centroid of the culvert entrance to the water surface at the entrance. The coefficient of discharge, C_d, is frequently taken as 0.62 for square-edged entrances and 1.0 for well-rounded entrances (Franzini and Finnemore, 2002). However, original data from USGS (Bodhaine, 1976) indicates that, for square-edged entrances, C_d depends on H/D, with $C_d = 0.44$ when $H/D = 1.4$ and $C_d = 0.59$ when $H/D = 5$. Beveling or rounding culvert entrances generally reduces the flow contraction at the inlet and results in higher values of C_d than for square entrances, and a 45° bevel is recommended for ease of construction (U.S. Federal Highway Administration, 1985). In cases where the culvert entrance acts like an orifice, the downstream water level does not influence the flow through the culvert and the flow is said to be under *inlet control*. According to ASCE (1992), Equation 3.230 is applicable only when $H/D \geq 2$, but Franzini and Finnemore (2002) state that the error in Equation 3.230 is less than 2% when $H/D \geq 1.2$. The occurrence of Type 3 flow requires a relatively square entrance that will cause contraction of the flow area to less than the area of the culvert barrel. In addition, the combination of barrel length, roughness, and bed slope must be such that the contracted flow will not expand to the full area of the barrel. If the water surface of the expanded flow comes into contact with the top of the culvert, Type 2 flow will occur because the passage of air to the culvert will be sealed off, causing the culvert to flow full throughout its length. In such cases, the culvert is called *hydraulically long*; otherwise, the culvert is *hydraulically short*. Bodhaine (1976) reported that a culvert has to meet the criterion that $L \leq 10D$ to allow Type 3 flow on a mild slope, and this rule can be used to differentiate between hydraulically short and hydraulically long culverts.

Based on experiments conducted by the U.S. National Bureau of Standards, the following best-fit power relationship has been developed to facilitate the description of Type 3 flow through culverts:

$$\boxed{\frac{H}{D} = c\left[\frac{Q}{AD^{0.5}}\right]^2 + Y - 0.5S} \tag{3.231}$$

where A is the full cross-sectional area of the culvert, S is the slope of the culvert, and c and Y are empirical constants that depend on the inlet type and are given in Table 3.12. Equation 3.231 is an empirical relation based on U.S. Customary units with H and D in ft, A in ft^2, and Q in ft^3/s. Equation 3.231 applies for $Q/AD^{0.5} \geq 4.0$ and is the basis for the culvert-design nomographs developed by the U.S. Federal Highway Administration (1985). When applying Equation 3.231 to mitered inlets, a slope correction factor of $+0.7S$ should be used instead of $-0.5S$.

Whereas Types 1 to 3 flows include most design circumstances when the culvert entrance is submerged, it is important to keep in mind that other submerged-entrance

TABLE 3.12: Constants for Empirical Culvert-Design Equations

Shape and material	Inlet shape	c	Y	Form (Type 5)	K or K'	M or M'
Circular concrete	Square edge with headwall	0.0398	0.67	1	0.0098	2.0
	Groove end with headwall	0.0292	0.74		0.0018	2.0
	Groove end projecting	0.0317	0.69		0.0045	2.0
Circular CMP*	Headwall	0.0078	0.69	1	0.0078	2.0
	Mitered to slope	0.0463	0.75		0.0210	1.33
	Projecting	0.0553	0.54		0.0340	1.5
Circular	Beveled ring, 45°	0.0300	0.74	1	0.0018	2.5
	Beveled ring, 33.7°	0.0243	0.83		0.0018	2.5
Rectangular box	30°–75° wingwall flares	0.0347	0.86	1	0.026	1.0
	30° and 75° wingwall flares	0.0400	0.80		0.061	0.75
	0° wingwall flares	0.0423	0.82		0.061	0.75
Rectangular box	45° wingwall flare, $w/D = 0.043$	0.0309	0.80	2	0.510	0.667
	18° to 33.7° wingwall flare, $w/D = 0.083$	0.0249	0.83		0.486	0.667
	90° headwall, 19-mm chamfers	0.0375	0.79		0.515	0.667
	90° headwall, 45° bevels	0.0314	0.82		0.495	0.667
	90° headwall, 33.7° bevels	0.0252	0.865		0.486	0.667
	19-mm chamfers, 45° skewed headwall	0.04505	0.68		0.545	0.667
	19-mm chamfers, 30° skewed headwall	0.0425	0.705		0.533	0.667
	19-mm chamfers, 15° skewed headwall	0.0402	0.73		0.522	0.667
	45° bevels, 10°–45° skewed headwall	0.0327	0.75		0.498	0.667
	19-mm chamfers, 45° wingwall flare, nonoffset	0.0339	0.803		0.497	0.667
	19-mm chamfers, 18.4° wingwall flare, nonoffset	0.0361	0.806		0.493	0.667
	19-mm chamfers, 18.4° wingwall flare, nonoffset, 30° skew	0.0368	0.71		0.495	0.667
	Top bevels, 45° wingwall flare, offset	0.0302	0.835		0.497	0.667
	Top bevels, 33.7° wingwall flare, offset	0.0252	0.881		0.495	0.667
	Top bevels, 18.4° wingwall flare, offset	0.0227	0.897		0.493	0.667
Box CM	90° headwall	0.0379	0.69	1	0.0083	2.0
	Thick wall projecting	0.0419	0.64		0.0145	1.75
	Thin wall projecting	0.0496	0.57		0.0340	1.5
Ellipse concrete	Horizontal ellipse, square edge with headwall	0.0398	0.67	1	0.0100	2.0
	Horizontal ellipse, groove end with headwall	0.0292	0.74		0.0018	2.5
	Horizontal ellipse, groove end projecting	0.0317	0.69		0.0045	2.0
	Vertical ellipse, square edge with headwall	0.0398	0.67		0.010	2.0
	Vertical ellipse, groove end with headwall	0.0292	0.74		0.0018	2.5
	Vertical ellipse, groove end projecting	0.0317	0.69		0.0095	2.0
Arch CM*	46-cm corner radius, 90° headwall	0.0379	0.69	1	0.0083	2.0
	46-cm corner radius, mitered to slope	0.0463	0.75		0.0300	1.0
	46-cm corner radius, projecting	0.0496	0.57		0.0340	1.5
	46-cm corner radius, projecting	0.0496	0.57		0.0300	1.5
	46-cm corner radius, no bevels	0.0368	0.68		0.0088	2.0
	46-cm corner radius, 33.7° bevels	0.0269	0.77		0.0030	2.0
	79-cm corner radius, projecting	0.0496	0.57		0.0300	1.5
	79-cm corner radius, no bevels	0.0368	0.68		0.0088	2.0
	79-cm corner radius, 33.7° bevels	0.0269	0.77		0.0030	2.0
	90° headwall	0.0379	0.69		0.0083	2.0
	Mitered to slope	0.0463	0.75		0.0300	1.0
	Thin-wall projecting	0.0496	0.57		0.0340	1.5
Circular	Smooth-tapered inlet throat	0.0196	0.90	2	0.534	0.555
	Rough-tapered inlet throat	0.0210	0.90		0.519	0.64
Elliptical inlet face	Tapered inlet, beveled edges	0.0368	0.83	2	0.536	0.622
	Tapered inlet, square edges	0.0478	0.80		0.5035	0.719
	Tapered inlet, thin edge projecting	0.0598	0.75		0.547	0.80
Rectangular	Tapered inlet throat	0.0179	0.97	2	0.475	0.667
Rectangular concrete	Side tapered, less favorable edge	0.0446	0.85	2	0.56	0.667
	Side tapered, more favorable edge	0.0378	0.87		0.56	0.667
Rectangular concrete	Side tapered, less favorable edge	0.0446	0.65	2	0.50	0.667
	Side tapered, more favorable edge	0.0378	0.71		0.50	0.667

Source: U.S. Federal Highway Administration (1985).
*Notes: CMP = corrugated metal pipe; CM = corrugated metal.

flow regimes are possible. For example, a scenario in which the culvert entrance is fully submerged, the tailwater is low, and the culvert flows partially full at the exit is certainly possible, even in cases where the normal depth of flow in the culvert exceeds the culvert diameter. In these cases, the culvert flow can be estimated by assuming that the hydraulic grade line at the culvert exit is at a point halfway between the critical depth and the crown of the culvert, and then using Equation 3.228 with Δh equal to the difference between the assumed hydraulic grade line elevation at the exit and the headwater elevation (USFHWA, 1985). In cases where the actual tailwater elevation exceeds the assumed hydraulic grade line elevation, the actual tailwater depth should be used in this approximation. In reality, a more conservative estimate of the culvert capacity is achieved by assuming Type 2 flow versus this flow condition.

Submerged entrances usually lead to greater flows through the culvert than unsubmerged entrances. In some cases, however, culverts must be designed so that the entrances are not submerged. Such cases include those in which the top of the culvert forms the base of a roadway. In the case of an unsubmerged entrance, Figure 3.42 shows three possible flow regimes: (a) The culvert has a mild slope and a low tailwater, in which case the critical depth occurs somewhere near the exit of the culvert (Type 4 flow); (b) the culvert has a steep slope and a low tailwater, in which case the critical depth occurs somewhere near the entrance of the culvert, at approximately $1.4y_c$ downstream from the entrance, and the flow approaches normal depth at the outlet end (Type 5 flow); and (c) the culvert has a mild slope and the tailwater submerges y_c (Type 6 flow).

Type 4 flow.   For Type 4 flow (mild slope, low tailwater), the critical flow depth occurs at the exit of the culvert. Applying the energy equation between the headwater and the culvert exit gives

$$\Delta h + \frac{V_1^2}{2g} - \frac{V^2}{2g} = h_i + h_f \tag{3.232}$$

where Δh is the difference between the headwater elevation and the elevation of the (critical) water surface at the exit of the culvert, V_1 is the headwater velocity, h_i is the entrance loss given by Equation 3.224, and h_f is the friction loss in the culvert given by Equation 3.223. Equation 3.232 assumes that the culvert slope is mild, and the velocity of the headwater is *not* neglected, as in the case of a ponded headwater where $H/D > 1.2$. Equation 3.232 yields the following expression for the discharge, Q, through the culvert:

$$\boxed{Q = A_c \sqrt{2g\left(\Delta h + \frac{V_1^2}{2g} - h_i - h_f\right)}} \tag{3.233}$$

Where A_c is the flow area at the critical flow section at the exit of the culvert. Equation 3.233 is an implicit expression for the discharge, since the critical depth (a component of Δh), headwater velocity, V_1, entrance loss, h_i, and the friction loss, h_f, all depend on the discharge, Q.

Type 5 flow.   For Type 5 flow (steep slope, low tailwater), the critical flow depth occurs at the entrance of the culvert. Applying the energy equation between the headwater and the culvert entrance gives

$$\Delta h + \frac{V_1^2}{2g} - \frac{V^2}{2g} = h_i \tag{3.234}$$

where Δh is the difference between the headwater elevation and the elevation of the (critical) water surface at the entrance of the culvert. Equation 3.234 leads to the following expression for the discharge, Q, through the culvert:

$$Q = A_c \sqrt{2g\left(\Delta h + \frac{V_1^2}{2g} - h_i\right)} \tag{3.235}$$

Equation 3.235 is an implicit expression for the discharge, since the critical depth (a component of Δh), headwater velocity, V_1, and entrance loss, h_i, all depend on the discharge, Q. In some cases, the approach velocity head, $V_1^2/2g$, is neglected, the entrance loss is given by Equation 3.224, and Equation 3.235 is put in the form (Sturm, 2001)

$$Q = C_d A_c \sqrt{2g(H - y_c)} \tag{3.236}$$

where C_d is defined by the relation

$$C_d = \frac{1}{\sqrt{1 + k_e}} \tag{3.237}$$

and H is the headwater depth. The USGS (Bodhaine, 1976) developed values for C_d as a function of the headwater depth-to-diameter ratio, H/D. For circular culverts set flush in a vertical headwall, $C_d = 0.93$ for $H/D < 0.4$, and it decreases to 0.80 at $H/D = 1.5$ when the entrance becomes submerged. For box culverts set flush in a vertical headwall, C_d can be taken as 0.95.

Based on experiments conducted by the U.S. National Bureau of Standards, the following best-fit power relationships have been developed to facilitate the description of Type 5 flow through culverts:

$$\frac{H}{D} = \frac{E_c}{D} + K\left[\frac{Q}{AD^{0.5}}\right]^M - 0.5S \tag{3.238}$$

$$\frac{H}{D} = K'\left[\frac{Q}{AD^{0.5}}\right]^{M'} \tag{3.239}$$

Where E_c is the specific energy under critical flow conditions at the culvert entrance, A is the full cross-sectional area of the culvert, S is the slope of the culvert, and K, K', M, and M' are empirical constants that depend on the inlet type and are given in Table 3.12. Equation 3.238 is the preferred equation, commonly referred to as "Form 1," and Equation 3.239 is used more easily and is commonly referred to as "Form 2." Equations 3.238 and 3.239 are empirical relations based on U.S. Customary units with H, D, and E_c in ft, A in ft^2, and Q in ft^3/s. Equations 3.238 and 3.239 are applicable for values of $Q/AD^{0.5} \leq 3.5$ and are the bases for the culvert-design nomographs developed by the U.S. Federal Highway Administration (1985). When applying Form 1 (Equation 3.238) to mitered inlets, a slope correction factor of $+0.7S$ should be used instead of $-0.5S$.

Type 6 flow.   For Type 6 flow (mild slope, tailwater submerges y_c), the water surface at the culvert exit is approximately equal to the tailwater elevation. Applying the energy equation between the headwater and the culvert exit gives

$$\Delta h + \frac{V_1^2}{2g} - \frac{V^2}{2g} = h_i + h_f \qquad (3.240)$$

where Δh is the difference between the headwater elevation and the tailwater elevation at the exit of the culvert. Equation 3.240 leads to the following expression for the discharge, Q, through the culvert:

$$Q = A \sqrt{2g\left(\Delta h + \frac{V_1^2}{2g} - h_i - h_f\right)} \qquad (3.241)$$

where A is the flow area at the exit of the culvert. Equation 3.241 is an implicit expression for the discharge, since the headwater velocity, V_1, entrance loss, h_i, and the friction loss, h_f, all depend on the discharge, Q.

Determination of culvert capacity.   The culvert capacity is determined by the headwater, tailwater, and culvert dimensions. The headwater elevation is usually specified based on the elevation of the roadway under which the culvert passes and whether overtopping will be allowed; the tailwater elevation is usually specified as either normal flow conditions in the downstream channel, or a constant (ponded) elevation, or derived from a given depth-flowrate relation for the downstream channel (called a *rating curve*). Based on the allowable headwater depth, H, and the height of the culvert opening, D, it is first determined whether the culvert entrance is submerged. If $H/D > 1.2$, the culvert entrance is submerged; if not, the entrance is not submerged. Once the submergence condition is determined, the discharge capacity of the culvert is calculated as follows:

Submerged entrance.   When the culvert entrance is submerged, the flow is either Type 1, 2, or 3. The flow type and the associated flowrate are determined by the following procedure:

1. If the culvert exit is submerged, then the flow is Type 1 and the discharge is given by Equation 3.228.

2. If the culvert exit is not submerged, the flow is either Type 2 or 3. Assume that the flow is Type 2 and calculate the discharge using Equation 3.228, with the appropriate definition of Δh.

3. Use the flowrate calculated in step 2 to determine the normal depth of flow in the culvert.

4. If the normal depth of flow calculated in step 3 is greater than the height of the culvert, then the flow is Type 2 and the discharge calculated in step 2 is correct.

5. If the normal depth of flow calculated in step 3 is less than the height of the culvert, then the flow is probably Type 3. Calculate the discharge using Equation 3.230 and verify that the normal depth of flow in the culvert is less than the culvert height. If the normal depth is less than the culvert height, then Type 3 flow is confirmed.

6. If neither Type 2 nor Type 3 flow can be confirmed, then some intermediate flow regime is occurring and the capacity of the culvert should be taken as the lesser of the two calculated discharges.

It is useful to note that circular culverts flow full when the discharge rate exceeds 1.07 Q_{full}, where Q_{full} is the full-flow discharge calculated using the Manning equation (Brater et al., 1996).

Unsubmerged entrance. When the culvert entrance is unsubmerged, the flow is possibly Type 4, 5, or 6. The flow type and associated flowrate are determined by the following procedure:

1. Assume that the flow is Type 4; use Equation 3.233, along with the critical flow equation (Equation 3.95), to calculate the discharge, Q, and the critical depth, y_c. Use Q in the Manning equation to calculate the normal depth, y_n. If $y_n > y_c$ and the tailwater depth is less than y_c, then Type 4 flow is verified and the calculated discharge is correct.

2. Assume that the flow is Type 5; use Equation 3.235, along with the critical flow equation (Equation 3.95), to calculate the discharge, Q, and the critical depth, y_c. Use Q in the Manning equation to calculate the normal depth, y_n. If $y_n < y_c$ and the tailwater depth is less than y_c, then Type 5 flow is verified and the calculated discharge is correct.

3. Assume that the flow is Type 6; use Equation 3.241 to calculate the discharge, Q, and use Q to calculate the normal depth, y_n, and critical depth, y_c. If $y_n > y_c$ and the tailwater depth is greater than y_c, then Type 6 flow is verified and the calculated discharge is correct.

In calculating the capacities of culverts it is important to keep in mind that there are flow regimes other than the Type 1 to Type 6 flows identified here (USFHWA, 1985). However, Type 1 to Type 6 conditions are representative of most culvert flows, and they can generally be confirmed by using the calculated flowrates to verify the assumed flow conditions.

EXAMPLE 3.22

What is the capacity of a 1.22-m by 1.22-m concrete box culvert ($n = 0.013$) with a rounded entrance ($k_e = 0.05$, $C_d = 0.95$) if the culvert slope is 0.5%, the length is 36.6 m, and the headwater level is 1.83 m above the culvert invert? Consider the following cases: (a) free-outlet conditions, and (b) tailwater elevation 0.304 m above the top of the box at the outlet. What must the headwater elevation be in case (b) for the culvert to pass the flow that exists in case (a)?

Solution Since the headwater depth exceeds 1.2 times the height of the culvert opening, the culvert entrance is *submerged*. An elevation view of the culvert is shown in Figure 3.43.

(a) For free-outlet conditions, two types of flow are possible: the normal depth of flow is greater than the culvert height (Type 2), or the normal depth of flow is less than the culvert height (Type 3). To determine the flow type, assume a certain type of flow, calculate the discharge and depth of flow, and see if the assumption is confirmed. If the assumption is not confirmed, then the initial assumption is

FIGURE 3.43: Elevation view of culvert

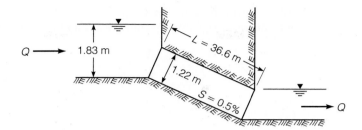

incorrect. Assuming Type 2 flow, then the flowrate equation, Equation 3.228, is given by

$$Q = A \sqrt{\frac{2g\,\Delta h}{\frac{2gn^2 L}{R^{\frac{4}{3}}} + k_e + 1}} \qquad (3.242)$$

where Δh is the difference in water levels between the entrance and exit of the culvert $[= 1.83 + 0.005(36.6) - 1.22 = 0.793$ m$]$, n is the roughness coefficient $(=0.013)$, L is the length of the culvert $(=36.6$ m$)$, and R is the hydraulic radius given by

$$R = \frac{A}{P}$$

where A is the cross-sectional area of the culvert and P is the wetted perimeter of the culvert,

$$A = (1.22)(1.22) = 1.49 \text{ m}^2$$
$$P = 4(1.22) = 4.88 \text{ m}$$

and therefore

$$R = \frac{1.49}{4.88} = 0.305 \text{ m}$$

Substituting known values of the culvert parameters into Equation 3.242 gives

$$Q = (1.49) \sqrt{\frac{2(9.81)(0.793)}{\frac{2(9.81)(0.013)^2(36.6)}{(0.305)^{\frac{4}{3}}} + 0.05 + 1}}$$

which simplifies to

$$Q = 4.59 \text{ m}^3/\text{s}$$

The next step is to calculate the normal depth at a discharge of 4.59 m^3/s using the Manning equation, where

$$Q = \frac{1}{n} A R^{\frac{2}{3}} S_0^{\frac{1}{2}}$$

and S_0 is the slope of the culvert ($= 0.005$). If the normal depth of flow is y_n, then the area, A, wetted perimeter, P, and hydraulic radius, R, are given by

$$A = by_n = 1.22y_n$$

$$P = b + 2y_n = 1.22 + 2y_n$$

$$R = \frac{A}{P} = \frac{1.22y_n}{1.22 + 2y_n}$$

The Manning equation gives

$$4.59 = \frac{1}{0.013} \frac{(1.22y_n)^{\frac{5}{3}}}{(1.22 + 2y_n)^{\frac{2}{3}}} (0.005)^{\frac{1}{2}}$$

which simplifies to

$$\frac{(1.22y_n)^{\frac{5}{3}}}{(1.22 + 2y_n)^{\frac{2}{3}}} = 0.844$$

which yields

$$y_n = 1.25 \text{ m}$$

Therefore, the initial assumption that the normal depth is greater than the height of the culvert ($= 1.22$ m) is verified, and Type 2 flow is confirmed. The flow through the culvert is equal to 4.59 m³/s.

(b) In this case, the tailwater is 0.304 m above the culvert outlet, and therefore the difference in water levels between the inlet and outlet, Δh, decreases by 0.304 m to 0.793 m -0.304 m $= 0.489$ m. The flow equation in this case, Type 1 flow, is the same as Equation 3.242, with $\Delta h = 0.489$ m, which gives

$$Q = (1.49) \sqrt{\frac{2(9.81)(0.489)}{\frac{2(9.81)(0.013)^2(36.6)}{(0.305)^{\frac{4}{3}}} + 0.05 + 1}}$$

which simplifies to

$$Q = 3.60 \text{ m}^3/\text{s}$$

Therefore, when the tailwater depth rises to 0.305 m above the culvert exit, the discharge decreases from 4.59 m³/s to 3.60 m³/s.

When the headwater is at a height x above the culvert inlet and the tailwater is 0.305 m above the outlet, the flow through the culvert is 4.59 m³/s. The difference between the headwater and tailwater elevations, Δh, is given by

$$\Delta h = [x + 1.22 + 0.005(36.6)] - [1.22 + 0.305] = x - 0.122$$

The flow equation for Type 1 flow (Equation 3.242) requires that

$$4.59 = (1.49) \sqrt{\frac{2(9.81)(x - 0.122)}{\frac{2(9.81)(0.013)^2(36.6)}{(0.305)^{\frac{4}{3}}} + 0.05 + 1}}$$

which leads to

$$x = 0.915 \text{ m}$$

Therefore, the headwater depth at the entrance of the culvert for a flow of 4.59 m^3/s is 1.22 m + 0.915 m = 2.14 m.

In many cases it is instructive to calculate the headwater elevation versus the flowrate through the culvert, and such relations are called *culvert performance curves* (Tuncock and Mays, 1999). This relationship is particularly useful in identifying the flow conditions under which roadways will be overtopped. In cases where the culvert headwater elevation exceeds the roadway crest elevation (i.e., the roadway is overtopped) the flow must be partitioned between flow through the culvert and flow over the roadway. Under overtopping conditions, the roadway is typically assumed to perform like a rectangular weir, in which case the flow over the roadway, Q_r, is given by

$$\boxed{Q_r = C_d L H_r^{3/2}} \tag{3.243}$$

where C_d is the discharge coefficient, L is the roadway length over the culvert, and H_r is the head of water over the crest of the roadway. The discharge coefficient, C_d, can be estimated from the head of water over the roadway (H_r), the width of the roadway (L_r), and the submergence depth downstream of the roadway (y_d), using the relations shown in Figure 3.44, where the discharge coefficient is expressed in the form

$$C_d = k_r C_r \tag{3.244}$$

where C_r is derived from H_r using either Figure 3.44(a) for $H_r/L_r > 0.15$ or Figure 3.44(b) for $H_r/L_r \leq 0.15$, and k_r is derived from Figure 3.44(c) for a given value of y_d/H_r. Values of C_d given in Figure 3.44 are in U.S. Customary units, applicable for Q_r in ft^3/s, L in ft and H_r in ft.

Design considerations.   It is generally necessary to take into account regulatory requirements in designing culverts. Typical design criteria are as follows (Atlanta Regional Commission-Georgia Stormwater Management Manual, 2003; Debo and Reese, 1995):

Flow.   Several flows are typically considered in culvert design. A minimum flow is considered to ensure a self-cleansing velocity, a design flow is considered to ensure that nuisance flooding does not occur, and a maximum flow is considered to ensure that the culvert does not cause major flooding during extreme runoff events. Minimum (self-cleansing) flows typically are taken as 2-year events,[||] design flows as 10-year or 25-year events, and maximum flows as 100-year events. Roadway overtopping may be allowed for maximum-flow events.

Headwater Elevation.   The allowable headwater depth is usually the primary basis for sizing a culvert. The allowable headwater elevation is taken as the elevation above which damage may be caused to adjacent property and/or the roadway and is determined from an evaluation of land use upstream of the culvert and the proposed or existing roadway elevation. Typically a 45-cm (18-in.) freeboard is required.

[||] An *n*-year event is equalled or exceeded once every *n* years.

FIGURE 3.44: Discharge coefficient for roadway overtopping.

Source: USFHWA (1985).

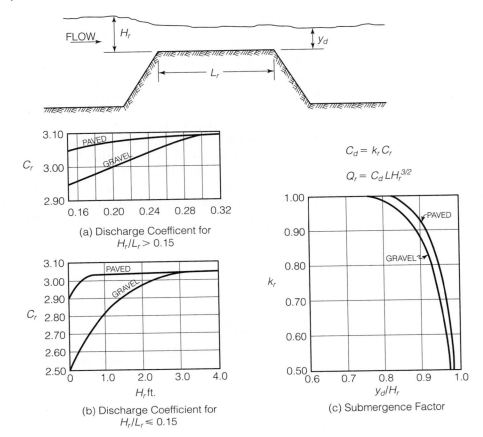

(a) Discharge Coefficent for $H_r/L_r > 0.15$

(b) Discharge Coefficient for $H_r/L_r \leqslant 0.15$

(c) Submergence Factor

$$C_d = k_r C_r$$

$$Q_r = C_d L H_r^{3/2}$$

Slope. The culvert slope should approximate the existing topography, and the maximum slope is typically 10% for concrete pipe and 15% for corrugated metal pipe (CMP) without using a pipe restraint.

Size. Local drainage regulations often require a minimum culvert size (usually in the 30- to 60-cm range). Debris potential is an important consideration in determining the minimum acceptable size of the culvert, and some localities require that the engineer assume 25% debris blockage in computing the required size of any culvert.

Allowable Velocities. Both minimum and maximum velocities must be considered in designing a culvert. A minimum velocity in a culvert of 0.6 to 0.9 m/s (2 to 3 ft/s) at the 2-year flow is frequently required to assure self-cleansing. The maximum allowable velocity for corrugated metal pipe is 3 to 5 m/s (10 to 15 ft/s), and usually there is no specified maximum allowable velocity for reinforced concrete pipe, although velocities greater than 4 to 5 m/s (12 to 15 ft/s) are rarely used because of potential problems with scour.

Material. The most common culvert materials are concrete (reinforced and nonreinforced), corrugated aluminum, and corrugated steel. The selection of a culvert material depends on the required structural strength, hydraulic roughness, durability, and corrosion and abrasion resistance. In general, corrugated culverts have significantly higher frictional resistance than concrete culverts, and most cities require the use of concrete pipe for culverts placed in critical areas or within the public right-of-way. Corrosion is generally

TABLE 3.13: Manning n in Culverts

Type of conduit	Wall and joint description	n
Concrete pipe	Good joints, smooth walls	0.011–0.013
	Good joints, rough walls	0.014–0.016
	Poor joints, rough walls	0.016–0.017
	Badly spalled	0.015–0.020
Concrete box	Good joints, smooth finished walls	0.012–0.015
	Poor joints, rough, unfinished walls	0.014–0.018
Spiral rib metal pipe	19-mm × 19-mm recesses at 30-cm spacing, good joints	0.012–0.013
Corrugated metal pipe, pipe arch, and box	68-mm × 13-mm annular corrugations	0.022–0.027
	68-mm × 13-mm helical corrugations	0.011–0.023
	150-mm × 25-mm helical corrugations	0.022–0.025
	125-mm × 25-mm corrugations	0.025–0.026
	75-mm × 25-mm corrugations	0.027–0.028
	150-mm × 50-mm structural plate	0.033–0.035
	230-mm × 64-mm structural plate	0.033–0.037
Corrugated polyethylene	Corrugated	0.018–0.025
PVC	Smooth	0.009–0.011

Source: U.S. Federal Highway Administration (1985).

a concern with corrugated metal culverts. Recommended Manning n values for culvert design are given in Table 3.13.

Inlet. The geometry of a culvert entrance is an important aspect of culvert design, since the culvert entrance exerts a significant influence on the hydraulic characteristics of a culvert. The four standard inlet types are: (1) flush setting in a vertical headwall, (2) wingwall entrance, (3) projecting entrance, and (4) mitered entrance set flush with a sloping embankment. Structural stability, aesthetics, and erosion control are among the factors that influence the selection of the inlet configuration. Headwalls increase the efficiency of an inlet, provide embankment stability and protection against erosion, and shorten the length of the required structure. Headwalls are usually required for all metal culverts and where buoyancy protection is necessary. Wingwalls are used where the side slopes of the channel adjacent to the inlet are unstable or where the culvert is skewed to the normal channel flow. The entrance loss coefficient, k_e, used to describe the entrance losses in most discharge formulae depends on the pipe material, shape, and entrance type, and can be estimated using the guidelines in Table 3.14. In some cases, the invert of the culvert entrance is depressed below the bottom of the incoming stream bed to increase the headwater depth and the capacity of the culvert. Clearly, this design will only be effective for certain types of flow, for example Type 3 flow. The difference in elevation between the stream bed and the invert of the culvert entrance is called the *fall*. If high headwater depths are to be encountered, or the approach velocity in the channel will cause scour, a short channel apron should be provided at the toe of the headwall.

TABLE 3.14: Culvert Entrance Loss Coefficients

Culvert type and entrance conditions	k_e
Pipe, concrete:	
Projecting from fill, socket end (groove end)	0.2
Projecting from fill, square-cut end	0.5
Headwall or headwall and wingwalls	
Socket end of pipe (groove end)	0.2
Square edge	0.5
Rounded (radius $= D/12$)	0.2
Mitered to conform to fill slope	0.7
End section conforming to fill slope	0.5
Beveled edges, 33.7° or 45° bevels	0.2
Side- or slope-tapered inlet	0.2
Pipe, or pipe arch, corrugated metal:	
Projecting from fill (no headwall)	0.9
Headwall or headwall and wingwalls, square edge	0.5
Mitered to conform to fill slope, paved or unpaved slope	0.7
End section conforming to fill slope	0.5
Beveled edges, 33.7° or 45° bevels	0.2
Side- or slope-tapered inlet	0.2
Box, reinforced concrete:	
Headwall parallel to embankment (no wingwalls)	
Square edged on 3 edges	0.5
Rounded on 3 edges	0.2
Wingwalls at 30° to 75° to barrel	
Square edged at crown	0.4
Crown edge rounded	0.2
Wingwalls at 10° to 25° to barrel	
Square edged at crown	0.5
Wingwalls parallel (extension of sides)	
Square edged at crown	0.7
Side or slope-tapered inlet	0.2

Source: U.S. Federal Highway Administration (1985).

Outlet. Outlet protection should be provided where discharge velocities will cause erosion problems. Protection against erosion at culvert outlets varies from limited riprap placement to complex and expensive energy dissipation devices such as hydraulic jump basins, impact basins, drop structures, and stilling wells (USFHWA, 1985). Outlet protection is typically designed for a 25-year flow.

Debris Control. Usage of smooth well-designed inlets, avoidance of multiple barrels, and alignment of culverts with natural drainage channels will help pass most floating debris. In cases where debris blockage is unavoidable, debris control structures may be required (USFHWA, 1985).

EXAMPLE 3.23

A culvert under a roadway is to be designed to accommodate a 100-year peak flow of 2.49 m³/s. The invert elevation at the culvert inlet is 289.56 m, the invert elevation at the outlet is 288.65 m, and the length of the culvert is to be 22.9 m. The channel downstream of the culvert has a rectangular cross section with a bottom width of 1.5 m, a slope of 4%, and a Manning's n of 0.045. The paved roadway crossing the culvert has a length of 15.2 m, an elevation of 291.08 m, and a width of 18.3 m. Considering a circular reinforced concrete pipe (RCP) culvert with a diameter of 610 mm and a conventional square-edge inlet and headwall, determine the depth of water flowing over the roadway, the flow over the roadway, and the flow through the culvert.

Solution   For the given design flowrate, the tailwater elevation can be derived from the normal-flow condition in the downstream channel. Characteristics of the rectangular downstream channel are given as: $b = 1.5$ m, $S_0 = 0.04$, and $n = 0.045$. Taking $Q = 2.49$ m³/s, the Manning equation gives

$$Q = \frac{1}{n}AR^{2/3}S_0^{1/2}$$

$$2.49 = \frac{1}{0.045}(1.5y_n)\left(\frac{1.5y_n}{1.5 + 2y_n}\right)^{2/3}(0.04)^{1/2}$$

which yields a normal-flow depth, $y_n = 0.73$ m. Since the invert elevation of the downstream channel at the culvert outlet is 288.65 m, the tailwater elevation, TW, under the design condition is given by

$$\text{TW} = 288.65 \text{ m} + 0.73 \text{ m} = 289.38 \text{ m}$$

Since the diameter of the culvert is 0.61 m and the tailwater depth is 0.73 m, the culvert outlet is submerged; and since the roadway elevation is 291.08 m and the tailwater elevation is 289.38 m, the tailwater is below the roadway. Assuming that roadway overtopping (by the headwater) occurs under the design condition, the design flow is equal to the sum of the flow through the culvert and the flow over the roadway such that

$$Q = A\sqrt{\frac{2g\Delta h}{\frac{2gn^2L}{R^{\frac{4}{3}}} + k_e + 1}} + C_dL_RH_r^{3/2} \tag{3.245}$$

where Type 1 flow through the culvert exists (see Equation 3.228). From the given data: $Q = 2.49$ m³/s, $D = 0.61$ m, $A = \pi D^2/4 = 0.292$ m², $n = 0.012$ (Table 3.13 for concrete pipe, good joints, smooth walls), $L = 22.9$ m, $R = D/4 = 0.153$ m, $k_e = 0.5$ (Table 3.14 for headwall, square edge), $L_R = 15.2$ m, and

$$\Delta h = (\text{Roadway Elevation} + H_r) - \text{Tailwater Elevation}$$

$$= (291.08 + H_r) - 289.38$$

$$= 1.70 + H_r \tag{3.246}$$

Combining Equations 3.245 and 3.246 with the given data yields

$$2.49 = 0.292\sqrt{\frac{2(9.81)(1.70 + H_r)}{\frac{2(9.81)(0.012)^2(22.9)}{(0.153)^{\frac{4}{3}}} + 0.5 + 1}} + C_d(15.2)H_r^{3/2}$$

which simplifies to

$$2.49 = 0.855\sqrt{1.70 + H_r} + 15.2 C_d H_r^{3/2} \tag{3.247}$$

The discharge coefficient, C_d, depends on the head over the roadway, H_r, via the graphical relations in Figure 3.44. Taking $L_r = 18.3$ m and $y_d = 0$ (since the tailwater is below the roadway), the simultaneous solution of Equation 3.247 and the graphical relations in Figure 3.44 is done by iteration in the following table:

(1) H_r (m)	(2) H_r (ft)	(3) H_r/L_r	(4) C_r	(5) y_d/H_r	(6) k_r	(7) $C'_d = k_r C_r$	(8) C_d	(9) H_r (m)
1.00	3.28	0.055	3.04	0.00	1.00	3.04	1.68	0.14
0.14	0.46	0.008	3.00	0.00	1.00	3.00	1.66	0.14

Column 1 is the assumed H_r in meters, column 2 is the assumed H_r in feet, column 4 is C_r derived from H_r/L_r (in column 3) and H_r (in ft) using Figure 3.44, column 6 is k_r derived from y_d/H_r (in column 5) using Figure 3.44, column 7 is C_d in U.S. Customary units obtained by multiplying columns 4 and 6, column 8 is C_d in SI units obtained by multiplying column 7 by the conversion factor of 0.552, and column 9 is obtained by substituting C_d in column 8 into Equation 3.247 and solving for H_r. The iterations indicate that $C_d = 1.66$, $H_r = 0.14$ m, and the flow over the roadway, Q_r, is given by

$$Q_r = C_d L_R H_r^{3/2} = (1.66)(15.2)(0.14)^{3/2} = 1.32 \text{ m}^3/\text{s}$$

The corresponding flow through the culvert is equal to 2.49 m³/s − 1.32 m³/s = 1.17 m³/s. Therefore, a culvert diameter of 610 mm will result in roadway overtopping, with a flow of 1.17 m³/s passing through the culvert, 1.32 m³/s passing over the roadway, and a depth of flow over the roadway equal to 14 cm. A larger culvert diameter could be explored if less roadway overtopping at the design flowrate is desired.

3.5 Design of Open Channels

Constructed open channels frequently serve as major drainageways in urban storm-water-management systems, and are also used to transport water in water-supply and irrigation projects. Constructed open channels are generally classified as either *unlined* or *lined*. Unlined channels are simply excavated channels in the ground through which water flows, and lined channels are excavated channels on the perimeter of which various lining materials are placed to provide stability and prevent erosion. Lining materials are classified as either rigid or flexible. Rigid linings include concrete or asphaltic concrete pavement, and a variety of precast interlocking blocks and articulated mats. Flexible linings include such materials as loose stone (riprap), vegetation, manufactured mats of lightweight materials, fabrics, or combinations of these materials.

3.5.1 Basic Principles

The objective in designing an open channel is to determine the shape, dimensions, and lining material that will safely accommodate the design flow at a reasonable cost and limit the erosion and deposition of materials in the channel.

Best hydraulic section.   The *best hydraulic section* can be determined by requiring that the flow area of the channel be minimized while maintaining the hydraulic capacity. Consider the Manning equation given by

$$Q = \frac{1}{n} A R^{\frac{2}{3}} S_0^{1/2} \tag{3.248}$$

where Q is the flow in the channel, A is the cross-sectional area, R is the hydraulic radius ($= A/P$), P is the wetted perimeter, and S_0 is the slope of the channel. Equation 3.248 can be rearranged into the form

$$A = \left(\frac{Qn}{S_0^{1/2}} \right)^{\frac{3}{5}} P^{\frac{2}{5}} \tag{3.249}$$

which demonstrates that for given values of Q, n, and S_0, the area, A, is proportional to the wetted perimeter, P, and minimizing A also minimizes P. The best hydraulic section is defined as the section that minimizes the flow area for given values of Q, n, and S_0.

To illustrate the process of determining the best hydraulic section, consider the trapezoidal section shown in Figure 3.45, where the shape parameters are the bottom width, b, and the side slope $m : 1$ (H:V). The flow area, A, and the wetted perimeter, P, are given by

$$A = by + my^2 \tag{3.250}$$

$$P = b + 2y\sqrt{m^2 + 1} \tag{3.251}$$

Eliminating b from Equations 3.250 and 3.251 leads to

$$A = (P - 2y\sqrt{m^2 + 1})y + my^2 \tag{3.252}$$

and eliminating A from Equations 3.249 and 3.252 yields

$$(P - 2y\sqrt{m^2 + 1})y + my^2 = cP^{\frac{2}{5}} \tag{3.253}$$

FIGURE 3.45: Trapezoidal section

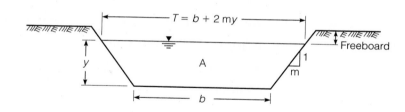

where c is a constant given by

$$c = \left(\frac{Qn}{S_0^{1/2}} \right)^{\frac{3}{5}} \qquad (3.254)$$

Holding m constant in Equation 3.253, differentiating with respect to y, and setting $\partial P/\partial y$ equal to zero leads to

$$P = 4y\sqrt{1 + m^2} - 2my \qquad (3.255)$$

Similarly, holding y constant in Equation 3.253, differentiating with respect to m, and setting $\partial P/\partial m$ equal to zero leads to

$$m = \frac{\sqrt{3}}{3} \approx 0.577 \qquad (3.256)$$

On the basis of Equations 3.255 and 3.256, the best hydraulic section for a trapezoid is one having the following geometric characteristics:

$$P = 2\sqrt{3}y, \qquad b = 2\frac{\sqrt{3}}{3}y, \qquad A = \sqrt{3}y^2 \qquad (3.257)$$

where $P = 3b$, indicating that the sides of the channel have the same length as the bottom. The best side slope, indicated by Equation 3.256, makes an angle of $60°$ with the horizontal. In cases where the side slopes are controlled by the angle of repose of the soil surrounding the channel, Equation 3.255 is used with a specified side slope, m, to find the best bottom width-to-depth ratio. This procedure is usually necessary, since the recommended side slopes for excavated channels are usually less than $1.5:1$ (H:V) or $33.7°$.

The area, A, wetted perimeter, P, and top width, T, of the best hydraulic sections for a variety of channel shapes are given in Table 3.15. Although the best hydraulic section appears to be the most economical in terms of excavation and channel lining, it is important to note that this section may not always be the most economical one, since: (a) the flow area does not include freeboard, and therefore is not the total area to be excavated; (b) it may not be possible to excavate a stable best hydraulic section in the available natural material; (c) for lined channels, the cost of lining may be comparable to the excavation costs; and (d) other factors such as the ease of access to the site and the cost of disposing of removed material may affect the economics of the channel design.

TABLE 3.15: Geometric Characteristics of Best Hydraulic Sections

Shape	Best geometry	A	P	T
Trapezoid	Half of a hexagon	$\sqrt{3}y^2$	$2\sqrt{3}y$	$\frac{4}{3}\sqrt{3}y$
Rectangle	Half of a square	$2y^2$	$4y$	$2y$
Triangle	Half of a square	y^2	$2\sqrt{2}y$	$2y$
Semicircle	—	$\frac{\pi}{2}y^2$	πy	$2y$
Parabola	—	$\frac{4}{3}\sqrt{2}y^2$	$\frac{8}{3}\sqrt{2}y$	$2\sqrt{2}y$

Minimum permissible velocity. The minimum permissible velocity is the lowest velocity that will prevent both sedimentation and vegetative growth in the channel. A velocity of 0.6 to 0.9 m/s (2 to 3 ft/s) will prevent sedimentation when the silt load is low, and a velocity of 0.75 m/s (2.5 ft/s) is usually sufficient to prevent vegetative growth (French, 1985). These estimates of minimum permissible velocities are only approximate values and can vary significantly in practice. Canals carrying turbid waters are seldom bothered by plant growth, while in channels transporting clear water some plant species flourish at velocities that are significantly in excess of the velocity that will cause erosion in the channel (Fortier and Scobey, 1926). A velocity of 0.6 m/s (2 ft/s) is sufficient to move a 15-mm diameter organic or a 2-mm sand particle (ASCE, 1982).

Channel slopes. The relevant slopes in channel design are the longitudinal slope and the side slope of the channel. Longitudinal slopes are constrained by the ground slope, the minimum permissible velocity, and the maximum allowable shear stress on the channel lining. Excavation is usually minimized by laying the channel on a slope equal to the slope of the ground surface. However, if the resulting flow velocity is less than the minimum permissible velocity, then a steeper slope that produces a higher velocity must be used, within the limits of the allowable shear stress on the channel lining. The side slopes of excavated channels are influenced by the material in which the channel is excavated. Suitable side slopes for channels excavated in various types of materials are shown in Table 3.16. These values are recommended for preliminary design. In deep cuts, side slopes are often steeper above the water surface than below the water surface, and in small drainage ditches, the side slopes are often steeper than they would be in an irrigation channel excavated in the same material (French, 1985). If concrete is the lining material, then side slopes greater than 1:1 usually require the use of forms, and for side slopes greater than 0.75:1 (H:V) the linings must be designed to withstand earth pressures. The U.S. Bureau of Reclamation recommends a 1.5:1 (H:V) slope for the usual sizes of concrete-lined canals (USBR, 1978), and the U.S. Federal Highway Administration recommends that side slopes in roadside and median channels not exceed 3:1 (USFHWA, 1988).

Freeboard. The *freeboard* is defined as the vertical distance between the water surface and the top of the channel when the channel is carrying the design flow at normal depth. Freeboard is provided to account for the uncertainty in the design, construction, and operation of the channel. As a minimum, the freeboard should be sufficient to prevent waves or fluctuations in the water surface from overflowing the

TABLE 3.16: Recommended Side Slopes in Various Types of Material

Material	Side slope (H:V)
Rock	Nearly vertical
Muck and peat soils	0.25:1
Stiff clay	0.5:1 to 1:1
Firm compacted clay or soils having clay, silt, and sand mixtures	1.5:1
Silt, loam, and sandy soils	2:1
Sandy loam, porous clay, and fine sands	3:1

Source: Chow (1959).

FIGURE 3.46: Shear stress distribution in a trapezoidal channel

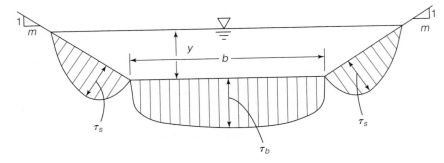

sides. A recommendation for the required freeboard in unlined channels is given by the U.S. Bureau of Reclamation as (Chow, 1959; Wurbs and James, 2002)

$$F = 0.55\sqrt{Cy}$$

(3.258)

where F is the required freeboard in meters, y is the design flow depth in meters, and C is a coefficient that varies from 1.5 at a flow of 0.57 m³/s to 2.5 for a flow of 85 m³/s or more. Additional freeboard must be provided to accommodate the superelevation of the water surface that occurs at channel bends. Applying the momentum equation to flow around a channel bend, the superelevation, h_s, of the water surface can be estimated by (Finnemore and Franzini, 2002)

$$h_s = \frac{V^2 T}{g r_c}$$

(3.259)

where V is the average velocity in the channel, T is the top-width of the channel, and r_c is the radius of curvature of the centerline of the channel. Equation 3.259 is valid only for subcritical flow conditions, in which case the elevation of the water surface at the outer channel bank will be $h_s/2$ higher than the centerline water-surface elevation, and the elevation of the water surface at the inner channel bank will be $h_s/2$ lower than the centerline water elevation. Equation 3.259 is a theoretical relation derived from the momentum equation (normal to the flow direction) and assumes a uniform velocity and constant curvature across the stream. If the effects of nonuniform velocity distribution and variation in curvature across the stream are taken into account, the superelevation, h_s, may be as much as 20% higher than given by Equation 3.259 (Finnemore and Franzini, 2002). The additional freeboard to accommodate the superelevation of the water surface around bends need only be provided in the vicinity of the bend. In order to minimize flow disturbances around bends, it is also recommended that the radius of curvature be at least three times the channel width (USACE, 1995). The minimum freeboard is usually 30 cm (ASCE, 1992).

3.5.2   Unlined Channels

The perimeter of an unlined channel consists of the native material in which the channel is excavated. The primary constraints in designing unlined channels are to prevent deposition of suspended sediment and to prevent scour of the perimeter material. Recall that the average shear stress, τ_0, around the perimeter of a channel is given by (see Section 3.2.2)

$$\tau_0 = \gamma R S_f \tag{3.260}$$

where γ is the specific weight of the fluid, R is the hydraulic radius of the channel, and S_f is the slope of the energy grade line. This average shear stress, τ_0, is also called the *unit tractive force*. In reality, the boundary shear stress is not uniformly distributed around the perimeter of the channel, and the distribution of shear stress around the perimeter of a trapezoidal section is shown in Figure 3.46. In trapezoidal sections, which is the shape most commonly used for unlined channels, the maximum shear stress on the bottom of the channel, τ_b, occurs at the center of the channel bottom and can be approximated by (Lane, 1955)

$$\boxed{\tau_b = \gamma y S_f} \tag{3.261}$$

and the maximum shear stress, τ_s, on the sides of a trapezoidal channel occurs approximately two-thirds of the way down the sides of the channel and can be approximated by

$$\boxed{\tau_s = K_s \gamma y S_f} \tag{3.262}$$

where K_s is a factor depending on the bottom-width to depth ratio, b/y, and the side slope, m, as given in Figure 3.47 (Anderson et al., 1970).

A particle of (submerged) weight w_p on the bottom of the channel resists the shear force of the flowing fluid by the friction force between the particle and surrounding particles on the bottom of the channel, which is given by $\mu_p w_p$, where μ_p is the coefficient of friction between particles on the bottom of the channel. Therefore, when particle motion on the bottom of the channel is incipient, the shear stress on the bottom of the channel is τ_b', and

$$A_p \tau_b' = \mu_p w_p \tag{3.263}$$

FIGURE 3.47: Side shear stress factor in trapezoidal channels

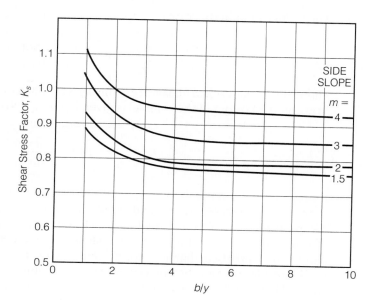

where A_p is the effective surface area of a particle on the bottom of the channel. The coefficient of friction between particles on the bottom of the channel can be related to the *angle of repose*, α, of the particle material by

$$\mu_p = \tan \alpha \tag{3.264}$$

Combining Equations 3.263 and 3.264 leads to the following expression for the critical bottom shear stress under incipient motion:

$$\tau_b' = \frac{w_p}{A_p} \tan \alpha \tag{3.265}$$

For particles on the side of the channel, the force on each particle consists of the shear force exerted by the fluid, which acts in the direction of flow, plus the component of the particle weight that acts down the side of the channel. Therefore the total force, F_p, tending to move a particle on the side of the channel is given by

$$F_p = \sqrt{(\tau_s A_p)^2 + (w_p \sin \theta)^2} \tag{3.266}$$

where θ is the angle that the side slope makes with the horizontal. When motion is incipient, the force tending to move the particle is equal to the frictional force, F_f, which is equal to the component of the particle weight normal to the side of the channel, multiplied by the coefficient of friction ($\tan \alpha$), in which case

$$F_f = w_p \cos \theta \tan \alpha \tag{3.267}$$

When motion is incipient, $F_p = F_f$, and the shear stress on the side of the channel is τ_s'. Equations 3.266 and 3.267 combine to yield

$$w_p \cos \theta \tan \alpha = \sqrt{(\tau_s' A_p)^2 + (w_p \sin \theta)^2}$$

which can be put in the form

$$\tau_s' = \frac{w_p}{A_p} \cos \theta \tan \alpha \sqrt{1 - \frac{\tan^2 \theta}{\tan^2 \alpha}} \tag{3.268}$$

The ratio of the critical shear stress on the side of the channel to the critical shear stress on the bottom of the channel is given by the *tractive force ratio*, K, where

$$\boxed{K = \frac{\tau_s'}{\tau_b'} = \sqrt{1 - \frac{\sin^2 \theta}{\sin^2 \alpha}}} \tag{3.269}$$

Equation 3.269 is generally applicable to cohesionless lining materials, where the angle of repose depends on the size and angularity of the particles as shown in Figure 3.48. The particle sizes to be used with Figure 3.48 are the 75-percentile diameters by weight. The relation between the permissible shear stress on the bottom of a channel, τ_b', and the mean particle diameter for channels excavated in cohesionless soils is shown in Figure 3.49(a), and the corresponding permissible shear stress on the side of the channel is equal to $K\tau_b'$, where K is the tractive force ratio given by Equation 3.269. In

FIGURE 3.48: Angles of repose of noncohesive material
Source: Lane (1955).

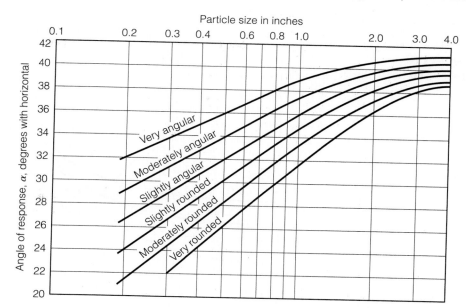

channels excavated in cohesive materials, the permissible shear stress on the bottom of a channel, τ_b', is shown in Figure 3.49(b).

It has been observed that sinuous canals scour more easily than straight channels, and the permissible shear stresses must be adjusted based on the degree of sinuousness of the channel. Correction factors for the permissible shear stress in sinuous channels are given in Table 3.17.

For channels excavated in cohesionless materials, the allowable shear stress on the sides of the channel usually governs the design, while in channels excavated in cohesive materials the shear stress on the bottom of the channel is usually critical. The following procedure is suggested for designing unlined earthen channels in cohesionless soils:

1. Estimate the roughness coefficient, n, based on the perimeter characteristics of the channel, and select the freeboard coefficient, C, based on the design flowrate in the channel. The roughness coefficient in the channel can be estimated using Table 3.18.

2. Estimate the angle of repose of the channel material from Figure 3.48.

3. Estimate the channel sinuousness and correction factor for permissible shear stress from Table 3.17.

4. Specify a side-slope angle based on the guidelines in Table 3.16.

5. Estimate the tractive force ratio from Equation 3.269.

6. Estimate the permissible shear stress on the bottom and sides of the channel from Figure 3.49 and the tractive force ratio. Correct for sinuousness.

7. Assume that the permissible shear stress on the side of the channel is the limiting factor in the channel design, and determine the normal depth of flow, y.

8. Calculate the required bottom width, b, of the channel using the Manning equation and y from step 7.

FIGURE 3.49: Permissible unit tractive force for channels in (a) noncohesive material, and (b) cohesive material

Source: USFHWA (1988).

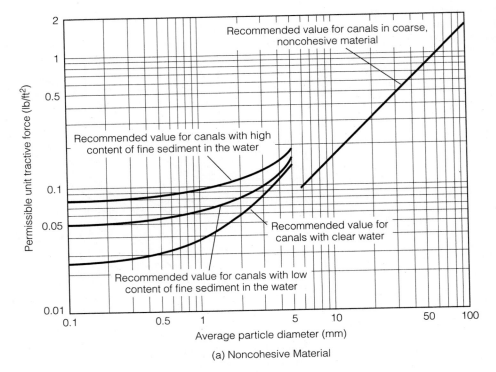

(a) Noncohesive Material

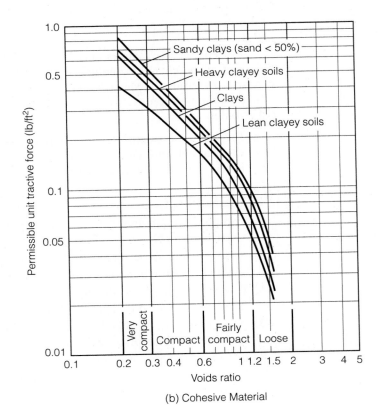

(b) Cohesive Material

TABLE 3.17: Correction Factors for Degree of Sinuousness

Degree of sinuousness	Correction factor
Straight channels	1.00
Slightly sinuous channels	0.90
Moderately sinuous channels	0.75
Very sinuous channels	0.60

Source: Lane (1955).

TABLE 3.18: Roughness Coefficients in Excavated Open Channels

Type	Characteristics	Minimum n	Normal n	Maximum n
Earth, straight, and uniform	Clean, recently completed	0.016	0.018	0.020
	Clean, after weathering	0.018	0.022	0.025
	Gravel, uniform section, clean	0.022	0.025	0.030
	with short grass, few weeds	0.022	0.027	0.033
Earth, winding, and sluggish	No vegetation	0.023	0.025	0.030
	Grass, some weeds	0.025	0.030	0.033
	Dense weeds or aquatic plants in deep channels	0.030	0.035	0.040
	Earth bottom and rubble sides	0.028	0.030	0.035
	Stony bottom and weedy banks	0.025	0.035	0.040
	Cobble bottom and clean sides	0.030	0.040	0.050
Dragline-excavated or dredged	No vegetation	0.025	0.028	0.033
	Light brush on banks	0.035	0.050	0.060
Rock cuts	Smooth and uniform	0.025	0.035	0.040
	Jagged and irregular	0.035	0.040	0.050
Channels not maintained, weeds and brush uncut	Dense weeds high as flow depth	0.050	0.080	0.120
	Clean bottom, brush on sides	0.040	0.050	0.080
	Same, highest stage of flow	0.045	0.070	0.110
	Dense brush, high stage	0.080	0.100	0.140

Source: Chow (1959).

9. Verify that the maximum shear stress on the bottom of the channel ($= \gamma y S_0$) is less than or equal to the permissible shear stress on the bottom of the channel.

10. Compare the design velocity with the minimum permissible velocity and calculate the Froude number of the design flow. Verify that the flow is subcritical.

11. Estimate the required freeboard in the channel using Equation 3.258.

This design procedure is illustrated in the following example.

EXAMPLE 3.24

Design a trapezoidal channel to carry 12 m³/s through a slightly sinuous route on a slope of 0.15%. The channel is to have side slopes of 2:1 (H:V), be excavated in coarse alluvium with a 75-percentile diameter of 2 cm, and particles on the perimeter of the channel are expected to be moderately rounded.

Solution

Step 1. The first step is to estimate the Manning roughness coefficient, n. Based on the data given in Table 3.18, an estimated n value for a uniform section lined with gravel is 0.025. This can be compared with the estimate given by Equation 3.45, which states that

$$n = 0.038d^{\frac{1}{6}}$$

where d is the characteristic particle diameter on the boundary (in meters). Using the 75-percentile diameter, d_{75}, for d yields

$$n = 0.038(d_{75})^{\frac{1}{6}} = 0.038(0.02)^{\frac{1}{6}} = 0.020$$

This compares with $n = 0.025$ estimated from Table 3.18, and the more conservative value of 0.025 will be used in the design.

Step 2. The angle of repose of the channel material is estimated from Figure 3.48, where

$$d_{75} = 2 \text{ cm} = 0.8 \text{ in.}$$

Therefore, the angle of repose, α, is equal to $32°$ (based on moderately rounded material).

Step 3. Since the channel is slightly sinuous, the correction factor, C_s, for the maximum shear stress is given by Table 3.17 as $C_s = 0.90$.

Step 4. Specify a channel side slope of $2:1$ (H:V). The angle that the side slope makes with the horizontal, θ, is given by

$$\theta = \tan^{-1}\left(\frac{1}{2}\right) = 26.6°$$

Step 5. The tractive force ratio, K, is given by Equation 3.269 as

$$K = \frac{\tau_s'}{\tau_b'} = \sqrt{1 - \frac{\sin^2 \theta}{\sin^2 \alpha}} = \sqrt{1 - \frac{\sin^2 26.6°}{\sin^2 32°}} = 0.53$$

Step 6. The permissible shear stress on the bottom of the channel is estimated from Figure 3.49 as 0.33 lb/ft^2 or 15.9 N/m^2 for a mean particle size of 20 mm. Correcting this permissible shear stress for sinuousness leads to an allowable shear stress on the bottom of the channel, τ_b', equal to

$$\tau_b' = C_s(15.9) = 0.9(15.9) = 14.3 \text{ N/m}^2$$

and the permissible shear stress on the side of the channel, τ_s', is therefore given by

$$\tau_s' = K\tau_b' = 0.53(14.3) = 7.6 \text{ N/m}^2$$

Step 7. The normal depth of flow, y, can now be estimated by assuming that particle motion is incipient on the side of the channel. For a side slope of $2:1$, Figure 3.47 gives $K_s = 0.79$ (assuming $b/y \geq 4$), and since the maximum shear stress on the side of the channel is given by Equation 3.262 as $K_s \tau_b (= K_s \gamma R S_0)$, then

$$0.79 \gamma y S_0 = 7.6 \text{ N/m}^2$$

which leads to

$$y = \frac{7.6}{0.79 \gamma S_0} = \frac{7.6}{0.79(9790)(0.0015)} = 0.66 \text{ m}$$

Step 8. The Manning equation requires that

$$Q = \frac{1}{n} A R^{\frac{2}{3}} S_0^{\frac{1}{2}} = \frac{1}{n} \frac{A^{\frac{5}{3}}}{P^{\frac{2}{3}}} S_0^{\frac{1}{2}}$$

Substituting known quantities gives

$$12 = \frac{1}{0.025} \frac{A^{\frac{5}{3}}}{P^{\frac{2}{3}}} (0.0015)^{\frac{1}{2}}$$

which simplifies to

$$\frac{A^{\frac{5}{3}}}{P^{\frac{2}{3}}} = 7.75$$

Since

$$A = [b + my]y = [b + 2(0.66)](0.66) = 0.66[b + 1.32]$$
$$P = b + 2y\sqrt{1 + m^2} = b + 2(0.66)\sqrt{1 + 2^2} = b + 2.95$$

the Manning equation can be written as

$$\frac{(b + 1.32)^{\frac{5}{3}}}{(b + 2.95)^{\frac{2}{3}}} = 15.5$$

Solving for b gives

$$b = 15.2 \text{ m}$$

Since $b/y = 15.2/0.66 = 23$, the initial assumption in Step 7 that $b/y \geq 4$ is verified. Therefore, the channel should have a width of 15.2 m, side slopes of 2:1, and a normal depth of flow at 12 m³/s of 0.66 m.

Step 9. The maximum shear stress on the bottom of the channel, τ_b, is given by

$$\tau_b = \gamma y S_0 = (9790)(0.66)(0.0015) = 9.7 \text{ N/m}^2$$

Since this is less than the permissible shear stress on the bottom (14.2 N/m²), then the channel design is acceptable from the viewpoint of permissible shear stress.

Step 10. The flow area, A, is given by

$$A = [b + my]y = [15.2 + (2)(0.66)](0.66) = 10.9 \text{ m}^2$$

and the average velocity in the channel, V, is given by

$$V = \frac{Q}{A} = \frac{12}{10.9} = 1.1 \text{ m/s}$$

This velocity should be sufficient to prevent sedimentation and vegetative growth. The Froude number, Fr, can be estimated using the relation

$$\text{Fr} = \frac{V}{\sqrt{gD}}$$

where D is the hydraulic depth given by

$$D = \frac{A}{T} = \frac{A}{b + 2my} = \frac{10.9}{15.2 + 2(2)(0.66)} = 0.61 \text{ m}$$

Therefore, the Froude number is

$$\text{Fr} = \frac{1.1}{\sqrt{(9.81)(0.61)}} = 0.45$$

The flow is subcritical and acceptable (Fr < 1).

Step 11. The required freeboard, F, can be estimated using Equation 3.258,

$$F = 0.55\sqrt{Cy}$$

where F is the freeboard in meters, y is the design flow depth in meters, and C is a coefficient that varies from 1.5 at a flow of 0.57 m³/s to 2.5 for a flow of 85 m³/s. Based on a flow of 12 m³/s, C can be interpolated as 1.6 and therefore the required freeboard is

$$F = 0.55\sqrt{(1.6)(0.66)} = 0.57 \text{ m}$$

The total depth of the channel to be excavated is therefore equal to the normal depth plus the freeboard, 0.66 m + 0.57 m = 1.23 m. The channel is to have a bottom width of 15.2 m, and side slopes of 2:1 (H:V).

From the viewpoint of practical construction, the dimensions of earth channels should be specified to the nearest 0.1 m (Kay, 1998).

The design of channels in cohesive soils is similar to the procedure in the previous example, with the exception that the shear stress on the bottom of the channel is usually assumed to govern the design.

EXAMPLE 3.25

Show that the permissible shear stresses on the sides of trapezoidal channels govern the channel design whenever $K/K_s < 1$. Verify this result using the previous example.

Solution The permissible shear stress on the side of a channel, τ'_s, is related to the permissible shear stress on the bottom of the channel, τ'_b, by Equation 3.269, where

$$\tau'_s = K\tau'_b \tag{3.270}$$

The maximum shear stress on the side of a channel, τ_s, is related to the maximum shear stress on the bottom of the channel, τ_b, by Equation 3.262, where

$$\tau_s = K_s \tau_b \tag{3.271}$$

If the maximum shear stress on the side of the channel is equal to the permissible shear stress on the side of the channel,

$$\tau_s = \tau_s' \tag{3.272}$$

Substituting Equations 3.270 and 3.271 into Equation 3.272 gives

$$K\tau_b' = K_s \tau_b$$

which yields

$$\tau_b = \left(\frac{K}{K_s}\right)\tau_b'$$

Therefore, whenever $K/K_s < 1$, then $\tau_b < \tau_b'$, and the permissible shear stress on the side of the channel governs the design.

In the previous example, $K = 0.53$, $K_s = 0.79$, and

$$\frac{K}{K_s} = \frac{0.53}{0.77} = 0.67 < 1$$

This result confirms that the permissible shear stress on the sides of the channel governs the design.

3.5.3 Lined Channels

Channel linings are used to control erosion in excavated channels, and these linings are generally classified as either *rigid* or *flexible*. Rigid linings include cast-in-place concrete, stone masonry, soil cement, and grouted riprap, while flexible linings include riprap, vegetation, and gravel. Rigid linings are called "rigid" since they tend to crack when deflected, and flexible linings are called "flexible" because they are able to conform to changes in channel shape while maintaining the overall integrity of the channel lining. Channels with rigid linings are called *rigid-boundary* channels, and channels with flexible linings are called *flexible-boundary* channels.

Rigid-boundary channels are preferred for a variety of purposes, such as to (1) transport water at high velocities to reduce construction and excavation costs, (2) decrease seepage losses, (3) decrease operation and maintenance costs, and (4) ensure the stability of the channel section. According to ASCE (1992), all channels carrying supercritical flow should be lined with concrete and continuously reinforced both longitudinally and laterally. Since channels with rigid linings are capable of high conveyance and high-velocity flow, flood-control channels with rigid linings are often used to reduce the amount of land required for a surface-drainage system. When land is costly or unavailable because of restrictions, use of rigid-channel linings is preferred (Cotton, 2001). A concrete-lined channel under construction using prefabricated concrete panels is shown in Figure 3.50. Rigid-boundary channels are highly susceptible

FIGURE 3.50: Concrete-lined canal

to failure from structural instability caused by freeze-thaw, swelling, and excessive soil pore-water pressures, and rigid linings tend to fail when a portion of the lining is damaged. Construction of rigid linings requires specialized equipment using relatively costly material. As a result, the cost of rigid linings is high. Prefabricated linings can be a less expensive alternative if shipping distances are not excessive.

In contrast to rigid-boundary channels, flexible-boundary channels are generally less expensive, permit infiltration and exfiltration, filter out contaminants, provide better habitat opportunities for local flora and fauna, and have a natural appearance. However, flexible-boundary channels have the disadvantage of being limited in the magnitude of the erosive force that they can sustain without damage to either the channel or the lining (USFHWA, 1988). Flexible linings are widely used as temporary channel linings for control of erosion during construction or reclamation of disturbed areas.

3.5.3.1 Rigid-boundary channels

A recommended design procedure for rigid-boundary channels is as follows (French, 1985):

1. Estimate the roughness coefficient, n, and freeboard coefficient, C, for the specified lining material and design flowrate, Q. Manning roughness coefficients for rigid-boundary channels are listed in Table 3.19. ASCE (1992) has recommended that open-channel designs not use a roughness coefficient lower than 0.013 for well-troweled concrete, and other finishes should have proportionally higher n values assigned to them.

2. Compute the normal depth of flow, y, using the Manning equation

$$Q = \frac{1}{n}AR^{\frac{2}{3}}S_0^{1/2} \tag{3.273}$$

where A is the flow area, R is the hydraulic radius, S_0 is the (given) longitudinal slope of the channel, and the shape of the cross section is specified by the designer. If appropriate, the relative dimensions of the best hydraulic section may be specified.

TABLE 3.19: Roughness Coefficients in Rigid-Boundary Open Channels

Type	Characteristics	Minimum n	Normal n	Maximum n
Cement	Neat surface	0.010	0.011	0.013
	Mortar	0.011	0.013	0.015
Concrete	Trowel finish	0.011	0.013	0.015
	Float finish	0.013	0.015	0.016
	Finished, with gravel on bottom	0.015	0.017	0.020
	Unfinished	0.014	0.017	0.020
	Gunite, good section	0.016	0.019	0.023
	Gunite, wavy section	0.018	0.022	0.025
	On good excavated rock	0.017	0.020	—
	On irregular excavated rock	0.022	0.027	—
Concrete bottom float finished with sides of:	Dressed stone in mortar	0.015	0.017	0.020
	Random stone in mortar	0.017	0.020	0.024
	Cement rubble masonry, plastered	0.016	0.020	0.024
	Cement rubble masonry	0.020	0.025	0.030
	Dry rubble or riprap	0.020	0.030	0.035
Brick	Glazed	0.011	0.013	0.015
	In cement mortar	0.012	0.015	0.018
Masonry	Cemented rubble	0.017	0.025	0.030
	Dry rubble	0.023	0.032	0.035
Dressed ashlar	—	0.013	0.015	0.017
Asphalt	Smooth	0.013	0.013	—

Source: Chow (1959).

3. Check the minimum permissible velocity and the Froude number. Repeat steps 2 and 3 if necessary to meet the minimum-velocity and subcritical-flow requirements.

4. Calculate the required freeboard using Equation 3.258, and increase the freeboard on channel bends by one-half the superelevation height given by Equation 3.259.

As an additional constraint in designing concrete-lined channels, ASCE (1992) recommends that flow velocities not exceed 2.1 m/s or result in a Froude number greater than 0.8 for nonreinforced linings, and that flow velocities not exceed 5.5 m/s for reinforced linings.

EXAMPLE 3.26

Design a lined trapezoidal channel to carry 20 m³/s on a longitudinal slope of 0.0015. The lining of the channel is to be float-finished concrete. Consider: (a) the best hydraulic section, and (b) a section with side slopes of 1.5 : 1 (H : V).

Solution

(a) According to Table 3.19, $n = 0.015$. Using the best (trapezoidal) hydraulic section, the bottom width, b, and side slope, m, are given by Equations 3.256 and 3.257 as

$$b = 1.15y, \qquad m = 0.58 \quad (= 60° \text{ angle})$$

and, according to Table 3.15,

$$A = 1.73y^2, \qquad P = 3.46y, \qquad T = 2.31y, \qquad R = \frac{A}{P} = 0.5y$$

Substituting the geometric characteristics of the channel into the Manning equation yields

$$20 = \frac{1}{0.015}(1.73y^2)(0.5y)^{\frac{2}{3}}(0.0015)^{\frac{1}{2}}$$

or

$$y^{8/3} = 7.12$$

which leads to

$$y = 2.09 \text{ m}$$

and hence the bottom-width, b, of the channel is given by

$$b = 1.15y = 1.15(2.09) = 2.40 \text{ m}$$

The flow area is therefore given by

$$A = 1.73y^2 = 1.73(2.09)^2 = 7.6 \text{ m}^2$$

and the average velocity, V, is

$$V = \frac{Q}{A} = \frac{20}{7.6} = 2.6 \text{ m/s}$$

This velocity should be sufficient to prevent sedimentation and vegetative growth. Since the velocity is greater than 2.1 m/s, the lining should be reinforced. The hydraulic depth, D, is given by

$$D = \frac{A}{T} = \frac{1.73y^2}{2.31y} = 0.749y = 0.749(2.09) = 1.6 \text{ m}$$

and the Froude number, Fr, is

$$\text{Fr} = \frac{V}{\sqrt{gD}} = \frac{2.6}{\sqrt{(9.81)(1.6)}} = 0.66$$

The flow is therefore subcritical (Fr < 1).
The required freeboard, F, can be estimated using Equation 3.258,

$$F = 0.55\sqrt{Cy}$$

where F is the freeboard in meters, y is the design flow depth in meters, and C is a coefficient that varies from 1.5 at a flow of 0.57 m^3/s to 2.5 for a flow of

85 m^3/s. Based on a flow of 20 m^3/s, C can be interpolated as 1.7. The required freeboard is

$$F = 0.55\sqrt{(1.7)(2.09)} = 1.04 \text{ m}$$

The total depth of the channel to be excavated and lined is equal to the normal depth plus the freeboard, 2.09 m + 1.04 m = 3.13 m. The channel is to have a bottom width of 2.40 m, and side slopes of 0.58 : 1 (H : V).

(b) If the channel side slope is 1.5 : 1, then $m = 1.5$ and Equation 3.255 gives the perimeter, P, as

$$P = 4y\sqrt{1 + m^2} - 2my = 4y\sqrt{1 + 1.5^2} - 2(1.5)y = 4.21y$$

Also, since

$$P = b + 2y\sqrt{1 + m^2}$$

then

$$b = P - 2y\sqrt{1 + m^2} = 4.21y - 2y\sqrt{1 + 1.5^2} = 0.60y$$

Also,

$$T = b + 2my = 0.60y + 2(1.5)y = 3.6y$$

$$A = (b + my)y = (0.60y + 1.5y)y = 2.10y^2$$

$$R = \frac{A}{P} = \frac{2.10y^2}{4.21y} = 0.499y$$

Substituting the geometric characteristics of the channel into the Manning equation gives

$$20 = \frac{1}{0.015}(2.10y^2)(0.499y)^{\frac{2}{3}}(0.0015)^{\frac{1}{2}}$$

or

$$y^{8/3} = 5.86$$

which leads to

$$y = 1.94 \text{ m}$$

and hence the bottom width, b, of the channel is given by

$$b = 0.60y = 0.60(1.94) = 1.16 \text{ m}$$

The flow area is therefore given by

$$A = 2.10y^2 = 2.10(1.94)^2 = 7.90 \text{ m}^2$$

and the average velocity, V, is

$$V = \frac{Q}{A} = \frac{20}{7.90} = 2.53 \text{ m/s}$$

This velocity should be sufficient to prevent sedimentation and vegetative growth. Since the velocity is greater than 2.1 m/s, the lining should be reinforced.

The hydraulic depth, D, is given by

$$D = \frac{A}{T} = \frac{2.10y^2}{3.6y} = 0.58y = 0.58(1.94) = 1.13 \text{ m}$$

and the Froude number, Fr, is

$$\text{Fr} = \frac{V}{\sqrt{gD}} = \frac{2.53}{\sqrt{(9.81)(1.13)}} = 0.76$$

The flow is therefore subcritical (Fr < 1).

The required freeboard, F, can be estimated using Equation 3.258:

$$F = 0.55\sqrt{Cy}$$

Based on a flow of 20 m³/s, C can be interpolated as 1.7 and the required freeboard is

$$F = 0.55\sqrt{(1.7)(1.94)} = 1.00 \text{ m}$$

The total depth of the channel to be excavated and lined is equal to the normal depth plus the freeboard, 1.94 m + 1.00 m = 2.94 m. The channel is to have a bottom width of 1.16 m and side slopes of 1.5 : 1 (H : V).

From the viewpoint of practical construction, the dimensions of lined channels should be specified to the nearest 0.05 m (Kay, 1998).

3.5.3.2   Flexible-boundary channels

In flexible-boundary channels the perimeter of the channel is covered with a moveable lining such as grass or riprap. The erosion resistance of the channel is determined by properties of the lining rather than the properties of the native material in which the channel is excavated. Flexible-boundary channels are designed such that the shear stress on the perimeter of the channel under maximum-flow conditions is less than the permissible shear stress on the lining material. Permissible shear stresses for various flexible linings are shown in Table 3.20, where vegetative linings are classified by their *retardance* as shown in Table 3.21. Channel shapes commonly used for grass-lined channels are trapezoidal, triangular, and parabolic, with the latter two shapes being the most popular. In designing grass-lined channels, the ability of tractors, or other farm-type machinery, to cross the channels during periods of no flow is an important consideration, and this may require that side slopes of a channel be designed to allow tractors to cross, rather than for hydraulic efficiency or stability. The freeboard requirement, F (in meters), for grass-lined channels can be estimated using the relation (ASCE, 1992)

$$\boxed{F = 0.152 + \frac{V^2}{2g}} \tag{3.274}$$

where V is the average velocity in the channel, under design conditions. The minimum freeboard should normally be at least 30 cm at the maximum-design water-surface

TABLE 3.20: Permissible Shear Stresses for Lining Materials (USFHWA, 1988)

Lining category	Lining type	Permissible shear stress (Pa)
Temporary*	Woven paper net	7.2
	Jute net	21.6
	Fiberglass roving:	
	Single	28.7
	Double	40.7
	Straw with net	69.5
	Curled wood mat	74.3
	Synthetic mat	95.7
Gravel riprap	2.5-cm (1-inch)	15.7
	5-cm (2-inch)	31.4
Rock riprap	15-cm (6-inch)	95.7
	30-cm (12-inch)	191.5
Vegetative	Class A	177.2
	Class B	100.6
	Class C	47.9
	Class D	28.7
	Class E	16.8

*Some "temporary" linings become permanent when buried.

elevation and an additional freeboard equal to the superelevation of the water surface should be provided around bends.

The Manning roughness coefficient for nonvegetative flexible-lining materials varies with flow depth as shown in Table 3.22. The relative channel roughness is higher for shallow flow depths and lower for large flow depths, and the range of flow depths from 15 cm to 60 cm is typical of highway drainage channels. The Manning roughness coefficient for vegetative linings varies significantly depending on the amount of submergence of the vegetation and the flow force exerted on the channel bed. Values of the Manning roughness coefficient, n, as a function of the hydraulic radius, R, and the channel slope, S_0, are given in Table 3.23 for various retardances (USFHWA, 1988; Kouwen et al., 1980). It is widely recognized that formulating the flow resistance in vegetated channels in terms of a Manning roughness coefficient is fundamentally inappropriate, since vegetation creates resistance by both frictional energy losses and volume displacement. Attempts to model the two resistance components separately have not been successful because of the difficulty in isolating their effects (Green, 2005; Fisher, 1992). New methods of estimating flow resistance in vegetated channels are still being developed (e.g., Carollo et al., 2005).

Bends.   Flow around a bend generates secondary currents, which impose higher shear stresses on the perimeter of a channel compared with straight sections. The maximum shear stress, τ_r, on the perimeter of a channel section in a bend is given by

$$\tau_r = K_r \tau_b \tag{3.275}$$

TABLE 3.21: Retardance in Grass-Lined Channels (Coyle, 1975; USFHWA, 1996)

Retardance	Cover	Condition
A	Reed canary grass	Excellent stand, tall (average 90 cm)
	Yellow bluestem *Ischaemum*	Excellent stand, tall (average 90 cm)
	Weeping lovegrass	Excellent stand, tall (average 75 cm)
B	Smooth bromegrass	Good stand, mowed (average 30–40 cm)
	Bermuda grass	Good stand, tall (average 30 cm)
	Native grass mixture (little bluestem, blue grama, and other long and short Midwest grasses)	Good stand, unmowed
	Tall fescue	Good stand, unmowed (average 45 cm)
	Lespedeza sericea	Good stand, not woody, tall (average 50 cm)
	Grass-legume mixture—timothy smooth	Good stand, uncut (average 50 cm)
	Tall fescue, with bird's foot Trefoil or Iodino	Good stand, uncut (average 45 cm)
	Blue grama	Good stand, uncut (average 35 cm)
	Kudzu	Very dense growth, uncut
		Dense growth, uncut
	Weeping lovegrass	Good stand, tall (average 60 cm)
		Good stand, unmowed (average 35 cm)
	Alfalfa	Good stand, uncut (average 30 cm)
C	Bahia	Good stand, uncut (15–18 cm)
	Bermuda grass	Good stand, mowed (average 15 cm)
	Redtop	Good stand, headed (40–60 cm)
	Grass-legume mixture—summer	Good stand, uncut (15–20 cm)
	Centipede grass	Very dense cover (average 15 cm)
	Kentucky bluegrass	Good stand, headed (15–30 cm)
	Crabgrass	Fair stand, uncut (average 25 cm to 120 cm)
	Common *lespedeza*	Good stand, uncut (average 30 cm)
D	Bermuda grass	Good stand, cut to 6-cm height
	Red fescue	Good stand, headed (30 to 45 cm)
	Buffalo grass	Good stand, uncut (8 to 15 cm)
	Grass-legume mixture—fall, spring	Good stand, uncut (10 to 13 cm)
	Lespedeza sericea	After cutting to 5-cm height; very good stand before cutting
	Common *lespedeza*	Excellent stand, uncut (average 10 cm)
E	Bermuda grass	Good stand, cut to 4-cm height
	Bermuda grass	Burned stubble

TABLE 3.22: Manning n for Nonvegetative Flexible-Lining Materials (USFHWA, 1988)

Lining category	Lining type	Depth range		
		0–15 cm	15–60 cm	>60 cm
Temporary	Jute net	0.028	0.022	0.019
	Fiberglass roving	0.028	0.021	0.019
	Straw with net	0.065	0.033	0.025
	Curled wood mat	0.066	0.035	0.028
	Synthetic mat	0.036	0.025	0.021
Gravel riprap	2.5-cm d_{50}	0.044	0.033	0.030
	5-cm d_{50}	0.066	0.041	0.034
Rock riprap	15-cm d_{50}	0.104	0.069	0.035
	30-cm d_{50}	–	0.078	0.040

TABLE 3.23: Manning n for Vegetative Linings (Kouwen et al., 1980)

Retardance	Formula for n^*
A	$n = \dfrac{1.22R^{1/6}}{30.2 + 19.97\log(R^{1.4}S_o^{0.4})}$
B	$n = \dfrac{1.22R^{1/6}}{37.4 + 19.97\log(R^{1.4}S_o^{0.4})}$
C	$n = \dfrac{1.22R^{1/6}}{44.6 + 19.97\log(R^{1.4}S_o^{0.4})}$
D	$n = \dfrac{1.22R^{1/6}}{49.0 + 19.97\log(R^{1.4}S_o^{0.4})}$
E	$n = \dfrac{1.22R^{1/6}}{52.1 + 19.97\log(R^{1.4}S_o^{0.4})}$

$^*R$ in meters.

where K_r is a factor that depends on the ratio of the channel curvature, r_c, to the bottom width, b, as shown in Figure 3.51 (USFHWA, 1988; Lane, 1955), and τ_b is the maximum shear stress on the bottom of the channel given by Equation 3.261. The increased shear stress caused by a bend persists downstream of the bend for a distance L_p given by (Nouh and Townsend, 1979)

$$\frac{L_p}{R} = 0.604\left(\frac{R^{1/6}}{n_b}\right) \tag{3.276}$$

where R is the hydraulic radius of the channel flow, and n_b is the Manning roughness coefficient in the bend.

Riprap lining.   *Riprap* consists of broken rock, cobbles, or boulders placed on the perimeter of a channel to protect against the erosive action of water. The two most important considerations in designing riprap linings are: (1) riprap gradation and thickness; and (2) use of filter material under riprap. According to the U.S. Federal Highway Administration (USFHWA, 1988), riprap gradation should follow a smooth size distribution curve. Most riprap gradations will fall into the range of d_{100}/d_{50} and d_{50}/d_{20} between 1.5 and 3.0, which is acceptable. The most important criterion

FIGURE 3.51: Shear stress factor for channel bends

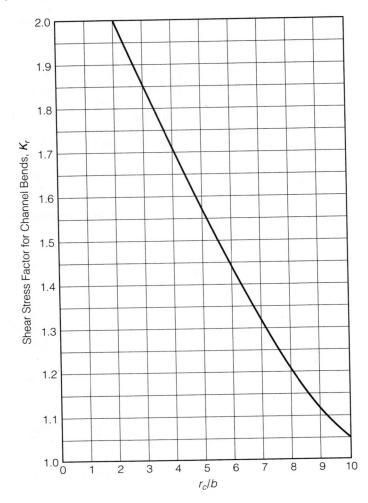

is the proper distribution of sizes in the gradation so that intersticies formed by larger stones are filled with smaller sizes, preventing the formation of open pockets. In general, riprap constructed with angular stones has the best performance. The thickness of a riprap lining should equal the diameter of the largest rock size in the gradation. For most gradations, this will mean a thickness of from 1.5 to 3.0 times the mean riprap diameter (USFHWA, 1988). Channels with gravel or riprap on side slopes less than 3:1 are generally assumed to be stable as long as the maximum shear stress on the bottom of the channel is less than the permissible shear stress (USFHWA, 1988).

Design procedure. Channels with flexible linings on longitudinal slopes less than 10% and side slopes less than or equal to 3:1 (H:V) can be designed using the following steps (USFHWA, 1988):

1. Select a flexible lining and determine the permissible shear stress, τ_p, from Table 3.20.

2. Select the channel shape, slope, and design discharge.
3. Determine the appropriate expression for Manning n:

- For nonvegetative linings use Table 3.22.
- For vegetative linings use Table 3.23.

4. Calculate the normal flow depth, y, in the channel using the Manning equation.
5. Calculate the maximum shear stress, τ_b, on the bottom of the channel using Equation 3.261. If the maximum shear stress, τ_b, on the bottom of the channel is greater than the permissible shear stress, τ_p, determined in step 1, then the lining is not acceptable; repeat steps 1 through 5.

   For riprap or gravel linings on steep side slopes (steeper than 3 : 1), the permissible shear stress on the sides of the channel must be greater than the maximum shear stress on the sides of the channel. The permissible shear stress on the sides of a channel is equal to the product of the tractive force ratio (Equation 3.269) and the permissible shear stress on the bottom of the channel.

6. For channel bends:

   (a) Determine the factor for maximum shear stress on channel bends, K_r, from Figure 3.51.
   (b) Calculate the shear stress in the bend, τ_r, using Equation 3.275. If $\tau_r > \tau_p$, the lining is not acceptable in the bend section; repeat steps 1 through 6.
   (c) If a lining material with a higher permissible shear stress is used in the bend section, calculate the length of protection, L_p, downstream of the bend using Equation 3.276.
   (d) Calculate the required freeboard using Equation 3.274 and add the superelevation, $h_s/2$, where h_s is given by Equation 3.259.

For channels on a slope greater than 10%, special design procedures are necessary (USFHWA, 1988).

EXAMPLE 3.27

A roadway-median channel is to be lined with Bermuda grass that has an average height of 15 cm, and the channel is to carry a design flow of 0.60 m³/s on a slope of 0.7%. The channel has a trapezoidal cross section with a bottom width of 1.5 m and side slopes of 3 : 1 (H : V). Determine whether the channel lining is acceptable, and find the depth of flow in the channel. What freeboard is required? Will the channel lining be adequate around a bend with a radius of curvature of 10 m?

Solution

   Step 1. The retardance of the grass lining can be classified as "C" according to Table 3.21, and the permissible shear stress, τ_p, for Class C vegetative lining is given by Table 3.20 as 47.9 Pa.

   Step 2. From the given data, $Q = 0.60$ m³/s, $S_0 = 0.007$, $b = 1.5$ m, and $m = 3$.

Step 3. The area, A, wetted perimeter, P, and hydraulic radius, R, are given by

$$A = (b + my)y = 1.5y + 3y^2$$

$$P = b + 2y\sqrt{1 + m^2} = 1.5 + 2y\sqrt{1 + 3^2} = 1.5 + 6.325y$$

$$R = \frac{A}{P} = \frac{1.5y + 3y^2}{1.5 + 6.325y}$$

The Manning n for a channel with Class C vegetative lining is given by Table 3.23 as

$$n = \frac{1.22R^{1/6}}{44.6 + 19.97\log(R^{1.4}S_0^{0.4})}$$

$$= \frac{1.22R^{1/6}}{44.6 + 19.97\log(R^{1.4}) + 19.97\log(0.007^{0.4})}$$

$$n = \frac{1.22R^{1/6}}{27.39 + 27.96\log R}$$

Step 4. From the Manning equation

$$Q = \frac{1}{n}\frac{A^{5/3}}{P^{2/3}}S_0^{1/2}$$

$$0.60 = \frac{27.39 + 27.96\log\left(\frac{1.5y+3y^2}{1.5+6.32y}\right)}{1.22\left(\frac{1.5y+3y^2}{1.5+6.32y}\right)^{1/6}}\frac{(1.5y + 3y^2)^{5/3}}{(1.5 + 6.32y)^{2/3}}(0.007)^{1/2}$$

which yields

$$y = 0.45 \text{ m}$$

Step 5. For a flow depth of 0.45 m, the maximum shear stress on the bottom of the channel, τ_b, is given by Equation 3.261 as

$$\tau_b = \gamma y S_0 = (9.79)(0.45)(0.007) = 0.031 \text{ kPa} = 31 \text{ Pa}$$

The maximum shear stress on the bottom of the channel (31 Pa) is less than the permissible shear stress on the channel lining (= 47.9 Pa) and therefore the channel lining is acceptable. (Since the channel lining is neither gravel nor riprap, it is not necessary to check the permissible shear stress on the sides of the channel.)

The velocity, V, of flow in the channel is given by

$$V = \frac{Q}{A} = \frac{0.60}{1.28} = 0.47 \text{ m/s}$$

and hence the required freeboard, F, is given by Equation 3.274 as

$$F = 0.152 + \frac{V^2}{2g} = 0.152 + \frac{0.47^2}{2(9.81)} = 0.16 \text{ m}$$

This calculated freeboard (= 16 cm) is less than the minimum freeboard of 30 cm recommended by ASCE (1992), therefore the design freeboard in the channel should be at least 30 cm.

If the radius of curvature, r_c, is 10 m, then $r_c/b = 10/1.5 = 6.67$ and the factor K_r for the maximum shear stress on the channel bend is given by Figure 3.51 as $K_r = 1.35$. The maximum shear stress in the bend, τ_r, is therefore given by Equation 3.275 as

$$\tau_r = K_r \tau_b = 1.35(0.031) = 0.041 \text{ kPa} = 41 \text{ Pa}$$

Since the maximum shear stress in the bend (=41 Pa) is less than the permissible shear stress on the channel lining (=47.9 Pa), the lining will be adequate around the bend.

Composite linings. In cases where different lining materials are used for different parts of the channel perimeter, the lining is called a *composite lining*. Typically, the lining material used on the side slopes has a lower permissible shear stress than that used on the bottom of the channel. This arrangement is acceptable since, in trapezoidal channels, the shear stresses on the sides of the channel are usually significantly less than on the bottom of the channel. Recall that the maximum shear stress on the side of the channels is given by Equation 3.262 as

$$\tau_s = K_s \tau_b$$

where K_s is given by Figure 3.47, and for slopes commonly encountered in practice ($m \geq 1.5$) it is clear that $K_s \leq 1$ and hence $\tau_b \geq \tau_s$. Aside from considerations of allowable stress, in channels lined with vegetation, frequent low flows will kill the submerged vegetation and a nonvegetative low-flow lining of concrete or riprap must be used. In composite channels, the effective Manning n is obtained by dividing the channel into two parts: the low-flow bottom section and the channel sides. Assigning a Manning n to each of these sections, and assuming that the flow in each section has the same mean velocity, then the effective Manning n for the composite section is given by

$$n = \left[\frac{P_L}{P} + \left(1 - \frac{P_L}{P} \right) \left(\frac{n_2}{n_1} \right)^{3/2} \right]^{2/3} n_1 \qquad (3.277)$$

where n_1 is the roughness of the low-flow (bottom) lining, n_2 is the roughness of the high-flow (side) lining, P_L is the wetted perimeter of the low-flow channel, and P is the wetted perimeter of the entire channel.

Channels with composite linings can be designed using the following procedure (USFHWA, 1988):

1. Determine the permissible shear stress, τ_p, for both lining types in the channel (see Table 3.20).
2. Estimate the depth of flow in the channel.
3. Estimate Manning n for each lining type (Table 3.22 for nonvegetative linings and Table 3.23 for vegetative linings).
4. Compute the ratio of high-flow to low-flow Manning n values, n_2/n_1.

5. Compute the ratio of low-flow channel wetted perimeter, P_L, to total wetted perimeter, P, (P_L/P).
6. Determine the effective Manning n of the section using Equation 3.277.
7. Determine the normal flow depth, y, using the effective Manning n calculated in step 6.
8. Compare the estimated flow depth in step 2 with the calculated flow depth in step 7. Repeat steps 2 through 7 until the estimated flow depth in step 2 is equal to the calculated flow depth in step 7.
9. Determine the maximum shear stress on the bottom of the channel, τ_b, using Equation 3.261, and the maximum shear stress on the sides of the channel, τ_s, using Equation 3.262.
10. Compare the maximum shear stresses on the bottom and sides of the channel, τ_b and τ_s, to the permissible shear stress, τ_p, for each of the channel linings. If τ_b or τ_s is greater than τ_p for the respective lining, a different combination of linings should be evaluated.

EXAMPLE 3.28

A roadside channel is to be constructed with a low-flow concrete bottom and grass sides. To further channelize low flows, the concrete bottom is made triangular with side slopes 12:1 (H:V). The top-width of the concrete-bottom section is 1 m, the longitudinal slope of the channel is 2.5%, and the sides of the channel have a slope of 3:1 (H:V) and are lined with Bermuda grass cut to 6 cm. For a design flow of 0.4 m^3/s, assess the adequacy of the channel lining and determine the depth of flow in the channel.

Solution

Step 1. The retardance of the grass lining can be taken as "D" according to Table 3.21, and the permissible shear stress, τ_p, for Class D vegetative lining is given by Table 3.20 as 28.7 Pa.

Step 2. Assume a depth of flow, $y = 1.00$ m.

Step 3. For a concrete lining, Table 3.19 gives $n = 0.013$ (trowel finish). For the grass lining, Manning n is given by Table 3.23 for Class D retardance as

$$n = \frac{1.22R^{1/6}}{49.0 + 19.97\log(R^{1.4}S_0^{0.4})} \tag{3.278}$$

For a depth of flow, y, expressions for the area, A, and wetted perimeter, P, are given by

$$A = \frac{T_b}{2}\left(\frac{T_b}{2m_1}\right) + \left[T_b + \left(y - \frac{T_b}{2m_1}\right)m_2\right]\left(y - \frac{T_b}{2m_1}\right) \tag{3.279}$$

$$P = 2\sqrt{1 + m_1^2}\left(\frac{T_b}{2m_1}\right) + 2\sqrt{1 + m_2^2}\left(y - \frac{T_b}{2m_1}\right) \tag{3.280}$$

where m_1 is the side slope of the triangular concrete-bottom section, m_2 is the slope of the grass sides, T_b is the top width of the concrete bottom, and y is the flow depth. From the given data, $T_b = 1$ m, $m_1 = 12$, $m_2 = 3$, and assuming $y = 1.00$ m gives

$$A = \frac{1}{2}\left(\frac{1}{2(12)}\right) + \left[1 + \left(1.00 - \frac{1}{2(12)}\right)(3)\right]\left(1.00 - \frac{1}{2(12)}\right)$$

$$= 3.734 \text{ m}^2$$

$$P = 2\sqrt{1 + 12^2}\left(\frac{1}{2(12)}\right) + 2\sqrt{1 + 3^2}\left(1.00 - \frac{1}{2(12)}\right) = 7.064 \text{ m}$$

$$R = \frac{A}{P} = \frac{3.734}{7.064} = 0.529 \text{ m}$$

Substituting $R = 0.529$ m and $S_0 = 0.025$ into Equation 3.278 gives

$$n = 0.0385$$

Step 4. The low-flow Manning roughness coefficient, n_1, is 0.013 and the Manning roughness coefficient of the side lining (grass), n_2, is 0.0385. Therefore $n_2/n_1 = 0.0385/0.013 = 2.96$.

Step 5. The low-flow wetted perimeter, P_L, is given by

$$P_L = 2\sqrt{1 + m_1^2}\left(\frac{T_b}{2m_1}\right) = 2\sqrt{1 + 12^2}\left(\frac{1}{2(12)}\right) = 1.003 \text{ m}$$

For $y = 1.00$ m, $P = 7.064$ m and therefore

$$\frac{P_L}{P} = \frac{1.003}{7.064} = 0.142$$

Step 6. Equation 3.277 gives the effective Manning n of the channel section as

$$n = \left[\frac{P_L}{P} + \left(1 - \frac{P_L}{P}\right)\left(\frac{n_2}{n_1}\right)^{3/2}\right]^{2/3} n_1$$

$$= \left[0.142 + (1 - 0.142)(2.96)^{3/2}\right]^{2/3}(0.013) = 0.0355$$

Step 7. From the Manning equation

$$Q = \frac{1}{n}AR^{2/3}S_0^{1/2} = \frac{1}{n}\frac{A^{5/3}}{P^{2/3}}S_0^{1/2}$$

which can be put in the form

$$\frac{A^5}{P^2} = \left[\frac{Qn}{S_0^{1/2}}\right]^3$$

From the given data: $Q = 0.4 \text{ m}^3/\text{s}$, $n = 0.0355$, and $S_0 = 0.025$, therefore

$$\frac{A^5}{P^2} = \left[\frac{(0.4)(0.0847)}{0.025^{1/2}}\right]^3 = 0.000724 \qquad (3.281)$$

Both A and P are functions of the flow depth, y, as given by Equations 3.279 and 3.280. Combining Equations 3.279 to 3.281 and solving for y gives

$$y = 0.234 \text{ m}$$

Step 8. The calculated flow depth ($= 0.234$ m) is not equal to the assumed flow depth ($= 1.00$ m), therefore steps 2 to 7 must be repeated until the assumed flow depth in step 2 is equal to the calculated flow depth in step 7. These calculations are summarized in the following table:

Assumed y (m)	A (m^2)	P (m)	R (m)	n_2	n_2/n_1	P_L/P	n	Calculated y (m)
1.000	3.734	7.06	0.529	0.0385	2.96	0.142	0.0355	0.234
0.400	0.764	3.27	0.234	0.0516	3.97	0.307	0.0419	0.253
0.265	0.394	2.42	0.163	0.0636	4.89	0.415	0.0464	0.265

Hence the flow depth in the channel is 0.265 m.

Step 9. For a flow depth of 0.265 m, the shear stress on the bottom of the channel is given by Equation 3.261 as

$$\tau_b = \gamma y S_0 = (9.79)(0.265)(0.025) = 0.065 \text{ kPa} = 65 \text{ Pa}$$

For side slopes of 3:1, the side shear stress factor, K_s, is given by Figure 3.47 as $K_s = 0.85$, and the maximum shear stress on the sides of the channel, τ_s, is given by

$$\tau_s = K_s \tau_b = 0.85(0.065) = 0.055 \text{ kPa} = 55 \text{ Pa}$$

Step 10. The maximum shear stress on the sides of the channel (55 Pa) is greater than the permissible shear stress on the grass lining ($= 28.7$ Pa) estimated in step 1. Therefore, the grass lining on the channel sides (6-cm high Bermuda grass) is not acceptable.

3.5.4 Levees

Levees are earth embankments constructed nearly parallel to streams and rivers to prevent inundation of adjacent lands during periods of high flows. Design considerations for levees are as follows (Prakash, 2004):

Location: Usually the distance between levees on two sides of a river or the distance of the levee from the centerline of the river depends on the availability of land. If land is available or can be acquired without significant environmental, socioeconomic, or political problems, the levee may be located on natural high grounds or ridges such that sufficient waterway (or floodway) is available to pass the design flood. Otherwise, other measures of flood control may have to be adopted. In some cases, alignment of levees has to be adjusted to protect or avoid existing historic, archeological, or other significant features.

Slope Stability: Levees are designed as earth dams so that the side slopes are stable under dry conditions and also when water is impounded up to the design-flood elevation on the river side with dry conditions or low water levels on the landward side. Usual side slopes of levees vary from 2:1 to 4:1 (H:V). Flatter slopes must be provided for embankment materials that have smaller angles of repose.

Seepage: During high river stages, seepage through the levee section may result in piping unless appropriate measures for seepage control and/or piping control are implemented. It is desirable to incorporate such measures in the design of the levee. Commonly used methods include berms to minimize potential for piping, trenches to collect and transport seepage water to nearby tributaries, relief wells, and inverted filters.

Interior Drainage: Construction of levees along the main channel may obstruct tributaries and natural overland flow entering the river through various reaches of its banks. This, along with seepage from the main river channel, may result in impoundment of water on the landward side of the levee. Measures for drainage of this water must be incorporated in the levee design. This may include pumping of water over the levee and diversion or discharge of impounded water into downstream tributaries through ditches or drain tiles nearly parallel to the levee.

Top Width and Freeboard: To permit the movement of equipment for maintenance and inspection, the top width of the levee should be greater than 3 meters. However, smaller top widths may be used if other means of access are available (e.g., an existing road close to the levee). Freeboards of levees vary from 1.0 to 1.5 m, with 1.0 m being the commonly adopted value (U.S. Army Corps of Engineers, 1994; 1995).

Erosion and Scour Protection: Methods for erosion protection of levees are generally the same as for channel-bank protection.

Hydraulic analyses associated with levees include floodplain delineation and floodway determination using computer models such as HEC-RAS (U.S. Army Corps of Engineers, 1998) and interior-drainage analysis using models such as HEC-IFH (U.S. Army Corps of Engineers, 1992). The design of levees is closely linked to floodplain-management practices discussed in Section 7.6.

3.6　Design of Sanitary Sewers

Sanitary sewers are an important class of subsurface open channels (partially full pipes) that are used to transport domestic, commercial, and industrial wastewater, ground-water infiltration from surrounding soils, and extraneous inflows from sources such as roof leaders, basement drains, and submerged manhole covers. The *design life* or *design period* of sewer systems is usually on the order of 50 years (ASCE, 1982). In cases where there might be significant additional development beyond the design period, it is generally prudent to secure easements and rights-of-way for future expansions. Pump stations in sewer systems can be designed with shorter design periods since they are relatively short lived and easier to update. Sewage treatment plants, which are also relatively easy to update, are generally designed for periods of about 20 years.

3.6.1　Design Flow

Sewer flows consist of two components: flows contributed through service connections and flows contributed by infiltration and inflow (I/I).

3.6.1.1 Service flow

Flows contributed through service connections are usually related to present and projected populations in the service area multiplied by per-capita wastewater production rates. Methods of population projection are described in Section 2.5.1, and the uncertainty in these projections resulting from economic and social variables should always be considered. Design population densities in residential areas are typically taken as the saturation densities, and typical values are listed in Table 3.24. The average daily per-capita domestic wastewater flows vary considerably within the United States, and several design per-capita flowrates are listed in Table 3.25. Local regulatory agencies usually specify the per-capita domestic wastewater flowrates to be used in their jurisdiction. Wastewater flows from commercial and industrial areas must be added to the domestic flowrates shown in Table 3.25. Typical flowrates associated with commercial and industrial areas are shown in Table 3.26.

It is fairly common to assume that the average rate of sewage flow, including domestic, commercial, and industrial flows, is equal to the average rate of water consumption, but this assumption should be made only after careful consideration of the nature of the community. For example, sewage flows might be only approximately equal to water consumption during dry weather and, in arid regions, evaporation and other losses may lead to significantly less sewage flows than water-use rates. It is

TABLE 3.24: Typical Saturation Densities

Type of area	Density (persons/ha)
Large lots	5–7
Small lots, single-family	75
Small lots, two-family	125
Multistory apartments	2500

Source: American Concrete Pipe Association (1981).

TABLE 3.25: Average Per-Capita Wastewater Domestic Flowrates

City	Flowrate (L/d/person)	Comments
Berkeley, CA	350	
Boston, MA	380	Includes infiltration; multiply by 3 when sewer is flowing full
Des Moines, IA	$380 \times$ factor	Factor $= \dfrac{18 + \sqrt{P}}{4 + \sqrt{P}}$ where P is the population in thousands
Detroit, MI	980	
Las Vegas, NV	950	
Little Rock, AR	380	
Milwaukee, WI	1000	Plus additional flow for inflow/infiltration
Orlando, FL	950	
Shreveport, LA	570	Plus 5600 L/d/ha infiltration

Source: ASCE (1982).

TABLE 3.26: Commercial and Industrial Wastewater Flowrates

City	Commercial	Industrial
Grand Rapids, MI	150–190 L/person/day, office buildings 1500–1900 L/day/room, hotels 750 L/d/bed, hospitals 750–1150 L/d/room, schools	2.3×10^6 L/d/ha
Kansas City, MO	47,000 L/d/ha	94,000 L/d/ha
Memphis, TN	19,000 L/d/ha	19,000 L/d/ha
Santa Monica, CA	91,000 L/d/ha	127,000 L/d/ha

Source: American Concrete Pipe Association (1981).

generally reported that 60%–80% of the total water supplied to a community becomes wastewater, with low ratios generally applicable to semi-arid regions (Viessman and Hammer, 2005). Ordinarily, the annual variation in the ratio of sewage to water supply in a city is not great, and thus the amount of water used by a city is a good indicator of the amount of sewage that will be generated.

3.6.1.2 Inflow and infiltration (I/I)

Inflow is defined as surface water entering the sewer via flooded sewer vents, leaky manholes, illicitly connected storm drains, basement drains, and other means than ground water. Inflow is usually the result of rain and/or snowmelt events. Inflows are difficult to predict and are usually lumped with the infiltration, which is then called the infiltration and inflow (I/I). In many existing systems, inflow is the largest flow component in sanitary sewers on rainy days and is often responsible for the backup of wastewater into basements and homes, or the bypassing of untreated wastewater to streams and other watercourses. Inflow rates are typically in the range of 600–12,000 m^3/d/km on a maximum-hour basis (Lagvankar and Velon, 2000).

Infiltration is defined as water that enters the sewer from ground water, and sources of infiltration flows include defective pipes, pipe joints, connections, or manhole walls. Regulatory agencies in most states specify maximum allowances for infiltration. For sanitary sewers up to 600 mm in diameter, it is common to allow 71 m^3/d/km for the total length of main sewers, laterals, and house connections. However, it should be noted that infiltration rates as high as 140 m^3/d/km have been recorded for sewers submerged in ground water, with rates up to 2350 m^3/d/km in isolated segments. As a typical example, Miami-Dade county (Florida) requires a design maximum infiltration allowance of 4.6 m^3/d/km per 100 mm of sewer diameter when there is a 60-cm static head above the crown of the pipe (Miami-Dade Public Works Department, 2001). Based on a survey of 128 cities, ASCE (1982) reported an average design infiltration allowance of 3.9 m^3/d/km per 100 mm of sewer diameter. In spite of these regulations on design inflow rates, McGhee (1991) reported that sewer size apparently has little effect on infiltration inflow, since larger sewers tend to have better joint workmanship, offsetting the increased size of the potential infiltration area. The actual amount of infiltration depends on the quality of sewer installation, the height of the ground water table, and the properties of the soil. Expansive soils tend to pull joints apart, while granular soils permit water to move easily through joints and breaks. The use of pressure sewers will reduce I/I to near zero.

3.6.1.3 Combined flow

Wastewater flows, exclusive of I/I, vary continuously, with very low flows during the early-morning hours and peak flows occurring at various times during the daylight hours. The I/I component remains fairly constant, except during and immediately following periods of heavy rainfall, when significant increases usually occur. Based on data reported in ASCE (1982), the ratio of the peak flow, Q_{peak}, to the average daily wastewater flow, Q_{ave}, can be estimated by the relation

$$\boxed{\frac{Q_{peak}}{Q_{ave}} = \frac{5.5}{P^{0.18}}} \tag{3.282}$$

where P is the population of the service area in thousands and the peak flow is defined as the maximum flow occurring during a 15-minute period for any 12-month period (American Concrete Pipe Association, 1985). Other peak-flow factors that are sometimes used in practice are shown in Table 3.27 (Haestad Methods, Inc., 2002). The ratio of the minimum flow, Q_{min}, to the average daily wastewater flow can be estimated by the relation

$$\boxed{\frac{Q_{min}}{Q_{ave}} = 0.2P^{0.16}} \tag{3.283}$$

Equations 3.282 and 3.283 indicate that the ratio of peak flow to minimum flow varies from less than 3 : 1 for sewers serving large populations to more than 20 : 1 for sewers serving smaller populations. Flow fluctuations are less for higher populations, since peak loads from different sources are more likely to occur in a staggered manner.

The flowrates commonly used in sewer design are the peak and minimum flowrates generated by the projected population of the service area. To estimate these flows, the average daily flow is first estimated by the product of the population projection and the average domestic per-capita flowrate. This flow is added to the

TABLE 3.27: Peak-Flow Factors for Sanitary-Sewer Flows

Q_{peak}/Q_{ave}	Reference
$\dfrac{5.5}{P^{0.18}}$	ASCE (1982)
$\dfrac{5.0}{P^{0.2}}$	Babbitt and Baumann (1958)
$\dfrac{18 + P^{0.5}}{4 + P^{0.5}}$	Great Lakes Upper Mississippi River Board (1978)
$1.0 + \dfrac{14.0}{4.0 + P^{0.5}}$	Harmon (1918)
$\dfrac{2.69}{Q_{ave}}$	Jakovlev et al. (1975)[†]

[†]Formula commonly referred to as the Federov formula.

commercial and industrial contributions to estimate the average wastewater flowrate. The peak and minimum flowrates are then calculated from the average flowrate using the factors given by Equations 3.282 and 3.283, and the (constant) I/I is added to the calculated flowrates to yield the overall peak and minimum flowrates to be accommodated by the sewer system. When commercial, institutional, or industrial wastewaters make up a significant portion of the average flows (such as 25 percent or more of all flows, exclusive of infiltration), peaking factors for the various flow components should be estimated separately (Metcalf & Eddy, Inc., 1981).

EXAMPLE 3.29

A trunk sewer is to be sized for a 25-km² city in which the master plan calls for 60% residential, 30% commercial, and 10% industrial development. The residential area is to be 40% large lots, 55% small single-family lots, and 5% multistory apartments. The average domestic wastewater flowrate is 800 L/d/person, the average commercial flowrate is 25,000 L/d/ha, and the average industrial flowrate is 40,000 L/d/ha. Infiltration and inflow is 1000 L/d/ha for the entire area. Estimate the peak and minimum flows to be handled by the trunk sewer.

Solution   From the given data, the total area of the city is 25 km² = 2500 ha and the residential area is 60% of 2500 ha = 1500 ha. The residential saturation densities can be taken as those in Table 3.24. Taking the per-capita flowrate as 800 L/d/person ($=9.26 \times 10^{-6}$ m³/s/person) gives the wastewater flows in the following table:

Type	Area (ha)	Density (persons/ha)	Population	Flow (m³/s)
Large lots	0.40(1500) = 600	6	3600	0.03
Small single-family lots	0.55(1500) = 825	75	61,875	0.57
Multistory apartments	0.05(1500) = 75	2500	187,500	1.74
Total			252,975	2.34

The commercial sector of the city covers 30% of 2500 ha = 750 ha, with a flowrate per unit area of 25,000 L/d/ha = 2.89×10^{-4} m³/s/ha. Hence the average flow from the commercial sector is $(2.89 \times 10^{-4})(750) = 0.22$ m³/s.

The industrial sector of the city covers 10% of 2500 ha = 250 ha, with a flowrate per unit area of 40,000 L/d/ha = 4.63×10^{-4} m³/s/ha. Hence the average flow from the commercial sector is $(4.63 \times 10^{-4})(250) = 0.12$ m³/s.

The infiltration and inflow from the entire area is 1000 L/ha × 2500 ha = 2.5×10^{6} L/d = 0.03 m³/s.

On the basis of these calculations, the average daily wastewater flow (excluding I/I) is 2.34 + 0.22 + 0.12 = 2.68 m³/s. Assume that the total population, P, of the city is equal to the residential population of 252,975, then the peak and minimum flow ratios can be estimated by Equations 3.282 and 3.283 as

$$\frac{Q_{peak}}{Q_{ave}} = \frac{5.5}{P^{0.18}} = \frac{5.5}{(252.975)^{0.18}} = 2.0$$

$$\frac{Q_{min}}{Q_{ave}} = 0.2P^{0.16} = 0.2(252.975)^{0.16} = 0.48$$

The peak and minimum flows are estimated by multiplying the average wastewater flows by these factors and adding the I/I. Thus

$$\text{Peak flow} = 2.0(2.68) + 0.03 = 5.39 \text{ m}^3/\text{s}$$

$$\text{Minimum flow} = 0.48(2.68) + 0.03 = 1.32 \text{ m}^3/\text{s}$$

The trunk sewer must be of sufficient size to accommodate the peak flow (without flowing full), and must also maintain a minimum scour velocity to prevent solids accumulation. These design requirements are covered in the following section.

3.6.2 Hydraulics of Sewers

Design guidelines published by the American Society of Civil Engineers (ASCE, 1982) and the Water Environment Federation (formerly the Water Pollution Control Federation; WPCF, 1982) state that sanitary sewers through 375 mm (15 in.) in diameter be designed to flow half full at the design flowrate, with larger sewers designed to flow three-fourths full. These guidelines ensure proper ventilation in sewers and also reflect the fact that smaller wastewater flows are much more uncertain than larger flows. Since sanitary sewers transport a significant amount of suspended solids, the prevention of solids deposition by specifying minimum permissible velocities (under minimum-flow conditions) is an important aspect of the hydraulic design of sanitary sewers. Minimum permissible velocities are sometimes called *self-cleansing* velocities. The specification of maximum permissible velocities is also important to prevent excessive scouring of the sewer pipe. ASCE (1982) recommends that flow velocities in sanitary sewers should not be less than 0.60 m/s (2 ft/s) or greater than 3.5 m/s (10 ft/s). Ideally, self-cleansing conditions should be achieved at least once per day (Butler and Davies, 2000); however, in cul-de-sacs served by short dead-end sections, it will often be impossible to meet minimum-velocity requirements, and such pipes may require periodic flushing to scour the accumulated sediments. A flow velocity in excess of 1 m/s is required to scour juvenile consolidated material deposited in sanitary sewers (Fan et al., 2001).

In accordance with the aforementioned design requirements, it is usually necessary to calculate the depth of flow and average velocity in circular pipes for given values of the pipe diameter, D, flowrate, Q, and pipe slope, S_0, where, in uniform flow, the slope of the channel, S_0, is equal to the slope of the energy grade line, S_f. Consider the circular-pipe cross section in Figure 3.52, where h is the depth of flow and θ is the water-surface angle. The depth of flow, h, cross-sectional area, A, and

FIGURE 3.52: Flow in partially filled pipe

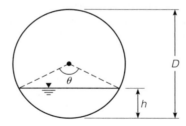

wetted perimeter, P, can be expressed in terms of θ by the geometric relations

$$h = \frac{D}{2}\left[1 - \cos\left(\frac{\theta}{2}\right)\right] \tag{3.284}$$

$$A = \left(\frac{\theta - \sin\theta}{8}\right)D^2 \tag{3.285}$$

$$P = \frac{D\theta}{2} \tag{3.286}$$

The Manning equation is commonly applied to describe the flow in sanitary sewers and can be written in the form

$$Q = \frac{1}{n}\frac{A^{5/3}}{P^{2/3}}S_0^{1/2} \tag{3.287}$$

where n is the Manning roughness coefficient. Combining Equations 3.285 to 3.287 leads to the following form of the Manning equation:

$$\boxed{\theta^{-\frac{2}{3}}(\theta - \sin\theta)^{\frac{5}{3}} - 20.16nQD^{-\frac{8}{3}}S_0^{-\frac{1}{2}} = 0} \tag{3.288}$$

where the only unknown is θ and the quantities n, Q, D, and S_0 are assumed to be given for this analysis. The solution of Equation 3.288 (i.e., θ) is substituted into Equation 3.284 to obtain the depth of flow in the sewer pipe. The average flow velocity is obtained by first calculating the flow area, A, from Equation 3.285, and then calculating the velocity, V, by

$$V = \frac{Q}{A} \tag{3.289}$$

The only obstacle to solving Equation 3.288 for θ, and hence obtaining h and V, is that Equation 3.288 must be solved numerically. A variety of techniques, including Newton's method, are available for this purpose. Combining Equations 3.288 and 3.284 gives a nonlinear relationship between the flowrate, Q, and the depth of flow, h, and it can be shown (see Problem 3.133) that the maximum flowrate occurs when $h/D \approx 0.94$. This condition manifests itself as a flow instability when the pipe is flowing almost full, and there is a tendency for the pipe to run temporarily full at irregular intervals (Henderson, 1966; Hager, 1999). This condition, sometimes referred to as *slugging* (Sturm, 2001), results in streaming air pockets at the crown of the pipe and pulsations that could damage pipe joints and cause undesirable fluctuations in discharge. This condition is avoided in practice by designing pipes such that $h/D \leq 0.75$. Another interesting feature of open-channel flow in circular pipes is that the velocity is the same whether the pipe flows half full or completely full (see Problem 3.134). According to the Manning equation (Equation 3.288), the velocity increases with depth of flow until it reaches a maximum at $h/D \approx 0.94$; the velocity then decreases with increasing depth and becomes equal to the half-full velocity when the pipe flows full (Gribbin, 1997).

EXAMPLE 3.30

Water flows at a rate of 4 m³/s in a circular concrete sewer of diameter 1500 mm and a Manning n of 0.015. If the slope of the sewer is 1%, calculate the depth of flow and velocity in the sewer.

Solution According to Equation 3.288,

$$\theta^{-\frac{2}{3}}(\theta - \sin\theta)^{\frac{5}{3}} - 20.16nQD^{-\frac{8}{3}}S_0^{-\frac{1}{2}} = 0$$

where in this case $n = 0.015$, $Q = 4$ m³/s, $S_0 = 0.01$, and $D = 1.5$ m. Therefore

$$\theta^{-\frac{2}{3}}(\theta - \sin\theta)^{\frac{5}{3}} - 20.16(0.015)(4)(1.5)^{-\frac{8}{3}}(0.01)^{-\frac{1}{2}} = 0$$

which leads to

$$\theta^{-\frac{2}{3}}(\theta - \sin\theta)^{\frac{5}{3}} - 4.10 = 0$$

Solving for θ yields

$$\theta = 3.50 \text{ radians}$$

Therefore, the normal flow depth, h, and area, A, are given by Equations 3.284 and 3.285 as

$$h = \frac{D}{2}\left[1 - \cos\left(\frac{\theta}{2}\right)\right] = \frac{1.5}{2}\left[1 - \cos\left(\frac{3.50}{2}\right)\right] = 0.88 \text{ m}$$

$$A = \left(\frac{\theta - \sin\theta}{8}\right)D^2 = \left(\frac{3.50 - \sin 3.50}{8}\right)(1.5)^2 = 1.08 \text{ m}^2$$

The average flow velocity, V, in the sewer is given by

$$V = \frac{Q}{A} = \frac{4}{1.08} = 3.70 \text{ m/s}$$

The value of the Manning n in sanitary sewers will generally approach a constant which is not a function of the pipe material, but is determined by the grit accumulation and slime build-up on the pipe walls. According to ASCE (1982), this n value is usually on the order of 0.013. A higher n value should be used in cases where additional roughness sources are known or anticipated. Metcalf and Eddy (1981) suggest that an n value of 0.015 be used to analyze most older existing sewers. Some experiments have shown that the value of n for concrete and vitrified clay pipes flowing partly full is greater than for the same pipes flowing full (Yarnell and Woodward, 1920; Wilcox, 1924). These results have been summarized by Camp (1946), where the values of n/n_{full} as a function of the relative flow depth, h/D, are given in Table 3.28; n_{full} is the n value when the pipe is flowing full.

Table 3.28 indicates that the n value for pipes flowing partially full can be as much as 29% higher than the full-flow n value. Similar variations in n value with depth have not been found for PVC pipes (Neale and Price, 1964). ASCE (1982) recommends that until more information and better analyses are available, the decision to use a constant or variable n must be left to the engineer. However, it seems prudent to either incorporate the variation of n with flow depth or apply a factor of safety on the order of 1.3 to the selected constant n value to accommodate the uncertainty of n with depth. Head losses in bends are primarily associated with large-diameter pipes (>915 mm), are difficult to quantify, and are usually accounted for by increasing the Manning n in the bend by 25% to 40% (McGhee, 1991).

TABLE 3.28: Variation of the Manning n With Depth

h/D	n/n_{full}
0	1.00
0.1	1.22
0.2	1.28
0.3	1.29
0.4	1.28
0.5	1.25
0.6	1.22
0.7	1.18
0.8	1.14
0.9	1.08
1.0	1.00

Source: Camp (1946).

In engineering practice in the United Kingdom, the Darcy–Weisbach equation is generally used to describe flow in sewers, and roughness heights in the range of 0.6 mm to 1.5 mm are recommended, depending on the design self-cleansing velocity (Butler and Davies, 2000).

3.6.3   Sewer-Pipe Material

A variety of pipe materials are used in practice, including concrete, vitrified clay, cast iron, ductile iron, and various thermoplastic materials including PVC. Pipes are broadly classified as either *rigid pipes* or *flexible pipes*. Rigid pipes derive a substantial part of their load-carrying capacity from the structural strength inherent in the pipe wall, while flexible pipes derive their load-carrying capacity from the interaction of the pipe and the embedment soils affected by the deflection of the pipe to the equilibrium point under load. Pipe materials classified as rigid and flexible are listed in Table 3.29, and the advantages and disadvantages to using various pipe materials, along with typical n values, are given in Table 3.30.

The type of pipe material to be used in any particular case is dictated by several factors, including the type of wastewater to be transported (residential, industrial, or combination), scour and abrasion conditions, installation requirements, type of soil, trench-load conditions, bedding and initial backfill material available, infiltration/exfiltration requirements, and cost effectiveness. Nonreinforced and reinforced concrete sewer pipe is used frequently in sanitary-sewer systems, and the commercially available diameters of concrete pipe are given in Appendix D, Table D.3 (American Concrete Pipe Association, 1985). It is useful to keep in mind that the size of reinforced

TABLE 3.29: Rigid and Flexible Pipe Materials

Rigid pipe	Flexible pipe
Concrete	Ductile iron
Cast iron	Steel
Vitrified clay	Thermoplastic (e.g., PVC)

TABLE 3.30: Sewer-Pipe Materials

Material	Advantages	Disadvantages	Applications	Manning n	Available diameters
Concrete	Readily available in most localities Wide range of structural and pressure strengths Wide range of nominal diameters Wide range of laying lengths	High weight Subject to corrosion on interior if atmosphere over wastewater contains hydrogen sulfide Subject to corrosion on exterior if buried in an acid or high-sulfate environment	Widely used for gravity sewers	0.011–0.015*	300 mm–3050 mm[‡]
Vitrified clay	High resistance to chemical corrosion Not susceptible to damage from hydrogen sulfide High resistance to abrasion Wide range of fittings available Manufactured in standard and extra-strength classifications	Limited range of sizes available High weight Subject to shear and beam breakage when improperly bedded Brittle	Widely used in small and medium-sized sewers	0.010–0.015	100 mm–610 mm
Cast iron	Long laying lengths High pressure and load-bearing capacity	Subject to corrosion where acids are present Subject to chemical attack in corrosive soils Subject to shear and beam breakage when improperly bedded High weight	No longer used, replaced by ductile-iron pipe	0.011–0.015[†]	50 mm–1220 mm
Ductile iron	Long laying lengths High pressure and load-bearing capacity High impact strength High beam strength	Subject to corrosion where acids are present Susceptible to hydrogen sulfide attack Subject to chemical attack in corrosive soils High weight	Often used for river crossings and where the pipe must support unusually high load, where an unusually leakproof sewer is required, or where unusual root problems are likely to develop. Should not be used where the ground water is brackish, unless suitable protective measures are taken. Ductile-iron pipe is employed in sewerage primarily for force mains and for piping in and around buildings.	0.011–0.015[†]	100 mm–910 mm
Plastic[§]	Light weight Long laying lengths High impact strength Ease in field cutting and tapping Corrosion resistant Low friction	Subject to attack by certain organic chemicals Subject to excessive deflection when improperly bedded and haunched Limited range of sizes available Subject to surface changes effected by long-term ultraviolet exposure	Used as an alternative to vitrified-clay pipe	0.010–0.015	100 mm–1220 mm

*The n values shown are for new pipes.
[†] Cement-lined and seal-coated.
[‡] Diameters are for reinforced-concrete pipe (ASTM C76, C361). Nonreinforced concrete pipe is available in nominal diameters from 150 mm to 600 mm (ASTM C14).
[§] Plastic pipe used in sewage systems includes PVC (polyvinyl chloride), ABS (acrylonitrile-butadiene-styrene), and PE (polyethylene) (ASTM D1785, D2241, D2729, F679, F794).

concrete pipe varies in 75-mm (3-in.) increments between 305 mm (12 in.) and 915 mm (36 in.), and in 150-mm increments between 915 mm (36 in.) and 2745 mm (108 in.). Numerous test programs have established values for the roughness coefficient of concrete pipe from 0.009 to 0.011; however, higher values of 0.012 and 0.013 are typically used in design to account for the possibility of slime or grease buildup in sanitary sewers (American Concrete Pipe Association, 1985).

3.6.4 System Layout

The layout of a sanitary-sewer system begins with the selection of an outlet, the determination of the tributary area, the location of the trunk or main sewer, the determination of whether there is a need for, and the location of, pumping stations and force mains; the location of underground rock formations; the location of water and gas lines, electrical, telephone, and television wires, and other underground utilities. The *main sewer* refers to the sewer pipeline that receives many tributary branches and serves a large territory; it is the principal sewer to which branch sewers are tributary. In larger systems, the main sewer is also called the *trunk sewer*. The selected system outlet depends on the scope and objectives of the particular project and may consist of a pumping station, an existing trunk or main sewer, or a treatment plant.

Since the flows in sewer systems are driven by gravity, preliminary layouts are generally made using topographic maps. Trunk and main sewers are located at the lower elevations of the service area, although existing roadways and the availability of rights-of-way may affect the exact locations. In developed areas, sanitary sewers are commonly located at or near the centers of roadways and alleys. In very wide streets, however, it may be more economical to install sanitary sewers on both sides of the street. When sanitary sewers are in close proximity to public water supplies, it is common practice to use pressure-type sewer pipe (i.e., force main), concrete encasement of the sewer pipe, or sewer pipe with joints that meet stringent infiltration/exfiltration requirements, or at least to put water pipes and sewer pipes on opposite sides of the street. Most building codes prohibit sanitary-sewer installation in the same trench as water mains, require that sewers be at least 3 m horizontally from water mains and, where they cross, at least 450 mm vertical separation between sewers and water mains (Lagvankar and Velon, 2000)

Manholes provide access to the sewer system for preventive maintenance and emergency service and are generally located at the junctions of sanitary sewers, at changes of grade or alignment, and at the beginning of the sewer system. A typical manhole is illustrated in Figure 3.53(a), indicating that manhole covers are typically 0.61 m (2 ft) in diameter, with a working space in the manhole typically 1.2 m (4 ft) in diameter. Manholes are typically precast and delivered to the site ready for installation, and a typical precast manhole is shown in Figure 3.54. The invert of the manhole should conform to the shape and slope of the incoming and outgoing sewer lines. In some cases, a section of pipe is laid through the manhole and the upper portion is sawed or broken off. More commonly, U-shaped channels are specified. The spacing between manholes varies with local conditions and methods of sewer maintenance and is often in the range of 90 to 150 m, with spacings of 150 to 300 m used for larger sewers that a person can walk through. A commonly used design criterion in some jurisdictions is that sewers with diameters of 600 mm or less should be straight between manholes (Lagvankar and Velon, 2000). In cases where a sewer pipe enters a manhole at an

FIGURE 3.53: Typical manholes
Source: ASCE (1992).

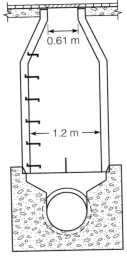

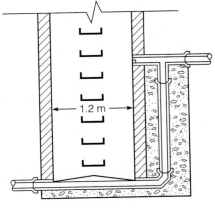

(a) Typical Manhole

(b) Drop Manhole

FIGURE 3.54: Precast manhole

elevation considerably higher than the outgoing pipe, it is generally not acceptable to let the incoming wastewater simply pour into the manhole, since this does not provide an acceptable workspace for maintenance and repair. Under these conditions, a *drop manhole*, illustrated in Figure 3.53(b), is used. Drop manholes are typically specified when the invert of the inflow pipe is more than 0.6 m above the elevation that would be obtained by matching the crowns of the inflow and outflow pipes.

Sanitary sewers should be buried to a sufficient depth that they can receive the contributed flow from the tributary area by gravity flow. Deep basements and buildings on land substantially below street level may require individual pumping facilities. In northern states, a cover of 3 m is typically required to prevent freezing, while in southern states the minimum cover is dictated by traffic loads, and ranges upward from 0.75 m, depending on the pipe size and anticipated loads (McGhee,

1991). The depth of sanitary sewers is such that they pass under all other utilities, with the possible exception of storm sewers. Sanitary sewers typically have a minimum diameter of 205 mm (8 in.), and it is common practice to lay service connections at a slope of 2%, with a minimum slope of 1%. In some developments where the houses are set far back from the street, the length and slope of the house connections may determine the minimum sewer depths. Service connections typically use 150- or 205-mm diameter pipe (Corbitt, 1990), although connections using 100-mm diameter pipe have been used successfully in some areas (Metcalf & Eddy, 1981). The diameters of sanitary sewers should never be allowed to decrease in the downstream direction, since this can cause sediment accumulation and blockage where this reduction occurs.

Pump stations, also known as *lift stations*, are frequently necessary in flat terrain to raise the wastewater to a higher elevation so that gravity flow can continue at reasonable slopes and depths. Pump stations typically have a *wet well* and a *dry well*, where the wet well receives the wastewater flow and is sized for a 10- to 30-min detention time with a bottom slope of 2 : 1, and the dry well is the pump and motor area. Ventilation, humidity control (dry well), and standby power supply are important design considerations. For smaller stations (<3800 m^3/d), at least two pumps should be provided; for larger pump stations, three or more pumps. In both cases, the pump capacities should be sufficient to pump at the maximum wastewater flowrate if any one pump is out of service (Corbitt, 1990). To avoid clogging, the associated suction and discharge piping should be at least 100 mm in diameter and the pump should be capable of passing 75-mm diameter solids. In cases where pumping directly into a gravity-flow sanitary sewer is not practical, wastewater can be pumped from the wet well through a pressurized sewer pipe over some obstruction or hill to another gravity sewer, or directly to a wastewater-treatment facility. Pressurized sewer pipes are called *force mains*, which are described in more detail in Section 3.6.7.

3.6.5 Hydrogen Sulfide Generation

Hydrogen-sulfide (H_2S) generation is a common problem in sanitary sewers. Among the problems associated with H_2S generation are odor, health hazard to maintenance crews, and corrosion of unprotected sewer pipes produced from cementitious materials and metals. The design of sanitary sewers seeks to avoid septic conditions, and to provide an environment that is relatively free of H_2S. Generation of H_2S in sanitary sewers results primarily from the action of sulfate-reducing bacteria (SRB) on the pipe floor which convert sulfates to hydrogen sulfide under anaerobic (reducing) conditions. Compounds such as sulfite, thiosulfate, free sulfur, and other inorganic sulfur compounds occasionally found in wastewater can also be reduced to sulfide. Corrosive conditions occur when sulfuric acid (H_2SO_4) is derived through the oxidation of hydrogen sulfide by aerobic bacteria (*Thiobacillus thiooxidans*) and fungi that reside on the exposed sewer pipe wall. This process can only take place where there is an adequate supply of H_2S (>2 ppm), high relative humidity, and atmospheric oxygen. The effect of H_2SO_4 on concrete surfaces exposed to the sewer environment can be devastating. Concrete pipes, asbestos-cement pipes, and mortar linings on ferrous pipes experience surface reactions in which the surface material is converted to an expanding, pasty mass which may fall away and expose new surfaces to corrosive attack. The color of corroded concrete surfaces can be various shades of yellow caused by the direct oxidation of H_2S to elemental sulfur. Entire pump stations have been known to collapse due to loss of structural stability from corrosion. Hydrogen sulfide

TABLE 3.31: Sulfide Generation Based on Z Values

Z Values	Sulfide condition
$Z < 5000$	Sulfide rarely generated
$5000 \leq Z \leq 10,000$	Marginal condition for sulfide generation
$Z > 10,000$	Sulfide generation common

Source: ASCE (1982).

gas is extremely toxic and can cause death at concentrations as low as 300 ppm (0.03%) in air. A person who ignores the first odor of the gas quickly loses the ability to smell the gas, eliminating further warning and leading to deadly consequences.

Significant factors that contribute to H_2S generation are high wastewater temperatures and low flow velocities. The potential for sulfide generation can be assessed using the formula (Pomeroy and Parkhurst, 1977)

$$Z = 0.308 \frac{\text{EBOD}}{S_0^{0.50} Q^{0.33}} \times \frac{P}{B} \qquad (3.290)$$

where EBOD is the *effective biochemical oxygen demand* in mg/L, defined by the relation

$$\text{EBOD} = \text{BOD} \times 1.07^{T-20} \qquad (3.291)$$

where BOD is the average five-day *biochemical oxygen demand* (mg/L at 20°C) during the highest 6-hour flow period of the day, T is the temperature of the wastewater in the sewer (°C), S_0 is the slope of the sewer, Q is the flowrate in the sewer (m³/s), P is the wetted perimeter (m), and B is the top-width (m) of the sewer flow. The relationship given by Equation 3.290 is commonly referred to as the Z *formula*, and the relationship between the calculated Z value and the potential for sulfide generation is given in Table 3.31.

It is usually impractical or impossible to design a sulfide-free sewer system, and engineers endeavor to minimize sulfide generation and use corrosion-resistant materials to the extent possible. Increased turbulence within sewers will increase the rate at which H_2S is released from wastewater, and structures causing avoidable turbulence should be identified and retrofitted to produce a more streamlined flow. Wastewater-treatment plants are usually located at the terminus of sewer systems, and chlorination with either elemental chlorine or hypochlorite quickly destroys sulfide and odorous organic sulfur compounds. Chlorination in sanitary sewers is generally considered impractical. The dissolving of air or oxygen in the wastewater as it moves through the sewer system, addition of chemicals such as iron and nitrate salts, and periodic sewer flushing are effective sulfide-control measures.

EXAMPLE 3.31

A 915-mm diameter concrete sewer is laid on a slope of 0.9% and is to carry 1.7 m³/s of domestic wastewater. If the five-day BOD of the wastewater at 20°C is expected to be 300 mg/L, determine the potential for sulfide generation when the wastewater temperature is 25°C.

Solution   From the given data, $Q = 1.7 \text{ m}^3/\text{s}$, $D = 915 \text{ mm} = 0.915 \text{ m}$, $S_0 = 0.009$, and Equation 3.288 gives the flow angle θ by the relation

$$\theta^{-\frac{2}{3}}(\theta - \sin\theta)^{\frac{5}{3}} - 20.16nQD^{-\frac{8}{3}}S_0^{-\frac{1}{2}} = 0$$

Taking the Manning n as 0.013, Equation 3.288 gives

$$\theta^{-\frac{2}{3}}(\theta - \sin\theta)^{\frac{5}{3}} - 20.16(0.013)(1.7)(0.915)^{-\frac{8}{3}}(0.009)^{-\frac{1}{2}} = 0$$

which simplifies to

$$\theta^{-\frac{2}{3}}(\theta - \sin\theta)^{\frac{5}{3}} = 5.95$$

and yields

$$\theta = 4.3 \text{ radians}$$

The flow perimeter, P, is given by Equation 3.286 as

$$P = \frac{D\theta}{2}$$

and the top-width, B, can be inferred from Figure 3.52 as

$$B = 2\left(\frac{D}{2}\right)\sin\left(\frac{\theta}{2}\right) = D\sin\left(\frac{\theta}{2}\right)$$

Hence

$$\frac{P}{B} = \frac{D\theta/2}{D\sin(\theta/2)} = \frac{\theta}{2\sin(\theta/2)}$$

and in this case

$$\frac{P}{B} = \frac{4.3}{2\sin(4.3/2)} = 2.57$$

The effective BOD, EBOD at $T = 25°\text{C}$, is given by Equation 3.291 as

$$\text{EBOD} = \text{BOD} \times 1.07^{T-20} = 300 \times 1.07^{25-20} = 421 \text{ mg/L}$$

According to the Z formula, Equation 3.290,

$$Z = 0.308\frac{\text{EBOD}}{S_0^{0.50}Q^{0.33}} \times \frac{P}{B} = 0.308\frac{421}{(0.009)^{0.50}(1.7)^{0.33}} \times 2.57 = 2948$$

Therefore, according to Table 3.31, hydrogen sulfide will be rarely generated.

3.6.6 Design Computations

Basic information required prior to computing the sizes and slopes of the sewer pipes includes: (1) a topographic map showing the proposed locations of the sewer lines, (2) the tributary areas to each line, (3) the (final) ground-surface elevations along each line, (4) the elevations of the basements of low-lying houses and other buildings, and (5) the elevations of existing sanitary sewers which must be intercepted.

Line no. (1)	Location (2)	Manhole no. From (3)	Manhole no. To (4)	Length (m) (5)	Area Increment (ha) (6)	Area Total (ha) (7)	Maximum flow I/I (m³/s) (8)	Maximum flow Sewage (m³/s) (9)	Maximum flow Total (m³/s) (10)	Minimum flow I/I (m³/s) (11)	Minimum flow Sewage (m³/s) (12)	Minimum flow Total (m³/s) (13)	Slope of sewer (14)	Diam (mm) (15)	Min velocity (m/s) (16)	Max velocity (m/s) (17)	Max depth (mm) (18)	Manhole invert drop (m) (19)	Fall in sewer (m) (20)	Sewer invert elevation Upper end (m) (21)	Sewer invert elevation Lower end (m) (22)	Ground surface elevation Upper end (m) (23)	Ground surface elevation Lower end (m) (24)

FIGURE 3.55: Typical computation form for design of sanitary sewers

After the sewer layout has been developed, design computations are performed using the form shown in Figure 3.55 (ASCE, 1982). Design computations begin with the characteristics of the sewer-pipe configuration in columns 1 to 5 and lead to the computation of the sewer slopes in column 14, diameters in column 15, and sewer invert elevations in columns 21 and 22. The steps to be followed in using Figure 3.55 in the design of sanitary sewers are as follows:

1. Computations begin with the uppermost pipe in the sewer system.

2. List the pipeline number in column 1 (usually starting from the number 1), list the street location of the pipe in column 2, list the beginning and ending manhole numbers in columns 3 and 4 (the manhole numbers usually start from 1 at the uppermost manhole), and list the length of the sewer pipe in column 5. The ground-surface elevations at the upstream and downstream manhole locations are listed in columns 23 and 24. These ground-surface elevations are used as reference elevations to ensure that the cover depth is acceptable and to compare the pipe slope with the ground slope.

3. In column 6, list the land area that will contribute wastewater flow to the sewer line. The contributing land area can be estimated using a topographic map and the proposed development plan.

4. In column 7, list the total area contributing wastewater flow to the sewer pipe. This total contributing area is the sum of the area that contributes directly to the pipe (listed in column 6) and the area that contributes flow to the upstream pipes that feed the sewer pipe.

5. In column 8, the contribution of infiltration and inflow (I/I) to the pipe flow is listed. I/I is usually calculated by multiplying the length of the pipe by a design inflow rate in $m^3/d/km$ and adding this inflow to the I/I contribution from all upstream connected pipes.

6. In column 9, the maximum sewage flow is calculated by multiplying the contributing area listed in column 7 by the average wastewater flowrate (usually given in $m^3/d/ha$) and the peaking factor (derived using a relation similar to Equation 3.282).

7. In column 10, the peak design flow is calculated as the sum of I/I (column 8) and the peak wastewater flow (column 9).

8. In columns 11, 12, and 13, the minimum design flow is calculated using a similar procedure to that used in calculating the peak design flow. The I/I contribution (column 11) is the same as calculated in step 5 (column 8); the minimum wastewater flow (column 12) is the average wastewater flow multiplied by the minimum-flow factor (derived using a relation similar to Equation 3.283); and the minimum design flow (column 13) is the sum of I/I (column 11) and the minimum wastewater flow (column 12).

9. For a given pipe diameter, the slope of the sewer (column 14) is equal to the steeper of the ground slope and the slope that yields the minimum permissible velocity (0.60 m/s) at the minimum design flow. The relationship between the pipe slope and the flow velocity is given by Equations 3.285 to 3.289. Once the slope is determined (using the minimum flowrate), the flow depth and velocity at the maximum flowrate (column 10) are calculated. If the flow depth under maximum-flow conditions exceeds the acceptable limit ($0.5D$ or $0.75D$), or the maximum-flow velocity exceeds an acceptable limit (3.5 m/s), then the pipe

diameter is increased to the next commercial size (e.g., see Table D.1), and the slope (column 14) is recalculated. This iteration may need to be repeated until the minimum permissible velocity, acceptable flow depth, and maximum-velocity criteria are all met. After this iteration, the diameter of the sewer pipe is listed in column 15, the minimum velocity in column 16, the maximum velocity in column 17, and the maximum-flow depth in column 18. In cul-de-sacs and other dead-end street sections, it is often impossible to satisfy the minimum-velocity requirements. Such pipes will usually require periodic flushing to scour accumulated sediments. The minimum practicable slope for construction is 0.08%, and therefore pipe slopes less than this value should not be specified (Metcalf & Eddy, Inc., 1981).

10. A drop in the sewer invert at the manhole upstream of a sewer pipe is necessary when either the sewer line changes direction or the diameter of the sewer is increased (over the upstream pipe). If a change in direction occurs, it is advisable to drop the pipe invert by 30 mm to compensate for the energy losses. If the diameter of the sewer pipe leaving a manhole is larger than the diameter of the sewer pipe entering the manhole, head losses are accounted for either by matching the crown elevations of the entering and leaving pipes or by aligning points that are 80% of the diameter from the pipe inverts. It is usually preferable to align the crowns of the pipes because the associated drop in the invert will always exceed 30 mm, losses associated with a change in direction can be neglected, and aligning the crowns of the pipes ensures that the smaller sewer will not flow full as a result of the backwater from the larger pipe unless the larger pipe also flows full. The drop in the sewer invert resulting from either a change in pipe direction or a change in pipe diameter is listed in column 19.

11. The fall in the sewer line is equal to the product of the slope (column 14) and the length of the sewer (column 5) and is listed in column 20.

12. The invert elevations of the upper and lower ends of the sewer line are listed in columns 21 and 22. The difference in these elevations is equal to the fall in the sewer calculated in column 20, and the invert elevation at the upper end of the sewer differs from the invert elevation at the lower end of the upstream pipe by the manhole invert drop listed in column 19.

13. Repeat steps 3 to 12 for all connected pipes, proceeding downstream until the sewer main is reached. Then repeat steps 3 to 12 for the sewer main until the connection to the next lateral is reached.

14. Repeat steps 3 to 13, beginning with the outermost pipe in the next sewer lateral. Continue designing the sewer main until the outlet point of the sewer system is reached.

The design procedure described here is usually automated to some degree and is most easily implemented using a spreadsheet program. The design procedure is illustrated by the following example.

EXAMPLE 3.32

A sewer system is to be designed to service the residential area shown in Figure 3.56. The average per-capita wastewater flowrate is estimated to be 800 L/d/person, and the infiltration and inflow (I/I) is estimated to be 70 m^3/d/km. The sewer system is to join an existing main sewer at manhole (MH) 5, where the average wastewater flow is 0.37 m³/s, representing the contribution of approximately 100,000 people. The I/I contribution to the flow in the main sewer at MH 5 is negligible. The existing sewer main at MH 5 is 1065 mm in diameter, has an invert elevation of 55.35 m, and is laid on a slope of 0.9%. The layout of the proposed sewer system shown in Figure 3.56 is based on the topography of the area. Pipe lengths, contributing areas, and ground-surface elevations are given in the following table:

Line no. (1)	Location (2)	Manhole no. From (3)	Manhole no. To (4)	Length (m) (5)	Contributing area (ha) (6)	Ground surface elevation Upper end (m) (7)	Ground surface elevation Lower end (m) (8)
0	Main Street	—	5	—	—	—	60.04
1	A Street	1	2	53	0.47	65.00	63.80
2	A Street	2	3	91	0.50	63.80	62.40
3	A Street	3	5	100	0.44	62.40	60.04
4	A Street	4	5	89	0.90	61.88	60.04
5	Main Street	5	12	69	0.17	60.04	60.04
6	B Street	6	8	58	0.43	65.08	63.20
7	P Avenue	7	8	50	0.48	63.60	63.20
8	B Street	8	10	91	0.39	63.20	62.04
9	Q Avenue	9	10	56	0.88	62.72	62.04
10	B Street	10	12	97	0.45	62.04	60.04
11	B Street	11	12	125	0.90	61.88	60.04
12	Main Street	12	19	75	0.28	60.04	60.20
13	C Street	13	15	57	0.60	64.40	62.84
14	P Avenue	14	15	53	0.76	63.24	62.84
15	C Street	15	17	97	0.51	62.84	61.60
16	Q Avenue	16	17	63	0.94	62.12	61.60
17	C Street	17	19	100	0.46	61.60	60.20
18	C Street	18	19	138	1.41	61.92	60.20
19	Main Street	19	26	78	0.30	60.20	60.08

Design the sewer system between A Street and C Street for a saturation density of 130 persons/ha. Local municipal guidelines require that the sewer pipes have a minimum cover of 2 m, a minimum slope of 0.08%, a peak-flow factor of 3.0, a minimum-flow factor of 0.5, and a minimum allowable pipe diameter of 150 mm.

Solution   From the given data, the average wastewater flow is 800 L/d/person × 130 persons/ha = 104000 L/d/ha = 0.0722 m³/min/ha. The infiltration and inflow (I/I) is 70 m³/d/km = 4.86 × 10⁻⁵ m³/min/m.

The results of the design computations are shown in Table 3.57. The computations begin with Line 0, which is the existing sewer main that must be extended to accommodate the sewer lines in the proposed residential development. The average flow in the sewer main is 0.37 m³/s = 22.2 m³/min, hence the maximum flow is 3.0 × 22.2 = 66.6 m³/min (column 10) and the minimum flow is 0.5 × 22.2 = 11.1 m³/min (column 13). With a slope of 0.009 (column 14) and a diameter of 1065 mm (column 15), the velocity at the minimum flowrate is calculated using Equations 3.285 to 3.289 as 1.75 m/s (column 16). The velocity at the maximum flowrate is 2.88 m/s (column 17) with a maximum depth of flow, calculated using Equation 3.284 as 476 mm,

FIGURE 3.56: Residential sewer project

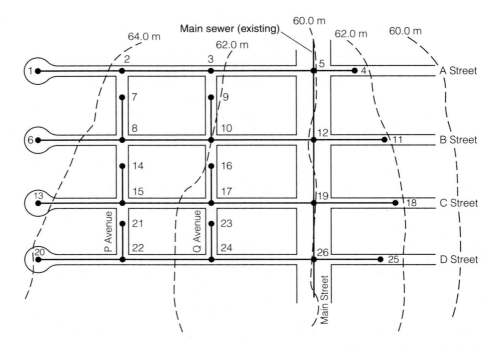

which is 45% of the pipe diameter. The invert elevation of the main sewer at MH 5 is 55.35 m (column 22) and the ground-surface elevation at MH 5 is 60.04 m (column 24).

The design of the sewer system begins with Line 1 on A Street, which goes from MH 1 to MH 2 and is 53 m long. The area contributing wastewater flow is 0.47 ha (column 7), hence the average wastewater flow in Line 1 is 0.0722 m³/min/ha × 0.47 ha = 0.0339 m³/min. The I/I contribution is 4.86 × 10⁻⁵ m³/min/m × 53 m = 0.0026 m³/min (columns 8 and 11), the peak wastewater flow is 3.0 × 0.0339 = 0.102 m³/min (column 9), giving a total peak flow of 0.102 + 0.0026 = 0.105 m³/min (column 10). Similarly, the minimum wastewater flow is 0.5 × 0.0339 = 0.0170 m³/min (column 12), and the minimum total flow is 0.0170 + 0.0026 = 0.0196 m³/min (column 13). Using the minimum allowable pipe diameter of 150 mm and a slope of 0.047 (column 14) yields a velocity of 0.60 m/s (column 16), which is the minimum permissible velocity. At the peak flow of 0.105 m³/s, the velocity is 0.99 m/s (column 17) and the depth of flow is 23 mm (column 18). The peak velocity is less than the maximum allowable velocity of 3.5 m/s, and the depth of flow is (much) less than the maximum desirable depth of flow of 75 mm (half full), and so the sewer line is acceptable. With a slope of 0.047 and a length of 53 m, the fall in the sewer is 0.047 × 53 m = 2.49 m (column 20). The ground elevation at the upstream end of the sewer is 65.00 m (column 23). With a cover of 2.11 m (slightly above the minimum cover of 2 m), the invert elevation of the upstream end of the pipe is taken as 65.00 − 2.11 − 0.15 = 62.74 m (column 21), and therefore the invert elevation of the downstream end of the pipe is equal to the upstream invert minus the fall in the sewer, which is 62.74 − 2.49 = 60.25 m (column 22).

The design of Lines 2 and 3 follows the same sequence as for Line 1, with the exception that the wastewater flows in each pipe are derived from the sum of the contributing areas of all upstream pipes plus the pipe being designed and that the

Line no.	Location	Manhole no.		Length (m)	Area		Maximum flow			Minimum flow			Slope of sewer	Diam (mm)	Min velocity (m/s)	Max velocity (m/s)	Max depth (mm)	Invert drop (m)	Fall in sewer (m)	Sewer invert elevation		Ground surface elevation	
		From	To		Increment (ha)	Total (ha)	I/I (m^3/min)	Sewage (m^3/min)	Total (m^3/min)	I/I (m^3/min)	Sewage (m^3/min)	Total (m^3/min)								Upper end (m)	Lower end (m)	Upper end (m)	Lower end (m)
(1)	(2)	(3)	(4)	(5)	(6)	(7)	(8)	(9)	(10)	(11)	(12)	(13)	(14)	(15)	(16)	(17)	(18)	(19)	(20)	(21)	(22)	(23)	(24)
0	Main Street	–	5	–	–	–	–	–	66.6	–	–	11.1	0.009	1065	1.75	2.88	476	–	–	–	55.35	–	60.04
1	A Street	1	2	53	0.47	0.47	0.0026	0.102	0.105	0.0026	0.0170	0.0196	0.047	150	0.60	0.99	23	–	2.49	62.74	60.25	65.00	63.80
2	A Street	2	3	91	0.50	0.97	0.0070	0.210	0.217	0.0070	0.0350	0.0420	0.024	150	0.60	0.97	40	–	2.18	60.25	58.07	63.80	62.40
3	A Street	3	5	100	0.44	1.41	0.0120	0.305	0.317	0.0120	0.0509	0.0629	0.018	150	0.61	0.97	52	–	1.80	58.07	56.27	62.40	60.04
4	A Street	4	5	89	0.90	0.90	0.0043	0.195	0.199	0.0043	0.0325	0.0368	0.027	150	0.60	0.98	37	–	2.40	58.67	56.27	61.88	60.04
5	Main Street	5	12	69	0.17	309.96	0.0197	67.14	67.16	0.0197	11.19	11.21	0.001	1220	0.78	1.24	879	0.155	0.07	55.20	55.13	60.04	60.04
6	B Street	6	8	58	0.43	0.43	0.0028	0.0932	0.0960	0.0028	0.0155	0.0183	0.050	150	0.60	0.99	22	–	2.90	62.90	60.00	65.08	63.20
7	P Avenue	7	8	50	0.48	0.48	0.0024	0.104	0.106	0.0024	0.0173	0.0197	0.048	150	0.60	1.00	23	–	2.40	61.34	58.99	63.60	63.20
8	B Street	8	10	91	0.39	1.30	0.0097	0.282	0.292	0.0097	0.0469	0.0566	0.019	150	0.60	0.97	49	–	1.73	58.99	57.26	63.20	62.04
9	Q Avenue	9	10	56	0.88	0.88	0.0027	0.191	0.194	0.0027	0.0318	0.0345	0.029	150	0.60	1.00	36	–	1.62	60.44	58.82	62.72	62.04
10	B Street	10	12	97	0.45	2.67	0.0171	0.578	0.595	0.0171	0.0964	0.114	0.011	205	0.61	0.95	86	0.055	1.07	57.21	56.14	62.04	60.04
11	B Street	11	12	125	0.90	0.90	0.0061	0.195	0.201	0.0061	0.0325	0.0386	0.026	150	0.60	0.97	37	–	3.25	59.45	56.20	61.88	60.04
12	Main Street	12	19	75	0.28	313.81	0.0465	67.97	68.02	0.0465	11.33	11.38	0.001	1220	0.79	1.24	887	–	0.08	55.13	55.06	60.04	60.20
13	C Street	13	15	57	0.60	0.60	0.0028	0.130	0.133	0.0028	0.0217	0.0245	0.040	150	0.60	1.00	27	–	2.28	62.20	59.92	64.40	62.84
14	P Avenue	14	15	53	0.76	0.76	0.0026	0.165	0.168	0.0026	0.0274	0.0300	0.034	150	0.61	1.02	32	–	1.80	60.38	58.58	63.24	62.84
15	C Street	15	17	97	0.51	1.87	0.0101	0.405	0.415	0.0101	0.0675	0.0776	0.015	150	0.61	0.98	63	–	1.46	58.58	57.12	62.84	61.60
16	Q Avenue	16	17	63	0.94	0.94	0.0031	0.204	0.207	0.0031	0.0339	0.0370	0.028	150	0.60	1.01	37	–	1.76	59.90	58.21	62.12	61.60
17	C Street	17	19	100	0.48	3.27	0.0180	0.708	0.726	0.0180	0.1180	0.1360	0.010	205	0.60	0.96	83	0.055	1.00	57.07	56.07	61.60	60.20
18	C Street	18	19	138	1.41	1.41	0.0067	0.305	0.312	0.0067	0.0509	0.0576	0.019	150	0.60	0.99	51	–	2.62	58.75	56.13	61.92	60.20
19	Main Street	19	26	78	0.30	318.79	0.0750	69.05	69.13	0.0750	11.51	11.59	0.001	1220	0.79	1.25	900	–	0.08	55.06	54.98	60.20	60.08

FIGURE 3.57: Sewer design calculations

I/I flow in each pipe is the sum of all upstream I/I flows plus the I/I contribution to the pipe being designed. Using this approach, the invert elevation at the end of Line 3 is 56.27 m (column 22), where the sewer lateral joins the main sewer. The crown elevation of the 150 mm lateral (Line 3) is $56.27 + 0.15 = 56.42$ m, which matches the crown elevation of the sewer main of $55.35 + 1.065 = 56.42$ m. Line 4 is designed from the other side of Main Street and joins the main sewer at MH 5 with an invert elevation of 56.27 m, which also aligns the crown of the sewer with that of the main.

The main sewer leaving MH 5 (Line 5) is designed next. The tributary area to Line 5 is the sum of the contributing areas of all contributing sewer laterals (Lines 1 to 4, $1.41 + 0.9 = 2.31$ ha) plus the equivalent area of the average flow in the main sewer upstream of MH 5 (0.37 m³/s ÷ 0.0722 m³/min/ha = 307.48 ha) plus the area that contributes directly to Line 5 (0.17 ha). Hence the total contributing area is $2.31 + 307.48 + 0.17 = 309.96$ ha (column 7). The I/I contribution to Line 5 is the sum of the I/I contributions to all upstream laterals ($0.012 + 0.0043 = 0.0163$ m³/min) plus the I/I contribution directly to Line 5 (69 m × 4.86 × 10^{-5} m³/min/m = 0.0034 m³/min) for a total I/I contribution of $0.0163 + 0.0034 = 0.0197$ m³/min (column 8). The peak flow (67.16 m³/min) and the minimum flow (11.21 m³/min) are calculated using the flow factors as previously described. Using a pipe diameter of 1065 mm (the diameter of the upstream sewer section) on a slope that yields the minimum permissible velocity (0.60 m/s) is not acceptable since the capacity of the sewer pipe will be exceeded under peak-flow conditions. Using the next larger commercial size of 1220 mm (column 15) on a slope of 0.001 (column 14) yields flow conditions where the pipe flows slightly less than three-fourths full (879 mm), and the minimum and maximum velocity criteria are met. Using the smallest slope (0.001) that meets the depth-of-flow and velocity criteria is a good choice since the ground surface is flat, minimizing excavation. Aligning the crowns of the incoming and outgoing sewer pipes at MH 5 requires a drop of $1.220 - 1.065 = 0.155$ m (column 19), hence the sewer invert leaving MH 5 is at elevation $55.35 - 0.155 = 55.20$ m (column 21). The drop in the sewer line is $0.001 × 69$ m = 0.074 m (column 20), and the invert elevation at the end of Line 5 is $55.20 - 0.07 = 55.13$ m.

The design of the other sewer laterals and the main sewer line follows the same procedure. In cases where laterals intersect with crown elevations more than 0.6 m apart (vertically), a drop manhole must be used (see Figure 3.53b). Drop manholes are required at MH 8, MH 10, MH 15, and MH 17. All laterals intersecting the main sewer have crown elevations that match those of the main sewer. All pipes in the sewer system are made of concrete; the 150-mm and 205-mm pipes are nonreinforced, and the sewer main consists entirely of reinforced concrete pipe (RCP).

Sanitary sewers are designed to function properly for the duration of the design period; however, special consideration should also be given to conditions that may develop at minimum flow during the first few years of sewer operation. In these early years, frequent cleanout could be necessary and maintenance costs excessive. In some cases it may be more economical to use smaller pipes in some parts of the sewer network, and plan for pipe replacement within the design period.

3.6.7 Force Mains

Where pumping of sewage is required, such as in flat terrains, *force mains* (pressure conduits) must be designed to carry the flows. In these cases, the costs of pumping and associated equipment are important considerations. Force mains must be able to transport sewage at velocities sufficient to avoid deposition and yet not so high as to create pipe erosion problems. Force mains are generally 200 mm (8 in.) in diameter or greater, but for small pumping stations smaller pipes may sometimes be acceptable. Velocities encountered in force-main operations are in the range of 0.6 to 3 m/s (USEPA, 2000e). In designing force mains, high points in lines should be avoided if possible, since this eliminates the need for air-relief valves. Good design practice dictates that the hydraulic gradient should lie above the force main at all points during periods of minimum-flow pumping.

3.7 Computer Models

Several good computer models are available for simulating flow in open channels, and many such models are parts of more comprehensive models that simulate surface runoff from rainfall (see Chapter 5). In engineering practice, the use of computer models to apply the fundamental principles covered in this chapter is sometimes essential. There are usually a variety of models to choose from for a particular application, but in doing work to be reviewed by regulatory agencies, models developed and maintained by agencies of the U.S. government have the greatest credibility and, perhaps more important, are almost universally acceptable in supporting permit applications and defending design protocols. A secondary guideline in choosing a model is that the simplest model that will accomplish the design objectives should be given the most serious consideration. Several of the more widely used models developed and endorsed by agencies of the U.S. government are described briefly here.

HEC-RAS. The HEC-RAS (River Analysis System) code was developed and is maintained by the U.S. Army Corps of Engineers Hydrologic Engineering Center (HEC) (U.S. Army Corps of Engineers, 2002). This is a widely used open-channel hydraulics program that simulates flow in single reaches or complex networks of natural or man-made channels. The model can perform hydraulic computations through bridges, culverts, side-flow weirs, floodplain encroachments, and split flows. HEC-RAS can simulate both steady and unsteady flows. Steady-flow water-surface profile computations use the standard-step method to solve the energy equation, while unsteady-flow simulations are based on an implicit finite-difference solution of one-dimensional differential forms of the continuity and momentum equations, called the St. Venant equations. The methods for modeling the effects of hydraulic structures are essentially the same with either steady- or unsteady-flow simulations. The user interacts with the model through a graphical user interface (GUI) for file management, data entry and editing, hydraulic analyses, graphical displays of input and output, and report preparation. The input and output can be viewed and printed graphically, and the computed results can be exported to GIS and CADD files.

WSP2. This code was developed and is maintained by National Resources Conservation Service (NRCS), an agency within the U.S. Department of Agriculture. WSP2 calculates the water-surface profile in one-dimensional, gradually varied steady flow. Calculations are made using the standard-step method, and this code has the capability to handle flow through hydraulic structures, including flow through bridges and culverts.

WSPRO. This code was developed and is maintained by the U.S. Geological Survey and the Federal Highway Administration (USFHWA, 1998). WSPRO is easy to use for highway design, floodplain mapping, flood-insurance studies, and stage-discharge relations. WSPRO can analyze all types of flows through bridges, culverts, roadway crossings with multiple openings, and embankment overflows.

HY8. This code was developed and is maintained by the U.S. Federal Highway Administration. HY8 is a widely used program for the design and analysis of culvert systems, and has several options to account for different culvert shapes, sizes, materials, inlet, and outlet conditions.

Summary

The fundamental equations governing flow in open channels are the continuity, momentum, and energy equations; these are presented in forms that are most applicable to steady open-channel flow. In cases of uniform flow, the momentum equation can be approximated by either the Darcy–Weisbach or Manning equation, although the Darcy–Weisbach equation is generally preferable. In cases of nonuniform flow, the energy equation is used to calculate the water-surface profile using either the direct-integration, direct-step, or standard-step method. The concept of specific energy is used to define critical, subcritical, and supercritical flow conditions, with transitions from supercritical to subcritical flow accomplished via a hydraulic jump. Hydraulic structures commonly used in engineering practice include weirs and Parshall flumes for measuring flowrates; spillways, gates, and stilling basins for regulating flows; and culverts for passing water under roadways. In weirs and Parshall flumes, the flowrate and depth have a unique relationship, allowing the flowrate passing the structure to be estimated from upstream and downstream depth measurements. Flows over spillways are similar to flows over weirs. Stilling basins are designed to localize hydraulic jumps and are sized based on the Froude number and velocity of the incoming flow. The flowrate through gates is expressed in terms of the water depths upstream and downstream of the gate. Culverts are widely used as drainage structures under roadways, and the flowrate through a culvert can depend on either the headwater depth or both the headwater and tailwater depths. Open channels can be categorized as either unlined or lined. For each type of channel, design considerations include selecting the shape of the channel section that minimizes excavation, ensuring a minimum permissible velocity to prevent deposition of suspended material, selecting a stable side slope, and allowing for adequate freeboard. The primary constraint in designing both unlined and lined channels is ensuring that the shear stress exerted by flowing water on the channel perimeter is less than the allowable shear stress, and a special consideration in grass-lined channels is accounting for the dependence of the Manning roughness coefficient on the flow velocity and hydraulic radius of the channel. The principles of open-channel flow and associated design guidelines are particularly useful in sizing major stormwater-drainage channels. Sanitary sewers convey wastewater to treatment plants or disposal locations and, since sewer pipes are designed to flow partially full, the principles of open-channel flow are applicable. Design criteria for sanitary sewers generally incorporate regulations specifying maximum and minimum allowable velocities and maximum allowable depths of flow.

Problems

3.1. An open channel has a trapezoidal cross section with a bottom-width of 5 m and side slopes of 2:1 (H:V). If the depth of flow is 2 m and the average velocity in the channel is 1 m/s, calculate the discharge in the channel.

3.2. Water flows at 8 m^3/s through a rectangular channel 4 m wide and 3 m deep. If the flow velocity is 1 m/s, calculate the depth of flow in the channel. If this channel expands (downstream) to a width of 5 m and the depth decreases by 0.5 m from the upstream depth, then what is the flow velocity in the expanded section?

3.3. Show that for circular pipes of diameter D the hydraulic radius, R, is related to the pipe diameter by $4R = D$.

3.4. A trapezoidal channel is to be excavated at a site where permit restrictions require that the channel have a bottom-width of 5 m, side slopes of 1.5:1 (H:V), and a depth of flow of 1.8 m. If the soil material erodes when the shear stress on the perimeter of the channel exceeds 3.5 Pa, determine the appropriate slope and flow capacity of the channel. Use the Darcy–Weisbach equation and assume that the excavated channel has an equivalent sand roughness of 3 mm.

3.5. Water flows in a 8-m wide rectangular channel that has a longitudinal slope of 0.0001. The channel has an equivalent sand roughness of 2 mm. Calculate the uniform flow depth in the channel when the flowrate is 15 m^3/s. Use the Darcy–Weisbach equation.

3.6. Derive Equation 3.45 from Equations 3.38, 3.39, and 3.43.

3.7. Given that hydraulically rough flow conditions occur in open channels when $u_* k_s / \nu \geq 100$, show that this condition can be expressed in terms of Manning parameters as

$$n^6 \sqrt{R S_0} \geq 9.6 \times 10^{-14}$$

If a concrete-lined rectangular channel with a bottom width of 5 m is constructed on a slope of 0.05%, determine the minimum flow depth for hydraulically rough flow conditions to exist.

3.8. Water flows at a depth of 2.20 m in a trapezoidal, concrete-lined section ($k_s = 2$ mm) with a bottom-width of 3.6 m and side slopes of 2:1 (H:V). The slope of the channel is 0.0006 and the water temperature is 20°C. Assuming uniform-flow conditions, estimate the average velocity and flowrate in the channel. Use both the Darcy–Weisbach and Manning equations and compare your answers.

3.9. Water flows in a trapezoidal channel that has a bottom width of 5 m, side slopes of 2:1 (H:V), and a longitudinal slope of 0.0001. The channel has an equivalent sand roughness of 1 mm. Calculate the uniform flow depth in the channel when the flowrate is 18 m^3/s. Is the flow hydraulically rough, smooth, or in transition? Would the Manning equation be valid in this case?

3.10. Show that the Manning n, can be expressed in terms of the Darcy friction factor, f, by the following relation:

$$n = \frac{f^{\frac{1}{2}} R^{\frac{1}{6}}}{8.86}$$

where R is the hydraulic radius of the flow. Does this relationship conclusively show that n is a function of the flow depth?

3.11. Water flows at 20 m^3/s in a trapezoidal channel that has a bottom width of 2.8 m, side slope of 2:1 (H:V), longitudinal slope of 0.01, and a Manning n of 0.015. (a) Use the Manning equation to find the normal depth of flow, and (b) determine the equivalent sand roughness of the channel. Assume the flow is fully turbulent.

3.12. It has been shown that in fully turbulent flow the Manning n can be related to the height, d, of the roughness projections by the relation

$$n = 0.039d^{\frac{1}{6}}$$

If the estimated roughness height, d, in a channel is 30 mm, then determine the percentage error in n resulting from a 70% error in estimating d.

3.13. Show that the minimum value of $n/k_s^{1/6}$ given by Equation 3.50 is

$$\frac{n}{k_s^{1/6}} = 0.039$$

Determine the range of R/k_s in which $n/k_s^{1/6}$ does not deviate by more than 5% from the minimum value. [*Hint*: You may need to use the relation $\log x = 0.4343 \ln x$.]

3.14. Show that the turbulence condition $u_* k_s/\nu > 70$ can be put in the form

$$k_s\sqrt{RS_0} > 2.2 \times 10^{-5}$$

A trapezoidal concrete channel with a bottom-width of 3 m and side slopes of $2:1$ (H:V) is estimated to have an equivalent sand roughness of 3 mm, and is laid on a slope of 0.1%. Determine the minimum flow depth for fully turbulent conditions to exist. Can the Manning equation be used at this flow depth? If turbulent-flow conditions do not exist at a given flow depth, give the equation that should be used to relate the flowrate to the depth of flow.

3.15. Water flows at a depth of 4.00 m in a trapezoidal, concrete-lined section with a bottom-width of 4 m and side slopes of $3:1$ (H:V). The slope of the channel is 0.0001 and the water temperature is $20°$ C. Assess the validity of using the Manning equation, assuming $n = 0.013$.

3.16. A trapezoidal irrigation channel is to be excavated to supply water to a farm. The design flowrate is 1.8 m^3/s, the side slopes are $2:1$ (H:V), the longitudinal slope of the channel is 0.1%, Manning's n is 0.025, and the geometry of the channel is to be such that the length of each channel side is equal to the bottom width. (a) Specify the dimensions of the channel required to accommodate the design flow under normal conditions; and (b) if the channel lining can resist an average shear stress of up to 4 Pa, under what flow conditions is the channel lining stable?

3.17. Derive the equation for equivalent Manning roughness in a compound channel proposed by Horton (1933) and Einstein (1934).

3.18. Derive the equation for equivalent Manning roughness in a compound channel proposed by Lotter (1933).

3.19. The sections in the floodplain shown in Figure 3.5 have the following values of the Manning n:

Section	n
1	0.040
2	0.030
3	0.015
4	0.013
5	0.017
6	0.035
7	0.060

If the depth of flow in the floodplain is 5 m, use the formulae in Table 3.2 to estimate the composite roughness.

FIGURE 3.58: Flow in an open channel

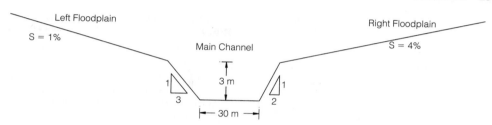

3.20. Consider the drainage channel and adjacent floodplains shown in Figure 3.58. The Manning roughness coefficients are given by:

Section	n
Left floodplain	0.040
Main channel	0.016
Right floodplain	0.050

and the longitudinal slope is 0.5%. Field tests have shown that the Horton (1933) equation best describes the composite roughness in the channel. Find the capacity of the main channel and the lateral extent of the floodplains for a 100-year flow of 1590 m³/s.

3.21. Use Equation 3.53 to show that the velocity in an open channel is equal to the depth-averaged velocity at a distance of 0.368d from the bottom of an open channel, where d is the depth of flow.

3.22. Use Equation 3.53 to show that the depth-averaged velocity in an open channel of depth d can be estimated by averaging the velocities at 0.2d and 0.8d from the bottom of the channel.

3.23. Water flows at 8.4 m³/s in a trapezoidal channel with a bottom-width of 2 m and side slopes of 2:1 (H:V). Over a distance of 100 m, the bottom-width expands to 2.5 m, with the side slopes remaining constant at 2:1. If the depth of flow at both of these sections is 1 m and the channel slope is 0.001, calculate the head loss between the sections. What is the power in kilowatts that is dissipated?

3.24. Use the Darcy–Weisbach equation to show that the head loss per unit length, S, between any two sections in an open channel can be estimated by the relation

$$S = \frac{\overline{f}}{4\overline{R}} \frac{\overline{V}^2}{2g}$$

where $\overline{f}$, $\overline{R}$, and $\overline{V}$ are the average friction factor, hydraulic radius, and flow velocity, respectively, between the upstream and downstream sections.

3.25. Determine the critical depth for 30 m³/s flowing in a rectangular channel with width 5 m. If the depth of flow is equal to 3 m, is the flow supercritical or subcritical?

3.26. Determine the critical depth for 50 m³/s flowing in a trapezoidal channel with bottom-width 4 m and side slopes of 1.5:1 (H:V). If the depth of flow is 3 m, calculate the Froude number and state whether the flow is subcritical or supercritical.

3.27. A rectangular channel 2 m wide carries 3 m³/s of water at a depth of 1.2 m. If an obstruction 40 cm wide is placed in the middle of this channel, find the elevation of the water surface at the constriction. What is the minimum width of the constriction that will not cause a rise in the water surface upstream?

3.28. Water flows at 1 m³/s in a rectangular channel of width 1 m and depth 1 m. What is the maximum contraction of the channel that will not choke the flow?

3.29. A lined rectangular concrete drainage channel is 10.0 m wide and carries a flow of 8 m³/s. In order to pass the flow under a roadway, the channel is contracted to a width of 6 m. Under design conditions, the depth of flow just upstream of the contraction is 1.00 m, and the contraction takes place over a distance of 7 m. (a) If the energy loss in the contraction is equal to $V_1^2/2g$, where V_1 is the average velocity upstream of the contraction, what is the depth of flow in the constriction? (b) Does consideration of energy losses have a significant effect on the depth of flow in the constriction? (c) If the width of the constriction is reduced to 4.50 m and a flow of 8 m³/s is maintained, determine the depth of flow within the constriction (include energy losses). (d) If reducing the width of the constriction to 4.50 m influences the upstream depth, determine the new upstream depth.

3.30. A rectangular channel 3 m wide carries 4 m³/s of water at a depth of 1.5 m. If an obstruction 15 cm high is placed across the channel, calculate the elevation of the water surface over the obstruction. What is the maximum height of the obstruction that will not cause a rise in the water surface upstream?

3.31. Show that the critical step height required to choke the flow in a rectangular open channel is given by

$$\frac{\Delta z_c}{y_1} = 1 + \frac{\text{Fr}_1^2}{2} - \frac{3}{2}\text{Fr}_1^{2/3}$$

where z_c is the critical step height, y_1 is the flow depth upstream of the step, and Fr_1 is the Froude number upstream of the step. Use this equation to verify your answer to Problem 3.30.

3.32. Water flows at 4.3 m³/s in a rectangular channel of width 3 m and depth of flow 1 m. If the channel width is decreased by 0.75 m and the bottom of the channel is raised by 0.25 m, what is the depth of flow in the constriction?

3.33. Water flows at 18 m³/s in a trapezoidal channel with a bottom-width of 5 m and side slopes of 2:1 (H:V). The depth of flow in the channel is 2 m. If a bridge pier of width 50 cm is placed in the middle of the channel, what is the depth of flow adjacent to the pier? What is the maximum width of a pier that will not cause a rise in the water surface upstream of the pier?

3.34. Water flows at 15 m³/s in a trapezoidal channel with a bottom-width of 4.5 m and side slopes of 1.5:1 (H:V). The depth of flow in the channel is 1.9 m. If a step of height 15 cm is placed in the channel, what is the depth of flow over the step? What is the maximum height of the step that will not cause a rise in the water surface upstream of the step?

3.35. Water flows at 20 m³/s with a uniform depth of 3 m in a trapezoidal channel of base width 3 m and side slopes 1:1. If a channel transition restricts the flow locally by raising the side walls to the vertical position, calculate the depth of water in the rectangular constriction. What is the minimum allowable width of the constriction to prevent choking?

3.36. Water flows at 36 m³/s in a rectangular channel of width 10 m and a Manning n of 0.030. If the depth of flow at a channel section is 3 m and the slope of the channel is 0.001, classify the water-surface profile. What is the slope of the water surface at the observed section? Would the shape of the water-surface profile be much different if the depth of flow were equal to 2 m?

3.37. Water flows at 30 m³/s in a rectangular channel of width 8 m. The Manning n of the channel is 0.035. Determine the range of channel slopes that would be classified as steep and the range that would be classified as mild.

3.38. Water flows in a trapezoidal channel where the bottom-width is 6 m and side slopes are 2:1 (H:V). The channel lining has an estimated Manning n of 0.045, and the slope of the channel is 1.5%. When the flowrate is 80 m³/s, the depth of flow at a gaging station is 5 m. Classify the water-surface profile, state whether the depth increases or decreases in the downstream direction, and calculate the slope of the water surface at the gaging station. On the basis of this water-surface slope, estimate the depth of flow 100 m downstream and upstream of the gaging station.

3.39. Derive an expression relating the conjugate depths in a hydraulic jump where the slope of the channel is equal to S_0. [*Hint*: Assume that the length of the jump is equal to $5y_2$ and that the shape of the jump between the upstream and downstream depths can be approximated by a trapezoid.]

3.40. If 100 m³/s of water flows in a channel 8 m wide at a depth of 0.9 m, calculate the downstream depth required to form a hydraulic jump and the fraction of the initial energy lost in the jump.

3.41. The head loss, h_L, across a hydraulic jump is described by the equation:

$$y_1 + \frac{V_1^2}{2g} = y_2 + \frac{V_2^2}{2g} + h_L$$

where the subscripts 1 and 2 refer to the upstream and downstream locations, respectively. Show that the dimensionless head loss, h_L/y_1, is given by

$$\frac{h_L}{y_1} = 1 - \frac{y_2}{y_1} + \frac{\mathrm{Fr}_1^2}{2}\left[1 - \left(\frac{y_1}{y_2}\right)^2\right]$$

where Fr_1 is the upstream Froude number.

3.42. Show that the hydraulic jump equation for a trapezoidal channel is given by

$$\frac{by_1^2}{2} + \frac{my_1^3}{3} + \frac{Q^2}{gy_1(b + my_1)} = \frac{by_2^2}{2} + \frac{my_2^3}{3} + \frac{Q^2}{gy_2(b + my_2)}$$

where b is the bottom-width of the channel, m is the side slope of the channel, y_1 and y_2 are the conjugate depths, and Q is the volumetric flowrate.

3.43. Water flows in a horizontal trapezoidal channel at 21 m³/s, where the bottom-width of the channel is 2 m, side slopes are 1:1, and depth of flow is 1 m. Calculate the downstream depth required for a hydraulic jump to form at this location. What would be the energy loss in the jump?

3.44. A flume with a triangular cross section and side slopes of 2:1 (H:V) contains water flowing at 0.30 m³/s at a depth of 15 cm. Verify that the flow is supercritical and calculate the conjugate depth.

3.45. Water flows at 10 m³/s in a rectangular channel of width 5.5 m. The slope of the channel is 0.15% and the Manning roughness coefficient is 0.038. Estimate the depth 100 m upstream of a section where the flow depth is 2.2 m using: (a) the direct-integration method, and (b) the standard-step method. Approximately how far upstream of this section would you expect to find uniform flow?

3.46. Water flows at 5 m³/s in a 4-m wide rectangular channel that is laid on a slope of 4%. If the channel has a Manning n of 0.05 and the depth at a given section is 1.5 m, how far upstream or downstream is the depth equal to 1 m?

3.47. If the depth of flow in the channel described in Problem 3.11 is measured as 1.4 m, find the location where the depth is 1.6 m.

3.48. A trapezoidal canal has a longitudinal slope of 1%, side slopes of 3:1 (H:V), a bottom width of 3.00 m, a Manning n of 0.015, and carries a flow of 20 m³/s. The depth of flow at a gaging station is observed to be 1.00 m. Answer the following questions:

(a) What is the normal depth of flow in the channel?

(b) What is the critical depth of flow in the channel?

(c) Classify the slope of the channel and the water surface profile at the gaging station.

(d) How far from the gaging station is the depth of flow equal to 1.1 m? Does this depth occur upstream or downstream of the gaging station?

(e) If the bottom of the channel just downstream of the gaging station is raised by 0.20 m, determine the resulting depth of flow at the downstream section. The bottom width of the channel remains constant at 3 m.

3.49. A rectangular channel 6 m wide carries a discharge of 0.8 m³/s. At a certain section the channel roughness changes from rough to smooth. The normal depths in the rough and smooth reaches are 0.9 and 0.7 m, respectively. The channel slope is 0.5%. Using a single step of the direct-step method, estimate the length of the reach of nonuniform flow.

3.50. Water flows at 11 m³/s in a rectangular channel of width 5 m. The slope of the channel is 0.1% and the Manning roughness coefficient is equal to 0.035. If the depth of flow at a selected section is 2 m, then calculate the upstream depths at 20-m intervals along the channel until the depth of flow is within 5% of the uniform depth.

3.51. Water flows in an open channel whose slope is 0.05%. The Manning roughness coefficient of the channel lining is estimated to be 0.040 when the flowrate is 250 m³/s. At a given section of the channel, the cross section is trapezoidal with a bottom-width of 12 m, side slopes of 2:1 (H:V), and a depth of flow of 8 m. Use the standard-step method to calculate the depth of flow 100 m upstream from this section where the cross section is trapezoidal with a bottom-width of 16 m and side slopes of 3:1 (H:V).

3.52. Show that the energy equation for open channel flow between stations B (upstream) and A (downstream) can be written in the form

$$\left[Z + \alpha \frac{V^2}{2g} \right]_A^B = \overline{S}_f \Delta L$$

where Z is the water surface elevation, α is the energy coefficient, V is the average velocity, $\overline{S}_f$ is the average friction slope between stations A and B, and ΔL is the distance between stations A and B.

A backwater curve is being computed in the stream between two sections A and B, 140 m apart. The hydraulic properties of the cross-sections are as follows:

| Water surface | Area (m²) | | Wetted perimeter (m) | |
elevation (m)	Section A	Section B	Section A	Section B
518.5	—	118.45	—	36.27
518.2	181.86	108.42	52.21	35.05
517.9	166.81	98.66	50.84	33.83
517.6	152.13	86.96	48.77	32.77
517.2	137.82	78.18	47.40	31.70
516.9	123.88	—	46.02	—

For a flow in the channel of 280 m³/s and a Manning coefficient of 0.040, the water elevation at Station A is 517.4 m. Compute the surface elevation at Station B.

Just upstream of Station B, the flow is partially obstructed by a large bridge pier in the channel, presenting an obstruction 2.50 m wide normal to the direction of flow. The channel at this location can be considered roughly rectangular in cross-section, with the bottom at Station B at elevation 515.10 m. Compute the water surface elevation adjacent to the pier.

3.53. A trapezoidal drainage channel of bottom-width 5 m, side slopes 2:1 (H:V), Manning's n of 0.018, and longitudinal slope of 0.1% terminates at a gate where the relationship between the flow through the gate, Q, headwater elevation (HW), tailwater elevation (TW), and gate opening (h) is given by

$$Q = 13.3h\sqrt{HW - TW}$$

where Q is in m³/s, and HW, TW, and h are in meters. If the elevation of the bottom of the channel at the gate location is 0.00 m, the tailwater elevation is 1.00 m and the flow in the channel is 20 m³/s, estimate the minimum gate opening such that the water surface elevation 100 m upstream of the gate does not exceed 2.20 m. [Note: The depth of flow 100 m upstream of the gate is *not* 2.20 m.]

3.54. Water flows at 10 m³/s in a 5-m wide channel. What is the height of a suppressed rectangular (sharp-crested) weir that will cause the depth of flow in the channel to be 2 m?

3.55. How does a sill differ from a weir? Show that Equation 3.162 gives a reasonable approximation to the discharge coefficient for both weirs and sills.

3.56. A sharp-crested rectangular weir is to be installed to measure flows that are expected to be on the order of 1 to 3 m³/s. If the depth of the channel is such that the maximum depth of water that can be accommodated at the weir is 3.0 m and the width of the channel is 5 m, determine the height of a suppressed weir that can be used to measure the flowrate.

3.57. The stage in a river is to be maintained at an elevation of 20.00 m behind a sharp-crested rectangular weir. If the river is 10 m wide and 1 m deep and the flowrate in the river is 5 m³/s, what is the crest elevation of the suppressed weir that must be used?

3.58. A contracted sharp-crested rectangular weir is to be used to measure the flowrate in an open channel that is 5 m wide and 2 m deep. It is desired that the depth of water over the crest of the weir be no less than 0.5 m when the flowrate is 1 m³/s and that the water level be at least 30 cm below the top of the channel. What should the crest length of the weir be, and how far below the top of the channel should the crest of the weir be located? Sketch the weir.

3.59. Water flows at a rate of 5 m³/s in a rectangular channel of width 10 m. The normal depth of flow in the channel is 5 m. If a sharp-crested (suppressed) weir of height 5 m is installed across the channel, determine the new flow depth in the channel for the same flowrate. What would the flow depth be if the rectangular weir were contracted (on both sides) to a width of 7 m? What would the flow depth be if a 7-m wide Cipolletti weir were used?

3.60. A Cipolletti weir is to be designed to measure the flow in a channel. Under design conditions, the normal depth of flow is 2 m and the flow rate is 5 m³/s. Design the weir with a top width of 10 m.

3.61. A 0.8-m high suppressed rectangular weir is placed in a 1-m wide rectangular channel. If the head over the weir is observed to be 20 cm, estimate the flowrate in the channel. If the downstream water depth rises to 5 cm above the crest of the weir and the upstream water depth remains at 20 cm above the crest, estimate the new flowrate in the channel.

3.62. A suppressed rectangular weir is being used to measure flows in an irrigation canal. The weir is 5 m wide and 2 m high and, under flood conditions, the upstream and downstream depths are measured as 2.5 m and 2.3 m, respectively. Use all applicable formulae to estimate the range of possible flows over the weir under these conditions. Assess the reliability of your flow estimate. How could a more precise estimate of the flowrate be obtained?

3.63. A small irrigation channel has a trapezoidal cross section with a slope of 0.1%, bottom-width of 2 m, side slopes of 2 : 1 (H : V), and a total channel depth of 2 m. The maximum flow expected in the channel is 3 m^3/s, and the maximum flow depth at a flow-measurement structure is not to exceed 1.5 m. (a) Design an unsuppressed rectangular weir of width 2 m that can be placed in the channel to measure the flow. (b) If the elevation of the water surface downstream of the weir is 20 cm below the water surface upstream of the weir, by how much does this affect the free-flow weir capacity (in m^3/s)? (c) What range of elevation differences between the headwater and tailwater will limit the reduction in weir capacity to less than 10%?

3.64. Equations 3.168 and 3.169 have been proposed for estimating the flowrates over submerged rectangular weirs. Compare the predicted discharges given by these formulae and assess whether either one is preferable.

3.65. A V-notch weir is to be used to measure a maximum flow of 12 L/s. If a notch angle of 30° is selected, what should the depth of the notch be?

3.66. If a V-notch weir has a notch angle of 50° and a height of 50 cm, estimate the capacity of the weir.

3.67. A 90° V-notch weir is to be used to measure flows as low as 1 L/min. Estimate the depth of flow over the weir, accounting for viscous and surface-tension effects if necessary.

3.68. A 70° V-notch weir is observed to have a 40-mm head. Estimate the flowrate over the weir.

3.69. A V-notch weir is to be designed to measure the flowrate in an irrigation channel. Under design conditions, the flowrate in the channel is 0.30 m^3/s and the depth of flow is 1 m. Design the weir with a top width of 1 m.

3.70. Determine the crest length of a compound weir that has a notch angle of 90°, a notch height of 20 cm, and a discharge of 1.5 m^3/s when the water depth is 1.20 m. The apex of the notch is placed at ground level.

3.71. A 25-cm high broad-crested weir is placed in a 1.5-m wide channel. If the maximum depth of water that can be measured upstream of the weir is equal to 75 cm, what is the maximum flowrate that can be measured by the weir?

3.72. A broad-crested rectangular weir of length 1 m, width 1 m, and height 30 cm is being considered to measure the flow in a canal. For what range of flows would this weir length be adequate?

3.73. A broad-crested weir is to be used to measure the flow in an irrigation channel. The design section upstream of the weir is rectangular with a width of 3 m, and the depth of flow is 4 m at a flowrate of 5 m^3/s. Design the height and length of the weir.

3.74. A broad-crested weir is to be constructed to measure the flowrate in a 3-m wide channel. If the flowrate in the channel is expected to be in the range of 0.8–1.5 m^3/s and the depth of the channel is 2.5 m, select a height and length of weir such that the water level in the channel does not rise any higher than 2 m above the bottom of the channel.

3.75. Flow is being measured by a Parshall flume that has a throat width of 0.91 m. Determine the flowrate through the flume when the water depth in the converging section is 45.7 cm and the depth in the throat section is 32.0 cm.

3.76. A Parshall flume has a throat width of 6.1 m, and the water depth in the converging section is 1.22 m and in the throat section is 1.10 m. Estimate the discharge. What would the discharge through the flume be if the water depth in the throat section were equal to 61.0 cm?

3.77. A Parshall flume is to be designed to measure the discharge in a channel. If the design flowrate is 1 m^3/s, determine the width of the flume to be used such that the depth of flow immediately upstream of the throat under design conditions is 1 m. By what percentage is the capacity of the flume reduced if the downstream depth is 0.85 m?

3.78. Show that Equation 3.199 can be derived from Equation 3.198. Explain your approach. [*Hint*: $dy/dx = \alpha$ at $x = X_{DT}$.]

3.79. Design a 40-m long overflow spillway that will discharge a design flood of 1600 m^3/s at a maximum allowable pool elevation of 200 m. The bottom elevation behind the spillway is 170 m, the upstream face of the spillway is has a 1:1 slope, and the spillway chute is to have a slope of 1:1.5 (H:V).

3.80. An overflow spillway is to be designed to discharge 500 m^3/s when the maximum pool elevation is at 200 m. The bottom elevation of the reservoir behind the dam is 150 m, and the spillway is to have an upstream face with a 1:1 slope. To accommodate three Tainter gates, the spillway will need two 1-m wide round-nosed piers between two square abutments. Determine the required spillway lengths for crest elevations of 195, 190, 185, and 180 m. Determine the minimum spillway length and corresponding crest elevation.

3.81. A spillway is 10 m wide and discharges 120 m^3/s under design conditions. The reservoir level behind the spillway is at elevation 100.00 m, and the water-surface elevation of the receiving water downstream of the stilling basin is 80.00 m. Assuming a 15% energy loss in the flow down the spillway, find the invert elevation of the floor of the stilling basin and design the stilling basin.

3.82. A 10-m wide emergency spillway with well-rounded abutments has a crest height 5.00 m above the bottom of a storage reservoir in which the maximum-pool elevation is 6.00 m. The emergency spillway terminates with a stilling basing in a 10-m wide rectangular concrete channel that has an invert elevation equal to the elevation of the bottom of the storage reservoir. If energy losses down the spillway are neglected, what type of stilling basin should be used?

3.83. A vertical gate is installed at the end of a canal that is 7 m wide; the depth of flow is 3.5 m. If the gate has a width of 3 m, estimate the flow through the gate when the gate is raised 0.5 m. Neglect the effect of the downstream water depth.

3.84. Explain why the contraction coefficient of a vertical gate with a rounded edge is greater than that of a vertical gate with a sharp edge.

3.85. Derive Equation 3.215.

3.86. If a vertical gate has an opening of 70 cm and discharges 16 m^3/s into a 5-m wide channel, estimate the maximum depth of water downstream of the gate for which the discharge under the gate will not be "drowned." [*Hint*: As long as a hydraulic jump can form, the discharge will not be drowned.]

3.87. Water flows under a vertical gate and into a 3-m wide rectangular channel where the flowrate and depth of flow are 4 m^3/s and 1.1 m, respectively. Estimate the minimum gate opening to prevent the formation of a hydraulic jump. What energy loss (in kilowatts) can be expected in a hydraulic jump?

3.88. Water is released from under a vertical gate into a 2-m wide lined rectangular channel. The gate opening is 50 cm, and the flowrate into the channel is 10 m^3/s. The channel is lined with reinforced concrete and has a Manning roughness coefficient of 0.015, a horizontal slope, and merges with a river where the depth of flow is 3.5 m. Does a hydraulic jump occur in the lined channel? If so, how far downstream of the gate does the jump occur, and what could be done to ensure that a hydraulic jump does not occur?

3.89. A vertical sluice gate is opened 20 cm and the depth of water behind the gate is 1.7 m. If the width of the channel is 1.5 m, estimate the free-flow discharge through the gate and the distinguishing condition for the tailwater depth.

3.90. Derive Equation 3.219.

3.91. Water is ponded behind a vertical gate to a height of 5 m in a rectangular channel of width 8 m. If the tailwater depth is 4.0 m and the gate opening is 1.40 m, what is the discharge through the gate?

3.92. Water is ponded behind a vertical gate to a height of 5 m in a rectangular channel of width 8 m. Calculate the gate opening that will release 50 m^3/s through the gate. How would this discharge be affected by a downstream flow depth of 4 m?

3.93. The contraction coefficient for a radial gate is given by

$$C_c = 1 - 0.75\left(\frac{\theta}{90}\right) + 0.36\left(\frac{\theta}{90}\right)^2$$

Find the value of θ that gives the minimum value of C_c.

3.94. A 500-mm diameter concrete drainage culvert ($n = 0.013$) is to be placed under a roadway. During the design storm, it is expected that water will pond behind the culvert to a height 20 cm above the top of the culvert. If the culvert entrance is to be well rounded ($k_e = 0.05$, $C_d = 0.95$), the slope of the culvert is 2%, the length of the culvert is 20 m, and the exit is not submerged, estimate the discharge through the culvert.

3.95. A culvert is to be designed to pass water under a roadway into a recreational lake, where the design elevation of the lake surface is 10.00 m and the bottom elevation of the lake is 9.50 m. The culvert is to be 380-mm diameter corrugated metal pipe (CMP) with 75-mm × 25-mm corrugations, 8 m long, and have a horizontal slope and a mitered entrance. Calculate the culvert performance curve from the no-flow condition up to a headwater elevation of 12 m. If the culvert is required to pass 0.30 m^3/s under design conditions, what is the minimum safe elevation of the roadway? If the lake elevation drops to 9.75 m, determine the percentage change in the culvert capacity.

3.96. What is the capacity of a 1.5-m by 1.5-m concrete box culvert ($n = 0.013$) with a rounded entrance ($k_e = 0.05$, $C_d = 0.95$) if the culvert slope is 0.7%, the length is 40 m, and the headwater level is 2 m above the culvert invert? Consider the following cases: (a) free-outlet conditions, and (b) tailwater elevation 0.5 m above the top of the box at the outlet. What must the headwater elevation be in case (b) for the culvert to pass the flow that exists in case (a)?

3.97. A concrete pipe culvert ($n = 0.013$) is to be sized to pass 0.80 m^3/s when the headwater depth is 1 m. Site conditions require that the longitudinal slope of the culvert be 2% and the length of the culvert be 10 m. If the culvert entrance conditions are such that $k_e = 0.5$, $C_d \approx 1$, and the discharge is free, determine the required diameter of the culvert.

3.98. A circular concrete culvert is designed to pass water under a roadway. Under design conditions, the headwater depth is 2 m, the tailwater depth is 1 m, the flow through the culvert is 1 m^3/s, the length of the culvert is 15 m, and the slope is 1.5%. Determine the required diameter of the culvert.

3.99. A 2-m × 2-m concrete box culvert is to handle a design flow of 4 m^3/s. The culvert is to be laid on a slope of 0.1, be 25 m long, and have a low tailwater elevation. Assuming an entrance loss coefficient of 0.1, determine the headwater depth under design conditions.

3.100. A drainage canal has a trapezoidal cross section with a bottom-width of 5 m, side slopes of 2:1 (H:V), a longitudinal slope of 0.5%, and a Manning roughness coefficient of 0.022. A 10-m long, 2-m × 2-m box culvert is to be placed in the canal to permit drainage under a roadway that crosses the canal. When the flow through the culvert is 10 m^3/s, what is the depth of flow at the entrance and exit of the culvert? Does the culvert have an adequate capacity to pass the flow without the entrance being submerged? [*Hint:* (1) Assume that the culvert tailwater depth is equal to the normal depth of flow in the downstream channel; and (2) calculate the water-surface profile in the culvert to determine the depth of flow at the culvert entrance.]

3.101. Is it possible to have a condition in which a culvert entrance is not submerged while its exit is submerged? If so, sketch the flow regime within the culvert and describe how you would calculate the capacity of the culvert.

3.102. A 450-mm × 450-mm concrete box culvert is being proposed to provide drainage across a rural driveway. The culvert is to be placed in a vertical headwall with 45° wingwall flares, 4 m long, and have a slope of 3%. Use the U.S. Federal Highway Administration (USFHWA) equations (Equations 3.231 and 3.238) to calculate the culvert performance curve for flowrates up to 0.6 m^3/s. State your assumptions in using these equations, and confirm your results for a flowrate of 0.6 m^3/s using the USFHWA nomograph shown in Figure 3.59.

3.103. A culvert under a roadway is to be designed for a 100-year peak surface runoff of 3.00 m^3/s. The invert elevation at the culvert inlet is 12.11 m, the invert elevation at the outlet is 11.71 m, and the length of the culvert is to be 20 m. The channel downstream of the culvert has a trapezoidal cross section with a bottom width of 1 m, side slopes of 2:1 (H:V), a longitudinal slope of 2%, and a Manning's n of 0.040. The gravel roadway crossing the culvert has a length of 15.0 m, an elevation of 13.50 m, and a width of 16.2 m. Considering a circular reinforced concrete pipe (RCP) culvert with a diameter of 760 mm and an entrance that is mitered to conform to the fill slope, determine the depth of water flowing over the roadway, the flow over the roadway, and the flow through the culvert.

3.104. Show that the best hydraulic section for a rectangular-shaped section is one in which the bottom-width is equal to twice the flow depth.

3.105. Show that the best hydraulic section for a triangular-shaped section is one in which the top-width is equal to twice the flow depth.

3.106. How do the side slopes in the best trapezoidal section compare with the side slopes in the best triangular section?

3.107. Water flows in a concrete-lined trapezoidal channel on a slope of 0.5%. The channel has a bottom-width of 10 m, side slopes of 2:1 (H:V), and a flowrate of 150 m^3/s. If the temperature of the water is 20°C: (a) Calculate the critical depth. (b) Calculate the normal depth. (c) Do you expect the velocity in the channel to prevent sedimentation and the growth of vegetation? (d) What freeboard would be appropriate for this channel? (e) If the actual depth of flow is 3 m, classify the water-surface profile.

3.108. If the best trapezoidal section is to be excavated in loose sandy earth, what side slope would you use? How would this compare with a trapezoidal channel excavated in stiff clay?

3.109. A trapezoidal channel has been designed with a bottom-width of 10 m, side slopes of 2:1 (H:V), and a longitudinal slope of 0.053%. The Manning n of the channel material is estimated to be 0.030. If the design flowrate is 28 m^3/s and the radius of curvature of the channel is 100 m, determine the design depth of flow in the channel and the minimum freeboard that must be provided. Is this channel design acceptable from the viewpoint of minimum permissible velocity?

FIGURE 3.59: USFHWA culvert chart.

Source: USFHWA (1985).

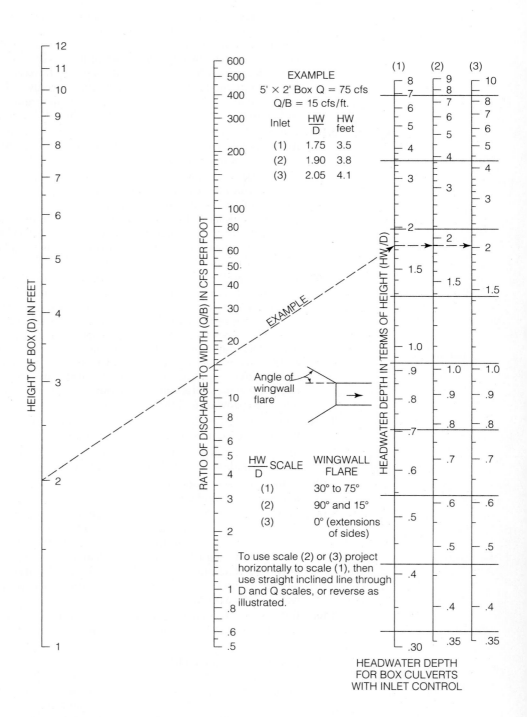

RATIO OF DISCHARGE TO WIDTH (Q/B) IN CFS PER FOOT

HEIGHT OF BOX (D) IN FEET

HEADWATER DEPTH IN TERMS OF HEIGHT (HW/D)

EXAMPLE

5' × 2' Box Q = 75 cfs

Q/B = 15 cfs/ft.

Inlet	$\frac{HW}{D}$	HW feet
(1)	1.75	3.5
(2)	1.90	3.8
(3)	2.05	4.1

Angle of wingwall flare

$\frac{HW}{D}$ SCALE	WINGWALL FLARE
(1)	30° to 75°
(2)	90° and 15°
(3)	0° (extensions of sides)

To use scale (2) or (3) project horizontally to scale (1), then use straight inclined line through D and Q scales, or reverse as illustrated.

HEADWATER DEPTH
FOR BOX CULVERTS
WITH INLET CONTROL

3.110. Design a trapezoidal channel to carry 30 m³/s through a slightly sinuous channel on a slope of 0.002. The channel is to be excavated in coarse alluvium with a 75-percentile diameter of 2.5 cm, with the particles on the perimeter of the channel moderately rounded.

3.111. Design a trapezoidal channel to carry 10 m³/s through a straight channel on a slope of 0.2%. The channel will be excavated in coarse alluvium with a 75-percentile diameter of 1 cm, with the particles on the channel perimeter moderately rounded.

3.112. A straight trapezoidal channel is to be excavated on a slope of 0.1% and have a design discharge of 27 m³/s. Right-of-way constraints require the top-width of the channel to be 17 m, and the channel is expected to consist of clean gravel (moderately angular) with a 75-percentile diameter of 3 cm. If freeboard is taken to be zero, determine the required bottom-width and side slopes of the channel.

3.113. The theoretical tractive-force ratio, K, in trapezoidal channels is given by Equation 3.269, and the bottom and side shear stresses are given by Equations 3.261 and 3.262. Use these relations to identify the side slopes under which the critical shear stress will be on the side of an unlined (trapezoidal) channel. If the side slopes of a particular channel are 25°, and the angle of repose is 30°, will the critical shear stress on the sides be reached before the critical shear stress on the bottom is exceeded?

3.114. Design a (straight) lined trapezoidal channel to carry 30 m³/s on a longitudinal slope of 0.002. The lining of the channel is to be float-finished concrete.

3.115. A lined rectangular channel is to be constructed on a slope of 0.042% to handle a design flowrate of 4.5 m³/s. The lining of the channel is to be unfinished concrete, and the maximum radius of curvature is 150 m. Determine the depth and width of the channel to be excavated.

3.116. A lined (straight) triangular channel is to be constructed on a slope of 0.032% to handle a design flowrate of 4.4 m³/s. The lining of the channel is to be smooth asphalt. Determine the dimensions of the channel to be excavated.

3.117. A parabolic-shaped channel is to be excavated and lined with mortar. If the channel is to carry 6 m³/s on a slope of 0.05%, determine the depth of channel to be excavated. Give the equation of the channel side.

3.118. Derive Equation 3.269 from Equations 3.265 and 3.268.

3.119. Consider a flexible-boundary channel in which the lining material has a permissible average shear stress of τ_p. Use the Manning equation to show that the permissible velocity, v_p, in the channel is given by

$$v_p = \frac{0.010}{n}\left(\frac{R^4}{y^3}\right)^{1/6} \tau_p^{1/2}$$

where R is the hydraulic radius and y is the depth of flow. Explain why it is better to design a channel based on maximum permissible shear stress rather than maximum permissible velocity.

3.120. A roadside channel is to be lined with Bermuda grass and regularly mowed to a height of 4 cm. The channel is to be designed for a flow of 1.1 m³/s and a longitudinal slope of 1.4%, and it is to have a triangular cross section with side slopes of 3:1 (H:V). Determine the depth of flow in the channel, and assess the adequacy of the channel lining.

3.121. Design a triangular grass-lined channel to handle an intermittent flow of 5 m³/s. The channel is to have side slopes of 4:1 (H:V), be excavated in an erosion-resistant soil, on a longitudinal slope of 0.0025, and lined with Bermuda grass. The retardance can vary between E and B.

3.122. Design a triangular grass-lined channel to carry 2 m³/s on a slope of 1%. The retardance can vary between A and C, and the grass is similar to Kentucky bluegrass.

3.123. A trapezoidal drainage channel on a golf course is lined with Kentucky bluegrass and is expected to convey 10 m³/s under heavy-rainfall conditions. The slope of the drainage channel is 0.1%. If the channel has a bottom-width of 3 m, side slopes of 2:1 (H:V), and a total depth of 2 m, assess the adequacy of the existing channel.

3.124. A median channel is to have a low-flow concrete bottom and grass sides. The bottom-width of the channel is 1.2 m, the longitudinal slope of the channel is 1.5%, and the sides of the channel have a slope of 4:1 (H:V) and are lined with Class B vegetation. For a design flow of 0.7 m³/s, assess the adequacy of the channel lining.

3.125. Show that the area, A, and wetted perimeter, P, of a channel with a triangular bottom section are given by

$$A = \frac{T_b}{2}\left(\frac{T_b}{2m_1}\right) + \left[T_b + \left(y - \frac{T_b}{2m_1}\right)m_2\right]\left(y - \frac{T_b}{2m_1}\right) \tag{3.292}$$

$$P = 2\sqrt{1 + m_1^2}\left(\frac{T_b}{2m_1}\right) + 2\sqrt{1 + m_2^2}\left(y - \frac{T_b}{2m_1}\right) \tag{3.293}$$

where T_b is the top-width of the bottom section, m_1 is the side slope of the bottom section, y is the depth of flow, and m_2 is the side slope of the channel.

3.126. Estimate the maximum and minimum wastewater flowrates from a 65-ha residential development that will consist of 10% large lots, 75% small single-family lots, and 15% small two-family lots. Assume an average per-capita flowrate of 350 L/d/person.

3.127. The master plan of a 45-km² city calls for land usages that are 65% residential, 25% commercial, and 10% industrial. The residential development is to be 15% large lots, 75% small single-family lots, and 10% multistory apartments. The average domestic wastewater flowrate is 500 L/d/person, the average commercial flowrate is 50,000 L/d/ha, and the average industrial flowrate is 90,000 L/d/ha. Infiltration and inflow is 1500 L/d/ha for the entire area. Estimate the peak and minimum flows to be handled by the trunk sewer.

3.128. Show that the Manning equation can be written in the form given by Equation 3.288.

3.129. Water flows at a rate of 7 m³/s in a circular concrete sewer of diameter 1600 mm. If the slope of the sewer is 0.01, calculate the depth of flow and velocity in the sewer. What diameter of pipe would be required for the pipe to flow three-quarters full?

3.130. You are designing the sanitary sewer lines for a 10-ha subdivision in Berkeley, California. The subdivision is to contain small single-family houses, and approximately 1 km of sewer line. Estimate the maximum flowrate to be handled by the sewer system. What diameter and pipe material would you use if the sewers are to be laid on an average slope of 3%?

3.131. Water flows at 3 m³/s in a circular concrete sewer laid on a slope of 0.005. If a velocity of 2 m/s is desired in the sewer, calculate the diameter of sewer pipe that should be used. What would the depth of flow be in the sewer pipe?

3.132. Water flows at 3.5 m³/s in a 1400-mm diameter concrete sewer ($n = 0.015$). Find the slope at which the sewer flows half full.

3.133. Show that the maximum flowrate in a circular pipe occurs when $h/D \approx 0.94$, where h is the depth of flow and D is the diameter of the pipe. Assume that the Manning n is constant with depth.

3.134. Show that the flow velocity in a circular pipe is the same whether the pipe is flowing half full or completely full.

3.135. A minimum wastewater flow of 0.15 m³/s is to be transported in a 915-mm diameter concrete pipe. Find the slope of the pipe that will give a minimum velocity of 0.6 m/s. If the maximum flowrate is 0.29 m³/s, assess the adequacy of the pipe in terms of maximum flow depth and maximum allowable velocity. Assuming the pipe is to carry an average flow of 0.25 m³/s, assess the potential for hydrogen sulfide generation if the temperature of the wastewater is 25°C and the five-day BOD is 250 mg/L at 20°C.

3.136. A 1220-mm diameter concrete sewer is to carry 0.50 m³/s of domestic wastewater at a temperature of 23°C on a slope of 0.009. Estimate the maximum five-day BOD in order that hydrogen sulfide generation not be a problem.

3.137. A sanitary sewer is to transport wastewater with a five-day BOD of 300 mg/L at 20°C. The flowrate of the wastewater is 0.01 m³/s, the diameter of the pipe is 205 mm, and the slope is 0.1%. Assess the potential for hydrogen sulfide generation.

3.138. A previously designed sewer line is made of 535-mm diameter reinforced concrete pipe on a slope of 0.001 and enters a manhole with a crown elevation of 2.50 m. The (average) design flowrate of the sewer line is 0.130 m³/s. It is expected that an additional flow of 0.02 m³/s will enter at the manhole, and the downstream sewer line leaving the manhole will carry an average flow of 0.15 m³/s. The downstream sewer line is to be designed for a peak-flow factor of 2.0, a minimum-flow factor of 0.5, and a maximum allowable flow depth equal to 75% of the pipe diameter. If the ground-surface elevation upstream of the new sewer line is 5.00 m, the ground-surface elevation at the manhole downstream of the new sewer line is 4.50 m, and the new sewer line is 150 m long, determine: (a) the diameter and slope of the new sewer line, and (b) the crown and invert elevations at the upstream and downstream ends of the new sewer line.

3.139. A sewer system is to be designed to service the area shown in Figure 3.56. The average per-capita wastewater flowrate is estimated to be 1000 L/d/person, and the infiltration and inflow (I/I) is estimated to be 100 m³/d/km. The sewer system is to join an existing main sewer at manhole (MH) 5, where the average wastewater flow is 0.40 m³/s, representing the contribution of approximately 120,000 people. The I/I contribution to the flow in the main sewer at MH 5 is negligible, and the main sewer at MH 5 is 1220 mm in diameter, has an invert elevation of 55.00 m, and is laid on a slope of 0.8%. The layout of the sewer system shown in Figure 3.56 is based on the topography of the area, and the pipe lengths, contributing areas, and ground-surface elevations are shown in the following table:

Line no. (1)	Location (2)	Manhole no. From (3)	Manhole no. To (4)	Length (m) (5)	Contributing area (ha) (6)	Ground-surface elevation Upper end (m) (7)	Ground-surface elevation Lower end (m) (8)
0	Main Street	—	5	—	—	—	60.04
1	A Street	1	2	55	0.47	65.00	63.80
2	A Street	2	3	90	0.50	63.80	62.40
3	A Street	3	5	100	0.44	62.40	60.04
4	A Street	4	5	90	0.90	61.88	60.04
5	Main Street	5	12	70	0.17	60.04	60.04

Line no. (1)	Location (2)	Manhole no.		Length (m) (5)	Contributing area (ha) (6)	Ground-surface elevation	
		From (3)	To (4)			Upper end (m) (7)	Lower end (m) (8)
6	B Street	6	8	60	0.43	65.08	63.20
7	P Avenue	7	8	50	0.48	63.60	63.20
8	B Street	8	10	90	0.39	63.20	62.04
9	Q Avenue	9	10	55	0.88	62.72	62.04
10	B Street	10	12	95	0.45	62.04	60.04
11	B Street	11	12	125	0.90	61.88	60.04
12	Main Street	12	19	75	0.28	60.04	60.20
13	C Street	13	15	55	0.60	64.40	62.84
14	P Avenue	14	15	55	0.76	63.24	62.84
15	C Street	15	17	95	0.51	62.84	61.60
16	Q Avenue	16	17	65	0.94	62.12	61.60
17	C Street	17	19	100	0.46	61.60	60.20
18	C Street	18	19	140	1.41	61.92	60.20
19	Main Street	19	26	80	0.30	60.20	60.08

Design the sewer system between A Street and C Street for a saturation density of 150 persons/ha. Municipal guidelines require that the sewer pipes have a minimum cover of 2 m, a minimum slope of 0.08%, a peak flow factor of 2.5, a minimum flow factor of 0.7, and a minimum allowable pipe diameter of 150 mm.

CHAPTER 4

Probability and Statistics in Water-Resources Engineering

4.1 Introduction

All hydrologic phenomena obey the laws of nature and can, in theory, be described by the solution of the relevant fundamental equations. In most cases, however, the uncertainty and amount of the detail that needs to be known about the physical system, combined with the limitations of analytical and numerical techniques for solving systems of equations, preclude an exact description of most natural phenomena. Processes that can be specified with certainty are called *deterministic*, while those that cannot are called *stochastic*. The two types of stochastic processes commonly encountered in engineering practice are: (1) processes where the outcome is uncertain because of the lack of information on the process or parameters affecting the process, called *epistemic uncertainty*, and (2) processes where the outcome is uncertain because the process generating the outcome is fundamentally random, called *natural uncertainty*. In the first case, epistemic uncertainty can be reduced, or even eliminated, by additional measurements of the variables affecting the process or by the use of more accurate models. In the second case, natural randomness is intrinsic to the process and uncertainty in the outcome might not be reduced by additional measurements. In many cases, recorded data includes both epistemic and natural uncertainty, and separation of these components is a challenging task (Merz and Thieken, 2005). Whether uncertainty is epistemic, natural, or a combination of both, an uncertain outcome is generally defined by a probability distribution that describes the relative likelihood of all possible outcomes. In practice, uncertain variables are collectively referred to as *random variables*.

A common application of probability theory in water-resources engineering involves the assignment of an exceedance probability, P_e, to hydrologic events. A system designed to handle a particular hydrologic event can be expected to fail with a probability equal to the probability of exceedance, P_e, of the design event. This probability of system failure is called the *risk* of failure, and the probability that the system will not fail, $1 - P_e$, is called the *reliability* of the system. Exceedance probabilities of hydrologic events are usually stated in terms of the probability that a given event is exceeded in any given year. The average number of years between exceedances is called the *return period*, T, which is related to the exceedance probability by

$$T = \frac{1}{P_e} \qquad (4.1)$$

The return period, T, is sometimes called the *recurrence interval* and is usually a criterion for selecting design events in water-resource systems. In cases where water-resource systems are to be in place for only a short period (less than a year), then it might be appropriate to calculate return periods of design events based on their occurrence within a specific season of the year (McCuen and Beighley, 2003).

4.2 Probability Distributions

A *probability function* defines the relationship between the outcome of a random process and the probability of occurrence of that outcome. If the sample space, S, contains discrete elements, then the sample space is called a *discrete sample space*; if S is a continuum, the sample space is called a *continuous sample space*. The sample space of a random variable is commonly denoted by an upper-case letter (e.g., X), and the corresponding lower-case letter denotes an element of the sample space (e.g., x). *Discrete probability distributions* describe the probability of outcomes in discrete sample spaces, while *continuous probability distributions* describe the probability of outcomes in continuous sample spaces.

4.2.1 Discrete Probability Distributions

If X is the sample space of a discrete random variable, with outcomes $x_1, x_2, \ldots, x_N$, then the probability of occurrence of each element, x_n, can be written as a *probability function*, $f(x_n)$, sometimes referred to as the *probability distribution function*. If $f(x_n)$ is a valid probability function, then it must necessarily have the following properties:

$$f(x_n) \geq 0, \qquad \forall\, x_n \tag{4.2}$$

$$\sum_{n=1}^{N} f(x_n) = 1 \tag{4.3}$$

where N is the number of elements in X. In many engineering applications, the quantity of interest is the probability that a random process will generate an outcome that is greater than or equal to some outcome x_n, where x_n is a design variable and exceedance of x_n will result in system failure.

Based on the definition of the probability distribution function, the probability that an outcome is less than or equal to x_n is given by

$$P(x_i \leq x_n) = \sum_{x_i \leq x_n} f(x_i) \tag{4.4}$$

and the probability that the outcome is greater than x_n is given by

$$P(x_i > x_n) = 1 - P(x_i \leq x_n) \tag{4.5}$$

The function $P(x_i \leq x_n)$ is called the *cumulative distribution function* and is commonly written as $F(x_n)$, where

$$F(x_n) = P(x_i \leq x_n) \tag{4.6}$$

These definitions of the probability distribution function, $f(x_n)$, and cumulative distribution function, $F(x_n)$, can be expanded to describe several random variables simultaneously. As an illustration, consider the case of two discrete random variables, X and Y, with elements x_p and y_q, where $p \in [1, N]$ and $q \in [1, M]$. The *joint probability distribution function* of X and Y, $f(x_p, y_q)$, is the probability of x_p and y_q occurring simultaneously. Such distributions are called *bivariate probability distributions*. If more than two variables are involved, the joint probability distribution is referred to as a *multivariate probability distribution*. Based on the definition of the bivariate probability distribution, $f(x_p, y_q)$, the following conditions must necessarily be satisfied

$$f(x_p, y_q) \geq 0, \qquad \forall x_p, y_q \tag{4.7}$$

$$\sum_{p=1}^{N} \sum_{q=1}^{M} f(x_p, y_q) = 1 \tag{4.8}$$

As in the case of a single random variable, the cumulative distribution function, $F(x_p, y_q)$, is defined as the probability that the outcomes are simultaneously less than or equal to x_p and y_q, respectively. Hence $F(x_p, y_q)$ is related to the joint probability distribution as follows:

$$F(x_p, y_q) = \sum_{x_i \leq x_p} \sum_{y_j \leq y_q} f(x_i, y_j) \tag{4.9}$$

In multivariate analysis, the probability distribution of any single random variate is called the *marginal probability distribution function*. In the case of a bivariate probability function, $f(x_p, y_q)$, the marginal probability of x_p, $g(x_p)$, is given by the relation

$$g(x_p) = \sum_{q=1}^{M} f(x_p, y_q) \tag{4.10}$$

Clearly, several probability functions can be derived from the basic function describing the probability of occurrence of discrete events. The particular function of interest in any given case (e.g., cumulative or joint) depends on the question being asked.

4.2.2 Continuous Probability Distributions

If X is a random variable with a continuous sample space, then there are an infinite number of possible outcomes and the probability of occurrence of any single value of X is actually zero. This problem is addressed by defining the probability of an outcome being in the range $[x, x + \Delta x]$ as $f(x)\Delta x$, where $f(x)$ is the *probability density function*. Based on this definition, any valid probability density function must satisfy the following conditions:

$$f(x) \geq 0, \qquad \forall x \tag{4.11}$$

$$\int_{-\infty}^{+\infty} f(x')\, dx' = 1 \tag{4.12}$$

The cumulative distribution function, $F(x)$, describes the probability that the outcome of a random process will be less than or equal to x and is related to the probability density function by the equation

$$F(x) = \int_{-\infty}^{x} f(x')\, dx' \tag{4.13}$$

which can also be written as

$$f(x) = \frac{dF(x)}{dx} \tag{4.14}$$

In describing the probability distribution of more than one random variable, the *joint probability density function* is used. In the case of two variables, X and Y, the probability that x will be in the range $[x, x + \Delta x]$ and y will be in the range $[y, y + \Delta y]$ is approximated by $f(x, y) \Delta x \Delta y$, where $f(x, y)$ is the joint probability density function of x and y. As in the case of discrete random variables, the bivariate cumulative distribution function, $F(x, y)$, and marginal probability distributions, $g(x)$ and $h(y)$, are defined as

$$F(x, y) = \int_{-\infty}^{x} \int_{-\infty}^{y} f(x', y') \, dx' \, dy' \tag{4.15}$$

$$g(x) = \int_{-\infty}^{\infty} f(x, y') \, dy' \tag{4.16}$$

$$h(y) = \int_{-\infty}^{\infty} f(x', y) \, dx' \tag{4.17}$$

4.2.3   Mathematical Expectation and Moments

Assuming that x_i is an outcome of the discrete random variable X, $f(x_i)$ is the probability distribution function of X, and $g(x_i)$ is an arbitrary function of x_i, then the expected value of the function g, represented by $\langle g \rangle$, is defined by the equation

$$\langle g \rangle = \sum_{i=1}^{N} g(x_i) f(x_i) \tag{4.18}$$

where N is the number of possible outcomes in the sample space, X. The expected value of a random function is the *arithmetic average* of the function calculated from an infinite number of random outcomes, and the analogous result for a function of a continuous random variable is given by

$$\langle g \rangle = \int_{-\infty}^{+\infty} g(x') f(x') \, dx' \tag{4.19}$$

where $g(x)$ is a continuous random function and $f(x)$ is the probability density function of x.

Several random functions are particularly useful in characterizing the distribution of random variables. The first is simply

$$g(x) = x \tag{4.20}$$

In this case $\langle g \rangle = \langle x \rangle$ corresponds to the arithmetic average of the outcomes over an infinite number of realizations. The quantity $\langle x \rangle$ is called the *mean* of the random variable and is usually denoted by μ_x. According to Equations 4.18 and 4.19, μ_x is defined for both discrete and continuous random variables by

$$\mu_x = \begin{cases} \sum_{i=1}^{N} x_i f(x_i) & \text{(discrete)} \\[2ex] \int_{-\infty}^{+\infty} x' f(x') \, dx' & \text{(continuous)} \end{cases} \tag{4.21}$$

A second random function that is frequently used is

$$g(x) = (x - \mu_x)^2 \tag{4.22}$$

which equals the square of the deviation of a random outcome from its mean. The expected value of this quantity is referred to as the *variance* of the random variable and is usually denoted by σ_x^2. According to Equations 4.18 and 4.19, the variance of discrete and continuous random variables are given by the relations

$$\sigma_x^2 = \begin{cases} \sum_{i=1}^N (x_i - \mu_x)^2 f(x_i) & \text{(discrete)} \\ \int_{-\infty}^{+\infty} (x' - \mu_x)^2 f(x') \, dx' & \text{(continuous)} \end{cases} \tag{4.23}$$

The square root of the variance, σ_x, is called the *standard deviation* of x and measures the average magnitude of the deviation of the random variable from its mean. Random outcomes occur that are either less than or greater than the mean, μ_x, and the symmetry of these outcomes about μ_x is measured by the *skewness* or *skewness coefficient*, which is the expected value of the function

$$g(x) = \frac{(x - \mu_x)^3}{\sigma_x^3} \tag{4.24}$$

If the random outcomes are symmetrical about the mean, the skewness is equal to zero; otherwise, a nonsymmetric distribution will have a positive or negative skew depending on the location of the tail of the distribution. There is no universal symbol that is used to represent the skewness, but in this text skewness will be represented by g_x. For discrete and continuous random variables, we have

$$g_x = \begin{cases} \frac{1}{\sigma_x^3} \sum_{i=1}^N (x_i - \mu_x)^3 f(x_i) & \text{(discrete)} \\ \frac{1}{\sigma_x^3} \int_{-\infty}^{+\infty} (x' - \mu_x)^3 f(x') \, dx' & \text{(continuous)} \end{cases} \tag{4.25}$$

The variables μ_x, σ_x, and g_x are measures of the average, variability about the average, and the symmetry about the average, respectively.

EXAMPLE 4.1

A water-resource system is designed such that the probability, $f(x_i)$, that the system capacity is exceeded x_i times during the 50-year design life is given by the following discrete probability distribution:

x_i	$f(x_i)$
0	0.13
1	0.27
2	0.28
3	0.18
4	0.09
5	0.03
6	0.02
>6	0.00

What is the mean number of system failures expected in 50 years? What are the variance and skewness of the number of failures?

Solution The mean number of failures, μ_x, is defined by Equation 4.21, where

$$\mu_x = \sum_{i=1}^{N} x_i f(x_i)$$

$$= (0)(0.13) + (1)(0.27) + (2)(0.28) + (3)(0.18) + (4)(0.09)$$

$$+(5)(0.03) + (6)(0.02)$$

$$= 2$$

The variance of the number of failures, σ_x^2, is defined by Equation 4.23 as

$$\sigma_x^2 = \sum_{i=1}^{N} (x_i - \mu_x)^2 f(x_i)$$

$$= (0 - 2)^2(0.13) + (1 - 2)^2(0.27) + (2 - 2)^2(0.28) + (3 - 2)^2(0.18)$$

$$+(4 - 2)^2(0.09) + (5 - 2)^2(0.03) + (6 - 2)^2(0.02)$$

$$= 1.92$$

The skewness coefficient, g_x, is defined by Equation 4.25 as

$$g_x = \frac{1}{\sigma_x^3} \sum_{i=1}^{N} (x_i - \mu_x)^3 f(x_i)$$

$$= \frac{1}{(1.92)^{\frac{3}{2}}} \Big[(0 - 2)^3(0.13) + (1 - 2)^3(0.27)$$

$$+(2 - 2)^3(0.28) + (3 - 2)^3(0.18)$$

$$+(4 - 2)^3(0.09) + (5 - 2)^3(0.03) + (6 - 2)^3(0.02) \Big]$$

$$= 0.631$$

EXAMPLE 4.2

The probability density function, $f(t)$, of the time between storms during the summer in Miami is estimated as

$$f(t) = \begin{cases} 0.014e^{-0.014t}, & t > 0 \\ 0, & t \leq 0 \end{cases}$$

where t is the time interval between storms in hours. Estimate the mean, standard deviation, and skewness of t.

Solution   The mean interstorm time, μ_t, is given by Equation 4.21 as

$$\mu_t = \int_0^\infty t' f(t')\, dt'$$

$$= \int_0^\infty t' (0.014 e^{-0.014 t'})\, dt'$$

$$= 0.014 \left(\int_0^\infty t' e^{-0.014 t'}\, dt' \right)$$

The quantity in parentheses can be conveniently integrated by using the result (Dwight, 1961)

$$\int_0^\infty x e^{-ax}\, dx = \frac{1}{a^2}$$

Hence, the expression for μ_t can be written as

$$\mu_t = 0.014 \left(\frac{1}{0.014^2} \right) = 71 \text{ hours}$$

The variance of t, σ_t^2, is given by Equation 4.23 as

$$\sigma_t^2 = \int_0^\infty (t' - \mu_t)^2 f(t')\, dt'$$

$$= \int_0^\infty (t' - 71)^2 0.014 e^{-0.014 t'}\, dt'$$

$$= 0.014 \int_0^\infty (t'^2 - 142 t' + 5041) e^{-0.014 t'}\, dt'$$

Using the integration results (Dwight, 1961),

$$\int_0^\infty x^2 e^{-ax}\, dx = \frac{2}{a^3}; \quad \int_0^\infty x e^{-ax}\, dx = \frac{1}{a^2}; \quad \int_0^\infty e^{-ax}\, dx = \frac{1}{a}$$

the expression for σ_t^2 can be written as

$$\sigma_t^2 = 0.014 \left[\frac{2}{0.014^3} - 142 \left(\frac{1}{0.014^2} \right) + 5041 \left(\frac{1}{0.014} \right) \right] = 5102 \text{ h}^2$$

The standard deviation, σ_t, of the storm interevent time is therefore equal to $\sqrt{5102} = 71$ h.

The skewness of t, g_t, is given by Equation 4.25 as

$$g_t = \frac{1}{\sigma_t^3} \int_0^\infty (t' - \mu_t)^3 f(t')\, dt'$$

$$= \frac{1}{(71)^3} \int_0^\infty (t' - 71)^3 0.014 e^{-0.014 t'}\, dt'$$

$$= \frac{0.014}{(71)^3} \int_0^\infty (t'^3 - 213 t'^2 + 15{,}123 t' - 357{,}911) e^{-0.014 t'}\, dt'$$

Using the previously cited integration formulae plus (Dwight, 1961)

$$\int_0^\infty x^3 e^{-ax}\, dx = \frac{6}{a^4}$$

gives

$$g_t = 3.91 \times 10^{-8} \left(\frac{6}{0.014^4} - 213 \frac{2}{0.014^3} \right.$$

$$\left. + 15{,}123 \frac{1}{0.014^2} - 357{,}911 \frac{1}{0.014} \right) = 2.1$$

The positive skewness ($=2.1$) indicates that the probability distribution of t has a long tail to the right of μ_t (toward higher values of t).

4.2.4 Return Period

Consider a random variable X, with the outcome having a return period T given by x_T. Let p be the probability that $X \geq x_T$ in any observation, or $p = P(X \geq x_T)$. For each observation, there are two possible outcomes, either $X \geq x_T$ with probability p, or $X < x_T$ with probability $1 - p$. Assuming that all observations are independent, then the probability of a return period τ is the probability of $\tau - 1$ observations where $X < x_T$ followed by an observation where $X \geq x_T$. The expected value of τ, $\langle \tau \rangle$, is therefore given by

$$\langle \tau \rangle = \sum_{\tau=1}^{\infty} \tau (1 - p)^{\tau-1} p$$

$$= p + 2(1 - p)p + 3(1 - p)^2 p + 4(1 - p)^3 p + \cdots$$

$$= p[1 + 2(1 - p) + 3(1 - p)^2 + 4(1 - p)^3 + \cdots] \qquad (4.26)$$

The expression within the brackets has the form of the power series expansion

$$(1 + x)^n = 1 + nx + \frac{n(n-1)}{2}x^2 + \frac{n(n-1)(n-2)}{6}x^3 + \cdots \qquad (4.27)$$

with $x = -(1 - p)$ and $n = -2$. Equation 4.26 can therefore be written in the form

$$\langle \tau \rangle = \frac{p}{[1 - (1 - p)]^2} = \frac{1}{p} \qquad (4.28)$$

Since $T = \langle \tau \rangle$, Equation 4.28 can be expressed as

$$T = \frac{1}{p} \qquad (4.29)$$

or

$$\boxed{T = \frac{1}{P(X \geq x_T)}} \qquad (4.30)$$

which can also be written as

$$P(X \geq x_T) = \frac{1}{T} \tag{4.31}$$

In engineering practice it is more common to describe an event by its return period than its exceedance probability. For example, floodplains are usually delineated for the "100-year flood," which has an exceedance probability of 1% in any given year. However, because of the misconception that the 100-year flood occurs once every 100 years, ASCE (1996a) recommends that the reporting of return periods should be avoided, and annual exceedance probabilities reported instead.

EXAMPLE 4.3

Analyses of the maximum-annual floods over the past 150 years in a small river indicates the following cumulative distribution:

n	Flow, x_n (m^3/s)	$P(X < x_n)$
1	0	0
2	25	0.19
3	50	0.35
4	75	0.52
5	100	0.62
6	125	0.69
7	150	0.88
8	175	0.92
9	200	0.95
10	225	0.98
11	250	1.00

Estimate the magnitudes of the floods with return periods of 10, 50, and 100 years.

Solution   According to Equation 4.31, floods with return periods, T, of 10, 50, and 100 years have exceedance probabilities of $1/10 = 0.10$, $1/50 = 0.02$, and $1/100 = 0.01$. These exceedance probabilities correspond to cumulative probabilities of $1 - 0.1 = 0.9$, $1 - 0.02 = 0.98$, and $1 - 0.01 = 0.99$, respectively. Interpolating from the given cumulative probability distribution gives the following results:

Return period (years)	Flow (m^3/s)
10	163
50	225
100	238

4.2.5 Common Probability Functions

There are an infinite number of valid probability distributions. Engineering analyses, however, are typically confined to well-defined ones that are relatively simple, associated with identifiable processes, and can be fully characterized by a relatively small number of parameters. Probability functions that can be expressed analytically are the functions of choice in engineering applications. In using theoretical probability distribution functions to describe observed phenomena, it is important to understand the fundamental processes that lead to these distributions, since there is an implicit assumption that the theoretical and observed processes are the same. Probability distribution functions are either discrete or continuous. Commonly encountered discrete and continuous probability distribution functions, and their associated processes, are described in the following sections.

Binomial distribution.   The *binomial distribution*, also called the *Bernoulli distribution*, is a discrete probability distribution that describes the probability of outcomes as either successes or failures. For example, in studying the probability of the annual maximum flow in a river exceeding a certain value, the maximum flow in any year may be deemed either a success (the flow exceeds a specified value) or a failure (the flow does not exceed the specified given value). Since many engineered systems are designed to operate below certain threshold (design) conditions, the analysis in terms of success and failure is fundamental to the assessment of system reliability. Specifically, the binomial distribution describes the probability of n successes in N trials, given that the outcome in any one trial is independent of the outcome in any other trial and the probability of a success in any one trial is p. Such trials are called *Bernoulli trials*, and the process generating Bernoulli trials is called a *Bernoulli process*. The binomial (Bernoulli) probability distribution, $f(n)$, is given by

$$f(n) = \frac{N!}{n!(N - n)!} p^n (1 - p)^{N-n} \qquad (4.32)$$

This discrete probability distribution follows directly from permutation and combination analysis and can also be written in the form

$$f(n) = \binom{N}{n} p^n (1 - p)^{N-n} \qquad (4.33)$$

where

$$\binom{N}{n} = \frac{N!}{n!(N - n)!} \qquad (4.34)$$

is commonly referred to as the *binomial coefficient*. The parameters of the binomial distribution are the total number of outcomes, N, and the probability of success, p, in each outcome, and this distribution is illustrated in Figure 4.1. The mean, variance, and skewness coefficient of the binomial distribution are given by

$$\mu_n = Np, \quad \sigma_n^2 = Np(1 - p), \quad g = \frac{1 - 2p}{\sqrt{Np(1 - p)}} \qquad (4.35)$$

The Bernoulli probability distribution, $f(n)$, is symmetric for $p = 0.5$, skewed to the right for $p < 0.5$, and skewed to the left for $p > 0.5$.

EXAMPLE 4.4

The capacity of a stormwater-management system is designed to accommodate a storm with a return period of 10 years. What is the probability that the stormwater

FIGURE 4.1: Binomial probability distribution.

Source: Benjamin and Cornell (1970)

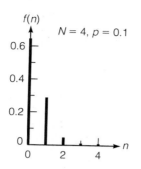

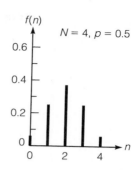

system will fail once in 20 years? What is the probability that the system fails *at least* once in 20 years?

Solution   The stormwater-management system fails when the magnitude of the design storm is exceeded. The probability, p, of the 10-year storm being exceeded in any one year is $1/10 = 0.1$. The probability of the 10-year storm being exceeded once in 20 years is given by the binomial distribution, Equation 4.32, as

$$f(n) = \frac{N!}{n!(N-n)!}p^n(1-p)^{N-n}$$

where $n = 1$, $N = 20$, and $p = 0.1$. Substituting these values gives

$$f(1) = \frac{20!}{1!(20-1)!}(0.1)^1(1-0.1)^{20-1} = 0.27$$

Therefore, the probability that the stormwater-management system fails once in 20 years is 27%.

The probability, P, of the 10-year storm being equalled or exceeded *at least* once in 20 years is given by

$$P = \sum_{i=1}^{20} f(i) = 1 - f(0) = 1 - \frac{20!}{0!(20-0)!}(0.1)^0(1-0.1)^{20-0} = 1 - 0.12 = 0.88$$

The probability that the design event will be exceeded (at least once) is referred to as the *risk of failure*, and the probability that the design event will not be exceeded is called the *reliability* of the system. In this example, the risk of failure is 88% over 20 years and the reliability of the system over the same period is 12%.

Geometric distribution.   The *geometric distribution* is a discrete probability distribution function that describes the number of Bernoulli trials, n, up to and including the one in which the first success occurs. The probability that the first success in a Bernoulli trial occurs on the nth trial is found by noting that for the first exceedance to be on the nth trial, there must be $n - 1$ preceding trials without success followed by a success. The probability of this sequence of events is $(1 - p)^{n-1}p$, and therefore the (geometric) probability distribution of the first success being in the nth trial, $f(n)$, is given by

$$\boxed{f(n) = p(1-p)^{n-1}} \tag{4.36}$$

FIGURE 4.2: Geometric
probability distribution

Source: Benjamin and Cornell
(1970)

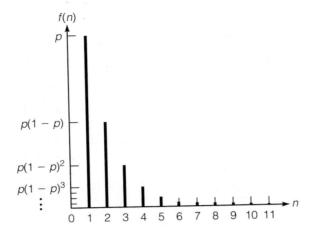

The parameter of the geometric distribution is the probability of success, p, and
this distribution is illustrated in Figure 4.2. The mean and variance of the geometric
distribution are given by

$$\mu_n = \frac{1}{p}, \quad \sigma_n^2 = \frac{1 - p}{p^2} \tag{4.37}$$

The probability distribution of the number of Bernoulli trials between successes
can be found from the geometric distribution by noting that the probability that n
trials elapse between successes is the same as the probability that the first success
is in the $n + 1$st trial. Hence the probability n trials between successes is equal to
$p(1 - p)^n$.

EXAMPLE 4.5

A stormwater-management system is designed for a 10-year storm. Assuming that
exceedance of the 10-year storm is a Bernoulli process, what is the probability that the
capacity of the stormwater-management system will be exceeded for the first time in
the fifth year after the system is constructed? What is the probability that the system
capacity will be exceeded within the first 5 years?

Solution   From the given data, $n = 5$, $p = 1/10 = 0.1$, and the probability that
the 10-year storm will be exceeded for the first time in the fifth year is given by the
geometric distribution (Equation 4.36) as

$$f(5) = (0.1)(1 - 0.1)^4 = 0.066 = 6.6\%$$

The probability, P, that the 10-year storm will be exceeded within the first 5 years is
given by

$$P = \sum_{n=1}^{5} p(1 - p)^{n-1} = \sum_{n=1}^{5} (0.1)(1 - 0.1)^{n-1} = \sum_{n=1}^{5} (0.1)(0.9)^{n-1}$$

$$= 0.1 + 0.09 + 0.081 + 0.073 + 0.066$$

$$= 0.41 = 41\%$$

Poisson distribution.   The *Poisson process* is a limiting case of an infinite number of Bernoulli trials, where the number of successes, n, becomes large and the probability of success in each trial, p, becomes small in such a way that the expected number of successes, Np, remains constant. Under these circumstances it can be shown that (Thiébaux, 1994; Benjamin and Cornell, 1970)

$$\lim_{N \to \infty, \, p \to 0} \frac{N!}{n!(N - n)!} \, p^n (1 - p)^{N-n} = \frac{(Np)^n e^{-Np}}{n!} \tag{4.38}$$

This approximation can be safely applied when $N \geq 100$, $p \leq 0.01$, and $Np \leq 20$ (Devore, 2000). Denoting the expected number of successes, Np, by λ, the probability distribution of the number of successes, n, in an interval in which the expected number of successes is given by λ is described by the *Poisson distribution*:

$$\boxed{f(n) = \frac{\lambda^n e^{-\lambda}}{n!}} \tag{4.39}$$

Haan (1977) describes the Poisson process as corresponding to the case where, for a given interval of time, the measurement increments decrease, resulting in a corresponding decrease in the event probability within the measurement increment, in such a way that the total number of events expected in the overall time interval remains constant and equal to λ. The only parameter in the Poisson distribution is λ. The mean, variance, and skewness of the discrete Poisson distribution are given by

$$\mu_n = \lambda, \quad \sigma_n^2 = \lambda, \quad g_n = \lambda^{-\frac{1}{2}} \tag{4.40}$$

Poisson distributions for several values of λ are illustrated in Figure 4.3. The Poisson process for a continuous time is defined in a similar way to the Poisson process on a discrete time scale. The assumptions underlying the Poisson process on a continuous time are: (1) the probability of an event in a short interval between t and $t + \Delta t$ is $\lambda \, \Delta t$ (proportional to the length of the interval) for all values of t; (2) the probability of more than one event in any short interval t to $t + \Delta t$ is negligible in comparison to $\lambda \Delta t$; and (3) the number of events in any interval of time is independent of the number of events in any other nonoverlapping interval of time.

Extending the result for the discrete Poisson process given by Equation 4.39 to the continuous Poisson process, the probability distribution of the number of events, n, in time t for a Poisson process is given by

$$\boxed{f(n) = \frac{(\lambda t)^n e^{-\lambda t}}{n!}} \tag{4.41}$$

The parameter of the Poisson distribution for a continuous process is λt, where λ is the average rate of occurrence of the event. The mean, variance, and skewness of the continuous Poisson distribution are given by

$$\mu_n = \lambda t, \quad \sigma_n^2 = \lambda t, \quad g_n = (\lambda t)^{-\frac{1}{2}} \tag{4.42}$$

The occurrences of storms and major floods have both been successfully described as Poisson processes (Borgman, 1963; Shane and Lynn, 1964).

FIGURE 4.3: Poisson probability distribution.

Source: Benjamin and Cornell (1970)

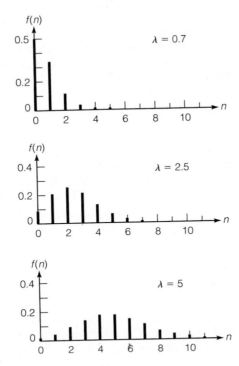

EXAMPLE 4.6

A flood-control system is designed for a runoff event with a 50-year return period. Assuming that exceedance of the 50-year runoff event is a Poisson process, what is the probability that the design event will be exceeded twice in the first 10 years of the system? What is the probability that the design event will be exceeded more than twice in the first 10 years?

Solution The probability distribution of the number of exceedances is given by Equation 4.39 (since the events are discrete). Over a period of 10 years, the expected number of exceedances, λ, of the 50-year event is given by

$$\lambda = Np = (10)\left(\frac{1}{50}\right) = 0.2$$

and the probability of the design event being exceeded twice is given by Equation 4.39 as

$$f(2) = \frac{0.2^2 e^{-0.2}}{2!} = 0.016 = 1.6\%$$

The probability, P, that the design event is exceeded more than twice in 10 years can be calculated using the relation

$$P = 1 - [f(0) + f(1) + f(2)]$$

$$= 1 - \left[\frac{0.2^0 e^{-0.2}}{0!} + \frac{0.2^1 e^{-0.2}}{1!} + \frac{0.2^2 e^{-0.2}}{2!} \right]$$

$$= 1 - [0.819 + 0.164 + 0.016] = 0.001$$

Therefore, there is only a 0.1% chance that the 50-year design event will be exceeded more than twice in any 10-year interval.

Exponential distribution.   The *exponential distribution* describes the probability distribution of the time between occurrences of an event in a continuous Poisson process. For any time duration, t, the probability of zero events during time t is given by Equation 4.41 as $e^{-\lambda t}$. Therefore, the probability that the time between events is less than t is equal to $1 - e^{-\lambda t}$. This is equal to the cumulative probability distribution, $F(t)$, of the time, t, between occurrences, which is equal to the integral of the probability density function, $f(t)$. The probability density function can therefore be derived by differentiating the cumulative distribution function; hence,

$$f(t) = \frac{d}{dt} F(t)$$

$$= \frac{d}{dt} (1 - e^{-\lambda t}) \tag{4.43}$$

which simplifies to

$$\boxed{f(t) = \lambda e^{-\lambda t}} \tag{4.44}$$

This distribution is called the exponential probability distribution. The mean, variance, and skewness coefficient of the exponential distribution are given by

$$\mu_t = \frac{1}{\lambda}, \quad \sigma_t^2 = \frac{1}{\lambda^2}, \quad g_t = 2 \tag{4.45}$$

The positive and constant value of the skewness indicates that the distribution is skewed to the right for all values of λ. The exponential probability distribution is illustrated in Figure 4.4. The cumulative distribution function, $F(t)$, of the exponential distribution function is given by

$$F(t) = \int_0^t \lambda e^{-\lambda \tau} d\tau \tag{4.46}$$

which leads to

$$\boxed{F(t) = 1 - e^{-\lambda t}} \tag{4.47}$$

In hydrologic applications, the exponential distribution is sometimes used to describe the interarrival times of random shocks to hydrologic systems, such as slugs of polluted runoff entering streams (Chow et al., 1988) and the arrival of storm events (Bedient and Huber, 1992).

FIGURE 4.4: Exponential probability distribution

Source: Benjamin and Cornell (1970).

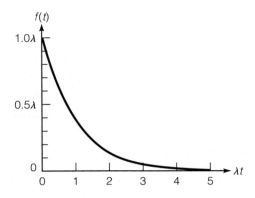

EXAMPLE 4.7

In the 86-year period between 1903 and 1988, Florida was hit by 56 hurricanes (Winsberg, 1990). Assuming that the time interval between hurricanes is given by the exponential distribution, what is the probability that there will be less than 1 year between hurricanes? What is the probability that there will be less than 2 years between hurricanes?

Solution The mean time between hurricanes, μ_t, is estimated as

$$\mu_t \approx \frac{86}{56} = 1.54 \text{ y}$$

The mean, μ_t, is related to the parameter, λ, of the exponential distribution by Equation 4.45, which gives

$$\lambda = \frac{1}{\mu_t} = \frac{1}{1.54} = 0.649 \text{ y}^{-1}$$

The cumulative probability distribution, $F(t)$, of the time between hurricanes is given by Equation 4.47 as

$$F(t) = 1 - e^{-\lambda t} = 1 - e^{-0.649t}$$

The probability that there will be less than one year between hurricanes is given by

$$F(1) = 1 - e^{-0.649(1)} = 0.48 = 48\%$$

and the probability that there will be less than two years between hurricanes is

$$F(2) = 1 - e^{-0.649(2)} = 0.73 = 73\%$$

Gamma/Pearson Type III distribution. The *gamma distribution* describes the probability of the time to the nth occurrence of an event in a Poisson process. This probability distribution is derived by finding the probability distribution of $t = t_1 + t_2 + \cdots + t_n$, where t_i is the time from the $i - 1$th to the ith event. This is simply equal to the sum of time intervals between events. The probability distribution of the time, t, to the nth event can be derived from the exponential distribution to yield

$$f(t) = \frac{\lambda^n t^{n-1} e^{-\lambda t}}{(n-1)!}$$

(4.48)

where $f(t)$ is the gamma probability distribution, illustrated in Figure 4.5 for several values of n. The mean, variance, and skewness coefficient of the gamma distribution are given by

$$\mu_t = \frac{n}{\lambda}, \quad \sigma_t^2 = \frac{n}{\lambda^2}, \quad g_t = \frac{2}{n}$$

(4.49)

In cases where n is not an integer, the gamma distribution, Equation 4.48, can be written in the more general form

$$f(t) = \frac{\lambda^n t^{n-1} e^{-\lambda t}}{\Gamma(n)}$$

(4.50)

where $\Gamma(n)$ is the gamma function defined by the equation

$$\Gamma(n) = \int_0^\infty t^{n-1} e^{-t}\, dt$$

(4.51)

which has the property that

$$\Gamma(n) = (n-1)!, \quad n = 1, 2, \ldots$$

(4.52)

In hydrologic applications, the gamma distribution is useful for describing skewed variables without the need for transformation and has been used to describe the distribution of the depth of precipitation in individual storms. On the basis of the fundamental process leading to the gamma distribution, it should be clear why the gamma distribution is equal to the exponential distribution when $n = 1$. The gamma

FIGURE 4.5: Gamma probability distribution

Source: Benjamin and Cornell (1970)

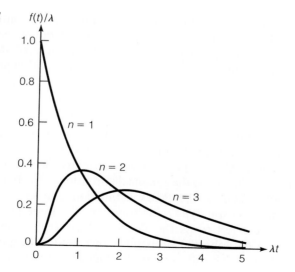

distribution given by Equation 4.50 can also be written in the form

$$f(t) = \frac{\lambda(\lambda t)^{n-1} e^{-\lambda t}}{\Gamma(n)} \tag{4.53}$$

Using the transformation

$$x = \lambda t \tag{4.54}$$

the probability density function of x, $p(x)$, is related to $f(t)$ by the relation

$$p(x) = f(t)\left|\frac{dt}{dx}\right| = f(t)\frac{1}{\lambda} \tag{4.55}$$

Combining Equations 4.53 to 4.55 and replacing n by α yields

$$p(x) = \frac{1}{\Gamma(\alpha)} x^{\alpha-1} e^{-x}, \quad x \geq 0 \tag{4.56}$$

This distribution is commonly referred to as the *one-parameter gamma distribution* (Yevjevich, 1972). The mean, variance, and skewness coefficient of the one-parameter gamma distribution are given by

$$\mu_x = \alpha, \quad \sigma_x^2 = \alpha, \quad g_x = \frac{2}{\sqrt{\alpha}} \tag{4.57}$$

Application of the one-parameter gamma distribution in hydrology is quite limited, primarily because of the difficulty in fitting a one-parameter distribution to observed hydrologic events (Yevjevich, 1972). A more general gamma distribution that does not have a lower bound of zero can be obtained by replacing x in Equation 4.56 by $(x - \gamma)/\beta$ to yield the following probability density function:

$$p(x) = \frac{1}{\beta^\alpha \Gamma(\alpha)}(x - \gamma)^{\alpha-1} e^{-(x-\gamma)/\beta}, \quad x \geq \gamma \tag{4.58}$$

where the parameter γ is the lower bound of the distribution, β is a scale parameter, and α is a shape parameter. The distribution function given by Equation 4.58 is called the *three-parameter gamma distribution*. The mean, variance, and skewness coefficient of the three-parameter (α, β, and γ) gamma distribution are given by

$$\mu_x = \gamma + \alpha\beta, \quad \sigma_x^2 = \alpha, \quad g_x = \frac{2}{\sqrt{\alpha}} \tag{4.59}$$

The three-parameter gamma distribution has the desirable properties (from a hydrologic viewpoint) of being bounded on the left and a positive skewness.

Pearson (1930) proposed a general equation for a distribution that fits many distributions—including the gamma distribution—by choosing appropriate parameter values, and a form of the Pearson function, similar to the gamma distribution, is known as the *Pearson Type III distribution*. The three-parameter gamma distribution, sometimes referred to as the Pearson Type III distribution, was first applied in hydrology to describe the probability distribution of annual-maximum flood peaks (Foster, 1924). When the measured data are very positively skewed, the data are usually log-transformed (i.e., x is the logarithm of a variable), and the distribution is called the *log–Pearson Type III distribution*. This distribution is widely used in hydrology, primarily because it has been (officially) recommended for application to peak flood flows by the U.S. Interagency Advisory Committee on Water Data (1982). An example of frequency analysis using the log–Pearson Type III distribution is given in Section 4.3.3.

EXAMPLE 4.8

A commercial area experiences significant flooding whenever a storm yields more than 13 cm of rainfall in 24 hours. In a typical year, there are four such storms. Assuming that these rainfall events are a Poisson process, what is the probability that it will take more than two years to have four flood events?

Solution   The probability density function of the time for four flood events is given by the gamma distribution, Equation 4.48. From the given data, $n = 4$, $\mu_t = 365$ d ($=1$ year), and

$$\lambda = \frac{n}{\mu_t} = \frac{4}{365} = 0.01096 \text{ d}^{-1}$$

The probability of up to t_o days elapsing before the fourth flood event is given by the cumulative distribution function

$$F(t_o) = \int_0^{t_o} f(t)\, dt = \int_0^{t_o} \frac{\lambda^n t^{n-1} e^{-\lambda t}}{(n-1)!}\, dt$$

$$= \frac{(0.01096)^4}{(4-1)!} \int_0^{t_o} t^{4-1} e^{-0.01096 t}\, dt = 2.405 \times 10^{-9} \int_0^{t_o} t^3 e^{-0.01096 t}\, dt$$

According to Dwight (1961),

$$\int t^3 e^{at}\, dt = e^{at}\left(\frac{t^3}{a} - \frac{3t^2}{a^2} + \frac{6t}{a^3} - \frac{6}{a^4} \right)$$

Using this relation to evaluate $F(t_o)$ gives

$$F(t_o) = -2.405 \times 10^{-9} \left[e^{-0.01096 t}\left(\frac{t^3}{0.01096} + \frac{3t^2}{0.01096^2} + \frac{6t}{0.01096^3} + \frac{6}{0.01096^4} \right) \right]_0^{t_o}$$

Taking $t_o = 730$ d ($=2$ years) gives

$$F(730) = -2.405 \times 10^{-9} \left[e^{-0.01096 t}\left(\frac{t^3}{0.01096} + \frac{3t^2}{0.01096^2} + \frac{6t}{0.01096^3} + \frac{6}{0.01096^4} \right) \right]_0^{730}$$

$$= -2.405 \times 10^{-9} \{ [e^{-8}(3.549 \times 10^{10} + 1.331 \times 10^{10}$$

$$+ 0.333 \times 10^{10} + 0.042 \times 10^{10})] - [1(4.158 \times 10^8)] \}$$

$$= 0.958$$

Hence, the probability that four flood events will occur in less than two years is 0.958, and the probability that it will take more than two years to have four flood events is $1 - 0.958 = 0.042$ or 4.2%.

Normal distribution. The *normal distribution*, also called the *Gaussian distribution*, is a symmetrical bell-shaped curve describing the probability density of a continuous random variable. The functional form of the normal distribution is given by

$$f(x) = \frac{1}{\sigma_x \sqrt{2\pi}} \exp\left[-\frac{1}{2}\left(\frac{x - \mu_x}{\sigma_x}\right)^2 \right] \tag{4.60}$$

where the parameters μ_x and σ_x are equal to the mean and standard deviation of x, respectively. Normally distributed random variables are commonly described by the shorthand notation $N(\mu, \sigma^2)$, and the shape of the normal distribution is illustrated in Figure 4.6. Most hydrologic variables cannot (theoretically) be normally distributed, since the range of normally distributed random variables is from $-\infty$ to ∞, and negative values of many hydrologic variables do not make sense. However, if the mean of a random variable is more than three or four times greater than its standard deviation, errors in the normal-distribution assumption can, in many cases, be neglected (Haan, 1977). Annual rainfall on the Greek island of Crete has be shown to mostly follow a normal distribution (Naoum and Tsanis, 2003).

It is sometimes convenient to work with the *standard normal deviate, z,* which is defined by

$$z = \frac{x - \mu_x}{\sigma_x} \tag{4.61}$$

where x is normally distributed. The probability density function of z is therefore given by

$$f(z) = \frac{1}{\sqrt{2\pi}}e^{-z^2/2} \tag{4.62}$$

Equation 4.61 guarantees that z is normally distributed with a mean of zero and a variance of unity, and is therefore an $N(0, 1)$ variate. The cumulative distribution, $F(z)$, of the standard normal deviate is given by

$$F(z) = \int_{-\infty}^{z} f(z')\,dz' = \frac{1}{\sqrt{2\pi}} \int_{-\infty}^{z} e^{-z'^2/2}\,dz' \tag{4.63}$$

where $F(z)$ is sometimes referred to as the area under the standard normal curve. These values are tabulated in Appendix C.1. Values of $F(z)$ can be approximated by

FIGURE 4.6: Normal probability distribution.

Source: Benjamin and Cornell (1970)

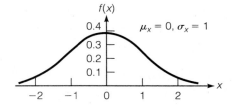
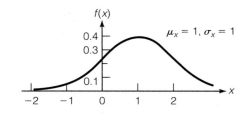

the analytic relation (Abramowitz and Stegun, 1965)

$$F(z) \approx \begin{cases} B, & z < 0 \\ 1 - B, & z > 0 \end{cases} \tag{4.64}$$

where

$$B = \frac{1}{2}\Big[1 + 0.196854|z| + 0.115194|z|^2 + 0.000344|z|^3 + 0.019527|z|^4\Big]^{-4} \tag{4.65}$$

and the error in $F(z)$ using Equation 4.65 is less than 0.00025. Conditions under which any random variable can be expected to follow a normal distribution are specified by the *central limit theorem*:

Central limit theorem.   *If S_n is the sum of n independently and identically distributed random variables X_i, each having a mean μ and variance σ^2, then in the limit as n approaches infinity, the distribution of S_n approaches a normal distribution with mean $n\mu$ and variance $n\sigma^2$.*

Unfortunately, not many hydrologic variables are the sum of independent and identically distributed random variables. Under some very general conditions, however, it can be shown that if $X_1, X_2, \ldots, X_n$ are n random variables with $\mu_{X_i} = \mu_i$ and $\sigma^2_{X_i} = \sigma_i^2$, then the sum, S_n, defined by

$$S_n = X_1 + X_2 + \cdots + X_n \tag{4.66}$$

is a random variable whose probability distribution approaches a normal distribution with mean, μ, and variance, σ^2 given by

$$\mu = \sum_{i=1}^{n} \mu_i \tag{4.67}$$

$$\sigma^2 = \sum_{i=1}^{n} \sigma_i^2 \tag{4.68}$$

The application of this result generally requires a large number of independent variables to be included in the sum S_n, and the probability distribution of each X_i has negligible influence on the distribution of S_n. In hydrologic applications, annual averages of variables such as precipitation and evaporation are assumed to be calculated from the sum of independent identically distributed measurements, and they tend to follow normal distributions (Chow et al., 1988).

EXAMPLE 4.9

The annual rainfall in the Upper Kissimmee River basin has been estimated to have a mean of 130.0 and a standard deviation of 15.6 cm (Chin, 1993b). Assuming that the annual rainfall is normally distributed, what is the probability of having an annual rainfall of less than 101.6 cm?

Solution From the given data: $\mu_x = 130.0$ cm, and $\sigma_x = 15.6$ cm. For $x = 101.6$ cm, the standard normal deviate, z, is given by Equation 4.61 as

$$z = \frac{x - \mu_x}{\sigma_x} = \frac{101.6 - 130.0}{15.6} = -1.82$$

The probability that $z \leq -1.82$, $F(-1.82)$, is given by Equation 4.64, where

$$B = \frac{1}{2}\left[1 + 0.196854|z| + 0.115194|z|^2 + 0.000344|z|^3 + 0.019527|z|^4\right]^{-4}$$

$$= \frac{1}{2}\left[1 + 0.196854|-1.82| + 0.115194|-1.82|^2 + 0.000344|-1.82|^3\right.$$

$$\left.+0.019527|-1.82|^4\right]^{-4}$$

$$= 0.034$$

and therefore

$$F(-1.82) = 0.034 = 3.4\%$$

There is a 3.4% probability of having an annual rainfall less than 101.6 cm.

Log-normal distribution. In cases where the random variable, X, is equal to the product of n random variables $X_1, X_2, \ldots, X_n$, the logarithm of X is equal to the sum of n random variables, where

$$\ln X = \ln X_1 + \ln X_2 + \cdots + \ln X_n \tag{4.69}$$

Therefore, according to the central limit theorem, $\ln X$ will be asymptotically normally distributed and X is said to have a *log-normal distribution*. Defining the random variable, Y, by the relation

$$Y = \ln X \tag{4.70}$$

then if Y is normally distributed, the theory of random functions can be used to show that the probability distribution of X, the log-normal distribution, is given by

$$f(x) = \frac{1}{x\sigma_y \sqrt{2\pi}} \exp\left[-\frac{(\ln x - \mu_y)^2}{2\sigma_y^2}\right], \quad x > 0 \tag{4.71}$$

where μ_y and σ_y^2 are the mean and variance of Y, respectively. The mean, variance, and skewness of a log-normally distributed variable, X, are given by

$$\mu_x = \exp(\mu_y + \sigma_y^2/2), \quad \sigma_x^2 = \mu_x^2[\exp(\sigma_y^2) - 1], \quad g_x = 3C_v + C_v^3 \tag{4.72}$$

where C_v is the *coefficient of variation* defined as

$$C_v = \frac{\sigma_x}{\mu_x} \tag{4.73}$$

If Y is defined by the relation

$$Y = \log_{10} X \qquad (4.74)$$

then Equation 4.71 still describes the probability density of X, with $\ln x$ replaced by $\log x$, and the moments of X are related to the moments of Y by

$$\mu_x = 10^{(\mu_y + \sigma_y^2/2)}, \quad \sigma_x^2 = \mu_x^2\left[10^{(\sigma_y^2)} - 1\right], \quad g_x = 3C_v + C_v^3 \qquad (4.75)$$

Many hydrologic variables exhibit a marked skewness, largely because they cannot be negative. Whereas the normal distribution allows the random variable to range without limit from negative infinity to positive infinity, the log-normal distribution has a lower limit of zero. The log-normal distribution has been found to reasonably describe such variables as daily precipitation depths, daily peak discharge rates, and the distribution of hydraulic conductivity in porous media (Viessman and Lewis, 1996). The log-normal distribution is commonly used in China to describe annual flood extremes (Singh and Strupczewski, 2002b).

EXAMPLE 4.10

Annual-maximum discharges in the Guadalupe River near Victoria, Texas, show a mean of 801 m³/s and a standard deviation of 851 m³/s. If the capacity of the river channel is 900 m³/s, and the flow is assumed to follow a log-normal distribution, what is the probability that the maximum discharge will exceed the channel capacity?

Solution From the given data: $\mu_x = 801$ m³/s; and $\sigma_x = 851$ m³/s. Equation 4.72 gives

$$\exp(\mu_y + \sigma_y^2/2) = 801 \qquad (= \mu_x)$$

and

$$(801)^2\left[\exp(\sigma_y^2) - 1\right] = (851)^2 \quad (= \sigma_x^2)$$

which are solved to yield

$$\mu_y = 6.31 \quad \text{and} \quad \sigma_y = 0.870$$

When $x = 900$ m³/s, the log-transformed variable, y, is given by $y = \ln 900 = 6.80$ and the normalized random variate, z, is given by

$$z = \frac{y - \mu_y}{\sigma_y} = \frac{6.80 - 6.31}{0.870} = 0.563$$

The probability that $z \le 0.563$, $F(0.563)$, is a standard normal distribution is given by Equation 4.64, where

$$B = \frac{1}{2}\left[1 + 0.196854|z| + 0.115194|z|^2 + 0.000344|z|^3 + 0.019527|z|^4\right]^{-4}$$

$$= \frac{1}{2}\left[1 + 0.196854|0.563| + 0.115194|0.563|^2 + 0.000344|0.563|^3\right.$$

$$\left. + 0.019527|0.563|^4\right]^{-4}$$

$$= 0.286$$

which yields

$$F(0.563) = 1 - 0.286 = 0.714 = 71.4\%$$

The probability that the maximum discharge in the river is greater than the channel capacity of 900 m³/s is therefore equal to $1 - 0.714 = 0.286 = 28.6\%$. This is the probability of flooding in any given year.

Uniform distribution.   The *uniform distribution* describes the behaviour of a random variable in which all possible outcomes are equally likely within the range $[a, b]$. The uniform distribution can be applied to either discrete or continuous random variables. For a continuous random variable, x, the uniform probability density function, $f(x)$, is given by

$$f(x) = \frac{1}{b - a}, \qquad a \leq x \leq b \tag{4.76}$$

where the parameters a and b define the range of the random variable. The mean, μ_x, and variance, σ_x^2, of a uniformly distributed random variable are given by

$$\mu_x = \frac{1}{2}(a + b), \qquad \sigma_x^2 = \frac{1}{12}(b - a)^2 \tag{4.77}$$

EXAMPLE 4.11

Conflicting data from several remote rain gages in a region of the Amazon basin indicate that the annual rainfall in 1997 was between 810 mm and 1080 mm. Assuming that the uncertainty can be described by a uniform probability distribution, what is the probability that the rainfall is greater than 1000 mm?

Solution   In this case, $a = 810$ mm, $b = 1080$ mm, and the uniform probability distribution of the 1997 rainfall is given by Equation 4.76 as

$$f(x) = \frac{1}{1080 - 810} = 0.00370 \quad \text{for} \quad 810 \text{ mm} \leq x \leq 1080 \text{ mm}$$

The probability, P, that $x \geq 1000$ mm is therefore given by

$$P = 0.00370 \times (1080 - 1000) = 0.296 = 29.6\%$$

Extreme-value distributions.   Extreme values are either maxima or minima of random variables. Consider the sample set $X_1, X_2, \ldots, X_n$, and let Y be the largest of the sample values. If $F(y)$ is the probability that $Y < y$ and $P_{X_i}(x_i)$ is the probability that $X_i < x_i$, then the probability that $Y < y$ is equal to the probability that all the x_is are less than y, which means that

$$F(y) = P_{X_1}(y)P_{X_2}(y)\ldots P_{X_n}(y) = [P_X(y)]^n \tag{4.78}$$

This equation gives the cumulative probability distribution of the extreme value, Y, in terms of the cumulative probability distribution of each outcome, which is also called the *parent distribution*. The probability density function, $f(y)$ of the extreme value, Y, is therefore given by

$$f(y) = \frac{d}{dy}F(y) = n[P_X(y)]^{n-1}p_X(y)$$

(4.79)

where $p_X(y)$ is the probability density function of each outcome. Equation 4.79 indicates that the probability distribution of extreme values, $f(y)$, depends on both the sample size and the parent distribution. Equation 4.79 was derived for maximum values, and a similar result can be derived for minimum values (see Problem 4.19).

Distributions of extreme values selected from large samples of many probability distributions have been shown to converge to one of three types of extreme-value distribution (Fisher and Tippett, 1928): Type I, Type II, or Type III. In Type I distributions, the parent distribution is unbounded in the direction of the desired extreme, and all moments of the distribution exist. In Type II distributions, also called *Frechet distributions*, the parent distribution is unbounded in the direction of the desired extreme and all moments of the distribution do not exist. In Type III distributions, the parent distribution is bounded in the direction of the desired extreme. The distributions of maxima of hydrologic variables are typically of Type I, since most hydrologic variables are (quasi-) unbounded to the right; Type II distributions are seldom used in hydrologic applications; and the distributions of minima are typically of Type III, since many hydrologic variables are bounded on the left by zero.

Extreme-value Type I (Gumbel) distribution.   The *extreme-value Type I distribution* requires that the parent distribution be unbounded in the direction of the extreme value. Specifically, Type I distributions require that the parent distribution falls off in an exponential manner, such that the upper tail of the cumulative distribution function, $P_X(x)$, can be expressed in the form

$$P_X(x) = 1 - e^{-g(x)}$$

with $g(x)$ an increasing function of x (Benjamin and Cornell, 1970). In using the Type I distribution to estimate maxima, parent distributions with the unbounded property include the normal, log-normal, exponential, and gamma distributions. Using the normal distribution as the parent distribution, the probability density function for the Type I extreme-value distribution is given by (Gumbel, 1958)

$$f(x) = \frac{1}{a}\exp\left\{\pm\frac{x-b}{a} - \exp\left[\pm\frac{x-b}{a}\right]\right\}, \quad -\infty < x < \infty,$$
$$-\infty < b < \infty, \quad a > 0$$

(4.80)

where a and b are scale and location parameters, b is the mode of the distribution and the minus of the $\pm$ used for maximum values. The plus of the $\pm$ is used for minimum values. The Type I distribution for maximum values is illustrated in Figure 4.7. The Type I extreme-value distribution is sometimes referred to as the *Gumbel extreme-value* distribution, the *Fisher–Tippett Type I* distribution, or the *double-exponential* distribution. The mean, variance, and skewness coefficient for the extreme-value

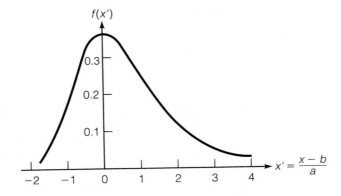

Type I distribution applied to maximum values are

$$\mu_x = b + 0.577a, \quad \sigma_x^2 = 1.645a^2, \quad g_x = 1.1396 \qquad (4.81)$$

and for the Type I distribution applied to minimum values,

$$\mu_x = b - 0.577a, \quad \sigma_x^2 = 1.645a^2, \quad g_x = -1.1396 \qquad (4.82)$$

By using the transformation

$$y = \frac{x - b}{a} \qquad (4.83)$$

the extreme-value Type I distribution can be written in the form

$$f(y) = \exp[\pm y - \exp(\pm y)] \qquad (4.84)$$

which yields the following cumulative distribution functions:

$$\boxed{F(y) = \exp[-\exp(-y)] \qquad \text{(maxima)}} \qquad (4.85)$$

$$\boxed{F(y) = 1 - \exp[-\exp(y)] \qquad \text{(minima)}} \qquad (4.86)$$

These cumulative distributions are most useful in determining the return periods of extreme events, such as flood flows (annual-maximum flows), maximum rainfall, and maximum wind speed (Gumbel, 1954). The extreme-value Type I (Gumbel) distribution is used extensively in flood studies in the United Kingdom and in many other parts of the world (Cunnane, 1988) and has been applied by the U.S. National Weather Service in analyzing annual-maximum precipitation amounts for given durations.

EXAMPLE 4.12

The annual-maximum discharges in the Guadalupe River near Victoria, Texas, between 1935 and 1978 show a mean of 811 m³/s and a standard deviation of 851 m³/s. Assuming that the annual-maximum flows are described by an extreme-value Type I

(Gumbel) distribution, estimate the annual-maximum flowrate with a return period of 100 years.

Solution   From the given data: $\mu_x = 811$ m^3/s, and $\sigma_x = 851$ m^3/s. According to Equation 4.81, the scale and location parameters, a and b, are derived from μ_x and σ_x by

$$a = \frac{\sigma_x}{\sqrt{1.645}} = \frac{851}{\sqrt{1.645}} = 664 \text{ m}^3/\text{s}$$

$$b = \mu_x - 0.577a = 811 - 0.577(664) = 428 \text{ m}^3/\text{s}$$

These parameters are used to transform the annual maxima, X, to the normalized variable, Y, where

$$Y = \frac{X - b}{a} = \frac{X - 428}{664}$$

For a return period of 100 years, the exceedance probability is $1/100 = 0.01$ and the cumulative probability, $F(y)$, is $1 - 0.01 = 0.99$. The extreme-value Type I probability distribution given by Equation 4.85 yields

$$0.99 = \exp\left[-\exp\left(-y_{100}\right)\right]$$

where y_{100} is the event with a return period of 100 years. Solving for y_{100} gives

$$y_{100} = -\ln\left(-\ln 0.99\right) = 4.60$$

The annual-maximum flow with a return period of 100 years, x_{100}, is therefore given by

$$y_{100} = \frac{x_{100} - 428}{664}$$

which leads to

$$x_{100} = 428 + 664y_{100} = 428 + 664(4.60) = 3482 \text{ m}^3/\text{s}$$

Extreme-value Type III (Weibull) distribution.   The extreme-value Type III distribution requires that the parent distribution be bounded in the direction of the extreme value. Specifically, Type III distributions (for minima) require that the lefthand tail of the cumulative distribution function, $P_X(x)$, rises from zero for values of $x \geq c$ such that

$$P_X(x) = \alpha(x - c)^a, \qquad x \geq c$$

where c is the lower limit of x (Benjamin and Cornell, 1970). The extreme-value Type III distribution has mostly been used in hydrology to estimate low streamflows, which are bounded on the left by zero. The extreme-value Type III distribution for minimum values is commonly called the *Weibull distribution* (after Weibull, 1939) and for $c = 0$ is given by

$$f(x) = ax^{a-1}b^{-a} \exp\left[-\left(\frac{x}{b}\right)^a\right], \qquad x \geq 0, \quad a, b > 0 \qquad (4.87)$$

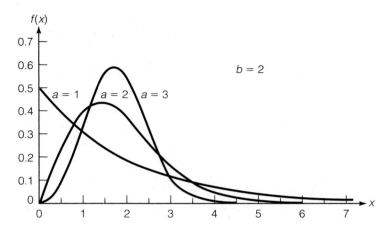

The Weibull distribution is illustrated in Figure 4.8. The mean and variance of the Weibull distribution are given by

$$\mu_x = b\Gamma\left(1 + \frac{1}{a}\right), \quad \sigma_x^2 = b^2\left[\Gamma\left(1 + \frac{2}{a}\right) - \Gamma^2\left(1 + \frac{1}{a}\right)\right] \tag{4.88}$$

and the skewness coefficient is

$$g_x = \frac{\Gamma(1 + 3/a) - 3\Gamma(1 + 2/a)\Gamma(1 + 1/a) + 2\Gamma^3(1 + 1/a)}{[\Gamma(1 + 2/a) - \Gamma^2(1 + 1/a)]^{\frac{3}{2}}} \tag{4.89}$$

where $\Gamma(n)$ is the gamma function defined by Equation 4.51. The cumulative Weibull distribution is given by

$$F(x) = 1 - \exp\left[-\left(\frac{x}{b}\right)^a\right] \tag{4.90}$$

which is useful in determining the return period of minimum events. If the lower bound of the parent distribution is nonzero, then a displacement parameter, c, must be added to the Type III extreme-value distributions for minima, and the probability density function becomes (Haan, 1977)

$$f(x) = a(x - c)^{a-1}(b - c)^{-a}\exp\left[-\left(\frac{x - c}{b - c}\right)^a\right] \tag{4.91}$$

and the cumulative distribution then becomes

$$F(x) = 1 - \exp\left[-\left(\frac{x - c}{b - c}\right)^a\right] \tag{4.92}$$

Equation 4.91 is sometimes called the *three-parameter Weibull distribution* or the *bounded exponential distribution*. The mean and variance of the three-parameter

Weibull distribution are given by

$$\mu_x = c + (b - c)\Gamma\left(1 + \frac{1}{a}\right), \quad \sigma_x^2 = (b - c)^2\left[\Gamma\left(1 + \frac{2}{a}\right) - \Gamma^2\left(1 + \frac{1}{a}\right)\right] \quad (4.93)$$

and the skewness coefficient of the three-parameter Weibull distribution is given by Equation 4.89.

EXAMPLE 4.13

The annual-minimum flows in a river at the location of a water-supply intake have a mean of 123 m^3/s and a standard deviation of 37 m^3/s. Assuming that the annual low flows have a lower bound of 0 m^3/s and are described by an extreme-value Type III distribution, what is the probability that an annual-low flow will be less than 80 m^3/s?

Solution   From the given data: $\mu_x = 123$ m^3/s, and $\sigma_x = 37$ m^3/s. According to Equation 4.88, the parameters of the probability distribution, a and b, must satisfy the relations

$$b\Gamma\left(1 + \frac{1}{a}\right) = 123 \quad (= \mu_x) \quad (4.94)$$

and

$$b^2\left[\Gamma\left(1 + \frac{2}{a}\right) - \Gamma^2\left(1 + \frac{1}{a}\right)\right] = (37)^2 \quad (= \sigma_x^2)$$

Combining these equations to eliminate b yields

$$\frac{123^2}{\Gamma^2\left(1 + \frac{1}{a}\right)}\left[\Gamma\left(1 + \frac{2}{a}\right) - \Gamma^2\left(1 + \frac{1}{a}\right)\right] = 37^2$$

or

$$\frac{\Gamma\left(1 + \frac{2}{a}\right) - \Gamma^2\left(1 + \frac{1}{a}\right)}{\Gamma^2\left(1 + \frac{1}{a}\right)} = 0.0905$$

This equation can be solved iteratively for a by defining $f(a)$ as

$$f(a) = \frac{\Gamma\left(1 + \frac{2}{a}\right) - \Gamma^2\left(1 + \frac{1}{a}\right)}{\Gamma^2\left(1 + \frac{1}{a}\right)} = \frac{\Gamma\left(1 + \frac{2}{a}\right)}{\Gamma^2\left(1 + \frac{1}{a}\right)} - 1$$

and selecting different values for a until $f(a) = 0.0905$. Using the tabulated values of the gamma function in Appendix D.3:

a	$\Gamma\left(1 + \frac{2}{a}\right)$	$\Gamma\left(1 + \frac{1}{a}\right)$	$f(a)$
2.00	1.000	0.886	0.273
3.00	0.903	0.893	0.132
4.00	0.886	0.906	0.079
3.79	0.888	0.904	0.0866
3.69	0.888	0.903	0.0890
3.65	0.889	0.903	0.0903

Therefore, $a = 3.65$ and Equation 4.94 gives

$$b = \frac{123}{\Gamma\left(1 + \frac{1}{a}\right)} = \frac{123}{\Gamma\left(1 + \frac{1}{3.65}\right)} = \frac{123}{0.903} = 136$$

The cumulative (Weibull) distribution of the annual-low flows, X, is given by Equation 4.90 as

$$F(x) = 1 - \exp\left[-\left(\frac{x}{b}\right)^a\right] = 1 - \exp\left[-\left(\frac{x}{136}\right)^{3.65}\right]$$

and the probability that $X \leq 80$ m³/s is given by

$$F(80) = 1 - \exp\left[-\left(\frac{80}{136}\right)^{3.65}\right] = 0.134 = 13.4\%$$

There is a 13.4% probability that the annual-low flow is less than 80 m³/s.

General extreme-value distribution. The *general extreme-value* (GEV) distribution incorporates the extreme-value Type I, II, and III distributions for maxima. The GEV distribution was first proposed by Jenkinson (1955) to describe the distribution of the largest values of meteorological data when the limiting form of the extreme-value distribution is unknown. The GEV distribution has to date been used to model a wide variety of natural extremes, including floods, rainfall, wind speeds, wave heights, and other maxima (Martins and Stedinger, 2000). The GEV cumulative distribution function is given by

$$F(x) = \begin{cases} \exp\left\{-\left[1 - \frac{c(x-a)}{b}\right]^{1/c}\right\} & c \neq 0 \\ \\ \exp\left\{-\exp\left[-\frac{(x-a)}{b}\right]\right\} & c = 0 \end{cases} \tag{4.95}$$

where a, b, and c are location, scale, and shape parameters respectively. For $c = 0$, the GEV distribution is the same as the extreme-value Type I distribution; for $c < 0$ the GEV distribution corresponds to the Type II distribution for maxima that have a finite lower bound of $a + b/c$; and for $c > 0$ the GEV distribution corresponds to the Type III distribution for maxima that have a finite upper bound at $a + b/c$. The mean, variance, and skewness coefficient for the GEV distribution are given by (Kottegoda and Rosso, 1997)

$$\mu_x = a + \frac{b}{c}[1 - \Gamma(1 + c)], \qquad\qquad c > -1 \tag{4.96}$$

$$\sigma_x^2 = \left(\frac{b}{c}\right)^2 [\Gamma(1 + 2c) - \Gamma^2(1 + c)], \qquad\qquad c > -0.5 \tag{4.97}$$

$$g_x = \text{sign}\,(c)\frac{-\Gamma(1 + 3c) + 3\Gamma(1 + c)\Gamma(1 + 2c) - 2\Gamma^3(1 + c)}{[\Gamma(1 + 2c) - \Gamma^2(1 + c)]^{3/2}} - 1, \quad c > -1/3 \tag{4.98}$$

where it is noted that the moments only exist for certain values of c. The generalized extreme-value distribution is commonly used in Great Britain to describe annual flood extremes (Singh and Strupczewski, 2002b).

EXAMPLE 4.14

The annual-maximum 24-hour rainfall amounts in a subtropical area have a mean of 15.4 cm, a standard deviation of 3.42 cm, and a skewness coefficient of 1.24. Assuming that the annual-maximum 24-hour rainfall is described by a general extreme-value (GEV) distribution, estimate the annual-maximum 24-hour rainfall with a return period of 100 years.

Solution   From the given data: μ_x = 15.4 cm, σ_x = 3.42 cm, and g_x = 1.24. In order to find the GEV cumulative distribution function, the parameters a, b, and c must first be determined. The shape parameter, c, can be determined directly from the skewness coefficient, g_x, using Equation 4.98, where

$$g_x = 1.24 = \text{sign}\,(c)\frac{-\Gamma(1\,+\,3c)\,+\,3\Gamma(1\,+\,c)\Gamma(1\,+\,2c)\,-\,2\Gamma^3(1\,+\,c)}{[\Gamma(1\,+\,2c)\,-\,\Gamma^2(1\,+\,c)]^{3/2}}\,-\,1$$

which can be solved to give $c = -0.129$. Rearranging Equation 4.97 gives b as

$$b = \sqrt{\frac{c^2\sigma_x^2}{\Gamma(1\,+\,2c)\,-\,\Gamma^2(1\,+\,c)}} = \sqrt{\frac{(-0.129)^2(3.42)^2}{1.236\,-\,1.093^2}} = 2.17 \text{ cm}$$

and rearranging Equation 4.96 gives a as

$$a = \mu_x\,-\,\frac{b}{c}[1\,-\,\Gamma(1\,+\,c)] = 15.4\,-\,\frac{2.17}{-0.129}[1\,-\,1.093] = 13.8 \text{ cm}$$

Hence the cumulative distribution function, $F(x)$, is given by Equation 4.95 as

$$F(x) = \exp\left\{-\left[1\,-\,\frac{c(x\,-\,a)}{b}\right]^{1/c}\right\} = \exp\left\{-\left[1\,-\,\frac{-0.129(x\,-\,13.8)}{2.17}\right]^{1/-0.129}\right\}$$

which yields

$$F(x) = \exp\left\{-\left[1\,+\,0.0594(x\,-\,13.8)\right]^{-7.75}\right\}$$

For a return period of 100 years, $F(x_{100}) = 1\,-\,1/100 = 0.99$ and

$$F(x_{100}) = 0.99 = \exp\left\{-\left[1\,+\,0.0594(x_{100}\,-\,13.8)\right]^{-7.75}\right\}$$

which yields

$$x_{100} = 27.4 \text{ cm}$$

Therefore, an annual-maximum 24-hour rainfall of 27.4 cm has a return period of 100 years.

FIGURE 4.9: Chi-square probability distribution.

Source: Benjamin and Cornell (1970)

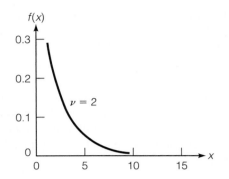
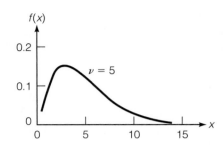

Chi-square distribution. Unlike the previously described probability distributions, the chi-square distribution is not used to describe a hydrologic process but is more commonly used in testing how well observed outcomes of hydrologic processes are described by theoretical probability distributions. The *chi-square distribution* is the probability distribution of a variable obtained by adding the squares of ν normally distributed random variables, all of which have a mean of zero and a variance of 1. That is, if $X_1, X_2, \ldots, X_\nu$ are normally distributed random variables with mean of zero and variance of 1 and a new random variable χ^2 is defined such that

$$\chi^2 = \sum_{i=1}^{\nu} X_i^2 \tag{4.99}$$

then the probability density function of χ^2 is defined as the chi-square distribution and is given by

$$f(x) = \frac{x^{-(1-\nu/2)} e^{x/2}}{2^{\nu/2} \Gamma(\nu/2)}, \qquad x, \nu > 0 \tag{4.100}$$

where $x = \chi^2$ and ν is called the number of *degrees of freedom*. The chi-square distribution is illustrated in Figure 4.9. The mean and variance of the chi-square distribution are given by

$$\mu_{\chi^2} = \nu, \quad \sigma_{\chi^2}^2 = 2\nu \tag{4.101}$$

The chi-square distribution is commonly used in determining the confidence intervals of various sample statistics. The cumulative chi-square distribution is given in Appendix C.3.

EXAMPLE 4.15

A random variable is calculated as the sum of the squares of 10 normally distributed variables with a mean of zero and a standard deviation of 1. What is the probability that the sum is greater than 20? What is the expected value of the sum?

Solution The random variable cited here is a χ^2 variate with 10 degrees of freedom. The cumulative distribution of the χ^2 variate is given in Appendix C.3 as a function of the number of degrees of freedom, ν, and the exceedance probability, α. In this case, for $\nu = 10$, and $\chi^2 = 20$, Appendix C.3 gives (by interpolation)

$$\alpha = 0.031$$

Therefore, the probability that the sum of the squares of 10 N(0,1) variables exceeds 20 is 0.031 or 3.1%.

The expected value of the sum is, by definition, equal to the mean, μ_{χ^2}, which is given by Equation 4.101 as

$$\mu_{\chi^2} = \nu = 10$$

4.3   Analysis of Hydrologic Data

4.3.1   Estimation of Population Distribution

Utilization of hydrologic data in engineering practice is typically a two-step procedure. In the first step, the cumulative probability distribution of the measured data is compared with a variety of theoretical distribution functions, and the theoretical distribution function that best fits the probability distribution of the measured data is taken as the population distribution. In the second step, the exceedance probabilities of selected events are estimated using the theoretical population distribution. Some federal agencies require the use of specific probability distributions, and in some cases they also specify the minimum size of the data set to be used in estimating the parameters of the distribution.

In estimating probability distributions from measured data, the measurements are usually assumed to be independent and drawn from identical probability distributions. In cases where the observations are correlated, then alternative methods such as *time-series analysis* are used in lieu of probability theory. The most common methods of estimating population distributions from measured data are: (1) visually comparing the probability distribution of the measured data with various theoretical distributions, and picking the closest distribution that is consistent with the underlying process generating the sample data, and (2) using hypothesis-testing methods to assess whether various probability distributions are consistent with the probability distribution of the measured data.

4.3.1.1   Probability distribution of observed data

To assist in identifying a theoretical probability distribution that adequately describes observed outcomes, it is generally useful to graphically compare the probability distribution of observed outcomes with various theoretical probability distributions. The first step in plotting the probability distribution of observed data is to rank the data, such that for N observations a rank of 1 is assigned to the observation with the largest magnitude and a rank of N is assigned to the observation with the lowest magnitude. The exceedance probability of the m-ranked observation, x_m, denoted by

$P_X(X > x_m)$, is commonly estimated by the relation

$$P_X(X > x_m) = \frac{m}{N + 1}, \qquad m = 1, \ldots, N \tag{4.102}$$

or as a cumulative distribution function

$$P_X(X < x_m) = 1 - \frac{m}{N + 1}, \qquad m = 1, \ldots, N \tag{4.103}$$

Equation 4.102 is called the *Weibull formula* (Weibull, 1939) and is widely used in practice. The main drawback of the Weibull formula for estimating the cumulative probability distribution from measured data is that it is asymptotically exact (as the number of observations approaches infinity) only for a population with an underlying uniform distribution, which is relatively rare in nature. To address this shortcoming, Gringorten (1963) proposed that the exceedance probability of observed data be estimated using the relation

$$P_X(X > x_m) = \frac{m - a}{N + 1 - 2a}, \qquad m = 1, \ldots, N \tag{4.104}$$

where a is a parameter that depends on the population distribution. For a normal (or log-normal) distribution, $a = 0.375$; for a Gumbel distribution, $a = 0.44$. Bedient and Huber (2002) suggest that $a = 0.40$ is a good compromise for the usual situation in which the exact distribution is unknown.

EXAMPLE 4.16

The annual peak flows in the Guadalupe River near Victoria, Texas, between 1965 and 1978 are as follows:

Year	Peak flow (ft^3/s)
1965	15,000
1966	9790
1967	70,000
1968	44,300
1969	15,200
1970	9190
1971	9740
1972	58,500
1973	33,100
1974	25,200
1975	30,200
1976	14,100
1977	54,500
1978	12,700

Use the Weibull and Gringorten formulae to estimate the cumulative probability distribution of annual peak flow and compare the results.

Solution   The ranking of flows between 1965 and 1978 are shown in columns 1 and 2 of the following table:

(1) Rank, m	(2) Flow, q_m (ft^3/s)	(3) Weibull $P(Q > q_m)$	(4) Gringorten $P(Q > q_m)$
1	70,000	0.067	0.042
2	58,500	0.133	0.113
3	54,500	0.200	0.183
4	44,300	0.267	0.253
5	33,100	0.333	0.324
6	30,200	0.400	0.394
7	25,200	0.467	0.465
8	15,200	0.533	0.535
9	15,000	0.600	0.606
10	14,100	0.667	0.676
11	12,700	0.733	0.746
12	9790	0.800	0.817
13	9740	0.867	0.887
14	9190	0.933	0.958

In this case, $N = 14$ and the Weibull exceedance probabilities are given by Equation 4.102 as

$$P(Q > q_m) = \frac{m}{14 + 1} = \frac{m}{15}$$

These probabilities are tabulated in column 3. The Gringorten exceedance probabilities are given by (assuming $a = 0.40$)

$$P(Q > q_m) = \frac{m - a}{N + 1 - 2a} = \frac{m - 0.40}{14 + 1 - 2(0.40)} = \frac{m - 0.40}{14.2}$$

and these probabilities are tabulated in column 4. The probability functions estimated using the Weibull and Gringorten formulae are compared in Figure 4.10, where the distributions appear very similar. At high flows, the Gringorten formula assigns a lower exceedance probability and a correspondingly higher return period.

Visual comparison of the probability distribution calculated from observed data with various theoretical population distributions is facilitated by using special probability graph paper for each theoretical distribution. For example, on normal probability paper, the cumulative normal distribution plots as a straight line and the normality of the distribution of the sample data is assessed on the basis of how well this observed distribution fits a straight line. The construction of probability graph paper for various probability distributions is described by Haan (1977).

4.3.1.2 Hypothesis tests

As an alternative to the visual comparison of observed probability distributions with various theoretical probability distributions, a quantitative comparison can be made using a hypothesis test. The two most common hypothesis tests for assessing whether an observed probability distribution can be approximated by a given (theoretical) population distribution are the *chi-square test* and the *Kolmogorov–Smirnov test*.

FIGURE 4.10: Comparison between Weibull and Gringorten probability distributions

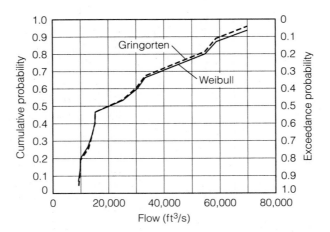

The chi-square test.   Based on sampling theory, it is known that if the N outcomes are divided into M classes, with X_m being the number of outcomes in class m, and p_m being the theoretical probability of an outcome being in class m, then the random variable,

$$\chi^2 = \sum_{m=1}^{M} \frac{(X_m - Np_m)^2}{Np_m} \tag{4.105}$$

has a chi-square distribution. The number of degrees of freedom is $M - 1$ if the expected frequencies can be computed without having to estimate the population parameters from the sample statistics, while the number of degrees of freedom is $M - 1 - n$ if the expected frequencies are computed by estimating n population parameters from sample statistics. In applying the chi-square goodness of fit test, the null hypothesis is taken as H_0: The samples are drawn from the proposed probability distribution. The null hypothesis is accepted at the α significance level if $\chi^2 \in [0, \chi_\alpha^2]$ and rejected otherwise.

EXAMPLE 4.17

Analysis of a 47-year record of annual rainfall indicates the following frequency distribution:

Range (mm)	Number of outcomes	Range (mm)	Number of outcomes
<1000	2	1250–1300	7
1000–1050	3	1300–1350	5
1050–1100	4	1350–1400	3
1100–1150	5	1400–1450	2
1150–1200	6	1450–1500	2
1200–1250	7	>1500	1

The measured data also indicate a mean of 1225 mm and a standard deviation of 151 mm. Using a 5% significance level, assess the hypothesis that the annual rainfall is drawn from a normal distribution.

Solution   The first step in the analysis is to derive the theoretical frequency distribution. Appendix C.1 gives the cumulative probability distribution of the standard normal deviate, z, which is defined as

$$z = \frac{x - \mu_x}{\sigma_x} = \frac{x - 1225}{151}$$

where x is the annual rainfall. Converting the annual rainfall amounts into standard normal deviates, z, yields:

Rainfall (mm)	z	$P(Z < z)$
1000	−1.49	0.07
1050	−1.16	0.12
1100	−0.83	0.20
1150	−0.50	0.31
1200	−0.17	0.43
1250	0.17	0.57
1300	0.50	0.69
1350	0.83	0.80
1400	1.16	0.88
1450	1.49	0.93
1500	1.82	0.97

and therefore the theoretical frequencies are given by

Range (mm)	Theoretical probability, p_m	Theoretical outcomes, Np_m
< 1000	0.07	3.29
1000–1050	0.05	2.35
1050–1100	0.08	3.76
1100–1150	0.11	5.17
1150–1200	0.12	5.64
1200–1250	0.14	6.58
1250–1300	0.12	5.64
1300–1350	0.11	5.17
1350–1400	0.08	3.76
1400–1450	0.05	2.35
1450–1500	0.04	1.88
> 1500	0.03	1.41

where the total number of observations, N, is equal to 47. Based on the observed and theoretical frequency distributions, the chi-square statistic is given by Equation 4.105 as

$$\chi^2 = \frac{(2 - 3.29)^2}{3.29} + \frac{(3 - 2.35)^2}{2.35} + \frac{(4 - 3.76)^2}{3.76} + \frac{(5 - 5.17)^2}{5.17} + \frac{(6 - 5.64)^2}{5.64}$$

$$+ \frac{(7 - 6.58)^2}{6.58} + \frac{(7 - 5.64)^2}{5.64} + \frac{(5 - 5.17)^2}{5.17} + \frac{(3 - 3.76)^2}{3.76}$$

$$+ \frac{(2 - 2.35)^2}{2.35} + \frac{(2 - 1.88)^2}{1.88} + \frac{(1 - 1.41)^2}{1.41}$$

$$= 1.42$$

Since both the mean and standard deviation were estimated from the measured data, the χ^2 statistic has $M - 1 - n$ degrees of freedom, where $M = 12$ ($=$ number of intervals), and $n = 2$ ($=$ number of population parameters estimated from measured data), hence $12 - 1 - 2 = 9$ degrees of freedom. Using a 5% significance level, the hypothesis that the observations are drawn from a normal distribution is accepted if

$$0 \leq 1.42 \leq \chi^2_{0.05}$$

Appendix C.3 gives that for 9 degrees of freedom, $\chi^2_{0.05} = 16.919$. Since $0 \leq 1.42 \leq 16.919$, the hypothesis that the annual rainfall is drawn from a normal distribution is *accepted* at the 5% level.

The effectiveness of the chi-square test is diminished if both the number of data intervals, called cells, is less than 5 and the expected frequency in any cell is less than 5 (Haldar and Mahadevan, 2000; McCuen, 2002a).

Kolmogorov–Smirnov test.   As an alternative to the chi-square goodness of fit test (to assess the hypothesis that observations are drawn from a population with a given theoretical probability distribution), the Kolmogorov–Smirnov test can be used. This test differs from the chi-square test in that no parameters from the theoretical probability distribution need to be estimated from the observed data. In this sense, the Kolmogorov–Smirnov test is called a *nonparametric* test. The procedure for implementing the Kolmogorov–Smirnov test is as follows (Haan, 1977):

1. Let $P_X(x)$ be the specified theoretical cumulative distribution function under the null hypothesis.
2. Let $S_N(x)$ be the sample cumulative distribution function based on N observations. For any observed x, $S_N(x) = k/N$, where k is the number of observations less than or equal to x.
3. Determine the maximum deviation, D, defined by

$$\boxed{D = \max |P_X(x) - S_N(x)|} \tag{4.106}$$

4. If, for the chosen significance level, the observed value of D is greater than or equal to the critical value of the Kolmogorov–Smirnov statistic tabulated in Appendix C.4, the hypothesis is rejected.

An advantage of the Kolmogorov–Smirnov test over the chi-square test is that it is not necessary to divide the data into intervals; thus any error of judgment associated with the number or size of the intervals is avoided (Haldar and Mahadevan, 2000).

EXAMPLE 4.18

Use the Kolmogorov–Smirnov test at the 10% significance level to assess the hypothesis that the data in Example 4.17 are drawn from a normal distribution.

Solution   Based on the given data, the measured and theoretical probability distributions of the annual rainfall are as follows:

(1) Rainfall x (mm)	(2) Normalized rainfall $(x - \mu_x)/\sigma_x$	(3) $S_N(x)$	(4) $P_X(x)$	(5) $|P_X(x) - S_N(x)|$
1000	−1.490	2/47 = 0.043	0.068	0.025
1050	−1.159	5/47 = 0.106	0.123	0.017
1100	−0.828	9/47 = 0.191	0.204	0.013
1150	−0.497	14/47 = 0.298	0.310	0.012
1200	−0.166	20/47 = 0.426	0.434	0.008
1250	0.166	27/47 = 0.574	0.566	0.008
1300	0.497	34/47 = 0.723	0.690	0.033
1350	0.828	39/47 = 0.830	0.796	0.034
1400	1.159	42/47 = 0.894	0.877	0.017
1450	1.490	44/47 = 0.936	0.932	0.004
1500	1.821	46/47 = 0.979	0.966	0.013

The normalized rainfall amounts in column 2 are calculated using the rainfall amount, x, in column 1; mean, μ_x, of 1225 mm; and standard deviation, σ_x, of 151 mm. The sample cumulative distribution function, $S_N(x)$, in column 3 is calculated using the relation

$$S_N(x) = \frac{k}{N}$$

where k is the number of measurements less than or equal to the given rainfall amount, x, and $N = 47$ is the total number of measurements. The theoretical cumulative distribution function, $P_X(x)$, in column 4 is calculated using the normalized rainfall (column 2) and the standard normal distribution function given in Appendix C.1. The absolute value of the difference between the theoretical and sample distribution functions is given in column 5. The maximum difference, D, between the theoretical and sample distribution functions is equal to 0.034 and occurs at an annual rainfall of 1350 mm. For a sample size of 11 and a significance level of 10%, Appendix C.4 gives the critical value of the Kolmogorov–Smirnov statistic as 0.352. Since $0.034 < 0.352$, the hypothesis that the measured data are from a normal distribution is accepted at the 10% significance level.

The Kolmogorov–Smirnov test is generally more efficient than the chi-square test when the sample size is small (McCuen, 2002a). However, neither of these tests is very powerful in the sense that the probability of accepting a false hypothesis is quite high, especially for small samples (Haan, 1977). In spite of this cautionary note, both the chi-square and the Kolmogorov–Smirnov tests are widely used in engineering applications.

4.3.2   Estimation of Population Parameters

The probability distribution of a random variable X is typically written in the form $f_X(x)$. However, this distribution could also be written in the form $f_X(x \mid \theta_1, \theta_2, \ldots, \theta_m)$ to indicate that the probability distribution of the random variable also depends on the values of the m parameters $\theta_1, \theta_2, \ldots, \theta_m$. This explicit expression for the probability distribution is particularly relevant when the population parameters are estimated using sample statistics, which are themselves random variables. The most common methods for estimating the parameters from measured data are: the method of moments, the maximum-likelihood method, and the method of L-moments.

4.3.2.1   Method of moments

The *method of moments* is based on the observation that the parameters of a probability distribution can usually be expressed in terms of the first few moments of the distribution. These moments can be estimated using sample statistics, and then the parameters of the distribution can be calculated using the relationship between the population parameters and the moments. The three moments most often used in hydrology are the mean, standard deviation, and skewness, defined by Equations 4.21, 4.23, and 4.25 for both discrete and continuous probability distributions. In practical applications, these moments must be estimated from finite samples, and the following equations usually provide the best estimates:

$$\hat{\mu}_x = \frac{1}{N} \sum_{i=1}^{N} x_i \tag{4.107}$$

$$\hat{\sigma}_x^2 = \frac{1}{N-1} \sum_{i=1}^{N} (x_i - \hat{\mu})^2 \tag{4.108}$$

$$\hat{g}_x = \frac{N}{(N-1)(N-2)} \frac{\sum_{i=1}^{N} (x_i - \hat{\mu})^3}{\hat{\sigma}^3} \tag{4.109}$$

where $\hat{\mu}_x$, $\hat{\sigma}_x^2$, and $\hat{g}_x$ are unbiased estimates of the mean (μ_x), variance (σ_x^2), and skewness (g_x) of the population from which the N samples of x, denoted by x_i, are drawn (Benjamin and Cornell, 1970). Unbiased estimates of the variance, $\hat{\sigma}_x^2$, and skewness, $\hat{g}_x$, are generally dependent on the underlying population distribution. The accuracy of the estimated skewness, $\hat{g}_x$, is usually of most concern, since it involves the summation of the cubes of deviations from the mean and is therefore subject to larger errors in its computation. Although Equation 4.109 is most often used to estimate the skewness, other estimates have been proposed (Bobée and Robitaille, 1975; Tasker and Stedinger, 1986).

EXAMPLE 4.19

The number of hurricanes hitting Florida each year between 1903 and 1988 is given in the following table (Winsberg, 1990):

	00	01	02	03	04	05	06	07	08	09
1900				1	1	0	2	0	0	1
1910	1	1	0	0	0	1	2	1	0	1
1920	0	1	0	0	2	1	2	0	2	1
1930	0	0	1	2	0	2	1	0	0	1
1940	0	1	0	0	1	2	1	2	2	1
1950	2	0	0	1	0	0	1	0	0	0
1960	1	0	0	3	0	1	2	0	1	0
1970	0	0	1	0	0	1	0	0	0	1
1980	0	0	0	0	0	2	0	1	0	

If the number of hurricanes per year is described by a Poisson distribution, estimate the parameters of the probability distribution. What is the probability of three hurricanes hitting Florida in one year?

Solution   The Poisson probability distribution, $f(n)$, is given by

$$f(n) = \frac{(\lambda t)^n e^{-\lambda t}}{n!}$$

where n is the number of occurrences per year and the parameter λt is related to the mean, μ_n, and standard deviation, σ_n, of n by

$$\mu_n = \lambda t, \qquad \sigma_n = \sqrt{\lambda t}$$

Based on the 86 years of data (1903–1988), the mean number of occurrences per year, μ_n, is estimated by the first-order moment, $\overline{N}$, as

$$\mu_n \approx \overline{N} = \frac{1}{86} \sum n_i = \frac{1}{86}(56) = 0.65$$

Hence $\lambda t \approx 0.65$ and the probability distribution of the number of hurricanes per year in Florida is given by

$$f(n) = \frac{0.65^n e^{-0.65}}{n!}$$

Putting $n = 3$ gives the probability of three hurricanes in one year as

$$f(3) = \frac{0.65^3 e^{-0.65}}{3!} = 0.02 = 2\%$$

It is interesting to note that the probability of at least one hurricane hitting Florida in any year is given by $1 - f(0) = 1 - e^{-0.65} = 0.48 = 48\%$.

Estimates of the mean, variance, and skewness given by Equations 4.107 to 4.109 are based on samples of a random variable and are therefore random variables themselves. The standard deviation of an estimated parameter is commonly called the *standard error*, and the standard errors of the estimated mean, variance, and skewness are given by the following relations

$$S_{\hat{\mu}_x} = \frac{\hat{\sigma}_x}{\sqrt{N}} \tag{4.110}$$

$$S_{\hat{\sigma}_x} = \hat{\sigma}_x \sqrt{\frac{1 + 0.75\hat{g}_x}{2N}} \tag{4.111}$$

$$S_{\hat{g}_x} = \left[10^{A - B \log_{10}(N/10)} \right]^{0.5} \tag{4.112}$$

where A and B are given by

$$A = \begin{cases} -0.33 + 0.08|\hat{g}_x| & \text{if } |\hat{g}_x| \le 0.90 \\ -0.52 + 0.30|\hat{g}_x| & \text{if } |\hat{g}_x| > 0.90 \end{cases} \tag{4.113}$$

and

$$B = \begin{cases} 0.94 - 0.26|\hat{g}_x| & \text{if } |\hat{g}_x| \leq 1.50 \\ 0.55 & \text{if } |\hat{g}_x| > 1.50 \end{cases} \qquad (4.114)$$

where $S_{\hat{\mu}_x}$, $S_{\hat{\sigma}_x}$, and $S_{\hat{g}_x}$ are the standard errors of $\hat{\mu}_x$, $\hat{\sigma}_x$, and $\hat{g}_x$ respectively. Combining Equations 4.107 to 4.114 shows that, for a given sample size, the relative accuracy of the skewness is much less than both the mean and standard deviation, especially for small sample sizes.

EXAMPLE 4.20

Annual-maximum peak flows in the Rocky River from 1960 to 1980 show a mean of 52 m^3/s, standard deviation of 21 m^3/s, and skewness of 0.8. Calculate the standard errors, and assess the relative accuracies of the estimated parameters.

Solution   From the given data: $N = 21$ (1960 to 1980), $\hat{\mu} = 52$ m^3/s, $\hat{\sigma} = 21$ m^3/s, and $\hat{g} = 0.8$. Substituting these values into Equations 4.110 to 4.112 yields

$$S_{\hat{\mu}} = \frac{\hat{\sigma}}{\sqrt{N}} = \frac{21}{\sqrt{21}} = 4.6 \text{ m}^3/\text{s}$$

$$S_{\hat{\sigma}} = \hat{\sigma}\sqrt{\frac{1 + 0.75\hat{g}}{2N}} = 21\sqrt{\frac{1 + 0.75(0.8)}{2(21)}} = 4.1 \text{ m}^3/\text{s}$$

$$S_{\hat{g}} = \left[10^{A - B\log_{10}(N/10)}\right]^{0.5}$$

where Equations 4.113 and 4.114 give

$$A = -0.33 + 0.08|0.8| = -0.27$$

$$B = 0.94 - 0.26|0.8| = 0.73$$

and therefore

$$S_{\hat{g}} = \left[10^{-0.27 - 0.73\log_{10}(21/10)}\right]^{0.5} = 0.56$$

Hence, the standard errors of the mean, standard deviation, and skewness are 4.6 m^3/s, 4.1 m^3/s, and 0.56 respectively.

Using the ratio of the standard error to the estimated value as a measure of the relative error of the estimated value, then

$$\text{relative error of mean} = \frac{S_{\hat{\mu}}}{\hat{\mu}} \times 100 = \frac{4.6}{52} \times 100 = 8.8\%$$

$$\text{relative error of standard deviation} = \frac{S_{\hat{\sigma}}}{\hat{\sigma}} \times 100 = \frac{4.1}{21} \times 100 = 20\%$$

$$\text{relative error of skewness} = \frac{S_{\hat{g}}}{\hat{g}} \times 100 = \frac{0.56}{0.8} \times 100 = 70\%$$

Based on these results, it is clear that the accuracies of the estimated parameters deteriorate as the order of the moments associated with the parameters increases.

4.3.2.2   Maximum likelihood method

The *maximum likelihood method* selects the population parameters that maximize the likelihood of the observed outcomes. Consider the case of n independent outcomes $x_1, x_2, \ldots, x_n$, where the probability of any outcome, x_i, is given by $p_X(x_i \mid \theta_1, \theta_2, \ldots, \theta_m)$, where $\theta_1, \theta_2, \ldots, \theta_m$ are the population parameters. The probability of the n observed (independent) outcomes is then given by the product of the probabilities of each of the outcomes. This product is called the *likelihood function*, $L(\theta_1, \theta_2, \ldots, \theta_m)$, where

$$L(\theta_1, \theta_2, \ldots, \theta_m) = \prod_{i=1}^{n} p_X(x_i \mid \theta_1, \theta_2, \ldots, \theta_m) \tag{4.115}$$

The values of the parameters that maximize the value of L are called the *maximum likelihood estimates* of the parameters. Since the form of the probability function, $p_X(x \mid \theta_1, \theta_2, \ldots, \theta_m)$, is assumed to be known, the maximum likelihood estimates can be derived from Equation 4.115 by equating the partial derivatives of L with respect to each of the parameters, θ_i, to zero. This leads to the following m equations

$$\frac{\partial L}{\partial \theta_i} = 0, \qquad i = 1, \ldots, m \tag{4.116}$$

This set of m equations can then be solved simultaneously to yield the m maximum likelihood parameters $\hat{\theta}_1, \hat{\theta}_2, \ldots, \hat{\theta}_m$. In some cases, it is more convenient to maximize the natural logarithm of the likelihood function than the likelihood function itself. This approach is particularly convenient when the probability distribution function involves an exponential term. It should be noted that since the logarithmic function is monotonic, values of the estimated parameters that maximize the logarithm of the likelihood function also maximize the likelihood function.

EXAMPLE 4.21

Use the maximum likelihood method to estimate the parameter in the Poisson distribution of hurricane hits described in the previous example. Compare this result with the parameter estimate obtained using the method of moments.

Solution   The Poisson probability distribution can be written as

$$f(n) = \frac{\theta^n e^{-\theta}}{n!}$$

where $\theta = \lambda t$ is the population parameter of the Poisson distribution. Define the likelihood function, L', for the 86 years of data as

$$L' = \prod_{i=1}^{86} f(n_i)$$

where n_i is the number of occurrences in year i. It is more convenient to work with the log-likelihood function, L, which is defined as

$$L = \ln L' = \sum_{i=1}^{86} \ln f(n_i)$$

$$= \sum_{i=1}^{86} \left[\ln\left(\frac{1}{n_i!}\right) + n_i \ln\theta - \theta \right]$$

$$= \sum_{i=1}^{86} \ln\left(\frac{1}{n_i!}\right) + \ln\theta \sum_{i=1}^{86} n_i - 86\theta$$

Taking the derivative with respect to θ and putting $\partial L/\partial\theta = 0$ gives

$$\frac{\partial L}{\partial \theta} = \frac{1}{\theta} \sum_{i=1}^{86} n_i - 86 = 0$$

which leads to

$$\theta = \frac{1}{86} \sum_{i=1}^{86} n_i$$

This is the same estimate of θ that was derived in the previous example using the method of moments.

The method of moments and the maximum likelihood method do not always yield the same estimates of the population parameters. The maximum likelihood method is generally preferred over the method of moments, particularly for large samples (Haan, 1977). The method of moments is severely affected if the data contain errors in the tails of the distribution, where the moment arms are long (Chow, 1954), and is particularly severe in highly skewed distributions (Haan, 1977). In contrast, the relative asymptotic bias of upper quartiles (high-return-period floods) is smallest for the method of moments and is largest for the maximum likelihood method when the true distribution is either the log-normal distribution or the gamma distribution and another distribution is fitted to it (Strupczewski et al., 2002). Several useful relationships between measured data and maximum likelihood parameter estimates for a variety of distributions can be found in Bury (1999).

4.3.2.3 Method of L-moments

The typically small sample sizes available for characterizing hydrologic time series yield estimates of the third and higher moments that are usually very uncertain. This has led to the use of an alternative system for estimating the parameters of probability distributions called *L-moments*. The rth probability-weighted moment, β_r, is defined by the relation

$$\beta_r = \int_{-\infty}^{+\infty} x[F_X(x)]^r f_X(x)\,dx \tag{4.117}$$

where F_X and f_X are the cumulative distribution function and probability density function of x, respectively. The L-moments, λ_r, are linear combinations of the probability-weighted moments, β_r, and the first four L-moments are computed as

$$\lambda_1 = \beta_0 \tag{4.118}$$

$$\lambda_2 = 2\beta_1 - \beta_0 \tag{4.119}$$

$$\lambda_3 = 6\beta_2 - 6\beta_1 + \beta_0 \tag{4.120}$$

$$\lambda_4 = 20\beta_3 - 30\beta_2 + 12\beta_1 - \beta_0 \tag{4.121}$$

Since the probability-weighted moments involve raising values of F_X rather than x to powers, and because $F_X \leq 1$, then estimates of probability-weighted moments and L-moments are much less susceptible to the influences of a few large or small values in the sample. Hence, L-moments are generally preferable to product moments for estimating the parameters of probability distributions of hydrologic variables.

Consider a sample of N measured values of a random variable X. To estimate the L-moments of the probability distribution from which the sample was taken, first rank the values as $x_1 \leq x_2 \leq x_3 \leq \cdots \leq x_N$, and estimate the probability-weighted moments as follows (Hosking and Wallis, 1997):

$$b_0 = \frac{1}{N} \sum_{i=1}^{N} x_i \tag{4.122}$$

$$b_1 = \frac{1}{N(N-1)} \sum_{i=2}^{N} (i-1)x_i \tag{4.123}$$

$$b_2 = \frac{1}{N(N-1)(N-2)} \sum_{i=3}^{N} (i-1)(i-2)x_i \tag{4.124}$$

$$b_3 = \frac{1}{N(N-1)(N-2)(N-3)} \sum_{i=4}^{N} (i-1)(i-2)(i-3)x_i \tag{4.125}$$

The sample estimates of the first four L-moments, denoted by L_1 to L_4, are then calculated by substituting b_0 to b_3 for β_0 to β_3 respectively in Equations 4.118 to 4.121. The L-moments of various probability distributions are given in terms of the parameters of the distributions in Table 4.1. Equating the sample L-moments to the theoretical L-moments of a distribution and then solving for the distribution parameters constitutes the method of L-moments.

EXAMPLE 4.22

The annual-maximum flows for the period 1946 to 1970 in Dry–Gulch Creek are as follows:

Year	Maximum flow (m³/s)	Year	Maximum flow (m³/s)	Year	Maximum flow (m³/s)
1946	126	1955	3	1964	11
1947	178	1956	2	1965	1122
1948	251	1957	141	1966	2
1949	35	1958	282	1967	1259
1950	71	1959	112	1968	158
1951	501	1960	40	1969	126
1952	891	1961	63	1970	35
1953	18	1962	398		
1954	2239	1963	708		

TABLE 4.1: Moments and L-Moments of Common Probability Distributions

Distribution	Parameters	Moments	L-Moments
Normal	μ_X, σ_X	$\mu_X = \mu_X$ $\sigma_X = \sigma_X$	$\lambda_1 = \mu_X$ $\lambda_2 = \frac{\sigma_X}{\pi^{1/2}}$
Log-normal $(Y = \ln X)$	μ_Y, σ_Y	$\mu_Y = \mu_Y$ $\sigma_Y = \sigma_Y$	$\lambda_1 = \exp\left(\mu_Y + \frac{\sigma_Y^2}{2}\right)$ $\lambda_2 = \exp\left(\mu_Y + \frac{\sigma_Y^2}{2}\right)$ $\mathrm{erf}\left(\frac{\sigma_Y}{2}\right)$
Exponential	ξ, η	$\mu_X = \xi + \frac{1}{\eta}$ $\sigma_X = \frac{1}{\eta^2}$	$\lambda_1 = \xi + \frac{1}{\eta}$ $\lambda_2 = \frac{1}{2\eta}$
Gumbel	ξ, α	$\mu_X = \xi + 0.5772\alpha$ $\sigma_X^2 = 1.645\alpha^2$	$\lambda_1 = \xi + 0.5772\alpha$ $\lambda_2 = 0.6931\alpha$
Generalized extreme value	ξ, α, κ	$\mu_X = \xi + \frac{\alpha}{\kappa}[1 - \Gamma(1 + \kappa)]$ $\sigma_X^2 = \left(\frac{\alpha}{\kappa}\right)^2 \{\Gamma(1 + 2\kappa)$ $-[\Gamma(1 + \kappa)]^2\}$	$\lambda_1 = \xi + \frac{\alpha}{\kappa}[1 - \Gamma(1 + \kappa)]$ $\lambda_2 = \frac{\alpha}{\kappa}(1 - 2^{-\kappa})\Gamma(1 + \kappa)$

Sources: Dingman (2002), Hosking and Wallis (1997).

A histogram of these data indicates that they are likely drawn from a log-normal distribution. Use the method of L-moments to estimate the mean and standard deviation of the log-normal distribution, and compare these parameters with those estimated using the method of moments.

Solution For the $N = 25$ years of data, the rank (i) and corresponding flow (Q_i) are given in columns 1 and 2, respectively, in the following table:

(1)	(2)	(3)	(4)	(5)
i	Q_i (m³/s)	$(1 - i)Q_i$	$\ln Q_i$	$(\ln Q_i - \mu_Y)^2$
1	2		0.693	15.314
2	2	2	0.693	15.314

(1)	(2)	(3)	(4)	(5)
i	$Q_i(\text{m}^3/\text{s})$	$(1 - i)Q_i$	$\ln Q_i$	$(\ln Q_i - \mu_Y)^2$
3	3	6	1.099	12.305
4	11	33	2.398	4.878
5	18	72	2.890	2.945
6	35	175	3.555	1.105
7	35	210	3.555	1.105
8	40	280	3.689	0.842
9	63	504	4.143	0.215
10	71	639	4.263	0.118
11	112	1120	4.718	0.013
12	126	1386	4.836	0.053
13	126	1512	4.836	0.053
14	141	1833	4.949	0.117
15	158	2212	5.063	0.208
16	178	2670	5.182	0.331
17	251	4016	5.525	0.844
18	282	4794	5.642	1.072
19	398	7164	5.986	1.904
20	501	9519	6.217	2.592
21	708	14,160	6.562	3.826
22	891	18,711	6.792	4.778
23	1122	24,684	7.023	5.839
24	1259	28,957	7.138	6.409
25	2239	53,736	7.714	9.655
Sum	8772	178,395	115.163	91.836

The first probability-weighted moment, b_0, is given by Equation 4.122 as

$$b_0 = \frac{1}{N} \sum_{i=1}^{N} Q_i = \frac{1}{25}(8772) = 350.9$$

The product $(1 - i)Q_i$ is shown in column 3, and the second probability-weighted moment, b_1 is given by Equation 4.123 as

$$b_1 = \frac{1}{N(N - 1)} \sum_{i=2}^{N}(i - 1)Q_i = \frac{1}{25(25 - 1)}(178,395) = 297.3$$

The first L-moment, λ_1, is given by Equation 4.118 as

$$\lambda_1 = b_0 = 350.9$$

and the second L-moment, λ_2, is given by Equation 4.119 as

$$\lambda_2 = 2b_1 - b_0 = 2(297.3) - 350.9 = 243.8$$

The relationship between the L-moments, λ_1 and λ_2, and the parameters of the log-normal distribution, μ_Y and σ_Y, (where $Y = \ln Q$) is given in Table 4.1, which yields

$$\lambda_1 = 350.9 = \exp\left(\mu_Y + \frac{\sigma_Y^2}{2}\right)$$

$$\lambda_2 = 243.8 = \exp\left(\mu_Y + \frac{\sigma_Y^2}{2}\right)\operatorname{erf}\left(\frac{\sigma_Y}{2}\right)$$

and solving for μ_Y and σ_Y gives

$$\mu_Y = 4.81$$

$$\sigma_Y = 1.45$$

Hence, using the L-moment method, the estimated mean and standard deviation of the log-normal distribution are 4.81 and 1.45 respectively, where the flows are given in m^3/s.

Column 4 in the preceeding table gives $\ln Q_i$, and column 5 gives $(\ln Q_i - \mu_Y)^2$, and the method of moments gives the mean (μ_Y) and standard deviation (σ_Y) of the log-normal distribution as

$$\mu_Y = \frac{1}{N}\sum_{i=1}^{N}\ln Q_i = \frac{1}{25}(115.163) = 4.61$$

$$\sigma_Y = \sqrt{\frac{1}{N-1}\sum_{i=1}^{N}(\ln Q_i - \mu_Y)^2} = \sqrt{\frac{1}{25-1}\,91.836} = 1.96$$

Hence, using the moment method, the estimated mean and standard deviation of the log-normal distribution are 4.61 and 1.96, respectively.

The difference between the mean and standard deviation (μ_Y, σ_Y) of the log-normal distribution estimated using the moment method (4.61, 1.96) compared with the mean and standard deviation estimated using the L-moment method (4.81, 1.45) is accounted for by the few outliers in the short period of record, which arguably tend to make the L-moment parameter estimates more reliable than the moment parameter estimates.

4.3.3 Frequency Analysis

Frequency analysis is concerned with estimating the relationship between an event, x, and the return period, T, of that event. Recalling that T is related to the exceedance probability of x, $P_X(X \geq x)$, by the relation

$$T = \frac{1}{P_X(X \geq x)} \tag{4.126}$$

then T is related to the cumulative distribution function of x, $P_X(X < x)$, by the relation

$$T = \frac{1}{1 - P_X(X < x)} \tag{4.127}$$

Many cumulative distribution functions cannot be expressed analytically and are tabulated as functions of normalized variables, X', where

$$X' = \frac{X - \mu_x}{\sigma_x} \tag{4.128}$$

and μ_x and σ_x are the mean and standard deviation of the population, respectively. The cumulative probability distribution of X', $P_{X'}(X' < x')$, for many distributions depends only on x' and the skewness coefficient, g_x, of the population and can be readily tabulated in statistical tables. Denoting the value of X' with return period T by x'_T, for many distributions

$$x'_T = K_T(T, g_x) \tag{4.129}$$

where $K(T, g_x)$ is derived from the cumulative distribution function of X' and is commonly called the *frequency factor*. Combining Equations 4.128 and 4.129 leads to

$$\boxed{x_T = \mu_x + K_T \sigma_x} \tag{4.130}$$

where x_T is the realization of X with return period T. The frequency factor is applicable to many, but not all, probability distributions. The frequency factors for a few probability distributions that are commonly used in practice are described here.

Normal distribution.   In the case of a normal distribution, the variable X' is equal to the standard normal deviate, and has a $N(0, 1)$ distribution. The cumulative distribution function of the standard normal deviate is given in Appendix C.1, and the frequency factor is equal to the standard normal deviate corresponding to a given exceedance probability or return period. The frequency factor (= standard normal deviate), K_T, can also be approximated by the empirical relation (Abramowitz and Stegun, 1965)

$$\boxed{K_T = w - \frac{2.515517 + 0.802853w + 0.010328w^2}{1 + 1.432788w + 0.189269w^2 + 0.001308w^3}} \tag{4.131}$$

where

$$w = \left[\ln\left(\frac{1}{p^2}\right) \right]^{\frac{1}{2}}, \qquad 0 < p \le 0.5 \tag{4.132}$$

and p is the exceedance probability $(= 1/T)$. When $p > 0.5$, $1 - p$ is substituted for p in Equation 4.132 and the value of K_T is computed using Equation 4.131 is given a negative sign. According to Abramowitz and Stegun (1965), the error in using Equation 4.131 to estimate the frequency factor is less than 0.00045.

EXAMPLE 4.23

Annual rainfall at a given location is normally distributed with a mean of 127 cm and a standard deviation of 19 cm. Estimate the magnitude of the 50-year annual rainfall.

Solution   The 50-year rainfall, x_{50}, can be written in terms of the frequency factor, K_{50}, as

$$x_{50} = \mu_x + K_{50}\sigma_x$$

From the given data, $\mu_x = 127$ cm, $\sigma_x = 19$ cm, and the exceedance probability, p, is given by

$$p = \frac{1}{T} = \frac{1}{50} = 0.02$$

The intermediate variable, w, is given by Equation 4.132 as

$$w = \left[\ln\left(\frac{1}{p^2}\right)\right]^{\frac{1}{2}} = \left[\ln\left(\frac{1}{0.02^2}\right)\right]^{\frac{1}{2}} = 2.797$$

Equation 4.131 gives the frequency factor, K_{50}, as

$$K_{50} = w - \frac{2.515517 + 0.802853w + 0.010328w^2}{1 + 1.432788w + 0.189269w^2 + 0.001308w^3}$$

$$= 2.797 - \frac{2.515517 + 0.802853(2.797) + 0.010328(2.797)^2}{1 + 1.432788(2.797) + 0.189269(2.797)^2 + 0.001308(2.797)^3}$$

$$= 2.054$$

The 50-year rainfall is therefore given by

$$x_{50} = \mu_x + K_{50}\sigma_x = 127 + 2.054(19) = 166 \text{ cm}$$

Log-normal distribution. In the case of a log-normal distribution, the random variable is first transformed using the relation

$$Y = \ln X \tag{4.133}$$

and the value of Y with return period T, y_T, is given by

$$y_T = \mu_y + K_T\sigma_y \tag{4.134}$$

where μ_y and σ_y are the mean and standard deviation of Y and K_T is the frequency factor of the standard normal deviate with return period T. The value of the original variable, X, with return period T, x_T, is then given by

$$x_T = \ln^{-1} y_T = e^{y_T} \tag{4.135}$$

EXAMPLE 4.24

The annual rainfall at a given location has a log-normal distribution with a mean of 127 cm and a standard deviation of 19 cm. Estimate the magnitude of the 50-year annual rainfall.

Solution From the given data, $\mu_x = 127$ cm, $\sigma_x = 19$ cm, and the mean, μ_y, and standard deviation, σ_y, of the log-transformed variable, $Y = \ln X$, are related to μ_x and σ_x by Equation 4.72, where

$$\mu_x = \exp\left(\mu_y + \frac{\sigma_y^2}{2}\right), \quad \sigma_x^2 = \mu_x^2[\exp(\sigma_y^2) - 1]$$

These equations can be put in the form

$$\mu_y = \frac{1}{2} \ln\left(\frac{\mu_x^4}{\mu_x^2 + \sigma_x^2}\right), \quad \sigma_y = \sqrt{\ln\frac{\sigma_x^2 + \mu_x^2}{\mu_x^2}}$$

which lead to

$$\mu_y = \frac{1}{2} \ln\left(\frac{127^4}{127^2 + 19^2}\right) = 4.83$$

and

$$\sigma_y = \sqrt{\ln\frac{19^2 + 127^2}{127^2}} = 0.149$$

Since the rainfall, X, is log-normally distributed, Y $(= \ln X)$ is normally distributed and the value of Y corresponding to a 50-year return period, y_{50}, is given by

$$y_{50} = \mu_y + K_{50}\sigma_y$$

where K_{50} is the frequency factor for a normal distribution and a return period of 50 years. In the previous example it was shown that $K_{50} = 2.054$, therefore

$$y_{50} = 4.83 + 2.054(0.149) = 5.14$$

and the corresponding 50-year rainfall, x_{50}, is given by

$$x_{50} = e^{y_{50}} = e^{5.14} = 171 \text{ cm}$$

Gamma/Pearson Type III distribution.   In the case of the Pearson Type III distribution, also called the three-parameter gamma distribution, the frequency factor depends on both the return period, T, and the skewness coefficient, g_x. If the skewness coefficient falls between -1 and $+1$, approximate values of the frequency factor for the gamma/Pearson Type III distribution, K_T, can be estimated using the relation (Kite, 1977; Viessman and Lewis, 2003)

$$K_T = \frac{1}{3k}\left\{[(x'_T - k)k + 1]^3 - 1\right\} \tag{4.136}$$

where x'_T is the standard normal deviate corresponding to the return period T and k is related to the skewness coefficient by

$$k = \frac{g_x}{6} \tag{4.137}$$

When the skewness, g_x, is equal to zero, the expanded form of Equation 4.136 indicates that $K_T = x'_T$, and the gamma/Pearson Type III distribution is identical to the normal distribution. In cases where the skewness coefficient falls outside the -1 to $+1$ range, the frequency factors given in Appendix C.2 should be used.

EXAMPLE 4.25

The annual-maximum discharges in a river show a mean of 811 m^3/s, a standard deviation of 851 m^3/s, and a skewness of 0.94. Assuming a Pearson Type III distribution, estimate the 100-year discharge. If subsequent measurements indicate that the skewness coefficient is actually equal to 1.52, how does this affect the estimated 100-year discharge?

Solution From the given data: $\mu_x = 811$ m^3/s, $\sigma_x = 851$ m^3/s, and $g_x = 0.94$. The standard normal deviate corresponding to a 100-year return period, x'_{100}, is 2.33 (Appendix C.1), $k = g_x/6 = 0.94/6 = 0.157$, and the frequency factor K_{100} is given by Equation 4.136 as

$$K_{100} = \frac{1}{3k}\left\{[(x'_{100} - k)k + 1]^3 - 1\right\}$$

$$= \frac{1}{3(0.157)}\left\{[(2.33 - 0.157)(0.157) + 1]^3 - 1\right\}$$

$$= 3.00$$

The 100-year discharge, x_{100}, is therefore given by

$$x_{100} = \mu_x + K_{100}\sigma_x = 811 + (3.00)(851) = 3364 \text{ m}^3/\text{s}$$

If the skewness is equal to 1.52, this is outside the $[-1, +1]$ range of applicability of Equation 4.136, and the frequency factor must be obtained from the tabulated frequency factors in Appendix C.2. For a return period of 100 years and a skewness coefficient of 1.52, Appendix C.2 gives $K_{100} = 3.342$, and the 100-year discharge is given by

$$x_{100} = \mu_x + K_{100}\sigma_x = 811 + (3.342)(851) = 3655 \text{ m}^3/\text{s}$$

Therefore the 62% change in estimated skewness from 0.94 to 1.52 causes a 9% change in the 100-year discharge from 3364 m^3/s to 3655 m^3/s.

Log–Pearson Type III distribution. In the case of a log–Pearson Type III distribution, the random variable is first transformed using the relation

$$Y = \ln X \tag{4.138}$$

The value of Y with return period T, y_T, is given by

$$y_T = \mu_y + K_T\sigma_y \tag{4.139}$$

where μ_y and σ_y are the mean and standard deviation of Y and K_T is the frequency factor of the Pearson Type III distribution with return period T and skewness coefficient g_y. The value of the original variable, X, with return period T, x_T, is then given by

$$x_T = \ln^{-1} y_T = e^{y_T} \tag{4.140}$$

It should be clear that whenever the skewness of $Y (= \ln X)$ is equal to zero, the log–Pearson Type III distribution is identical to the log-normal distribution.

EXAMPLE 4.26

The natural logarithms of annual-maximum discharges (m^3/s) in a river show a mean of 6.33, a standard deviation of 0.862, and a skewness of -0.833. Assuming a log–Pearson Type III distribution, estimate the 100-year discharge.

Solution   From the given data, $\mu_y = 6.33$, $\sigma_y = 0.862$, and $g_y = -0.833$, which yields $k = g_y/6 = -0.833/6 = -0.139$. The standard normal deviate corresponding to a 100-year return period, x'_{100}, is 2.33 (Appendix C.1), and the frequency factor for a 100-year return period, K_{100}, is given by Equation 4.136 as

$$K_{100} = \frac{1}{3k}\left\{[(x'_{100} - k)k + 1]^3 - 1\right\}$$

$$= \frac{1}{3(-0.139)}\left\{[(2.33 + 0.139)(-0.139) + 1]^3 - 1\right\}$$

$$= 1.72$$

The 100-year log-discharge, y_{100}, is therefore given by

$$y_{100} = \mu_y + K_{100}\sigma_y = 6.33 + (1.72)(0.862) = 7.81$$

and the 100-year discharge, x_{100}, is

$$x_{100} = e^{y_{100}} = e^{7.81} = 2465 \text{ m}^3/\text{s}$$

The U.S. Interagency Advisory Committee on Water Data published an important document called Bulletin 17B (USIAC, 1982), which provides detailed guidance on the application of the log–Pearson Type III distribution to annual-maximum flood flows. The flood-frequency analysis procedures outlined in Bulletin 17B are reiterated in Report EM 1110-2-1415 (U.S. Army Corps of Engineers, 1993) and coded into the computer program HEC-FFA (Flood Frequency Analysis) developed and maintained by the Hydrologic Engineering Center (HEC, 1992).

Of particular concern in applying the log–Pearson Type III distribution to observed data is that skew coefficients estimated from small samples might be highly inaccurate. In cases where this is a concern, Bulletin 17B outlines a procedure for developing regional skew coefficients based on at least 40 gaging stations located within a 160-km radius of the site of concern, with each having at least 25 years of data. If this procedure is not feasible, the generalized skew map shown in Figure 4.11 is provided as an easier but less accurate alternative. This map of generalized logarithmic skew coefficients for peak annual flows was developed from skew coefficients computed for 2972 gaging stations, all having at least 25 years of record, following the procedures outlined in Bulletin 17B. Depending on the number of years of record, regionalized skew coefficients are used either in lieu of or in combination with values computed from observed flows at the particular stream gage of concern. Equation 4.141 allows a weighted skew coefficient G_W to be computed by combining a regionalized skew coefficient G_R and station skew coefficient G,

$$G_W = \frac{(MSE_R)(G) + (MSE_S)(G_R)}{MSE_R + MSE_S} \tag{4.141}$$

FIGURE 4.11: Generalized skew coefficients of the logarithms of annual-maximum streamflow

Source: Interagency Advisory Committee on Water Data (Bulletin 17B, 1982).

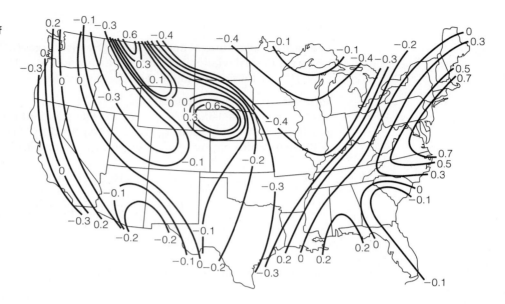

where the station skew G is computed from observed flows at the station of interest, the regional skew, G_R, is either developed from multiple stations following procedures outlined in Bulletin 17B or read from Figure 4.11, MSE_S denotes the mean-square error of the station skew, which can be estimated by (Wallis et al., 1974)

$$MSE_S = 10^{[A - B(\log_{10}(N/10))]} \qquad (4.142)$$

where

$$A = \begin{cases} -0.33 + 0.08|G| & \text{if } |G| \leq 0.90 \\ -0.52 + 0.30|G| & \text{if } |G| > 0.90 \end{cases}$$

$$B = \begin{cases} 0.94 - 0.26|G| & \text{if } |G| \leq 1.50 \\ 0.55 & \text{if } |G| > 1.50 \end{cases}$$

and MSE_R is the mean-square error of the regional skew. If G_R is taken from Figure 4.11, then MSE_R is equal to 0.302.

EXAMPLE 4.27

Observations of annual-maximum flood flows in the Ocmulgee River near Macon, Georgia, from 1980–2000 yield a skew coefficient of the log-transformed data equal to -0.309. Determine the (weighted) skew coefficient that should be used for flood-frequency analysis.

Solution   In central Georgia, where Macon is located, Figure 4.11 gives a generalized skew coefficient G_R of -0.1, and MSE_R is equal to 0.302. From the observed annual-maximum flood flows, $G = -0.309$, and for the 21-year period of record (1980–2000),

$N = 21$. By definition

$$A = -0.33 + 0.08|G| = -0.33 + 0.08|-0.309| = -0.305$$

$$B = 0.94 - 0.26|G| = 0.94 - 0.26|-0.309| = 0.860$$

$$MSE_S = 10^{[A-B(\log_{10}(N/10))]} = 10^{[-0.305-0.860(\log_{10}(21/10))]} = 0.262$$

and the weighted skew coefficient is given by

$$G_W = \frac{(MSE_R)(G) + (MSE_S)(G_R)}{MSE_R + MSE_S}$$

$$= \frac{(0.302)(-0.309) + (0.262)(-0.1)}{0.302 + 0.262} = -0.212$$

Therefore, a skew coefficient of -0.212 is more appropriate than the measured skew coefficient of -0.309 for frequency analyses of flood flows in the Ocmulgee River at Macon, Georgia. As the period of record increases, it is expected that the measured skew coefficient will approach the generalized skew coefficient derived from Figure 4.11.

Extreme-value Type I distribution.   In the case of an extreme-value Type I distribution, the frequency factor, K_T, can be written as (Chow, 1953)

$$K_T = -\frac{\sqrt{6}}{\pi}\left\{0.5772 + \ln\left[\ln\left(\frac{T}{T-1}\right)\right]\right\} \tag{4.143}$$

This expression for K_T is valid only in the limit as the number of samples approaches infinity. For a finite sample size K_T varies with the sample size as shown in Table 4.2

TABLE 4.2: Gumbel Extreme-Value Frequency Factors

Sample size	Return period								
	2.33	5	10	20	25	50	75	100	1000
15	0.065	0.967	1.703	2.410	2.632	3.321	3.721	4.005	6.265
20	0.052	0.919	1.625	2.302	2.517	3.179	3.563	3.386	6.006
25	0.044	0.888	1.575	2.235	2.444	3.088	3.463	3.729	5.842
30	0.038	0.866	1.541	2.188	2.393	3.026	3.393	3.653	5.727
40	0.031	0.838	1.495	2.126	2.326	2.943	3.301	3.554	5.476
50	0.026	0.820	1.466	2.086	2.283	2.889	3.241	3.491	5.478
60	0.023	0.807	1.446	2.059	2.253	2.852	3.200	3.446	5.410
70	0.020	0.797	1.430	2.038	2.230	2.824	3.169	3.413	5.359
75	0.019	0.794	1.423	2.029	2.220	2.812	3.155	3.400	5.338
100	0.015	0.779	1.401	1.998	2.187	2.770	3.109	3.340	5.261
∞	−0.067	0.720	1.305	1.866	2.044	2.592	2.911	3.137	4.900

Source: Viessman and Lewis (2003).

EXAMPLE 4.28

Based on 30 years of data, the annual-maximum discharges in a river have a mean of 811 m^3/s and a standard deviation of 851 m^3/s. Assuming that the annual maxima are described by an extreme-value Type I distribution, estimate the 100-year discharge. What would be the 100-year discharge if the statistics were based on 60 years of data? Compare your results with the prediction made using Equation 4.143.

Solution   From the given data: $\mu_x = 811$ m^3/s, $\sigma_x = 851$ m^3/s, and the 100-year discharge, x_{100}, is given by

$$x_{100} = \mu_x + K_{100}\sigma_x = 811 + K_{100}851$$

For 30 years of data, the 100-year frequency factor is given by Table 4.2 as $K_{100} = 3.653$, and the 100-year discharge is given by

$$x_{100} = 811 + (3.653)851 = 3920 \text{ m}^3/\text{s}$$

For 60 years of data, the 100-year frequency factor is given by Table 4.2 as $K_{100} = 3.446$, and the 100-year discharge is given by

$$x_{100} = 811 + (3.446)851 = 3740 \text{ m}^3/\text{s}$$

For an infinite number of years of data, the 100-year frequency factor is given by Equation 4.143 as

$$K_{100} = -\frac{\sqrt{6}}{\pi}\left\{0.5772 + \ln\left[\ln\left(\frac{100}{100-1}\right)\right]\right\}$$

$$= 3.137$$

which gives

$$x_{100} = 811 + (3.137)851 = 3480 \text{ m}^3/\text{s}$$

Hence, for 30, 60, and an infinite number of years of data, the estimated 100-year discharge is 3920 m^3/s, 3740 m^3/s, and 3480 m^3/s, respectively. Based on these results, it is clear that, for the same statistics, the estimated 100-year discharge decreases as the number of years of data increases. In fact there is approximately 13% difference between the 100-year discharge based on 30 years of data and the 100-year discharge based on an infinite number of years of data.

General extreme-value (GEV) distribution.   In the case of the general extreme-value (GEV) distribution, the value of the random variable, x_T, with a return period T is given by (Kottegoda and Rosso, 1997)

$$x_T = a + \frac{b}{c}\left[1 - \left(\ln\frac{T}{T-1}\right)^c\right] \tag{4.144}$$

where a, b, and c, are the location, scale, and shape parameters of the GEV distribution.

EXAMPLE 4.29

The annual-maximum 24-hour rainfall in a subtropical area is described by a general extreme-value distribution with parameters: $a = 13.8$ cm, $b = 2.17$ cm, and $c = -0.129$. Determine the annual-maximum 24-hour rainfall with a return period of 20 years.

Solution   From the given data: $T = 20$ years, $a = 13.8$ cm, $b = 2.17$ cm, and $c = -0.129$. The annual-maximum 24-hour rainfall with a return period of 20 years is given by Equation 4.144 as

$$x_{20} = a + \frac{b}{c}\left[1 - \left(\ln\frac{T}{T-1}\right)^c\right] = 13.8 + \frac{2.17}{-0.129}\left[1 - \left(\ln\frac{20}{20-1}\right)^{-0.129}\right]$$
$$= 21.7 \text{ cm}$$

Therefore, an annual-maximum 24-hour rainfall of 21.7 cm has a return period of 20 years.

Frequency analyses are generally based on assumed (population) probability distributions and sample estimates of the population parameters. The uncertainty of sample estimates of parameters requires caution when using data records shorter than 10 years and when estimating variables with recurrence intervals longer than twice the record length (Viessman and Lewis, 1966). Some hydrologists recommend that extrapolations to average recurrence intervals that are more than twice the length of the data set should generally be avoided (Davie, 2002).

Summary

Many design variables in water-resources engineering are either intrinsically random or uncertain, and such variables can be defined only by using probability distributions. The parameters of probability distributions can usually be expressed in terms of the mean, variance, and skewness of the outcomes, and these properties are defined in terms of mathematical expectation and moments. Discrete random processes in which the outcomes fall into only two categories can be approximated as Bernoulli processes, in which cases the binomial and geometric distributions are applicable. Continuous random processes can be approximated as Poisson processes, where the Poisson, exponential, and gamma distributions applicable.

Outcomes derived from the sum of a large number of independent and identically distributed random variables can be approximated by the normal distribution, and outcomes derived from the product of such variables can be approximated by the log-normal distribution. In cases where only extreme values are considered, the Gumbel (for maxima), Weibull (for minima), and general extreme-value distributions are appropriate.

The parameters of probability distributions can be estimated using either the method of moments, the maximum likelihood method, or the method of L-moments; and the agreement between a hypothetical probability distribution and observations can be measured by using either the chi-square test or the Kolmogorov–Smirnov test. The relationship between the outcome of a hydrologic process and its return period can be estimated using frequency factors, and these factors are readily available for the normal, log-normal, Pearson Type III, log-Pearson Type III, extreme-value Type I, and general extreme-value distributions.

Problems

4.1. A flood-control system is designed such that the probability that the system capacity is exceeded X times in 30 years is given by the following discrete probability distribution:

x_i	$f(x_i)$
0	0.04
1	0.14
2	0.23
3	0.24
4	0.18
5	0.10
6	0.05
7	0.02
8	0.01
> 9	0.00

What is the mean number of system failures expected in 30 years? What are the variance and skewness of the number of failures?

4.2. The probability distribution of the time between flooding in a residential area is given by

$$f(t) = \begin{cases} 0.143e^{-0.143t}, & t > 0 \\ 0, & t \le 0 \end{cases}$$

where t is the time interval between flood events in days. Estimate the mean, standard deviation, and skewness of t.

4.3. The annual-maximum flows in a river are known to vary between 4 m^3/s and 10 m^3/s and have a probability distribution of the form

$$f(x) = \frac{\alpha}{x^2}$$

where x is the annual-maximum flow. Determine (a) the value of α, (b) the return period of a maximum flow of 7 m^3/s, and (c) the mean and standard deviation of the annual-maximum flows.

4.4. The flows during the summer in a regulated channel vary between 1 m^3/s and 2 m^3/s, and all flows between these values are equally likely. Derive an expression for the probability density function of the river flow, and use this distribution to calculate the mean, variance, and skewness of the streamflows. Identify the flowrate that has a return period of 50 years. [*Hint:* Your derived expression must satisfy the relations: $f(Q) \ge 0$ and $\int_{-\infty}^{\infty} f(Q)\,dQ = 1$.]

4.5. The average rainfall over a certain area is estimated to be 1524 mm, and the rainfall is as likely to be above average in any year (wet year) as it is to be below average. For any 20-year interval, what is the probability of experiencing the following events: (a) one wet year with 19 dry years, and (b) 10 wet years with 10 dry years?

4.6. Use the theory of combinations to show that if the probability of an event occurring in any trial is p, then if all trials are independent, the probability of the event occurring n times in N trials, $f(n)$, is given by

$$f(n) = \frac{N!}{n!(N-n)!}\, p^n (1-p)^{N-n}$$

4.7. A flood-control system is designed for a rainfall event with a return period of 25 years. What is the probability that the design storm will be exceeded five times in 10 years? What is the probability that the design rainfall will be equalled or exceeded at least once in a 10-year period?

4.8. A water-resource system is designed for a hydrologic event with a 25-year return period. Assuming that the hydrologic events can be taken as a Bernoulli process, what is the probability that the design event will be exceeded once in the first five years of system operation? More than once in five years?

4.9. A stormwater-management system is being designed for a (design) life of 10 years, and a (design) runoff event is to be selected such that there is only a 1% chance that the system capacity will be exceeded during the design life. Determine the return period of the design runoff event. If the cumulative distribution function of the runoff, Q (in m^3/s), is given by

$$F(Q) = \exp\left[-\exp\left(-Q\right)\right]$$

what is the design Q?

4.10. A water-resource system is designed for a 50-year storm. Assuming that the occurrence of the 50-year storm is a Bernoulli process, what is the probability that the 50-year storm will be equalled or exceeded one year after the system is constructed? Within the first six years?

4.11. South Florida experiences a six-month wet season between May and October. Within the wet season there are, on average, 35 days with more than 2.5 cm of rainfall. Assuming that the time between 2.5-cm rainfall events is exponentially distributed, what is the probability that there is less than one week between 2.5-cm rainfall events during the wet season?

4.12. A drainage system is designed for a storm with a return period of 10 years. What is the average interval between floods? What is the probability that there will be more than six months between floods?

4.13. Explain why a gamma distribution with $n = 1$ is the same as an exponential distribution.

4.14. A residential development in the southern United States floods whenever more than 10.2 cm of rain falls in 24 hours. In a typical year, there are three such storms. Assuming that these rainfall events are a Poisson process, what is the probability that it will take more than one year to have three flood events?

4.15. The annual rainfall in Everglades National Park has been estimated to have a mean of 141.2 cm and a standard deviation of 28.2 cm (Chin, 1993a). Assuming that the annual rainfall is normally distributed, what is the probability of having (a) an annual rainfall of less than 127.0 cm, (b) an annual rainfall of less than 152.4 cm, and (c) an annual rainfall between 127.0 and 152.4 cm?

4.16. Annual-maximum discharges in a river show a mean of 620 m^3/s and a standard deviation of 311 m^3/s. If the capacity of the river is 780 m^3/s and the flow can be assumed to follow a log-normal distribution, what is the probability that the maximum discharge will exceed the channel capacity?

4.17. The annual rainfall in a rural town is shown to be log-normally distributed with a mean of 114 cm and a standard deviation of 22 cm. Estimate the rainfall having a return period of 100 years. How would your estimate differ if the rainfall were normally distributed?

4.18. Streamflow data collected from several sources indicate that the peak flow in a drainage channel at a given location is somewhere in the range between 95 and 115 m^3/s. Assuming that the uncertainty can be described by a uniform distribution, what is the probability that the peak flow is greater than 100 m^3/s?

4.19. The probability distribution for the maximum values in a sample set is given by Equation 4.79. Derive a similar expression for the minimum values in a sample set.

4.20. The annual-maximum discharges in a river show a mean of 480 m^3/s and a standard deviation of 320 m^3/s. Assuming that the annual-maximum flows are described by an extreme-value Type I (Gumbel) distribution, use Equation 4.85 to estimate the annual-maximum flowrates corresponding to return periods of 50 and 100 years.

4.21. The annual-minimum rainfall amounts have been tabulated for a region and show a mean of 710 mm and a standard deviation of 112 mm. Assuming that these minima are described by an extreme-value Type I distribution, estimate the average interval between years where the minimum rainfall exceeds 600 mm.

4.22. The annual-low flows in a drainage channel have a mean of 43 m^3/s and a standard deviation of 12 m^3/s. Assuming that the annual-low flows have a lower bound of 0 m^3/s and are described by an extreme-value Type III distribution, what is the probability that an annual-low flow will be less than 10 m^3/s?

4.23. Measurements of annual-minimum flows in a river indicate that the minimum annual-minimum is 9 m^3/s, an annual-minimum flow of 34 m^3/s has a return of 10 years, and an annual-minimum flow of 76 m^3/s has a return period of 50 years. If these data follow an extreme-value Type III distribution, what would the 100-year annual-minimum flow be?

4.24. The annual-maximum streamflows in a river have a mean of 63.2 m^3/s, a standard deviation of 13.7 m^3/s, and a skewness coefficient of 1.86. Assuming that the annual-maximum streamflows are described by a general extreme-value (GEV) distribution, estimate the annual-maximum streamflow with a return period of 50 years.

4.25. A random variable, X, is equal to the sum of the squares of 15 normally distributed variables with a mean of zero and a standard deviation of 1. What is the probability that X is greater than 25? What is the expected value of X?

4.26. Exceedance probabilities of measured data can be estimated using the Gringorten formula

$$P_X(X > x_m) = \frac{m - a}{N + 1 - 2a}$$

where x_m is a measurement with rank m, N is the total number of measurements, and a is a constant that depends on the population probability distribution. Values of a are typically in the range of 0.375 to 0.44. Determine the formula for the return period of x_m. What is the maximum error in the return period that can result from uncertainty in a?

4.27. The annual peak flows in a river between 1980 and 1996 are as follows:

Year	Peak flow (ft^3/s)
1980	8000
1981	5550
1982	3390
1983	6390
1984	5889
1985	7182
1986	10,584
1987	11,586
1988	8293
1989	9193
1990	5142
1991	7884
1992	4132
1993	12,136
1994	5129
1995	7236
1996	6222

Use the Weibull and Gringorten formulae to estimate the cumulative probability distribution of peak flow and compare the results.

4.28. Analysis of a 50-year record of annual rainfall indicates the following frequency distribution:

Range (mm)	Number of outcomes
< 960	4
960–1064	5
1064–1138	6
1138–1201	6
1201–1260	5
1260–1318	7
1318–1382	4
1382–1455	3
1455–1560	6
> 1560	4

The measured data also indicate a mean of 1260 mm and a standard deviation of 234 mm. Using the chi-square test with a 5% significance level, assess the hypothesis that the annual rainfall is drawn from a normal distribution.

4.29. Use the Kolmogorov–Smirnov test at the 10% significance level to assess the hypothesis that the data in Problem 4.28 are drawn from a normal distribution.

4.30. A sample of a random variable has a mean of 1.6, and the random variable is assumed to have an exponential distribution. Use the method of moments to determine the parameter (λ) in the exponential distribution.

4.31. A sample of a random variable has a mean of 1.6 and a standard deviation of 1.2. If the random variable is assumed to have a log-normal distribution, apply the method of moments to estimate the parameters in the log-normal distribution.

4.32. Annual-maximum peak flows in the Davidian River from 1940 to 1980 show a mean of 35 m^3/s, standard deviation of 15 m^3/s, and skewness of 0.4. Calculate the standard errors, and assess the relative accuracies or the estimated parameters.

4.33. A sample of random data is to be fitted to the exponential probability distribution given by Equation 4.44. Derive the maximum likelihood estimate of the distribution parameter λ. Compare your result with the method of moments estimate.

4.34. Derive the maximum-likelihood estimator of the parameters in the normal distribution.

4.35. The annual-maximum flows for the period 1970 to 1996 in the Bucking Bronco River are as follows:

Year	Maximum flow (m^3/s)	Year	Maximum flow (m^3/s)	Year	Maximum flow (m^3/s)
1970	189	1979	5	1988	17
1971	267	1980	3	1989	1683
1972	377	1981	212	1990	4
1973	53	1982	423	1991	1889
1974	107	1983	168	1992	237
1975	752	1984	60	1993	189
1976	1337	1985	95	1994	53
1977	27	1986	597	1995	550
1978	3359	1987	1062	1996	320

A histogram of these data indicate that they are likely drawn from a log-normal distribution. Use the method of L-moments to estimate the mean and standard deviation of the log-normal distribution, and compare these parameters with those estimated using the method of moments.

4.36. The annual rainfall at a given location is normally distributed with a mean of 152 cm and a standard deviation of 30 cm. Estimate the magnitude of the 10-year and 100-year annual rainfall amounts.

4.37. The annual rainfall at a given location has a log-normal distribution with a mean of 152 cm and a standard deviation of 30 cm. Estimate the magnitude of the 10-year and 100-year annual rainfall amounts. Compare your results with those obtained in Problem 4.36.

4.38. The annual-maximum discharges in a river show a mean of 480 m^3/s, a standard deviation of 320 m^3/s, and a skewness of 0.87. Assuming a Pearson Type III distribution, use the frequency factor to estimate the 50-year and 100-year annual-maximum flows.

4.39. Compare the Pearson Type III frequency factor for a 100-year event estimated using Equation 4.136 with the same frequency factor estimated using Appendix C.2, for skewness coefficients of 3.0, 2.0, 1.0, 0.5, and 0.0. How do these results support the assertion that Equation 4.136 should be used only when the skewness coefficient is in the range $[-1, +1]$?

4.40. The annual-maximum discharges in a river show a mean of 480 m^3/s, a standard deviation of 320 m^3/s, and a log-transformed skewness of 0.1. Assuming a log–Pearson Type III distribution, estimate the 100-year annual-maximum flow.

4.41. Observations of annual-maximum flood flows in the Mississippi River near Baton Rouge, Louisiana, from 1985–2000 yield a skew coefficient of the log-transformed data equal to -0.210. Determine the (weighted) skew coefficient that should be used for flood-frequency analysis.

4.42. Based on 25 years of data, the annual-maximum discharges in a river have a mean of 480 m^3/s and a standard deviation of 320 m^3/s. Assuming that the annual maxima are described by an extreme-value Type I distribution, estimate the 100-year discharge. Compare your result with the prediction made using Equation 4.143, and with the result obtained in Problem 4.20.

4.43. Consider the series of annual-maximum streamflow measurements given in the table below. (a) Estimate the return period of a 30-m^3/s (annual-maximum) flow using the Weibull formula; and (b) estimate the return period of a 30-m^3/s (annual-maximum) flow using the Gringorten formula. It is postulated that the observed data follow the extreme-value Type I distribution. (c) Use the method of moments to estimate the parameters (a and b) of the extreme-value Type I distribution; (d) use the chi-square test with six classes to assess whether the data fit an extreme-value Type I distribution; and (e) assuming that the data are adequately described by an extreme-value Type I distribution, use the frequency factor to determine the annual-maximum streamflow with a 100-year return period.

Year	Flow (m^3/s)
1970	20
1971	23
1972	13
1973	28
1974	35
1975	19

Year	Flow (m³/s)
1976	14
1977	10
1978	31
1979	25
1980	9
1981	40
1982	22
1983	19
1984	17

4.44. The annual-maximum streamflows in a river are described by a general extreme-value distribution with parameters: $a = 55.2$ m³/s, $b = 8.68$ m³/s, and $c = -0.163$. Determine the annual-maximum streamflow with a return period of 100 years.

CHAPTER 5

Surface-Water Hydrology

5.1 Introduction

Surface-water hydrology deals with the distribution, movement, and properties of water above the surface of the earth. Applications of surface-water hydrology in engineering practice include modeling rainfall events and predicting the quantity and quality of the resulting surface runoff. The temporal distribution of rainfall at a given location is called a *hyetograph*, the temporal distribution of surface runoff at any location is called a *hydrograph*, and the temporal distribution of pollutant concentration in the runoff is called a *pollutograph*. The estimation of hydrographs and pollutographs from hyetographs and the design of systems to control the quantity and quality of surface runoff are the responsibility of a water-resources engineer. Agricultural production is a significant component of most societies, and the water supplied to agricultural crops by rainfall must usually be supplemented by irrigation water derived from various terrestrial sources. To determine the irrigation requirements, water-resources engineers must be able to predict the amount of water consumed by *evapotranspiration*, and ensure that water resources are made available to sustain important agricultural activities.

The land area that can contribute to the runoff at any particular location is determined by the shape and topography of the surrounding region. The potential contributing area is called the *watershed*; the area within a watershed over which rainfall occurs is called the *catchment* area. In most engineering applications, the watershed and catchment areas are taken to be the same, and are sometimes referred to as *drainage basins* or *drainage areas*. The characteristics of the catchment area determine the quantity, quality, timing, and distribution of the surface runoff for a given rainfall event. Pollutants contained in surface runoff generally depend on the land uses within the catchment area, and classes of pollutants contained in surface-water runoff that are typically of interest include suspended solids, heavy metals, nutrients, organics, oxygen-demanding substances, and bacteria (ASCE, 1992). These pollutants may be in solution or suspension, or attached to particles of sediment.

5.2 Rainfall

The precipitation of water vapor from the atmosphere occurs in many forms, the most important of which are rain and snow. Hail and sleet are less frequent forms of precipitation. Engineered drainage systems in most urban communities are designed primarily to control the runoff from rainfall. The formation of precipitation usually results from the lifting of moist air masses within the atmosphere, which results in the cooling and condensation of moisture. The four conditions that must be present for the production of precipitation are: (1) cooling of the air mass, (2) condensation of water droplets onto nuclei, (3) growth of water droplets, and (4) mechanisms to cause a sufficient density of the droplets.

Cloud droplets generally form on condensation nuclei, which are usually less than 1 micron in diameter and typically consist of sea salts, dust, or combustion by-products. These particles are called *aerosols*. In pure air, condensation of water vapor to form liquid water droplets occurs only when the air is supersaturated. Typically, once condensation begins, water droplets and ice crystals are both formed. However, the greater saturation vapor pressure over water compared to over ice results in a vapor pressure gradient that causes the growth of ice crystals at the expense of water droplets. This is called the *Bergeron process* or *ice-crystal process*. When the condensed moisture droplet is large enough, the precipitation falls to the ground. Moisture droplets larger than about 0.1 mm are large enough to fall, and these drops grow as they collide and coalesce to form larger droplets. Rain drops falling to the ground are typically in the size range of 0.5–3 mm, while drizzle is a subset of rain with droplet sizes less than 0.5 mm.

The main mechanisms of air-mass lifting are frontal lifting, orographic lifting, and convective lifting. In *frontal lifting*, warm air is lifted over cooler air by frontal passage. The resulting precipitation events are called *cyclonic* or *frontal* storms, and the zone where the warm and cold air masses meet is called a *front*. Frontal precipitation is the dominant type of precipitation in the north-central United States and other continental areas (Elliot, 1995). In a *warm front*, warm air advances over a colder air mass with a relatively slow rate of ascent (0.3%–1% slope) causing a large area of precipitation in advance of the front, typically 300 to 500 km (200 to 300 mi) ahead of the front. In a *cold front*, warm air is pushed upward at a relatively steep slope (0.7%–2%) by advancing cold air, leading to smaller precipitation areas in advance of the cold front. Precipitation rates are generally higher in advance of cold fronts than in advance of warm fronts. In *orographic lifting*, warm air rises as it flows over hills or mountains, and the resulting precipitation events are called *orographic storms*. An example of the orographic effect can be seen in the northwestern United States, where the westerly air flow results in higher precipitation and cooler temperatures to the west of the Cascade mountains (e.g., Seattle, Washington) than to the east (e.g., Boise, Idaho). Orographic precipitation is a major factor in most mountainous areas, and the amount of precipitation typically shows a strong correlation with elevation (Naoum and Tsanis, 2003). In *convective lifting*, air rises by virtue of being warmer and less dense than the surrounding air, and the resulting precipitation events are called *convective storms* or, more commonly, *thunderstorms*. A typical thunderstorm is illustrated in Figure 5.1, where the localized nature of such storms is clearly apparent. Convective precipitation is common during the summer months in the central United States and other continental climates with moist summers. Convective storms are typically of short duration, and usually occur on hot midsummer days as late-afternoon storms.

5.2.1 Local Rainfall

Records of rainfall have been collected for more than 2000 years (Ward and Robinson, 1999). Rainfall is typically measured using rain gages operated by government agencies

such as the National Weather Service and local drainage districts. Rainfall amounts are described by the volume of rain falling per unit area and are given as a depth of water. Gages for measuring rainfall are categorized as either nonrecording (manual) or recording.

In the United States, many of the rainfall data reported by the National Weather Service (NWS) are collected manually using a standard rain gage that consists of a 20.3-cm (8-in.) diameter funnel that passes water into a cylindrical measuring tube, the whole assembly being placed within an overflow can. The measuring tube has a cross-sectional area one-tenth that of the collector funnel; therefore, a 2.5-mm (0.1-in.) rainfall will occupy a depth of 25 mm (1 in.) in the collector tube. The capacity of the standard collector tube is 50 mm (2 in.), and rainfall in excess of this amount collects in the overflow can. The manual NWS gage is primarily used for collecting daily rainfall amounts.

Automatic-recording gages are usually used for measuring rainfall at intervals less than one day, and for collecting data in remote locations. Recording gages use either a tipping bucket, weighing mechanism, or float device. The tipping-bucket rain gage was invented by Sir Christopher Wren in about 1662, and now accounts for nearly half of all recording rain gages worldwide (Upton and Rahimi, 2003). A tipping-bucket gage is based on funneling the collected rain to a small bucket that tilts and empties each time it fills, generating an electronic pulse with each tilt. The number of bucket-tilts per time interval provides a basis for determining the precipitation depth over time. Typical problems with tipping-bucket rain gages include blockages, wetting and evaporation losses of typically 0.05 mm per rainfall event (Niemczynowicz, 1986), rain missed during the tipping process which typically takes about half a second (Marsalek, 1981), wind effect, position, and shelter. A weighing-type gage is based on continuously recording the weight of the accumulated precipitation, and float-type recording gages operate by catching rainfall in a tube containing a float whose rise is recorded with time. The accuracy of both manual and automatic-recording rain gages is typically on the order of 0.25 mm (0.01 in.) per rainfall event.

Rain-gage measurements are actually point measurements of rainfall and are only representative of a small area surrounding the rain gage. Areas on the order of

FIGURE 5.2: Weather radar station (WSR-88D)

25 km^2 (10 mi^2) have been taken as characteristic of rain-gage measurements (Gupta, 1989; Ponce, 1989), although considerably smaller characteristic areas can be expected in regions where convection storms are common.

In the United States, the National Weather Service makes extensive use of *Weather Surveillance Radar 1988 Doppler* (WSR-88D), commonly known as next-generation weather radar (*NEXRAD*). A typical WSR-88D tower is shown in Figure 5.2, and each WSR-88D station measures weather activity over a 230-km-radius circular area with a 10-cm wavelength signal. This radar system provides estimates of rainfall from the reflectivity of the S-band signal (1.55 GHz to 3.9 GHz) within cells that are approximately 4 km × 4 km. Since radar systems measure only the movement of water in the atmosphere, and not the volume of water falling to the ground, radar-estimated rainfall should generally be corrected to correlate with field measurements on the ground. Comparisons of rainfall predictions using WSR-88D with rainfall measurements using dense networks of rain gages have shown that WSR-88D estimates tend to be about 5% to 10% lower than rain-gage estimates (Johnson et al., 1999); however, radar provides the advantage of covering large areas with high spatial and temporal resolution.

Rainfall measurements are seldom used directly in design applications, but rather the statistics of the rainfall measurements are typically used. Rainfall statistics are most commonly presented in the form of *intensity-duration-frequency* (IDF) curves, which express the relationship between the average intensity in a rainstorm and the averaging time (= duration), with the average intensity having a given probability. Such curves are also called *intensity-frequency duration* (IFD) curves (Gyasi-Agyei, 2005). To fully understand the meaning and application of the IDF (or IFD) curve, it is best to review how this curve is calculated from raw rainfall measurements. The data required to calculate the IDF curve are a record of rainfall measurements in the form of the depth of rainfall during fixed intervals of time, Δt, typically on the order of 5 minutes. Local rainfall data are usually in the form of daily totals for nonrecording gages, with smaller time increments used in recording gages. For a rainfall record containing several years of data, the following computations lead to the IDF curve:

1. For a given duration of time (= averaging period), starting with Δt, determine the maximum rainfall amount for this duration in *each* year.

2. The precipitation amounts, one for each year, are rank-ordered, and the return period, T, for each precipitation amount is estimated using the Weibull formula

$$T = \frac{n + 1}{m} \tag{5.1}$$

where n is the number of years of data and m is the rank of the data corresponding to the event with return period T. As an alternative to using the Weibull formula to estimate the return periods of the precipitation amounts, the extreme-value Type I distribution may be assumed. In this case, the return periods of the precipitation amounts are derived from the mean and standard deviation of the annual-maximum depths (Akan, 1993).

3. Steps 1 and 2 are repeated, with the duration increased by Δt. A maximum duration needs to be specified, and for urban-drainage applications is typically on the order of 1 to 2 h.

4. For each return period, T, the precipitation amount versus duration can be plotted. This relationship is called the depth-duration-frequency curve. Dividing the precipitation amount by the corresponding duration yields the average intensity, which is plotted versus the duration, for each return period, to yield the IDF curve.

This procedure is illustrated in the following example.

EXAMPLE 5.1

A rainfall record contains 32 years of rainfall measurements at 5-minute intervals. The maximum rainfall amounts for intervals of 5 min, 10 min, 15 min, 20 min, 25 min, and 30 min have been calculated and ranked. The top three rainfall amounts, in millimeters, for each time increment are given in the following table.

Rank	Δt in minutes					
	5	10	15	20	25	30
1	12.1	18.5	24.2	28.3	29.5	31.5
2	11.0	17.9	22.1	26.0	28.4	30.2
3	10.7	17.5	21.9	25.2	27.6	29.9

Calculate the IDF curve for a return period of 20 years.

Solution For each time interval, Δt, there are $n = 32$ ranked rainfall amounts of annual maxima. The relationship between the rank, m, and the return period, T, is given by Equation 5.1 as

$$T = \frac{n + 1}{m} = \frac{32 + 1}{m} = \frac{33}{m}$$

The return period can therefore be used in lieu of the rankings, and the given data can be put in the form:

Return period, T (years)	Δt in minutes					
	5	10	15	20	25	30
33	12.1	18.5	24.2	28.3	29.5	31.5
16.5	11.0	17.9	22.1	26.0	28.4	30.2
11	10.7	17.5	21.9	25.2	27.6	29.9

The rainfall increments with a return period, T, of 20 years can be linearly interpolated between the rainfall increments corresponding to $T = 33$ years and $T = 16.5$ years to yield:

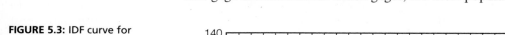

Duration (min)	5	10	15	20	25	30
Rainfall (mm)	11.2	18.0	22.5	26.5	28.6	30.5

and the average intensities for each duration are obtained by dividing the rainfall amounts by the corresponding duration to yield:

Duration (min)	5	10	15	20	25	30
Intensity (mm/h)	134	108	90	79	69	61

These points define the IDF curve for a return period of 20 years, and this IDF curve is plotted in Figure 5.3.

Many of the data used in deriving IDF curves are derived from tipping-bucket rain-gage measurements. These gages, the most popular type of rain gage employed

FIGURE 5.3: IDF curve for 20-year return period

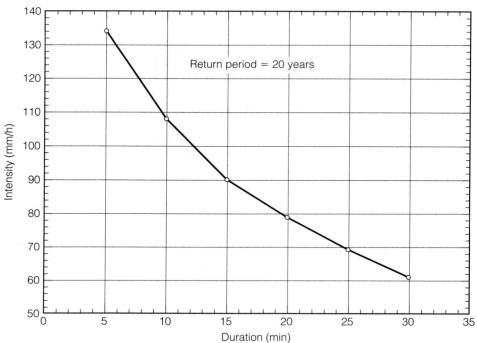

worldwide, are known to underestimate rainfall at high intensities because of the rain missed during the tipping of the bucket. Unless the rain gage used in deriving the IDF curve is dynamically calibrated, or the rain gage is self-calibrating, the derived IDF curve can lead to significant underdesign of urban drainage systems (Molini et al., 2005).

EXAMPLE 5.2

How many years of rainfall data are required to derive the IDF curve for a return period of 10 years?

Solution   The return period, T, is related to the number of years of data, n, and the ranking, m, by

$$T = \frac{n + 1}{m}$$

For $T = 10$ years and $m = 1$,

$$n = mT - 1 = (1)(10) - 1 = 9 \text{ years}$$

Therefore a minimum of 9 years of data are required to estimate the IDF curve for a 10-year return period.

The previous example has illustrated the derivation of IDF curves from n annual maxima of rainfall measurements, where the series of annual maxima is simply called the *annual series*. As an alternative to using an n-year annual series to derive IDF curves, a *partial-duration series* is sometimes used in which the largest n rainfall amounts in an n-year record are selected for each duration, regardless of the year in which the rainfall amounts occur. In this case, the return period, T, assigned to each rainfall amount is still calculated using Equation 5.1. The frequency distribution of rainfall amounts derived using the partial-duration series differs from that derived using the annual series. However, a rainfall amount with a return period T derived from a partial-duration series can be converted to a corresponding rainfall amount for an annual series by using the empirical factors given in Table 5.1. These factors are applicable for return periods greater than 2 years. Partial-duration and annual series compare favorably at the larger recurrence intervals, but for smaller recurrence intervals the partial-duration series will normally indicate events of greater magnitude.

TABLE 5.1: Factors for Converting Partial-Duration Series to Annual Series

Return period (years)	Factor
2	0.88
5	0.96
10	0.99
25	1.00
>25	1.00

Source: Frederick et al. (1977).

EXAMPLE 5.3

A 10-year rainfall record measures the rainfall increments at 5-minute intervals. The top six rainfall increments derived from the partial-duration series are as follows:

Rank	1	2	3	4	5	6
5-min rainfall (mm)	22.1	21.9	21.4	20.7	20.3	19.8

Estimate the frequency distribution of the annual maxima with return periods greater than 2 years.

Solution   The ranked data were derived from a 10-year partial-duration series ($n = 10$), where the return period, T, is given by

$$T = \frac{n + 1}{m} = \frac{10 + 1}{m} = \frac{11}{m}$$

Using this (Weibull) relation, the frequency distribution of the 5-min rainfall increments is given by

Return period (y)	11	5.5	3.7	2.8	2.2	1.8
5-min rainfall (mm)	22.1	21.9	21.4	20.7	20.3	19.8

The calculated rainfall amounts for the partial-duration series can be converted to corresponding rainfall amounts for the annual series of maxima by using the factors in Table 5.1. Applying linear interpolation in Table 5.1 leads to the following conversion factors:

Return period (y)	11	5.5	3.7	2.8	2.2	1.8
Factor	0.99	0.96	0.93	0.90	0.89	—

Applying these factors to the partial-duration frequency distribution leads to the following frequency distribution of annual-maxima 5-min rainfall increments:

Return period (y)	11	5.5	3.7	2.8	2.2
5-min rainfall (mm)	21.9	21.0	19.9	18.6	18.1

This analysis can be repeated for different durations in order to determine the IDF curves from a partial-duration series.

The frequency distributions of local rainfall in the United States have been published by Hershfield (1961) for storm durations from 30 min to 24 hours, and return periods from 1 to 100 years. Hershfield's paper is commonly referred to as TP-40 (an acronym for the U.S. Weather Bureau* *Technical Paper Number 40*, published by Hershfield in 1961) or simply the *Rainfall Frequency Atlas*. The frequency distributions in TP-40 were derived from data at approximately 4000 stations by assuming a Gumbel distribution. The data in TP-40 for the 11 western states have been updated by Miller and colleagues (1973). These 11 western states are Montana, Wyoming, Colorado, New Mexico, Idaho, Utah, Nevada, Arizona, Washington, Oregon, and California.

*The United States Weather Bureau is now called the National Weather Service.

FIGURE 5.4:
Intensity-duration-frequency
(IDF) curve

Source: Florida Department of
Transportation (2000).

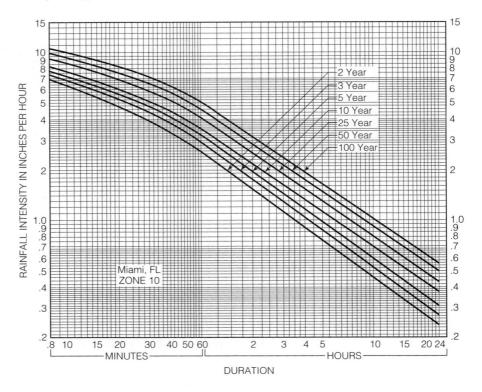

Frequency distributions for local rainfall in Alaska can be found in TP-47 (Miller, 1963), for Hawaii in TP-43 (U.S. Department of Commerce, 1962), and for Puerto Rico and the Virgin Islands in TP-42 (U.S. Department of Commerce, 1961). The U.S. Weather Bureau has also published rainfall frequency maps for storm durations from 2 to 10 days (USWB, 1964) in TP-49. In designing urban drainage systems, rainfall durations of less than 60 min and sometimes as brief as 5 min are commonly used (Wenzel, 1982). The rainfall statistics for durations from 5 to 60 min have been published by Frederick and colleagues (1977) for the eastern and central United States. This paper is commonly referred to as HYDRO-35 (an acronym for the NOAA *Technical Memorandum Number NWS HYDRO-35*), and these results partially supersede those in TP-40.

A typical IDF curve (for Miami, Florida) is illustrated in Figure 5.4. Although IDF curves are available for many locations within the United States, usually from a local water-management or drainage district, there are some locations for which IDF curves are not available. In these cases, IDF curves can be estimated from the published National Weather Service frequency distributions in TP-40 using a methodology proposed by Chen (1983). This method is based on three rainfall depths derived from TP-40: the 10-year 1-hour rainfall (R_1^{10}) shown in Figure 5.5, the 10-year 24-hour rainfall (R_{24}^{10}) shown in Figure 5.6, and the 100-year 1-hour rainfall (R_1^{100}) shown in Figure 5.7. The IDF curve can then be expressed as a relationship between the average intensity, i (in./h), for a rainfall duration t (min) by

$$i = \frac{a}{(t + b_1)^{c_1}} \qquad (5.2)$$

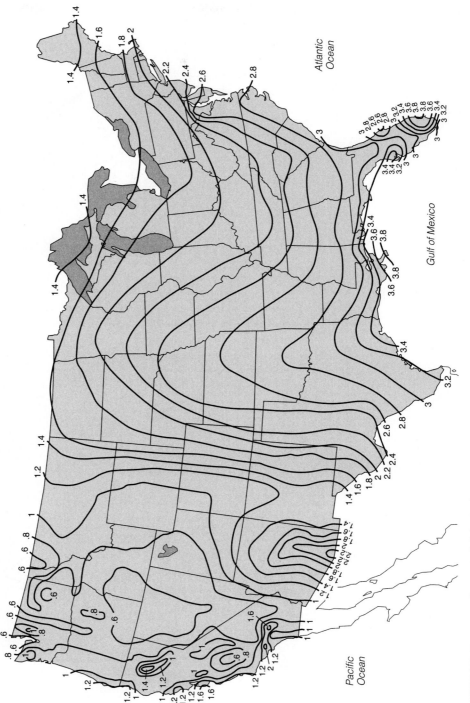

FIGURE 5.5: Ten-year 1-hour rainfall (inches)

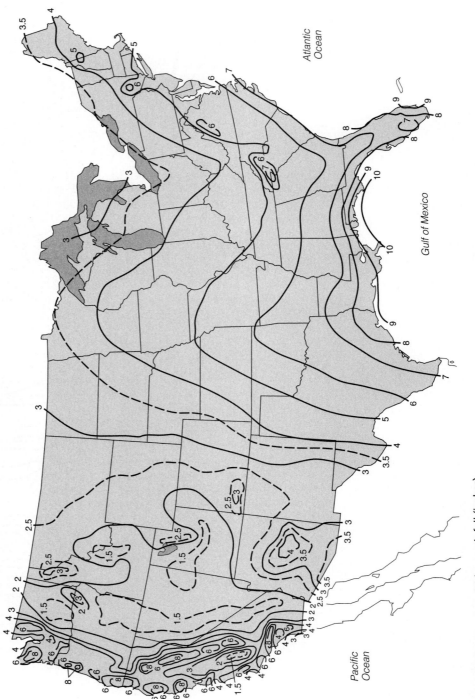

FIGURE 5.6: Ten-year 24-hour rainfall (inches)

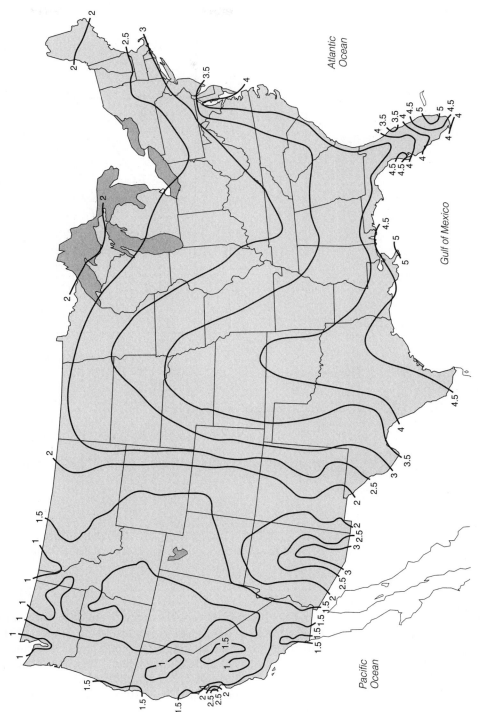

FIGURE 5.7: One-hundred-year 1-hour rainfall (inches).

Source: Chen (1983).

FIGURE 5.8: Constants in Chen IDF curve.

Source: Chen (1983).

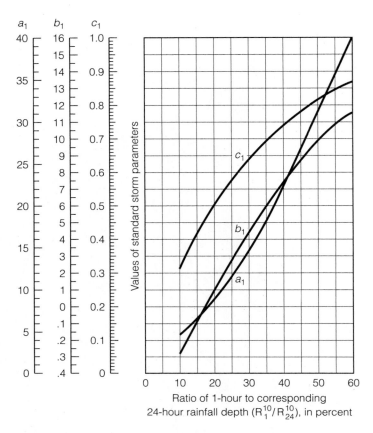

Ratio of 1-hour to corresponding 24-hour rainfall depth (R_1^{10}/R_{24}^{10}), in percent

where a is a constant given by

$$a = a_1 R_1^{10}[(x - 1)\log(T_p/10) + 1] \qquad (5.3)$$

a_1, b_1, and c_1 are empirical functions of R_1^{10}/R_{24}^{10} derived from Figure 5.8, x is defined by

$$x = \frac{R_1^{100}}{R_1^{10}} \qquad (5.4)$$

and T_p is the return period for the partial-duration series, which is assumed to be related to return period, T, for the annual-maximum series by the relation

$$T_p = -\frac{1}{\ln(1 - 1/T)} \qquad (5.5)$$

For $T > 10$ years there is not a significant difference between T and T_p. An IDF curve can be developed for any location using the Chen (1983) method, which is appropriate for storm durations ranging from 5 min to 24 hours and return periods greater than or equal to 1 year. In applying the Chen method within the United States, it is important to note that values of R_1^{10}, R_{24}^{10}, interpolated from Figures 5.5 and 5.6, respectively, have standard errors of at least 10%, and values of R_1^{100} interpolated from Figure 5.7 have standard errors of at least 20% (Hershfield, 1961).

The Chen method has been widely applied both within and outside the United States; however, in countries such as Canada, India, and Italy alternative generic IDF curves have proven to be more appropriate than the Chen method (Alila, 2000). Urbanization can have a significant effect on local rainfall characteristics, and the stationarity of IDF curves derived from historical data is sometimes a concern. For example, predevelopment rainfall characteristics in the cities of St. Louis, Phoenix, and Houston are demonstrably different from current rainfall characteristics (Burian and Shepherd, 2005; Balling and Brazel, 1987; Huff and Vogel, 1978).

EXAMPLE 5.4

Estimate the IDF curve for 50-year storms in Miami using the Chen method. What is the average intensity of a 50-year 1-hour storm?

Solution For Miami, $R_1^{10} = 3.6$ in. (Figure 5.5), $R_{24}^{10} = 9$ in. (Figure 5.6), and $R_1^{100} = 4.7$ in. (Figure 5.7). For a return period, T, equal to 50 years,

$$T_p = -\frac{1}{\ln(1 - 1/T)} = -\frac{1}{\ln(1 - 1/50)} = 49.5 \text{ years}$$

$$x = \frac{R_1^{100}}{R_1^{10}} = \frac{4.7}{3.6} = 1.31$$

Since $R_1^{10}/R_{24}^{10} = 3.6/9 = 0.4 = 40\%$, then Figure 5.8 gives $a_1 = 22.8$, $b_1 = 7.5$, and $c_1 = 0.74$. Equation 5.3 gives

$$a = a_1 R_1^{10}[(x - 1)\log(T_p/10) + 1] = (22.8)(3.6)[(1.31 - 1)\log(49.5/10) + 1] = 99.8$$

and therefore the IDF curve is given by

$$i = \frac{a}{(t + b_1)^{c_1}} = \frac{99.8}{(t + 7.5)^{0.74}}$$

For a storm of duration, t, equal to 1 hour ($= 60$ min),

$$i = \frac{99.8}{(60 + 7.5)^{0.74}} = 4.4 \text{ in./h}$$

A 50-year 1-hour storm in Miami therefore has an average intensity of 4.4 in./h.

Wenzel (1982) developed empirical IDF curves for several large cities in the United States using the Chen (1983) form of the IDF curves given by Equation 5.2. Values of a, b_1, and c_1 for the IDF curves corresponding to a 10-year return period are given in Table 5.2. These IDF curves are particularly useful to engineers, since most drainage systems are designed for rainfall events with return periods on the order of 10 years.

Functional forms of the IDF curve other than that given by Equation 5.2 have been proposed and are widely used in practice. Two common examples are

$$i = \frac{aT^m}{(b + t)^n}, \qquad i = a + b(\ln t) + c(\ln t)^2 + d(\ln t)^3 \qquad (5.6)$$

TABLE 5.2: Ten-Year IDF Constants for Major U.S. Cities

City	a	b_1	c_1
Atlanta, Georgia	64.1	8.16	0.76
Chicago, Illinois	60.9	9.56	0.81
Cleveland, Ohio	47.6	8.86	0.79
Denver, Colorado	50.8	10.50	0.84
Helena, Montana	30.8	9.56	0.81
Houston, Texas	98.3	9.30	0.80
Los Angeles, California	10.9	1.15	0.51
Miami, Florida	79.9	7.24	0.73
New York, New York	51.4	7.85	0.75
Olympia, Washington	6.3	0.60	0.40
Santa Fe, New Mexico	32.2	8.54	0.76
St. Louis, Missouri	61.0	8.96	0.78

Source: Wenzel (1982).

where i is the average rainfall intensity, t is the duration, T is the return period, and a, b, c, d, m, and n are locally calibrated constants.

5.2.2   Spatially Averaged Rainfall

Surface runoff from catchment areas is typically estimated using rainfall amounts that are spatially averaged over the catchment, and spatially averaged rainfall is sometimes referred to as *mean areal precipitation* (U.S. Army Corps of Engineers, 2000). Rainfall is never uniformly distributed in space, and spatially averaged rainfall tends to be scale dependent and not statistically homogeneous (in space). The scale dependence of spatially averaged rainfall can be derived from rain-gage measurements using numerical interpolation schemes such as kriging (Journel, 1989).

Consider the averaging area and distribution of rain gages illustrated in Figure 5.9. The objective is to estimate the average rainfall over the given area based on rainfall measurements at the individual rain gages. The basis of the estimation scheme is an interpolation function that estimates the rainfall at any point in the given area, usually as a weighted average of the rainfall measurements at the

FIGURE 5.9: Estimation of spatially averaged rainfall

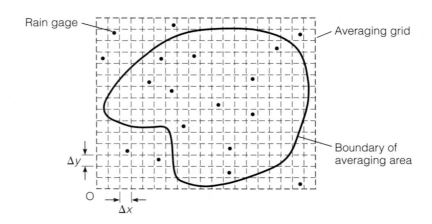

individual rain gages. A linear interpolation function has the form

$$\hat{P}(\mathbf{x}) = \sum_{i=1}^{N} \lambda_i P(\mathbf{x}_i) \tag{5.7}$$

where $\hat{P}(\mathbf{x})$ is the (interpolated) rainfall at location $\mathbf{x}$, $P(\mathbf{x}_i)$ is the measured precipitation at rain gage i that is located at $\mathbf{x}_i$, λ_i is the weight given to measurements at station i, and these weights generally satisfy the relation

$$\sum_{i=1}^{N} \lambda_i = 1 \tag{5.8}$$

There are a variety of ways to estimate the weights, λ_i, depending on the underlying assumptions about the spatial distribution of the rainfall. Some of the more common assumptions are as follows:

1. The rainfall is uniformly distributed in space. Under this assumption, equal weight is assigned to each station

$$\lambda_i = \frac{1}{N} \tag{5.9}$$

   The estimated rainfall at any point is simply equal to the arithmetic average of the measured data.

2. The weights are calculated from the kriging weights (Journel, 1989), using either the covariance function or the variogram of the rainfall. This method is equivalent estimating the point rainfall from the contours of equal rainfall (isohyets) derived from the measured data.

3. The rainfall at any point is estimated by the rainfall at the nearest station. Under this assumption, $\lambda_i = 1$ for the nearest station, and $\lambda_i = 0$ for all other stations. This methodology is the discrete equivalent of the graphical *Thiessen polygon method* (Thiessen, 1911) that has been used for many years in hydrology. This approach can reasonably be called the *discrete Thiessen method*. The Thiessen method should not be used to estimate rainfall depths of mountainous watersheds, since elevation is also a strong factor influencing the areal distribution (Soil Conservation Service, 1993).

4. The weight assigned to each measurement station is inversely proportional to the distance from the estimation point to the measurement station. This approach is frequently referred to as the *reciprocal-distance approach* (Simanton and Osborn, 1980; Wei and McGuinness, 1973). A particular example of the reciprocal-distance approach is the *inverse-distance-squared method* in which the station weights are given by

$$\lambda_i = \frac{\frac{1}{d_i^2}}{\sum_{i=1}^{N} \frac{1}{d_i^2}} \tag{5.10}$$

   where d_i is the distance to station i, and N is the number of stations within some defined radius of where the rainfall is to be estimated. In applying the inverse-distance-squared method, the U.S. Army Corps of Engineers (USACE, 2000) recommends centering North/South and East/West reference axes at the location where the rainfall is to be estimated, and then using the nearest measurement within each of the four quadrants to estimate the rainfall.

After specifying the station weights in the rainfall interpolation formula, the next step is to numerically discretize the averaging area by an averaging grid as indicated in Figure 5.9. The definition of the averaging grid requires specification of the origin, O; discretizations in the x- and y-directions, Δx and Δy; and the number of cells in each of the coordinate directions. The rainfall, $\hat{P}(\mathbf{x}_j)$, at the center, $\mathbf{x}_j$, of each cell is then calculated using the interpolation formula (Equation 5.7) with specified weights, and the average rainfall over the entire area, $\overline{P}$, is given by

$$\overline{P} = \frac{1}{A} \sum_{j=1}^{J} \hat{P}(\mathbf{x}_j) A_j \qquad (5.11)$$

where A is the averaging area, J is the number of cells that contain a portion of the averaging area, and A_j is the amount of the averaging area contained in cell j. This method is well suited to spreadsheet calculation, or to computer programs written specifically for this task.

EXAMPLE 5.5

The spatially averaged rainfall is to be calculated for the catchment area shown in Figure 5.10. There are five rain gages in close proximity to the catchment area, and the Cartesian coordinates of these gages are as follows:

Gage	x (km)	y (km)
A	1.3	7.0
B	1.0	3.7
C	4.2	4.9
D	3.5	1.4
E	2.1	−1.0

FIGURE 5.10: Catchment area

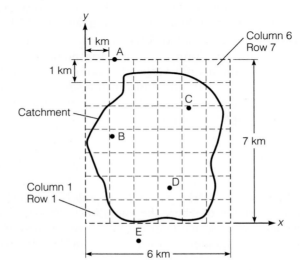

The rainfall measured at each of the gages during a 1-hour interval is 60 mm at A, 90 mm at B, 65 mm at C, 35 mm at D, and 20 mm at E. Use the discrete Thiessen method with the 1-km × 1-km grid shown in Figure 5.10 to estimate the average rainfall over the catchment area during the 1-hour interval.

Solution   Using the discrete Thiessen method the weights, λ_i, are assigned to each cell in Figure 5.10 using the convention that the nearest station has a weight of 1, and all other stations have weights of zero. Computations are summarized in Table 5.3, where the row numbers increase from the bottom to the top and the column numbers increase from left to right in Figure 5.10. The station weights assigned to each cell, based on the rain gage closest to the center of the cell, are shown in columns 3 to 7 in Table 5.3, and the area, A_i, of the catchment contained in each cell, i, is shown in column 8. The rainfall amount, P_i, assigned to each cell is equal to the rainfall at the nearest station (according to the discrete Thiessen method) and is given in column 9. The total area, A, of the catchment is obtained by summing the values in column 8 and is equal to 28.91 km². The average rainfall over the catchment, $\overline{P}$, is given by

$$\overline{P} = \frac{1}{A} \sum P_i A_i = \frac{1}{28.91}(1707.2) = 59 \text{ mm}$$

The average rainfall over the entire catchment is therefore equal to 59 mm.

Station weights are used in an operational setting where it is not practical to perform more detailed analysis of precipitation fields. For example, the National Weather Service River Forecast System (NWSRFS), which is the tool employed to generate hydrologic forecasts in real time throughout the United States by the National Weather Service (NWS), uses station weights to generate mean areal precipitation (MAP) time series for model calibration and to compute MAP for operational forecasting on large rivers at (typically) 6-hour intervals (NWS, 2002). The two primary ways currently used by the National Weather Service to compute station weights are the Thiessen and inverse-distance-squared weighting methodologies (NWS, 2002); however, it is recognized that alternative approaches might be superior under certain circumstances (Fiedler, 2003).

It is clear that the accuracy of estimated mean areal precipitation over a catchment is directly related to the density of rain gages within the area and to the spatial characteristics of storms occurring within the catchment. Accordingly, the National Weather Service recommends that the minimum number of rain gages, N, for a local flood-warning network within a catchment of area, A, is given by

$$N = 0.73A^{0.33} \tag{5.12}$$

where A is in square kilometers. However, even if more than the minimum number of gages are used, not all storms will be adequately gaged. It is interesting to note that precipitation gages are typically 20 to 30 cm in diameter, and these measurements are routinely used to estimate the average rainfall over areas exceeding 1 km². Clearly, isolated storms may not be measured well if storm cells are located over areas that are not gaged, or if the distribution of rainfall within storm cells is very nonuniform over the catchment area.

TABLE 5.3: Calculation of Catchment-Averaged Rainfall

(1)	(2)	(3)	(4)	(5)	(6)	(7)	(8)	(9)	(10)
		\multicolumn{5}{}{Weights, λ_i}					Area, A_i	Rainfall, P_i	$P \times A$
Row	Column	A	B	C	D	E	(km^2)	(mm)	(km$^2 \cdot$mm)
1	1					1	0.01	20	0.2
	2					1	0.7	20	14
	3				1		0.95	35	33.25
	4				1		0.9	35	31.5
	5				1		0.8	35	28
	6				1		0.1	35	3.5
2	1					1	0.15	20	3
	2				1		1	35	35
	3				1		1	35	35
	4				1		1	35	35
	5				1		1	35	35
	6				1		0.3	35	10.5
3	1		1				0.7	90	63
	2		1				1	90	90
	3				1		1	35	35
	4				1		1	35	35
	5				1		1	35	35
	6				1		0.2	35	7
4	1		1				0.95	90	85.5
	2		1				1	90	90
	3		1				1	90	90
	4			1			1	65	65
	5			1			1	65	65
	6			1			0.4	65	26
5	1		1				0.6	90	54
	2		1				1	90	90
	3		1				1	90	90
	4			1			1	65	65
	5			1			1	65	65
	6			1			0.75	65	48.75
6	1	1					0	60	0
	2	1					0.7	60	42
	3			1			1	65	65
	4			1			1	65	65
	5			1			1	65	65
	6			1			0.35	65	22.75
7	1	1					0	60	0
	2	1					0.15	60	9
	3	1					0.55	60	33
	4			1			0.45	65	29.25
	5			1			0.2	65	13
	6			1			0	65	0
Total							28.91		1707.2

5.2.3 Design Rainfall

A hypothetical rainfall event corresponding to a specified return period is usually the basis for the design and analysis of stormwater-management systems. However, it should be noted that the return period of a rainfall event is not equal to the return period of the resulting runoff, and therefore the reliability of surface-water management systems will depend on such factors as the antecedent moisture conditions in the catchment. In contrast to the single-event design-storm approach, a continuous-simulation approach is sometimes used where a historical rainfall record is used as input to a rainfall-runoff model; the resulting runoff is analyzed to determine the hydrograph corresponding to a given return period. The design-storm approach is more widely used in engineering practice than the continuous-simulation approach. Design storms can be either synthetic or actual (historic) design storms, with synthetic storms defined from historical rainfall statistics.

Synthetic design storms are characterized by their return period, duration, depth, temporal distribution, and spatial distribution. The selection of these quantities for design purposes is described in the following sections.

5.2.3.1 Return period

The return period of a design rainfall should be selected on the basis of economic efficiency (ASCE, 1992). In practice, however, the return period is usually selected on the basis of level of protection. Typical return periods are given in Table 5.4, although longer return periods are sometimes used. For example, return periods of 10 to 30 years are commonly used for designing storm sewers in commercial and high-value districts (Burton, 1996). In selecting the return period for a particular project, local drainage regulations should be reviewed and followed. An implicit assumption in designing drainage systems for a given return period of rainfall is that the return period of the resulting runoff is equal to the return period of the design rainfall. Based on this assumption, the risk of failure of the drainage system is taken to be the same as the exceedance probability of the design rainfall.

5.2.3.2 Rainfall duration

The design duration of a storm is usually selected on the basis of the time-response characteristics of the catchment. The time response of a catchment is measured by

TABLE 5.4: Typical Return Periods

Land use	Design storm return period (years)
Minor drainage systems:	
Residential	2–5
High-value general commercial area	2–10
Airports (terminals, roads, aprons)	2–10
High-value downtown business areas	5–10
Major drainage-system elements	up to 100 years

Source: ASCE (1992).

the travel time of surface runoff from the most remote point of the catchment to the catchment outlet and is called the *time of concentration*. The duration of design storms used to design water-control systems must generally equal or exceed the time of concentration of the area covered by the control system. On small urban catchments (< 40 ha), current practice is to select the duration of the design rainfall as equal to the time of concentration. This approach usually leads to the maximum peak runoff for a given return period. For the design of detention basins, however, the duration causing the largest detention volume is most critical, and several different storm durations may need to be tried to identify the most critical design-storm duration (Akan, 1993). For catchments with high infiltration losses, the duration of the critical rainfall associated with the maximum peak runoff may be shorter than the time of concentration (Chen and Wong, 1993). Many drainage districts require that the performance of drainage systems be analyzed for a standard 24-hour storm with a specified return period, typically on the order of 25 years. It has been shown that 24 hours is a good design storm duration for watersheds in Maryland with areas in the range of 5 to 130 km^2 (Levy and McCuen, 1999).

5.2.3.3   Rainfall depth

The design-rainfall depth for a selected return period and duration is obtained directly from the intensity-duration-frequency (IDF) curve of the catchment. The IDF curve can be estimated from rainfall measurements, derived from National Weather Service (NWS) publications such as TP-40 (Hershfield, 1961), or obtained from regulatory manuals that govern local drainage designs.

5.2.3.4   Temporal distribution

Realistic temporal distributions of rainfall within design storms are best determined from historical rainfall measurements. In many cases, however, either the data are not available or such a detailed analysis cannot be justified. Under these conditions, the designer must resort to empirical distributions. Frequently used methods for estimating the rainfall distribution in storms are the triangular method, alternating-block method, and the NRCS 24-hour hyetograph.

Triangular method.   A common approximation for the rainfall distribution is the triangular distribution shown in Figure 5.11. In this distribution, the time to peak, t_p, is related to the (intensity-weighted) mean of the rainfall distribution, $\bar{i}$, by the relation

$$\boxed{\frac{t_p}{t_d} = 3\frac{\bar{i}}{t_d} - 1} \tag{5.13}$$

FIGURE 5.11: Triangular rainfall distribution

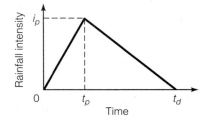

where t_d is the duration of the storm. Using this equation, the nondimensional time to peak, t_p/t_d, can be related to the nondimensional mean of the rainfall distribution. Yen and Chow (1980) and El-Jabi and Sarraf (1991) studied a wide variety of storms and found that t_p/t_d was typically in the range of 0.32 to 0.51, while McEnroe (1986) found much shorter time to peak by considering only 1-hour storms with return periods of two years or greater.

EXAMPLE 5.6

The IDF curve for 10-year storms in Houston, Texas, is given by

$$i = \frac{2497}{(t_d + 9.30)^{0.80}}$$

where i is the average intensity in mm/h and t_d is the duration in minutes. Assuming that the mean of the rainfall distribution is equal to 40% of the rainfall duration, estimate the triangular hyetograph for a 1-hour storm. [This example is adapted from ASCE (1992).]

Solution For a 1-hour storm, $t_d = 60$ min and the average intensity, i, is given by

$$i = \frac{2497}{(t_d + 9.30)^{0.80}} = \frac{2497}{(60 + 9.30)^{0.80}} = 84.1 \text{ mm/h}$$

The rainfall amount, P, in 1 hour is therefore given by

$$P = 84.1 \text{ mm/h} \times 1 \text{ h} = 84.1 \text{ mm}$$

The peak of the triangular hyetograph occurs at t_p, where Equation 5.13 gives

$$\frac{t_p}{t_d} = 3\frac{\bar{i}}{t_d} - 1$$

From the given data, $\bar{i}/t_d = 0.4$, in which case

$$\frac{t_p}{t_d} = 3(0.4) - 1 = 0.2$$

and

$$t_p = 0.2t_d = 0.2(1) = 0.2 \text{ h}$$

The triangular hyetograph has a peak equal to i_p, a base of t_d, and an area under the hyetograph of 84.1 mm. Therefore, using the formula for the area of a triangle

$$\frac{1}{2} i_p t_d = 84.1$$

leads to

$$i_p = \frac{84.1}{\frac{1}{2}t_d} = \frac{84.1}{\frac{1}{2}(1)} = 168 \text{ mm/h}$$

The triangular hyetograph for this 10-year 1-hour storm is illustrated in Figure 5.12.

FIGURE 5.12: Estimated 10-year 1-hour rainfall hyetograph in Houston, Texas

Adapted from ASCE (1992).

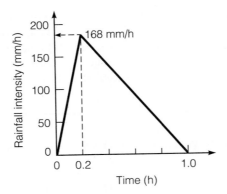

Alternating-block method. A hyetograph can be derived from the IDF curve for a selected storm duration and return period using the *alternating-block method*. The hyetograph generated by the alternating-block method describes the rainfall in n time intervals of duration Δt, for a total storm duration of $n\Delta t$. The procedure for determining the alternating-block hyetograph from the IDF curve is as follows:

1. Select a return period and duration for the storm.
2. Read the average intensity from the IDF curve for storms of duration $\Delta t, 2\Delta t, \ldots,$ $n\Delta t$. Determine the corresponding precipitation depths by multiplying each intensity by the corresponding duration.
3. Calculate the difference between successive precipitation depth values. These differences are equal to the amount of precipitation during each additional unit of time Δt.
4. Reorder the precipitation increments into a time sequence with the maximum intensity occurring at the center of the storm, and the remaining precipitation increments arranged in descending order alternately to the right and left of the central block.

The alternating-block method assumes that the maximum rainfall for any duration less than or equal to the total storm duration has the same return period (ASCE, 1992). In these cases, the IDF curve is said to be nested within the design hyetograph. Field data confirm that this assumption is very conservative, particularly for longer storms.

EXAMPLE 5.7

The IDF curve for 10-year storms in Houston, Texas, is given by

$$i = \frac{2497}{(t + 9.30)^{0.80}} \text{ mm/h}$$

where t is the duration in minutes. Use the alternating-block method to calculate the hyetograph for a 10-year 1-hour storm using 9 time intervals. Compare this result with the triangular hyetograph determined in Example 5.6.

Solution For a 60-minute storm with 9 time intervals, the time increment, Δt, is given by

$$\Delta t = \frac{60}{9} = 6.67 \text{ min}$$

The average intensities for storm durations equal to multiples of Δt are derived from the IDF curve using $t = \Delta t, 2\Delta t, \ldots, 9\Delta t$ and the results are given in column 3 of the following table:

(1) Increment	(2) t (min)	(3) i (mm/h)	(4) it (mm)	(5) Rainfall amount (mm)	(6) Intensity (mm/h)
1	6.67	272	30.2	30.2	272
2	13.33	206	45.8	15.6	140
3	20.00	167	55.7	9.90	89.1
4	26.67	142	63.1	7.40	66.6
5	33.33	124	68.9	5.80	52.2
6	40.00	110	73.3	4.40	39.6
7	46.67	99.8	77.6	4.30	38.7
8	53.33	91.2	81.1	3.50	31.5
9	60.00	84.1	84.1	3.00	27.0

The precipitation for each rainfall duration, t, is given in column 4 ($=$ col. 2 $\times$ col. 3), the rainfall increments corresponding to the duration increments are given in column 5, and the corresponding intensities are given in column 6. In accordance with the alternating-block method, the maximum intensity ($=272$ mm/h) is placed at the center of the storm, and the other intensities are arranged in descending order alternately to the right and left of the center block. The alternating-block hyetograph is therefore given by:

Time (min)	Average intensity (mm/h)
0–6.67	27.0
6.67–13.33	38.7
13.43–20.00	52.2
20.00–26.67	89.1
26.67–33.33	272
33.33–40.00	140
40.00–46.67	66.6
46.67–53.33	39.6
53.33–60.00	31.5

In the triangular hyetograph derived in Example 5.6, the location of the peak intensity can be varied, while in the alternating-block method the peak intensity always occurs at 50% of the storm duration. Also, the triangular hyetograph derived in the previous example resulted in a maximum intensity of 168 mm/h, while the alternating-block method resulted in a much higher maximum intensity of 272 mm/h. This difference is a result of the short duration used to calculate the peak intensity and reflects the conservative nature of the alternating-block method.

The alternating-block method is the recommended approach for constructing frequency-based hypothetical storms; however, the durations of storms constructed using this method should generally be less than 10 days (U.S. Army Corps of Engineers, 2000).

FIGURE 5.13: Geographic boundaries of NRCS rainfall distributions

Rainfall distribution

Type I

Type IA

Type II

Type III

NRCS 24-hour hyetographs.   The Natural Resources Conservation Service (formerly the Soil Conservation Service) has developed 24-hour rainfall distributions for four geographic regions in the United States (SCS, 1986). These rainfall distributions are approximately consistent with local IDF curves (constructed in a way similar to the alternating-block method), and the geographic boundaries corresponding to these rainfall distributions are shown in Figure 5.13. Types I and IA rainfall distributions are characteristic of a Pacific maritime climate with wet winters and dry summers. The Type I distribution is applicable to Hawaii, the coastal side of the Sierra Nevada in southern California, and the interior regions of Alaska. Type IA represents areas on the coastal side of the Sierra Nevada and the Cascade Mountains in Oregon, Washington, northern California, and the coastal regions of Alaska. Type II rainfall is characteristic of most regions in the continental United States, with the exceptions of the Gulf coast regions of Texas, Louisiana, Alabama, south Florida, and most of the Atlantic coastline, which are characterized by Type III rainfall, where tropical storms are prevalent and produce large 24-hour rainfall amounts. Puerto Rico and the Virgin Islands are characterized by Type II rainfall. The 24-hour rainfall hyetographs are illustrated in Figure 5.14, where the abscissa is the time in hours, and the ordinate gives the dimensionless precipitation, P/P_T, where P is the cumulative rainfall (a function of time) and P_T is the total 24-hour rainfall amount. The coordinates of the rainfall distributions are given in Table 5.5. The peak rainfall intensity occurs at the time when the slope of the cumulative rainfall distribution is steepest, which for the NRCS 24-hour hyetographs are 8.2 h (Type IA), 10.0 h (Type I), 12.0 h (Type II), and 12.2 h (Type III). Comparing the NRCS 24-hour rainfall distributions indicates that Type IA yields the least intense storms, Type II the most intense storms, and Type II and Type III distributions are very similar to each other.

EXAMPLE 5.8

The precipitation resulting from a 10-year 24-hour storm on the Gulf coast of Texas is estimated to be 180 mm. Calculate the NRCS 24-h hyetograph.

FIGURE 5.14: NRCS 24-hour rainfall distributions

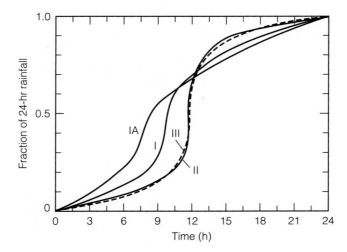

TABLE 5.5: NRCS 24-Hour Rainfall Distributions

Time (h)	Type I P/P_T	Type IA P/P_T	Type II P/P_T	Type III P/P_T
0.0	0.000	0.000	0.000	0.000
0.5	0.008	0.010	0.005	0.005
1.0	0.017	0.022	0.011	0.010
1.5	0.026	0.036	0.017	0.015
2.0	0.035	0.051	0.023	0.020
2.5	0.045	0.067	0.029	0.026
3.0	0.055	0.083	0.035	0.032
3.5	0.065	0.099	0.041	0.037
4.0	0.076	0.116	0.048	0.043
4.5	0.087	0.135	0.056	0.050
5.0	0.099	0.156	0.064	0.057
5.5	0.112	0.179	0.072	0.065
6.0	0.126	0.204	0.080	0.072
6.5	0.140	0.233	0.090	0.081
7.0	0.156	0.268	0.100	0.089
7.5	0.174	0.310	0.110	0.102
8.0	0.194	0.425	0.120	0.115
8.5	0.219	0.480	0.133	0.130
9.0	0.254	0.520	0.147	0.148
9.5	0.303	0.550	0.163	0.167
10.0	0.515	0.577	0.181	0.189
10.5	0.583	0.601	0.203	0.216
11.0	0.624	0.623	0.236	0.250
11.5	0.655	0.644	0.283	0.298
12.0	0.682	0.664	0.663	0.500
12.5	0.706	0.683	0.735	0.702

TABLE 5.5: (*Continued*)

Time (h)	Type I P/P_T	Type IA P/P_T	Type II P/P_T	Type III P/P_T
13.0	0.728	0.701	0.776	0.751
13.5	0.748	0.719	0.804	0.785
14.0	0.766	0.736	0.825	0.811
14.5	0.783	0.753	0.842	0.830
15.0	0.799	0.769	0.856	0.848
15.5	0.815	0.785	0.869	0.867
16.0	0.830	0.800	0.881	0.886
16.5	0.844	0.815	0.893	0.895
17.0	0.857	0.830	0.903	0.904
17.5	0.870	0.844	0.913	0.913
18.0	0.882	0.858	0.922	0.922
18.5	0.893	0.871	0.930	0.930
19.0	0.905	0.884	0.938	0.939
19.5	0.916	0.896	0.946	0.948
20.0	0.926	0.908	0.953	0.957
20.5	0.936	0.920	0.959	0.962
21.0	0.946	0.932	0.965	0.968
21.5	0.956	0.944	0.971	0.973
22.0	0.965	0.956	0.977	0.979
22.5	0.974	0.967	0.983	0.984
23.0	0.983	0.978	0.989	0.989
23.5	0.992	0.989	0.995	0.995
24.0	1.000	1.000	1.000	1.000

Source: SCS (1986).

Solution   On the Gulf coast of Texas, 24-hour storms are characterized by Type III rainfall. The hyetograph is determined by multiplying the ordinates of the Type III hyetograph in Table 5.5 by $P_T = 180$ mm to yield the following select points (this example is for illustrative purposes only and not all hyetograph points are shown):

Time (h)	P/P_T	Cumulative precipitation, P (mm)
0	0	0
3	0.032	6
6	0.072	13
9	0.148	27
12	0.500	90
15	0.848	153
18	0.922	166
21	0.968	174
24	1.000	180

5.2.3.5   Spatial distribution

The spatial distribution of storms is usually important in calculating the runoff from large catchments. For any given return period and duration, the spatially averaged rainfall depth over an area is generally less than the maximum point-rainfall depth. The *areal-reduction factor* (ARF) is defined as the ratio of the areal-average rainfall to the point rainfall depth and, in the absence of local data, can be estimated using Figure 5.15 (Hershfield, 1961; U.S. Weather Bureau, 1957). Areal-averaged rainfall is usually assumed to be uniformly distributed over the catchment, and for areas less than 25 km² (10 mi²), areal-reduction factors are not recommended (World Meteorological Organization, 1983; Jens, 1979). The areal-reduction factor, F, given in Figure 5.15 can be approximated algebraically by the relation (Leclerc and Schaake, 1972; Eagleson, 1972)

$$F = 1 - \exp\left(-1.1t_d^{\frac{1}{4}}\right) + \exp\left(-1.1t_d^{\frac{1}{4}} - 0.01A\right)$$ (5.14)

where t_d is the rainfall duration in hours, and A is the catchment area in square miles. Although Equation 5.14 was based on areal reduction factors (ARFs) derived from a limited data set many years ago, inclusion of more recent data and re-estimation of the ARFs leads to approximately the same relation (Allen and DeGaetano, 2005). Caution should be exercised in using generalized areal-reduction factors, since local and regional effects may lead to significantly different reduction factors (Huff, 1995).

EXAMPLE 5.9

The local rainfall resulting from a 10-year 24-hour storm on the Gulf coast of Texas is 180 mm. Estimate the average rainfall on a 100 km² catchment.

Solution   From the given data: $t_d = 24$ h, $A = 100$ km² $= 40$ mi², and the areal-reduction factor is given by Equation 5.14 as

$$F = 1 - \exp\left[-1.1t_d^{\frac{1}{4}}\right] + \exp\left[-1.1t_d^{\frac{1}{4}} - 0.01A\right]$$

FIGURE 5.15: Areal-reduction factor vs. catchment area and storm duration

Source: Hershfield (1961).

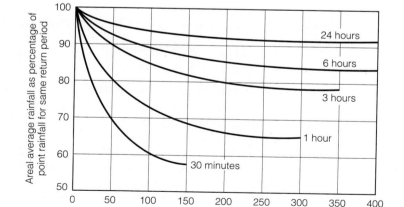

$$= 1 - \exp[-1.1(24)^{\frac{1}{4}}] + \exp[-1.1(24)^{\frac{1}{4}} - 0.01(40)]$$
$$= 0.97$$

Therefore, for a local-average rainfall of 180 mm, the average precipitation over a 100-km^2 catchment is expected to be 0.97×180 mm $= 175$ mm.

Areal-reduction factors are particularly useful in converting temporal rainfall distributions derived from point measurements to area-averaged temporal rainfall distributions. The conversion process is to simply multiply the ordinates of the point-rainfall distribution by the areal-reduction factor corresponding to the area of the catchment under consideration. The exact pattern of spatial variation in storm depth is normally disregarded, except in major-structure designs that are based on extreme rainfall events, such as the probable maximum precipitation.

5.2.4 Extreme Rainfall

In large water-resource projects it is frequently necessary to consider the consequences of extreme rainfall events. Such rainfall events include the probable maximum storm and the standard project storm.

5.2.4.1 Probable maximum precipitation

The *probable maximum precipitation* (PMP) is the theoretically greatest depth of precipitation for a given duration that is physically possible over a given size storm area at a particular geographical location at a certain time of the year (U.S. National Weather Service, 1982). Estimates of PMP are required for design situations in which failure could result in catastrophic consequences. For example, a flood overtopping an embankment dam could breach the dam and result in severe downstream flooding that is much worse than if the dam did not exist. The most common methods used to estimate the PMP are the rational estimation method and the statistical estimation method.

Rational estimation method. The rational estimation method is based on the relation

$$\text{PMP} = \left(\frac{\text{precipitation}}{\text{moisture}} \right)_{\text{max}} \times (\text{moisture supply})_{\text{max}} \qquad (5.15)$$

where the maximum ratio of precipitation to atmospheric moisture (first term) is obtained from historical rainfall and meteorological records, and the maximum moisture supply (second term) is obtained from meteorological tables of *effective precipitable water* based on the maximum persisting dew point for a given drainage basin. Studies of PMPs in the United States using the rational estimation method have been conducted by the U.S. National Weather Service for the entire country, and these studies are contained in several hydrometeorological reports (HMRs). The HMR reports corresponding to various areas in the United States are shown in Figure 5.16. A major dividing line for the HMRs is the 105th meridian (105° longitude), which separates the relatively mountainous western portion of the United

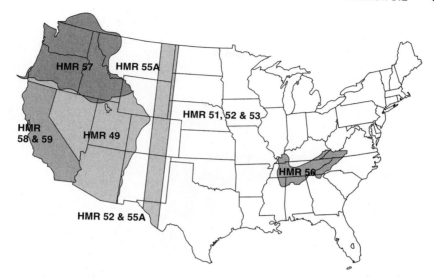

FIGURE 5.16: Regions of the United States covered by generalized PMP studies

States from the more gently sloping eastern portion. The 105th meridian approximately coincides with the slopes of the Rocky Mountains. PMPs are characterized in HMRs by isohyetal maps of rainfall for given storm areas and durations. Standard storm areas covered by HMR maps are: 26 km² (10 mi²), 518 km² (200 mi²), 2590 km² (1000 mi²), 12,950 km² (5000 mi²), 25,900 km² (10,000 mi²), and 51,800 km² (20,000 mi²); and standard storm durations covered by HMR maps are: 6 h, 12 h, 24 h, 48 h, and 72 h. Hydrometeorological reports (HMRs) shown in Figure 5.16 can be downloaded from the U.S. National Weather Service at www.nws.noaa.gov, and the PMP at any location for any storm area and storm duration can be obtained from these reports. An example of the format in which the data are presented is given in Figure 5.17, which shows a PMP map for a storm duration of 24-hours covering an area of 26 km² (10 mi²) derived from HMR 51 (U.S. National Weather Service, 1978). The stippled areas indicate mountainous regions where assumptions regarding moisture variations with elevation are somewhat questionable. These stippled areas include the Appalachian Mountains (which extend from Georgia to Maine) and the slopes of the Rocky Mountains between the 103rd and 105th meridian. To determine the PMP for a drainage basin in one of the Gulf Coast states south of the last PMP isohyet shown on a PMP map (for example, a basin in Florida), the PMP values given by the southernmost isohyet should be used. The World Meteorological Organization (1986) addresses PMP estimates for regions throughout the world.

Statistical estimation method.   The most widely used statistical method for estimating the PMP was proposed by Hershfield (1961, 1965), and has become one of the standard methods suggested by the World Meteorological Organization (WMO, 1986) for estimating the PMP at any location. The procedure is based on the equation

$$P_{\mathrm{m}} = \overline{P} + k_{\mathrm{m}}\sigma_{\mathrm{P}}$$

(5.16)

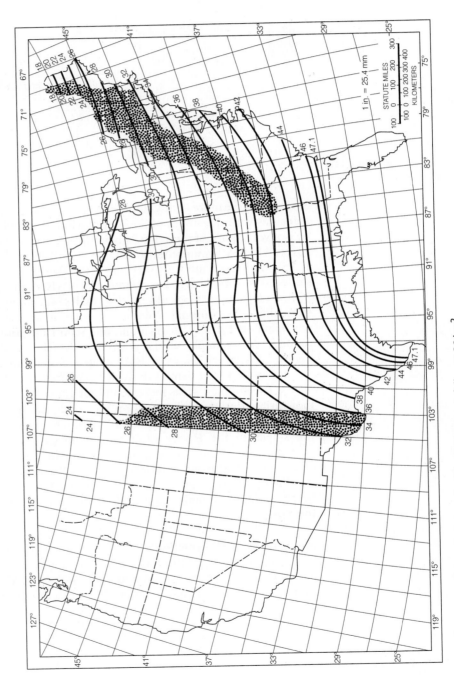

FIGURE 5.17: All season PMP (in inches) for 24-hour rainfall over 26 km²

where P_m is the annual-maximum rainfall amount in a given duration, $\overline{P}$ and σ_P are the mean and standard deviation of the annual-maximum rainfall amounts for a given duration, and k_m is a factor that depends on the storm duration and $\overline{P}$. Hershfield (1961) analyzed 24-hour rainfall records from 2600 stations with a total of 95,000 station-years of data and recommended taking k_m as 15; while some later investigations have recommended taking k_m as 20 for 24-hour rainfall amounts (National Research Council, 1985).

EXAMPLE 5.10

The annual-maximum 24-hour rainfall amounts in a tropical city have a mean of 15.2 cm and a standard deviation of 4.8 cm. Estimate the probable maximum precipitation (PMP) at this location.

Solution   From the given data: $\overline{P} = 15.2$ cm, and $\sigma_P = 4.8$ cm. Taking $k_m = 20$ gives the probable maximum precipitation, P_m, as

$$P_m = \overline{P} + k_m \sigma_P = 15.2 + (20)(4.8) = 111 \text{ cm}$$

Therefore, the PMP at this location is 111 cm.

The important difference between the two methods, from an engineering viewpoint, is that the rational estimation method implies zero risk of exceeding the PMP, while the statistical method implies that any rainfall amount has a finite risk of being exceeded. A study by Koutsoyiannis (1999) indicates that the PMP amounts calculated using the rational method have return periods on the order of 60,000 years, and the factor k_m in Equation 5.16 should be taken as a random variable with a generalized extreme value (GEV) distribution given by

$$F(k_m) = \exp\left\{-\left[1 + \frac{0.13(k_m - 0.44)}{0.60}\right]^{-7.69}\right\} \tag{5.17}$$

where $F(k_m)$ is the cumulative distribution function of the frequency factor k_m in Equation 5.16. The value of k_m given by Equation 5.17, for a specified exceedance probability, yields the precipitation depth for a 24-hour duration. Estimation of the rainfall depth for durations other than 24 hours, with the same return period, can be derived from the 24-hour rainfall amount using the intensity-duration-frequency (IDF) curve for the particular geographic area. IDF curves can generally be expressed in the form

$$i = \frac{f(T)}{g(t)} \tag{5.18}$$

where i is the average rainfall intensity in a storm of duration t and return period T, $f(T)$ is a function of the return period, T, and $g(t)$ is a function of the storm duration, t. For extreme storms with a return period T, Equation 5.18 gives the following relation between the rainfall amount in 24 hours and the rainfall in duration t:

$$f(T) = \frac{P_{24}}{24} g(24) = \frac{P_t}{t} g(t) \tag{5.19}$$

where P_t and P_{24} are the rainfall amounts in duration t and 24 hours, respectively. Equation 5.19 simplifies to the more useful form

$$P_t = \frac{t}{24}\frac{g(24)}{g(t)}P_{24} \tag{5.20}$$

The utility of this equation in estimating the rainfall in extreme storms of duration t is illustrated in the following example.

EXAMPLE 5.11

The annual-maximum 24-hour rainfall amounts on a Pacific island have a mean of 32.0 cm and a standard deviation of 15.6 cm. The IDF curve for the 10-year storm on the island is given by the Chen method as

$$i = \frac{2030}{(t + 7.25)^{0.75}} \text{ mm/h}$$

where i is the intensity in mm/h and t is the duration in minutes. Find the annual-maximum 6-hour rainfall with a return period of 100,000 years.

Solution   From the given data, the annual-maximum 24-hour rainfall series gives: $\overline{P} = 32.0$ cm, and $\sigma_P = 15.6$ cm. For an extreme storm with a return period, T, of 100,000 years, Equation 5.17 gives

$$F(k_m) = 1 - \frac{1}{100,000} = \exp\left\{-\left[1 + \frac{0.13(k_m - 0.44)}{0.60}\right]^{-7.69}\right\} \tag{5.21}$$

where k_m is the frequency factor with an exceedance probability of 100,000 years. Solving Equation 5.21 gives $k_m = 16.45$, and therefore the 24-hour rainfall extreme, P_{24}, with a return period of 100,000 years is given by Equation 5.16 as

$$P_{24} = \overline{P} + k_m\sigma_P = 32.0 + 16.45(15.6) = 288.6 \text{ cm} \tag{5.22}$$

The IDF curve for the island is given by the Chen method in the form

$$i = \frac{f(T)}{g(t)} = \frac{f(T)}{(t + 7.25)^{0.75}}$$

where $g(t)$ is given as

$$g(t) = (60t + 7.25)^{0.75} \tag{5.23}$$

where t is in hours. Combining Equations 5.20, 5.22, and 5.23 gives the 6-hour rainfall as

$$P_6 = \frac{6}{24}\frac{g(24)}{g(6)}P_{24} = \frac{6}{24}\frac{(60 \times 24 + 7.25)^{0.75}}{(60 \times 6 + 7.25)^{0.75}}(288.6) = 202 \text{ cm}$$

Therefore, the 6-hour annual-maximum rainfall with a return period of 100,000 years is approximately 202 cm.

It is important to keep in mind that return periods associated with PMP events are generally large, on the order of tens of thousands of years, and cannot be verified in practice (Jothityangkoon and Sivapalan, 2003). In fact, given that the time scale of climate change is also on the order of tens of thousands of years, estimates of PMP return periods must be considered highly uncertain.

World-record precipitation amounts.   The maximum observed rainfall amounts on the Earth for given durations are approximated by the following relation (World Meteorological Organization, 1983)

$$P_{m,obs} = 422t_d^{0.475}$$ (5.24)

where $P_{m,obs}$ is the maximum observed point-rainfall in millimeters and t_d is the storm duration in hours. Several of the world's maximum observed point-rainfalls as a function of duration are listed in Table 5.6.

EXAMPLE 5.12

Estimate the maximum amount of rainfall that has been observed in 24 h.

Solution   According to Equation 5.24, for a 24-h storm ($t_d = 24$ h) the maximum observed rainfall amount, P_m, is given by

$$P_{m,obs} = 422t_d^{0.475} = 422(24)^{0.475} = 1910 \text{ mm} = 191 \text{ cm}$$

This is slightly higher than the observed 24-h maximum of 187 cm given in Table 5.6. It is noteworthy that this 24-h rainfall is significantly greater than the average annual rainfall in Miami, Florida (152 cm).

5.2.4.2   Probable maximum storm

The spatial and temporal distribution of the probable maximum precipitation (PMP) that generates the most severe runoff conditions is the *probable maximum storm* (PMS), and the most severe runoff condition, corresponding to the PMS, is called the *probable maximum flood* (PMF). The duration of the PMS that causes the largest flood at a site of interest is the *critical duration* for that drainage basin. The critical duration is determined by routing runoff hydrographs resulting from PMP rainfall of various durations, and selecting the duration causing the maximum flood. In general, the critical duration should never be less than the time of concentration of the drainage basin.

TABLE 5.6: Maximum Observed Point Rainfalls

Duration	Precipitation (cm)	Location	Date
1 min	3.8	Barot, Guadeloupe (West Indies)	26 November 1970
1 hour	40.1	Shangdi, Inner Mongolia, China	3 July 1975
1 day	187.0	Cilaos, Réunion (Indian Ocean)	15–16 March 1952
1 week	465.3	Commerson, Réunion	21–27 January 1980
1 month	930.0	Cherrapunji, India	1–31 July 1861
1 year	2646.1	Cherrapunji, India	1 August 1860–31 July 1861

References: National Weather Service, www.nws.noaa.gov; Ward and Trimble (2004); Dingman (2002).

The U.S. National Weather Service (NWS, 1982) recommends using 6-h increments in rainfall amounts for sequencing the PMP. Normally, the temporal distribution is defined by the order in which 6-h incremental rainfall amounts are arranged in a 3-day (72-h) sequence. For a given storm-area, the 6-h incremental rainfall amounts are calculated by first determining the PMPs for 6, 12, 18, 24, ..., 72 h; then successive subtraction of the PMP for each of these durations from that of the duration 6-h longer gives 6-h increments of the PMP. As the next step, the NWS recommends arranging the 6-h incremental rainfall amounts such that they decrease progressively to either side of the greatest 6-h increment. The four greatest 6-h increments can be placed at any position in the sequence, except within the first 24 h of the storm. An example of an allowable temporal sequence in a 72-h storm is shown in Figure 5.18; however, other arrangements in which the rainfall increments decrease progressively on both sides of the greatest 6-h increment are also used. For example, increments ranked 1 to 6 can be ordered as 6, 4, 2, 1, 3, 5 or 5, 3, 1, 2, 4, 6 (Prakash, 2004). According to the U.S. National Weather Service (USNWS, 1978), allowing the PMP for all durations (6 to 72 h) to occur in a single storm is not an undue maximization. In cases where the duration of the PMS is less than 6 h, the U.S. Army Corps of Engineers (USACE, 1979) recommends the breakup of 6 h according to the percentages given in Table 5.7.

The spatial distribution of the PMS must be specified in addition to the temporal distribution described previously. The standard isohyetal pattern recommended for spatial distribution of PMS east of the 105th meridian is shown in Figure 5.19. This pattern contains 19 isohyets (A through S) providing coverage of drainage areas up to about 155,400 km^2 (60,000 mi^2). The 5 isohyets not shown in Figure 5.19 cover areas between 25,900 km^2 (10,000 mi^2) and 155,000 km^2 (60,000 mi^2), and the 26-km^2 (10-mi^2) isohyet is taken to be the same as point rainfall. The equation of an elliptical isohyet encompassing an area A is given by

$$r^2 = \frac{a^2 b^2}{a^2 \sin^2 \theta + b^2 \cos^2 \theta} \tag{5.25}$$

where r and θ are the radial coordinates of any point on the ellipse, and a and b are the lengths of the semimajor and semiminor axes that are related to the enclosed area A by the relations

$$a = 2.5b \tag{5.26}$$

$$b = \left(\frac{A}{2.5\pi} \right)^{1/2} \tag{5.27}$$

Equations 5.25 and 5.26 indicate that PMS isohyetal pattern is characterized by concentric ellipses in which the ratio of the lengths of the major and minor axes is 2.5 to 1. The U.S. National Weather Service (NWS, 1982) recommends that the rainfall distribution shown in Figure 5.19 be applied only to the three greatest 6-h increments of PMS (18-h PMS). For the nine remaining 6-h increments in the 3-day storm, the NWS recommends a uniform distribution of rainfall throughout the area of the PMS. For each of the three greatest 6-h rainfall increments in the PMS, the rainfall within each isohyet is estimated from a nomogram developed by the U.S. National Weather Service (NWS, 1982), and the nomogram for the first (highest) 6-h rainfall increment in the PMS is shown in Figure 5.20.

FIGURE 5.18: Example of temporal sequence allowed for 6-h increments of PMP

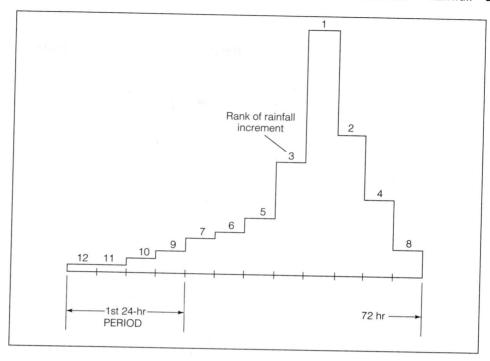

FIGURE 5.18: Example of temporal sequence allowed for 6-h increments of PMP

TABLE 5.7: Hyetograph of 6-h PMP

Sequence (h)	Percent of 6-h depth
First hour	10
Second hour	12
Third hour	15
Fourth hour	38
Fifth hour	14
Sixth hour	11

Source: U.S. Army Corps of Engineers (1979).

To illustrate the use of Figure 5.20, if a PMS covers a 259-km^2 (100-mi^2) catchment and the highest 6-h rainfall amount is 50.8 cm (20 in.), then, according to Figure 5.20, there is 1.12(50.8) = 57 cm of rainfall within the A (10-mi^2) isohyet, 1.04(50.8) = 52.8 cm within the B (25-mi^2) isohyet, and so on up to the D isohyet.

The U.S. National Weather Service (NWS, 1982) recommends centering the isohyetal pattern shown in Figure 5.19 over the catchment area to obtain the hydrologically most critical runoff volume. However, hydrologic trials with varying storm-center locations might be necessary to determine the critical location that produces the most runoff volume. It is necessary to use only as many of the isohyets in Figure 5.19 as needed to cover the catchment area. The orientation of the PMS isohyets derived from historical observations east of the 105th meridian are shown in Figure 5.21. The orientation shown in Figure 5.21 should be compared with the orientation of the

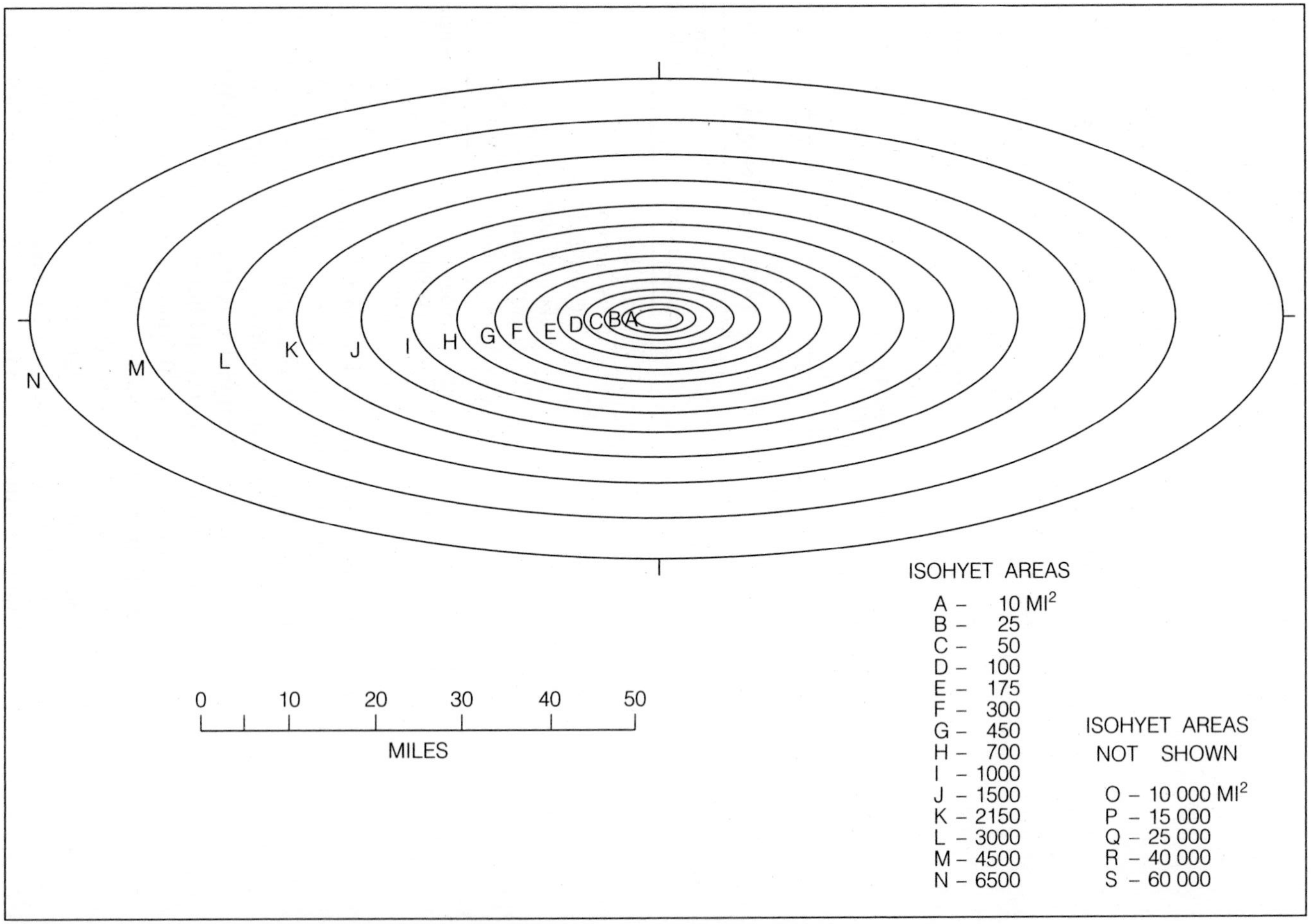

FIGURE 5.19: Standard isohyetal pattern recommended for spatial distribution of PMP East of the 105th Meridian

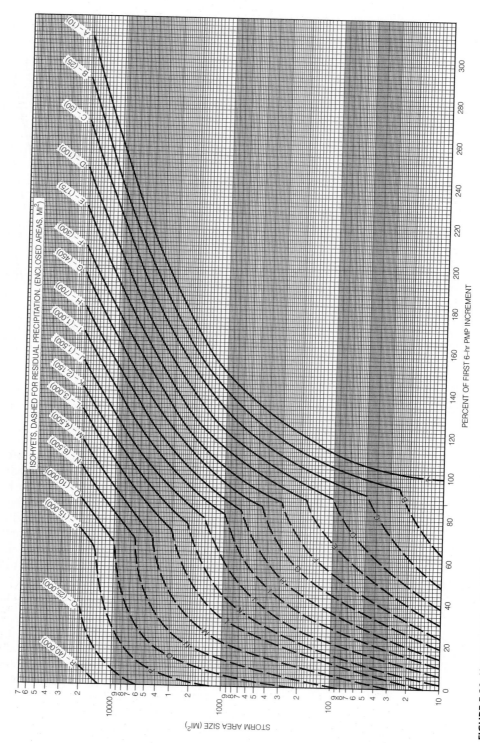

FIGURE 5.20: Nomogram for first 6-h increment for standard isohyet area sizes between 10 and 40,000 mi²

371

FIGURE 5.21: Recommended orientations for PMS within ±40°, East of the 105th Meridian

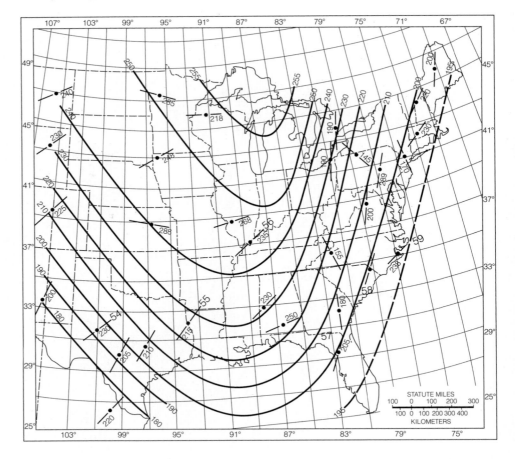

FIGURE 5.21: Recommended orientations for PMS within ±40°, East of the 105th Meridian

longitudinal axis of the catchment area. It is usually preferable to align the longitudinal axis of the PMS with the longitudinal axis of the catchment area. If the orientation of the catchment and the storm orientation given in Figure 5.21 differ by less than ±40°, then the standard isohyetal pattern should be aligned with the longitudinal axis of the catchment area. If the catchment orientation and the storm orientation given in Figure 5.21 differ by more than ±40°, then the standard isohyetal pattern should be aligned with the longitudinal axis of the catchment; however, the isohyet values for each of the 6-h increments of the PMP are to be reduced in accordance with the factors in Figure 5.22.

EXAMPLE 5.13

The South-Dade Watershed in Miami, Florida, has a drainage area of 1036 km² (400 mi²) and is approximately rectangular in shape with dimensions 19.3 km × 53.6 km. The longitudinal axis of the watershed is oriented in the North/South direction. Determine the appropriate storm area to be used in estimating the PMS. If using this storm area in HMR 51 yields a 72-h PMP of 105 cm (41.4 in.), and a maximum incremental 6-h rainfall of 65.5 cm (25.8 in.), determine the isohyetal rainfall pattern to be used in constructing the PMS.

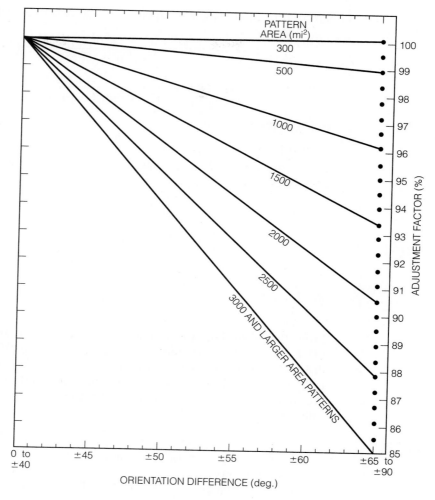

FIGURE 5.22: Adjustment factor for Isohyet values when basin orientation differs from PMS orientation by more than ±40°

Solution   The isolines shown in Figure 5.21 indicate that the recommended orientation of the principal axes of the PMS isohyets is approximately 195°, which is within 15° of the North/South direction (180°). Since the recommended orientation of the PMS is within ±40° of the basin orientation, then the principal axis of the PMS isohyets can be taken to coincide with the longitudinal axis of the watershed (North/South direction).

The outer isohyet of the PMS is an ellipse centered at the center of the watershed and passing through all corners of the rectangular watershed area. Since the dimensions of the watershed are 19.3 km × 53.6 km, then the radial coordinates of a point on the outer ellipse are given by

$$r = \sqrt{\left(\frac{19.3}{2}\right)^2 + \left(\frac{53.6}{2}\right)^2} = 28.5 \text{ km}$$

$$\theta = \tan^{-1}\left(\frac{19.3}{53.6}\right) = 19.8°$$

Substituting these data into Equation 5.25 and invoking Equation 5.26 yields

$$r^2 = \frac{a^2 b^2}{a^2 \sin^2 \theta + b^2 \cos^2 \theta}$$

$$28.5^2 = \frac{(2.5b)^2 b^2}{(2.5b)^2 \sin^2(19.8) + b^2 \cos^2(19.8)}$$

which yields

$$b = 14.4 \text{ km}$$

Using Equation 5.27 to calculate the area, A, covered by the outer isohyet of the PMS gives

$$b = \left(\frac{A}{2.5\pi}\right)^{1/2}$$

$$14.4 = \left(\frac{A}{2.5\pi}\right)^{1/2}$$

which yields

$$A = 1629 \text{ km}^2$$

From the given data, the 72-h PMS over an area of 1629 km^2 (629 mi^2) is 105 cm (41.4 in.), and the maximum 6-h increment is 65.5 cm (25.8 in.). The isohyetal rainfall distribution in the PMS is expressed in the form of the average rainfall depth between isohyets, and isohyets A through H are appropriate for this drainage area. Using the adjustment factors in Figure 5.20 yields the rainfall distribution given in the following table:

(1) Isohyet contour	(2) Area (km^2)	(3) Adjustment factor (Figure 5.20)	(4) Isohyetal precipitation depth (cm)	(5) Average depth between isohyets (cm)
A	26	1.40	91.7	91.7
B	65	1.31	85.8	88.8
C	129	1.22	79.9	82.8
D	259	1.14	74.7	77.3
E	453	1.06	69.4	72.1
F	776	0.97	63.5	66.5
G	1165	0.91	59.6	61.6
H	1813	0.81	53.0	56.3

Columns 1 and 2 are the standard PMS isohyets, column 3 is the adjustment factor derived from Figure 5.20, column 4 is obtained by multiplying column 3 by the 6-h rainfall depth of 65.5 cm, and column 5 is the average 6-h rainfall between PMS isohyets. The actual rainfall over the drainage basin is determined by superimposing the PMS standard isohyetal pattern over the drainage basin, and partitioning the rainfall in column 5 between rain that falls within the drainage basin and rain that falls outside the drainage basin.

The U.S. Army Corps of Engineers Hydrologic Engineering Center (HEC, 1984) has a computer program called HMR 52, which computes the basin-average rainfall for the PMS based on the PMP estimate from HMR 51.

5.2.4.3 Standard project storm

The *standard project storm* (SPS) is a temporal distribution of precipitation that is reasonably characteristic of large storms that have occurred or could occur in the locality of concern (U.S. Army Corps of Engineers, 1989). For drainage areas east of the 105th meridian (east of the Rocky Mountains), detailed guidance on the estimation of the SPS has been published by the U.S. Army Corps of Engineers (1965) as generalized regional relationships for depth, duration, and area of precipitation. For areas west of the 105th meridian, special studies are made to develop appropriate SPS estimates. Procedures for estimating the SPS are different for small basins with areas less than 2600 km^2 (1000 mi^2), than for large basins with areas greater than 2600 km^2 (1000 mi^2). For small drainage basins ($<$2600 km^2), the SPS can be reasonably estimated as 50% of the probable maximum storm (PMS), while for large drainage basins ($>$2600 km^2) the spatial and temporal distribution of rainfall are of particular importance. For example, the total rainfall over the Kansas River basin during the period 25–31 May 1903, which produced an estimated peak discharge of 7400 m^3/s on the Kansas River at Kansas City, was almost identical with the total precipitation that occurred over the basin on 9–12 July 1951 to produce a peak discharge of 14,000 m^3/s. This vividly illustrates why estimation of SPS on large drainage basins is a difficult proposition.

5.3 Rainfall Abstractions

The processes of interception, infiltration, and depression storage are commonly referred to as *rainfall abstractions*. These processes must generally be accounted for in estimating the surface runoff resulting from a given rainfall event.

5.3.1 Interception

Interception is the process by which rainfall is abstracted prior to reaching the ground. The wetting of surface vegetation is typically the primary form of interception, although rainfall is also intercepted by buildings and other above-ground structures. In urban areas, the density of vegetation is usually not sufficient to cause an appreciable amount of interception. However, in areas where there is a significant amount of vegetation, such as forested areas, interception can significantly reduce the amount of rainfall that reaches the ground. Therefore, in projects that involve the clearing of wooded areas, engineers must be prepared to account for the increased runoff that will occur as a result of reduced interception.

Methods used for estimating interception are mostly empirical, where the amount of interception is expressed either as a fraction of the amount of precipitation or as an empirical function of the rainfall amount. The interception percentages over seasonal and annual time scales for several types of vegetation have been summarized by Woodall (1984) and are given in Table 5.8. These data indicate that, on an annual basis, tree interception can abstract as much as 48% of rainfall amount (*Picea abies*) and grasses on the order of 13%. Caution should be exercised expressing interception simply as a percentage of rainfall, since the storm characteristics (intensity and duration), local climate, and the age and density of the vegetation have a significant influence on the interception percentages. Where possible, local data should be used to estimate interception losses.

Many interception functions are similar to that originally suggested by Horton (1919), where the interception, I, for a single storm, is related to the rainfall amount, P, by an equation of the form

$$I = a + bP^n \qquad (5.28)$$

where a and b are constants. When I is measured in millimeters, typical values are $n = 1$ (for most vegetative covers); a between 0.02 mm for shrubs and 0.05 mm for pine woods; and b between 0.18 and 0.20 for orchards and woods and 0.40 for shrubs. Surface vegetation generally has a finite interception capacity, which should not be exceeded by the estimated interception amount. The interception storage capacity of surface vegetation can range from less than 0.3 mm to 13 mm, with a typical value for turf grass of 1.3 mm.

More sophisticated interception functions have been suggested to account for the limited storage capacity of surface vegetation and evaporation during the storm (Meriam, 1960; Gray, 1973; Brooks et al., 1991), where the interception, I, is expressed in the form

$$I = S(1 - e^{-\frac{P}{S}}) + KEt \qquad (5.29)$$

where S is the available storage, P is the amount of rainfall during the storm, K is the ratio of the surface area of one side of the leaves to the projection of the vegetation on the ground (called the *leaf area index*), E is the evaporation rate during the storm, and t is the duration of the storm. Available storage, S, is typically in the range of 3–4 mm for fully developed pine trees; 7 mm for spruce, fir, and hemlock; 3 mm for leafed-out hardwoods; and 1 mm for bare hardwoods (Helvey, 1971). More sophisticated models of canopy interception, particularly in forested areas, are still being developed (Hall, 2003).

Interception by forest litter is much smaller than canopy interception. The amount of litter interception is largely dependent on the thickness of the litter, water-holding capacity, frequency of wetting, and evaporation rate. Studies have shown that it is only a few millimeters in depth in most cases (Chang, 2002). Typically, about 1% to 5% of annual precipitation and less than 50 mm/year are lost to litter interception (Helvey and Patric, 1965).

TABLE 5.8: Interception Percentages in Selected Studies

Cover type	Season	Interception (%)	Reference
Conifers			
Picea abies	Year	48	Leyton et al. (1967)
Tsuga canadensis	Summer	33	Voigt (1960)
Pseudotsuga	Year	36	Aussenac and Boulangeat (1980)
Pseudotsuga	Summer	24	Rothacher (1963)
	Winter	14	
Pinus caribaea	Year	18	Waterloo et al. (1999)
Pinus radiata	Year	26	Feller (1981)
Pinus radiata	Year	19	Smith (1974)
Pinus resinosa	Summer	19	Voigt (1960)
Pinus strobus	Year	16	Helvey (1967)
Pinus taeda	Year	14	Swank et al. (1972)
Evergreen hardwoods			
Notofagus sp.	Year	33	Aldridge and Jackson (1973)
Notofagus/Podocarpus	Summer	30	Rowe (1979)
	Winter	21	
Acacia	Year	19	Beard (1962)
Eucalyptus regnans	Year	19	Feller (1981)
Melaleuca quinquenervia	Summer	19	Woodall (1984)
Moist tropical forest	Summer	16	Jackson (1971)
Mixed eucalypts	Year	11	Smith (1974)
Deciduous hardwoods			
Carpinus sp.	Year	36	Leyton et al. (1967)
Fagus grandifolia	Summer	25	Voigt (1960)
Fagus silvatica	Summer	21	Aussenac and Boulangeat (1980)
	Winter	6	
Liriodendron	Year	10	Helvey (1964)
Grasses			
Themeda sp.	Year	13	Beard (1962)
Cymbopogon sp.	Year	13	Beard (1962)
Soil cover			
Hardwood litter	Year	3	Helvey (1964)
Pinus strobus litter	Year	3	Helvey (1967)
Oak litter	Year	2	Blow (1955)
Pinus caribaea litter	Year	8	Waterloo et al. (1999)
Pinus taeda litter	Year	4	Swank et al. (1972)

Source: Woodall (1984).

EXAMPLE 5.14

A pine forest is to be cleared for a commercial development in which all the trees on the site will be removed. The IDF curve for a 20-year rainfall is given by the relation

$$i = \frac{2819}{t + 16}$$

where i is the rainfall intensity in mm/h and t is the duration in minutes. The storage capacity of the trees in the forest is estimated as 6 mm, the leaf area index is 7, and the evaporation rate during the storm is estimated as 0.2 mm/h. (a) Determine the increase in precipitation reaching the ground during a 20-min storm that will result from clearing the site; and (b) compare your result with the interception predicted by the Horton-type empirical equation of the form $I = a + bP^n$, where a and b are constants and P is the precipitation amount.

Solution

(a) For a 20-min storm, the average intensity, i, is given by the IDF equation as

$$i = \frac{2819}{t + 16} = \frac{2819}{20 + 16} = 78 \text{ mm/h}$$

and the precipitation amount, P, is given by

$$P = it = (78)\left(\frac{20}{60}\right) = 26 \text{ mm}$$

The interception, I, of the wooded area can be estimated by Equation 5.29, where $S = 6$ mm, $P = 26$ mm, $K = 7$, $E = 0.2$ mm/h, and $t = 20/60$ h $= 0.33$ h. Hence

$$I = S(1 - e^{-\frac{P}{S}}) + KEt = 6(1 - e^{-\frac{26}{6}}) + (7)(0.2)(0.33) = 5.9 + 0.5 = 6.4 \text{ mm}$$

The wooded area intercepts approximately 6.4 mm of the 26 mm that falls on the wooded area. Prior to clearing the wooded area, the rainfall reaching the ground in a 20-min storm is $26 - 6.4 = 19.6$ mm. After clearing the wooded area, the rainfall reaching the ground is expected to be 26 mm, an increase of 33% over the incident rainfall prior to clearing the wooded area. These calculations also show that evaporation contributes only 0.5 mm of the 6.4 mm intercepted, indicating that evaporation during a storm contributes relatively little to interception.

(b) Using the interception formula $I = a + bP^n$ for pine woods, it can be assumed that $n = 1$, $a = 0.05$ mm, and $b = 0.19$. Hence

$$I = a + bP^n = 0.05 + 0.19(26) = 5 \text{ mm}$$

The pine woods are estimated to intercept 5 mm of rainfall, in which case the predevelopment rainfall reaching the ground is $26 \text{ mm} - 5 \text{ mm} = 21$ mm and the postdevelopment rainfall reaching the ground is 26 mm, an increase of 24% over predevelopment conditions.

5.3.2 Depression Storage

Water that accumulates in surface depressions during a storm is called *depression storage*. This portion of rainfall does not contribute to surface runoff; it either infiltrates or evaporates following the rainfall event. Depression storage is generally expressed as an average depth over the catchment area, and typical depths of depression storage are given in Table 5.9. These typical values are for moderate slopes; the values would

TABLE 5.9: Typical Values of Depression Storage

Surface Type	Depression storage (mm)	Reference
Pavement:		
Steep	0.5	Pecher (1969), Viessman et al. (1977)
Flat	1.5, 3.5	Pecher (1969), Viessman et al. (1977)
Impervious areas	1.3–2.5	Tholin and Kiefer (1960)
Lawns	2.5–5.1	Hicks (1944)
Pasture	5.1	ASCE (1992)
Flat roofs	2.5–7.5	Butler and Davies (2000)
Forest litter	7.6	ASCE (1992)

be larger for flat slopes, and smaller for steep slopes (Butler and Davies, 2000). In estimating surface runoff from rainfall, depression storage is usually deducted from the initial rainfall.

EXAMPLE 5.15

A 10-min storm produces 12 mm of rainfall on an impervious parking lot. Estimate the fraction of this rainfall that becomes surface runoff.

Solution On the impervious parking lot, water trapped in depression storage forms puddles and does not contribute to runoff. Depression storage is typically in the range of 1.3–2.5 mm, with an average value of 1.9 mm. The fraction of runoff, C, is therefore estimated by

$$C = \frac{12 - 1.9}{12} = 0.84$$

This indicates an abstraction of 16% for a 12-mm storm, which corresponds to 84% runoff. Clearly, the fraction of runoff will increase for higher precipitation amounts.

5.3.3 Infiltration

The process by which water seeps into the ground through the soil surface is called *infiltration* and is usually the dominant rainfall abstraction process. Infiltration capacity is determined primarily by the surface cover and the properties of the underlying soil. Soils covered with grass or other vegetation tend to have significantly higher infiltration capacities than bare soils. Bare-soil infiltration rates are considered high when they are greater than 25 mm/h, low when they are less than 2.5 mm/h, and grass cover tends to increase these values by a factor between 3 and 7.5 (Viessman and Lewis, 2003).

Soils are typically classified by their particle-size distribution, and the U.S. Department of Agriculture (USDA) soil-classification system based on particle size is shown in Table 5.10. *Soil texture* is determined by the proportions (by weight) of clay, silt and sand, after particles larger than sand (>2 mm) are removed. The USDA scheme for defining soil texture is shown in Figure 5.23, which is commonly called the USDA soil texture triangle. If a significant proportion of the soil (>15%) is larger than sand, an adjective such as "gravelly" or "stony" is added to the soil texture specification.

TABLE 5.10: USDA Soil Classification System

Soil	Particle sizes
Clay	<0.002 mm
Silt	0.002–0.05 mm
Sand	0.05–2 mm
Gravel	>2 mm

FIGURE 5.23: USDA soil texture triangle

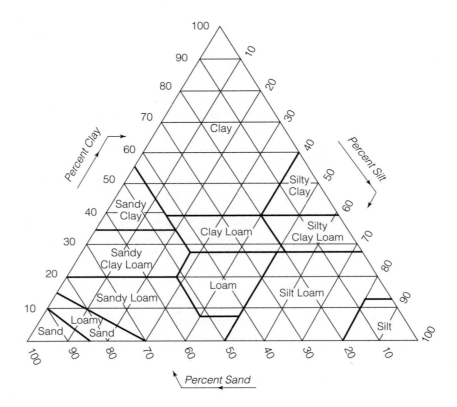

The vertical soil profile is generally categorized into horizons that are distinguished by the proportion of organic material and the degree to which material has been removed (eluviated) or deposited (illuviated) by chemical and physical processes. Soil scientists identify soil horizons on the basis of color, texture, and structure, and these horizons are generally designated by letters as follows:

O-Horizon. Surface litter consisting primarily of organic matter.

A-Horizon. Topsoil consisting of decaying organic matter and inorganic minerals.

E-Horizon. The zone of leaching where percolating water dissolves water-soluble matter.

B-Horizon. The subsoil below the A- and E-horizons that contain minerals and humic compounds. Usually contains more clay than the A-horizon.

C-Horizon. A zone consisting primarily of undecomposed mineral particles and rock fragments. This is the soil parent material from which the A- and B-horizons were formed.

R-Horizon. Bedrock, an impenetrable layer.

Not all of the soil horizons listed above are present in all soils, and the boundaries between layers are commonly gradational.

Quantitative estimation of runoff from rainfall generally requires an analytic description of the infiltration process. A variety of models are used to describe the infiltration process, with no one model being best for all cases. To appreciate the assumptions and limitations of the various models, it is important to look first at the fundamental process of infiltration.

5.3.3.1   The infiltration process

Infiltration describes the entry of water into the soil through the soil surface, and *percolation* describes the movement of water within the soil. The infiltration rate is equal to the percolation rate just below the ground surface, and the percolation rate, q_0, is given by Darcy's law as:

$$q_0 = -K(\theta) \frac{\partial h}{\partial z} \tag{5.30}$$

where $K(\theta)$ is the vertical hydraulic conductivity expressed as a function of the moisture content (= the volume of water per unit volume of the soil), θ; and h is the piezometric head of the pore water defined by

$$h = \frac{p}{\gamma} + z \tag{5.31}$$

where p is the pore-water pressure, γ is the specific weight of water, and z is the vertical coordinate (positive upward). A negative pore pressure indicates that the pressure is below atmospheric pressure. In the unsaturated soil beneath the ground surface, the soil moisture is usually under tension, with negative pore pressures.

Laboratory and field experiments indicate that there is a fairly stable relationship between the pore pressure, p, and moisture content, θ, that is unique to each soil. A typical relationship between $-p/\gamma$ and θ is illustrated in Figure 5.24(a), where the

FIGURE 5.24: Typical moisture retention curve and $C(\theta)$

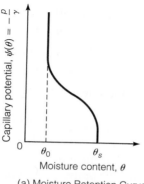

(a) Moisture Retention Curve

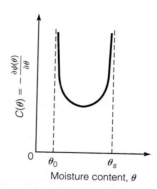

(b) Derived Function, $C(\theta)$

capillary potential, $\psi(\theta)$, defined by

$$\psi(\theta) = -\frac{p}{\gamma} \tag{5.32}$$

is commonly used in lieu of the pressure head and is closely related to the *matric potential*, which is defined as p/γ. The typical moisture retention curve, Figure 5.24(a), indicates that when the pores are filled with water at atmospheric pressure, the moisture content is equal to the saturated moisture content, θ_s, and the pore pressure and capillary potential are both equal to zero. The saturated moisture content, θ_s, is numerically equal to the porosity of the soil. As the moisture content is reduced, the pore pressure decreases and the capillary potential increases in accordance with Equation 5.32. This trend continues until the moisture content is equal to θ_0, at which point the pore water becomes discontinuous and further reductions in the pore pressure do not result in a decreased moisture content.

Combining Darcy's law, Equation 5.30, with the definition of the piezometric head, Equation 5.31, yields the following expression for the vertical percolation rate (just below the ground surface) in terms of the capillary potential:

$$q_0 = -K(\theta)\frac{\partial h}{\partial z}$$

$$= -K(\theta)\frac{\partial}{\partial z}\left(\frac{p}{\gamma} + z\right) = -K(\theta)\frac{\partial}{\partial z}\left[-\psi(\theta) + z\right]$$

$$= K(\theta)\left[\frac{\partial\psi(\theta)}{\partial z} - 1\right] \tag{5.33}$$

The chain rule of differentiation guarantees that the *suction gradient*, $\partial\psi/\partial z$, is given by

$$\frac{\partial\psi(\theta)}{\partial z} = \frac{\partial\psi(\theta)}{\partial\theta}\frac{\partial\theta}{\partial z} \tag{5.34}$$

where $\partial\psi/\partial\theta$ is a soil property derived from the moisture retention curve, Figure 5.24(a), and $\partial\theta/\partial z$ is called the *wetness gradient*. Combining Equations 5.34 and 5.33 yields the following relationship between the percolation rate, q_0, and the moisture content, θ:

$$q_0 = K(\theta)\left[\frac{\partial\psi(\theta)}{\partial\theta}\frac{\partial\theta}{\partial z} - 1\right]$$

$$= -K(\theta)\left[C(\theta)\frac{\partial\theta}{\partial z} + 1\right] \tag{5.35}$$

where $C(\theta)$ is a soil property defined by

$$C(\theta) = -\frac{\partial\psi(\theta)}{\partial\theta} \tag{5.36}$$

and is derived directly from the moisture retention curve for the soil. The inverse of $C(\theta)$ is sometimes called the *specific moisture capacity* (Kemblowki and Urroz, 1999) or the *specific water capacity* (Mays, 2001). The functional form of $C(\theta)$ corresponding to a typical moisture retention curve is illustrated in Figure 5.24(b), where it should be noted that $C(\theta)$ is always positive. Although the percolation rate, q_0, just below the ground surface given by Equation 5.35 is numerically equal to the infiltration rate, f, Equation 5.35 requires that q_0 is positive in the upward direction, while f is conventionally taken as positive in the downward direction. Consequently, the infiltration rate, f, is simply equal to negative q_0, in which case Equation 5.35 yields the following theoretical expression for the infiltration rate:

$$f = K(\theta)\left[C(\theta)\frac{\partial\theta}{\partial z} + 1\right] \tag{5.37}$$

Practical application of Equation 5.37 requires specification of the hydraulic conductivity function $K(\theta)$. According to Bear (1979), field experiments indicate that $K(\theta)$ can be adequately described by

$$K(\theta) = K_0\left(\frac{\theta - \theta_0}{\theta_s - \theta_0}\right)^3 \tag{5.38}$$

where K_0 is the hydraulic conductivity at saturation. According to Equation 5.38, the hydraulic conductivity, $K(\theta)$, increases monotonically from zero to K_0 as θ increases from θ_0 to θ_s. Combining Equations 5.37 and 5.38 yields the infiltration equation

$$\boxed{f = K_0\left(\frac{\theta - \theta_0}{\theta_s - \theta_0}\right)^3\left[C(\theta)\frac{\partial\theta}{\partial z} + 1\right]} \tag{5.39}$$

The fundamental infiltration process is illustrated in Figure 5.25 for the case where water is ponded above the ground surface. The initial moisture distribution between the ground surface and the water table (before ponding) approximates an equilibrium distribution where the conditions at the surface are described by $\theta = \theta_0$ and $\partial\theta/\partial z = 0$. Under these conditions, Equation 5.39 indicates that

$$f = 0 \tag{5.40}$$

Immediately after infiltration begins, the soil just below the ground surface becomes saturated, but it is still unsaturated further down in the soil column, leading to a sharp

FIGURE 5.25: Infiltration process

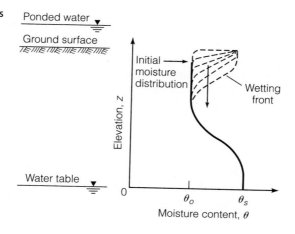

moisture gradient near the surface. Under these circumstances, $\theta = \theta_s$ and $\partial\theta/\partial z > 0$ at the ground surface. The infiltration rate is given by Equation 5.39 as

$$f = K_0 \left[C(\theta_s)\frac{\partial\theta}{\partial z} + 1 \right] \tag{5.41}$$

As infiltration proceeds, K_0 and $C(\theta_s)$ remain constant, and the moisture content gradient, $\partial\theta/\partial z$, gradually decreases to zero as the wetting front penetrates the soil column (see Figure 5.25). Therefore, the infiltration rate gradually decreases from its maximum value given by Equation 5.41 ($\partial\theta/\partial z > 0$), at the beginning of the infiltration process, to the asymptotic minimum infiltration rate when $\partial\theta/\partial z = 0$ and

$$f = K_0 \tag{5.42}$$

The minimum infiltration rate is therefore equal to the (vertical) saturated hydraulic conductivity of the soil. As the ponded water continues to infiltrate into the soil, conditions near the ground surface remain approximately constant, with infiltration proceeding at the minimum rate. Eventually, the entire soil column becomes saturated and the recharge rate at the water table equals the infiltration rate. In reality, soil is typically stratified, with the infiltration rate being eventually limited by the rate of percolation through the least pervious subsoil layer.

Several models are commonly used to estimate infiltration, and the validity of each of these models should be viewed relative to their consistency with the theoretical infiltration process described here. The models most frequently used in engineering practice are the Horton, Green–Ampt, and Natural Resources Conservation Service (NRCS) models. No single approach works best for all situations, and in most cases the methods are limited by knowledge of their site-specific parameters. Many of these models distinguish between the actual infiltration rate, f, and the *potential infiltration rate*, f_p, which is equal to the infiltration rate when water is ponded at the ground surface.

5.3.3.2 Horton model

Horton (1939, 1940) proposed the following empirical equation to describe the decline in the potential infiltration rate, f_p, as a function of time

$$\boxed{f_p = f_c + (f_0 - f_c)e^{-kt}} \tag{5.43}$$

where f_0 is the initial (maximum) infiltration rate, f_c is the asymptotic (minimum) infiltration rate ($t \to \infty$), and k is a decay constant. It has been shown previously that the asymptotic minimum infiltration rate must be equal to the saturated hydraulic conductivity of the soil. Equation 5.43 describes an infiltration capacity that decreases exponentially with time, ultimately approaching a constant value, and assumes an infiltration process described by the relation

$$\frac{df_p}{dt} = -k(f_p - f_c) \tag{5.44}$$

The Horton model fits well with experimental data (Singh, 1989). Typical values of f_0, f_c, and k are given in Table 5.11 (rank-ordered by f_c), and Singh (1992) recommends that f_0/f_c be on the order of 5. The variability of the infiltration parameters in the Horton model reflects the condition that infiltration depends on several factors that

TABLE 5.11: Typical Values of Horton Infiltration Parameters

Soil type	f_0 (mm/h)	f_c (mm/h)	k (min^{-1})
General[†]			
Coarse-textured soils	250	25	0.03
Medium-textured soils	200	12	0.03
Fine-textured soils	125	6	0.03
Clays/paved areas	75	3	0.03
By USDA soil texture*			
Sand	—	210	—
Loamy sand	—	61	—
Sandy loam	—	26	—
Loam	—	13	—
Silt loam	—	7	—
Sandy clay	—	4	—
Clay loam	—	2	—
Silty clay loam	—	1	—
Sandy clay	—	1	—
Silty clay	—	1	—
Clay	—	0.5	—
By specific soil[‡]			
Dothan loamy sand	88	67	0.02
Fuquay pebbly loamy sand	158	61	0.08
Tooup sand	584	46	0.55
Carnegie sandy loam	375	45	0.33
Leefield loamy sand	288	44	0.13
Alphalpha loamy sand	483	36	0.64

Sources: [†]Butler and Davies (2000); *Schueler (1987); [‡]Rawls et al. (1976).

are not explicitly accounted for in Table 5.11, such as the initial moisture content and organic content of the soil, vegetative cover, and season (Linsley et al., 1982).

The temporal variation in infiltration rate given by Equation 5.43 is applicable when the water is continuously ponded above the soil column, and the functional form of this equation is illustrated in Figure 5.26. In cases where water is not continuously ponded above the soil column, the potential infiltration, f_p, can be expressed in terms of the cumulative infiltration, F, by an implicit relationship. The cumulative infiltration

FIGURE 5.26: Horton infiltration model

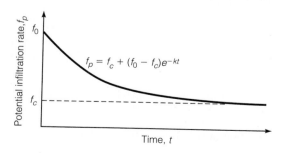

$$f_p = f_c + (f_0 - f_c)e^{-kt}$$

as a function of time corresponding to Equation 5.43 is given by

$$F(t) = \int_0^t f_p(\tau)\,d\tau$$

$$= \int_0^t \left[f_c + (f_0 - f_c)e^{-k\tau} \right] d\tau$$

$$= f_c t + \frac{f_0 - f_c}{k}(1 - e^{-kt}) \tag{5.45}$$

Equations 5.43 and 5.45 form an implicit relationship between the cumulative infiltration, F, and the potential infiltration rate, f_p, where t is simply a parameter in the relationship. Hence, at any time during a rainfall event, if the cumulative infiltration is known, it can be substituted into Equation 5.45 to obtain the value of the parameter t, which is then substituted into Equation 5.43 to obtain the corresponding infiltration capacity. Alternatively, Equations 5.43 and 5.45 can be combined to yield the following direct relationship between F and f_p:

$$F(t) = \left[\frac{f_c}{k} \ln(f_0 - f_c) + \frac{f_0}{k} \right] - \frac{f_c}{k} \ln(f_p - f_c) - \frac{f_p}{k} \tag{5.46}$$

EXAMPLE 5.16

A catchment soil has Horton infiltration parameters: $f_0 = 100$ mm/h, $f_c = 20$ mm/h, and $k = 2$ min^{-1}. What rainfall rate would result in ponding from the beginning of the storm? If this rainfall rate is maintained for 40 minutes, describe the infiltration as a function of time during the storm.

Solution   According to the Horton model of infiltration, the potential infiltration rate varies between a maximum of 100 mm/h ($= f_0$) and an (asymptotic) minimum of 20 mm/h ($= f_c$). Any storm in which the rainfall rate exceeds 100 mm/h during the entire storm will cause ponding from the beginning of the storm. Under these circumstances, the infiltration rate, f, as a function of time is given by Equation 5.43 as

$$f = f_c + (f_0 - f_c)e^{-kt} = 20 + (100 - 20)e^{-2t}$$

$$= 20 + 80e^{-2t}, \qquad 0 \le t \le 40 \text{ min}$$

EXAMPLE 5.17

A catchment soil is found to have the following Horton infiltration parameters: $f_0 = 100$ mm/h, $f_c = 20$ mm/h, and $k = 2$ min^{-1}. The design storm is given by the following hyetograph:

Interval (min)	Average rainfall (mm/h)
0–10	10
10–20	20
20–30	80
30–40	100
40–50	80
50–60	10

Estimate the time at which ponding begins.

Solution   From the given data: $f_0 = 100$ mm/h, $f_c = 20$ mm/h, and $k = 2$ min^{-1} = 120 h^{-1}. According to Equation 5.46, the cumulative infiltration, F, is related to the infiltration capacity, f_p, by

$$F = \left[\frac{f_c}{k} \ln(f_0 - f_c) + \frac{f_0}{k} \right] - \frac{f_c}{k} \ln(f_p - f_c) - \frac{f_p}{k}$$

$$= \left[\frac{20}{120} \ln(100 - 20) + \frac{100}{120} \right] - \frac{20}{120} \ln(f_p - 20) - \frac{f_p}{120}$$

which yields

$$F = 1.564 - 0.167 \ln(f_p - 20) - 0.00833 f_p \qquad (5.47)$$

where f_p is in mm/h. Each 10-minute increment of the storm will now be taken sequentially.

$t = 0{-}10$ min: During this period the rainfall intensity, i ($=10$ mm/h), is less than the minimum infiltration rate, f_c ($=20$ mm/h), and no ponding occurs. The cumulative infiltration, F, after 10 min is given by

$$F = i\,\Delta t = 10\left(\frac{10}{60} \right) = 1.67 \text{ mm}$$

$t = 10{-}20$ min: During this period, the rainfall intensity, i ($=20$ mm/h), is equal to the minimum infiltration rate, f_c ($=20$ mm/h), and no ponding occurs. The cumulative infiltration, F, after 20 min is given by

$$F = i\,\Delta t + 1.67 = 20\left(\frac{10}{60} \right) + 1.67 = 5.01 \text{ mm}$$

$t = 20{-}30$ min: During this period, the rainfall intensity, i ($=80$ mm/h), exceeds the minimum infiltration rate, f_c ($=20$ mm/h), and therefore ponding is possible. At $t = 20$ min, $F = 5.01$ mm and Equation 5.47 gives

$$5.01 = 1.564 - 0.167 \ln(f_p - 20) - 0.00833 f_p$$

which yields an infiltration capacity, f_p, of

$$f_p = 20 \text{ mm/h}$$

Since the rainfall rate (80 mm/h) exceeds the infiltration capacity (20 mm/h) from the beginning of the time interval, ponding starts at the beginning of the time interval, at $t = 20$ min.

5.3.3.3   Green–Ampt model

This physically based semi-empirical model was first proposed by Green and Ampt (1911) and was put on a firm physical basis by Philip (1954). The Green–Ampt model, sometimes called the *delta function model* (Salvucci and Entekhabi, 1994; Philip, 1993), is today one of the most realistic models of infiltration available to the engineer. A typical vertical section of soil is shown in Figure 5.27, where it is assumed that water is ponded to a depth H on the ground surface and that there is a sharp interface between the wet soil and the dry soil. This interface is called the *wetting front*, and

FIGURE 5.27: Green–Ampt soil column

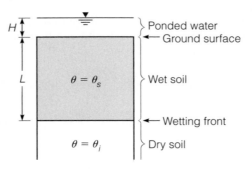

as the ponded water infiltrates into the soil, the wetting front moves downward. Flow through saturated porous media is described by Darcy's law, which can be written in the form

$$q_s = -K_s \frac{dh}{ds} \tag{5.48}$$

where q_s is the flow per unit area or *specific discharge* in the s-direction, K_s is the hydraulic conductivity of the saturated soil, h is the piezometric head, and dh/ds is the gradient of the piezometric head in the s-direction. In reality, K_s is usually less than the hydraulic conductivity at saturation, because of entrapped air which prevents complete saturation. Using a finite-difference approximation to the Darcy equation over the depth of the wet soil, L, leads to

$$f_p = -K_s \frac{(-\Phi_f) - (H + L)}{L} \tag{5.49}$$

where f_p is the potential infiltration rate (equal to the specific discharge when water is ponded above the ground surface), $-\Phi_f$ is the head at the wetting front, and $H + L$ is the head at the top of the wet soil. The suction head Φ_f is defined as $-p/\gamma$, where p is the pressure head in the water at the wetting front and p is generally negative (i.e., below atmospheric pressure). The suction head, Φ_f, is equal to the capillary potential under initial soil moisture conditions. Assuming that the soil was initially dry, then the total volume of water infiltrated, F, is given by

$$F = (n - \theta_i)L \tag{5.50}$$

where n is the porosity of the soil (equal to volumetric water content at saturation) and θ_i is the initial volumetric water content of the dry soil. The volumetric water content of a soil is the volume of water in a soil sample divided by the volume of the sample. In reality, the value of n used in Equation 5.50 should be reduced to account for the entrapment of air. Assuming that $H \ll (L + \Phi_f)$, then Equations 5.49 and 5.50 can be combined to yield

$$f_p = K_s + \frac{K_s(n - \theta_i)\Phi_f}{F} \tag{5.51}$$

Recognizing that the potential infiltration rate, f_p, and the cumulative infiltrated amount, F, are related by

$$f_p = \frac{dF}{dt} \tag{5.52}$$

then Equations 5.51 and 5.52 can be combined to yield the following differential equation for F:

$$\frac{dF}{dt} = K_s + \frac{K_s(n - \theta_i)\Phi_f}{F} \tag{5.53}$$

Separation of variables allows this equation to be written as

$$\int_0^F \frac{F'}{K_s F' + K_s(n - \theta_i)\Phi_f} \, dF' = \int_0^t dt' \tag{5.54}$$

which incorporates the initial condition that $F = 0$ when $t = 0$. Integrating Equation 5.54 leads to the following equation for the cumulative infiltration versus time:

$$K_s t = F - (n - \theta_i)\Phi_f \ln\left[1 + \frac{F}{(n - \theta_i)\Phi_f}\right] \tag{5.55}$$

This equation assumes that water has been continuously ponded above the soil column from time $t = 0$, which will happen only if the rainfall intensity exceeds the infiltration capacity from the beginning of the storm. If the rainfall intensity, i, is initially less than the infiltration capacity, then the actual infiltration, f, will be equal to the rainfall intensity until the infiltration capacity, f_p, (which continually decreases) becomes equal to the rainfall intensity. Assuming that this happens at time t_p, then

$$f = \begin{cases} i, & t < t_p \\ f_p = i, & t = t_p \end{cases} \tag{5.56}$$

Combining Equations 5.56 and 5.51 leads to the following expression for the total infiltrated volume, F_p, at $t = t_p$:

$$F_p = \frac{\Phi_f(n - \theta_i)}{i/K_s - 1} \tag{5.57}$$

Since the infiltration rate has been equal to i up to this point, then the time, t_p, at which the infiltrated volume becomes equal to F_p is given by

$$t_p = \frac{F_p}{i} \tag{5.58}$$

For $t > t_p$, the rainfall intensity exceeds the potential infiltration rate, and therefore infiltration continues at the potential rate given by

$$f = f_p = K_s + \frac{K_s(n - \theta_i)\Phi_f}{F}, \qquad t > t_p \tag{5.59}$$

and the cumulative infiltration as a function of time can be written in the form (Mein and Larson, 1973)

$$K_s(t - t_p + t_p') = F - (n - \theta_i)\Phi_f \ln\left[1 + \frac{F}{(n - \theta_i)\Phi_f}\right] \tag{5.60}$$

where t_p' is the equivalent time to infiltrate F_p under the condition of surface ponding from $t = 0$.

TABLE 5.12: Typical Values of Green–Ampt Parameters

USDA soil-texture class	Hydraulic conductivity K_s (mm/h)	Wetting front suction head Φ_f (mm)	Porosity n	Field capacity θ_0	Wilting point θ_w
Sand	120	49	0.437	0.062	0.024
Loamy sand	30	61	0.437	0.105	0.047
Sandy loam	11	110	0.453	0.190	0.085
Loam	3	89	0.463	0.232	0.116
Silt loam	7	170	0.501	0.284	0.135
Sandy clay loam	2	220	0.398	0.244	0.136
Clay loam	1	210	0.464	0.310	0.187
Silty clay loam	1	270	0.471	0.342	0.210
Sandy clay	1	240	0.430	0.321	0.221
Silty clay	1	290	0.479	0.371	0.251
Clay	0.3	320	0.475	0.378	0.265

Source: Rawls et al. (1983).

Typical values of the Green–Ampt parameters for several soil types are given in Table 5.12. These values should not be used for bare soils with crusted surfaces. Methods for estimating Green–Ampt soil parameters from readily available soil characteristics have been presented by Rawls and Brakensiek (1982) and Rawls and colleagues (1983). The field capacity, θ_0, and wilting point, θ_w, give some indication of the dry-soil moisture content, θ_i, where the field capacity is the residual water content after gravity drainage, and the wilting point is the limiting water content below which plants cannot extract water for transpiration. ASCE (1996b) recommends caution in applying the Green–Ampt method in forested areas, where the basic assumptions of the method may not be valid. According to Ward and Dorsey (1995), the Green–Ampt method best describes the infiltration process in soils that exhibit a sharp wetting front, such as coarse-textured soils with uniform pore shapes. However, in many agricultural soils sharp wetting fronts are not the norm (Thomas and Phillips, 1979; Quisenberry and Phillips, 1976).

EXAMPLE 5.18

A catchment area consists almost entirely of loamy sand, which typically has a saturated hydraulic conductivity of 30 mm/h, average suction head of 61 mm, porosity of 0.44, field capacity of 0.105, wilting point of 0.047, and depression storage of 5 mm. You are in the process of designing a stormwater-management system to handle the runoff from this area, and have selected the following 1-hour storm as the basis of your design:

Interval (min)	Ave. rainfall (mm/h)
0–10	10
10–20	20
20–30	80
30–40	100
40–50	80
50–60	10

Determine the runoff versus time for average initial moisture conditions, and contrast the amount of rainfall with the amount of runoff. Use the Green–Ampt method for your calculations and assume that the initial moisture conditions are midway between the field capacity and wilting point.

Solution   The basic equations of the Green–Ampt model are Equations 5.59 and 5.60. In the present case: $K_s = 30$ mm/h; $\theta_i = \frac{1}{2}(0.105 + 0.047) = 0.076$; $n = 0.44$; and $\Phi_f = 61$ mm. The infiltration capacity, f_p, as a function of the cumulative infiltration, F, is

$$f_p = K_s + \frac{K_s(n - \theta_i)\Phi_f}{F}$$

$$= 30 + \frac{30(0.44 - 0.076)(61)}{F}$$

$$= 30 + \frac{666.12}{F} \tag{5.61}$$

The cumulative infiltration as a function of time is given by

$$K_s(t - t_p + t'_p) = F - (n - \theta_i)\Phi_f \ln\left[1 + \frac{F}{(n - \theta_i)\Phi_f}\right]$$

$$30(t - t_p + t'_p) = F - (0.44 - 0.076)(61)\ln\left[1 + \frac{F}{(0.44 - 0.076)(61)}\right]$$

$$= F - 22.2\ln(1 + 0.0450F) \tag{5.62}$$

If ponding occurs from $t = 0$, then

$$30t = F - 22.2\ln(1 + 0.0450F) \tag{5.63}$$

Each 10-minute increment in the storm will now be taken sequentially, and the computation of the runoff is summarized in Table 5.13.

$t = 0$–10 min: During this period, the rainfall intensity, i ($= 10$ mm/h) is less than the saturated hydraulic conductivity, K_s ($= 30$ mm/h), and therefore no ponding occurs. The entire rainfall amount of $10 \times (10/60) = 1.7$ mm is infiltrated.

$t = 10$–20 min: During this period, the rainfall rate (20 mm/h) is still less than the minimum infiltration capacity (30 mm/h) and all the rainfall infiltrates. The rainfall amount during this period is $20 \times (10/60) = 3.3$ mm, and the cumulative infiltration after 20 minutes is 1.7 mm $+ 3.3$ mm $= 5.0$ mm.

$t = 20$–30 min: During this period the rainfall rate ($= 80$ mm/h) exceeds the minimum infiltration capacity ($= 30$ mm/h), and therefore ponding is possible. Use Equation 5.61 to determine the cumulative infiltration, F_{80}, corresponding to a potential infiltration rate of 80 mm/h:

$$80 = 30 + \frac{666.12}{F_{80}}$$

which leads to

$$F_{80} = 13.3 \text{ mm}$$

TABLE 5.13: Computation of Rainfall Excess Using Green–Ampt Model

Time (min)	F (mm)	Δt (h)	ΔF (mm)	i (mm/h)	$i\,\Delta t$ (mm)	Depression storage, ΔS (mm)	Total storage (mm)	Runoff (mm)
0	0.0							
		0.167	1.7	10	1.7	0	0	0
10	1.7							
		0.167	3.3	20	3.3	0	0	0
20	5.0							
		0.167	12.9	80	13.4	0.5	0.5	0
30	17.9							
		0.167	9.9	100	16.7	4.5	5.0	2.3
40	27.8							
		0.167	8.5	80	13.4	0	5.0	4.9
50	36.3							
		0.167	6.7	10	1.7	−5.0	0	0
60	43.0							
Total:					50.2			7.2

After 20 min (0.334 h) the total infiltration was 5.0 mm, therefore if $0.334 + t'$ is the time when the cumulative infiltration is 13.3 mm, then

$$5.0 + 80t' = 13.3$$

which leads to

$$t' = 0.104 \text{ h } (=6.2 \text{ min})$$

Therefore, the time at which ponding occurs, t_p, is

$$t_p = 0.334 \text{ h } + 0.104 \text{ h}$$
$$= 0.438 \text{ h } (=26.3 \text{ min})$$

The next step is to find the time, t_p', that it would take for 13.3 mm to infiltrate, if infiltration occurs at the potential rate from $t = 0$. Infiltration as a function of time is given by Equation 5.63, therefore

$$30t_p' = 13.3 - 22.2 \ln[1 + (0.0450)(13.3)]$$

which leads to

$$t_p' = 0.096 \text{ h } (=5.8 \text{ min})$$

Therefore, the equation for the cumulative infiltration as a function of time after $t = 0.438$ h is given by Equation 5.62, which can be written as

$$30(t - 0.438 + 0.096) = F - 22.2 \ln(1 + 0.0450F)$$

or

$$30(t - 0.342) = F - 22.2 \ln(1 + 0.0450F) \tag{5.64}$$

At the end of the current time period, $t = 3(0.167) = 0.501$ h, and substituting this value into Equation 5.64 leads to

$$F = 17.9 \text{ mm}$$

Since the rainfall during this period is 13.4 mm, and the cumulative infiltration (= cumulative rainfall) up to the beginning of this period is 5 mm, then the amount of rainfall that does not infiltrate is equal to 5 mm + 13.4 mm − 17.9 mm = 0.5 mm. This excess amount goes toward filling up the depression storage, which has a maximum capacity of 5 mm. The available depression storage at the end of this time period is 5 mm − 0.5 mm = 4.5 mm. Since the depression storage is not filled, there is no runoff.

t = **30−40 min:** Since the rainfall rate is now higher than during the previous time interval, ponding continues to occur. The time at the end of this period is 0.668 h. Substituting this value for *t* into Equation 5.64 gives the cumulative infiltration at the end of the time period as

$$F = 27.8 \text{ mm}$$

Since the cumulative infiltration up to the beginning of this period is 17.9 mm, then the infiltrated amount during this period is 27.8 mm − 17.9 mm = 9.9 mm. The rainfall during this period is 16.7 mm; therefore, the amount of rain that does not infiltrate is 16.7 mm − 9.9 mm = 6.8 mm. Since there is 4.5 mm in available depression storage, then the amount of runoff is 6.8 mm − 4.5 mm = 2.3 mm. The depression storage is filled at the end of this time period.

t = **40−50 min:** The rainfall rate (80 mm/h) is still higher than the infiltration capacity and ponding continues. The time at the end of this period is 0.835 h, and Equation 5.64 gives the cumulative infiltration at the end of this time period as

$$F = 36.3 \text{ mm}$$

The cumulative infiltration up to the beginning of this period is 27.8 mm; therefore, the infiltration during this time interval is 36.3 mm − 27.8 mm = 8.5 mm. The rainfall during this period is 13.4 mm, therefore the amount of rain that does not infiltrate is 13.4 mm − 8.5 mm = 4.9 mm. Since the depression storage is full, then all 4.9 mm is contributed to runoff.

t = **50−60 min:** The rainfall rate (10 mm/h) is below the minimum infiltration capacity (30 mm/h). Under these circumstances, the ponded water will infiltrate and, if there is sufficient infiltration capacity, some of the rainfall may also infiltrate. The time at the end of this period is 1 h, and if infiltration continues at the potential rate, then the cumulative infiltration at the end of the period is given by Equation 5.64 as

$$F = 44 \text{ mm}$$

Because the cumulative infiltration up to the beginning of this period is 36.3 mm, the potential infiltration during this period is 44 mm − 36.3 mm = 7.7 mm. The rainfall during this period is 1.7 mm and there is 5 mm of depression storage to infiltrate. Since the rainfall plus depression storage is equal to 6.7 mm, then all the rainfall and depression storage is infiltrated during this time interval. The cumulative infiltration at the end of this period is 36.3 mm + 6.7 mm = 43 mm.

The total rainfall during this storm is 50.2 mm and the total runoff is 7.2 mm. The runoff is therefore equal to 14% of the rainfall.

5.3.3.4 NRCS curve-number model

The *curve-number model* was originally developed by the Natural Resources Conservation Service* (NRCS), within the U.S. Department of Agriculture. This model was published in 1954 in the first edition of the National Engineering Handbook, which has subsequently been revised several times (e.g., SCS, 1993). This empirical method is the most widely used method for estimating *rainfall excess* (= rainfall minus abstractions) in the United States, and the popularity of this method is attributed to its ease of application, lack of serious competition, and extensive database of parameters. The curve-number model was originally developed for calculating daily runoff as affected by land-use practices in small agricultural watersheds; then, because of its overwhelming success, it was subsequently adapted to urban catchments.

The NRCS curve-number model separates the rainfall into three components: *rainfall excess*, Q, *initial abstraction*, I_a, and *retention*, F. These components are illustrated graphically in Figure 5.28. The initial abstraction includes the rainfall that is stored in the catchment area before runoff begins. This includes interception, infiltration, and depression storage. If the amount of rainfall is less than the initial abstraction, then runoff does not occur. The retention, F, is the portion of the rainfall reaching the ground that is retained by the catchment and consists primarily of infiltrated water. The basic assumption of the NRCS model is that for any rainfall event the precipitation, P, runoff, Q, retention, F, and initial abstraction, I_a, are related by

$$\frac{F}{S} = \frac{Q}{P - I_a} \tag{5.65}$$

where S is the *potential maximum retention* and measures the retention capacity of the soil. The maximum retention, S, does not include I_a. The rationale for Equation 5.65 is that for any rainfall event, the portion of available storage (= S) that is filled, F, is equal to the portion of available rainfall (= $P - I_a$) that appears as runoff, Q. Equation 5.65 is, of course, applicable only when $P > I_a$. Conservation of mass requires that

$$F = P - Q - I_a \tag{5.66}$$

Eliminating F from Equations 5.65 and 5.66 yields

$$Q = \frac{(P - I_a)^2}{(P - I_a) + S}, \qquad P > I_a \tag{5.67}$$

FIGURE 5.28: Components in the NRCS curve-number model

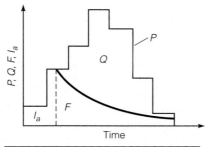

---

*The Natural Resources Conservation Service (NRCS) was formerly called the Soil Conservation Service (SCS).

Empirical data indicate that the initial abstraction, I_a, is directly related to the maximum retention, S, and the following relation is commonly assumed:

$$I_a = 0.2\,S \tag{5.68}$$

Recent research has indicated that the factor of 0.2 is probably adequate for large storms in rural areas, but it is likely an overestimate for small to medium storms and is probably too high for urban areas (Schneider and McCuen, 2005; Singh, 1992). However, since the storage capacity, S, of many catchments has been determined based on Equation 5.68, it is recommended that the 0.2 factor be retained when using storage estimates calibrated from field measurements of rainfall and runoff.

Combining Equations 5.67 and 5.68 leads to

$$\boxed{Q = \frac{(P - 0.2S)^2}{P + 0.8S}, \qquad P > 0.2S} \tag{5.69}$$

This equation is the basis for estimating the volume of runoff, Q, from a volume of rainfall, P, given the maximum retention, S. It should be noted, however, that the NRCS method was originally developed as a runoff index for 24-h rainfall amounts and should be used with caution in attempting to analyze incremental runoff amounts during the course of a storm (Kibler, 1982) or runoff from durations other than 24 hours. This is elucidated by noting that the curve-number model predicts the same runoff for a given rainfall amount, regardless of duration. This is obviously not correct, since long-duration low-intensity storms will have a smaller runoff amount than short-duration high-intensity storms, when both storms have the same total rainfall. The curve-number model is less accurate when the runoff is less than 10 mm, and in these cases another method should be used to determine the runoff (SCS, 1986). In applications, the curve-number model is quite satisfactory when used for its intended purpose, which is to evaluate the effects of land-use changes and conservation practices on direct runoff. Since it was not developed to reproduce individual historical events, only limited success has been achieved in using it for that purpose.

The total infiltration for a rainfall event is given by the combination of Equations 5.66, 5.68, and 5.69, which yields

$$F = \frac{(P - 0.2S)S}{P + 0.8S}, \qquad P > 0.2S \tag{5.70}$$

The infiltration rate, f, can be derived from Equation 5.70 by differentiation, where

$$\boxed{f = \frac{dF}{dt} = \frac{S^2 i}{(P + 0.8S)^2}} \tag{5.71}$$

and i is the rainfall intensity given by

$$i = \frac{dP}{dt} \tag{5.72}$$

The functional form of the infiltration rate formula given by Equation 5.71 is not physically realistic, since it requires that the infiltration rate be dependent on the rainfall intensity (Morel-Seytoux and Verdin, 1981). In spite of this limitation, the NRCS curve-number model is widely used in practice.

Instead of specifying S directly, a *curve number*, CN, is usually specified where CN is related to S by

$$CN = \frac{1000}{10 + 0.0394S} \tag{5.73}$$

where S is given in millimeters. Equation 5.73 is modified from the original formula, which required S in inches. Clearly, in the absence of available storage ($S = 0$, impervious surface) the curve number is equal to 100, and for an infinite amount of storage the curve number is equal to zero. The curve number therefore varies between 0 and 100. The utilization of the curve number, CN, in the rainfall-runoff relation is the basis for the naming of the curve-number model. In fact, the NRCS adopted the term "curve number" because the rainfall-runoff equation may be expressed graphically with curves for different values of CN.

In practical applications, the curve number is considered to be a function of several factors including hydrologic soil group, cover type, treatment (management practice), hydrologic condition, antecedent runoff condition, and impervious area in the catchment. Curve numbers corresponding to a variety of urban land uses are given in Table 5.14. Soils are classified into four hydrologic soil groups: A, B, C, and D; descriptions of these groups are given in Table 5.15. Soils are grouped based on profile characteristics that include depth, texture, organic matter content, structure, and degree of swelling when saturated. Minimum infiltration rates associated with the hydrologic soil groups are given in Table 5.15, where it should be noted that these rates are for bare soil after prolonged wetting. The NRCS has classified more than 5000 soils into these four groups (Rawls et al., 1982). Local NRCS offices can usually provide information on local soils and their associated soil groups, but it should be noted that activities such as the operation of heavy equipment can substantially change the local soil characteristics (Haan et al., 1994). There are a variety of methods for determining cover type, the most common ones being field reconnaissance, aerial photographs, and land-use maps. In agricultural practice, *treatment* is a cover-type modifier used to describe the management of cultivated lands. It includes mechanical practices such as contouring and terracing, and management practices such as crop rotations and reduced or no tillage. Hydrologic condition indicates the effects of cover type and treatment on infiltration and runoff, and is generally estimated from the density of plant and residue cover. Good hydrologic condition indicates that the soil has a low runoff potential for that specific hydrologic soil group, cover type, and treatment. The percentage of impervious area and the means of conveying runoff from impervious areas to the drainage systems should generally be considered in estimating the curve number for urban areas. An impervious area is considered directly connected if runoff from the area flows directly into the drainage system. The impervious area is not directly connected if runoff from the impervious area flows over a pervious area and then into a drainage system.

The *antecedent runoff condition* (ARC) is a measure of the actual available storage relative to the average available storage at the beginning of the rainfall event. The antecedent runoff condition is closely related to the antecedent moisture content of the soil and is grouped into three categories: ARC I, ARC II, and ARC III. The average curve numbers normally cited for a particular land area correspond to ARC II conditions, and these curve numbers can be adjusted for drier than normal conditions (ARC I) or wetter than normal conditions (ARC III) by using Table 5.16.

TABLE 5.14: Curve Numbers for Various Urban Land Uses

Cover type and hydrologic condition	Curve numbers for hydrologic soil group			
	A	B	C	D
Lawns, open spaces, parks, golf courses:				
Good condition: grass cover on 75% or more of the area	39	61	74	80
Fair condition: grass cover on 50% to 75% of the area	49	69	79	84
Poor condition: grass cover on 50% or less of the area	68	79	86	89
Paved parking lots, roofs, driveways, etc.	98	98	98	98
Streets and roads:				
Paved with curbs and storm sewers	98	98	98	98
Gravel	76	85	89	91
Dirt	72	82	87	89
Paved with open ditches	83	89	92	93
Commercial and business areas (85% impervious*)	89	92	94	95
Industrial districts (72% impervious*)	81	88	91	93
Row houses, town houses, and residential with lot sizes 1/8 ac or less (65% impervious*)	77	85	90	92
Residential average lot size:				
1/8 ac or less (town houses) (65% impervious*)	77	85	90	92
1/4 ac (38% impervious*)	61	75	83	87
1/3 ac (30% impervious*)	57	72	81	86
1/2 ac (25% impervious*)	54	70	80	85
1 ac (20% impervious*)	51	68	79	84
2 ac (12% impervious*)	46	65	77	82

*The impervious area is assumed to be directly connected to the drainage system, with the impervious area having a CN of 98, and the pervious area taken as equivalent to open space in good hydrologic condition.

TABLE 5.15: Description of NRCS Soil Groups

Group	Description	Minimum infiltration rate (mm/h)
A	Deep sand; deep loess; aggregated silts	> 7.6
B	Shallow loess; sandy loam	3.8–7.6
C	Clay loams; shallow sandy loam; soils low in organic content; soils usually high in clay	1.3–3.8
D	Soils that swell significantly when wet; heavy plastic clays; certain saline soils	0–1.3

TABLE 5.16: Antecedent Runoff Condition Adjustments

CN for ARC II	Corresponding CN for condition	
	ARC I	ARC III
100	100	100
95	87	99
90	78	98
85	70	97
80	63	94
75	57	91
70	51	87
65	45	83
60	40	79
55	35	75
50	31	70
45	27	65
40	23	60
35	19	55
30	15	50
25	12	45
20	9	39
15	7	33
10	4	26
5	2	17
0	0	0

The curve-number adjustments in Table 5.16 can be approximated by the relations (Hawkins et al., 1985; Chow et al., 1988)

$$CN_I = \frac{CN_{II}}{2.3 - 0.013CN_{II}} \tag{5.74}$$

and

$$CN_{III} = \frac{CN_{II}}{0.43 + 0.0057CN_{II}} \tag{5.75}$$

where CN_I, CN_{II}, and CN_{III} are the curve numbers under AMC I, AMC II, and AMC III conditions, respectively.

The guidelines for selecting curve numbers given in Tables 5.14 and 5.16 are useful in cases where site-specific data on the maximum retention, S, either are not available or cannot be reasonably estimated. Parameters of the curve-number model given in Table 5.14 were estimated based on data from small (< 4 ha) agricultural watersheds in the midwestern United States. Current empirical evidence suggests that hydrological systems are overdesigned when using curve numbers from Tables 5.14 and 5.16 (Schneider and McCuen, 2005). Whenever S is available for a catchment, then the curve number should be estimated directly using Equation 5.73, and, if necessary, adjusted using Table 5.16. As a precautionary note, the NRCS does not recommend the use of the curve-number model when CN is less than 40.

EXAMPLE 5.19

The drainage facilities of a catchment are to be designed for a rainfall of return period 25 years and duration 2 hours, where the IDF curve for 25-year storms is given by

$$i = \frac{830}{t + 33}$$

where i is the rainfall intensity in cm/h, and t is the storm duration in minutes. A double-ring infiltrometer test on the soil shows that the minimum infiltration rate is on the order of 5 mm/h. The urban area being developed consists of mostly open space with less than 50% grass cover. Use the NRCS curve-number method to estimate the total amount of runoff (in cm), assuming the soil is in average condition at the beginning of the design storm. Estimate the percentage increase in runoff that would occur if heavy rainfall occurred within the previous five days and the soil was saturated.

Solution   The amount of rainfall can be estimated using the IDF curve. For a 2-hour 25-year storm, the average intensity is given by

$$i = \frac{830}{120 + 33} = 5.42 \text{ cm/h}$$

and the total amount of rainfall, P, in the 2-h storm is equal to $(5.42)(2) = 10.8$ cm.

Since the minimum infiltration rate of the soil is 5 mm/h, then according to Table 5.15 it can be inferred that the soil is in Group B. The description of the area, open space with less than 50% grass and Group B soil, is cited in Table 5.14 to have a curve number equal to 79. The curve number, CN, and soil storage, S (in mm), are related by Equation 5.73 as

$$CN = \frac{1000}{10 + 0.0394S}$$

and therefore the storage, S, in this case is given as

$$S = \frac{1}{0.0394}\left(\frac{1000}{CN} - 10\right) = \frac{1}{0.0394}\left(\frac{1000}{79} - 10\right) = 67.5 \text{ mm}$$

The runoff amount, Q, can be calculated from the rainfall amount, P ($=10.8$ cm), and the maximum storage, S ($=6.75$ cm) using Equation 5.69, where

$$Q = \frac{(P - 0.2S)^2}{P + 0.8S}, \qquad P > 0.2S$$

Since P ($=10.8$ cm) $> 0.2S$ ($=1.35$ cm), then this equation is valid and

$$Q = \frac{[10.8 - 0.2(6.75)]^2}{10.8 + 0.8(6.75)} = 5.51 \text{ cm}$$

Hence, there is 5.51 cm of runoff from the storm with a rainfall amount of 10.8 cm.

When the soil is saturated, Table 5.16 indicates that the curve number, CN, increases to 94 and the maximum available soil storage is given by

$$S = \frac{1}{0.0394}\left(\frac{1000}{CN} - 10\right) = \frac{1}{0.0394}\left(\frac{1000}{94} - 10\right) = 16.2 \text{ mm}$$

and the runoff amount, Q, is given by

$$Q = \frac{[10.8 - 0.2(1.62)]^2}{10.8 + 0.8(1.62)} = 9.07 \text{ cm}$$

Therefore, under saturated conditions, the runoff amount increases from 5.51 cm to 9.07 cm, which corresponds to a 64.6% increase.

Soil type and soil properties at most locations in the United States can be obtained from the USDA Natural Resources Conservation Service Soil Survey Geographic (SSURGO) database. The digital SSURGO maps duplicate the original soil-survey maps that were compiled to 1:24,0000 scale, 7.5-min orthophoto quadrangles. This database generally includes the hydrologic soil group and permeability range of each soil, and the permeability is commonly taken as being equal to the saturated hydraulic conductivity and the infiltration capacity of the soil. It should be noted that some investigations have shown soil saturated hydraulic conductivity is often much higher than soil permeability reported in soil surveys (Troch et al., 1993; Rossing, 1996); however, soil-survey permeabilities have sometimes been used as worst-case infiltration capacities in rainfall-runoff models (Walter et al., 2003).

5.3.3.5   Comparison of infiltration models

The Horton, Green–Ampt, and curve-number models are all commonly used in engineering practice and are justified by their inclusion in the ASCE Manual of Practice on the Design and Construction of Urban Stormwater Management Systems (ASCE, 1992). These models are representative of the three classes of infiltration models: physically based, semi-empirical, and empirical models. The Green–Ampt model is a physically based model, the Horton model is a semi-empirical model, and the curve-number model is an empirical model. In comparing the performance of various infiltration models, the size of the area is an important consideration. The scales at which infiltration models are applied include the local (soil column) scale, the agricultural field scale, and the catchment scale. An infiltration model that performs better at one scale might not necessarily perform better at another scale.

Some comparative studies have indicated that the Green–Ampt type models perform better than the curve-number model in predicting runoff volumes from catchments, and hence peak runoff rates (Chahinian et al., 2005; Van Mullem, 1991; Hjelmfelt, 1991). This result is not unexpected, given the relatively poor performance of the curve-number model when applied to individual storms (Willeke, 1997), the general recognition that the curve-number method best represents a long-term expected-value relationship between rainfall and runoff (Smith, 1997), the recognition that the relation of soil and land-use conditions to curve number may vary regionally (Miller and Cronshey, 1989), and the recognition that the curve-number model does not properly account for soil moisture (Michel et al., 2005). Mishra et al. (2003) compared the performance of fourteen physically based, semi-empirical, and empirical infiltration models, including the Green–Ampt and Horton models (not including the curve-number model), and concluded that semi-empirical models tend to perform better than other physically based models, which are apparently better for laboratory-tested soils than for field soils. The results reported by Mishra et al. (2003) indicated that the Horton model performs significantly better than the Green–Ampt model.

   The measurable nature of the Green–Ampt parameters and the physical basis of the model are appealing features; however, both the Green–Ampt and Horton models suffer from being unbounded (Ponce and Hawkins, 1996), in that they allow an unlimited amount of infiltration and they do not account for the effects of spatial variability in parameters such as the saturated hydraulic conductivity (Woolhiser et al., 1996). A pragmatic reason for using the curve-number model is that it is a method supported by a U.S. government agency (the Department of Agriculture), which gives its users basic protection in litigation (Smith, 1997). The authoritative origin of the curve-number model apparently qualifies as a defense in legal proceedings that an engineer has performed a hydrologic analysis in accordance with generally accepted standards (Willeke, 1997).

5.3.4   Rainfall Excess on Composite Areas

In many cases, a catchment can be delineated into several subcatchments with different abstraction characteristics. The runoff from such composite catchments depends on how the subcatchments are connected, and this is illustrated in Figure 5.29 for the case of pervious and impervious subcatchments within a larger catchment. The subcatchments consist of a pervious area and two separate impervious ones, with only one of the impervious areas directly connected to the catchment outlet. The runoff from the composite catchment is equal to the runoff from the impervious area, A_1, that is directly connected to the catchment outlet, plus the runoff from the pervious area, A_2, which assimilates a portion of the runoff from the impervious area, A_3, that is not directly connected to the catchment outlet. To calculate the rainfall excess on the composite catchment, the rainfall excesses on the each of the subcatchments are first calculated using methods such as the NRCS curve-number method, and then these rainfall excesses are routed to the outlet of the composite catchment.
   Rainfall abstractions in each subcatchment can be defined by a functional relation such as

$$Q_i = f_i(P), \qquad i = [1,3] \tag{5.76}$$

FIGURE 5.29: Runoff from composite catchment

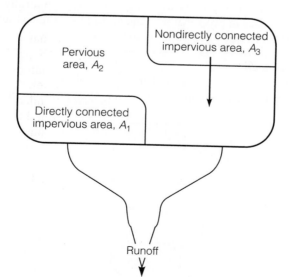

where Q_i is the rainfall excess on subcatchment i, and f_i is an abstraction function that relates Q_i to the precipitation, P, on the subcatchment. If the NRCS curve-number method is used to calculate the rainfall excess on subcatchment i, then f_i is given by

$$f_i(P) = \begin{cases} \frac{(P-0.2S_i)^2}{P+0.8S_i} & \text{if } P > 0.2S_i \\ 0 & \text{if } P \leq 0.2S_i \end{cases} \tag{5.77}$$

where S_i is the available storage in the subcatchment i. It can usually be assumed that the precipitation depth, P, on each subcatchment is the same. Since the rainfall excess, Q_3, from the nondirectly connected impervious area, A_3, drains into the pervious area, A_2, the effective precipitation, P_{eff}, on the pervious area is given by

$$P_{\text{eff}} = P + Q_3 \frac{A_3}{A_2} \tag{5.78}$$

The total rainfall excess, Q, from the composite catchment is equal to the sum of the rainfall excess on the directly connected impervious area plus the rainfall excess on the (directly connected) pervious area. Therefore

$$Q = f_1(P) \frac{A_1}{A} + f_2(P_{\text{eff}}) \frac{A_2}{A}$$

$$= f_1(P) \frac{A_1}{A} + f_2\left(P + Q_3 \frac{A_3}{A_2}\right) \frac{A_2}{A} \tag{5.79}$$

where A is the total area of the composite catchment given by

$$A = \sum_{i=1}^{3} A_i \tag{5.80}$$

This routing methodology for estimating the rainfall excess on composite catchments can be extended to any arrangement of subcatchments, provided that the flow paths to the catchment outlet are clearly defined. It is generally recognized that urban-area runoff increases in proportion to the amount of impervious area and the way this area is connected to outflow points by the drainage system.

EXAMPLE 5.20

The commercial site illustrated in Figure 5.30 covers 25 ha, of which the parking lot covers 7.5 ha, the building covers 6.3 ha, open grass covers 10 ha, and the site is graded

FIGURE 5.30: Commercial site

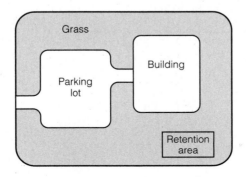

such that all of the runoff is routed to a grass retention area that covers 1.2 ha. Both the roof of the building and the parking lot are impervious (CN = 98) and drain directly onto the grassed area, which contains Type B soil and is in good condition. For a storm with a precipitation of 200 mm, calculate the surface runoff that enters the retention area.

Solution   In this case, both the parking lot and the roof of the building are not directly connected to the retention area, and therefore the runoff must be routed to the retention area through the grass. For the parking lot and building, CN = 98, and the available storage, S_1, can be derived from Equation 5.73 as

$$S_1 = \frac{1}{0.0394}\left(\frac{1000}{CN} - 10\right) = \frac{1}{0.0394}\left(\frac{1000}{98} - 10\right) = 5 \text{ mm}$$

For a precipitation, P, of 200 mm, the rainfall excess, Q_1, from the parking lot and building is given by

$$Q_1 = \frac{(P - 0.2S_1)^2}{P + 0.8S_1} = \frac{[200 - (0.2)(5)]^2}{200 + (0.8)(5)} = 194 \text{ mm}$$

The total area of the parking lot and building is equal to 7.5 ha + 6.3 ha = 13.8 ha, and the grassed area (including the retention area) is 10 ha + 1.2 ha = 11.2 ha. The effective rainfall, P_{eff}, on the grassed area is therefore given by

$$P_{eff} = 200 + 194\frac{13.8}{11.2} = 439 \text{ mm}$$

The grassed area contains Type B soil and is in good condition, and Table 5.14 gives CN = 61. The available storage, S_2, is derived from Equation 5.73 as

$$S_2 = \frac{1}{0.0394}\left(\frac{1000}{CN} - 10\right) = \frac{1}{0.0394}\left(\frac{1000}{61} - 10\right) = 162 \text{ mm}$$

The rainfall excess from the 11.2-ha grassed area, Q_2, is given by

$$Q_2 = \frac{(P_{eff} - 0.2S_2)^2}{P_{eff} + 0.8S_2} = \frac{[439 - (0.2)(162)]^2}{439 + (0.8)(162)} = 291 \text{ mm}$$

and the rainfall excess, Q, from the entire 25-ha composite catchment is

$$Q = 291\frac{11.2}{25} = 130 \text{ mm}$$

Therefore, a rainfall of 200 mm on the composite catchment will result in 130 mm of runoff, which will be disposed of via infiltration in the retention area.

A commonly used approximation in estimating the rainfall excess from composite areas is to apply the NRCS curve-number method with an area-weighted curve number. In a composite catchment with n subcatchments, the area-weighted curve number, CN_{eff}, is defined as

$$CN_{eff} = \frac{1}{A}\sum_{i=1}^{n} CN_i A_i \tag{5.81}$$

where CN_i and A_i are the curve number and area of subcatchment i and A is the total area of the composite catchment. If the flow paths in the composite catchment are known, then it is generally preferable that the rainfall excesses be routed through the catchment, as described previously.

EXAMPLE 5.21

The commercial site illustrated in Figure 5.30 covers 25 ha, of which the parking lot covers 7.5 ha, the building covers 6.3 ha, open grass covers 10 ha, and the site is graded such that all of the runoff is routed to a grass retention area that covers 1.2 ha. Both the roof of the building and the parking lot are impervious ($CN = 98$) and drain directly onto the grassed area ($CN = 61$). For a storm with a precipitation of 200 mm, use an area-weighted curve number to calculate the rainfall excess that enters the retention area. Contrast your result with that obtained using the routing method in Example 5.20.

Solution   The area-weighted curve number, CN_{eff}, is given by Equation 5.81 as

$$CN_{eff} = \frac{1}{A} \sum_{i=1}^{n} CN_i A_i$$

$$= \frac{1}{25}[(98)(7.5) + (98)(6.3) + (10)(61) + (1.2)(61)]$$

$$= 81$$

The available storage, S, corresponding to CN_{eff}, is

$$S = \frac{1}{0.0394}\left(\frac{1000}{CN_{eff}} - 10\right) = \frac{1}{0.0394}\left(\frac{1000}{81} - 10\right) = 60 \text{ mm}$$

and the rainfall excess, Q, resulting from a rainfall, P, of 200 mm is given by

$$Q = \frac{(P - 0.2S)^2}{P + 0.8S} = \frac{[200 - (0.2)(60)]^2}{200 + (0.8)(60)} = 143 \text{ mm}$$

This estimated runoff of 143 mm is 10% higher than the 130 mm estimated by routing the rainfall excess from the subcatchments.

5.4 Runoff Models

Runoff models predict the temporal distribution of runoff at a catchment outlet based on the temporal distribution of effective rainfall and the catchment characteristics. The *effective rainfall* is defined as the incident rainfall minus the abstractions and is sometimes referred to as the *rainfall excess*. The most important abstractions are usually infiltration and depression storage, and the catchment characteristics that are usually most important in translating the effective rainfall distribution to a runoff distribution at the catchment outlet are those related to the topography and surface cover of the catchment. Runoff models are classified as either *distributed-parameter* models or *lumped-parameter* models. Distributed-parameter models account for runoff processes on scales smaller than the size of the catchment, such as accounting for the runoff from every roof, over every lawn, and in every street gutter, while lumped-parameter models consider the entire catchment as a single hydrologic element, with the runoff characteristics described by one or more (lumped) parameters.

In cases where the surface runoff flows into an unlined drainage channel that penetrates the saturated zone of an aquifer, the flow in the drainage channel consists of both surface-water runoff and ground-water inflow to the channel, and these flow components must generally be modeled separately. The flow resulting from surface runoff is called *direct runoff*, and the flow resulting from ground-water inflow is subdivided into *base flow* and *interflow*, which is sometimes referred to as *throughflow*. Base flow is typically (quasi-) independent of the rainfall event, is equal to the flow of ground water from the saturated zone into the drainage channel, and depends on the difference between the water-table elevation in the surrounding aquifer and the water-surface elevation in the drainage channel (Chin, 1991). Interflow is the inflow of ground water to the drainage channel that occurs between the ground surface and the water table and is typically caused by a low-permeability subsurface layer that impedes the vertical infiltration of rainwater along with large pores left by rotting tree roots and burrowing animals. Interflow contributions to river flows can be significant in forested areas. Various methods have been adopted for computationally separating the direct runoff, interflow and base flow components of observed hydrographs (Singh, 1992; Tallaksen, 1995; McCuen, 1998). The simplest approach is to assume a constant base flow equal to the discharge just before the rain begins (Wurbs and James, 2002). The direct runoff resulting from a storm event is added to the base flow and interflow to yield the flow hydrograph in the drainage channel. Base flow contributed by infiltrating ground water is a relatively slow process compared with overland flow and interflow to a channel and, as a consequence, overland flow and interflow are sometimes collectively referred to as *quickflow* (Fitts, 2002).

A fundamental hypothesis that was originally made by Horton (1933b; 1945) is that overland flow occurs when the rainfall rate exceeds the infiltration capacity of the soil. This type of overland flow is commonly referred to as *Hortonian overland flow*. In current hydrologic practice, it is generally recognized that rainfall rates rarely exceed the infiltration capacities of soils, and runoff does occur when rainfall rates are less than the soil infiltration capacities. Betson (1964) proposed that, whereas surface runoff may be generated by a Hortonian mechanism, within a catchment there are only limited areas that contribute overland flow to a runoff hydrograph, and this is referred to as the *partial-area concept*. Hewlett and Hibbert (1967) were the first to suggest that there may be a process other than the Hortonian process that is responsible for the generation of overland flow. In modern engineering practice, it is generally recognized that the two primary hydrological mechanisms that generate overland flow are *infiltration excess* and *saturation excess*. Saturation excess is fundamentally different from infiltration excess in that overland flow is generated at locations where the soil is saturated at the surface. Unlike Hortonian runoff, where soil type and land use play a controlling role in runoff generation, landscape position, local topography, and soil depth are some of the primary controls in saturation-excess runoff. Saturation excess is the basis of the concept of *variable source-area* hydrology that acknowledges that the spatial extent of saturation will vary seasonally, depending on the relative rates of rainfall and evapotranspiration. In the saturation-excess process, rainfall causes a thin layer of soil on some parts of the basin to saturate upward from some restricting boundary to the ground surface, especially in zones of shallow, wet, or less-permeable soil. This process occurs frequently on the footslopes of hills, bottoms of valleys, swamps, and shallow soils. Determining which process dominates, infiltration excess or saturation excess, is fundamental to identifying appropriate methods for describing the rainfall-runoff relation. Unfortunately, it is all too common in hydrologic practice to assume, without evidence, that the runoff mechanism is Hortonian and apply models that may not reflect reality. In cases where significant surface runoff is observed while estimated infiltration capacities and rainfall intensities indicate that significant Hortonian runoff should not occur, saturation-excess runoff is likely to be the primary runoff mechanism (Walter et al., 2003).

A wide variety of models are available for calculating runoff from rainfall, and their applicability must be assessed in light of the fundamental rainfall-runoff process. The applicability of various runoff models can be broadly associated with the *scale* of the catchment, which can be classified as *small*, *midsize*, or *large* (Ponce, 1989). In small catchments, the response to rainfall events is sufficiently rapid and the catchment is sufficiently small that runoff during a relatively short time interval can be adequately modeled by assuming a constant rainfall in space and time. The *rational method* is the most widely used runoff model in small catchments. In midsize catchments, the slower response requires that the temporal distribution of rainfall be accounted for; however, the catchment is still smaller than the characteristic storm scale, and the rainfall can be assumed to be uniform over the catchment. *Unit hydrograph* models are the most widely used runoff models in midsize catchments. In large catchments, both the spatial and temporal variations in precipitation events must be incorporated in the runoff model, and models that explicitly incorporate routing methodologies are the most appropriate. Runoff regimes within a catchment vary from overland flow at the smallest scales to river flow at the largest scales, and runoff models must necessarily accommodate this scale effect. Small catchments have predominantly overland-flow runoff, while large catchments typically have a significant amount of runoff in identifiable river or drainage channels. As a consequence, the channel storage characteristics increase significantly from small catchments to large catchments.

5.4.1 Time of Concentration

The parameter most often used to characterize the response of a catchment to a rainfall event is the *time of concentration*, defined as the time to equilibrium of a catchment under a steady rainfall excess, or sometimes as the longest travel time that it takes surface runoff to reach the discharge point of a catchment (Wanielista et al., 1997). Most equations for estimating the time of concentration, t_c, express t_c as function of the rainfall intensity, i, catchment length scale, L, average catchment slope, S_0, and a parameter that describes the catchment surface, C, hence the equations for t_c typically have the functional form

$$t_c = f(i, L, S_0, C) \tag{5.82}$$

The time of concentration of a catchment includes the time of overland flow and the travel time in drainage channels leading to the catchment outlet.

5.4.1.1 Overland flow

Several equations are commonly used to estimate the time of concentration for overland flow. The most popular ones are described here.

Kinematic-wave equation.   A fundamental expression for the time of concentration in overland flow can be derived by considering the one-dimensional approximation of the surface-runoff process illustrated in Figure 5.31. The boundary of the catchment area is at $x = 0$, i_e is the rainfall-excess rate, y is the runoff flow depth, and q is the volumetric flow rate per unit width of the catchment area. Within the control volume of length Δx, the law of conservation of mass requires that the net mass inflow is equal to the rate of change of mass within the control volume. This law can be stated mathematically by the relation

$$\left[(\rho q) - \frac{\partial(\rho q)}{\partial x}\frac{\Delta x}{2}\right] + \left[i_e \rho\, \Delta x\right] - \left[(\rho q) + \frac{\partial(\rho q)}{\partial x}\frac{\Delta x}{2}\right] = \frac{\partial y}{\partial t}\rho\, \Delta x \qquad (5.83)$$

where the first term in square brackets is the inflow into the control volume, the second term is the rainfall excess entering the control volume, the third term is the outflow, and the righthand side of Equation 5.83 is equal to the rate of change of fluid mass within the control volume. Taking the density, ρ, as being constant and simplifying yields

$$\frac{\partial y}{\partial t} + \frac{\partial q}{\partial x} = i_e \qquad (5.84)$$

This equation contains two unknowns, q and y, and a second relationship between these variables is needed to solve this equation. Normally, the second equation is the momentum equation; however, a unique relationship between the flow rate, q, and the flow depth, y, can be assumed to have the form

$$q = \alpha y^m \qquad (5.85)$$

where α is a proportionality constant. The assumption of a relationship such as Equation 5.85 is justified in that equations describing steady-state flow in open channels, such as the Manning and Darcy–Weisbach equations, can be put in the form of Equation 5.85. Combining Equations 5.84 and 5.85 leads to the following differential equation for y:

$$\frac{\partial y}{\partial t} + \alpha m y^{m-1}\frac{\partial y}{\partial x} = i_e \qquad (5.86)$$

FIGURE 5.31: One-dimensional approximation of surface-runoff process

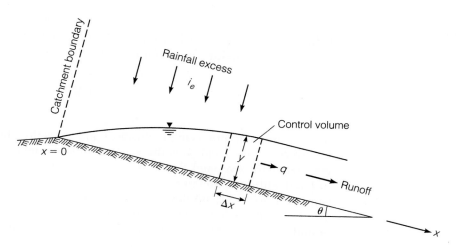

Solution of this equation can be obtained by comparing it with

$$\frac{dy}{dt} = \frac{\partial y}{\partial t} + \frac{dx}{dt}\frac{\partial y}{\partial x} = i_e \tag{5.87}$$

which gives the rate of change of y with respect to t observed by moving at a velocity dx/dt. Consequently, Equation 5.86 is equivalent to the following pair of equations:

$$\frac{dx}{dt} = \alpha m y^{m-1} \tag{5.88}$$

$$\frac{dy}{dt} = i_e \tag{5.89}$$

where dx/dt is called the *wave speed*, and Equation 5.89 is called the *kinematic-wave equation*. Solution of Equation 5.89 subject to the boundary condition that $y = 0$ at $t = 0$ yields

$$y = i_e t \tag{5.90}$$

Substituting this result into Equation 5.88 and integrating subject to the boundary condition that $x = 0$ at $t = 0$ yields

$$x = \alpha i_e^{m-1} t^m \tag{5.91}$$

Equations 5.90 and 5.91 are parametric equations describing the water surface illustrated in Figure 5.31, and the discharge at any location along the catchment area can be obtained by combining Equations 5.85 and 5.90 to yield

$$q = \alpha (i_e t)^m \tag{5.92}$$

Defining the time of concentration, t_c, of a catchment as the time required for a kinematic wave to travel the distance L from the catchment boundary to the catchment outlet, then Equation 5.91 gives the time of concentration as

$$t_c = \left(\frac{L}{\alpha i_e^{m-1}}\right)^{\frac{1}{m}} \tag{5.93}$$

If the Manning equation is used to relate the runoff rate to the depth, then Equation 5.93 can be written in the form (ASCE, 1992)

$$t_c = 6.99 \frac{(nL)^{0.6}}{i_e^{0.4} S_0^{0.3}} \tag{5.94}$$

where t_c is in minutes, i_e in mm/h, and L in m, n is the Manning roughness coefficient for overland flow, and S_0 is the ground slope. Estimates of the Manning roughness coefficient for overland flow are given in Table 5.17, where the surface types are ordered with increasing roughness. On the basis of Equation 5.94, the time of concentration for overland flow should be regarded as a function of the rainfall-excess rate (i_e), the catchment-surface roughness (n), the flow length from the catchment boundary to the outlet (L), and the slope of the flow path (S_0).

TABLE 5.17: Manning's n for Overland Flow

Surface type	Manning n	Range
Smooth concrete	0.011	0.01–0.014
Bare sand	0.01	0.01–0.016
Graveled surface	0.012	0.010–0.018
Asphalt	0.012	0.010–0.018
Bare clay	0.012	0.010–0.016
Smooth earth	0.018	0.015–0.021
Bare clay-loam (eroded)	0.02	0.012–0.033
Bare smooth soil	0.10	—
Range (natural)	0.13	0.01–0.32
Sparse vegetation	0.15	—
Short grass	0.15	0.10–0.25
Light turf	0.20	—
Woods, no underbrush	0.20	0.1–0.3
Dense grass	0.24	0.15–0.35
Lawns	0.25	0.20–0.30
Dense turf	0.35	0.30–0.35
Pasture	0.35	0.30–0.40
Dense shrubbery and forest litter	0.40	—
Woods, light underbrush	0.40	0.3–0.5
Bermuda grass	0.41	0.30–0.50
Bluegrass sod	0.45	0.39–0.63
Woods, dense underbrush	0.80	0.6–0.95

Sources: ASCE (1992); Wurbs and James (2002); Crawford and Linsley (1966); Engman (1986); McCuen et al. (1996); McCuen (2005).

Equation 5.94 assumes that the surface runoff is described by the Manning equation, which is valid only for turbulent flows; however, at least a portion of the surface runoff will be in the laminar and transition regimes (Wong and Chen, 1997). This limitation associated with using the Manning equation can be addressed by using the Darcy–Weisbach equation, which yields

$$\alpha = \left(\frac{8gS_0}{C\nu^k} \right)^{\frac{1}{(2-k)}} \quad \text{and} \quad m = \frac{3}{2-k} \tag{5.95}$$

where ν is the kinematic viscosity of water, and C and k are parameters relating the Darcy–Weisbach friction factor, f, to the Reynolds number, Re:

$$f = \frac{C}{\text{Re}^k} \tag{5.96}$$

where $k = 0$ for turbulent flow, $k = 1$ for laminar flow, $0 < k < 1$ for transitional flow, and the Reynolds number is defined by

$$\text{Re} = \frac{q}{\nu} \tag{5.97}$$

Laminar flow typically occurs where $\text{Re} < 200$, turbulent flow where $\text{Re} > 2,000$, and transition flow where $200 < \text{Re} < 2,000$. Values of C for overland flow have not been widely published, but Radojkovic and Maksimovic (1987) and Wenzel (1970) indicate that for concrete surfaces C values of 41.8, 2, and 0.04 are appropriate for laminar, transition, and turbulent flow regimes, respectively. Combining the kinematic-wave expression for t_c, Equation 5.93, with the Darcy–Weisbach equation, Equations 5.95 to 5.97, yields

$$t_c = \left[\frac{0.21(3.6 \times 10^6 \nu)^k CL^{2-k}}{S_0 i_e^{1+k}} \right]^{\frac{1}{3}}$$

(5.98)

where t_c is in minutes, ν is in m²/s, L is in meters, and i_e is in mm/h. Equation 5.98, called the Chen and Wong formula (Wong, 2005), can be used to account for various flow regimes in overland flow, and it has been shown that assuming a single flow regime (laminar, transition, or turbulent) will tend to underestimate the time of concentration (Wong and Chen, 1997). Indications are that overland flow is predominantly in the transition regime and that Equation 5.98 may be most applicable using $k \approx 0.5$.

NRCS method.   NRCS (SCS, 1986) proposed that overland flow consists of two sequential flow regimes: *sheet flow* and *shallow concentrated flow*. Sheet flow is characterized by runoff that occurs as a continuous sheet of water flowing over the land surface, while shallow concentrated flow is characterized by flow in isolated rills and then gullies of increasing proportions. Ultimately, most surface runoff enters open channels and pipes, which is the third regime of surface runoff included in the time of concentration. In many cases, the time of concentration of a catchment is expressed as the sum of the travel time as sheet flow plus the travel time as shallow concentrated flow plus the travel time as open-channel flow. The flow characteristics of sheet flow are sufficiently different from shallow concentrated flow that separate equations are recommended. The flow length of the sheet flow regime should generally be less than 100 m, and the travel time, t_f (in hours), over a flow length, L (in m), is estimated by (SCS, 1986)

$$t_f = 0.0288 \frac{(nL)^{0.8}}{P_2^{0.5} S_0^{0.4}}$$

(5.99)

where n is the Manning roughness coefficient for overland flow (Table 5.17), S_0 is the land slope, and P_2 is the two-year 24-hour rainfall (in cm). Equation 5.99 was developed from the kinematic-wave equation (Equation 5.94) by Overton and Meadows (1976) using the following assumptions: (1) The flow is steady and uniform with a depth of about 3 cm; (2) the rainfall intensity is uniform over the catchment; (3) the rainfall duration is 24 hours; (4) infiltration is neglected; and (5) the maximum flow length is 100 m. In considering the validity of these assumptions, it should be noted that overland flow depth may be significantly different from 3 cm in many areas, the rainfall duration may differ from 24 hours, and the actual travel time can increase if there is a significant amount of infiltration in the catchment. By limiting the maximum flow length to 100 m, the catchment is necessarily small, and the assumption of a spatially uniform rainfall distribution is reasonable. After a maximum distance of 100 m, sheet flow usually becomes shallow concentrated flow, and the average velocity, V_{sc}, is taken to be a function of the slope of the flow path and the type of land surface in accordance with the Manning equation

$$V_{sc} = \frac{1}{n} R^{\frac{2}{3}} S_0^{\frac{1}{2}}$$

(5.100)

where n is the roughness coefficient, R is the hydraulic radius, and S_0 is the slope of the flow path. For unpaved areas, it is commonly assumed that $n = 0.05$ and $R = 12$ cm; and for paved areas, $n = 0.025$ and $R = 6$ cm. Equation 5.100 can also be expressed in the form

$$V_{sc} = kS_0^{\frac{1}{2}}$$

(5.101)

where k ($= R^{2/3}/n$) is called the *intercept coefficient*, and several suggested values of k are given in Table 5.18. In addition to intercept coefficients for shallow concentrated flow, Table 5.18 also gives bulk intercept coefficients for the entire overland flow, including sheet flow and shallow concentrated flow regimes. The average velocity, V_{sc}, derived from Equation 5.101 is then combined with the flow length, L_{sc}, of shallow concentrated flow to yield the flow time, t_{sc}, as

$$t_{sc} = \frac{L_{sc}}{V_{sc}}$$

(5.102)

The total time of concentration, t_c, of overland flow is taken as the sum of the sheet flow time, t_f, given by Equation 5.99, and the shallow concentrated flow time, t_{sc}, given by Equation 5.102. The overland flow time of concentration is added to the channel flow time to obtain the time of concentration of the entire catchment.

Kirpich equation.   An empirical time of concentration formula that is especially popular is the *Kirpich formula* (Kirpich, 1940) given by

$$t_c = 0.019 \frac{L^{0.77}}{S_0^{0.385}}$$

(5.103)

TABLE 5.18: Intercept Coefficient for Overland-Flow Velocity vs. Slope

Land cover/flow regime	k (m/s)
Forest with heavy ground litter; hay meadow (overland flow)	0.76
Trash fallow or minimum tillage cultivation; contour or strip cropped; woodland (overland flow)	1.52
Short grass pasture (overland flow)	2.13
Cultivated straight row (overland flow)	2.74
Nearly bare and untilled (overland flow); alluvial fans in western mountain regions	3.05
Grassed waterway (shallow concentrated flow)	4.57
Unpaved (shallow concentrated flow)	4.91
Paved area (shallow concentrated flow); small upland gullies	6.19

Source: U.S. Federal Highway Administration (1996).

where t_c is the time of concentration in minutes, L is the flow length in meters, and S_0 is the average slope along the flow path. Equation 5.103 was originally developed and calibrated from NRCS data reported by Ramser (1927) on seven partially wooded agricultural catchments in Tennessee, ranging in size from 0.4 to 45 ha, with slopes varying from 3% to 10%; it has found widespread use in urban applications to estimate both overland flow and channel flow times. Equation 5.103 is most applicable for natural basins with well-defined channels, bare-earth overland flow, and flow in mowed channels including roadside ditches (Debo and Reese, 1995). Rossmiller (1980) reviewed field applications of the Kirpich equation and suggested that for overland flow on concrete or asphalt surfaces, t_c should be multiplied by 0.4; for concrete channels, multiply t_c by 0.2; and for general overland flow and flow in natural grass channels, multiply t_c by 2. According to Prakash (1987), the Kirpich equation yields relatively low estimates of the time of concentration. The Kirpich formula is usually considered applicable to small agricultural watersheds with drainage areas less than 80 ha.

Izzard equation.   The Izzard equation (Izzard, 1944; 1946) was derived from laboratory experiments on pavements and turf where overland flow was dominant. The Izzard equation is given by

$$t_c = \frac{530KL^{1/3}}{i_e^{2/3}}, \qquad \text{where } i_e L < 3.9 \text{ m}^2/\text{h} \tag{5.104}$$

where t_c is the time of concentration in minutes, L is the overland flow distance in meters, i_e is the effective rainfall intensity in mm/h, and K is a constant given by

$$K = \frac{2.8 \times 10^{-6} i_e + c_r}{S_0^{1/3}} \tag{5.105}$$

where c_r is a retardance coefficient that is determined by the catchment surface as given in Table 5.19 and S_0 is the catchment slope.

Kerby equation.   The Kerby equation (Kerby, 1959) is given by

$$t_c = 1.44 \left(\frac{Lr}{\sqrt{S_0}} \right)^{0.467}, \qquad \text{where } L < 365 \text{ m} \tag{5.106}$$

TABLE 5.19: Values of c_r in the Izzard Equation

Surface	c_r
Very smooth asphalt	0.0070
Tar and sand pavement	0.0075
Crushed-slate roof	0.0082
Concrete	0.012
Closely clipped sod	0.016
Tar and gravel pavement	0.017
Dense bluegrass	0.060

Source: Izzard (1944; 1946).

TABLE 5.20: Values of r in the Kerby Equation

Surface	r
Smooth pavements	0.02
Asphalt/concrete	0.05–0.15
Smooth bare packed soil, free of stones	0.10
Light turf	0.20
Poor grass on moderately rough ground	0.20
Average grass	0.40
Dense turf	0.17–0.80
Dense grass	0.17–0.30
Bermuda grass	0.30–0.48
Deciduous timberland	0.60
Conifer timberland, dense grass	0.60

Sources: Kerby (1959); Westphal (2001).

where t_c is the time of concentration in minutes, L is the length of flow in meters, r is a retardance roughness coefficient given in Table 5.20, and S_0 is the slope of the catchment. The Kerby equation is an empirical relation developed by Kerby (1959) using published research on airport drainage done by Hathaway (1945), consequently, it is sometimes referred to as the Kerby–Hathaway equation. Catchments with areas less than 4 ha, slopes less than 1%, and retardance coefficients less than 0.8 were used in calibrating the Kerby equation, and application of this equation should also be limited to this range. In addition, since the Kerby equation applies to the overland/sheet flow regime, it is recommended that the length of flow, L, be less than 100 m (Westphal, 2001).

EXAMPLE 5.22

An urban catchment with an asphalt surface has an average slope of 0.5%, and the distance from the catchment boundary to the outlet is 90 m. For a 20-min storm with an effective rainfall rate of 75 mm/h, estimate the time of concentration using: (a) the kinematic-wave equation, (b) the NRCS method, (c) the Kirpich equation, (d) the Izzard equation, and (e) the Kerby equation.

Solution

(a) *Kinematic-wave equation:* If the overland flow is assumed to be fully turbulent, the Manning form of the kinematic-wave equation (Equation 5.94) can be used. From the given data: $L = 90$ m, $i_e = 75$ mm/h, $S_0 = 0.005$, and for an asphalt surface Table 5.17 gives $n = 0.012$. According to Equation 5.94

$$t_c = 6.99 \frac{(nL)^{0.6}}{i_e^{0.4} S_0^{0.3}} = 6.99 \frac{(0.012 \times 90)^{0.6}}{(75)^{0.4}(0.005)^{0.3}} = 6 \text{ min}$$

If the overland flow is assumed to be in the transition range (which is more probable), then the Darcy–Weisbach form of the kinematic-wave equation (Equation 5.98) must be used. Substituting $\nu = 10^{-6}$ m²/s, $k = 0.5$, and $C = 2$

into Equation 5.98 (Chen and Wong formula) gives

$$t_c = \left[\frac{0.21(3.6 \times 10^6 \nu)^k CL^{2-k}}{S_0 i_e^{1+k}} \right]^{\frac{1}{3}}$$

$$= \left[\frac{0.21(3.6 \times 10^6 \times 10^{-6})^{0.5}(2)(90)^{2-0.5}}{(0.005)(75)^{1+0.5}} \right]^{\frac{1}{3}} = 6 \text{ min}$$

Therefore, both forms of the kinematic-wave equation yield the same result (to the nearest minute).

(b) *NRCS method:* Use $n = 0.012$, $L = 90$ m, $S_0 = 0.005$, and take P_2 as $(20/60) \times 75 = 25$ mm $= 2.5$ cm. According to the NRCS approach, assuming the sheet-flow regime ($L \le 100$ m) yields the following estimate of the time of concentration:

$$t_c = 0.0288 \frac{(nL)^{0.8}}{P_2^{0.5} S_0^{0.4}} = 0.0288 \frac{(0.012 \times 90)^{0.8}}{(2.5)^{0.5}(0.005)^{0.4}} = 0.16 \text{ h} = 10 \text{ min}$$

This is an overestimate of t_c, since P_2 should be the 24-h rainfall rather than the 20-min rainfall.

(c) *Kirpich equation:* Use $L = 90$ m and $S_0 = 0.005$. According to the Kirpich equation, with a factor of 0.4 to account for the asphalt surface,

$$t_c = 0.4(0.019) \frac{L^{0.77}}{S_0^{0.385}} = 0.4(0.019) \frac{(90)^{0.77}}{(0.005)^{0.385}} = 2 \text{ min}$$

(d) *Izzard equation:* Use $S_0 = 0.005$, $L = 90$ m, $i_e = 75$ mm/h, and a retardance coefficient, c_r, given by Table 5.19 as 0.007. The constant, K, is given by

$$K = \frac{2.8 \times 10^{-6} i_e + c_r}{S_0^{1/3}} = \frac{2.8 \times 10^{-6} \times 75 + 0.007}{(0.005)^{1/3}} = 0.0422$$

The Izzard equation gives the time of concentration as

$$t_c = \frac{530 K L^{1/3}}{i_e^{2/3}} = \frac{530(0.0422)(90)^{1/3}}{(75)^{2/3}} = 6 \text{ min}$$

In this case, $i_e L = (0.075)(90) = 6.75$ m^2/h. Therefore, since $i_e L > 3.9$ m^2/h, the Izzard equation is not strictly applicable.

(e) *Kerby equation:* Use $L = 90$ m, $S_0 = 0.005$, and a retardance coefficient, r, given by Table 5.20 as 0.02. According to the Kerby equation

$$t_c = 1.44 \left(\frac{Lr}{\sqrt{S_0}} \right)^{0.467} = 1.44 \left(\frac{90 \times 0.02}{\sqrt{0.005}} \right)^{0.467} = 7 \text{ min}$$

The computed times of concentration are summarized in the following table:

Equation	t_c (min)
Kinematic wave	6
NRCS	10
Kirpich	2
Izzard	6
Kerby	7

Noting that the NRCS method overestimates t_c and that the Kirpich equation would give $t_c = 5$ min if the Rosmiller factor of 0.4 were not applied, the (overland flow) time of concentration of the catchment can be taken to be on the order of 6 min.

5.4.1.2   Channel flow

Channel flow elements in flow paths include street gutters, roadside swales, storm sewers, drainage channels, and small streams. In these cases, it is recommended that velocity-based equations such as Manning and Darcy–Weisbach be used to estimate the flow time in each segment (ASCE, 1992). Here the flow time, t_0, is estimated by the relation

$$t_0 = \frac{L}{V_0}$$

(5.107)

where L is the length of the flow path in the channel and V_0 is the estimated velocity of flow. The time of concentration of the entire catchment is equal to the sum of the time of concentration for overland flow and the channel flow time, t_0.

EXAMPLE 5.23

A catchment consists of an asphalt pavement that drains into a rectangular concrete channel. The asphalt surface has an average slope of 0.6%, and the distance from the catchment boundary to the drain is 50 m. The drainage channel is 30 m long, 25 cm wide, 20 cm deep, and has a slope of 0.8%. For an effective rainfall rate of 60 mm/h, the flowrate in the channel is estimated to be 0.025 m³/s. Estimate the time of concentration of the catchment.

Solution   The flow consists of both overland flow and channel flow. Use the kinematic-wave equation to estimate the time of concentration of overland flow, t_1, where $L = 50$ m, $i_e = 60$ mm/h, $S_0 = 0.006$, and for an asphalt surface Table 5.17 gives $n = 0.012$. The kinematic-wave equation (Equation 5.94) gives

$$t_1 = 6.99 \frac{(nL)^{0.6}}{i_e^{0.4} S_0^{0.3}} = 6.99 \frac{(0.012 \times 50)^{0.6}}{(60)^{0.4}(0.006)^{0.3}} = 5 \text{ min}$$

The flow area, A, in the drainage channel can be calculated using the Manning equation

$$Q = \frac{1}{n} A R^{2/3} S_0^{1/2} = \frac{1}{n} A \left(\frac{A}{P}\right)^{2/3} S_0^{1/2} = \frac{1}{n} \frac{A^{5/3}}{P^{2/3}} S_0^{1/2}$$

where $Q = 0.025$ m³/s, $n = 0.013$ (for concrete), and $S_0 = 0.008$. The area, A, and wetted perimeter, P, can be written in terms of the depth of flow, d, in the drainage channel as $A = 0.25d$ and $P = 2d + 0.25$. Therefore, the Manning equation can be put in the form

$$0.025 = \frac{1}{0.013} \frac{(0.25d)^{5/3}}{(2d + 0.25)^{2/3}}(0.008)^{1/2}$$

which yields

$$d = 0.10 \text{ m} = 10 \text{ cm}$$

The flow area, A, in the drainage channel is given by

$$A = 0.25d = (0.25)(0.10) = 0.025 \text{ m}^2$$

Therefore, the flow velocity, V_0, is given by

$$V_0 = \frac{Q}{A} = \frac{0.025}{0.025} = 1.0 \text{ m/s}$$

Since the length, L, of the drainage channel is 30 m, the flow time, t_2, in the channel is given by

$$t_2 = \frac{L}{V_0} = \frac{30}{1.0} = 30 \text{ s} = 0.5 \text{ min}$$

The time of concentration, t_c, of the entire catchment area is equal to the overland flow time, t_1, plus the channel flow time, t_2. Hence

$$t_c = t_1 + t_2 = 5 + 0.5 = 5.5 \text{ min}$$

Because of uncertainties in estimating t_c, the time of concentration of the catchment can reasonably be taken as 6 minutes.

5.4.1.3   Accuracy of estimates

Wong (2005) compared 9 equations for estimating t_c in experimental plots having concrete and grass surfaces and concluded that t_c equations that do not account for rainfall intensity are valid for only a limited range of rainfall intensities. The equations used in the Wong (2005) study included, among others, the Izzard equation (Equation 5.104), the Kerby equation (Equation 5.106), and the Chen and Wong equation (Equation 5.98). The results of this study showed that best agreement with experimental data for both concrete and grass surfaces was obtained using the Chen and Wong equation.

McCuen and colleagues (1984) compared 11 equations for estimating t_c in 48 urban catchments in the United States. The catchments used in the study all had areas less than 1600 ha (4000 ac), average impervious areas of approximately 29%, and times of concentration from 0.21 to 6.14 h, with an average time of concentration of 1.5 h. The equations used in these studies included, among others, the kinematic-wave equation (Equation 5.94), the Kirpich equation (Equation 5.103), and the Kerby equation (Equation 5.106). The results of this study indicated that the error in the estimated value of t_c exceeded 0.5 h for more than 50% of the catchments, with the standard deviation of the errors ranging from 0.37 to 2.27 h. These results indicate that relatively large errors can be expected in estimating t_c from commonly used equations, and Singh (1989) has noted that this can lead to significant errors in design discharges.

It is considered good practice to use at least three different methods to estimate the time of concentration and, within the range of these estimates, the final value should be selected by judgment (Prakash, 2004). Estimates of travel time in drainage channels are usually much more accurate than estimates of overland flow; therefore, in catchments where flow in drainage channels constitutes a significant portion of the travel time, the time of concentration can be estimated more accurately.

5.4.2 Peak-Runoff Models

Peak-runoff models estimate only the peak runoff, not the entire runoff hydrograph. Peak flows are required for the hydraulic design of bridges and culverts, for evaluation of flooding potential, and for the design of stormwater conveyance structures such as sewer pipes.

5.4.2.1 The rational method

The *rational method* is the most widely used peak-runoff method in urban hydrology. This method has been used by engineers since the nineteenth century (Mulvaney, 1850; Kuichling, 1889). The rational method relates the peak-runoff rate, Q_p, to the rainfall intensity, i, by the relation

$$Q_p = CiA \tag{5.108}$$

where C is the runoff coefficient and A is the area of the catchment. Application of the rational method assumes that: (1) the entire catchment area is contributing to the runoff, in which case the duration of the storm must equal or exceed the time of concentration of the catchment; (2) the rainfall is distributed uniformly over the catchment area; and (3) all catchment losses are incorporated into the runoff coefficient, C. The runoff mechanism assumed by the rational method is illustrated in Figure 5.32 (for a constant rainfall), where the runoff from a storm gradually increases until an equilibrium is reached in which the runoff rate, Q, is equal to the effective rainfall rate over the entire catchment area. This condition occurs at the time of concentration, t_c. Under these circumstances, the duration of the rainfall does not affect the peak runoff as long as the duration equals or exceeds t_c. Since the storm duration must equal or exceed the time of concentration for the rational method to be applicable, and the average intensity of a storm is inversely proportional to the duration of the storm, then the rainfall intensity used in the rational method to generate the largest peak runoff must correspond to a storm whose duration is equal

FIGURE 5.32: Runoff mechanism assumed by rational method

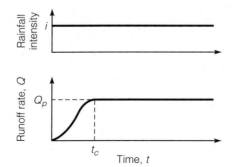

to the time of concentration. The rational method implicitly assumes that the rainfall excess, i_e, is related to the rainfall rate, i, by

$$i_e = Ci \tag{5.109}$$

which means that the rainfall-excess rate is a constant fraction, C, of the rainfall rate. The assumption that C is a constant that is independent of the rainfall rate and antecedent moisture conditions makes the application of the rational method questionable from a physical viewpoint, but this approximation improves as the imperviousness of the catchment increases. In addition to this flaw in the underlying theory of the rational method, the assumed uniformity of the rainfall distribution over the catchment area and the assumption of equilibrium conditions indicate that the rational method should be limited to areas smaller than about 80 ha (200 ac) (ASCE, 1992), which typically have times of concentration less than 20 min (Wanielista et al., 1997). Debo and Reese (1995) recommend that the rational method be limited to areas smaller than 40 ha (100 ac), some jurisdictions limit the use of the rational method to areas smaller than 8 ha (20 ac) (South Carolina, 1992), and in some cases local characteristics may limit the application of the rational method to areas smaller than 4 ha (10 ac) (Haestad Methods, Inc., 2002). Use of the rational method is not recommended in any catchment where ponding of stormwater might affect the peak discharge, or where the design and operation of large drainage facilities are to be undertaken.

Application of the rational method requires a simultaneous solution of the IDF equation and the time-of-concentration expression to determine the rainfall intensity and time of concentration. Typically, the IDF curve is of the form

$$i = f(t_c) \tag{5.110}$$

and an expression for the time of concentration is of the form

$$t_c = g(i) \tag{5.111}$$

where f and g are known functions; Equation 5.110 implicitly assumes that the storm duration is equal to the time of concentration. Typical runoff coefficients for rainfall intensities with 2- to 10-year return periods are shown in Table 5.21. Higher coefficients should be used for less frequent storms having higher return periods, and the runoff coefficients can be reasonably increased by 10%, 20%, and 25% for 20-, 50-, and 100-year storms, respectively (Akan, 1993). Obviously the increased runoff coefficients must not be greater than 1.0. It is common practice to use the impervious fraction of a catchment as a preliminary estimate of the runoff coefficient (Rodriguez et al., 2003); this approach is similar to the Lloyd-Davies variation of the rational method which is widely used in the United Kingdom and assumes that 100% of runoff comes from impervious surfaces that are directly connected to the drainage system, and 0% of the runoff from pervious surfaces (Lloyd-Davies, 1906; Butler and Davies, 2000).

EXAMPLE 5.24

A new 1.2-ha suburban residential development is to be drained by a storm sewer that connects to the municipal drainage system. The development is characterized by an average runoff coefficient of 0.4, a Manning n for overland flow of 0.20, an average overland flow length of 70 m, and an average slope of 0.7%. The time of concentration can be estimated by the kinematic-wave equation. Local drainage regulations require

TABLE 5.21: Typical Runoff Coefficients (2-year to 10-year Return Periods)

Description of area	Runoff coefficient
Business:	
Downtown areas	0.70–0.95
Neighborhood areas	0.50–0.70
Residential:	
Single-family areas	0.30–0.50
Multiunits, detached	0.40–0.60
Multiunits, attached	0.60–0.75
Residential, suburban	0.25–0.40
Apartment dwelling areas	0.50–0.70
Industrial:	
Light areas	0.50–0.80
Heavy areas	0.60–0.90
Parks, cemeteries	0.10–0.25
Railroad yard areas	0.20–0.35
Unimproved areas	0.10–0.30
Pavement:	
Asphalt or concrete	0.70–0.95
Brick	0.70–0.85
Roofs	0.75–0.95
Lawns, sandy soil:	
Flat, 2%	0.05–0.10
Average, 2%–7%	0.10–0.15
Steep, 7% or more	0.15–0.20
Lawns, heavy soil:	
Flat, 2%	0.13–0.17
Average, 2%–7%	0.18–0.22
Steep, 7% or more	0.25–0.35

Source: ASCE (1992).

the sewer pipe to be sized for the peak runoff resulting from a 10-year rainfall event. The 10-year IDF curve is given by

$$i = \frac{315.5}{t^{0.81} + 6.19}$$

where i is the rainfall intensity in cm/h and t is the duration in minutes. Local drainage regulations further require a minimum time of concentration of 5 minutes. Determine the peak runoff rate to be handled by the storm sewer.

Solution   The time of concentration, t_c, is estimated by the kinematic-wave equation (Equation 5.94) as

$$t_c = 6.99 \frac{(nL)^{0.6}}{i_e^{0.4} S_0^{0.3}}$$

where $n = 0.20$, $L = 70$ m, and $S_0 = 0.007$, and therefore

$$t_c = 6.99 \frac{(0.20 \times 70)^{0.6}}{i_e^{0.4}(0.007)^{0.3}} = \frac{151}{i_e^{0.4}} \text{ min}$$

where i_e is in mm/h. Equating the storm duration to the time of concentration, the effective rainfall rate, i_e, for a 10-year storm is given by the IDF relation as

$$i_e = Ci = \frac{3155C}{t_c^{0.81} + 6.19} \text{ mm/h}$$

Combining the latter two equations with $C = 0.4$ yields

$$i_e = \frac{3155(0.4)}{\left(\dfrac{151}{i_e^{0.4}}\right)^{0.81} + 6.19} = \frac{1262 i_e^{0.324}}{58.2 + 6.19 i_e^{0.324}}$$

Solving this equation yields $i_e = 58$ mm/h and a corresponding time of concentration of 30 min, which exceeds the minimum allowable time of concentration of 5 minutes. The peak runoff, Q_p, from the residential development is given by the rational formula as

$$Q_p = CiA = i_e A$$

where $i_e = 58$ mm/h $= 1.61 \times 10^{-5}$ m/s, and $A = 1.2$ ha $= 1.2 \times 10^4$ m^2. Therefore

$$Q_p = (1.61 \times 10^{-5})(1.2 \times 10^4) = 0.19 \text{ m}^3/\text{s}$$

The storm sewer serving the residential area should be sized to accommodate 0.19 m^3/s when flowing full.

It is important to keep in mind that the assumptions of the rational method are valid only for small catchments, with areas typically less than 80 ha (200 ac) and times of concentration typically less than 20 minutes. For larger catchments, the peak runoff will likely result from storms with durations considerably longer than the time of concentration of the catchment (Levy and McCuen, 1999).

In using the rational method to calculate the peak discharge from a composite catchment area, the rational formula should generally be applied in two ways (ASCE, 1992): (1) using the entire drainage area, and (2) using the most densely developed directly connected area. In using the entire drainage area, the rainfall duration is equated to the time of concentration of the entire catchment, the average intensity corresponding to a duration equal to the time of concentration is derived from the IDF curve, the average runoff coefficient is an area-weighted average, and the resulting peak runoff is calculated by substituting these values into the rational formula, Equation 5.108. Using only the portion of the catchment that is most densely developed and directly connected to the catchment outlet, the smaller (sub)catchment area results in a shorter time of concentration and higher average rainfall intensity, and a higher average runoff coefficient because of the more impervious nature of the subcatchment; the peak runoff from this area is obtained by substituting these values into Equation 5.108. If the increased average rainfall intensity and runoff coefficient are sufficient to offset the smaller catchment area, then the peak runoff from the densely developed directly connected area will be greater than the peak runoff from the entire catchment and will govern the design of the downstream drainage structures.

EXAMPLE 5.25

Consider the case where the residential development described in Example 5.24 contains 0.4 ha of impervious area that is directly connected to the storm sewer. If the runoff coefficient of the impervious area is 0.9, the Manning n for overland flow on the impervious surface is 0.03, the average flow length is 20 m, and the average slope is 0.1%, estimate the design runoff to be handled by the storm sewer.

Solution   The time of concentration, t_c, of the impervious area is given by

$$t_c = 6.99 \frac{(nL)^{0.6}}{i_e^{0.4} S_0^{0.3}}$$

where $n = 0.03$, $L = 20$ m, and $S_0 = 0.001$. Therefore

$$t_c = 6.99 \frac{(0.03 \times 20)^{0.6}}{i_e^{0.4}(0.001)^{0.3}} = \frac{40.9}{i_e^{0.4}} \text{ min}$$

where i_e is in mm/h. Taking the storm duration as t_c, the 10-year effective rainfall rate, i_e, is given by the IDF curve as

$$i_e = Ci = \frac{3155C}{t_c^{0.81} + 6.19} \text{ mm/h}$$

Combining the latter two equations with $C = 0.9$ yields

$$i_e = \frac{3155(0.9)}{\left(\dfrac{40.9}{i_e^{0.4}}\right)^{0.81} + 6.19} = \frac{2840 i_e^{0.324}}{20.2 + 6.19 i_e^{0.324}}$$

Solving this equation yields $i_e = 303$ mm/h and a corresponding time of concentration of 4.2 min, which is less than the minimum allowable time of concentration of 5 minutes. Taking $t_c = 5$ min, the IDF relation yields

$$i_e = \frac{3155(0.9)}{5^{0.81} + 6.19} = 288 \text{ mm/h}$$

The peak runoff, Q_p, from the impervious area is given by the rational formula as

$$Q_p = CiA = i_e A$$

where $i_e = 288$ mm/h $= 8 \times 10^{-5}$ m/s, and $A = 0.4$ ha $= 4000$ m^2; therefore

$$Q_p = (8 \times 10^{-5})(4000) = 0.32 \text{ m}^3/\text{s}$$

Example 5.24 indicated that the peak runoff from the entire composite catchment is 0.19 m^3/s, and this example shows that the peak runoff from the directly connected impervious area is 0.32 m^3/s. The storm sewer serving this residential development should therefore be designed to accommodate 0.32 m^3/s.

5.4.2.2   NRCS-TR55 method

The Natural Resources Conservation Service (NRCS) computed the runoff from many small and midsize catchments using the NRCS regional 24-h hyetographs to describe the rainfall distribution, the NRCS curve-number model to calculate the rainfall excess, and the NRCS unit-hydrograph method to calculate the runoff hydrograph.

These computations were performed using the TR-20 computer program (SCS, 1983). On the basis of these results, the NRCS proposed the *graphical peak-discharge method* or, more commonly, the *TR-55 method* for estimating peak runoff from small and midsize catchments, with times of concentration between 0.1h and 10 h. In applying the TR-55 method, it is important to note that the NRCS regional 24-h hyetographs are designed to contain the (average) intensity of any duration of rainfall for the frequency of the event chosen (i.e., NRCS hyetographs are consistent with local IDF curves) and therefore peak runoffs are generated from NRCS 24-h storms during an interval of maximum rainfall excess roughly equal to the time of concentration of the catchment. The TR-55 method, named after the technical report in which it is described (SCS, 1986), expresses the peak runoff, q_p, in m^3/s as

$$q_p = q_u A Q F_p \tag{5.112}$$

where q_u is the unit peak discharge in m^3/s per cm of runoff per km^2 of catchment area, A is the catchment area in km^2, Q is the runoff in centimeters from a 24-h storm with a given return period, and F_p is the pond and swamp adjustment factor (dimensionless). The runoff, Q, is derived directly from the NRCS curve-number model, Equation 5.69, using the 24-h precipitation. F_p is derived from Table 5.22, assuming that the ponds and/or swampy areas are distributed throughout the catchment, and the unit-peak discharge, q_u, is obtained using the empirical relation

$$\log(q_u) = C_0 + C_1 \log t_c + C_2 (\log t_c)^2 - 2.366 \tag{5.113}$$

where C_0, C_1, and C_2 are obtained from Table 5.23, and t_c is in hours. Values of t_c in Equation 5.113 must be between 0.1 h and 10 h (calculated using the NRCS method described in Section 5.4.1.1). Values of I_a in Table 5.23 are derived using

$$I_a = 0.2\, S \tag{5.114}$$

where S is obtained from the curve number in accordance with Equation 5.73. If $I_a/P < 0.1$, where P corresponds to the 24-h precipitation for the given return period, values of C_0, C_1, and C_2 corresponding to $I_a/P = 0.1$ should be used, and if $I_a/P > 0.5$, values of C_0, C_1, and C_2 corresponding to $I_a/P = 0.5$ should

TABLE 5.22: Pond and Swamp Adjustment Factor, F_p

Percentage of pond and swamp areas	F_p
0	1.00
0.2	0.97
1.0	0.87
3.0	0.75
5.0*	0.72

*If the percentage of pond and swamp areas exceeds 5%, then consideration should be given to routing the runoff through these areas.

TABLE 5.23: Parameters Used to Estimate Unit Peak Discharge, q_u

Rainfall type	I_a/P	C_0	C_1	C_2
I	0.10	2.30550	−0.51429	−0.11750
	0.20	2.23537	−0.50387	−0.08929
	0.25	2.18219	−0.48488	−0.06589
	0.30	2.10624	−0.45695	−0.02835
	0.35	2.00303	−0.40769	0.01983
	0.40	1.87733	−0.32274	0.05754
	0.45	1.76312	−0.15644	0.00453
	0.50	1.67889	−0.06930	0.0
IA	0.10	2.03250	−0.31583	−0.13748
	0.20	1.91978	−0.28215	−0.07020
	0.25	1.83842	−0.25543	−0.02597
	0.30	1.72657	−0.19826	0.02633
	0.50	1.63417	−0.09100	0.0
II	0.10	2.55323	−0.61512	−0.16403
	0.30	2.46532	−0.62257	−0.11657
	0.35	2.41896	−0.61594	−0.08820
	0.40	2.36409	−0.59857	−0.05621
	0.45	2.29238	−0.57005	−0.02281
	0.50	2.20282	−0.51599	−0.01259
III	0.10	2.47317	−0.51848	−0.17083
	0.30	2.39628	−0.51202	−0.13245
	0.35	2.35477	−0.49735	−0.11985
	0.40	2.30726	−0.46541	−0.11094
	0.45	2.24876	−0.41314	−0.11508
	0.50	2.17772	−0.36803	−0.09525

be used. These approximations result in reduced accuracy of the peak-discharge estimates (SCS, 1986), and McCuen and Okunola (2002) have noted that for times of concentration less than 0.3 h, the TR-55 graphical method may underestimate the peak discharge, relative to the TR-20 model, by as much as 15% for values of I_a/P near the lower limit of 0.1. The Federal Highway Administration (USFHWA, 1995) and the NRCS (SCS, 1986) have both recommended that the NRCS-TR55 method be used only with homogenous catchments, where the curve numbers vary within ± 5 between zones, CN of the catchment should be greater than 40, t_c should be between 0.1 and 10 h, and t_c should be approximately the same for all main channels.

EXAMPLE 5.26

A 2.25-km^2 catchment with 0.2% pond area is estimated to have a curve number of 85, a time of concentration of 2.4 h, and a 24-h Type III precipitation of 13 cm. Estimate the peak runoff from the catchment.

Solution   For CN = 85, the storage, S, is given by

$$S = \frac{1}{0.0394}\left(\frac{1000}{\text{CN}} - 10\right) = \frac{1}{0.0394}\left(\frac{1000}{85} - 10\right) = 45 \text{ mm}$$

For $P = 130$ mm, the runoff, Q, is given by

$$Q = \frac{[P - 0.2S]^2}{P + 0.8S} = \frac{[130 - 0.2(45)]^2}{130 + 0.8(45)} = 88 \text{ mm} = 8.8 \text{ cm}$$

From Table 5.22, $F_p = 0.97$. By definition

$$\frac{I_a}{P} = \frac{0.2S}{P} = \frac{(0.2)(45)}{130} = 0.069$$

Since $I_a/P < 0.1$, use $I_a/P = 0.1$ in Table 5.23, which gives $C_0 = 2.47317$, $C_1 = -0.51848$, $C_2 = -0.17083$, and since $t_c = 2.4$ h, Equation 5.113 gives

$$\log(q_u) = C_0 + C_1 \log t_c + C_2 (\log t_c)^2 - 2.366$$

$$= 2.47317 - 0.51848 \log(2.4) - 0.17083(\log 2.4)^2 - 2.366$$

$$= -0.116$$

which leads to

$$q_u = 10^{-0.116} = 0.765 \ (\text{m}^3/\text{s})/\text{cm}/\text{km}^2$$

Therefore, according to the TR-55 method, the peak discharge, q_p, is given by Equation 5.112 as

$$q_p = q_u A Q F_p = 0.765(2.25)(8.8)(0.97) = 14.7 \ \text{m}^3/\text{s}$$

It is important to recognize that hydrologic models used to estimate peak-discharge rates are not highly accurate. Calibrated peak-discharge models often have standard errors of 25% or more, with uncalibrated models having significantly higher standard errors (McCuen, 2001).

5.4.2.3 USGS regional regression equations

In areas where similar hydrologic conditions exist, it is possible to relate measured peak-runoff rates to catchment characteristics using regression techniques. Peak-runoff regression equations are typically in the form

$$Q_T = a x_1^b x_2^c x_3^d \tag{5.115}$$

where Q_T is the peak discharge associated with the recurrence interval T; x_1, x_2, and x_3 are predictor variables that characterize the catchment area; and a, b, c, and d are parameters determined from a regression analysis. The drainage area is always included as a predictor variable, with other variables typically including catchment slope or channel slope, percent imperviousness, percent of catchment covered by lakes and ponds, and mean annual precipitation or precipitation for a specified recurrence interval and duration.

Since 1973, the U.S. Geological Survey (USGS) in cooperation with state highway departments and other entities has developed and published regional regression equations for estimating flood-flow frequency relations for rural, unregulated catchments in every state in the United States, Puerto Rico, American Samoa, and a number of metropolitan areas in the United States. The USGS, in cooperation with the Federal Highway Administration (FHWA) and Federal Emergency Management Agency (FEMA), compiled the regression equations into a computer program called the National Flood Frequency (NFF) Program and accompanying report (Jennings et al., 1994). The NFF program includes about 2065 regression equations for 289 flood regions in the United States, and the summary report accompanying the NFF program presents the equations for each state with references to reports documenting the original statewide frequency and regression analyses. The typical procedure in developing these regression equations is to apply the log–Pearson Type III distribution to measured annual-maximum flood flows, identify flood flows with return periods of 2, 5, 10, 25, 50, 100, and 500 years, and then perform regression analyses to estimate the relationship between the peak flows and the catchment and local climatic parameters. Due to the relatively short record length of the stream-gage data used to develop many of the equations, estimation errors on the order of 15% to 100% are typical of peak-discharge rates estimated using the USGS regression equations.

To illustrate the formulation and use of the USGS regression equations, consider the case of Florida, which is divided into the four hydrologic regions shown in Figure 5.33, in one of which floods are undefined (Bridges, 1982). The peak-discharge regression equations developed for these regions have return periods that range from 2 to 500 years and are given as follows (Bridges, 1982):

Region A:

$$Q_2 = 93.4A^{0.756}S^{0.268}(L + 3)^{-0.803} \tag{5.116}$$

$$Q_5 = 192A^{0.722}S^{0.255}(L + 3)^{-0.759} \tag{5.117}$$

$$Q_{10} = 274A^{0.708}S^{0.248}(L + 3)^{-0.738} \tag{5.118}$$

$$Q_{25} = 395A^{0.696}S^{0.240}(L + 3)^{-0.717} \tag{5.119}$$

$$Q_{50} = 496A^{0.690}S^{0.234}(L + 3)^{-0.705} \tag{5.120}$$

$$Q_{100} = 609A^{0.685}S^{0.227}(L + 3)^{-0.695} \tag{5.121}$$

$$Q_{500} = 985A^{0.668}S^{0.196}(L + 3)^{-0.687} \tag{5.122}$$

Region B:

$$Q_2 = 44.2A^{0.658}(L + 0.6)^{-0.561} \tag{5.123}$$

$$Q_5 = 113A^{0.0.614}(L + 0.6)^{-0.573} \tag{5.124}$$

$$Q_{10} = 182A^{0.592}(L + 0.6)^{-0.580} \tag{5.125}$$

$$Q_{25} = 298A^{0.570}(L + 0.6)^{-0.585} \tag{5.126}$$

$$Q_{50} = 410A^{0.556}(L + 0.6)^{-0.589} \tag{5.127}$$

$$Q_{100} = 584A^{0.543}(L + 0.6)^{-0.591} \tag{5.128}$$

$$Q_{500} = 936A^{0.521}(L + 0.6)^{-0.594} \tag{5.129}$$

FIGURE 5.33: Flood-frequency region map for Florida.

Source: Bridges (1982).

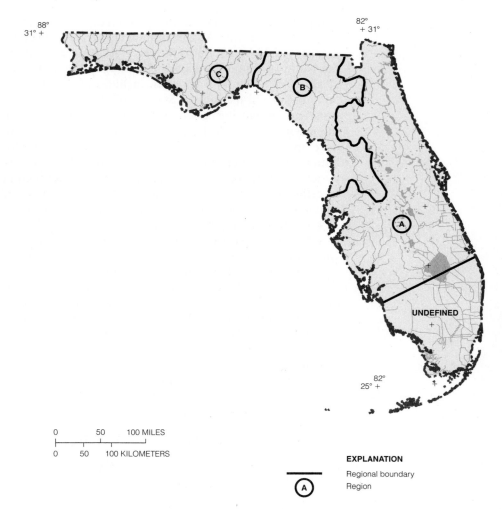

Region C:

$$Q_2 = 58.9A^{0.824}S^{0.387}(L + 3)^{-0.785} \qquad (5.130)$$

$$Q_5 = 117A^{0.844}S^{0.482}(L + 3)^{-1.06} \qquad (5.131)$$

$$Q_{10} = 164A^{0.860}S^{0.534}(L + 3)^{-1.21} \qquad (5.132)$$

$$Q_{25} = 234A^{0.882}S^{0.586}(L + 3)^{-1.37} \qquad (5.133)$$

$$Q_{50} = 291A^{0.900}S^{0.626}(L + 3)^{-1.48} \qquad (5.134)$$

$$Q_{100} = 351A^{0.918}S^{0.658}(L + 3)^{-1.58} \qquad (5.135)$$

$$Q_{500} = 507A^{0.960}S^{0.725}(L + 3)^{-1.79} \qquad (5.136)$$

The predictor variables used in these regression equations are: drainage area (A) in mi^2, channel slope (S) in ft/mi, and percentage of the drainage area covered by lakes and ponds (L) in %. The regression equations were developed from peak-discharge records at 182 gaging stations and are applicable to natural-flow streams. The standard errors of peak-flow estimates derived from the regression equations range from 40%–60% for Region A, 60%–65% for Region B, and 44%–76% for Region C. In addition to these regression equations, which apply to natural-flow streams, regression equations have also been developed for the urban areas of Tampa Bay and Leon County in Florida (Lopez and Woodham, 1982; Franklin and Losey, 1984). In the urbanized areas of Tampa Bay, the predictor variables are the drainage area, basin development factor, main-channel slope, and detention storage area; while in the urban areas of Leon County the predictor variables are the drainage area, and the impervious area as a percentage of the drainage area.

5.4.3 Continuous-Runoff Models

Continuous-runoff models estimate the entire runoff hydrograph from the rainfall excess remaining after initial abstraction, infiltration, and depression storage have been taken into account. A good example of a method for calculating the rainfall excess is the Green–Ampt model discussed earlier. Since surface conditions tend to be spatially variable within a catchment area, the rainfall excess will also be variable in both space and time, and the runoff hydrograph at the catchment outlet will depend on such factors as the topography and the surface roughness of the catchment. Four types of models commonly used in engineering practice to estimate the runoff hydrograph from the rainfall excess are: (1) unit-hydrograph models, (2) time-area models, (3) kinematic-wave models, and (4) nonlinear reservoir models. These models are not normally used to estimate the peak runoff, since the accuracy of the estimated peak runoff depends on the temporal resolution of the rainfall excess (Aronica et al., 2005b). Continuous-runoff models are mostly applicable to designing storage reservoirs in stormwater-management systems.

5.4.3.1 Unit-hydrograph theory

The idea of using a unit hydrograph to describe the response of a catchment to rainfall excess was first introduced by Sherman (1932), and is currently the most widely used method of estimating runoff hydrographs (Viessman and Lewis, 1996). A *unit hydrograph* is defined as the temporal distribution of runoff resulting from a unit depth (1 cm or 1 in.) of rainfall excess occurring over a given duration, and distributed uniformly in time and space over the catchment area.

   To demonstrate the application of unit-hydrograph models, consider the unit hydrograph, $u_{\Delta t}(t)$, for a given catchment and rainfall-excess duration, Δt, illustrated in Figure 5.34(a). This is commonly referred to as the Δt-unit hydrograph. Assuming that the catchment response is linear, then the runoff hydrograph, $Q(t)$, for a rainfall excess $P_{\Delta t}$ occurring over a duration Δt is given by

$$\boxed{Q(t) = P_{\Delta t} u_{\Delta t}(t)} \tag{5.137}$$

where the ordinates of $Q(t)$ are equal to $P_{\Delta t}$ times the ordinates of the unit hydrograph, $u_{\Delta t}(t)$. This is illustrated in Figure 5.34(b). If the rainfall excess, P, occurs over a duration equal to an integral multiple of Δt, say $n\Delta t$, and assuming the catchment response is linear, the response of the catchment is equal to that of n storms occurring sequentially, with the rainfall excess in each storm equal to $P_{n\Delta t}/n$. The runoff

FIGURE 5.34: Applications of the unit hydrograph

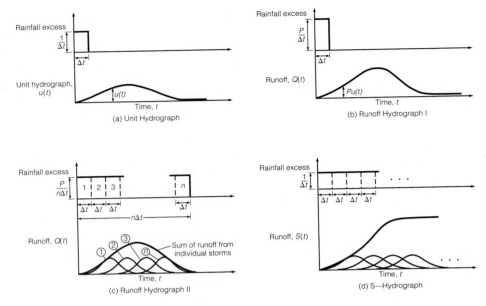

(a) Unit Hydrograph

(b) Runoff Hydrograph I

(c) Runoff Hydrograph II

(d) S—Hydrograph

hydrograph, $Q(t)$, is then given by

$$Q(t) = \frac{P_{n\Delta t}}{n} \sum_{i=0}^{n-1} u_{\Delta t}(t - i\,\Delta t) \tag{5.138}$$

where the response of the catchment is equal to the summation of the responses to the incremental rainfall excesses of duration Δt. This is illustrated in Figure 5.34(c).

EXAMPLE 5.27

The 10-min unit hydrograph for a 2.25-km^2 urban catchment is given by

Time (min)	0	30	60	90	120	150	180	210	240	270	300	330	360	390
Runoff (m^3/s)	0	1.2	2.8	1.7	1.4	1.2	1.1	0.91	0.74	0.61	0.50	0.28	0.17	0

(a) Verify that the unit hydrograph is consistent with a 1-cm rainfall excess, (b) estimate the runoff hydrograph for a 10-min rainfall excess of 3.5 cm, and (c) estimate the runoff hydrograph for a 20-min rainfall excess of 8.5 cm.

Solution

(a) The area, A_h, under the unit hydrograph can be estimated by numerical integration as

$$A_h = (1800)(1.2 + 2.8 + 1.7 + 1.4 + 1.2 + 1.1 + 0.91 + 0.74 + 0.61$$
$$+ 0.50 + 0.28 + 0.17)$$
$$= 22{,}698 \text{ m}^3$$

where the time increment between the hydrograph ordinates is 30 min = 1800 s. Since the area of the catchment is 2.25 km^2 = 2.25 × 10^6 m^2, the depth, h, of rainfall excess is given by

$$h = \frac{22,698}{2.25 \times 10^6} = 0.01 \text{ m} = 1 \text{ cm}$$

Since the depth of rainfall excess is 1 cm, the given hydrograph qualifies as a unit hydrograph.

(b) For a 10-min rainfall excess of 3.5 cm, the runoff hydrograph is estimated by multiplying the ordinates of the unit hydrograph by 3.5. This yields the following runoff hydrograph:

Time (min)	0	30	60	90	120	150	180	210	240	270	300	330	360	390
Runoff (m^3/s)	0	4.2	9.8	6.0	4.9	4.2	3.9	3.2	2.6	2.1	1.8	0.98	0.60	0

(c) The runoff hydrograph for a 20-min rainfall excess of 8.5 cm is calculated by adding the runoff hydrographs from two consecutive 10-min events, with each event corresponding to a rainfall excess of 4.25 cm. The unit hydrograph is first interpolated for 10-minute intervals and multiplied by 4.25 to give the runoff from a 10-min 4.25-cm event. This hydrograph is then added to the same hydrograph shifted forward by 10 minutes. The computations are summarized in Table 5.24, where Runoff 1 and Runoff 2 are the runoff hydrographs of the two 10-min 4.25-cm events, respectively.

For storms in which the duration of the rainfall excess is not an integral multiple of Δt, the runoff is derived from the *S-hydrograph*, $S_{\Delta t}(t)$, which is defined as the response to a storm of infinite duration and intensity $1/\Delta t$ as illustrated in Figure 5.34(d). The S-hydrograph, which was originally proposed by Morgan and Hulinghorst (1939), is related to the Δt-unit hydrograph, $u_{\Delta t}(t)$, by

$$S_{\Delta t}(t) = \sum_{i=0}^{t/\Delta t} u_{\Delta t}(t - i\,\Delta t) \tag{5.139}$$

The S-hydrograph defined by Equation 5.139 converges quickly, since the ordinates of the individual-storm unit hydrographs, $u_{\Delta t}(t)$, are only nonzero for a finite interval of time. The S-hydrograph is the response of a catchment to a rainfall excess of intensity $1/\Delta t$ beginning at $t = 0$ and continuing to $t = \infty$. If the rainfall excess were to begin at $t = \tau$, the response of the catchment would be identical to the response to a rainfall excess beginning at $t = 0$, except that the S-hydrograph would begin at time $t = \tau$ instead of $t = 0$. In fact, linearity of the catchment response requires that the difference between the runoff hydrograph resulting from an infinite rainfall excess of intensity $1/\Delta t$ beginning at $t = 0$ and the runoff hydrograph resulting from an infinite rainfall excess of intensity $1/\Delta t$ beginning at $t = \tau$ must be equal to the runoff hydrograph resulting from a rainfall excess of intensity $1/\Delta t$ beginning at $t = 0$

TABLE 5.24: Hydrograph Computation

Time (min)	Runoff 1 (m^3/s)	Runoff 2 (m^3/s)	Total runoff (m^3/s)
0	0	0	0
10	1.70	0	1.70
20	3.40	1.70	5.10
30	5.10	3.40	8.50
40	7.35	5.10	12.45
50	9.65	7.35	17.00
60	11.90	9.65	21.55
70	10.33	11.90	22.23
80	8.80	10.33	19.13
90	7.23	8.80	16.03
100	6.80	7.23	14.03
110	6.38	6.80	13.18
120	5.95	6.38	12.33
130	5.65	5.95	11.60
140	5.40	5.65	11.05
150	5.10	5.40	10.50
160	4.97	5.10	10.07
170	4.80	4.97	9.77
180	4.68	4.80	9.48
190	4.42	4.68	9.10
200	4.12	4.42	8.54
210	3.87	4.12	7.99
220	3.61	3.87	7.48
230	3.40	3.61	7.01
240	3.15	3.40	6.55
250	2.98	3.15	6.13
260	2.76	2.98	5.74
270	2.59	2.76	5.35
280	2.42	2.59	5.01
290	2.30	2.42	4.72
300	2.13	2.30	4.43
310	1.83	2.13	3.96
320	1.49	1.83	3.32
330	1.19	1.49	2.68
340	1.02	1.19	2.21
350	0.89	1.02	1.91
360	0.72	0.89	1.61
370	0.47	0.72	1.19
380	0.26	0.47	0.73
390	0	0.26	0.26
400	0	0	0

and ending at $t = \tau$. This is illustrated in Figure 5.35. The τ-unit hydrograph, denoted by $u_\tau(t)$, is derived from the S-hydrograph by

$$u_\tau(t) = \frac{\Delta t}{\tau}[S_{\Delta t}(t) - S_{\Delta t}(t - \tau)] \tag{5.140}$$

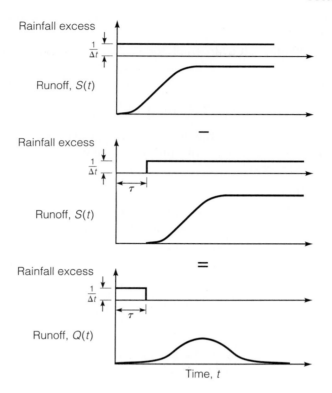

and the catchment response, $Q(t)$, to a rainfall excess P_τ with duration τ is

$$Q(t) = P_\tau u_\tau = \frac{P_\tau \Delta t}{\tau}[S(t) - S(t - \tau)]$$ (5.141)

The basic assumptions of the unit-hydrograph model are: (1) The rainfall excess has a constant intensity within the effective duration; (2) the rainfall excess is uniformly distributed throughout the catchment; (3) the *base time* (= duration) of the runoff hydrograph resulting from a rainfall excess of given duration is constant; (4) the ordinates of all the runoff hydrographs from the catchment are directly proportional to the amount of rainfall excess; and (5) for a given catchment, the hydrograph resulting from a given rainfall excess reflects the unchanging characteristics of the catchment. The last assumption is sometimes called the *principle of invariance*. The assumption of a uniformly distributed rainfall excess in space and time limits application of unit-hydrograph methods to relatively small catchment areas, typically in the range of 0.5 ha to 25 km^2 (1 ac to 10 mi^2), although even smaller catchment limitations, on the order of 2.5 km^2 (1 mi^2), are frequently used. If the catchment area is too large for the unit-hydrograph model to be applied over the entire catchment, then the area should be divided into smaller areas that are analyzed separately. The assumption of linearity has been found to be satisfactory in many practical cases (Chow et al., 1988); however, unit-hydrograph models are applicable only when channel conditions remain unchanged and catchments do not have appreciable storage.

EXAMPLE 5.28

The 30-min unit hydrograph for a 2.25-km^2 catchment is given by

Time (min)	0	30	60	90	120	150	180	210	240	270	300	330	360	390
Runoff (m^3/s)	0	1.2	2.8	1.7	1.4	1.2	1.1	0.91	0.74	0.61	0.50	0.28	0.17	0

(a) Determine the S-hydrograph, (b) calculate the 40-min unit hydrograph for the catchment, and (c) verify that the 40-min unit hydrograph corresponds to a 1-cm rainfall excess.

Solution

(a) The S-hydrograph is calculated by summing the lagged unit hydrographs (Equation 5.139), and these computations are summarized in the following table:

Runoff* (m^3/s)	0	30	60	90	120	150	180	210	240	270	300	330	360	390	420
$u_{\Delta t}(t)$	0	1.2	2.8	1.7	1.4	1.2	1.1	0.91	0.74	0.61	0.50	0.28	0.17	0	0
$u_{\Delta t}(t - \Delta t)$	0	0	1.2	2.8	1.7	1.4	1.2	1.1	0.91	0.74	0.61	0.50	0.28	0.17	0
$u_{\Delta t}(t - 2\Delta t)$	0	0	0	1.2	2.8	1.7	1.4	1.2	1.1	0.91	0.74	0.61	0.50	0.28	0.17
$u_{\Delta t}(t - 3\Delta t)$	0	0	0	0	1.2	2.8	1.7	1.4	1.2	1.1	0.91	0.74	0.61	0.50	0.28
$u_{\Delta t}(t - 4\Delta t)$	0	0	0	0	0	1.2	2.8	1.7	1.4	1.2	1.1	0.91	0.74	0.61	0.50
$u_{\Delta t}(t - 5\Delta t)$	0	0	0	0	0	0	1.2	2.8	1.7	1.4	1.2	1.1	0.91	0.74	0.61
$u_{\Delta t}(t - 6\Delta t)$	0	0	0	0	0	0	0	1.2	2.8	1.7	1.4	1.2	1.1	0.91	0.74
$u_{\Delta t}(t - 7\Delta t)$	0	0	0	0	0	0	0	0	1.2	2.8	1.7	1.4	1.2	1.1	0.91
$u_{\Delta t}(t - 8\Delta t)$	0	0	0	0	0	0	0	0	0	1.2	2.8	1.7	1.4	1.2	1.1
$u_{\Delta t}(t - 9\Delta t)$	0	0	0	0	0	0	0	0	0	0	1.2	2.8	1.7	1.4	1.2
$u_{\Delta t}(t - 10\Delta t)$	0	0	0	0	0	0	0	0	0	0	0	1.2	2.8	1.7	1.4
$u_{\Delta t}(t - 11\Delta t)$	0	0	0	0	0	0	0	0	0	0	0	0	1.2	2.8	1.7
$u_{\Delta t}(t - 12\Delta t)$	0	0	0	0	0	0	0	0	0	0	0	0	0	1.2	2.8
$u_{\Delta t}(t - 13\Delta t)$	0	0	0	0	0	0	0	0	0	0	0	0	0	0	1.2
$u_{\Delta t}(t - 14\Delta t)$	0	0	0	0	0	0	0	0	0	0	0	0	0	0	0
S-hydrograph	0	1.2	4.0	5.7	7.1	8.3	9.4	10.3	11.1	11.7	12.2	12.4	12.6	12.6	12.6

*$\Delta t = 30$ min

(b) To estimate the 40-min unit hydrograph, the S-hydrograph must be shifted by 40 minutes and subtracted from the original S-hydrograph (Equation 5.140). These computations are summarized in columns 2 to 4 in the following table.

(1) t (min)	(2) $S(t)$ (m^3/s)	(3) $S(t - 40)$ (m^3/s)	(4) $S(t) - S(t - 40)$ (m^3/s)	(5) 40-min UH (m^3/s)
0	0	0	0	0
30	1.2	0	1.2	0.9
60	4.0	0.8	3.2	2.4
90	5.7	3.1	2.6	2.0
120	7.1	5.1	2.0	1.5
150	8.3	6.6	1.7	1.3
180	9.4	7.9	1.5	1.1
210	10.3	9.0	1.3	1.0
240	11.1	10.0	1.1	0.8
270	11.7	10.8	0.9	0.7
300	12.2	11.5	0.7	0.5
330	12.4	12.0	0.4	0.3
360	12.6	12.3	0.3	0.2
390	12.6	12.5	0.1	0.1
420	12.6	12.6	0	0

Note that values of $S(t - 40)$ must be interpolated. From the given data, $\Delta t = 30$ min and $\tau = 40$ min, therefore $\Delta t/\tau = 30/40 = 0.75$, and the 40-min unit hydrograph is obtained by multiplying column 4 by 0.75 (see Equation 5.140). The 40-min unit hydrograph is given in column 5.

(c) The area under the computed 40-min unit hydrograph is given by

$$\text{Area} = (30)(60)(0.9 + 2.4 + 2.0 + 1.5 + 1.3 + 1.1 + 1.0 + 0.8 + 0.7$$
$$+ 0.5 + 0.3 + 0.2 + 0.1)$$
$$= 23,040 \text{ m}^3$$

Since the catchment area is 2.25 km^2 = 2.25 $\times$ 10^6 m^2, the depth of rainfall is given by

$$\text{Depth of rainfall} = \frac{23,040}{2.25 \times 10^6} = 0.01 \text{ m} = 1 \text{ cm}$$

The depth of rainfall is equal to 1 cm, and therefore the 40-min unit hydrograph calculated in (b) is a valid unit hydrograph.

Unit hydrographs can also be used to determine the runoff in cases where the rainfall excess is nonuniform in time. In these cases, the rainfall excess is discretized into uniform events over time intervals Δt, and the runoffs from each of these discrete events are added together to give the runoff from the entire event. Hence, if the rainfall excess is divided into N discrete events, with rainfall excesses $P_i, i = 1,\ldots,N$, then the runoff, $Q(t)$, from the entire event is given by

$$Q(t) = \sum_{i=1}^{N} P_i u_{\Delta t}(t - i \Delta t + \Delta t) \tag{5.142}$$

where $u_{\Delta t}(t)$ is the Δt-unit hydrograph.

EXAMPLE 5.29

The 30-min unit hydrograph for a catchment is given by

Time (min)	0	30	60	90	120	150	180	210	240	270	300	330	360	390
Runoff (m^3/s)	0	1.2	2.8	1.7	1.4	1.2	1.1	0.91	0.74	0.61	0.50	0.28	0.17	0

Estimate the runoff resulting from the following 90-min storm:

Time (min)	Rainfall Excess (cm)
0–30	3.1
30–60	2.5
60–90	1.7

Solution   The given 90-min storm can be viewed as three consecutive 30-min storms with rainfall excess amounts of 3.1 cm, 2.5 cm, and 1.7 cm. The runoff from each storm is estimated by multiplying the 30-min unit hydrograph by 3.1, 2.5, and 1.7, respectively. The hydrographs are then lagged by 30 minutes and summed to estimate the total runoff from the storm (see Equation 5.142). These computations are summarized in the following table:

Time (min)	UH × 3.1 (m³/s)	UH × 2.5 (m³/s)	UH × 1.7 (m³/s)	Total runoff (m³/s)
0	0	0	0	0
30	3.7	0	0	3.7
60	8.7	3.0	0	11.7
90	5.3	7.0	2.0	14.3
120	4.3	4.3	4.8	13.4
150	3.7	3.5	2.9	10.1
180	3.4	3.0	2.4	8.8
210	2.8	2.8	2.0	7.6
240	2.3	2.3	1.9	6.4
270	1.9	1.9	1.5	5.3
300	1.6	1.5	1.3	4.4
330	0.9	1.3	1.0	3.2
360	0.5	0.7	0.9	2.1
390	0	0.4	0.5	0.9
420	0	0	0.3	0.3
450	0	0	0	0

Instantaneous unit hydrograph. The *instantaneous unit hydrograph* (IUH) is the runoff hydrograph that would result if a unit depth of water were instantaneously deposited uniformly over an area and then allowed to run off. To develop an IUH, the S-hydrograph derived from a Δt-unit hydrograph must first be obtained. The resulting S-hydrograph is then differenced by the time τ, and the difference scaled to develop the τ-unit hydrograph, which becomes the IUH in the limit as τ approaches zero. According to this definition, and using Equation 5.140, the ordinates of the IUH, $Q_{IUH}(t)$, are given by

$$Q_{IUH}(t) = \lim_{\tau \to 0} u_\tau(t) \tag{5.143}$$

$$= \lim_{\tau \to 0} \left\{ \frac{\Delta t}{\tau} [S_{\Delta t}(t) - S_{\Delta t}(t - \tau)] \right\} \tag{5.144}$$

$$= \Delta t \lim_{\tau \to 0} \left\{ \frac{S_{\Delta t}(t) - S_{\Delta t}(t - \tau)}{\tau} \right\} \tag{5.145}$$

which yields

$$\boxed{Q_{IUH}(t) = \Delta t \frac{dS_{\Delta t}}{dt}} \tag{5.146}$$

This relation shows that the ordinates of the IUH can be determined directly from the slope of the S-hydrograph. The S-hydrograph can be expressed in terms of the instantaneous unit hydrograph by integrating Equation 5.146, which yields

$$S_{\Delta t}(t) = \frac{1}{\Delta t} \int_0^t Q_{\text{IUH}} \, dt \tag{5.147}$$

Substituting Equation 5.147 into Equation 5.140 gives the following relation between the τ-unit hydrograph, $u_\tau(t)$, and the instantaneous unit hydrograph, $Q_{\text{IUH}}(t)$:

$$u_\tau(t) = \frac{\Delta t}{\tau}[S_{\Delta t}(t) - S_{\Delta t}(t - \tau)] \tag{5.148}$$

$$= \frac{\Delta t}{\tau}\left[\frac{1}{\Delta t}\int_0^t Q_{\text{IUH}}\, dt - \frac{1}{\Delta t}\int_0^{t-\tau} Q_{\text{IUH}}\, dt\right] \tag{5.149}$$

which yields

$$\boxed{u_\tau(t) = \frac{1}{\tau}\int_{t-\tau}^t Q_{\text{IUH}}\, dt} \tag{5.150}$$

This relation shows that the instantaneous unit hydrograph, $Q_{\text{IUH}}(t)$, is fundamental to determining the unit hydrograph for any duration τ. For a finite time interval, τ, Equation 5.150 can be approximated by

$$u_\tau(t) \approx \frac{1}{\tau}\left[\frac{Q_{\text{IUH}}(t) + Q_{\text{IUH}}(t - \tau)}{2}\tau\right] \tag{5.151}$$

which simplifies to

$$\boxed{u_\tau(t) \approx \frac{1}{2}\left[Q_{\text{IUH}}(t) + Q_{\text{IUH}}(t - \tau)\right]} \tag{5.152}$$

Use of this approximate equation is allowed for small values of τ and permits direct calculation of a unit hydrograph from an IUH, bypassing the normal S-hydrograph procedure.

5.4.3.2 Unit-hydrograph models

Unit hydrographs are estimated in practice either from contemporaneous rainfall and runoff measurements or by using empirical relationships. The estimation of unit hydrographs from measured rainfall and runoff hydrographs is seldom an option in urban hydrology, where runoff is to be estimated from planned developments. The most common method of estimating the unit hydrograph in urban applications is to use empirical relationships to construct a *synthetic unit hydrograph*, where the shape of the unit hydrograph is related to the catchment characteristics and the duration of the rainfall excess. The shape parameters are usually the peak runoff, time to peak, and time base of the unit hydrograph. A wide variety of methods to estimate synthetic unit hydrographs have been developed, but only some of these methods are appropriate for use in designing urban stormwater systems (ASCE, 1992). Several of the more popular empirical unit hydrographs are described below.

Espey–Altman 10-min unit hydrograph. Espey and Altman (1978) and Espey et al. (1977) reported on unit hydrographs resulting from 10-min rainfall excesses over 41 watersheds ranging in size from 0.036 to 39 km^2 (0.014 to 15 mi^2), and impervious fractions ranging from 2% to 100%. Of the 41 watersheds studied, 16 were in

FIGURE 5.36: Catchment conveyance factor.

Source: Espey and Altman (1978).

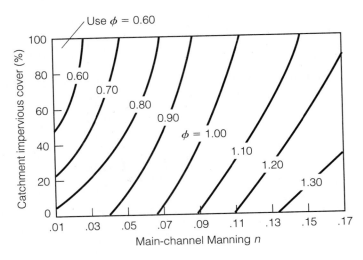

Texas, nine in North Carolina, six in Kentucky, four in Indiana, two in Colorado, two in Mississippi, one in Tennessee, and one in Pennsylvania. The characteristics of the catchment areas were measured by the following five parameters: (1) size of the drainage area, A; (2) main-channel length, L, defined as the total distance along the main channel from the catchment outlet to the upstream boundary of the catchment; (3) main-channel slope, S, defined as $H/0.8L$, where H is the difference in elevation between a point on the channel bottom at a distance of $0.2L$ downstream from the upstream catchment boundary and a point on the channel bottom at the catchment outlet; (4) impervious cover, I (%), which is assumed to be 5% for undeveloped catchments; and (5) dimensionless catchment conveyance factor, ϕ, which is a function of the impervious cover, I, and the Manning n of the main channel, as shown in Figure 5.36. The variables A, L, S, and I can usually be determined from existing topographic maps and aerial photographs, while the Manning n must be based on field inspections.

The observed 10-min unit hydrographs were described by the following five parameters: (1) time of rise to the peak of the unit hydrograph, T_p, measured from the beginning of the runoff; (2) peak flow of the unit hydrograph, Q_p; (3) time base of the unit hydrograph, T_B; (4) width of the hydrograph at 50% of Q_p, W_{50}; and (5) width of the unit hydrograph at 75% of Q_p, W_{75}. The empirical relationships between these five unit-hydrograph parameters and the catchment characteristics are given by Equations 5.153 to 5.157

$$T_p = 4.1 \frac{L^{0.23}\phi^{1.57}}{S^{0.25}I^{0.18}} \tag{5.153}$$

$$Q_p = 359 \frac{A^{0.96}}{T_p^{1.07}} \tag{5.154}$$

$$T_B = 1645 \frac{A}{Q_p^{0.95}} \tag{5.155}$$

$$W_{50} = 252 \frac{A^{0.93}}{Q_p^{0.92}} \tag{5.156}$$

$$W_{75} = 95 \frac{A^{0.79}}{Q_p^{0.78}} \tag{5.157}$$

where L is in meters, I is in %, A is in km^2, T_p, T_B, W_{50}, and W_{75} are in minutes, Q_p is in m^3/s, and S and ϕ are dimensionless. The application of Equations 5.153 to 5.157 is illustrated in Figure 5.37, where the widths W_{50} and W_{75} are normally positioned such that one-third lies on the rising side and two-thirds on the recession side of the unit hydrograph. In the data reported by Espey and Altman (1978), Equation 5.154 explains approximately 94% of the observed variance in the peak runoff, and Equations 5.153 to 5.157 collectively specify seven points through which the unit hydrograph must pass. Some minor adjustment to the shape of the hydrograph is generally necessary to ensure that it is indeed a unit hydrograph.

Of all the unit-hydrograph models covered in this section, the Espey–Altman model is the only one developed specifically for urban applications.

EXAMPLE 5.30

A 2.25-km^2 urban catchment has a main channel with a slope of 0.5% and Manning n of 0.06, the catchment is 40% impervious, and the distance along the main channel from the catchment boundary to the outlet is 1680 m. Calculate the 10-min unit hydrograph using the Espey–Altman method.

Solution   From the given data: $A = 2.25$ km^2, $S = 0.005$, $I = 40$%, and $L = 1680$ m. For $n = 0.06$, Figure 5.36 gives $\phi = 0.85$. Substituting these data into Equations 5.153 to 5.157 yields

$$T_p = 4.1 \frac{L^{0.23}\phi^{1.57}}{S^{0.25}I^{0.18}} = 4.1 \frac{(1680)^{0.23}(0.85)^{1.57}}{(0.005)^{0.25}(40)^{0.18}} = 33.9 \text{ min} = 0.565 \text{ h}$$

$$Q_p = 359 \frac{A^{0.96}}{T_p^{1.07}} = 359 \frac{(2.25)^{0.96}}{(33.9)^{1.07}} = 18.0 \text{ m}^3/\text{s}$$

$$T_B = 1645 \frac{A}{Q_p^{0.95}} = 1645 \frac{(2.25)}{(18.0)^{0.95}} = 238 \text{ min} = 3.96 \text{ h}$$

FIGURE 5.37: Espey–Altman unit hydrograph

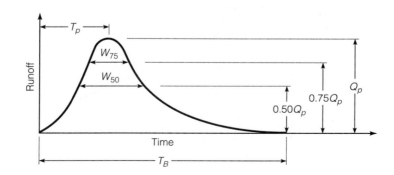

$$W_{50} = 252 \frac{A^{0.93}}{Q_p^{0.92}} = 252 \frac{(2.25)^{0.93}}{(18.0)^{0.92}} = 37.5 \text{ min} = 0.625 \text{ h}$$

$$W_{75} = 95 \frac{A^{0.79}}{Q_p^{0.78}} = 95 \frac{(2.25)^{0.79}}{(18.0)^{0.78}} = 18.9 \text{ min} = 0.315 \text{ h}$$

The hydrograph corresponding to these parameters is shown in Figure 5.38, where all intermediate points have been linearly interpolated. The hydrograph coordinates are:

Time (h)	0	0.357	0.460	0.565	0.775	0.982	3.96
Runoff (m^3/s)	0	9.0	13.5	18.0	13.5	9.0	0.0

For the estimated hydrograph to qualify as a unit hydrograph, it must correspond to a rainfall excess of 1 cm. This can be verified by dividing the area under the estimated hydrograph (= volume of runoff) by the catchment area. The area under the estimated hydrograph is given by (see Figure 5.38):

$$\text{Hydrograph area} = 3600 \left[\frac{1}{2}(0.357)(9.0) + \frac{1}{2}(9.0 + 13.5)(0.460 + 0.357) \right.$$

$$+ \frac{1}{2}(13.5 + 18.0)(0.565 - 0.460)$$

$$+ \frac{1}{2}(18.0 + 13.5)(0.775 - 0.565)$$

$$+ \frac{1}{2}(13.5 + 9.0)(0.982 - 0.775)$$

$$\left. + \frac{1}{2}(9.0 + 0.0)(3.96 - 0.982) \right]$$

$$= 84,443 \text{ m}^3$$

FIGURE 5.38: Espey–Altman runoff hydrograph

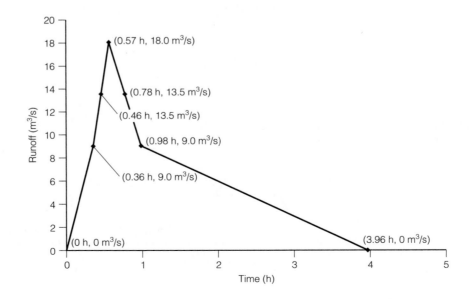

The catchment area is 2.25 km² = 2.25 × 10⁶ m², and therefore the runoff depth is given by

$$\text{Runoff depth} = \frac{8,4443 \text{ m}^3}{2.25 \times 10^6 \text{ m}^2} = 0.0375 \text{ m} = 3.75 \text{ cm}$$

Since the estimated hydrograph corresponds to a runoff depth of 3.75 cm and a unit hydrograph corresponds to a depth of 1 cm, the estimated hydrograph coordinates must be adjusted by a factor of $1/3.75 = 0.267$. Applying this factor to the estimated hydrograph coordinates yields the following Espey–Altman 10-min unit hydrograph:

Time (h)	0	0.357	0.460	0.565	0.775	0.982	3.96
Runoff (m³/s)	0	2.40	3.60	4.81	3.60	2.40	0.0

Snyder unit-hydrograph model.   The Snyder unit-hydrograph model is an empirical model that was developed based on studies of 20 watersheds located primarily in the Appalachian Highlands in the United States (Snyder, 1938). Snyder collected rainfall and runoff data from gaged watersheds with areas ranging from 26 to 26,000 km², derived the unit hydrographs, and related the hydrograph parameters to measurable watershed characteristics. Although the methodology proposed by Snyder was developed several decades ago, it is still widely used, and is part of the computer code HEC-HMS that is used by the U.S. Army Corps of Engineers for hydrologic studies (USACE, 2000).

Snyder's observations indicated that the peak runoff per unit (1 cm) of rainfall excess, Q_p (m³/s), is given by

$$Q_p = 2.75 \frac{C_p A}{t_l} \tag{5.158}$$

where t_l is the time lag in hours, A is the catchment area in km², and C_p is the peaking coefficient. The time lag, t_l, is defined as the time from the centroid of the rainfall excess to the peak runoff. The duration of rainfall excess, t_r, is related to the time lag, t_l, by

$$t_r = \frac{t_l}{5.5} \tag{5.159}$$

Based on Snyder's observations, the time lag, t_l, is related empirically to the catchment characteristics by

$$t_l = 0.75 C_t (LL_c)^{0.3} \tag{5.160}$$

where C_t is the basin coefficient that accounts for the slope, land use, and associated storage characteristics of the river basin, L is the length of the main stream from the outlet to the catchment boundary in kilometers, and L_c is the length along the main stream from the outlet to a point nearest to the catchment centroid in kilometers. The term $(LL_c)^{0.3}$ is sometimes called the *shape factor* of the catchment (McCuen, 2005). The parameters C_p in Equation 5.158 and C_t in Equation 5.160 are best found via calibration, as they are not physically based. Snyder (1938) reported that C_t

TABLE 5.25: Variability in Snyder Unit-Hydrograph Parameters

C_t	C_p	Location	Reference
1.01–4.33	0.26–0.67	27 watersheds in Pennsylvania	Miller et al. (1983)
0.40–2.26	0.31–1.22	13 watersheds (1.28–192 km^2) in central Texas	Hudlow and Clark (1969)
0.3–0.7	0.35–0.59	Northwestern United States (C_t); Sacramento and lower San Joaquin rivers	Linsley (1943)
0.4–2.4	0.4–1.1	12 watersheds (0.05–635 km^2) in eastern New South Wales, Australia	Cordery (1968)

typically ranges from 1.35 to 1.65, with a mean of 1.5; however, values of C_t have been found to vary significantly outside this range in very mountainous or very flat terrain. Bedient and Huber (1992) report that C_p ranges from 0.4 to 0.8, where the larger values of C_p are associated with smaller values of C_t. The variability in C_t and C_p is evident from the range of values reported in Table 5.25. If the duration of the desired unit hydrograph for the watershed of interest is significantly different from that specified in Equation 5.159, the following relationship can be used to adjust the time lag, t_l:

$$t_{lR} = t_l + 0.25(t_R - t_r) \tag{5.161}$$

where t_R is the desired duration of the rainfall excess, and t_{lR} is the lag of the desired unit hydrograph. For durations other than the standard duration, the peak runoff, Q_{pR}, is given by

$$Q_{pR} = 2.75 \frac{C_p A}{t_{lR}} \tag{5.162}$$

The U.S. Army Corps of Engineers developed the following empirical equations to help define the shape of the unit hydrograph:

$$W_{50} = \frac{2.14}{(Q_{pR}/A)^{1.08}} \tag{5.163}$$

$$W_{75} = \frac{1.22}{(Q_{pR}/A)^{1.08}} \tag{5.164}$$

where W_{50} and W_{75} are the widths in hours of the unit hydrograph at 50% and 75% of Q_{pR}, where Q_{pR} is the peak discharge in m^3/s, and A is the catchment area in km^2. These time widths are proportioned such that one-third is before the peak and two-thirds after the peak. The time to peak discharge, T_p, is computed from the duration of rainfall excess, t_R, and time lag, t_{lR}, using the relation

$$T_p = \frac{1}{2} t_R + t_{lR} \tag{5.165}$$

The time base of the unit hydrograph, T_B, is computed such that the unit hydrograph represents 1 cm of runoff. A rough estimate of T_B in small watersheds is three to five time T_p.

EXAMPLE 5.31

Derive the 2-h Snyder unit hydrograph for a 60-km² watershed where the main stream is 11 km long and the distance from the watershed outlet to the point on the stream nearest to the centroid of the watershed is 4 km. Previous investigations in similar terrain indicate that $C_p = 0.6$ and $C_t = 1.5$.

Solution   From the given data: $A = 60$ km², $L = 11$ km, $L_c = 4$ km, $C_p = 0.6$, and $C_t = 1.5$. According to Equation 5.160, the time lag, t_l, is given by

$$t_l = 0.75 C_t (LL_c)^{0.3} = 0.75(1.5)(11 \times 4)^{0.3} = 3.50 \text{ h}$$

This time lag corresponds to a rainfall-excess duration, t_r, given by Equation 5.159 as

$$t_r = \frac{t_l}{5.5} = \frac{3.50}{5.5} = 0.636 \text{ h}$$

Since a 2-h unit hydrograph is required, $t_R = 2$ h, and the adjusted time lag, t_{lR}, is given by Equation 5.161 as

$$t_{lR} = t_l + 0.25(t_R - t_r) = 3.50 + 0.25(2.00 - 0.636) = 3.84 \text{ h}$$

This gives a time to peak, T_p, as

$$T_p = \frac{t_R}{2} + t_{lR} = \frac{2.00}{2} + 3.84 = 4.84 \text{ h}$$

and the peak flow, Q_{pR}, is given by Equation 5.162 as

$$Q_{pR} = 2.75 \frac{C_p A}{t_{lR}} = 2.75 \frac{(0.6)(60)}{3.84} = 25.8 \text{ m}^3/\text{s}$$

The hydrograph shape parameters, W_{50} and W_{75}, are given by Equations 5.163 and 5.164 as

$$W_{50} = \frac{2.14}{(Q_{pR}/A)^{1.08}} = \frac{2.14}{(25.8/60)^{1.08}} = 5.32 \text{ h}$$

$$W_{75} = \frac{1.22}{(Q_{pR}/A)^{1.08}} = \frac{1.22}{(25.8/60)^{1.08}} = 3.04 \text{ h}$$

Putting one-third of W_{50} and W_{75} before the peak and two-thirds of W_{50} and W_{75} after the peak leads to the following points on the unit hydrograph:

t (h)	Q (m³/s)
0.0	0.0
3.07	12.9
3.83	19.4
4.84	25.8
6.87	19.4
8.39	12.9
T_B	0.0

Since the runoff depth is 1 cm and the area of the catchment is 60×10^6 m²,

$$1 \text{ cm} = \left[\frac{1}{2}(3.07)(12.9) + \frac{1}{2}(12.9 + 19.4)(3.83 - 3.07) + \frac{1}{2}(19.4 + 25.8) \right.$$

$$\times (4.84 - 3.83) + \frac{1}{2}(25.8 + 19.4)(6.87 - 4.84) + \frac{1}{2}(19.4 + 12.9)$$

$$\left. \times (8.39 - 6.87) + \frac{1}{2}(T_B - 8.39)(12.9)\right] \frac{(3600)(100)}{60 \times 10^6}$$

which yields $T_B = 14.80$ h. As expected, T_B is in the range of three to five times T_p. This completes specification of the final point on the Snyder unit hydrograph, which is plotted in Figure 5.39.

NRCS dimensionless unit hydrograph.   The Natural Resources Conservation Service (SCS, 1985; SCS, 1986) developed a dimensionless unit hydrograph that represents the average shape of a large number of unit hydrographs from small agricultural watersheds throughout the United States. This dimensionless unit hydrograph is frequently applied to urban catchments. The NRCS dimensionless unit hydrograph is illustrated in Figure 5.40(a) and expresses the normalized runoff, Q/Q_p, as a function of the normalized time, t/T_p, where Q_p is the peak runoff and T_p is the time to the peak of the hydrograph from the beginning of the rainfall excess. The coordinates of the NRCS dimensionless unit hydrograph are given in Table 5.26.

The NRCS dimensionless unit hydrograph can be converted to an actual hydrograph by multiplying the abscissa by T_p and the ordinate by Q_p. The time, T_p, is estimated using

$$T_p = \frac{1}{2}t_r + t_l \tag{5.166}$$

where t_r is the duration of the rainfall excess and t_l is the time lag from the centroid of the rainfall excess to the peak of the runoff hydrograph. NRCS recommends that the specified value of t_r not exceed two-tenths of t_c or three-tenths of T_p for the NRCS

FIGURE 5.39: Snyder 2-h unit hydrograph

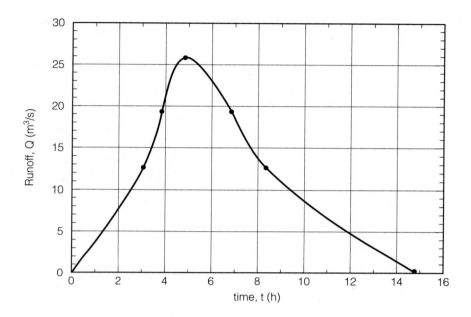

FIGURE 5.40: NRCS unit hydrographs

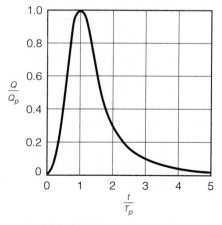

(a) Dimensionless Unit Hydrograph

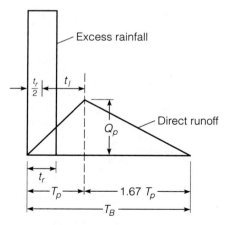

(b) Triangular Unit Hydrograph

TABLE 5.26: NRCS Dimensionless
Unit Hydrograph

t/T_p	Q/Q_p	t/T_p	Q/Q_p
0.0	0.000	2.6	0.107
0.2	0.100	2.8	0.077
0.4	0.310	3.0	0.055
0.6	0.660	3.2	0.040
0.8	0.930	3.4	0.029
1.0	1.000	3.6	0.021
1.2	0.930	3.8	0.015
1.4	0.780	4.0	0.011
1.6	0.560	4.2	0.008
1.8	0.390	4.4	0.006
2.0	0.280	4.6	0.004
2.2	0.207	4.8	0.002
2.4	0.147	5.0	0.000

dimensionless unit hydrograph to be valid. The time lag, t_l, can be estimated by

$$t_l = 0.6t_c \tag{5.167}$$

for rural catchments and, if justified by available data, can also be estimated using the more detailed relation

$$t_l = \frac{L^{0.8}(S + 1)^{0.7}}{1900Y^{0.5}} \tag{5.168}$$

where t_l is in hours, L is the length to the catchment divide (ft), and S is the potential maximum retention (in.) given by

$$S = \frac{1000}{CN} - 10 \tag{5.169}$$

where CN is the curve number and Y is the average catchment slope (%). Care should be taken in estimating Y from digital elevation models (DEMs), since the value of Y can be sensitive to the DEM grid size (Hill and Neary, 2005). The curve number in Equation 5.169 should be based on antecedent runoff condition II (AMCII), since it is being used as a measure of the surface roughness and not runoff potential (Haan et al., 1994). Equation 5.168 is applicable to curve numbers between 50 and 95, and catchment areas less than 8 km^2 (2000 ac). Otherwise, Equation 5.167 should be used to estimate the basin lag, t_l (Ponce, 1989). If the watershed is in an urban area, then t_l given by Equation 5.168 is adjusted for imperviousness and/or improved water courses by the factor M, where

$$M = 1 - I(-6.8 \times 10^{-3} + 3.4 \times 10^{-4}CN - 4.3 \times 10^{-7}CN^2 - 2.2 \times 10^{-8}CN^3) \tag{5.170}$$

and I is either the percentage impervious or the percentage of the main watercourse that is hydraulically improved from natural conditions. If part of the area is impervious and part of the channel is improved, then two values of M are determined and both are multiplied by t_l. Once t_l is determined, the time to peak, T_p, is estimated using Equation 5.166. The time base of the unit hydrograph, T_B, is equal to $5T_p$, and both T_p and T_B should be rounded to the nearest whole-number multiple of t_r.

The NRCS dimensionless unit hydrograph can be approximated by the triangular unit hydrograph illustrated in Figure 5.40(b). It incorporates the following key properties of the dimensionless unit hydrograph: (1) the total volume under the dimensionless unit hydrograph is the same, (2) the volume under the rising limb is the same, and (3) the peak discharge is the same. Taking Q_p and T_p to be the same in both the NRCS dimensionless unit hydrograph (Figure 5.40(a)) and the triangular approximation (Figure 5.40(b)), then the time base of the triangular unit hydrograph must be equal to $2.67T_p$. For a runoff depth, h, from a catchment area, A, the triangular unit hydrograph requires that

$$A \cdot h = \frac{1}{2} Q_p(2.67T_p) \tag{5.171}$$

If the runoff depth, h, is 1 cm, A is in km^2, Q_p is in (m^3/s)/cm, and T_p is in hours, then Equation 5.171 yields

$$\boxed{Q_p = 2.08 \frac{A}{T_p}} \tag{5.172}$$

This estimate of Q_p provides the peak runoff in both the curvilinear unit hydrograph (Figure 5.40(a)) and the triangular approximation (Figure 5.40(b)). The coefficient of 2.08 in Equation 5.172 is appropriate for the average rural experimental watersheds used in calibrating the formula, but it should be increased by about 20% for steep mountainous conditions and decreased by about 30% for flat swampy conditions. Flat swampy conditions are typically associated with areas having an average slope less than 0.5% (Lin and Perkins, 1989). Adjustment of the coefficient in Equation 5.172 must necessarily be accompanied by an adjustment to the time base of the unit hydrograph in order to maintain a runoff depth of 1 cm.

According to Debo and Reese (1995), the triangular unit-hydrograph approximation produces results that are sufficiently accurate for most stormwater-management facility designs, including curbs, gutters, storm drains, channels, ditches, and culverts.

EXAMPLE 5.32

A 2.25-km^2 urban catchment is estimated to have an average curve number of 70, an average slope of 0.5%, and a flow length from the catchment boundary to the outlet of 1680 m. If the imperviousness of the catchment is estimated at 40%, determine the NRCS unit hydrograph for a 30-minute rainfall excess. Determine the approximate NRCS triangular unit hydrograph, and verify that it corresponds to a rainfall excess of 1 cm.

Solution   The time lag of the catchment can be estimated using Equation 5.168, where $L = 1680$ m $= 5512$ ft, $Y = 0.5\%$, and S is derived from CN $= 70$ using Equation 5.169:

$$S = \frac{1000}{\text{CN}} - 10 = \frac{1000}{70} - 10 = 4.3 \text{ in.}$$

The time lag, t_l, is therefore given by

$$t_l = \frac{L^{0.8}(S + 1)^{0.7}}{1900Y^{0.5}} = \frac{(5512)^{0.8}(4.3 + 1)^{0.7}}{1900(0.5)^{0.5}} = 2.35 \text{ h}$$

For a 40% impervious catchment, $I = 40\%$ and $CN = 70$, Equation 5.170 gives the correction factor, M, as

$$M = 1 - I[-6.8 \times 10^{-3} + 3.4 \times 10^{-4}CN - 4.3 \times 10^{-7}CN^2 - 2.2 \times 10^{-8}CN^3]$$

$$= 1 - (40)[-6.8 \times 10^{-3} + 3.4 \times 10^{-4}(70) - 4.3 \times 10^{-7}(70)^2 - 2.2 \times 10^{-8}(70)^3]$$

$$= 0.706$$

Applying the correction factor, M, to calculate the adjusted time lag yields

$$t_l = 0.706 \times 2.35 \text{ h} = 1.66 \text{ h}$$

For a 30-minute rainfall excess, $t_r = 0.5$ h and the time to peak, T_p, is given by Equation 5.166 as

$$T_p = \frac{1}{2}t_r + t_l = \frac{1}{2}(0.5) + 1.66 = 1.91 \text{ h}$$

The NRCS unit hydrograph is limited to cases where $t_r/T_p \leq 0.3$; in this case, $t_r/T_p = 0.5/1.91 = 0.26$. The NRCS unit hydrograph is therefore applicable. The peak of the unit hydrograph, Q_p, is estimated using Equation 5.172, where $A = 2.25 \text{ km}^2$ and

$$Q_p = 2.08\frac{A}{T_p} = 2.08\frac{2.25}{1.91} = 2.45 \text{ (m}^3\text{/s)/cm}$$

With $T_p = 1.91$ h and $Q_p = 2.45$ (m^3/s)/cm, the NRCS 30-min unit hydrograph is obtained by multiplying t/T_p in Table 5.26 by 1.91 h and Q/Q_p by 2.45 (m^3/s)/cm to yield

t (h)	Q [(m^3/s)/cm]	t (h)	Q [(m^3/s)/cm]
0.0	0.0	3.82	0.69
0.38	0.25	4.20	0.51
0.76	0.76	4.58	0.36
1.15	1.62	4.97	0.26
1.53	2.28	5.35	0.19
1.91	2.45	5.73	0.14
2.29	2.28	6.49	0.07
2.67	1.91	8.02	0.03
3.06	1.37	8.79	0.01
3.44	0.96	9.55	0.00

For the approximate triangular unit hydrograph, T_p and Q_p are the same, but the base of the hydrograph is $2.67T_p = 2.67 \times 1.91$ h $= 5.10$ h ($=18,360$ s). The area under the triangular hydrograph ($=$ volume of runoff) is given by

$$\text{Area} = \frac{1}{2}\text{ base} \times \text{height} = \frac{1}{2}18,360 \times 2.45 = 22,500 \text{ m}^3$$

Since the catchment area is 2.25 km² = 2.25 × 10⁶ m², the depth of rainfall is equal to the volume of runoff divided by the area of the catchment. Hence

$$\text{Depth of rainfall} = \frac{22{,}500 \text{ m}^3}{2.25 \times 10^6 \text{ m}^2} = 0.01 \text{ m} = 1 \text{ cm}$$

Since the runoff hydrograph corresponds to a unit depth (1 cm) of rainfall, it is a valid unit hydrograph.

Accuracy of unit-hydrograph models.   According to Dunne and Leopold (1978), the unit-hydrograph method gives estimates of flood peaks that are usually within 25% of their true value; this is close enough for most planning purposes and about as close as can be expected, given the usual lack of detailed information about catchment processes and states. Errors larger than 25% can be expected if synthetic unit hydrographs are used without verification for the region of application (Dingman, 2002).

5.4.3.3   Time-area models

Time-area models describe the relationship between the travel time to the catchment outlet and the location within the catchment. This relationship is based on the estimated velocity of direct runoff and neglects storage effects. The time-area model can be illustrated by considering the runoff isochrones illustrated in Figure 5.41, and the corresponding histogram of contributing area vs. time shown in Figure 5.42. These figures indicate that the incremental area contributing to runoff varies throughout a storm, with full contribution from all areas for $t \geq t_5$. If the rainfall excess is expressed as the average rainfall-excess intensity over the time increments in the time-area

FIGURE 5.41: Runoff isochrones

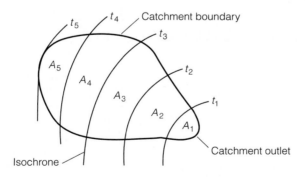

FIGURE 5.42: Time-area histogram

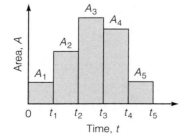

histogram, then the runoff at the end of time increment j, Q_j, can be expressed in terms of the contributing areas by the relation

$$Q_j = i_j A_1 + i_{j-1} A_2 + \cdots + i_1 A_j$$

or

$$Q_j = \sum_{k=1}^{j} i_{j-k+1} A_k \qquad (5.173)$$

where i_j and A_j are the average rainfall-excess intensity and contributing area, respectively, during time increment j. The difficulty in estimating isochronal lines and the need to account for storage effects make the time-area model difficult to apply in practice. The time-area model uses a simple routing procedure that does not consider attenuation of the runoff hydrograph associated with storage effects, and is therefore limited to small watersheds with areas less than 200 ha (500 ac) (Ward, 2004). As a direct result of neglecting attenuation, peak flows will likely be overestimated if the time-area model is applied to large watersheds.

EXAMPLE 5.33

A 100-ha catchment is estimated to have the following time-area relationship:

Time (min)	Contributing area (ha)
0	0
5	3
10	9
15	25
20	51
25	91
30	100

If the rainfall-excess distribution is given by

Time (min)	Average intensity (mm/h)
0–5	132
5–10	84
10–15	60
15–20	36

estimate the runoff hydrograph using the time-area model.

Solution   First, tabulate the contributing area for each of the given time intervals:

Time interval (min)	Contributing area (ha)
0–5	3
5–10	6
10–15	16
15–20	26
20–25	40
25–30	9

The runoff hydrograph is calculated using Equation 5.173, and the computations are given in the following table (a factor of 0.002778 is used to convert ha·mm/h to m³/s):

Time (min)		Runoff hydrograph (m³/s)
0		0
5	[132(3)](0.002778) =	1.1
10	[84(3) + 132(6)](0.002778) =	4.0
15	[60(3) + 84(6) + 132(16)](0.002778) =	7.8
20	[36(3) + 60(6) + 84(16) + 132(26)](0.002778) =	14.6
25	[36(6) + 60(16) + 84(26) + 132(40)](0.002778) =	24.0
30	[36(16) + 60(26) + 84(40) + 132(9)](0.002778) =	18.6
35	[36(26) + 60(40) + 84(9)](0.002778) =	11.4
40	[36(40) + 60(9)](0.002778) =	5.5
45	[36(9)](0.002778) =	0.9
50	0 =	0

The expected runoff hydrograph in the above table has a time base of 50 min and a peak runoff rate of 24.0 m³/s.

The Hydrologic Engineering Center (1998, 2000) has developed the following synthetic time-area relationship based on empirical analyses of several catchments

$$\frac{A}{A_c} = 1.414 \left(\frac{t}{t_c} \right)^{1.5} \quad \text{for} \quad 0 \leq \frac{t}{t_c} \leq 0.5 \tag{5.174}$$

$$\frac{A}{A_c} = 1 - 1.414 \left(1 - \frac{t}{t_c} \right)^{1.5} \quad \text{for} \quad 0.5 \leq \frac{t}{t_c} \leq 1.0 \tag{5.175}$$

where A is the contributing area at time t, A_c is the catchment area, and t_c is the time of concentration of the catchment. To account for storage effects in the catchment, the Hydrologic Engineering Center (HEC) utilizes the continuity equation in the form

$$\bar{I} - \left(\frac{O_1 + O_2}{2} \right) = \frac{S_2 - S_1}{\Delta t} \tag{5.176}$$

where $\bar{I}$ is the mean inflow during time interval Δt, O is the outflow, and S is the storage. The subscripts 1 and 2 refer to the beginning and end of the time interval Δt. In applying Equation 5.176, $\bar{I}$ is the mean runoff during Δt accounting for translation effects only (as done in the previous example); and O_1 and O_2 are the outflows at the beginning and end of the time interval when storage effects are considered. Assuming that the storage S is linearly proportional to the outflow, O, commonly referred to as the *linear-reservoir assumption*, then

$$S = RO \tag{5.177}$$

where R is a proportionality constant that is sometimes called the *storage coefficient* and has units of time. Equations 5.176 and 5.177 combine to yield the relation

$$\boxed{O_2 = C\bar{I} + (1 - C)O_1} \tag{5.178}$$

where

$$C = \frac{2\Delta t}{2R + \Delta t} \tag{5.179}$$

Equation 5.178 is the basis for the *Clark Method* of estimating the unit hydrograph from time-area curves (Clark, 1945), and this method is still widely used in current engineering practice (HEC, 2000; HEC, 2001). In the Clark Method, the runoff resulting from the translation of an instantaneous unit depth of rainfall is first calculated using Equation 5.173, and then this runoff hydrograph is routed through a linear reservoir using Equation 5.178. The resulting runoff hydrograph is, by definition, an instantaneous unit hydrograph (IUH). The unit hydrograph for any duration, τ, can then be determined from this IUH using Equation 5.152.

EXAMPLE 5.34

A catchment has an area of 95 km² and a time of concentration of 7.5 h, and the Clark storage coefficient is estimated to be 3.4 hours. Estimate the 2-hour unit hydrograph for the catchment.

Solution From the given data: $A_c = 95 \text{ km}^2$, $t_c = 7.5$ h, and $R = 3.4$ h. The computations are presented in the following table, and the sequence of calculations is described below.

(1) t (h)	(2) t/t_c	(3) A/A_c	(4) A (km²)	(5) ΔA (km²)	(6) $Q(t)$ (m³/s)	(7) IUH (m³/s)	(8) 2-h UH (m³/s)
0.0	0.000	0.000	0.0			0	0
				2.3	12.8		
0.5	0.067	0.024	2.3			1.8	0.9
				4.2	23.5		
1.0	0.133	0.069	6.5			4.7	2.4
				5.5	30.4		
1.5	0.200	0.126	12.0			8.3	4.1
				6.5	36.0		
2.0	0.267	0.195	18.5			12.1	6.0
				7.4	40.9		
2.5	0.333	0.272	25.9			16.0	8.9
				8.1	45.2		
3.0	0.400	0.358	34.0			20.0	12.4
				8.8	49.1		
3.5	0.467	0.451	42.8			24.0	16.1
				9.4	52.0		
4.0	0.533	0.549	52.2			27.8	19.9
				8.8	49.1		
4.5	0.600	0.642	61.0			30.7	23.4
				8.1	45.2		
5.0	0.667	0.728	69.1			32.7	26.4
				7.4	40.9		
5.5	0.733	0.805	76.5			33.8	28.9
				6.5	36.0		

(1) t (h)	(2) t/t_c	(3) A/A_c	(4) A (km^2)	(5) ΔA (km^2)	(6) $Q(t)$ (m^3/s)	(7) IUH (m^3/s)	(8) 2-h UH (m^3/s)
6.0	0.800	0.874	83.0			34.1	31.0
				5.5	30.4		
6.5	0.867	0.931	88.5			33.6	32.2
				4.2	23.5		
7.0	0.933	0.976	92.7			32.2	32.5
				2.3	12.8		
7.5	1.000	1.000	95.0			29.6	31.7
				0.0	0.0		
8.0			95.0			25.5	29.8
8.5						22.0	27.8
9.0						19.0	25.6
9.5						16.4	23.0
.						.	.
.						.	.
.						.	.
29.5						0.0	0.1
30.0						0.0	0.1
30.5						0.0	0.0

- Column 2 ($=t/t_c$) is derived from column 1 ($=t$) by dividing column 1 by t_c ($=7.5$ h); column 3 ($=A/A_c$) is derived from column 2 using either Equation 5.174 ($t/t_c \leq 0.5$) or Equation 5.175 ($t/t_c \geq 0.5$); and column 4 ($=A$) is derived from column 3 by multiplying column 3 by A_c ($=95$ km^2).

- The incremental area, ΔA, between isochrones is calculated by subtracting the areas included by adjacent isochrones in column 4, and the results are shown in column 5.

- Assuming pure translation of the rainfall excess, the average runoff rate, $Q(t)$, during each time increment in response to an instantaneous rainfall excess of 1 cm ($=0.01$ m) is given by

$$Q(t) = \frac{(0.01 \text{ m})(\Delta A \times 10^6 \text{ m}^2/\text{km}^2)}{(\Delta t \times 3600 \text{ s/h})} \text{ m}^3/\text{s}$$

where ΔA is the incremental area in column 5 in km^2, and Δt is the time increment in hours. The calculated runoff rate, $Q(t)$, is given in column 6. The runoff hydrograph in column 6 is the instantaneous unit hydrograph with storage effects neglected (i.e., pure translation).

- Storage effects are accounted for by using the linear-reservoir routing relation given by Equation 5.178, where

$$O_2 = C\bar{I} + (1 - C)O_1$$

and

$$C = \frac{2\Delta t}{2R + \Delta t}$$

In this case, $\Delta t = 0.5$ h, $R = 3.4$ h,

$$C = \frac{2(0.5)}{2(3.4) + 0.5} = 0.137$$

and the linear-reservoir routing relation is given by

$$O_2 = 0.137\bar{I} + (1 - 0.137)O_1 = 0.137\bar{I} + 0.863O_1$$

The results of applying this routing relation are given in column 7, where O_1 and O_2 are the routed runoff rates at the beginning and end of the time increment, respectively (in column 7), and $\bar{I}$ is the average translated runoff during the time increment, given in column 6. After applying the routing relation, the hydrograph shown in column 7 is the instantaneous unit hydrograph with storage effects taken into account.

- The 2-h unit hydrograph, $u_2(t)$, is derived from the instantaneous unit hydrograph in column 7 using Equation 5.152,

$$u_2(t) \approx \frac{1}{2}[Q_{\text{IUH}}(t) + Q_{\text{IUH}}(t - 2)]$$

where $Q_{\text{IUH}}(t)$ and $Q_{\text{IUH}}(t - 2)$ are the instantaneous unit hydrograph runoff rates tabulated in column 7. Applying this equation to the data in column 7 yields the 2-h unit hydrograph given in column 8, with the corresponding times given in column 1.

5.4.3.4 Kinematic-wave model

The kinematic-wave model describes runoff by solving the one-dimensional continuity equation

$$\frac{\partial y}{\partial t} + \frac{\partial q}{\partial x} = i_e \tag{5.180}$$

and a momentum equation of the form

$$q = \alpha y^m \tag{5.181}$$

where y is the flow depth, q is the flow per unit width of the catchment, and i_e is the effective rainfall. Combining Equations 5.180 and 5.181 yields the following kinematic-wave equation for overland flow:

$$\boxed{\frac{\partial y}{\partial t} + \alpha m y^{m-1}\frac{\partial y}{\partial x} = i_e} \tag{5.182}$$

The kinematic-wave model solves Equation 5.182 for the flow depth, y, as a function of x and t for given initial and boundary conditions. This solution is then substituted into Equation 5.181 to determine the runoff, q, as a function of space and time. In cases where the rainfall excess is independent of x and t, then the solution was derived previously and is given by Equation 5.92. In the more general case with variable effective rainfall, a numerical solution of Equation 5.182 is required. The kinematic-wave model is useful in developing surface-runoff hydrographs for subwatersheds that contribute lateral flows to streams; for example, overland flow to a stream from the left and right sides of its bank may be computed using kinematic-wave models.

In cases where overland flow enters a drainage channel, flow in the channel can be described by

$$\frac{\partial A}{\partial t} + \frac{\partial Q}{\partial x} = q_0 \tag{5.183}$$

and

$$Q = \alpha A^m \tag{5.184}$$

where A is the cross-sectional flow area, Q is the flowrate in the channel, and q_0 is the overland inflow entering the channel per unit length of the channel. Combining Equations 5.183 and 5.184 yields the following kinematic-wave equation for channel flow:

$$\boxed{\frac{\partial A}{\partial t} + \alpha m A^{m-1} \frac{\partial A}{\partial x} = q_0} \tag{5.185}$$

Because this equation does not allow for hydrograph diffusion, the application of the kinematic-wave equation is limited to flow conditions that do not demonstrate appreciable hydrograph attenuation. Accordingly, the kinematic-wave approximation works best when applied to short, well-defined channel reaches as found in urban drainage applications (ASCE, 1996b). A typical kinematic-wave model divides the catchment into overland flow planes that feed collector channels. The kinematic-wave model is used only in numerical models, and the documentation accompanying commercial software packages that implement the kinematic-wave model generally provides detailed descriptions of the numerical procedures used to solve the kinematic-wave equation.

5.4.3.5 Nonlinear-reservoir model

The nonlinear-reservoir model views the catchment as a very shallow reservoir, where the inflow is equal to the rainfall excess, the outflow is a (nonlinear) function of the depth of flow over the catchment, and the difference between the inflow and outflow is equal to the rate of change of storage within the catchment. The nonlinear-reservoir model can be stated as

$$A\frac{dy}{dt} = Ai_e - Q \tag{5.186}$$

where A is the surface area of the catchment and Q is the surface runoff at the catchment outlet. The model assumes uniform overland flow at the catchment outlet at a depth equal to the difference between the water depth, y, and the (constant) depression storage, y_d; therefore, Q can be expressed as a function of y. Equation 5.186 is an ordinary differential equation in y that can be solved, given an initial condition within the catchment and the variation of i_e as a function of time.

A particular application of the nonlinear-reservoir model is described by Huber and Dickinson (1988), where the runoff at the catchment outlet is given by the Manning relation

$$Q = \frac{CW}{n}(y - y_d)^{5/3}S_0^{1/2} \tag{5.187}$$

where C is a constant to account for the units ($C = 1$ for SI units, $C = 1.49$ for U.S. Customary units), W is a representative width of the catchment, n is an average value of the Manning roughness coefficient for the catchment, and S_0 is the average slope of the catchment. Estimates of the Manning n for overland flow are given in Table 5.17. The combination of Equations 5.186 and 5.187 can be put in the simple finite-difference form

$$\frac{y_2 - y_1}{\Delta t} = \bar{i}_e - \frac{CWS_0^{1/2}}{An}\left(\frac{y_1 + y_2}{2} - y_d\right)^{5/3} \tag{5.188}$$

where Δt is the time step, y_1 and y_2 are the water depths at the beginning and end of the time step, and $\bar{i}_e$ is the average effective rainfall over the time step. Equation 5.188 must be solved for y_2, which is then substituted into Equation 5.187 to determine the runoff, Q.

EXAMPLE 5.35

A 1-ha catchment consists of mostly light turf with an average slope of 0.8%. The width of the catchment is approximately 100 m, and the depression storage is estimated to be 5 mm. Use the nonlinear-reservoir model to calculate the runoff from the following 15-min rainfall excess:

Time (min)	Effective rainfall (mm/h)
0–5	120
5–10	70
10–15	50

Solution   The nonlinear-reservoir model is given by Equation 5.188, where $C = 1$, $W = 100$ m, $S_0 = 0.008$, $A = 1$ ha $= 10^4$ m^2, $n = 0.20$ (from Table 5.17), and $y_d = 5$ mm $= 0.005$ m. Taking $\Delta t = 5$ min $= 300$ s and substituting the given parameters into Equation 5.188 gives

$$\frac{y_2 - y_1}{300} = \bar{i}_e - \frac{(1)(100)(0.008)^{1/2}}{(10^4)(0.20)}\left(\frac{y_1 + y_2}{2} - 0.005\right)^{5/3}$$

which simplifies to

$$y_2 = y_1 + 300\left[2.78 \times 10^{-7}\bar{i}_e - 0.00447\left(\frac{y_1 + y_2}{2} - 0.005\right)^{5/3}\right]$$

where the factor 2.78×10^{-7} is introduced to convert $\bar{i}_e$ in mm/h to m/s. For given values of y_1, this equation is solved for $y = y_2$ as a function of time. The corresponding runoff, Q, is given by Equation 5.187 as

$$Q = \frac{CW}{n}(y - y_d)^{5/3}S_0^{1/2} = \frac{(1)(100)}{0.20}(y - 0.005)^{5/3}(0.008)^{1/2} = 44.7(y - 0.005)^{5/3}$$

Starting with $y_1 = 0$ m at $t = 0$, the runoff computations are tabulated for $t = 0$ to 60 min as follows:

t (min)	$\bar{i}_e$ (mm/h)	y (m)	Q (m^3/s)
0		0	
	120		
5		0.0100	0.00654
	70		
10		0.0154	0.0221
	50		
15		0.0187	0.0351
	0		
20		0.0177	0.0309
	0		
25		0.0168	0.0273
	0		
30		0.0160	0.0243
	0		
35		0.0153	0.0218
	0		
40		0.0147	0.0197
	0		
45		0.0141	0.0177
	0		
50		0.0136	0.0161
	0		
55		0.0131	0.0146
	0		
60		0.0127	0.0134

Beyond $t = 60$ min, the runoff, Q, decreases gradually to zero.

5.4.3.6 Santa Barbara Urban Hydrograph model

The Santa Barbara Urban Hydrograph (SBUH) model was developed for the Santa Barbara County Flood Control and Water Conservation District in California (Stubchaer, 1975) and has been adopted by other water-management agencies in the United States (for example, the South Florida Water Management District, 1994). In the SBUH model, the impervious portion of the catchment is assumed to be directly connected to the drainage system, abstractions from rain falling on impervious surfaces are neglected, and abstractions from pervious areas are accounted for by using either the NRCS curve-number method or a similar technique. The SBUH combines the runoff from impervious and pervious areas to develop a runoff hydrograph that is routed through an imaginary reservoir that causes a time delay equal to the time of concentration of the catchment. The computations proceed through consecutive time intervals, Δt, during which the instantaneous runoff, I, is calculated using the relation

$$I = [ix + i_e(1.0 - x)]A \qquad (5.189)$$

where i is the rainfall rate, x is the fraction of the catchment that is impervious, i_e is the effective rainfall rate, and A is the area of the catchment. Over the finite interval,

Δt, the law of conservation of mass requires that the rate of runoff excess, I, minus the outflow rate, Q, from the catchment is equal to the rate of change of catchment storage, S; and this requirement can be written in the finite difference form

$$\left(\frac{I_j + I_{j+1}}{2}\right) - \left(\frac{Q_j + Q_{j+1}}{2}\right) = \frac{S_{j+1} - S_j}{\Delta t} \tag{5.190}$$

where j is the time step index. Approximating the catchment as a linear reservoir assumes that the catchment storage volume, S, is linearly proportional to the catchment outflow rate, Q, such that

$$S_j = KQ_j \quad \text{and} \quad S_{j+1} = KQ_{j+1} \tag{5.191}$$

where K is a proportionality constant. Also, since the time for the entire catchment to drain is, by definition, equal to the time of concentration, t_c, the SBUH model takes K equal to t_c, and Equation 5.191 becomes

$$S_j = t_cQ_j \quad \text{and} \quad S_{j+1} = t_cQ_{j+1} \tag{5.192}$$

Substituting Equation 5.192 into Equation 5.190 yields the following routing equation:

$$\boxed{Q_j = Q_{j-1} + K_r(I_{j-1} + I_j - 2Q_{j-1})} \tag{5.193}$$

where K_r is a routing constant given by

$$\boxed{K_r = \frac{\Delta t}{2t_c + \Delta t}} \tag{5.194}$$

An important limitation of the SBUH method is that the calculated peak discharge cannot occur after precipitation ceases. In reality, for short-duration storms over flat and large watersheds, the peak discharge can occur after rainfall ends. The SBUH method has been reported to greatly overpredict peak-flow rates for pasture conditions, and slightly overpredict flows for forest conditions (Jackson et al., 2001).

EXAMPLE 5.36

A 2.25-km^2 catchment is estimated to have a time of concentration of 45 minutes, and 45% of the catchment is impervious. Estimate the runoff hydrograph for the following rainfall event:

Time (min)	Rainfall (mm/h)	Rainfall excess (mm/h)
0	0	0
10	210	150
20	126	102
30	78	66

Use a time increment of 10 minutes to calculate the runoff hydrograph.

Solution From the given data, $\Delta t = 10$ min and $t_c = 45$ min, therefore Equation 5.194 gives

$$K_r = \frac{\Delta t}{2t_c + \Delta t} = \frac{10}{2(45) + 10} = 0.10$$

Also, from the given data, $x = 0.45$, $A = 2.25$ km^2, and Equation 5.189 gives the instantaneous runoff, I, as

$$I = [ix + i_e(1.0 - x)]A = [i(0.45) + i_e(1.0 - 0.45)](2.25)(0.278)$$

$$= 0.625[0.45i + 0.55i_e] \tag{5.195}$$

where the factor 0.278 is required to give I in m^3/s. The catchment runoff is given by Equation 5.193 as

$$Q_j = Q_{j-1} + K_r(I_{j-1} + I_j - 2Q_{j-1}) = Q_{j-1} + (0.10)(I_{j-1} + I_j - 2Q_{j-1}) \tag{5.196}$$

Beginning with $I_1 = Q_1 = 0$, Equation 5.195 is applied to calculate I at each time step, and Equation 5.196 is applied to calculate the runoff hydrograph. These calculations are given in the following table:

t (min)	i (mm/h)	i_e (mm/h)	I (m^3/s)	Q (m^3/s)
0	0	0	0	0
10	210	150	111	11.1
20	126	102	70.5	27.0
30	78	66	44.6	33.1
40	0	0	0	30.9
50	0	0	0	24.7
60	0	0	0	19.8
70	0	0	0	15.8
80	0	0	0	12.7
90	0	0	0	10.1
100	0	0	0	8.1
110	0	0	0	6.5
120	0	0	0	5.2
130	0	0	0	4.1
140	0	0	0	3.3
150	0	0	0	2.7
160	0	0	0	2.1
170	0	0	0	1.7
180	0	0	0	1.4
190	0	0	0	1.1
200	0	0	0	0.9

The runoff, Q, as a function of time reaches a peak at the time the rainfall ends and decreases exponentially thereafter.

5.4.3.7 Extreme runoff events

The *probable maximum flood* (PMF) is the flood discharge that may be expected from the most severe combination of meteorologic and hydrologic conditions that are possible in a region (U.S. Army Corps of Engineers, 1965). Practical applications of such floods are usually confined to the determination of spillway requirements for high dams, but in unusual cases may constitute the design flood for local protection works where an exceptionally high degree of protection is advisable and economically obtainable.

The *standard project flood* (SPF) is the most severe flood considered reasonably characteristic of a specific region. The SPF is intended as a practicable expression of the degree of protection that should be sought as a general rule in the design of flood-control works for communities where protection of human life and unusually high-valued property is involved. The standard project flood excludes extremely rare conditions and, typically, the peak discharge of an SPF is about 40% to 60% of that of a PMF for the same basin and has a return period of about 500 years. The commonly used methodology for estimating the SPF in the United States has been documented by the U.S. Army Corps of Engineers (USACE, 1965). In most cases, the SPF is the runoff hydrograph from the *standard project storm* (SPS) described in Section 5.2.4.3. In some cases, particularly in very large drainage basins, the SPF estimate may be based on a study of actual hydrographs or stages of record, or on other procedures not directly involving an SPS estimate. In cases where floods are predominantly the result of melting snow, the SPF estimate is based on estimates of the most critical combinations of snow, temperature and water losses considered reasonably characteristic of the region. Since the statistical probability of SPF occurrence varies with the size of the drainage area and other hydrometeorological factors, it is not considered feasible to assign specific frequency estimates to SPF determinations in general.

The standard project flood should not be confused with the design flood of a project. The term *design flood* refers to the flood hydrograph or peak discharge value adopted as the basis for design of a particular project after full consideration has been given to flood characteristics, frequencies, and the economic and other practical considerations entering into selection of the design discharge criteria. The design flood for a particular project may be either greater or less than the standard project flood.

5.5 Routing Models

Routing is the process of determining the spatial and temporal distribution of flow rate and flow depth along a watercourse such as a river or storm sewer. Routing is sometimes called *flow routing* or *flood routing*, and routing models are generally classified as either *hydrologic-routing* models or *hydraulic-routing* models. Hydrologic-routing models are based on the simultaneous solution of the continuity equation and a second equation which usually expresses the storage volume within a channel reach as a function of inflow and outflow. Hydraulic-routing models are based on the simultaneous solution of the continuity equation and momentum equation for open channels. Although hydraulic-routing models are generally considered to be more accurate than hydrologic-routing models, the simplicity and acceptable accuracy of hydrologic-routing models in some circumstances make them appealing, particularly in the design of storage ponds and reservoirs.

5.5.1 Hydrologic Routing

The basic equation used in hydrologic routing is the continuity equation

$$\frac{dS}{dt} = I(t) - O(t) \tag{5.197}$$

where S is the storage between upstream and downstream sections, t is time, $I(t)$ is the inflow at the upstream section, and $O(t)$ is the outflow at the downstream section. Hydrologic routing is frequently applied to storage reservoirs and stormwater-detention basins, where $I(t)$ is the inflow to the storage reservoir, typically from rivers, streams, or drainage channels; $O(t)$ is the outflow from the reservoir, typically over spillways, weirs, or orifice-type outlets; and S is the storage in the reservoir. Although the application of hydrologic routing to rivers and drainage channels is also well established, in many of these cases hydraulic routing is preferable. The procedure for hydrologic routing depends on the particular system being modeled. In the case of storage reservoirs, the storage, S, is typically a function of the outflow, $O(t)$, and the *modified Puls method* (Puls, 1928),* is preferred. In the case of channel routing, S is typically related to both the upstream inflow, $I(t)$, and downstream outflow, $O(t)$, and other methods such as the *Muskingum method* are preferred.

5.5.1.1 Modified Puls method

Over the finite interval of time between t and $t + \Delta t$, Equation 5.197 can be written in the finite difference form

$$\frac{S_2 - S_1}{\Delta t} = \left(\frac{I_1 + I_2}{2} \right) - \left(\frac{O_1 + O_2}{2} \right) \tag{5.198}$$

where the subscripts 1 and 2 refer to the values of the variables at times t and $t + \Delta t$, respectively. For computational convenience, Equation 5.198 can be put in the form

$$\boxed{(I_1 + I_2) + \left(\frac{2S_1}{\Delta t} - O_1 \right) = \left(\frac{2S_2}{\Delta t} + O_2 \right)} \tag{5.199}$$

In using this form of the continuity equation to route the hydrograph through a storage reservoir, it is important to recognize what is known and what is to be determined. Since the inflow hydrograph is generally a given, then I_1 and I_2 are known at every time step. For ungated spillways, orifice-type outlets, and weir-type outlets, the discharge, O, is a known function of the water-surface elevation (stage) in the reservoir. Since the storage, S, in the reservoir is typically a known function of the stage in the reservoir, then $2S/\Delta t + O$ is typically a known function of the stage in the reservoir, and the following procedure can be used to route the inflow hydrograph through the reservoir:

1. Substitute known values of I_1, I_2, and $2S_1/\Delta t - O_1$ into the lefthand side of Equation 5.199. This gives the value of $2S_2/\Delta t + O_2$.
2. From the reservoir characteristics, determine the discharge, O_2, corresponding to the calculated value of $2S_2/\Delta t + O_2$.
3. Subtract $2O_2$ from $2S_2/\Delta t + O_2$ to yield $2S_2/\Delta t - O_2$ at the end of this time step.
4. Repeat steps 1 to 3 until the entire outflow hydrograph, $O(t)$, is calculated.

In performing these computations, it is recommended to select a time step, Δt, such that there are a minimum of five or six points on the rising side of the inflow hydrograph, one of which coincides with the inflow peak.

The *storage-indication method* is very similar to the modified Puls method, with the exception that the continuity equation is expressed in the form

$$\left(\frac{I_1 + I_2}{2} \right) \Delta t + \left(S_1 - \frac{O_1 \Delta t}{2} \right) = \left(S_2 + \frac{O_2 \Delta t}{2} \right) \tag{5.200}$$

---

*The method proposed by Puls (1928) was modified by the U.S. Bureau of Reclamation (1949), hence the name *modified Puls method*.

rather than in the form given by Equation 5.199 (McCuen, 2005). The primary difference between the modified Puls method and the storage-indication method is that the modified Puls method is based on the relation between $2S/\Delta t + O$ and O, while the storage-indication method is based on the relation between $S + O\Delta t/2$ and O; otherwise, the calculation procedures are almost identical. It is fairly common practice to refer to the modified Puls and storage-indication methods as being the same, although they are technically different.

For irregular-shaped detention basins, the surface area, A_s, versus elevation, h, is determined from contour maps of the detention-basin site, and the storage-versus-elevation relation is subsequently calculated using the equation

$$S_2 = S_1 + (h_2 - h_1)\frac{A_{s1} + A_{s2}}{2} \tag{5.201}$$

where S_1 and A_{s1} correspond to elevation h_1, and S_2 and A_{s2} correspond to elevation h_2. A more accurate relationship proposed by Paine and Akan (2001) is given by

$$S_2 = S_1 + \frac{h_2 - h_1}{3}(A_{s1} + A_{s2} + \sqrt{A_{s1}A_{s2}}) \tag{5.202}$$

The difference between Equations 5.201 and 5.202 decreases as the height increment decreases.

EXAMPLE 5.37

A stormwater-detention basin is estimated to have the following storage characteristics:

Stage (m)	Storage (m^3)
5.0	0
5.5	694
6.0	1525
6.5	2507
7.0	3652
7.5	4973
8.0	6484

The discharge weir from the detention basin has a crest elevation of 5.5 m, and the weir discharge, Q, is given by

$$Q = 1.83H^{\frac{3}{2}}$$

where Q is in m^3/s and H is the height of the water surface above the crest of the weir in meters. The catchment runoff hydrograph is given by

Time (min)	0	30	60	90	120	150	180	210	240	270	300	330	360	390
Runoff (m^3/s)	0	2.4	5.6	3.4	2.8	2.4	2.2	1.8	1.5	1.2	1.0	0.56	0.34	0

If the prestorm stage in the detention basin is 5.0 m, estimate the discharge hydrograph from the detention basin.

Solution   This problem requires that the runoff hydrograph be routed through the detention basin. Using $\Delta t = 30$ min, the storage and outflow characteristics can be put in the convenient tabular form:

Stage (m)	S (m^3)	O (m^3/s)	$2S/\Delta t + O$ (m^3/s)
5.0	0	0	0
5.5	694	0	0.771
6.0	1525	0.647	2.34
6.5	2507	1.83	4.62
7.0	3652	3.36	7.42
7.5	4973	5.18	10.7
8.0	6484	7.23	14.4

and the computations in the routing procedure are summarized in the following table:

(1) Time (min)	(2) I (m^3/s)	(3) $2S/\Delta t - O$ (m^3/s)	(4) $2S/\Delta t + O$ (m^3/s)	(5) O (m^3/s)
0	0	0	0	0
30	2.4	1.04	2.4	0.68
60	5.6	0.52	9.04	4.26
90	3.4	0.46	9.52	4.53
120	2.8	0.78	6.66	2.94
150	2.4	0.84	5.98	2.57
180	2.2	0.88	5.44	2.28
210	1.8	0.94	4.88	1.97
240	1.5	0.98	4.24	1.63
270	1.2	1.00	3.68	1.34
300	1.0	1.02	3.20	1.09
330	0.56	1.04	2.58	0.77
360	0.34	0.98	1.94	0.48
390	0	0.86	1.32	0.23
420	0	0.78	0.86	0.04
450	0	0.78	0.78	0.00
480	0	0.78	0.78	0.00

The sequence of computations begins with tabulating the inflow hydrograph in columns 1 and 2, and the first row of the table ($t = 0$ min) is given by the initial conditions. At $t = 30$ min, $I_1 = 0$ m^3/s, $I_2 = 2.4$ m^3/s, $2S_1/\Delta t - O_1 = 0$ m^3/s, and Equation 5.199 gives

$$\frac{2S_2}{\Delta t} + O_2 = (I_1 + I_2) + \left(\frac{2S_1}{\Delta t} - O_1\right) = 0 + 2.4 + 0 = 2.4 \text{ m}^3/\text{s}$$

This value of $2S/\Delta t + O$ is written in column 4 (at $t = 30$ min), the corresponding outflow, O, from the detention basin is interpolated from the reservoir properties as 0.68 m^3/s, and this value is written in column 5 (at $t = 30$ min), and the corresponding value of $2S/\Delta t - O$ is given by

$$\frac{2S}{\Delta t} - O = \left(\frac{2S}{\Delta t} + O\right) - 2O = 2.4 - 2(0.68) = 1.04 \text{ m}^3/\text{s}$$

which is written in column 3 (at $t = 30$ min). The variables at $t = 30$ min are now all known, and form a new set of initial conditions to repeat the computation procedure for subsequent time steps.

The outflow hydrograph, given in column 5, indicates a peak discharge of 4.53 m³/s from the detention basin. This is a reduction of 19% from the peak catchment runoff ($=$ basin inflow) of 5.6 m³/s.

It is interesting to note that the peak outflow from the reservoir must occur at the intersection of the inflow and outflow hydrographs, since before this point inflows exceed outflows and the water level in the reservoir must be increasing; while beyond this point outflows exceed inflows and the water level in the reservoir and outflow rate must be decreasing.

5.5.1.2 Muskingum method

The Muskingum method for flood routing was originally developed by McCarthy (1938) for flood-control studies in the Muskingum River basin in Ohio. It is primarily used for routing flows in drainage channels, including rivers and streams, and is the most widely used method of hydrologic stream-channel routing (Ponce, 1989). The Muskingum method approximates the storage volume in a channel by a combination of *prism storage* and *wedge storage*, as illustrated in Figure 5.43 for the case in which the inflow exceeds the outflow. The prism storage is a volume of constant cross section corresponding to uniform flow in a prismatic channel; the wedge storage is generated by the passage of the flow hydrograph. A *negative wedge* is produced when the outflow exceeds the inflow, and it occurs as the water level recedes in the channel.

Assuming that the flow area is directly proportional to the channel flow, the volume of prism storage can be expressed as KO, where K is the travel time through the reach and O is the flow through the prism. The wedge storage can therefore be approximated by $KX(I - O)$, where X is a weighting factor in the range $0 \leq X \leq 0.5$. For reservoir-type storage, $X = 0$; for a full wedge, $X = 0.5$. The total storage, S, between the inflow and outflow sections is therefore given by

$$S = KO + KX(I - O) \tag{5.203}$$

FIGURE 5.43: Muskingum storage approximation.

Source: Chow et al. (1988).

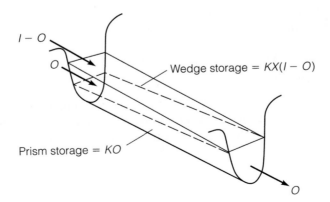

or

$$S = K[XI + (1 - X)O] \tag{5.204}$$

Applying Equation 5.204 at time increments of Δt, the storage, S, in the channel between the inflow and outflow sections at times $j\Delta t$ and $(j + 1)\Delta t$ can be written as

$$S_j = K[XI_j + (1 - X)O_j] \tag{5.205}$$

and

$$S_{j+1} = K[XI_{j+1} + (1 - X)O_{j+1}] \tag{5.206}$$

respectively, and the change in storage over Δt is therefore given by

$$S_{j+1} - S_j = K\{[XI_{j+1} + (1 - X)O_{j+1}] - [XI_j + (1 - X)O_j]\} \tag{5.207}$$

The discretized form of the continuity equation, Equation 5.198, can be written as

$$S_{j+1} - S_j = \frac{(I_j + I_{j+1})}{2} \Delta t - \frac{(O_j + O_{j+1})}{2} \Delta t \tag{5.208}$$

Combining Equations 5.207 and 5.208 yields the routing expression

$$\boxed{O_{j+1} = C_1 I_{j+1} + C_2 I_j + C_3 O_j} \tag{5.209}$$

where C_1, C_2, and C_3 are given by

$$C_1 = \frac{\Delta t - 2KX}{2K(1 - X) + \Delta t} \tag{5.210}$$

$$C_2 = \frac{\Delta t + 2KX}{2K(1 - X) + \Delta t} \tag{5.211}$$

$$C_3 = \frac{2K(1 - X) - \Delta t}{2K(1 - X) + \Delta t} \tag{5.212}$$

and it is apparent that

$$C_1 + C_2 + C_3 = 1 \tag{5.213}$$

The routing equation, Equation 5.209, is applied to a given inflow hydrograph, I_j $(j = 1, J)$, and initial outflow, O_1, to calculate the outflow hydrograph, O_j $(j = 2, J)$, at a downstream section. The constants in the routing equation, C_1, C_2, and C_3 are expressed in terms of the routing time step, Δt, and the channel parameters K and X. Ideally, subreaches selected for channel routing should be such that the travel time, K, through each subreach is equal to the time step, Δt. If this is not possible, then to avoid negative flows, Δt should be selected such that (Hjelmfelt, 1985)

$$2KX \leq \Delta t \leq 2K(1 - X) \tag{5.214}$$

where this condition ensures that the routing coefficients C_1 to C_3 are all positive. It is frequently recommended that Δt be assigned any convenient value between $K/3$ and K (ASCE, 1996; Viessman and Lewis, 2003).

EXAMPLE 5.38

The flow hydrograph at a channel section is given by:

Time (min)	Flow (m^3/s)
0	10.0
30	10.0
60	25.0
90	45.0
120	31.3
150	27.5
180	25.0
210	23.8
240	21.3
270	19.4
300	17.5
330	16.3
360	13.5
390	12.1
420	10.0
450	10.0
480	10.0

Use the Muskingum method to estimate the hydrograph 1200 m downstream from the channel section. Assume that $X = 0.2$ and $K = 40$ min.

Solution   In accordance with Equation 5.214, select Δt such that

$$2KX \le \Delta t \le 2K(1 - X) \Longrightarrow 2(40)(0.2) \le \Delta t \le 2(40)(1 - 0.2) \Longrightarrow 16\,\text{min} \le \Delta t \le 64\,\text{min}$$

and it is frequently recommended that

$$\frac{K}{3} \le \Delta t \le K$$

or

$$\frac{40}{3}\,\text{min} \le \Delta t \le 40\,\text{min} \Longrightarrow 13.3\,\text{min} \le \Delta t \le 40\,\text{min}$$

Taking $\Delta t = 30$ min, the Muskingum constants are given by Equations 5.210 to 5.212 as

$$C_1 = \frac{\Delta t - 2KX}{2K(1 - X) + \Delta t} = \frac{30 - 2(40)(0.2)}{2(40)(1 - 0.2) + 30} = 0.149$$

$$C_2 = \frac{\Delta t + 2KX}{2K(1 - X) + \Delta t} = \frac{30 + 2(40)(0.2)}{2(40)(1 - 0.2) + 30} = 0.489$$

$$C_3 = \frac{2K(1 - X) - \Delta t}{2K(1 - X) + \Delta t} = \frac{2(40)(1 - 0.2) - 30}{2(40)(1 - 0.2) + 30} = 0.362$$

These results can be verified by taking $C_1 + C_2 + C_3 = 0.149 + 0.489 + 0.362 = 1$. The Muskingum routing equation, Equation 5.209, is therefore given by

$$O_{j+1} = C_1 I_{j+1} + C_2 I_j + C_3 O_j = 0.149 I_{j+1} + 0.489 I_j + 0.362 O_j$$

This routing equation is applied repeatedly to the given inflow hydrograph, and the results are as follows:

(1) Time (min)	(2) I (m^3/s)	(3) O (m^3/s)
0	10.0	10.0
30	10.0	10.0
60	25.0	12.2
90	45.0	23.4
120	31.3	35.1
150	27.5	32.1
180	25.0	28.8
210	23.8	26.2
240	21.3	24.3
270	19.4	22.1
300	17.5	20.1
330	16.3	18.3
360	13.5	16.6
390	12.1	14.4
420	10.0	12.6
450	10.0	11.0
480	10.0	10.3
510	10.0	10.1
540	10.0	10.1
570	10.0	10.0
600	10.0	10.0

The computations begin with the inflow hydrograph in columns 1 and 2 and the initial outflow, O, at $t = 0$ min. At $t = 30$ min, $I_1 = 10$ m^3/s, $I_2 = 10$ m^3/s, $O_1 = 10$ m^3/s, and the Muskingum method (for j = 1) gives the outflow, O_2, at $t = 30$ min as

$$O_2 = 0.149I_2 + 0.489I_1 + 0.362O_1 = 0.149(10) + 0.489(10) + 0.362(10) = 10 \, m^3/s$$

This procedure is repeated for subsequent times, and the outflow hydrograph is shown in column 3.

The proportionality factor, K, is a measure of the time of travel through the channel reach, and X in natural streams is typically between 0 and 0.3, with a mean value near 0.2 (Chow et al., 1988). Experience has shown that for channels with mild slopes and over-bank flow, the parameter X will approach 0.0; while for steeper streams, with well-defined channels that do not have flows going out of the bank, X will be closer to 0.5. (Hydrologic Engineering Center, 2000). Values of X in excess of 0.5 produce hydrograph amplification, which is unrealistic. If measured inflow and outflow hydrographs are available for the channel reaches, then K and X can be estimated by selecting the values of these parameters that given the best fit between the measured and predicted outflow hydrographs.

Comparison between a computed and observed hydrograph is generally done using a goodness of fit index, which is typically called an *objective function*. Two of the most commonly used objective functions are the sum of absolute errors and the sum of squared residuals, defined by

$$\text{sum of absolute errors} = \sum_{j=1}^{N} |O_j^{\text{meas}} - O_j^{\text{obs}}| \qquad (5.215)$$

$$\text{sum of squared residuals} = \sum_{j=1}^{N} (O_j^{\text{meas}} - O_j^{\text{obs}})^2 \qquad (5.216)$$

where O_j^{meas} and O_j^{obs} are the measured and observed outflow hydrographs, respectively. The sum of absolute errors objective function gives equal weight to each difference between the measured and observed hydrograph, while the sum of squared residuals gives more weight to the larger differences.

Since K is defined as the travel time through the channel reach, K can be estimated by the observed time for the hydrograph peak to move through the channel reach. In the absence of measured data, K is usually taken as the estimated mean travel time between the inflow and outflow section and X is taken as 0.2. According to Chow et al. (1988), great accuracy in determining X may not be necessary, because the results of the Muskingum method are relatively insensitive to this parameter.

EXAMPLE 5.39

Measured flows at an upstream and downstream section of a drainage channel are as follows:

Time (h)	Upstream flow (m³/s)	Downstream flow (m³/s)
0.0	42.4	52.0
0.5	46.3	51.0
1.0	51.8	51.4
1.5	56.2	52.8
2.0	57.6	54.7
2.5	65.1	58.1
3.0	80.3	65.0
3.5	90.6	75.6
4.0	103.2	87.2
4.5	81.5	93.3
5.0	78.3	85.3
5.5	65.8	77.5
6.0	59.6	68.8
6.5	56.3	61.7
7.0	50.8	56.1
7.5	53.1	52.8
8.0	50.7	51.1

Estimate the Muskingum coefficients X and K that should be used in routing flows through this section of the drainage channel.

Solution   From the given data: $\Delta t = 0.5$ h, and the Muskingum routing equation is given by the combination of Equations 5.209 to 5.212 as

$$O_{j+1} = \left[\frac{\Delta t - 2KX}{2K(1 - X) + \Delta t} \right] I_{j+1} + \left[\frac{\Delta t + 2KX}{2K(1 - X) + \Delta t} \right] I_j$$

$$+ \left[\frac{2K(1 - X) - \Delta t}{2K(1 - X) + \Delta t} \right] O_j \tag{5.217}$$

For given values of Δt, X, and K, Equation 5.217 is used to calculate the outflow hydrograph corresponding to the given inflow hydrograph, and the sum of absolute errors between the calculated and measured hydrograph is given by Equation 5.215. For various assumed values of X and K, the sum of absolute errors is given in the following table:

(1) X	(2) K (h)	(3) Sum of absolute errors (m^6/s^2)
0.5	0.645	43.1
0.4	0.638	32.2
0.3	0.618	24.3
0.2	0.624	19.8
0.1	0.610	19.3
0.05	0.610	20.4
0.0	0.639	21.8
0.15	0.619	18.6
0.14	0.622	18.4
0.13	0.614	18.6

For each given value of X in column 1, the value of K that minimizes the sum of absolute errors is given in column 2, and the corresponding sum of absolute errors is given in column 3. Based on the tabulated results, the values of X and K that minimize the sum of absolute errors are $X = 0.14$ and $K = 0.622$ h.

The sum of squared residuals between a calculated and measured hydrograph is given by Equation 5.216, and the sum of squared residuals for various values of X and K is given in the following table:

(1) X	(2) K (h)	(3) Sum of squared residuals (m^6/s^2)
0.5	0.597	245.9
0.4	0.619	141.6
0.3	0.631	83.2
0.2	0.636	53.4
0.1	0.636	42.5
0.05	0.635	42.0
0.0	0.633	44.3
0.07	0.636	41.87
0.08	0.636	41.94
0.06	0.636	41.90
0.15	0.636	46.0

For each value of X in column 1, the value of K that minimizes the sum of squared residuals is given in column 2, and the corresponding sum of squared residuals is given in column 3. Based on the tabulated results, the values of X and K that minimize the sum of squared residuals are $X = 0.07$ and $K = 0.636$ h.

The optimal values of X and K based on the sum of absolute errors and the sum of squared residuals are different, and the calculated outflow hydrographs for each optimal (X, K) combination are compared with the measured hydrograph in Figure 5.44. Clearly, both (X, K) combinations give a good fit to the measured outflow hydrograph, with the value of X and K based on minimizing the sum of absolute errors $(0.14, 0.622$ h$)$ giving a better fit to the hydrograph peak flow.

In cases where flow measurements are not available, a method for estimating K and X was proposed by Cunge (1969), where K is estimated by the relation

$$K = \frac{\Delta x}{c} \tag{5.218}$$

where Δx is the distance between the inflow and outflow sections and c is the wave celerity, which can be taken as

$$c = \frac{5}{3}v \tag{5.219}$$

where v is the average velocity at the bankfull discharge. The coefficient $\frac{5}{3}$ in Equation 5.219 is derived from the Manning equation applied to wide rectangular channels (Dingman, 1984); in triangular channels the coefficient is $\frac{4}{3}$, and in wide parabolic channels it is $\frac{11}{9}$ (Viessman and Lewis, 2003). The value of X suggested by

FIGURE 5.44: Calibrated versus measured outflow hydrograph

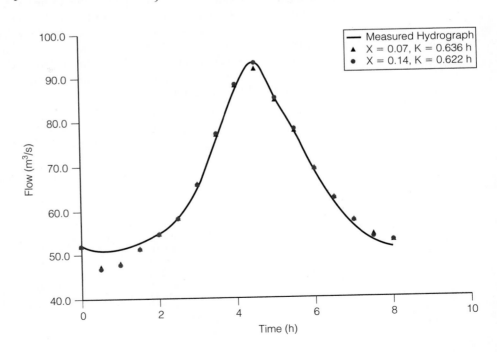

Cunge (1969) is

$$X = \frac{1}{2}\left(1 - \frac{q_0}{S_0 c \, \Delta x}\right) \tag{5.220}$$

where q_0 is the flow per unit top width, calculated at the average flow rate (midway between the base flow and the peak flow of the inflow hydrograph) and S_0 is the slope of the channel. In cases where the Muskingum method is applied using values of K and X estimated from Equations 5.218 to 5.220, the approach is referred to as the *Muskingum–Cunge method*. The accuracy of this method is derived from the fact that taking K and X as given by Equations 5.218 and 5.220 makes the Muskingum equation equal to the finite difference form of the diffusion-wave formulation of the momentum equation, which is in the class of more accurate hydraulic-routing methods. When using the Muskingum–Cunge method, the values of Δx and Δt should be selected to assure that the flood-wave details are properly routed. Nominally, the time to peak of inflow is broken into 5 or 10 time steps Δt. To give both temporal and spatial resolution, the total reach length, L, can be divided into several increments of length Δx, and outflow from each is treated as inflow to the next.

In cases where there are significant backwater effects and where the channel is either very steep or very flat, dynamic effects may be significant and hydraulic routing is preferred over hydrologic routing.

5.5.2 Hydraulic Routing

In hydraulic routing, the one-dimensional continuity and momentum equations are solved along the channel continuum. This approach is in contrast to hydrologic routing, which is based on the finite difference approximation to the continuity equation plus a second equation that relates the channel storage to the flowrate. The one-dimensional continuity and momentum equations in open channels, first presented by Barre de Saint-Venant (1871), are commonly called the Saint-Venant equations, and can be written as

$$\boxed{\frac{\partial Q}{\partial x} + \frac{\partial A}{\partial t} = 0 \qquad \text{(1-D continuity)}} \tag{5.221}$$

$$\boxed{\frac{1}{A}\frac{\partial Q}{\partial t} + \frac{1}{A}\frac{\partial}{\partial x}\left(\frac{Q^2}{A}\right) + g\frac{\partial y}{\partial x} - g(S_0 - S_f) = 0 \qquad \text{(1-D momentum)}} \tag{5.222}$$

where Q is the volumetric flowrate, x is the distance along the channel, t is time, A is the cross-sectional area, g is gravity, y is the depth of flow, S_0 is the slope of the channel, and S_f is the slope of the energy grade line, which can be estimated using the Manning equation. The assumptions inherent in the Saint-Venant equations are: (1) The flow is one-dimensional (i.e., the depth and velocity vary only in the longitudinal direction of the channel); (2) the flow is gradually varied along the channel (i.e., the vertical pressure distribution is hydrostatic); (3) the longitudinal axis of the channel is straight; (4) the bottom slope of the channel is small and the channel bed is fixed (i.e., the effects of scour and deposition are negligible); and (5) the friction coefficients for steady uniform flow are applicable. With the exception of a few simple cases, the Saint-Venant equations, given by Equations 5.221 and 5.222, cannot be solved analytically; they are usually solved with numerical models that use implicit or explicit finite difference algorithms (Amein and Fang, 1969) or the method of characteristics (Amein, 1966).

In many cases, some terms in the momentum equation are small compared with others in the same equation, and the momentum equation can be simplified by neglecting some of these terms. When the full momentum equation is used, this is commonly called the *dynamic model* and is given by Equation 5.222. The first two terms in the dynamic model (Equation 5.222) are the local and convective acceleration, also called the *inertial terms*, and neglecting these terms leads to the following *diffusion model*:

$$g\frac{\partial y}{\partial x} - g(S_0 - S_f) = 0 \qquad \text{(diffusion model)} \tag{5.223}$$

In cases where both the inertial terms and spatial variations in depth-of-flow are small relative to other terms, the momentum equation leads to the following *kinematic model*:

$$(S_0 - S_f) = 0 \qquad \text{(kinematic model)} \tag{5.224}$$

which includes only gravity and flow-resistance terms. The kinematic model is frequently used for overland-flow routing.

EXAMPLE 5.40

Write a simple finite difference model for solving the Saint-Venant equations.

Solution   The dependent variables in the Saint-Venant equations are the flowrate, Q, and the depth of flow, y; and the independent variables are the distance along the channel, x, and the time, t. In a finite difference approximation, the solution is calculated at discrete intervals, Δx, along the x-axis, and at discrete intervals, Δt, in time. Denoting the distance index by i, and the time index by j, the following convention for specifying variables at index locations is adopted:

$$Q_{i,j} = Q(i\Delta x, j\Delta t)$$

$$y_{i,j} = y(i\Delta x, j\Delta t)$$

$$A_{i,j} = A(y_{i,j})$$

where $A_{i,j}$ is the flow area corresponding to $y_{i,j}$.

The (explicit) finite difference approximation to the one-dimensional continuity equation (Equation 5.221) can be written as

$$\frac{Q_{i+1,j} - Q_{i,j}}{\Delta x} + \frac{A_{i,j+1} - A_{i,j}}{\Delta t} = 0$$

and the (explicit) finite-difference approximation to the one-dimensional momentum equation (Equation 5.222) can be written as

$$\frac{1}{A_{i,j}}\frac{Q_{i,j+1} - Q_{i,j}}{\Delta t} + \frac{1}{A_{i,j}}\frac{1}{\Delta x}\left(\frac{Q_{i+1,j}^2}{A_{i+1,j}} - \frac{Q_{i,j}^2}{A_{i,j}}\right) + g\frac{y_{i+1,j} - y_{i,j}}{\Delta x} - g(S_0 - S_{i,j}) = 0$$

where $S_{i,j}$ is the friction slope calculated using the Manning equation with $Q_{i,j}$ and $y_{i,j}$. These finite-difference equations, along with given initial and boundary conditions, can be used to solve for Q and y at the discrete points $i\Delta x$ and $j\Delta t$ along the x- and t-axes, respectively.

It should be noted that this numerical scheme requires some constraints on the choice of the discrete intervals Δx and Δt in order to produce accurate and stable solutions (Ames, 1977).

5.6 Water-Quality Models

Pollutants in stormwater originate from a variety of sources, including oil from motor vehicles, suspended sediment from construction activities, chemicals from lawns, and fecal droppings from animals. Urban areas are of special concern with respect to their impact on water quality. While urban areas might be small compared with agricultural areas, they produce high concentrations of pollutants per unit area, and they lack the natural buffering that often exists in agricultural areas (McCuen, 2005). Field studies have demonstrated that urban stormwater runoff is frequently a major source of pollution in receiving waters, particularly for bacteria and some heavy metals (Field et al., 2000). For this reason, pretreatment of stormwater runoff prior to discharge into receiving waters is usually required by local regulations.

The water quality in storm sewers and other drainage conduits is influenced by both runoff-related sources and nonstormwater sources. Nonstormwater sources of pollution in storm sewers include illicit connections, interactions with sewage systems, improper disposal, spills, malfunctioning septic tanks, and infiltration of contaminated water. The removal of nonstormwater sources of pollutants can result in a dramatic improvement in the quality of water discharged from storm sewers. Runoff-related sources of pollution in storm sewers originate from land surfaces and catch basins. On land surfaces, pollutants accumulate much more rapidly on impervious areas than they do on pervious areas, and the primary areas of accumulation are streets and gutters.

Initial planning estimates of pollutant levels in stormwater effluents can be estimated using either event-mean concentrations or regression equations. The most widely used regression equations for pollutant loads are incorporated in the USGS and EPA models.

5.6.1 Event-Mean Concentrations

The majority of stormwater pollutants fall into the following categories: solids, oxygen-demanding substrates, nutrients, and heavy metals. Among the toxic heavy metals detected in stormwater runoff, lead, zinc, and copper are the most abundant and detected most frequently. Typical concentrations of various pollutants in urban stormwater runoff are shown in Table 5.27. In many cases, a typical runoff concentration is assigned to a pollutant and this is called the *event-mean concentration* (EMC). Most pollutants in stormwater runoff have a strong affinity to suspended solids (SS), and the removal of SS often removes many other pollutants found in urban stormwater (Urbonas and Stahre, 1993). According to ASCE (1992), for general-purpose planning, the concentrations of pollutants in runoff from large residential and commercial areas can be assumed to be in the range of the concentrations shown in Table 5.27, with central business districts usually having the highest pollutant loadings per unit area.

TABLE 5.27: Typical Water Quality of Runoff from Residential and Commercial Areas

Category	Sources	Deterious effects	Measure	Typical concentrations
Sediment	Eroding rock, building sites, streets, lawns	Clogs waterways, smothers bottom-living aquatic organisms and increases turbidity	Total suspended solids (TSS)	180–548 mg/L
Oxygen-demanding substrates	Decaying organic matter	Consume oxygen in water, sometimes creating an oxygen deficit that leads to fish kills	Biochemical oxygen demand (BOD) Chemical oxygen demand (COD)	12–19 mg/L 82–178 mg/L
Nutrients	Nitrogen and phosphorus from landscape runoff, atmospheric deposition, and faulty septic tanks	Cause unwanted and uncontrolled growth of algae and aquatic weeds	Total phosphorus, as P (TP) Soluble phosphorus, as P (SP) Total Kjeldahl nitrogen, as N (TKN) Nitrite + nitrate, as N (NO2+NO3)	0.42–0.88 mg/L 0.15–0.28 mg/L 1.90–4.18 mg/L 0.86–2.2 mg/L
Heavy metals	Vehicles, highway materials, atmospheric deposition, industry	Can disrupt the reproduction of fish and shellfish and accumulate in fish tissues	Total copper (Cu) Total lead (Pb) Total zinc (Zn)	43–118 μg/L 182–443 μg/L 202–633 μg/L

Source: USEPA (1983).

Whalen and Cullum (1988) compared the quality of runoff from residential, commercial, light industrial, roadway, and mixed-urban land uses. Their results indicated that higher nutrient loads are generated by residential land uses, metal contamination is more widespread from commercial and roadway areas, and there are no discernible trends for suspended solids in runoff as a function of land use. Heavy metals found in urban runoff are 10–1000 times the concentration of metals found in sanitary sewage (Wanielista, 1978).

The quality of urban runoff also depends on the stage of development. During the initial phase of urbanization, the dominant source of pollution will be sediments from bare-soil areas at construction sites. During the intermediate phase of urbanization, sediments from construction sites will decline, but sediments from stream-bank erosion will increase because of the increased runoff rate and volume. During the mature phase of urbanization (when the stream channels have stabilized and there is limited new construction), the primary source of pollution will be from washoff of accumulated deposits on impervious surfaces.

EXAMPLE 5.41

Plans are being considered to rezone a 100-ha rural area for commercial and residential use. Annual rainfall in the area is 150 cm, the existing (rural) land is approximately 5% impervious, and the rezoned land is expected to be 40% impervious. Estimate the current and future amounts of suspended solids in the runoff.

Solution   Since the annual rainfall is 150 cm and the existing land is 5% impervious,

$$\text{existing annual runoff} = 0.05(150 \text{ cm}) = 7.5 \text{ cm}$$

and since the rezoned land is 40% impervious,

$$\text{expected annual runoff} = 0.40(150 \text{ cm}) = 60 \text{ cm}$$

According to Table 5.27, the EMC of suspended solids for residential and commercial areas is typically in the range of 180–548 mg/L with a midrange value of 364 mg/L. In the absence of any local EMC data, the EMC of suspended solids can be taken as 364 mg/L ($= 0.384 \text{ kg/m}^3$). Since the area of the catchment is 100 ha ($= 10^6 \text{ m}^2$), the total suspended solids in 7.5 cm ($= 0.075$ m) of annual runoff from the rural land is given by

$$\text{existing suspended solids} = (0.075 \text{ m})(10^6 \text{ m}^2)(0.364 \text{ kg/m}^3) = 27{,}300 \text{ kg}$$

and the suspended solids contained in 60 cm (0.60 m) of annual runoff from the rezoned/developed land is given by

$$\text{expected suspended solids} = (0.60 \text{ m})(10^6 \text{ m}^2)(0.364 \text{ kg/m}^3) = 218{,}400 \text{ kg}$$

These results indicate that an order-of-magnitude increase in suspended solids contained in the surface runoff is to be expected, increasing from approximately 27,300 kg/year to 218,400 kg/year. An adequate stormwater-management system should be put in place to remove some of the anticipated suspended sediment load prior the runoff being discharged from the area.

5.6.2 USGS Model

The U.S. Geological Survey analyzed the pollutant loads resulting from 2813 storms at 173 urban stations in 30 metropolitan areas and developed empirical equations to estimate the annual pollutant loads in terms of the rainfall and catchment characteristics (Driver and Tasker, 1988; 1990). Pollutant loads refer to the amount of pollutant per unit time, usually a year, and water managers are usually more interested in pollutant loads than runoff concentrations in individual storms (Black, 1996). The USGS regression equations for annual load have the form

$$Y = 0.454(N)(BCF)10^{[a+b\sqrt{(DA)}+c(IA)+d(MAR)+e(MJT)+f(X2)]} \tag{5.225}$$

where Y is the pollutant load (kg) for the pollutants listed in Table 5.28, N is the average number of storms in a year, BCF is a bias correction factor, DA is the total contributing drainage area (ha), IA is the impervious area as a percentage of the total contributing area (%), MAR is the mean annual rainfall (cm), MJT is the mean minimum January temperature (°C), and $X2$ is an indicator variable that is equal to 1.0 if commercial land use plus industrial land use exceeds 75% of the total contributing drainage area and is zero otherwise. The regression constants in Equation 5.225 depend on the type of pollutant and are given in Table 5.29. A storm is defined as a rainfall event with at least 1.3 mm (0.05 in.) of rain, and storms are separated by at least 6 consecutive hours of zero rainfall. In accordance with this definition, the mean number of storms per year in several cities within the United States is given in Table 5.30, where the cities have been rank-ordered by the number of storms. As in the case of most empirical relationships, application of Equation 5.225 is not recommended beyond the ranges of variables used in developing the equation, which are given in Table 5.31.

EXAMPLE 5.42

The pollutant load from a 100-ha planned development in Fort Lauderdale, Florida, is to be assessed. The development will be 40% impervious, with 50% commercial and industrial use. According to Winsberg (1990), there are typically 84 storms per year in Fort Lauderdale, the mean annual rainfall is 147 cm, and the mean minimum January

TABLE 5.28: Pollutants in USGS Formula

Y	Pollutant
COD	Chemical oxygen demand
SS	Total suspended solids
DS	Dissolved solids
TN	Total nitrogen
AN	Total ammonia plus organic nitrogen
TP	Total phosphorus
DP	Dissolved phosphorus
CU	Total recoverable copper
PB	Total recoverable lead
ZN	Total recoverable zinc

TABLE 5.29: Regression Constants for USGS Pollutant Load Equation

Y	a	b	c	d	e	f	BCF
COD	1.1174	0.1427	0.0051	—	—	—	1.298
SS	0.5926	0.0988	—	0.0104	−0.0535	—	1.521
DS	1.1025	0.1583	—	—	−0.0418	—	1.251
TN	−0.2433	0.1018	0.0061	—	—	−0.4442	1.345
AN	−1.4002	0.1002	0.0064	0.00890	−0.0378	−0.4345	1.277
TP	−2.0700	0.1294	—	0.00921	−0.0383	—	1.314
DP	−1.3661	0.0867	—	—	—	—	1.469
CU	−1.9336	0.1136	—	—	−0.0254	—	1.403
PB	−1.9679	0.1183	0.0070	0.00504	—	—	1.365
ZN	−1.6302	0.1267	0.0072	—	—	—	1.322

Source: Driver and Tasker (1990).

TABLE 5.30: Mean Number of Storms per Year

City	Mean number of storms
Miami, FL	100
Cleveland, OH	100
Bellevue, WA	98
Seattle, WA	97
Portland, OR	96
Knoxville, TN	92
Chicago, IL	87
Boston, MA	84
New York, NY	83
New Orleans, LA	83
Tampa, FL	79
Winston–Salem, NC	77
Philadelphia, PA	77
St. Louis, MO	74
Houston, TX	70
Denver, CO	57
Austin, TX	54
Dallas, TX	53
San Francisco, CA	46
Fresno, CA	41
Los Angeles, CA	32
Phoenix, AZ	31

Sources: Driver and Tasker (1988); McCuen (2005).

temperature is 14.4°C. Estimate the annual load of total phosphorus contained in the runoff.

Solution   From the given data: $N = 84$ storms, $BCF = 1.314$ (Table 5.29 for TP), $DA = 100$ ha, $IA = 40\%$, $MAR = 147$ cm, $MJT = 14.4°C$, and $X2 = 0$ (since

TABLE 5.31: Ranges of Variables Used in Developing USGS Equation

Y	DA (ha)	IA (%)	MAR (cm)	MJT °C
COD	4.9–183	4–100	21.3–157.5	−16.0–14.8
SS	4.9–183	4–100	21.3–125.4	−16.0–10.1
DS	5.2–117	19–99	26.0–95.5	−11.4–2.1
TN	4.9–215	4–100	30.0–157.5	−16.0–14.8
AN	4.9–183	4–100	21.3–157.5	−16.0–14.8
TP	4.9–215	4–100	21.3–157.5	−16.0–14.8
DP	5.2–183	5–99	21.3–117.3	−11.8–2.1
CU	3.6–215	6–99	21.3–157.5	−9.3–14.8
PB	4.9–215	4–100	21.3–157.5	−16.0–14.8
ZN	4.9–215	13–100	21.3–157.5	−11.4–14.8

Source: Driver and Tasker (1990).

commercial plus industrial use is less than 75%). These variables are within the ranges given in Table 5.31, and therefore the USGS regression equation, Equation 5.225, can be used. The regression constants taken from Table 5.29 (for TP) are $a = -2.0700$, $b = 0.1294$, $d = 0.00921$, $e = -0.0383$, and $c = f = 0$. Substituting data into Equation 5.225 gives

$$Y = 0.454(N)(BCF)10^{[a+b\sqrt{(DA)}+c(IA)+d(MAR)+e(MJT)+f(X2)]}$$

$$= 0.454(84)(1.314)10^{[-2.0700+0.1294\sqrt{(100)}+0.00921(147)-0.0383(14.4)]}$$

$$= 53 \text{ kg/year}$$

The annual load of total phosphorus contained in the runoff from the planned development is expected to be on the order of 53 kg.

5.6.3   EPA Model

The U.S. Environmental Protection Agency has also developed a set of empirical formulae that can be used to estimate the average annual pollutant loads in urban stormwater runoff (Heany et al., 1977). The empirical equation for urban areas having separate storm-sewer systems is given by

$$M_s = 0.0442\alpha Pfs \tag{5.226}$$

where M_s is the amount of pollutant (kg) generated per hectare of land per year, α is a pollutant loading factor given in Table 5.32 for various pollutants (BOD_5 = five-day biochemical oxygen demand, SS = suspended solids, VS = volatile solids, PO_4 = phosphate, and N = nitrogen), P is the precipitation (cm/year), f is a population density function, and s is a street-sweeping factor. The population density function, f, for residential areas is given by

$$f = 0.142 + 0.134D^{0.54} \tag{5.227}$$

TABLE 5.32: Pollutant Loading Factor, α

Land Use	BOD_5	SS	VS	PO_4	N
Residential	0.799	16.3	9.4	0.0336	0.131
Commercial	3.200	22.2	14.0	0.0757	0.296
Industrial	1.210	29.1	14.3	0.0705	0.277
Other	0.113	2.7	2.6	0.0099	0.060

Source: Heany et al. (1977).

where D is the population density in persons per hectare. For commercial and industrial areas, the population density function, f, is equal to 1.0; for other types of developed areas, such as parks, cemeteries, and schools, f is taken as equal to 0.142. The street-sweeping factor, s, depends on the sweeping interval, N_s (days); if $N_s > 20$ days, then $s = 1.0$, and if $N_s \leq 20$ days, then s is given by

$$\boxed{s = \frac{N_s}{20}} \tag{5.228}$$

Disposal of street-sweeping wastes may pose a problem because of possible high levels of lead, copper, zinc, and other wastes from automobile traffic (Dodson, 1998). The average annual pollutant concentration can be derived from the annual pollutant load by dividing the annual pollutant load by the annual runoff. Heany and colleagues (1977) suggested that the annual runoff, R (cm), can be estimated using the formula

$$\boxed{R = \left[0.15 + 0.75\left(\frac{I}{100}\right)\right] P - 3.004 d^{0.5957}} \tag{5.229}$$

where I is the imperviousness of the catchment (%), P is the annual rainfall (cm), and d is the depression storage (cm), which can be estimated using the relation

$$\boxed{d = 0.64 - 0.476\left(\frac{I}{100}\right)} \tag{5.230}$$

EXAMPLE 5.43

A 100-ha residential development has a population density of 15 persons per hectare, streets are swept every two weeks, and the area is 40% impervious. The average annual rainfall is 147 cm. Estimate the annual load of phosphate (PO_4) expected in the runoff.

Solution From the given data: $\alpha = 0.0336$ (Table 5.32: PO_4, Residential), $P = 147$ cm, $D = 15$ persons/ha, $N_s = 14$ days, Equation 5.227 gives

$$f = 0.142 + 0.134 D^{0.54} = 0.142 + 0.134(15)^{0.54} = 0.72$$

and Equation 5.228 gives

$$s = \frac{N_s}{20} = \frac{14}{20} = 0.7$$

Substituting these data into the EPA model, Equation 5.226, gives

$$M_s = 0.0442\alpha Pfs = 0.0442(0.0336)(147)(0.72)(0.7) = 0.11 \text{ kg/ha}$$

For a 100-ha development, the annual phosphate load is 100 ha $\times$ 0.11 kg/ha = 11 kg. The average concentration in the runoff is obtained by dividing the annual load ($= 11$ kg) by the annual runoff. Equation 5.230 gives the depression storage, d, as

$$d = 0.64 - 0.476 \left(\frac{I}{100} \right)$$

$$= 0.64 - 0.476 \left(\frac{40}{100} \right)$$

$$= 0.45 \text{ cm}$$

and Equation 5.229 gives the runoff, R, as

$$R = \left[0.15 + 0.75 \left(\frac{I}{100} \right) \right] P - 3.004 d^{0.5957}$$

$$= \left[0.15 + 0.75 \left(\frac{40}{100} \right) \right] (147) - 3.004(0.45)^{0.5957}$$

$$= 64 \text{ cm}$$

Since the catchment area is 100 ha = 10^6 m^2, then the volume, V, of annual runoff is 0.64×10^6 m^3 and the average concentration, c, of PO$_4$ is given by

$$c = \frac{11 \text{ kg}}{0.64 \times 10^6 \text{ m}^3} = 1.7 \times 10^{-5} \text{ kg/m}^3 = 0.017 \text{ mg/L}$$

The USGS and EPA regression equations are useful in estimating the annual pollutant loads on receiving waterbodies. In contrast to estimating annual pollutant loads, individual storm-event pollutant loads are typically estimated using either regression equations or process-based water-quality models. The regression equations empirically relate the event loads to the storm and catchment characteristics (e.g., Jewell and Adrian, 1981, 1982; Driver and Lystrom, 1986; Driver and Tasker, 1988; Driver, 1990). The process-based water-quality models typically simulate the dry-weather accumulation of pollutants on the catchment surface and the subsequent washoff caused by surface runoff (e.g., Heany et al., 1976; Geiger and Dorsch, 1980; Johanson et al., 1984; Hemain, 1986; Huber and Dickinson, 1988). A comparative evaluation of urban stormwater-quality models concluded that, once calibrated, both regression equations and process-based models can estimate event pollutant loads satisfactorily (Vaze and Chiew, 2003). Therefore, if only estimates of event loads are required, versus contaminant concentrations during the runoff events, regression models should be used because they are simpler and require less data compared to process-based models. A detailed review of approaches for controlling the quality of surface runoff from urban catchments can be found in Chin (2006).

5.7 Design of Stormwater-Management Systems

Stormwater-management systems are designed to control the quantity, quality, timing, and distribution of runoff resulting from storm events. Other objectives in the design of stormwater-management systems include: erosion control, reuse storage, and ground-water recharge. A typical urban stormwater-management system has two distinct subsystems: a minor and a major one. The minor system consists of storm sewers that route the design runoff to receiving waters, and it is typically designed to handle runoff events with return periods of 2 to 10 years. Typical return periods for various types of service areas are given in Table 5.4. The major system consists of the above-ground conveyance routes that transport stormwater from larger runoff events with return periods from 25 to 100 years. Major urban conveyance systems that are covered by the National Flood Insurance Program (in the United States) are typically designed for a runoff with a 100-year return period.

5.7.1 Minor System

Most minor stormwater-management systems are designed for urban environments, and the principal hydraulic elements of the minor system are shown in Figure 5.45. The (minor) stormwater-management system collects surface runoff via inlets, and the surface runoff is routed to a treatment unit and/or receiving waterbody, usually through underground pipes called *storm sewers*. In some cases, the surface runoff is discharged directly into a receiving body of water such as a drainage canal. In some older U.S. and European cities, storm and sanitary sewers are combined into a single system; these are called *combined-sewer systems*.

5.7.1.1 Storm sewers

Storm sewers are typically located a short distance behind the curb, or in the roadway near the curb. These sewers should be straight between manholes (where possible); where curves are necessary to conform to street layout, the radius of curvature should not be less than 30 m. There should be at least 0.9 m (3 ft) of cover over the crowns of the sewer pipes to prevent excessive loading on the pipe, and crossings with underground utilities should be avoided whenever possible, but, if necessary, should be at an angle greater than 45°. Manholes, also called *clean-out structures* (ASCE, 1992) or *access holes* (USFHWA, 1996), are placed along the sewer pipeline to provide convenient access for inspection, maintenance, and repair of storm-drainage systems; they are normally located at the junctions of sewers, changes in grade or alignment, and where there are changes in pipe size. Manhole spacings depend on the pipe sizes, and typical maximum spacings are given in Table 5.33. Drop manholes (see Figure 3.53(b)) are provided for sewers entering a manhole at an elevation of 0.6 m or more above the manhole invert.

  The rational method is commonly used to determine the peak flows to be handled by storm sewers. The flow calculations proceed from the most upstream pipe in the system and, with each new inlet (inflow), the pipe immediately downstream of the inlet is expected to carry the runoff from a storm of duration equal to the time of concentration of the contributing area. Two separate contributing areas must

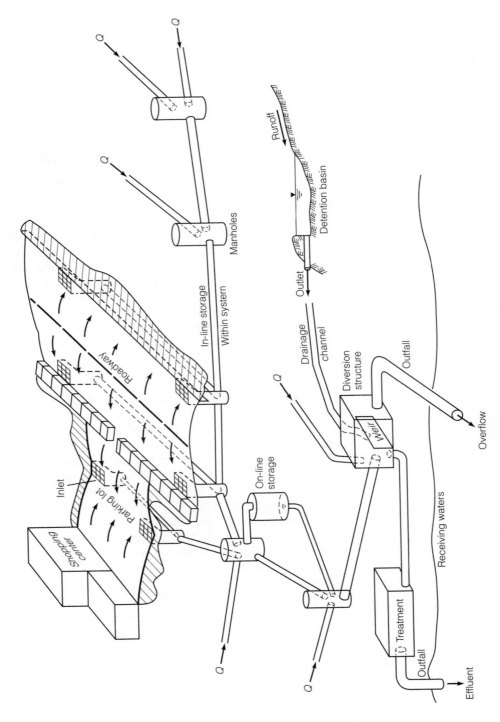

FIGURE 5.45: Principal hydraulic elements in (minor) stormwater management system.

Source: ASCE (1992).

TABLE 5.33: Typical Manhole Spacings

Pipe size	Maximum spacing
38 cm or less	122 m
46 cm to 91 cm	152 m
107 cm or greater	183 m

Source: Boulder County (1984).

be considered: (1) the entire contributing upstream area, and (2) the impervious upstream area directly connected to the inlets. Directly connected impervious area (DCIA) must be considered separately, since it typically has a considerably shorter time of concentration than the entire upstream contributing area, resulting in a higher design-rainfall intensity and possibly a higher peak runoff rate than the entire catchment. A minimum time of concentration such as 5 min is generally adopted to preclude unrealistically high design rainfall intensities. Most impervious areas are transportation related, mainly roads, driveways, and parking lots.

The minimization of directly connected impervious area is by far the most effective method of controlling the quality of surface runoff (ASCE, 1992) and, in many cases, DCIA is a key indicator of urbanization's effect on the quantity and quality of surface runoff (Lee and Heany, 2003; Jones et al., 2005). Typically, runoff from non-DCIA areas occurs only for larger storms, and the accurate estimation of DCIA is an important component of the cost-effective design and adequacy of roadway drainage systems (Aronica and Lanza, 2005a). Portions of roadways contributing to DCIA generally have curbs and gutters, while portions drained by roadside swales are usually not associated with DCIA. The procedure for calculating the design flows in storm sewers is illustrated in the following example.

EXAMPLE 5.44

Consider the two inlets and two pipes shown in Figure 5.46. Catchment A has an area of 1 ha and is 50% impervious; catchment B has an area of 2 ha and is 10% impervious.

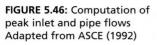

FIGURE 5.46: Computation of peak inlet and pipe flows
Adapted from ASCE (1992)

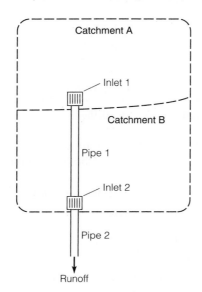

TABLE 5.34: Catchment Characteristics

Catchment	Surface	C	L (m)	n	S_0
A	Pervious	0.2	80	0.2	0.01
	Impervious	0.9	60	0.1	0.01
B	Pervious	0.2	140	0.2	0.01
	Impervious	0.9	65	0.1	0.01

All impervious areas are directly connected to the sewer inlets. The runoff coefficient, C; length of overland flow, L; roughness coefficient, n; and average slope, S_0, of the pervious and impervious surfaces in both catchments are given in Table 5.34. The design storm has a return period of 10 years, and the 10-year IDF curve can be approximated by

$$i = \frac{7620}{t + 36}$$

where i is the average rainfall intensity in mm/h and t is the duration of the storm in minutes. Calculate the peak flows to be handled by the inlets and pipes.

Solution   Using the given IDF curve, the effective rainfall rate, i_e, is given by the rational formula as

$$i_e = Ci = C\frac{7620}{t_c + 36} \tag{5.231}$$

where C is the runoff coefficient. The duration, t, of the design storm is taken to be equal to the time of concentration, t_c, given by Equation 5.94 as

$$t_c = 6.99 \frac{(nL)^{0.6}}{i_e^{0.4} S_0^{0.3}} \tag{5.232}$$

Simultaneous solution of Equations 5.231 and 5.232 using the catchment characteristics in Table 5.34 leads to the following times of concentration, t_c:

Catchment	Surface	t_c (min)
A	Pervious area	46
	Impervious area	11
B	Pervious	71
	Impervious	12

Consider now the flows at specific locations.

Inlet 1 and Pipe 1: When the entire catchment A is contributing, the time of concentration is 46 min (this is the time for both pervious and impervious areas to be fully contributing), the average rainfall rate, i, from the IDF curve is 92.9 mm/h ($= 2.58 \times 10^{-5}$ m/s), and the weighted-average runoff coefficient, $\overline{C}$, is given by

$$\overline{C} = 0.5(0.9) + 0.5(0.2) = 0.55$$

Since the area of the catchment is 1 ha ($= 10{,}000$ m^2), the peak runoff rate, Q_p, from the catchment is given by the rational formula as

$$Q_p = \overline{C}iA = (0.55)(2.58 \times 10^{-5})(10{,}000) = 0.142 \text{ m}^3/\text{s}$$

Considering only the impervious portion of the catchment, the time of concentration is 11 min, the average rainfall rate, i, from the IDF curve is 162 mm/h ($= 4.50 \times 10^{-5}$ m/s), the runoff coefficient, C, is 0.9, the contributing area is 0.5 ha ($= 5000$ m^2), and the peak runoff rate, Q_p, is given by

$$Q_p = CiA = (0.9)(4.50 \times 10^{-5})(5000) = 0.203 \text{ m}^3/\text{s}$$

The calculated peak runoff from the directly connected impervious area is greater than the calculated runoff from the entire area, and therefore the design flow to be handled by Inlet 1 and Pipe 1 is controlled by the directly connected impervious area and is equal to 0.203 m^3/s.

Inlet 2: When the entire catchment B is contributing, the time of concentration is 71 min, the average rainfall rate, i, from the IDF curve is 71.2 mm/h ($= 1.98 \times 10^{-5}$ m/s), and the weighted average runoff coefficient, $\overline{C}$, is given by

$$\overline{C} = 0.1(0.9) + 0.9(0.2) = 0.27$$

Since the area of the catchment is 2 ha ($= 20,000$ m^2), then the peak runoff rate, Q_p, from the catchment is given by the rational formula as

$$Q_p = \overline{C}iA = (0.27)(1.98 \times 10^{-5})(20,000) = 0.107 \text{ m}^3/\text{s}$$

Considering only the impervious portion of the catchment, the time of concentration is 12 min, the average rainfall rate, i, from the IDF curve is 159 mm/h ($= 4.41 \times 10^{-5}$ m/s), the runoff coefficient, C, is 0.9, the contributing area is 0.2 ha ($= 2000$ m^2), and the peak runoff rate, Q_p, is given by

$$Q_p = CiA = (0.9)(4.41 \times 10^{-5})(2000) = 0.079 \text{ m}^3/\text{s}$$

The calculated peak runoff from the directly connected impervious area is less than the calculated runoff from the entire area, and therefore the design flow to be handled by Inlet 2 is controlled by the entire catchment and is equal to 0.107 m^3/s.

Pipe 2: First consider the case where the entire tributary area of 3 ha ($= 30,000$ m^2) is contributing runoff to pipe 2. The time of concentration of catchment A is equal to 46 min plus the time of flow in pipe 1, which, in lieu of hydraulic calculations, can be taken as 2 min. Therefore, the time of concentration of catchment A is 48 min. The time of concentration of catchment B is 71 min, and therefore the time of concentration of the entire tributary area to pipe 2 (including both catchments A and B) is equal to 71 min. The average rainfall intensity corresponding to this duration (from the IDF curve) is 71.2 mm/h ($= 1.98 \times 10^{-5}$ m/s); the area-weighted runoff coefficient, $\overline{C}$, is given by

$$\overline{C} = \frac{1}{3}[(0.5 + 0.2)(0.9) + (0.5 + 1.8)(0.2)] = 0.36$$

and the rational formula gives the peak runoff rate, Q_p, as

$$Q_p = \overline{C}iA = (0.36)(1.98 \times 10^{-5})(30,000) = 0.214 \text{ m}^3/\text{s}$$

Considering only the impervious portions of catchments A and B, the contributing area is 0.7 ha ($= 7000$ m^2), the time of concentration is 13 min (equal to the time of concentration for Inlet 1 plus travel time of 2 min in pipe), the corresponding average rainfall intensity from the IDF curve is 156 mm/h ($= 4.32 \times 10^{-5}$), the runoff coefficient is 0.9, and the rational formula gives a peak runoff, Q_p, of

$$Q_p = CiA = (0.9)(4.32 \times 10^{-5})(7000) = 0.272 \text{ m}^3/\text{s}$$

Therefore, the peak runoff rate calculated by using the entire catchment is less than the peak runoff rate calculated by considering only the directly connected impervious portion of the tributary area. The design flow to be handled by pipe 2 is therefore controlled by the directly connected impervious area and is equal to 0.272 m³/s.

Most storm sewers are sized to flow full at the design discharge, although where ground elevations are sufficient, a limited surcharge above the pipe crown may be permitted (ASCE, 1992). To prevent deposition of suspended materials, the minimum slope of the sewer should produce a velocity of at least 60 to 90 cm/s (2 to 3 ft/s) when the sewer is flowing full; to prevent scouring, the velocity should be less than 3 to 4.5 m/s (10 to 15 ft/s). The appropriate flow equation for sizing storm sewers is the Darcy–Weisbach equation, but it is also common practice to use the Manning equation. Caution should be exercised when using the Manning equation, which is valid only for hydraulically rough (fully turbulent) flow and is appropriate only when the following condition is satisfied (French, 1985):

$$n^6 \sqrt{RS_0} \geq 9.6 \times 10^{-14} \tag{5.233}$$

where n is the Manning roughness coefficient, R is the hydraulic radius of the pipe (in meters), and S_0 is the slope of the pipe. Manning roughness coefficients recommended for closed-conduit flow are given in Table 5.35. In cases where the Darcy–Weisbach equation is used, the pipe roughness is commonly assumed to be independent of the pipe material (due to the accumulation of slime) and a value of 0.6 mm is typically

TABLE 5.35: Manning Coefficient in Closed Conduits

Material	n
Asbestos-cement pipe	0.011–0.015
Brick	0.013–0.017
Cast-iron pipe (cement lined and seal coated)	0.011–0.015
Concrete (monolithic):	
Smooth forms	0.012–0.014
Rough forms	0.015–0.017
Concrete pipe	0.011–0.015
Corrugated metal pipe (1.3-cm × 6.4-cm corrugations):	
Plain	0.022–0.026
Paved invert	0.018–0.022
Spun asphalt lined	0.011–0.015
Plastic pipe (smooth)	0.011–0.015
Vitrified clay:	
Pipes	0.011–0.015
Liner plates	0.013–0.017

Source: ASCE (1982).

recommended (Butler and Davies, 2000). In cases where the pipe material has a larger roughness height than 0.6 mm, the roughness height of the pipe material should be used.

In the hydraulic design of sewer pipes, the basic objective is to calculate the size and slope of the pipes that will carry the design flows at velocities that are within a specified range and with flow depths that are less than or equal to the diameter of the pipes. In most situations, it can be assumed that the flow is uniform and any losses other than pipe friction can be accounted for by assuming point losses at each manhole. In calculating the diameter, D, of storm sewers, the Manning equation can be put in the convenient form

$$D = \left(\frac{3.21Qn}{\sqrt{S_0}} \right)^{\frac{3}{8}} \qquad (5.234)$$

where Q is the design flowrate in m³/s and D is in meters. If the Darcy–Weisbach equation is used, the convenient form is

$$D = \left(\frac{0.811fQ^2}{gS_0} \right)^{\frac{1}{5}} \qquad (5.235)$$

where f is the friction factor. The actual size of the pipe to be used should be the next larger commercial size than calculated using either Equation 5.234 or 5.235. Concrete, asbestos-cement, and clay pipes are commonly used for diameters between 10 cm (4 in.) and 60 cm (25 in.), with reinforced or prestressed concrete pipes commonly used for diameters larger than 60 cm (Novotny et al., 1989). It is generally recommended to choose a pipe diameter larger than 30 cm (12 in.) to prevent clogging and facilitate maintenance (Gribbin, 1997).

Junction and manhole losses usually have a significant effect on flows in sewers. Junctions are locations where two or more pipes join together and enter another pipe or channel, and these transitions need to be smooth to avoid high head losses. Conditions that promote turbulent flow and associated high head losses include a large angle between the incoming pipes ($>60°$), a large vertical distance between the pipes (>15 cm between the two inverts), and the absence of a semicircular channel at the bottom of the manhole (ASCE, 1992). Manholes are generally placed at sewer junctions, as well as at other locations, to permit access to the entire pipeline system. The head loss, h_m, in manholes can be estimated using the equation

$$h_m = K_c \frac{V^2}{2g} \qquad (5.236)$$

where K_c is a head-loss coefficient and V is the average velocity in the inflow pipe. Head-loss coefficients for manholes with single inflow and outflow pipes aligned opposite to each other vary between 0.12 and 0.32, while the loss coefficients vary between 1.0 and 1.8 for inflow and outflow pipes at $90°$ to each other (ASCE, 1992). To maintain the specific energy of the flow, a minimum of 3 to 6 cm drop in the sewer invert at manholes is advisable. Under no conditions should the crown of the upstream pipe be lower than the crown of the downstream pipe. In cases where the pipe diameter increases, it is recommended that the crowns of the upstream and downstream pipes be aligned.

EXAMPLE 5.45

A concrete sewer pipe is to be laid parallel to the ground surface on a slope of 0.5% and is to be designed to carry 0.43 m³/s of stormwater runoff. Estimate the required pipe diameter using (a) the Manning equation, and (b) the Darcy–Weisbach equation. If service manholes are placed along the pipeline, estimate the head loss at each manhole.

Solution   From the given data: $S_0 = 0.005$, $Q = 0.43$ m³/s, and $n = 0.013$ (Table 5.35, average for concrete pipe).

(a) The Manning equation (Equation 5.234) gives

$$D = \left[\frac{3.21Qn}{\sqrt{S_0}} \right]^{\frac{3}{8}} = \left[\frac{3.21(0.43)(0.013)}{\sqrt{0.005}} \right]^{\frac{3}{8}} = 0.60 \text{ m}$$

Use the limitation given by Equation 5.233 to check whether the Manning equation is valid:

$$n^6\sqrt{RS_0} = (0.013)^6\sqrt{(0.6/4)(0.005)} = 1.3 \times 10^{-13} \geq 9.6 \times 10^{-14}$$

Therefore, the Manning equation is valid. Using a 60-cm pipe, the flow velocity, V, is given by

$$V = \frac{Q}{A} = \frac{0.43}{\frac{\pi}{4}(0.6)^2} = 1.52 \text{ m/s}$$

This velocity exceeds the minimum velocity to prevent sedimentation (0.60 to 0.90 m/s) and is less than the maximum velocity to prevent excess scour (3 to 4.5 m/s). According to the Manning equation, the pipe should have a diameter of 60 cm.

(b) The Darcy–Weisbach equation (Equation 5.235) gives

$$D = \left[\frac{0.811fQ^2}{gS_0} \right]^{\frac{1}{5}} = \left[\frac{0.811f(0.43)^2}{(9.81)(0.005)} \right]^{\frac{1}{5}} = 1.250f^{\frac{1}{5}}$$

which can be put in the form

$$f = 0.328D^5 \tag{5.237}$$

The friction factor, f, also depends on D via the Colebrook equation (Equation 2.35) which is given by

$$\frac{1}{\sqrt{f}} = -2\log\left(\frac{k_s/D}{3.7} + \frac{2.51}{Re\sqrt{f}} \right) \tag{5.238}$$

The equivalent sand roughness, k_s, of concrete is in the range 0.3–3.0 mm (Table 2.1) and can be taken as $k_s = 1.7$ mm. Assuming that the temperature of the water is 20°C, the kinematic viscosity, ν, is equal to 1.00×10^{-6} m/s^2, and the Reynolds number, Re, is given by

$$\text{Re} = \frac{VD}{\nu} = \frac{4Q}{\pi D\nu} = \frac{4(0.43)}{\pi D(1.00 \times 10^{-6})} = \frac{5.48 \times 10^5}{D} \tag{5.239}$$

Combining Equations 5.237 to 5.239 gives

$$\frac{1}{\sqrt{0.328D^5}} = -2\log\left(\frac{0.0017/D}{3.7} + \frac{2.51}{\frac{5.48\times10^5}{D}\sqrt{0.328D^5}}\right)$$

which simplifies to

$$1.75D^{-5/2} = -2\log\left(4.59 \times 10^{-4}D^{-1} + 8.00 \times 10^{-6}D^{-3/2}\right)$$

and yields

$$D = 0.60 \text{ m} = 60 \text{ cm}$$

Therefore, the Darcy–Weisbach equation requires that the sewer pipe be at least 60 cm in diameter.

The service manholes placed along the pipe will each cause a head loss, h_m, where

$$h_m = K_c \frac{V^2}{2g}$$

For inflow and outflow pipes aligned opposite to each other, K_c is between 0.12 and 0.32 and can be assigned an average value of $K_c = 0.22$. Since $V = 1.52$ m/s, the head loss, h_m, is therefore given by

$$h_m = 0.22 \frac{(1.52)^2}{2(9.81)} = 0.026 \text{ m} = 0.26 \text{ cm}$$

This head loss must be accounted for in computing the energy grade line in the sewer system.

5.7.1.2 Street gutters and inlets

Surface runoff from urban streets are typically routed to sewer pipes through street gutters and inlets. To facilitate drainage, urban roadways are designed with both cross-slopes and longitudinal slopes. The cross slope directs the surface runoff to the sides of the roadway, where the pavement intersects the curb and forms an open channel called a *gutter*. Longitudinal slopes direct the flow in the gutters to stormwater *inlets* that direct the flow into sewer pipes. Typical cross slopes on urban roadways are in the range of 1.5% to 6%, depending on the type of pavement surface (Easa, 1995), and typical longitudinal slopes are in the range of 0.5% to 5%, depending on the topography. The spacing between stormwater inlets depends on several criteria, but it is usually controlled by the allowable water spread toward the crown of the street.

FIGURE 5.47: Triangular curb gutter

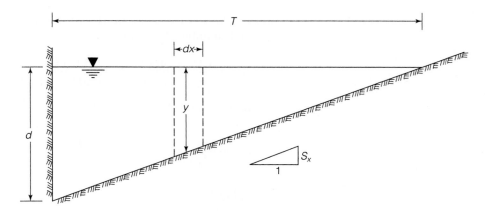

Flow in a triangular curb gutter is illustrated in Figure 5.47. The incremental flow, dQ, through any gutter width dx can be estimated using the relation

$$dQ = Vy\,dx \tag{5.240}$$

where V is the average velocity and y is the depth of flow within an elemental flow area of width dx. Using the Manning equation, the average velocity within the elemental flow area can be estimated by

$$V = \frac{1}{n}\,y^{2/3}S_0^{1/2} \tag{5.241}$$

where n is the Manning roughness coefficient, and S_0 is the longitudinal slope of the gutter. If S_x is the cross slope of the gutter, then

$$\frac{dy}{dx} = S_x \tag{5.242}$$

Combining Equations 5.240 to 5.242 gives

$$dQ = \frac{1}{n}\,y^{5/3}\,\frac{S_0^{1/2}}{S_x}\,dy \tag{5.243}$$

Since the flow depth, y, varies from 0 to d across the gutter, the total flow, Q, in the gutter is given by

$$Q = \int_0^d \frac{1}{n}\,y^{5/3}\,\frac{S_0^{1/2}}{S_x}\,dy \tag{5.244}$$

which yields (ASCE, 1992; USFHWA, 1996)

$$\boxed{Q = 0.375\left(\frac{1}{nS_x}\right)d^{8/3}S_0^{1/2}} \tag{5.245}$$

Equation 5.245 is viewed as preferable to the direct application of the Manning equation to the gutter, since the hydraulic radius does not adequately describe the gutter cross section, particularly when the top-width T exceeds 40 times the depth at the curb (ASCE, 1992). A limitation of Equation 5.245 is that it neglects the resistance

TABLE 5.36: Typical Manning n Values for Street and Pavement Gutters

Type of gutter or pavement	Manning n
Concrete gutter, troweled finish	0.012
Asphalt pavement:	
Smooth texture	0.013
Rough texture	0.016
Concrete gutter with asphalt pavement:	
Smooth	0.013
Rough	0.015
Concrete pavement:	
Float finish	0.014
Broom finish	0.016

Source: USFHWA (1984b).

of the curb face, which is negligible if the cross slope is 10% or less (Johnson and Chang, 1984). The depth and top-width of gutter flow are related by

$$d = TS_x \tag{5.246}$$

Typical Manning n values for street and pavement gutters are given in Table 5.36 (USFHWA, 1984b), and a Manning n of 0.016 is recommended for most applications (Guo, 2000). To facilitate proper drainage, it is recommended that the gutter grade exceed 0.4% and the street cross slope exceed 2% (ASCE, 1992). A gutter cross slope may be the same as that of the pavement or may be designed to be steeper. The gutter, together with a curb, should be at least 15 cm (6 in.) deep and 60 cm (2 ft) wide, with the deepest portion adjacent to the curb. The maximum allowable width of street flooding depends on the type of street and is usually specified separately for minor and major design-runoff events. Typical regulatory requirements for allowable pavement encroachment are given in Table 5.37, and these typically correspond to rainfall events with a return period of 10 years. Allowable pavement encroachment and design return period are the bases for computing the street drainage capacity using the modified Manning equation (Equation 5.245). The flowrate corresponding to the allowable pavement encroachment is commonly called the *street hydraulic conveyance capacity*.

EXAMPLE 5.46

A four-lane collector roadway is to be constructed with 3.66-m (12-ft) lanes, a cross slope of 2%, a longitudinal slope of 0.5%, and pavement made of rough asphalt. The (minor) roadway-drainage system consists of curbs and gutters and is to be designed for a rainfall intensity of 150 mm/h. Determine the spacing of the inlets.

Solution For a collector street, Table 5.37 indicates that at least one lane must be free of water. However, since the roadway has four lanes, the drainage system must necessarily be designed to leave two lanes free of water (one on each side of the crown). Since each lane is 3.66 m wide, the allowable top-width, T, is 3.66 m, with $n = 0.016$ (rough asphalt), $S_x = 0.02$, and $S_0 = 0.005$. The maximum allowable depth of flow, d, at the curb is given by

$$d = TS_x = (3.66)(0.02) = 0.0732 \text{ m} = 7.32 \text{ cm}$$

TABLE 5.37: Typical Regulatory Requirements for Pavement Encroachment

Street type	Minor storm runoff	Major storm runoff
Local*	No curb overtopping;[†] flow may spread to crown of street	Residential dwellings, public, commercial and industrial buildings shall not be inundated at the ground line, unless buildings are floodproofed. The depth of water over the gutter flow line shall not exceed an amount specified by local regulation, often 30 cm (12 in.).
Collector[‡]	No curb overtopping;[†] flow spread must leave at least one lane free of water	Same as for local streets
Arterial[§]	No curb overtopping;[†] flow spread must leave at least one lane free of water in each direction	Residential dwellings, public, commercial, and industrial buildings shall not be inundated at the ground line, unless buildings are floodproofed. Depth of water at the street crown shall not exceed 15 cm (6 in.) to allow operation of emergency vehicles. The depth of water over the gutter flow line shall not exceed a locally prescribed amount.
Freeway**	No encroachment allowed on any traffic lanes	Same as for arterial streets

Source: Denver Regional Urban Storm Drainage Criteria Manual (1984).
*A local street is a minor traffic carrier within a neighborhood characterized by one or two moving lanes and parking along curbs. Traffic control may be by stop or yield signs.
[†]Where no curb exists, encroachment onto adjacent property should not be permitted.
[‡]A collector street collects and distributes traffic between arterial and local streets. There may be two or four moving traffic lanes and parking may be allowed adjacent to curbs.
[§]An arterial street permits rapid and relatively unimpeded traffic movement. There may be four to six lanes of traffic, and parking adjacent to curbs may be prohibited. The arterial traffic normally has the right-of-way over collector streets. An arterial street will often include a median strip with traffic channelization and signals at numerous intersections.
**Freeways permit rapid and unimpeded movement of traffic through and around a city. Access is normally controlled by interchanges at major arterial streets. There may be eight or more traffic lanes, frequently separated by a median strip.

and the maximum allowable flowrate, Q, in the gutter is given by the Manning equation (Equation 5.245) as

$$Q = 0.375 \left[\frac{1}{nS_x} \right] d^{8/3} S_0^{1/2} = 0.375 \left[\frac{1}{(0.016)(0.02)} \right] (0.0732)^{8/3} (0.005)^{1/2}$$

$$= 0.0777 \text{ m}^3/\text{s}$$

Since the design-rainfall intensity is 150 mm/h = 4.17×10^{-5} m/s, the contributing area, A, required to produce a runoff of 0.0777 m³/s is given by

$$A = \frac{0.0777}{4.17 \times 10^{-5}} = 1863 \text{ m}^2$$

The roadway has two lanes contributing runoff to each gutter. Therefore, the width of the contributing area is $2 \times 3.66 = 7.32$ m, and the length, L, of roadway required for a contributing area of 1863 m^2 is given by

$$L = \frac{1863}{7.32} = 255 \text{ m}$$

The required spacing of inlets is therefore 255 m. This is a rather large spacing, and assumes that 100% of the gutter flow is intercepted by each inlet. In reality, the inlets will probably not be designed to intercept all the gutter flow, resulting in some carryover and required spacings less than 255 m. The requirement that inlets be placed at immediately upgrade of intersections, at pedestrian cross walks, upstream of bridges, and at vertical sag locations may also affect the actual spacing of inlets in the gutter.

There are a number of locations where inlets are required, regardless of the contributing drainage area. These locations are called *geometric controls*, since they are determined by the geometry of the roadway system, and must generally be marked on drainage plans before any computations of runoff, roadway encroachment, and inlet capacity. Examples of geometric-control locations are (Young and Stein, 1999):

- At all sag points in the gutter grade
- Immediately upstream of median breaks, entrance/exit ramps, crosswalks, and street intersections (i.e., at any location where water could flow onto the roadway)
- Immediately upgrade of bridges (to prevent pavement drainage from flowing onto bridge decks)
- Immediately downstream of bridges (to intercept bridge deck drainage)
- Immediately upgrade of cross-slope reversals
- At the end of channels in cut sections
- On side streets immediately upgrade from intersections
- Behind curbs, shoulders, or sidewalks to drain low areas

In addition to these areas, stormwater from cut slopes and adjacent regions draining toward the roadway should be intercepted before it reaches the pavement. In practice, inlet locations are frequently dictated by street-geometrical conditions rather than spread-of-water computations.

Stormwater inlets can take many forms but are usually classified as either curb inlets, grate inlets, combination inlets, or slotted drains. The various inlet types are illustrated in Figure 5.48. Municipalities sometimes specify the manufacturer and specific inlet types that are acceptable within their jurisdiction. Since it is usually uneconomical to make stormwater inlets as wide as the design spread on roadways, inlets on continuous grades typically intercept only a portion of the gutter flow, and the fraction of flow intercepted under design conditions is called the *inlet efficiency*, E, and is given by

$$E = \frac{Q_i}{Q} \tag{5.247}$$

FIGURE 5.48: Stormwater inlets.
Source: USFHWA (1984a).

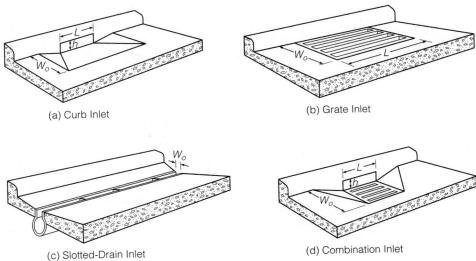

(a) Curb Inlet

(b) Grate Inlet

(c) Slotted-Drain Inlet

(d) Combination Inlet

where Q_i is the intercepted flow, and Q is the total gutter flow. The flow that is not intercepted by an inlet is called the *bypass flow* or *carryover flow*. The efficiency of inlets in passing debris is critical in sag locations, since all runoff entering the sag must be passed through the inlet. For this reason, grate inlets are not recommended in sag locations, because of their tendency to become clogged, and curb-opening or combination inlets are recommended for use in these locations.

Curb inlets. Curb inlets are vertical openings in the curb covered by a top slab. The capacity of a curb inlet depends on the size of the inlet, the amount of debris blockage, whether the inlet is depressed, and whether deflectors are used. Details on the performance of curb inlets can be found in regulatory manuals such as the Federal Highway Administration Urban Drainage Design Manual (USFHWA, 1996) and the Denver Regional Urban Storm Drainage Criteria Manual (1984). Curb-opening heights vary in dimension, with typical opening heights in the range of 10 cm to 15 cm (4 in. to 6 in.). Opening heights should not exceed 15 cm (6 in.) to reduce the risk of children entering the inlet. Curb inlets are most effective on flatter slopes (less than 3%), in sags, and with gutter flows that carry significant amounts of floating debris (USFHWA, 1996).

On continuous grades, the length, L_T, of a curb inlet required for total interception of gutter flow on a pavement section with a uniform cross slope is given by (USFHWA, 1996)

$$L_T = 0.817 Q^{0.42} S_0^{0.3} \left(\frac{1}{n S_x} \right)^{0.6} \tag{5.248}$$

where L_T is in meters, Q is the gutter flow in m^3/s, S_0 is the longitudinal slope of the gutter, n is the Manning roughness coefficient, and S_x is the cross slope. For curb inlets shorter than L_T, the ratio, R, of the intercepted flow to the gutter flow is given

FIGURE 5.49: Depressed curb inlet

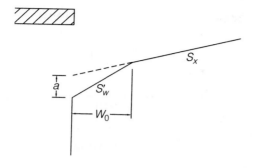

by (USFHWA, 1996)

$$R = 1 - \left(1 - \frac{L}{L_T}\right)^{1.8}$$

(5.249)

where L is the length of the curb inlet in the same units as L_T. In cases of a depressed curb inlet, shown in Figure 5.49, Equations 5.248 and 5.249 can still be used to calculate the inlet capacity, with the cross slope, S_x, replaced by the *equivalent cross slope*, S_e, given by

$$S_e = S_x + S'_w R_w$$

(5.250)

where S_x is the cross slope of the pavement shown in Figure 5.49, and S'_w is the cross slope of the gutter measured from the cross slope of the pavement, given by

$$S'_w = \frac{a}{W_0}$$

(5.251)

where a is the gutter depression, and W_0 is the width of the depressed gutter, and R_w is the ratio of the frontal flow to the depressed section to the total gutter flow given by (USFHWA, 1996)

$$R_w = 1 - \left(1 - \frac{W_0}{T}\right)^{8/3}$$

(5.252)

EXAMPLE 5.47

A smooth-asphalt roadway has a cross slope of 3%, a longitudinal slope of 2%, a curb height of 15 cm, and a 90-cm wide concrete gutter. If the gutter flow is 0.08 m³/s, determine the length of a 12-cm high curb inlet that is required to remove all the water from the gutter. Consider the cases where: (a) there is an inlet depression of 25 mm; and (b) there is no inlet depression. What inlet length would remove 80% of the gutter flow?

Solution   From the given data: $Q = 0.08 \text{ m}^3/\text{s}$, $S_x = 3\% = 0.03$, and $S_0 = 2\% = 0.02$. Assuming that a significant portion of the gutter flow extends onto the smooth-asphalt pavement, Table 5.36 gives $n = 0.013$. Using the Manning equation

(Equation 5.245) to find the depth, d, at the curb gives

$$Q = 0.375 \left(\frac{1}{nS_x} \right) d^{8/3} S_0^{1/2}$$

$$0.08 = 0.375 \left(\frac{1}{0.013 \times 0.03} \right) d^{8/3} (0.02)^{1/2}$$

and solving for d leads to

$$d = 0.061 \text{ m} = 6.1 \text{ cm}$$

This flow depth is less than the curb height of 15 cm, and therefore the flow is constrained within the roadway. The top-width, T, of the gutter flow, is given by

$$T = \frac{d}{S_x} = \frac{0.061}{0.03} = 2.03 \text{ m}$$

The spread of 2.03 m is much larger than the gutter width of 90 cm, verifying the assumption that a significant portion of the gutter flow extends onto the smooth-asphalt pavement, and hence justifying the assumed n value of 0.013. For an inlet width, W_0, of 0.90 m, the ratio, R_w, of frontal flow to the total gutter flow is given by Equation 5.252 as

$$R_w = 1 - \left(1 - \frac{W_0}{T} \right)^{8/3}$$

$$= 1 - \left(1 - \frac{0.90}{2.03} \right)^{8/3} = 0.79$$

(a) For a gutter depression, a, of 25 mm (= 0.025 m), the cross slope, S_w', of the gutter relative to the pavement slope is given by Equation 5.251 as

$$S_w' = \frac{a}{W_0} = \frac{0.025}{0.90} = 0.028$$

Hence the equivalent cross slope, S_e, of the depressed inlet is given by Equation 5.250 as

$$S_e = S_x + S_w' R_w$$

$$= 0.03 + (0.028)(0.79) = 0.052$$

The length, L_T, of the curb inlet required to intercept all of the gutter flow is given by Equation 5.248, with S_e replacing S_x, as

$$L_T = 0.817 Q^{0.42} S_0^{0.3} \left(\frac{1}{nS_e} \right)^{0.6}$$

$$= 0.817 (0.08)^{0.42} (0.02)^{0.3} \left(\frac{1}{0.013 \times 0.052} \right)^{0.6}$$

$$= 6.98 \text{ m}$$

(b) If there is no inlet depression, then $S_e = S_x = 0.03$ and the length, L_T, of the curb inlet required to intercept all of the gutter flow is given by Equation 5.248 as

$$L_T = 0.817 Q^{0.42} S_0^{0.3} \left(\frac{1}{nS_x} \right)^{0.6}$$

$$= 0.817(0.08)^{0.42}(0.02)^{0.3} \left(\frac{1}{0.013 \times 0.03} \right)^{0.6}$$

$$= 9.71 \text{ m}$$

It is interesting to note that the required length of the curb inlet without a depression ($= 9.71$ m) is 39% longer than the required length with a depressed inlet ($= 6.98$ m).

The fraction, R, of gutter flow removed by an inlet of length L is given by Equation 5.249. When $R = 80\% = 0.80$, Equation 5.249 gives

$$0.80 = 1 - \left(1 - \frac{L}{L_T} \right)^{1.8}$$

Solving for L/L_T gives

$$\frac{L}{L_T} = 0.59$$

Therefore, the length of the curb inlet required to intercept 80% of the flow is $0.59(6.98 \text{ m}) = 4.12$ m for a depressed inlet, and $0.59(9.71 \text{ m}) = 5.73$ m for an undepressed inlet.

In sag vertical-curve locations, curb inlets act as weirs up to a depth equal to the opening height, and the inlet operates as an orifice when the water depth is greater than 1.4 times the opening height. Between these depths, transition between weir and orifice flow occurs. The weir flow equation gives the flowrate, Q_i (m^3/s), into the curb inlet as (USFHWA, 1996)

$$\boxed{Q_i = 1.25(L + 1.8W_0)d^{1.5}} \tag{5.253}$$

where L is the length of the curb opening (m), W_0 is the width of the inlet depression (m), and d is the depth of flow at the curb upstream of the gutter depression. The weir equation, Equation 5.253, is valid when

$$d \leq h + a \tag{5.254}$$

where h is the height of the curb opening, and a is the depth of the gutter depression. Without a depressed gutter, the inflow to a curb inlet is given by

$$\boxed{Q_i = 1.60Ld^{1.5}, \quad d \leq h} \tag{5.255}$$

For curb-opening lengths greater than 3.6 m, Equation 5.255 for nondepressed inlets gives inlet capacities greater than those calculated using Equation 5.253 for depressed inlets. Since depressed inlets will perform at least as well as nondepressed inlets of the same length, Equation 5.255 should be used for all curb-opening inlets having lengths greater than 3.6 m (USFHWA, 1996). When the flow depth, d, exceeds 1.4 times the opening height, h, the inflow to the curb inlet is given by the orifice equation

$$Q_i = 0.67A \left[2g \left(d - \frac{h}{2} \right) \right]^{\frac{1}{2}} \tag{5.256}$$

where A is the area of the curb opening ($= hL$) and d is the flow depth in the depressed gutter.

EXAMPLE 5.48

A roadway has a flow depth of 8 cm in a 60-cm wide gutter, and a corresponding flowrate of 0.08 m³/s. If the gutter flow drains into a curb inlet in a sag location, determine the length of a 15-cm (6-in.) high inlet that is required to remove all the water from the gutter. Consider the cases where (a) the inlet is depressed, and (b) the inlet is not depressed.

Solution

(a) Since the flow depth (8 cm) is less than the height of the inlet (15 cm), the curb inlet acts as a weir. In this case, $Q_i = 0.08$ m³/s, $W_0 = 0.6$ m, $d = 0.08$ m, and the weir equation (Equation 5.253) can be put in the form

$$L = \frac{Q_i}{1.25d^{1.5}} - 1.8W_0$$

The required weir length, L, is therefore given by

$$L = \frac{0.08}{1.25(0.08)^{1.5}} - 1.8(0.6) = 1.75 \text{ m}$$

With an inlet depression, the length of the curb opening should be at least 1.75 m.

(b) In the case of no inlet depression, the inlet equation (Equation 5.255) can be put in the form

$$L = \frac{Q_i}{1.60d^{1.5}}$$

and the required weir length is given by

$$L = \frac{0.08}{1.60(0.08)^{1.5}} = 2.21 \text{ m}$$

The presence of an inlet depression reduces the required curb length from 2.21 m to 1.75 m, a reduction of 21%.

Grate inlets. Grate inlets consist of an opening in the gutter covered by one or more grates. The main advantage of grate inlets are that they are installed along the roadway where water is flowing, and their main disadvantage is interference with bicycles and the tendency for debris blockage. If clogging due to debris is not expected, then a grate or grate/curb combination type inlet will provide more capacity than a curb

TABLE 5.38: Grate Inlets

Name	Description
P–50	Parallel-bar grate with bar spacing 48 mm on center.
P–50 × 100	Parallel-bar grate with bar spacing 48 mm on center and 10-mm diameter lateral rods spaced at 102 mm on center.
P–30	Parallel-bar grate with 29-mm-on-center bar spacing.
Curved vane	Curved-vane grate with 83-mm longitudinal bar and 108-mm transverse-bar spacing on center.
45°–60 Tilt bar	45° tilt-bar grate with 57-mm longitudinal bar and 102-mm transverse-bar spacing on center.
45°–85 Tilt bar	45° tilt-bar grate with 83-mm longitudinal bar and 102-mm transverse-bar spacing on center.
30°–85 Tilt bar	30° tilt-bar grate with 83-mm longitudinal bar and 102-mm transverse-bar spacing on center.
Reticuline	"Honeycomb" pattern of lateral bars and longitudinal bearing bars.

inlet. Grates typically consist of longitudinal and/or transverse bars oriented parallel and perpendicular to the gutter flow, respectively, and design procedures have been developed for the grates listed in Table 5.38 (USFHWA, 1996). The P–50 grate has been found to be unsafe for bicycle traffic (Burgi, 1978; Nicklow, 2001). Grates are typically available with longitudinal dimensions in the range 610 mm to 1220 mm (2 ft to 4 ft) and transverse dimensions in the range 381 mm to 914 mm (1 ft to 3 ft). Typical P–50×100 and reticuline grates are shown in Figure 5.50.

The type of grate to be used in any area is usually specified by local municipal codes. In cases where there is some flexibility in grate selection, it is instructive to note that the top five grates for debris-handling efficiency are, in rank order (USFHWA, 1996): curved vane, 30°–85 tilt bar, 45°–85 tilt bar, P–50, and P–50×100.

Grated inlets on continuous grades intercept a portion of the frontal flow and a portion of the side flow. *Splash-over* occurs when a portion of the frontal flow passes directly over the inlet, and the ratio, R_f, of the frontal flow intercepted by the inlet to the total frontal flow is called the *frontal-flow interception efficiency*, which can be estimated using the relation (USFHWA, 1996)

$$R_f = 1 - 0.295(V - V_0) \tag{5.257}$$

where V is the gutter flow velocity, and V_0 is the critical gutter velocity where splash-over first occurs. The *splash-over velocity*, V_0, depends on the length and type of inlet in accordance with the experimental relations given in Figure 5.51. The ratio, R_s, of the side flow intercepted to the total side flow is called the *side-flow interception efficiency*, and can be estimated using

$$R_s = \left[1 + \frac{0.0828V^{1.8}}{S_x L^{2.3}} \right]^{-1} \tag{5.258}$$

where S_x is the cross slope of the gutter, and L is the length of the grate inlet. The Manning equation (Equation 5.245) gives the ratio of frontal flow to total gutter flow, R_w, by

$$R_w = 1 - \left(1 - \frac{W_0}{T} \right)^{8/3} \tag{5.259}$$

FIGURE 5.50: Types of grates

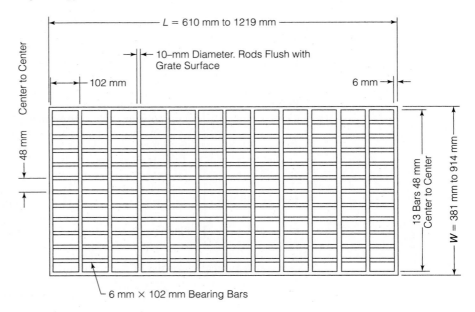

(a) P–50 × 100 Grate

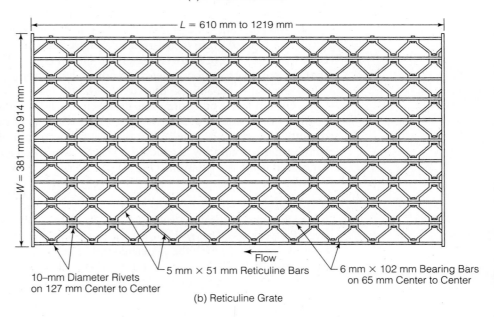

(b) Reticuline Grate

where W_0 is the frontal width of the grate inlet and T is the top-width of flow in the gutter. This is the same as Equation 5.252 that was used to calculate the frontal flow over a width W_0. Combining Equations 5.257 to 5.259 gives the ratio, R, of the flow intercepted by the grate inlet to the total gutter flow as

$$R = R_f R_w + R_s(1 - R_w)$$

(5.260)

FIGURE 5.51: Splash-over velocities at grate inlets

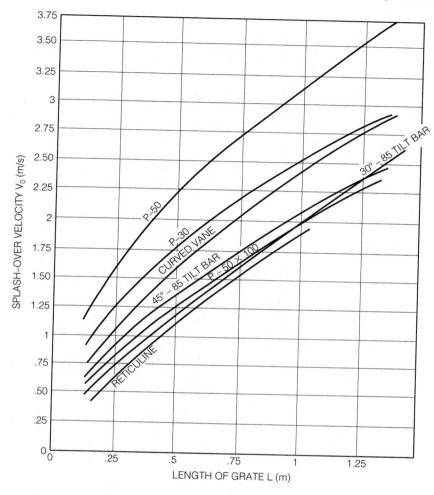

Therefore, if the flow in the gutter is Q, the amount of flow intercepted by the grate inlet is RQ.

EXAMPLE 5.49

A smooth-asphalt roadway has a cross slope of 3%, a longitudinal slope of 2%, a curb height of 8 cm, and a 90-cm wide concrete gutter. If the gutter flow is estimated to be 0.08 m³/s, determine the size and interception capacity of a P–50 × 100 grate that should be used to intercept as much of the flow as possible.

Solution From the given data: $Q = 0.08 \text{ m}^3/\text{s}$, $S_x = 3\% = 0.03$, and $S_0 = 2\% = 0.02$. Assuming that a significant portion of the gutter flow extends onto the smooth-asphalt pavement, Table 5.36 gives $n = 0.013$. Using the Manning equation (Equation 5.245) to find the depth, d, at the curb gives

$$Q = 0.375 \left(\frac{1}{nS_x} \right) d^{8/3} S_0^{1/2}$$

$$0.08 = 0.375 \left(\frac{1}{0.013 \times 0.03} \right) d^{8/3} (0.02)^{1/2}$$

and solving for d leads to

$$d = 0.061 \text{ m} = 6.1 \text{ cm}$$

This flow depth is less than the curb height of 8 cm, and therefore the flow is constrained within the roadway. The top-width, T, of the gutter flow, the flow area, A, and the flow velocity, V, are given by

$$T = \frac{d}{S_x} = \frac{0.061}{0.03} = 2.03 \text{ m}$$

$$A = \frac{1}{2} dT = \frac{1}{2}(0.061)(2.03) = 0.0619 \text{ m}^2$$

$$V = \frac{Q}{A} = \frac{0.08}{0.0619} = 1.29 \text{ m/s}$$

The spread of 2.03 m is much wider than the gutter width of 0.90 m, verifying the assumption that a significant portion of the gutter flow extends onto the smooth-asphalt pavement, and hence justifying the assumed n value of 0.013. Since the spread ($T = 2.03$ m) exceeds the maximum transverse dimension of typical grates (= 914 mm), use a grate with a transverse dimension, W_0, of 914 mm. According to Figure 5.51, the length of a P–50×100 grate corresponding to a splash-over velocity of 1.29 m/s is approximately 50 cm. Therefore, any grate longer than 50 cm will intercept 100% of the frontal flow, and have a frontal-flow efficiency, R_f, equal to 1.0. To maximize side-flow interception, use a (typical) maximum available grate length of 1220 mm. The selected grate therefore has dimensions of 914 mm × 1220 mm (= 3 ft × 4 ft).

The ratio, R_w, of frontal flow to total gutter flow is given by Equation 5.259 as

$$R_w = 1 - \left(1 - \frac{W_0}{T} \right)^{8/3}$$

$$= 1 - \left(1 - \frac{0.914}{2.03} \right)^{8/3} = 0.80$$

The ratio, R_s, of the side flow intercepted to the total side flow is given by Equation 5.258 as

$$R_s = \left[1 + \frac{0.0828 V^{1.8}}{S_x L^{2.3}} \right]^{-1}$$

$$= \left[1 + \frac{0.0828(1.29)^{1.8}}{(0.03)(1.22)^{2.3}} \right]^{-1} = 0.27$$

Therefore, the ratio, R, of the intercepted flow to the total gutter flow is given by Equation 5.260 as

$$R = R_f R_w + R_s (1 - R_w)$$

$$= (1.0)(0.80) + (0.27)(1 - 0.80) = 0.85$$

Based on this result, the grate intercepts $0.85(0.08 \text{ m}^3/\text{s}) = 0.068 \text{ m}^3/\text{s}$, and $0.08 \text{ m}^3/\text{s} - 0.068 \text{ m}^3/\text{s} = 0.012 \text{ m}^3/\text{s}$ bypasses the grate inlet.

Grate inlets in sag vertical curves operate as weirs for shallow ponding depths and as orifices at greater depths. The depths at which grates operate as weirs or orifices depend on the bar configuration and size of the grate, and grates of larger dimension will operate as weirs to greater depths than smaller grates or grates with less opening area (USFHWA, 1996). Typically, for depths of water not exceeding 12 cm, the following weir equation can be used to calculate the capacity, Q_i (m^3/s), of a grate inlet

$$Q_i = 1.66Pd^{1.5}$$

(5.261)

where P is the perimeter of the grate opening (m) and d is the depth of flow above the grate (m). If the grate is adjacent to a curb, then that side of the grate is not counted in the perimeter. Typically, if the flow depth over a grate exceeds 43 cm, then the following orifice equation is used to compute the capacity, Q_i, of the grate inlet

$$Q_i = 0.67A\sqrt{2gd}$$

(5.262)

where A is the open area in the grate and d is the depth of flow above the grate. For depths of flow typically between 12 cm and 43 cm, the capacity of the grate is somewhere between that calculated by Equations 5.261 and 5.262. Grates alone are not typically recommended for installation in sags because of their tendency to clog and cause flooding during severe weather. A combination inlet (see next section) is usually a better choice in sag locations.

EXAMPLE 5.50

A roadway has a cross slope of 2%, a flow depth at the curb of 8 cm, and a corresponding flowrate in the gutter of 0.08 m^3/s. The gutter flow is to be removed in a vertical sag by a grate inlet that is mounted flush with the curb. Calculate the minimum dimensions of the grate inlet. Assume that the grate opening is 50% clogged with debris, which covers 25% of the perimeter of the grate.

Solution   Since the depth of flow is less than 12 cm, the inflow to the inlet is probably given by the weir equation (Equation 5.261), which can be put in the form

$$P = \frac{Q_i}{1.66d^{1.5}}$$

where P is the grate-inlet perimeter not including the side adjacent to the curb, $Q_i = 0.08 \text{ m}^3/\text{s}$, $d = 0.08 \text{ m}$, and

$$P = \frac{0.08}{1.66(0.08)^{1.5}} = 2.13 \text{ m}$$

Since the grate is flush-mounted with the curb, the length plus twice the width must equal 213 cm. Since 25% of the grate perimeter is blocked by debris, the grate must have a perimeter that exceeds 213/0.75 = 284 cm. Hence a 914 mm × 1219 mm grate

FIGURE 5.52: Grate inlet in parking lot and uncurbed roadway

with perimeter of 305 cm would have some reserve capacity to accommodate debris blockage.

In addition to draining roadways with curbs and gutters, grate inlets are also used to drain parking lots and roadways without curbs and gutters. Examples of these grate inlets are shown in Figure 5.52. In draining uncurbed roadways, the grate inlet is typically placed in a rectangular paved area that is directly connected to the roadway pavement, as illustrated in the righthand picture. The capacities of the grate inlets shown in Figure 5.52 are calculated using the weir equation, Equation 5.261, for depths less than 12 cm; the orifice equation, Equation 5.262, for depths greater than 43 cm; and an interpolated weir/orifice capacity for intermediate depths.

Combination inlets. Combination inlets consist of combined curb and grate inlets. Typically, the curb opening is upstream of the grate, intercepting debris, reducing the spread on the roadway pavement, and increasing the efficiency of the grate inlet. A combination inlet in which the curb opening is upstream of the grate is called a *sweeper configuration*, and the upstream curb inlet is called a *sweeper inlet*. In sag locations the grate is best placed at the center of the curb opening. A combination inlet with the sweeper configuration is shown in Figure 5.53, where the precast unit, prior to installation, is also shown. It should be clear that this combination inlet is positioned in a sag location. In most combination inlets the grate is located in a depressed gutter. The interception capacity of a combination inlet is equal to the sum of the capacity

FIGURE 5.53: Combination inlet and precast unit

of the curb opening upstream of the grate plus the grate capacity. The three-step procedure to calculate the capacity of a combination inlet is:

1. Calculate the capacity of the curb inlet upstream of the grate inlet using the procedure described previously for curb inlets. Subtract the curb-inlet capacity from the gutter flow to obtain the gutter flow approaching the grate inlet.

2. Calculate the capacity of the grate inlet. The gutter cross section is typically depressed over a width, W_0, equal to the width of the grate. In this case, the ratio, R_w, of flow over width W_0 to flow over the top width T is given by

$$R_w = \left\{ 1 + \frac{S_w/S_x}{\left[1 + \frac{S_w/S_x}{\frac{T}{W_0}-1}\right]^{8/3} - 1} \right\}^{-1} \tag{5.263}$$

where S_w is the slope of the depressed gutter, and S_x is the slope of the undepressed gutter. The side-flow interception capacity, R_s, is calculated using Equation 5.258, the frontal-flow interception, R_f, is calculated using Equation 5.257, and the fraction of gutter flow intercepted by the grate is given by Equation 5.260.

3. Add the capacity of the curb opening calculated in step 1 to the capacity of the grate inlet calculated in step 2 to obtain the capacity of the combination inlet.

EXAMPLE 5.51

A combination inlet consists of a 2-m curb inlet and a 0.6-m by 0.6-m P-30 grate inlet adjacent to the downstream 0.6 m of the curb opening. The gutter section has a width of 0.6 m, a longitudinal slope of 2%, a cross slope of 3%, and a gutter depression of 25 mm in front of the curb opening. The roadway pavement consists of smooth asphalt. For a gutter flow of 0.07 m³/s, calculate the interception capacity of the combination inlet.

Solution

Step 1. Calculate the interception capacity of the curb opening upstream of the grate inlet. From the given data, $Q = 0.07$ m³/s, $S_0 = 2\% = 0.02$, $n = 0.013$ (Table 5.36), $S_x = 3\% = 0.03$, $a = 25$ mm $= 0.025$ m, and $W_0 = 0.6$ m. The fraction, R_w, of gutter flow over the depressed section in front of the curb opening is given by Equation 5.263, where

$$R_w = \left\{ 1 + \frac{S_w/S_x}{\left[1 + \frac{S_w/S_x}{\frac{T}{W_0}-1}\right]^{8/3} - 1} \right\}^{-1} \tag{5.264}$$

In this case,

$$S_w = S_x + \frac{a}{W_0} = 0.03 + \frac{0.025}{0.6} = 0.072$$

and Equation 5.264 can be written as

$$\frac{Q_w}{Q} = \left\{ 1 + \frac{0.072/0.03}{\left[1 + \frac{0.072/0.03}{\frac{T}{0.6}-1} \right]^{8/3} - 1} \right\}^{-1}$$

$$= \left\{ 1 + \frac{2.4}{\left[1 + \frac{2.4}{1.67T-1} \right]^{8/3} - 1} \right\}^{-1} \qquad (5.265)$$

where Q_w is the flow over width W_0, and Q is the total gutter flow given as 0.07 m³/s. Hence, Equation 5.265 can be written as

$$Q_w = 0.07 \left\{ 1 + \frac{2.4}{\left[1 + \frac{2.4}{1.67T-1} \right]^{8/3} - 1} \right\}^{-1} \qquad (5.266)$$

It is convenient for subsequent analyses to work with the flow, Q_s, over the section outside of the depressed section, in which case

$$Q_s = Q - Q_w = 0.07 - 0.07 \left\{ 1 + \frac{2.4}{\left[1 + \frac{2.4}{1.67T-1} \right]^{8/3} - 1} \right\}^{-1}$$

$$(5.267)$$

This equation must be solved simultaneously with the Manning equation, Equation 5.245, which can be written as

$$Q_s = \frac{0.375}{n} S_x^{5/3} S_0^{1/2} (T - W_0)^{8/3} \qquad (5.268)$$

which, in this case gives

$$Q_s = \frac{0.375}{0.013} (0.03)^{5/3} (0.02)^{1/2} (T - 0.6)^{8/3}$$

which simplifies to

$$Q_s = 0.0118(T - 0.6)^{8/3} \qquad (5.269)$$

Simultaneous solution of Equations 5.267 and 5.269 gives $Q_s = 0.0162$ m³/s and $T = 1.72$ m. The flow ratio, R_w, over width W_0 is therefore given by

$$R_w = \frac{Q - Q_s}{Q} = \frac{0.07 - 0.0162}{0.07} = 0.77$$

The equivalent cross slope, S_e, in the gutter depression is given by Equation 5.250 as

$$S_e = S_x + S'_w R_w = S_x + \left(\frac{a}{W_0} \right) R_w = 0.03 + \left(\frac{0.025}{0.6} \right) 0.77 = 0.062$$

Equation 5.248 gives the length, L_T, of curb opening for 100% interception as

$$L_T = 0.817 Q^{0.42} S_0^{0.3} \left(\frac{1}{nS_e} \right)^{0.6}$$

$$= 0.817(0.07)^{0.42}(0.02)^{0.3} \left(\frac{1}{0.013 \times 0.062} \right)^{0.6} = 5.94 \text{ m}$$

Since the length, L, of curb opening upstream of the grate is 2 m − 0.6 m = 1.4 m, the efficiency of the curb opening is given by Equation 5.249 as

$$R = 1 - \left(1 - \frac{L}{L_T} \right)^{1.8} = 1 - \left(1 - \frac{1.4}{5.94} \right)^{1.8} = 0.384$$

The flow, Q_{ic}, intercepted by the curb opening is therefore given by

$$Q_{ic} = 0.384(0.07 \text{ m}^3/\text{s}) = 0.027 \text{ m}^3/\text{s}$$

Step 2. Calculate the interception capacity of the grate inlet. The gutter flow immediately upstream of the grate, Q_g, is given by

$$Q_g = Q - Q_{ic} = 0.07 - 0.027 = 0.043 \text{ m}^3/\text{s}$$

The ratio of flow over the grate, Q_w, to the gutter flow upstream of the grate, Q_g, is given by Equation 5.265 (with Q_g replacing Q) and, similar to Equation 5.267, the side flow, Q_s, is given by

$$Q_s = 0.043 - 0.043 \left\{ 1 + \frac{2.4}{\left[1 + \frac{2.4}{1.67T-1} \right]^{8/3} - 1} \right\}^{-1} \tag{5.270}$$

The Manning equation is given by Equation 5.268, which leads to Equation 5.269. Solving Equations 5.269 and 5.270 simultaneously gives $Q_s = 0.00556 \text{ m}^3/\text{s}$ and $T = 1.35 \text{ m}$. The ratio, R_w, of flow over W_0 to the total gutter flow is therefore given by

$$R_w = \frac{Q_g - Q_s}{Q_g} = \frac{0.043 - 0.00556}{0.043} = 0.87$$

Next, calculate the frontal-flow interception efficiency, R_f. The total flow area, A, in the gutter is given by

$$A = \frac{1}{2}[T^2 S_x + aW_0] = \frac{1}{2}[(1.35)^2(0.03) + (0.025)(0.6)] = 0.0348 \text{ m}^2$$

and hence the average velocity, V, in the gutter is given by

$$V = \frac{Q_g}{A} = \frac{0.043}{0.0348} = 1.24 \text{ m/s}$$

For a 0.6-m long P-30 grate, the splash-over velocity is approximately 2 m/s (see Figure 5.51), and since the average gutter velocity is 1.24 m/s, the frontal-flow interception efficiency is 100% and hence

$$R_f = 1.0$$

The side-flow interception efficiency, R_s, is given by Equation 5.258 as

$$R_s = \left[1 + \frac{0.0828V^{1.8}}{S_xL^{2.3}}\right]^{-1}$$

$$= \left[1 + \frac{0.0828(1.24)^{1.8}}{(0.03)(0.6)^{2.3}}\right]^{-1} = 0.071$$

The flow intercepted by the grate, Q_{ig}, is given by Equation 5.260 as

$$Q_{ig} = Q_g[R_fR_w + R_s(1 - R_w)] = 0.043[(1.0)(0.87)$$
$$+ (0.071)(1 - 0.87)] = 0.038 \text{ m}^3/\text{s} \qquad (5.271)$$

Step 3.   Calculate the total interception capacity of the combination inlet. The interception capacity of the combination inlet, Q_i, is the sum of the curb opening capacity, Q_{ic}, and the grate capacity, Q_{ig}, hence

$$Q_i = Q_{ic} + Q_{ig} = 0.027 + 0.038 = 0.065 \text{ m}^3/\text{s}$$

Since the gutter flow upstream of the combination inlet is 0.07 m³/s, then 0.07 m³/s − 0.065 m³/s = 0.005 m³/s bypasses the combination inlet.

In sag vertical curve locations, the capacity of a combination inlet is equal to the capacity of the sweeper portion of the curb inlet plus the grate capacity. For flow depths less than the opening height of the sweeper inlet, the capacity of the sweeper inlet is given by the weir equation, Equation 5.253 (depressed gutter) or Equation 5.255 (undepressed gutter), and for flow depths greater than 1.4 times the opening height the capacity of the sweeper inlet is given by the orifice equation, Equation 5.256. For flow depths less than about 12 cm, the capacity of a grate inlet is given by the weir equation, Equation 5.261, and for depths greater than about 43 cm, the capacity of a grate inlet is given by Equation 5.262. For flow depths in between weir and orifice conditions, inlet capacities can usually be interpolated. Combination inlets in sag locations are frequently designed assuming complete clogging of the grate (Nicklow, 2001).

EXAMPLE 5.52

A combination inlet is to be placed at a sag location of a roadway to remove a gutter flow of 0.15 m³/s. The combination inlet consists of a 15-cm high by 2.5-m long curb opening with a 0.6-m by 1.2-m P-50 grate centered in front of the curb opening. If the roadway has a cross slope of 3% and the perimeter of the grate is 30% clogged, determine the spread of water on the roadway.

Solution   Assuming that the flow depth, d, in the gutter is less than the height of the curb opening (15 cm), the curb opening acts like a weir, and the inflow to the curb opening, Q_c, is given by Equation 5.255 as

$$Q_c = 1.60Ld^{1.5} \tag{5.272}$$

where $L = 2.5\,\text{m} - 1.2\,\text{m} = 1.3\,\text{m}$. Assuming that the flow depth over the grate is less than 12 cm, then the grate inlet acts like a weir, and the inflow to the grate inlet, Q_g, is given by Equation 5.261 as

$$Q_g = 1.66Pd^{1.5} \tag{5.273}$$

where, for 30% clogging, $P = 0.7(L + 2W_0) = 0.7(1.2 + 2 \times 0.6) = 1.68\,\text{m}$. Combining Equations 5.272 and 5.273 gives the capacity of the combination inlet, Q, as

$$Q = Q_c + Q_g \tag{5.274}$$

Taking $Q = 0.15\,\text{m}^3/\text{s}$, combining Equations 5.272 to 5.274, and substituting the given data yields

$$0.15 = 1.60(1.30)d^{1.5} + 1.66(1.68)d^{1.5}$$

which gives

$$d = 0.098\,\text{m} = 9.8\,\text{cm}$$

This depth of flow at the curb verifies weir-flow assumptions for the curb opening and the grate inlet. The spread, T, on the roadway is given by

$$T = \frac{d}{S_x} = \frac{0.098}{0.03} = 3.27\,\text{m}$$

Slotted-drain inlets.   Slotted inlets are used in areas where it is desirable to intercept sheet flow before it crosses onto a section of roadway. Typical isometric and elevation views of a slotted drain are illustrated in Figure 5.54, and an operational slotted drain and typical corrugated metal pipe (CMP) used in constructing slotted drains are shown in Figure 5.55. The main advantage of slotted drains is their ability to intercept flow over a wide section of roadway, and their main disadvantage is that they are very susceptible to clogging from sediments and debris. Sediment deposition in the pipe

FIGURE 5.54: Slotted-drain inlet

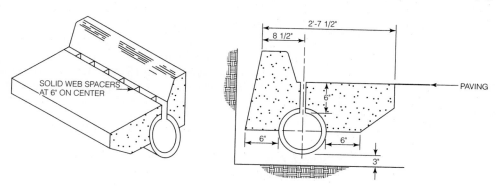

FIGURE 5.55: Operational slotted drain and typical slotted-drain pipe

is the most frequently encountered problem (USFHWA, 1996). Slotted-drain inlets function like curb inlets when they are placed parallel to the flow, and like grate inlets when placed perpendicular to the flow.

For slotted-drain inlets installed parallel to the gutter flow on continuous grades, and having slot widths greater than 4.45 cm (1.75 in.), the length, L_T, of drain required to intercept a flow Q_i is given by Equation 5.248, which is also used to calculate the length of curb inlets for 100% flow interception, and is repeated here for convenience as

$$L_T = 0.817 Q^{0.42} S_0^{0.3} \left(\frac{1}{nS_x} \right)^{0.6} \tag{5.275}$$

where S_0 is the longitudinal slope of the drain, n is the Manning roughness coefficient of the gutter, and S_x is the cross slope of the gutter. In cases where the slotted-drain inlet is shorter than L_T, the ratio, R, of the intercepted flow to the gutter flow is given by

$$R = 1 - \left(1 - \frac{L}{L_T} \right)^{1.8} \tag{5.276}$$

It is common practice to use slotted-drain lengths that are much longer than lengths used for curb inlets (Loganathan et al., 1996).

Slotted-drain inlets installed perpendicular to the flow direction perform like short grate inlets. Assuming a splash-over velocity of 0.3 m/s and no side flow, the grate-inlet efficiency equations can be used to calculate the interception capacity of slotted-drain inlets (Young and Stein, 1999).

In sag locations, slotted drains operate as weirs for depths below approximately 5 cm (2 in.) and as orifices in locations where the depth at the upstream edge of the slot is greater than about 12 cm (5 in.). Between these depths transition flow occurs (USFHWA, 1996). The capacity of a slotted inlet operating as a weir can be computed using the relation (Young and Stein, 1999)

$$Q_i = 1.4 L d^{0.5} \tag{5.277}$$

where L is the length of the slot (m), and d is the depth of flow adjacent to the inlet (m). The capacity of a slotted inlet operating as an orifice can be computed using the relation (Young and Stein, 1999)

$$Q_i = 0.8Lw\sqrt{2gd} \tag{5.278}$$

where w is the width of the slot. Slotted drains are usually not recommended in sag locations because of potential problems with clogging from debris.

EXAMPLE 5.53

A roadway has a cross slope of 2.5% and a longitudinal slope of 1.5%, and the flowrate in the gutter is 0.1 m³/s. The flow is to be removed by a slotted drain with a slot width of 5 cm, and the Manning n of the roadway is 0.015. Estimate the minimum length of slotted drain that can be used.

Solution   From the given data, $Q_i = 0.1$ m³/s, $S_0 = 0.015$, $n = 0.015$, $S_x = 0.025$, and Equation 5.275 gives the length of the drain as

$$L_T = 0.817Q_i^{0.42}S_0^{0.3}\left[\frac{1}{nS_x}\right]^{0.6} = 0.817(0.1)^{0.42}(0.015)^{0.3}\left[\frac{1}{(0.015)(0.025)}\right]^{0.6} = 10.0 \text{ m}$$

Hence the slotted drain must be at least 10.0 m long to remove the gutter flow.

5.7.1.3 Roadside and median channels

Roadside channels are commonly used with uncurbed roadway sections to convey runoff from the roadway pavement and from areas which drain toward the roadway. Right-of-way constraints limit the use of roadside channels on most urban roadways. These channels are normally trapezoidal or triangular in cross section, lined with grass or other protective lining, and outlet to a storm-drain piping system via a drop inlet, to a detention or retention basin, or to an outfall channel. A typical drop inlet and median channel are illustrated in Figure 5.56. Drop inlets are similar to grate inlets used in pavement drainage and should be placed flush with the channel bottom and constructed of traffic-safe bar grates. In addition, paving around the inlet perimeter can help prevent erosion and might slightly increase the interception capacity of the inlet by accelerating the flow. Small dikes are often placed downstream of drop inlets to impede bypass flow. The height of dike required for 100% interception is not large and can be determined by calculating the ponding height required for interception of the flow by a sag grate. Roadside channels are typically designed as grass-lined channels following the design protocol described in Section 3.5.3.2. Design flows typically have return periods of 5 to 10 years. It is recommended that the channel side slopes not exceed 3:1 and, in areas where traffic safety may be of concern, channel side slopes should be 4:1 or flatter (USFHWA, 1996). Longitudinal slopes are generally dictated by the road profile; however, channels with gradients greater than 2% tend to flow in a supercritical state (USFHWA, 1988) and may require the use of flexible linings to maintain stability. Most flexible lining materials are suitable for protecting channel gradients up to 10% (USFHWA, 1988). In roadside channels a freeboard of 15 cm under design flow conditions is generally considered adequate.

FIGURE 5.56: Typical median inlet

Source: USFHWA (1988).

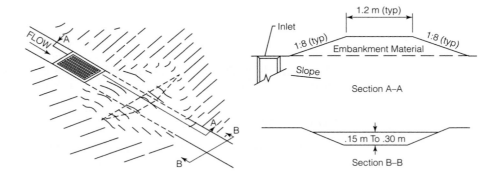

5.7.2 Runoff Controls

Urban stormwater-management systems are designed to control both the quantity and quality of stormwater runoff. Quantity control usually requires that peak postdevelopment runoff rates do not exceed peak predevelopment runoff rates (for a design rainfall), and quality control usually requires a defined level of treatment, such as a specified detention time in a sedimentation basin or the retention of a specified volume of initial runoff. The runoff events used to design flood-control and water-quality control systems are generally different. Flood-control systems are designed for large, infrequent runoff events with return periods of 10 to 100 years, while quality-control systems are designed for small frequent events with return periods of less than one year. On a long-term basis, most of the pollutant load in stormwater runoff is contained in the smaller, more frequent storms.

Runoff controls can be either on-site or regional controls. *On-site controls* handle runoff from individual developments, while *regional controls* handle runoff from several developments. The main advantage of on-site facilities is that developers can be required to build them, while the major disadvantage is the larger overall land area that is required compared with regional controls. The main advantage of regional facilities is that they provide more storage and can be designed for longer release periods, while their major disadvantages are the complex arrangements that are necessary to collect funds from developers and to use those funds efficiently for the intended stormwater-management facilities. The minimization of directly connected impervious areas remains one of the most effective source controls that can be implemented to reduce the quantity and improve the quality of runoff at the source, and it can significantly reduce the capacity requirements in other runoff controls. Under ideal conditions, the minimization of directly connected impervious areas can virtually eliminate surface runoff from storms with less than 13 mm (0.5 in.) of precipitation (Urbonas and Stahre, 1993).

5.7.2.1 Stormwater impoundments

Stormwater impoundments are facilities (basins) that collect surface runoff and release it either at a reduced rate through an outlet or by infiltration into the ground. These impoundments are frequently used for both flood-control and water-quality control purposes. The two major types of impoundments are detention and retention impoundments (basins). *Detention basins* are water-storage areas where the stored water is released gradually through an uncontrolled outlet, and *retention basins* are water-storage areas where there is either no outlet or the impounded water is stored for a prolonged period. Infiltration basins and ponds that maintain water permanently, with freeboard provided for flood storage, are the most common types of retention basins (ASCE, 1992). Storage impoundments where a permanent water body forms the base of the storage area are called *wet* basins; impoundments where the ground surface forms the base of the storage area are called *dry* basins. Wet-detention basins are commonly referred to as *detention ponds*.

Most detention basins are constructed by a combination of cut and fill and must have at least one service outlet and an emergency spillway. In some cases, multiple outlets at different elevations are used to facilitate the discharge from the basin under multiple design storms with return periods between 20 and 50 years (ASCE, 1992). An emergency spillway provides for controlled overflow during large storm events, typically the 100-year storm (Yen and Akan, 1999). A schematic diagram of a (dry) detention basin is given in Figure 5.57, and a schematic diagram of a detention pond is illustrated in Figure 5.58. Dry-detention basins are typically open areas characterized by low-level outlets that can discharge any accumulated basin inflows, while wet-detention basins (detention ponds) have high-level outlets that discharge when the water level in the pond rises above the permanent pool. Dry-detention basins generally empty after a storm, while detention ponds retain the water much longer above a permanent pool of water. In modern practice, detention ponds are sometimes referred to as retention ponds (because they retain pollutants); however, the latter term should be reserved for ponds without outlets.

The most common types of outlets from detention ponds are orifice-type and weir-type outlets, and these outlets are typically part of a *single-stage riser* as shown in

FIGURE 5.57: Dry detention basin

Source: Urbonas and Roesner (1993).

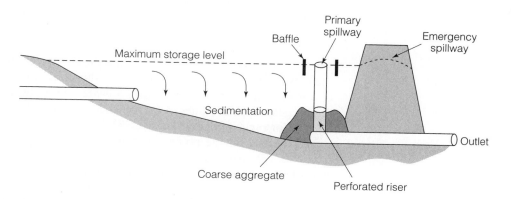

FIGURE 5.58: Detention pond

Source: Urbonas and Roesner, 1993.

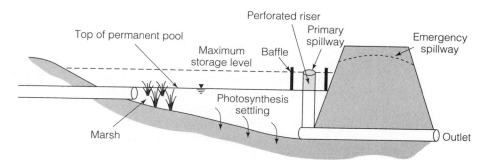

FIGURE 5.59: Single-stage riser

Source: McCuen (2005).

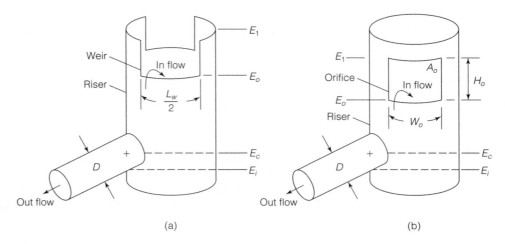

(a)   (b)

Figure 5.59. The water stored in the detention pond above the permanent pool flows through the orifice or over the weir into the riser and out of the pipe leading from the riser. For weir flow, the discharge over the weir, Q_w, is given by

$$Q_w = C_w L_w h^{3/2} \qquad (5.279)$$

where C_w is the weir coefficient, L_w is the length of the weir, and h is the elevation of the water surface above the crest of the weir. A typical value of C_w is 1.83 when Q_w is in m^3/s, and L_w and h are in meters. For orifice flow, the discharge through the orifice, Q_0, is given by

$$Q_0 = C_d A_0 \sqrt{2gh} \qquad (5.280)$$

where C_d is a discharge coefficient, A_0 is the cross-sectional area of the orifice, and h is the elevation of the water surface above the center of the orifice (for free discharges) or h is the difference between the headwater and tailwater elevations (for submerged discharges). Typical values of C_d are 0.6 for square-edge uniform-entrance conditions, and 0.4 for ragged-edge orifices (USFHWA, 1996).

Where drainage policies require control of flow rates of two exceedance frequencies, the *two-stage riser* is an alternative for control. Its structure is similar to that of the single-stage riser, except that it includes either two weirs or a weir and an orifice as shown in Figure 5.60. For the commonly used weir/orifice structure, the orifice is used to control the more frequent event, and the larger event is controlled using the weir. The runoff from the smaller and larger events are also referred to as the low-stage and high-stage events. According to McCuen (2005), there have been few theoretical or empirical studies on the hydraulics of two-stage risers; and a number of procedures have been proposed for routing flows through these structures. Some proposed procedures have assumed that the flow through the orifice ceases when weir flow begins, while other proposed procedures have assumed that the flow through the orifice is independent of the flow over the weir. It appears to be more realistic to assume that the two flows are not independent and that the interdependence between weir flow and orifice flow increases as the elevation of the water surface above the weir increases. The general form of the stage-discharge relationship is typically

FIGURE 5.60: Two-stage riser

Source: McCuen (2005).

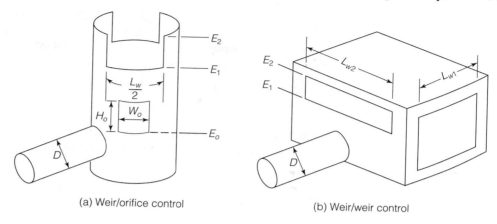

(a) Weir/orifice control   (b) Weir/weir control

given by

$$
Q = \begin{cases}
0, & E \le E_0 \\
C_w L_w (E - E_0)^{1.5}, & E_0 \le E \le E_0 + H_0 \\
C_d A_0 \sqrt{2g(E - E_0)}, & E_0 + H_0 \le E \le E_1 \\
C_w L_w (E - E_1)^{1.5} + C_d A_0 \sqrt{2g(E - E_1)}, & E_1 < E
\end{cases}
$$

(5.281)

where E is the water-surface elevation in the pond surrounding the riser, E_0 is the elevation of the bottom of the orifice, H_0 is the maximum height of the orifice, and E_1 is the elevation of the weir. If the barrel diameter of the riser is large enough to assume that the orifice and weir flows are independent, then the orifice part of Equation 5.281 would be $C_d A_0 \sqrt{2g(E - E_0)}$ rather than $C_d A_0 \sqrt{2g(E - E_1)}$.

EXAMPLE 5.54

An outlet structure from a detention pond is to be designed for both water-quality and water-quantity control purposes. For water-quality control, the outlet structure is to be sized such that the first 2.5 cm of runoff from the catchment is discharged in not less than 24 hours. For water-quantity control, the outlet structure is to be sized such that the peak runoff is less than or equal to the predevelopment peak of 0.92 m³/s. The catchment area is 46 ha, and the average wet-season pool elevation in the detention pond is 102.50 m. Storage of the water-quality volume causes the water-surface elevation in the detention pond to rise to 102.85 m, and under design-flood conditions the maximum allowable pool elevation in the detention pond is 104.00 m. The outlet structure is to consist of a 1.21-m diameter vertical circular pipe (riser), with the water-quality volume discharged through an orifice in the riser, and the flood flow discharged over a high-level weir at the top of the riser. The high-level discharge weir must have a crest length not greater than 50% of the riser diameter. Determine the required dimensions and elevations of the orifice and weir components of the two-stage riser outlet.

Solution   The water-quality volume will be discharged through an unsubmerged orifice (which acts like a weir), where the bottom side of the orifice is at the average wet-season pool elevation of 102.50 m and the top of the orifice is at the pool elevation when the water-quality volume is stored, which is 102.85 m. Since the catchment area is 46 ha = 4.6×10^5 m^2 and the water-quality depth is 2.5 cm = 0.025 m, then

$$\text{water-quality volume} = 4.6 \times 10^5 \text{ m}^2 \times 0.025 \text{ m} = 11,500 \text{ m}^3$$

If this water-quality volume is to be detained in the pond for one day (= 86,400 seconds), then the average discharge rate through the orifice/weir, Q_0, is given by

$$Q_0 = \frac{\text{water-quality volume}}{1 \text{ day}} = \frac{11,500 \text{ m}^3}{86,400 \text{ s}} = 0.133 \text{ m}^3/\text{s}$$

Specifying the maximum orifice discharge rate as equal to the average discharge rate for 24-h detention (which guarantees a detention time greater than 24 h) yields, according to Equation 5.281,

$$Q_0 = C_w L_w (E - E_0)^{1.5}$$

where C_w can be taken as 1.83, $E = 102.85$ m, and $E_0 = 102.50$ m. Substituting into the above equation yields

$$0.133 = 1.83 L_w (102.85 - 102.50)^{1.5}$$

which gives

$$L_w = 0.35 \text{ m}$$

The required height of the orifice, H_0, is $102.85 - 102.50$ m = 0.35 m, and therefore the dimensions of the orifice required to detain the water-quality volume for at least 24 hours ($L_w \times H_0$) is 0.35 m $\times$ 0.35 m.

The riser diameter, D, is equal to 1.21 m and the maximum length of the flood-discharge weir is $0.5D = 0.5(1.21$ m$) = 0.605$ m. The discharge from the outlet structure when the pool elevation exceeds the crest elevation of the flood-discharge weir is given by Equation 5.281 as

$$Q = C_w L_w (E - E_1)^{1.5} + C_d A_0 \sqrt{2g(E - E_1)}$$

When the pool elevation is at its maximum allowable level, $Q = 0.92$ m^3/s, $C_w = 1.83$, $L_w = 0.605$ m, $E = 104.00$ m, $C_d = 0.6$, and $A_0 = 0.35$ m $\times$ 0.35 m $= 0.1225$ m^2. Substituting into the above equation yields

$$0.92 = 1.83(0.605)(104.00 - E_1)^{1.5} + 0.6(0.1225)\sqrt{2(9.81)(104.00 - E_1)}$$

which gives

$$E_1 = 103.30 \text{ m}$$

Therefore, the bottom of the flood discharge weir should be at elevation 103.30 m to give a design-flood discharge equal to the predevelopment value of 0.92 m^3/s.

In summary, the required two-stage riser should have a 0.35-m × 0.35-m orifice with top and bottom elevations of 102.85 m and 102.50 m, and a 0.605-m long flood-discharge weir with a crest elevation of 103.30 m. The riser will meet both the water-quality and water-quantity (flood-discharge) objectives of the detention pond.

By their very nature, flood-control basins tend to be wet, soggy, and soft. It should never be assumed that a concrete outlet structure can be adequately supported in this environment without piles or other geotechnical treatment. Without proper support, the outlet structure can settle and separate from the outfall pipe, creating a guaranteed failure scenario, or at least an expensive maintenance situation (Paine and Akan, 2001). As the height of the outlet riser increases, conditions become more unstable, and, in order to increase stability, the outlet structure should be located in the embankment rather than in the open bottom of the basin.

5.7.2.2 Flood control

Both detention and retention basins are used for flood control. The design of retention basins for flood control consists first of providing sufficient freeboard in an impoundment area to store the runoff volume resulting from the design runoff event and then of verifying that the infiltration capacity of the impoundment is sufficient to remove the stored water in a reasonable amount of time, usually a typical interstorm period. The design runoff event is usually equal to the runoff resulting from a storm of specified duration and frequency, such as a 72-hour storm with a 25-year return period. The design of detention basins for flood control consists of the following steps: (1) select a drainage-basin configuration; (2) select an outlet structure; (3) route the design-runoff hydrograph through the detention basin and determine the peak discharge from the detention basin and the maximum water elevation in the basin; and (4) repeat steps 1 to 3 until the peak discharge and maximum water-surface elevation are acceptable. An acceptable peak discharge is usually one that is less than or equal to the predevelopment-peak discharge, and an acceptable maximum water-surface elevation maintains the stored runoff within the confines of the detention basin. Most drainage ordinances require that postdevelopment discharges not exceed predevelopment discharges for multiple events, such as the 2-year, 10-year, 25-year, and 50-year storms (ASCE, 1996a). Detention facilities should be designed to drain within the typical interstorm period, usually on the order of 72 hours (Debo and Reese, 1995). Some guidelines for determining the required size of a detention basin for flood control are given below.

- **Make a preliminary selection of a drainage basin.** A preliminary estimate of the required drainage-basin volume can be obtained by subtracting the predevelopment runoff volume from the postdevelopment runoff volume. This volume represents the approximate storage requirement to ensure that the postdevelopment peak discharge rate is less than or equal to the predevelopment peak runoff rate.

- **Make a preliminary selection of an outlet structure.** Use the required volume of the detention basin estimated in step 1 with the storage-elevation function for the site to estimate the maximum headwater elevation at the outlet location.

Select an outlet structure that will pass the maximum allowable outflow at this headwater elevation.

- **Route the runoff hydrograph through the detention basin.** The runoff hydrograph is routed through the detention basin using the modified Puls method described in Section 5.5.1.1. This procedure yields the discharge hydrograph from the detention basin.

- **Assess the performance of the detention basin.** Determine whether the peak discharge from the detention basin is less than or equal to the predevelopment peak discharge. If it is, the required detention volume corresponds to the maximum stage in the detention basin. If the peak discharge from the detention basin exceeds the predevelopment peak discharge, then adjust the outlet structure to discharge the predevelopment peak at the maximum stage in the detention basin, and repeat the previous two steps.

EXAMPLE 5.55

The estimated runoff hydrographs from a site before and after development are as follows:

Time (min)	0	30	60	90	120	150	180	210	240	270	300	330	360	390
Before (m^3/s)	0	1.2	1.7	2.8	1.4	1.2	1.1	0.91	0.74	0.61	0.50	0.28	0.17	0
After (m^3/s)	0	2.2	7.7	1.9	1.1	0.80	0.70	0.58	0.38	0.22	0.11	0	0	0

The postdevelopment detention basin is to be a detention pond drained by an outflow weir. The elevation versus storage in the detention pond is

Elevation (m)	Storage (m^3)
0	0
0.5	5,544
1.0	12,200
1.5	20,056

where the weir crest is at elevation 0 m, which is also the initial elevation of the water in the detention pond prior to runoff. The performance of the weir is given by

$$Q = 1.83bh^{\frac{3}{2}}$$

where Q is the overflow rate (m^3/s), b is the crest length (m), and h is the head on the weir (m). Determine the required crest length of the weir for the detention pond to perform its desired function. What is the maximum water-surface elevation expected in the detention pond?

Solution The required detention-pond volume is first estimated by subtracting the predevelopment runoff volume, V_1, from the postdevelopment runoff volume, V_2. From the given hydrographs:

$$V_1 = (30)(60)[1.2 + 1.7 + 2.8 + 1.4 + 1.2 + 1.1 + 0.91 + 0.74 + 0.61$$
$$+ 0.50 + 0.28 + 0.17]$$
$$= 22,698 \text{ m}^3$$

and

$$V_2 = (30)(60)[2.2 + 7.7 + 1.9 + 1.1 + 0.80 + 0.70 + 0.58 + 0.38$$
$$+ 0.22 + 0.11]$$
$$= 28{,}242 \text{ m}^3$$

A preliminary estimate of the required volume, V, of the detention pond above the normal pool elevation is

$$V = V_2 - V_1 = 28{,}242 - 22{,}698 = 5544 \text{ m}^3$$

From the storage-elevation function, the head h corresponding to a storage volume of 5544 m³ is 0.50 m. The maximum predevelopment runoff, Q, is 2.8 m³/s, and the weir equation gives

$$Q = 1.83bh^{\frac{3}{2}}$$

or

$$b = \frac{Q}{1.83h^{3/2}} = \frac{2.8}{1.83(0.50)^{3/2}} = 4.33 \text{ m}$$

Based on this preliminary estimate of the crest length, use a trial length of 4.25 m. The corresponding weir-discharge equation is

$$Q = 1.83bh^{\frac{3}{2}} = 1.83(4.25)h^{\frac{3}{2}} = 7.78h^{\frac{3}{2}}$$

The postdevelopment-runoff hydrograph can be routed through the detention pond using $\Delta t = 30$ min. The storage and outflow characteristics of the detention pond can be put in the following form

Elevation (m)	Storage, S (m³)	Outflow, O (m³/s)	$2S/\Delta t + O$ (m³/s)
0	0	0	0
0.5	5,544	2.75	8.91
1.0	12,200	7.78	21.34
1.5	20,056	14.29	36.58

The routing computations (using the modified Puls method) are summarized in the following table:

Time (min)	Inflow, I (m³/s)	$2S/\Delta t - O$ (m³/s)	$2S/\Delta t + O$ (m³/s)	O (m³/s)
0	0	0	0	0
30	2.2	0.84	2.2	0.68
60	7.7	3.76	10.74	3.49
90	1.9	4.26	13.36	4.55
120	1.1	2.78	7.26	2.24

From these results, it is already clear that the maximum outflow from the detention pond is 4.55 m³/s, which is higher than the predevelopment peak of 2.8 m³/s and is therefore unacceptable. Decreasing the crest length (=4.25 m) of the weir by the factor 2.8/4.55 = 0.62 gives a new crest length, b, of 0.62(4.25 m) = 2.64 m. Using a rounded number of 2.50 m, the revised weir equation is

$$Q = 1.83bh^{\frac{3}{2}} = 1.83(2.5)h^{\frac{3}{2}} = 4.58h^{\frac{3}{2}}$$

The revised storage-outflow characteristics of the detention basin are:

Elevation (m)	Storage, S (m^3)	Outflow, O (m^3/s)	$2S/\Delta t + O$ (m^3/s)
0	0	0	0
0.5	5,544	1.62	7.78
1.0	12,200	4.58	18.14
1.5	20,056	8.41	30.70

The routing computations are summarized in the following table:

Time (min)	Inflow, I (m^3/s)	$2S/\Delta t - O$ (m^3/s)	$2S/\Delta t + O$ (m^3/s)	O (m^3/s)
0	0	0	0	0
30	2.2	1.28	2.2	0.46
60	7.7	6.00	11.18	2.59
90	1.9	7.90	15.60	3.85
120	1.1	5.88	10.90	2.51

From these results, it is already clear that the maximum outflow from the detention basin is 3.85 m^3/s, which is higher than the predevelopment peak of 2.8 m^3/s and is therefore unacceptable. The crest length must be further decreased until the maximum weir discharge is less than or equal to 2.8 m^3/s. This occurs when the crest length, b, is decreased to 1.30 m. The revised weir-discharge equation is then given by

$$Q = 1.83bh^{\frac{3}{2}} = 1.83(1.30)h^{\frac{3}{2}} = 2.38h^{\frac{3}{2}}$$

and the revised storage-outflow characteristics of the reservoir are:

Elevation (m)	Storage, S (m^3/s)	Outflow, O (m^3/s)	$2S/\Delta t + O$ (m^3/s)
0	0	0	0
0.5	5,544	0.84	7.00
1.0	12,200	2.38	15.94
1.5	20,056	4.37	26.66

The routing computations are summarized in the following table:

Time (min)	Inflow, I (m^3/s)	$2S/\Delta t - O$ (m^3/s)	$2S/\Delta t + O$ (m^3/s)	O (m^3/s)
0	0	0	0	0
30	2.2	1.68	2.2	0.26
60	7.7	8.32	11.58	1.63
90	1.9	12.42	17.92	2.75
120	1.1	10.84	15.42	2.29
150	0.80	9.08	12.74	1.83
180	0.70	7.66	10.58	1.46
210	0.58	6.60	8.94	1.17
240	0.38	5.68	7.56	0.94
270	0.22	4.78	6.28	0.75
300	0.11	3.89	5.11	0.61
330	0	3.04	4.00	0.48

Time (min)	Inflow, I (m^3/s)	$2S/\Delta t - O$ (m^3/s)	$2S/\Delta t + O$ (m^3/s)	O (m^3/s)
360	0	2.32	3.04	0.36
390	0	1.76	2.32	0.28
420	0	1.34	1.76	0.21
450	0	1.02	1.34	0.16
480	0	0.78	1.02	0.12
510	0	0.60	0.78	0.09
540	0	0.46	0.60	0.07
570	0	0.34	0.46	0.06
600	0	0.26	0.34	0.04
630	0	0.20	0.26	0.03
660	0	0.16	0.20	0.02
690	0	0.12	0.16	0.02
720	0	0.10	0.12	0.01

For a crest length of 1.30 m, the maximum postdevelopment discharge is 2.75 m^3/s and is therefore acceptable. The maximum water level in the detention pond corresponds to a weir overflow rate of 2.75 m^3/s; from the weir-discharge equation, this corresponds to $h = 1.10$ m. If this water elevation is excessive, the engineer could consider expanding the proposed detention basin.

It is interesting to note that for any pond with an uncontrolled outlet the peak water-surface elevation in the pond will always occur at the time when the outflow hydrograph intersects the receding limb of the inflow hydrograph. Prior to this intersection, inflow exceeds outflow and the water level is rising, and beyond this intersection outflow exceeds inflow and the water level is falling.

Flood-control systems are primarily designed to ensure that postdevelopment peak-discharge rates do not exceed predevelopment peak-discharge rates. Using detention basins to accomplish this goal generally results in a postdevelopment-discharge hydrograph that is shifted in time and has an overall greater volume compared to the predevelopment-discharge hydrograph. Consequently, the postdevelopment-discharge hydrograph generally has higher off-peak discharge rates than the predevelopment hydrograph. Effects of increased runoff volumes from developed areas include (1) prolonged rise in the water surface downstream of the development, which might affect the slope and stability of channels; (2) the increase in runoff volume represents the amount of ground-water recharge that is no longer being absorbed on-site; and (3) an increased volume of water is released into downstream detention ponds (Haestad Methods, Inc., 1997a).

In cases where there is not ample surface area available to meet storage requirements, *underground detention* might be necessary. Underground detention may consist of a series of large pipes or prefabricated custom chambers manufactured specifically for underground detention.

5.7.2.3 Water-quality control

The quality of urban runoff is determined principally by nonstructural controls and structural controls. *Nonstructural controls* are practices that reduce the accumulation and generation of potential pollutants at or near their source, while *structural controls* are practices that involve an engineered facility to control the quality of runoff. Examples of nonstructural controls include land-use planning and management, floodplain protection, public education, fertilizer and pesticide application control, street sweeping, household hazardous waste recycling programs, and erosion control at construction sites. Examples of structural controls include stormwater-detention basins and infiltration basins.

Water-quality control regulations usually require a defined level of treatment, such as a specified detention time in a sedimentation basin and/or the retention of a specified volume of initial runoff called the *water-quality volume*. The most effective stormwater-management systems are designed to satisfy both detention and retention criteria. Detention basins are commonly used for sedimentation purposes, and infiltration basins or underground exfiltration trenches are used for retention purposes. The specified retention volume (i.e., the water-quality volume) is usually less than or equal to the runoff volume in at least 90% of the annual runoff events, and typically corresponds to a runoff depth on the order of 1.3 cm (0.5 in.).

Detention Systems

The design of detention basins for water-quality control is fundamentally different than for flood control because flood-control detention basins are designed to attenuate peak runoff rates from large storms (with long return periods), whereas water-quality detention basins are designed to provide sufficient detention time for the sedimentation of pollutant loads in smaller, more frequent storms that usually contain much higher concentrations of pollutants than runoff from large-rainfall events. Processes such as natural die-off of bacteria and plant uptake of soluble nitrogen and phosphorus also occur in detention basins, but sedimentation is the principal process of pollutant removal. A common practice is to use *dual-purpose basins* designed to control both peak discharges and pollution from stormwater runoff.

The main design criteria for water-quality detention basins are (Akan, 1993): (1) detain the design runoff long enough to provide the targeted level of treatment, and (2) evacuate the design runoff soon enough to provide available storage for the next runoff event. The required detention time is determined by the settling velocities of the pollutants in the runoff. A mean detention time of about 18 hours is usually sufficient to settle out 60% of total suspended solids, lead, and hydrocarbons, and 45% of total BOD, copper, and phosphates from urban storm runoff (Whipple and Randall, 1983). The required evacuation time for a water-quality detention basin is based on the average time between design runoff events and is usually specified by local regulatory agencies. According to the Environmental Protection Agency (USEPA, 1986a), the average interval between storms in most parts of the United States is between 73 and 108 hours, while the average time between storms in the southwestern part of the United States is much higher, on the order of 277 hours (USEPA, 1986a). Florida requires that detention basins empty within 72 hours after a storm event, and Delaware requires that 90% of the runoff be evacuated in 36 hours or in 18 hours for residential areas (Akan, 1993).

The design of a detention basin for given detention and evacuation times is a reservoir routing problem. *Wet-detention basins* (detention ponds) contain a permanent pool of water, and the detention time, t_d, is estimated from the outflow hydrograph by the relation

$$\int_0^{t_d} O(t) \, dt = V \tag{5.282}$$

where $O(t)$ is the outflow from the detention basin as a function of time and V is the average volume of the detention basin. The evacuation time in wet-detention basins is typically taken as the time from the peak of the outflow hydrograph (when the storage is a maximum) to the time that 95% of the surcharge storage has been evacuated (Wurbs and James, 2002). Detention basins without a permanent pool of water are called *dry-detention basins*. The detention time in dry-detention basins is typically taken as the time difference between the centroids of the inflow and outflow hydrographs (Haan et al., 1994; Wurbs and James, 2002). The evacuation time is taken as the interval between when inflow first enters the detention basin and when outflow from the basin ceases.

EXAMPLE 5.56

A runoff hydrograph is routed through a detention pond, and the results are given in the following table:

Time (min)	Inflow (m^3/s)	Storage (m^3/s)	Outflow (m^3/s)
0	0	10,000	0
30	2.2	11,746	0.26
60	7.7	18,955	1.63
90	1.9	23,653	2.75
120	1.1	21,817	2.29
150	0.8	19,819	1.83
180	0.7	18,208	1.46
210	0.58	16,993	1.17
240	0.38	15,958	0.94
270	0.22	14,977	0.75
300	0.11	14,050	0.61
330	0	13,168	0.48
360	0	12,412	0.36
390	0	11,836	0.28
420	0	11,395	0.21
450	0	11,062	0.16
480	0	10,810	0.12
510	0	10,621	0.09
540	0	10,477	0.07
570	0	10,360	0.06
600	0	10,270	0.04
630	0	10,207	0.03
660	0	10,162	0.02
690	0	10,126	0.02
720	0	10,099	0.01

Estimate the detention time and evacuation time of the detention pond.

Solution   The average volume, V, of the detention pond is determined by averaging the storage over the duration of the discharge (0 to 720 min), which yields

$V = 13{,}567$ m^3. The detention time, t_d, is defined by Equation 5.282, and t_d is the time when the cumulative outflow is equal to V. The cumulative outflow as a function of time is tabulated as follows:

Time (min)	Outflow (m^3/s)	Cumulative outflow (m^3)	Time (min)	Outflow (m^3/s)	Cumulative outflow (m^3/s)
0	0	0	390	0.28	26,406
30	0.26	234	420	0.21	26,847
60	1.63	1,935	450	0.16	27,180
90	2.75	5,877	480	0.12	27,432
120	2.29	10,413	510	0.09	27,621
150	1.83	14,121	540	0.07	27,765
180	1.46	17,082	570	0.06	27,882
210	1.17	19,449	600	0.04	27,972
240	0.94	21,348	630	0.03	28,035
270	0.75	22,869	660	0.02	28,080
300	0.61	24,093	690	0.02	28,116
330	0.48	25,074	720	0.01	28,143
360	0.36	25,830			

From these results, $t_d = 146$ min $= 2.4$ h. The peak of the outflow hydrograph (2.75 m^3/s) occurs at $t = 90$ min and corresponds to a surcharge of 23,653 m^3 − 10,000 m^3 = 13,653 m^3. When 95% of the surcharge has been evacuated, the remaining storage is 10,000 m^3 + 0.05(13,653 m^3) = 10,682 m^3, and from the given data this storage occurs at t = 500 min. The evacuation time is therefore equal to 500 min − 90 min = 410 min = 6.8 h.

EXAMPLE 5.57

If the inflow and outflow hydrographs given in the previous example were from a dry detention basin, estimate the detention time and evacuation time.

Solution The detention time can be approximated by the difference between the centroids of the inflow and outflow hydrographs. Denoting the points on the inflow hydrograph by $I_i = I(t_i)$, and the outflow hydrograph by $O_i = O(t_i)$, the centroids of the inflow and outflow hydrographs, t_I and t_O, are defined as

$$t_I = \frac{\sum_{i=1}^{N_I} t_i I_i}{\sum_{i=1}^{N_I} I_i}$$

and

$$t_O = \frac{\sum_{i=1}^{N_O} t_i O_i}{\sum_{i=1}^{N_O} O_i}$$

where N_I and N_O are the number of points on the inflow and outflow hydrographs, respectively. The computations of t_I and t_O are summarized in the following table:

t_i (min)	I_i (m^3/s)	$t_i I_i$ (m^3)	O_i (m^3/s)	$t_i O_i$ (m^3)
0	0	0	0	0
30	2.2	3,960	0.26	468
60	7.7	27,720	1.63	5,868

t_i (min)	I_i (m^3/s)	$t_i I_i$ (m^3)	O_i (m^3/s)	$t_i O_i$ (m^3)
90	1.9	10,260	2.75	14,850
120	1.1	7,920	2.29	16,488
150	0.80	7,200	1.83	16,470
180	0.70	7,560	1.46	15,768
210	0.58	7,308	1.17	14,742
240	0.38	5,472	0.94	13,536
270	0.22	3,564	0.75	12,150
300	0.11	1,980	0.61	10,980
330	0	0	0.48	9,504
360	0	0	0.36	7,776
390	0	0	0.28	6,552
420	0	0	0.21	5,292
450	0	0	0.16	4,320
480	0	0	0.12	3,456
510	0	0	0.09	2,754
540	0	0	0.07	2,268
570	0	0	0.06	2,052
600	0	0	0.04	1,440
630	0	0	0.03	1,134
660	0	0	0.02	792
690	0	0	0.02	828
720	0	0	0.01	432
Total	15.69	82,944	15.64	169,920

Based on these results:

$$t_I = \frac{82,944}{15.69} = 88 \text{ min}$$

$$t_O = \frac{169,920}{15.64} = 181 \text{ min}$$

The detention time, t_d, is therefore given by

$$t_d = t_O - t_I = 181 - 88 = 93 \text{ min} = 1.6 \text{ h}$$

The evacuation time is equal to the time from when inflow first enters the detention basin to when outflow from the basin ceases. Inflow begins at $t = 0$ min and ceases at about $t = 720$ min. Therefore, the evacuation time is equal to 720 min or 12 h.

Wet-detention basins.   Wet-detention basins are the most common type of detention basin used in stormwater management. *Wet-detention basins*, also called *wet-detention ponds* or simply *detention ponds*, are designed to maintain a permanent pool of water and temporarily store runoff until it is released at a controlled rate. Detention ponds remove pollutants by physical, chemical, and biological processes. In addition to sedimentation, chemical flocculation occurs when heavier sediment particles overtake and coalesce with smaller (lighter) particles; biological removal is accomplished by uptake of pollutants by aquatic plants and metabolism by phytoplankton and microorganisms. Detention ponds are typically used for drainage areas of 4 ha or more, and the runoff volume to be treated, called the *water-quality volume*, typically corresponds to 1.3 cm of runoff from the drainage area. The removal of dissolved pollutants primarily occurs between storms. In cases where there is no pond outlet, detention ponds are called *retention ponds*, although the term "retention pond" is frequently used (incorrectly) to describe wet-detention ponds in general. Detention and retention ponds are sometimes called *amenity lakes*, and can be included in the design of golf courses and the landscaping of parks and open spaces.

The layout of a typical detention pond is illustrated in Figure 5.61. The three most important factors in determining the removal efficiency of detention ponds are: (1) the volume of the permanent pool; (2) the depth of the permanent pool; and (3) the presence of a shallow littoral zone. The volume of the permanent pool should be sufficient to provide two to four weeks of detention time so that algae can grow, and the ratio of the volume of the detention pond to the detained volume for water-quality treatment should be at least 4 to achieve total suspended-sediment removal rates of 80% to 90% (ASCE, 1998). The depth of the permanent pool should be greater than 1 to 2 meters, to prevent wind-generated waves from resuspending accumulated bottom sediments and to reduce bottom-weed growth by minimizing sunlight penetration to the bottom of the pond. The depth of the pond should be less than 3 to 5 meters so that the water remains well mixed and the bottom sediment remains aerobic. An anaerobic condition in the bottom of the pond will mobilize nutrients and metals into the water column and significantly reduce the effectiveness of the detention pond. In Florida, detention ponds up to 9 meters deep have been used successfully when excavated in high-ground-water areas, probably because of the improved circulation at the bottom of the pond as a result of ground water moving through the pond (ASCE, 1998). The presence of a littoral zone is essential to the proper performance of a detention

FIGURE 5.61: Layout of detention pond.

Source: Schueler (1987).

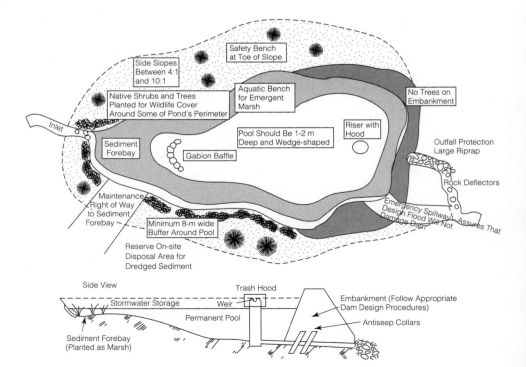

pond, since the aquatic plants in the littoral zone provide much of the biological assimilation of the dissolved stormwater pollutants. The littoral zone should cover 25% to 50% of the surface area of the detention pond (ASCE, 1998) and have a slope of 6:1 (H:V) or less to a depth of 60 cm (2 ft) below the permanent-pond elevation. This is sometimes called an *aquatic bench* (Paine and Akan, 2001). Small side slopes provide a measure of public safety, especially for children. It is recommended that the flow length in the detention pond be extended as much as possible between the inlet and outlet structures and that the outlet structure from the detention pond be designed such that an average-annual runoff event, captured as a surcharge above the permanent pool, be drained in approximately 72 hours, or whatever is the local interstorm duration (Urbonas and Stahre, 1993). The length-to-width ratio of detention ponds is sometimes recommended to be greater than 4:1 (Yousef and Wanielista, 1989), but ASCE (1998) suggests that a length-to-width ratio of 3:1 is preferable. This length-to-width requirement is intended to minimize short circuiting, enhance sedimentation, and prevent vertical stratification within the permanent pool (Hartigan, 1989). Overflow from detention ponds is allowed for larger-rainfall events, with appropriate restrictions for flood control.

Dry-detention basins.   *Dry-detention basins* are areas that are normally dry, but function as detention reservoirs during runoff events. Dry-detention basins remove pollutants primarily by sedimentation. The removal efficiency of these basins is regarded as poor for detention times shorter than 12 hours, good for detention times longer than 24 hours, and excellent for a detention times of around 48 hours (USFHWA, 1996). The design of dry-detention basins for pollutant removal is much less scientific than for detention ponds. Although sedimentation is still the primary pollutant-removal process, the estimation of basin performance is derived mostly from empirical results. Dry-detention basins should have a volume at least equal to the average runoff event during the year (Grizzard et al., 1986), and this volume be drained in no less than 40 hours (Urbonas and Stahre, 1993). However, dry-detention basins with long drain times tend to be breeding grounds for mosquitoes, have "boggy" bottoms with wetland vegetation, and are usually difficult to maintain and clean. Typical removal rates for properly designed dry-detention basins are given in Table 5.39.

Dry-detention basins are typically used where the drainage area exceeds 4 ha; however, because they take up large areas, dry-detention basins are generally not well suited for high-density residential developments. In many cases, dry-detention

TABLE 5.39: Typical Removal Rates in Dry-Detention Basins

Pollutant	Removal Rate
TSS	50%–70%
TP	10%–20%
Nitrogen	10%–20%
Organic matter	20%–40%
Pb	75%–90%
Zn	30%–60%
Hydrocarbons	50%–70%
Bacteria	50%–90%

Source: Urbonas and Stahre (1993).

FIGURE 5.62: Layout of extended (dry) detention basin.

Source: Schueler (1987).

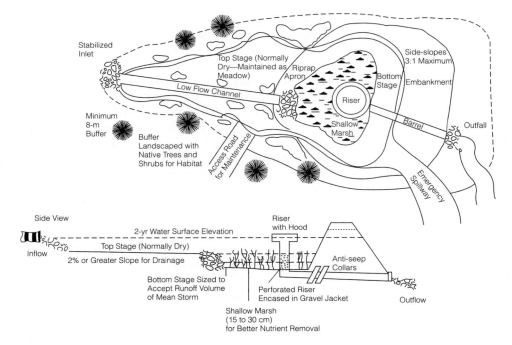

basins have a dual purpose in both quality and quantity (peak-flow) control. For dual-purpose dry-detention basins, the bottom portion of the basin is designed with a its own low-level outlet structure for water-quality control, while the basin as a whole is designed to control the peak discharge. The bottom portion of the basin used for water-quality control typically has a detention time much longer than the basin as a whole and this (bottom) portion of the basin is sometimes called an *extended dry-detention basin* or simply *extended detention basin* and is illustrated in Figure 5.62. The water-quality volume in the dry-detention basin typically corresponds to 1.3 cm of runoff from the drainage area. In designing dry-detention basins, the volume of the basin can be taken as equal to the runoff volume to be treated, provided the catchment area is less than 100 ha. If the catchment area is larger than 100 ha, reservoir routing is necessary to determine the volume of the basin. The calculated volume should be increased by 20% to account for sediment accumulation (ASCE, 1998). The outlet structure, such as a V-notch weir or perforated riser, should be designed to drain the basin in the specified design period. To ensure that small-runoff events will be adequately detained, ASCE (1998) recommends that the outlet empty less than 50% of the design volume in the first one-third of the design emptying period. The shape of dry-detention basins should be such that they gradually expand from the inlet and contract toward the outlet to reduce short circuiting. Riprap or other methods of stabilization should be provided within a low-flow channel and at the outflow channel to resist erodible velocities. A length-to-width ratio of 2:1 or greater, preferably up to a ratio of 4:1, is recommended (ASCE, 1998). Basin side slopes of 4:1 (H:V) or greater provide for facility maintenance and safety concerns, and a forebay with a volume equal to approximately 10% of the total design volume can help with the maintenance of the basin by facilitating sediment deposition near the inflow, thereby extending the service life of the remainder of the basin. Embankments for small on-site basins should be protected from at least the 100-year flood; whenever possible, dry-detention basins should be incorporated within larger flood-control facilities. Some dry-detention basins are hardly noticed by the public, since they may be implemented as multiple-use facilities that function as parks and recreation areas, and are filled only during exceptional storms. Maintenance of dry-detention basins is both essential and costly, with the general objectives being to prevent clogging, prevent standing water, and prevent the growth of weeds and wetland plants.

EXAMPLE 5.58

A dry-detention basin is to be designed for a 60-ha (150-ac) residential development. The runoff coefficient of the area is estimated as 0.3, and the average rainfall depth during the year is 3.3 cm (1.3 in.). Design the detention basin and estimate the suspended-solids removal efficiency.

Solution   The average runoff volume, V, is given by

$$V = CAD$$

where C is the runoff coefficient ($= 0.3$), A is the catchment area ($=60$ ha), and D is the average rainfall depth ($=3.3$ cm). Hence,

$$V = (0.30)(60 \times 10^4)(3.3 \times 10^{-2}) = 5940 \text{ m}^3$$

The required volume of the dry-detention basin should be increased by 20% to account for sediment accumulation and is therefore equal to $1.2(5940 \text{ m}^3) = 7130 \text{ m}^3$. The outlet between the bottom of the detention basin and the elevation corresponding to a basin storage of 7130 m^3 should be designed to discharge 7130 m^3 in approximately 40 hours. An appropriate outlet structure could be a perforated riser or a V-notch weir, and overflow from the detention basin should be accommodated when the storage volume exceeds 7130 m^3. In accordance with Table 5.39, the suspended-solids removal in the basin is expected to be in the range of 50% to 70%.

Wet- vs. dry-detention basins.   In deciding whether to use a wet- or dry-detention basin as a water-quality control, an important consideration is whether nutrient removal (nitrogen and phosphorus) is an important requirement. This is particularly the case when the quality of the receiving water is sensitive to nutrient loadings. Properly designed wet-detention basins (detention ponds) generally provide much better nutrient removal than dry-detention basins, since many of the nutrients in surface runoff are in dissolved form and are not significantly affected by the sedimentation process in dry-detention basins (Hartigan, 1989). This functional advantage of wet-detention basins must be balanced against their greater land requirements. For example, the permanent pool of a wet-detention basin can require anywhere from two to seven times more storage than the alternative dry-detention basin.

Retention systems

Stormwater retention is the most effective quality-control method, but it can be used only in situations where the captured volume of water can infiltrate into the ground before the next storm. The most common retention systems are infiltration basins, swales, and below-ground exfiltration trenches. In many cases, these systems closely reproduce the predevelopment water balance (Dodson, 1998).

Infiltration basins.   *Infiltration basins* are excavated areas that impound stormwater runoff, which then infiltrates into the ground. Infiltration basins, also called *dry-retention basins*, are similar in appearance and construction to dry-detention basins, except that the detained stormwater runoff is exfiltrated through permeable soils beneath the basin. The layout of a typical infiltration basin is illustrated in Figure 5.63. Infiltration basins typically serve areas ranging from front yards to 20 ha (ASCE, 1992). Infiltration basins are classified as either on-line or off-line. *On-line infiltration basins* retain a specified water-quality volume; when a larger runoff occurs, it overflows the basin, which then acts as a detention pond for the larger event. Some drainage systems divert the water-quality volume out of the normal drainage path and into *off-line infiltration basins* that hold it for later treatment.

Infiltration basins must be located in soils that allow the runoff to infiltrate within 72 hours, or within 24 to 36 hours for infiltration areas that are planted with grass. The seasonal high-water table should be at least 1.2 m (4 ft) below the ground surface in the infiltration basin to assure that the pollutants in the runoff are removed by the vegetation, soil, and microbes before reaching the water table (Urbonas and Stahre, 1993). Infiltration rates can be calculated using the Green–Ampt or similar models (see Section 5.3.3.3), but soils with saturated infiltration rates less than 8 mm/h are not suitable for infiltration basins (ASCE, 1998). Design guidelines suggested by ASCE (1998) are that water ponding in infiltration basins be less than 0.3 m during the design storm and that the design infiltration rate be limited to a maximum of 50 mm/h to account for clogging. The bottom of an infiltration basin should be graded as flat as

FIGURE 5.63: Layout of infiltration basin

Source: Guo (2001a).

Top-view

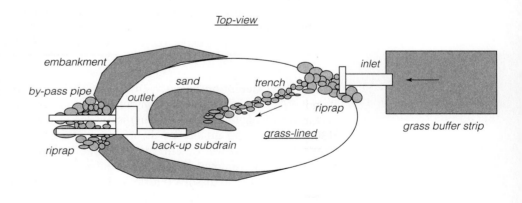

Side Profile

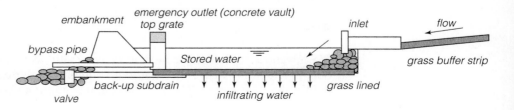

possible to allow for uniform ponding and infiltration, and the side slope of the basin should be less than 3 : 1 (H : V) to allow for easier mowing and better bank stabilization. Water-tolerant turf such as reed canary grass or tall fescue should be used in the basin to promote better infiltration and pollutant filtering. In urban settings, infiltration basins are commonly integrated into recreational areas, greenbelts, neighborhood parks, and open spaces, while in highway drainage they may be located in rights-of-way or in open space within freeway interchange loops. Infiltration basins are susceptible to clogging and sedimentation, and can require large land areas. Although the two main problems typically associated with infiltration basins are clogging and contamination of underlying soil and ground water, data collected at several infiltration basins have indicated minimal clogging and soil-contamination depths less than 50 cm after about 20 years of operation (Dechesne et al., 2005). During routine operation, standing water in infiltration basins can create problems of security and insect breeding.

EXAMPLE 5.59

An infiltration basin is to be designed to retain the first 1.3 cm of runoff from a 10-ha catchment. The area to be used for the infiltration basin is turfed, and field measurements indicate that the native soil has a minimum infiltration rate of 150 mm/h. If the retained runoff is to infiltrate within 24 hours, determine the surface area that must be set aside for the basin.

Solution   The volume, V, of the runoff corresponding to a depth of 1.3 cm = 0.013 m on an area of 10 ha = 10^5 m^2 is given by

$$V = (0.013)(10^5) = 1300 \text{ m}^3$$

The native soil has a minimum infiltration rate of 150 mm/h. To account for clogging, however, the design infiltration rate will be taken as 50 mm/h = 0.05 m/h. The area, A, of infiltration basin required to infiltrate 1300 m^3 in 24 h is given by

$$A = \frac{1300}{(0.05)(24)} = 1080 \text{ m}^2 = 0.11 \text{ ha}$$

If runoff is not to pond to more than 0.3 m, then the maximum volume that can be handled by a 0.11-ha infiltration basin is 0.3 m × 0.11 ha = 330 m^3. Since the runoff volume to be handled is 1300 m^3, the area of the infiltration basin should be at least 1300/330 × 0.11 ha = 0.43 ha.

In areas where the water table is shallow, particular care should be taken to assess the mounding of the water table below the basin. In cases where the infiltrated water causes the water table to rise to the ground elevation, ponding will occur and the infiltration basin will not function properly. A methodology for assessing the mounding beneath circular infiltration basins can be found in Guo (2001).

Swales.   Swales are shallow vegetated (grass) open channels with small longitudinal and side slopes that transport and infiltrate runoff from adjacent land areas. Swales are commonly used in highway medians and for roadside drainage on rural roads. Typically, swales are designed as free-flowing open channels with inclined slopes, but they can also be designed to be nonflowing, with all the surface runoff retained and

infiltrated into the ground. Nonflowing swales are sometimes called *retention swales* (Field et al., 2000).

Retention swales can be designed either to retain a fixed volume of runoff or to retain the entire runoff from a design storm. Retention swales used for water-quality control are typically designed to retain a fixed depth of runoff and are sized based on their stage-storage relation; elevated outlets, such as those shown in Figure 5.64, are used for excess runoff above the water-quality volume. Retention swales used for water-quantity control are designed to retain the entire runoff from a design storm, and are sized such that the infiltration rate is equal to the peak runoff rate. In this case, if the peak runoff rate is Q_p, then the length, L, of swale required for the infiltration rate to be equal to the runoff rate is given by

$$L = \frac{Q_p}{fP} \tag{5.283}$$

where f is equal to the infiltration capacity of the soil and P is the wetted perimeter of the swale. For any cross-sectional shape, the wetted perimeter can be expressed in terms of the runoff rate via the Manning equation. For triangular-shaped swales, Equation 5.283 can be combined with the Manning equation to yield (Wanielista and colleagues, 1997)

$$L = \frac{151,400 Q_p^{5/8} m^{5/8} S^{3/16}}{n^{3/8}(1 + m^2)^{5/8} f} \tag{5.284}$$

where L is the swale length in meters, Q_p is the peak runoff rate in m³/s, m is the side slope, S is the longitudinal slope, n is the Manning roughness coefficient, and f is the infiltration capacity in cm/h. Typical values of the Manning roughness coefficient are $n = 0.20$ for routinely mowed swales and $n = 0.24$ for infrequently mowed swales (ASCE, 1998). In the case of trapezoidal sections

$$L = \frac{360,000 Q}{\left\{ b + 2.38 \left[\frac{Qn}{(2\sqrt{1+m^2}-m)S^{1/2}} \right]^{3/8} \sqrt{1 + m^2} \right\} f} \tag{5.285}$$

where b is the bottom width of the swale in meters. Equation 5.285 applies to the *best trapezoidal section*, where the perimeter, P, is related to the flow depth, y, and the side slope m by

$$P = 4y\sqrt{1 + m^2} - 2my \tag{5.286}$$

For swales to be effective, the infiltration rate of the underlying soil should be greater than 13 mm/h, the longitudinal grade should be set as flat as possible to promote infiltration, never steeper than 3%–5% (Urbonas and Stahre, 1993; Yu et al., 1994), and side slopes should be flatter than 4:1 (H:V) to maximize the contact area. It is sometimes recommended that swales be designed for 2-year storms and that the runoff volume in nonflowing or slow-moving swales be infiltrated within 36 hours (Urbonas and Stahre, 1993). In cases where the length of the swale required to infiltrate the runoff is excessive, then a *check dam* can be used to store the runoff within the swale, in which case the depth in the swale should not exceed 0.5 m. Check dams usually consist of crushed rock or pretreated timber up to 0.6 m in height. Grassed swales should be mowed to stimulate vegetative growth, control weeds, and maintain the capacity of the system.

EXAMPLE 5.60

A triangular-shaped swale is to retain the runoff from a catchment with design peak-runoff rate of 0.02 m³/s. The longitudinal slope of the swale is to be 3% with side slopes of 4:1 (H:V). If the grassed swale has a Manning n of 0.24 (infrequently mowed grass) and a minimum infiltration rate of 150 mm/h, determine the length of swale required.

Solution   From the given data: $Q = 0.02$ m³/s, $m = 4$, $S = 0.03$, $n = 0.24$, and $f = 150$ mm/h $= 15$ cm/h. Equation 5.284 gives the required swale length, L, as

$$L = 151,400 \, \frac{Q^{5/8} m^{5/8} S^{3/16}}{n^{3/8}(1 + m^2)^{5/8} f}$$

$$= 151,400 \, \frac{(0.02)^{5/8}(4)^{5/8}(0.03)^{3/16}}{(0.24)^{3/8}(1 + 4^2)^{5/8} 15}$$

$$= 314 \text{ m}$$

This length of swale is quite long. A ponded infiltration basin would require less area and would probably be more cost effective.

In the previous example, the swale was designed to retain and infiltrate the entire surface runoff. In cases where the swale is to provide treatment of the surface runoff, primarily through trapping a portion of the sediment and organic biosolids in the vegetative cover, the swale is designed as a *biofilter* and is called a *biofiltration swale*. For the swale to perform adequately as a biofilter, the following design criteria are recommended (ASCE, 1998):

- Minimum hydraulic residence time of 5 minutes
- Maximum flow velocity of 0.3 m/s

- Maximum bottom width of 2.4 m
- Minimum bottom width of 0.6 m
- Maximum depth of flow no greater than one-third of the gross or emergent vegetation height for infrequently mowed swales, or no greater than one-half of the vegetation height for regularly mowed swales, up to a maximum height of approximately 75 mm for grass and approximately 50 mm below the normal height of the shortest plant species
- Minimum length of 30 m

For biofiltration swales, ASCE (1998) recommends longitudinal slopes of 1%–2%, with a minimum of 0.5% and a maximum of 6%. When the longitudinal slope is less than 1%–2%, perforated underdrains should be installed or, if there is adequate moisture, wetland species should be established. If the slope is greater than 2%, check dams should be used to reduce the effective slope to approximately 2%. Using these guidelines, the following design procedure is proposed for biofiltration swales (updated from ASCE, 1998):

1. Estimate the runoff rate for the design event and limit the discharge to approximately 0.03 m³/s by dividing the flow among several swales, installing upstream detention to control release rates, or reducing the developed surface area to reduce the runoff coefficient and gain space for biofiltration.
2. Establish the slope of the swale.
3. Select a vegetation cover suitable for the site.
4. Estimate the height of vegetation that is expected to occur during the storm runoff season. The design flow depth should be at least 50 mm less than this vegetation height and a maximum of approximately 75 mm in biofiltration swales and 25 mm in filter strips.
5. Typically, biofiltration swales are designed as trapezoidal channels (skip this step for filter-strip design). When using a rectangular section, provide reinforced vertical walls.
6. For a trapezoidal cross section, select a side slope that is no steeper than 3 : 1 (H : V), with 4 : 1 (H : V) or flatter preferred.
7. Compute the bottom-width and flow velocity. Limit the design velocity to less than 0.3 m/s.
8. Compute the swale length using the design velocity from step 7 and an assumed hydraulic detention time, preferably greater than 5 minutes. If the computed swale length is less than 30 m, increase the swale length to 30 m and adjust the bottom-width.

Biofiltration swales are low-cost stormwater best-management practices (BMP) that have proven effective for controlling runoff pollution from land surfaces, especially highways and agricultural lands. Biofiltration swales are attractive options for agencies such as departments of transportation, since they are easily incorporated into the landscape, such as highway medians. The primary mechanisms for pollutant removal in swales are filtration by vegetation, settling of particulates, and infiltration into the subsurface zone. In general, biofiltration swales show good performance for removal of suspended solids, but are not considered efficient for removal of nutrients (Yu et al., 2001). Recent studies indicate that the most effective biofiltration swales have a minimum length of 75 m and a maximum longitudinal slope of 3% (Yu et al., 2001).

EXAMPLE 5.61

A biofiltration swale is to be constructed on a 1% slope to handle a design runoff of $0.03 \ \text{m}^3/\text{s}$. During the wet season, the swale is expected to be covered with grass having an average height of 130 mm with Class E retardance. Design the biofiltration swale.

Solution From the given data: $Q = 0.03 \ \text{m}^3/\text{s}$, $S_0 = 0.01$, and the average height of the vegetation is 130 mm. This given data covers the specifications in steps 1 to 3 of the design procedure.

Step 4. The design depth in the biofiltration swale should be at least 50 mm below the height of the vegetation (130 mm − 50 mm = 80 mm), with a maximum height of 75 mm. Therefore, in this case, the design flow depth is taken as 75 mm.

Step 5. Use a trapezoidal section for the swale.

Step 6. Use side slopes of 4:1 ($m = 4$), a bottom-width b, and a depth $y = 75$ mm (=0.075 m). The flow area, A, wetted perimeter, P, and hydraulic radius, R, are given by

$$A = by + my^2 = b(0.075) + (4)(0.075)^2 = 0.075b + 0.0225$$

$$P = b + 2\sqrt{1 + m^2}\,y = b + 2\sqrt{1 + 4^2}(0.075) = b + 0.618$$

$$R = \frac{A}{P} = \frac{0.075b + 0.0225}{b + 0.618} \tag{5.287}$$

where $0.6 \ \text{m} < b < 2.4 \ \text{m}$.

Step 7. The Manning equation requires that

$$Q = \frac{1}{n}AR^{2/3}S_0^{1/2} = \frac{1}{n}\frac{A^{5/3}}{P^{2/3}}S_0^{1/2}$$

In this case,

$$0.03 = \frac{1}{n}\frac{(0.075b + 0.0225)^{5/3}}{(b + 0.618)^{2/3}}(0.01)^{1/2}$$

or

$$\frac{1}{n^3}\frac{(0.075b + 0.0225)^5}{(b + 0.618)^2} = 0.027 \tag{5.288}$$

For Class E retardance, Manning's n is given by Table 3.23 as

$$n = \frac{1.22R^{1/6}}{52.1 + 19.97\log(R^{1.4}S_0^{0.4})} \tag{5.289}$$

Simultaneous solution of Equations 5.287, 5.288, and 5.289 gives

$$b = 4.44 \ \text{m}$$

which exceeds the maximum bottom-width (for a uniform flow distribution) of 2.4 m. Repeated solution of Equations 5.287, 5.288, and 5.289

shows that a flow of 0.014 m³/s will require a bottom-width, b of 2.4 m, with a flow velocity ($= Q/A$) of 0.067 m/s. Since the flow velocity is less than the maximum velocity of 0.3 m/s, two swales each having a bottom-width of 2.4 m and handling half the flow should be used.

Step 8. Using a detention time of 5 minutes with the design velocity of 0.067 m/s gives the length, L, of the swale as

$$L = Vt = (0.067)(5 \times 60) = 20.1 \text{ m}$$

which is shorter than the minimum length of 30 m. Therefore use a length of 30 m.

In summary, two biofiltration swales are required, each with a trapezoidal cross-section with a bottom width of 2.4 m, side slopes of 4 : 1, and 30 m long.

Vegetated filter strips.   *Vegetated filter strips* (also called *buffer strips*) are mildly sloping vegetated surfaces, usually grass, that are located between impervious surfaces (pollutant sources) and water-quality control areas. Vegetated filter strips are designed to slow the runoff velocity from the impervious area, reducing the peak-runoff rate and increasing the opportunities for infiltration, sedimentation, and trapping of the pollutants. Vegetated filter strips are often used as pretreatment (to remove sediments) for other structural practices such as dry-detention basins and exfiltration trenches. These areas are designed to receive overland sheet flow, and they provide little treatment for concentrated flows. The design procedure for filter strips is the same as that for biofiltration swales, with the additional constraints that the average depth of flow be no more than 25 mm and the hydraulic radius be taken equal to the flow depth (ASCE, 1998). The width of the filter strip should be sufficiently limited to achieve a uniform-flow distribution. Grassed filter strips may develop a berm of sediment at the upper edge that must be periodically removed. Mowing will maintain a thicker vegetative cover, providing better sediment retention. Recommended areas for use are in agriculture and low-density developments. Although studies indicate highly varying effectiveness, trees in strips can be more effective than grass strips alone because of their greater uptake and long-term retention of plant nutrients. Properly constructed forested and grass fiter strips can be expected to remove more than 60% of the particulates and perhaps as much as 40% of plant nutrients in urban runoff (Dodson, 1998). In arid and semiarid climates, grass buffer strips need to be irrigated (Field et al., 2000).

Exfiltration trenches.   Exfiltration trenches, also called *infiltration trenches*, *percolation trenches*, and *french drains*, are common in urban areas with large impervious areas and high land costs. An exfiltration trench typically consists of a long narrow excavation, ranging from 1 to 4 m in depth, backfilled with gravel aggregate (2.5 to 7.6 cm) and surrounded by a filter fabric to prevent the migration of fine soil particles into the trench, which can cause clogging of the gravel aggregate (Harrington, 1989). The maximum trench depth is limited by trench-wall stability, seasonal high ground-water levels, and the depth to any impervious soil layer. Exfiltration trenches 1 m wide and 1 to 2 m deep seem to be most efficient (ASCE, 1998). Exfiltration trenches can have their top elevation either at the ground surface or below ground.

FIGURE 5.65: Layout of surface
exfiltration trench
Source: Guo (2001a).

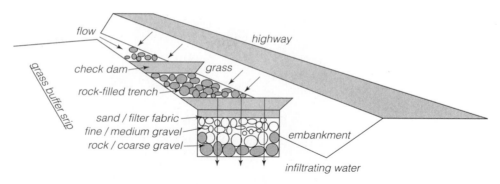

FIGURE 5.65: Layout of surface
exfiltration trench
Source: Guo (2001a).

FIGURE 5.66: Picture of surface
exfiltration trench
Source: Guo (2001a).

Surface trenches receive sheet flow runoff directly from adjacent areas after it has been filtered by a grass buffer. The layout of a typical surface exfiltration trench is shown in Figure 5.65 and pictured in Figure 5.66. Surface exfiltration trenches are typically used in residential areas where smaller loads of sediment and oil can be trapped by grassed filter strips that are at least 6 m wide (Harrington, 1989). For below-ground exfiltration trenches, runoff is collected by a stormwater inlet on top of a catch basin, and water collected in the catch basin is delivered to the below-ground exfiltration trench by a perforated pipe. A below-ground exfiltration trench under construction is shown in Figure 5.67, and a close-up view of the perforated pipe used in the trench is shown in Figure 5.68. The exfiltration trench shown in Figure 5.67 consists of a catch basin in the center, perforated pipe extending from both sides into the trench, and a partial fill of rockfill aggregate which is surrounded by a filter fabric. For below-ground exfiltration trenches, a minimum of 0.3 m of soil cover should be provided for the establishment of vegetation. Adequate ground-water (quality) protection is generally obtained by providing at least 1.2 m separation between the bottom of the trench and the seasonal high water table.

Exfiltration trenches are frequently used for highway median strips and parking lot "islands" (depressions between two lots or adjacent sides of one lot). To prevent clogging, inlets to underground exfiltration trenches should include trash racks, catch basins, and baffles to reduce sediment, leaves, other debris, and oils and greases. Exfiltration trenches are especially vulnerable to clogging during construction. Once

FIGURE 5.67: Below-ground exfiltration trench under construction

FIGURE 5.68: Perforated pipe used in below-ground exfiltration trench

operational, routine maintenance consists of vacuuming debris from the catch basin inlets and, if needed, using high-pressure hoses to wash clogging materials out of the pipe. However, once trenches are clogged, rehabilitation is difficult. Surface trenches are more susceptible to sediment accumulations than underground trenches, but their accessibility makes them easier to maintain. About 50% of exfiltration trenches constructed in the eastern United States have failed (Schueler et al., 1991). Although the nature and reason for these failures are not typically reported, clogging within the trench and of its infiltrating surfaces is suspected (Field et al., 2000).

Exfiltration trenches are considered off-line systems, and their purpose is to store and exfiltrate the runoff from frequent storms into the ground. Pollutant-removal mechanisms in exfiltration trenches include adsorption, filtering, and microbial decomposition below the trench. Exfiltration trenches are typically designed to serve single-family residential areas up to 4 ha (10 ac) and commercial areas up to 2 ha (5 ac) in size, and are particularly suited for rights-of-way, parking lots, easements, and other areas with limited space. Due to concerns about soil and ground-water contamination, exfiltration trenches should not be used at industrial and commercial sites that may be susceptible to spillage of soluble pollutants, such as gasoline, oils, and solvents.

Exfiltration trenches can be used only at sites with porous soils, favorable site geology, and proper ground-water conditions. Site conditions that are favorable to exfiltration trenches are (Stahre and Urbonas, 1989; Harrington, 1989; ASCE, 1998):

1. The hydraulic conductivity surrounding the trench exceeds 2 m/d.
2. The distance between the bottom of the trench and the seasonal high water table or bedrock exceeds 1.2 m.
3. Water-supply wells are more than 30 m from the trench (to prevent possible contamination).
4. The trench is located at least 6 m from building foundations (to avoid possible hydrostatic pressures on foundations or basements).
5. The ground slope downstream of the trench does not exceed 20%, which would increase the chance of downstream seepage and slope failure.

Exfiltration trenches are generally designed to retain a volume equal to the difference between the runoff volume and the volume of water exfiltrated during a storm (ASCE, 1998). Assuming that the water in the trench percolates through one-half of the trench height, saturated-flow conditions exist between the trench and the water table, and there is negligible outflow from the bottom of the trench (due to clogging), the total outflow rate, Q_{out}, from the two (long) sides of the trench is given by Darcy's law (see Chapter 6, Section 6.2.1) as

$$Q_{out} = 2\left(K_t \frac{H}{2} L\right) = K_t HL \tag{5.290}$$

where K_t is the trench hydraulic conductivity, H is the height of the trench, L is the length of the trench, and the hydraulic gradient is assumed equal to unity. In cases where saturated-flow conditions do not exist between the trench and the water table, the assumption of a unit hydraulic gradient gives a conservative estimate of the trench outflow (Duchene et al., 1994). Taking t as the duration of the storm, the volume exfiltrated in time t, V_{out}, is given by

$$V_{out} = Q_{out}t = K_t HLt \tag{5.291}$$

The runoff volume into the trench during time t, V_{in}, is given by the rational formula (see Section 5.4.2.1) as

$$V_{in} = CiAt \tag{5.292}$$

where C is the runoff coefficient, i is the average intensity of a storm with duration t, and A is the area of the catchment contributing flow to the exfiltration trench. The storage capacity of the trench, V_{stor}, is given by

$$V_{stor} = nWHL \tag{5.293}$$

where n is the porosity in the trench, typically taken as 40% (ASCE, 1998). The trench dimensions must be such that

$$V_{\text{in}} = V_{\text{stor}} + V_{\text{out}} \tag{5.294}$$

Combining Equations 5.291 to 5.294 yields

$$CiAt = nWHL + K_tHLt \tag{5.295}$$

Solving for the trench length, L, gives

$$\boxed{L = \frac{CiAt}{(nW + K_t t)H}} \tag{5.296}$$

The rainfall intensity, i, is related to the storm duration, t, by an intensity-duration-frequency (IDF) curve that typically has the form

$$i = \frac{a}{(t + b_1)^{c_1}} \tag{5.297}$$

where a, b_1, and c_1 are constants (see Section 5.2.1). The combination of Equations 5.296 and 5.297 indicates that the required trench length, L, varies as a function of the storm duration, t, and the design length should be chosen as the maximum length required for any storm duration (with a given return period). In some cases, the runoff from a specified rainfall depth is required to be handled by the trench, and in these cases ASCE (1998) recommends that the trench be designed for an IDF curve with a return period in which the specified rainfall depth falls in 1 hour. The contributing area to an exfiltration trench is usually less than 4 ha due to storage requirements for peak-runoff control.

EXAMPLE 5.62

An exfiltration trench is to be designed to handle the runoff from a 1-ha commercial area with an average runoff coefficient of 0.7. The IDF curve for the design rainfall is

$$i = \frac{548}{(t + 7.24)^{0.73}}$$

where i is the average rainfall intensity in mm/h and t is the storm duration in minutes. The trench hydraulic conductivity estimated from field tests is 15 m/d, the seasonal high water table is 5 m below the ground surface, and local regulations require that a safety factor of 2 be applied to the trench hydraulic conductivity to account for clogging. Design the exfiltration trench.

Solution   From the given data: $C = 0.7$, $A = 1\,\text{ha} = 10{,}000\,\text{m}^2$, and $K_t = 15/2 = 7.5\,\text{m/d} = 0.31\,\text{m/h}$ (using a safety factor of 2). According to ASCE (1998) guidelines, the porosity, n, of the gravel pack can be taken as 40% ($n = 0.4$), and a trench width, W, of 1 m and height, H, of 2 m can be expected to perform efficiently. Substituting these values into Equation 5.296 gives

$$L = \frac{CiAt}{(nW + K_t t)H} = \frac{(0.7)i(10{,}000)t}{(0.4 \times 1 + 0.31t)(2)} = \frac{7000it}{0.8 + 0.62t} \tag{5.298}$$

where i in m/h, and t in hours are related by

$$i = \frac{0.548}{(60t + 7.24)^{0.73}} \text{ m/h} \tag{5.299}$$

Combining Equations 5.298 and 5.299 gives the required trench length, L, as a function of the storm duration, t, as

$$L = \frac{3836t}{(0.8 + 0.62t)(60t + 7.24)^{0.73}} \tag{5.300}$$

Taking the derivative with respect to t gives

$$\frac{dL}{dt} = \frac{[(0.8 + 0.62t)(60t + 7.24)^{0.73}]3836 - 3836t[(0.8 + 0.62t)0.73(60t + 7.24)^{-0.27}60 + (60t + 7.24)^{0.73}0.62]}{[(0.8 + 0.62t)(60t + 7.24)^{0.73}]^2}$$

$$= \frac{3836(0.8 + 0.62t)(60t + 7.24)^{0.73} - 168,000t(0.8 + 0.62t)(60t + 7.24)^{-0.27} - 2378t(60t + 7.24)^{0.73}}{[(0.8 + 0.62t)(60t + 7.24)^{0.73}]^2}$$

and the maximum-value criterion, $dL/dt = 0$, yields

$$3836(0.8 + 0.62t)(60t + 7.24)^{0.73} - 168,000t(0.8 + 0.62t)(60t + 7.24)^{-0.27}$$
$$- 2378t(60t + 7.24)^{0.73} = 0$$

which gives $t = 0.759$ h, and substituting into Equation 5.300 yields $L = 127$ m. On the basis of this result, a trench length of 127 m gives the trench volume required to handle the design storm without causing surface ponding. Since the seasonal high-water table is 5 m below the ground surface and the minimum allowable spacing between the bottom of the trench and the water table is 1.2 m, a (maximum) trench height of 5 m − 1.2 m = 3.8 m would still be adequate and would yield the shortest possible trench length (for the specified trench width of 1 m). Since the trench length is inversely proportional to the trench height (see Equation 5.296), using a trench height, H, of 3.8 m would give a required trench length, L, of

$$L = 127 \text{ m} \times \frac{2}{3.8} = 67 \text{ m}$$

Hence, a trench capable of handling the design storm is 1 m wide, 3.8 m high, and 67 m long. If the (vertical) side slopes are not stable for a trench of this depth, then the trench height and length can be adjusted according to their inverse proportionality.

In designing an exfiltration trench there must be reasonable assurance that the aquifer beneath it can conduct water away from the trench as fast as water exfiltrates from the trench to the water table. Under normal circumstances, the water table beneath the trench will mound until there is sufficient induced gradient to conduct the exfiltrated water away. The trench will fail if the water-table mound rises above the trench bottom to interfere with the operation of the trench. A method to predict the mounding depth under exfiltration trenches is presented in Chapter 6, Section 6.8.8.

5.7.2.4   Best-management practices

A stormwater *best-management practice* (BMP) is a method or combination of methods found to be the most effective and feasible means of preventing or reducing the amount of pollution generated by nonpoint sources to a level compatible with water-quality goals (South Florida Water Management District, 2002). Methods of controlling pollutants in stormwater runoff are categorized as nonstructural or structural practices. Nonstructural BMPs are practices that improve water quality by reducing the accumulation and generation of potential pollutants at or near their source. Structural BMPs involve building an engineered facility for controlling quantity and quality of urban runoff.

Nonstructural BMPs generally fall into the following four categories: planning and regulatory tools; source controls; maintenance and operational procedures; and educational and outreach programs. Planning and regulatory tools include hazardous materials codes, zoning, land-development and land-use regulations, water-shortage and conservation policies, and controls on types of flow allowed to drain into municipal storm-sewer systems. Source controls include erosion and sediment control during construction, collection and proper disposal of waste materials, and modified use of chemicals such as fertilizer, pesticides, and herbicides. Maintenance and operational procedures include turf and landscape management, street cleaning, catch-basin cleaning, road maintenance, and canal/ditch maintenance. Educational and outreach programs include distributing toxics checklists for meeting household hazardous-waste regulations, producing displays and exhibits for school programs, distributing free seedlings for erosion control, and creating volunteer opportunities such as water-quality monitoring. Prevention practices such as planning and zoning tools to ensure setback, buffers, and open-space requirements can be implemented with ease at the planning stage of any development with a high degree of success

Structural BMPs for controlling stormwater runoff generally fall into two main categories: retention systems and detention systems. Retention BMP systems include exfiltration trenches, infiltration basins, grassed swales, vegetated filter strips, and retention ponds. Detention BMP systems include dry- and wet-detention basins. Important factors to be considered in selecting a BMP are the amount of runoff to be handled by the system and the soil type. Guidance for selecting best-management practices is given in Table 5.40. Generally, well-designed BMPs listed in Table 5.40 can provide high pollutant removal for nonsoluble particulate pollutants, such as suspended sediment and trace metals. Much lower removal rates are achieved for soluble pollutants such as phosphorus and nitrogen. Each site has its natural attributes that influence the type and configuration of the stormwater-management system. For

TABLE 5.40: Water-Quality Control System Selection Criteria (USFHWA, 1996)

Best management practices	Contributing Area (ha)					Acceptable NRCS soil types
	0–2	2–4	4–12	12–20	>20	
Exfiltration trenches	•	•				A, B
Infiltration basins		•	•	•		A, B
Grassed swales	•	•	•			A, B, C
Filter strips	•					A, B, C
Retention ponds			•	•	•	B, C, D
Detention ponds			•	•	•	A, B, C, D

example, a site with sandy soils would suggest the use of infiltration practices such as retention areas integrated into a development's open space and landscaping, while natural low areas offer opportunities for detention systems.

A comprehensive stormwater-management program should include a combination of structural and nonstructural components that are properly selected, designed, implemented, inspected, and regularly maintained.

5.7.3 Major System

The major drainage system includes features such as natural and constructed open channels, streets, and drainage easements such as floodplains. The major drainage system handles runoff events that exceed the capacity of the minor drainage system, typically the 100-year storm, and must be planned concurrently with the design of the minor drainage system. The design of open channels has been discussed extensively in Chapter 4, Section 3.5. In major urban drainage systems, concrete- and grass-lined channels are the most common, with grass lining usually preferred for aesthetic reasons. Concrete-lined channels have smaller roughness coefficients, require smaller flow areas, and are used when hydraulic, topographic, and right-of-way needs are important considerations.

5.8 Evapotranspiration

Evaporation is the process by which water is transformed from the liquid phase to the vapor phase, and *transpiration* is the process by which water moves through plants and evaporates through leaf stomatae, which are small openings in the leaves. In cases where the ground surface is covered by vegetation, it is usually not easy to differentiate between evaporation from the ground surface and transpiration through plants; the combined process of evaporation and transpiration is called *evapotranspiration*. The relative contribution of evaporation and transpiration to evapotranspiration varies. For example, when crops are in their initial stages of development and cover less than 10% of the ground area, water is predominantly lost by soil evaporation, but once the crop is well developed and completely covers the soil, transpiration becomes the dominant process. Evaporation from soil is affected by such factors as soil water content, type, the presence or absence of surface mulches, and environmental conditions imposed on the soil (Burt et al., 2005); transpiration is affected by such factors as the type of vegetation and the stage of development.

Evapotranspiration does not contribute significantly to the water budget over time scales of individual storms, but over longer time periods (weeks and months) it is a major component in the terrestrial water budget. On an annual basis, approximately 70% of the rainfall in the United States is returned to the atmosphere via evapotranspiration, and predicting evapotranspiration is of primary interest in the design of irrigation and surface-water storage systems.

Three standard evapotranspiration rates are commonly used in practice: (1) potential evapotranspiration, (2) reference-crop evapotranspiration, and (3) actual evapotranspiration. *Potential evapotranspiration* is used synonymously with the term *potential evaporation*, which is defined as the quantity of water evaporated per unit area, per unit time, from an idealized, extensive free water surface under existing atmospheric conditions. Potential evaporation is typically used as a measure of the meteorological control on evaporation from an open water surface (e.g., lake,

TABLE 5.41: Grass-Reference
Evapotranspiration Rates

Range (mm/d)	Classification
0–2.5	Low
2.5–5.0	Moderate
5.0–7.5	High
>7.5	Very high

Source: Allen et al. (1988).

reservoir) or bare saturated soil. *Reference-crop evapotranspiration* is the rate of evapotranspiration from an area planted with a specific (reference) crop, where water availability is not a limiting factor. Reference-crop evapotranspiration is used as a measure of evapotranspiration from a standard vegetated surface. Grass and alfalfa are by far the most commonly used reference crops in hydrologic practice. Grass is characteristic of short vegetation (8–15 cm tall) and alfalfa is characteristic of tall vegetation (>50 cm tall). Grass-reference evapotranspiration rates are a measure of the meteorological influence on evapotranspiration, and typical values are given in Table 5.41. The *actual evapotranspiration* is the evapotranspiration that occurs under actual soil, ground-cover, and water-availability conditions. Typically, the actual evapotranspiration is taken as the reference-crop evapotranspiration multiplied by a *crop coefficient*.

Over the past several decades, hundreds of empirical and semi-empirical methods of estimating evapotranspiration have evolved. These can be broadly classified as *radiation*, *temperature*, *combination*, or *evaporation-pan* methods. Radiation and temperature methods relate evapotranspiration solely to net radiation (solar plus longwave) and air temperature, respectively. Combination methods are the most comprehensive and account for both energy utilization and the processes required to remove the vapor, once it has evaporated into the atmosphere. In contrast to the combination methods, the radiation and temperature methods neglect the wind velocity and specific-humidity gradient above the evaporating surface, both of which influence the ability of the air to transport water vapor away from the evaporative surface. Combination methods are usually most accurate in estimating actual and reference-crop evapotranspiration, with radiation methods providing reasonable estimates in most cases, and better estimates in some cases (Sumner and Jacobs, 2005; Irmak et al., 2003c). Most temperature methods perform poorly because they do not account for the net radiation, vapor-pressure deficit, or sunshine percentage, which play important roles in determining the evapotranspiration, particularly in humid regions where the variation in evapotranspiration is more often due to variations in these factors than to variations in temperature (Trajkovic, 2005; Irmak et al., 2003c). Evaporation-pan methods are empirical formulations that relate reference-crop evapotranspiration to measured water evaporation from standardized open pans. Pan-evaporation estimates can be poor and erratic over daily time scales, and are better suited to longer-term estimates (weekly, monthly). Several evapotranspiration methods of various types are listed in Table 5.42. There seems to be a consensus among present-day hydrologists that the physically based Penman–Monteith equation provides the best description of the evapotranspiration process. This assertion is supported by several quantitative comparisons between measured and predicted evapotranspiration at several test sites around the world (ASCE, 1990).

TABLE 5.42: Methods for Estimating Evapotranspiration

Classification	Method	References
Radiation	Jensen–Haise	Jensen and Haise (1963); Jensen et al. (1971)
	FAO-24 Radiation	Doorenbos and Pruitt (1975; 1977)
	Priestly–Taylor	Priestly and Taylor (1972)
	Turc	Turc (1961); Jensen (1966)
Temperature	NRCS Blaney–Criddle	USDA (1970)
	FAO-24 Blaney–Criddle	Doorenbos and Pruitt (1977); Allen and Pruitt (1986)
	Hargreaves	Hargreaves et al. (1985); Hargreaves and Samani (1985)
	Thornthwaite	Thornthwaite (1948); Thornthwaite and Mather (1955)
Combination	Penman–Monteith	Monteith (1965; 1981); Allen (1986); Allen et al. (1989)
	Penman	Penman (1963)
	1972 Kimberly–Penman	Wright and Jensen (1972)
	1982 Kimberly–Penman	Wright (1982)
	FAO-24 Penman	Doorenbos and Pruitt (1975; 1977)
	FAO-PPP-17 Penman	Frére and Popov (1979)
	Businger-van Bavel	Businger (1956); van Bavel (1966)
Evaporation-pan	Christiansen	Christiansen (1968); Christiansen and Hargreaves (1969)
	FAO-24 Pan	Doorenbos and Pruitt (1977)

Source: ASCE (1990).

Satellite-based measurement, used in association with energy-balance models, is an evolving method to estimating evapotranspiration and crop coefficients over field and catchment scales (Tasumi et al., 2005). The commonly used Surface Energy Balance Algorithm for Land (SEBAL), which has been applied in more than 30 countries worldwide, is estimated to have a field-scale accuracy of 85% for daily time scales and 95% for seasonal time scales (Bastiaanssen et al., 2005).

5.8.1 ASCE Penman–Monteith Method

The ASCE Penman–Monteith (ASCE-PM) equation is based on the Penman–Monteith form of the Penman combination equation (Monteith, 1981) and is widely accepted as the best-performing method for estimating evapotranspiration (ET) from meteorological data (Todorovic, 1999). The Penman–Monteith method has been recommended as the primary method for defining grass reference ET (Smith et al., 1991; Allen et al., 1998). The basic hypothesis of the Penman–Monteith approach is that transpiration of water through leaves is composed of three serial processes: the transport of water through the surface of the leaves against a surface or canopy resistance, r_s; molecular diffusion against a molecular boundary-layer resistance, r_b; and turbulent transport against an aerodynamic resistance, r_a, between the layer in the immediate vicinity of the canopy surface and the planetary boundary layer. The boundary-layer resistance, r_b, for water vapor is usually much smaller than the aerodynamic resistance, r_a, or the surface resistance, r_s, and can be ignored in comparison with both other resistances. The ASCE Penman–Monteith equation or *ASCE-PM equation* for estimating the *crop evapotranspiration*, ET_c, from vegetated surfaces where availability of water is not a limiting factor is given by

$$ET_c = \frac{1}{\rho_w \lambda} \left[\frac{\Delta(R_n - G) + \rho_a c_p \frac{e_s - e_a}{r_a}}{\Delta + \gamma \left(1 + \frac{r_s}{r_a}\right)} \right] \tag{5.301}$$

where ρ_w is the density of water; λ is the latent heat of vaporization of water; Δ is the gradient of the saturated vapor pressure versus temperature curve; R_n is the net radiation (solar plus long wave); G is the soil heat flux; ρ_a is the density of moist air; c_p is the specific heat of moist air ($= 1.013$ kJ/(kg°C)); e_s is the saturation vapor pressure; e_a is the ambient vapor pressure; r_a is the aerodynamic resistance to vapor and heat diffusion; γ is the psychrometric constant; and r_s is the bulk surface resistance. Equation 5.301 is dimensionally homogeneous, and any consistent set of units can be used. The ASCE-PM equation (Equation 5.301) is commonly referred to as simply the Penman–Monteith equation, with the ASCE designation primarily associated with the methodology used to estimate the parameters in the equation. It is important to keep in mind that Equation 5.301 is applicable when the availability of water is not a limiting factor; hence, the evapotranspiration estimated using the Penman–Monteith equation (Equation 5.301) depends on weather-related and crop parameters only. Several studies have shown relatively good and consistent agreement of the Penman–Monteith equation with measured ET values in various climates (Allen et al., 1989, 1994a, 1998; Howell et al., 2000; Wright et al., 2000; Itenfisu et al., 2003). The Penman–Monteith equation can also be used to estimate the evaporation from large open bodies of water, and recommended methods to be used in estimating the parameters in the Penman–Monteith equation are described below.

Aerodynamic resistance, r_a.   The aerodynamic resistance, r_a, measures the resistance to vapor flow from air flowing over vegetated surfaces. In the absence of significant thermal stratification, r_a can be estimated using the relation (Pereira et al., 1999)

$$r_a = \frac{\ln\left[\frac{z_m - d}{z_{0m}}\right] \ln\left[\frac{z_h - d}{z_{0h}}\right]}{k^2 u_z} \tag{5.302}$$

where z_m is the height of wind measurements, d is the zero-plane displacement height, z_h is the height of air temperature and humidity measurements, z_{0m} is the roughness length governing momentum transfer, z_{0h} is the roughness length governing heat and vapor transfers, k is the von Kármán constant, and u_z is the wind speed measured at height z_m. Typically, for full-cover uniform crops, d and z_{0m} are related to the crop height, h, by (Brutsaert, 1982; ASCE, 1990)

$$d = \frac{2}{3}h \quad \text{and} \quad z_{0m} = 0.123h \tag{5.303}$$

Since the momentum transfer governs the heat and vapor transfer, the roughness height z_{0h} is assumed to be a function of z_{0m}, where

$$z_{0h} = az_{0m} \tag{5.304}$$

For tall and partially covering crops $a = 1$ and for fully covering crops $a = 0.1$ (Monteith, 1973; Campbell, 1977; Thom and Oliver, 1977; Brutsaert, 1982; Allen et al., 1989; ASCE, 1990). Three-cup and propeller *anemometers* are both widely used to measure wind speed, and measurements by both types are reliable, provided that the mechanical parts are functioning properly. Anemometers are typically placed at a standard height of either 5 m or 10 m above ground in meteorologic applications and 2 m or 3 m in agrometeorologic applications; for evapotranspiration applications, wind speeds at 2 m above the ground surface are generally required. To adjust wind speed collected at elevations other than the standard height of 2 m, the following logarithmic wind-speed profile can be used above a short grassed surface to adjust measurements (Allen et al., 1998)

wind profile over short grassed surface:

$$u_2 = \frac{4.87}{\ln{(67.8z - 5.42)}} u_z \tag{5.305}$$

where u_2 is the wind speed 2 m above the ground surface, and u_z is the wind speed measured z m above the ground surface. Care should be taken that measured wind speeds are characteristic of the local area. In cases where the wind speed is not measured over a short grassed surface, the following equation should be used to adjust the wind speed:

general wind profile:

$$u_2 = \frac{\ln{\left(\dfrac{2-d}{z_{0m}}\right)}}{\ln{\left(\dfrac{z-d}{z_{0m}}\right)}} u_z \tag{5.306}$$

where d and z_{0m} are given by Equation 5.303. It is important to note that Equation 5.305 is derived from Equation 5.306 by assuming a crop height, h, of 0.12 meters. The standard grass reference crop has a height, h, of 0.12 m. Using the standardized height for wind speed, temperature, and humidity as 2 m ($z_m = z_h = 2$ m), assuming a fully covering crop ($a = 0.1$), and taking the von Kármán constant, k, as 0.41, Equations 5.302 to 5.304 give

$$r_a = \frac{\ln{\left[\dfrac{2 - \frac{2}{3}(0.12)}{0.123(0.12)}\right]} \ln{\left[\dfrac{2 - \frac{2}{3}(0.12)}{(0.1)0.123(0.12)}\right]}}{(0.41)^2 u_2} = \frac{208}{u_2} \tag{5.307}$$

This equation is commonly used as a starting point to estimate the aerodynamic resistance of a standard grass reference crop.

The aerodynamic resistance of open water can be estimated using the equation (Thom and Oliver, 1977)

$$r_a = \frac{4.72\left[\ln\frac{z_m}{z_0}\right]^2}{1 + 0.536u_2}\ \text{s/m} \tag{5.308}$$

where z_m is the measurement height of meteorological variables above the water surface, and z_0 is the aerodynamic roughness of the surface. Taking $z_m = 2$ m and $z_0 = 1.37$ mm (Thom and Oliver, 1977), the aerodynamic resistance is determined by the wind speed. The aerodynamic resistance of open water is generally higher than found for either short vegetation or forest and reflects a greater resistance to vapor transport due to lower levels of turbulence.

(Bulk) surface resistance, r_s. The surface resistance, r_s, describes the resistance to vapor flow through leaf stomatae, and is sometimes referred to as the *canopy resistance* or *stomatal resistance*. The surface resistance, r_s, varies as leaf stomatae open and close in response to various micrometeorological conditions and is dependent on the particular plant species. It is generally assumed that the surface resistances of trees are greater than those of shorter vegetation, since trees tend to have stomatal control, while shorter vegetation does not (Olmsted, 1978). An acceptable approximation for estimating the surface resistance, r_s, of dense full-cover vegetation is (Allen et al., 1998)

$$r_s = \frac{r_l}{\text{LAI}_{\text{active}}} \tag{5.309}$$

where r_l is the bulk stomatal resistance of a well-illuminated leaf (s/m), and $\text{LAI}_{\text{active}}$ is the active (sunlit) leaf-area index (dimensionless). The bulk stomatal resistance, r_l, of a single well-illuminated leaf typically has a value on the order of 100 s/m, and the *leaf area index*, LAI, is defined as the ratio of the surface area of the leaves (upper side only) to the projection of the vegetation on the ground surface. The *active leaf area index* includes only the leaf area that actively contributes to the surface heat and vapor transfer, typically the upper, sunlit-portion of a dense canopy. LAI values for various crops differ widely, but values of 3–5 are common for many mature crops. The LAI for grass and alfalfa reference crops can be estimated using the following relations (Allen et al., 1998):

For clipped grass:
$$\text{LAI} = 24h \tag{5.310}$$

For alfalfa:
$$\text{LAI} = 5.5 + 1.5\ln h \tag{5.311}$$

where h is the height of the vegetation in meters. For various reference crops (grass, alfalfa), a general equation for estimating $\text{LAI}_{\text{active}}$ is (Allen et al., 1998)

$$\text{LAI}_{\text{active}} = 0.5\text{LAI} \tag{5.312}$$

For a standard grass reference surface with a height of 0.12 m and a stomatal resistance of 100 s/m, the bulk surface resistance, r_s, is given by

$$r_s = \frac{100}{0.5(24)(0.12)} = 70\ \text{s/m} \tag{5.313}$$

The bulk surface resistance, r_s, for large open-water bodies such as lakes is zero by definition, since there are no restrictions to water moving between the liquid and vapor phases at the surface of the lake.

Net radiation, R_n.   The net radiation, R_n, is approximately equal to the net solar (shortwave) radiation, S_n, plus the net longwave radiation, L_n, hence

$$\boxed{R_n = S_n + L_n}$$

(5.314)

The net radiation, R_n, can be measured directly using *pyradiometers* or *net radiometers*, which sense both short- and longwave radiation using upward- and downward-facing sensors. Net radiation is difficult to measure because net radiometers are hard to maintain and calibrate, resulting in systematic biases (EWRI, 2002). As a consequence, the net radiation, R_n, is often predicted using empirical equations. According to EWRI (2002), such predictions are routine and highly accurate. The net radiation, R_n, is normally positive during the day and negative during the night, and the total daily value of R_n is almost always positive, except in extreme conditions at high latitudes (Allen et al., 1998).

The net shortwave radiation, S_n, is equal to the fraction of the incoming solar radiation that is not reflected by the ground cover and is given by

$$S_n = (1 - \alpha)R_s$$

(5.315)

where α is the *albedo* or *canopy reflection coefficient*, defined as the fraction of shortwave radiation reflected at the surface, and R_s is the total incoming solar radiation. Typical albedos for various surfaces are given in Table 5.43, and a more detailed list of albedos can be found in Ponce and colleagues (1997). Total incoming solar radiation, R_s, is commonly measured using *pyranometers*, which measure the incoming shortwave radiation to a solid angle in the shape of a hemisphere oriented upward. When a pyranometer is oriented downward it measures the reflected shortwave radiation and is called an *albedometer*. When two pyranometers are used together, one oriented upward and the other downward, the net shortwave radiation, S_n, is measured directly and the instrument is called a *net pyranometer*.

The availability of measurements of R_s and/or R_n varies, and where they are not available, empirical estimation methods are required. Empirical formulae relating R_s directly to measurements of maximum and minimum temperature, $T_{\max}$ and $T_{\min}$, are very convenient for engineers, agronomists, climatologists, and others who work with the U.S. National Weather Service climatological data, where only $T_{\max}$, $T_{\min}$, and

TABLE 5.43: Typical Albedos

Land cover	Albedo, α
Open water	0.08
Tall forest	0.11–0.16
Tall farm crops	0.15–0.20
Cereal crops	0.20–0.26
Short farm crops	0.20–0.26
Grass and pasture	0.20–0.26
Bare soil	0.10 (wet)–0.35 (dry)

Source: Shuttleworth and Maidment (1993).

daily rainfall are recorded on a regular basis. In the absence of direct measurements, R_s can be estimated using the relation (Hargreaves and Samani, 1982)

$$R_s = (KT)(T_{max} - T_{min})^{0.5} S_0 \tag{5.316}$$

where KT is an empirical coefficient, S_0 is the extraterrestrial radiation (MJ m^{-2}d^{-1}), T_{max} is the maximum daily temperature (°C), and T_{min} is the minimum daily temperature (°C). Hargreaves (1994) recommended using KT = 0.19 for coastal regions, and KT = 0.162 for interior regions. Several other methods for estimating KT have been proposed (Allen, 1997; Samani, 2000); however, all of these methods give similar estimates of KT at lower land elevations. The conventional method for estimating the total incoming solar radiation, R_s, is the Angstrom formula, which relates the solar radiation to extraterrestrial radiation and relative sunshine duration by

$$R_s = \left(a_s + b_s \frac{n}{N} \right) S_0 \tag{5.317}$$

where n is the number of bright-sunshine hours per day (h), N is the total number of daylight hours in the day (h), S_0 is the extraterrestrial radiation (MJ m^{-2}d^{-1}), a_s is a regression constant expressing the fraction of extraterrestrial radiation reaching the earth on overcast days ($n = 0$), and $a_s + b_s$ is the fraction of extraterrestrial radiation reaching the earth on clear days ($n = N$). Depending on atmospheric conditions (humidity, dust) and solar declination (latitude and month), the Angstrom values a_s and b_s will vary. Where no actual solar radiation data are available and no calibration has been carried out to estimate a_s and b_s, the values $a_s = 0.25$ and $b_s = 0.50$ are recommended. Sunshine duration, n, is commonly recorded using a *Campbell–Stokes heliograph* in which a glass globe focuses the radiation beam to a special recording paper and a trace is burned on the paper as the sun is moving. No trace is recorded in the absence of bright sunshine. The extraterrestrial solar radiation, S_0, in Equation 5.317 can be estimated using the following equation (Duffie and Beckman, 1980):

$$S_0 = \frac{24(60)}{\pi} G_{sc} d_r (\omega_s \sin \phi \sin \delta + \cos \phi \cos \delta \sin \omega_s) \tag{5.318}$$

where S_0 is in MJ/(m^{2}d), G_{sc} is the solar constant equal to 0.0820 MJ/(m^2·min), d_r is the inverse relative distance between the earth and the sun (dimensionless), ω_s is the sunset-hour angle in radians, ϕ is the latitude in radians, and δ is the solar declination in radians. Variables in Equation 5.318 can be estimated using the following relations:

$$d_r = 1 + 0.033 \cos \left(\frac{2\pi}{365} J \right) \tag{5.319}$$

$$\delta = 0.4093 \sin \left(\frac{2\pi}{365} J - 1.405 \right) \tag{5.320}$$

$$\omega_s = \cos^{-1}(- \tan \phi \tan \delta) \tag{5.321}$$

where J is the number of the day in the year between 1 (1 January) and 365 or 366 (31 December), and J is commonly referred to as the *Julian day*. Equation 5.318 is appropriate for calculating the extraterrestrial solar radiation, S_0, over daily time scales; however, for hourly or shorter durations the following modified equation is recommended (Allen et al., 1998):

$$S_0 = \frac{24(60)}{\pi} G_{sc} d_r [(\omega_2 - \omega_1) \sin\phi \sin\delta + \cos\phi \cos\delta (\sin\omega_2 - \sin\omega_1)] \quad (5.322)$$

where ω_1 and ω_2 are the solar angles at the beginning and end of the time interval. The number of daylight hours, N, is directly related to the sunset-hour angle, ω_s, and is given by (Duffie and Beckman, 1980)

$$N = \frac{24}{\pi} \omega_s \quad (5.323)$$

Cloudy skies generally reduce the number of hours of bright sunshine from N to n. Combining Equations 5.315, 5.317, and 5.318 gives the following combined relation for estimating the daily-averaged net shortwave radiation:

$$S_n = (1 - \alpha) \left[a_s + b_s \frac{n}{N} \right] \left[\frac{1440}{\pi} G_{sc} d_r (\omega_s \sin\phi \sin\delta + \cos\phi \cos\delta \sin\omega_s) \right]$$
$$(5.324)$$

In addition to the shortwave (0.3–3 μm) solar energy that is added to the surface vegetation, there is also longwave radiation (3–100 μm) that is emitted by both the atmosphere and the ground. The rate of longwave emission from any body is proportional to the absolute temperature of the surface of the body raised to the fourth power. This relation is expressed quantitatively by the Stefan–Boltzmann law. The net flux of longwave radiation leaving the earth's surface is less than that given by the Stefan–Boltzmann law due to the absorption and downward radiation from the sky. Water vapor, clouds, carbon dioxide, and dust in the atmosphere are absorbers and emitters of longwave radiation. As humidity and cloudiness play an important role, the Stefan–Boltzmann law is corrected by these two factors when estimating the net flux of longwave radiation at the ground surface. The net incoming longwave radiation, L_n, can be estimated using the equation

$$L_n = -\sigma \left(\frac{T_{max,K}^4 + T_{min,K}^4}{2} \right) \varepsilon' f \quad (5.325)$$

where L_n is in MJ m^{-2}d^{-1}, σ is the Stefan–Boltzmann constant (= 4.903 ×10^{-9} MJ m^{-2}K^{-4}d^{-1}), $T_{max,K}$ is the daily maximum absolute temperature (K), and $T_{min,K}$ is the daily minimum absolute temperature (K), ε' is the net emissivity between the atmosphere and the ground (dimensionless), and f is the *cloudiness factor* (dimensionless). The net emissivity, ε', can be estimated using the equation (Brunt, 1932; Doorenbos and Pruitt, 1975; 1977; Allen et al., 1989)

$$\varepsilon' = 0.34 - 0.14 \sqrt{e_a} \quad (5.326)$$

where e_a is the vapor pressure (kPa) of water in the atmosphere. The cloudiness factor, f, can be estimated using the relation

$$f = 1.35 \frac{R_s}{R_{s0}} - 0.35 \quad (5.327)$$

where R_{s0} is the clear-sky solar radiation that can be related to the extraterrestrial radiation, S_0, and the elevation, z using the relation (Doorenbos and Pruitt, 1977)

$$R_{s0} = (0.75 + 2 \times 10^{-5}z)S_0 \tag{5.328}$$

where z is the land-surface elevation in meters. In cases where measurements of R_s are not available, the cloudiness factor, f, can be estimated by the equation

$$f = 0.22\frac{n}{N} \tag{5.329}$$

Combining Equations 5.325, 5.326, and 5.327 gives the following relation to be used in estimating the net longwave radiation:

$$L_n = -\sigma\left(\frac{T_{max,K}^4 + T_{min,K}^4}{2}\right)(0.34 - 0.14\sqrt{e_a})\left(1.35\frac{R_s}{R_{s0}} - 0.35\right) \tag{5.330}$$

The relative solar radiation, R_s/R_{s0}, indicates the relative cloudiness and must be limited to $0.25 \le R_s/R_{s0} \le 1.0$.

Soil heat flux, G.   The soil heat flux, G, is the energy utilized in heating the soil, and is positive when the soil is warming and negative when the soil is cooling. The soil heat flux, G, can be estimated using the relation

$$G = c_s\frac{T_i - T_{i-1}}{\Delta t}\Delta z \tag{5.331}$$

where c_s is the soil heat capacity, T_{i-1} and T_i are the soil temperatures at the beginning and end of time step Δt, respectively, and Δz is the effective soil depth. It is commonly assumed that the soil temperature is equal to the air temperature; however, the soil temperature generally lags the air temperature and therefore time intervals, Δt, exceeding one day are recommended (Allen et al., 1998). The penetration depth of the temperature wave is determined by the length of the time interval. The effective soil depth, Δz, is only 0.10–0.20 m for a time interval of one or a few days but might be 2 m or more for monthly periods. The soil heat capacity, c_s, is related to the mineral composition and water content of the soil. Averaged over one day, G is typically small, but becomes more significant for hourly or monthly time periods. For hourly (or shorter) time intervals, G beneath a dense cover of grass or alfalfa does not correlate well with air temperature; however, hourly G does correlate well with net radiation and can be approximated as a fraction of R_n according to the following relations:

Short reference crop (grass):

$$G_{hour} = \begin{cases} 0.1R_n & \text{(day)} \\ 0.5R_n & \text{(night)} \end{cases}$$

Tall reference crop (alfalfa):

$$G_{hour} = \begin{cases} 0.04R_n & \text{(day)} \\ 0.2R_n & \text{(night)} \end{cases}$$

For daily time intervals beneath a grass reference surface, it can be assumed that

$$G_{\text{day}} = 0 \tag{5.332}$$

For monthly time intervals, assuming a constant heat capacity of 2.1 MJ/(m³°C) and a soil depth of about 2 m, Equation 5.331 can be used to derive G for monthly time intervals as

$$G_{\text{month}} = 0.07(T_{i+1} - T_{i-1}) \tag{5.333}$$

where T_{i-1} and T_{i+1} are the mean temperatures in the previous and following months, respectively. If T_{i+1} is unknown, then the following equation can be used in lieu of Equation 5.333:

$$G_{\text{month}} = 0.14(T_i - T_{i-1}) \tag{5.334}$$

where T_i is the mean temperature in the current month. The soil heat flux, G, can be measured directly using soil heat flux plates and thermocouples or thermistors.

Latent heat of vaporization, λ. The latent heat of vaporization, λ, is the energy required to change a unit mass of water from liquid to vapor at a constant temperature and pressure. The latent heat of vaporization, λ, can be expressed as a function of the water-surface temperature, T_s, using the empirical equation (Harrison, 1963)

$$\boxed{\lambda = 2.501 - 0.002361 T_s} \tag{5.335}$$

where λ is in MJ/kg and T_s is in °C. Since λ varies only slightly over normal temperature ranges, a constant value of 2.45 MJ/kg corresponding to a temperature of 20°C is commonly assumed.

Psychrometric constant, γ. The psychrometric constant, γ, depends on the atmospheric pressure, p, and the latent heat of vaporization, λ, and is defined as

$$\boxed{\gamma = \frac{c_p p}{\varepsilon \lambda} = 0.0016286 \frac{p}{\lambda} \text{ kPa/°C}} \tag{5.336}$$

where the specific heat of moist air, c_p, is taken as 1.013 kJ/(kg°C), and ε is the ratio of the molecular weight of water vapor to the molecular weight of dry air ($= 0.622$). The atmospheric pressure, p, at an elevation z above sea level can be estimated using the relation (Allen et al., 1998)

$$p = 101.3 \left(\frac{293 - 0.0065z}{293} \right)^{5.26} \tag{5.337}$$

where p is in kPa and z is in meters. Assuming an atmospheric pressure of 101.32 kPa, and a latent heat of vaporization of 2.444 MJ/kg, then the psychrometric constant, γ, given by Equation 5.336 is equal to 0.06752 kPa/°C.

Saturation vapor pressure, e_s. The saturation vapor pressure, e_s, can be estimated from the ambient temperature, T, using the relation

$$e_s(T) = 0.6108 \exp \left[\frac{17.27T}{T + 237.3} \right] \tag{5.338}$$

where e_s is in kPa and T is in $°C$. For hourly time intervals, T can be taken as the mean air temperature during the hourly period, while for estimates of daily-averaged ET, the saturation vapor pressure, e_s, should be estimated using the relation

$$e_s = \frac{e_s(T_{max}) + e_s(T_{min})}{2} \tag{5.339}$$

Combining Equations 5.338 and 5.339 gives the following expression for estimating the saturation vapor pressure from temperature measurements:

$$e_s = 0.3054 \left[\exp\left(\frac{17.27 T_{max}}{T_{max} + 237.3} \right) + \exp\left(\frac{17.27 T_{min}}{T_{min} + 237.3} \right) \right] \tag{5.340}$$

Vapor-pressure gradient, Δ. The gradient of the saturated vapor-pressure versus temperature curve, Δ, can be estimated using the equation

$$\Delta = \frac{4098 e_s}{(T + 237.3)^2} \tag{5.341}$$

where Δ is in kPa/$°C$, e_s is the saturation vapor pressure in kPa, and T is the air temperature in $°C$. Since e_s is determined by the air temperature, T, via Equation 5.338, then Δ can be calculated directly from the air temperature using the relation

$$\Delta = \frac{4098 \left[0.6108 \exp\left(\frac{17.27 T}{T + 237.3} \right) \right]}{(T + 237.3)^2} \tag{5.342}$$

For daily ET estimates, the air temperature, T, is taken as the average of the maximum and minimum air temperatures.

Actual vapor pressure, e_a. The actual vapor pressure, e_a, can be estimated directly from the measured dew point temperature, T_{dew}, using Equation 5.338, which yields

$$e_a = e_s(T_{dew}) = 0.6108 \exp\left[\frac{17.27 T_{dew}}{T_{dew} + 237.3} \right] \tag{5.343}$$

where T_{dew} is in $°C$. Equation 5.343 follows directly from the definition of the dew-point temperature, which is the temperature to which air needs to be cooled to make the air saturated. The dew-point temperature, T_{dew}, is often measured with a mirrorlike metallic surface that is artificially cooled, and when dew forms on the surface, its temperature is sensed as T_{dew}. Other dew-sensor systems use chemical or electric properties of certain materials that are altered when absorbing water vapor. Instruments for measuring dew-point temperature require careful operation and maintenance and are seldom available at weather stations. As an alternative to Equation 5.343, the actual vapor pressure, e_a, can be estimated using measured temperatures and relative humidities using the relation

$$e_a = \frac{e_s(T_{max}) \dfrac{RH_{max}}{100} + e_s(T_{min}) \dfrac{RH_{min}}{100}}{2} \tag{5.344}$$

where $e_s(T_{max})$ and $e_s(T_{min})$ are the saturation vapor pressures at the maximum and minimum temperatures, respectively, and RH_{max} and RH_{min} are the maximum and minimum relative humidities over the course of one day. Relative humidity is measured using a *hygrometer*. Modern hygrometers use a film from a dielectric polymer that changes its dielectric constant with changes in surface moisture, thus inducing a variation of the capacity of a condenser using that dielectric. Equation 5.344 can also be used to estimate the average vapor pressure over periods longer than one day. For periods of a week, ten days, or a month, RH_{max} and RH_{min} can be taken as the average of daily maxima and minima (respectively) during the averaging period. When using equipment where errors in estimating RH_{min} can be large, or when RH data integrity is in doubt, then it is recommended to use only RH_{max} in the following equation (Allen et al., 1998):

$$e_a = e_s(T_{min}) \frac{RH_{max}}{100} \tag{5.345}$$

In the absence of RH_{max} and RH_{min}, the following equation can be used:

$$e_a = \frac{RH_{mean}}{100} \left[\frac{e_s(T_{max}) + e_s(T_{min})}{2} \right] \tag{5.346}$$

where RH_{mean} is the mean relative humidity, defined as the average between RH_{max} and RH_{min}. However, Equation 5.346 is less desirable than Equations 5.344 or 5.345.

Air density, ρ_a. The density of moist air, ρ_a, is a function of the temperature and pressure of the air, and can be estimated using the ideal-gas law

$$\rho_a = \frac{p}{T_{Kv} R} \tag{5.347}$$

where p is atmospheric pressure in kPa, T_{Kv} is the virtual temperature in K given by (Allen et al., 1998)

$$T_{Kv} = 1.01(T + 273) \tag{5.348}$$

where T is the air temperature in $°C$, and R is the specific gas constant, equal to 0.287 kJ/(kg·K). Combining Equations 5.347 and 5.348 and taking R equal to 0.287 kJ/(kg·K) yields

$$\rho_a = 3.450 \frac{p}{T + 273} \text{ kg/m}^3 \tag{5.349}$$

where the atmospheric pressure, p, can be assumed to equal 101.32 kPa.

EXAMPLE 5.63

A constructed cattail (*Typha latifolia*) marsh is located in south Florida at $23°$ $38'$ N and $80°$ $25'$ W, and measurements during the month of February indicate the following conditions (Abtew, 1996): Maximum temperature = $26.0°$C, minimum temperature = $16.9°$C, mean temperature = $21.3°$C, maximum humidity = 98.2%, minimum humidity = 63.2%, and average wind speed at 10 m height = 3.96 m/s. The temperature and humidity measurements were collected at 4 m above the ground surface, typical cattail height in the marsh is 0.6 m, the cattails only partially cover the wetland, the surface resistance of cattails has been shown to be approximately 90 s/m (Abtew and Obeysekera, 1995), the average amount of sunshine is typically 69% of the possible amount (Winsberg, 1990), and the mean temperatures in previous and following months (January and March) are $18.7°$C and $21.3°$C, respectively. Assuming an albedo of 0.20, use these data to estimate the evapotranspiration of cattails during the month of February. Subsequent direct measurements reported by Abtew (1996) indicate an average vapor-pressure deficit of 0.67 kPa, average incoming solar radiation of 12.95 MJ/(m²·d), and average net radiation of 8.83 MJ/(m²·d) for February. How would these measurements affect your predictions?

Solution   From the given data: $\phi = 23°38' = 23.63° = 0.412$ rad, $T_{max} = 26.0°$C, $T_{min} = 16.9°$C, $RH_{max} = 98.2\%$, $RH_{min} = 63.2\%$, $z_m = 10$ m, $u_{10} = 3.96$ m/s, $z_h = 4$ m, $h = 0.6$ m, $r_s = 90$ s/m $= 1.04 \times 10^{-3}$ d/m, sunshine fraction $(n/N) = 0.69$, $T_{i-1} = 18.7°$C, $T_{i+1} = 21.3°$C, and $\alpha = 0.20$. The parameters in the Penman–Monteith equation are estimated as follows:

r_a: The aerodynamic resistance, r_a, is given by Equation 5.302 as

$$r_a = \frac{\ln\left[\frac{z_m - d}{z_{0m}}\right] \ln\left[\frac{z_h - d}{z_{0h}}\right]}{k^2 u_z}$$

where d and z_{0m} are given by Equation 5.303 as

$$d = \frac{2}{3}h = \frac{2}{3}(0.6) = 0.4 \text{ m}$$

$$z_{0m} = 0.123h = 0.123(0.6) = 0.0738 \text{ m}$$

and, since the cattails only partially cover the wetland, Equation 5.304 gives

$$z_{0h} = (1)z_{0m} = (1)(0.0738) = 0.0738 \text{ m}$$

Taking the von Kármán constant, k, as 0.41 and substituting the given and derived data into the expression for r_a yields

$$r_a = \frac{\ln\left[\frac{10 - 0.4}{0.0738}\right]\ln\left[\frac{4 - 0.4}{0.0738}\right]}{(0.41)^2 3.96} = 28.4 \text{ s/m} = 3.29 \times 10^{-4} \text{ d/m}$$

Δ: The slope of the vapor-pressure versus temperature curve, Δ, is given by Equation 5.342 as

$$\Delta = \frac{4098\left[0.6108\exp\left(\frac{17.27T}{T + 237.3}\right)\right]}{(T + 237.3)^2}$$

where Δ is in kPa/°C, and T is the average temperature in °C. When estimating daily-averaged evapotranspiration rates, T is taken as the average of the maximum and minimum daily temperature; hence

$$T = \frac{T_{max} + T_{min}}{2} = \frac{26.0 + 16.9}{2} = 21.5°C$$

Substituting into the expression for Δ yields

$$\Delta = \frac{4098\left[0.6108\exp\left(\frac{17.27(21.5)}{21.5+237.3}\right)\right]}{(21.5 + 237.3)^2} = 0.157 \text{ kPa/°C}$$

R_n: The net radiation, R_n, is equal to the sum of the net shortwave and longwave radiation according to Equation 5.314, where

$$R_n = S_n + L_n$$

The net shortwave radiation can be estimated using Equation 5.324 as

$$S_n = (1 - \alpha)\left[a_s + b_s\frac{n}{N}\right]\left[\frac{1440}{\pi}G_{sc}d_r(\omega_s \sin\phi \sin\delta + \cos\phi\cos\delta\sin\omega_s)\right]$$

$$(5.350)$$

where a_s and b_s can be taken as 0.25 and 0.50, respectively (since there are no calibrated values); the solar constant, G_{sc} is equal to 0.0820 MJ/(m²·min); the Julian day, J, for mid-February is 45, d_r is the relative distance between the earth and the sun given by Equation 5.319 as

$$d_r = 1 + 0.033\cos\left(\frac{2\pi}{365}J\right) = 1 + 0.033\cos\left(\frac{2\pi}{365}45\right) = 1.024$$

the solar declination, δ, is given by Equation 5.320 as

$$\delta = 0.4093\sin\left(\frac{2\pi}{365}J - 1.405\right) = 0.4093\sin\left(\frac{2\pi}{365}45 - 1.405\right)$$

$$= -0.241 \text{ radians}$$

and the sunset-hour angle, ω_s, is given by Equation 5.321 as

$$\omega_s = \cos^{-1}[-\tan\phi\tan\delta] = \cos^{-1}[-\tan(0.412)\tan(-0.241)] = 1.463 \text{ radians}$$

Substituting the given and derived values of α, a_s, b_s, n/N (= sunshine fraction), G_{sc}, d_r, ω_s, ϕ, and δ into Equation 5.350 yields

$$S_n = (1 - 0.20)[0.25 + 0.50(0.69)]$$

$$\left\{\frac{1440}{\pi}(0.0820)(1.024)[1.463\sin(0.412)\sin(-0.241)\right.$$

$$\left.+ \cos(0.412)\cos(-0.241)\sin(1.463)]\right\}$$

$$= 13.6 \text{ MJ/(m²·d)}$$

The net longwave radiation, L_n, is given by Equation 5.330 as

$$L_n = -\sigma \left(\frac{T_{max,K}^4 + T_{min,K}^4}{2} \right) (0.34 - 0.14\sqrt{e_a}) \left(1.35\frac{R_s}{R_{s0}} - 0.35 \right) \quad (5.351)$$

where $\sigma = 4.903 \times 10^{-9}$ MJ m^{-2}K^{-4}d^{-1}; $T_{max,K}$ and $T_{min,K}$ are the maximum and minimum daily temperatures given as 299.2 K (= 26.0°C) and 290.1 K (= 16.9°C), respectively; e_a is the actual vapor pressure given by Equation 5.344

$$e_a = \frac{e_s(T_{max})\frac{RH_{max}}{100} + e_s(T_{min})\frac{RH_{min}}{100}}{2}$$

where

$$e_s(T_{max}) = 0.6108 \left[\exp\left(\frac{17.27 T_{max}}{T_{max} + 237.3} \right) \right] = 0.6108 \left[\exp\left(\frac{17.27(26.0)}{26.0 + 237.3} \right) \right]$$

$$= 3.36 \text{ kPa}$$

$$e_s(T_{min}) = 0.6108 \left[\exp\left(\frac{17.27 T_{min}}{T_{min} + 237.3} \right) \right] = 0.6108 \left[\exp\left(\frac{17.27(16.9)}{16.9 + 237.3} \right) \right]$$

$$= 1.93 \text{ kPa}$$

$$e_s(T) = 0.6108 \left[\exp\left(\frac{17.27 T}{T + 237.3} \right) \right] = 0.6108 \left[\exp\left(\frac{17.27(21.5)}{21.5 + 237.3} \right) \right]$$

$$= 2.64 \text{ kPa}$$

Substituting the given and derived values of $e_s(T_{max})$, RH_{max}, $e_s(T_{min})$, and RH_{min} into the expression for e_a yields

$$e_a = \frac{3.36\frac{98.2}{100} + 1.93\frac{63.2}{100}}{2} = 2.26 \text{ kPa}$$

According to Equation 5.317, the incoming shortwave (solar) radiation, R_s, can be estimated by the relation

$$R_s = \left(a_s + b_s\frac{n}{N} \right) S_0 = (0.25 + 0.50 \times 0.69) S_0 = 0.595 S_0$$

where S_0 is the extraterrestrial radiation, and the clear-sky solar radiation, R_{s0}, can be estimated by Equation 5.328 (taking $z = 0$) as

$$R_{s0} = [0.75 + 2 \times 10^{-5}(0)]S_0 = 0.75 S_0$$

Substituting the known values of σ, $T_{max,K}$, $T_{min,K}$, e_a, R_s, and R_{s0} into Equation 5.361 yields

$$L_n = -4.903 \times 10^{-9} \left(\frac{299.2^4 + 290.1^4}{2} \right) (0.34 - 0.14\sqrt{2.26})$$

$$\times \left(1.35\frac{0.595 S_0}{0.75 S_0} - 0.35 \right) = -3.46 \text{ MJ/(m}^2\cdot\text{d)}$$

where the negative value indicates that the net longwave radiation in February is away from the earth. The total available energy, R_n, in February is then equal to the sum of S_n and L_n and is given by

$$R_n = S_n + L_n = 13.6 - 3.46 = 10.1 \text{ MJ/(m}^2\text{d)}$$

G: According to Equation 5.333,

$$G_{\text{month}} = 0.07(T_{i+1} - T_{i-1}) = 0.07(21.3 - 18.7) = 0.182 \text{ MJ/(m}^2\text{d)}$$

γ: The psychrometric constant, γ, can be estimated using Equation 5.336 as

$$\gamma = 0.0016286 \frac{p}{\lambda} \text{ kPa/}^\circ\text{C}$$

where the atmospheric pressure, p, can be taken as 101.32 kPa, and the latent heat of vaporization, λ, can be taken as 2.45 MJ/kg. Therefore, γ is given by

$$\gamma = 0.0016286 \frac{101.32}{2.45} = 0.0674 \text{ kPa/}^\circ\text{C}$$

Since the pressure, p, and latent heat of vaporization, λ, remain approximately constant throughout the year, γ remains approximately constant.

ρ_a: The density of air, ρ_a, can be estimated using Equation 5.349 as

$$\rho_a = 3.450 \frac{p}{T + 273} = 3.450 \frac{101.32}{21.5 + 273} = 1.187 \text{ kg/m}^3$$

Taking the density of water, ρ_w, as 998.2 kg/m³; the latent heat of vaporization, λ, as 2.45 MJ/kg; the specific heat of moist air, c_p, as 1.013 kJ/(kg°C) = 1.013×10⁻³ MJ/(kg°C), and substituting into the Penman–Monteith equation, Equation 5.301 yields

$$ET_c = \frac{1}{\rho_w \lambda} \left[\frac{\Delta(R_n - G) + \rho_a c_p \frac{e_s - e_a}{r_a}}{\Delta + \gamma\left(1 + \frac{r_s}{r_a}\right)} \right]$$

$$= \frac{1}{(998.2)(2.45)} \left[\frac{0.157(10.1 - 0.182) + (1.187)(1.013 \times 10^{-3})\frac{2.64 - 2.26}{3.29 \times 10^{-4}}}{0.157 + 0.0674\left(1 + \frac{1.04 \times 10^{-3}}{3.29 \times 10^{-4}}\right)} \right]$$

$$= 0.0028 \text{ m/d} = 2.8 \text{ mm/d}$$

Therefore the average evapotranspiration during February is estimated to be 2.8 mm/d. Subsequent direct measurements indicate that $e_s - e_a = 0.67$ kPa and $R_n = 8.83$ MJ/(m²·d), and using these values in the Penman–Monteith equation gives

$$ET_c = \frac{1}{(998.2)(2.45)} \left[\frac{0.157(8.83 - 0.182) + (1.187)(1.013 \times 10^{-3})\frac{0.67}{3.29 \times 10^{-4}}}{0.157 + 0.0674\left(1 + \frac{1.04 \times 10^{-3}}{3.29 \times 10^{-4}}\right)} \right]$$

$$= 0.0036 \text{ m/d} = 3.6 \text{ mm/d}$$

Therefore, using direct measurements of net radiation and vapor pressure deficit change the estimated evaporation rate by 29 % (3.6 mm/d vs. 2.8 mm/d). A close inspection of this result indicates that the increased evapotranspiration is primarily caused by the measured vapor-pressure deficit being much higher than the estimated vapor-pressure deficit.

5.8.2   Potential Evapotranspiration

The potential evapotranspiration, introduced by Penman (1948), is defined as "the amount of water transpired in a given time by a short green crop, completely shading the ground, of uniform height and with adequate water status in the soil profile." According to this definition, the potential evapotranspiration rate is not related to a specific crop, and many types of vegetation can be described as a "short green crop." Several equations have been proposed for estimating the potential evapotranspiration. In many cases, the preferred method is that suggested by Penman (1956), which can be derived from the Penman–Monteith equation, Equation 5.301, given by

$$\text{ET}_c = \frac{1}{\rho_w \lambda} \left[\frac{\Delta(R_n - G) + \rho_a c_p \frac{e_s - e_a}{r_a}}{\Delta + \gamma \left(1 + \frac{r_s}{r_a} \right)} \right] \tag{5.352}$$

The Penman (1956) equation for estimating the potential evapotranspiration, ET_p, is derived directly from this equation by taking $r_s = 0$, which gives the Penman equation as

$$\text{ET}_p = \frac{1}{\rho_w \lambda} \left[\frac{\Delta(R_n - G) + \rho_a c_p \frac{e_s - e_a}{r_a}}{\Delta + \gamma} \right] \tag{5.353}$$

This equation has been used in several regional studies to estimate the distribution of potential evapotranspiration (e.g., Bidlake et al., 1996), with open water and wetland systems commonly assumed to evaporate at the potential rate (Abtew et al., 2003).

EXAMPLE 5.64

Estimate the potential evapotranspiration at the location of the cattail marsh described in the previous example. Comment on the difference between the actual and potential evapotranspiration at the site.

Solution   Potential evapotranspiration is calculated using the same Penman–Monteith equation as for actual evapotranspiration, with the exception that $r_s = 0$ and r_a is the aerodynamic resistance of open water, which is given by Equation 5.308 as

$$r_a = \frac{4.72 \left[\ln \frac{z_m}{z_0} \right]^2}{1 + 0.536 u_2} \text{ s/m}$$

where $z_m = 2$ m (standard for open-water), $z_0 = 1.37$ mm $= 0.00137$ m, and the wind speed at 2 m height is derived from the given wind speed at 10 m height

$(u_{10} = 3.96$ m/s) using Equation 5.305, which yields

$$u_2 = \frac{4.87}{\ln(67.8z - 5.42)}u_z$$

$$= \frac{4.87}{\ln(67.8(10) - 5.42)}3.96$$

$$= 2.96 \text{ m/s}$$

Substituting the known and derived values of z_m, z_0, and u_2 into the expression for r_a gives

$$r_a = \frac{4.72\left[\ln \frac{2}{0.00137}\right]^2}{1 + 0.536(2.96)} = 97 \text{ s/m} = 1.12 \times 10^{-3} \text{ d/m}$$

For open water, the albedo is 0.08, and since the albedo used in the previous example is 0.20, the previously calculated net solar radiation, S_n [$= 13.6$ MJ/(m²·d)], becomes

$$S_n = 13.6 \frac{(1 - 0.08)}{(1 - 0.20)} = 15.6 \text{ MJ/(m}^2\text{d)}$$

Since $L_n = -3.46$ MJ/(m²d), the net radiation, R_n, over open water is given by

$$R_n = S_n + L_n = 15.6 + (-3.46) = 12.1 \text{ MJ/(m}^2\text{d)}$$

Substituting into the Penman–Monteith equation, using applicable data from the previous example, yields

$$\text{ET}_c = \frac{1}{\rho_w \lambda} \left[\frac{\Delta(R_n - G) + \rho_a c_p \frac{e_s - e_a}{r_a}}{\Delta + \gamma\left(1 + \frac{r_s}{r_a}\right)} \right]$$

$$= \frac{1}{(998.2)(2.45)} \left[\frac{0.157(12.1 - 0.182) + (1.187)(1.013 \times 10^{-3})\frac{2.64 - 2.26}{1.12 \times 10^{-3}}}{0.157 + 0.0674\left(1 + \frac{0}{1.12 \times 10^{-3}}\right)} \right]$$

$$= 0.0042 \text{ m/d} = 4.2 \text{ mm/d}$$

Therefore, the (theoretical) potential evapotranspiration of 4.2 mm/d at the cattail marsh is significantly higher than the (theoretical) actual transpiration of 2.8 mm/d. This difference is due to the reduced surface resistance and increased absorption of solar energy associated with open water versus cattail vegetation.

5.8.3 Reference Evapotranspiration

The reference-evapotranspiration concept was introduced in the late 1970s and early 1980s to avoid the ambiguity in surface vegetation associated with the definition of potential evapotranspiration. The reference evapotranspiration is frequently defined as "the rate of evapotranspiration from a hypothetical reference crop with an assumed crop height of 0.12 m, a fixed surface resistance of $70\,\text{s m}^{-1}$, and an albedo of 0.23, closely resembling the evapotranspiration from an extensive surface of green grass of uniform height, actively growing, well-watered, and completely shading the ground." In the literature, the terms "reference evapotranspiration" and "reference-crop evapotranspiration" are used synonymously.

The most widely used and recommended methods for estimating reference evapotranspiration are the FAO56 and ASCE Penman–Monteith equations, with evaporation pans and empirical methods also being used. In California, the California Irrigation Management Information System (CIMIS) Penman equation is widely used. All Penman equations give similar estimates of the reference evapotranspiration (Temesgen et al., 2005).

5.8.3.1 FAO56-Penman–Monteith method

The International Commission for Irrigation and Drainage (ICID) and the Food and Agriculture Organization of the United Nations (FAO) have recommended that the FAO56-Penman–Monteith (FAO56-PM) method be used as the standard method to estimate reference-crop evapotranspiration. For grass and alfalfa reference crops, r_a can be approximated by

$$r_a = \begin{cases} \dfrac{208}{u_2}\ \text{s/m} & \text{(grass reference crop)} \\[2ex] \dfrac{110}{u_2}\ \text{s/m} & \text{(alfalfa reference crop)} \end{cases} \tag{5.354}$$

where u_2 is the wind speed in m/s at a standardized height of 2 m above the ground (Allen et al., 1989). The surface resistance, r_s, of grass and alfalfa reference crops are given by

$$r_s = \begin{cases} 70\ \text{s/m} & \text{(grass reference crop)} \\[2ex] 45\ \text{s/m} & \text{(alfalfa reference crop)} \end{cases} \tag{5.355}$$

The FAO56-PM equation as given by FAO Irrigation and Drainage Paper No. 56 gives an explicit expression for the grass reference-crop evapotranspiration. According to the Penman–Monteith equation (Equation 5.301), the evapotranspiration rate of any vegetated surface is given by

$$\text{ET}_c = \frac{1}{\rho_w \lambda}\left[\frac{\Delta(R_n - G) + \rho_a c_p \dfrac{e_s - e_a}{r_a}}{\Delta + \gamma\left(1 + \dfrac{r_s}{r_a}\right)}\right] \tag{5.356}$$

The FAO56-PM method assumes that the density of water is equal to $1000\ \text{kg/m}^3$, and since values of λ vary only slightly over normal temperature ranges, a constant value of 2.45 MJ/kg corresponding to a temperature of $20°\text{C}$ is assumed. Equation 5.336 can be used to estimate the specific heat of water, c_p, Equation 5.349 can be used to estimate the density of air, and the aerodynamic and surface resistances of a grass reference crop are given by

$$r_a = \frac{208}{u_2} \text{ s/m} = \frac{2.407 \times 10^{-3}}{u_2} \text{d/m} \tag{5.357}$$

$$r_s = 70 \text{ s/m} = 8.102 \times 10^{-4} \text{d/m} \tag{5.358}$$

Substituting into the Penman–Monteith equation, Equation 5.356 yields

$$\text{ET}_c = \frac{1}{(1000)(2.45)} \left[\frac{\Delta(R_n - G) + \left(3.450\frac{p}{T_{\text{mean}}+273}\right)\left(\frac{\gamma\varepsilon\lambda}{p}\right)\frac{e_s - e_a}{\frac{2.407\times10^{-3}}{u_2}}}{\Delta + \gamma\left(1 + \frac{8.102\times10^{-4}}{\frac{2.407\times10^{-3}}{u_2}}\right)} \right] \text{m/d}$$

$$= \frac{1000}{(1000)(2.45)} \left[\frac{\Delta(R_n - G) + \left(3.450\frac{p}{T_{\text{mean}}+273}\right)\left(\frac{\gamma(0.622)(2.45)}{p}\right)\frac{e_s - e_a}{\frac{2.407\times10^{-3}}{u_2}}}{\Delta + \gamma\left(1 + \frac{8.102\times10^{-4}}{\frac{2.407\times10^{-3}}{u_2}}\right)} \right] \text{mm/d}$$

which simplifies into the conventional FAO56-PM equation given by

$$\boxed{\text{ET}_0 = \frac{0.408\Delta(R_n - G) + \gamma\frac{900}{T_{\text{mean}}+273}u_2(e_s - e_a)}{\Delta + \gamma(1 + 0.34u_2)}} \tag{5.359}$$

where ET_0 is the grass reference evapotranspiration (mm/d).

The FAO56-PM equation (Equation 5.359) requires temperature, humidity, radiation, and wind-speed data as listed in Table 5.44. It is generally recommended that the climate data used to estimate reference evapotranspiration be collected in an environment similar to that of the reference crop (grass). Therefore, weather stations used to collect data to estimate reference evapotranspiration should be located in well-irrigated and well-maintained grass areas, with the surrounding grass area exceeding 2 ha. In many cases, especially in developing countries, the quality of data and the difficulties in gathering all the necessary weather parameters can present serious limitations to using the FAO56-PM method (Irmak et al., 2003a). In particular, wind speed and humidity data measured at airports and hill locations can introduce errors in ET_0 estimates, since these locations have different microclimates than irrigated agricultural areas for which ET_0 estimations are made. The FAO56-PM method recommends using the equations described in Section 5.8.1 to estimate the weather parameters in the FAO56-PM equation. Although alternative equations are available in the technical literature to estimate these weather parameters, strict adherence to the FAO56-recommended equations should be followed, since alternative equations can lead to a lack of consistency in estimating the FAO56-PM reference evapotranspiration (Nandagiri and Kovoor, 2005).

TABLE 5.44: Data Required for FAO56-PM Equation

Parameter	Data required
Location	Altitude above sea level and latitude of the location should be specified. These data are needed to adjust some weather parameters for the local average value of atmospheric pressure, to compute extraterrestrial radiation (R_a) and, in some cases, daylight hours.
Temperature	Daily maximum and minimum air temperatures are required. Where only mean daily temperatures are available, the calculations can still be executed but some underestimation of ET_0 will probably occur due to the nonlinearity of the saturation vapor-pressure versus temperature relationship. Using mean air temperature instead of maximum and minimum air temperatures yields a lower saturation vapor pressure, e_s, and hence a lower vapor-pressure difference ($e_s - e_a$), and a lower ET_0 estimate.
Humidity	The average daily vapor pressure, e_a, is required. The actual vapor pressure, where not available, can be derived from maximum and minimum relative humidity, psychrometric data (i.e., dry- and wet-bulb temperatures), or dewpoint temperature.
Radiation	The daily net radiation is required. These data are not commonly available but can be derived from the average shortwave radiation measured with a pyranometer or from the daily actual duration of bright sunshine measured with a sunshine recorder.
Wind speed	The average daily wind speed measured at 2 m above ground level is required. It is important to verify the height at which wind speed is measured, as wind speeds measured at different heights above the ground surface differ.

The accuracy of ET_0 estimation models should ideally be measured relative to data derived from lysimeter measurements. However, it is widely known that environmental and management requirements for lysimeter experiments are very demanding (Allen et al., 1994b), and it has been suggested that the FAO56-PM equation be considered superior to most lysimeter-measured ET_0, and that the FAO56-PM equation can be used in calibrating other ET_0 models.

In areas where substantial changes in wind speed, dew point, or cloudiness occur during the day, calculation of ET_0 using hourly time steps is generally better than using 24-h time steps. However, the hourly-averaged surface resistance, r_s, of grass can deviate significantly from the daily-averaged value. Under most conditions, application of the FAO56-PM equation with 24-h time steps described here produces accurate results.

EXAMPLE 5.65

Weather conditions on typical days in Miami during the months of January, April, July, and October are shown in the following table (Winsberg, 1990):

Month	Minimum humidity (%)	Maximum humidity (%)	Minimum temperature (°C)	Maximum temperature (°C)	Wind speed (m/s)	Daylight fraction
January	59	84	14.4	23.9	4.5	0.69
April	54	79	20.0	28.3	4.9	0.78
July	63	85	24.4	31.7	3.6	0.76
October	64	86	21.7	28.9	4.0	0.72

Given that Miami is located at approximately 26°N latitude, compute the grass reference evapotranspiration for the given months using the FAO56-PM method. Typical monthly rainfall amounts for January, April, July, and October are 5.0 cm, 8.2 cm, 15.2 cm, and 19.1 cm (Henry et al., 1994). Compare the estimated monthly reference evapotranspiration with the monthly rainfall, and assess the implications of this differential on available water resources.

Solution   The FAO56-PM equation is given by Equation 5.359, and the parameters of this equation either are known or can be estimated from the given data. The computational procedure will be illustrated for the month of January. The FAO56-PM equation gives the grass reference evapotranspiration, ET_0, as

$$ET_0 = \frac{0.408\Delta(R_n - G) + \gamma\frac{900}{T+273}u_2(e_s - e_a)}{\Delta + \gamma(1 + 0.34u_2)}$$

and the computation of the parameters in this equation for January is as follows:

Δ: The slope of the vapor-pressure versus temperature curve, Δ, is given by Equation 5.342 as

$$\Delta = \frac{4098\left[0.6108\exp\left(\frac{17.27T}{T+237.3}\right)\right]}{(T + 237.3)^2}$$

where Δ is in kPa/°C, and T is the average temperature in °C. When estimating daily-averaged evapotranspiration rates, T is taken as the average of the maximum and minimum daily temperature, and for a typical day in January the given data indicate that

$$T = \frac{14.4 + 23.9}{2} = 19.2°C$$

and therefore

$$\Delta = \frac{4098\left[0.6108\exp\left(\frac{17.27(19.2)}{19.2+237.3}\right)\right]}{(19.2 + 237.3)^2} = 0.139\text{ kPa/°C}$$

R_n: The net radiation, R_n, is equal to the sum of the net shortwave and longwave radiation according to Equation 5.314, where

$$R_n = S_n + L_n$$

The net shortwave radiation can be estimated using Equation 5.324 as

$$S_n = (1 - \alpha)\left[a_s + b_s\frac{n}{N}\right]\left[\frac{1440}{\pi}G_{sc}d_r(\omega_s\sin\phi\sin\delta + \cos\phi\cos\delta\sin\omega_s)\right]$$

$$(5.360)$$

where the albedo, α, is equal to 0.23 for the grass reference crop; a_s and b_s can be taken as 0.25 and 0.50, respectively (since there are no calibrated values); the fraction of bright sunshine during daylight hours for January is given as 0.69; the solar constant, G_{sc}, is equal to 0.0820 MJ/(m²·min); d_r is the relative distance between the earth and the sun, given by Equation 5.319 as

$$d_r = 1 + 0.033\cos\left(\frac{2\pi}{365}J\right) = 1 + 0.033\cos\left(\frac{2\pi}{365}15\right) = 1.032$$

where the mean Julian day, J, for January is taken as 15; the solar declination, δ, is given by Equation 5.320 as

$$\delta = 0.4093 \sin\left(\frac{2\pi}{365}J - 1.405\right) = 0.4093 \sin\left(\frac{2\pi}{365}15 - 1.405\right)$$

$$= -0.373 \text{ radians}$$

and the sunset-hour angle, ω_s, is given by Equation 5.321 as

$$\omega_s = \cos^{-1}[-\tan\phi\tan\delta] = \cos^{-1}[-\tan(26°)\tan(-0.373)] = 1.38 \text{ radians}$$

where the latitude, ϕ, is given as 26° (= 0.454 rad). Substituting the values of α, a_s, b_s, n/N, G_{sc}, d_r, ω_s, ϕ, and δ into Equation 5.360 yields

$$S_n = (1 - 0.23)[0.25 + 0.50(0.69)]\left\{\frac{1440}{\pi}(0.0820)(1.032)[1.38\sin(0.454)\right.$$

$$\left. \sin(-0.373) + \cos(0.454)\cos(-0.373)\sin(1.38)]\right\}$$

$$= 10.68 \text{ MJ/(m}^2\cdot\text{d)}$$

The net longwave radiation, L_n, is given by Equation 5.330 as

$$L_n = -\sigma\left(\frac{T_{max,K}^4 + T_{min,K}^4}{2}\right)(0.34 - 0.14\sqrt{e_a})\left(1.35\frac{R_s}{R_{so}} - 0.35\right) \quad (5.361)$$

where $\sigma = 4.903 \times 10^{-9}$ MJ m^{-2}K^{-4}d^{-1}; $T_{max,K}$ and $T_{min,K}$ are the maximum and minimum daily temperatures given as 297.1 K (= 23.9°C) and 287.6 K (= 14.4°C), respectively; e_a is the actual vapor pressure given by Equation 5.344

$$e_a = \frac{e_s(T_{max})\frac{RH_{max}}{100} + e_s(T_{min})\frac{RH_{min}}{100}}{2}$$

where RH_{max} and RH_{min} are the maximum and minimum relative humidities given as 84% and 59%, respectively, and

$$e_s(T) = 0.6108\left[\exp\left(\frac{17.27T}{T + 237.3}\right)\right] = 0.6108\left[\exp\left(\frac{17.27(19.2)}{19.2 + 237.3}\right)\right]$$

$$= 2.22 \text{ kPa}$$

$$e_s(T_{max}) = 0.6108\left[\exp\left(\frac{17.27T_{max}}{T_{max} + 237.3}\right)\right] = 0.6108\left[\exp\left(\frac{17.27(23.9)}{23.9 + 237.3}\right)\right]$$

$$= 2.97 \text{ kPa}$$

$$e_s(T_{min}) = 0.6108\left[\exp\left(\frac{17.27T_{min}}{T_{min} + 237.3}\right)\right] = 0.6108\left[\exp\left(\frac{17.27(14.4)}{14.4 + 237.3}\right)\right]$$

$$= 1.64 \text{ kPa}$$

which yields

$$e_a = \frac{2.97\frac{84}{100} + 1.64\frac{59}{100}}{2} = 1.73 \text{ kPa}$$

According to Equation 5.317, the net shortwave (solar) radiation, R_s, can be estimated by the relation

$$R_s = \left(a_s + b_s\frac{n}{N}\right)S_0 = (0.25 + 0.50 \times 0.69)\,S_0 = 0.595S_o$$

and the clear-sky solar radiation, R_{s0}, can be estimated by Equation 5.328 (taking $z = 0$) as

$$R_{s0} = [0.75 + 2 \times 10^{-5}(0)]S_0 = 0.75S_0$$

Substituting the known values of σ, $T_{\max,K}$, $T_{\min,K}$, e_a, R_s, and R_{s0} into Equation 5.361 yields

$$L_n = -4.903 \times 10^{-9}\left(\frac{297.1^4 + 287.6^4}{2}\right)\left(0.34 - 0.14\sqrt{1.73}\right)$$

$$\times \left(1.35\frac{0.595S_o}{0.75S_o} - 0.35\right) = -4.03 \text{ MJ/(m}^2\cdot\text{d)}$$

where the negative value indicates that the net longwave radiation in January is away from the earth. The total available energy, R_n, in January is then equal to the sum of S_n and L_n and is given by

$$R_n = S_n + L_n = 10.68 - 4.03 = 6.65 \text{ MJ/(m}^2\text{d)}$$

G: According to Equation 5.332, $G = 0$ MJ/(m²d) when averaged over one day.

γ: The psychrometric constant, γ, can be estimated using Equation 5.336 as

$$\gamma = 0.0016286\frac{p}{\lambda} \text{ kPa/}^\circ\text{C}$$

where the atmospheric pressure, p, can be taken as 101.32 kPa, and the latent heat of vaporization, λ, can be taken as 2.45 MJ/kg. Therefore, γ is given by

$$\gamma = 0.0016286\frac{101.32}{2.45} = 0.0674 \text{ kPa/}^\circ\text{C}$$

Since the pressure, p, and latent heat of vaporization, λ, remain approximately constant throughout the year, γ remains approximately constant.

Based on the given and calculated parameters for the month of January, the grass reference-crop evapotranspiration estimated by the Penman equation is given by

$$\text{ET}_0 = \frac{0.408\Delta(R_n - G) + \gamma\frac{900}{T+273}u_2(e_s - e_a)}{\Delta + \gamma(1 + 0.34u_2)}$$

$$= \frac{0.408(0.139)(6.65 - 0) + (0.0674)\frac{900}{19.2+273}4.5(2.22 - 1.73)}{0.139 + 0.0674(1 + 0.34 \times 4.5)}$$

$$= 2.7 \text{ mm/d}$$

The computations illustrated here for January are repeated for other months of the year, and the results are tabulated below:

Month	Δ (kPa/°C)	R_n (MJ/m²d)	ET (mm/d)	ET (mm)	Rain (mm)
Jan	0.138	6.7	2.7	83	50
Apr	0.181	14.3	5.3	159	82
Jul	0.221	16.4	5.6	174	152
Oct	0.192	10.4	3.9	121	191

These results indicate that the grass reference evapotranspiration exceeds rainfall in January, April, and July, which indicates that grass irrigation could be necessary during these months.

5.8.3.2   ASCE Penman–Monteith method

ASCE has developed equations to estimate reference evapotranspiration for two types of vegetated surfaces: one consists of a short crop with an approximate height of 0.12 m (similar to clipped grass), and the other is a tall crop with an approximate height of 0.50 m (similar to full-cover alfalfa). Both estimation equations are derived from the ASCE Penman–Monteith equation (Equation 5.301) and are expressed in the generalized form (Walter et al., 2000)

$$\text{ET}_{\text{ref}} = \frac{0.408\Delta(R_n - G) + \gamma \frac{C_n}{T+273} u_2(e_s - e_a)}{\Delta + \gamma(1 + C_d u_2)} \tag{5.362}$$

where ET_{ref} is the standardized reference crop for short (ET_0) or tall (ET_r) surfaces, C_n is a numerator constant that changes with reference type and calculation time step, and C_d is a denominator constant that changes with reference type and calculation time step. Values of C_n and C_d to be used in the ASCE standardized method are given in Table 5.45. The ASCE standardized equation, Equation 5.362, is derived from the ASCE Penman–Monteith, Equation 5.301, by substituting the parameter values shown in Table 5.46. The ASCE standardized equation recognizes that surface resistance is significantly higher during the nighttime than during the daytime, which is due to stomatal closure during the nighttime hours, and nighttime is considered to occur whenever the net radiation, R_n, is negative. On reviewing the assumptions in Table 5.46, it is clear that for daily time intervals for short reference crops the assumed parameters are the same as those for the FAO56-PM equation, and the ASCE standardized equation is the same as the FAO56-PM equation in this case. The ASCE standardized method goes beyond the FAO56-PM method, which applies only to a short (grass) reference crop, in that it is applied also to a tall reference crop (alfalfa), and it is also applied to hourly time steps with adjustment of the surface resistance to reflect daytime or nighttime hours. The calculation of ET_0 and ET_r generally assumes a constant albedo of 0.23 throughout the daytime and nighttime hours; however, it is also recognized that, in reality, albedo varies somewhat with time of day and with time of season and latitude due to change in sun angle (EWRI, 2002).

Yoder et al. (2005) investigated the sensitivity of the reference ET calculated using the standardized ASCE-PM equation to the expressions used to calculate

TABLE 5.45: Values of C_n and C_d in ASCE Standardized Method

Calculation time step	Short reference, ET_0		Tall reference, ET_r		Units for ET_0, ET_r	Units for R_n, G
	C_n	C_d	C_n	C_d		
Daily	900	0.34	1600	0.38	mm/d	MJ/(m^2·d)
Hourly (in daytime)	37	0.24	66	0.25	mm/h	MJ/(m^2·h)
Hourly (in nighttime)	37	0.96	66	1.7	mm/h	MJ/(m^2·h)

TABLE 5.46: Assumed Parameters in ASCE Standardized Method

Term	ET_0	ET_r
Reference vegetation height, h	0.12 m	0.50 m
Height of air temperature and humidity measurements, z_h	1.5–2.5 m	1.5–2.5 m
Height corresponding to wind speed, z_w	2.0 m	2.0 m
Zero plane displacement height	0.08 m	0.08 m
Latent heat of vaporization, λ	2.45 MJ/kg	2.45 MJ/kg
Surface resistance, r_s (daily)	70 s/m	45 s/m
Surface resistance, r_s (daytime)	50 s/m	30 s/m
Surface resistance, r_s (nighttime)	200 s/m	200 s/m

the actual vapor pressure (e_a), the saturation vapor pressure (e_s), and clear-sky solar radiation (R_{s0}). It was found that using Equations 5.339, 5.344, and 5.328 to estimate e_s, e_a, and R_{s0}, respectively, provided the most accurate estimates of ET_0. For consistency in estimating the ASCE-PM reference evapotranspiration, it is recommended that the equations given in Section 5.8.1 be strictly adhered to in estimating the weather parameters in the ASCE-PM equation. These are the same equations recommended for use in the FAO56-PM equation.

The ASCE-PM method for a grass reference crop, which is included in the ASCE standardized method, was adopted by NRCS into Chapter 2 of the NRCS Irrigation Guide (Martin and Gilley, 1993).

5.8.3.3 Evaporation pans

Evaporation pans provide a direct measure of evaporation in the field and are used to estimate the reference-crop evapotranspiration, ET_0, by multiplying pan-evaporation measurements by *pan coefficients*. Pans provide a measurement of the integrated effect of radiation, wind, temperature, and humidity on the evaporation from an open water surface. The most commonly used pan design is the U.S. Weather Bureau Class A pan, which is 120.7 cm (4 ft) in diameter, 25 cm (10 in.) deep, and made of either galvanized iron (22 gauge) or Monel^{TM} metal (0.8 mm). If galvanized, the pan should be painted annually with aluminum paint. The pan is mounted on a wooden frame (slatted platform) 15 cm above ground level, the soil is built up to within 5 cm of the bottom of the pan, and the pan should be surrounded by short grass turf 4 to 10 cm high that is well irrigated during the dry season. The pan is filled with water to within 5 cm of the rim. The water level should not fall more than 7.5 cm below the rim, and the water should be regularly renewed, at least weekly, to eliminate excessive turbidity. A typical Class A pan station is shown in Figure 5.69. A Class A pan station

FIGURE 5.69: Class A evaporation pan

generally includes an anemometer mounted 15 cm (6 in.) above the pan rim and, in some cases, a mesh cover to prevent animals from entering the pan or drinking the water. A 12.5-mm (0.5-in.) mesh screen is commonly used; however, the mesh lowers the measured evaporation by 5%–20% (Stanhill, 1962; Dagg, 1968). Screens over the pan are not standard equipment and preferably should not be used. Pans should be protected by fences to keep animals from drinking. Pan-evaporation rates are generally greater than evaporation rates from large bodies of water. Numerous studies have shown that large water bodies have evaporation rates far higher near the edge of the water than toward the center, where the air is more saturated and able to absorb less water vapor. The small size of an evaporation pan means that the whole pan is effectively an 'edge' and will have higher evaporation rates than much larger bodies of water (Davie, 2002). A second, smaller, problem is that sides of the pan and the water inside will absorb radiation and warm up quicker than a much larger body of water, providing an extra energy source and greater evaporation rate. Reference-crop evapotranspiration, ET_0, is estimated from pan measurements by multiplicative factors called *pan coefficients*. Therefore

$$\boxed{ET_0 = k_p E_p} \tag{5.363}$$

where k_p is the pan coefficient and E_p is the measured pan evaporation.

Pan coefficients vary seasonally, are typically in the range of 0.35 to 0.85 (Doorenbos and Pruitt, 1975; 1977), and representative pan coefficients for various site conditions are given in Table 5.47. Key factors affecting the pan coefficient are the average wind speed, upwind fetch characteristics, and ambient humidity. The estimation of evapotranspiration using the pan coefficients in Table 5.47 is commonly called the *FAO-24 pan evaporation method*. This method was published by the Food and Agriculture Organization (FAO) in paper number 24, hence the name. The pan coefficients given in Table 5.47 are described by the following regression equations (Allen and Pruitt, 1991; Allen et al., 1988)

TABLE 5.47: Pan Coefficients for Various Site Conditions

Wind	Upwind fetch of green crop, m	Case A: Pan surrounded by short green crop			Upwind fetch of dry fallow, m	Case B: Pan surrounded by dry, bare area		
		Mean relative humidity, %				Mean relative humidity, %		
		Low <40	Med 40–70	High >70		Low <40	Med 40–70	High >70
Light (<2 m/s)	1	0.55	0.65	0.75	1	0.70	0.80	0.85
	10	0.65	0.75	0.85	10	0.60	0.70	0.80
	100	0.70	0.80	0.85	100	0.55	0.65	0.75
	1000	0.75	0.85	0.85	1000	0.50	0.60	0.70
Moderate (2–5 m/s)	1	0.50	0.60	0.65	1	0.65	0.75	0.80
	10	0.60	0.70	0.75	10	0.55	0.65	0.70
	100	0.65	0.75	0.80	100	0.50	0.60	0.65
	1000	0.70	0.80	0.80	1000	0.45	0.55	0.60
Strong (5–8 m/s)	1	0.45	0.50	0.60	1	0.60	0.65	0.70
	10	0.55	0.60	0.65	10	0.50	0.55	0.65
	100	0.60	0.65	0.70	100	0.45	0.50	0.60
	1000	0.65	0.70	0.75	1000	0.40	0.45	0.55
Very strong (>8 m/s)	1	0.40	0.45	0.50	1	0.50	0.60	0.65
	10	0.45	0.55	0.60	10	0.45	0.50	0.55
	100	0.50	0.60	0.65	100	0.40	0.45	0.50
	1000	0.55	0.60	0.65	1000	0.35	0.40	0.45

Source: Doorenbos and Pruitt (1977).

Surrounded by short green crop:

$$k_p = 0.108 - 0.0286u_2 + 0.0422\ln(\text{FET}) + 0.1434\ln(\text{RH}_{\text{mean}})$$
$$- 0.000631[\ln(\text{FET})]^2 \ln(\text{RH}_{\text{mean}}) \tag{5.364}$$

Surrounded by dry area:

$$k_p = 0.61 + 0.00341\text{RH}_{\text{mean}} - 0.000162u_2\text{RH}_{\text{mean}} - 0.00000959u_2\text{FET}$$
$$+ 0.00327u_2 \ln(\text{FET}) - 0.00289u_2 \ln(86.4u_2) - 0.0106\ln(86.4u_2)\ln(\text{FET})$$
$$+ 0.00063[\ln(\text{FET})]^2 \ln(86.4u_2) \tag{5.365}$$

where u_2 is the average daily wind speed at 2 m height (m/s), RH_{mean} is the average daily relative humidity (%) calculated by averaging the maximum and minimum relative humidity over the course of a day, and FET is the upwind fetch (m). In applying Equations 5.364 and 5.365 the following data ranges must be strictly observed (Allen et al., 1998):

$$1 \text{ m} \leq \text{FET} \leq 1000 \text{ m} \tag{5.366}$$

$$30\% \leq \text{RH}_{\text{mean}} \leq 84\% \tag{5.367}$$

$$1 \text{ m/s} \leq u_2 \leq 8 \text{ m/s} \tag{5.368}$$

The pan coefficients given in Table 5.47 are applicable to short, irrigated grass turf. For taller and aerodynamically rougher crops, the values of k_p would be higher and vary less with differences in weather conditions (ASCE, 1990). It is recommended that the pan be installed inside a short green cropped area with a size of at least 15 m by 15 m, and at least 10 m from the green crop edge in the general upwind direction (Allen et al., 1998).

Where a standard pan environment is maintained and where strong dry-wind conditions occur only occasionally, mean monthly evapotranspiration for a well-watered short grass should be predictable to within ±10% or better for most climates. Pan-based estimates of ET_0 using Equations 5.363 and 5.364 have been shown to be in close agreement with daily ET_0 estimates provided by the California Irrigation Management System (CIMIS) (Snyder et al., 2005). The accuracy of the pan-evaporation method is dependent on the quality of station maintenance and is susceptible to the microclimatic conditions under which the pan operates.

EXAMPLE 5.66

Evaporation-pan measurements indicate the daily evaporation depths over the course of one week during the summer: 8.2 mm, 7.5 mm, 7.6 mm, 6.8 mm, 7.6 mm, 8.9 mm, and 8.5 mm. During this period the average wind speed is 1.9 m/s and the average daily relative humidity is 70%. If the evaporation pan is installed in a green area with an upstream fetch of up to 1 km, determine the 7-day average reference evapotranspiration.

Solution From the given data, $u_2 = 1.9$ m/s, $\text{RH}_{\text{mean}} = 70\%$, and FET = 1000 m. Based on these data, Table 5.47 and Equation 5.364 give $k_p = 0.85$. The 7-day average pan evaporation rate, E_p, is given by

$$E_p = \frac{8.2 + 7.5 + 7.6 + 6.8 + 7.6 + 8.9 + 8.5}{7} = 7.9 \text{ mm/d}$$

and hence the reference evapotranspiration, ET_0, is given by Equation 5.363 as

$$\text{ET}_0 = k_p E_p = (0.85)(7.9) = 6.7 \text{ mm/d}$$

Therefore, a 7-day average grass-reference evapotranspiration of 6.7 mm/d is indicated by the pan data.

5.8.3.4 Empirical methods

The preferred FAO56-PM equation requires solar radiation, wind speed, air temperature, and humidity data to calculate ET_0. However, all of these input variables are not always available for a given location. Although the temperature and humidity data are routinely measured, and solar radiation data can be estimated with sufficient accuracy, wind-speed data are rarely available, and there are no widely accepted and reliable methods to predict the wind speeds with sufficient accuracy. In such cases, simplified radiation-based or temperature methods, which require fewer input parameters, can be used to estimate the grass-reference evapotranspiration (ET_0) or alfalfa-reference evapotranspiration (ET_r).

In a recent study, Irmak et al. (2003a) evaluated the performance of 21 ET_0 and ET_r methods in Florida (humid climate) and reported that all methods produced significantly different daily ET_0 values than the FAO56-PM method. To accommodate limitations on the availability of weather data and the limitations of several empirical ET_0 estimation methods, Irmak et al. (2003b) developed a set of empirical equations that are in closer agreement with the FAO56-PM equation than any of the other commonly used methods to estimate ET_0 in humid climates in the southeastern United States. These equations have been calibrated to match the FAO56-PM equation and are given by

$$ET_0 = -0.611 + 0.149R_s + 0.079T_{\text{mean}} \tag{5.369}$$

$$ET_0 = 0.489 + 0.289R_n + 0.023T_{\text{mean}} \tag{5.370}$$

where ET_0 is the reference evapotranspiration, in mm/d, R_s is the incoming solar radiation in MJ/(m²·d), T_{mean} is the mean daily air temperature in °C computed as the average of daily maximum and minimum air temperature, and R_n is the net radiation in MJ/(m²·d). Equation 5.369 should be used when measurements or estimates of R_s and T_{mean} are available, and Equation 5.370 when R_n and T_{mean} are available.

5.8.4 Actual Evapotranspiration

Crop evapotranspiration, ET_c, under standard conditions can be derived directly from meteorological and crop data using the general Penman–Monteith equation given by Equation 5.301. However, this approach is seldom taken because of the difficulty in estimating such crop parameters as albedo, α, aerodynamic resistance, r_a, and surface resistance, r_s. The more common approach is to first calculate a reference evapotranspiration, assuming either standardized grass or alfalfa cover, and then multiply the reference evapotranspiration by a crop-specific coefficient to estimate the actual crop evapotranspiration under *standard conditions*, which exist in large fields under excellent agronomic and soil–water conditions. Crop evapotranspiration, ET_c, differs distinctly from grass-reference crop evapotranspiration, ET_0 (grass) or ET_r (alfalfa), as the ground cover, canopy properties, and aerodynamic resistance of the crop are different from those of the reference crop.

5.8.4.1 Index-of-dryness method

It is generally accepted that the mean annual evapotranspiration, ET, from a catchment must be less than the mean annual precipitation, P, on the catchment. Furthermore, it has been shown that the ratio of mean annual potential evapotranspiration, ET_p, to precipitation, P, can be used to provide fairly accurate estimates of the mean annual evapotranspiration. This ratio (ET_p/P) is commonly referred to as the *index of dryness*, and ET approaches P in regions where the index of dryness is much greater than one, and the ET approaches the ET_p where the index of dryness is much less than one. This relationship can be estimated by the equation (Zhang et al., 2004)

$$\frac{\text{ET}}{\text{ET}_p} = 1 + \frac{P}{\text{ET}_p} - \left[1 + \left(\frac{P}{\text{ET}_p} \right)^w \right]^{1/w} \tag{5.371}$$

where the parameter w represents the integrated effects of catchment characteristics on annual evapotranspiration. For forested catchments w can be taken as 2.84, and for grassed catchments as 2.55. These relative values of w indicate higher evapotranspiration from forested areas than grassed areas under the same climatic conditions.

EXAMPLE 5.67

In western Kansas the mean annual potential evapotranspiration is 125 cm, and the mean annual precipitation is 80 cm. Estimate the index of dryness and the actual evapotranspiration for grassed catchments in western Kansas.

Solution   From the given data: $\text{ET}_p = 125$ cm, $P = 80$ cm, and the index of dryness is given by

$$\text{index of dryness} = \frac{\text{ET}_p}{P} = \frac{125}{80} = 1.56$$

Using Equation 5.371 with $w = 2.55$ (for grassed catchments) gives

$$\frac{\text{ET}}{\text{ET}_p} = 1 + \frac{P}{\text{ET}_p} - \left[1 + \left(\frac{P}{\text{ET}_p} \right)^w \right]^{1/w}$$

$$= 1 + \frac{1}{1.56} - \left[1 + \left(\frac{1}{1.56} \right)^{2.55} \right]^{1/2.55} = 0.525$$

Therefore, the actual evapotranspiration, ET, in western Kansas can be estimated by

$$\text{ET} = 0.525\,\text{ET}_p = 0.525(125 \text{ cm}) = 66 \text{ cm}$$

The actual annual evapotranspiration (66 cm) is much closer to the annual precipitation (80 cm) compared with the potential evapotranspiration (125 cm). It is also notable that the actual evapotranspiration is less than the precipitation, while the potential evapotranspiration is greater than the precipitation.

5.8.4.2   Crop-coefficient method

In the crop-coefficient approach, crop evapotranspiration, ET_c, is estimated by the relation

$$\boxed{\text{ET}_c = k_c\text{ET}_0} \tag{5.372}$$

where k_c is a *crop coefficient*, which varies primarily with the characteristics of a specific crop and only to a limited extent with climate. The crop-coefficient approach enables the transfer of standard values of k_c between locations and between climates. The crop coefficient predicts ET_c under standard conditions, where no limitations are placed on crop growth or evapotranspiration due to water shortage, crop density, disease, weed, insect, or salinity pressures. The crop coefficient, k_c, represents an integration of the effects of four primary characteristics that distinguish the crop from reference grass: crop height, albedo, surface resistance, and evaporation from the underlying exposed soil. Changes in these characteristics over the growing season generally affect the crop coefficient. The crop-coefficient approach is particularly useful when short-term estimates of ET are needed, such as in irrigation scheduling for individual agricultural fields (Allen et al., 2005b).

Calculation steps to be followed in estimating the crop evapotranspiration using crop coefficients were recommended in FAO-56 and are given by (Allen et al., 1998):

1. Identify the crop growth stages, determine their lengths, and select the corresponding k_c coefficients;
2. Adjust the selected k_c coefficients for frequency of wetting or climatic conditions during the stage;
3. Construct the crop coefficient curve; and
4. Calculate ET_c using Equation 5.372.

This approach is called the *FAO-56 crop coefficient approach*, and guidance for each calculation step is given below:

Step 1: Select crop coefficients. As crops develop, the ground cover, crop height, and leaf area change, and these changes directly influence the crop evapotranspiration. The growing period can be divided into four distinct growth stages: initial, crop-development, mid-season, and late-season.

Initial Stage: The initial stage runs from planting to approximately 10% ground cover. The leaf area is small, and evapotranspiration is predominantly from soil evaporation. The crop coefficient during the initial stage ($k_{c\ ini}$) is larger when the soil is wet from irrigation and rainfall, and smaller when the soil surface is dry.

Crop-Development Stage: The crop-development stage runs from 10% ground cover to effective full cover. As the crop develops and shades more of the ground, evaporation becomes more restricted and transpiration gradually becomes the major process.

Mid-Season Stage: The mid-season stage runs from effective full cover to the start of maturity, which is often indicated by the beginning of aging, yellowing or senescence of leaves, leaf drop, or the browning of fruit to the degree that the crop evapotranspiration is reduced. This is the longest stage for perennials and many annuals, but it may be relatively short for vegetable crops that are harvested fresh for their green vegetation. At the mid-season stage, the crop coefficient reaches its maximum value ($k_{c\ mid}$) and is relatively constant for most growing and cultural conditions.

Late-Season Stage: The late-season stage runs from the start of maturity to harvest or full senescence. The calculation of evapotranspiration is presumed to end when the crop is harvested, dries out naturally, reaches full senescence, or experiences

TABLE 5.48: Lengths of Growth Stages for Various Crops (days)

Crop	Initial (L_{ini})	Dev. (L_{dev})	Mid (L_{mid})	Late (L_{late})	Total	Plant date
Broccoli	35	45	40	15	135	September
Cabbage	40	60	50	15	165	September
Carrots	30	50	90	30	200	October
Cauliflower	35	50	40	15	140	September
Onion (green)	30	55	55	40	180	March
Tomato	35	40	50	30	155	April/May
Cantaloupe	30	45	35	10	120	January
	10	60	25	25	120	August
Potato	30	35	50	25	140	December
Beans (green)	20	30	30	10	90	February/March
	15	25	25	10	75	August/September
Grapes	20	40	120	60	240	April
	20	50	75	60	205	March

Source: Allen et al. (1988).

leaf drop. The value of the crop coefficient at the end of the late-season stage ($k_{c\ end}$) depends on crop- and water-management practices.

General lengths of the four distinct growth stages and total growing period for various types of climates and locations can be found in FAO Irrigation and Drainage Paper No. 24, and selected lengths for various crops in California are shown in Table 5.48. The values in Table 5.48 are useful only as a general guide and for comparison purposes, and local observations of the specific plant-stage development should be used, wherever possible, to incorporate effects of plant variety, climate, and cultural practices. Local information can be obtained by interviewing farmers, ranchers, agricultural extension agents, and local researchers. Typical crop coefficients for various stages of development are shown in Table 5.49.

Step 2: Adjust selected crop coefficients. Values of $k_{c\ ini}$ in Table 5.49 are for well-watered crops, and are therefore significantly influenced by irrigation practices during the initial growth phase. Values of $k_{c\ mid}$ and $k_{c\ end}$ are more affected by crop height, local wind speed, and humidity; and various adjustment factors have been developed to modify the values of $k_{c\ mid}$ and $k_{c\ end}$ in Table 5.49. For $k_{c\ ini}$, adjustment factors

TABLE 5.49: Typical Crop Coefficients for Various Growth Stages

Crop class	$k_{c\ ini}$	$k_{c\ mid}$	$k_{c\ end}$
Small vegetables	0.7	1.05	0.95
Vegetables—solanum family	0.6	1.15	0.80
Vegetables—cucumber family	0.5	1.00	0.80
Roots and tubers	0.5	1.10	0.95
Legumes	0.4	1.15	0.55
Perennial vegetables	0.5	1.00	0.80
Oil crops	0.35	1.15	0.35
Cereals	0.3	1.15	0.4
Sugar cane	0.40	1.25	0.75

Source: Allen et al. (1988).

FIGURE 5.70: Adjustments to $k_{c\ ini}$ for small infiltration depths

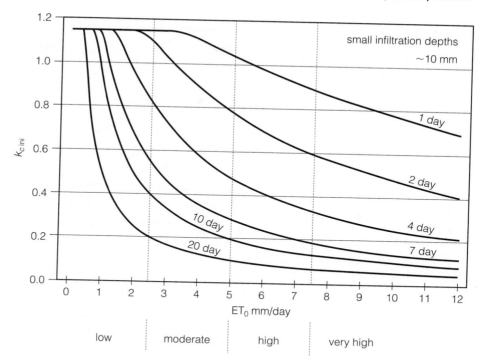

related to the reference evapotranspiration, ET_0, and the interval between irrigation or significant rainfall events are given in Figures 5.70 and 5.71. In cases where the wetting events are light to medium (3–10 mm), Figure 5.70 is applicable, and in cases where the wetting events are large (> 40 mm), adjustment factors are given in Figure 5.71 for coarse-, medium-, and fine-textured soils (Allen et al., 1998). In cases where wetting-event depth is between 10 mm and 40 mm, the adjustment factor can be linearly interpolated between Figures 5.70 and 5.71 for given values of ET_0 and time interval between wetting events. Empirical equations that closely match Figures 5.70 and 5.71 have been proposed by Allen et al. (2005c).

The relative impact of climate on $k_{c\ mid}$ is shown in Figure 5.72, where the adjustments to the values in Table 5.49 are given for various types of climates, mean daily wind speeds, and crop heights. Values of $k_{c\ end}$ given in Table 5.49 are typical values under standard climatic conditions. More arid climates and conditions of greater wind speed will have higher values of $k_{c\ end}$, and more humid climates and conditions of lower wind speed will have lower values of $k_{c\ end}$. For specific adjustments at locations where RH_{min} differs from 45% or where u_2 is larger or smaller than 2.0 m/s, the following equation can be used:

$$k_{c\ end} = k'_{c\ end} + [0.04(u_2 - 2) - 0.004(RH_{min} - 45)]\left(\frac{h}{3}\right)^{0.3} \qquad (5.373)$$

where $k'_{c\ end}$ is the value of $k_{c\ end}$ given by Table 5.49, u_2 is the mean daily wind speed at 2 m height over grass during the late-season growth stage (1 m/s $\leq u_2 \leq$ 6 m/s), RH_{min} is the mean daily minimum relative humidity during the late-season stage (20% $\leq RH_{min} \leq$ 80%), and h is the mean plant height during the late-season stage

FIGURE 5.71: Adjustments to $k_{c\,ini}$ for large infiltration depths

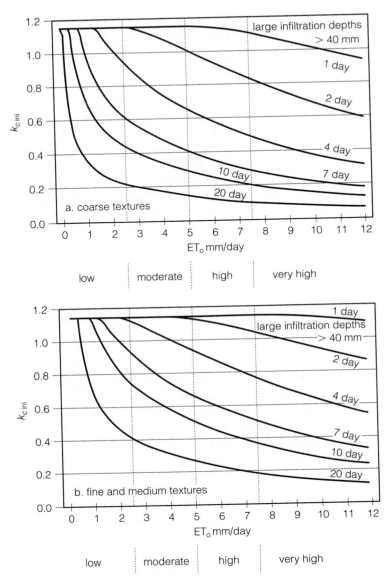

(0.1 m $\leq h \leq$ 10 m). Equation 5.373 is applied only when $k'_{c\,end}$ is greater than 0.45, and in cases where $k'_{c\,end} < 0.45$, no adjustment is made to $k_{c\,end}$.

Step 3: Construct the crop-coefficient curve. A typical crop-coefficient curve is shown in Figure 5.73. To construct this curve, the growing period is first divided into the four general growth stages that describe crop phenology (initial, crop development, mid-season, and late-season stage), and the three k_c values that correspond to $k_{c\,ini}$, $k_{c\,mid}$, and $k_{c\,end}$ are determined as described previously. The crop-coefficient curve is constructed by connecting straight-line segments through each of the four growth stages. Horizontal lines are drawn through $k_{c\,ini}$ in the initial stage and through $k_{c\,mid}$.

FIGURE 5.72: Adjustments to $k_{c\ mid}$

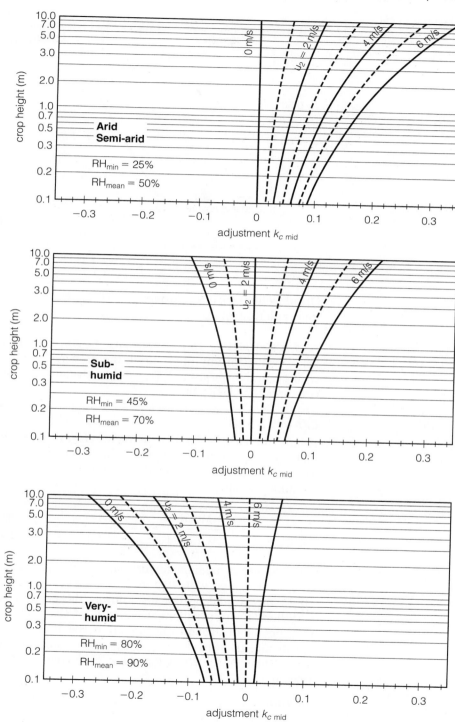

FIGURE 5.73: Crop coefficient curve

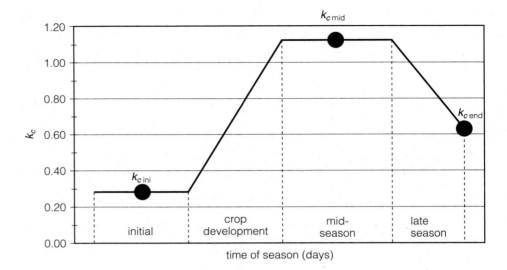

Diagonal lines are drawn from $k_{c\,ini}$ to $k_{c\,mid}$ within the course of the crop-development stage and from $k_{c\,mid}$ to $k_{c\,end}$ within the course of the late-season stage.

Step 4: Calculating crop evapotranspiration, ET_c. From the crop-coefficient curve, the k_c value for any time during the growing period can be determined. Then the crop evapotranspiration, ET_c, can be calculated by multiplying the k_c values by the corresponding ET_0 values.

EXAMPLE 5.68

A large tomato farm in Santa Rosa, California, is developing an irrigation schedule to match the evapotranspiration and rainfall characteristics on the farm. The reference-evapotranspiration rates in the region have been estimated using the FAO56-PM equation as follows:

Month	May	June	July	August	September
ET_0 (mm/d)	4.5	5.4	5.3	4.9	3.9

The soil has a medium texture and, during the initial stage of growth, the tomatoes are irrigated with 20 mm of water every 4 days. The average wind speed 2 m above the ground on the farm during the growing season is 4 m/s, conditions are very humid with a minimum humidity of around 80%, and the tomato plants grow to a height of around one meter. If the tomatoes are typically planted on May 1, estimate the variation of actual evapotranspiration throughout the growing season. Tomatoes are fruits that have crop coefficients very similar to those of small vegetables.

Solution The durations of the various growth stages for the tomato crop are given in Table 5.48 and, since the tomato crop is planted on May 1, the start and end dates are as follows:

Phase	Duration (days)	Start	End
Initial	35	May 1	June 4
Crop development	40	June 4	July 13
Mid-season	50	July 13	August 31
Late Season	30	August 31	September 29

The crop coefficients during the various growth phases can be estimated from Table 5.49 (for small vegetables) as $k_{c\ ini} = 0.7$, $k_{c\ mid} = 1.05$, and $k_{c\ end} = 0.95$. Adjustments to these crop coefficients are given below:

$k_{c\ ini}$: The 35-day initial-growth phase goes from May 1 to June 4, and the average reference evapotranspiration during this time, ET_i, can be taken as the day-weighted average evapotranspiration in May (4.5 mm/d) and June (5.4 mm/d). Therefore for the initial-growth phase the average evapotranspiration is given by

$$ET_i = \frac{(31)(4.5) + (4)(5.4)}{35} = 4.6 \text{ mm/d}$$

For 10 mm of irrigation every 4 days, Figure 5.70 gives a correction factor of 0.52, and for 40 mm of irrigation every 4 days (for medium-textured soils), Figure 5.71 gives a correction factor of 0.96. Therefore, for 20 mm of irrigation every 4 days, the correction factor can be interpolated as

$$\text{correction factor} = 0.52 + \frac{0.96 - 0.52}{40 - 10}(20 - 10) = 0.67$$

The corrected value of $k_{c\ ini}$ is therefore given by

$$k_{c\ ini} = 0.67(0.7) = 0.47$$

$k_{c\ mid}$: The mid-season growth phase goes from July 13 to August 31, the average wind speed during this time is 4 m/s, the crop height is 1 m, and for very humid conditions Figure 5.72 gives an adjustment of -0.03. The adjusted value of $k_{c\ mid}$ is therefore given by

$$k_{c\ mid} = 1.05 - 0.03 = 1.02$$

$k_{c\ end}$: The late-season growth phase ends on September 29, at which time the average wind speed at 2-m height, u_2, is 4 m/s, the minimum humidity, RH_{min}, is 80%, and the adjusted value of $k_{c\ end}$ is given by Equation 5.373 as

$$k_{c\ end} = k'_{c\ end} + [0.04(u_2 - 2) - 0.004(RH_{min} - 45)]\left(\frac{h}{3}\right)^{0.3}$$

$$= 0.95 + [0.04(4 - 2) - 0.004(80 - 45)]\left(\frac{1}{3}\right)^{0.3}$$

$$= 0.91$$

Combining the adjusted crop coefficients with the given reference evapotranspiration rate gives the following points on the evapotranspiration curve:

From	To	k_c	ET_0 (mm/d)	ET_c (mm/d)
May 1	May 31	0.47	4.5	2.1
June 1	June 4	0.47	5.4	2.5
July 13	July 31	1.02	5.3	5.4
August 1	August 31	1.02	4.9	5.0
	September 29	0.91	3.9	3.5

The evapotranspiration curve is plotted in Figure 5.74. As expected, the evapotranspiration rates are lowest in the initial stages of growth, are at a maximum during the mid-season stage, and decrease as the crop matures towards harvest.

The accuracy of the FAO-56 crop-coefficient approach in the Imperial Irrigation District in the western United States was assessed by Allen et al. (2005d). The results of this investigation demonstrated that the accuracy of the estimated annual crop evapotranspiration was, on average, within 8% of that determined by a water balance; on a monthly basis the average difference was on the order of 15%.

As two reference-crop definitions (grass and alfalfa) are in use in various parts of the world, two families of k_c curves for agricultural crops have been developed. These are the alfalfa-based k_c curves by Wright (1981; 1982) and grass-based k_c curves by Doorenbos and Pruitt (1977), ASCE (1990), and Allen et al. (1998). Caution must be exercised to avoid using grass-based k_c values with alfalfa-reference evapotranspiration, ET_r, or using alfalfa-based k_c values with grass-reference evapotranspiration, ET_0. Usually, alfalfa-based k_c can be converted to grass-based k_c by multiplying by a factor ranging from about 1.0 to 1.3, depending on the climate (1.05 for humid, calm

FIGURE 5.74: ET curve for tomato farm

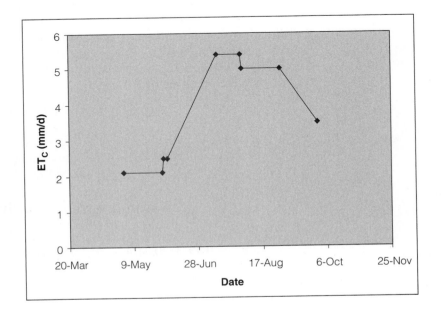

conditions; 1.2 for semi-arid, moderately windy conditions; and 1.35 for arid, windy conditions). A reference conversion ratio, k_{ratio}, can be established for any climate using the following relation (Allen et al., 1998):

$$k_{ratio} = 1.2 + [0.04(u_2 - 2) - 0.004(RH_{min} - 45)]\left(\frac{h}{3}\right)^{0.3} \qquad (5.374)$$

where u_2 is the mean daily wind speed at 2 m height over grass, RH_{min} is the mean daily minimum relative humidity, and h is the standard height of the alfalfa reference crop, which is equal to 0.5 m. Using k_{ratio} derived from Equation 5.374, the crop coefficient, $k_{c\ grass}$, based on a grass reference-crop evapotranspiration, is related to the alfalfa crop coefficient, $k_{c\ alfalfa}$, by

$$k_{c\ grass} = k_{ratio}k_{c\ alfalfa} \qquad (5.375)$$

EXAMPLE 5.69

In Kimberly, Idaho, during an average summer the minimum relative humidity is 30% and the mean wind speed is 2.2 m/s. Potato farmers have been using old maps showing the alfalfa-reference evapotranspiration, and during the summer growth season the alfalfa crop coefficient for potatoes is typically taken as 1.4. New maps have been generated showing the distribution of grass-reference evapotranspiration. What crop coefficient for potatoes should be used with these data?

Solution   From the given data: $RH_{min} = 30\%$, $u_2 = 2.2$ m/s, and for alfalfa $h = 0.5$ m. Substituting into Equation 5.374 gives

$$k_{ratio} = 1.2 + [0.04(u_2 - 2) - 0.004(RH_{min} - 45)]\left(\frac{h}{3}\right)^{0.3}$$

$$= 1.2 + [0.04(2.2 - 2) - 0.004(30 - 45)]\left(\frac{0.5}{3}\right)^{0.3} = 1.24$$

Therefore, the ratio of grass to alfalfa crop coefficient is 1.24. Since $k_{c\ alfalfa} = 1.4$, then Equation 5.375 gives

$$k_{c\ grass} = k_{ratio}k_{c\ alfalfa}$$

$$= (1.24)(1.4) = 1.7$$

Hence the grass crop coefficient for potatoes during the summer is 1.7. The fact that the grass crop coefficient is greater than the alfalfa crop coefficient indicates that the grass-reference ET is less than the alfalfa-reference ET.

It is very important to keep in mind that crop coefficients are comparable only to the extent that they are based on the same equation used to estimate the reference evapotranspiration. In an effort to put all crop coefficients on a common basis, there is a significant movement in the United States to refer all crop coefficients to the ASCE standardized equation for either grass or alfalfa, where it is noted that the ASCE standardized equation is the same as the FAO56-PM equation for a grass reference crop using daily time steps.

5.8.5 Selection of ET Estimation Method

The preferred method of estimating actual evapotranspiration is either using the Penman–Monteith equation (Equation 5.301) directly or using crop coefficients combined with a reference evapotranspiration calculated using the FAO56-PM equation (Equation 5.359). The Penman equation (Equation 5.353) is the preferred method of estimating potential evapotranspiration, ET_p, although utilization of the potential evapotranspiration is discouraged in favor of the reference evapotranspiration, ET_0, as a measure of meteorological effects on evapotranspiration (ET). Considerations in using Penman-type equations to estimate ET are the complexity of the model, lack of sufficient meteorological data to estimate the model parameters, and a recognition that in many cases evapotranspiration is dominated by solar radiation and therefore the use of simpler methods with fewer parameters is justified. For example, Abtew (1996) has noted that in south Florida most of the variance in daily evapotranspiration is explained by solar radiation, with the effect of humidity and wind speed relatively minimal.

5.9 Computer Models

Several good computer models are available for simulating surface-water hydrologic processes in both urban and rural catchments, and the use of computer models to apply the fundamental principles covered in this chapter is commonplace in engineering practice. Computer models are currently undergoing a rapid evolution to incorporate the growing databases associated with the widespread use of Geographic Information Systems (GIS) in urban planning and management. Typically, there are a variety of hydrologic models to choose from for a particular application. However, in doing work that is to be reviewed by regulatory agencies, models developed and maintained by the U.S. government have the greatest credibility and, perhaps more important, are almost universally acceptable in supporting permit applications and defending design protocols on liability issues. A secondary guideline in choosing a hydrologic model is that the simplest model that will accomplish the design objectives should be given the most serious consideration. Several of the more widely used surface-water hydrologic models that have been developed and endorsed by U.S. government agencies are described briefly here.

SWMM. In the United States, SWMM is regarded as the standard modeling tool for urban drainage-system evaluation. The basic SWMM code was developed and is maintained by the U.S. Environmental Protection Agency (USEPA). SWMM is primarily used to simulate the runoff from a continuous rainfall record and consists principally of a Runoff Module, a Transport Module, and a Storage/Treatment Module. The catchment is separated into subareas that are connected to small gutter/pipe elements leading to an outfall location. The Runoff Module simulates runoff by modeling each subarea as a nonlinear reservoir with precipitation as input and infiltration, evaporation, and surface runoff as outflows. Infiltration is modeled by either the Green–Ampt or Horton equations; and infiltrated water is routed through upper and lower subsurface zones and may contribute to runoff. The Transport Module receives subarea runoff hydrographs and uses either kinematic or dynamic routing to calculate sewer flows and the hydraulic head at key junctions. The Storage/Treatment Module simulates the effects of detention basins and combined-sewer overflows.

Quality constituents modeled by SWMM include suspended solids, settleable solids, biochemical oxygen demand, nitrogen, phosphorus, and grease. A variety of options are provided for determining surface-runoff loads, including constant concentration, regression relationships of load versus flow, and buildup-washoff. Many commercial versions of SWMM have added attractive output presentations, including elaborate graphics, with the basic USEPA SWMM as the main computing engine.

TR 20. This program was developed and is maintained by the Natural Resources Conservation Service, U.S. Department of Agriculture. TR 20 represents the catchment as several subbasins with homogeneous properties. The NRCS curve-number and unit-hydrograph methods are used to calculate the runoff hydrographs resulting from a single storm; the runoff hydrographs from the subareas are routed through drainage channels and reservoirs using the kinematic method (channel routing) and the storage-indication method (reservoir routing). The catchment-outflow hydrograph is the sum of the routed hydrographs at the catchment outlet. This model is designed to perform multiple analyses in a single run.

HEC-HMS. The HEC-HMS (Hydrologic Modeling System) program was developed and is maintained by the U.S. Army Corps of Engineers Hydrologic Engineering Center (HEC). This model is probably the most extensively applied of all watershed computer programs (Wurbs and James, 2002). HEC-HMS is an interactive model that computes surface-runoff hydrographs for given rainfall and snowmelt events for given watershed characteristics, combines runoff hydrographs from subwatersheds at prescribed locations, and routes them through reservoirs and channels in a given network. This model can also perform simplified dam-break analysis and economic analysis for flood damages. The HEC-HMS model includes the following features:

- A distributed runoff model which can be used with distributed precipitation data such as data available from weather radar.
- The program will develop a design storm for a specified exceedance frequency from an input IDF curve using the alternating-block method.
- Precipitation is separated into losses and runoff using either of the following options: NRCS Curve Number method, Green–Ampt, Holtan, Horton, or initial/uniform infiltration rate.
- Runoff hydrographs from each subbasin are developed with either the unit-hydrograph or kinematic-wave approach. The unit hydrograph can be either input or synthesized by the model using the NRCS dimensionless-unit hydrograph, Snyder, or Clark methods.
- The storage-outflow method is used for reservoir routing. Stream-routing options include storage-outflow, Muskingum, Muskingum–Cunge, and kinematic model.
- A continuous soil-moisture accounting model is used to simulate the long-term response of a watershed to wetting and drying.
- An automatic-calibration package can estimate certain model parameters and initial conditions, given observations of hydrologic conditions in the watershed.

- A link to a database-management system permits data storage, retrieval, and connectivity with other analysis tools available from the U.S. Army Corps of Engineers Hydrologic Engineering Center and other sources.

Only a few of the more widely used computer models have been cited here, and many other good models are capable of performing the same tasks.

Summary

The material covered in this chapter is particularly applicable to the design of urban storm-water-management systems and provides the fundamental background necessary to design such systems in accordance with the ASCE Manual of Practice (ASCE, 1992). The chapter begins with the quantitative specifications of design rainfall events, which include: return period, duration, depth, temporal distribution, and spatial distribution. Extreme-rainfall events such as the probable maximum storm and the standard project storm are also covered. After specifying the design rainfall event, abstraction models are used to account for interception (Equation 5.29), depression storage (Table 5.9), and infiltration (Horton, Green–Ampt, or NRCS models). Removal of the abstractions from the design rainfall yields the temporal distribution of runoff depth, which is used to calculate either the peak-runoff or continuous-runoff hydrograph. In small catchments, peak-runoff rates are adequately calculated using the rational method, with the NRCS-TR55 or USGS Regional Regression methods being suitable alternatives in some cases. Peak-runoff rates are used in sizing gutters, inlets, and storm sewers. Complete runoff hydrographs are used to size storage basins and are typically computed using a unit-hydrograph model, although the time-area, kinematic-wave, nonlinear-reservoir, and Santa Barbara models are also used in practice. The quality of stormwater runoff can be estimated using either event mean concentrations, USGS, or USEPA water-quality models.

Stormwater-management systems consist of both a minor and a major system. Components of minor stormwater systems include street gutters, inlets, roadside and median channels, storm sewers, and storage basins, while the major stormwater system includes drainage pathways when the capacity of the minor system is exceeded, such as floodways and constructed drainage canals. Procedures for designing storm sewers and roadway-drainage systems are all covered in detail. Stormwater impoundments are designed so that peak postdevelopment runoff is less than or equal to peak predevelopment runoff and so that sufficient detention time is provided for water-quality control. Procedures for designing stormwater impoundments for both flood control and water-quality control are covered in detail, and guidance is provided for selecting the best-management practices for a variety of catchment conditions.

Water consumption by evapotranspiration is of primary importance in designing irrigation systems in agricultural areas and in computing long-term water balances. The Penman–Monteith equation is widely regarded as one of the most accurate and technically sound methods for estimating evapotranspiration, and the application of this method is presented in detail. Methods used in practice to estimate potential, reference, and actual evapotranspiration are all covered, and guidance is provided for selecting the appropriate evapotranspiration-estimation method for a variety of site conditions.

In engineering practice, many of the techniques presented in this chapter are implemented using readily available computer programs. A few of the more widely used programs that have been developed by U.S. government agencies are SWMM, TR 20, and HEC-HMS.

Problems

5.1. Show that the maximum return period that can be investigated when using the Weibull formula to derive the IDF curve from an n-year annual rainfall series is equal to $n + 1$.

5.2. A rainfall record contains 50 years of measurements at 5-minute intervals. The annual-maximum rainfall amounts for intervals of 5 min, 10 min, 15 min, 20 min, 25 min, and 30 min are ranked as follows:

Rank	Δt in minutes					
	5	10	15	20	25	30
1	26.2	45.8	60.5	72.4	81.8	89.7
2	25.3	44.0	58.1	69.6	78.6	86.3
3	24.2	42.2	55.8	66.8	75.5	82.8

where the rainfall amounts are in millimeters. Calculate the IDF curve for a return period of 40 years.

5.3. An eight-year rainfall record measures the rainfall increments at 10-minute intervals, and the top five rainfall increments are as follows:

Rank	1	2	3	4	5
10-min rainfall (mm)	39.6	39.4	38.5	37.3	36.5

Estimate the frequency distribution of the annual maxima with return periods greater than two years.

5.4. Assuming that the ranked rainfall increments in Problem 5.2 were derived from a partial-duration series, calculate the annual-series IDF curve for a 40-year return period.

5.5. The rainfall data in Table 5.50 were compiled by determining the maximum amount of rainfall that occurred for a given time period during a given year. Use these data to develop IDF curves for 5-, 10-, and 25-year return periods. If these IDF curves are to be fitted to a function of the form

$$i = \frac{a}{t + b}$$

estimate the values of a and b for a 5-year return period.

5.6. Use the Chen method to estimate the 10-year IDF curve for Atlanta, Georgia.

5.7. Use the Chen method to estimate the 10-year IDF curves for New York City and Los Angeles. Compare your results with those obtained by Wenzel (1982) and listed in Table 5.2.

5.8. The spatially averaged rainfall is to be calculated for the catchment area shown in Figure 5.10, where the coordinates of the nearby rain gages are as follows:

Gage	x (km)	y (km)
A	1.3	7.0
B	1.0	3.7
C	4.2	4.9
D	3.5	1.4
E	2.1	−1.0

TABLE 5.50: Rainfall Maxima (in cm) for Given Durations

Year	Duration (h)						
	1	2	4	6	10	12	24
1946	1.7	2.9	4.6	4.8	4.8	4.8	5.1
1947	1.9	2.6	2.6	3.0	3.7	3.7	4.8
1948	1.9	2.3	2.7	2.9	3.1	3.1	3.2
1949	1.9	2.4	2.8	4.0	4.7	4.9	5.7
1950	1.9	1.9	2.2	2.5	2.5	2.5	3.0
1951	2.1	2.4	3.0	3.5	4.0	4.2	5.1
1952	1.7	2.5	2.7	2.8	3.0	3.0	4.0
1953	1.7	3.6	3.6	3.6	4.3	4.3	4.7
1954	1.9	2.1	2.8	3.0	3.8	3.8	3.8
1955	2.4	3.7	4.5	4.5	4.6	4.6	5.4
1956	1.0	1.3	1.6	1.7	1.7	1.7	1.8
1957	2.7	2.8	3.2	3.7	3.7	3.8	4.4
1958	1.3	1.5	2.0	2.6	3.5	3.8	4.1
1959	2.8	3.2	3.4	4.1	5.2	5.7	6.0
1960	2.0	2.1	2.8	4.1	5.6	6.0	7.9
1961	1.8	2.7	2.8	2.8	2.8	2.8	2.8
1962	1.5	2.6	2.8	3.1	3.8	4.5	6.4
1963	1.4	2.5	3.1	3.7	4.2	4.3	4.5
1964	1.6	1.8	2.2	2.6	3.6	3.6	4.1
1965	1.2	1.7	2.4	2.7	3.0	3.1	3.3
1966	1.4	1.5	1.5	2.3	2.5	2.7	2.8
1967	1.7	1.8	2.4	2.7	2.9	2.9	4.4
1968	1.9	2.6	4.1	5.1	5.1	5.2	5.4
1969	1.8	3.2	3.8	3.9	3.9	3.9	3.9
1970	1.3	1.6	1.7	1.7	1.7	1.7	2.7
1971	1.8	2.1	3.3	3.4	3.7	4.2	4.9
1972	1.6	2.0	2.3	2.6	2.7	2.7	2.7
1973	1.9	2.8	2.8	2.9	2.9	2.9	2.9
1974	3.1	3.4	3.6	3.6	3.6	3.6	3.8
1975	1.5	1.6	2.1	2.9	3.6	3.6	4.2
1976	2.9	3.4	4.0	4.1	4.4	4.5	4.9
1977	1.4	2.5	2.8	2.8	2.8	2.8	2.9
1978	1.6	2.1	2.2	2.2	2.2	2.2	2.7
1979	2.2	2.3	4.5	5.2	7.6	7.7	7.8
1980	2.5	3.0	3.2	3.2	4.3	4.6	4.7
1981	2.1	3.0	3.1	3.1	3.1	3.1	5.0
1982	1.6	2.3	2.8	3.2	3.8	4.1	5.0
1983	1.2	1.8	2.1	2.5	2.5	2.5	2.6
1984	1.5	4.2	4.2	4.2	4.2	4.2	4.2
1985	1.7	2.7	2.7	2.7	2.8	2.9	3.5
1986	3.7	6.7	9.7	10.4	10.4	10.7	11.7
1987	2.7	4.8	4.8	4.8	6.2	6.2	6.7
1988	2.0	6.9	6.9	6.9	6.9	6.9	10.1
1989	1.9	2.9	3.0	3.0	3.0	3.0	3.2
1990	1.2	1.4	1.6	1.7	2.1	2.1	2.5

and the coordinates of the grid origin are $(0, 0)$. The measured rainfall at each of the gages during a 1-hour interval is 60 mm at A, 90 mm at B, 65 mm at C, 35 mm at D, and 20 mm at E. Use the inverse-distance-squared method with the 1-km × 1-km grid shown in Figure 5.10 to estimate the average rainfall over the catchment during the 1-hour interval.

5.9. Repeat Problem 5.8 by assigning equal weight to each rain gage. Compare your result with that obtained using the inverse-distance-squared method. Which result do you think is more accurate?

5.10. The IDF curve for 10-year storms in Atlanta, Georgia, is given by

$$i = \frac{1628}{(t_d + 8.16)^{0.76}}$$

where i is the average intensity in mm/h and t_d is the duration in minutes. Assuming that the mean of the rainfall distribution is equal to 38% of the rainfall duration, estimate the triangular hyetograph for a 40-minute storm.

5.11. The IDF curve for 10-year storms in Santa Fe, New Mexico, is given by

$$i = \frac{818}{(t_d + 8.54)^{0.76}}$$

where i is the average intensity in mm/h and t_d is the duration in minutes. Assuming that the mean of the rainfall distribution is equal to 44% of the rainfall duration, estimate the triangular hyetograph for a 50-minute storm.

5.12. Use the alternating-block method and the IDF curve given in Problem 5.11 to calculate the hyetograph for a 10-year 1-hour storm using 11 time intervals.

5.13. Derive the IDF curve for Boston, Massachusetts, using the Chen method. Use the derived IDF curve with the alternating-block method to determine the hyetograph for a storm with a return period of 10 years and a duration of 40 minutes. Use nine time intervals to construct the hyetograph.

5.14. Use the 10-year IDF curve for Atlanta in Table 5.2 given by Wenzel (1982) and the alternating-block method to estimate the rainfall hyetograph for a 10-year 50-minute storm in Atlanta. Use seven time intervals.

5.15. The alternating-block method assumes that the maximum rainfall for any duration less than or equal to the total storm duration has the same return period. Discuss why this is a conservative assumption.

5.16. The precipitation resulting from a 25-year 24-hour storm in Miami, Florida, is estimated to be 260 mm. Calculate the NRCS 24-hour hyetograph.

5.17. The precipitation resulting from a 25-year 24-hour storm in Atlanta, Georgia, is estimated to be 175 mm. Calculate the NRCS 24-hour hyetograph, and compare it with the 25-year 24-hour hyetograph for Miami that is calculated in Problem 5.16.

5.18. The rainfall measured at a rain gage during a 10-hour storm is 193 mm. Estimate the average rainfall over a catchment containing the rain gage, where the area of the catchment is 200 km^2.

5.19. Use Equation 5.14 to show that for long-duration storms the areal-reduction factor becomes independent of the storm duration. Under these asymptotic conditions, determine the relationship between the areal-reduction factor and the catchment area.

5.20. The annual-maximum 24-hour rainfall amounts in a South-American city have a mean of 202 mm and a standard deviation of 65 mm. Estimate the probable maximum precipitation (PMP) using the Hershfield method, and compare this value to the world's largest observed 24-hour rainfall.

5.21. Assess whether the Chen method always leads to an IDF curve of the form

$$i = \frac{f(T)}{g(t)}$$

where i is the average rainfall intensity for a storm of duration t and return period T, $f(T)$ is a term that depends only on the return period, T, of the storm, and $g(t)$ is a term that depends only on the duration, t, of the storm.

5.22. The annual-maximum 24-hour rainfall amounts on a tropical island have a mean of 320 mm and a standard deviation of 156 mm. The IDF curve for the 50-year storm on the island is given by

$$i = \frac{6240}{(t + 6.87)^{0.65}} \text{ mm/h}$$

where i is the intensity in mm/h and t is the duration in minutes. Find the annual-maximum 8-hour rainfall with a return period of 50,000 years.

5.23. Estimate the maximum amount of rainfall that can be expected in one hour.

5.24. Maximum precipitation amounts can be estimated using Equation 5.24, and observed precipitation maxima are listed in Table 5.6. Compare the predictions given by Equation 5.24 with the observations in Table 5.6.

5.25. The Everglades Agricultural Area in central Florida has an area of approximately 2600 km^2 and is approximately square. Determine the appropriate storm area to be used in estimating the PMP. If using this storm area in HMR 51 yields a 72-h PMP of 107 cm (42 in.) and a maximum 6-h rainfall of 72 cm (28.3 in.), determine the isohyetal rainfall pattern to be used in constructing the PMS.

5.26. A pine forest is to be cleared for a development in which all the trees on the site will be removed. The IDF curve for the area is given by

$$i = \frac{2819}{t + 16}$$

where i is the average rainfall intensity in mm/h and t is the duration in minutes. The storage capacity of the trees in the forest is estimated as 5 mm, the leaf area index is 6, and the evaporation rate during the storm is estimated as 0.3 mm/h. Determine the increase in precipitation reaching the ground during a 30-min storm that will result from clearing the site.

5.27. The storage capacity of a forest canopy covering a catchment is 9 mm, the leaf-area index is 8, and the evaporation rate during a storm is 0.5 mm/h. If a storm has the hyetograph calculated in Problem 5.14, estimate the amount of precipitation intercepted during each of the seven time intervals of the hyetograph. Determine the hyetograph of the rainfall reaching the ground.

5.28. For the pine forest described in Problem 5.26, estimate the interception using a Horton-type empirical equation of the form $I = a + bP^n$, where a and b are constants and P is the precipitation amount.

5.29. A 25-min storm produces 30 mm of rainfall on the following surfaces: steep pavement, flat pavement, impervious surface, lawn, pasture, and forest litter. Assuming that depression storage is the dominant abstraction process, for each surface, estimate the fraction of rainfall that becomes surface runoff. On which surface is depression storage most significant, and on which surface is it least significant?

5.30. A soil has the Horton infiltration parameters: $f_0 = 200$ mm/h, $f_c = 60$ mm/h, and $k = 4$ min^{-1}. If rainfall in excess of 200 mm/h is maintained for 50 min, estimate the infiltration as a function of time. What is the infiltration rate at the end of 50 min? How would this rate be affected if the rainfall rate were less than 200 mm/h?

5.31. Show that the Horton infiltration model given by Equation 5.43 satisfies the differential equation

$$\frac{df_p}{dt} = -k(f_p - f_c)$$

5.32. A catchment has the Horton infiltration parameters: $f_0 = 150$ mm/h, $f_c = 50$ mm/h, and $k = 3$ min^{-1}. The design storm is:

Interval (min)	Average rainfall (mm/h)
0–10	20
10–20	40
20–30	80
30–40	170
40–50	90
50–60	20

Estimate the time when ponding begins.

5.33. An alternating-block analysis indicated the following rainfall distribution for a 70-min storm:

Interval	Rainfall (mm/h)
1	7
2	25
3	45
4	189
5	97
6	44
7	18

where each interval corresponds to 10 minutes. If infiltration can be described by the Horton parameters: $f_0 = 600$ mm/h, $f_c = 30$ mm/h, and $k = 0.5$ min^{-1}, and the depression storage is 4 mm, use the Horton method to determine the distribution of runoff, and hence the total runoff.

5.34. An area consists almost entirely of sandy loam, which typically has a saturated hydraulic conductivity of 11 mm/h, average suction head of 110 mm, porosity of 0.45, field capacity of 0.190, wilting point of 0.085, and depression storage of 4 mm. The design rainfall is given as:

Interval (min)	Average rainfall (mm/h)
0–10	20
10–20	40
20–30	60
30–40	110
40–50	60
50–60	20

Use the Green–Ampt method to determine the runoff versus time for average initial moisture conditions, and contrast the depth of rainfall with the depth of runoff. Assume that the initial moisture conditions are midway between the field capacity and wilting point.

5.35. Repeat Problem 5.34 for a sandy clay soil with a depression storage of 9 mm. [*Hint*: Use Table 5.12 to estimate the soil properties.]

5.36. Derive the NRCS curve-number model for the infiltration rate given by Equation 5.71. Explain why this infiltration model is unrealistic.

5.37. Drainage facilities are to be designed for a rainfall of return period 10 years and duration 1 hour. The IDF curve is given by

$$i = \frac{203}{(t + 7.24)^{0.73}}$$

where i is the average intensity in cm/h and t is the storm duration in minutes. The minimum infiltration rate is 10 mm/h, and the area to be drained is primarily residential with lot sizes on the order of 0.2 ha (0.5 ac). Use the NRCS method to estimate the total amount of runoff (in cm), assuming the soil is in average condition at the beginning of the design storm.

5.38. Repeat Problem 5.37 for the case in which heavy rainfall occurs within the previous five days and the soil is saturated.

5.39. Use Equation 5.71 to calculate the average infiltration rate during the storm described in Problem 5.37. Compare this calculated infiltration rate with the given minimum infiltration rate of 10 mm/h.

5.40. Data from a double-ring infiltrometer indicate that a soil in a catchment has the following Horton parameters: $f_0 = 250$ mm/h, $f_c = 44$ mm/h, and $k = 0.13$ min^{-1}. Observations also indicate that the catchment has an average depression storage of 6 mm. If the catchment is located where the 10-year 24-hour rainfall is 229 mm and can be described by the NRCS Type II distribution, estimate the curve number for the site. Use hourly time increments in your analysis. Would the curve number be different for a 20-year 24 hour rainfall? Is the dependence of the curve number on the rainfall amount physically reasonable?

5.41. An undeveloped parcel of land in south Florida has a water-table elevation 1.22 m below land surface, and an estimated cumulative water storage of 21 cm is recommended for use in the curve-number method. Consider the 1-day and 3-day storm events given in Tables 5.51 and 5.52, both of which have a total rainfall amount of 31.1 cm.

Assuming that the infiltration capacity of the soil is a constant, plot the relationship between curve number and infiltration capacity for both the 1-day and 3-day storms. Use these results to determine the infiltration capacity corresponding to the 21 cm of available storage. If the actual infiltration capacity of the soil is 4 cm/h, estimate the curve numbers that should be used for the 1-day and 3-day storms.

5.42. A proposed 20-ha development includes 5 ha of parking lots, 10 ha of buildings, and 5 ha of grassed area. The runoff from the parking lots and buildings are both routed directly to grassed areas. If the grassed areas contain Type A soil in good condition, estimate the runoff from the site for a 180-mm rainfall event.

5.43. Repeat Problem 5.42 for the case where the runoff from the buildings, specifically the roofs, is discharged directly onto the parking lots. Based on this result, what can you infer about the importance of directing roof drains to pervious areas?

5.44. Repeat Problem 5.42 using an area-weighted curve number. How would your result change if the roof drains were directly connected to the parking lot?

5.45. Discuss why it is preferable to route rainfall excesses on composite areas rather than using weighted-average curve numbers.

5.46. Consider a site that is I percent impervious in which the curve number of the impervious area is 98, and the curve number of the pervious area is CN_p. If the all of the impervious area is directly connected to the drainage system, show that the

TABLE 5.51: 1-Day Storm Event

Time (h)	Rainfall (cm)	Time (h)	Rainfall (cm)	Time (h)	Rainfall (cm)
0	0.0				
0.5	0.2	8.5	4.8	16.5	27.6
1.0	0.3	9.0	5.3	17.0	27.9
1.5	0.5	9.5	5.9	17.5	28.2
2.0	0.6	10.0	6.6	18.0	28.5
2.5	0.8	10.5	7.4	18.5	28.8
3.0	1.0	11.0	8.4	19.0	29.0
3.5	1.2	11.5	9.9	19.5	29.3
4.0	1.4	12.0	20.4	20.0	29.6
4.5	1.6	12.5	22.7	20.5	29.8
5.0	1.9	13.0	23.9	21.0	30.0
5.5	2.2	13.5	24.7	21.5	30.2
6.0	2.6	14.0	25.4	22.0	30.4
6.5	3.0	14.5	26.0	22.5	30.5
7.0	3.4	15.0	26.4	23.0	30.7
7.5	3.8	15.5	26.9	23.5	30.9
8.0	4.3	16.0	27.4	24.0	31.1

TABLE 5.52: 3-Day Storm Event

Time (h)	Rainfall (cm)	Time (h)	Rainfall (cm)	Time (h)	Rainfall (cm)
0	0.0				
1	0.2	25	3.54	49	8.44
2	0.27	26	3.75	50	8.66
3	0.41	27	3.95	51	8.93
4	0.55	28	4.16	52	9.24
5	0.69	29	4.34	53	9.62
6	0.82	30	4.55	54	10.1
7	0.98	31	4.75	55	10.7
8	1.1	32	4.96	56	11.3
9	1.3	33	5.17	57	12.1
10	1.4	34	5.37	58	13.1
11	1.5	35	5.58	59	14.4
12	1.7	36	5.76	60	23.2
13	1.8	37	5.97	61	25.7
14	1.9	38	6.17	62	26.9
15	2.1	39	6.38	63	27.6
16	2.2	40	6.58	64	28.3
17	2.35	41	6.79	65	28.7
18	2.51	42	7.00	66	29.1
19	2.65	43	7.20	67	29.6
20	2.79	44	7.41	68	30.0
21	2.93	45	7.59	69	30.2
22	3.06	46	7.80	70	30.5
23	3.20	47	8.00	71	30.8
24	3.34	48	8.21	72	31.1

composite curve number of the site, CN_c, is given by

$$CN_c = CN_p + \frac{I}{100}(98 - CN_p)$$

5.47. Consider the site described in Problem 5.46, with the exception that only a portion of the impervious area is directly connected to the drainage system. Show that the composite curve number of the site, CN_c, is then given by

$$CN_c = CN_p + \frac{I}{100}(98 - CN_p)(1 - 0.5R)$$

where R is the ratio of the unconnected impervious area to the total impervious area.

5.48. A catchment with a grass surface has an average slope of 0.8%, and the distance from the catchment boundary to the outlet is 80 m. For a 30-min storm with an effective rainfall rate of 70 mm/h, estimate the time of concentration using: (a) the kinematic-wave equation, (b) the NRCS method, (c) the Kirpich equation, (d) the Izzard equation, and (e) the Kerby equation.

5.49. What is the maximum flow distance that should be described by overland flow?

5.50. Find α and m in the kinematic-wave model (Equation 5.85) corresponding to: (a) the Manning equation, and (b) the Darcy–Weisbach equation.

5.51. An asphalt pavement drains into a rectangular concrete channel. The catchment surface has an average slope of 1.0%, and the distance from the catchment boundary to the drain is 30 m. The drainage channel is 60 m long, 20 cm wide, and 25 cm deep, and has a slope of 0.6%. For an effective rainfall rate of 50 mm/h, the flowrate in the channel is estimated to be 0.02 m³/s. Estimate the time of concentration of the catchment.

5.52. The surface of a 2-ha catchment is characterized by a runoff coefficient of 0.5, a Manning n for overland flow of 0.25, an average overland flow length of 60 m, and an average slope of 0.5%. Calculate the time of concentration using the kinematic-wave equation. The drainage channel is to be sized for the peak runoff resulting from a 10-year rainfall event, and the 10-year IDF curve is given by

$$i = \frac{150}{(t + 8.96)^{0.78}}$$

where i is the average rainfall intensity in cm/h and t is the duration in minutes. The minimum time of concentration is 5 minutes. Determine the peak runoff rate.

5.53. A 20-ha townhouse development is to be drained by a single drainage inlet. The average runoff coefficient for the site is estimated to be 0.7, Manning's n for overland flow is 0.25, the average overland flow length is 100 m, and the average slope of the site toward the inlet is 0.6%. The site is located in an area with an IDF curve given by

$$i = \frac{1020}{(t + 8.7)^{0.75}}$$

where i is the rainfall intensity in mm/h and t is the duration in minutes. What is the peak runoff rate expected at the drainage inlet? If an error of 10% is possible in each of the assumed site parameters, determine which parameter gives the highest peak runoff when the 10% error is included, and which gives the lowest peak runoff when the error is included. If the 10% error occurs simultaneously in all the drainage parameters, what would be the design peak discharge at the inlet?

5.54. Explain why higher runoff coefficients should be used for storms with longer return periods.

5.55. Suppose that the catchment described in Problem 5.52 contains 0.5 ha of impervious area that is directly connected to the storm sewer. If the runoff coefficient of the impervious area is 0.9, the Manning n for overland flow on the impervious surface is 0.035, the average flow length is 30 m, and the average slope is 0.5%, estimate the peak runoff.

5.56. A 4.2-km^2 catchment with 0.5% pond area has a curve number of 79, a time of concentration of 3 h, and a 24-h Type II precipitation of 10 cm. Estimate the peak runoff.

5.57. A 1-km^2 catchment with 3% pond area has a curve number of 70, a time of concentration of 1.7 h, and a 24-h Type I precipitation of 13 cm. Estimate the peak runoff.

5.58. A 10-ha single-family residential development in Atlanta is estimated to have a curve number of 70 and a time of concentration of 15 minutes. There are no ponds or swamps on the site, and the 10-year IDF curve is given by

$$i = \frac{2029}{(t + 7.24)^{0.73}}$$

where i is the rainfall intensity in mm/h, and t is the storm duration in minutes. Estimate the peak runoff from the site, and determine the equivalent runoff coefficient that could be used with the rational method to estimate the peak runoff.

5.59. The 15-min unit hydrograph for a 2.1-km^2 urban catchment is given by

Time (min)	Runoff (m^3/s)	Time (min)	Runoff (m^3/s)
0	0	210	0.66
30	1.4	240	0.49
60	3.2	270	0.36
90	1.5	300	0.28
120	1.2	330	0.25
150	1.1	360	0.17
180	1.0	390	0

(a) Verify that the unit hydrograph is consistent with a 1-cm rainfall excess; (b) estimate the runoff hydrograph for a 15-min rainfall excess of 2.8 cm; and (c) estimate the runoff hydrograph for a 30-min rainfall excess of 10.3 cm.

5.60. The 15-min unit hydrograph for a 2.1-km^2 catchment is given in Problem 5.59. Determine the S-hydrograph, calculate the 50-min unit hydrograph for the catchment, and verify that your calculated unit hydrograph corresponds to a 1-cm rainfall excess.

5.61. The 15-min unit hydrograph for a 2.1-km^2 catchment is given in Problem 5.59. Estimate the runoff resulting from the following 120-min storm:

Time (min)	Rainfall (cm)
0–30	2.4
30–60	4.5
60–90	2.1
90–120	0.8

5.62. In a 315-km^2 watershed the measured rainfall and resulting runoff in a small river leaving the catchment are as given in Table 5.53.

TABLE 5.53: Measured Rainfall and Runoff

Time (h)	Rainfall (cm)	River flow (m^3/s)
0		100
	0.5	
1		100
	2.5	
2		300
	2.5	
3		700
	0.5	
4		1000
5		800
6		600
7		400
8		300
9		200
10		100
11		100

Explain why the flow in the drainage channel is nonzero even when there is no rain. What is the name given to this no-rain flow? Separate the flow in the drainage channel into the component that results from rainfall (i.e., direct runoff) and the component that does not result from rainfall. Determine the unit hydrograph based on the given data. Compare the total rainfall depth with the total (direct) runoff depth to determine the total losses during the storm. If this loss is distributed uniformly over the duration of the storm event (4 h), estimate the duration of rainfall excess to be associated with the unit hydrograph. Use this derived unit hydrograph to estimate the runoff from the following storm event:

Time (h)	Effective rainfall (cm/h)
0–2	0.5
2–4	1.5
4–6	2.0
6–8	1.0

5.63. Calculate the Espey–Altman 10-minute unit hydrograph for a 3-km^2 catchment. The main channel has a slope of 0.9% and a Manning n of 0.10, the catchment is 50% impervious, and the distance along the main channel from the catchment boundary to the outlet is 1100 m.

5.64. If the rainfall excess versus time from a 2.25-km² catchment is given by

Time (min)	Rainfall excess (mm)
0–10	0
10–20	10
20–30	15
30–40	0
40–50	0

and the 10-min unit hydrograph derived using the Espey–Altman method is given by

Time (min)	Runoff (m³/s)
0	0
22	3.12
28	4.67
34	6.21
47	4.67
59	3.12
250	0

estimate the runoff hydrograph of the catchment.

5.65. Derive the 3-h Snyder unit hydrograph for a 100-km² watershed where the main stream is 15 km long and the distance from the watershed outlet to the point on the stream nearest to the centroid of the watershed is 7 km. Assume $C_p = 0.7$ and $C_t = 1.8$.

5.66. A 3-km² catchment has an average curve number of 60, an average slope of 0.8%, and a flow length from the catchment boundary to the outlet of 1000 m. If the catchment is 40% impervious, determine the NRCS unit hydrograph for a 20-minute storm.

5.67. A 16-km² rural catchment is estimated to have a time of concentration of 14.4 h. Estimate the runoff hydrograph resulting from a storm with the following rainfall excess:

Time (min)	Rainfall excess (cm)
1–30	2.9
30–60	1.5

5.68. Determine the approximate NRCS triangular unit hydrograph for the catchment described in Problem 5.66, and verify that it corresponds to a rainfall excess of 1 cm.

5.69. Show that any triangular unit hydrograph that has a peak runoff, Q_p, in (m³/s)/cm, given by

$$Q_p = 2.08 \frac{A}{T_p}$$

must necessarily have a runoff duration equal to $2.67T_p$, where T_p is the time to peak in hours, and A is the catchment area in km².

5.70. A 75-ha catchment has the following time-area relation:

Time (min)	Contributing area (ha)
0	0
5	2
10	7
15	19
20	38
25	68
30	75

Estimate the runoff hydrograph using the time-area model for a 30-min rainfall-excess distribution given by

Time (min)	Average intensity (mm/h)
0–5	120
5–10	70
10–15	50
15–20	30
20–25	20
25–30	10

5.71. A catchment has an area of 80 km^2 and a time of concentration of 7.0 h, and the Clark storage coefficient is estimated to be 3.0 hours. Estimate the 1-hour unit hydrograph for the catchment.

5.72. A 2-ha catchment consists of pastures with an average slope of 0.5%. The width of the catchment is approximately 250 m, and the depression storage is estimated to be 10 mm. Use the nonlinear reservoir model to calculate the runoff from the following 20-min rainfall excess:

Time (min)	Effective rainfall (mm/h)
0–5	110
5–10	50
10–15	40
15–20	10

5.73. A 1-km^2 catchment has a time of concentration of 25 minutes and is 45% impervious. Use the Santa Barbara urban hydrograph method to estimate the runoff hydrograph for the following rainfall event:

Time (min)	Rainfall (mm/h)	Rainfall excess (mm/h)
0	0	0
10	50	0
20	200	130
30	103	91
40	52	11

Use a time increment of 10 minutes to calculate the runoff hydrograph.

5.74. A stormwater detention basin has the following storage characteristics:

Stage (m)	Storage (m^3)
8.0	0
8.5	1041
9.0	2288
9.5	3761
10.0	5478
10.5	7460
11.0	9726

The discharge weir from the detention basin has a crest elevation of 8.0 m, and the weir discharge, Q (m^3/s), is given by

$$Q = 3.29 \, h^{\frac{3}{2}}$$

where h is the height of the water surface above the crest of the weir in meters. The catchment runoff hydrograph is given by:

Time (min)	Runoff (m^3/s)
0	0
30	3.6
60	8.4
90	5.1
120	4.2
150	3.6
180	3.3
210	2.7
240	2.3
270	1.8
300	1.5
330	0.84
360	0.51
390	0

If the prestorm stage in the detention basin is 8.0 m, estimate the discharge hydrograph from the detention basin.

5.75. The flow hydrograph at a channel section is given by:

Time (min)	Flow (m^3/s)
0	0.0
30	12.5
60	22.1
90	15.4
120	13.6
150	12.4
180	11.7
210	10.8
240	9.9
270	8.4

Time (min)	Flow (m³/s)
300	8.1
330	7.5
360	4.2
390	0.0

Use the Muskingum method to estimate the hydrograph 1000 m downstream from the channel section. Assume that $X = 0.3$ and $K = 35$ min.

5.76. Measured flows at an upstream and downstream section of a river are as follows:

Time (h)	Upstream flow (m³/s)	Downstream flow (m³/s)
0	10.0	10.0
30	10.0	10.0
60	25.0	12.2
90	45.0	23.4
120	31.3	35.1
150	27.5	32.1
180	25.0	28.8
210	23.8	26.2
240	21.3	24.3
270	19.4	22.1
300	17.5	20.1
330	16.3	18.3
360	13.5	16.6
390	12.1	14.4
420	10.0	12.6
450	10.0	11.0
480	10.0	10.3
510	10.0	10.1
540	10.0	10.1
570	10.0	10.0
600	10.0	10.0

Estimate the Muskingum coefficients X and K that should be used in routing flows through this section of the river.

5.77. Write a simple finite-difference model to solve the kinematic-wave equation for channel routing. [*Hint*: The kinematic-wave equation is obtained by combining Equations 5.221 and 5.224.]

5.78. Plans are being considered to develop a 60-ha area for commercial and residential use. Annual rainfall in the area is 120 cm, the existing (undeveloped) land is approximately 3% impervious, and the developed land is expected to be 30% impervious. Estimate the current and future amounts of lead in the runoff.

5.79. A 70-ha catchment is 50% impervious, with 60% commercial and industrial use. The site is located in a city where there are typically 84 storms per year, the mean annual rainfall is 95 cm, and the mean minimum January temperature is 8.1°C. Estimate the annual load of suspended solids contained in the runoff.

5.80. Calculate the annual load of lead in the runoff for the site described in Problem 5.79.

5.81. A 80-ha residential development has a population density of 15 persons per hectare, streets are swept every two weeks, and the area is 25% impervious. If the average annual rainfall is 98 cm, estimate the annual load of phosphate (PO_4) expected in the runoff.

5.82. A stormwater-management system is to be designed for a new 10-ha residential development. The site is located in a region where there are 62 storms per year with at least 1.3 mm of rain, and the IDF curve is given by

$$i = \frac{6000}{t + 20} \text{ mm/h} \tag{5.376}$$

where i is the average rainfall intensity and t is the duration of the storm in minutes. The mean annual rainfall in the region is 98.5 cm, and the mean January temperature is $9.6°$C. The developed site is to have an impervious area of 65%, an estimated depression storage of 1.5 mm, an estimated population density of 25 persons/hectare, and no street sweeping. Estimate the average concentration in mg/L of suspended sediments in the runoff using: (a) the USGS model, and (b) the EPA model.

5.83. Consider the two inlets and two pipes shown in Figure 5.46. Catchment A has an area of 0.5 ha and is 60% impervious, and catchment B has an area of 1 ha and is 15% impervious. The runoff coefficient, C; length of overland flow, L; roughness coefficient, n; and average slope, S_0, of the pervious and impervious surfaces in both catchments are given in the following table:

Catchment	Surface	C	L (m)	n	S_0
A	pervious	0.2	80	0.2	0.01
	impervious	0.9	60	0.1	0.01
B	pervious	0.2	140	0.2	0.01
	impervious	0.9	65	0.1	0.01

The IDF curve of the design storm is given by

$$i = \frac{8000}{t + 40}$$

where i is the average rainfall intensity in mm/h and t is the duration of the storm in minutes. Calculate the peak flows in the inlets and pipes.

5.84. Within the storm-sewer system for the site described in Problem 5.82, pipe I and pipe II intersect (at a manhole) and flow into Pipe III. All pipes have a slope of 2%. Pipe I and pipe II both drain 1-ha areas that are 65% impervious, with the impervious area directly connected to the inlet to pipe I, but with the inlet to pipe II not directly connected to any impervious area. The runoff coefficients can be taken as 0.3 and 0.9 for the pervious and impervious areas, respectively. The times of concentration of the catchments contributing to pipes I and II are:

Pipe	Area	t_c (min)
I	All	25
	DCIA*	12
II	All	30
	DCIA*	—

*DCIA means Directly Connected Impervious Area.

If the flow times in pipe I and pipe II are 3 min (in each pipe), estimate the design flows in pipes I, II, and III. What diameter of concrete pipe would you use for pipe III?

5.85. A concrete sewer pipe is to be laid on a slope of 0.90% and is to be designed to carry 0.50 m^3/s of stormwater runoff. Estimate the required pipe diameter using the Manning equation.

5.86. Repeat Problem 5.85 using the Darcy–Weisbach equation.

5.87. If service manholes are placed along the pipeline in Problem 5.85, estimate the head loss at each manhole.

5.88. A four-lane collector roadway is to be constructed with 3.66-m (12-ft) lanes, a cross slope of 1.5%, a longitudinal slope of 0.8%, and pavement of smooth asphalt. If the roadway drainage system is to be designed for a rainfall intensity of 120 mm/h, determine the spacing of the inlets.

5.89. A roadway with a rough-asphalt pavement has a cross slope of 2.5%, a longitudinal slope of 1.5%, a curb height of 15 cm, and a 90-cm wide concrete gutter. If the gutter flow is 0.09 m^3/s, determine the length of an 11-cm high curb inlet that is required to remove all the gutter flow. Consider the cases where (a) there is an inlet depression of 30 mm over a width of 70 cm; and (b) there is no inlet depression. What inlet length would remove 70% of the gutter flow?

5.90. A roadway has a flow depth at the curb of 9 cm and a corresponding flowrate in the gutter of 0.1 m^3/s. If the gutter flow is to be removed in a vertical sag by a 15-cm high curb inlet, determine the length of the inlet. Consider the cases where (a) the width of the inlet depression is 0.3 m, and (b) there is no inlet depression.

5.91. Use the form of the Manning equation given by Equation 5.245 to show that the ratio, R_w, of flow over width W_0 to flow over the top-width T in a triangular curb gutter is given by

$$R_w = 1 - \left(1 - \frac{W_0}{T}\right)^{8/3}$$

Explain the importance of this equation in the design of grate inlets.

5.92. A roadway with a rough-asphalt pavement has a cross slope of 2%, a longitudinal slope of 2.5%, a curb height of 8 cm, and a 90-cm wide concrete gutter. If flow in the gutter is 0.07 m^3/s, determine the size and interception capacity of a reticuline grate that should be used to intercept as much of the flow as possible.

5.93. A roadway has a cross slope of 1.5%, a flow depth at the curb of 9 cm, and a corresponding flowrate in the gutter of 0.1 m^3/s. The gutter flow is to be removed in a vertical sag by a grate inlet that is mounted flush with the curb. Calculate the minimum dimensions of the grate inlet.

5.94. Explain why the interception capacity of a combination inlet with a grate placed alongside a curb opening on a grade does not differ materially from the interception capacity of the grate only.

5.95. Show that the ratio, R_w, of flow over width W_0 to flow over the top width T in a depressed triangular curb gutter is given by

$$R_w = \left\{1 + \frac{S_w/S_x}{\left[1 + \dfrac{S_w/S_x}{\frac{T}{W_0}-1}\right]^{8/3} - 1}\right\}^{-1}$$

where W_0 is the top width of the flow in the depressed gutter, S_w is the slope of the depressed gutter, and S_x is the slope of the undepressed gutter. Explain the importance of this equation in the design of combination inlets.

5.96. A combination inlet consists of a 3-m curb inlet and a 0.6-m by 1.2-m reticuline grate inlet adjacent to the downstream 1.2 m of the curb opening. The gutter section has a width of 0.6 m, a longitudinal slope of 1.5%, a cross slope of 3.5%, and a gutter depression of 30 mm. The roadway pavement consists of rough asphalt. For a gutter flow of 0.09 m³/s, calculate the interception capacity of the inlet.

5.97. A combination inlet in a sag location of a roadway is to remove a gutter flow of 0.20 m³/s. The inlet consists of a 12-cm high by 3-m long curb opening with a 0.6-m by 0.9-m P-30 grate centered in front of the curb opening. If the roadway has a cross slope of 3.5% and the perimeter of the grate is 20% clogged, determine the spread of water on the roadway at the sag location.

5.98. A roadway has a cross slope of 1.5%, a longitudinal slope of 0.5%, and a flowrate in the gutter of 0.2 m³/s. The flow is to be removed by a slotted drain with a slot width of 5 cm, and the Manning n of the roadway is estimated as 0.017. Estimate the minimum length of slotted drain that can be used.

5.99. An outlet from a detention pond consists of a 1.50-m diameter riser with a 0.40 m × 0.40 m square orifice between elevations 1.00 m and 1.40 m. Flood flows discharge over the top of the riser (as weir flow), and the elevation of the top of the riser is 2.00 m. Determine the outlet discharges for pond elevations between 1.00 m and 2.40 m in 0.20-m intervals.

5.100. Show that if the bottom of a storage reservoir is a rectangle of dimensions $L \times W$, the longitudinal cross section is a trapezoid with base W and side slope angle α, and the transverse cross section is rectangular with base L, then the stage-storage relation is given by

$$ S = \frac{L}{\tan \alpha} h^2 + (LW)h $$

where S is the storage, and h is the depth.

5.101. The runoff hydrographs from a site before and after development are as follows:

Time (min)	Before (m³/s)	After (m³/s)
0	0	0
30	2.0	3.5
60	7.5	10.6
90	1.7	7.5
120	0.90	5.1
150	0.75	3.0
180	0.62	1.5
210	0.49	0.98
240	0.30	0.75
270	0.18	0.62
300	0.50	0.51
330	0	0.25
360	0	0.12
390	0	0

The postdevelopment detention basin is to be a wet-detention reservoir drained by an outflow weir. The elevation versus storage in the detention basin is

Elevation (m)	Storage (m³)
	0
0.5	11,022
1.0	24,683
1.5	41,522

where the weir crest is at elevation 0 m, which is also the initial elevation of the detention basin prior to runoff. The performance of the weir is given by

$$Q = 1.83bh^{\frac{3}{2}}$$

where Q is the overflow rate in m^3/s, b is the crest length in m, and h is the head on the weir in m. Determine the crest length of the weir for the detention basin to perform its desired function. What is the maximum water-surface elevation expected in the detention basin?

5.102. The runoff from the site described in Problem 5.82 is to be routed to a dual-purpose dry-detention reservoir for both flood control and water-quality control. The predevelopment and postdevelopment runoff volumes can be estimated using the USEPA empirical formula (Equation 5.229), where the predevelopment imperviousness and depression storage are 10% and 2.5 mm, respectively. The peak predevelopment runoff is 3 m^3/s, and the stage-storage relation of the detention reservoir is:

Elevation (m)	Storage (m^3)
0.0	0
0.5	300
1.0	600
1.5	900

The basin is to be drained by a circular orifice, where the discharge, Q, is related to the area of the opening, A, and the stage, h, by

$$Q = 0.65A\sqrt{2gh} \qquad (5.377)$$

Make a preliminary estimate the required diameter of the orifice opening.

5.103. A runoff hydrograph is routed through a wet-detention basin, and the results are given in the following table:

Time (min)	Inflow (m^3/s)	Storage (m^3)	Outflow (m^3/s)
0	0	5000	0
30	1.1	5873	0.13
60	3.8	9478	0.81
90	0.95	11,827	1.37
120	0.55	10,909	1.15
150	0.40	9910	0.91
180	0.35	9104	0.73
210	0.29	8497	0.59
240	0.19	7979	0.47
270	0.11	7489	0.37
300	0	7025	0.30
330	0	6584	0.24
360	0	6206	0.17
390	0	5918	0.14
420	0	5698	0.11
450	0	5531	0.08

Time (min)	Inflow (m³/s)	Storage (m³)	Outflow (m³/s)
480	0	5405	0.07
510	0	5311	0.06
540	0	5239	0.05
570	0	5180	0.04
600	0	5135	0.03
630	0	5104	0.02
660	0	5081	0.01
690	0	5063	0.01
720	0	5050	0.01

Estimate the detention time and evacuation time of the detention basin.

5.104. If the inflow and outflow hydrographs given in Problem 5.103 are for a dry-detention basin, estimate the detention time and evacuation time in the basin.

5.105. Determine the required volume of a dry-detention basin for a 25-ha residential development, where the runoff coefficient is estimated as 0.4 and the average rainfall depth during the year is 4.2 cm. Estimate the suspended-solids removal in the basin.

5.106. A retention area is to be designed to accommodate the runoff from a 10-year 45-minute storm. The average rainfall intensities in 5-minute intervals are as follows: 24.1, 30.4, 41.6, 67.4, 169, 97.3, 51.6, 36.2, and 26.8 mm/h. If the catchment area is 5 ha, and the infiltration capacity of the soil remains constant at 50 mm/h during the storm, estimate the storage volume in the retention area required to store the excess runoff. If the retention area covers 700 m², what depth of ponding would you expect to occur during the design storm?

5.107. The IDF curve for Miami, Florida, is given by the relation

$$i = \frac{7836}{48.6T^{-0.11} + t(0.5895 + T^{-0.67})}$$

where i is the average rainfall intensity in mm/h, T is the return period in years, and t is the rainfall duration in minutes. A 4-ha residential development consists primarily of single-family residences on 0.2-ha (1/2-acre) lots, the native soil is classified as Group B, the maximum distance from the catchment boundary to the drainage outlet is 200 m, and the average slope along the drainage path is 0.5%. The drainage outlet for the development is located adjacent to the roadway and has 0.8-ha of directly connected impervious area, consisting of asphalt pavement with a maximum drainage length of 80 m and an average slope of 0.7%. If the minimum allowable time of concentration is 5 minutes, determine the peak runoff from the catchment for a 25-year return period. Local drainage regulations require that the stormwater-management system be designed for a 25-year 3-day rainfall event. Estimate the volume of runoff from the site to be stored in a retention pond (lake), where the control (prestorm) elevation in the lake is 4.00 m NGVD, and when the water in the lake is at the control elevation, the dimensions of the lake at the water level are 15 m × 15 m, and the side slopes are 6:1. Determine the lake setback distance required to completely retain the 25-year 3-day runoff in the lake.

5.108. An infiltration basin is to retain the first 2.5 cm of runoff from a 20-ha catchment. The area to be used for the infiltration basin is turfed, and the soil has a minimum infiltration rate of 100 mm/h. If the retained runoff is to infiltrate within 36 hours, determine the surface area to be set aside for the basin.

5.109. The IDF curve for rainfall in Miami-Dade county (Florida) is given by

$$i = \frac{7836}{48.6T^{-0.11} + t(0.5895 + T^{-0.67})}$$

where i is the rainfall intensity in mm/h, T is the return period in years, and t is the rainfall duration in minutes. Estimate the maximum rainfall amount for all storms with a return period of 10 years. If a development site is characterized by a runoff coefficient of 0.4 and has a retention capacity for 50 mm of runoff, will any 10-year runoff be discharged from the site?

5.110. Derive Equation 5.284 for a swale with a triangular cross section.

5.111. Derive Equation 5.285 for a swale with a trapezoidal cross section.

5.112. A triangular-shaped swale is to retain the runoff from a catchment with a design runoff rate of 0.01 m³/s. The longitudinal slope of the swale is to be 1.5% with side slopes of 5:1 (H:V). If the grassed swale has a Manning n of 0.030 and a minimum infiltration rate of 200 mm/h, determine the length of swale required.

5.113. Repeat Problem 5.112 for a trapezoidal swale with a bottom width of 1 m.

5.114. A biofiltration swale is to be laid on a 2% slope to handle a portion of the design runoff equal to 0.002 m³/s. During the wet season, the swale is expected to be covered with grass having an average height of 100 mm with Class E retardance. Design the biofiltration swale.

5.115. An exfiltration trench is to be designed to handle the runoff from a 3-ha residential area with a runoff coefficient of 0.5. The IDF curve for the design rainfall is given by

$$i = \frac{403}{(t + 8.16)^{0.69}} \text{ mm/h}$$

where i is the average rainfall intensity, and t is the storm duration in minutes. The trench hydraulic conductivity is 35 m/d, and the seasonal high-water table is 4.6 m below the ground surface. Assuming a safety factor of 3 for the trench hydraulic conductivity, design the exfiltration trench.

5.116. An exfiltration trench is to be used to drain a residential area in Coral Gables, Florida. The trench will be adjacent to a main roadway and designed to remove runoff from all 10-year storms, for which the IDF curve is given by the local Public Works Department as

$$i = \frac{7836}{37.7 + 0.803t}$$

where i is the average rainfall intensity in mm/h, and t is the rainfall duration in minutes. The trench has a catchment area of 0.03 ha, of which 35% is directly connected impervious area ($C = 0.9$), and 65% is effectively pervious area ($C = 0.4$). If the trench hydraulic conductivity is measured at 5 m/d, the width of the trench is 1 m, and the height of the trench is 3 m, determine the required length of the trench. What is the total runoff depth that will be retained by the trench?

5.117. An exfiltration trench is to be designed to handle runoff from a 0.5-ha residential area with an average runoff coefficient of 0.6. The IDF curves for the trench location are given by

$$i = \frac{7836}{48.6T^{-0.11} + t(0.5895 + T^{-0.67})}$$

where i is the average rainfall intensity in mm/h, T is the return period in years, and t is the rainfall duration in minutes. Slug tests indicate that the trench hydraulic conductivity is 8 m/d, the seasonal high-water table is 3.5 m below the ground surface, and a safety factor of 2 is to be used in the design. If the trench is required to prevent flooding caused by any 10-year storm, and is required to retain the first 25 mm of runoff, design the trench. [*Hint*: In cases where a specified depth of runoff is to be retained by a trench, ASCE (1998) recommends that the trench be designed using the IDF curve with at least a return period in which the corresponding rainfall depth occurs in one hour.]

5.118. Determine the soil heat flux in April in Algiers (Algeria). The mean monthly temperatures in March, April, and May are 14.1°C, 16.1°C, and 18.8°C. What is the equivalent evaporation rate?

5.119. Show that the wind speed profile over a grassed surface (Equation 5.305) can be derived from the general wind-speed profile (Equation 5.306) by assuming a vegetation height of 0.12 meters.

5.120. The Greeley, Colorado, weather station is located at latitude 40.41° N and longitude 104.78° W at an elevation of 1462.4 meters. The anemometer is located 3 m above ground, and instruments measuring air temperature and relative humidity are located 1.68 m above ground. The weather station is surrounded by irrigated grass with a height of 0.12 meters. Daily measurements during a one-week period in July are given in the following table:

Day	T_{max} (°C)	T_{min} (°C)	e_a (kPa)	R_s (MJ/m$^2 \cdot$ d)	u_3 (m/s)
1	32.4	10.9	1.27	22.4	1.94
2	33.6	12.2	1.19	26.8	2.14
3	32.6	14.8	1.40	23.3	2.06
4	33.8	11.8	1.18	29.0	1.97
5	32.7	15.9	1.59	27.9	2.98
6	36.3	15.8	1.58	29.2	2.37
7	35.5	16.7	1.13	23.2	2.43

where T_{max} and T_{min} are the maximum and minimum temperatures in the day, e_a is the actual vapor pressure, R_s is the incoming solar radiation, and u_3 is the wind speed 3 m above the ground. Determine the grass and alfalfa reference evapotranspiration rates for each day of the week, and assess the degree of variability.

5.121. Hourly measurements at the Greeley weather station described in Problem 5.120 during a 24-hour period in July are given in the following table:

Hour	T_{hr} (°C)	e_a (kPa)	R_s (MJ/m$^2 \cdot$ h)	u_3 (m/s)
1	16.5	1.26	0.00	0.50
2	15.4	1.34	0.00	1.00
3	15.5	1.31	0.00	0.68
4	13.5	1.26	0.00	0.69
5	13.2	1.24	0.03	0.29
6	16.2	1.31	0.46	1.24
7	20.0	1.36	1.09	1.28
8	22.9	1.39	1.74	0.88
9	26.4	1.25	2.34	0.72
10	28.2	1.17	2.84	1.52
11	29.8	1.03	3.25	1.97
12	30.9	1.02	3.21	2.07
13	31.8	0.98	3.34	2.76
14	32.5	0.87	2.96	2.90

Hour	T_{hr} (°C)	e_a (kPa)	R_s (MJ/m^2·h)	u_3 (m/s)
15	32.9	0.86	2.25	3.10
16	32.4	0.93	1.35	2.77
17	30.2	1.14	0.88	3.41
18	30.6	1.27	0.79	2.78
19	28.3	1.27	0.27	2.95
20	25.9	1.17	0.03	3.27
21	23.9	1.20	0.00	2.86
22	20.1	1.35	0.00	0.58
23	19.9	1.35	0.00	0.95
24	18.4	1.32	0.00	0.30

where T_{hr} is the mean hourly temperature, e_a is the actual vapor pressure, R_s is the incoming solar radiation, and u_3 is the wind speed 3 m above the ground. Determine the grass and alfalfa reference evapotranspiration rates for each hour, and assess the degree of variability.

5.122. An evaporation pan measures a monthly evaporation of 152 mm. The pan is surrounded by a short green crop for approximately 300 m in the upwind direction, the average wind speed is 4 m/s, and the mean relative humidity is 80%. Estimate the reference evapotranspiration for short irrigated grass turf.

5.123. Show that Equations 5.364 and 5.365 are adequate representations of the data in Table 5.47.

5.124. In central Florida the mean annual potential evapotranspiration is 125 cm, and the mean annual precipitation is 150 cm. Estimate the index of dryness and the actual evapotranspiration for grassed catchments in central Florida.

CHAPTER 6

Ground-Water Hydrology

6.1 Introduction

Ground-water hydrology is the science dealing with the quantity, quality, movement, and distribution of water below the surface of the earth. The field of ground-water hydrology is sometimes called *geohydrology* or *hydrogeology*, where the former term is used in the context of engineering practice and the latter term in the context of geologic practice. A major application of the principles of ground-water hydrology is in the development of water supplies by means of wells and infiltration galleries. Other important applications include the evaluation, mitigation, and remediation of contaminated ground water; the storage of surface waters in underground reservoirs; and the lowering of ground-water levels to permit crop growth. Ground water accounts for approximately 30% of all the fresh water on earth, which is second only to the polar ice (69%), and two orders of magnitude greater than the amount of fresh water in lakes and rivers (0.3%).

The subsurface environment consists of a porous medium in which the void spaces have varying degrees of water saturation. Regions where the void spaces are completely filled with water are called *zones of saturation*, and regions where the void spaces are not completely filled with water are called *zones of aeration*. Water in a zone of aeration is sometimes called *vadose* water*, and the zone of aeration is sometimes called the *vadose zone*. Typically, the zone of aeration (vadose zone) lies above the zone of saturation, and the upper boundary of the zone of saturation is called the *phreatic surface†* or *water table*. At the water table, the pressure is equal to atmospheric pressure, and this condition is illustrated in Figure 6.1. Within the zone of aeration are three subzones: the soil-water, intermediate, and capillary zones. The *soil-water zone* is the region containing the roots of surface vegetation, voids left by decayed roots of earlier vegetation, and animal and worm burrows. The maximum moisture content of the soil in the soil-water zone corresponds to the maximum moisture that can be held by the soil against the force of gravity, regardless of the depth of the water table below ground surface. The maximum moisture content in the soil-water zone is called the *field capacity*, and the thickness of the soil-water zone is typically on the order of 1 to 3 m (Raudkivi and Callander, 1976; Tindall and Kunkel, 1999). Beneath the soil-water zone is the *intermediate zone*, which extends from the bottom of the soil-water zone to the upper limit of the capillary zone. The *capillary zone* extends from the water table up to the limit of the capillary rise of the water from the zone of saturation (through the pores of the porous formation). The thickness of the capillary zone depends on the pore size, and can vary from 1 cm to several meters.

---

*Vadose is a derivative of the Latin word *vadosus*, which means "shallow."
†*Phreatic* is a derivative of the Greek word *phreatos*, which means "well." In this context, a saturated zone is encountered when a well is dug.

FIGURE 6.1: Unconfined aquifer

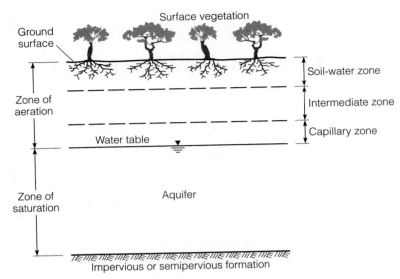

An *aquifer*[‡] is a geologic formation containing water that can be withdrawn in significant amounts. A common misconception is that an aquifer is always completely saturated with water; however, it is important to keep in mind that an aquifer consists of the entire water-bearing geologic unit (or entire group of water bearing units) and not just its saturated portion. *Aquicludes* contain water but are incapable of transmitting it in significant quantities, and *aquifuges* neither contain nor transmit water. A clay layer is an example of an aquiclude; solid rock is an example of an aquifuge; and, for most practical purposes, aquicludes can be taken as impervious formations. Aquifers are classified as either unconfined or confined. *Unconfined aquifers* are open to the atmosphere, as illustrated in Figure 6.1, and are also called *phreatic aquifers* or *water-table aquifers*. In *confined aquifers*, water in the saturated zone is bounded above by either impervious or semipervious formations. A typical configuration of a confined aquifer is shown in Figure 6.2, where the water in the confined aquifer is recharged by inflows at A (usually from rainfall).

Land surfaces that supply water to aquifers are called *recharge areas*, and maintaining an adequate recharge area (and recharge-water supply) is particularly important in urban areas where ground water is a major source of drinking water. The primary ground-water recharge mechanism is the infiltration of rainfall. *Piezometers* are observation wells with very short screened openings that are used to measure the piezometric head, ϕ, which for an incompressible fluid is given by

$$\phi = \frac{p}{\gamma} + z \tag{6.1}$$

where p and z are the pressure and elevation at the opening of the observation well (piezometer) and γ is the specific weight of the ground water. If the confined aquifer in Figure 6.2 is penetrated by piezometers at B and C, then the water levels in these piezometers rise to levels equal to the piezometric heads at B and C, respectively. At B, the piezometric head rises above the top confining layer of the (confined) aquifer,

---

[‡]*Aquifer* is a derivative of the Latin words *aqua* ("water") and *ferre* ("to bear").

FIGURE 6.2: Confined, unconfined, and artesian aquifers

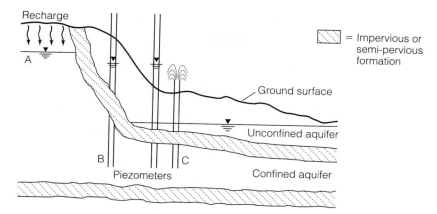

indicating that the water pressure at the top of the confined aquifer is greater than atmospheric pressure. This condition can be contrasted with the top of an unconfined aquifer, where the pressure is equal to atmospheric pressure. The piezometer at C behaves similarly to the piezometer at B, the difference being that the water level in the piezometer at C rises above the ground surface. In practical terms, this means that if a well were extended from the ground surface down into the confined aquifer at C, then the water would flow continuously from the well until the piezometric head was reduced to the elevation at the top of the well. An aquifer that produces flowing water when penetrated by a well from the ground surface is called an *artesian aquifer*.[§] As indicated in Figure 6.2, a confined aquifer can be an artesian aquifer at some locations, such as at C, and an unconfined aquifer at other locations, such as at A.

The presence of artesian aquifers and the distribution of pressures in aquifers is frequently identified by plotting the areal distribution of piezometric head, and such plots are commonly referred to as *piezometric surfaces* or *potentiometric surfaces*. In current practice, the term "potentiometric surface" is preferred. In cases where artesian aquifers intersect the ground surface, concentrated flows of ground water called *springs* are formed. Ground-water inflows into surface-water channels are a common source of perennial discharge in streams, which is commonly referred to as the *base flow* of the stream. Stream flows mostly consist of the base flow plus the flow resulting from stormwater runoff. Both unconfined and confined aquifers can be bounded by semipervious formations called *aquitards*, which are significantly less permeable than the aquifer but are not impervious. Of course, unconfined aquifers can only be bounded by impervious or semipervious layers on the bottom, while confined aquifers are bounded on both the top and bottom by either impervious or semipervious layers. Aquifers bounded by semipervious formations are called *leaky*, and terms such as *leaky-unconfined aquifer* and *leaky-confined aquifer* are used.

A microscopic view of the flow through porous media is illustrated in Figure 6.3, where water flows through the *void space* and around the *solid matrix* within the porous medium. It is difficult to describe the details of the flow field within the void spaces, since this would necessarily require a detailed knowledge of the geometry of the void space within the porous medium. To deal with this problem, it is convenient to work with spatially averaged variables rather than variables at a point (which is a

---

[§]The name "artesian" is derived from the name of the northern French city of Artois, where wells penetrating artesian aquifers are common.

FIGURE 6.3: Microscopic view of flow through porous media

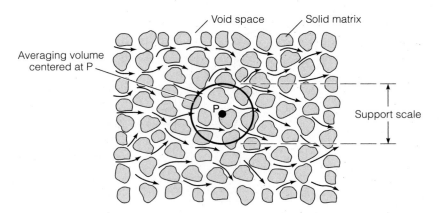

spatial average over a very small volume). Referring to Figure 6.3, a property of the porous medium at P can be taken as the average value of that property within a volume centered at P. The scale of the averaging volume is called the *support scale*. Almost all properties of a porous medium that are relevant to ground-water engineering are associated with a support scale, and in most cases the value of the averaged quantity is independent of the size of the support scale. A case in point is the *porosity*, n, of a porous medium, which is defined by the relation

$$n = \frac{\text{volume of voids}}{\text{sample volume}} \tag{6.2}$$

where the sample volume corresponds to the spherical volume with radius equal to the support scale. A typical relationship between the porosity and support scale is shown in Figure 6.4. When the support scale is very small, the porosity is sensitive to the location and size of the sample volume. Clearly, for sample volumes of the order of the size of the void space, it makes a significant difference whether the sample volume is located within a void space or within the solid matrix. As the support scale gets larger, the porosity becomes less sensitive to the location and size of the sample volume and approaches a constant value that is independent of the support scale. The porosity remains independent of the support scale until the averaging volume becomes so large that it encompasses portions of the porous medium that have significantly different characteristics; under these circumstances, the porosity again becomes dependent on the size of the support scale. The range within which the

FIGURE 6.4: Porosity versus support scale

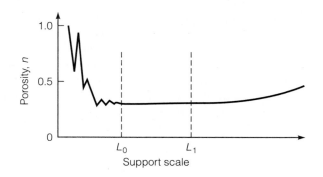

porosity is independent of the support scale is given by the interval between the scales L_0 and L_1 shown in Figure 6.4. Therefore, as long as the support scale is between L_0 and L_1, the porosity need not be associated with any particular support scale. The sample volume associated with the support scale L_0 is commonly referred to as the *representative elementary volume* (REV). For sandy soils, an averaging volume with a radius on the order of 10 to 20 grain diameters appears to be adequate for obtaining a stable average (Charbeneau, 2000). The relationship between the porosity and support scale is typical of the relationship between many other hydrogeologic parameters and support scales, although the REV of other parameters may be different. In general, there is no guarantee that a REV exists for any hydrogeologic parameter, and in the absence of a REV the value of the parameter must be associated with a support scale.

All earth materials are collectively known as *rocks*; and the three main categories of rocks are igneous, sedimentary, and metamorphic rocks. *Igneous rocks*, such as basalt and granite, are formed from molten or partially molten rock (magma) formed deep within the earth; *sedimentary rocks*, such as sand, gravel, sandstone, and limestone, are formed by the erosion of previously existing rocks and/or the deposition of marine sediment; and *metamorphic rocks*, such as schist and shale, are formed through the alteration of igneous or sedimentary rock by extreme heat or pressure or both. Typically, igneous and metamorphic rocks have less pore space and fewer passageways for water than sedimentary deposits. Rocks are further classified as either *consolidated* or *unconsolidated*. Solid masses of rock are referred to as consolidated, while rocks consisting of loose granular material are termed unconsolidated. In consolidated formations, *original porosity* or *primary porosity* is associated with pore spaces created during the formation of the rock, while *secondary porosity* is associated with pore spaces created after rock formation. Examples of secondary porosity in consolidated formations include fractures, and solution cavities in limestone. Representative values of porosity in consolidated formations are given in Table 6.1.

Between 60% and 90% of all developed aquifers consist of granular unconsolidated rocks (Lehr et al., 1988; Todd, 1980), where the porosities are associated with the intergranular spaces determined by the particle-size distribution. In general, granular material is classified by particle-size distribution, and many different organizations

TABLE 6.1: Representative Hydrologic Properties in Consolidated Formations

Material	Porosity	Specific yield	Hydraulic conductivity (m/d)
Sandstone	0.05–0.50	0.01–0.41	10^{-5}–4
Limestone	0–0.56	0–0.36	10^{-4}–2000
Schist	0.01–0.50	0.20–0.35	10^{-4}–0.2
Siltstone	0.20–0.48	0.01–0.35	10^{-6}–0.001
Claystone	0.41–0.45	—	—
Shale	0–0.10	0.01–0.05	10^{-8}–0.04
Till	0.22–0.45	0.01–0.34	10^{-5}–30
Basalt	0.01–0.50	—	10^{-6}–2000
Pumice	0.80–0.90	—	—
Tuff	0.10–0.55	0.01–0.47	—

TABLE 6.2: USDA Classification and Representative Hydrologic Properties in Unconsolidated Formations

Material*	Particle size* (mm)	Porosity	Specific yield	Hydraulic conductivity (m/d)
Very coarse gravel	32.0–64.0	—	—	—
Coarse gravel	16.0–32.0	0.24–0.40	0.10–0.26	860–8600
Medium gravel	8.0–16.0	0.24–0.44	0.13–0.45	20–1000
Fine gravel	4.0–8.0	0.25–0.40	0.15–0.40	—
Very fine gravel	2.0–4.0	—	—	—
Very coarse sand	1.0–2.0	—	—	—
Coarse sand	0.5–1.0	0.20–0.50	0.15–0.45	0.08–860
Medium sand	0.25–0.5	0.29–0.49	0.15–0.46	0.08–50
Fine sand	0.10–0.25	0.25–0.55	0.01–0.46	0.01–40
Very fine sand	0.05–0.125	—	—	—
Silt	0.002–0.05	0.34–0.70	0.01–0.40	10^{-5}–2
Clay	<0.002	0.33–0.70	0–0.20	$<10^{-2}$

*USDA Soil Classification System, gravel classification given by Morris and Johnson (1967).

have established classification standards for use in various disciplines. The United States Department of Agriculture (USDA) soil classification system is one of the most widely used in water-resources engineering and is given in Table 6.2, along with corresponding values of porosity. The porosities of granular materials tend to decrease with increasing particle size; however, this does not mean that water flows with more resistance through aquifers composed of larger particle sizes. In fact, the opposite is true. Porosities are considered small when $n < 0.05$, medium when $0.05 \leq n \leq 0.20$, and large when $n > 0.20$ (Kashef, 1986).

The most common aquifer materials are unconsolidated sands and gravels, which occur in alluvial valleys, coastal plains, dunes, and glacial deposits (Bouwer, 1978). Consolidated formations that make good aquifers are sandstones, limestones with solution channels, and heavily fractured volcanic and crystalline rocks. Clays, shales, and dense crystalline rocks are the most common materials found in aquitards. Aquifers range in thickness from less than 1 m to several hundred meters. They may be long and narrow, as in small alluvial valleys, or they may extend over millions of square kilometers and underlie major portions of states (Bouwer, 1984). The depth from the ground surface to the top of the saturated zone of an aquifer may range from 1 m to more than several hundred meters.

Geologic perspective. Geologic characterization of the subsurface environment is regarded as an essential component of most ground-water studies (Stone, 1999). Although geologic characterizations are typically performed by professional geologists, engineers are expected to have sufficient understanding of geologic principles and nomenclature to collaborate with geologists and understand the relationship between the geology and hydrology in the subsurface environment. The science of this relationship is called *hydrogeology*. The most fundamental aspect of the geologic setting in a subsurface environment is the *stratigraphy*, which describes the *geologic units* present and their relationship to each other. The description of each geologic unit (also called *stratigraphic unit*) includes the name, lithology, thickness, extent, and nature of contact with the underlying unit. Formal stratigraphic names consist of two parts: a unique geographic name, and a stratigraphic rank. The geographic name is usually derived from the location where the unit is best exposed or was first described, while the rank is based on the composition of the geologic unit. The most fundamental rank is the *formation*, which consists of a distinct group of rocks or deposits having a common lithologic character or origin. For example, the "Fort Thomson Formation" in Florida consists of interbedded shell beds and limestones (Florida Geological Survey, 1991). In some cases, the major rock type is used instead of the word "formation"—for example, "Key Largo Limestone" and "Pensacola Clay." Two or more closely related formations may be designated as a *group*; for example, the "Alum Bluff Group" found in the Florida panhandle includes the Chipola Formation, Oak Grove Sand, Shoal River Formation, and the Choctawhatchee Formation. It is vitally important that an engineer use the correct stratigraphic names in reporting on the subsurface environment, since failure to do so may lead others to question the credibility of the engineer's report. The *lithology* of a stratigraphic unit is described by its rock or deposit type, texture (grain size, shape, sorting), and mineral composition. Geologic materials in the stratigraphic column can be grouped into *hydrostratigraphic units* according to their hydrologic behavior or function. An aquifer is an example of a hydrostratigraphic unit, and can include more than one stratigraphic unit. The name of an aquifer, or other hydrostratigraphic unit, is usually derived from either its stratigraphy, depth, or lithology. If the aquifer coincides exactly with a stratigraphic unit, then the aquifer usually takes the name of the stratigraphic unit—for example,

FIGURE 6.5: Injection well and subsurface geology

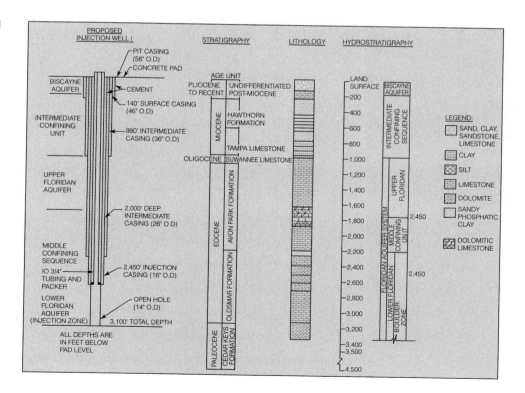

the Ogallala aquifer in the Ogallala Formation. In areas where multiple hydrostrati-graphic units exist, depth classifications lead to names like "shallow aquifer," "surficial aquifer," and "intermediate confining bed." In cases where the aquifer is classified by lithology, the name of the aquifer includes the predominant sediment or rock type in the aquifer, and names such as the "sandstone aquifer" and the "limestone aquifer" are used.

In engineering practice, knowledge and understanding of the stratigraphy, lithology, and hydrostratigraphy of the subsurface environment are essential to the design of such water-resource systems as water-supply wells, injection wells, and ground-water monitoring wells. For these systems, it is generally good engineering practice to present the final design on the same page as the subsurface stratigraphy, lithology, and hydrostratigraphy. An example of such a presentation is shown in Figure 6.5. This figure illustrates an injection well that extends approximately 944 m (3100 ft) below land surface, where the well is designed to inject effluent into the Boulder Zone, a highly transmissive layer within the Floridan Aquifer System. The injection well is encapsulated in a cement lining to prevent leakage of the effluent into the formations overlying the Boulder Zone.

6.2 Basic Equations of Ground-Water Flow

6.2.1 Darcy's Law

The study of flow through porous media was pioneered by Darcy (1856) using the experimental setup shown in Figure 6.6.* In this experiment, Darcy investigated the flowrate of water through a column of sand with cross-sectional area A and length L. Darcy (1856) found that the flowrate, Q, of water through the sand column could be described by the relation

$$Q = KA \frac{h_1 - h_2}{L} \tag{6.3}$$

FIGURE 6.6: Darcy's experimental setup

Source: Bear (1979).

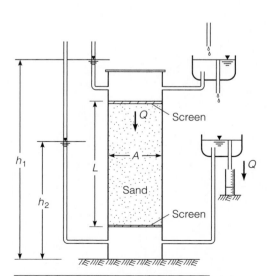

---

*The motivation for Darcy's experiments was to study the performance of the sand filters in the water-supply system for the city of Dijon, France.

where K is a proportionality constant, and h_1 and h_2 are the piezometric heads at the entrance and exit of the sand column, respectively. Recall that the piezometric head, h, of an incompressible fluid is given by $h = p/\gamma + z$. The piezometric head is sometimes called the *hydraulic head*. Defining the gradient in the piezometric head or *hydraulic gradient*, J, across the sand column by

$$J = \frac{h_2 - h_1}{L} \tag{6.4}$$

and defining the *specific discharge*, q, through the sand column by

$$q = \frac{Q}{A} \tag{6.5}$$

then Equation 6.3 can be written in the form

$$\boxed{q = -KJ} \tag{6.6}$$

which is commonly known as *Darcy's law*. The specific discharge, q, is sometimes called the *filtration velocity* or *Darcy's velocity*. The experimentally validated (phenomenological) relationship given by Equation 6.6 states that the specific discharge through a porous medium is linearly proportional to the gradient in the piezometric head in the direction of flow. The proportionality constant, K, is called the *hydraulic conductivity* of the porous medium. In the field of geotechnical engineering, the term *coefficient of permeability* is commonly used instead of hydraulic conductivity, but both terms refer to the same quantity.

Before proceeding to study the applications of Darcy's law, several important points must be noted. First of all, the specific discharge, q, is defined as the volumetric flowrate per unit cross section of the porous medium and, since the flow only occurs within the pores, the actual velocity of flow through the pores is necessarily higher than the specific discharge. The flowrate through the pores is called the *seepage velocity*, v, and is defined by

$$v = \frac{Q}{A_p} \tag{6.7}$$

where A_p is the area of the pores (normal to the flow direction). Comparing Equations 6.5 and 6.7, the seepage velocity, v, is related to the specific discharge, q, by

$$v = q\,\frac{A}{A_p} \tag{6.8}$$

The ratio of the pore area, A_p, to the bulk cross-sectional area, A, is defined as the *areal porosity*, and is equal to the volumetric porosity, n, defined by Equation 6.2 (Bear, 1972). In reality, not all of the pore space is connected and available for fluid flow, and therefore the effective areal porosity is less than the volumetric porosity, n. The ratio of the effective flow area to the bulk cross-sectional area is defined as the *effective porosity*, n_e, where

$$n_e = \frac{A_e}{A} \tag{6.9}$$

where A_e is the effective flow area through the pores. For unconsolidated porous media, the effective porosity is approximately equal to the volumetric porosity (Todd, 1980), while in many consolidated formations the effective porosity can be over an order of magnitude smaller than the total porosity, with the greatest difference occurring in fractured rocks (Domenico and Schwartz, 1998). Combining Equations 6.8 and 6.9, and taking A_p equal to A_e, yields the following relationship between the seepage velocity, v, and the specific discharge, q, given by Darcy's law:

$$v = \frac{q}{n_e} = -\frac{K}{n_e} J \qquad (6.10)$$

In practice, the effective porosity is sometimes denoted by n rather than n_e, and it is differentiated from the volumetric porosity by the stated definition of n.

EXAMPLE 6.1

Water flows through a sand aquifer with a piezometric head gradient of 0.01. (a) If the hydraulic conductivity and effective porosity of the aquifer are 2 m/d and 0.3, respectively, estimate the specific discharge and seepage velocity in the aquifer; (b) estimate the volumetric flowrate of the ground water if the aquifer is 15 m deep and 1 km wide. (c) How long does it take the ground water to move 100 m?

Solution

(a) From the given data: $J = -0.01$; $K = 2$ m/d; $n_e = 0.3$; and, according to Darcy's law, the specific discharge, q, is given by

$$q = -KJ = -(2)(-0.01) = 0.02 \text{ m/d}$$

The corresponding seepage velocity, v, is

$$v = \frac{q}{n_e} = \frac{0.02}{0.3} = 0.067 \text{ m/d}$$

(b) The volumetric flowrate, Q, of ground water across an area $A = 15$ m $\times$ 1 km $=$ 15,000 m^2 is

$$Q = qA = (0.02)(15,000) = 300 \text{ m}^3/\text{d}$$

(c) Ground water flows with the seepage velocity, v, and therefore the time, t, to travel 100 m is given by

$$t = \frac{100 \text{ m}}{v} = \frac{100 \text{ m}}{0.067 \text{ m/d}} = 1490 \text{ d} = 4.09 \text{ years}$$

This result is indicative of the slow movement of most ground waters.

The hydraulic conductivity, K, that appears in Darcy's law is a function of both the fluid properties and the geometry of the solid matrix. This point is intuitively obvious if one considers the water in Darcy's experiment being replaced by oil. Clearly there would be less flow than for water under the same piezometric gradient, thereby indicating a smaller hydraulic conductivity for oil in sand than for water in sand. Also, if clay were used instead of sand in Darcy's experiment, then there would be less flow than for sand under the same hydraulic gradient, indicating a smaller hydraulic conductivity for water in clay than for water in sand. The functional relationship between the hydraulic conductivity and the fluid and solid-matrix properties can be extracted using dimensional analysis. If the fluid properties are characterized by the specific weight, γ, and dynamic viscosity, μ, and the solid matrix is characterized by the length scale of the pores, d, then the hydraulic conductivity, K, is related to the fluid and solid matrix properties by the following functional relationship:

$$K = f_1(\gamma, \mu, d) \tag{6.11}$$

where f_1 is an undetermined function. Using the Buckingham pi theorem, this functional relationship between four variables containing three dimensions (mass, length, and time) can be expressed in the following form:

$$f_2\left(\frac{K\mu}{\gamma d^2}\right) = 0 \tag{6.12}$$

where f_2 is an undetermined function. The relationship given by Equation 6.12 can theoretically be solved for the dimensionless quantity, $K\mu/\gamma d^2$, to yield

$$\frac{K\mu}{\gamma d^2} = \alpha \tag{6.13}$$

where α is a dimensionless constant that incorporates the secondary structural characteristics of the porous medium that are not simply characterized by the pore scale, such as the distribution and shape of the pore sizes. Solving for K in Equation 6.13 yields

$$K = \alpha \frac{\gamma}{\mu} d^2 \tag{6.14}$$

This equation clearly separates the hydraulic conductivity into a term related to the fluid properties, γ/μ, and a term related to the geometry of the solid matrix, αd^2. The *intrinsic permeability*, k, is defined by the relation

$$\boxed{k = \alpha d^2} \tag{6.15}$$

The intrinsic permeability should not be confused with the "coefficient of permeability," a term used by geotechnical engineers to refer to the hydraulic conductivity. There have been a number of investigations to estimate the magnitude of k in Equation 6.15, and several of these results are summarized in Table 6.3. The characteristic pore size, d, is most often taken as the 10-percentile grain size, d_{10}, which is also called the *effective grain diameter*, and the proportionality constant, α, is commonly considered to decrease as the porosity, n, decreases (Venkataraman and Rao, 1998).

A comparative study by Sperry and Pierce (1995) indicates that the Hazen (1911) equation for intrinsic permeability will usually provide a good estimate of the intrinsic permeability, except for irregularly shaped particles. Others have suggested that the Kozeny–Carman equation is preferable and has a sounder theoretical basis (Aubertin et al., 2005; Barr, 2005). Using the definition of the intrinsic permeability given by

TABLE 6.3: Equations for Estimating Intrinsic Permeability

Equation	Reference	Comments
$k = 0.617 \times 10^{-3}d^2$	Krumbein and Monk (1942)	d is a measure of the pore size.
$k = 1.02 \times 10^{-3}d_{10}^2$	Hazen (1911)	Restricted to uniformity coefficient $(d_{60}/d_{10}) < 5$, and $0.1\text{mm} < d_{10} < 3$ mm.
$k = 0.654 \times 10^{-3}d_{10}^2$	Harleman et al. (1963)	—
$k = 0.750 \times 10^{-3}d^2e^{-1.31\sigma}$	Krumbein and Monk (1943)	d is the geometric mean diameter, and σ is the log standard deviation of the size distribution.
$k = C_s \dfrac{n^3}{1-n}D_s^2$	Taylor (1948)	C_s is a shape factor, n is the porosity, and D_s is the equivalent (uniform) spherical diameter of the porous medium $(D_s \approx d_{10})$.
$k = \dfrac{n^3}{180(1-n)^2}d^2$	Kozeny–Carman equation, (Kozeny, 1927; Carman 1937, 1956)	d is the median grain size, d_{50}.
$k = \dfrac{1}{m}\left[\dfrac{(1-n)^2}{n^3}\left(\dfrac{a}{100}\sum\dfrac{P}{d_g}\right)^2\right]^{-1}$	Fair and Hatch (1933)	m is a packing factor (≈ 5 by experiment), a is a sand shape factor which ranges from 6.0 for spherical grains to 7.7 for angular grains, P is the percentage of grains passing one sieve and held on the next, and d_g is the geometric mean of these sieve sizes.

Equation 6.15, the hydraulic conductivity, K, given by Equation 6.14, can be written in the form

$$K = k\frac{\gamma}{\mu} \tag{6.16}$$

On the basis of Equation 6.16, it is appropriate to associate the intrinsic permeability with the solid matrix, and associate the hydraulic conductivity with the fluid/matrix combination. The fluid properties, γ and μ, and hence the hydraulic conductivity, are significantly influenced by the temperature of the ground water. The depth to nearly constant temperature occurs at about 10 m in the tropics and at about 20 m in polar regions (Todd, 1980) and, above these depths, significant seasonal fluctuations in temperature can occur. The ground-water temperature typically increases with depth by about 2 to 3.5°C/100 m (Warner and Lehr, 1981; Todd, 1980), and the rate of increase in water temperature with depth is commonly referred to as the *geothermal gradient*. In the contiguous United States, ground-water temperatures vary from about 4°C in the north (10°C in the northwest) to around 20°C in the south (Miller et al., 1962). The standard value of hydraulic conductivity is defined for pure water at a temperature of 15.6°C (Fetter, 2001). The intrinsic permeability, k, in Equation 6.16 is generally a very small number when expressed in m^2 and is commonly expressed in darcys, where 1 darcy is equal to 0.987×10^{-12} m^2.

EXAMPLE 6.2

Laboratory analysis of an aquifer material indicates a porosity of 0.40 and a grain-size distribution as follows:

Grain size (mm)	Percent finer (%)
4.760	96.0
2.000	80.0
0.840	52.0
0.420	38.0
0.250	25.0
0.149	12.0
0.074	5.0

Estimate the hydraulic conductivity of the aquifer using all available empirical equations and compare the results. Assume a temperature of 20°C.

Solution   For water at 20°C, $\gamma = 9.789$ kN/m^3, $\mu = 1.002 \times 10^{-3}$ N·s/m^2, and the hydraulic conductivity, K, is related to the intrinsic permeability, k, by

$$K = k\frac{\gamma}{\mu}$$

$$= k\frac{9.789 \times 10^3}{1.002 \times 10^{-3}} = 9.969 \times 10^6 \, k$$

where k is in m^2 and K is in m/s. From the given grain-size distribution:

$$d_{10} = 0.128 \text{ mm}$$
$$d_{50} = 0.780 \text{ mm}$$
$$d_{60} = 1.171 \text{ mm}$$
$$U_c = \frac{d_{60}}{d_{10}} = \frac{1.171}{0.128} = 9.1 \, (= \text{uniformity coefficient})$$

The Fair and Hatch grain-size parameter, $\sum P/d_g$, can be easily derived by putting the data in the following form:

Size (mm)	Geometric mean of adjacent sieve sizes, d_i (mm)	Percent finer (%)	Percentage held between adjacent sieve sizes, P_i (%)	$\dfrac{P_i}{d_i}$ (mm^{-1})	$\log d_i$ (log mm)
4.760		96.0			
	3.085		16.0	5.19	0.489
2.000		80.0			
	1.296		28.0	21.61	0.113
0.840		52.0			
	0.594		14.0	23.57	−0.226
0.420		38.0			
	0.324		13.0	40.12	−0.489
0.250		25.0			
	0.193		13.0	67.36	−0.703
0.149		12.0			
	0.105		7.0	66.67	−0.979
0.074		5.0			
			Total:	224.52	

Therefore, the Fair and Hatch grain-size parameter is given by

$$\sum_{i=1}^{6} \frac{P_i}{d_i} = 224.52 \text{ mm}^{-1} = 2.245 \times 10^5 \text{ m}^{-1}$$

The mean, μ, and variance, σ^2, of the logarithms of the grain sizes can be estimated by

$$\mu = \sum_{i=1}^{6} \left(\frac{P_i}{100}\right) \log d_i$$

$$= (0.16)(0.489) + (0.28)(0.113) + (0.14)(-0.226) + (0.13)(-0.489)$$
$$+ (0.13)(-0.703) + (0.07)(-0.979)$$

$$= -0.145$$

$$\sigma^2 = \sum_{i=1}^{6} \left(\frac{P_i}{100}\right) (\log d_i - \mu)^2$$

$$= (0.16)(0.489 + 0.145)^2 + (0.28)(0.113 + 0.145)^2 + (0.14)(-0.226 + 0.145)^2$$
$$+ (0.13)(-0.489 + 0.145)^2 + (0.13)(-0.703 + 0.145)^2$$
$$+ (0.07)(-0.979 + 0.145)^2$$

$$= 0.188$$

Therefore, the geometric mean grain diameter, d_g, is given by

$$d_g = 10^\mu = 10^{-0.145} = 0.716 \text{ mm}$$

and the log standard deviation, σ, is calculated as

$$\sigma = \sqrt{0.188} = 0.434$$

The derived parameters of the grain-size distribution yield the following hydraulic conductivities:

Reference	Equation	Parameters	K (m/d)
Krumbein and Monk (1942)	$k = 0.617 \times 10^{-3} d^2$	$d = d_{10} = 1.28 \times 10^{-4}$ m	9
Hazen (1911)	$k = 1.02 \times 10^{-3} d_{10}^2$	$U_c = 9.1$	*
		$d_{10} = 1.28 \times 10^{-4}$ m	14
Harleman et al. (1963)	$k = 0.654 \times 10^{-3} d_{10}^2$	$d_{10} = 1.28 \times 10^{-4}$ m	9
Krumbein and Monk (1943)	$k = 0.750 \times 10^{-3} d^2 e^{-1.31\sigma}$	$d_g = 7.16 \times 10^{-4}$ m	184
		$\sigma = 0.434$	
Taylor (1948)	$k = C_s \dfrac{n^3}{1-n} D_s^2$	$C_s = 1$	1475
		$n = 0.40$	
		$D_s = d_{10} = 1.28 \times 10^{-4}$ m	
Kozeny–Carman	$k = \dfrac{n^3}{180(1-n)^2} d^2$	$n = 0.40$	503
		$d = d_{50} = 7.8 \times 10^{-4}$ m	
Fair and Hatch (1933)	$k = \dfrac{1}{m}\left[\dfrac{(1-n)^2}{n^3}\left(\dfrac{a}{100}\sum\dfrac{P}{d_g}\right)^2\right]^{-1}$	$m = 5$	128
		$n = 0.40$	
		$a = 6.9$ (average)	
		$\sum P/d_g = 2.24 \times 10^5$ m^{-1}	
		Average:	332

*Since $U_c > 5$, the Hazen equation is not applicable.

These results show that the hydraulic conductivities estimated using available empirical equations vary over two orders of magnitude (9–1475 m/d), with an average value of 332 m/d. This reflects the great uncertainty associated with estimating hydraulic conductivities from grain-size distributions.

Hydraulic conductivities of natural porous media cover a wide range of values, and typical magnitudes in various consolidated and unconsolidated formations are shown in Tables 6.1 and 6.2. These data indicate that hydraulic conductivities can range over several orders of magnitude for the same type of rock or sediment. The hydraulic conductivity of any porous formation can be classified as ranging from "very high" to "very low," according to the U.S. Bureau of Reclamation classification shown in Table 6.4 (USBR, 1977).

Darcy's law indicates a linear relationship between the flowrate and the gradient in the piezometric head measured in the direction of flow. This linear relationship is the same as for flow in pipes under low Reynolds number (Hagen–Poiseuille flow) and is symptomatic of the dominance of viscous forces as water flows through the pore spaces (Munson et al., 1994). As the Reynolds number increases, viscous forces

TABLE 6.4: Classification of Hydraulic Conductivities

Hydraulic conductivity, K (m/d)	Class	Unconsolidated deposits	Consolidated rocks
>1000	Very high	Clean gravel	Vesicular and scoriaceous basalt and cavernous limestone and dolomite
10–1000	High	Clean sand, and sand and gravel	Clean sandstone and fractured igneous and metamorphic rocks
0.01–10	Moderate	Fine sand	Laminated sandstone, shale, mudstone
0.0001–0.01	Low	Silt, clay, and mixtures of sand, silt, and clay	Massive igneous and metamorphic rocks
<0.0001	Very low	Massive clay	

Source: USBR (1977).

become less dominant, and the flowrate deviates from being linearly proportional to the piezometric head gradient. The characteristic Reynolds number, Re, can be defined in terms of the specific discharge, q; pore scale, d; and kinematic viscosity of water, ν, by the relation

$$\text{Re} = \frac{qd}{\nu} \tag{6.17}$$

Experiments indicate that deviations from Darcy's law begin to occur for values of Re > 1, but serious deviations do not occur up to Re = 10 (Ahmed and Sunada, 1969). Such conditions are routinely found in the immediate vicinity of large water-supply wells, and in fractured rock formations. Darcy's law has also been found to be invalid at very small seepage velocities in compact clays and other very-low-permeability materials (Bolt and Groenevelt, 1969), and it is not theoretically valid under some transient conditions (de Marsily, 1986). However, the effect of transient flows on the validity of Darcy's law can be taken as negligible in most cases of practical interest. Limitations related to moderate Reynolds numbers and transient conditions can be addressed by using the Forchheimer or Forchheimer–Dupuit equation, which, for an isotropic medium, is given by

$$-\nabla h = a\mathbf{q} + b\mathbf{q}|\mathbf{q}| \tag{6.18}$$

where ∇ is the gradient operator, $\mathbf{q}$ is the specific discharge vector, and a and b are constants. Although ground-water flow is adequately described by the Darcy equation in most practical cases, some analytical solutions to the ground-water flow equation using the Forchheimer equation are available (Moutsopoulos and Tsihrintzis, 2005).

EXAMPLE 6.3

A sand aquifer has a 10-percentile particle size of 0.4 mm and an effective porosity of 0.3. If the temperature of the water in the aquifer is 20°C, estimate the range of seepage velocities for which Darcy's law is valid.

Solution   Darcy's law can be taken to be valid when Re < 10,

$$\frac{qd}{\nu} < 10$$

which can be put in the form

$$q < \frac{10\nu}{d}$$

or

$$v < \frac{10\nu}{n_e d}$$

where the specific discharge, q, is related to the seepage velocity, v, and the effective porosity, n_e, by $v = q/n_e$. From the given data, $n_e = 0.3$, $d \approx d_{10} = 0.4\,\text{mm} = 4 \times 10^{-4}\,\text{m}$, and at $20°\text{C}$, $\nu = 1.00 \times 10^{-6}\,\text{m}^2/\text{s}$. Hence

$$v < \frac{10(1.00 \times 10^{-6})}{(0.3)(4 \times 10^{-4})} = 0.0833\,\text{m/s} = 7200\,\text{m/d}$$

Darcy's law can be applied in the aquifer whenever the seepage velocity is less than 7200 m/d.

Darcy's law relates the specific discharge to the piezometric gradient in the direction of flow by Equation 6.6. In the case of three-dimensional flow, the components of the specific-discharge vector are related to the corresponding components of the head-gradient vector by the relationship

$$q_i = -KJ_i \tag{6.19}$$

where q_i and J_i are the i-components of the specific discharge and head gradient respectively. This formulation assumes that the hydraulic conductivity is *isotropic*, in which case it does not depend on the flow direction. In cases where the hydraulic conductivity depends on the flow direction, the porous medium is called *anisotropic*, the hydraulic conductivity is a tensor, and the specific discharge is related to the head gradient by

$$\boxed{q_i = -K_{ij}J_j} \tag{6.20}$$

where the Einstein summation convention is used. The hydraulic conductivity is a symmetric second-rank tensor with nine components (K_{ij}, $i = 1,3; j = 1,3$), of which six are independent.

Anisotropy can be caused by a variety of factors, such as solution cavities in carbonate rocks preferentially forming in the horizontal flow direction, or the deposition of flat granular material (tilted slightly upward in the direction of flow) in the formation of alluvial aquifers. In cases where the coordinate axes are chosen to coincide with the *principal axes* of the hydraulic conductivity tensor, Darcy's law can be written in the simple form

$$q_i' = -K_{ii}'J_i' \tag{6.21}$$

where the primed quantities indicate components in the principal directions of the hydraulic conductivity tensor, and the Einstein summation is not used. In other words, if the coordinate axes coincide with the principal axes of the hydraulic conductivity tensor, then the components of the seepage velocity in the principal directions are given by

$$q_1' = -K_{11}'J_1'$$

$$q_2' = -K_{22}'J_2' \tag{6.22}$$

$$q_3' = -K_{33}'J_3'$$

If coordinate axes other than the principal axes are used, then the components of the hydraulic conductivity tensor can be computed using the relationship (Bear, 1972)

$$\boxed{K_{ij} = K_{11}'\alpha_{i1}\alpha_{j1} + K_{22}'\alpha_{i2}\alpha_{j2} + K_{33}'\alpha_{i3}\alpha_{j3}} \tag{6.23}$$

where α_{ij} is the cosine of the angle between the x_i- and x_j'-axes. Primed coordinates refer to the principal axes. In two-dimensional aquifers, Equation 6.23 can be written in the form

$$K_{11} = \frac{K_{11}' + K_{22}'}{2} + \frac{K_{11}' - K_{22}'}{2}\cos 2\theta \tag{6.24}$$

$$K_{22} = \frac{K_{11}' + K_{22}'}{2} - \frac{K_{11}' - K_{22}'}{2}\cos 2\theta \tag{6.25}$$

$$K_{12} = K_{21} = -\frac{K_{11}' - K_{22}'}{2}\sin 2\theta \tag{6.26}$$

where θ is the angle between the x_1- and x_1'-axes, measured counterclockwise from the x_1'-axis. These equations can also be used to estimate the principal components of the hydraulic conductivity and the orientation of the principal axes of the hydraulic conductivity tensor, K_{ij}.

In most cases, the principal axes are such that the x_1'- and x_2'-axes are in the horizontal plane and the x_3'-axis is vertically upward. Another common approximation is that the hydraulic conductivity is isotropic in the horizontal plane (i.e., $K_{11} = K_{22}$), and the ratio of the vertical hydraulic conductivity, K_{33}, to the horizontal hydraulic conductivity, K_{11} or K_{22}, is defined as the *anisotropy ratio*. Typical values of the anisotropy ratio are in the range of 0.1 to 0.5 for alluvial aquifers, with values as low as 0.01 where clay layers exist (Bedient et al., 1999). The primary cause of anisotropy on a small scale is the orientation of clay minerals in sedimentary rocks and unconsolidated sediments. In nongranular rocks, the size, shape, orientation, and spacing of fractures and other voids (such as solution cavities) are the primary causes of anisotropy. Porous formations are called *homogeneous* if the hydraulic conductivity is independent of location and *nonhomogeneous* if the hydraulic conductivity varies spatially.

EXAMPLE 6.4

The piezometric heads are measured at three locations in an aquifer. Point A is located at (0 km, 0 km), Point B at (1 km, −0.5 km), and Point C at (0.5 km, −1.2 km), and the piezometric heads at A, B, and C are 2.157 m, 1.752 m, and 1.629 m. (a) Determine the head gradient in the aquifer; (b) if the coordinate locations are measured relative to the principal axes (x', y'), and the principal components of the hydraulic conductivity tensor are $K_{xx}' = 15$ m/d and $K_{yy}' = 5$ m/d, calculate the magnitude and direction of the specific discharge; (c) if the coordinate axes are rotated 30° clockwise from the principal axes, calculate the components of the hydraulic conductivity tensor relative to the new coordinate axes; and (d) verify that the magnitude and direction of the specific discharge calculated in part (b) are not affected by axis rotation.

Solution

(a) The piezometric head distribution, $h(x', y')$, in the triangular region ABC can be assumed to be planar and given by

$$h(x', y') = ax' + by' + c$$

where a, b, and c are constants, and (x', y') are the (principal) coordinate locations. Applying this equation to points A, B, and C (with all linear dimensions in meters) yields

$$2.157 = a(0) + b(0) + c$$

$$1.752 = a(1000) + b(-500) + c$$

$$1.629 = a(500) + b(-1200) + c$$

The solution of these equations is $a = -0.0002337$, $b = 0.0003426$, and $c = 2.157$. From the planar head distribution, it is clear that the components of the head gradient are given by

$$\frac{\partial h}{\partial x'} = a \quad \text{and} \quad \frac{\partial h}{\partial y'} = b$$

Therefore, in this case the components of the head gradient are

$$\frac{\partial h}{\partial x'} = -0.0002337 \quad \text{and} \quad \frac{\partial h}{\partial y'} = 0.0003426$$

which can be written in vector notation as

$$\nabla' h = -0.0002337 i' + 0.0003426 j'$$

where i' and j' are unit vectors in the principal directions.

(b) The components of the specific discharge vector relative to the principal axes are given by

$$q'_x = -K'_{xx} J'_x$$

$$q'_y = -K'_{yy} J'_y$$

where $J'_x = \partial h / \partial x' = -0.0002337$, $J'_y = \partial h / \partial y' = 0.0003426$, $K'_{xx} = 15$ m/d, and $K_{yy} = 5$ m/d. Substituting these values yields

$$q'_x = -15(-0.0002337) = 0.00351 \text{ m/d}$$

$$q'_y = -5(0.0003426) = -0.00171 \text{ m/d}$$

The magnitude of the specific discharge, q', is given by

$$q' = \sqrt{(q'_x)^2 + (q'_y)^2} = \sqrt{0.00351^2 + (-0.00171)^2} = 0.00390 \text{ m/d}$$

and the direction of flow is at an angle η measured clockwise from the principal axes, where

$$\eta = \tan^{-1}\left(\frac{0.00171}{0.00351}\right) = 26.0°$$

(c) If the reference axes are rotated $30°$ clockwise from the principal axes, then, according to Equations 6.24 to 6.26, the components of the hydraulic conductivity tensor become

$$K_{xx} = \frac{K'_{xx} + K'_{yy}}{2} + \frac{K'_{xx} - K'_{yy}}{2} \cos 2\theta$$

$$= \frac{15 + 5}{2} + \frac{15 - 5}{2} \cos 2(-30°) = 12.5 \text{ m/d}$$

$$K_{yy} = \frac{K'_{xx} + K'_{yy}}{2} - \frac{K'_{xx} - K'_{yy}}{2} \cos 2\theta$$

$$= \frac{15 + 5}{2} - \frac{15 - 5}{2} \cos 2(-30°) = 7.5 \text{ m/d}$$

$$K_{xy} = K_{yx} = -\frac{K'_{xx} - K'_{yy}}{2} \sin 2\theta = -\frac{15 - 5}{2} \sin 2(-30°) = 4.33 \text{ m/d}$$

The components of the piezometric head gradient also change with axis rotation, and the new components, J_x and J_y, are given by

$$J_x = J'_x \cos 30° - J'_y \sin 30°$$

$$= (-0.0002337) \cos 30° - (0.0003426) \sin 30° = -0.0003738$$

$$J_y = J'_x \sin 30° + J'_y \cos 30°$$

$$= (-0.0002337) \sin 30° + (0.0003426) \cos 30° = 0.0001799$$

The components of the specific discharge relative to the new coordinate axes, q_x and q_y, are therefore

$$q_x = -K_{xx}J_x - K_{xy}J_y = -(12.5)(-0.0003738) - (4.33)(0.0001799)$$

$$= 0.00389 \text{ m/d}$$

$$q_y = -K_{yx}J_x - K_{yy}J_y = -(4.33)(-0.0003738) - (7.5)(0.0001799)$$

$$= 0.000269 \text{ m/d}$$

The magnitude of the specific discharge, q, is

$$q = \sqrt{q_x^2 + q_y^2} = \sqrt{0.00389^2 + (0.000269)^2} = 0.00390 \text{ m/d}$$

and the specific discharge vector is oriented at an angle ξ measured anticlockwise from the rotated x-axis, where

$$\xi = \tan^{-1}\left(\frac{0.000269}{0.00389}\right) = 4.0°$$

(d) The magnitude and direction of the specific discharge vector calculated for the rotated axes in (c) is exactly the same as the specific discharge vector calculated in (b). Hence, as expected, the specific discharge vector does not depend on orientation of the coordinate axes.

As with most hydrogeologic parameters, the hydraulic conductivity is associated with a support scale, which corresponds to the volume over which the hydraulic conductivity describes the proportionality between the average specific discharge and the average hydraulic gradient. Hydraulic conductivities averaged over a support scale tend to vary in space and can be described as a *random space function* (RSF). The probability distribution of the hydraulic conductivity at any point in space has been found to be log-normally distributed in most cases. Defining the natural logarithm of the hydraulic conductivity, Y, by

$$Y = \ln K \tag{6.27}$$

where K is the hydraulic conductivity, then the log-normal probability distribution of K can be described by the mean and variance of Y, $\langle Y \rangle$ and σ_Y^2, respectively. Alternatively, the mean of $\ln K$ can be expressed in terms of the *geometric mean* of the hydraulic conductivity, K_G, where

$$K_G = e^{\langle Y \rangle} \tag{6.28}$$

Besides the mean and variance of $\ln K$ at any point in an aquifer, the spatial variability in $\ln K$ is described by its spatial correlation, which is defined as the correlation between $\ln K$ measured at any two points in space. The spatial correlation of $\ln K$, $\rho_Y(\mathbf{r})$, in a statistically homogeneous porous medium is defined by the relation

$$\rho_Y(\mathbf{r}) = \frac{\langle (Y(\mathbf{x}) - \langle Y \rangle)(Y(\mathbf{x} + \mathbf{r}) - \langle Y \rangle) \rangle}{\sigma_Y^2} \tag{6.29}$$

where $\mathbf{r}$ is the separation vector between the measurements of Y, and $\mathbf{x}$ is the position vector indicating the location of the reference measurement. Statistical homogeneity of the porous medium guarantees that the spatial correlation calculated by Equation 6.29 is independent of the reference location, $\mathbf{x}$. In practice, several empirical correlation functions can be used to describe $\rho_Y(\mathbf{r})$ (Dagan, 1989), the most common of which is the *exponential correlation function* given by

$$\rho_Y(\mathbf{r}) = \exp\left[-\left(\frac{r_1^2}{\lambda_1^2} + \frac{r_2^2}{\lambda_2^2} + \frac{r_3^2}{\lambda_3^2} \right)^{\frac{1}{2}} \right] \tag{6.30}$$

where (r_1, r_2, r_3) are the components of the separation vector, $\mathbf{r}$, and $(\lambda_1, \lambda_2, \lambda_3)$ are the *correlation length scales* in each of the coordinate directions. The correlation length scales are empirical parameters that measure the separation distances at which the correlation between hydraulic conductivities is equal to e^{-1} ($=0.368$). The correlation between hydraulic conductivities at zero separation is equal to one, indicating perfect correlation.

The statistical characterization of the hydraulic conductivity provides the most realistic description of the distribution of hydraulic conductivities in porous media. However, the relationship between the statistical properties of the hydraulic conductivity and the support scale must not be overlooked. For example, whenever hydraulic conductivities are measured at small support scales, such as in permeameters in the laboratory where the support scale is on the order of a few centimeters, it is expected that individual measurements of the log-hydraulic conductivity may deviate significantly from the mean, $\langle Y \rangle$, the variance of the measurements, σ_Y^2, will be high, and the correlation length scales $(\lambda_1, \lambda_2, \lambda_3)$ will be small, and on the order of the support scale. If, in the same porous medium, the hydraulic conductivities are measured over a much larger support scale using aquifer tests in the field, where the support scale may be on the order of 100 m, then it is expected that individual measurements at different locations in the aquifer will be much closer to the ensemble mean, $\langle Y \rangle$, the variance of the measurements, σ_Y^2, will be small, and the correlation length scale will be large, and on the order of the support scale. The sensitivity of the statistical properties of the hydraulic conductivity to the support scale of the measurements is called the *scale effect*. At any given scale, the mean hydraulic conductivity, K_G, can be used in Darcy's law to yield the average specific discharge over the support scale, and on scales smaller than the support scale, the specific discharge may be expected to deviate from the averaged value. The average hydraulic conductivities over scales much larger than the support scale of individual measurements can be estimated by assuming a log-normal distribution of the hydraulic conductivity, with an exponential correlation function, and for small variances in hydraulic conductivity ($\sigma_Y < 1$) in isotropic porous media, the mean values of the hydraulic conductivity, $\langle K \rangle$, can be estimated by (Dagan, 1989)

$$\text{1-D Flow:} \quad \langle K \rangle = K_G \left(1 - \frac{\sigma_Y^2}{2} \right) \tag{6.31}$$

$$\text{2-D Flow:} \quad \langle K \rangle = K_G \tag{6.32}$$

$$\text{3-D Flow:} \quad \langle K \rangle = K_G \left(1 + \frac{\sigma_Y^2}{6} \right) \tag{6.33}$$

Large-scale averaged hydraulic conductivities, $\langle K \rangle$, are commonly referred to as *macroscopic hydraulic conductivities*.

EXAMPLE 6.5

Ten aquifer tests have been performed at different locations in an aquifer, and the calculated hydraulic conductivities are found to be: 15, 20, 30, 10, 8, 55, 17, 86, 12, and 35 m/d. The support scale for each of these measurements is around 100 m, and the geologic logs indicate that the aquifer can be considered statistically homogeneous. Estimate the macroscopic hydraulic conductivity (applied over scales of several hundreds of meters) that is appropriate for simulating three-dimensional flows in the aquifer.

Solution   The natural logs of the hydraulic conductivities are given in the following table:

K	$Y (= \ln K)$	Y^2
15	2.71	7.33
20	3.00	8.97
30	3.40	11.57
10	2.30	5.30
8	2.08	4.32
55	4.01	16.06
17	2.83	8.03
86	4.45	19.84
12	2.48	6.17
35	3.56	12.64
Sum	30.82	100.23

The mean and variance of the log-hydraulic conductivity, $\langle Y \rangle$ and σ_Y^2, can be estimated by $\overline{Y}$ and S_Y^2, respectively, where

$$\overline{Y} = \frac{1}{N} \sum_{i=1}^{N} Y_i$$

and

$$S_Y^2 = \frac{1}{N-1} \sum_{i=1}^{N} (Y_i - \overline{Y})^2$$

$$= \frac{1}{N-1} \sum_{i=1}^{N} Y_i^2 - \frac{N}{N-1} \overline{Y}^2$$

where N is the number of measurements, and Y_i are the log-hydraulic conductivity measurements. From the given data,

$$\overline{Y} = \frac{1}{10}(30.82) = 3.082$$

and

$$S_Y^2 = \frac{1}{10-1}100.23 - \frac{10}{9}3.082^2 = 0.583$$

For three-dimensional flows, the mean (large-scale) hydraulic conductivity is given by Equation 6.33 as

$$\langle K \rangle = K_G\left(1 + \frac{\sigma_Y^2}{6}\right)$$

where K_G is the geometric hydraulic conductivity given by

$$K_G = e^{\langle Y \rangle} \approx e^{\overline{Y}} = e^{3.082} = 22 \text{ m/d}$$

Substituting $K_G = 22$ m/d, and $\sigma_Y^2 = 0.583$ into Equation 6.33 yields the macroscopic hydraulic conductivity

$$\langle K \rangle = 22\left(1 + \frac{0.583}{6}\right) = 24 \text{ m/d}$$

In concluding the discussion of Darcy's law, it should be noted that there are some porous media where Darcy's law is not applicable, notably in fractured media. In these formations, the flow is either modeled in individual fractures or, if the fractures are sufficiently dense, modeled as an *equivalent* porous medium. Techniques for describing flow in fractured media can be found in de Marsily (1986).

6.2.2   General Flow Equation

Consider the control volume shown in Figure 6.7. The net mass inflow rate to the control volume is given by

$$
\text{Net mass inflow rate} = \left[(\rho q_x) - \frac{\partial(\rho q_x)}{\partial x}\frac{\Delta x}{2}\right]\Delta y\,\Delta z - \left[(\rho q_x) + \frac{\partial(\rho q_x)}{\partial x}\frac{\Delta x}{2}\right]\Delta y\,\Delta z
$$

$$
+ \left[(\rho q_y) - \frac{\partial(\rho q_y)}{\partial y}\frac{\Delta y}{2}\right]\Delta x\,\Delta z - \left[(\rho q_y) + \frac{\partial(\rho q_y)}{\partial y}\frac{\Delta y}{2}\right]\Delta x\,\Delta z
$$

$$
+ \left[(\rho q_z) - \frac{\partial(\rho q_z)}{\partial z}\frac{\Delta z}{2}\right]\Delta x\,\Delta y - \left[(\rho q_z) + \frac{\partial(\rho q_z)}{\partial z}\frac{\Delta z}{2}\right]\Delta x\,\Delta y
$$

$$(6.34)$$

where ρ is the density of the fluid. Combining terms in Equation 6.34 and simplifying leads to

$$
\text{Net mass inflow rate} = -\left[\frac{\partial(\rho q_x)}{\partial x} + \frac{\partial(\rho q_y)}{\partial y} + \frac{\partial(\rho q_z)}{\partial z}\right]\Delta x\,\Delta y\,\Delta z \qquad (6.35)
$$

In accordance with the law of conservation of mass, the net mass inflow rate into the control volume is equal to rate of change of mass within the control volume, which is given by

$$
\text{Rate of change of mass} = \frac{\partial(n\rho)}{\partial t}\Delta x\,\Delta y\,\Delta z \qquad (6.36)
$$

FIGURE 6.7: Control volume in a porous medium

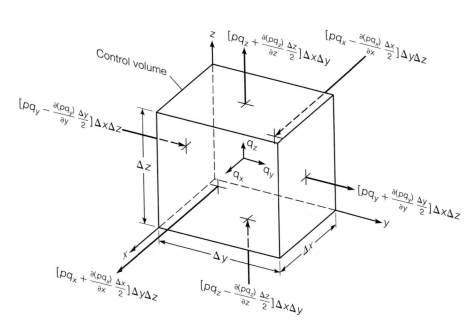

where n is the porosity of the porous medium, and t is the time. Combining Equations 6.35 and 6.36 yields the following relation:

$$-\left[\frac{\partial(\rho q_x)}{\partial x} + \frac{\partial(\rho q_y)}{\partial y} + \frac{\partial(\rho q_z)}{\partial z}\right] = \frac{\partial(n\rho)}{\partial t} \tag{6.37}$$

This equation can be further reduced for the cases where the density, ρ, is independent of space and time. Under this condition, Equation 6.37 reduces to

$$-\left(\frac{\partial q_x}{\partial x} + \frac{\partial q_y}{\partial y} + \frac{\partial q_z}{\partial z}\right) = \frac{\partial n}{\partial t} \tag{6.38}$$

The *specific storage*, S_s, of a porous medium is defined as the volume of water released from storage per unit volume of the porous medium per unit decline in piezometric head, and typical values of S_s in unconsolidated materials range from 10^{-4} to 10^{-9} m^{-1} (Roscoe Moss Company, 1990; Cheng, 2000). Water released from storage due to declining piezometric head (i.e., declining pore pressure) is associated with the net effect of: (1) expansion of water due to compressibility; and (2) reduction in void space associated with the expansion of the solid grains and the compressibility of the solid matrix. To account for these compressibility effects, the specific storage, S_s, can be expressed in terms of the elastic properties of water and the porous medium using the following equation (Bear, 1979):

$$\boxed{S_s = \gamma(nE_w + \alpha)} \tag{6.39}$$

where γ is the specific weight of water, E_w is the compressibility coefficient[†] of water, and α is the compressibility coefficient of the porous matrix. The compressibility of pure water, E_w, is equal to 4.59×10^{-7} m^2/kN at 20°C, and typical values of n, α, and S_s for several aquifer materials are given in Table 6.5. Considering the control volume shown in Figure 6.7, the time rate of change of fluid volume within the control volume can be expressed in terms of the specific storage, S_s, by the following relation:

$$\frac{\partial n}{\partial t} \Delta x \, \Delta y \, \Delta z = S_s \frac{\partial \phi}{\partial t} \Delta x \, \Delta y \, \Delta z \tag{6.40}$$

where ϕ is the piezometric head, defined as $p/\gamma + z$. Equation 6.40 simplifies into

$$\frac{\partial n}{\partial t} = S_s \frac{\partial \phi}{\partial t} \tag{6.41}$$

and combining Equations 6.38 and 6.41 yields

$$-\left(\frac{\partial q_x}{\partial x} + \frac{\partial q_y}{\partial y} + \frac{\partial q_z}{\partial z}\right) = S_s \frac{\partial \phi}{\partial t} \tag{6.42}$$

[†]The *compressibility coefficient* is defined as the inverse of the bulk modulus of elasticity

TABLE 6.5: Specific Storage Coefficients in Aquifers

Aquifer material	n	α (m^2/N)	S_s (m^{-1})
Plastic clay	0.5	3×10^{-7}—2×10^{-6}	9.8×10^{-3}
Stiff clay	0.4	7×10^{-8}—3×10^{-7}	2.0×10^{-3}
Medium-hard clay	0.4	7×10^{-8}—1×10^{-7}	9.8×10^{-4}
Loose sand	0.4	5×10^{-8}—1×10^{-7}	7.9×10^{-4}
Dense sand	0.3	5×10^{-9}—2×10^{-8}	2.0×10^{-4}
Dense sandy gravel	0.3	5×10^{-9}—1×10^{-8}	8.0×10^{-5}
Fissured jointed rock	0.1	3×10^{-10}—7×10^{-9}	5.4×10^{-6}
Boise sandstone	0.26	1×10^{-10}	2.2×10^{-6}
Weber sandstone	0.06	3×10^{-11}	5.9×10^{-7}

Sources: Cheng (2000); Domenico and Mifflin (1965).

According to Darcy's law, the components of the specific discharge vector, q_x, q_y, and q_z, can be written in terms of the piezometric head gradients, where

$$q_x = -K_{xx} \frac{\partial \phi}{\partial x}$$

$$q_y = -K_{yy} \frac{\partial \phi}{\partial y} \qquad (6.43)$$

$$q_z = -K_{zz} \frac{\partial \phi}{\partial z}$$

and it is assumed that the coordinate axes are in the directions of the principal axes of the hydraulic conductivity. Combining Equations 6.42 and 6.43 leads to

$$\boxed{\frac{\partial}{\partial x} \left(K_{xx} \frac{\partial \phi}{\partial x} \right) + \frac{\partial}{\partial y} \left(K_{yy} \frac{\partial \phi}{\partial y} \right) + \frac{\partial}{\partial z} \left(K_{zz} \frac{\partial \phi}{\partial z} \right) = S_s \frac{\partial \phi}{\partial t}} \qquad (6.44)$$

This general equation is applicable to both isotropic and anisotropic formations. In cases where the aquifer is homogeneous and anisotropic, Equation 6.44 becomes

$$\boxed{K_{xx} \frac{\partial^2 \phi}{\partial x^2} + K_{yy} \frac{\partial^2 \phi}{\partial y^2} + K_{zz} \frac{\partial^2 \phi}{\partial z^2} = S_s \frac{\partial \phi}{\partial t}} \quad \text{homogeneous, anisotropic} \qquad (6.45)$$

In cases where the hydraulic conductivity is isotropic, the hydraulic conductivity, K, is independent of the coordinate direction, which means that

$$K_{xx} = K_{yy} = K_{zz} = K \qquad (6.46)$$

and Equation 6.45 can be written as

$$\boxed{\frac{\partial^2 \phi}{\partial x^2} + \frac{\partial^2 \phi}{\partial y^2} + \frac{\partial^2 \phi}{\partial z^2} = \frac{S_s}{K} \frac{\partial \phi}{\partial t}} \quad \text{homogeneous, isotropic} \qquad (6.47)$$

or in the more convenient vector form as

$$\nabla^2 \phi = \frac{S_s}{K} \frac{\partial \phi}{\partial t} \qquad \text{homogeneous, isotropic} \qquad (6.48)$$

where $\nabla^2()$ is the Laplacian operator defined by

$$\nabla^2 f = \frac{\partial^2 f}{\partial x^2} + \frac{\partial^2 f}{\partial y^2} + \frac{\partial^2 f}{\partial z^2} \qquad (6.49)$$

where f is a scalar function. In cases where there is significant radial symmetry, it is convenient to use cylindrical coordinates, and the Laplacian in Equation 6.48 is given by

$$\nabla^2 f = \frac{\partial^2 f}{\partial r^2} + \frac{1}{r} \frac{\partial f}{\partial r} + \frac{1}{r^2} \frac{\partial^2 f}{\partial \theta^2} + \frac{\partial^2 f}{\partial z^2} \qquad (6.50)$$

In summary, the governing equation for the flow of a homogeneous fluid in anisotropic nonhomogeneous porous media is given Equation 6.44, which simplifies to Equation 6.45 for homogeneous media and to Equation 6.47 for media that are both homogeneous and isotropic.

EXAMPLE 6.6

Give the equation describing the piezometric head distribution in a homogeneous anisotropic aquifer in which vertical variations in the piezometric head are negligible. Explain why (in this case) the vertical hydraulic conductivity does not influence the head distribution. Is the assumption of negligible vertical variation in piezometric head reasonable in some cases?

Solution The piezometric head distribution, $\phi(x, y, z, t)$, in a homogeneous anisotropic aquifer is given by Equation 6.45 as

$$K_{xx} \frac{\partial^2 \phi}{\partial x^2} + K_{yy} \frac{\partial^2 \phi}{\partial y^2} + K_{zz} \frac{\partial^2 \phi}{\partial z^2} = S_s \frac{\partial \phi}{\partial t}$$

If vertical variations in piezometric head are negligible, then ϕ does not depend on z, which means that $\partial^2 \phi / \partial z^2 = 0$, and therefore Equation 6.45 becomes

$$K_{xx} \frac{\partial^2 \phi}{\partial x^2} + K_{yy} \frac{\partial^2 \phi}{\partial y^2} = S_s \frac{\partial \phi}{\partial t}$$

Since the vertical hydraulic conductivity, K_{zz}, does not appear in this equation, then K_{zz} has no effect on the head distribution, $\phi(x, y, t)$.

The assumption of negligible vertical variation in the piezometric head is exact when the vertical pressure distribution is hydrostatic. This is certainly reasonable for horizontal ground-water flow.

Equivalent anisotropic/isotropic media. Consider a porous medium that is homogeneous and anisotropic, with a given hydraulic conductivity (K_{xx}, K_{yy}, K_{zz}) and specific storage, S_s. The spatial features of the anisotropic domain, described relative to the

(x, y, z) coordinates, can be transformed into a new domain, described by (x', y', z') coordinates, where

$$x' = \sqrt{\frac{K}{K_{xx}}}\, x, \qquad y' = \sqrt{\frac{K}{K_{yy}}}\, y, \qquad z' = \sqrt{\frac{K}{K_{zz}}}\, z \qquad (6.51)$$

and K is an arbitrary coefficient with dimensions of a hydraulic conductivity. Applying these relationships between (x, y, z) and (x', y', z') yields

$$\frac{\partial \phi}{\partial x'} = \frac{\partial \phi}{\partial x}\frac{dx}{dx'} = \sqrt{\frac{K_{xx}}{K}}\frac{\partial \phi}{\partial x} \qquad (6.52)$$

$$\frac{\partial^2 \phi}{\partial x'^2} = \frac{\partial}{\partial x}\left(\frac{\partial \phi}{\partial x'}\right)\frac{dx}{dx'} = \frac{K_{xx}}{K}\frac{\partial^2 \phi}{\partial x^2} \qquad (6.53)$$

$$\frac{\partial \phi}{\partial y'} = \frac{\partial \phi}{\partial y}\frac{dy}{dy'} = \sqrt{\frac{K_{yy}}{K}}\frac{\partial \phi}{\partial y} \qquad (6.54)$$

$$\frac{\partial^2 \phi}{\partial y'^2} = \frac{\partial}{\partial y}\left(\frac{\partial \phi}{\partial y'}\right)\frac{dy}{dy'} = \frac{K_{yy}}{K}\frac{\partial^2 \phi}{\partial y^2} \qquad (6.55)$$

$$\frac{\partial \phi}{\partial z'} = \frac{\partial \phi}{\partial z}\frac{dz}{dz'} = \sqrt{\frac{K_{zz}}{K}}\frac{\partial \phi}{\partial z} \qquad (6.56)$$

$$\frac{\partial^2 \phi}{\partial z'^2} = \frac{\partial}{\partial z}\left(\frac{\partial \phi}{\partial z'}\right)\frac{dz}{dz'} = \frac{K_{zz}}{K}\frac{\partial^2 \phi}{\partial z^2} \qquad (6.57)$$

Substituting Equations 6.52 to 6.57 into Equation 6.45 gives the following equation for the head distribution in the (x', y', z') domain:

$$\frac{\partial^2 \phi}{\partial x'^2} + \frac{\partial^2 \phi}{\partial y'^2} + \frac{\partial^2 \phi}{\partial z'^2} = \frac{S_s}{K}\frac{\partial \phi}{\partial t} \qquad (6.58)$$

which indicates that the head distribution can be calculated by taking the hydraulic conductivity as homogeneous and isotropic with a value of K in the transformed domain. By further requiring that the calculated flowrates across any boundary in the transformed domain be equal to the flowrate across the corresponding boundary in the real (x, y, z) domain, it can be shown that K should be taken as (de Marsily, 1986)

$$K = \sqrt[3]{K_{xx} K_{yy} K_{zz}} \qquad (6.59)$$

This result is extremely useful in practical applications, since it means that solutions to the flow equation in isotropic homogeneous formations can be applied to any anisotropic homogeneous formation by simply specifying the hydraulic conductivity components (K_{xx}, K_{yy}, K_{zz}), and transforming from (x', y', z') coordinates to (x, y, z) coordinates using Equation 6.51.

EXAMPLE 6.7

The steady-state head distribution in a large homogeneous isotropic aquifer caused by pumping at a rate Q from the location (x_0', y_0', z_0') is given by

$$\phi(x', y', z') = -\frac{Q}{4\pi} \frac{1}{\sqrt{(x' - x_0')^2 + (y' - y_0')^2 + (z' - z_0')^2}}$$

Determine the head distribution caused by pumping at a rate Q in a homogeneous anisotropic aquifer where $K_{xx} = 10$ m/d, $K_{yy} = 5$ m/d, and $K_{zz} = 1$ m/d.

Solution   The hydraulic conductivity, K, of the equivalent isotropic aquifer is given by Equation 6.59 as

$$K = \sqrt[3]{K_{xx} K_{yy} K_{zz}} = \sqrt[3]{(10)(5)(1)} = 3.68 \text{ m/d}$$

and the relationships between the coordinates in the anisotropic aquifer and equivalent isotropic aquifer are given by Equation 6.51 as

$$x' = \sqrt{\frac{K}{K_{xx}}} x = \sqrt{\frac{3.68}{10}} x = 0.607x$$

$$y' = \sqrt{\frac{K}{K_{yy}}} y = \sqrt{\frac{3.68}{5}} y = 0.858y$$

$$z' = \sqrt{\frac{K}{K_{zz}}} z = \sqrt{\frac{3.68}{1}} z = 1.92z$$

where the primed coordinates apply to the equivalent isotropic aquifer. Applying the coordinate transformation to the head distribution in the equivalent isotropic aquifer gives the head distribution resulting from pumping in an anisotropic aquifer as

$$\phi(x, y, z) = -\frac{Q}{4\pi} \frac{1}{\sqrt{(0.607)^2(x - x_0)^2 + (0.858)^2(y - y_0)^2 + (1.92)^2(z - z_0)^2}}$$

$$= -\frac{Q}{4\pi} \frac{1}{\sqrt{0.368(x - x_0)^2 + 0.736(y - y_0)^2 + 3.69(z - z_0)^2}}$$

where (x_0, y_0, z_0) is the pumping location in the anisotropic aquifer.

The governing equation for flow in porous media, Equation 6.44, contains only one unknown quantity, the piezometric head, ϕ; and Equation 6.44 along with the initial and boundary conditions for the piezometric head represents a complete statement of the problem of flow in porous media (Bear, 1979). Solving Equation 6.44 with associated boundary and initial conditions yields the piezometric head distribution, which is also called the piezometric surface or potentiometric surface. The distribution of seepage velocity and specific discharge in the porous medium can be derived directly from the potentiometric surface and hydraulic-conductivity distribution using Darcy's law.

6.2.3　Two-Dimensional Approximations

6.2.3.1　Unconfined aquifers

In the case of unconfined aquifers, the general flow equation, Equation 6.44, can be simplified using an approximation first suggested by Dupuit (1863). Consider the typical flow conditions in an unconfined aquifer shown in Figure 6.8. The surface streamline is coincident with the phreatic surface (water table), and the streamlines become more horizontal with depth below the phreatic surface. The streamlines eventually become horizontal at the base of the aquifer (which is assumed to be horizontal). Since the equipotential lines are necessarily perpendicular to the streamlines, the equipotential lines are curved, as illustrated in Figure 6.8(a). The Dupuit (1863) approximation is that the equipotential lines can be assumed to be vertical, making the flowlines horizontal, and the specific discharge components constant over the depth, in which case

$$
\begin{aligned}
q_x &= -K_{xx}\frac{\partial \phi}{\partial x} \approx -K_{xx}\frac{\partial h}{\partial x} \\
q_y &= -K_{yy}\frac{\partial \phi}{\partial y} \approx -K_{yy}\frac{\partial h}{\partial y}
\end{aligned}
\tag{6.60}
$$

where h is the depth of the saturated zone. The Dupuit approximation is equivalent to assuming that the vertical pressure distribution in the aquifer is hydrostatic, leading to two-dimensional flow in the horizontal plane. The Dupuit approximation was independently proposed by Forchheimer (1930) and is sometimes called the *Dupuit–Forchheimer* approximation.[‡] Clearly, for the Dupuit approximation to be valid, the actual equipotential lines must be nearly vertical, and therefore the slope of the phreatic surface must be nearly horizontal. These conditions are met by most unconfined aquifers, which typically have (phreatic) surface slopes in the range of 0.1% to 1% (Bear, 1979). Analysis of ground-water flow based on the Dupuit approximation is commonly called the *hydraulic approach*.

　　A general equation for flow in unconfined aquifers can now be derived using the Dupuit approximation. Consider the control volume shown in Figure 6.9, which is bounded on the bottom by the base of the aquifer and on the top by the phreatic

FIGURE 6.8: Dupuit approximation in an unconfined aquifer

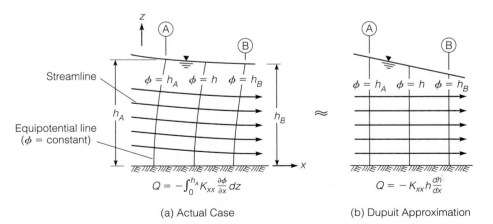

(a) Actual Case　　　　　　　　(b) Dupuit Approximation

[‡] Arsène Dupuit (1804–1866) was an Italian-born engineer who practiced mostly in France, and Philipp Forchheimer (1852–1933) was an Austrian hydraulics professor at the University of Graz in Austria.

FIGURE 6.9: Control volume in an unconfined aquifer

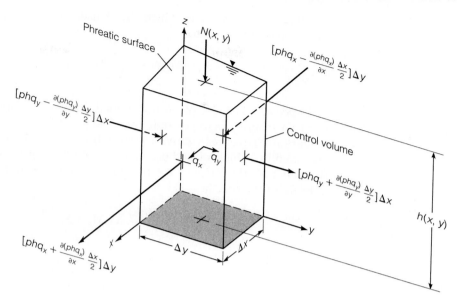

surface. In addition to the inflows and outflows from the lateral boundaries of the control volume, there is also a recharge, $N(x, y)$, which accounts for the net inflow from above the aquifer, and is typically associated with infiltrated water from the ground surface that penetrates the vadose zone. The recharge, $N(x, y)$, has units of volume inflow per unit area per unit time [L/T]. The net mass inflow rate to the control volume is given by

$$\text{Net mass inflow rate} \approx \left[(\rho q_x h) - \frac{\partial(\rho q_x h)}{\partial x} \frac{\Delta x}{2} \right] \Delta y - \left[(\rho q_x h) + \frac{\partial(\rho q_x h)}{\partial x} \frac{\Delta x}{2} \right] \Delta y$$

$$+ \left[(\rho q_y h) - \frac{\partial(\rho q_y h)}{\partial y} \frac{\Delta y}{2} \right] \Delta x - \left[(\rho q_y h) + \frac{\partial(\rho q_y h)}{\partial y} \frac{\Delta y}{2} \right]$$

$$\times \Delta x + \rho N \Delta x \Delta y \tag{6.61}$$

where the net mass inflow rate is regarded as approximate, since vertical flow in the aquifer is neglected. Combining terms in Equation 6.61 and simplifying yields

$$\text{Net mass inflow rate} \approx - \left[\frac{\partial(\rho q_x h)}{\partial x} + \frac{\partial(\rho q_y h)}{\partial y} \right] \Delta x \Delta y + \rho N \Delta x \Delta y \tag{6.62}$$

The net mass inflow rate to the control volume is equal to rate of change of mass within the control volume, which is given by

$$\text{Rate of change of mass} = \frac{\partial(nh \Delta x \Delta y \rho)}{\partial t} \tag{6.63}$$

Combining Equations 6.62 and 6.63 leads to the Dupuit approximation to the flow equation as

$$- \left[\frac{\partial(\rho q_x h)}{\partial x} + \frac{\partial(\rho q_y h)}{\partial y} \right] + \rho N = \frac{\partial(nh\rho)}{\partial t} \tag{6.64}$$

For the common case in which the density, ρ, is independent of space and time, Equation 6.64 reduces to

$$-\left[\frac{\partial(q_x h)}{\partial x} + \frac{\partial(q_y h)}{\partial y}\right] + N = \frac{\partial(nh)}{\partial t} \tag{6.65}$$

The *specific yield*, S_y, of an unconfined aquifer is defined as the volume of water released from storage per unit (plan) area per unit decline in the phreatic surface (water table). Representative values of the specific yield for several aquifer materials are given in Tables 6.1 and 6.2. Considering the control volume shown in Figure 6.9, the time rate of change of water volume within the control volume can be expressed in terms of the specific yield by

$$\frac{\partial(nh)}{\partial t}\Delta x\,\Delta y = S_y\frac{\partial h}{\partial t}\Delta x\,\Delta y \tag{6.66}$$

which simplifies into

$$\frac{\partial(nh)}{\partial t} = S_y\frac{\partial h}{\partial t} \tag{6.67}$$

Combining Equations 6.65 and 6.67 yields

$$-\left[\frac{\partial(q_x h)}{\partial x} + \frac{\partial(q_y h)}{\partial y}\right] + N = S_y\frac{\partial h}{\partial t} \tag{6.68}$$

Substituting the approximate Darcy relations between the specific discharge and the depth of the saturated zone, h, Equation 6.60, leads to

$$\boxed{\frac{\partial}{\partial x}\left(K_{xx}\,h\,\frac{\partial h}{\partial x}\right) + \frac{\partial}{\partial y}\left(K_{yy}\,h\,\frac{\partial h}{\partial y}\right) + N = S_y\frac{\partial h}{\partial t}} \tag{6.69}$$

This equation is applicable to two-dimensional flow in anisotropic nonhomogeneous porous formations where the Dupuit approximation is valid.

Consider the case of a stratified aquifer shown in Figure 6.10, where the aquifer has n layers, and layer i ($i = 1,\ldots,n$) has a thickness Δz_i, and hydraulic conductivity (principal) components K_{xx}^i and K_{yy}^i. The volumetric flowrate in the x-direction, Q_x, is equal to the sum of the volumetric flowrates (in the x-direction) in each of the layers. Therefore

$$Q_x = -\sum_{i=1}^{n} K_{xx}^i\frac{\partial h}{\partial x}\Delta z_i\,\Delta y \tag{6.70}$$

and the effective x-component of the hydraulic conductivity, $\overline{K}_{xx}$, is defined by the relation

$$Q_x = -\overline{K}_{xx}\frac{\partial h}{\partial x}h\,\Delta y \tag{6.71}$$

where

$$\overline{K}_{xx} = \frac{1}{h}\sum_{i=1}^{n} K_{xx}^i\,\Delta z_i \tag{6.72}$$

FIGURE 6.10: Control volume in a stratified unconfined aquifer

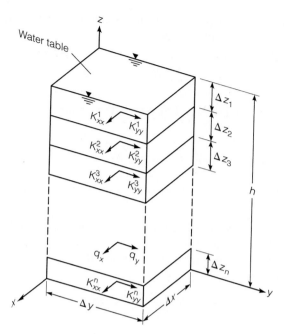

Similarly, the effective hydraulic conductivity in the y-direction is given by

$$\overline{K}_{yy} = \frac{1}{h} \sum_{i=1}^{n} K_{yy}^{i} \, \Delta z_i \tag{6.73}$$

Equations 6.72 and 6.73 yield effective hydraulic conductivities that can be substituted into the Dupuit approximation of the flow equation, Equation 6.69, to determine the two-dimensional distribution of the water table in stratified phreatic aquifers. It should be clear that the effective hydraulic conductivities given by Equations 6.72 and 6.73 are both nonhomogeneous in space, since the effective hydraulic conductivities are a function of the saturated depth, h, which is necessarily nonhomogeneous.

EXAMPLE 6.8

The hydraulic conductivity distribution in a 30-m thick stratified surficial aquifer is given in the following table:

Depth (m)	K_{xx} (m/d)	K_{yy} (m/d)
0–5	25	30
5–10	30	33
10–15	40	37
15–20	32	28
20–25	22	19
25–30	13	11

Estimate the effective hydraulic conductivity when the water table is 4 m below the ground surface. Would the effective hydraulic conductivity be the same at a location where the water table is 5 m below the ground surface?

Solution The effective hydraulic conductivity components, $\overline{K}_{xx}$ and $\overline{K}_{yy}$, are given by Equations 6.72 and 6.73 as

$$\overline{K}_{xx} = \frac{1}{h} \sum_{i=1}^{n} K_{xx}^{i} \Delta z_i \qquad \text{and} \qquad \overline{K}_{yy} = \frac{1}{h} \sum_{i=1}^{n} K_{yy}^{i} \Delta z_i$$

When the water table is 4 m below the ground surface, $h = 30 - 4 = 26$ m, and therefore

$$\overline{K}_{xx} = \frac{1}{26} (25 \times 1 + 30 \times 5 + 40 \times 5 + 32 \times 5 + 22 \times 5 + 13 \times 5) = 27.3 \text{ m/d}$$

$$\overline{K}_{yy} = \frac{1}{26} (30 \times 1 + 33 \times 5 + 37 \times 5 + 28 \times 5 + 19 \times 5 + 11 \times 5) = 25.8 \text{ m/d}$$

These calculations account for the fact that the ground water flows through only the top 1 m of the upper layer.

When the water table is 5 m below the ground surface, $h = 30 - 5 = 25$ m, and therefore

$$\overline{K}_{xx} = \frac{1}{25} (30 \times 5 + 40 \times 5 + 32 \times 5 + 22 \times 5 + 13 \times 5) = 27.4 \text{ m/d}$$

$$\overline{K}_{yy} = \frac{1}{25} (33 \times 5 + 37 \times 5 + 28 \times 5 + 19 \times 5 + 11 \times 5) = 25.6 \text{ m/d}$$

Hence, $\overline{K}_{xx}$ increases slightly and $\overline{K}_{yy}$ decreases slightly when the water table falls from 4 m to 5 m below the ground surface.

In cases where the hydraulic conductivity of the aquifer is homogeneous, such as when the aquifer is composed of a single homogeneous layer, Equation 6.69 describes the distribution of saturated thickness, h, by

$$\boxed{K_{xx} \frac{\partial}{\partial x} \left(h \frac{\partial h}{\partial x} \right) + K_{yy} \frac{\partial}{\partial y} \left(h \frac{\partial h}{\partial y} \right) + N = S_y \frac{\partial h}{\partial t} \qquad \text{homogeneous, anisotropic}}$$

(6.74)

Noting the identities

$$h \frac{\partial h}{\partial x} = \frac{1}{2} \frac{\partial h^2}{\partial x}, \qquad h \frac{\partial h}{\partial y} = \frac{1}{2} \frac{\partial h^2}{\partial y} \qquad (6.75)$$

then Equation 6.74 can be written in the more compact form

$$K_{xx} \frac{\partial^2 h^2}{\partial x^2} + K_{yy} \frac{\partial^2 h^2}{\partial y^2} + 2N = 2S_y \frac{\partial h}{\partial t} \qquad (6.76)$$

A common assumption is that the hydraulic conductivity is isotropic in the horizontal plane, in which case

$$K_{xx} = K_{yy} = K \qquad (6.77)$$

and Equation 6.76 can be written as

$$K\frac{\partial^2 h^2}{\partial x^2} + K\frac{\partial^2 h^2}{\partial y^2} + 2N = 2S_y\frac{\partial h}{\partial t} \qquad \text{homogeneous, isotropic} \qquad (6.78)$$

or in the more compact vector form as

$$K\nabla^2(h^2) + 2N = 2S_y\frac{\partial h}{\partial t} \qquad \text{homogeneous, isotropic} \qquad (6.79)$$

where $\nabla^2()$ is the Laplacian operator. Equation 6.79, commonly called the *Boussinesq equation*, is an approximate governing equation for two-dimensional flow in unconfined aquifers, derived by invoking the Dupuit approximation. The Boussinesq equation (Equation 6.79) is a nonlinear partial differential equation in one unknown: the saturated thickness, h. The statement of this equation, along with the initial and boundary conditions for the saturated thickness, represents a complete statement of the problem of flow in unconfined aquifers. Solving this problem for the distribution of saturated thickness, h, also yields the distribution of specific discharge in the aquifer, since the specific discharge can be derived directly from the saturated thickness according to Equation 6.60.

There are many problems in ground-water engineering that relate to conditions induced by pumping (or recharging) wells. In these cases, the induced stress on the aquifer is symmetric around the well and it is therefore preferable to work with radial coordinates rather than Cartesian coordinates. The Laplacian operator in radial coordinates is given by

$$\nabla^2() = \frac{\partial^2()}{\partial r^2} + \frac{1}{r}\frac{\partial()}{\partial r} + \frac{1}{r^2}\frac{\partial^2()}{\partial\theta^2} \qquad (6.80)$$

Combining Equations 6.79 and 6.80, the ground-water flow equation in radial coordinates for an isotropic hydraulic conductivity is given by

$$\frac{\partial^2(h^2)}{\partial r^2} + \frac{1}{r}\frac{\partial(h^2)}{\partial r} + \frac{1}{r^2}\frac{\partial^2(h^2)}{\partial\theta^2} + \frac{2N}{K} = 2\frac{S_y}{K}\frac{\partial h}{\partial t} \qquad \text{homogeneous, isotropic}$$

$$(6.81)$$

EXAMPLE 6.9

Write an equation for the saturated thickness surrounding a single pumping well in an unconfined isotropic homogeneous aquifer. Assume that the influence of the well is radially symmetric. How would this equation be simplified for steady-state conditions without recharge?

Solution   In this case, it is appropriate to use radial coordinates, and the saturated thickness, h, can be described by Equation 6.81. Since the influence of the well is radially symmetric around the well, the origin of the radial coordinate system should be taken as the well location and, since h is independent of θ, $\partial^2(h^2)/\partial\theta^2 = 0$. The governing equation, Equation 6.81, then simplifies to

$$\frac{\partial^2(h^2)}{\partial r^2} + \frac{1}{r}\frac{\partial(h^2)}{\partial r} + \frac{2N}{K} = 2\frac{S_y}{K}\frac{\partial h}{\partial t}$$

Under steady-state conditions, $\partial h / \partial t = 0$, and in the absence of recharge $N = 0$. Hence, for the steady-state no-recharge case, the saturated thickness, h, is described by

$$\frac{\partial^2 (h^2)}{\partial r^2} + \frac{1}{r}\frac{\partial (h^2)}{\partial r} = 0 \qquad (6.82)$$

This equation is considerably simpler than the general equation for the saturated thickness given by Equation 6.81.

6.2.3.2   Confined aquifers

Consider the case of a confined aquifer of thickness b illustrated in Figure 6.11. Invoking the Dupuit approximation that the streamlines are horizontal and the seepage velocities are vertically uniform, the mass inflow rate to the control volume is given by

$$\text{Net mass inflow rate} = \left[\rho q_x - \frac{\partial(\rho q_x)}{\partial x}\frac{\Delta x}{2}\right] b\,\Delta y - \left[\rho q_x + \frac{\partial(\rho q_x)}{\partial x}\frac{\Delta x}{2}\right] b\,\Delta y$$

$$+ \left[\rho q_y - \frac{\partial(\rho q_y)}{\partial y}\frac{\Delta y}{2}\right] b\,\Delta x - \left[\rho q_y + \frac{\partial(\rho q_y)}{\partial y}\frac{\Delta y}{2}\right] b\,\Delta x$$

$$+ \rho N\,\Delta x\,\Delta y \qquad (6.83)$$

Combining terms in Equation 6.83 and simplifying yields

$$\text{Net mass inflow rate} = -\left[\frac{\partial(\rho q_x)}{\partial x} + \frac{\partial(\rho q_y)}{\partial y}\right] b\,\Delta x\,\Delta y + \rho N\,\Delta x\,\Delta y \qquad (6.84)$$

FIGURE 6.11: Control volume in a confined aquifer

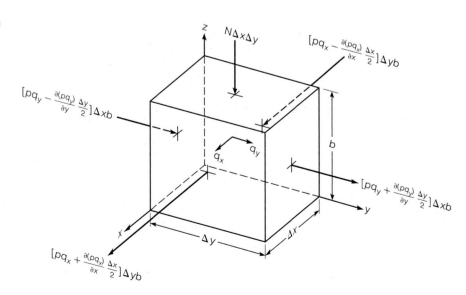

The net mass inflow rate to the control volume is equal to rate of change of mass within the control volume, which is given by

$$\text{Rate of change of mass} = S \, \Delta x \, \Delta y \rho \, \frac{\partial \phi}{\partial t} \tag{6.85}$$

where S is the *storage coefficient* of the confined aquifer, which is defined as the volume of water released from storage per unit surface area of the aquifer per unit change in piezometric head. The storage coefficient is also called the *storativity*, and typically has magnitudes in the range $10^{-6} < S < 10^{-2}$ (Şen, 1995), which are much smaller than the specific yields in phreatic aquifers. In fact, storage coefficients are typically 1000 to 10,000 times smaller than corresponding specific yields in unconfined aquifers (de Marsily, 1986). This indicates that, for the same volume of withdrawal or recharge, changes in piezometric surface elevations are much larger in a confined aquifer than in an unconfined aquifer of the same material. In cases where land subsidence is negligible, water released from storage due to declining piezometric heads in confined aquifers is associated with the net effect of the elastic expansion of the water and the elastic expansion of the solid matrix. Consequently, the storage coefficient can be expressed in terms of the elastic properties of the water and porous medium using the following equation (Bear, 1979)

$$S = \gamma n b \left(E_w + \frac{\alpha}{n} \right) \tag{6.86}$$

where γ is the specific weight of water, b is the thickness of the aquifer, E_w is the compressibility coefficient of water, and α is the compressibility coefficient of the solid matrix. Typically, 40% of the magnitude of S is contributed by the compressibility of water, and 60% by the compressibility of the solid matrix. Lohman (1979) has proposed the following relationship for estimating the storage coefficient:

$$S \approx 3 \times 10^{-6} b \tag{6.87}$$

where b is the aquifer thickness in meters. Combining Equations 6.84 and 6.85 leads to

$$-\left[\frac{\partial(\rho q_x)}{\partial x} + \frac{\partial(\rho q_y)}{\partial y} \right] + \frac{\rho N}{b} = \frac{S}{b} \rho \frac{\partial \phi}{\partial t} \tag{6.88}$$

For the common case in which the density, ρ, is independent of space and time, Equation 6.88 reduces to

$$-\left(\frac{\partial q_x}{\partial x} + \frac{\partial q_y}{\partial y} \right) + \frac{\rho N}{b} = \frac{S}{b} \frac{\partial \phi}{\partial t} \tag{6.89}$$

According to Darcy's law

$$q_x = -K_{xx} \frac{\partial \phi}{\partial x} \tag{6.90}$$

$$q_y = -K_{yy} \frac{\partial \phi}{\partial y} \tag{6.91}$$

and substituting these relations into Equation 6.89 leads to

$$\frac{\partial}{\partial x}\left(K_{xx}\frac{\partial \phi}{\partial x}\right) + \frac{\partial}{\partial y}\left(K_{yy}\frac{\partial \phi}{\partial y}\right) + \frac{N}{b} = \frac{S}{b}\frac{\partial \phi}{\partial t} \qquad (6.92)$$

This equation is applicable to anisotropic nonhomogeneous formations. In cases where the effective hydraulic conductivity of the aquifer is homogeneous, such as when the aquifer is composed of a single homogeneous layer, Equation 6.92 becomes

$$K_{xx}\frac{\partial^2 \phi}{\partial x^2} + K_{yy}\frac{\partial^2 \phi}{\partial y^2} + \frac{N}{b} = \frac{S}{b}\frac{\partial \phi}{\partial t} \qquad \text{homogeneous, anisotropic} \qquad (6.93)$$

A common assumption is that the hydraulic conductivity is isotropic in the horizontal plane, in which case

$$K_{xx} = K_{yy} = K \qquad (6.94)$$

and Equation 6.93 can be written as

$$K\frac{\partial^2 \phi}{\partial x^2} + K\frac{\partial^2 \phi}{\partial y^2} + \frac{N}{b} = \frac{S}{b}\frac{\partial \phi}{\partial t} \qquad \text{homogeneous, isotropic} \qquad (6.95)$$

or in the more compact form as

$$K\nabla^2 \phi + \frac{N}{b} = \frac{S}{b}\frac{\partial \phi}{\partial t} \qquad \text{homogeneous, isotropic} \qquad (6.96)$$

where $\nabla^2()$ is the two-dimensional Laplacian operator.

EXAMPLE 6.10

If the storage coefficient in a 20-m thick homogeneous isotropic confined aquifer can be estimated using Equation 6.87, determine the volume of water released per m^2 of aquifer when the piezometric head drops by 2 m. Where is this water "released" from? If the piezometric head distribution reaches a steady state without recharge, state the equation describing the head distribution.

Solution   From the given data: $b = 20$ m, and Equation 6.87 gives the storage coefficient as

$$S \approx 3 \times 10^{-6}b = 3 \times 10^{-6}(20) = 6 \times 10^{-5}$$

This gives the volume of water (in m^3) released per m^2 of aquifer per meter drop in the piezometric head. Hence, for a 2-m drop in piezometric head, the volume of water released per m^2 is $2 \times (6 \times 10^{-5}) = 1.2 \times 10^{-4}$ m^3 per m^2. This water is "released" by the volumetric expansion of water and the expansion of the solid matrix (= reduction in pore volume) caused by a decrease in pore pressure.

For a homogeneous isotropic confined aquifer, the piezometric head distribution is given by Equation 6.95 as

$$K\frac{\partial^2 \phi}{\partial x^2} + K\frac{\partial^2 \phi}{\partial y^2} + \frac{N}{b} = \frac{S}{b}\frac{\partial \phi}{\partial t}$$

Under steady-state conditions, $\partial\phi/\partial t = 0$, and without recharge, $N = 0$. In this case, Equation 6.95 becomes

$$K\frac{\partial^2\phi}{\partial x^2} + K\frac{\partial^2\phi}{\partial y^2} = 0$$

or

$$\frac{\partial^2\phi}{\partial x^2} + \frac{\partial^2\phi}{\partial y^2} = 0$$

which is considerably simpler than the general equation for confined homogeneous aquifers given by Equation 6.95.

Basic equations in terms of transmissivity. The transmissivity (T_{xx}, T_{yy}) of a (confined) aquifer is defined as the product of the hydraulic conductivity (K_{xx}, K_{yy}) and the aquifer thickness, b, where

$$T_{xx} = K_{xx}b, \qquad T_{yy} = K_{yy}b \tag{6.97}$$

Hence, the general flow equation applicable to anisotropic nonhomogeneous formations (Equation 6.92) can be written as

$$\frac{\partial}{\partial x}\left(T_{xx}\frac{\partial\phi}{\partial x}\right) + \frac{\partial}{\partial y}\left(T_{yy}\frac{\partial\phi}{\partial y}\right) + N = S\frac{\partial\phi}{\partial t} \tag{6.98}$$

For homogeneous anisotropic formations

$$T_{xx}\frac{\partial^2\phi}{\partial x^2} + T_{yy}\frac{\partial^2\phi}{\partial y^2} + N = S\frac{\partial\phi}{\partial t} \qquad \text{homogeneous, anisotropic} \tag{6.99}$$

and for homogeneous isotropic formations

$$T\frac{\partial^2\phi}{\partial x^2} + T\frac{\partial^2\phi}{\partial y^2} + N = S\frac{\partial\phi}{\partial t} \qquad \text{homogeneous, isotropic} \tag{6.100}$$

Many analyses of ground-water flow in confined aquifers are conducted using transmissivities rather than hydraulic conductivities, and in these cases Equations 6.98 to 6.100 are the appropriate governing equations.

EXAMPLE 6.11

A homogeneous anisotropic confined aquifer is 25 m thick and has principal hydraulic conductivities of $K_{xx} = 34$ m/d and $K_{yy} = 15$ m/d. Determine the principal transmissivities and state the differential equation describing the piezometric head distribution in the absence of recharge. Would the governing equation be any different if the aquifer were 50-m thick, $K_{xx} = 17$ m/d, and $K_{yy} = 7.5$ m/d?

Solution From the given data: $b = 25$ m, and the principal transmissivities are given by Equation 6.97 as

$$T_{xx} = K_{xx}b = (34)(25) = 850 \text{ m}^2/\text{d}$$

$$T_{yy} = K_{yy}b = (15)(25) = 375 \text{ m}^2/\text{d}$$

In the absence of recharge, $N = 0$, the differential equation describing the piezometric head distribution is derived from Equation 6.99 as

$$T_{xx}\frac{\partial^2\phi}{\partial x^2} + T_{yy}\frac{\partial^2\phi}{\partial y^2} = S\frac{\partial\phi}{\partial t}$$

or

$$850\frac{\partial^2\phi}{\partial x^2} + 375\frac{\partial^2\phi}{\partial y^2} = S\frac{\partial\phi}{\partial t}$$

If the aquifer were 50 m thick, $K_{xx} = 17$ m/d, and $K_{yy} = 7.5$ m/d, then $T_{xx} = (17)(50) = 850 \text{ m}^2/\text{d}$ and $T_{yy} = (7.5)(50) = 375 \text{ m}^2/\text{d}$. Therefore, the governing differential equation would be no different.

Equivalent anisotropic/isotropic media. Consider a porous medium that is homogeneous and anisotropic, with transmissivities (T_{xx}, T_{yy}), storage coefficient, S, and spatial coordinates (x, y). The spatial features of the anisotropic domain, described by the (x, y) coordinates, can be transformed into a new domain, described by (x', y') coordinates, where

$$x' = \sqrt{\frac{T}{T_{xx}}}\, x, \qquad y' = \sqrt{\frac{T}{T_{yy}}}\, y \tag{6.101}$$

and T is defined as an arbitrary coefficient with dimensions of a transmissivity $[L^2/T]$. Applying these relationships between (x, y) and (x', y') yields

$$\frac{\partial\phi}{\partial x'} = \frac{\partial\phi}{\partial x}\frac{dx}{dx'} = \sqrt{\frac{T_{xx}}{T}}\frac{\partial\phi}{\partial x} \tag{6.102}$$

$$\frac{\partial^2\phi}{\partial x'^2} = \frac{\partial}{\partial x}\left(\frac{\partial\phi}{\partial x'}\right)\frac{dx}{dx'} = \frac{T_{xx}}{T}\frac{\partial^2\phi}{\partial x^2} \tag{6.103}$$

$$\frac{\partial\phi}{\partial y'} = \frac{\partial\phi}{\partial y}\frac{dy}{dy'} = \sqrt{\frac{T_{yy}}{T}}\frac{\partial\phi}{\partial y} \tag{6.104}$$

$$\frac{\partial^2\phi}{\partial y'^2} = \frac{\partial}{\partial y}\left(\frac{\partial\phi}{\partial y'}\right)\frac{dy}{dy'} = \frac{T_{yy}}{T}\frac{\partial^2\phi}{\partial y^2} \tag{6.105}$$

Substituting Equations 6.102 to 6.105 into Equation 6.99 gives the following equation for the piezometric head distribution in the (x', y') domain:

$$T\frac{\partial^2\phi}{\partial x'^2} + T\frac{\partial^2\phi}{\partial y'^2} = S\frac{\partial\phi}{\partial t} \tag{6.106}$$

which indicates that the head distribution can be calculated by taking the transmissivity as homogeneous and isotropic with a value of T in the transformed domain. By further requiring that the calculated flowrate across any boundary in the transformed domain be equal to the flowrate across the corresponding boundary in the real (x, y) domain, de Marsily (1986) has shown that T should be taken as

$$T = \sqrt{T_{xx} T_{yy}} \tag{6.107}$$

This result is extremely useful in practical applications, since it means that solutions to the flow equation in isotropic homogeneous formations can be applied to any anisotropic homogeneous formation by simply specifying the transmissivity components (T_{xx}, T_{yy}) and transforming from (x', y') coordinates to (x, y) coordinates using Equation 6.101.

EXAMPLE 6.12

The analytic solution of the ground-water flow equation in an extensive isotropic homogeneous two-dimensional confined aquifer for the case of a single pumping well can be approximated by

$$\phi(x', y') = \phi_0 + \frac{Q_w}{2\pi T} \left[0.5772 + \ln \frac{(x'^2 + y'^2)S}{4Tt} \right] \tag{6.108}$$

where $\phi(x', y')$ is the head distribution, x' and y' are the spatial coordinates, ϕ_0 is the initial head distribution in the aquifer (prior to pumping), Q_w is the pumping rate from the well, T is the transmissivity, S is the storage coefficient, and t is the time since pumping began. What will be the piezometric head distribution if the aquifer is anisotropic with transmissivities T_{xx} and T_{yy}?

Solution   Application of the head distribution in an isotropic medium to an anisotropic medium requires a scaling of the spatial coordinates and the transmissivity in accordance with Equations 6.101 and 6.107. These equations yield

$$T = \sqrt{T_{xx} T_{yy}} \Rightarrow \frac{T}{T_{xx}} = \sqrt{\frac{T_{yy}}{T_{xx}}}, \qquad \frac{T}{T_{yy}} = \sqrt{\frac{T_{xx}}{T_{yy}}}$$

$$x'^2 = \frac{T}{T_{xx}} x^2 = \sqrt{\frac{T_{yy}}{T_{xx}}} x^2$$

$$y'^2 = \frac{T}{T_{yy}} y^2 = \sqrt{\frac{T_{xx}}{T_{yy}}} y^2$$

Substituting these relationships into Equation 6.108 gives the piezometric head distribution in an anisotropic homogeneous aquifer as

$$\phi(x, y) = \phi_0 + \frac{Q_w}{2\pi \sqrt{T_{xx} T_{yy}}} \left[0.5772 + \ln \frac{(\sqrt{T_{yy}/T_{xx}} x^2 + \sqrt{T_{xx}/T_{yy}} y^2)S}{4\sqrt{T_{xx} T_{yy}} t} \right]$$

The previous example illustrates the application of an analytic solution to the ground-water flow equation in an isotropic medium to an anisotropic medium. Implicit within this approach is that the initial and boundary conditions in both the isotropic and anisotropic domains are also related by the coordinate transformations. In the following section, several solutions to the ground-water flow equation in isotropic formations are presented for a variety of initial and boundary conditions. These results can be extended to anisotropic formations by using the approach described in this section.

6.3 Solutions of the Ground-Water Flow Equation

Most engineering applications of ground-water hydrology involve a particular solution of the ground-water flow equation, and these solutions can be either numerical or analytic. Numerical models typically solve the three-dimensional flow equation given by Equation 6.44 and can easily accommodate complex initial and boundary conditions along with complex hydraulic conductivity and specific storage distributions. A complete description of a widely used numerical model is given by McDonald and Harbaugh (1988) and Harbaugh and McDonald (1996). Analytic models typically solve the two-dimensional (Dupuit) approximations to the flow equations and are appropriate whenever the characteristics of the porous formation, as well as the boundary and initial conditions, are particularly simple. Common applications of analytic models are for matching observed aquifer responses with analytic solutions in order to determine aquifer properties. These applications are emphasized in the following sections.

6.3.1 Steady Unconfined Flow Between Two Reservoirs

Consider the case of unconfined flow between two reservoirs shown in Figure 6.12. The flow is from a reservoir with water depth h_L to a reservoir with water depth h_R, through a distance L of stratified phreatic aquifer with hydraulic conductivities K_1 and K_2 over thicknesses b_1 and b_2, respectively. Invoking the Dupuit approximation, the governing flow equation is given by Equation 6.69, which is rewritten here for convenient reference:

$$\frac{\partial}{\partial x}\left(\overline{K}_{xx}h\frac{\partial h}{\partial x}\right) + \frac{\partial}{\partial y}\left(\overline{K}_{yy}h\frac{\partial h}{\partial y}\right) + N = S_y\frac{\partial h}{\partial t} \qquad (6.109)$$

FIGURE 6.12: Unconfined flow between two reservoirs

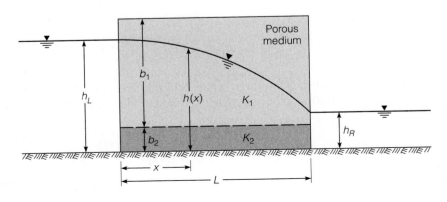

Since conditions are uniform in the y-direction (perpendicular to the page), conditions are at steady state, and there is no recharge, then the partial derivatives with respect to y and t are equal to zero, N is equal to zero, and Equation 6.109 reduces to

$$\frac{d}{dx}\left(\overline{K}_{xx}h\,\frac{dh}{dx}\right) = 0 \tag{6.110}$$

where the partial derivatives have been replaced by total derivatives, since the saturated-zone thickness, h, depends only on x. The effective hydraulic conductivity, $\overline{K}_{xx}$, derived from Equation 6.72, is given by

$$\overline{K}_{xx} = \frac{1}{h}\left[K_1(h - b_2) + K_2 b_2\right] \tag{6.111}$$

and substituting Equation 6.111 into Equation 6.110 yields

$$\frac{d}{dx}\left\{\left[K_1(h - b_2) + K_2 b_2\right]\frac{dh}{dx}\right\} = 0 \tag{6.112}$$

Integrating Equation 6.112 once leads to

$$\left[K_1(h - b_2) + K_2 b_2\right]\frac{dh}{dx} = C_1 \tag{6.113}$$

where C_1 is a constant. Separating terms in Equation 6.113 leads to

$$K_1 h\,\frac{dh}{dx} + (K_2 - K_1)b_2\,\frac{dh}{dx} = C_1 \tag{6.114}$$

which can also be written in the form

$$\frac{K_1}{2}\frac{d(h^2)}{dx} + (K_2 - K_1)b_2\,\frac{dh}{dx} = C_1 \tag{6.115}$$

which integrates directly to yield

$$\boxed{\frac{K_1}{2}h^2 + (K_2 - K_1)b_2 h = C_1 x + C_2} \tag{6.116}$$

The constants C_1 and C_2 can be determined by requiring $h(x)$ in Equation 6.116 to satisfy the boundary conditions: $h(0) = h_L$, and $h(L) = h_R$. Invoking these boundary conditions leads to

$$C_1 = \frac{K_1}{2L}\left(h_R^2 - h_L^2\right) + (K_2 - K_1)\frac{b_2}{L}\left(h_R - h_L\right) \tag{6.117}$$

$$C_2 = \frac{K_1}{2}h_L^2 + (K_2 - K_1)b_2 h_L \tag{6.118}$$

Hence the complete solution to the problem of unconfined flow between two reservoirs is given by Equation 6.116, where the constants C_1 and C_2 are given by Equations 6.117 and 6.118. Once the distribution of saturated-zone thickness, h, is calculated, then the flow between the two reservoirs, Q, can be calculated directly from Darcy's law and the Dupuit approximation using the relation

$$Q = -K_1(h - b_2)\frac{dh}{dx} - K_2 b_2\,\frac{dh}{dx} \tag{6.119}$$

where dh/dx is determined from the saturated-zone thickness, h, described by Equation 6.116. Combining Equations 6.116 and 6.119 yields the simple result

$$Q = -C_1 \qquad (6.120)$$

The approach used here can be extended to cases where there are more than two layers, and also to cases where a recharge, N, is present.

EXAMPLE 6.13

The Biscayne aquifer, one of the most permeable in the world, consists principally of two layers: the Miami Limestone Formation and the Fort Thomson Formation. In one particular area, the Miami Limestone Formation extends from ground surface at 2.44 m NGVD to -3.00 m NGVD, and the Fort Thomson Formation extends from -3.00 m NGVD to -15.24 m NGVD. The hydraulic conductivity of the Miami Limestone Formation can be taken as 1500 m/d, and the hydraulic conductivity of the Fort Thompson Formation as 12,000 m/d. Calculate the shape of the phreatic surface and the flowrate between two fully penetrating canals 1 km apart, when the water elevations in the two canals are 1.07 m NGVD and 1.00 m NGVD.

Solution A schematic diagram of the Biscayne aquifer between two fully penetrating canals is shown in Figure 6.13. The phreatic surface is described by Equations 6.116, 6.117, and 6.118, where $h_L = 1.07$ m $- (-15.24$ m$) = 16.31$ m, $h_R = 1.00$ m $- (-15.24$ m$) = 16.24$ m, $K_1 = 1500$ m/d, $K_2 = 12,000$ m/d, $b_2 = -3.00$ m $- (-15.24$ m$) = 12.24$ m, and $L = 1000$ m. Substituting these values into Equations 6.116, 6.117, and 6.118 yields

$$750h^2 + 128,520h = C_1 x + C_2 \qquad (6.121)$$

where

$$C_1 = -10.7 \text{ m}^2/\text{d}, \qquad C_2 = 2{,}296{,}000 \text{ m}^3/\text{d} \qquad (6.122)$$

Combining these results gives the following (implicit) equation for the phreatic surface:

$$h^2 + 171.4h = -0.01428x + 3061 \qquad (6.123)$$

The flowrate, Q, between the two reservoirs is given by Equation 6.120, where

$$Q = -C_1 = -(-10.7 \text{ m}^2/\text{d}) = 10.7 \text{ m}^2/\text{d}$$

FIGURE 6.13: Unconfined flow in the Biscayne aquifer

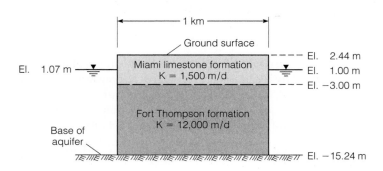

The case of flow between two fully penetrating water bodies can also be applied to the cases where the water bodies do not fully penetrate the aquifer. Bear (1979) indicates that beyond a distance on the order of two aquifer depths from a partially penetrating water body, the streamlines are usually very near to being horizontal and the effect of partial penetration is negligible. A practical application of this approximation is in estimating the leakage from partially penetrating open channels. A detailed investigation of this problem by Chin (1991) indicated that a distance of 10 aquifer depths from a partially penetrating open channel might be necessary in order to neglect the effect of partial penetration on leakage.

6.3.2   Steady Flow to a Well in a Confined Aquifer

In isotropic and homogeneous confined aquifers, the distribution of piezometric head for two-dimensional flows is given by Equation 6.96, which can be written in the form

$$\frac{\partial^2 \phi}{\partial r^2} + \frac{1}{r}\frac{\partial \phi}{\partial r} + \frac{1}{r^2}\frac{\partial^2 \phi}{\partial \theta^2} + \frac{N}{Kb} = \frac{S}{Kb}\frac{\partial \phi}{\partial t} \tag{6.124}$$

If a well fully penetrates the confined aquifer, and withdraws water at a constant rate uniformly over the aquifer thickness, b, as illustrated in Figure 6.14, then the piezometric head, ϕ, is radially symmetric (independent of θ) and independent of time. Consequently, all derivatives of $\phi(r)$ with respect to θ, and t are zero. If there is no recharge of the confined aquifer, then $N = 0$ and Equation 6.124 reduces to

$$\frac{d^2 \phi}{dr^2} + \frac{1}{r}\frac{d\phi}{dr} = 0 \tag{6.125}$$

where partial derivatives of ϕ are replaced by total derivatives, since ϕ is only a function of the radial distance from the well, r. Multiplying Equation 6.125 by r yields

$$r\frac{d^2 \phi}{dr^2} + \frac{d\phi}{dr} = 0$$

which can be written as

$$\frac{d}{dr}\left(r\frac{d\phi}{dr}\right) = 0 \tag{6.126}$$

FIGURE 6.14: Fully penetrating well in a confined aquifer

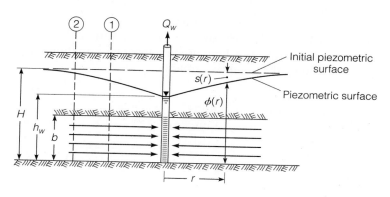

This equation is to be solved with the boundary conditions

$$2\pi r_w b K \frac{d\phi}{dr}\bigg|_{r=r_w} = Q_w \tag{6.127}$$

$$\phi(r_w) = h_w \tag{6.128}$$

where r_w is the radius of the well, K is the hydraulic conductivity of the aquifer, Q_w is the flowrate out of the well, and h_w is the depth of the water in the well, which is also equal to the piezometric head at the well ($r = r_w$). Integrating Equation 6.126 with respect to r yields

$$r\frac{d\phi}{dr} = A \tag{6.129}$$

and integrating again with respect to r yields

$$\phi = A \ln r + B \tag{6.130}$$

where A and B are constants to be determined from the boundary conditions, Equations 6.127 and 6.128. Applying the boundary conditions yields

$$A = \frac{Q_w}{2\pi b K}, \qquad B = h_w - A \ln r_w \tag{6.131}$$

Substituting Equation 6.131 into Equation 6.130 yields the following expression for the piezometric head distribution induced by a fully penetrating pumping well in a confined aquifer:

$$\phi(r) = h_w + \frac{Q_w}{2\pi K b} \ln\left(\frac{r}{r_w}\right) \tag{6.132}$$

Although this solution is exact for the flow equation and defined boundary conditions, there is an apparent paradox in that Equation 6.132 indicates that the piezometric head, ϕ, increases without bound as the distance from the well, r, increases. Clearly, this is not realistic, since the piezometric surface should be asymptotic to the initial piezometric surface before pumping—that is, the piezometric surface corresponding to $Q_w = 0$. This is illustrated in Figure 6.14. The reason that an exact solution to the steady-state flow equation does not give a realistic result is that a steady state for this flow condition is in fact impossible, since the water being pumped is constantly being drawn from storage and should result in a continuous decline and expansion of the piezometric surface. In spite of the limitation of the steady-state approximation, Equation 6.132 provides a reasonably accurate description of the piezometric surface, as long as $\phi < H$, where H is the elevation of the piezometric surface in the absence of pumping. This limitation in applying Equation 6.132 is frequently made explicit by defining the boundary condition

$$\phi(R) = H \tag{6.133}$$

in lieu of Equation 6.128, where R is commonly referred to as the *radius of influence* of the well. Using Equation 6.133 as a boundary condition, the head distribution is given by

$$\boxed{\phi(r) = H - \frac{Q_w}{2\pi K b} \ln\left(\frac{R}{r}\right)} \tag{6.134}$$

This equation, originally derived by Thiem (1906), is commonly known as the *Thiem equation*. From a theoretical viewpoint, R must necessarily be transient and cannot be taken as a constant; however, since $\phi(r)$ depends on the logarithm of R, then $\phi(r)$ is not very sensitive to errors in R. The most commonly used (empirical) equation for estimating R (in meters) was originally proposed by Sichard (1927) as

$$R = 3000 s_w \sqrt{K} \tag{6.135}$$

where $s_w = H - \phi(r_w)$ is the drawdown at the well in meters, and K is the hydraulic conductivity in meters per second. Other semiempirical and empirical equations used to estimate the radius of influence are (Bear, 1979)

$$R = (1.9 \text{ to } 2.45) \sqrt{\frac{HKt}{S}} \tag{6.136}$$

and

$$R = 575 s_w \sqrt{HK} \tag{6.137}$$

where H is in meters, and t is the time in seconds.

In many cases, it is convenient to work with the *drawdown*, $s(r)$, instead of the piezometric head, $\phi(r)$, where

$$s(r) = H - \phi(r) \tag{6.138}$$

Combining Equations 6.138 and 6.132 gives

$$s(r) = s_w - \frac{Q_w}{2\pi T} \ln\left(\frac{r}{r_w}\right), \qquad s(r) \geq 0 \tag{6.139}$$

where s_w is the drawdown at the well and T is the *transmissivity* of the confined aquifer defined by

$$T = Kb \tag{6.140}$$

Equation 6.139 describes the drawdown surface surrounding the well, commonly referred to as the *cone of depression*. The theoretical drawdown distribution given by Equation 6.139 is used in *aquifer pump tests* to determine the transmissivity of the aquifer by measuring the drawdowns resulting from pumpage, and then solving for the transmissivity. Applying either Equation 6.134 or Equation 6.139 to estimate the drawdowns at $r = r_1$ and $r = r_2$ and subtracting the result yields

$$s_1 - s_2 = \frac{Q_w}{2\pi T} \ln\left(\frac{r_2}{r_1}\right) \tag{6.141}$$

where s_1 and s_2 are the drawdowns at r_1 and r_2, respectively. Equation 6.141 is sometimes referred to as the Thiem equation (Hermance, 1999), and can be put in the more useful form

$$\boxed{T = \frac{Q_w}{2\pi(s_1 - s_2)} \ln\left(\frac{r_2}{r_1}\right)} \tag{6.142}$$

Hence, by measuring the pumping rate, Q_w, and drawdowns, s_1 and s_2, at two monitoring wells located at $r = r_1$ and $r = r_2$, the transmissivity of the aquifer, T, can be estimated directly using Equation 6.142. It is interesting to note that, in reality, even though the steady-state approximation inherent in Equation 6.142 is not strictly correct, the hydraulic gradient, $s_1 - s_2$, in the area surrounding the well approaches a pseudosteady state, and Equation 6.142 provides reasonably accurate estimates of the transmissivity.

EXAMPLE 6.14

An aquifer pump test was conducted in a confined aquifer where the initial piezometric surface was at elevation 14.385 m, and well logs indicate that the thickness of the aquifer is 25 m. The well was pumped at 31.54 L/s, and after one day the piezometric levels at 50 m and 100 m from the pumping well were measured as 13.585 m and 14.015 m, respectively. Assuming steady-state conditions, estimate the transmissivity and hydraulic conductivity of the aquifer.

Solution   The equation to be used to estimate the transmissivity is given by Equation 6.142 as

$$T = \frac{Q_w}{2\pi(s_1 - s_2)} \ln\left(\frac{r_2}{r_1}\right)$$

where $Q_w = 31.54$ L/s $= 2725$ m³/d, $r_1 = 50$ m, $r_2 = 100$ m, $s_1 = 14.385$ m $-$ 13.585 m $= 0.800$ m, and $s_2 = 14.385$ m $-$ 14.015 m $= 0.370$ m. Substituting these values yields

$$T = \frac{2725}{2\pi(0.800 - 0.370)} \ln\left(\frac{100}{50}\right) = 699 \text{ m}^2/\text{d}$$

The transmissivity, T, is related to the hydraulic conductivity, K, by

$$T = Kb$$

where b is the aquifer thickness. Since $b = 25$ m,

$$K = \frac{T}{b} = \frac{699}{25} = 28.0 \text{ m/d}$$

Therefore, the results of the aquifer pump test indicate that the transmissivity of the aquifer is 699 m²/d, and the hydraulic conductivity is 28.0 m/d.

As a reference point, aquifers with transmissivity values less than 10 m²/d can supply only enough water for domestic wells and other low-yield uses (Lehr et al., 1988).

6.3.3   Steady Flow to a Well in an Unconfined Aquifer

In isotropic and homogeneous unconfined aquifers, the distribution of saturated-zone thickness, h, can be estimated by assuming two-dimensional flow (Dupuit approximation) and describing the distribution of saturated-zone thickness by Equation 6.81. The case of a fully penetrating pumping well in an unconfined aquifer is illustrated in Figure 6.15. Since the aquifer response to pumping is radially symmetric around the well, derivatives with respect to θ are equal to zero and Equation 6.81 can be written as

$$\frac{d^2(h^2)}{dr^2} + \frac{1}{r}\frac{d(h^2)}{dr} = 0 \tag{6.143}$$

FIGURE 6.15: Fully penetrating well in an unconfined aquifer

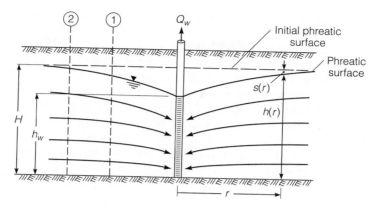

where partial derivatives have been replaced by total derivatives (since h is a function of r only), the recharge, N, has been set equal to zero, and the derivative of h with respect to time has been set equal to zero (since steady state is assumed). Equation 6.143 can be written as

$$\frac{d}{dr}\left(r\,\frac{dh^2}{dr}\right) = 0 \qquad (6.144)$$

This equation is to be solved with the boundary conditions

$$2\pi r_w h\, K\frac{dh}{dr}\bigg|_{r=r_w} = Q_w \qquad (6.145)$$

$$h(r_w) = h_w \qquad (6.146)$$

where r_w is the radius of the well, K is the hydraulic conductivity of the aquifer, Q_w is the pumping rate, and h_w is the depth of the water in the well. Integrating Equation 6.144 twice with respect to r yields

$$h^2 = A\ln r + B \qquad (6.147)$$

where A and B are constants to be determined using the boundary conditions. Applying the boundary conditions yields

$$A = \frac{Q_w}{\pi K}, \qquad B = h_w^2 - A\ln r_w \qquad (6.148)$$

Substituting Equation 6.148 into Equation 6.147 yields the following expression for the saturated thickness in an unconfined aquifer surrounding a pumping well:

$$h^2 = h_w^2 + \frac{Q_w}{\pi K}\ln\left(\frac{r}{r_w}\right) \qquad (6.149)$$

This equation is sometimes referred to as the *Dupuit equation* for radial flow (Charbeneau, 2000), and it plays the same role for unconfined aquifers that the Thiem equation plays for confined aquifers. As was the case for confined aquifers, Equation 6.149 is apparently exact for the assumed flow equation and defined boundary conditions. However, there is an apparent paradox in that Equation 6.149 indicates that the saturated thickness, h, increases without bound as the distance from the well, r, increases. Clearly, this is not realistic, since the water table should be asymptotic to the water table prior to pumping—that is, when the pumping rate, Q_w, is equal to zero. The reason that an exact solution to the steady-state flow equation does not give a realistic result is that steady state for this flow condition is impossible. As in the case of a confined aquifer, the pumpage must continuously be drawn from aquifer storage, and therefore there can be no steady state. In spite of this paradox, Equation 6.149 provides a reasonably accurate description of the phreatic surface, as long as $h < H$, where H is the saturated thickness in the absence of pumping. The *drawdown*, s, is defined by

$$s(r) = H - h(r) \tag{6.150}$$

in which case Equation 6.149 can be written as

$$(H - s)^2 = (H - s_w)^2 + \frac{Q_w}{\pi K} \ln\left(\frac{r}{r_w}\right) \tag{6.151}$$

where s_w is the drawdown at the well. The drawdown distribution given by Equation 6.151 is commonly used in aquifer pump tests to determine the hydraulic conductivity of the aquifer by measuring the drawdowns resulting from pumping water out of the well, and then solving for the hydraulic conductivity. Applying Equation 6.151 to estimate the drawdowns at $r = r_1$ and $r = r_2$ and subtracting the result yields

$$(H - s_2)^2 - (H - s_1)^2 = \frac{Q_w}{\pi K} \ln\left(\frac{r_2}{r_1}\right)$$

which simplifies to

$$\left(s_1 - \frac{s_1^2}{2H}\right) - \left(s_2 - \frac{s_2^2}{2H}\right) = \frac{Q_w}{2\pi KH} \ln\left(\frac{r_2}{r_1}\right) \tag{6.152}$$

Defining the *modified drawdowns*, s_1' and s_2' by the relations

$$s_1' = s_1 - \frac{s_1^2}{2H}, \qquad s_2' = s_2 - \frac{s_2^2}{2H} \tag{6.153}$$

then Equation 6.152 can be written as

$$s_1' - s_2' = \frac{Q_w}{2\pi T} \ln\left(\frac{r_2}{r_1}\right) \tag{6.154}$$

where T is the transmissivity of the unconfined aquifer defined by

$$T = KH \tag{6.155}$$

Making T the subject of the formula in Equation 6.154 yields the useful equation

$$\boxed{T = \frac{Q_w}{2\pi(s_1' - s_2')} \ln\left(\frac{r_2}{r_1}\right)} \tag{6.156}$$

which is exactly the same result that was derived for a confined aquifer (see Equation 6.142), except that modified drawdowns are used. It should be noted, however, that in the common case where the drawdowns are small compared with the aquifer saturated thickness, or specifically

$$s_1^2 \ll 2H \quad \text{and} \quad s_2^2 \ll 2H$$

the modified drawdowns, s_1' and s_2', can be replaced by the actual drawdowns, s_1 and s_2. On the basis of Equation 6.156, the transmissivity of the aquifer can be estimated from measurements of the pumping rate, Q_w, and drawdowns, s_1 and s_2, at two monitoring wells located at $r = r_1$ and $r = r_2$. The hydraulic conductivity can then be estimated using Equation 6.155.

EXAMPLE 6.15

An aquifer pump test has been conducted in an unconfined aquifer of saturated thickness 15 m. The well was pumped at a rate of 100 L/s, and the drawdowns at 50 m and 100 m from the pumping well after one day of pumping were 0.412 m and 0.251 m, respectively. Estimate the transmissivity and hydraulic conductivity of the aquifer. Assume steady state.

Solution The steady-state estimate of the transmissivity, T, is given by Equation 6.156 as

$$T = \frac{Q_w}{2\pi(s_1' - s_2')} \ln\left(\frac{r_2}{r_1}\right)$$

where $Q_w = 100$ L/s $= 8640$ m^3/d, $r_1 = 50$ m, $r_2 = 100$ m, $s_1 = 0.412$ m, $s_2 = 0.251$ m, $H = 15$ m, and s_1' and s_2' are given by Equation 6.153 as

$$s_1' = s_1 - \frac{s_1^2}{2H} = 0.412 - \frac{0.412^2}{2(15)} = 0.406 \text{ m}$$

and

$$s_2' = s_2 - \frac{s_2^2}{2H} = 0.251 - \frac{0.251^2}{2(15)} = 0.249 \text{ m}$$

The transmissivity is therefore estimated as

$$T = \frac{8640}{2\pi(0.406 - 0.249)} \ln\left(\frac{100}{50}\right) = 6070 \text{ m}^2/\text{d}$$

The hydraulic conductivity, K, is given by

$$K = \frac{T}{H} = \frac{6070}{15} = 405 \text{ m/d}$$

Therefore, the transmissivity and hydraulic conductivity of the aquifer are estimated from the pump-test data to be 6070 m^2/d and 405 m/d, respectively.

6.3.4   Steady Flow to a Well in a Leaky Confined Aquifer

In isotropic and homogeneous semiconfined (leaky) aquifers, the distribution of piezometric head for two-dimensional flows is given by Equation 6.96, which can be written in the form

$$\frac{\partial^2 \phi}{\partial r^2} + \frac{1}{r}\frac{\partial \phi}{\partial r} + \frac{1}{r^2}\frac{\partial^2 \phi}{\partial \theta^2} + \frac{N}{Kb} = \frac{S}{Kb}\frac{\partial \phi}{\partial t} \tag{6.157}$$

If a well fully penetrates the semiconfined aquifer and withdraws water at a constant rate uniformly over the aquifer thickness, b, as illustrated in Figure 6.16, then the piezometric head, ϕ, is radially symmetric (independent of θ) and, under steady-state conditions, is independent of time. Therefore, all derivatives with respect to θ and t are zero, and Equation 6.157 can be written in the form

$$\frac{d^2 \phi}{dr^2} + \frac{1}{r}\frac{d\phi}{dr} + \frac{N}{Kb} = 0 \tag{6.158}$$

where partial derivatives have replaced total derivatives, since ϕ depends only on r. Leakage into the semiconfined aquifer can be calculated by applying Darcy's law across the semipermeable layer, in which case

$$N = -K'\frac{\phi - \phi_0}{b'} \tag{6.159}$$

where K' is the hydraulic conductivity of the semipermeable layer, b' is the thickness of the semipermeable layer, and ϕ_0 is the piezometric head above the semipermeable layer, assumed to be constant. Combining Equations 6.158 and 6.159 leads to

$$\frac{d^2 \phi}{dr^2} + \frac{1}{r}\frac{d\phi}{dr} + \frac{K'(\phi_0 - \phi)}{Kbb'} = 0 \tag{6.160}$$

Defining the parameter λ by the expression

$$\boxed{\lambda^2 = \frac{Kbb'}{K'}} \tag{6.161}$$

FIGURE 6.16: Fully penetrating well in a semiconfined aquifer

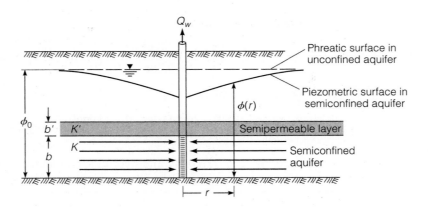

then Equation 6.160 can be written as

$$\frac{d^2\phi}{dr^2} + \frac{1}{r}\frac{d\phi}{dr} + \frac{\phi_0 - \phi}{\lambda^2} = 0 \tag{6.162}$$

The quantity λ defined by Equation 6.161 is commonly used to parameterize the leakage in semiconfined aquifers, and is called the *leakage factor*. Defining new variables ϕ' and r' by

$$\phi' = \phi_0 - \phi \tag{6.163}$$

$$r' = \frac{r}{\lambda} \tag{6.164}$$

then Equation 6.162 can be written in the form

$$r'^2 \frac{d^2\phi'}{dr'^2} + r'\frac{d\phi'}{dr'} - r'^2\phi' = 0 \tag{6.165}$$

This equation is a modified Bessel's equation of order zero, discussed in Appendix D.2, and has the solution

$$\phi' = AI_0(r') + BK_0(r') \tag{6.166}$$

or

$$\phi_0 - \phi(r) = AI_0(r/\lambda) + BK_0(r/\lambda) \tag{6.167}$$

where $I_0(x)$ and $K_0(x)$ are modified Bessel functions of the first and second kind (respectively) of order zero, and A and B are constants to be determined from the boundary conditions. In this case, the boundary conditions are

$$\phi(\infty) = \phi_0 \tag{6.168}$$

and

$$2\pi r_w bK \frac{d\phi}{dr}\bigg|_{r=r_w} = Q_w \tag{6.169}$$

Noting the following Bessel function properties

$$I_0(\infty) = \infty \tag{6.170}$$

and

$$K_0(\infty) = 0 \tag{6.171}$$

then the boundary condition given by Equation 6.168 requires that $A = 0$ in Equation 6.167, which now becomes

$$\phi_0 - \phi(r) = BK_0(r/\lambda) \tag{6.172}$$

Application of the second boundary condition, Equation 6.169, requires using the Bessel function identity

$$\frac{d}{dr'}K_0(r') = -K_1(r') \tag{6.173}$$

where $K_1(r')$ is the modified Bessel function of the second kind of order one. Differentiating Equation 6.172 and applying the boundary condition given by Equation 6.169

leads to the following expression for B:

$$B = \frac{Q_w}{2\pi T(r_w/\lambda)K_1(r_w/\lambda)} \tag{6.174}$$

where $T(= Kb)$ is the transmissivity of the semiconfined aquifer. Substituting Equation 6.174 into Equation 6.172 yields the following expression for the piezometric head distribution:

$$\phi_0 - \phi(r) = \frac{Q_w}{2\pi T} \frac{K_0(r/\lambda)}{(r_w/\lambda)K_1(r_w/\lambda)} \tag{6.175}$$

Defining the drawdown, $s(r)$, by

$$s(r) = \phi_0 - \phi(r) \tag{6.176}$$

then Equations 6.175 and 6.176 yield the following expression for the drawdown:

$$s(r) = \frac{Q_w}{2\pi T} \frac{K_0(r/\lambda)}{(r_w/\lambda)K_1(r_w/\lambda)} \tag{6.177}$$

In this case, there are two aquifer properties: transmissivity, T, and leakage factor, λ. These aquifer properties can be estimated using Equation 6.177 with drawdown measurements at two locations. Denoting the drawdown measurements by s_1 and s_2, at r_1 and r_2, respectively, then Equation 6.177 yields

$$\boxed{\frac{K_0(r_1/\lambda)}{K_0(r_2/\lambda)} = \frac{s_1}{s_2}} \tag{6.178}$$

and

$$\boxed{T = \frac{Q_w}{2\pi s_1} \frac{K_0(r_1/\lambda)}{(r_w/\lambda)K_1(r_w/\lambda)}} \tag{6.179}$$

Equation 6.178 can be used to estimate the leakage factor from the drawdown measurements. This leakage factor can be used in Equation 6.179 to calculate the transmissivity of the semiconfined aquifer.

EXAMPLE 6.16

An aquifer pump test is conducted in a leaky confined aquifer in which the semiconfining layer is estimated to have a thickness of 2 m, and the aquifer has a thickness of 20 m. The pumping rate is 50 L/s from a well of radius 0.5 m, and the steady-state drawdowns at 50 m and 100 m from the well are 0.3 m and 0.1 m, respectively. Estimate the transmissivity, hydraulic conductivity, and leakage factor of the aquifer, and the hydraulic conductivity of the semiconfining layer.

Solution   The leakage factor, λ, can be estimated using Equation 6.178, where

$$\frac{K_0(r_1/\lambda)}{K_0(r_2/\lambda)} = \frac{s_1}{s_2}$$

Substituting $r_1 = 50$ m, $r_2 = 100$ m, $s_1 = 0.3$ m, and $s_2 = 0.1$ m leads to

$$\frac{K_0(50/\lambda)}{K_0(100/\lambda)} = \frac{0.3}{0.1} = 3$$

This equation is an implicit equation for λ which can be easily solved by trial and error using the (modified) Bessel functions built into Excel™ spreadsheets and tabulated in Appendix D.2. The result is

$$\lambda = 62.7 \text{ m}$$

The transmissivity of the semiconfined aquifer can be estimated using Equation 6.179, where

$$T = \frac{Q_w}{2\pi s_1} \frac{K_0(r_1/\lambda)}{(r_w/\lambda)K_1(r_w/\lambda)}$$

Substituting $Q_w = 50$ L/s $= 4320$ m³/d, and $r_w = 0.5$ m leads to

$$T = \frac{4320}{2\pi(0.3)} \frac{K_0(50/62.7)}{(0.5/62.7)K_1(0.5/62.7)} = 1300 \text{ m}^2/\text{d}$$

Since the thickness of the aquifer is 20 m, the hydraulic conductivity, K, is

$$K = \frac{1300}{20} = 65 \text{ m/d}$$

Recall the definition of the leakage factor, where

$$\lambda^2 = \frac{Kbb'}{K'}$$

where b is the thickness of the aquifer, b' is the thickness of the semiconfining layer, and K' is the hydraulic conductivity of the semiconfining layer. In the present case, $\lambda = 62.7$ m, $K = 65$ m/d, and $b' = 2$ m, therefore

$$K' = \frac{Kbb'}{\lambda^2} = \frac{(65)(20)(2)}{(62.7)^2} = 0.66 \text{ m/d}$$

An interesting approximation to the drawdown equation (Equation 6.177) is obtained when $r_w/\lambda \ll 1$, in which case $(r_w/\lambda)K_1(r_w/\lambda) \approx 1$ and Equation 6.177 can be approximated by

$$s(r) = \frac{Q_w}{2\pi T} K_0(r/\lambda) \tag{6.180}$$

which indicates that the drawdown distribution is independent of the radius of the well, r_w. According to Bear (1979), Equation 6.180 is accurate to within 1% of Equation 6.177 whenever $r_w/\lambda < 0.02$.

EXAMPLE 6.17

Repeat the previous problem using the approximation given by Equation 6.180.

Solution   Equation 6.180 gives the following relation for estimating the leakage factor, λ:

$$\frac{K_0(r_1/\lambda)}{K_0(r_2/\lambda)} = \frac{s_1}{s_2}$$

where $r_1 = 50$ m, $r_2 = 100$ m, $s_1 = 0.3$ m, and $s_2 = 0.1$ m. Hence

$$\frac{K_0(50/\lambda)}{K_0(100/\lambda)} = \frac{0.3}{0.1} = 3$$

This is the same expression for λ used in the previous example, and the result is given by

$$\lambda = 62.7 \text{ m}$$

This value of λ indicates that

$$\frac{r_w}{\lambda} = \frac{0.5}{62.7} = 0.008 \ll 0.02$$

and therefore using the approximation given by Equation 6.180 is justified. Equation 6.180 gives the transmissivity, T, as

$$T = \frac{Q_w}{2\pi s_1} K_0(r_1/\lambda)$$

where $Q_w = 50 \text{ L/s} = 4320 \text{ m}^3/\text{d}$, and therefore

$$T = \frac{4320}{2\pi(0.3)} K_0(50/62.7) = 1300 \text{ m}^2/\text{d}$$

This transmissivity is the same as was obtained in the previous example. Calculations of both the hydraulic conductivity of the aquifer and semiconfining layer are the same as in the previous example, yielding

$$K = 65 \text{ m/d} \quad \text{and} \quad K' = 0.66 \text{ m/d}$$

The results of this example support the assertion that Equation 6.180 is an adequate approximation to Equation 6.177 when $r_w/\lambda < 0.02$.

The derived leakage factor can be used to classify the "leakiness" of the semiconfining layer using the classification system in Table 6.6. Aquifers with leakage factors exceeding 10,000 m can be treated as confined aquifers.

The equations derived in this section assume that the flow is horizontal in the aquifer and vertical in the semiconfining (leaky) layer. Errors introduced by these assumptions are generally less than 5%, provided that the hydraulic conductivity of the aquifer is more than two orders of magnitude greater than that of the semiconfining layer (Neuman and Witherspoon, 1969a).

6.3.5 Steady Flow to a Well in an Unconfined Aquifer with Recharge

In isotropic and homogeneous unconfined aquifers, the distribution of saturated thickness can be estimated by assuming two-dimensional flow (Dupuit approximation).

TABLE 6.6: Classification of Leakage

Leakage factor, λ (m)	Condition
< 1000	High leakage
1000–5000	Moderate leakage
5000–10,000	Low leakage
> 10,000	Negligible leakage

Source: Şen (1995).

Under these circumstances, the flow is described by Equation 6.81, which is given by

$$\frac{\partial^2(h^2)}{\partial r^2} + \frac{1}{r}\frac{\partial(h^2)}{\partial r} + \frac{1}{r^2}\frac{\partial^2(h^2)}{\partial \theta^2} + \frac{2N}{K} = 2\frac{S_y}{K}\frac{\partial h}{\partial t} \tag{6.181}$$

Since the solution is radially symmetric and steady state, h is only a function of r, and Equation 6.181 becomes

$$\frac{d^2(h^2)}{dr^2} + \frac{1}{r}\frac{d(h^2)}{dr} + \frac{2N}{K} = 0 \tag{6.182}$$

where the partial derivatives have been replaced by total derivatives. Equation 6.182 can also be written in the more compact form

$$\frac{d}{dr}\left(r\frac{dh^2}{dr}\right) + r\frac{2N}{K} = 0 \tag{6.183}$$

This equation is to be solved using the boundary conditions

$$2\pi r_w hK \left.\frac{dh}{dr}\right|_{r=r_w} = Q_w \tag{6.184}$$

$$h(r_w) = h_w \tag{6.185}$$

where r_w is the radius of the well, and Q_w is the rate at which water is being pumped from the well. Since the governing differential equation is in terms of h^2, it is convenient to write the boundary conditions in terms of h^2 as

$$\pi r_w K \left.\frac{dh^2}{dr}\right|_{r=r_w} = Q_w \tag{6.186}$$

and

$$h^2(r_w) = h_w^2 \tag{6.187}$$

Integrating Equation 6.183 with respect to r yields

$$\frac{dh^2}{dr} = -\frac{Nr}{K} + \frac{A}{r} \tag{6.188}$$

and integrating again with respect to r yields

$$h^2 = -\frac{Nr^2}{2K} + A \ln r + B \tag{6.189}$$

where A and B are integration constants. Applying the boundary conditions yields the following expression for the distribution of aquifer thickness:

$$\boxed{h^2 = h_w^2 + \frac{N}{2K}(r_w^2 - r^2) + A \ln\left(\frac{r}{r_w}\right)} \tag{6.190}$$

where

$$A = \frac{Q_w}{\pi K} + \frac{Nr_w^2}{K} \tag{6.191}$$

EXAMPLE 6.18

Ground water is pumped from a well at the rate of 20 L/s, and the well is located in an unconfined aquifer with an average surficial recharge of 0.5 m/d. If the radius of the well is 0.5 m, the hydraulic conductivity of the aquifer is 20 m/d, and the saturated thickness 100 m from the well is 25 m, estimate the saturated thickness at the well. Determine the additional drawdown at the well in the absence of surficial recharge.

Solution From the given data: $Q_w = 20$ L/s $= 1728$ m³/d; $N = 0.5$ m/d; $r_w = 0.5$ m; $K = 20$ m/d; and at $r = 100$ m, $h = 25$ m. Equation 6.190 gives the saturated thickness at the well, h_w, as

$$h_w = \sqrt{h^2 - \frac{N}{2K}(r_w^2 - r^2) - A\ln\left(\frac{r}{r_w}\right)}$$

$$= \sqrt{25^2 - \frac{0.5}{2(20)}(0.5^2 - 100^2) - A\ln\left(\frac{100}{0.5}\right)} = \sqrt{750 - 5.30A}$$

The constant A is given by Equation 6.191 as

$$A = \frac{Q_w}{\pi K} + \frac{Nr_w^2}{K} = \frac{1728}{\pi(20)} + \frac{(0.5)(0.5)^2}{20} = 27.5 \text{ m}^2$$

Hence, the saturated thickness, h_w, at the well is given by

$$h_w = \sqrt{750 - 5.30(27.5)} = 24.6 \text{ m}$$

In the absence of surficial recharge, $N = 0$ m/d, the saturated thickness at the well, h_w, is given by Equation 6.190 as

$$h'_w = \sqrt{h^2 - A\ln\left(\frac{r}{r_w}\right)} = \sqrt{25^2 - A\ln\left(\frac{100}{0.5}\right)} = \sqrt{625 - 5.30A}$$

where

$$A = \frac{Q_w}{\pi K} = \frac{1728}{\pi(20)} = 27.5 \text{ m}^2$$

Therefore,

$$h'_w = \sqrt{625 - 5.30(27.5)} = 21.9 \text{ m}$$

and the additional drawdown, Δs, in the absence of recharge is given by

$$\Delta s = h_w - h'_w = 24.6 - 21.9 = 2.7 \text{ m}$$

6.3.6   Unsteady Flow to a Well in a Confined Aquifer

In isotropic and homogeneous confined aquifers, the distribution of piezometric head for two-dimensional flows is given by Equation 6.96, where

$$\frac{\partial^2 \phi}{\partial r^2} + \frac{1}{r}\frac{\partial \phi}{\partial r} + \frac{1}{r^2}\frac{\partial^2 \phi}{\partial \theta^2} + \frac{N}{Kb} = \frac{S}{Kb}\frac{\partial \phi}{\partial t} \tag{6.192}$$

If a well fully penetrates the confined aquifer, and withdraws water uniformly over the aquifer thickness, b, then the piezometric head, ϕ, is radially symmetric and therefore independent of θ. If the leakage, N, through the confining layers is equal to zero, then Equation 6.192 becomes

$$\frac{\partial^2 \phi}{\partial r^2} + \frac{1}{r}\frac{\partial \phi}{\partial r} = \frac{S}{T}\frac{\partial \phi}{\partial t} \tag{6.193}$$

where $T(= Kb)$ is the transmissivity of the aquifer. Defining the drawdown in the aquifer, $s(r, t)$, by the relation

$$s(r, t) = \phi_0 - \phi(r, t) \tag{6.194}$$

where ϕ_0 is the piezometric head in the confined aquifer prior to pumping, then Equation 6.193 can be written in terms of drawdown as

$$\frac{\partial^2 s}{\partial r^2} + \frac{1}{r}\frac{\partial s}{\partial r} = \frac{S}{T}\frac{\partial s}{\partial t} \tag{6.195}$$

This equation is to be solved subject to the following initial and boundary conditions:

$$s(r, 0) = 0 \tag{6.196}$$

$$s(\infty, t) = 0 \tag{6.197}$$

$$\lim_{r \to 0} r\frac{\partial s}{\partial r} = -\frac{Q_w}{2\pi T} \tag{6.198}$$

where Q_w is the pumping rate out of the well. Solution of this problem is facilitated by changing variables from (r, t) to (r, u), where

$$u = \frac{r^2 S}{4Tt} \tag{6.199}$$

and u is sometimes called the *Boltzman variable*. Using the chain rule on Equation 6.199, the following relationships can be derived:

$$\frac{\partial s}{\partial r} = \frac{2u}{r}\frac{\partial s}{\partial u} \tag{6.200}$$

$$\frac{\partial^2 s}{\partial r^2} = \frac{2u}{r}\left[\frac{2u}{r}\frac{\partial^2 s}{\partial u^2} + \frac{1}{r}\frac{\partial s}{\partial u}\right] \tag{6.201}$$

$$\frac{\partial s}{\partial t} = -\frac{u}{t}\frac{\partial s}{\partial u} \tag{6.202}$$

Substituting Equations 6.200 to 6.202 into Equation 6.195 yields

$$\frac{\partial^2 s}{\partial u^2} + \left(1 + \frac{1}{u}\right)\frac{\partial s}{\partial u} = 0 \tag{6.203}$$

The striking result here is that the variable t has dropped out, and the solution, s, of Equation 6.203 depends only on the Boltzman variable, u. Consequently, the partial derivatives can be replaced by total derivatives, and Equation 6.203 can be written as

$$\frac{d^2s}{du^2} + \left(1 + \frac{1}{u}\right)\frac{ds}{du} = 0 \tag{6.204}$$

The boundary conditions, with respect to u, can be derived from Equations 6.196 to 6.198, which yield

$$s(\infty) = 0 \tag{6.205}$$

$$\lim_{u \to 0} u \frac{ds}{du} = -\frac{Q_w}{4\pi T} \tag{6.206}$$

Hence, the problem of unsteady flow to a well in a confined aquifer is defined by Equation 6.204, subject to the boundary conditions given by Equations 6.205 and 6.206. To solve Equation 6.204, let

$$p = \frac{ds}{du} \tag{6.207}$$

then Equation 6.204 can be written as

$$\frac{dp}{du} + \left(1 + \frac{1}{u}\right)p = 0 \tag{6.208}$$

which can be written as

$$\frac{dp}{p} = -\left(1 + \frac{1}{u}\right)du \tag{6.209}$$

and integrating this equation gives

$$\ln p = -u - \ln u + A' \tag{6.210}$$

where A' is a constant to be determined from the boundary conditions. Rearranging Equation 6.210 leads to

$$up = e^{A'-u}$$

$$= Ae^{-u} \tag{6.211}$$

where A is a constant ($= e^{A'}$). Substituting the definition of p from Equation 6.207 leads to

$$u\frac{ds}{du} = Ae^{-u} \tag{6.212}$$

Applying the boundary condition given by Equation 6.206 to find A leads to

$$u\frac{ds}{du} = -\frac{Q_w}{4\pi T}e^{-u} \tag{6.213}$$

which can be put in the form

$$ds = -\frac{Q_w}{4\pi T}\frac{e^{-u}}{u}du \tag{6.214}$$

or

$$s(u) = -\frac{Q_w}{4\pi T} \int_a^u \frac{e^{-x}}{x} dx + B \tag{6.215}$$

where a and B are constants. Applying the boundary condition given in Equation 6.205 leads to

$$B = \frac{Q_w}{4\pi T} \int_a^\infty \frac{e^{-x}}{x} dx \tag{6.216}$$

Substituting Equation 6.216 into Equation 6.215 yields

$$\begin{aligned} s(u) &= \frac{Q_w}{4\pi T} \left[\int_a^\infty \frac{e^{-x}}{x} dx - \int_a^u \frac{e^{-x}}{x} dx \right] \\ &= \frac{Q_w}{4\pi T} \int_u^\infty \frac{e^{-x}}{x} dx \end{aligned} \tag{6.217}$$

This equation is sufficient to describe the transient drawdown in response to a fully penetrating well in a confined aquifer. The integral in Equation 6.217 is commonly referred to as the *well function*, $W(u)$, where

$$\boxed{W(u) = \int_u^\infty \frac{e^{-x}}{x} dx} \tag{6.218}$$

and Equation 6.217 can be written as

$$\boxed{s(u) = \frac{Q_w}{4\pi T} W(u)} \tag{6.219}$$

This equation is widely used in ground-water engineering and is called the *Theis equation*, after Theis (1935) who originally derived it. The well function is shown for several values of u in Table 6.7. The well function, $W(u)$, is closely related to the *exponential integral*, $\mathrm{Ei}(x)$, that is commonly used in mathematics,[*] and $W(u)$ can be expressed in terms of an infinite series as

$$W(u) = -\gamma - \ln u + u - \frac{u^2}{2 \cdot 2!} + \frac{u^3}{3 \cdot 3!} - \frac{u^4}{4 \cdot 4!} + \cdots \tag{6.220}$$

where $\gamma = 0.5772157\ldots$ is Euler's constant. The utility of this series expression is that for small values of u, the higher-order terms can be neglected, yielding an analytic expression for $W(u)$. In fact, for values of u less than or equal to 0.004, the first two terms in Equation 6.220 give values of $W(u)$ that are accurate to within 0.1%, and therefore the following approximation is appropriate:

$$\boxed{W(u) = -0.5772 - \ln u, \quad u \le 0.004} \tag{6.221}$$

This approximation was first applied to the analysis of ground-water flow by Cooper and Jacob (1946) and is commonly referred to as the *Cooper–Jacob approximation*. For $u < 0.03$, the error in Equation 6.221 is less than 1%, and for $u = 0.1$ the error is about 5%. For any $u \le 1$, Equation 6.220 converges and can be used in

---

[*]$W(u) = -\mathrm{Ei}(-u)$.

TABLE 6.7: Well Function, $W(u)$

u	1.0	2.0	3.0	4.0	5.0	6.0	7.0	8.0	9.0
$\times 1$	.2194	.0489	.0130	.0038	.0011	.0004	.0001	.0000	.0000
$\times 10^{-1}$	1.8229	1.2227	.9057	.7024	.5598	.4544	.3738	.3106	.2602
$\times 10^{-2}$	4.0379	3.3547	2.9591	2.6813	2.4679	2.2953	2.1508	2.0269	1.9187
$\times 10^{-3}$	6.3315	5.6394	5.2349	4.9482	4.7261	4.5448	4.3916	4.2591	4.1423
$\times 10^{-4}$	8.6332	7.9401	7.5348	7.2472	7.0242	6.8420	6.6879	6.5545	6.4368
$\times 10^{-5}$	10.9357	10.2428	9.8372	9.5494	9.3264	9.1440	8.9899	8.8564	8.7386

computer/calculator programs to calculate $W(u)$, while for $u > 1$, the following expression can be used (Abramowitz and Stegun, 1972):

$$W(u) = \frac{e^{-u}}{u} \frac{u^2 + a_1 u + a_2}{u^2 + b_1 u + b_2}, \qquad u > 1 \tag{6.222}$$

where

$$a_1 = 2.334733, \quad a_2 = 0.250621, \quad b_1 = 3.330657, \quad b_2 = 1.681534 \tag{6.223}$$

The relative error in using Equation 6.222 is on the order of 10^{-5} for all values of $u > 1$ (Cheng, 2000).

The utility of the Theis equation, Equation 6.219, and the approximation given by Equation 6.221, lies primarily in the determination of the aquifer properties, T and S, from aquifer pump tests. To use the Theis equation to estimate T and S, the relationship between the drawdown, $s(r,t)$, and r^2/t must be measured in the aquifer for a given pumping rate, Q_w. The relationship between s and r^2/t can be obtained from measurements of drawdown, s, versus time, t, at a single monitoring well at a known distance r from the pumping well, or the relationship between s and r^2/t can be obtained from drawdown measurements at several monitoring wells, at varying distances from the pumping well, at a single instance of time, t. To demonstrate how the aquifer properties can be derived from measured values of drawdown, s, versus r^2/t, consider the definition of the Boltzman variable, Equation 6.199, and the logarithm of the Theis equation (Equation 6.219), which yield

$$\ln\left(\frac{r^2}{t}\right) = \ln u + \alpha \tag{6.224}$$

$$\ln s = \ln W(u) + \beta \tag{6.225}$$

where α and β are constants given by

$$\alpha = \ln\left(\frac{4T}{S}\right) \tag{6.226}$$

$$\beta = \ln\left(\frac{Q_w}{4\pi T}\right) \tag{6.227}$$

Based on Equations 6.224 and 6.225, it is apparent that the theoretical relationship between $\ln s$ and $\ln(r^2/t)$ can be derived directly from the relationship between $\ln W(u)$ and $\ln u$, simply by adding α to $\ln u$ and β to $\ln W(u)$. This relationship is illustrated

FIGURE 6.17: Estimation of aquifer properties using Theis equation and well function

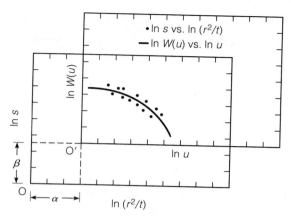

in Figure 6.17, where the plot of $\ln W(u)$ versus $\ln u$ is commonly referred to as a *type curve*.

To determine α and β from field measurements, $\ln s$ versus $\ln (r^2/t)$ is plotted on axes with the same origin and scale as a reference plot of $\ln W(u)$ versus $\ln u$. The plots are then superimposed on a computer monitor such that the axes of the plots are parallel, and the curve of $\ln W(u)$ versus $\ln u$ closely matches the curve of $\ln s$ versus $\ln (r^2/t)$. These matched curves should resemble those shown in Figure 6.17. After matching the curves, using a best-fit approach such as least-squares analysis, the constants α and β are derived from the coordinates of the origin of the $\ln W(u)$ versus $\ln u$ curve (O' in Figure 6.17) on the $\ln s$ versus $\ln (r^2/t)$ curve. Using the definition of β given by Equation 6.227, the transmissivity, T, of the aquifer is given by

$$T = \frac{Q_w}{4\pi} e^{-\beta}$$

(6.228)

and once T is determined, S can be derived from the definition of α given by Equation 6.226 as

$$S = 4Te^{-\alpha}$$

(6.229)

The value of the storage coefficient, S, can be used to confirm the assumption that the aquifer is confined. For example, if the calculated storage coefficient exceeds 0.1, the aquifer is most likely unconfined (Gorelick et al., 1993).

In cases where transient drawdowns are measured at several observation wells surrounding a single pumping well, the Theis equation can be used to estimate T and S for each record of drawdown measurements. The estimated values of T and S will generally differ due to the heterogeneity of the aquifer. The geometric mean of T can be used to estimate the effective transmissivity of the aquifer under parallel flow conditions, and the geometric mean of S can be used to estimate the storativity of the aquifer (Sánchez-Vila et al., 1999).

EXAMPLE 6.19

A confined aquifer of thickness 30 m is pumped at a rate of 75 L/s from a well of radius 0.3 m. The recorded drawdowns in a monitoring well located 100 m from the pumping well are given in Table 6.8. Use the Theis equation to estimate the hydraulic conductivity, transmissivity, and storage coefficient of the aquifer.

TABLE 6.8: Drawdowns in Monitoring Well

Time (s)	Drawdown (cm)
1	0.00
10	0.04
100	1.98
1000	62.80
10,000	183.20
100,000	313.80

Solution The Theis equation is used to extract the transmissivity and storage coefficient by matching the standard curve of $\ln W(u)$ versus $\ln u$ (i.e., type curve), to the measured data in the form of $\ln s$ versus $\ln (r^2/t)$, where s is the drawdown, r is the distance of the monitoring well from the pumping well, and t is the time since the beginning of pumping. In this case, $r = 100$ m, and the relationship between s and r^2/t is given in Table 6.9, where t is the time in days, and s is the drawdown in meters. The superposition of the $\ln W(u)$ versus $\ln u$ curve onto the $\ln s$ versus $\ln r^2/t$ curve is illustrated in Figure 6.18, and the axis displacement indicates that $\alpha = 15.1$ and $\beta = -0.557$. Since $Q_w = 75$ L/s $= 6480$ m^3/d, then Equation 6.228 yields

$$T = \frac{Q_w}{4\pi} e^{-\beta} = \frac{6480}{4\pi} e^{-(-0.557)} = 900 \text{ m}^2/\text{d}$$

TABLE 6.9: s Versus r^2/t

r^2/t (m^2/d)	s (m)
8.64×10^8	0.000
8.64×10^7	0.0004
8.64×10^6	0.0198
8.64×10^5	0.628
8.64×10^4	1.832
8.64×10^3	3.138

FIGURE 6.18: Application of theis equation and well function

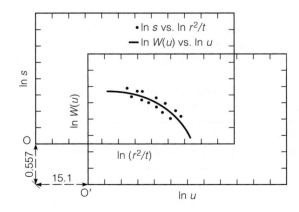

and Equation 6.229 yields

$$S = 4Te^{-\alpha} = 4(900)e^{-15.1} = 0.0010$$

The hydraulic conductivity, K, of the aquifer is given by

$$K = \frac{T}{b} = \frac{900}{30} = 30 \text{ m/d}$$

The Theis equation (Equation 6.219) can also be applied in anisotropic formations using the coordinate transformation technique described in Section 6.2.3.2, which also contains a specific example on the application of the Cooper–Jacob approximation to the Theis equation in anisotropic formations. An important limitation of the Theis equation is that it neglects storage in the well, by assuming that all pumpage is extracted from the aquifer. This assumption has been investigated by Papadopulos and Cooper (1967), who showed that well-storage effects are negligible when calculating the transient drawdown in an aquifer if

$$t > 2500 \frac{r_c^2}{T} \tag{6.230}$$

where t is the duration of pumping, r_c is the diameter of the well casing, and T is the transmissivity of the aquifer. When calculating the drawdown at a well ($r = r_w$), storage effects are negligible when

$$t > 250 \frac{r_c^2}{T} \tag{6.231}$$

In most practical cases, Equations 6.230 and 6.231 are satisfied and the Theis equation is applicable. In cases where well-storage effects are significant, the analytical formulation developed by Papadopulos and Cooper (1967) should be used in lieu of the Theis equation. A second limitation of the Theis equation relates to the assumption of an infinite aquifer. This assumption is justified as long as the cone of depression does not intersect any boundaries. Other assumptions implicit in the Theis equation include a horizontal aquifer with constant thickness, and a fully penetrating well. Care should be taken that the Theis equation is applied to field scenarios that meet the assumed conditions.

6.3.7   Unsteady Flow to a Well in an Unconfined Aquifer

In isotropic and homogeneous unconfined aquifers, the distribution of saturated thickness, h, can be estimated by invoking the Dupuit approximation of two-dimensional flow, in which case the distribution of saturated thickness is given by Equation 6.81. Since the drawdown induced by a single pumping well will be radially symmetric, and taking the recharge, N, equal to zero, Equation 6.81 can be written as

$$\frac{\partial^2(h^2)}{\partial r^2} + \frac{1}{r}\frac{\partial(h^2)}{\partial r} = 2\frac{S_y}{K}\frac{\partial h}{\partial t} \tag{6.232}$$

Defining the drawdown, $s(r,t)$, by the relation

$$s(r,t) = H - h(r,t) \tag{6.233}$$

where H is the saturated thickness prior to pumping, and substituting Equation 6.233 into 6.232 leads to

$$\frac{\partial^2}{\partial r^2}\left(s - \frac{s^2}{2H}\right) + \frac{1}{r}\frac{\partial}{\partial r}\left(s - \frac{s^2}{2H}\right) = \frac{S_y}{KH}\frac{\partial s}{\partial t} \qquad (6.234)$$

Noting that

$$s - \frac{s^2}{2H} = s\left(1 - \frac{s}{2H}\right) \qquad (6.235)$$

then whenever $s \ll 2H$,

$$s - \frac{s^2}{2H} \approx s \qquad (6.236)$$

and Equation 6.234 can be written as

$$\frac{\partial^2 s}{\partial r^2} + \frac{1}{r}\frac{\partial s}{\partial r} = \frac{S_y}{T}\frac{\partial s}{\partial t}, \qquad s \ll 2H \qquad (6.237)$$

where T is the transmissivity defined by

$$T = KH \qquad (6.238)$$

The initial and boundary conditions for unsteady flow to a well in an unconfined infinite aquifer are

$$s(r, 0) = 0 \qquad (6.239)$$

$$s(\infty, t) = 0 \qquad (6.240)$$

$$\lim_{r \to 0} r\frac{\partial s}{\partial r} = -\frac{Q_w}{2\pi T} \qquad (6.241)$$

Therefore, under the restriction that $s \ll 2H$, the governing equation and associated initial and boundary conditions are exactly the same as the governing equation and associated initial and boundary conditions that describe unsteady drawdown in a confined aquifer, with the exception that the storage coefficient, S, in the confined aquifer case is replaced by the specific yield, S_y, in the unconfined aquifer case. The small-drawdown approximation is justified when the drawdown, s, is less than 5% of the original saturated thickness, H (Boonstra, 1998a). The similarity between the confined and unconfined cases can be observed by comparing Equations 6.237 and 6.239 to 6.241 with Equations 6.195 to 6.198. Consequently, the Theis equation describing the drawdown as a function of time in confined aquifers, Equation 6.219, can also be used in unconfined aquifers, as can the curve-matching approach for determining the hydraulic properties of the aquifer. However, the curve-matching approach in unconfined aquifers produces the specific yield, S_y, instead of the storage coefficient, S, for confined aquifers.

EXAMPLE 6.20

An aquifer pump test is conducted at 80 L/s in an unconfined aquifer, and comparison of the drawdown curve with the well function indicates that the best match occurs when $\alpha = 9.9$ and $\beta = -0.60$. If the maximum drawdown during the aquifer pump test is

1.5 m and the saturated thickness prior to pumping is 20 m, estimate the transmissivity and specific yield of the aquifer.

Solution   Since the maximum drawdown is 1.5 m and the saturated thickness prior to pumping is 20 m, then

$$\frac{s}{2H} \le \frac{1.5}{2(20)} = 0.0375$$

Since $s/2H \ll 1$, the confined-aquifer approximation can be used to analyze the data from the aquifer pump test. From the given data: $Q_w = 80$ L/s $= 6912$ m^3/d, $\alpha = 9.9$, $\beta = -0.60$, and Equation 6.228 gives

$$T = \frac{Q_w}{4\pi} e^{-\beta} = \frac{6912}{4\pi} e^{-(-0.60)} = 1000 \text{ m}^2/\text{d}$$

and Equation 6.229 gives

$$S_y = 4Te^{-\alpha} = 4(1000)e^{-9.9} = 0.20$$

In pumping tests within unconfined aquifers, the aquifer response in many cases tends to be different for small and large times. Specifically, the storage coefficient determined from small-time data is significantly smaller than for large-time data. This phenomenon is typically associated with the fact that gravity drainage from the pores above the water table is not immediate, and initial release of water is associated (only) with the compressibility of the water and aquifer material. After some time, gravity drainage begins to contribute significantly to aquifer recharge, producing a *delayed yield*, which is noticeably present in finer-grained aquifer materials, which may take weeks to fully drain (Gorelick et al., 1993). Neuman (1975, 1973, 1972) derived the following solution for a pumping well in an unconfined aquifer with delayed yield:

$$\boxed{s = \frac{Q_w}{4\pi T} W(u, u_y, \Gamma)} \tag{6.242}$$

where $W(u, u_y, \Gamma)$ is called the *unconfined well function* or *Neuman's well function* (Batu, 1998) and

$$u = \frac{r^2 S}{4Tt} \tag{6.243}$$

$$u_y = \frac{r^2 S_y}{4Tt} \tag{6.244}$$

$$\Gamma = \frac{r^2 K_v}{H^2 K_h} \tag{6.245}$$

$$W(u, u_y, \Gamma) = \int_0^\infty 4yJ_0(y\Gamma^{1/2}) \sum_{n=0}^\infty a_n(y)\, dy \tag{6.246}$$

where K_h and K_v are the horizontal and vertical hydraulic conductivities, respectively, J_0 is the Bessel function of the first kind of order zero, and

$$a_o = \frac{\left\{1 - \exp\left[-\frac{\Gamma(y^2 - \gamma_0^2)}{4u}\right]\right\} \tanh \gamma_0}{\left[y^2 + (1 + \sigma)\gamma_0^2 - \frac{(y^2 - \gamma_0^2)^2}{\sigma}\right]\gamma_0} \tag{6.247}$$

$$a_n = \frac{\left\{1 - \exp\left[-\frac{\Gamma(y^2 + \gamma_n^2)}{4u}\right]\right\} \tanh \gamma_n}{\left[y^2 - (1 + \sigma)\gamma_n^2 - \frac{(y^2 + \gamma_n^2)^2}{\sigma}\right]\gamma_n} \tag{6.248}$$

where

$$\sigma = \frac{u}{u_y} = \frac{S}{S_y} \tag{6.249}$$

and γ_0 and γ_n are the roots of the characteristic equations

$$\sigma\gamma_o \sinh \gamma_0 - (y^2 - \gamma_0^2) \cosh \gamma_0 = 0; \qquad \gamma_0^2 < y^2 \tag{6.250}$$

$$\sigma\gamma_n \sin \gamma_n + (y^2 + \gamma_n^2) \cos \gamma_n = 0 \tag{6.251}$$

where

$$(2n - 1)\frac{\pi}{2} < \gamma_n < n\pi \tag{6.252}$$

Equations 6.242 to 6.251 collectively describe the drawdown in an unconfined aquifer with delayed yield. In most cases, $S_y \gg S$, which gives $\sigma \approx 0$ and allows Equation 6.242 to reduce to two asymptotic families of type curves, $W(u, \Gamma)$ and $W(u_y, \Gamma)$, respectively known as Type A and Type B curves (Neuman, 1975):

$$W(u, \Gamma) = \int_0^\infty 64yJ_0\left(y\Gamma^{1/2}\right) \sum_{n=1}^\infty \frac{1 - \exp\left\{-\frac{\Gamma}{16u}\left[4y^2 + (2n - 1)^2\pi^2\right]\right\}}{(2n - 1)^2\pi^2\left[4y^2 + (2n - 1)^2\pi^2\right]} dy \tag{6.253}$$

$$W(u_y, \Gamma) = \int_0^\infty 4yJ_0\left(y\Gamma^{1/2}\right)\left\{ \frac{\left[1 - \exp\left(-\frac{\Gamma y \tanh y}{4u_y}\right)\right]\tanh y}{2y^3}\right.$$

$$\left. + \sum_{n=1}^\infty \frac{16}{(2n - 1)^2\pi^2\left[4y^2 + (2n - 1)^2\pi^2\right]}\right\} dy \tag{6.254}$$

TABLE 6.10: Well Function $W(u, \Gamma)$ for Unconfined Aquifers

Γ \ $1/u$	0.001	0.01	0.06	0.2	0.6	1.0	2.0	4.0	6.0
4.0×10^{-1}	0.0234	0.0240	0.0230	0.0214	0.0188	0.0170	0.0138	0.00933	0.00639
8.0×10^{-1}	0.144	0.140	0.131	0.119	0.0988	0.0849	0.0603	0.0317	0.0174
1.4×10^{0}	0.358	0.345	0.318	0.279	0.217	0.175	0.107	0.0445	0.0210
2.4×10^{0}	0.662	0.633	0.570	0.483	0.343	0.256	0.133	0.0476	0.0214
4.0×10^{0}	1.02	0.963	0.849	0.688	0.438	0.300	0.140	0.0478	0.0215
8.0×10^{0}	1.57	1.46	1.23	0.918	0.497	0.317	0.141	0.0478	0.0215
1.4×10^{1}	2.05	1.88	1.51	1.03	0.507	0.317	0.141	0.0478	0.0215
2.4×10^{1}	2.52	2.27	1.73	1.07	0.507	0.317	0.141	0.0478	0.0215
4.0×10^{1}	2.97	2.61	1.85	1.08	0.507	0.317	0.141	0.0478	0.0215
8.0×10^{1}	3.56	3.00	1.92	1.08	0.507	0.317	0.141	0.0478	0.0215
1.4×10^{2}	4.01	3.23	1.93	1.08	0.507	0.317	0.141	0.0478	0.0215
2.4×10^{2}	4.42	3.37	1.94	1.08	0.507	0.317	0.141	0.0478	0.0215
4.0×10^{2}	4.77	3.43	1.94	1.08	0.507	0.317	0.141	0.0478	0.0215
8.0×10^{2}	5.16	3.45	1.94	1.08	0.507	0.317	0.141	0.0478	0.0215
1.4×10^{3}	5.40	3.46	1.94	1.08	0.507	0.317	0.141	0.0478	0.0215
2.4×10^{3}	5.54	3.46	1.94	1.08	0.507	0.317	0.141	0.0478	0.0215
4.0×10^{3}	5.59	3.46	1.94	1.08	0.507	0.317	0.141	0.0478	0.0215
8.0×10^{3}	5.62	3.46	1.94	1.08	0.507	0.317	0.141	0.0478	0.0215
1.4×10^{4}	5.62	3.46	1.94	1.08	0.507	0.317	0.141	0.0478	0.0215

These Type A and Type B curves are tabulated in Tables 6.10 and 6.11 and can be used to determine the aquifer parameters based on small-time and large-time pumping data using the relations

$$s = \frac{Q_w}{4\pi T} W(u, \Gamma) \qquad \text{for small time} \qquad (6.255)$$

$$s = \frac{Q_w}{4\pi T} W(u_y, \Gamma) \qquad \text{for large time} \qquad (6.256)$$

The application of Equations 6.255 and 6.256 is illustrated in Figure 6.19, where Type A curves merge out of the Theis curve shown on the left, and Type B curves merge into the Theis curve shown on the right. There are typically three distinct segments on the time-drawdown curve in unconfined aquifers, as shown in Figure 6.20 (Mays, 2001). The first segment occurs immediately after pumping begins, when water released from the aquifer is associated with the compressibility of the aquifer matrix and ground water. Such releases are parameterized by the storage coefficient of the aquifer. The second drawdown segment shown in Figure 6.20 illustrates the intermediate stage when the expansion of the cone of depression decreases because of gravity drainage, and the slope of the time-drawdown curve decreases, reflecting recharge. During the third segment shown in Figure 6.20, the time-drawdown curves conform to the Theis solution, with gravity drainage parameterized by the specific yield. This latter condition may start from several minutes to several days after pumping begins. Pump-drawdown data collected during the last segment of the drawdown curve can

TABLE 6.11: Well Function $W(u_y, \Gamma)$ for Unconfined Aquifers

$1/u_y$ \ Γ	0.001	0.01	0.06	0.2	0.6	1.0	2.0	4.0	6.0
4.0×10^{-4}	5.62	3.45	1.94	1.08	0.508	0.318	0.141	0.0479	0.0215
8.0×10^{-4}	5.62	3.45	1.94	1.08	0.508	0.318	0.141	0.0480	0.0216
1.4×10^{-3}	5.62	3.45	1.94	1.08	0.508	0.318	0.142	0.0481	0.0217
2.4×10^{-3}	5.62	3.45	1.94	1.09	0.508	0.318	0.142	0.0484	0.0219
4.0×10^{-3}	5.62	3.45	1.94	1.09	0.508	0.318	0.142	0.0488	0.0221
8.0×10^{-3}	5.62	3.45	1.94	1.09	0.509	0.319	0.143	0.0496	0.0228
1.4×10^{-2}	5.62	3.45	1.94	1.09	0.510	0.321	0.145	0.0509	0.0239
2.4×10^{-2}	5.62	3.45	1.94	1.09	0.512	0.323	0.147	0.0532	0.0257
4.0×10^{-2}	5.62	3.45	1.94	1.09	0.516	0.327	0.152	0.0568	0.0286
8.0×10^{-2}	5.62	3.46	1.94	1.09	0.524	0.337	0.162	0.0661	0.0362
1.4×10^{-1}	5.62	3.46	1.94	1.10	0.537	0.350	0.178	0.0806	0.0486
2.4×10^{-1}	5.62	3.46	1.95	1.11	0.557	0.374	0.205	0.106	0.0714
4.0×10^{-1}	5.62	3.46	1.96	1.13	0.589	0.412	0.248	0.149	0.113
8.0×10^{-1}	5.62	3.46	1.98	1.18	0.667	0.506	0.357	0.266	0.231
1.4×10^{0}	5.63	3.47	2.01	1.24	0.780	0.642	0.517	0.445	0.419
2.4×10^{0}	5.63	3.49	2.06	1.35	0.954	0.850	0.763	0.718	0.703
4.0×10^{0}	5.63	3.51	2.13	1.50	1.200	1.13	1.08	1.06	1.05
8.0×10^{0}	5.64	3.56	2.31	1.85	1.68	1.65	1.63	1.63	1.63
1.4×10^{1}	5.65	3.63	2.55	2.23	2.15	2.14	2.14	2.14	2.14
2.4×10^{1}	5.67	3.74	2.86	2.68	2.65	2.65	2.64	2.64	2.64
4.0×10^{1}	5.70	3.90	3.24	3.15	3.14	3.14	3.14	3.14	3.14
8.0×10^{1}	5.76	4.22	3.85	3.82	3.82	3.82	3.82	3.82	3.82
1.4×10^{2}	5.85	4.58	4.38	4.37	4.37	4.37	4.37	4.37	4.37
2.4×10^{2}	5.99	5.00	4.91	4.91	4.91	4.91	4.91	4.91	4.91
4.0×10^{2}	6.16	5.46	5.42	5.42	5.42	5.42	5.42	5.42	5.42
8.0×10^{2}	6.47	6.11	6.11	6.11	6.11	6.11	6.11	6.11	6.11
1.4×10^{3}	6.67	6.67	6.67	6.67	6.67	6.67	6.67	6.67	6.67
2.4×10^{3}	7.21	7.21	7.21	7.21	7.21	7.21	7.21	7.21	7.21
4.0×10^{3}	7.72	7.72	7.72	7.72	7.72	7.72	7.72	7.72	7.72
8.0×10^{3}	8.41	8.41	8.41	8.41	8.41	8.41	8.41	8.41	8.41
1.4×10^{4}	8.97	8.97	8.97	8.97	8.97	8.97	8.97	8.97	8.97
2.4×10^{4}	9.51	9.51	9.51	9.51	9.51	9.51	9.51	9.51	9.51
4.0×10^{4}	19.4	19.4	19.4	19.4	19.4	19.4	19.4	19.4	19.4

be used to estimate the transmissivity and specific yield of the aquifer via the Theis equation.

In analyzing pump-test data in unconfined aquifers using the above-described Neuman method, the following steps are suggested (Domenico and Schwartz, 1998):

- Match the early-time drawdown data to the Type A curves. This match yields a value of Γ, and the origin shift (α, β). To extract the storage coefficient, S, and the transmissivity, T, from the origin shift, one of the following methods should be used:

FIGURE 6.19: Unconfined aquifer well functions $W(u, \Gamma)$ and $W(u_y, \Gamma)$

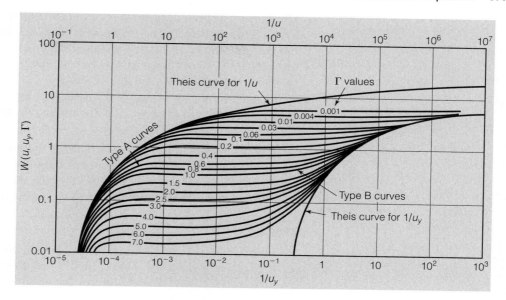

FIGURE 6.20: Three segments on time-drawdown curve in unconfined aquifers

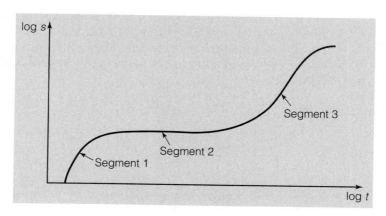

- Match the $\ln W(u, \Gamma)$ vs. $\ln u$ curve to the $\ln s$ vs. $\ln(r^2/t)$ curve, determine the origin shift (α, β), and calculate T and S using the following relations:

$$T = \frac{Q_w}{4\pi} e^{-\beta}$$

$$S = 4Te^{-\alpha}$$

- Match the $\ln W(u, \Gamma)$ vs. $\ln(1/u)$ curve to the $\ln s$ vs. $\ln(t/r^2)$ curve, determine the origin shift (α, β), and calculate T and S using the following relations:

$$T = \frac{Q_w}{4\pi} e^{-\beta}$$

$$S = 4Te^{\alpha}$$

- Match the late-time drawdown data to the Type B curve with the same Γ value as the Type A match. The origin shift (α, β), yields the specific yield, S_y, and the transmissivity, T, using the same relations as in the previous step. The transmissivity, T, calculated in this step should be close to the value obtained in the previous step.
- Using the estimated values of Γ, and horizontal hydraulic conductivity, K_h ($= T/H$), determine the vertical hydraulic conductivity of the aquifer using Equation 6.245.

In applying either the Theis or Neuman method to calculate the hydraulic parameters of an aquifer, it must generally be considered that these methods assume that the drawdown is sufficiently small such that the transmissivity can be assumed to be property of the aquifer that remains constant during the aquifer pump test.

EXAMPLE 6.21

An unconfined aquifer is composed primarily of medium-grained sand and gravel, with a clayey zone at shallow depths. The aquifer is underlain by marls having a relatively low hydraulic conductivity. The bottom of the aquifer is 13.75 m below the ground surface, and the water table is initially at a depth of 5.51 m. An aquifer pumping test is conducted using a well that is screened between the depths of 7.00 m and 13.75 m, and has a diameter of 320 mm. The pumping test lasts 48 hours and 50 minutes and the pumping rate varies between 14.2 L/s and 15.2 L/s and averages about 14.7 L/s. The observed drawdowns 10 m from the well are given in Table 6.12. Determine the aquifer properties using the Neuman type-curve method, and provide an analysis of the results.

Solution From the given data: $Q_w = 14.7$ L/s $= 1270$ m^3/d, and $r = 10$ m.

Step 1. Comparing the $\ln W(u, \Gamma)$ vs. $\ln u$ curve with the $\ln s$ vs. $\ln(r^2/t)$ curve to determine the origin shift (α, β) and curve parameter Γ yields the

TABLE 6.12: Drawdown Versus Time at 10 m from Pumping Well

Time (min)	Drawdown (m)	Time (min)	Drawdown (m)	Time (min)	Drawdown (m)
0.25	0.100	10.00	0.216	248.75	0.327
0.50	0.121	13.20	0.221	303.92	0.349
0.73	0.135	16.12	0.225	371.23	0.354
0.97	0.163	19.05	0.229	431.37	0.360
1.23	0.172	22.50	0.232	544.93	0.366
1.45	0.178	28.72	0.236	688.25	0.375
2.00	0.179	35.90	0.242	786.58	0.397
2.48	0.185	43.87	0.246	987.07	0.403
2.93	0.191	53.58	0.254	1154.65	0.401
3.53	0.201	73.08	0.265	1355.03	0.417
4.53	0.216	100.00	0.276	1585.43	0.430
5.53	0.218	133.33	0.298	1969.47	0.456
6.55	0.218	155.90	0.309	2529.58	0.477
7.87	0.218	190.48	0.319	2939.33	0.506

following values:

$$\Gamma = 0.01$$

$$\alpha = 15.28$$

$$\beta = -2.75$$

and hence the transmissivity, T, and storage coefficient, S, are given by

$$T = \frac{Q_w}{4\pi}e^{-\beta} = \frac{1270}{4\pi}e^{-(-2.75)} = 1581 \text{ m}^2/\text{d}$$

$$S = 4Te^{-\alpha} = 4(1581)e^{-15.28} = 0.00146$$

Step 2. Taking $\Gamma = 0.01$ and comparing the $\ln W(u_y, \Gamma)$ vs. $\ln u_y$ curve with the $\ln s$ vs. $\ln(r^2/t)$ curve to determine the origin shift (α, β) yields

$$\alpha = 11.60$$

$$\beta = -2.73$$

and hence the transmissivity, T, and specific yield, S_y, are given by

$$T = \frac{Q_w}{4\pi}e^{-\beta} = \frac{1270}{4\pi}e^{-(-2.73)} = 1550 \text{ m}^2/\text{d}$$

$$S_y = 4Te^{-\alpha} = 4(1550)e^{-11.60} = 0.0568$$

The values of the transmissivity calculated at the early time (1581 m²/d) and late time (1550 m³/d) are close, and therefore the transmissivity can be taken as the (rounded) average of 1570 m³/d.

Step 3. For a transmissivity, T, equal to 1570 m²/d and a saturated thickness, H, equal to 13.75 m −5.51 m = 8.24 m, the horizontal hydraulic conductivity, K_h, is given by

$$K_h = \frac{T}{H} = \frac{1570}{8.24} = 191 \text{ m/d}$$

Taking $\Gamma = 0.01$ and using the definition of Γ given by Equation 6.245 yields

$$\Gamma = \frac{r^2 K_v}{H^2 K_h}$$

$$0.01 = \frac{(10)^2 K_v}{(8.24)^2(191)}$$

which gives

$$K_v = 1.30 \text{ m/d}$$

In summary, the test data indicate that the aquifer has a horizontal hydraulic conductivity of 191 m/d, a vertical hydraulic conductivity of 1.30 m/d, a specific yield of 0.057, and a storage coefficient of 0.00146.

Analysis of Results:

The results show that the horizontal hydraulic conductivity (191 m/d) is more than 100 times greater than the vertical hydraulic conductivity. This may be attributed to the presence of a clayey formation in the upper part of the aquifer. The specific storage, $S_s = S/H = 0.00146/8.24 = 1.77 \times 10^{-4}$, is relatively high and gives an indication that the compressibility of the aquifer is greater than usually encountered in deep confined aquifers having similar aquifer materials.
[This example is adapted from Neuman (1975) and Batu (1998). The data for this example are taken from an aquifer pumping test performed in 1965 by the French Bureau de Recherches Géologiques et Minières ant Saint Pardon de Conques, located in the Vallée de la Garonne, Gironde, France.]

6.3.8 Unsteady Flow to a Well in a Leaky Confined Aquifer

In isotropic and homogeneous semiconfined (leaky) aquifers, the distribution of piezometric head for two-dimensional flows is given by Equation 6.96. If a well fully penetrates the semiconfined aquifer and withdraws water uniformly over the aquifer thickness, b, then the piezometric head, ϕ, is radially symmetric (independent of θ), all derivatives with respect to θ are zero, and Equation 6.96 can be written in the form

$$\frac{\partial^2 \phi}{\partial r^2} + \frac{1}{r}\frac{\partial \phi}{\partial r} + \frac{N}{Kb} = \frac{S}{T}\frac{\partial \phi}{\partial t} \tag{6.257}$$

The leakage into the semiconfined aquifer can be calculated by applying Darcy's law across the semiconfining layer, in which case

$$N = -K' \frac{\phi - \phi_0}{b'} \tag{6.258}$$

where K' is the hydraulic conductivity of the confining layer, b' is the thickness of the confining layer, and ϕ_0 is the piezometric head above the confining layer, assumed to be constant. Combining Equations 6.257 and 6.258 leads to

$$\frac{\partial^2 \phi}{\partial r^2} + \frac{1}{r}\frac{\partial \phi}{\partial r} + \frac{\phi_0 - \phi}{\lambda^2} = \frac{S}{T}\frac{\partial \phi}{\partial t} \tag{6.259}$$

where λ is the leakage factor defined by Equation 6.161. If the drawdown, $s(r,t)$, is defined by the relation

$$s(r,t) = \phi_0 - \phi(r,t) \tag{6.260}$$

then Equation 6.259 can be written in terms of drawdown, s, as

$$\frac{\partial^2 s}{\partial r^2} + \frac{1}{r}\frac{\partial s}{\partial r} - \frac{s}{\lambda^2} = \frac{S}{T}\frac{\partial s}{\partial t} \tag{6.261}$$

The initial and boundary conditions in the case of unsteady flow to a well in an infinite semiconfined aquifer are as follows:

$$s(r, 0) = 0 \tag{6.262}$$

$$s(\infty, t) = 0 \tag{6.263}$$

$$\lim_{r \to 0} r \frac{\partial s}{\partial r} = -\frac{Q_w}{2\pi T} \tag{6.264}$$

The solution to this problem has been derived by Hantush and Jacob (1955) as

$$\boxed{s(r, t) = \frac{Q_w}{4\pi T} W\left(u, \frac{r}{\lambda}\right)} \tag{6.265}$$

where $W(u, r/\lambda)$ is called the *Hantush and Jacob well function for leaky aquifers* (Hantush and Jacob, 1955), or simply the *leaky well function* (Freeze and Cherry, 1979), and is defined as

$$\boxed{W\left(u, \frac{r}{\lambda}\right) = \int_u^\infty \frac{1}{y} \exp\left(-y - \frac{r^2}{4\lambda^2 y}\right) dy} \tag{6.266}$$

Values of $W(u, r/\lambda)$ for several values of u and r/λ are tabulated in Table 6.13. For $r/\lambda = 0$, $W(u, r/\lambda)$ is equal to $W(u)$. The well function, $W(u, r/\lambda)$, in leaky confined aquifers is used in the same way as the well function, $W(u)$, in confined aquifers to determine the transmissivity, T, and storage coefficient, S, from aquifer pump tests. In the case of leaky confined aquifers, a curve of $\ln s$ versus $\ln(r^2/t)$ is overlain on a curve of $\ln W(u, r/\lambda)$ versus $\ln u$ (plotted for several different values of r/λ). The displacement of the origin of the $\ln W(u, r/\lambda)$ versus $\ln u$ curve relative to the $\ln s$ versus $\ln(r^2/t)$ curve yields the displacement coordinates (α', β'), and the shape of the $\ln s$ versus $\ln(r^2/t)$ curve yields the best-fit value of r/λ, and hence the leakage factor λ can be derived from the known distance, r, of the monitoring well from the pumping well. The transmissivity, T, and storage coefficient, S, can then be derived from the displacement coordinates using the relations

$$\boxed{T = \frac{Q_w}{4\pi} e^{-\beta'}} \tag{6.267}$$

and

$$\boxed{S = 4Te^{-\alpha'}} \tag{6.268}$$

EXAMPLE 6.22

A leaky confined aquifer of thickness 30 m is pumped at a rate of 75 L/s from a fully penetrating well of radius 0.3 m. The drawdown curve is measured at an observation well 10 m from the pumping well, and matched with the well function for a leaky confined aquifer. The best match is obtained for $\alpha' = 15$, $\beta' = -0.60$, and $r/\lambda = 0.04$. If the thickness of the semiconfining layer overlying the aquifer is 2 m, estimate the transmissivity and storage coefficient of the aquifer and the hydraulic conductivity of the semiconfining layer.

TABLE 6.13: Well Function for Leaky Aquifer, $W(u, r/\lambda)$

u \ r/λ	0.00	0.002	0.004	0.007	0.01	0.02	0.04	0.06	0.08	0.10
0.00		12.6611	11.2748	10.1557	9.4425	8.0569	6.6731	5.8456	5.2950	4.8541
1×10^{-6}	13.2383	12.4417	11.2711	10.1557						
2×10^{-6}	12.5451	12.1013	11.2259	10.1554						
5×10^{-6}	11.6289	11.4384	10.9642	10.1290	9.4425					
8×10^{-6}	11.1589	11.0377	10.7151	10.0602	9.4313					
1×10^{-5}	10.9357	10.8382	10.5725	10.0034	9.4176	8.0569				
2×10^{-5}	10.2426	10.1932	10.0522	9.7126	9.2961	8.0558				
5×10^{-5}	9.3263	9.3064	9.2480	9.0957	8.8827	8.0080	6.6730			
7×10^{-5}	8.9899	8.9756	8.9336	8.8224	8.6625	7.9456	6.6726			
1×10^{-4}	8.6332	8.6233	8.5937	8.5145	8.3983	7.8375	6.6693	5.8658	5.2950	4.8541
2×10^{-4}	7.9402	7.9352	7.9203	7.8800	7.8192	7.4472	6.6242	5.8637	5.2949	4.8541
5×10^{-4}	7.0242	7.0222	7.0163	6.9999	6.9750	6.8346	6.3626	5.8011	5.2848	4.8530
7×10^{-4}	6.6879	6.6865	6.6823	6.6706	6.6527	6.5508	6.1917	5.7274	5.2618	4.8478
1×10^{-3}	6.3315	6.3305	6.3276	6.3194	6.3069	6.2347	5.9711	5.6058	5.2087	4.8292
2×10^{-3}	5.6394	5.6389	5.6374	5.6334	5.6271	5.5907	5.4516	5.2411	4.9848	4.7079
5×10^{-3}	4.7261	4.7259	4.7253	4.7237	4.7212	4.7068	4.6499	4.5590	4.4389	4.2990
7×10^{-3}	4.3916	4.3915	4.3910	4.3899	4.3882	4.3779	4.3374	4.2719	4.1839	4.0771
1×10^{-2}	4.0379	4.0378	4.0375	4.0368	4.0351	4.0285	4.0003	3.9544	3.8920	3.8190
2×10^{-2}	3.3547	3.3547	3.3545	3.3542	3.3536	3.3502	3.3365	3.3141	3.2832	3.2442
5×10^{-2}	2.4679	2.4679	2.4678	2.4677	2.4675	2.4662	2.4613	2.4531	2.4416	2.4271
7×10^{-2}	2.1508	2.1508	2.1508	2.1507	2.1506	2.1497	2.1464	2.1408	2.1331	2.1232
1×10^{-1}	1.8229	1.8229	1.8229	1.8228	1.8227	1.8222	1.8220	1.8164	1.8114	1.8050
2×10^{-1}	1.2227	1.2226	1.2226	1.2226	1.2226	1.2224	1.2215	1.2201	1.2181	1.2155
5×10^{-1}	0.5598	0.5598	0.5598	0.5598	0.5598	0.5597	0.5595	0.5592	0.5587	0.5581
7×10^{-1}	0.3738	0.3738	0.3738	0.3738	0.3738	0.3737	0.3736	0.3734	0.3732	0.3729
1.00	0.2194	0.2194	0.2194	0.2194	0.2194	0.2194	0.2193	0.2192	0.2191	0.2190
2.00	0.0489	0.0489	0.0489	0.0489	0.0489	0.0489	0.0489	0.0489	0.0489	0.0488
5.00	0.0011	0.0011	0.0011	0.0011	0.0011	0.0011	0.0011	0.0011	0.0011	0.0011
7.00	0.0001	0.0001	0.0001	0.0001	0.0001	0.0001	0.0001	0.0001	0.0001	0.0001
8.00	0.0000	0.0000	0.0000	0.0000	0.0000	0.0000	0.0000	0.0000	0.0000	0.0000

Solution From the given data: $Q_w = 75$ L/s $= 6480$ m³/d, $\alpha' = 15$, $\beta' = -0.60$, and Equation 6.267 gives the transmissivity, T, of the aquifer as

$$T = \frac{Q_w}{4\pi} e^{-\beta'} = \frac{6480}{4\pi} e^{-(-0.60)} = 940 \text{ m}^2/\text{d}$$

Equation 6.268 gives the storage coefficient, S, of the aquifer as

$$S = 4Te^{-\alpha'} = 4(940)e^{-15} = 0.0012$$

Since the leaky well function that matches the drawdown data has the parameter $r/\lambda = 0.04$, and the distance, r, of the observation well from the pumping well is 10 m, then

$$\lambda = \frac{r}{0.04} = \frac{10}{0.04} = 250 \text{ m}$$

The leakage factor, λ, is defined by Equation 6.161 as

$$\lambda = \sqrt{\frac{Kbb'}{K'}} = \sqrt{\frac{Tb'}{K'}}$$

which can be put in the form

$$K' = \frac{Tb'}{\lambda^2}$$

where K' and b' are the hydraulic conductivity and thickness of the semiconfining layer, respectively. In this case, $T = 940$ m²/d, $b' = 2$ m, $\lambda = 250$ m, and therefore

$$K' = \frac{(940)(2)}{(250)^2} = 0.03 \text{ m/d}$$

This analysis of unsteady flow to a well in a leaky confined aquifer neglects the effect of storage in the semiconfining (leaky) layer, neglects the drawdown in the overlying aquifer, and neglects storage in the well. In some cases, these approximations can lead to significant errors. Neuman and Witherspoon (1969b) have shown that the following condition justifies neglecting the storage in the semiconfining layer:

$$\frac{r}{4b}\left(\frac{K'S'_s}{KS_s}\right) < 0.01 \tag{6.269}$$

where S_s is the specific storage of the aquifer, and S'_s is the specific storage of the semiconfining layer. Hantush (1960) also showed that storage in the semiconfining layer can be neglected if

$$t > 0.036 \frac{b'S'}{K'} \tag{6.270}$$

where S' is the storage coefficient of the semiconfining layer. If either Equation 6.269 or 6.270 is satisfied, then it is reasonable to neglect storage in the semiconfining layer. Neuman and Witherspoon (1969a) have shown that the drawdown in the overlying aquifer can be neglected when either of the following conditions are satisfied:

$$T_s > 100T \tag{6.271}$$

or

$$t < \frac{S'(b')^2}{10bK'} \tag{6.272}$$

where T_s is the transmissivity of the overlying aquifer. Storage in the pumping well can be neglected whenever (Fetter, 2001)

$$t > 30 \frac{r_w^2 S}{T}\left[1 - \left(\frac{10r_w}{b}\right)^2\right] \tag{6.273}$$

and

$$\frac{r_w}{(Tb'/K')^{1/2}} < 0.1 \tag{6.274}$$

Whenever storage effects in the semiconfining layer are significant and/or there is appreciable drawdown in the overlying aquifer, the analytical formulation developed by Neuman and Witherspoon (1968; 1969a; 1969b; 1972) should be applied to describe the aquifer response. In cases where the storage in the well is significant, the drawdown distribution in the aquifer can be estimated using the analytic relations described by Kabala (1993) and Lee (1999).

EXAMPLE 6.23

In the previous example, the thickness and the hydraulic conductivity of the overlying aquifer are 23 m and 40 m/d, respectively, the semiconfining layer has a storage coefficient of 0.0001, and the duration of the aquifer pump test was 1 day. Assess whether neglecting storage in the semiconfining layer, neglecting drawdown in the overlying aquifer, and neglecting storage in the pumping well are justified.

Solution   From the given data: $T_s = 40 \times 23 = 920 \text{ m}^2/\text{d}$, $S' = 0.0001$, $b' = 2 \text{ m}$, $b = 30 \text{ m}$, $r_w = 0.3 \text{ m}$, and $r = 10 \text{ m}$. Analysis of the pump-test data yields the following hydraulic properties of the semiconfined aquifer: $T = 940 \text{ m}^2/\text{d}$, $S = 0.0012$, $K' = 0.03 \text{ m/d}$, and $K = T/b = 940/30 = 31.3 \text{ m/d}$. The specific storage of the aquifer and the confining layer are given by

$$S_s = \frac{S}{b} = \frac{0.0012}{30} = 4 \times 10^{-5}$$

$$S'_s = \frac{S'}{b'} = \frac{0.0001}{2} = 5 \times 10^{-5}$$

Storage in the semiconfining layer can be neglected if

$$\frac{r}{4b}\left(\frac{K'S'_s}{KS_s}\right)^{1/2} < 0.01$$

$$\frac{10}{4 \times 30}\left(\frac{0.03 \times 5 \times 10^{-5}}{31.3 \times 4 \times 10^{-5}}\right)^{1/2} < 0.01$$

$$0.0029 < 0.01$$

which indicates that storage in the semiconfining layer can be neglected. The alternate criterion for neglecting storage in the semiconfining layer is

$$t > 0.036\,\frac{b'S'}{K'}$$

$$1\,\text{d} > 0.036\,\frac{(2)(0.0001)}{0.03}\,\text{d}$$

$$1\,\text{d} > 2 \times 10^{-4}\,\text{d}$$

which also indicates that storage in the semiconfining layer can be neglected. Satisfaction of either criterion for neglecting storage in the semiconfining layer is sufficient to neglect this effect. Since both criteria for neglecting storage in the semiconfining layer are satisfied, there is more than sufficient justification for neglecting this effect.

Drawdown in the overlying aquifer can be neglected if

$$T_s > 100T$$

$$920 \text{ m}^2/\text{d} > (100)(940) \text{ m}^2/\text{d}$$

$$920 \text{ m}^2/\text{d} > 94,000 \text{ m}^2/\text{d}$$

which indicates that drawdown in the overlying aquifer cannot be neglected. The alternate criterion for neglecting drawdown in the overlying aquifer is

$$t < \frac{S'(b')^2}{10bK'}$$

$$1 \text{ d} < \frac{0.0001(2)^2}{10(30)(0.03)} \text{ d}$$

$$1 \text{ d} < 4.4 \times 10^{-5} \text{ d}$$

which indicates that drawdown in the overlying aquifer cannot be neglected. Both criteria for neglecting drawdown in the overlying aquifer indicate that drawdown in the overlying aquifer should be considered in the analysis.

Storage in the pumping well can be neglected if

$$\frac{r_w}{(Tb'/K')^{1/2}} < 0.1$$

$$\frac{0.3}{(940 \times 2/0.03)^{1/2}} < 0.1$$

$$0.0012 < 0.1$$

which indicates that storage in the pumping well can be neglected. The additional criterion for neglecting storage in the pumping well is

$$t > 30 \frac{r_w^2 S}{T}\left[1 - \left(\frac{10r_w}{b}\right)^2\right]$$

$$1 \text{ d} > 30 \frac{0.3^2(0.0012)}{940}\left[1 - \left(\frac{10 \times 0.3}{30}\right)^2\right] \text{ d}$$

$$1 \text{ d} > 3.4 \times 10^{-6} \text{ d}$$

Since both criteria for neglecting storage in the pumping well are satisfied, this effect can be neglected.

The results of these analyses support the assumptions that storage in the semiconfining layer and storage in the well can both be neglected. However, drawdown in the overlying aquifer should be taken into account. Based on these results, the analysis of pump-test data using the Hantush–Jacob method and associated assumptions is not justified.

If a well in a semiconfined aquifer is pumped long enough, all the water will come from leakage across the semiconfining layer and none from elastic storage in the aquifer (Hantush and Jacob, 1954). This occurs when

$$t > 8 \frac{b'S}{K'} \tag{6.275}$$

and in this case the drawdown can be estimated using the relation

$$s = \frac{Q_w}{2\pi T} K_0(r/\lambda) \tag{6.276}$$

where K_0 is the zero-order modified Bessel function of the second kind.

6.3.9 Partially Penetrating Wells

The analytic solutions presented in the previous sections have all assumed that the pumping wells fully penetrate the aquifer and that the flow induced by the pumping well is approximately two dimensional. In cases where the pumping well does not fully penetrate the aquifer, in the immediate vicinity of the well the flow pattern is significantly three dimensional and the two-dimensional approximation is not valid. This effect is illustrated in Figure 6.21 for the case of a confined aquifer. The average flowpath induced by a partially penetrating well is longer than for a fully penetrating

FIGURE 6.21: Partially penetrating well in a confined aquifer

Source: Todd (1980).

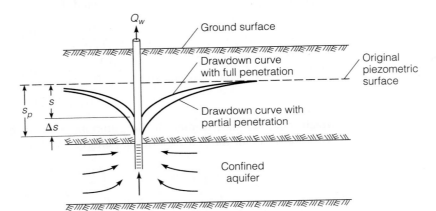

(a) Flow to a Partially Penetrating Well

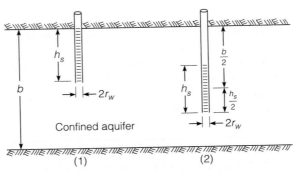

(b) Two Cases of Partial Penetration

well; consequently, for the same pumping rate, the drawdown at the well induced by a partially penetrating well is greater than for a fully penetrating well. This condition is illustrated in Figure 6.21(a), where the drawdown, s_p, at the partially penetrating well can be expressed in the form

$$s_p = s + \Delta s \tag{6.277}$$

where s is the drawdown induced if the well were fully penetrating and Δs is the additional drawdown caused by partial penetration. Figure 6.21(b) illustrates two cases of partial penetration: a case in which the well penetrates from the top of the aquifer and a case in which the well is centered within the aquifer.

In the case where the well penetrates from the top of the confined aquifer, the additional drawdown associated with partial penetration, Δs, can be estimated by (DeGlee, 1930; Bear, 1979; Prakash, 2004)

$$\Delta s = \frac{Q_w}{2\pi T} \frac{1-p}{p} \ln \frac{(1.2-p)h_s}{r_w} \tag{6.278}$$

where Q_w is the pumping rate, T is the transmissivity of the aquifer, p is the *penetration fraction* defined as

$$p = \frac{h_s}{b} \tag{6.279}$$

where h_s is the length of the screen, b is the thickness of the aquifer, and r_w is the radius of the well. An alternative semiempirical estimate of the drawdown at a partially penetrating well that extends from the top of a confined aquifer was proposed by Kozeny (1933) based on earlier theoretical work by Muskat (1932; 1937), and this relation is still recognized as a valid approach (Bear, 1979; Prakash, 2004). The Kozeny (1933) equation is given by

$$s_p = \frac{s}{p \left[1 + 7\sqrt{\frac{r_w}{2h_s}} \cos\left(\frac{\pi p}{2}\right) \right]} \tag{6.280}$$

where s_p is the drawdown at the partially penetrating (pumped) well, and s is the drawdown in the case that the well is fully penetrating and is given by

$$s = \frac{Q_w}{2\pi T} \ln\left(\frac{R}{r_w}\right) \tag{6.281}$$

where R is the radius of influence of the fully penetrating well.

In the cases where the well is screened in the center of the confined aquifer, then the drawdown correction, Δs, is given by (DeGlee, 1930; Bear, 1979; Prakash, 2004)

$$\Delta s = \frac{Q_w}{2\pi T} \frac{1-p}{p} \ln \frac{(1.2-p)h_s}{2r_w} \tag{6.282}$$

Corrections for other screen placements in a confined aquifer can be found in Huisman (1972), and Equations 6.278 and 6.282 are applicable for $10(r_w/b) \le p \le 0.8$.

EXAMPLE 6.24

A partially penetrating well with a screen length of 8 m is to be installed in a 20-m thick confined aquifer. The hydraulic conductivity of the aquifer is 100 m/d, the radius of the well is 0.3 m, the pumping rate is 40 L/s, and the radius of influence is 200 meters. Compare the drawdown at the partially penetrating well with the drawdown at a fully penetrating well at the same location, where: (a) the partially penetrating well is screened from the top of the aquifer, and (b) the partially penetrating well is screened in the center of the aquifer.

Solution

(a) From the given data: $Q_w = 40$ L/s $= 3456$ m^3/d; $K = 100$ m/d; $b = 20$ m; $h_s = 8$ m; $p = h_s/b = 8/20 = 0.4$; $r_w = 0.3$ m, and $R = 200$ m. For a fully penetrating well, the drawdown at the well is given by Equation 6.281 as

$$s_w = \frac{Q_w}{2\pi T} \ln\left(\frac{R}{r_w}\right)$$

$$= \frac{3456}{2\pi(100)(20)} \ln\left(\frac{200}{0.3}\right) = 1.79 \text{ m}$$

In the case of a partially penetrating well, $10(r_w/b) = 10(0.3/20) = 0.15$ and $p = 0.4$, hence the requirement that $10(r_w/b) \le p \le 0.8$ is satisfied and Equations 6.278 and 6.282 are applicable. Equation 6.278 gives the additional drawdown, Δs, when the well is screened from the top of the aquifer, as

$$\Delta s = \frac{Q_w}{2\pi T} \frac{1-p}{p} \ln \frac{(1.2 - p)h_s}{r_w}$$

$$= \frac{3456}{2\pi(100 \times 20)} \frac{1 - 0.4}{0.4} \ln \frac{(1.2 - 0.4)8}{0.3} = 1.26 \text{ m}$$

Therefore, according to Equation 6.278, the partially penetrating well produces a drawdown of 1.79 m + 1.26 m = 3.05 m at the partially penetrating well. The Kozeny equation (Equation 6.280) gives the drawdown at the partially penetrating well as

$$s_p = \frac{s}{p\left[1 + 7\sqrt{\frac{r_w}{2h_s}} \cos\left(\frac{\pi p}{2}\right)\right]}$$

$$= \frac{1.79}{(0.4)\left[1 + 7\sqrt{\frac{0.3}{2(8)}} \cos\left(\frac{\pi(0.4)}{2}\right)\right]} = 2.52 \text{ m}$$

Therefore, the Kozeny equation gives a drawdown at the partially penetrating well which is 17% less than predicted by Equation 6.278. Both drawdowns (3.05 m and 2.52 m) can be contrasted with the drawdown at a fully penetrating well equal to 1.79 m.

(b) In the case where the well is screened in the center of the aquifer, Equation 6.282 gives the additional drawdown as

$$\Delta s = \frac{Q_w}{2\pi T}\frac{1-p}{p}\ln\frac{(1.2-p)h_s}{2r_w}$$

$$= \frac{3456}{2\pi(100\times20)}\frac{1-0.4}{0.4}\ln\frac{(1.2-0.4)8}{2(0.3)} = 0.98\text{ m}$$

Therefore, the partially penetrating well produces a drawdown of 1.79 m + 0.98 m = 2.77 m, compared with the 1.79 m drawdown of the fully penetrating well.

The drawdown, s_p, at a partially penetrating well in an unconfined aquifer can be expressed in the form

$$\boxed{s_p^2 = s^2 + 2h_w\Delta s} \tag{6.283}$$

where s is the drawdown induced if the well were fully penetrating, h_w is the saturated thickness of the aquifer at the well, and Δs is a drawdown increment. For the case where the well penetrates the top of the unconfined aquifer

$$\Delta s = \frac{Q_w}{2\pi Kh_w}\frac{1-p}{p}\ln\frac{(1.2-p)h_s}{r_w} \tag{6.284}$$

and in the case where the well is centered in the unconfined aquifer

$$\Delta s = \frac{Q_w}{2\pi Kh_w}\frac{1-p}{p}\ln\frac{(1.2-p)h_s}{2r_w} \tag{6.285}$$

EXAMPLE 6.25

Repeat the previous example for the case of an unconfined aquifer, where the drawdown at the fully penetrating well is 1 m.

Solution

(a) In the case where the well penetrates the top of the aquifer, Equation 6.284 gives

$$\Delta s = \frac{Q_w}{2\pi Kh_w}\frac{1-p}{p}\ln\frac{(1.2-p)h_s}{r_w}$$

$$= \frac{3456}{2\pi(100)(19)}\frac{1-0.4}{0.4}\ln\frac{(1.2-0.4)8}{0.3} = 1.33\text{ m}$$

where $h_w = 20\text{ m} - 1\text{ m} = 19\text{ m}$ is the saturated thickness at the well under full penetration. According to Equation 6.283, the drawdown, s_p, under partially penetrating conditions is given by

$$s_p = \sqrt{s^2 + 2h_w\,\Delta s}$$

where $s = 1\text{ m}$, $h_w = 19\text{ m}$, and $\Delta s = 1.33\text{ m}$, hence

$$s_p = \sqrt{1^2 + 2(19)(1.33)} = 7.18\text{ m}$$

Therefore, the additional drawdown due to partial penetration is 7.18 m − 1 m = 6.18 m.

(b) In the case where the well is screened in the center of the aquifer, Equation 6.285 gives

$$\Delta s = \frac{Q_w}{2\pi K h_w} \frac{1-p}{p} \ln \frac{(1.2-p)h_s}{2r_w}$$

$$= \frac{3456}{2\pi(100)(19)} \frac{1-0.4}{0.4} \ln \frac{(1.2-0.4)8}{2(0.3)} = 1.03 \text{ m}$$

and the corresponding total drawdown, s_p, is

$$s_p = \sqrt{s^2 + 2h_w \Delta s} = \sqrt{1^2 + 2(19)(1.03)} = 6.34 \text{ m}$$

Hence the additional drawdown due to partial penetration is 6.34 m − 1 m = 5.34 m. Placing the well in the center of the aquifer rather than at the top of the aquifer reduces the pumping head by 7.18 m − 6.34 m = 0.84 m.

As the distance from a partially penetrating well increases, the flow becomes more two dimensional and the effect of partial penetration on the drawdown decreases. In homogeneous isotropic aquifers, this distance is estimated to be 1.5 to 2 times the saturated thickness of the aquifer. Hantush (1964) has shown that the effect of partial penetration in anisotropic aquifers is negligible at any location more than r' away from the pumping well, where r' is given by

$$r' = 1.5H\sqrt{\frac{K_h}{K_v}} \tag{6.286}$$

where H is the saturated thickness of the aquifer, and K_h and K_v are the horizontal and vertical hydraulic conductivities, respectively. Wells that penetrate more than 85% of the saturated thickness of an aquifer can usually be assumed to be fully penetrating (USBR, 1995).

6.4 Principle of Superposition

The principle of superposition can be stated as follows (Bear, 1979): If $s_1, s_2, \ldots, s_n$ are n general solutions of a homogeneous linear partial differential equation

$$L(s) = 0 \tag{6.287}$$

where L represents a linear operator, then

$$s = \sum_{i=1}^{n} C_i s_i \tag{6.288}$$

is also a solution of Equation 6.287, and $C_i, i = [1, n]$ are constants determined by the boundary conditions.

6.4.1 Multiple Wells

The principle of superposition can be elucidated by considering the case of two fully penetrating wells in an isotropic and homogeneous confined aquifer, with one well located at (x_1, y_1) and the other at (x_2, y_2). Assuming two-dimensional flow, the governing equation for the drawdown, $s(r, t)$ induced by these wells is given by Equation 6.195 as

$$\frac{\partial^2 s}{\partial r^2} + \frac{1}{r}\frac{\partial s}{\partial r} = \frac{S}{T}\frac{\partial s}{\partial t} \tag{6.289}$$

This equation can be written in the form

$$L(s) = 0 \tag{6.290}$$

where $L()$ is the linear operator

$$L() = \frac{\partial^2()}{\partial r^2} + \frac{1}{r}\frac{\partial()}{\partial r} - \frac{S}{T}\frac{\partial()}{\partial t} \tag{6.291}$$

In the case of unsteady flow to wells in an infinite confined aquifer, Equation 6.289 is to be solved subject to the following initial and boundary conditions:

$$s(r, 0) = 0 \tag{6.292}$$

$$s(\infty, t) = 0 \tag{6.293}$$

$$\lim_{r_1 \to 0} r_1 \frac{\partial s}{\partial r_1} = -\frac{Q_1}{2\pi T} \tag{6.294}$$

$$\lim_{r_2 \to 0} r_2 \frac{\partial s}{\partial r_2} = -\frac{Q_2}{2\pi T} \tag{6.295}$$

where r_1 and r_2 are radial coordinates measured from (x_1, y_1) and (x_2, y_2), respectively, and Q_1 and Q_2 are the respective pumping rates from the two wells. To justify solving this problem by superposition, it is noted that if $s_1(r, t)$ and $s_2(r, t)$ are the drawdowns induced by each well operating by itself, then the following equations must be satisfied:

$$L(s_1) = 0 \tag{6.296}$$

and

$$L(s_2) = 0 \tag{6.297}$$

with boundary conditions

$$s_1(r_1, 0) = 0 \tag{6.298}$$

$$s_1(\infty, t) = 0 \tag{6.299}$$

$$\lim_{r_1 \to 0} r_1 \frac{\partial s_1}{\partial r_1} = -\frac{Q_1}{2\pi T} \tag{6.300}$$

and

$$s_2(r_2, 0) = 0 \tag{6.301}$$

$$s_2(\infty, t) = 0 \tag{6.302}$$

$$\lim_{r_2 \to 0} r_2 \frac{\partial s_2}{\partial r_2} = -\frac{Q_2}{2\pi T} \tag{6.303}$$

Since both s_1 and s_2 satisfy the governing (homogeneous) differential equation, Equation 6.290, then by the principle of superposition

$$s = C_1 s_1 + C_2 s_2 \tag{6.304}$$

also satisfies the governing differential equation. The initial condition, Equation 6.292, and the boundary condition given by Equation 6.293 are both satisfied by Equation 6.304 by virtue of the fact that s_1 and s_2 satisfy Equations 6.298, 6.299, 6.301, and 6.302. The boundary condition given by Equation 6.294 requires that $C_1 = 1$, by virtue of Equation 6.300 and the fact that the continuity equation guarantees that

$$\lim_{r_1 \to 0} r_1 \frac{\partial s_2}{\partial r_1} = 0 \tag{6.305}$$

The boundary condition given by Equation 6.295 requires that $C_2 = 1$, by virtue of Equation 6.303 and the fact that the continuity equation guarantees that

$$\lim_{r_2 \to 0} r_2 \frac{\partial s_1}{\partial r_2} = 0 \tag{6.306}$$

Therefore, in the case of two pumping wells in an infinite confined aquifer, it has been demonstrated that the drawdown induced by the two wells, $s(r, t)$, is equal to the sum of the drawdowns induced by each of the wells operating individually, in which case

$$s(r, t) = s_1(r, t) + s_2(r, t) \tag{6.307}$$

This analysis can be extended to the case of multiple wells, in which case it can be demonstrated that the induced drawdown is simply the sum of the drawdowns induced by each of the wells. Although the example provided here extends only to multiple wells in confined aquifers of infinite extent, the principle of superposition applies also to steady flows in infinite unconfined aquifers, where the governing equation is linear in h^2, and also to unsteady flows in infinite unconfined aquifers, provided the governing equation is linearized by assuming that the drawdowns are small relative to the aquifer thickness. As a rule of thumb, linearization of the governing equation in unconfined aquifers is justified whenever drawdowns are less than 5% of the initial aquifer thickness (Boonstra, 1998a).

EXAMPLE 6.26

A municipal wellfield consists of five water-supply wells, each rated at 16 L/s. The wells are located 100 m apart along a north-south line and fully penetrate a 20-m thick confined aquifer that has a transmissivity of 1000 m^2/d and a storage coefficient of 6×10^{-5}. Estimate the drawdown 500 m west of the center well after one week, one month, one year, and one decade of wellfield operation.

Solution   In accordance with the principle of superposition, the drawdown s_P at a location, P, 500 m west of the center well is given by

$$s_P = \sum_{i=1}^{5} s_i$$

where s_i is the drawdown at P induced by well i. The drawdown in the aquifer induced by each well is given by the Theis equation (Equation 6.219), where

$$s_i = \frac{Q_w}{4\pi T} W(u_i)$$

and hence

$$s_P = \frac{Q_w}{4\pi T} \sum_{i=1}^{5} W(u_i)$$

where

$$u_i = \frac{r_i^2 S}{4Tt}$$

From the given data: $Q_w = 16$ L/s $= 1382$ m³/d, $T = 1000$ m²/d, $S = 0.00006$, $r_1 = 500$ m, $r_2 = r_3 = \sqrt{500^2 + 100^2} = 509.9$ m, $r_4 = r_5 = \sqrt{500^2 + 200^2} = 538.5$ m, and therefore

$$s_P = \frac{1382}{4\pi(1000)} \left[W\left(\frac{500^2 \times 0.00006}{4t(1000)} \right) + 2W\left(\frac{509.9^2 \times 0.00006}{4t(1000)} \right) \right.$$

$$\left. + 2W\left(\frac{538.5^2 \times 0.00006}{4t(1000)} \right) \right]$$

$$= 0.110 \left[W\left(\frac{0.00375}{t} \right) + 2W\left(\frac{0.00390}{t} \right) + 2W\left(\frac{0.00435}{t} \right) \right]$$

Applying this result for $t = 7$ days ($=1$ week), 30 days ($=1$ month), 365 days ($=1$ year), and 3650 days ($=10$ years) gives

t	s_P (m)
1 week	3.78
1 month	4.58
1 year	5.96
1 decade	7.22

6.4.2   Well in Uniform Flow

In the previous application, the drawdown in the aquifer is zero whenever the pumping rate is equal to zero. In many cases, there is a regional mean flow, which can

694

be characterized by a uniform seepage velocity, v_0. In this case, the drawdown, s', is related to v_0 by Darcy's law, and for confined aquifers it is given by

$$s' = \frac{nv_0}{K}x \qquad (6.308)$$

where n is the effective porosity, K is the hydraulic conductivity of the aquifer, and x is the distance downstream from where the drawdown is zero. This fundamental solution to the ground-water flow equation can be used with the principle of superposition in cases where a regional flow exists. The drawdown, $s''(x, y)$, induced by a single well in an infinite aquifer is given by Equation 6.139 as

$$s''(x, y) = s_w - \frac{Q_w}{2\pi T} \ln\left(\frac{\sqrt{(x - x_0)^2 + (y - y_0)^2}}{r_w}\right), \qquad s'(r) \geq 0 \qquad (6.309)$$

where s_w is the drawdown at the pumping well located at (x_0, y_0), Q_w is the pumping rate, T is the transmissivity, and r_w is the radius of the well. Since both Equations 6.308 and 6.309 satisfy the governing flow equation, then by the principle of superposition, the sum of these solutions also satisfies the governing flow equation. Second, the required boundary conditions are that

$$\lim_{r \to r_w} r\frac{\partial s}{\partial r} = -\frac{Q_w}{2\pi T} \qquad (6.310)$$

and that as $x \to \pm\infty$ the drawdown approaches the uniform-flow solution. If we consider the drawdown, $s(x, y)$, defined by

$$s(x, y) = s'(x, y) + s''(x, y) \qquad (6.311)$$

then it is clear that $s(x, y)$ satisfies both boundary conditions, and the drawdown induced by a pumping well in an aquifer with a regional flow is given by

$$s(x, y) = s_w + \frac{nv_0}{K}x - \frac{Q_w}{2\pi T} \ln\left(\frac{\sqrt{(x - x_0)^2 + (y - y_0)^2}}{r_w}\right) \qquad (6.312)$$

EXAMPLE 6.27

A 0.4-m diameter fully penetrating well pumps 15 L/s from a 25-m thick confined aquifer with a hydraulic conductivity of 30 m/d and a porosity of 0.2. The regional mean flow has a seepage velocity of 1 m/d. If the incremental drawdown at the well induced by pumping is 2 m, estimate the drawdown 50 m upstream of the well with and without the regional mean flow. At what distance downstream of the well is the seepage velocity equal to zero?

Solution   From the given data: $s_w = 2$ m, $n = 0.2$, $v_0 = 1$ m/d, $K = 30$ m/d, $Q_w = 15$ L/s $= 1300$ m³/d, $b = 25$ m, $T = Kb = (30)(25) = 750$ m²/d, and $r_w = 0.2$ m. Taking $(x_0, y_0) = (0$ m, 0 m$)$, then at 50 m upstream of the well $(x, y) = (-50$ m, 0 m$)$

and the drawdown, $s(-50, 0)$, is given by Equation 6.312 as

$$s(x, y) = s_w + \frac{nv_0}{K}x - \frac{Q_w}{2\pi T}\ln\left(\frac{\sqrt{(x - x_0)^2 + (y - y_0)^2}}{r_w}\right)$$

$$= 2 + \left[\frac{(0.2)(1)}{30}(-50)\right] - \frac{1300}{2\pi(750)}\ln\left(\frac{\sqrt{50^2 + 0^2}}{0.2}\right)$$

$$= 0.48 + [-0.33]$$

where the drawdown in square brackets represents the contribution of the regional mean flow. Therefore, in the absence of a regional mean flow, the drawdown 50 m upstream of the well is 0.48 m, and with the regional mean flow the drawdown is 0.48 m − 0.33 m = 0.15 m.

The seepage velocity is equal to zero where $ds/dx = 0$. Differentiating Equation 6.312 with respect to x and taking $y = y_0 = 0$ and $x_0 = 0$ gives

$$\frac{ds}{dx} = \frac{nv_0}{K} - \frac{Q_w}{2\pi Tx}$$

Taking $ds/dx = 0$ and rearranging yields

$$x = \frac{Q_w K}{2\pi Tnv_0}$$

Substituting parameter values gives

$$x = \frac{(1300)(30)}{2\pi(750)(0.2)(1)} = 41.4 \text{ m}$$

Therefore, at a distance 41.4 m downstream of the well the seepage velocity is equal to zero. At this location, the direction of the seepage velocity changes from toward the well to away from the well. A practical result is that contaminant sources farther than 41.4 m downstream of the well cannot impact the well.

This analysis of a single well in a uniform flow can be extended to multiple wells in a uniform flow by superimposing the drawdowns induced by additional wells. Well(s) in a uniform flow field are commonly used to intercept contaminant plumes in ground water. This situation is illustrated in Figure 6.22, where the region contributing flow to the well is commonly called the *capture zone*. The ground-water divide between the region that flows to the well and the region flowing by the well is given by

$$\frac{y}{x} = -\tan\left(\frac{2\pi KbJ}{Q}y\right)$$

where J is the original piezometric gradient. It can be shown that the boundary of the capture zone approaches $y = y_L$ as $x \to \infty$, where

$$y_L = \pm\frac{Q}{2KbJ}$$

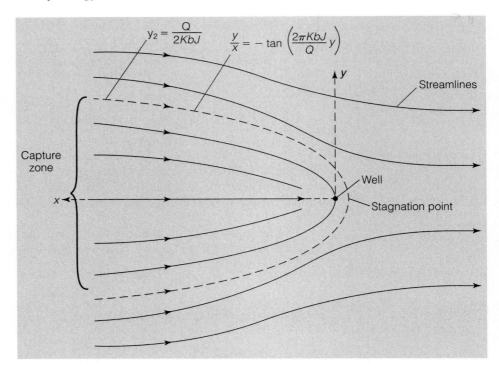

and the stagnation point, where the seepage velocity is equal to zero, occurs at

$$x_S = -\frac{Q}{2\pi KbJ}, \quad y = 0$$

6.5   Method of Images

The method of images is a practical application of the principle of superposition in which several fundamental solutions to the governing (linear) flow equation are combined to yield a solution to the flow equation that satisfies some prescribed boundary conditions. This approach is illustrated in the following sections.

6.5.1   Constant-Head Boundary

Consider the well shown in Figure 6.23, located at a distance L away from a constant-head boundary such as a fully penetrating reservoir. Under steady-state conditions, a solution is sought that satisfies the linearized flow equation given by

$$\nabla^2 s = 0 \tag{6.313}$$

where $s(x, y)$ is the drawdown defined by

$$s(x, y) = H - h(x, y) \tag{6.314}$$

FIGURE 6.23: Well near a constant-head boundary

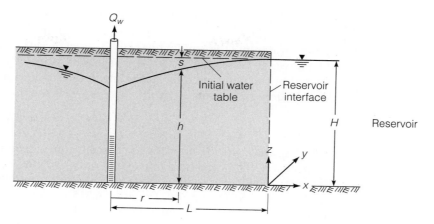

FIGURE 6.23: Well near a constant-head boundary

and $\nabla^2()$ is the Laplacian operator. The boundary conditions to be satisfied in this case are

$$s(0, y) = 0 \tag{6.315}$$

$$\lim_{x \to -\infty} s(x, y) = 0 \tag{6.316}$$

$$\lim_{r \to r_w} r \frac{ds}{dr} = -\frac{Q_w}{2\pi T} \tag{6.317}$$

where r is the radial coordinate originating from the location of the pumping well. If the location of the pumping well is (x_0, y_0), then r is related to the Cartesian coordinates by

$$r^2 = (x - x_0)^2 + (y - y_0)^2 \tag{6.318}$$

Although the solution of the governing equation (Equation 6.313) and associated boundary conditions (Equations 6.315 to 6.317) seems challenging, the solution is actually quite simple using the method of images. Consider the situation illustrated in Figure 6.24, where the fully penetrating reservoir is replaced by a well that is identical

FIGURE 6.24: Pumping well and image well relative to a constant-head boundary

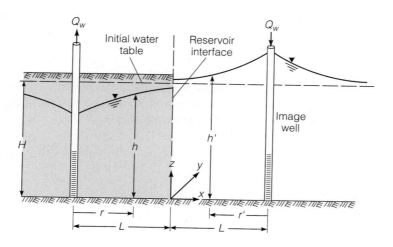

to the pumping well and placed symmetrically about the reservoir interface. This is called an *image well* and differs from the pumping well only in that the image well pumps water into the aquifer at a rate Q_w, and therefore the magnitude of the buildup of the water table caused by the image well is exactly the same as the drawdown caused by the pumping well. According to the superposition principle, the drawdown distribution, $s(x, y)$, caused by the pumping well and image well combined is given by

$$s(x, y) = s_p(x, y) + s_i(x, y) \tag{6.319}$$

where s_p and s_i are the drawdowns caused by the individual pumping and image wells, respectively. The combined drawdown $s(x, y)$ defined by Equation 6.319 satisfies the boundary conditions required for a well adjacent to a fully penetrating reservoir, given by Equations 6.315 to 6.317. Equation 6.315 is satisfied by virtue of the fact that the drawdown induced by the pumping well and the buildup induced by the recharge well along the reservoir boundary at $x = 0$ are equal, and therefore the total drawdown, s, given by Equation 6.319, is zero along $x = 0$. The second boundary condition, Equation 6.316, is met by virtue of the fact that both s_p and s_i approach zero as $x \to -\infty$, and the third boundary condition, Equation 6.317, is met by virtue of the fact that

$$\lim_{r \to r_w} r \frac{ds_p}{dr} = -\frac{Q_w}{2\pi T} \tag{6.320}$$

and

$$\lim_{r \to r_w} r \frac{ds_i}{dr} = 0 \quad \text{(continuity)} \tag{6.321}$$

and therefore

$$\lim_{r \to r_w} r \frac{d}{dr}(s_p + s_i) = -\frac{Q_w}{2\pi T} \tag{6.322}$$

which is exactly the requirement of Equation 6.317. The expression for the drawdown induced by a single pumping well in an infinite unconfined aquifer is

$$s_p = \frac{Q_w}{2\pi T} \ln\left(\frac{R}{r}\right) \tag{6.323}$$

and for the image (recharge) well the drawdown, s_i, is equal to

$$s_i = -\frac{Q_w}{2\pi T} \ln\left(\frac{R}{r'}\right) \tag{6.324}$$

where r' is measured relative to the recharge well. According to the analysis presented here, the drawdown distribution, s, induced by a pumping well located at a distance L from a fully penetrating reservoir is given by

$$s = s_p + s_i = \frac{Q_w}{2\pi T}\left[\ln\left(\frac{R}{r}\right) - \ln\left(\frac{R}{r'}\right)\right] \tag{6.325}$$

or

$$\boxed{s = \frac{Q_w}{2\pi T} \ln\left(\frac{r'}{r}\right)} \tag{6.326}$$

which can be written in Cartesian coordinates as

$$s = \frac{Q_w}{4\pi T} \ln \left[\frac{(x' - 2L)^2 + y'^2}{x'^2 + y'^2} \right] \tag{6.327}$$

where x' are y' are the coordinates relative to the pumping well location. The example presented here illustrates that the method of images simply consists of the superposition of fundamental solutions to the flow equation, where the fundamental solutions are combined in such a way that the prescribed boundary conditions are met.

EXAMPLE 6.28

A well is located 100 m west of a fully penetrating river that runs in a north-south direction. If the well pumping rate is 20 L/s, the saturated thickness of the aquifer is 20 m, and the hydraulic conductivity is 28 m/d, estimate the drawdowns 30 m north, south, east, and west of the well. What is the leakage out of the river per unit length of river at the section closest to the well?

Solution   From the given data: $Q_w = 20$ L/s $= 1728$ m^3/d, $K = 28$ m/d, $H = 20$ m, $T = KH = (28)(20) = 560$ m^2/d, and $L = 100$ m. The drawdown, $s(x,y)$, is given by Equation 6.327 as

$$
\begin{aligned}
s(x,y) &= \frac{Q_w}{4\pi T} \ln \left[\frac{(x' - 2L)^2 + y'^2}{x'^2 + y'^2} \right] \\
&= \frac{1728}{4\pi(560)} \ln \left[\frac{(x' - 2 \times 200)^2 + y'^2}{x'^2 + y'^2} \right] = 0.246 \ln \left[\frac{(x' - 200)^2 + y'^2}{x'^2 + y'^2} \right]
\end{aligned}
$$

and the drawdowns 30 m from the well are given in the following table:

Location	x' (m)	y' (m)	s (m)
North	0	30	0.94
South	0	-30	0.94
East	30	0	0.85
West	-30	0	1.00

The leakage out of the river per unit length (of river), q, is given by

$$
\begin{aligned}
q &= -KH \frac{ds}{dx'} \bigg|_{x'=100, \, y'=0} \\
&= -T \frac{d}{dx'} \left[\frac{Q_w}{4\pi T} \ln \frac{(x' - 200)^2}{x'^2} \right]_{x'=100} \\
&= -\frac{Q_w}{4\pi} \left[\frac{2x'(x' - 200) - 2(x' - 200)^2}{x'(x' - 200)^2} \right]_{x'=100} \\
&= -\frac{1728}{4\pi} \left[\frac{2(100)(100 - 200) - 2(100 - 200)^2}{100(100 - 200)^2} \right]_{x'=100} = 5.5 \text{ (m}^3\text{/d)/m}
\end{aligned}
$$

Hence the leakage rate out of the fully penetrating river is 5.5 (m³/d)/m at the section closest to the well. The leakage rates at other river sections are less than this value.

6.5.2 Impermeable Boundary

Consider the pumping well shown in Figure 6.25, located a distance L away from a fully penetrating impermeable boundary. The impermeable boundary could be due to a fault or could simply result from a lateral change in aquifer material. Under steady-state conditions, the equation to be solved is

$$\nabla^2 s = 0 \tag{6.328}$$

where $s(x, y)$ is the drawdown defined by

$$s(x, y) = H - h(x, y) \tag{6.329}$$

and the boundary conditions to be satisfied are

$$\left. \frac{\partial s}{\partial x} \right|_{x=0} = 0 \tag{6.330}$$

$$\lim_{x \to -\infty} s(x, y) = 0 \tag{6.331}$$

$$\lim_{r \to r_w} r \frac{ds}{dr} = -\frac{Q_w}{2\pi T} \tag{6.332}$$

Consider the superposition of the drawdowns induced by a pumping well and an identical image well located symmetrically about the impermeable boundary as illustrated in Figure 6.26. If $s_p(x, y)$ and $s_i(x, y)$ are the drawdowns caused by the real and image wells, respectively, in an infinite aquifer, then it can be shown that the sum of the drawdowns, $s(x, y)$, given by

$$s(x, y) = s_p(x, y) + s_i(x, y) \tag{6.333}$$

is the solution of Equation 6.328 and satisfies the boundary conditions given by Equations 6.330 to 6.332. This can be demonstrated by noting that both s_p and s_i satisfy Equation 6.328; therefore, the sum of s_p and s_i also satisfies Equation 6.328.

FIGURE 6.25: Well near an impermeable boundary

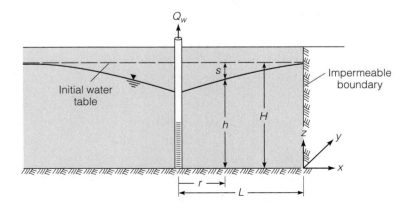

FIGURE 6.26: Pumping well and image well relative to an impermeable boundary

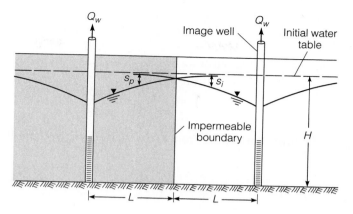

Since the drawdowns induced by the real and image wells are symmetric relative to the location of the impermeable boundary, then the sum of the drawdowns is also symmetric, and the slope of the drawdown curve is equal to zero at the impermeable boundary. Therefore, the boundary condition given by Equation 6.330 is satisfied by the sum of the drawdowns. Since the drawdowns induced by both the real and image well approach zero as the distance from the well increases, then the boundary condition given by Equation 6.331 is satisfied by the sum of the drawdowns. Finally, since s_p and s_i satisfy the equations

$$\lim_{r \to r_w} r \frac{ds_p}{dr} = -\frac{Q_w}{2\pi T} \tag{6.334}$$

and

$$\lim_{r \to r_w} r \frac{ds_i}{dr} = 0 \tag{6.335}$$

then

$$\lim_{r \to r_w} r \frac{d(s_p + s_i)}{dr} = -\frac{Q_w}{2\pi T} \tag{6.336}$$

which shows that the sum of the drawdowns also satisfies the boundary condition given by Equation 6.332. This completes the proof that the sum of the drawdowns induced by the real and image well satisfies both the governing equation and boundary conditions of the problem.

EXAMPLE 6.29

A well is located 100 m west of an impervious boundary that runs in the north-south direction. If the well pumps 20 L/s, the saturated thickness of the aquifer is 20 m, and the hydraulic conductivity is 28 m/d, estimate the drawdowns 30 m north, south, east, and west of the well. The radius of influence of the well can be taken as 600 m.

Solution   The drawdown distribution, $s(x, y)$, is the sum of the drawdowns induced by two pumping wells placed symmetrically about the impermeable boundary. Hence,

$$s(x, y) = s_p + s_i = \frac{Q_w}{2\pi T}\left[\ln\left(\frac{R}{r}\right) + \ln\left(\frac{R}{r'}\right)\right] = \frac{Q_w}{2\pi T}\ln\left(\frac{R^2}{rr'}\right)$$

where R is the radius of influence, r is the radial distance of (x, y) from the pumping well, and r' is the radial distance of (x, y) from the image well. From the given data, $Q_w = 20$ L/s $= 1728$ m³/d, $K = 28$ m/d, $H = 20$ m, $T = KH = (28)(20) = 560$ m²/d, and $R = 600$ m. Substituting the given data into the drawdown equation yields

$$s(x, y) = \frac{1728}{2\pi(560)} \ln\left(\frac{600^2}{rr'}\right) = 0.491(12.8 - \ln rr')$$

The drawdowns 30 m from the well are given in the following table:

Location	r (m)	r' (m)	s (m)
North	30	202	2.01
South	30	202	2.01
East	30	170	2.09
West	30	230	1.94

Impermeable barriers are frequently associated with the rising side of a buried valley. This situation is quite common in the northern, once-glaciated parts of the United States. Indeed, an aquifer is often cut off in two parallel directions by buried-valley walls (AWWA, 2003b).

6.5.3   Other Applications

The previous examples clearly demonstrate the fundamental reasons why the method of images works, and provide sufficient guidance to apply this method to other cases. The linearity and homogeneity of the governing differential equation guarantee that superimposed solutions will also satisfy the governing differential equation. The selection of the location(s) of image wells is controlled by the requirement that the superimposed drawdowns must meet the boundary conditions. In the case of a constant-head boundary, an image well is placed to ensure zero drawdown at the constant-head boundary; in the case of an impermeable boundary, an image well is placed to ensure that the slope of the drawdown curve is zero at the impermeable boundary.

6.6   Saltwater Intrusion

In coastal aquifers, a transition region exists where the water in the aquifer changes from fresh water to saltwater. However, because saltwater is denser than fresh water, the saltwater tends to form a wedge beneath the fresh water, as shown in Figure 6.27, for the case of an unconfined aquifer. This illustration is somewhat idealized, since in reality there is not a sharp interface between fresh water and saltwater, but rather a "blurred" interface resulting from diffusion and mixing caused by the relative movement of the fresh water and saltwater. This relative movement is usually associated with tides and temporal variations in aquifer stresses. The thickness of the transition zone between fresh water and saltwater can range from a few meters to over a hundred meters (Visher and Mink, 1964). The intrusion of saltwater into coastal aquifers is generally of concern because of the associated deterioration in ground-water

FIGURE 6.27: Saltwater interface in a coastal aquifer

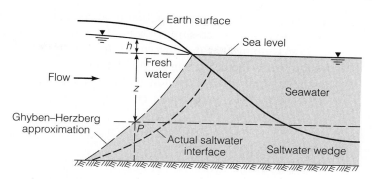

quality. Since the recommended maximum contaminant level (MCL) for chloride in drinking water is 250 mg/L and a typical chloride level in seawater is 14,000 mg/L, then mixing more than 1.8% seawater with nonsaline water renders the mixture nonpotable. This percentage is even less if the fresh water contains a nonzero chloride concentration. In the United States, saltwater intrusion has resulted in the degradation of aquifers in at least 20 of the coastal states (Newport, 1977) and has been primarily caused by overpumping in sensitive portions of the aquifers. The most seriously affected states are Florida, California, Texas, New York, and Hawaii (Rail, 1989).

An approximate method for determining the location of the saltwater interface was introduced independently by Badon-Ghyben (1888) and Herzberg (1901) and is called the *Ghyben–Herzberg approximation*. Under this approximation, the pressure distribution is assumed to be hydrostatic within any vertical section of the aquifer, which implicitly assumes that the streamlines are horizontal. Under this assumption, the hydrostatic pressure at point P in Figure 6.27 can be calculated from either the fresh-water head or the saltwater head, which means that

$$\gamma_f(h + z) = \gamma_s z \tag{6.337}$$

where γ_f is the specific weight of fresh water, γ_s is the specific weight of saltwater, h is the elevation of the water table above sea level, and z is the depth of the saltwater interface below sea level. Solving Equation 6.337 for z leads to

$$z = \frac{\gamma_f}{\gamma_s - \gamma_f} h \quad \text{or} \quad z = \frac{\rho_f}{\rho_s - \rho_f} h \tag{6.338}$$

where ρ_f is the density of fresh water and ρ_s is the density of saltwater. This is called the *Ghyben–Herzberg equation*. Under typical conditions, $\rho_f = 1000 \text{ kg/m}^3$ and $\rho_s = 1025 \text{ kg/m}^3$. Substituting these values into Equation 6.338 leads to

$$z \approx 40\,h \tag{6.339}$$

which means that the saltwater interface will typically be found at a distance below sea level equal to 40 times the elevation of the water table above sea level. The Ghyben–Herzberg approximation also means that the slope of the salt water interface is 40 times greater than the slope of the water table. Near the shore, the depth to the interface predicted by the Ghyben–Herzberg approximation tends to be less than the actual depth observed in the field (Fitts, 2002). In fact, at the shoreline the

Ghyben–Herzberg approximation predicts that the saltwater interface is at sea level, while there must necessarily be a nonzero thickness of fresh water there.

Assuming that the flow in the fresh-water portion of the aquifer is horizontal and towards the coast, neglecting direct surface recharge (such as from rainfall), and assuming that there is no flow within the saltwater wedge, the flowrate, Q, of fresh water toward the coast can be estimated using the Darcy equation

$$Q = K(h + z)\frac{dh}{dx} \tag{6.340}$$

where K is the hydraulic conductivity of the aquifer, and x is the distance inland from the shoreline. Equation 6.340 uses the Dupuit approximation, which assumes horizontal flow and equates the horizontal piezometric head gradient to the slope of the water table. Combining Equations 6.340 and 6.338 yields

$$Q = K\left(\frac{\gamma_s}{\gamma_s - \gamma_f}\right)h\frac{dh}{dx} \tag{6.341}$$

and integrating Equation 6.341 yields

$$Qx = \frac{K}{2}\left(\frac{\gamma_s}{\gamma_s - \gamma_f}\right)h^2 + C \tag{6.342}$$

where C is an integration constant. Applying the boundary condition that $h = 0$ at $x = 0$ yields $C = 0$, and applying the boundary condition that $h = h_L$ at $x = L$ yields

$$\boxed{Q = \frac{K}{2L}\left(\frac{\gamma_s}{\gamma_s - \gamma_f}\right)h_L^2} \tag{6.343}$$

This equation is particularly useful in estimating the flow of fresh water toward the coast, based on the elevation, h_L, of the water at a distance L from the coast. The water-table profile can be estimated by combining Equations 6.342 (with $C = 0$) and 6.343 to yield

$$\boxed{h = h_L\sqrt{\frac{x}{L}}} \tag{6.344}$$

The results presented here demonstrate that a small number of piezometric head measurements can be used to obtain an estimate of the fresh-water discharge of an aquifer and the location of the interface between fresh water and saltwater.

EXAMPLE 6.30

Measurements in a coastal aquifer indicate that the saltwater interface intercepts the bottom of the aquifer approximately 2 km from the shoreline. If the hydraulic conductivity of the aquifer is 50 m/d and the bottom of the aquifer is 60 m below sea level, estimate the fresh-water discharge per kilometer of shoreline.

Solution   From the given data: $K = 50$ m/d, $L = 2$ km $= 2000$ m, and $z = 60$ m. Assuming $\rho_f = 1000$ kg/m^3 and $\rho_s = 1025$ kg/m^3, then Equation 6.338 gives

$$z = \frac{\rho_f}{\rho_s - \rho_f} h_L$$

$$60 = \frac{1000}{1025 - 1000} h_L$$

which yields $h_L = 1.5$ m. Substituting given data into Equation 6.343 gives

$$Q = \frac{K}{2L}\left(\frac{\gamma_s}{\gamma_s - \gamma_f}\right) h_L^2$$

$$= \frac{50}{2(2000)}\left(\frac{1025}{1025 - 1000}\right)(1.5)^2$$

$$= 1.15 \text{ m}^2/\text{d}$$

Therefore, the fresh-water discharge per kilometer of shoreline is $1.15 \times 1000 = 1150$ (m^3/d)/km.

In applying the Ghyben–Herzberg approximation, Equation 6.338, it is useful to note that the assumption of horizontal flow produces acceptable results, except near the coastline where vertical flow components become significant, in which case the actual saltwater interface is expected to be found below the location predicted by the Ghyben–Herzberg equation (Bear, 1979). In the case of confined aquifers, the Ghyben–Herzberg approximation is also applicable, with the elevation of the water table replaced by the elevation of the piezometric surface. Bear and Dagan (1962) have shown that the length of saltwater intrusion into a horizontal confined aquifer of thickness b is predicted to within 5% by the Ghyben–Herzberg equation, provided that $\pi(\Delta\gamma/\gamma_f)Kb/Q > 8$, where Q is the rate of flow of fresh water per unit breadth of the aquifer, and $\Delta\gamma = \gamma_s - \gamma_f$.

Besides saltwater intrusion caused by the density difference between saltwater and fresh water, a second important mechanism for saltwater intrusion is associated with the construction of unregulated coastal drainage canals. These canals allow the inland penetration of saltwater via tidal inflow and subsequent leakage of saltwater from the canals into the aquifer. To prevent saltwater intrusion in coastal drainage canals, salinity-control gates are typically placed at the downstream end of the canal to maintain a fresh-water head (on the upstream side of the gate) over the sea elevation (on the downstream side of the gate). The fresh-water head should be sufficient to prevent saltwater intrusion in accordance with the Ghyben–Herzberg equation. During periods of high runoff and when the stages in the canals are above a prescribed level, then the canal gates are opened to permit drainage while maintaining a fresh-water head that is sufficient to prevent saltwater intrusion.

EXAMPLE 6.31

Consider the gated canal in a coastal aquifer illustrated in Figure 6.28. If the aquifer thickness below the canal is 30 m, and at high tide the depth of seawater on the

FIGURE 6.28: Gated canal

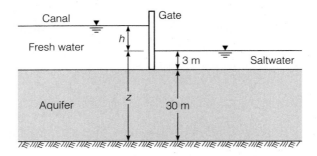

downstream side of the gate is 3 m, find the depth of fresh water on the upstream side of the gate that must be maintained to prevent saltwater intrusion.

Solution   The elevation of the fresh-water surface at the upstream side of the gate must be sufficient to maintain the saltwater interface at a depth of 33 m below sea level. According to the Ghyben–Herzberg equation (Equation 6.338), the height of the fresh-water surface above sea level, h, is given by

$$h = \frac{\rho_s - \rho_f}{\rho_f} z$$

where ρ_s and ρ_f are the densities of saltwater and fresh water, respectively, and z is the depth of the interface below sea level. Substituting $\rho_s = 1025$ kg/m^3, $\rho_f = 1000$ kg/m^3, and $z = 33$ m yields

$$h = \frac{1025 - 1000}{1000} 33$$
$$= 0.83 \text{ m}$$

Therefore, the fresh water on the upstream side of the gate must be held at 0.83 m above the sea level on the downstream side of the gate. The total depth of fresh water in the canal is 3 m + 0.83 m = 3.83 m.

In addition to salinity-control gates in coastal drainage channels, other methods of controlling saltwater intrusion include modification of pumping patterns, creation of fresh-water recharge areas, and installation of extraction and injection barriers. Extraction barriers are created by maintaining a continuous pumping trough with a line of wells adjacent to the sea, and injection barriers are created by injecting high-quality fresh water into a line of recharge wells to create a high-pressure ridge. In extraction barriers, seawater flows inland toward the extraction wells and fresh water flows seaward toward the extraction wells. The pumped water is brackish and is normally discharged to the sea.

Whenever water-supply wells are installed above the saltwater interface, the pumping rate from the wells must be controlled so as not to pull the saltwater up into the well. The process by which the saltwater interface rises in response to pumping is called *upconing*. This phenomenon is illustrated in Figure 6.29. Schmorak and Mercado (1969) proposed the following approximation of the rise height, z, of the

FIGURE 6.29: Upconing under a partially penetrating well

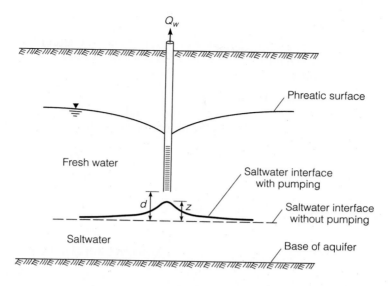

saltwater interface in response to pumping:

$$z = \frac{Q_w}{2\pi d K_x(\Delta\rho/\rho_f)} \qquad (6.345)$$

where Q_w is the pumping rate, d is the depth of the saltwater interface below the well before pumping, K_x is the horizontal hydraulic conductivity of the aquifer, ρ_f is the density of fresh water, and $\Delta\rho$ is defined by

$$\Delta\rho = \rho_s - \rho_f \qquad (6.346)$$

where ρ_s is the saltwater density. Equation 6.345 incorporates both the Dupuit and Ghyben–Herzberg approximations, and therefore care should be taken in cases where significant deviations from these approximations occur. Experiments have shown that whenever the rise height, z, exceeds a critical value, then the saltwater interface accelerates upward toward the well. This critical rise height has been estimated to be in the range $0.3d$ to $0.5d$ (Todd, 1980). Taking the maximum allowable rise height to be $0.3d$ in Equation 6.345 corresponds to a pumping rate, Q_{max}, given by

$$Q_{\text{max}} = 0.6\pi d^2 K_x \frac{\Delta\rho}{\rho_f} \qquad (6.347)$$

Therefore, as long as the pumping rate is less than or equal to Q_{max}, pumping of fresh water above a saltwater interface remains viable, although pumping rates must remain steady to avoid blurring the interface. For anisotropic aquifers in which the vertical component of the hydraulic conductivity is less than the horizontal component, a maximum well discharge larger than that given by Equation 6.347 is possible (Chandler and McWhorter, 1975).

EXAMPLE 6.32

A well pumps at 5 L/s in a 30-m thick coastal aquifer that has a hydraulic conductivity of 100 m/d. How close can the saltwater wedge approach the well before the quality of the pumped water is affected?

Solution   From the given data: $Q_w = 5$ L/s $= 432$ m^3/d, $K_x = 100$ m/d, $\rho_f = 1000$ kg/m^3, $\rho_s = 1025$ kg/m^3, and $\Delta\rho = \rho_s - \rho_f = 1025 - 1000 = 25$ kg/m^3. Equation 6.347 gives the minimum allowable distance of the saltwater wedge from the well as

$$d = \sqrt{\frac{Q_{max}}{0.6\pi K_x \, \Delta\rho/\rho_f}}$$

If $Q_{max} = 432$ m^3/d, then

$$d = \sqrt{\frac{432}{0.6\pi(100)(25)/1000}} = 9.6 \text{ m}$$

Therefore, the quality of pumped water will be impacted when the saltwater interface is located 9.6 m below the pumping well.

In cases where the pumping well fully penetrates the aquifer, the drawdown induced by the well must be limited to ensure that the toe of the saltwater wedge does not intersect the well (Mantoglou, 2003). In this case, the Ghyben–Herzberg equation can be used to estimate the limiting drawdown.

Saline ground water is a general term used to describe ground water containing more than 1000 mg/L of total dissolved solids. There are several classification schemes for ground water based on total dissolved solids, and a widely cited one, initially proposed by Carroll (1962), is given in Table 6.14. Intruded seawater has a total dissolved solids concentration of 35,000 mg/L and is classified as saline ground water. Other forms of saline ground water include *connate water** that was originally buried along with the aquifer material, water salinized by contact with soluble salts in the porous formation where it is situated, and water in regions with shallow water tables where evapotranspiration concentrates the salts in solution.

6.7   Ground-Water Flow in the Unsaturated Zone

Porous media in which the void spaces are not completely filled with water are called *unsaturated*, and an *unsaturated zone* is typically found between the ground surface and

TABLE 6.14: Classification of Saline Ground Water

Classification	Total dissolved solids (mg/L)
Fresh water	0–1000
Brackish water	1000–10,000
Saline water	10,000–100,000
Brine	> 100,000

Source: Carroll (1962).

*The word *connate* is derived from the latin word *connatus*, which means "born together."

the water table. The unsaturated zone is sometimes called the *zone of aeration* or the *vadose zone*. In some areas, such as parts of the eastern United States, the unsaturated zone can be quite thin, ranging from a few centimeters to several meters, while in other areas, such as in the western United States, the unsaturated zone is typically more than a hundred meters thick (Tindall and Kunkel, 1999). The fields of surface-water hydrology and ground-water hydrology are linked by processes that occur within the unsaturated zone. Understanding the movement of water in the unsaturated zone is important in describing the surface-infiltration process, the extraction of water for plant transpiration, and the movement of water and contaminants from the ground surface into the saturated zone of an aquifer.

Within the unsaturated zone three subzones can be delineated: the *soil-water zone*, *intermediate zone*, and *capillary zone*. The soil-water zone contains the roots of surface vegetation, and the maximum moisture content in the soil-water zone corresponds to the amount of water that can be held by the soil against the forces of gravity drainage, regardless of the depth of the water table below the ground surface. Water in the soil-water zone is called *soil water*, and the soil-water zone extends from the ground surface to the depth of plant roots. In agricultural fields, roots typically extend downward no more than 2 m, while some trees have tap roots that extend downward many meters. Transpiration, evaporation, and gravity drainage are important processes that control the moisture content within the soil-water zone. Beneath the soil-water zone, evaporation and transpiration are negligible, and an intermediate zone extends from the bottom of the soil-water zone to the upper limit of the capillary zone. Within the capillary zone, which is sometimes called the *capillary fringe*, water rises from the water table through the void spaces in much the same way that water rises in a capillary tube.

To further illustrate the dynamics of water movement in the capillary zone, consider the capillary tube shown in Figure 6.30, with the *surface tension* between tube material and water given by σ. Equilibrium at the water surface in the capillary tube requires that the weight of the water column be supported by the surface-tension force between the tube material and the water. Therefore

$$\gamma \frac{\pi D^2}{4} h_c = \pi D \sigma \cos \theta \tag{6.348}$$

FIGURE 6.30: Capillary rise

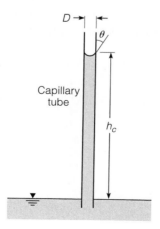

Capillary tube

h_c

where γ is the specific weight of water, D is the diameter of the capillary tube, h_c is the capillary rise, and θ is the *contact angle* between the water and the tube. Rearranging Equation 6.348 yields the following expression for the capillary rise, h_c:

$$h_c = \frac{4\sigma \cos\theta}{\gamma D} \qquad (6.349)$$

This relationship shows that the capillary rise is inversely proportional to the diameter of the tube. Since the pressure distribution within the capillary tube is hydrostatic, the pressure, p_c, at the water surface inside the capillary tube is given by

$$p_c = -\gamma h_c \qquad (6.350)$$

where the pressure is negative and therefore below atmospheric pressure.

Extending the behavior of capillary tubes to pore spaces above the saturated zone in ground water, it is easy to understand why the pressure is negative in the pore water above the water table, and why in fine media such as silts and clays, where the diameters of the void conduits are small, there are large capillary rises. Conversely, in coarse material such as sand and gravel, the void conduits have large diameters and the capillary rise is generally small. The capillary rises in several samples of unconsolidated materials having similar porosity (≈ 0.41) are given in Table 6.15. Although capillary rise through porous media is very much like the rise through capillary tubes, porous media differ from capillary tubes in that the void conduits are irregular, vary in size according to the gradation of the porous matrix, and sometimes contain "dead ends." Consequently, the moisture content in the porous medium, θ_c, defined by the relation

$$\theta_c = \frac{\text{volume of water in soil sample}}{\text{sample volume}} \qquad (6.351)$$

decreases with distance above the water table. The moisture content continues to decrease with distance above the water table, until the continuity of the capillary rise is broken. When this happens, the distribution of water is no longer analogous to the rise in a capillary tube but becomes discontinuous and held by the solid matrix. The maximum moisture content in this (discontinuous) zone is called the *field capacity*, θ_f.

The distribution of moisture capacity within the unsaturated zone is illustrated in Figure 6.31, which is commonly called the *retention curve*. Its exact shape depends on several factors, with pore-size distribution being the most important. Within the soil-zone and intermediate zone the moisture capacity is determined by surface tension

TABLE 6.15: Capillary Rise in Unconsolidated Materials

Material	Grain size (mm)	Capillary rise (cm)
Fine gravel	2–5	2.5
Very coarse sand	1–2	6.5
Coarse sand	0.5–1	13.5
Medium sand	0.2–0.5	24.6
Fine sand	0.1–0.2	42.8
Silt	0.05–0.1	105.5
Fine silt	0.02–0.05	200

Source: Lohman (1979).

FIGURE 6.31: Retention curve

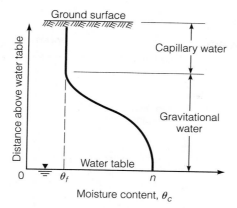

between the soil and water, and the maximum (equilibrium) moisture content is equal to the field capacity. Water clinging to soil particles that does not drain under the force of gravity is called *pendular water*. As the capillary zone is approached, the soil transitions from water being held (discontinuously) by the soil to where the water in the soil is continuous and rises above the water table in a manner analogous to the rise in a capillary tube. Within this region, the pressure distribution in the soil water is described by Equation 6.350, where h_c is the elevation above the water table. By definition, the water table is located where the pore-water pressure is equal to atmospheric pressure, and it is possible for the soil to be saturated above the water table. Therefore, the water table is only an approximate demarcation between the unsaturated and saturated zone. Water that is held discontinuously at the field capacity is called *capillary water*, while water that is continuously drawn from the water table is called *gravitational water*. The maximum moisture content in the retention curve must obviously be equal to the porosity, n.

In contrast to the equilibrium distribution of moisture content given by the retention curve, the actual distribution of water within the soil column is affected by several processes. First of all, plant transpiration and surface evaporation will extract water from the unsaturated zone. If this water is extracted from the capillary water, then, because this water is discontinuous, there is no mechanism to replenish it (other than rainfall or irrigation) and the moisture content will simply decrease below the field capacity. When the moisture content is reduced to the *wilting point*, the water is held so tightly by the soil that plants cannot extract any more water, and only evaporation can further reduce the moisture content. Under these conditions, capillary forces exceed the osmotic forces and plants growing in the soil are reduced to a wilted condition. The current measure of the wilting point is the water content corresponding to a capillary suction of 1.5 MPa (Brady, 1974). The soil can ultimately become dry as all the capillary water is removed by evaporation. At this point, a small amount of water called *hygroscopic water* is still held by the soil, and this water can be removed by oven drying the soil at a temperature of 105°C. The range of moisture contents is illustrated in Figure 6.32. Gravitational water extracted from the unsaturated zone behaves quite differently from capillary water. Since gravitational water is continuously connected to the water table, it is replenished by capillary rise, and therefore significant reductions in moisture content do not necessarily result from the extraction of gravitational water. Because of the relationship between pore

FIGURE 6.32: Range of moisture contents in porous media

Source: Bear (1979).

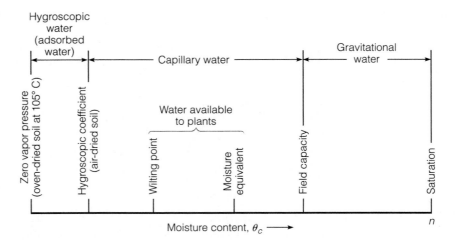

FIGURE 6.33: Specific yield in an unconfined aquifer

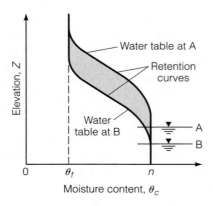

geometry and capillary rise, hysteresis effects are prevalent in the distribution of gravitational water. Clearly, the moisture distribution resulting from a falling water table will be different than that for a rising water table.

A useful application of the retention curve is to estimate the specific yield, S_y, of an unconfined aquifer, which is an important parameter in estimating transient effects in unconfined aquifers. Consider the case illustrated in Figure 6.33, where the water table is initially at A and falls to B. In this case, the depth of water released from storage is equal to the shaded area between the retention curves. Since the specific yield is defined as the volume of water released from storage per unit surface area per unit drop in the water table, the specific yield can be determined by dividing the depth of water released from storage by the magnitude of the drop in the water table (from A to B). A procedure for estimating the specific yield from the retention curve is illustrated in the following example.

EXAMPLE 6.33

The retention curve of a soil sample from the unsaturated zone is given in Table 6.16, where the soil's porosity is 0.2 and its field capacity is 0.05. Estimate the specific yield of the aquifer.

TABLE 6.16: Retention Curve

Elevation above water table (m)	Moisture content
0	0.20
0.5	0.19
1.0	0.18
1.5	0.16
2.0	0.13
2.5	0.11
3.0	0.08
3.5	0.06
4.0	0.05
4.5	0.05
5.0	0.05

Solution The initial depth of water in the unsaturated zone between the water table and 5 m above the water table, V_0, can be estimated by integrating the retention curve in Table 6.16. Therefore, using a trapezoidal rule to integrate over the 0.5-m spacing of data points yields

$$V_0 = (0.20)(0.25 \text{ m}) + (0.19 + 0.18 + 0.16 + 0.13 + 0.11 + 0.08 + 0.06$$
$$+ 0.05 + 0.05)(0.5 \text{ m}) + (0.05)(0.25 \text{ m})$$
$$= 0.5563 \text{ m}$$

The initial depth of water between 1 m below the initial water table and 5 m above the initial water table, V_1, is given by

$$V_1 = 0.5563 + (0.20)(1) = 0.7563 \text{ m}$$

If the water table falls by 1 m, then the depth of water between the new water table and 6 m above the water table, V_2, is given by

$$V_2 = 0.5563 + (0.05)(1) = 0.6063 \text{ m}$$

Therefore, when the water table falls 1 m, the depth of water released from storage, $V_1 - V_2$, is

$$V_1 - V_2 = 0.7563 - 0.6063 = 0.15 \text{ m}$$

and therefore the specific yield, S_y, is given by

$$S_y = \frac{0.15}{1} = 0.15$$

6.8   Engineered Systems

6.8.1   Design of Wellfields

The objective of wellfield design is to determine the number and location of wells required to supply water at a specified rate. A primary constraint in wellfield design is the maximum allowable drawdown, where the wells are to be arranged in such a way that the maximum allowable drawdown is not exceeded anywhere in the aquifer. Gupta (2001) has suggested a design procedure where each well in the wellfield is initially assumed to have a drawdown (at the well) equal to one-half the allowable drawdown, with the other half of the allowable drawdown caused by interference from other wells, boundary effects, and well losses. The pumping rate, Q_w, at each well that would cause a drawdown equal to one-half the allowable drawdown is determined using the Theis equation, with a time equal to one year. The number of wells required in the wellfield is then equated to the required water-supply rate, Q, divided by the pumping rate from each well, Q_w. This ratio is rounded upward to the nearest integer. Once the number of wells and the pumping rate from each well are established, the wells are arranged in a regular pattern and spaced such that the total drawdown at each well does not exceed the allowable drawdown. If an analytical approach is taken, then the total-drawdown field can be derived using the principle of superposition. Selecting the relative locations of the wells will depend on the individual site characteristics, such as property boundaries and existing pipe networks, but the wells should generally be aligned parallel to and as close as possible to surface-water recharge boundaries and as far away as possible from impermeable boundaries. Typically, production wells are spaced at least 75 m apart (Walton, 1991). A wellfield design is illustrated in the following example.

EXAMPLE 6.34

A small municipal wellfield is to be developed in an unconfined sand aquifer with a hydraulic conductivity of 50 m/d, saturated thickness of 30 m, and specific yield of 0.2. A service demand of 7000 m³/d is required from the wellfield, and the diameter of each well is to be 60 cm. If the drawdown in the aquifer is not to exceed 3 m when the wellfield is operational, develop a proposed layout for the wells. There are no nearby surface-water bodies.

Solution   The first step is to estimate the pumping rate from a single well that would cause a drawdown of 1.5 m, equal to one-half of the allowable drawdown of 3 m. The drawdown, s, resulting from the operation of each well can be estimated using the Theis equation, where

$$s = \frac{Q_w}{4\pi T} W(u)$$

where Q_w is the pumping rate from the well, T is the transmissivity of the aquifer, $W(u)$ is the well function, and u is defined by the relation

$$u = \frac{r^2 S_y}{4Tt}$$

where r is the distance from the well, S_y is the specific yield of the aquifer, and t is the time since the beginning of pumping. Since the drawdowns in the wellfield will vary between zero and 3 m, the average transmissivity, T, to be used in the wellfield design is given by

$$T = KH = 50 \times \left(30 - \frac{3}{2}\right) = 1425 \text{ m}^2/\text{d}$$

For a single well with $t = 365$ days,

$$u_w = \frac{r_w^2 S_y}{4Tt} = \frac{(0.3)^2(0.2)}{(4)(1425)(365)} = 8.652 \times 10^{-9}$$

Since $u_w \leq 0.004$, then with less than 0.1% error

$$W(u_w) = -0.5772 - \ln u_w$$

and therefore

$$W(u_w) = -0.5772 - \ln(8.652 \times 10^{-9}) = 17.99$$

According to the Theis equation, the pumping rate, Q_w, required to produce a drawdown, s_w, is given by

$$Q_w = \frac{4\pi T s_w}{W(u_w)}$$

For $s_w = 1.5$ m (one-half the allowable drawdown), then

$$Q_w = \frac{4\pi(1425)(1.5)}{17.99} = 1493 \text{ m}^3/\text{d}$$

Since the service demand is 7000 m^3/d, the number of wells required is given by

$$\text{number of wells} = \frac{7000}{1493} = 4.69 \approx 5 \text{ wells}$$

Consider the five-well wellfield with the arrangement shown in Figure 6.34. Each of the 5 wells is to operate at the same pumping rate, Q_w, given by

$$Q_w = \frac{7000}{5} = 1400 \text{ m}^3/\text{d}$$

and the arrangement consists of a central well (No. 1) surrounded by four wells (Nos. 2, 3, 4, 5) at a distance R away from the central well. Clearly, when all five wells are in operation, the drawdown will be greatest at the central well (No. 1). Denoting

FIGURE 6.34: Proposed wellfield

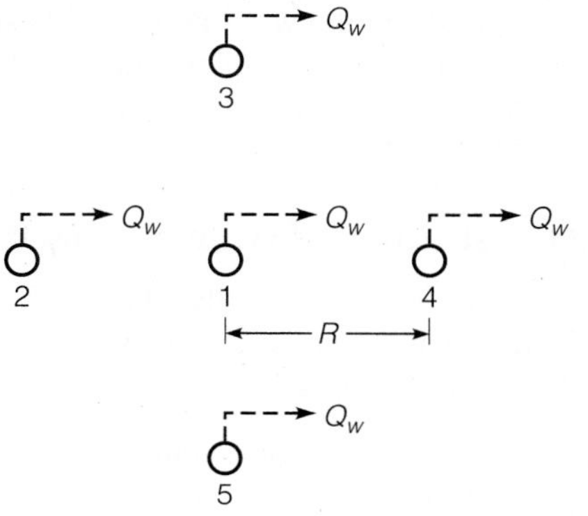

the drawdown at well No. 1 by s_1, then by the principle of superposition

$$s_1 = \frac{Q_w}{4\pi T} \sum_{i=1}^{5} W(u_i) \tag{6.352}$$

where u_i are given by

$$u_1 = \frac{r_w^2 S_y}{4Tt} = \frac{(0.3)^2(0.2)}{4(1425)(365)} = 8.652 \times 10^{-9}$$

and

$$u_2 = u_3 = u_4 = u_5 = \frac{R^2 S_y}{4Tt} = \frac{R^2(0.2)}{4(1425)(365)} = 9.613 \times 10^{-8} R^2 \tag{6.353}$$

and therefore,

$$W(u_1) = 17.99$$

$$W(u_2) = W(u_3) = W(u_4) = W(u_5)$$

Equating the drawdown at well No. 1 to the maximum allowable value of 3 m, then Equation 6.352 yields

$$3 = \frac{1400}{4\pi(1425)} \left[17.99 + 4W(u_2) \right]$$

which leads to

$$W(u_2) = 5.096$$

which, by the definition of the well function leads to

$$u_2 = 0.003437$$

Therefore, on the basis of Equation 6.353,

$$0.003437 = 9.613 \times 10^{-8} R^2$$

which leads to

$$R = 189 \text{ m}$$

Therefore, a wellfield consisting of 5 wells arranged as shown in Figure 6.34, with each well pumping 1400 m³/d, and spaced 189 m apart will yield the required amount of water and produce drawdowns that will not exceed the specified maximum of 3 m.

6.8.2 Design and Construction of Water-Supply Wells

Wells used for water supply are typically classified according to their method of construction and include: driven wells, drilled wells, and radial wells. These types of wells are described briefly as follows:

Driven Well. A *driven well* consists of a pointed screen, called a *drive point* or *well point*, and lengths of pipe attached to the top of the drive point. The drive point is a perforated pipe covered with woven wire mesh or a tubular brass jacket, or is similar to screens for drilled wells and is adaptable to driving. A pointed steel tip at the base of the drive point breaks through pebbles and thin layers of hard material and opens a passageway for the point. A driven well varies from 32 to 100 mm in diameter and can be installed to a maximum of about 9 to 12 m deep (AWWA, 2003b). For municipal water supplies, the driven well is used where thin deposits of sand and gravel are found at shallow depths.

Drilled Well. Major water-supply wells are typically constructed using rotary drilling methods. A *drilled well* is constructed using a drill rig to excavate or drill a hole, into which a casing is forced or placed to prevent it from collapsing. When a water-bearing formation of sufficient capacity is reached, a screen is set in place that allows water to flow into the casing and holds back fine materials in the formation. When the drilled well passes through rock, a screen is usually not used, unless the formation is fractured. A drilled well is commonly used for municipal water supply, with pipe diameters typically in the range of 50 to 1210 mm.

Radial Well. A *radial well* is a combination of a large-diameter central well and a series of horizontally driven wells projecting out of the vertical walls of the central well. The main well, or *central caisson*, serves as a collector for the water produced from the individual horizontal wells, called laterals. The laterals are installed in coarse formations, often in more than one layer or tier. The radial well, also called a *horizontal collector* well, is widely used because it can produce very large quantities of water. In many cases, a radial well is located along the shore of a lake or river because infiltration from the water body can recharge it.

Drilled wells are the most commonly used type of water-supply well, and guidelines for their design and construction are covered in this section.

The well design procedure starts with expectations of potential *water yield* from an aquifer, and a typical water-supply well is illustrated in Figure 6.35. A generally accepted, all-inclusive standard for designing water-supply wells is not available (USBR, 1995); however, the American Water Works Association (AWWA, 1990) and the National Water Well Association (Lehr et al., 1988) describe a number of commonly used design standards. The geology of the aquifer surrounding the well has a dominant influence on the well design, and the greater the certainty with which the geology surrounding the well is known, the better the well design. Wells drilled in consolidated formations such as sandstone or limestone generally do not require any casing, screen, or gravel pack, while wells drilled in unconsolidated formations such as sand or gravel will generally require a casing and screen and may require a gravel pack. For major water-supply wells with yields greater than 500 L/min, it is usually advisable to first drill a pilot hole at the proposed well location to collect the relevant geologic data; if this location proves to be acceptable, the expanded pilot hole is then used as the water-supply well. If the pilot hole indicates that the proposed site is inadequate, the site can be abandoned without the major cost of drilling a production well. For minor wells, well-drilling costs may be about the same as for a pilot hole. Consequently, smaller wells are drilled with a preliminary design developed on the basis of readily available information, and necessary adjustments and refinements in design are made during drilling and construction, as appropriate to maximize well yield (AWWA, 2001).

FIGURE 6.35: Typical water-supply well

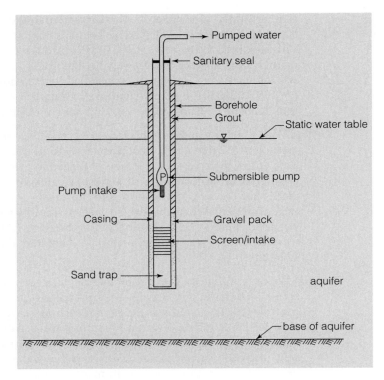

Several criteria for determining the optimal withdrawal rate from a well have been suggested. One of the most popular rules of thumb is that wells should be pumped at a rate that maintains a maximum allowable drawdown. A variety of criteria are used to establish the maximum allowable drawdown, including consideration of existing pump-intake elevations, allowable stage fluctuations in connected surface-water bodies, and the sometimes-used criteria that the maximum allowable drawdown should not exceed 50–60% of an unconfined aquifer's saturated thickness. Once the design pumping rate has been specified, with due consideration to the allowable drawdown, the structural components of the water-supply well are designed in accordance with the following guidelines:

Casing.   The well casing is simply a liner placed in the borehole to prevent the walls (of the borehole) from collapsing. The well casing typically consists of a solid pipe that is connected at its lower end to a screened interval, sometimes called the *well intake*. In production wells, the pump intake is typically placed within the casing; the diameter of the casing should be at least 5 cm larger than the nominal diameter of the pump bowls and the flow velocity in the casing should be less than 1 m/s to prevent excessive friction losses (Todd, 1980; Lehr et al., 1988). Recommended casing diameters based on pump sizes used in the water-well industry are given in Table 6.17 (Lehr et al., 1988; Johnson Division, 1966), however, in final design the manufacturer of the selected pump system should be consulted for advice. If the casing diameter based on pump size is less than the diameter required for a casing flow velocity of 1 m/s, then a casing diameter based on a 1 m/s flow velocity should be selected. Casing materials commonly

TABLE 6.17: Typical Casing Diameters Based on Pump Size

Pumping rate (L/min)	Optimum casing diameter (mm)	Minimum casing diameter (mm)
< 400	150 mm ID*	125 mm ID
300–660	200 mm ID	150 mm ID
570–1500	250 mm ID	200 mm ID
1300–2500	300 mm ID	250 mm ID
2300–3400	350 mm OD*	300 mm ID
3200–4900	400 mm OD	350 mm OD
4500–6800	500 mm OD	400 mm OD
6100–11,000	600 mm OD	500 mm OD

Source: Lehr et al. *Design and Construction of Water Wells*. 1988. ITP, Cincinnati.
*ID = inside diameter; OD = outside diameter.

used in practice include alloyed or unalloyed steel, fiberglass, and polyvinyl chloride (PVC). Fiberglass and PVC casings are used primarily in shallow, small-diameter wells where corrosion may be an issue, while steel casing is used in the majority of water-supply wells. Well casings should generally extend above the ground surface to prevent surface water from running down the hole and contaminating the well water. Joints for permanent steel casings should have threaded couplings or be welded to ensure watertightness from the bottom of the casing to above grade. Thermoplastic casing typically is bell-and-socket construction, joined by cementing, or else is joined with spline-lock fittings.

Grouting. Well *grouting* consists of placing a cement slurry (called *grout*) between the well casing and the borehole; it is generally necessary to prevent polluted surface water or low-quality water in overlying aquifers from seeping along the outside of the casing and contaminating the well or aquifer. Grouting also serves to anchor the casing inside the borehole, protect the casing from corrosion, and prevent the caving in of aquifer material surrounding the well. Depending on the method used for grouting, the borehole is drilled 50–150 mm larger than the outside diameter of the casing. The cement slurry used in grouting is usually a mix of about 50 L of water per 100 kg of cement, with additives used to provide special properties such as preventing the grout from cracking and accelerating or retarding the time of setting. Additives that are commonly used include bentonite clay, pozzolans, and perlite (Bouwer, 1978). Grout should be placed by pumping it through a pipe or hose (called a *tremie*) to the bottom of the casing and above the gravel pack. The tremie pipe is placed between the casing and the surrounding soil or rock. Pumping the grout from the bottom up ensures that no gaps or voids are left between the casing and the soil or rock. To prevent the penetration of the cement slurry into the gravel pack, a layer of clean sand (particle size 0.3 to 0.6 mm) is placed on top of the gravel pack. This layer of sand is commonly referred to as a *sand bridge*, and cement slurries generally do not penetrate the sand layer by more than a few centimeters. A minimum grout thickness of 50 mm is recommended and may even be required by some regulatory agencies.

FIGURE 6.36: Continuous-slot wire-wound well screen in unconsolidated foundation

Todd (1980), Courtesy Johnson Division, UOP, Inc.

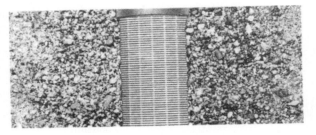

Screen intake. Wells completed in unconsolidated formations (such as sands and gravels) generally have screen intakes which keep the pump from being clogged with aquifer material. Screens are sometimes installed in fractured formations that could collapse into the borehole. A typical well screen in an unconsolidated formation is shown in Figure 6.36.

The primary design constraint in sizing screen intakes is an acceptable entrance velocity for flow through the screen openings. Entrance velocities through a screened intake need to be controlled for several reasons, most importantly: (1) to minimize head losses across the screen and prevent screen clogging; (2) to prevent *sand drive*, which is a compaction that occurs over time, decreasing the hydraulic conductivity around the screen and thereby decreasing the well efficiency; and (3) to prevent sand pumping, which quickly destroys the pump impellers and seals. The entrance velocity, v_s, can be expressed in the form (Todd, 1980)

$$v_s = \frac{Q_w}{c\pi d_s L_s P} \tag{6.354}$$

where Q_w is the pumping rate, c is the clogging coefficient (approximated as 0.5 by assuming that 50% of the open area of a screen is blocked by aquifer material), d_s is the screen diameter, L_s is the screen length, and P is the fraction of open area in the screen (which is part of the manufacturer's specifications). Walton (1962) has suggested that maximum desirable screen entrance velocities are related to the hydraulic conductivity of the aquifer material, which is fundamentally related to the grain-size distribution of the aquifer material. The maximum desirable screen entrance velocities for selected hydraulic conductivities are given in Table 6.18, and these limitations on entrance velocities are intended to prevent the larger grain sizes in the porous medium from being dislodged and clogging the screen. On the basis of Equation 6.354, it is clear that specification of a maximum screen entrance velocity leaves the designer the flexibility to select a variety of screen diameters and lengths. The U.S. Bureau of Reclamation (1995) recommends that the average entrance velocity generally be less than or equal to 1.8 m/min in order to limit turbulent flow and the associated head losses in the vicinity of screen intakes. To promote better and more efficient well construction, a design entrance velocity between 1.8 and 3.6 m/min has been adopted by many regulatory agencies (AWWA, 2003b). Given that screen entrance velocities tend to be nonuniform along the length of the screen intake (VonHofe and Helweg, 1998), it is recommended that a conservative average screen entrance velocity, such as 1.8 m/min, be used in design.

TABLE 6.18: Maximum Desirable Screen Entrance Velocities

Hydraulic conductivity of aquifer (m/d)	Maximum desirable screen entrance velocity (m/min)
> 250	3.7
250	3.4
200	3.0
160	2.7
120	2.4
100	2.1
80	1.8
60	1.5
40	1.2
20	0.9
< 20	0.6

Source: Walton (1962).

Well screen (and casing) diameters should be at least one pipe size larger than the largest diameter of the pumping equipment to be installed (AWWA, 2003b). This gap allows adequate space for pump installation and removal, efficient pump operation, and good hydraulic efficiency of the well. Analyses by VonHofe and Helweg (1998) have suggested that the pump intake diameter should be approximately 60% of the screen (and casing) diameter for efficient operation. For screens with a given length and diameter, head losses decrease rapidly with an increase in percentage of open area up to 15%, less rapidly up to about 25%, and relatively slowly between 25% and 60%. For practical purposes, a percentage of open area of about 15% ($P = 0.15$) is acceptable and easily obtained with many commercial screens.

The U.S. Bureau of Reclamation has recommended minimum diameters for various well capacities as shown in Table 6.19. These diameters may be increased

TABLE 6.19: Minimum Screen Diameters

Pumping rate (L/min)	Screen diameter (mm)
< 190	50
190–475	100
475–1330	150
1330–3040	200
3040–5320	250
5320–9500	300
9500–13,300	350
13,300–19,000	400
19,000–26,000	450
26,600–34,200	500

Source: USBR (1995).

to obtain acceptable entrance velocities, and smaller diameters are sometimes specified in the interest of economy.

An important consideration in the selection of well screens is the size of the screen opening, commonly referred to as the *slot size*, which must be sufficiently small to prevent the screen from being clogged by dislodged portions of the aquifer matrix. The appropriate screen size can be related to the *uniformity coefficient*, U_c, of the aquifer matrix, defined by

$$U_c = \frac{d_{60}}{d_{10}} \tag{6.355}$$

where d_{60} and d_{10} are 60-percentile and 10-percentile particle diameters, respectively, of the aquifer matrix. A uniformity coefficient (U_c) of 1 indicates that the aquifer material is of uniform size, $U_c \leq 5$ indicates a *poorly graded* material, and $U_c > 5$ indicates a *well-graded* material (Kashef, 1986). Aquifers where $U_c \geq 3$ and $d_{10} > 0.25$ mm are good candidates for being *developed naturally*, meaning that initial pumping of the well will remove the fines from the surrounding aquifer material, leaving the screen surrounded by a very permeable coarse-grained annular region.

In cases where the well can be developed naturally, Todd (1980) and Lehr et al. (1988) suggest the criteria in Table 6.20 for selecting the screen slot size. Within the range of slot sizes indicated in Table 6.20, sizes at the low end of the range are selected whenever the well is not overlaid with a firm layer of soil, such as clay or shale, and slot sizes at the high end of the range are selected whenever firm soil layers are present (Lehr et al., 1988). In cases where $d_{10} < 0.25$ mm, an artificial gravel pack should be used. By designing a screen slot size based on criteria in Table 6.20, the finer material in the aquifer surrounding the screen is removed during *well development*, which includes any mechanism that removes silt, fine sand, or other such material from the zone immediately surrounding a well intake. High-rate pumping and surge plungers are common methods used in well development. After well development, the screen intake is surrounded by a coarse layer with a hydraulic conductivity significantly higher than that of the undisturbed aquifer matrix. The coarse layer surrounding the screen is typically about 50-cm thick (Boonstra, 1998). If a screen is not available in the required slot size, the next smaller standard size should be selected. Slot openings in commercially available well screens typically range between 1 mm and 6 mm, and slot sizes are commonly specified in multiples of 0.025 mm (0.001 in.). Hence,

TABLE 6.20: Criteria for Screen Slot Sizes in Naturally Developed Wells

Aquifer properties	Screen slot size*
$U_c < 3$, $d_{10} > 0.25$ mm	$d_{40}-d_{60}$
$3 \leq U_c \leq 5$, $d_{10} > 0.25$ mm	$d_{40}-d_{70}$
$U_c > 5$, $d_{10} > 0.25$ mm	$d_{50}-d_{70}$

*Screen slot sizes are given as percentile sizes of aquifer material. For example, d_{40} is the 40-percentile size of the aquifer material.

a No. 40 slot has a 1.0 mm (0.040 in.) slot width. In cases where the grain-size distribution varies over the depth of the aquifer, the specification of screen sections with different slot sizes is common.

To maximize the pumping rate for a given drawdown, it is recommended that wells in confined aquifers be screened through the entire thickness of the aquifer (USBR, 1995), or at least through 70%–80% of the aquifer thickness (Wurbs and James, 2002). If the aquifer is homogeneous, the well screen is centered in the aquifer; otherwise the screen is set in the coarser part of the aquifer. To maximize the pumping rate for a given drawdown in unconfined aquifers, wells should be screened in the lower one-third to one-half of the saturated thickness, with the upper two-thirds of the saturated thickness reserved for drawdown. The screen-length specifications for confined and unconfined aquifers are not economically feasible in very deep and thick aquifers, and in these cases the usual practice is to penetrate a sufficient thickness of the aquifer to achieve the required discharge capacity at an acceptable drawdown. In cases where the screen does not fully penetrate the aquifer, drawdowns will be greater than estimated by assuming full penetration. This represents a decrease in well yield. A minimum of 1 to 2 m of screened intake is usually required for domestic wells (Fetter, 2001). Longer screen lengths generally result in higher pumping rates per unit drawdown.

The depth of the screen in unconsolidated formations is designed to protect the intake from surface contamination, and the minimum screen depth will vary with soil formations and surrounding conditions. In unconsolidated formations, screen depths of 8 to 9 m or more provide a reasonable level of protection from surface contamination (AWWA, 2003b).

Most screen sections are made in lengths ranging from 1.5 m to 6 m, and diameters ranging from 30 mm to 1500 mm. Screen sections are typically joined together by welding or couplings to give almost any length of screen. The least expensive and most commonly available screens are made of low-carbon steel. Screens made of nonferrous metals and alloys, plastics, and fiberglass are used in areas of aggressive corrosion and encrustation. Water-supply wells typically have stainless-steel screens with continuous slots that are made by winding wire around vertical rods.

Pump. Pumps are installed in water wells to lift the water in the well to the ground surface and deliver it to the point of use. The types of pumps used to remove water from wells are: vertical turbine pumps, submersible pumps, jet pumps, pneumatic pumps, airlift pumps, and positive displacement pumps. These types of pumps are described in Table 6.21. The vertical turbine pump is often the most suitable pump for ground-water applications, especially for moderate to large discharge rates (USBR, 1995). Submersible pumps are also used frequently and have the advantage of being able to lift water from deep wells where long shafts in crooked casings might prohibit the use of vertical turbine pumps. Both vertical turbine and submersible pumps are categorized as centrifugal pumps, since the rotating impellers induce water flow through centrifugal force. Typical vertical turbine and submersible pumps are illustrated in Figure 6.37. The sizes and performance characteristics of these pumps must meet the design/operational requirements of head (pressure) and flowrate to be delivered by the pump. The

TABLE 6.21: Water-Well Pumps

Pump Type	Description
Vertical turbine pumps	Vertical turbine pumps have the motor located on the discharge side at the ground surface and require a drive shaft extending down the well to the pump located below the water surface.
Submersible pumps	The term "submersible" is applied to turbine pumps when the motor is close-coupled beneath the bowl assembly of the pump and both are installed under water. This type of construction eliminates the surface motor, long drive shaft, shaft bearings, and lubrication system of the conventional turbine pump; however, the electrical connections are submerged. Submersible pumps are especially useful for high-head low-capacity applications such as domestic water supply.
Jet pumps	The jet pump combines two principles of pumping, that of the injector (jet) and that of the centrifugal pump. Jet pumps are inefficient when compared to ordinary centrifugal pumps, but are used in domestic wells because of other favorable characteristics.
Pneumatic pumps	Pneumatic pumps operate using air pressure and are generally used under special conditions such as contaminant cleanup and monitoring. They are used for purging, sampling, product-only pumping, and low- to moderate-flow pumping.
Airlift pumps	Airlift pumps release compressed air into the discharge pipe, and the reduced specific weight of the water column lifts the water to the surface. Airlift pumps are inefficient and expensive in comparison to other pumping methods and are rarely used.
Positive displacement pumps	Positive displacement pumps displace water through a pumping mechanism. There are several types of positive displacement pumps including: piston pumps, rotor peristaltic pumps, and Lemoineau-type pumps.
Suction pumps	Suction pumps are typically centrifugal pumps located above the ground surface, and their use is limited by the suction lift that can be developed. In practice, the suction lift is usually limited to where the water table is about 7 m below the pump intake.

performance characteristics (head versus discharge) provided by manufacturers are generally given per stage in the pump. For multiple-stage pumps, the performance characteristics are derived by adding all the heads (added at each stage) for a given discharge. An excellent review of water-well pump design can be found in USBR (1995).

Well-design specifications typically caution against installing the intake of a pump in the well screen, since such designs are assumed to increase screen

FIGURE 6.37: Water-supply pumps

Courtesy Goulds Pumps.

VERTICAL TURBINE PUMP   **SUBMERSIBLE PUMP**

entrance velocities, thereby decreasing well efficiencies (Driscoll, 1986; Roscoe Moss Company, 1990). However, contrary to this guideline, there is evidence that placing the pump intake within the well screen actually increases well efficiencies (Von Hofe and Helweg, 1998; Korom et al., 2003). If the pump is placed within the well screen, the optimum location of the pump intake within the well screen depends on the size of the aquifer matrix (Korom et al., 2003); however, placing the pump intake at 50% of the screen length would be a reasonable guideline based on available information. As mentioned in the context of screen design, the pump intake diameter should be approximately 60% of the screen diameter for efficient operation (VonHofe and Helweg, 1998). If the pump is located within the well casing, a minimum clearance of 25 mm around the pump bowls is recommended (USBR, 1995).

Gravel pack. In some cases, the aquifer material is so fine that selection of screen openings on the basis of its size and uniformity coefficient would yield openings so small that the entrance velocity would be unacceptably high. Under these circumstances, *gravel packs* or (equivalently) *filter packs* are placed in the annular region between the screen and the perimeter of the borehole. In this context, the term "gravel pack" refers to any filtering media that is placed around the well screen and is not limited to a coarse gravel material as the name implies. Fine to medium sand is commonly used as gravel-pack material. According to Ahren (1957), gravel packs in unconsolidated formations are usually justified when the uniformity coefficient, U_c, is less than 3 and the aquifer has a d_{10} less than 0.25 mm (0.010 in).

The thickness of the gravel pack is typically in the range of 8 to 23 cm (Todd, 1980; Ahren, 1957; Boonstra, 1998), with ideal thicknesses in the range of 10 to 15 cm (Lehr et al., 1988). It is difficult to develop a well through a gravel pack much thicker than 20 cm, since the seepage velocities induced during the development procedure must be able to penetrate the gravel pack to repair the damage done by drilling, break down any residual drilling fluid on the borehole wall, and remove finer particles near the borehole (USBR, 1995). The thickness of the gravel pack should not be less than 8 cm to ensure that a continuous pack will surround the entire screen (Boonstra, 1998). The specifications of the gravel pack are determined primarily by the aquifer matrix, and criteria for selecting the gravel pack and corresponding screen slot size are given in Table 6.22.

Typically, a gravel pack should have a uniformity coefficient between 1 and 2.5, with a median grain size between six and nine times the median grain size of the aquifer material. The corresponding slot size of the screen opening

TABLE 6.22: Criteria for Gravel Pack Selection

Uniformity coefficient of aquifer matrix, U_c	Gravel pack criteria	Screen slot size
< 2.5	(a) U_c between 1 and 2.5, with the 50% size not greater than six times the 50% size of the aquifer (preferable criteria)	5% to 10% passing size of the gravel pack
	(b) If (a) is not available, U_c between 2.5 and 5, with 50% size not greater than nine times the 50% size of the aquifer (alternative criteria)	
2.5–5	(a) U_c between 1 and 2.5, with the 50% size not greater than nine times the 50% size of the formation (preferable criteria)	5% to 10% passing size of the gravel pack
	(b) If (a) is not available, U_c between 2.5 and 5, with 50% size not greater than 12 times the 50% size of the aquifer (alternative criteria)	
> 5	Multiply the 30% passing size of the aquifer by 6 and 9 and locate the points on the grain-size distribution graph on the same horizontal line. Through these points draw two parallel lines representing materials with $U_c \leq 2.5$. Select gravel pack material that falls between the two lines.	5% to 10% passing size of the gravel pack

Source: USBR (1995).

should be between the 5- and 10-percentile grain size of the gravel pack, with the 10-percentile size usually being preferable to minimize entrance losses. The maximum grain size of a gravel pack should generally be around 10 mm, but less than 9 mm if placed through a nominal 100 mm (4 in.) tremie pipe (USBR, 1995), and the gravel pack should extend at least 1 m above the screen. Gravel packs usually consist of quartz-grained material that is sieved and washed to remove the finer material such as silt and clay. The grains of the gravel-pack material must be well rounded, to maximize the porosity; angular grains will tend to lock into adjacent grains, reducing the openings through which water and fine aquifer material can move. Gravel packs are usually considered essential in sandy aquifers.

In unconsolidated formations, whether a gravel pack is used or not, a properly developed well is surrounded by a coarse annular region that has a hydraulic conductivity much higher than the surrounding aquifer. This condition serves to increase the *effective radius* of the well, which is roughly equal to the radial extent of the coarse annular region surrounding the well screen, and drawdown calculations must be based on the effective radius.

Sand trap. A *sand trap* is a section of well casing installed below the screened intake section. Its function is to store sand and silt entering the well during pumping. The length of the sand trap is usually 2 to 6 m, and the diameter is typically the same as the screen.

Sanitary seal. At the surface, all wells should have a sanitary seal, which prevents contamination from entering the well casing. This seal is in addition to the grout that is placed between the borehole wall and the well casing to prevent surface contamination of water entering the well intake. The sanitary seal for a small well having a submersible pump is illustrated in Figure 6.38 and consists of a metal plate with a rubberized gasket around its perimeter that fits snugly into the top of the well casing. The sanitary seal has openings into the well for the discharge pipe and the pump power cable, and an air vent to let air into the casing as the water level drops.

EXAMPLE 6.35

A water-supply well is to be installed in an unconfined aquifer which has a saturated thickness of 25 m and a hydraulic conductivity of 30 m/d. A grain-size analysis of the aquifer indicates a uniformity coefficient of 2.7 and a 50-percentile grain size of 1 mm. If the well is to be pumped at 20 L/s, design the well screen and the gravel pack.

Solution   According to Table 6.22, since the uniformity coefficient (U_c) of the aquifer matrix is 2.7, the gravel pack should have a U_c between 1 and 2.5 and a 50% size not greater than nine times the 50-percentile size (d_{50}) of the aquifer. Hence, d_{50} for the gravel pack should be less than 9×1 mm $= 9$ mm. A check with a sand and gravel company will generally yield a commercial gravel pack material that satisfies these criteria. For example, you may find "5 mm $\times$ 9 mm" gravel, which contains only grain sizes between 5 mm and 9 mm. Assuming that the particle sizes are uniformly distributed between 5 mm and 9 mm, then $d_{10} \approx 5.4$ mm, $d_{60} \approx 7.4$ mm, and U_c of the gravel is approximately $7.4/5.4 = 1.4$. Hence, the 5 mm $\times$ 9 mm gravel is acceptable.

FIGURE 6.38: Sanitary seal in water-supply well

Source: AWWA (2003).

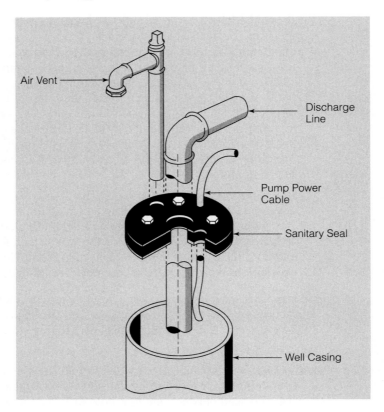

Since gravel packs should be between 8 cm and 23 cm thick, a reasonable thickness would be 10 cm. This is within the range of ideal thicknesses suggested by Lehr and colleagues (1988).

According to Table 6.22, the screen slot size should be between the 5% and 10% passing size of the gravel pack. In this case, the 10% passing size is estimated to be 5.4 mm, and it is reasonable to select a slot size of 5 mm. A review of manufacturers' literature on well screens will give the commercially available slot widths and associated fractions, P, of open area. Typically, $P = 0.10$. Since the screen length, L_S, in unconfined aquifers should be between 0.3 and 0.5 times the saturated thickness, select a screen length of 0.5 times the saturated thickness to maximize the pumping rate per unit drawdown, in which case $L_S = 0.5 \times 25$ m $= 12.5$ m. For a pumping rate of 20 L/s (= 1200 L/min), Table 6.17 indicates an optimum casing diameter of 250 mm, and Table 6.19 indicates a minimum screen diameter of 150 mm. Taking the screen and casing diameter as 250 mm, and assuming that 50% of the open area is clogged by the aquifer material, then the screen entrance velocity, v_s, is estimated by Equation 6.354 as

$$v_s = \frac{Q_w}{c\pi d_s L_s P}$$

where $Q_w = 20$ L/s $= 1728$ m³/d, $c = 0.5$, $d_s = 0.25$ m, $L_s = 12.5$ m, and $P = 0.1$, hence

$$v_s = \frac{1728}{(0.5)\pi(0.25)(12.5)(0.1)} = 3520 \text{ m/d} = 2.4 \text{ m/min}$$

According to Table 6.18, this screen entrance velocity is too high for an aquifer with a hydraulic conductivity of 30 m/d, where the desirable screen size would produce an entrance velocity on the order of 1.05 m/min. Taking $v_s = 1.05$ m/min ($= 1510$ m/d), the corresponding screen diameter, d_s, is given by

$$d_s = \frac{Q_w}{c\pi v_s L_s P} = \frac{1728}{(0.5)\pi(1510)(12.5)(0.1)} = 0.583 \text{ m} \approx 600 \text{ mm}$$

In summary, a screen length of 12.5 m, diameter of 600 mm, and slot size of 5 mm is acceptable. This screen should be surrounded by a 5 mm × 9 mm gravel pack with a thickness of about 10 cm; the diameter of the well casing should be 600 mm so that it can be easily joined to the screen.

6.8.3 Performance Assessment of Water-Supply Wells

The pumping rate corresponding to the maximum allowable drawdown can be estimated from the *specific capacity* of a well, which is defined as the well pumping rate per unit drawdown, where the drawdown is equal to the drawdown in the aquifer (at the boundary of the borehole) plus the well (head) loss resulting from flow through the gravel pack and well screen. The specific capacity is sometimes considered to be the most informative single factor in well performance (AWWA, 2003b). The specific capacity in the case of a fully penetrating well in a homogeneous isotropic confined aquifer with negligible head losses in the developed region surrounding the intake can be derived directly from the Thiem steady-state equation (Equation 6.134) as

$$\text{Specific capacity} = \frac{Q_w}{s_w} = \frac{2\pi T}{\ln(R/r_w)} \tag{6.356}$$

or from the Theis transient equation (Equation 6.219) as

$$\text{Specific capacity} = \frac{Q_w}{s_w} = \frac{4\pi T}{W(r_w^2 S/4Tt)} \tag{6.357}$$

where Q_w is the pumping rate, s_w is the drawdown at the well, T is the transmissivity of the aquifer, R is the radius of influence (see Equations 6.135 to 6.137), r_w is the effective radius of the well, $W(u)$ is the well function (Equation 6.218), S is the storage coefficient, and t is the time since the beginning of pumping. Although both Equations 6.356 and 6.357 are used to estimate the specific capacity of a well, Equation 6.357 is more realistic and indicates that the specific capacity gradually decreases with time. Indeed, a reduction of up to 40% in the specific capacity has been observed in one year in wells deriving water entirely from storage (Gupta, 2001). Classifications of well productivity in terms of specific capacity are given in Table 6.23. The specific capacity of water-supply wells is typically monitored over time to detect whether the well is becoming clogged. The specific capacity should be determined at least annually, and the analysis should occur only after a well is allowed to fully

TABLE 6.23: Classification of Well Productivity

Specific capacity, Q_w/s_w (L/min)/m	Productivity
< 0.3	Negligible
0.3–3	Very low
3–30	Low
30–300	Moderate
> 300	High

Source: Şen (1995).

recover. Then, the well should be pumped for an hour to determine the specific capacity. This result is then compared to the original data and plotted to show trends (AWWA, 2003b).

Rehabilitation of clogged wells should begin before the specific capacity falls below 85% of the original value (Thomas, 2002). The maximum rate at which water can be extracted from a well is defined as the *well yield*, which is equal to the specific capacity multiplied by the allowable drawdown. The well yield should not be confused with the *safe aquifer yield*, which is the maximum rate at which water can be withdrawn from an aquifer without depleting the water supply. Water managers are keenly aware that increases in ground-water withdrawals must generally be balanced by an increase in aquifer recharge, a decrease in aquifer discharge, and/or a loss of ground-water storage.

EXAMPLE 6.36

A water-supply well is to be constructed in a surficial aquifer where the saturated thickness is 25 m, and the maximum allowable drawdown in the surficial aquifer is 2 m. The well is to have an effective radius of 400 mm, and the hydraulic conductivity and specific yield of the aquifer are estimated as 20 m/d and 0.15, respectively. Estimate the specific capacity and the well yield after 3 years of service. Classify the productivity of the well.

Solution   From the given data: $H = 25$ m, $K = 20$ m/d, $T = KH = (20)(25) = 500$ m^2/d, $r_w = 400$ mm $= 0.4$ m, $S_y = 0.15$, and $t = 3$ years $= 3(365)$ days $= 1095$ days. Substituting these data into Equation 6.357 (using the small-drawdown assumption for surficial aquifers) gives

$$\text{Specific capacity} = \frac{4\pi T}{W(r_w^2 S_y/4Tt)} = \frac{4\pi(500)}{W(0.4^2 \times 0.15/4 \times 500 \times 1095)}$$

$$= 355 \text{ m}^2/\text{d} = 247 \text{ (L/min)/m}$$

Since the maximum allowable drawdown is 2 m, the well yield is given by

$$\text{well yield} = 247 \text{ (L/min)/m} \times 2 \text{ m} = 494 \text{ L/min}$$

and therefore the maximum allowable pumping rate is 494 L/min. According to Table 6.23, a well with a specific capacity of 247 (L/min)/m has a moderate productivity.

Head losses resulting from turbulent flow through the well screen and the coarse-grained material surrounding the well are collectively called *well losses*, and these well losses cause the water level inside the well to be less than the water level in the aquifer adjacent to the well. This is illustrated in Figure 6.39. Well losses are added to the *formation losses* to estimate the *total drawdown*. The well loss, h_w, is typically estimated by the relation

$$\boxed{h_w = \alpha Q_w^n} \tag{6.358}$$

where α is a constant, Q_w is the pumping rate, and n is an exponent that typically varies between 1.5 and 3 (Lennox, 1966), but is commonly assumed to be equal to 2. The assumption of $n = 2$ was originally proposed by Jacob (1947) and is still widely accepted. Values of α corresponding to various states of a well were suggested by Walton (1962) and are given in Table 6.24. These data indicate that for a properly designed and functional well, α should be less than 1800 s^2/m^5. Since *formation losses* are typically proportional to the well pumping rate, Q_w, and well losses are typically proportional to Q_w^2, then the drawdown in a well, s_w, can typically be expressed in the form

$$s_w = \beta Q_w + \alpha Q_w^2 \tag{6.359}$$

FIGURE 6.39: Well losses and formation losses

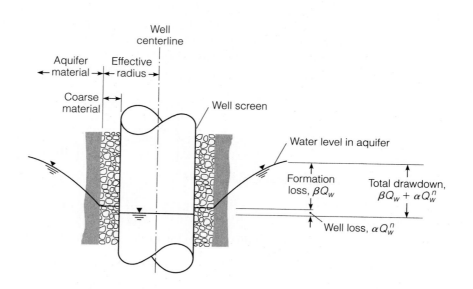

TABLE 6.24: Well-Loss Coefficients

Well loss coefficient, α (s^2/m^5)	Well condition
<1800	Properly designed and developed
1800–3600	Mild deterioration or clogging
3600–14,400	Severe deterioration or clogging
>14,400	Difficult to restore well to original capacity

Source: Walton (1962).

where β is the *formation-loss coefficient*. Dividing both sides of Equation 6.359 by Q_w yields

$$\frac{s_w}{Q_w} = \beta + \alpha Q_w \qquad (6.360)$$

which indicates a linear relationship between s_w/Q_w and Q_w. The coefficients α and β can be determined in the field using a *step-drawdown pumping test*, where the well pumping rate is increased in a series of discrete steps, with drawdown measurements taken at fixed time intervals. Typically, a step-drawdown test includes five to eight pumping steps, each lasting from one to two hours. The resulting data yield a relationship between s_w/Q_w and Q_w, which is fitted to the linear relationship given by Equation 6.360, and the slope and intercept of the best-fit line are taken as α and β, respectively. The *well efficiency*, e_w, is defined as the percentage of the total drawdown in the well caused by formation losses, and based on Equation 6.359 the well efficiency can be put in the form

$$e_w = \frac{\beta Q_w}{s_w} \times 100 \qquad (6.361)$$

EXAMPLE 6.37

A step-drawdown test is conducted at a well where a pumping rate of 864 m^3/d yields a drawdown in the well of 1.54 m, and a pumping rate of 1296 m^3/d yields a drawdown of 2.36 m. Assess the condition of the well, and estimate the well efficiency and specific capacity at a pumping rate of 2000 m^3/d.

Solution   The given data can be put in the following tabular form:

Q_w (m³/d)	s_w (m)	s_w/Q_w (d/m²)
864	1.54	0.00178
1296	2.36	0.00181

According to Equation 6.360,

$$\frac{s_w}{Q_w} = \beta + \alpha Q_w$$

and combining this equation with the step-drawdown data gives

$$0.00178 = \beta + 864\alpha$$

$$0.00181 = \beta + 1296\alpha$$

Solving these equations simultaneously yields

$$\alpha = 6.9 \times 10^{-8} \; d^2/m^5 = 515 \; s^2/m^5 \qquad \text{and} \qquad \beta = 0.00172 \; d/m^2$$

Comparing $\alpha = 515 \; s^2/m^5$ with the guidelines in Table 6.24 indicates that the well is in good condition.

At $Q_w = 2000$ m^3/d, the expected drawdown at the well, s_w, is given by Equation 6.359 as

$$s_w = \beta Q_w + \alpha Q_w^2 = (0.00172)(2000) + (6.9 \times 10^{-8})(2000)^2 = 3.72 \text{ m}$$

The drawdown associated with formation losses is $\beta Q_w = 0.00172(2000) = 3.44$ m, and hence the well efficiency, e_w, is given by Equation 6.361 as

$$e_w = \frac{\beta Q_w}{s_w} \times 100 = \frac{3.44}{3.72} \times 100 = 92\%$$

At $Q_w = 2000$ m³/d, the specific capacity of the well is given by

$$\text{Specific capacity} = \frac{Q_w}{s_w} = \frac{Q_w}{\beta Q_w + \alpha Q_w^2} = \frac{1}{\beta + \alpha Q_w}$$

$$= \frac{1}{0.00172 + (6.9 \times 10^{-8})(2000)} = 538 \text{ m}^2/\text{d} = 373 \text{ (L/min)/m}$$

Comparing this result with the classification system in Table 6.23 indicates that the well is highly productive at a pumping rate of 2000 m³/d.

This analysis assumes a priori that the well losses are proportional to Q_w^2, and this assumption is justified by the fact that the well-loss coefficient used to assess the condition of the well in Table 6.24 is based on this assumption. In the more general case,

$$s_w = \beta Q_w + \alpha Q_w^n \tag{6.362}$$

which can be put in the form

$$\log\left(\frac{s_w}{Q_w} - \beta\right) = \log \alpha + (n - 1)\log Q_w \tag{6.363}$$

which shows that the relation between $\log(s_w/Q_w - \beta)$ and $\log Q_w$ is linear with a slope of $n - 1$ and an intercept of $\log \alpha$. To apply Equation 6.363 to step-drawdown data, various values of β are tried until the relation between $\log(s_w/Q_w - \beta)$ and $\log Q_w$ becomes approximately linear, then values of n and α are obtained directly from the slope and intercept of the best-fit straight line. This approach, originally proposed by Rorabaugh (1953), is sometimes referred to as the *Rorabaugh method*.

EXAMPLE 6.38

The drawdowns, s_w (m), and corresponding pumping rates, Q_w (m³/d), at a well during a step-drawdown test were plotted on a graph of $\log(s_w/Q_w - \beta)$ versus $\log Q_w$ for various values of β. The plotted curve is most linear when $\beta = 6.36 \times 10^{-4}$ d/m², and the corresponding best-fit line through the data has a slope of 1.14 and an intercept of -6.72. Determine the formation and well-loss coefficients, and estimate the efficiency and productivity of the well when the pumping rate is 900 m³/d.

Solution From the given data, $\beta = 6.36 \times 10^{-4}$ d/m²; comparing the slope and intercept of the best-fit line with Equation 6.363 indicates that

$$\log \alpha = -6.72, \qquad (n - 1) = 1.14$$

which gives

$$\alpha = 1.91 \times 10^{-7} \text{ d}^2/\text{m}^5, \qquad n = 2.14$$

Hence, the formation and well-loss coefficients are $\beta = 6.36 \times 10^{-4}$ d/m^2 and $\alpha = 1.91 \times 10^{-7}$ d^2/m^5 ($= 1430$ s^2/m^5), respectively, and the drawdown, s_w, is given by

$$s_w = 6.36 \times 10^{-4} Q_w + 1.91 \times 10^{-7} Q_w^{2.14}$$

The well efficiency, e_w, is equal to the percentage of total drawdown caused by formation losses, and can be expressed in the form

$$e_w = \frac{\beta Q_w}{\beta Q_w + \alpha Q_w^n} \times 100 = \frac{\beta}{\beta + \alpha Q_w^{n-1}} \times 100$$

When $Q_w = 900$ m^3/d,

$$e_w = \frac{6.36 \times 10^{-4}}{6.36 \times 10^{-4} + (1.91 \times 10^{-7})(900)^{2.14-1}} \times 100 = 59\%$$

Since the well-loss coefficient, $\alpha = 1430$ s^2/m^5, indicates that the annular region surrounding the well is in fairly good condition (see Table 6.24), the low well efficiency of 59% can be attributed to formation losses being of comparable magnitude to the well losses. At $Q_w = 900$ m^3/d, the productivity of the well is measured by the specific capacity as

$$\text{Specific capacity} = \frac{Q_w}{s_w} = \frac{Q_w}{\beta Q_w + \alpha Q_w^n} = \frac{1}{\beta + \alpha Q_w^{n-1}}$$

$$= \frac{1}{6.36 \times 10^{-4} + (1.91 \times 10^{-7})(900)^{1.14}}$$

$$= 925 \text{ m}^2/\text{d} = 642 \text{ (L/min)/m}$$

Comparing this result with the classification system in Table 6.23 indicates that the well is highly productive at a pumping rate of 900 m^3/d.

6.8.4   Well Drilling

Numerous techniques and a wide variety of specialized equipment are used to drill water wells. These techniques can be broadly categorized as either *percussion* or *rotary* methods (Stone, 1999). A drill rig mounted on a truck, along with a large gasoline or diesel engine to drive the system, is used for each of these methods, and the appropriate drilling method is selected based on the local geology and the size and depth of the well to be installed.

Percussion methods.   Percussion drilling uses pounding or jetting to drill a hole in the ground. Three common percussion methods are: *cable-tool drilling*, *hammer drilling*, and *jetting*. In cable-tool drilling, a heavy drill string, supporting a sharpened tool or bit, is repeatedly lifted and dropped into the hole by a cable. The mixture of cuttings and water is periodically removed from the hole using a sand pump or bailer. The cable-tool method is mostly used in rock to depths of 600 meters. In the hammer method, compressed air is cycled to repeatedly pound the bit downward, a process much like that used in a jack-hammer. Cuttings are removed from the hole using compressed air. Jetting is similar to cable-tool drilling, with pressurized water on either side of a chisel-like bit, and water and cuttings flowing up the annulus between the drill string and the wall of the open hole.

Rotary methods.   Rotary drilling uses a rotating bit to drill a hole in the ground. The rotating drill bit has cutting teeth generally made of steel. For very hard geologic material, the teeth may be tipped with titanium or diamond. Rotary methods differ primarily in the type of drilling fluid used and the direction of circulation. In conventional mud- or air-rotary methods, fluid is pumped down the inside of the drill pipe down to the drill bit, and this fluid comes back up to the surface through the annulus between the pipe and the wall of the open hole. This fluid, sometimes called *drilling mud*, is recirculated during the drilling operation and, in addition to cooling the drill bit and forcing the tailings (geologic material) from the borehole, it also exerts pressure on the walls of the borehole and forms a clay lining on the wall of the well, preventing it from caving (in unconsolidated formations) during the drilling process. As the drill bit rotates and moves downward, creating a borehole, the drill crew adds additional lengths of 6-m pipe, and this string of pipes is increased in length until the desired well depth is reached. Drilling mud consists of a suspension of water, bentonite clay, and various organic additives.

6.8.5 Wellhead Protection

Public water-supply wells must generally be protected from contaminants introduced into the aquifer from areas surrounding the well. The most effective way of doing so is to control land uses in the surrounding areas by zoning regulations. In the United States, all states are required by law to develop comprehensive wellhead-protection (WHP) plans. Such plans must contain the following elements (USEPA, 1990a):

1. Specify the roles and duties of state agencies, local government entities, and public water suppliers in the development and implementation of WHP programs.
2. Delineate a *wellhead protection area* (WHPA) for each public water-supply well based on reasonably available hydrogeologic information on ground-water flow, recharge and discharge, and other information the state deems necessary to adequately determine the WHPA.
3. Identify sources of contaminants within each WHPA, including all potential anthropogenic sources that may have an adverse effect on health.
4. Develop management approaches that include, as appropriate, technical assistance, financial assistance, implementation of control measures, education, training, and demonstration projects that are used to protect the water supply within the WHPA from such contaminants.
5. Develop contingency plans indicating the location and provision of alternate drinking water supplies for each public water-supply system in the event of well contamination.
6. Site new wells properly to maximize yield and minimize potential contamination.
7. Ensure public participation by establishing procedures encouraging the public to participate in developing the WHP program elements.

Much thought and effort have been put into the development of WHP plans, in both the United States and Europe, for a variety of budgets and hydrogeologic settings, and several good WHP plans have been published (e.g., ASCE, 1997; Johnson, 1997; USEPA, 1987). Wellhead protection plans developed to date indicate that there is no universal approach to wellhead protection, given the variety of political, social, economic, and technical constraints that are encountered in practice. A WHP plan should generally contain the seven key elements listed above, be tailored to local conditions, and afford a defined level of protection to the public water supply.

Delineation of wellhead protection areas. The most important and challenging component of any WHP plan, from an engineering viewpoint, is the delineation of the wellhead protection area. A wellhead protection area (WHPA) is legally defined as "the surface and subsurface area surrounding a well or wellfield that supplies a public water system through which contaminants are likely to pass and eventually reach the water well or wellfield." A pumping well produces drawdowns in the aquifer surrounding the well, and the portion of the aquifer within which ground water moves toward the pumping well is called the *zone of contribution* (ZOC) of the well. Only pollutants introduced within the ZOC can contribute to contamination of the pumped water. Although regulators must be concerned with all pollutant sources within the ZOC, the level of protection necessarily varies with the distance from the well. In the immediate vicinity of the well, within tens of meters, protection is most stringent and all pollutant sources are typically prohibited; it is not unusual for this area to be fenced and have controlled access. This area is sometimes called the *remedial action zone*. Beyond this area, land uses that generate a limited amount of ground-water contamination, or potential for ground-water contamination, are typically permitted; however, sufficient distance to the well is provided such that the pollutant concentrations are attenuated to acceptably low levels by the time they reach the well. This area, where contaminant sources are controlled by regulatory permits, is commonly called the *attenuation zone*. Beyond the attenuation zone, but still within the ZOC, contaminant sources are unlikely to have any significant impact on the pumped water and regulatory burdens are reduced. This outer zone is commonly called the *wellfield management zone*. The primary factors to be considered in delineating WHPAs are the types and amounts of contaminants that could possibly be introduced into the ZOC and the attenuation characteristics of the contaminants in the subsurface environment.

The types of contaminants that are of most concern are organic chemicals and viruses. Organic chemicals are used as industrial solvents and pesticides, and they are the major component of gasoline, which is usually stored in underground storage tanks at gas stations. Viruses in ground water originate primarily from septic tank effluent, onsite domestic wastewater treatment systems, and broken sewer pipes. The attenuation of various contaminants in ground water occurs primarily via the processes of *dispersion*, *decay*, and *sorption*. Criteria commonly used for delineating WHPAs are: (1) distance, (2) drawdown, (3) time of travel, (4) flow boundaries, and (5) assimilative capacity. An assessment of these criteria, and typical threshold values, are given in Table 6.25. The specific criteria selected to delineate a WHPA depend on a variety of considerations, including overall protection goals, technical considerations, and policy considerations.

The most commonly used approach is the time of travel (TOT) approach, which relates locations within the WHPA to the times of travel from those locations to the water-supply well. To define the level of protection of the water supply from a particular contaminant, the time of travel is contrasted with the time for the

TABLE 6.25: WHPA Delineation Criteria

Criteria	Assessment	Typical thresholds
Distance	Does not directly incorporate contaminant fate and transport processes. Commonly an arbitrary policy decision.	300 m–3 km
Drawdown	Does not directly incorporate contaminant fate and transport processes. Defines areas where the influence of pumping is the same.	3 cm–30 cm
Time of travel	Incorporates advection and decay processes, neglects dispersion. Widely used criterion.	5 y–50 y
Flow boundaries	Highest level of protection, boundary of WHPA is taken as the ZOC boundary. Not practical in most cases.	Physical and hydrologic
Assimilative capacity	Incorporates all significant contaminant fate and transport processes. A rational approach to wellhead protection. Requires much technical expertise and is rarely used.	Requires drinking water standards to be met at well

Source: USEPA (1987).

contaminant to undergo an acceptable amount of attenuation. The time of travel is sometimes associated with the time that is available to clean up a spill within the WHPA. In cases where decay or biotransformation is the dominant mode of attenuation, reductions in contaminant concentration are commonly described by the first-order process

$$\frac{dc}{dt} = -\lambda c \tag{6.364}$$

where c is the contaminant concentration in ground water, t is the time since release of the contaminant, and λ is the decay parameter. Equation 6.364 can be solved to yield

$$\boxed{\frac{c}{c_0} = e^{-\lambda t}} \tag{6.365}$$

where c_0 is the initial concentration at $t = 0$. Equation 6.365 gives the attenuation of contaminant concentration as a function of travel time. Therefore, if a pollutant source is expected to contaminate the ground water at a concentration c_0 and the allowable concentration at the well contributed by this source is c, the minimum allowable time of travel to the well can be calculated using Equation 6.365. A useful parameter to characterize the time scale of decay is the *half-life*, T_{50}, which is the time for the concentration to decay to one-half of its original value. Equation 6.365 gives

$$T_{50} = \frac{0.693}{\lambda} \tag{6.366}$$

Time-of-travel contours surrounding single wells can be derived using the continuity relation

$$Q_w t = \pi r^2 b n \tag{6.367}$$

where Q_w is the pumping rate, t is time, r is the distance traveled in time t, b is the saturated thickness, and n is the porosity. Equation 6.367 can be put in the form

$$r = \sqrt{\frac{Q_w t}{\pi b n}} \tag{6.368}$$

which gives the radial distance from the well that has a travel time t. Equation 6.368 assumes that the aquifer properties are radially symmetric around the pumping well, a condition that is usually difficult to determine and subject to uncertainty. In the case of nonuniform aquifers and more complex multiwell scenarios, numerical ground-water models and probabilistic approaches can be used to calculate travel times (Feyen et al., 2003a).

EXAMPLE 6.39

A WHPA is to be delineated around a municipal water-supply well that pumps 0.35 m^3/s from an aquifer with a saturated thickness of 25 m and a porosity of 0.2. The contaminant of concern is viruses from residential septic tanks. Residential development is to be permitted on 2000-m^2 lots, and the viral concentration under each lot resulting from septic tank effluent is expected to be 50/L. A risk analysis indicates that the maximum allowable viral concentration in the pumped water is 0.01/L, and the decay constant, λ, for viruses can be taken as 0.5 d^{-1} (Yates, 1987). If there is to be 1 km^2 of residential development surrounding the water-supply well, estimate the boundary of the WHPA within which no residential development should be allowed.

Solution   The allowable viral concentration in the pumped water is 0.01/L. Since the lots are uniformly distributed around the water-supply well, when each lot contributes a concentration of 0.01/L to the pumped water, the average concentration in the pumped water is 0.01/L. Since $c_0 = 50$/L and $\lambda = 0.5$ d^{-1}, Equation 6.365 requires that

$$\frac{c}{c_0} = e^{-\lambda t}$$

$$\frac{0.01}{50} = e^{-0.5\,t}$$

which gives $t = 17.0$ days. The radial distance, r, corresponding to a travel time of 17.0 days ($= 1.47 \times 10^6$ seconds) is given by Equation 6.368 as

$$r = \sqrt{\frac{Q_w t}{\pi b n}}$$

$$= \sqrt{\frac{(0.35)(1.47 \times 10^6)}{\pi (25)(0.2)}} = 181 \text{ m}$$

Therefore, prohibiting residential development within a radius of 181 m of the well will provide a satisfactory level of protection from viral contamination. Ground-water monitoring is recommended to verify assumptions regarding viral concentrations in the ground water under septic tanks.

The delineation of WHPAs based on travel time is appropriate when decay is the dominant attenuation process, contaminant dispersion is negligible, and the aquifer and pollutant source parameters are known with a reasonable degree of certainty. In cases where these conditions are not met, more comprehensive WHP models are recommended (e.g., Chin and Chittaluru, 1994).

6.8.6   Design of Aquifer Pumping Tests

Aquifer pumping tests or *aquifer tests** are used to determine the hydraulic properties of aquifers. The methodology consists of matching field measurements of drawdowns caused by pumping wells with the corresponding theoretical drawdowns and determining the hydraulic properties of the aquifer that produce the best fit. The details of several analytic models for analyzing aquifer-test data in a variety of field scenarios have been presented in Section 6.3. Aside from analytic techniques for processing the measured data, several operational issues must be addressed to properly conduct aquifer tests. These operational procedures are described in detail by several standards of the American Society of Testing Materials (ASTM-D5092, 1990; ASTM-D4043, 1991a; ASTM-D4050, 1991b; ASTM-D4106, 1991c), and have also been described in detail by the U.S. Environmental Protection Agency (Osbourne, 1993). Aquifer tests are usually conducted using a pumping well and at least one observation well. The key elements in designing an aquifer test at any site are: (1) number and location of observation wells, (2) design of observation wells, (3) approximate duration of the test, and (4) discharge rate from the pumping well. Aquifer tests are usually quite expensive, and the installation of a pumping well and surrounding observation wells is typically justified only in cases where exploitation of the aquifer by water-supply production wells at the site is being contemplated. In most cases, the pumping well is subsequently utilized as a production well.

Prior to conducting an aquifer test, basic data on the aquifer must be collected. These data must include, if possible, the depth, thickness, areal extent, and lithology of the aquifer, the locations of aquifer discontinuities caused by changes in lithology, the locations of surface water bodies, preliminary estimates of the transmissivity and storage coefficient of the aquifer, and preliminary estimates of leakage coefficients if semiconfining layers are present. These data facilitate the design of the discharge rate from the pumping well as well as aid in the location of observation wells. Preliminary values of the transmissivity and storage coefficient of an aquifer can be estimated by conducting slug tests (see Section 6.8.7) on wells near the site, but such tests have no more than order-of-magnitude accuracy (Osbourne, 1993). It is advisable to use existing wells to conduct aquifer tests whenever possible; however, care should be taken that existing wells are properly constructed and developed and that these wells are screened in the same aquifer zone as the one being investigated.

---

*"Aquifer test" is the preferred terminology used by the United States Geological Survey (USGS).

6.8.6.1   Design of pumping well

The design of the pumping well includes consideration of: (1) well construction, (2) well development procedure, (3) well access for water-level measurements, (4) a reliable power source, (5) type of pump, (6) discharge-control and measurement equipment, and (7) method of waste disposal.

Well construction.   The diameter of the pumping well must be large enough to accommodate both the test pump and space for water-level measurement. Guidelines for casing diameter given in Table 6.17 are recommended. The well screen must have sufficient open area to minimize local well losses; guidelines given in Table 6.19 are recommended. If the well is located in an unconsolidated aquifer, a gravel pack should be placed in the annular region between the well screen and the perimeter of the borehole. The gravel pack should extend at least 30 cm above the top of the well screen and be designed in accordance with the guidelines given in Section 6.8.2. A seal of bentonite pellets should be placed on top of the gravel pack. A minimum of 1 m of pellets should be used, and an annulus seal of cement and/or bentonite grout should be placed on top of the bentonite pellets. The well casing should be protected at the surface with a concrete pad around the well to isolate the well bore from surface runoff.

Well development.   Pumping wells should be adequately developed to ensure that well losses are minimized. See Section 6.8.2 for a thorough discussion of well losses. If the well is suspected to be poorly developed or nothing is known, it is advisable to conduct a step-drawdown test to determine the magnitude of the well losses.

Water-level measurement access.   It must be possible to measure the depth of water in the pumping well before, during, and after pumping. Usually, electric-sounder or pressure-transducer systems are used.

Reliable power source.   Power must be continuously available to the pump during the test. A power failure during the test usually requires that the test be terminated and sufficient time permitted for water levels to stabilize. This can cause expensive and time-consuming delays.

Pump selection.   Electrically powered pumps produce the most constant discharge and are recommended for use during an aquifer test. The discharge of gasoline- or diesel-powered pumps may vary greatly over a 24-h period and requires more frequent monitoring during the aquifer test. According to Osbourne (1993), a diesel-powered turbine pump may have more than a 10% variation in discharge as a result of daily variations in temperature.

Discharge-control and measurement equipment.   Common methods of measuring well discharge include orifice plates and manometers, inline flow meters, inline calibrated pitot tubes, weirs or flumes, and (for low discharge rates) measuring the time taken to discharge a measured volume. The discharge of wells yielding less than 400 L/min can be readily measured with sufficient accuracy using a calibrated bucket or drum and a stopwatch (USBR, 1995). An important pump characteristic is that as the pump lift increases, the discharge decreases for a pump running at a constant speed. Therefore, the pump speed will usually need to be increased during the test to maintain a constant discharge.

Water disposal.   The volume of pumped water produced, the storage requirements, disposal alternatives, and any treatment needs must be assessed during the planning phase. Clearly, the pumped water cannot be allowed to infiltrate back into the aquifer during the aquifer test. If the aquifer is unconfined and the unsaturated zone overlying the aquifer is relatively permeable, the pumped water should be transported by pipeline to a location beyond the area of influence that will develop during the aquifer test. If the water table is more than 30 m below the ground surface and the unsaturated zone has a low hydraulic conductivity, an open ditch may be used to transport the pumped water (USBR, 1995).

6.8.6.2   Design of observation wells

If existing wells are to be used as observation wells, then it should be verified that these wells are screened in the aquifer being investigated and that the screens are not clogged due to the buildup of iron compounds, carbonate compounds, sulfate compounds, or bacterial growth. The response test (Stallman, 1971; Black and Kipp, 1977) is recommended if existing wells are to be used as observation wells. In addition, the following characteristics of observation wells should be considered:

Well diameter.   The well casing should be large enough to allow for accurate, rapid water-level measurements. Well casings 50 mm (2 in.) in diameter are usually adequate in aquifers less than 30 m deep, but they are often difficult to develop. Well casings 100 to 150 mm (4 to 6 in.) in diameter are easier to develop and should have a better aquifer response. If a water-depth recorder is to be used, then well casings of 100 to 150 mm will usually be required. Difficulties in drilling a straight hole usually dictate that a well over 60 m deep must be at least 100 mm in diameter. Wells with diameters larger than 150 mm may cause a lag in response time due to water held in casing storage, so smaller diameters are usually preferable (Lehr et al., 1988).

Well construction.   Observation wells should ideally have 1.5 to 6 m of perforated casing or screening near the bottom of the well. In addition, the placement of the gravel pack, bentonite pellets, cement or bentonite grout, and a concrete pad should follow the same guidelines as for the pumping well. After installation, observation wells should be developed by surging with a block, and/or submersible pump for a sufficient period (usually several hours) to meet a predetermined level of turbidity (Campbell and Lehr, 1972; Driscoll, 1986).

Distance from pumping well.   Single observation wells are usually located about three to five times the saturated thickness away from the pumping well, with observation wells in unconfined aquifers located closer to the pumping well than observation wells in confined aquifers (Lehr et al., 1988). This usually works out to a distance of 10 m to 100 m (Boonstra, 1998). If the pumping well is partially penetrating, observation wells should be located at a minimum distance equal to one-and-a-half to two times the saturated thickness from the pumping well (USBR, 1995). At least three observation wells at different distances from the pumping well are desirable, so that results can be averaged and obviously erroneous data can be disregarded. Whenever multiple observation wells are used, they are typically placed in a straight line or along perpendicular rays originating at the pumping well. If aquifer anisotropy is expected, then observation wells should be located in a pattern based on the suspected or known anisotropic conditions at the site. If the principal directions of anisotropy are not known, then at least three wells on different rays are required. If the aquifer is vertically anisotropic, then the minimum distance, r_{min}, of an observation well from the pumping well should be taken as (USBR, 1995)

$$r_{min} = 1.5b\left(\frac{K_h}{K_v}\right)^{\frac{1}{2}} \tag{6.369}$$

where b is the saturated thickness of the aquifer, K_h is the horizontal hydraulic conductivity, and K_v is the vertical hydraulic conductivity. Observation wells should be located far enough away from geologic and hydraulic boundaries to permit recognition of drawdown trends before the boundary conditions influence the drawdown readings.

6.8.6.3 Field procedures

Aside from the design of the pumping and observation wells, there are several operational guidelines to be considered.

Establishment of baseline conditions. Prior to the initiation of an aquifer test, it is essential to monitor the water levels in the pumping and observation wells in addition to wells adjacent to the site. These measurements will indicate whether a measurable trend exists in the ground-water levels. In addition, the influence and scheduling of offsite pumping on the aquifer test must be assessed, and controlled if necessary. As a general rule, at least one week of observations must be available prior to the initiation of the aquifer test. It is advisable to record barometric pressure, rainfall, and water levels in surface-water bodies within the anticipated cone of depression of the test well during the period of the aquifer test. It is relevant to note that a barometric pressure increase of 1 cm of mercury may result in a fall of about 8 cm in the water level in an observation well in a confined aquifer (Prakash, 2004).

Water-level measurements during aquifer test. Immediately before pumping is to begin, static water levels in all test wells should be recorded. The recommended time intervals for recording water levels during an aquifer test are given in Table 6.26 (after ASTM Committee D-18, D 4050). After pumping is terminated, recovery

TABLE 6.26: Recommended Time Intervals for Water-Level Measurement

Time since beginning of test	Measurement interval
0–3 min	Every 30 s
3–15 min	Every 1 min
15–60 min	Every 5 min
60–120 min	Every 10 min
120 min–10 h	Every 30 min
10 h–48 h	Every 4 h
48 h–shutdown	Every 24 h

measurements should be taken at the same time intervals as listed in Table 6.26, where the times are measured from the instant the pump is turned off. A check valve should be used to prevent backflow after pumping is terminated, and recovery measurements should continue until the prepumping state is recovered. The drawdown, s', after pumping is terminated in homogeneous isotropic aquifers can be approximated by (temporal) superposition of the Theis equation (Equation 6.219) as

$$s' = \frac{Q_w}{4\pi T} \left[W(u) - W(u') \right] \tag{6.370}$$

where Q_w is the constant pumping rate during the aquifer test, T is the aquifer transmissivity, $W(u)$ is the well function, and

$$u = \frac{r^2 S}{4Tt} \quad \text{and} \quad u' = \frac{r^2 S'}{4Tt'} \tag{6.371}$$

where r is the distance of the observation well from the pumping well, S is the storage coefficient during drawdown, S' is the storage coefficient during recovery, t is the time since pumping began, and t' is the time since pumping stopped. For small values of u and u' ($u, u' < 0.004$), the Cooper–Jacob approximation to the well function (Equation 6.221) can be applied to Equation 6.370, yielding

$$s' = \frac{Q_w}{4\pi T} \ln \frac{S't}{St'} \tag{6.372}$$

or

$$\boxed{s' = \frac{Q_w}{4\pi T} \left(\ln \frac{t}{t'} + \ln \frac{S'}{S} \right)} \tag{6.373}$$

This equation is a linear relationship between s' and $\ln t/t'$. Matching Equation 6.373 to the recovery data yields a slope of $Q_w/4\pi T$ and an intercept of $Q_w/4\pi T \ln S'/S$. Knowing the pumping rate, Q_w, and the storage coefficient during the drawdown test, S, then the aquifer transmissivity, T, and recovery storage coefficient, S', can be estimated from the slope and intercept of recovery data. The estimated value of T can be used to confirm the transmissivity determined from the pump-drawdown measurements; however, no independent confirmation of the drawdown storage coefficient is obtained from this analysis. In some cases, S' is assumed to be equal to S (e.g., de Marsily, 1986), in which case the transmissivity of the aquifer is all that is extracted from the recovery data.

Discharge-rate measurements.   During the initial hour of the aquifer test, the discharge from the pumping well should be measured as frequently as practical, and it is important to bring the discharge rate up to the design rate as quickly as possible. Because of the variety of environmental factors that can affect the discharge rate from diesel- and gasoline-driven pumps, the discharge should be checked four times per day, preferably early morning, mid-morning, mid-afternoon, and early evening. Osbourne (1993) has indicated that a 10% variation in discharge can result in a 100 percent variation in the estimate of aquifer transmissivity, and it is recommended that the discharge should never be allowed to vary by more than 5% during an aquifer test.

Length of test. The test should be of sufficient length to accurately identify the shape of the drawdown versus time curve from which the aquifer hydraulic properties are extracted. The drawdown curve should be plotted during the aquifer test. As a general guideline, when three or more drawdown readings taken at one-hour intervals at the most distant well fall on a straight line (on the log drawdown versus log time curve), the aquifer test can be terminated (Lehr et al., 1988). Aquifer tests are typically conducted for around 24 hours in confined aquifers and around 72 hours in unconfined aquifers.

6.8.7 Slug Test

In some field situations, the hydraulic conductivity of the porous medium may be too small or the diameter, depth, or yield of the well may be too small (e.g., in the range of 5 to 200 m³/day) to conduct an aquifer pumping test, or the scope of the investigation may not warrant an aquifer pumping test. In such cases, a *slug test* may be useful as a relatively quick and cost-effective method to estimate the aquifer hydraulic conductivity in the immediate vicinity of the well. This test is applicable to completely or partially penetrating wells in unconfined aquifers and fully penetrating wells in confined aquifers. The slug test is sometimes called a *bail-down test* (Fetter, 2001).

The field setup for a slug test is illustrated in Figure 6.40, where a well with casing radius r_c extends to a depth L_w below the water table in an unconfined aquifer of saturated thickness H. The well intake (screen) has a length L_e and effective radius r_w. The effective radius is generally greater than the actual radius of the intake and includes the high-permeability region surrounding the well caused by either well development or the placement of a gravel pack. The slug test is performed by instantaneously removing a volume V of water from the well (a "slug" of water), and relating the recovery of the water level in the well to the hydraulic conductivity of the surrounding aquifer. The instantaneous removal of water can be accomplished using a bailer (a type of bucket), or by submerging a closed cylinder in the well, letting the water reach equilibrium, and then quickly pulling the cylinder out. In some

FIGURE 6.40: Field setup for a slug test.

Source: Bouwer and Rice (1976).

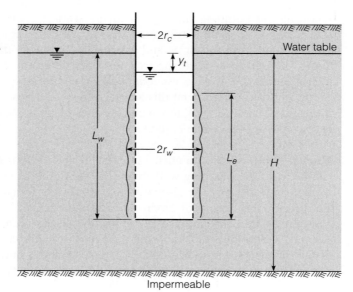

Impermeable

cases, compressed air (or partial vacuum) has been used to displace a slug of water in the well.

A variety of analytical techniques have been proposed to relate aquifer properties to the recovery data (Cooper et al., 1967; Hvorslev, 1951; Bouwer and Rice, 1976; Bouwer and Rice, 1989; Nguyen and Pinder, 1984; Van der Kamp, 1976), and a comparison of several of these methods can be found in Herzog (1994). The appropriate method of analysis depends primarily on the response characteristics of the water in the well. If the water level recovers to the initial static level in a smooth manner, then the response is called *overdamped*, while in cases where the water level in the well oscillates about the static water level, the response is called *underdamped*. An overdamped response is typical of most cases where the test well is screened in aquifers of moderate hydraulic conductivity, and methods proposed by Cooper et al.(1967), Hvorslev (1951), and Bouwer and Rice (1976, 1989) are appropriate in these cases. Overdamped responses typically occur in wells that are screened in high hydraulic conductivity aquifers, and the method proposed by Van der Kamp (1976) is appropriate for this case.

The most widely used slug-test procedure, applicable to underdamped responses in fully or partially penetrating wells in unconfined aquifers, was developed by Bouwer and Rice (1976) and subsequently updated by Bouwer (1989). The basis of the Bouwer and Rice method is the Thiem equation (Equation 6.134), which can be put in the form

$$y = \frac{Q_w}{2\pi K L_e} \ln\left(\frac{R_e}{r_w}\right) \tag{6.374}$$

where y is the drawdown at the well, Q_w is the rate at which ground water is removed from the aquifer, K is the hydraulic conductivity of the aquifer, and R_e is the effective radial distance over which the head y is dissipated. Equation 6.374 neglects well losses. Continuity requires that the flow into the well, Q_w, be equal to the rate at which the volume of water in the well increases, therefore

$$\frac{dy}{dt} = -\frac{Q_w}{\pi r_c^2} \tag{6.375}$$

Combining Equations 6.374 and 6.375 (eliminating Q_w), integrating, and solving for K yields

$$K = \frac{r_c^2 \ln (R_e/r_w)}{2L_e} \frac{1}{t} \ln \frac{y_0}{y_t} \tag{6.376}$$

where y_0 is the initial drawdown in the well and y_t is the drawdown at time t. Using Equation 6.376 to estimate the hydraulic conductivity, K, requires that r_c, r_w, and L_e be known from the dimensions of the well, and t, y_0, and y_t be measured during the recovery of the water level in the well. The effective radial distance, R_e, can be estimated using the following empirical relation (Bouwer and Rice, 1976):

$$\ln \frac{R_e}{r_w} = \left\{ \frac{1.1}{\ln (L_w/r_w)} + \frac{A + B \ln[(H - L_w)/r_w]}{(L_e/r_w)} \right\}^{-1} \quad \text{(partially penetrating well)}$$

$$\tag{6.377}$$

where A and B are dimensionless parameters that are related to L_e/r_w, as shown in Figure 6.41. Analyses by Bouwer and Rice (1976) have indicated that if $\ln[(H - L_w)/r_w] > 6$, a value of 6 should be used for this term in Equation 6.377. If $H = L_w$, the well is fully penetrating, the term $\ln[(H - L_w)/r_w]$ given in Equation 6.377 is indeterminate, and the following equation should be used to estimate R_e

$$\ln \frac{R_e}{r_w} = \left\{ \frac{1.1}{\ln (L_w/r_w)} + \frac{C}{(L_e/r_w)} \right\}^{-1} \qquad \text{(fully penetrating well)} \qquad (6.378)$$

where C is a dimensionless parameter related to L_e/r_w, as shown in Figure 6.41. Values of $\ln (R_e/r_w)$ are within 10% of experimental values when $L_e/L_w > 0.4$, and within 25% of experimental values when $L_e/L_w < 0.2$ (Bouwer, 1978). It should be noted that the analysis presented here applies equally well to cases in which a "slug" of water is instantaneously added to the well at the beginning of the test (Batu, 1998). Values of K derived from a slug test primarily measure the average horizontal component of the hydraulic conductivity over a cylindrical volume of aquifer with inner radius r_w, outer radius R_e, and height slightly larger than L_e. Because of the relatively small portion of the aquifer sampled by slug tests, it is not unusual for the hydraulic conductivity values derived from slug tests at closely spaced wells to differ by several orders of magnitude (Gorelick et al., 1993). Part of the reason is that hydraulic conductivities derived from slug tests are highly sensitive to minor variations in well-construction details, such as the screen position and the dimensions of the gravel pack. Averaging the hydraulic conductivities derived from spatially distributed slug tests to get a representative areal average should be done with caution, since significant underestimates of the areal average are frequently obtained (Bouwer, 1996). Slug tests are typically limited to aquifers with relatively low hydraulic conductivities, since rapid recovery in the test well yields poor data resolution (Lee, 1999). Slug tests are particularly useful in

FIGURE 6.41: Parameters in slug-test analysis.

Source: Bouwer and Rice (1976).

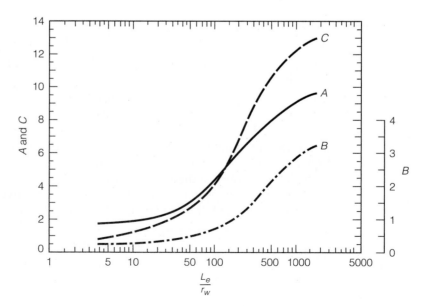

investigating localized flow conditions, which have an important effect on contaminant transport in ground water.

Data analysis.   Data collected during slug tests consist primarily of the drawdown, y_t, as a function of time, t, and a methodology for combining these data with Equation 6.376 to estimate the hydraulic conductivity is as follows:

1. Plot the measured data of $\ln y_t$ versus t.
2. According to Equation 6.376, the plot in step 1 should be a straight line. If some portion of the data does not show a linear relation, then only the linear portion is used in the analysis. In some cases, the plot of $\ln y_t$ versus t will yield a curve with two straight-line segments (Fetter, 2001). Such a situation occurs when the water in the gravel pack drains rapidly into the well. Once the water level in the gravel pack equals the water level in the well, then the second straight-line segment forms. This reflects the hydraulic conductivity of the undisturbed aquifer. If a double straight line forms, the second segment should be used in the analysis. If the water level in the well is not lowered to the level of the gravel pack or if the gravel pack is so permeable that it will drain at the same rate that the water table is lowered, then the double-straight-line effect should not develop. If the gravel pack is surrounded by a less permeable zone, the data may fit three straight lines, one at very small values of t, the second at intermediate values of t, and the third at large values of t (Prakash, 2004). Again, the last straight line is representative of the hydraulic conductivity of the undisturbed aquifer.
3. Fit a straight line through the linear portion of the curve identified in step 2. The equation of this line can be written as

$$\ln y_t = mt + \ln y_0 \qquad (6.379)$$

Determine the slope, m, of the fitted line. Equation 6.379 also indicates that this slope is given by

$$m = -\frac{1}{t} \ln \frac{y_0}{y_t} \qquad (6.380)$$

4. Based on the known values of L_w and H, select the equation to be used for $\ln(R_e/r_w)$ from Equations 6.377 and 6.378.
5. If Equation 6.377 is to be used, determine the values of A and B from Figure 6.41. If Equation 6.378 is to be used, determine the value of C from Figure 6.41.
6. Calculate the value of $\ln(R_e/r_w)$ using the results of step 5.
7. Calculate the hydraulic conductivity, K, using Equation 6.376 with the values of m, $\ln(R_e/r_w)$, r_c, and L_e, where

$$K = -\frac{r_c^2 \ln(R_e/r_w)}{2L_e} m \qquad (6.381)$$

The application of this analysis is illustrated in the following example.

EXAMPLE 6.40

A slug test is conducted in a monitoring well that penetrates 5 m below the water table in an unconfined aquifer with a saturated thickness of 15 m. The well casing has a 100-mm diameter, and the bottom 3 m of the well is screened and surrounded by a 100-mm thick gravel pack. The water level in the well is instantaneously drawn down 1500 mm, and the observed time-drawdown relation during recovery is given in the following table:

Time (s)	0	2	4	6	8	10	12	14	16
Drawdown (mm)	1500	645	372	195	130	67	44	23	15

Estimate the hydraulic conductivity and transmissivity of the aquifer.

Solution The measured data of $\ln y_t$ versus t is plotted in Figure 6.42. These data indicate that, except for the first data point ($t = 0$, $y_t = 1500$ mm), the relationship between $\ln y_t$ and t appears quite linear and can be fitted to the regression equation

$$\ln y_t = -0.270t + 6.976$$

which has a slope, m, of -0.270. From the given dimensions of the well and aquifer, $L_w = 5$ m, $H = 15$ m, $r_c = 50$ mm, $L_e = 3$ m, $r_w = r_c + 100$ mm $= 150$ mm, and $L_e/r_w = 3/0.15 = 20$. Equation 6.377 is appropriate for calculating $\ln(R_e/r_w)$ (since the well is partially penetrating), and the dimensionless parameters A and B are given by Figure 6.41 as

$$A = 2.1, \qquad B = 0.25$$

Substituting these data into Equation 6.377 yields

$$\ln \frac{R_e}{r_w} = \left\{ \frac{1.1}{\ln(L_w/r_w)} + \frac{A + B\ln[(H - L_w)/r_w]}{(L_e/r_w)} \right\}^{-1}$$

$$= \left\{ \frac{1.1}{\ln(5/0.15)} + \frac{2.1 + 0.25\ln[(15 - 5)/0.15]}{(3/0.15)} \right\}^{-1} = 2.12$$

and putting this result into Equation 6.381 gives

$$K = -\frac{r_c^2 \ln(R_e/r_w)}{2L_e}m$$

$$= -\frac{(0.05)^2(2.12)}{2(3)}(-0.270) = 2.39 \times 10^{-4} \text{ m/s} = 20.6 \text{ m/d}$$

FIGURE 6.42: Analysis of slug-test data

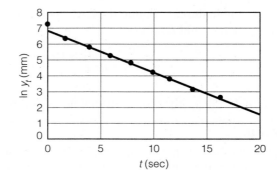

This result indicates an average (horizontal) hydraulic conductivity of 20.6 m/d in the immediate vicinity of the well. The transmissivity, T, of the aquifer can be estimated by

$$T = KH = (20.6)(15) = 309 \text{ m}^2/\text{d}$$

Although the Bouwer and Rice (1976) slug-test analysis was originally designed for unconfined aquifers, the test can also be used in confined or stratified aquifers if the top of the screen is some distance below the upper confining layer (Bedient et al., 1999).

Guidelines.   The Kansas Geological Survey (Butler et al., 1996) and Fetter (2001) have reviewed several slug-test methodologies and suggested the following guidelines for conducting slug tests:

1. **Three or more slug tests should be performed at a given well.** If the well has not been properly developed prior to slug testing, the slug test itself might cause additional development. This can be detected by a shift in the calculated transmissivity during the repeated slug tests. Also the slug test can mobilize fine material, which can result in a decrease in formation transmissivity.

2. **Two or more different initial displacements should be used during testing of a given well.** A series of tests with initial head displacement, y_0, which varies by a factor of at least 2, should be employed; however, the first and last tests should have the same y_0 so that a dynamic *skin effect* can be detected. The smearing of clay and silt on the surface of boreholes by augers sometimes results in a borehole skin that has a lower hydraulic conductivity than the surrounding aquifer. This skin effect is usually removed by well development.

3. **The slug should be introduced in a near-instantaneous fashion and a good estimate of the initial displacement should be obtained.** This is a basic assumption of the slug-test methodology. While it is easy to accomplish in a low-permeability system, it can be difficult to do in a system that has a very rapid response.

4. **Appropriate data-acquisition equipment should be used.** Manual methods of measurement might be satisfactory for a well in a low-permeability sediment that responds over a period of many minutes to hours. For more permeable formations, where the total period of response can be less than one minute, a data logger with a pressure transducer is mandatory. The pressure transducer should have the proper sensitivity for the planned head displacement.

It is important to keep in mind that slug tests measure the hydraulic conductivity in the immediate vicinity of the well, and therefore these tests may not be useful in estimating larger-scale aquifer hydraulic conductivities, such as those measured by aquifer pump tests. Comparison of hydraulic conductivities estimated by slug tests with values estimated by aquifer pump tests indicate that slug-test results are generally low (Prakash, 2004; Bromley et al., 2004). The likely reason for this discrepancy between small-scale and large-scale measurements of hydraulic conductivity is that slug tests do not account for preferential flow paths within aquifers.

6.8.8 Design of Exfiltration Trenches

Exfiltration trenches, sometimes called *french drains* or *percolation trenches*, are commonly used to discharge stormwater runoff into the subsurface and to return treated water from pump-and-treat systems into surficial aquifers. A typical exfiltration trench is illustrated in Figure 6.43. The operational characteristics of exfiltration trenches are influenced by the hydraulic conductivity of the aquifer material immediately surrounding the trench. This *trench hydraulic conductivity* is commonly determined from slug tests. The functional characteristics of an exfiltration trench depend on whether the trench is discharging water at a constant rate, such as in the case of a pump-and-treat system, or discharging surface-water runoff from a design storm, as in the case of stormwater-management applications. The design of a constant-flow exfiltration trench is described below, and the design of an exfiltration trench for stormwater-management applications is described in Chapter 5, Section 5.7.2.3.

Constant-flow design.   Constant-flow exfiltration trenches are commonly designed assuming that the water in the trench is exfiltrated through both the bottom and the long sides of the trench. The outflow from the bottom of the trench, Q_b, is given by Darcy's law as

$$Q_b = K_t W L \tag{6.382}$$

where K_t is the trench hydraulic conductivity, W is the width of the trench, L is the length of the trench, and the hydraulic gradient is assumed equal to unity. The outflow from each side of the trench, Q_s, is given by Darcy's law as

$$Q_s = K_t A_{\text{perc}} \tag{6.383}$$

FIGURE 6.43: Typical exfiltration trench

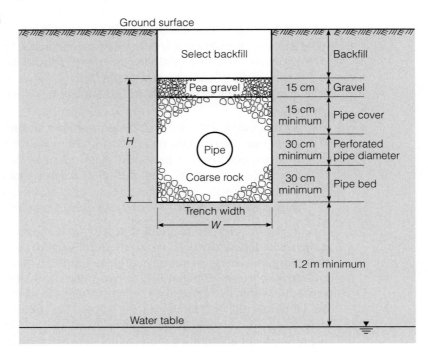

where A_{perc} is the side area through which water exfiltrates and the hydraulic gradient is taken as unity. It is common to assume that water exfiltrates through one-half of the trench height, H, in which case

$$A_{\text{perc}} = \frac{1}{2} LH \qquad (6.384)$$

and Equations 6.383 and 6.384 combine to give

$$Q_s = \frac{1}{2} K_t HL \qquad (6.385)$$

The total outflow rate from the exfiltration trench, Q, is equal to the sum of the flows out of the two (long) sides plus the flow out of the bottom. Therefore

$$Q = Q_b + 2Q_s$$

$$= K_t WL + 2\left(\frac{1}{2} K_t HL\right) = K_t L(W + H) \qquad (6.386)$$

Solving this equation for the trench length, L, yields

$$\boxed{L = \frac{Q}{K_t(W + H)}} \qquad (6.387)$$

The objective of the trench design is to select a length, L, width, W, and height, H, of the exfiltration trench, where the trench hydraulic conductivity, K_t, and the outflow rate, Q, are fixed by the site and design conditions. The maximum trench depth is limited by trench-wall stability, seasonal-high water-table elevation, and the depth to any impervious soil layer. Exfiltration trenches 1 m wide and 1 to 2 m deep are considered to be most efficient (ASCE, 1998). Adequate ground-water (quality) protection is generally obtained by providing at least 1.2 m separation between the bottom of the trench and the seasonal-high water table.

After selecting the geometry of the exfiltration trench, the aggregate mix to be placed in the trench surrounding the perforated pipe must be selected such that the hydraulic conductivity of the aggregate mix significantly exceeds the hydraulic conductivity of the porous matrix surrounding the trench. The hydraulic conductivities of various aggregate mixes have been measured by Cedergren and colleagues (1972), who showed that the hydraulic conductivity of any gravel mix can be expected to exceed 4000 m/d, which should be adequate for practically all exfiltration-trench designs. Consequently, any commercially available gradation of gravel can be used to surround the perforated pipe in the exfiltration trench. The perforated pipe used in exfiltration trenches usually consists of corrugated steel pipe with approximately 320 perforations per square meter; the diameter of the perforations is typically 0.95 mm (American Iron and Steel Institute, 1995).

After designing an exfiltration trench capable of injecting the design flow, Q, into the subsurface, the final step is to verify that the aquifer is capable of transporting the recharge water away from the trench at a sufficient rate to prevent the ground water from mounding to within a specified depth below the trench. The mounding of ground water under a rectangular recharge area of length L and width W has been investigated by Hantush (1967), who derived the following expression for the maximum saturated thickness, h_m, under a rectangular recharge area as a function of time:

$$h_m^2(t) = h_i^2 + \frac{2N}{K} \nu t S^* \left(\frac{W}{4\sqrt{\nu t}}, \frac{L}{4\sqrt{\nu t}} \right) \qquad (6.388)$$

where h_i is the saturated thickness without recharge; N is the recharge rate, which can be expressed in terms of the design flow, Q, by the relation

$$N = \frac{Q}{LW} \qquad (6.389)$$

and K (in Equation 6.388) is the hydraulic conductivity of the aquifer; ν is a combination of terms given by

$$\nu = \frac{Kb}{S_y} \qquad (6.390)$$

where b is a mean saturated thickness, typically taken to be the average of h_i and $h_m(t)$; S_y is the specific yield; t is the time since the beginning of recharge; and $S^*(\alpha, \beta)$ is a function defined by

$$S^*(\alpha, \beta) = \int_0^1 \text{erf}\left(\frac{\alpha}{\sqrt{\tau}} \right) \text{erf}\left(\frac{\beta}{\sqrt{\tau}} \right) d\tau \qquad (6.391)$$

where $\text{erf}(\xi)$ is the error function. General evaluation of the function $S^*(\alpha, \beta)$ requires numerical integration of Equation 6.391; however, values of the function for a range of α and β are presented in Table 6.27. For values of α and β beyond the range of Table 6.27, the following relations and approximations for $S^*(\alpha, \beta)$ are useful for

TABLE 6.27: S^* Function

$\alpha \setminus \beta$	0.0000	0.30	0.60	0.90	1.2	1.5	1.8	2.1	2.4	2.7	3.0
0.00	0.0000	0.0000	0.0000	0.0000	0.0000	0.0000	0.0000	0.0000	0.0000	0.0000	0.0000
0.10	0.0000	0.1290	0.1764	0.1957	0.2030	0.2052	0.2062	0.2065	0.2065	0.2065	0.2065
0.30	0.0000	0.3009	0.4314	0.4860	0.5070	0.5142	0.5164	0.5170	0.5171	0.5172	0.5172
0.60	0.0000	0.4314	0.6426	0.7360	0.7729	0.7857	0.7897	0.7907	0.7909	0.7910	0.7910
0.90	0.0000	0.4860	0.7360	0.8504	0.8966	0.9129	0.9180	0.9193	0.9196	0.9197	0.9197
1.20	0.0000	0.5070	0.7729	0.8966	0.9472	0.9653	0.9709	0.9724	0.9728	0.9728	0.9728
1.50	0.0000	0.5142	0.7857	0.9129	0.9653	0.9841	0.9900	0.9915	0.9919	0.9920	0.9920
1.80	0.0000	0.5164	0.7897	0.9180	0.9709	0.9900	0.9959	0.9975	0.9979	0.9979	0.9979
2.10	0.0000	0.5170	0.7907	0.9193	0.9724	0.9915	0.9975	0.9991	0.9995	0.9995	0.9995
2.40	0.0000	0.5171	0.7909	0.9196	0.9728	0.9919	0.9979	0.9995	0.9998	0.9999	0.9999
2.70	0.0000	0.5172	0.7910	0.9197	0.9728	0.9920	0.9979	0.9995	0.9999	1.0000	1.0000
3.00	0.0000	0.5172	0.7910	0.9197	0.9728	0.9920	0.9979	0.9995	0.9999	1.0000	1.0000

practical computations (Hantush, 1967):

$$S^*(\alpha, \beta) = S^*(\beta, \alpha) \tag{6.392}$$

$$S^*(0, \beta) = S^*(\alpha, 0) = 0 \tag{6.393}$$

$$S^*(\alpha, \beta) \simeq 1 - 4i^2 \text{erfc}(\beta), \quad \text{if } \alpha \geq 3.0 \tag{6.394}$$

$$S^*(\alpha, \beta) \simeq 1 - 4i^2 \text{erfc}(\alpha), \quad \text{if } \beta \geq 3.0 \tag{6.395}$$

$$S^*(\alpha, \beta) \simeq 1, \qquad\qquad \text{if } \alpha \geq 3.0 \text{ and } \beta \geq 3.0 \tag{6.396}$$

$$S^*(\alpha, \beta) \simeq \frac{4}{\pi} \alpha\beta \left\{ 3 + W(\alpha^2 + \beta^2) - \left[\frac{\alpha}{\beta} \tan^{-1} \frac{\beta}{\alpha} + \frac{\beta}{\alpha} \tan^{-1} \frac{\alpha}{\beta} \right] \right\},$$

$$\text{if } \alpha^2 + \beta^2 \leq 0.10 \tag{6.397}$$

where erfc(x) is the complementary error function, $i = \sqrt{-1}$, and $W(u)$ in Equation 6.397 is the Theis well function. Equation 6.388 gives the height of the ground-water mound under the trench as a function of time. Typically, operators are concerned with the long-term effect of recharge on ground water, such as the height of the ground-water mound 20, 50, or 100 years from initial operation (Bouwer et al., 1999). Ground-water levels that are too high will cause the ground to be waterlogged, and will threaten underground pipelines and basements. The design of an exfiltration trench is illustrated by the following example.

EXAMPLE 6.41

Design an exfiltration trench to inject 207 m³/d from a pump-and-treat system. Several slug tests at the site have-indicated a minimum trench hydraulic conductivity of 24 m/d. The depth from the ground surface to the seasonal-high water table is 4.35 m, and the aquifer is estimated to have a hydraulic conductivity of 107 m/d, a saturated thickness of 10.7 m, and a specific yield of 0.2. Local regulations require a factor of safety of 2 in the assumed trench hydraulic conductivity and a minimum backfill of 50 cm.

Solution   The length, L, of the trench is given by Equation 6.387 as

$$L = \frac{Q}{K_t(W + H)} \tag{6.398}$$

where the dimensions H and W are illustrated in Figure 6.43. Following ASCE (1998) guidelines, specify $W = 1$ m (a typical backhoe dimension), and $H = 2$ m (a typical depth for vertical slope stability). The design injection rate, Q, is 207 m³/d, and the design trench hydraulic conductivity, K_t, is 12 m/d, equal to the minimum measured value of 24 m/d divided by the factor of safety of 2. Substituting these values for Q, K_t, W, and H into Equation 6.398 yields

$$L = \frac{Q}{K_t(W + H)} = \frac{207}{12(1 + 2)} = 5.75 \text{ m} \tag{6.399}$$

A trench dimension of 6 m long by 1 m wide by 2 m deep, when filled with gravel, will be capable of transferring water into the aquifer at a rate of 207 m³/d. Since the seasonal-high water table is 4.35 m below the ground surface, there is sufficient room to install a 2-m deep trench covered with 0.5 m ($= 50$ cm) of backfill and still maintain a distance of at least 1.2 m between the bottom of the trench and the seasonal-high water table.

The next question is whether the aquifer will be able to transport the effluent away from the trench as fast as it is supplied, without causing the water table to rise to within 1.2 m of the bottom of the trench. The height of the water table above the base of the aquifer as a function of time is given by

$$h_m^2(t) = h_i^2 + \frac{2N}{K} \nu t S^* \left(\frac{W}{4\sqrt{\nu t}}, \frac{L}{4\sqrt{\nu t}} \right) \tag{6.400}$$

where $h_i = 10.7$ m and N is given by

$$N = \frac{Q}{LW} = \frac{207}{(6)(1)} = 34.5 \text{ m/d}$$

The hydraulic conductivity of the aquifer, K, is 107 m/d, and the parameter ν in Equation 6.400 is given by

$$\nu = \frac{Kb}{S_y} = \frac{K(h_i + h_m)/2}{S_y} = \frac{107(10.7 + h_m)/2}{0.2} = 267.5(10.7 + h_m)$$

Substituting the trench dimensions ($L = 6$ m, $W = 1$ m) and aquifer properties into Equation 6.400 yields

$$h_m^2 = 10.7^2 + \frac{2(34.5)}{107} [267.5(10.7 + h_m)]t S^* \left(\frac{1}{4\sqrt{(5725)t}}, \frac{6}{4\sqrt{(5725)t}} \right)$$

$$= 114 + 172.5(10.7 + h_m)t S^* \left(\frac{0.00330}{\sqrt{t}}, \frac{0.0198}{\sqrt{t}} \right) \tag{6.401}$$

This equation relates the maximum water surface elevation below the trench to the time since the trench began operation. Values of h_m for several values of t are shown in Table 6.28, where $S^*(\alpha, \beta)$ has been estimated using Equation 6.397. The values shown in Table 6.28 indicate that the water table under the trench steadily increases, and after 100 years the saturated thickness will be 10.87 m. This indicates that the water table below the exfiltration trench will rise (mound) by about 0.17 m ($= 17$ cm) and will remain at an acceptable depth (> 1.2 m) below the trench. The final trench design is illustrated in Figure 6.44.

TABLE 6.28: h_m Versus t

t (days)	h_m (m)
1	10.86
10	10.84
100	10.87
1000	10.91
10,000	10.94

FIGURE 6.44: Trench design

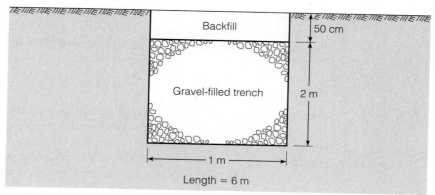

6.8.9   **Seepage Meters**

A conventional seepage meter is illustrated in Figure 6.45 and consists of a pan, formed from an inverted bucket or drum, connected to a collection bag made from thin plastic film. The pan is pushed into the bed of a lake or stream and the collection bag is weighed and attached to the pan. The bag is weighed again after some time has elapsed, and the flux of water into or out of the bed is readily calculated using the change in weight of the bag, the elapsed time, and the area of the pan. Interpretation of this measurement assumes that water flows into the pan at the same rate as it would if the seepage meter were absent. This assumption is intuitively appealing, because head losses in the pan and collection bag are expected to be negligible, although this is not always the case (Murdoch and Kelly, 2003). Some seepage meters use electronic flow meters to avoid the need for a collection bag. The seepage meter was initially developed to measure losses from irrigation canals (Israelsen and Reeve, 1944), and its application has been expanded to measure ground-water discharge into lakes and ponds, determine water budgets, obtain samples for chemical analyses, and study ground-water interactions with streams. Tests by Lee (1977), Erickson (1981), and Belanger and Montgomery (1992) suggest that seepage meters can yield measurements that are repeatable within 10%, while other studies (Isiorho and Meyer, 1999) are

FIGURE 6.45: Seepage meter

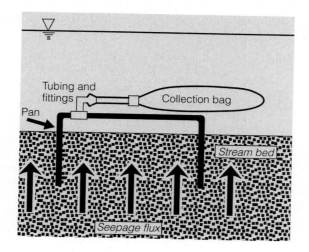

less encouraging, and show that repeat measurements in the laboratory can vary by more than a factor of 10. Seepage flux measured in the laboratory is typically less than the actual seepage rate, according to Erickson (1981) and Belanger and Montgomery (1992), who found the ratio of measured to actual seepage ranged from 0.66 to 0.77.

6.9 Computer Models

Several good computer codes are available for simulating ground-water flow. In their most general form, these codes provide numerical solutions to the combined Darcy and continuity equations, subject to prescribed initial and boundary conditions. Several practical applications of ground-water flow codes are listed in Table 6.29. In choosing a code for a particular application, there are usually a variety of codes to choose from; however, in doing work that is to be reviewed by regulatory agencies, codes developed and maintained by agencies of the U.S. government have the greatest credibility and, perhaps more important, are almost universally acceptable in supporting permit applications and defending design protocols. A secondary guideline in choosing a code is that the simplest code that will accomplish the design objectives should be given the most serious consideration.

By far the most widely used ground-water flow code is MODFLOW, which was developed and is maintained by the U.S. Geological Survey (McDonald and Harbaugh, 1988; Anderson and Woessner, 1992; Harbaugh and McDonald, 1996; USGS, 2000). MODFLOW is a three-dimensional finite-difference code that is very flexible, is applicable to numerous practical situations, has an input structure that allows for rapid and easy changes of input data, has a modular code that makes program modification simple, and is well documented. Flows can be steady or transient, aquifers confined or unconfined, and the aquifer properties can be either homogeneous or nonhomogeneous. MODFLOW has the capability of simulating a variety of aquifer stresses, including wells, areally distributed recharge, evapotranspiration, drains, and streams that penetrate the aquifer.

Other codes not developed by U.S. government agencies but widely used in ground-water engineering are DYNFLOW and PLASM. DYNFLOW is a three-dimensional finite-element code capable of handling irregular boundaries and has well-structured input files, and PLASM is a two-dimensional flow code that accounts for vertical leakance and has been widely taught at universities for many years. The PLASM code is simple, easy to modify and customize, and comes with a very helpful user manual.

TABLE 6.29: Practical Applications of Flow Models

No.	Application
1	Ground-water management and water supply
2	Aquifer-test evaluations
3	Wellhead-protection studies
4	Infiltration studies
5	Determination of velocity fields for contaminant-transport models
6	Determining barrier boundaries to stop saltwater intrusion

Source: ASCE (1996a).

Summary

Ground-water hydrology deals with the occurrence and movement of water below the surface of the earth. The basic equations that describe ground-water flows are the continuity and Darcy equations, where application of the Darcy equation is limited to low-Reynolds-number flows (Re < 10). Combining the continuity and Darcy equations yields a single governing equation in which the piezometric head is the dependent variable and space and time are the independent variables. Porous media and fluid properties are parameterized by the hydraulic conductivity tensor, K_{ij}, and the specific storage, S_s. In aquifers where vertical flows are negligible, ground-water movement can be approximated as two dimensional; the properties of the porous medium and fluid are parameterized by the transmissivity, T, and the specific yield, S_y (unconfined aquifer) or storage coefficient, S (confined aquifer). Solutions to the ground-water flow equation for a variety of initial and boundary conditions are presented. These solutions are frequently used in engineering practice, and include steady flow to a well in an unconfined aquifer with recharge, unsteady flow to a well in an unconfined aquifer, and cases of partially penetrating wells. Many of the solutions presented can be combined using the principle of superposition, where the validity of superposition is ensured by the linearity of the governing flow equation, boundary conditions, and initial conditions. Practical application of the principle of superposition is demonstrated by the method of images. In coastal areas, denser seawater intrudes into coastal aquifers (called salinity intrusion), and the application of the Ghyben–Herzberg equation to calculate the location of the saltwater interface in coastal aquifers is presented.

Engineered systems that are designed using the principles of ground-water hydrology include municipal wellfields and exfiltration trenches. In wellfield design, the objective is to design a system of wells that will produce a desired flowrate, within the constraints of available land area and allowable drawdown. A methodology for designing wellfields using fundamental solutions to the ground-water flow equation is presented as well as guidelines for designing individual water-supply wells and conducting aquifer pump tests. A methodology for obtaining localized estimates of the hydraulic conductivity using slug tests is also presented. In designing exfiltration trenches, the objective is to determine the dimensions and configuration of a trench that will exfiltrate water at a specified flowrate, within the constraint of allowable mounding of the water table. A methodology for designing exfiltration trenches and predicting mounding effects is presented. In many practical applications, a numerical ground-water flow code is required to handle complex geologies and boundary conditions. The most popular numerical code for calculating ground-water flows is MODFLOW, with other codes such as DYNFLOW and PLASM also in use.

Problems

6.1. Which property is a better measure of the productivity of an aquifer: porosity or hydraulic conductivity? Explain why.

6.2. Two piezometers are installed in a confined aquifer with an estimated hydraulic conductivity of 10 m/d and a porosity of 0.2. The reference piezometer is located at point A, and the second piezometer is located at point B, 1 km from point A at an angle of 45° clockwise from true north. The water level at B is 10 cm below A. Calculate the seepage velocity along line AB. Explain why this is not the actual seepage velocity in the aquifer. A third piezometer is located in the aquifer at point

C, 0.5 km from A at an angle of 140° clockwise from true north, and the water level at C is 8 cm below A. Calculate the seepage velocity and Darcy velocity (= specific discharge) in the aquifer.

6.3. If a contaminant is spilled in an aquifer, is the mean position of the contaminated cloud advected at the Darcy velocity or at the seepage velocity? Explain your answer. Why does a contaminant cloud "disperse" as it is advected through an aquifer?

6.4. An aquifer pump test indicates a hydraulic conductivity of 20 m/d when the water in the aquifer is at a temperature of 20°C. Determine the intrinsic permeability of the aquifer. If the fluid in the aquifer consists of spilled tetrachloroethylene (PERC), determine the hydraulic conductivity of the aquifer to PERC. Under the same piezometric gradient, which fluid moves faster? (Refer to Tables B.1 and B.2 for the fluid properties.)

6.5. A sand aquifer has a 10-percentile particle size of 0.5 mm. Estimate the intrinsic permeability and hydraulic conductivity at 20°C using the Harleman et al. (1963) equation. Compare your result with the representative hydraulic conductivity of medium sand given in Table 6.2.

6.6. Hazen (1911) developed the following empirical formula for the hydraulic conductivity, K (cm/s), in terms of the effective grain size, d_{10} (cm), of a porous medium:

$$K = cd_{10}^2$$

where c is approximately equal to 100. Show that this corresponds to using $\alpha = 1.02 \times 10^{-3}$ in Equation 6.15 to estimate the intrinsic permeability. Assume that the water temperature in the porous medium is 20°C.

6.7. An aquifer material has a porosity of 0.20 and a grain-size distribution as follows:

Grain size (mm)	Percent finer (%)
4.760	98.0
2.000	86.0
0.840	64.0
0.420	51.0
0.250	36.0
0.149	13.0
0.074	4.0

Estimate the hydraulic conductivity of the aquifer using all available empirical equations. Assume a temperature of 20°C.

6.8. Show that the seepage velocity, v, induced by a pumping well as a function of the distance, r, from the well is given by

$$v = \frac{Q_w}{2\pi r H n_e} \tag{6.402}$$

where Q_w is the well pumping rate, H is the saturated thickness, and n_e is the effective porosity.

6.9. Consider a well pumping at 0.4 m³/s in an aquifer with an effective porosity of 0.2, a saturated thickness of 24 m, and a hydraulic conductivity of 50 m/d. If the seepage velocity as a function of distance from the well is given by Equation 6.402, then calculate the extent of the circular area surrounding the well where Darcy's law is not valid. [*Hint:* Estimate the characteristic pore diameter by first computing the intrinsic permeability and then estimating the pore diameter using Equation 6.15.]

6.10. Derive Equations 6.24 to 6.26 from Equation 6.23.

6.11. Use Equations 6.24 to 6.26 to show that the principal components of the hydraulic conductivity tensor, K'_{11} and K'_{22}, can be expressed in terms of the general hydraulic conductivity tensor K_{ij} by the relations

$$K'_{11} = \frac{K_{11} + K_{22}}{2} + \left[\left(\frac{K_{11} - K_{22}}{2} \right)^2 + K_{12}^2 \right]^{\frac{1}{2}}$$

$$K'_{22} = \frac{K_{11} + K_{22}}{2} - \left[\left(\frac{K_{11} - K_{22}}{2} \right)^2 + K_{12}^2 \right]^{\frac{1}{2}}$$

6.12. Use Equations 6.24 to 6.26 to show that the angle of rotation needed to reach the principal axes from corresponding reference axes is given by

$$\theta = \frac{1}{2} \tan^{-1} \frac{2K_{12}}{K_{11} - K_{22}}$$

6.13. The aquifer described in Problem 6.2 is found to be anisotropic, with $K_{11} = 15$ m/d, $K_{22} = 5$ m/d, and $K_{12} = K_{21} = 0$ (i.e., the coordinate axes are the principal axes). If the x_1-axis is aligned with the east-west direction and the x_2-axis with the north-south direction, determine the seepage velocity and compare it with the seepage velocity obtained by assuming that the aquifer is homogeneous with a hydraulic conductivity of 10 m/d.

6.14. Ground-water flow in a regional aquifer has historically been toward the east, and the hydraulic conductivity tensor is estimated to be $K_{xx} = 100$ m/d, $K_{yy} = 10$ m/d, and $K_{xy} = 0$ m/d, where the x- and y-axes are along the east/west and north/south directions, respectively. The development of a new wellfield will cause the mean head gradient to shift from the east to the southeast (a rotation of $45°$). Compare the specific discharge in the direction of the head gradient for a gradient of 0.01 in the east direction with the specific discharge in the direction of the head gradient of 0.01 in the southeast direction. In either of these cases, does the ground water flow in the direction of the head gradient?

6.15. (a) A neutrally buoyant miscible contaminant is accidentally spilled into an unconfined aquifer. You have been asked to estimate the rate and direction of movement of the contaminant. Explain what field measurements and information you will need for your analysis, and how you would arrive at your estimate. State your assumptions. (b) If a surficial aquifer is composed mostly of sand, and a well driller hands you a bag of sand extracted from the aquifer, could you estimate the hydraulic conductivity of the aquifer? How? (c) If you have 10 measurements of hydraulic conductivity in an aquifer, how would you estimate an average hydraulic conductivity for the aquifer? Explain your rationale.

6.16. The hydraulic conductivities measured at 12 locations in a porous formation are 100, 200, 220, 250, 290, 50, 310, 400, 130, 190, 350, 500 m/d. Calculate the macroscopic hydraulic conductivities for both two- and three-dimensional flow.

6.17. If the hydraulic conductivity measurements given in Problem 6.16 were taken at 100-m intervals, then estimate the correlations between log-hydraulic conductivities at lags of 100 m and 200 m.

6.18. Write the expression for the exponential log-hydraulic conductivity correlation function along one of the Cartesian coordinate axes. [*Hint:* Take $r_1 \neq 0$, and $r_2 = r_3 = 0$.] Show that the correlation length scale in this direction is equal to the integral of the log-hydraulic conductivity correlation function in that direction.

6.19. The definition of the Laplacian operator in Cartesian coordinates is given by

$$\nabla^2 f = \frac{\partial^2 f}{\partial x^2} + \frac{\partial^2 f}{\partial y^2} + \frac{\partial^2 f}{\partial z^2}$$

If the relationship between cylindrical coordinates (r, θ, z) and Cartesian coordinates (x, y, z) is given by $x = r\cos\theta$, $y = r\sin\theta$, $z = z$, then show that the Laplacian operator in cylindrical coordinates is given by Equation 6.50. Under what circumstances would it be worthwhile to use cylindrical coordinates?

6.20. Consider the case of a three-dimensional homogeneous isotropic aquifer. Using cylindrical coordinates, give the differential equation describing the steady-state head distribution in the aquifer. Simplify this equation for a case where the head is radially symmetric.

6.21. A steady-state solution of the three-dimensional ground-water flow equation in an aquifer is found by making the simplifying assumption that the aquifer is isotropic and homogeneous, with a hydraulic conductivity of 25 m/d. The results of this simplified analysis indicate that the head, ϕ, at a location with Cartesian coordinates (100 m, 100 m, 10 m) is equal to 25 m. If the aquifer is actually anisotropic with $K_{xx} = 30$ m/d, $K_{yy} = 40$ m/d, and $K_{zz} = 13$ m/d, determine the actual location in the aquifer where the piezometric head is equal to 25 m. What is the water pressure (= pore pressure) at this location?

6.22. The hydraulic conductivity distribution in a 20-m thick stratified surficial aquifer is given by

Depth (m)	K_{xx} (m/d)	K_{yy} (m/d)
0–2	5	15
2–5	7	20
5–7	9	21
7–11	14	12
11–15	11	17
15–19	6	9
19–20	2	5

Determine the effective hydraulic conductivity when the water table is 2 m below the ground surface. Contrast this with the result when the water table is 3 m below the ground surface.

6.23. Consider the case of a two-dimensional unconfined homogeneous anisotropic aquifer. Using Cartesian coordinates, give the differential equation describing the steady-state saturated-zone thickness. Give a real-world example of a situation in which this equation would be applicable.

6.24. Consider the case of a two-dimensional confined homogeneous isotropic aquifer. Using Cartesian coordinates, give the differential equation describing the steady-state piezometric head distribution. Give a real-world example of a situation where this equation would be applicable.

6.25. A confined aquifer composed of a dense sandy-gravel matrix is 20 m thick and has a porosity of 0.3. If the compressibility of the aquifer material is estimated to be 8×10^{-9} m²/N, estimate the storage coefficient of the aquifer. Compare your result to the estimate given by the Lohman (1979) equation

$$S = 3 \times 10^{-6} b$$

where b is the aquifer thickness. Explain any reasons for a discrepancy. Assume that the temperature of the ground water is $15°C$. [*Note:* The compressibility is the reciprocal of the bulk modulus of elasticity.]

6.26. Explain why the storage coefficient is much smaller than the specific yield.

6.27. The steady-state head distribution caused by a single fully penetrating pumping well in an extensive (two-dimensional) isotropic confined aquifer can be estimated by

$$\phi(r) = \phi_0 - \frac{Q_w}{2\pi T} \ln\left(\frac{R}{r}\right), \qquad r < R$$

where ϕ is the piezometric head, r is the distance from the well, ϕ_0 is the head prior to pumping, Q_w is the pumping rate from the well, T is the transmissivity, and R is the radius of influence of the well. Determine the steady-state drawdown distribution for the case of an anisotropic aquifer with transmissivity components T_{xx} and T_{yy}.

6.28. Consider a two-layer stratified aquifer between two reservoirs. The water surfaces in the reservoirs are at elevations 5 and 4 m NGVD, respectively; the ground surface between the aquifers is at elevation 10 m NGVD; the top layer of the aquifer extends from ground surface down to -10 m NGVD, and the base of the aquifer (and reservoirs) is at -20 m NGVD. The hydraulic conductivity of the top layer is 50 m/d and that of the bottom layer is 100 m/d. If the reservoirs are 2 km apart, find the equation of the phreatic surface, and the flowrate between the reservoirs. Neglect surface recharge.

6.29. Show that the flow, Q, between two reservoirs separated by a two-layer aquifer can be expressed as

$$Q = \frac{K_1}{2L}(h_L^2 - h_R^2) + (K_1 - K_2)\frac{b_2}{L}(h_R - h_L) \qquad (6.403)$$

6.30. Consider the case of a fully penetrating canal shown in Figure 6.46 in which the drawdowns a distance L from the sides of the canal are s_L and s_R on the lefthand and righthand sides of the canal, respectively, and the effective hydraulic conductivity of the aquifer is K. Derive an expression for the leakage out of the canal per unit length of canal. Calculate the leakage when $K = 30$ m/d, $H = 20$ m, $L = 70$ m, and $s_L = s_R = 5$ cm.

6.31. Derive the general equation for the phreatic surface in a two-layer aquifer between two reservoirs when the recharge, $N(x)$, is not equal to zero. [*Hint:* An equation similar to Equation 6.116, but with an additional term to account for recharge.]

6.32. The equation describing the phreatic surface in a two-layer aquifer between two reservoirs has been shown to be

$$\frac{K_1}{2}h^2 + (K_2 - K_1)b_2 h + \frac{Nx^2}{2} = C_1 x + C_2$$

FIGURE 6.46: Leakage from a fully penetrating canal

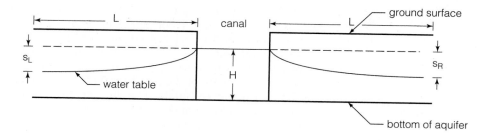

where C_1 and C_2 are constants given by

$$C_1 = \frac{K_1}{2L}(h_R^2 - h_L^2) + (K_2 - K_1)b_2\frac{(h_R - h_L)}{L} + \frac{NL}{2}$$

and

$$C_2 = \frac{K_1}{2}h_L^2 + (K_2 - K_1)b_2h_L$$

where N is the recharge rate between the two reservoirs. This equation describes a mounded phreatic surface, with flow to the left of the mound going toward the lefthand reservoir, and flow to the right of the mound going toward the righthand reservoir. Derive an expression for the location of the mound. It has been stated that a mound will always be located between the two reservoirs. Use your derived expression to determine whether this statement is true or false.

6.33. Derive the general equation for the phreatic surface in a three-layer aquifer between two reservoirs. Neglect surface recharge.

6.34. A well pumps at 0.4 m³/s from a confined aquifer whose thickness is 24 m. If the drawdown 50 m from the well is 1 m and the drawdown 100 m from the well is 0.5 m, then calculate the hydraulic conductivity and transmissivity of the aquifer. Do you expect the drawdowns at 50 m and 100 m from the well to approach a steady state? Explain your answer. If the radius of the pumping well is 0.5 m and the drawdown at the pumping well is measured to be 4 m, then calculate the radial distance to where the drawdown is equal to zero. Why is the steady-state drawdown equation not valid beyond this distance?

6.35. (a) For a confined aquifer, the Thiem equation can be used to express the drawdown surrounding a well in terms of well pumping rate, Q_w, transmissivity, T, and radius of influence, R. Explain why this equation does not accurately account for the drawdown at the well. Show that by selecting the radius of influence as

$$R = r_w \exp\left(\frac{2\pi T s_w}{Q_w}\right) \tag{6.404}$$

the Thiem equation can be made to accurately account for the drawdown at the well. What are the limitations of the Thiem equation and why does it have these limitations? Under what circumstances would you use the Thiem equation? (b) In deriving the Thiem equation, it is assumed that Darcy's law is applicable in the aquifer immediately surrounding the well intake. If a well has a pumping rate of 30 L/s, diameter of 10 cm, and a 1-m long intake, estimate the range of hydraulic conductivities for which the Thiem equation could be applied without violating Darcy's law.

6.36. Repeat Problem 6.34 for an unconfined aquifer, assuming that the saturated thickness is 24 m prior to pumping. If actual drawdowns, rather than the modified drawdowns, are used to calculate the transmissivity of the aquifer, then what would be the percentage difference in the calculated transmissivity?

6.37. A water-supply well is located at the center of a small circular island that has a radius of 1 km. The well has a diameter of 1 m and a pumping rate of 20,000 L/min, and the (phreatic) aquifer has a hydraulic conductivity of 40 m/d, a specific yield of 0.15, and a porosity of 0.2. If the water surrounding the island is at mean sea level and the base of the aquifer is 45 m below sea level, estimate the water-table elevation at the well intake. Use the seepage velocity halfway between the well and the coastline to estimate how long a contaminant would take to travel from the perimeter of the island to the well.

6.38. A pumping well has a radius of 0.5 m and extracts water at a rate of 0.4 m³/s from a semiconfined aquifer whose thickness is 24 m. The drawdown 50 m from the well is 1 m, and the drawdown 100 m from the well is 0.5 m. Use Equation 6.177 to calculate the leakage factor, transmissivity, and hydraulic conductivity of the aquifer. Compare the transmissivity with that obtained by neglecting leakage. (See Problem 6.34.)

6.39. Repeat Problem 6.38 using the approximate relation given by Equation 6.180. Verify that it is appropriate to use Equation 6.180.

6.40. If the piezometric surface in Problem 6.38 was initially 40 m above the base of the aquifer, then plot the leakage rate into the semiconfined aquifer as a function of the radial distance from the pumping well.

6.41. The transmissivity in a confined aquifer is estimated from steady-state drawdown measurements by Equation 6.142, and in a semiconfined aquifer by Equations 6.178 and 6.179. Identify the range of leakage factors for which the leakage can be neglected in the analysis of the drawdown data. Assume that leakage can be neglected whenever the neglect of leakage results in less than 0.1% error in the calculated transmissivity.

6.42. A well of radius 0.5 m in an unconfined aquifer pumps at a rate of 0.4 m³/s. If the hydraulic conductivity of the aquifer is 30 m/d, the saturated thickness of the aquifer at the well is 24 m, and the recharge rate to the aquifer is 500 mm/y, find an expression for the distribution of saturated thickness surrounding the well. Determine the distribution of seepage velocity surrounding the well.

6.43. Derive Equations 6.200 to 6.202 using the chain rule and Equation 6.199.

6.44. A well pumps water from a confined aquifer at a rate of 0.4 m³/s, where the radius of the well is 0.5 m and the thickness of the aquifer is 24 m. If the aquifer's storage coefficient is 0.0012 and its hydraulic conductivity is estimated to be 300 m/d, then calculate the drawdown at distances of 0.5 m, 50 m, and 100 m from the well as a function of time.

6.45. It has been widely asserted that the steady-state equation for estimating the transmissivity in confined aquifers given by Equation 6.142 can be used to estimate the transmissivity in spite of the unsteadiness in the measured drawdowns. Use the drawdown results derived in Problem 6.44 to estimate the time required for the transmissivity to be estimated within 1% accuracy by the steady-state equation.

6.46. A fully penetrating well pumps 0.2 m³/s from an unconfined aquifer where the specific yield is 0.15 and the hydraulic conductivity is 100 m/d. Prior to pumping, the saturated thickness was uniformly equal to 28 m. Calculate the drawdowns at 50 m and 100 m from the well after 1 sec, 1 minute, 1 hour, 1 day, 1 month, and 6 months. At each time, use the steady-state (Thiem) equation to estimate the transmissivity of the aquifer based on the calculated drawdowns at 50 and 100 m. On the basis of these results, what can you say about using a steady-state equation to estimate the transmissivity from pairs of drawdowns during the transient aquifer pump test?

6.47. A confined aquifer of thickness 24 m is pumped at a rate of 0.4 m³/s, and the recorded drawdowns in a monitoring well located 50 m from the pumping well are given in Table 6.30. Estimate the hydraulic conductivity, transmissivity, and storage coefficient of the aquifer.

6.48. Explain clearly how equations for isotropic formations can be utilized in anisotropic formations. Explain why some aquifers are anisotropic in the horizontal plane. Consider a case in which a confined aquifer is anisotropic with hydraulic conductivities given approximately by $K_{xx}=30$ m/d, $K_{yy}=15$ m/d, and $K_{xy}=10$ m/d. The thickness of the aquifer is 20 m, the storage coefficient is 0.005, and the porosity of the aquifer is 0.15. A pump test is to be conducted in the aquifer to confirm the hydraulic conductivities. If the pumping rate is 40 L/s, find the location of a monitoring well

TABLE 6.30: Drawdowns in Monitoring Well

Time (s)	Drawdown (cm)
1	0.00
10	9.54
100	76.39
1000	150.03
10,000	256.85
100,000	319.72

TABLE 6.31: Drawdowns in Monitoring Well

Time (s)	Drawdown (cm)
1	0.00
10	0.00
100	0.12
1000	12.21
10,000	36.89
100,000	68.64

such that a drawdown of 1 m would be expected after one day of pumping. How many possible locations for the monitoring well are there?

6.49. A well of radius 0.5 m pumps water out of an unconfined aquifer at a rate of $0.4 \text{ m}^3/\text{s}$. The saturated thickness prior to pumping is 24 m, and the measured drawdowns 50 m from the pumping well as a function of time are given in Table 6.31. Determine the hydraulic conductivity, transmissivity, and specific yield of the aquifer.

6.50. A confined aquifer of thickness 24 m is pumped at $0.27 \text{ m}^3/\text{s}$ and the recorded drawdowns in a monitoring well 50 m from the pumping well are given in Table 6.32. Use the Cooper–Jacob approximation of the well function to estimate the hydraulic conductivity and the storage coefficient of the aquifer. At what time does the Cooper–Jacob approximation become reasonable?

6.51. An aquifer pump test is conducted at a (pumping) rate of 22 L/s in an unconfined aquifer that has a saturated thickness of 30 m. Observed drawdowns 70 m from the well are as follows:

Time (min)	Drawdown (m)	Time (min)	Drawdown (m)	Time (min)	Drawdown (m)
0.01	0.001	1.88	0.036	1875.00	0.157
0.01	0.008	18.75	0.036	3330.00	0.187
0.02	0.013	59.40	0.046	5940.00	0.248
0.03	0.023	187.50	0.058	10,530.00	0.267
0.06	0.034	333.00	0.067	18,750.00	0.312
0.11	0.036	594.00	0.091		
0.19	0.036	1053.00	0.122		

Determine the aquifer properties using the Neuman type-curve method.

6.52. An aquifer pump test is conducted in an unconfined aquifer with a saturated thickness of 25 m. The pumping well is pumped at 40 L/s, and the drawdowns are

TABLE 6.32: Drawdowns in Monitoring Well

Time (s)	Drawdown (cm)
1	0.00
10	6.39
100	51.18
1000	100.52
10,000	172.09
100,000	214.21

measured at a monitoring well located 40 m away from the pumping well. The measured drawdowns as a function of time are given in the following table:

Time	Drawdown (cm)
18 s	0.8
37 s	4.4
1.08 min	10.2
1.85 min	17.7
3.07 min	25.2
6.15 min	33.7
10.8 min	37.8
18.4 min	39.3
1.54 h	43.3
2.69 h	45.5
4.61 h	49.5
7.68 h	55.1
15.4 h	67.9
1.12 d	82.8
1.92 d	98.4

Use a delayed-yield analysis to determine the storage coefficient, specific yield, horizontal hydraulic conductivity, and vertical hydraulic conductivity of the aquifer.

6.53. A confined anisotropic aquifer has a hydraulic conductivity of K_{xx} = 45 m/d, K_{yy} = 15 m/d, and K_{xy} = 0 m/d; a thickness of 14 m; and a storage coefficient of 10^{-4}. If a well is installed in the aquifer and pumped at 1000 L/min, estimate the drawdown at x = 100 m, y = 100 m after 1 week of pumping. How would this result change if the confining layer had a leakage factor of 2000 m?

6.54. A fully penetrating well of radius 0.5 m pumps water at 0.4 m³/s from a semiconfined aquifer of thickness 24 m. If the hydraulic conductivity of the aquifer is 300 m/d and the storage coefficient is 0.0012, estimate the drawdown as a function of time at a distance of 50 m from the well for leakage factors of 1 m, 10 m, and 100 m. Explain your results. If the thickness of the semiconfining layer is 5 m, then what are the hydraulic conductivities of the semiconfining layers corresponding to leakage factors of 10 m and 100 m?

6.55. An aquifer test was conducted in a 5-m thick shallow aquifer confined above by 4.5 m of clay. The 200-mm diameter well was pumped at the rate of 95 L/min for 20 hours, and drawdown was measured in a monitoring well located 20 m west of the pumping well. The data are shown in the table below. Estimate the transmissivity and storage coefficient of the aquifer, and the hydraulic conductivity of the confining layer.

Time (min)	Drawdown (m)	Time (min)	Drawdown (m)	Time (min)	Drawdown (m)
5	0.23	75	1.34	958	1.91
28	1.01	244	1.67	1129	1.95
41	1.09	493	1.82	1185	1.96
60	1.24	669	1.86		

6.56. In the aquifer described in Problem 6.55 the thickness and hydraulic conductivity of the overlying aquifer is 20 m and 25 m/d, respectively, and the semiconfining layer has a storage coefficient of 0.0015. Assess whether neglecting storage in the semiconfining layer, neglecting drawdown in the overlying aquifer, and neglecting storage in the pumping well are justified.

6.57. Consider the fully penetrating well described in Problem 6.34. What will be the drawdown in the well if only the top half of the aquifer is penetrated? Express the drawdown in the well as a function of the penetration factor. Can the drawdowns at 50 m and 100 m be reasonably estimated by assuming full penetration of the well, even if the well penetrates only the top half of the aquifer? Explain.

6.58. Consider the fully penetrating well described in Problem 6.34. Compare the additional drawdown in the well expected if the well penetrates only the top 40% of the aquifer with the additional drawdown expected if the well screen is centered in the aquifer and has a length equal to 40% of the aquifer thickness.

6.59. A 200-mm diameter water-supply well is to be installed in a confined aquifer of thickness 30 m, and the radius of influence of the well is expected to be 250 m. Estimate the screen length that would be required to limit the effect of partial penetration on drawdown at the well to less than 10 percent. Use both the Kozeny (1933) and DeGlee (1930) equations in your analysis.

6.60. Three wells are located in an infinite unconfined aquifer of saturated thickness 20 m, specific yield 0.2, and hydraulic conductivity 40 m/d. The planar coordinate locations of the wells are: point A (0 m, 0 m), point B (200 m, 200 m), and point C (200 m, −200 m). If all wells begin pumping at the same time and at a rate of 0.2 m^3/s, then calculate the drawdown as a function of time at a location (100 m, 100 m).

6.61. A well of radius 0.5 m pumps water at 0.4 m^3/s from a confined aquifer of thickness 24 m. The hydraulic conductivity of the aquifer is 300 m/d, the storage coefficient is 0.012, and the radius of influence can be taken as 1200 m. If the planar coordinates of the well are (0 m, 0 m), and a fully penetrating river runs along the line $x = 500$ m, then calculate the steady-state drawdowns at (100 m, 0 m) and (−100 m, 0 m).

6.62. A second well located at (0 m, 200 m) is added to the wellfield described in Problem 6.61. If this well also pumps at 0.4 m^3/s and the radius of influence is 1200 m, then calculate the drawdowns at (100 m, 0 m) and (−100 m, 0 m) and compare with those calculated in Problem 6.61. If this second well is to be placed parallel to the river and contribute no more than 1% of the total drawdown at the designated points, then determine the coordinates of the second well.

6.63. A well with an effective radius of 0.8 m pumps water at 0.5 m^3/s from an unconfined aquifer of saturated thickness 24 m, hydraulic conductivity 250 m/d, and specific yield 0.2. A canal that feeds the aquifer is located 500 m east of the well, and the canal runs in a north/south direction. Determine the fraction of the pumped water that originates in the canal. You may need to use the following integral relation to solve this problem:

$$\int \frac{dx}{a^2 + x^2} = \frac{1}{a} \tan^{-1} \left(\frac{x}{a} \right)$$

where a is any constant.

6.64. Repeat Problem 6.61, but with the fully penetrating river replaced by an impermeable barrier.

6.65. Repeat Problem 6.62, with the fully penetrating river replaced by an impermeable barrier.

6.66. A well of radius 0.5 m pumps water at 0.4 m^3/s from a confined aquifer of thickness 24 m, hydraulic conductivity 300 m/d, and storage coefficient 0.012. If a fully penetrating river is located 1 km east of the well and an impermeable boundary is located 1 km west of the well, then calculate the drawdown in the aquifer at points 200 m east of the well and 200 m west of the well. Assume that the radius of influence of the well is 1200 m.

6.67. Show that the drawdown distribution caused by a well in an infinite strip between two fully penetrating streams is given by

$$s = \frac{Q_w}{4\pi T} \sum_{n=-\infty}^{\infty} \ln \frac{(x + x_0 - 2nd)^2 + y^2}{(x - x_0 - 2nd)^2 + y^2} \tag{6.405}$$

where Q_w is the pumping rate, T is the aquifer transmissivity, x_0 is the distance of the well from one stream, and d is the distance between the streams.

6.68. Show the arrangement of image and real wells that is required to calculate the drawdown induced by a well in an aquifer quadrant, where the other three quadrants contain impermeable formations.

6.69. Show that the buildup caused by injecting water at a rate Q_w into a fully penetrating well in an unconfined infinite aquifer is exactly the same as the drawdown caused by withdrawing water at a rate Q_w from the same aquifer.

6.70. The water table at a given location in a coastal aquifer is 1 m above mean sea level, and the saturated zone in the aquifer is 24 m thick. Do you expect saltwater intrusion to be a problem at this location?

6.71. In a coastal aquifer the saltwater interface intercepts the bottom of the aquifer approximately 3 km from the shoreline. If the hydraulic conductivity of the aquifer is 40 m/d and the bottom of the aquifer is 50 m below sea level, estimate the freshwater discharge per kilometer of shoreline.

6.72. A drainage canal in a coastal area terminates in a gated structure, and the thickness of the aquifer beneath the canal is 24 m. On the downstream side of the gate, the seawater fluctuates between 30 cm above and below mean sea level. If the elevation of the bottom of the canal is 3 m below sea level, determine the minimum water elevation on the upstream side of the gate to prevent saltwater intrusion.

6.73. A water-supply well is being threatened by saltwater intrusion. The well is pumping 0.04 m^3/s from an aquifer that has a hydraulic conductivity of 500 m/d, saturated thickness of 50 m, porosity of 0.2, and specific yield of 0.15. If the well is screened to within 15 m of the bottom of the aquifer, what will be the thickness of the saltwater wedge under the well when the well becomes contaminated with saltwater?

6.74. Explain why a rising water table will not yield the same retention curve as a falling water table.

6.75. Consider a soil with the retention curve shown in Table 6.33. If the water table was initially 1.5 m below the ground surface and falls to 3 m below the ground surface, then determine the specific yield of the aquifer. Would the specific yield be any different if the water table had fallen from 3.5 m to 5 m below the ground surface?

6.76. A wellfield is to be developed to produce 4.44 m^3/s from an unconfined aquifer with a saturated thickness of 35 m, a hydraulic conductivity of 500 m/d, and a specific yield of 0.2. If the radius of each well is to be 0.5 m, the drawdown is not to exceed 2 m, and the wells are to be arranged along a straight line, then determine the number of wells, the pumping rate from each well, and the spacing between wells.

TABLE 6.33: Retention Curve

Elevation above water table (m)	Moisture content
0	0.18
0.5	0.16
1.0	0.15
1.5	0.14
2.0	0.11
2.5	0.08
3.0	0.06
3.5	0.05
4.0	0.05
4.5	0.05
5.0	0.05

6.77. A new wellfield is to be developed within a 500-m × 500-m block of land bounded on all sides by canals that are intended to feed the wellfield. The stages in all canals are to be maintained at 1.000 m NGVD, and the drawdown in the wellfield is not to exceed 2 m. The diameter of each well is to be 800 mm, and the wells are to be a minimum of 50 m apart and a minimum of 100 m from any canal. Field tests indicate that the aquifer has a saturated thickness of 20 m, a hydraulic conductivity of 75 m/d, and a specific yield of 0.26. Determine the number and location of the wells required to supply 6000 m³/d.

6.78. A wellfield is to be developed such that the wells are all parallel to a stream, with each well having a diameter of 0.1 m. The unconfined aquifer has a saturated thickness of 20 m, an effective porosity of 0.15 and a hydraulic conductivity of 60 m/d. If the wellfield is expected to deliver 5000 L/min, each well is to be 100 m from the stream, and the maximum allowable drawdown is 2 m, estimate the number of wells required, the pumping rate from each well, and the spacing between the wells.

6.79. A square-shaped 50-ha parcel of land has been identified as a possible site for a wellfield to supply a population of 50,000 people. The saturated thickness of the aquifer is 35 m, the hydraulic conductivity is 85 m/d, the porosity is 0.2, and the specific yield is 0.15. The per-capita demand of the population is 580 L/person/d. Design a wellfield that can be accommodated on the parcel of land. The radius of each well can be taken as 50 cm, and the drawdown must not exceed 6 m. If the wells are screened over the bottom 20 m of the aquifer and the screen has a diameter of 50 cm with 50% open area, assess whether the screen is adequate.

6.80. A well of radius 0.5 m is to be installed in an unconfined aquifer of hydraulic conductivity 300 m/d. What is the maximum allowable screen entrance velocity, and what is the corresponding maximum pumping rate? If the uniformity coefficient of the aquifer matrix is 3.2, then write the specifications for the gravel pack and the screen slot size.

6.81. A water-supply well is to be installed in a 30-m thick sand aquifer with a uniform grain-size distribution between 0.04 mm and 2.2 mm. The hydraulic conductivity of the aquifer is estimated to be 50 m/d, the porosity is 0.15, and the specific yield is 0.2. If the pumping rate is to be 800 L/min, design the casing, screen, and gravel pack.

6.82. Exploration of an unconfined aquifer at the site of a proposed water-supply well indicates a transmissivity of 4200 m²/d and a specific yield of 0.2. Drilling a pilot hole at the proposed well location indicates that the aquifer has a saturated thickness of 30 m, and a sieve analysis of the borehole cuttings indicates 10-percentile and 60-percentile particle sizes of 0.5 mm and 2.1 mm, respectively. The well is expected to yield 2500 L/min, and the turbine pump to be used has a bowl diameter of 300 mm. Available screens have slot sizes in the range of 30 to 100 in increments of 5, and all have 15% open area. Specify the following well dimensions: (1) casing diameter; (2) borehole diameter; (3) screen diameter; (4) screen length; (5) screen slot size; (6) gravel-pack specifications (if used); and (6) location of pump intake.

6.83. A 0.3-m diameter well is to be installed in a 20-m thick confined aquifer in which the allowable drawdown of the potentiometric surface at the well is 5 m. The hydraulic conductivity and storage coefficient of the aquifer are estimated as 30 m/d and 10^{-4}, respectively. Estimate the specific capacity and the well yield after 1 year, and classify the productivity of the well.

6.84. A step-drawdown test is conducted in an aquifer and yields the results shown in Table 6.34. Determine the well-loss coefficient, formation-loss coefficient, and the specific capacity of the well. Assess the condition and productivity of the well at a pumping rate of 5000 m^3/s.

6.85. The drawdowns, s_w (cm), and corresponding pumping rates, Q_w (L/min), collected at a well during a step-drawdown test were plotted on a graph of $\log(s_w/Q_w - \beta)$ versus $\log Q_w$ for various values of β. The plotted curve is most linear when $\beta = 55.2$ s/m^2, and the corresponding best-fit line through the data has a slope of 1.22. If the well efficiency is known to be 32% when the pumping rate is 620 L/min, estimate the well-loss coefficient. Assess the productivity of the well when pumping at 1000 L/min.

6.86. A step-drawdown test yields the following results:

Pumping rate (L/min)	Drawdown (cm)
40	15
60	23
80	33
100	41
120	53
140	62
160	73

Use the Rorabaugh method to analyze the measurements and determine the efficiency and specific capacity of the well when the pumping rate is 500 L/min.

TABLE 6.34: Step-Drawdown Results

Pumping rate (m^3/s)	Drawdown (cm)
0	0
0.05	20
0.10	53
0.15	86
0.20	147
0.25	210
0.30	283

Compare your result with the conventional approximation that well losses are proportional to the square of the pumping rate.

6.87. The following results are derived from a step-drawdown test in a deep aquifer:

Pumping rate (m^3/d)	Drawdown (m)
500	2.40
1000	5.38
1500	9.28
2000	14.36
2500	20.82
3000	28.87
3500	38.70

Use the Rorabaugh method to estimate an empirical relation between pumping rate and drawdown. Determine the well loss as a percentage of total drawdown for the pumping rates shown in the above table. [Adapted from Todd and Mays (2005).]

6.88. Show that the half-life of a contaminant in ground water can be described by Equation 6.366.

6.89. The risk, R, of illness from viruses in pumped water is related to the viral concentration, c, by the relation

$$R = 1 - (1 + 4.76c)^{-94.9}$$

where c is in #/L. If an illness rate of 1 in 10,000 people is acceptable ($R = 10^{-4}$), determine the allowable viral concentration in the pumped water. A municipal water-supply well is to be installed 100 m from a school that uses a septic tank for onsite wastewater disposal. The well is to have a rated capacity of $0.4 \ m^3/s$, the saturated thickness of the aquifer is 20 m, and the porosity of the aquifer is 0.17. If the allowable risk of viral infection from the pumped water is 10^{-4} and the decay constant for viruses in ground water is $0.3 \ d^{-1}$, estimate the maximum allowable virus concentration in the ground water below the school.

6.90. A 30-cm diameter water-supply well is to pump 50 L/s and be located 150 m from a fully penetrating river in an aquifer with a saturated thickness of 25 m. The hydraulic conductivity of the aquifer is 30 m/d, the porosity is 0.2, and the radius of influence of the well is 1 km. Calculate the shortest time of travel of the ground water from the river to the well intake. If the decay constant of a contaminant is $0.01 \ d^{-1}$, and the allowable concentration of the contaminant in the pumped water is $1 \ \mu g/L$, estimate the maximum allowable concentration of the contaminant in the river by assuming the shortest time of travel. Assess the validity of this assumption.

6.91. Give an example of how you would use estimates of aquifer properties prior to conducting an aquifer pump test to aid in the design of the pump test.

6.92. Use the principle of superposition to show that Equation 6.370 can be used to describe the recovery data in a pump test.

6.93. Apply the Cooper–Jacob approximation to Equation 6.370 to show that the recovery data in a pump test can be described by Equation 6.373.

6.94. An aquifer pump test is conducted at a rate of 2000 L/min, and pumping is terminated after 4 hours. Measurements of drawdown versus time after pumping is terminated are given in the following table:

Drawdown (m)	Time since end of pumping (min)
1.01	1
0.90	2
0.83	3
0.75	5
0.70	7
0.61	10
0.55	15
0.60	20
0.42	30
0.37	40
0.31	60
0.26	80
0.23	100
0.19	140
0.15	180

If the pump-drawdown test indicates a storage coefficient of 0.0001, estimate the aquifer transmissivity and the recovery storage coefficient.

6.95. It has been stated that a 10% variation in discharge during an aquifer pump test can result in a 100% variation in the estimate of aquifer transmissivity (Osbourne, 1993). Assess the validity of this statement.

6.96. A slug test was performed in alluvial deposits of the Salt River bed west of Phoenix, Arizona. The geometry of the aquifer is shown in Figure 6.47. A solid cylinder with a volume equivalent to a 0.32-m change in water level in the well was also placed below the water level. When the water level had returned to equilibrium, the cylinder was quickly removed. The measured values of y_t versus time (t) are given

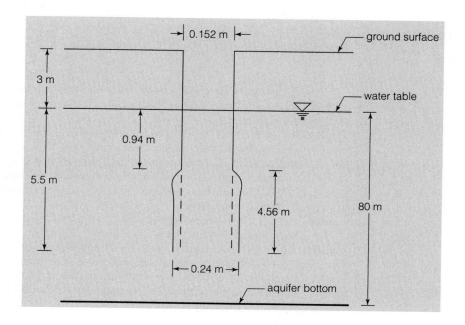

FIGURE 6.47: Slug test

in the following table:

t (s)	y_t (m)	y_t/y_0
0.0	0.290	1.000
0.7	0.238	0.821
1.4	0.190	0.655
2.8	0.150	0.517
3.7	0.117	0.403
6.0	0.072	0.248
8.8	0.030	0.103
12.8	0.012	0.041
18.6	0.005	0.017
28.6	0.002	0.007
38.6	0.001	0.003

The theoretical value of y_0 calculated from the displacement of the submerged cylinder is 0.32 m. Applying the Bouwer and Rice slug-test data analysis method, determine the horizontal hydraulic conductivity of the aquifer.

6.97. A monitoring well that penetrates 4 m below the water table in an unconfined aquifer with a saturated thickness of 12 m is used to conduct a slug test. The well casing has a 150 mm diameter, and the bottom 2 m of the well is screened and surrounded by a 110-mm thick gravel pack. The observed time-drawdown relation during the slug test is given in the following table:

Time (s)	Drawdown (mm)
0	700
3	392
6	260
9	137
12	91
15	47
18	31
21	16
24	11

Estimate the hydraulic conductivity and transmissivity of the aquifer.

6.98. Repeat Problem 6.97 for the case in which the monitoring well fully penetrates the aquifer.

6.99. An exfiltration trench is to be designed to inject 500 m³/d from a pump-and-treat system. The trench hydraulic conductivity is 35 m/d, and the depth from the ground surface to the seasonal-high water table is 5.22 m. The aquifer is estimated to have a hydraulic conductivity of 70 m/d, a saturated thickness of 15 m, and a specific yield of 0.22. Design the exfiltration trench assuming a factor of safety of 2.5 in the trench hydraulic conductivity.

6.100. Repeat Problem 6.99 assuming 1 m of backfill over the trench (so that surface vegetation can be planted).

6.101. An exfiltration trench is to be designed such that the water table remains at least 1.2 m below the trench after one year of operation. The trench hydraulic conductivity is 10 m/d, the width of the trench is to be 1 m, the depth is to be 1.5 m, and space limitations will allow a maximum trench length of 10 m. The wet-season water table is 3.1 m below ground surface, the hydraulic conductivity of the aquifer is 15 m/d, the porosity is 0.15, and the saturated thickness of the aquifer is 7 m. What is the maximum allowable flowrate that can be exfiltrated through the trench?

6.102. An exfiltration trench is being considered as a means to discharge 500 m³/d continuously for 20 years into an aquifer having a saturated thickness of 17 m, a hydraulic conductivity of 7 m/d, and a specific yield of 0.14. The trench hydraulic conductivity is estimated to be 20 m/d, and the depth from the ground surface to the water table is 4 m. A factor of safety of 2, and at least 30 cm of backfill above the trench is must be used. Determine the required trench dimensions, and assess the practicality of the trench.

6.103. A 100-m × 100-m pond is to be used as a recharge basin in which water is expected to infiltrate at the rate of 5 meters per day. The unconfined aquifer beneath the recharge basin has a hydraulic conductivity of 90 m/d, a specific yield of 0.2, and a wet-season saturated thickness of 35 meters. At the start of recharge operations, the wet-season water table is 20 m below the bottom of the pond. Based on a consideration of the mounding effect, determine whether the pond can be successfully operated for 20 years.

CHAPTER 7

Water-Resources Planning and Management

7.1 Introduction

Water-resources planning and management is a multidisciplinary field that addresses issues related to developing the water resources of a region. Areas covered by water-resources planning and management include flood protection, adequacy of water supplies, environmental impacts, recreation, economic development, social well-being, and politics. A major water-resource challenge today is drought management, especially in the western United States, and controlling the impact of drought conditions on municipal and industrial water supplies (Galloway, 2005). The fundamental issue facing water-resource planners is the need to balance environmental concerns with the needs of people for sustainable development.

There are two levels of water-resources planning: *comprehensive planning* and *functional planning*. Comprehensive planning coordinates a wide range of activities, sets overall direction, identifies priorities, and provides a basis for managing conflicts (Dzurik, 2003). Comprehensive planning generally considers interrelationships between water, society, and economic development. In contrast to comprehensive planning, functional planning is confined to one major area or resource of primary interest, such as water-resources, transportation, or land-use planning. Functional planning is typically more detailed and technical than comprehensive planning and tends to be the only type of planning that remains politically palatable over time.

Water-resources planning and management is closely related to *water-resources systems engineering*, which is defined as the formulation and evaluation of alternative plans to determine the system configuration or set of actions that will best accomplish the project objectives within the constraints of natural laws, engineering principles, economics, environmental protection, social and political pressures, legal restrictions, and institutional and financial capabilities (Wurbs and James, 2002). The primary factors and constraints to be considered in planning the development of water resources are: sustainability, technical feasibility, public policy, regulations, political realities, and economic factors.

The benefits obtained from good planning include improved quality of decisions, better understanding of problems and issues, more consistent policy decisions with respect to long-term goals, more focus on priority issues, consensus building, and more rationality in public decision making.

7.2 Planning Process

The planning process as applied to water resources is comparable to other types of planning. It is a logical series of steps, beginning with identification of needs, proceeding to recommendations for action, and culminating in implementation and monitoring. The water-resources planning process generally has the following components:

1. Problem Identification
2. Data Collection and Analysis
3. Development of Goals and Objectives
4. Clarification and Diagnosis of the Problem or Issues
5. Identification of Alternative Solutions
6. Analysis of Alternatives
7. Evaluation and Recommendation of Actions
8. Development of an Implementation Program
9. Surveillance and Monitoring

Steps 3, 5, 6, 7, and 8 are at the heart of the planning process and are commonly known as the *rational planning model* (Dzurik, 2003). Descriptions of all steps in the planning process are given below.

Problem Identification. Problem identification is derived from the expressed needs and concerns relating to the water resources of an area. Problem identification should reflect the concerns of different groups, both public and private, and should be described in sufficient detail to allow for adequate attention in the planning process. Public involvement as well as coordination of various agencies and groups should begin at this stage. Specific needs usually result either from a problem that has already been experienced, such as flooding, or from an anticipated problem, such as inadequate water supply to meet future needs. In some instances, water-resources planning studies may be more generalized and not deal with any single problem or need. River-basin studies and areawide water-quality studies are typically of this type.

Data Collection and Analysis. Following problem identification, the study area should be defined and existing information for the study area analyzed for its relevance to the problem under consideration. This analysis should include identification of relevant geophysical and biological features; social, demographic, and cultural characteristics; land use; and economic activity such as manufacturing, commerce, and agriculture. It is also important to determine anticipated future conditions as shown by existing planning documents, which should be reviewed and modified as needed. In the United States, extensive data on the water resources of an area is typically available from the U.S. Geological Survey (USGS), while state and local government entities are usually rich sources of relevant information. As part of the data collection process, forecasts should be made of appropriate variables to determine likely future conditions of the problem under investigation.

Goals and Objectives. Specifying relevant planning goals and objectives and defining their relative importance is one of the most difficult tasks in water-resources planning, since there are many divergent interests that often generate conflicting objectives. The most common conflict in water-resources planning is between economic and environmental objectives. Statements of goals and objectives indicate what the planning effort hopes to accomplish.

Goals are broad and general, such as the attainment of clean water or provision of adequate water supplies. Although they are stated in general terms, goal statements relate human values to natural resources and the environment. Objectives are more specific statements of plans to be developed, and often several objectives must be accomplished to attain a goal. Objectives provide the basis for evaluating alternatives. Examples of typical water-resources planning objectives are:

- Prevent continued water degradation by waterborne wastes
- Prevent or reduce flood damages
- Provide water-based recreation
- Provide for efficient reuse of treated wastewater
- Provide for efficient development and management of water supplies

It is difficult to plan for vague or unrealistic objectives. Therefore, objectives should be stated as clearly and specifically as possible in order to facilitate their achievement.

Problem Diagnosis. Problem diagnosis involves formulating a clear understanding of the problem so that alternative solutions will effectively address it. Most commonly, this analysis will indicate that there are numerous alternative solutions. For example, flooding problems may be solved by either structural or nonstructural solutions.

Formulation of Alternatives. Alternative plans must be formulated to help the decision maker identify how they relate to the objectives and understand the trade-offs among them. Formulation of alternatives begins with identification of measures that will address the defined problems. Public and interagency participation is important at this point to ensure that the full range of measures is considered. Whenever possible, the alternatives should range from capital-intensive structural measures to nonstructural management or policy solutions. Some objectives have economic efficiency as their primary goal, while other frequently cited objectives focus on preservation or restoration of the environment. There is no standard number of alternatives to be developed. Judgment must be exercised to determine which plans are appropriate and to decide which alternatives to carry forward for more detailed study.

Analysis of Alternatives. The two primary aspects of analyzing alternatives are economic evaluation and impact assessment. In both cases, measures are developed which quantify the changes resulting from the alternative plans. It is always useful to have a no-action plan as a baseline for comparison.

Economic evaluation usually requires a benefit-cost analysis, for which several methods have been developed. The simplest form of benefit-cost analysis compares the present value of all the costs with the present value of all the benefits. If the benefits exceed the costs, then the plan is *economically justified*. The present-value approach is appropriate if the economic lives of the alternatives are the same. If they differ, it is more appropriate to convert each time stream of net benefits to an equivalent average-annual benefit for purposes of comparison (Dzurik, 2003).

Impact assessment includes environmental, social, and economic impacts, but the major concern is usually the environmental impact. Impact assessment is an analysis of the potential positive and negative impacts, and significant changes that might result from selecting an alternative. Impacts are identified by comparing inputs, outputs, and facility requirements of an alternative to the base condition in the absence of the alternative. Although no single measure of environmental impact can be obtained (such as the benefit-cost ratio for economic analysis), environmental assessment of each alternative allows for a detailed comparison of potential impacts, which allows for more informed decision making.

Evaluation and Recommendations. Evaluation is the process of analyzing alternative plans and comparing their beneficial and adverse contributions for the purpose of recommending a plan. A simple approach to evaluation proceeds with the following steps:

1. Identify the issues and objectives to which each alternative is directed.
2. Determine the positive and negative impacts of the alternatives, using public input as well as professional judgment.
3. Display the results of the evaluation so that decision makers know how each alternative relates to local, regional, state, and national issues and policies. This display could include trade-offs and choices for each alternative.

Specific criteria used by the U.S. Army Corps of Engineers that are useful in evaluating plans and reducing the number of alternatives are:

- Acceptability: Assess the workability and viability of a plan in terms of its acceptance by affected parties and its accommodation of known institutional constraints.
- Effectiveness: Appraise a plan's technical performance and contribution to planning objectives.
- Efficiency: Assess the plan's ability to meet objectives functionally and in the least costly way.
- Completeness: Assess whether all necessary investments to fully attain a plan are included.
- Certainty: Analyze the likelihood of the plan's meeting planning objectives.
- Geographic Scope: Determine if the area is large enough to fully address the problem.
- Benefit-Cost Ratio: Determine the economic effectiveness of the plan.
- Reversibility: Measure the capability to restore a complete project to original condition.
- Stability: Analyze the sensitivity of the plan to potential future developments.

The significance of each of the above tests in comparing plans is a matter of judgment and will vary with the type of plan being developed.

Final plan selection is done by those decision makers with legal authority, based on comparisons and recommendations set forth by planners. The selection should be based on the best use of resources considering all effects, monetary and non-monetary. Usually only one plan is selected, although the planning process may be repeated to develop new alternatives or combinations of existing alternatives.

Implementation. Implementation means carrying out a selected plan or set of recommendations. At this stage, the plan is adopted and put forward for design and construction, or for adoption of laws, policies, and management procedures. Often considerable effort goes into developing plans that are never adopted. In some cases, a plan is approved and adopted but never carried out, or set aside for years. In the latter case, care must be taken not to implement a plan that was adopted years ago unless a thorough reevaluation is done. In like fashion, a recently adopted plan that is being implemented should be continuously reviewed.

Surveillance and Monitoring. After a water-resources plan is implemented, the project should be monitored to see how well it meets the original goals and objectives. Many water-resources projects require long-term investments, so it is likely that modifications will be required as conditions change. It is common for such modifications to be needed long before the useful life of the investment is completed.

7.3   Legal and Regulatory Issues

7.3.1   Water Rights

A *water right* is the legal right for an entity to use, store, regulate, and/or divert water; and *water law* codifies the allocation and administration of water rights. Water rights in the United States are established primarily at the state level, but federal laws govern the water rights for military installations and other federal lands such as national parks and Indian reservations. The laws defining water rights and the institutions involved in allocating water resources constitute the framework for managing the water resources of the United States.

Water rights vary significantly across the United States, reflecting differing traditions and climates. Water rights have traditionally been based on two distinct doctrines: the *riparian doctrine* and the *appropriation doctrine*. However, in the current regulatory environment, a more accurate picture is that there are three doctrines: (1) riparian doctrine; (2) *regulated riparian doctrine*; and (3) appropriation doctrine. The emerging "regulated riparian doctrine" lays a system of government permits and regulation by state agencies on top of the traditional court-made riparian doctrine (AWWA, 2001). Ground-water policy is usually a blend of the three doctrines. The water rights in the 50 states are summarized in Table 7.1.

7.3.1.1   Surface-water rights

A body of water flowing in a well-defined channel or watercourse, including water found in lakes, ponds, rivers, creeks, streams, and springs, is regarded as surface water. Rights to use surface water according to the three governing doctrines are as follows:

Riparian Rights. A riparian* right allows owners of property adjacent to a water body to use water from that body. Therefore, water rights are property rights

---

*Riparian is derived from the word *ripa* which means "riverbank" in Latin. Land adjacent to a stream is called *riparian land*.

TABLE 7.1: Water-Rights Doctrines in the United States

State	Surface water			Ground water
	Appropriation	Riparian	Regulated riparian	Permit required?
Alabama			★	N
Alaska	★			Y
Arizona	★			Y
Arkansas			★	N
California		Other†		N
Colorado	★			Y
Connecticut			★	Y
Delaware			★	Y
Florida			★	Y
Georgia			★	Y
Hawaii		Other†		Y
Idaho	★			Y
Illinois		★		N
Indiana			★	N
Iowa			★	Y
Kansas	★			Y
Kentucky			★	Y
Louisiana		★		N
Maine		★		N
Maryland			★	Y
Massachusetts			★	Y
Michigan		★		N
Minnesota			★	Y
Mississippi			★	Y
Missouri		★		N
Montana	★			Y
Nebraska	★			N
Nevada	★			Y
New Hampshire		★		N
New Jersey			★	Y
New Mexico	★			Y
New York			★	Y
North Carolina			★	Y
North Dakota	★			Y
Ohio		★		Y
Oklahoma	★			Y
Oregon	★			Y
Pennsylvania		★		N
Rhode Island		★		N
South Carolina		★		Y
South Dakota	★			Y
Tennessee		★		N
Texas	★			N
Utah	★			Y
Vermont		★		N
Virginia			★	Y
Washington	★			Y
West Virginia		★		N
Wisconsin			★	Y
Wyoming	★			Y

Source: AWWA (2001).

†Water rights system contains aspects of all three systems.

associated with the ownership of land adjacent to a watercourse. Riparian rights are the basis of rules used to allocate water in the water-rich eastern United States and many parts of the world, and these policies have evolved almost naturally in environments where water is generally plentiful and excessive government involvement is unwanted.

Riparian rights are founded on two basic principles (Cech, 2002): the reasonable-use principle and the correlative-rights principle. The *reasonable-use principle* means that the riparian property owner can use as much water as desired for any reasonable purpose, as long as this does not interfere with the reasonable use by other riparian owners. Upper riparian owners may not obstruct the water flow, and lower riparian owners are prohibited from back-flooding an upstream owner. If these restrictions are violated, the injured riparian owners are entitled to collect damages for their losses. Under the *correlative-rights principle*, riparian landowners must share the total flow of the water in a stream. The proportion of use allocated to each riparian user is based on the amount of waterfront property owned along a stream and creates equal rights for all riparian land owners. There is no priority of water use in this system. The correlative-rights principle provides a minimum reasonable amount of water to all water users along a stream, and the presence of a water shortage means that all riparian property owners suffer in proportion to the amount of riparian land, and each must accept a reduction in supply. The rights of an owner of land abutting a water source are not affected by how long the land has been owned or how much water the owner has used in the past.

Riparian rights represent a common-law doctrine, which means that they are part of a large body of civil law that, for the most part, was made by judges in court decisions in individual cases, rather than statutory law enacted by a legislative body. Because the possession of rights arises from ownership of land abutting a water body, all riparian owners are treated as equals by the courts. Court cases can usually be traced to the issue of what is reasonable use. In making this determination, many factors are considered, including

- purpose of the use
- its suitability to the water body
- its economic or social value
- the amount of harm caused by the use
- the amount of harm avoided by changing either party's use
- the protection of existing values

In the past, when the technical capacity to use water was limited and abstractions from streams were needed only for a few cattle or a small vegetable plot, the riparian doctrine functioned adequately. In recent years, problems of concentrated and massive water use have become more common and severe in humid regions. Aquatic stream habitats have been threatened by concentrated withdrawals as cities have expanded and irrigation, which is highly consumptive, has increased (Eheart and Lund, 2002).

Regulated Riparian Rights.   Many states have overlaid the traditional riparian system with permit systems for regulating water use. This has been described as a *regulated riparian* system. Increased demands, drought, and expanding water-use conflicts are fueling the move from pure riparian systems to regulated riparian systems. The most important feature of regulated riparian statutes is that direct users of water must have a permit from a state administrative agency to use water. Permits are allocated to riparian landowners based on use, need, and climate factors, and can be changed or revoked by the state.

Appropriation Rights. An appropriation right means that owners of property are entitled to use water allocated to them, and water users gain rights based on the priority of their beneficial use and have an exclusive right to the allotted water. Beneficial uses are defined by states' regulatory codes and commonly include irrigation, mining, stock watering, manufacturing, municipal uses, domestic uses, and recreational uses. Unless a state has a minimum-flow rule, appropriators can lawfully divert the entire quantity of streamflow if they can use it beneficially. Impoundment for later use is common. When available water is short, the holder of senior water rights may use his or her entire allocation, whereas holders of junior water rights have to live with the shortage. Water-resources engineering projects, which may jeopardize senior water rights, are not permitted without proper compensation. Appropriation rights are mostly used in the more arid western United States, where water is more scarce than in the eastern United States.

In the United States, the appropriation-doctrine concept started with the forty-niners in the gold fields of California. When there was too little water for all the mining operations, the principle applied was "first in time, first in right." This principle was given legal recognition by the courts and later made into law by western state legislatures. The strict form of the appropriation doctrine, called the Colorado Doctrine, is now used in Alaska, Arizona, Colorado, Idaho, Montana, Nevada, New Mexico, Utah, and Wyoming. A combination of the riparian and prior appropriation doctrine, called the California Doctrine, is used in California, Kansas, Nebraska, North Dakota, Oklahoma, Oregon, South Dakota, Texas, and Washington. Currently, the most important water use in the western United States is for irrigation, so, in many cases, the appropriation doctrine is irrigation law.

The appropriation doctrine rests on two basic principles: priority in time and beneficial use (AWWA, 2003a).

Priority in Time. When streamflow is less than demand, use is prioritized based on who has been using the water for the longest period of time. The most recent uses are discontinued in order to provide water for the earlier uses. There is no sharing of suffering, which is a hallmark of the riparian doctrine.

Beneficial Use. A user who has been appropriated water continues to be entitled to it only when it can be used beneficially. Waste is (theoretically) prohibited, and any available water beyond the amount that can be used by one appropriator is available to others. In addition, nonuse of the water for a long period may result in loss of the water right by forfeiture or abandonment. Beneficial use is seldom defined precisely. There are two different but related aspects: the type of use (such as irrigation, mining, or municipal use) and efficiency. Most use is judged beneficial, and the majority of court cases have been about inefficient or wasteful uses.

In most states that use the appropriation system, water may not be taken or used until the user has obtained a permit from a water administration official, usually the state engineer.

7.3.1.2 Ground-water rights

Water rights for ground water are based on either the absolute ownership doctrine, reasonable use doctrine, correlative rights doctrine, prior appropriation doctrine, or permit system. Descriptions of these approaches are as follows (AWWA, 2003a):

Absolute Ownership Doctrine. The rule of absolute ownership is the oldest of the doctrines applied to ground water, evolving from the principle that land ownership encompasses everything beneath the land to the center of the earth. According to the rule of absolute ownership, a landowner may use all ground water that can be captured from beneath the owner's land. There are almost no restrictions applied to this rule. The water can be used for the owner's purposes both on and off the land, and it can be sold to others. There is no liability if pumping reduces a neighbor's supply or even dries up a neighbor's well. Texas and several other states have historically adhered to the absolute ownership doctrine but are slowly changing.

Reasonable Use Doctrine. The rule of reasonable use holds that ground-water rights are an incident of land ownership; however, if the use of ground water interferes with ground water used by neighboring landowners, then the use of ground water can continue only if its use is reasonable. Conversely, an owner is liable if unreasonable use causes harm to others. This amounts to a qualified right rather than an absolute right to use ground water. Reasonable use is much easier to determine for ground water than for surface water under the riparian doctrine. In general, any use of water that is not wasteful for a purpose associated with the land from which the water is drawn (an overlying use) is reasonable. Conversely, any use off the property may be considered unreasonable if it interferes with the use of the ground-water source by others. The overlying-use criterion does not resolve disputes if all parties to the dispute are using the water for overlying purposes. In this case, all uses are reasonable, all can continue, and "reasonable" use becomes the rule of capture–in other words, the owner with the biggest pump gets the water. On the other hand, because a nonoverlying use is considered unreasonable if it interferes with an overlying use, such uses are forbidden no matter how beneficial. This doctrine is common in the eastern United States.

Correlative Rights Doctrine. The rule of correlative rights holds that the right to make an overlying use of water is not absolute but is relative to the rights of other overlying users. The rule of correlative rights is used primarily when the ground-water supply is insufficient to satisfy the needs of all overlying users and sharing is required. In some cases, sharing is accomplished by prorating the supply on the basis of overlying acreage. This doctrine is an extension of the Reasonable Use Doctrine and is primarily used in California.

Prior Appropriation Doctrine. Under the prior appropriation doctrine, ground water is allocated similarly to surface water with priorities assigned based on the dates that users first appropriate the water for beneficial use. This doctrine is common in the western United States.

Permit System. The permit system is used to prevent overdevelopment of an aquifer. The distinguishing feature of the permit system is administrative regulation and management of ground water. This contrasts greatly with the relatively unregulated nature of ground-water use under the doctrines previously discussed. Systems in which state agencies issue permits specifying the amounts and conditions of water use have been adopted in a number of western and eastern states. Other doctrines may be reflected in the water rights documented by permits.

There is much variety in how water rights are interpreted and managed in the United States. The evolution of water-resources management is more complete in the West, where water is frequently in short supply, than in the East, where supplies have been more plentiful.

7.3.1.3   Federal water rights

The federal government owns vast amounts of land in the United States, particularly in national parks, monuments, forests, and wildlife refuges in the West. Federal water rights consist of reserved rights, rights for federal lands, rights on navigable waters, and contractual rights.

Reserved Rights. The federal reserved water rights doctrine, frequently referred to as the *Winters Doctrine* or *Indian Reserved Rights*, after *Winters* v. *United States*, holds that when the United States sets aside or reserves a part of its lands for particular uses or purposes, it reserves by implication the right to enough of the unappropriated waters on or adjacent to the lands to meet its uses and purposes. This implied reservation usually takes priority as of the time the lands are reserved. Reserved rights are particularly used to ensure the sustainability of Indian reservations, national parks, forests, monuments, wildlife refuges, and military installations.

Water Rights for Federal Lands Not Reserved. Not all federal lands are reservations, and the Bureau of Land Management administers millions of acres of unreserved public lands under congressionally authorized programs. Water rights of unreserved lands are still a matter of debate. Some contend that the federal government must comply with state law in acquiring water rights for use on unreserved lands, while others contend that the federal government can establish the rights based on congressional mandate. The U.S. Supreme Court has directed the federal government to comply with state law, but only insofar as the state law is consistent with Congress's objectives.

Federal Rights on Navigable Waters. Under the commerce clause of the U.S. constitution, the United States retains control over all navigable waters in the interest of navigation. On this basis, the United States has a dominant servitude over such waters, and the beds and banks of lakes and streams that contain them. All proprietary rights of others related to such water and the land beneath it are subject to this dominant servitude or easement for navigation and may be impaired or extinguished by the United States without reason.

Contractual Water Rights. Federal contract rights usually relate to the U.S. Bureau of Reclamation (USBR) and the U.S. Army Corps of Engineers. Under the Reclamation Act of 1902, the Secretary of the Interior, acting through the

USBR, is authorized to construct, operate, and maintain water-storage and distribution facilities in the 17 western states. The USBR is also authorized to make water from such projects available to users for irrigation and other purposes. To acquire water rights, the USBR usually files an application to appropriate a specific amount of unappropriated water in a river and/or acquires vested water rights by purchase or condemnation. After constructing a dam and storage reservoir and, if necessary, the distribution works to deliver project water to individual users, the USBR executes after-delivery contracts with irrigation districts or other similar entities. The USBR may also enter into contracts to furnish water for municipal and industrial uses. These contracts make project water available in return for the users' agreeing to repay a portion of the construction, operation, and maintenance costs over a specified period. The mission of the U.S. Army Corps of Engineers is to maintain and improve navigability in U.S. harbors and navigable rivers, and control floods. This mission is typically executed by building dams, gates, and levees.

7.3.2   U.S. Federal Laws Relating to Water Resources

This section identifies some of the federal laws that must be followed, and permits/approvals that must be secured for projects with federal interests. State and local agencies grant similar permits and approvals.

7.3.2.1   National Environmental Policy Act (NEPA)

This act mandates an interdisciplinary planning approach to ensure that environmental considerations are incorporated into project proposals for federal funding. A key requirement of NEPA is that an environmental assessment be done for all federally funded projects. Results of environmental assessments typically result in either a Finding of No Significant Impact (FONSI) or the preparation of an Environmental Impact Statement (EIS).

7.3.2.2   Clean Water Act (CWA)

This act, approved by Congress in 1972, is intended to restore and maintain the chemical, physical, and biological integrity of waters of the United States. The two fundamental national goals of the CWA are to eliminate the discharge of pollutants into the waters of the United States and to achieve water-quality levels that support fishing and swimming. Under the CWA, all states are required to develop water-quality standards for waters of the United States within their boundaries. States are required to review their water-quality standards at least once every three years and, if appropriate, revise or adopt new standards. The results of this triennial review must be submitted to the U.S. Environmental Protection Agency (USEPA), and the USEPA must approve or disapprove of any new or revised standards. Key sections of the CWA are as follows:

Section 201. Requires plans for the development of cost-effective, environmentally sound, and implementable treatment works that will meet the objectives of the Clean Water Act. The purpose of these plans is to ensure that investments in treatment works will proceed in an orderly fashion and will achieve maximum protection and enhancement of water quality.

Section 208. Requires statewide and areawide pollution-control planning. The objective is to reduce pollution from diffuse sources such as urban runoff, surface mining, and construction. Areawide plans must control the deposition of all residual waste generated in areas where water quality could be affected, and must include a process to control disposal of pollutants on land and in subsurface excavations. This section is the Clean Water Act's most comprehensive vehicle for in-depth water-quality management on an areawide basis.

Section 209. Requires preparation of Level B plans for all basins in the Unites States. These are preliminary or reconnaissance-level water and related land-use plans prepared to resolve complex long-range problems identified in the broader framework plans.

Section 303(d). Requires states to submit lists of surface waters that do not meet applicable water-quality standards after implementation of technology-based limitations, and to establish Total Maximum Daily Loads (TMDLs) for these waters on a prioritized schedule which considers the severity of the pollution and the uses to be made of such waters.

Section 303(e). Requires a basin plan representing the first phase of a two-phase planning process. Whereas Phase I emphasizes surface and nonsurface water discharges, Phase II adds an intensive study of nonpoint sources. Section 303(e) basin plans also provide an information base for the development of Section 208 areawide management plans and Section 201 facilities plans.

Section 304(a). Authorizes USEPA to develop and publish criteria for water quality that reflect the latest scientific knowledge about the kind and extent of all identifiable effects of pollutants in water on health and the environment.

Section 305(b). Requires each state to prepare and submit an annual report on the water quality of all waters in the state. This evaluation of water-quality problems is the basis of a state's long-range plans and programs for water management.

Section 311. Prohibits the discharge of hazardous substances in harmful quantities. Once a substance is classified as hazardous, a discharger must report any spilling, leaking, pouring, emitting, emptying, or dumping of that substance and is liable for the costs incurred to clean up the spill.

Section 319. Provides a framework for funding state and local efforts to address pollutant sources not addressed by the NPDES program (i.e., nonpoint sources). To obtain funding, states are required to submit nonpoint-source assessment reports identifying state waters that could not reasonably be expected to attain or maintain applicable water-quality standards or the goals and requirements of the CWA without additional control of nonpoint sources of pollution.

Section 320. Established the National Estuary Program, which focused point and nonpoint pollution control on geographically targeted, high-priority estuarine waters.

Section 401: National Pollutant Discharge Elimination System (NPDES). This permit system is administered by the U.S. Environmental Protection Agency

(USEPA) and is intended to control discharges from all point sources of pollution. Individual states may assume responsibility for the program under agreements with the USEPA; in such cases, the permits are issued by the state agency. Even if a state does not assume responsibility from USEPA for administration of the NPDES program, it still plays an active role in the permitting process. In order for an applicant to be eligible for an NPDES permit from USEPA, the applicant must obtain state certification that the discharge would not violate state water-quality standards. NPDES permits are required for point sources, such as discharges from municipal and industrial wastewater treatment plants, stormwater sewer systems, construction activities, animal feedlots, and mining operations. Return flows from irrigated agriculture, agricultural stormwater runoff, and discharges from nonpoint silvicultural activities are exempt from NPDES permit requirements. NPDES permits typically specify limits or allowable ranges for specified chemicals, monitoring and reporting requirements, operating and maintenance procedures, spill response procedures, and methods for handling emergencies. Best-management plans may be required to minimize the amount of contaminant being released.

Section 404: Dredge and Fill Permits. This section gives the U.S. Army Corps of Engineers the authority to control alteration of wetlands and requires a property owner to obtain a permit from the U.S. Army Corps of Engineers before dredging and/or filling any navigable waters of the United States. Individual states may apply to administer the Section 404 permit system, in which case the dredge and fill criteria must be at least as strict as the Corps criteria. Responsibilities for the Section 404 program are shared by the U.S. Environmental Protection Agency and the U.S. Army Corps of Engineers.

Section 503. Regulates sewage-sludge disposal from treatment plants. These regulations may indirectly affect desalination plant reject-water disposal.

The United States Congress periodically reauthorizes the CWA, normally with a series of amendments intended update or clarify the rules.

7.3.2.3 Rivers and Harbors Act

This act, originally promulgated in 1899, provides authority for the U.S. Army, through the U.S. Army Corps of Engineers, to exercise control over all construction in navigable waters of the United States. This act prohibits discharging refuse matter of any kind into navigable waters of the United States, prohibits depositing material of any kind any place on the bank of any navigable water that can be washed off into the waters, and prohibits the obstruction or alteration of navigable waters of the United States without approval (a permit) from the U.S. Army Corps of Engineers.

7.3.2.4 Federal Power Act

This act addresses the licensing of hydroelectric facilities and addresses all aspects of water-power planning, including navigation, irrigation, flood control, and hydropower. Regulations are enforced by the Federal Energy Regulatory Commission (FERC), and the regulations require that FERC consider fish and wildlife habitat, recreational opportunities, and environmental quality as part of the licensing process.

7.3.2.5   National Flood Insurance Program (NFIP)

This program is directed by the Federal Emergency Management Agency (FEMA) and requires local governments to enact floodplain-management guidelines for private development in designated flood-prone areas. If these guidelines are followed, these developments are eligible for federally subsidized flood insurance.

7.3.2.6   Fish and Wildlife Coordination Act

This act established a policy that wildlife conservation is to receive equal consideration with other aspects of water-resources development. The primary objective of the act is to identify potentially adverse effects on wildlife while a project is still in the proposal stage. The act requires any federal agency or permittee proposing to impound, divert, control, or modify a natural water body to consult with the U.S. Fish and Wildlife Service (USFWS) and the appropriate wildlife-management agency of the state involved.

7.3.2.7   Migratory Waterfowl Act

Regulations under this statute conserve and protect migratory birds in accordance with treaties between the United States, Mexico, Canada, Japan, and the former Soviet Union. The migratory birds protected under this act are specified in the respective treaties. If the planning watershed is located on a major waterfowl flyway in the United States, the environmental-planning document should identify and describe impacts to the life stages of the waterfowl that use the watershed.

7.3.2.8   National Wildlife Refuge System Administration Act

This statute consolidates various categories of wildlife ranges and refuges for management under one program. If a watershed includes refuges, the environmental-planning document should consider the effects of changing adjacent land uses on refuges.

7.3.2.9   Wild and Scenic Rivers Act

This act is administered through the National Park Service and is intended to preserve, in a free-flowing condition, certain waters possessing outstanding scenic, recreational, geologic, fish and wildlife, historic, cultural, and other similar values. This act prohibits the Federal Energy Regulatory Commission (FERC) from granting a permit for water projects on or directly affecting rivers designated as wild and scenic. Under this act, a federal agency may not help construct water-resources projects that would have direct and adverse effects on the free-flowing, scenic, and natural values of a designated wild or scenic river. If the projects would affect the free-flowing characteristics of a designated river or unreasonably diminish the scenic, recreational, and fish and wildlife values in the area, such activities should be developed in consultation with the National Park Service in a manner that minimizes adverse impacts.

7.3.2.10   Federal Water Project Recreation Act

In planning any federal, navigation, flood-control, reclamation, or water-resources project, this act requires parties to fully consider opportunities that the project affords for outdoor recreation and to enhance fish and wildlife.

7.3.2.11 Endangered Species Act

This act prohibits direct harm to species that have been designated by USEPA as threatened or endangered. This protection includes the habitat of the endangered species. These protections are not absolute, and Congress authorized the creation of the Endangered Species Committee, composed of federal officials and state representatives, which can grant exemptions from the Endangered Species Act's (ESA's) requirements in special circumstances. The ESA requires federal agencies to consult with the U.S. Fish and Wildlife Service and/or the National Marine Fisheries Service on activities that may affect any species listed as threatened or endangered. Environmental impact documents must analyze the effects of various alternatives on listed species.

7.3.2.12 Safe Drinking Water Act (SDWA)

The SDWA is intended to ensure that public drinking-water supply systems meet national water-quality standards for the protection of public health. Since its inception, the SDWA has been amended, or reauthorized, several times. The SDWA includes the wellhead-protection program, where each state is required to have a plan to protect public water-supply wells from possible contamination. The SDWA also addresses ground-water protection through the sole-source aquifer protection program, which is designed to protect the recharge areas of aquifers that are the principal sources of drinking water. Several important provisions of the SDWA are as follows:

Surface Water Treatment Rule (SWTR). All utilities served by surface water or "ground water under the direct influence of surface water" must comply with this rule. Ground water under the direct influence of surface water is defined as "any water beneath the surface of the ground with significant occurrences of insects or other microorganisms, algae, or large-diameter pathogens such as *Giardia lamblia*, or significant and relatively rapid shifts in water characteristics such as turbidity, temperature, conductivity, or pH, which closely correlate to climatological or surface water conditions." The SWTR's key requirements are that water suppliers provide multibarrier treatment, including filtration and disinfection, and achieve an overall 99.9% (3-log) removal-inactivation of *Giardia* cysts and 99.99% (4-log) removal-inactivation of viruses. To determine the number of *Giardia* cysts and viruses physically removed by a treatment process, USEPA developed a credit system for various treatment technologies as follows (removal credits for *Giardia* and viruses in parentheses):

- Conventional treatment (2.5-log/2-log)—coagulation, flocculation, sedimentation, and filtration.
- Direct filtration (2-log/1-log)—coagulation (with or without flocculation) and filtration.
- Slow sand filtration (2-log/2-log)—passing raw water through a sand filter at low velocities, generally less than 35 cm/h.
- Diatomaceous earth filtration (2-log/1-log)—a precoat cake of diatomaceous earth filter media on a support membrane, plus continuous addition of filter media to the filter water.

These treatment processes must be supplemented by disinfection to provide a total treatment process that achieves the required 3-log/4-log reduction. Effective disinfection is measured by the product of the residual disinfectant concentration at the first customer of the distribution system and the contact time.

Total Coliform Rule. The Total Coliform Rule (TCR) is based on the presence-absence concept, rather than on the traditional determination of coliform concentration. Compliance is based on the number or percentage of samples testing positive for coliforms during a given month. Samples are taken monthly, with the number of samples a function of population served. Samples range from one sample per month for service populations of 25 to 1000, to as many as 480 samples per month for populations larger than 3,960,001. If any sample is determined to be coliform positive, three repeat samples must be collected and analyzed within 24 hours, one at the original sampling location and at least one within five service connections upstream and downstream of the original site.

Lead and Copper Rule. Typically, high levels of lead and copper are not present in the source water; rather, they are leached from customer service piping or compounds used to join water system piping. The Lead and Copper Rule specifies optimal corrosion-control techniques, along with treating levels of pH in raw water supplies, and replacing service connections made of lead pipe or lead-based solder.

Ground Water Disinfection Rule. The Ground Water Disinfection Rule requires that all public water systems using ground water disinfect the source water at each well unless natural disinfection requirements are met or the system qualifies for a variance.

Underground Injection Control (UIC) Program. The purpose of the UIC program is to protect the quality of underground sources of drinking water. This purpose is achieved through rules that govern the construction and operation of injection wells.

Sole-Source Aquifer Protection Program. One mechanism for protecting an aquifer is to have it designated as a sole-source aquifer. Under provisions of the SDWA, the U.S. Environmental Protection Agency can, on its own initiative or as the result of a petition, designate an aquifer as a sole source for an area. After a determination is published, no federal money may be used or granted for any purpose that could result in contamination of the aquifer. As of 2002 there were 71 aquifers designated as "sole source" in the United States.

Wellhead Protection Program. Under Section 1428 of the Safe Drinking Water Act, each state is required to develop a comprehensive wellhead protection program. This requires the delineation of the wellhead protection area (possibly based on travel-time criteria), a contaminant source inventory, and a wellhead protection area management plan. Delineation of a wellhead protection area is typically done through the use of computer modeling of travel time and pollutant transport.

Source Water Assessment Program. All states are required to implement Source Water Assessment Programs (SWAPs) to assess areas serving as sources of drinking water in order to identify potential threats and initiate protection efforts.

7.3.2.13   Resource Conservation and Recovery Act (RCRA)

This act regulates the manufacture, storage, transportation, use, treatment, and disposal of solid and hazardous substances. A paper trail to track hazardous substances is required from creation to disposal. This act is principally designed to prevent contaminants from leaching into ground water from municipal landfills, underground storage tanks, surface impoundments, and hazardous-waste facilities. RCRA impacts water supply and wastewater treatment by providing that sludge from treatment plants may be included under a hazardous solid waste category. States may administer the hazardous waste program instead of USEPA, in which case state regulations must be as strict as federal regulations, states must provide information on the program to the public, and states must process applications in a manner similar to USEPA.

7.3.2.14   Comprehensive Environmental Response, Compensation, and Liability Act (CERCLA)

This act, more commonly known as "Superfund," provides funds to facilitate the cleanup of hazardous waste disposal sites. The USEPA prioritizes these sites on the basis of potential harm to the public health and environment, with higher-priority sites placed on the National Priorities List (NPL) for immediate action. Parties responsible for the spill or release of hazardous substances are held liable for the costs incurred in remedying the damages.

7.3.2.15   Toxic Substance Control Act (TSCA)

This act is intended to protect the public health and the environment from unreasonable risks by regulating the commercial manufacture, use, and disposal of chemical substances that have the potential to leach into the ground water. As with RCRA, businesses must maintain records, send reports and notices to USEPA, and handle chemicals according to federal regulations.

7.3.2.16   Federal Insecticide, Fungicide, and Rodenticide Act (FIFRA)

This act is the primary pesticide law in the United States and requires pesticide manufacturers to register their products with the USEPA prior to marketing them. This act controls the availability of pesticides that have the ability to leach into ground water.

7.3.3   Documentation of Environmental Impacts

Environmental planning documents are used for two primary reasons: (1) by decision makers and the public to understand the potential impacts of alternatives; and (2) to obtain permits and approvals for constructing and operating facilities. In accordance with the National Environmental Policy Act (NEPA), environmental documentation must be developed for all federally sponsored projects and discretionary actions that may directly or indirectly affect environmental resources or land-use patterns. NEPA procedures ensure that environmental information about the project is available to public officials and citizens before decisions are made and before actions are taken. Each state has its own framework for assessing the environmental impacts of water-resources projects. Once a project plan is submitted to a regulatory agency, the agency prepares an environmental assessment (EA), which evaluates whether the proposed project is likely to have significant environmental impacts and whether an environmental impact statement (EIS) is needed. If the EA determines that the EIS is not needed, the agency issues a finding of no significant impact (FONSI) that briefly explains why the action will not have a significant impact on the environment. Many states have their own environmental documentation statutes. For example, in California it is the California Environmental Quality Act (CEQA), where the assessment process begins with an initial study that is equivalent to the federal environmental assessment, then either a "negative declaration"—equivalent to the federal FONSI—or an environmental impact report (EIR)—equivalent to the federal EIS—must be prepared. Detailed environmental assessments or impact statements are not usually undertaken until after the preferred alternative is identified. However, the planning process leading to that selection should include preliminary consideration of all issues likely to require detailed assessment. The environmental planning process must be implemented concurrently with the technical evaluation. Project alternatives are technically and environmentally assessed only so far as is needed to compare alternatives and identify the best one. Thereafter, the full suite of environmental permits is pursued for the selected alternative, which generally requires in-depth environmental analyses and studies. Once these permits are obtained, the technical aspects of the project are developed for implementation.

There are three basic types of environmental impact documents: exclusions or exemptions, environmental assessments, and environmental impact statements or reports.

Exclusions. Exclusions are used for projects that occur routinely with anticipated outcomes that do not result in significant adverse impacts. Examples include restoring or rehabilitating facilities to a level that does not increase the original design capacity, or changing ownership of facilities. The Lead Regulatory Agency summarizes the reasons for using an exclusion document and adopts the findings. That agency must notify the public of the decision and allow public comment before implementing the proposed action.

Environmental Assessments. Environmental assessments or preliminary studies are frequently used in water-resources planning. Through this process a preliminary report is prepared to describe the affected environment with respect to the biological, human, and physical resources. The level of detail presented is minimal and focuses on the elements that may be affected by the proposed action. Generally, the environmental assessment describes only the potential impacts and benefits of the proposed action. If the proposed action has an adverse impact, mitigation measures are described in the document. Alternatives to the proposed action are described and the reasons for not pursing the alternatives are summarized. If the Lead Regulatory Agency determines that there would be no significant adverse impacts from implementing the proposed action, including mitigation measures, the environmental assessment is adopted and recorded in a public forum. The public must be notified that the final environmental assessment has been adopted and allowed to comment on the decision before the proposed action is implemented. If the environmental assessment shows that significant impacts could occur, or if the public review requires that several alternatives be evaluated in equal detail, the information in the environmental assessment is used to prepare an environmental impact statement or report. Sometimes an environmental assessment is completed to focus the analysis in an environmental impact document.

Environmental Impact Statements. Environmental impact statements (EISs) or environmental impact reports (EIRs) are used when several alternatives need to be evaluated at an equal level of detail. This document requires a detailed description of the affected environment and potential impacts and benefits of alternatives. Frequently a definition of "significant impact" is given for each element considered in the report, and adverse impacts are measured against impacts that would occur without the project and defined as "significant" or "not significant." The level of analysis of the impact assessment is more detailed than under an environmental assessment and frequently includes results of numerical models, field investigations, and discussions with community representatives or experts. The environmental impact documentation also includes an evaluation of cumulative impacts of implementing the alternatives with other planned programs that are not specifically part of the alternatives. Mitigation measures are proposed for the alternatives to reduce adverse impacts to a level of less significance. The environmental impact document usually defines an "environmentally preferred alternative" and the "preferred alternative". Frequently these alternatives are identical. However, when they are different, the document must describe the reasons to select the preferred alternative with fewer benefits to the environment. For some complex environmental documents, especially for watershed planning, the draft environmental document frequently does not include a preferred alternative or final mitigation measures. Environmental impact statements have to be prepared using an interdisciplinary approach that ensures integrated use of the natural and social sciences and the environmental design arts. In addition, different states have developed guidelines or regulations for the preparation of EISs for dams and environmental permit requirements for other water-resources engineering projects in their jurisdictions.

The three key steps in preparing an environmental document are: (1) selecting the study area; (2) selecting the study period and the baseline conditions; and (3) determining the level of detail in the analysis. In many water-resources projects, the study area is defined by a watershed; however, in some cases, such as when stored water or potential hydropower is to be used outside the watershed, the study area may need to be expanded to identify direct economic benefits or impacts of operating the dam. In most cases (as specified under NEPA), the environmental analyses need to compare only impacts of the alternatives with the future No Action Alternative condition, which is associated with the established baseline conditions. However, many state regulations require that the alternatives to the current and future No Action Alternative conditions be compared. The level of detail in the environmental impact analysis should be at a level that is easily understandable by the public, with technical details in appendices. The documents must generally address the impacts on biological, human, and physical resources of an area; and the level of detail in the analyses must be appropriate to discern the differences between alternatives, and sufficient to identify potential impacts.

7.3.4 Permit Requirements

In addition to federal laws which require the submission of environmental impact documents for federally sponsored projects, state, county, and municipal permit requirements exist for a variety of activities related to water-resources engineering.

Responsible local- and state-government agencies must generally be consulted for their permit requirements, and all permits must be in effect prior to constructing a new facility. Typical permit requirements for water-resources projects are as follows:

Stormwater-Management Permit. Most states require developers to obtain permits from state and/or local governmental entities to construct and operate stormwater-management systems. Such permits are usually obtained from departments with the word "environmental" in their name. These permits typically require that some form of structural best-management practice (BMP) be applied to new developments or redevelopments. These structural BMPs may include storage practices (detention ponds), filtration practices (grassed swales, and filter strips), and infiltration practices (infiltration basins, infiltration trenches, and porous pavement). Nonstructural BMPs may also be covered in permit requirements, such as requirements to limit growth to identified areas, protect sensitive areas such as wetlands and riparian areas, minimize imperviousness, maintain open space, and minimize disturbance of soils and vegetation.

NPDES Permit. The flowchart shown in Figure 7.1 illustrates most of the decisions that must be made in order to determine whether a particular stormwater discharge requires a NPDES permit under the Clean Water Act. Such permits are required to be obtained by municipalities, industries, and construction contractors that discharge either to waters of the United States or to municipal separate storm sewer systems (MS4s). Most municipalities are required to obtain NPDES permits for their public drainage systems, which include combined sewer systems (CSSs) and MS4s. NPDES permits are required for most industrial stormwater discharges, and these permits are classified as either general permits or individual permits. General permits cover entire industries, and any industry can be covered by a general permit by simply preparing a stormwater pollution prevention plan (SWPPP) and indicating its willingness to abide by the terms of a general permit, which is done via submittal of a notice of intent (NOI) to the regulatory agency granting the NPDES permit. In the relatively rare case where an industry is not covered by a general permit, an application for an individual permit may be submitted. NPDES permits are required for construction sites where the disturbed land is 2 ha (5 ac) or more. In most states, general and individual NPDES permits are available for construction sites, and almost all construction activities that require NPDES permit coverage for stormwater discharges can be covered under a general permit. Detailed guidance on securing NPDES permits can be found in Dodson (1998).

Construction sites can pose special problems for municipal stormwater collection systems, since, in a short time, stormwater discharges from construction sites can contribute more pollutants to a receiving stream than had been deposited over several decades (Dodson and Maske, 2001). While much of the potential pollutants might be sand, silt, and colloidal sediments, stormwater runoff from construction sites can include pollutants other than sediment, such as phosphorus and nitrogen from fertilizers, pesticides, petroleum derivatives, construction chemicals, and solid wastes that may become mobilized when land surfaces are disturbed. Construction permits generally include, as a minimum, requirements for construction-site owners or operators to implement appropriate BMPs such a silt fences, temporary detention ponds, and hay bales.

FIGURE 7.1: Stormwater discharges requiring NPDES permits

Source: Dodson (1998).

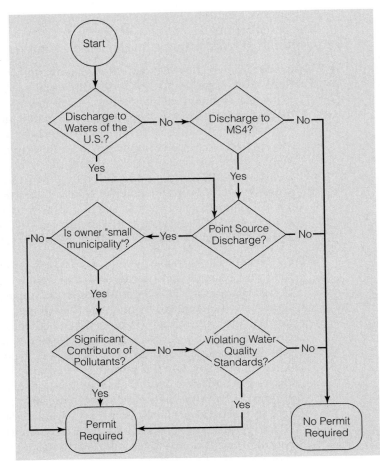

7.4   Economic Feasibility

Economic analysis is an essential component of evaluating alternatives in water-resources planning. In many cases, a proposed action may be politically attractive, technologically feasible, and environmentally acceptable, while economic analysis shows the action to cost more than it provides in benefits. Often complicating the economic analysis of a project, and particularly of water-resources projects, is determining the proper perspective from which to view the costs and benefits. For example, a project to develop water supplies may be viewed from the utility perspective, the ratepayer perspective, or the society perspective, and determining the benefits and costs for projects will vary according to the perspective from which the analysis takes place. The perspective that is appropriate for evaluation must be determined for each project, and no one perspective is appropriate for all cases. In water-supply projects, the society perspective provides the broadest coverage of costs, including those related to the environment, but the utility perspective should always be examined to make sure the program is affordable to the utility and its ratepayers. These types of projects should be evaluated from each perspective to assure the decision makers that significant, relevant impacts have not been overlooked.

7.4.1   Benefit-Cost Analysis

Benefit-cost analysis is the commonly used procedure for economic evaluation of public projects, and its success depends on the ability to assign monetary values to social and environmental costs and benefits. Benefit-cost analysis is most suitable for ranking or comparing alternatives designed to attain the same ends, rather than for testing the absolute desirability of one project. The boundaries of a benefit-cost analysis must always be specified at the outset; these are usually political boundaries (county, state), but may also be those of a corporate entity.

Costs.   Costs to an entity developing a water-resources project are generally classified as either direct or indirect. Direct costs are those borne by the entity itself, and typically include: design costs, construction costs, real-estate costs, rights-of-way costs, capital costs for equipment, and labor costs for operating and maintaining the system. Expenditures by those who receive direct benefits from a project in order to utilize it are included as direct costs. Indirect costs are those borne by parties not directly related to the project. Costs imposed on society by environmental degradation are prime examples of indirect costs. For overall economic efficiency, all costs should be accounted for, as under the society perspective, and this requires that both direct and indirect costs be included to the fullest extent possible.

Benefits.   Benefits are classified as either direct (internal), indirect (external), or intangible. Direct or internal benefits include additional revenue from such sources as water sales, power generation, recreation-use fees, impact fees from growth supported by a new water supply, additional tax revenues from growth in the service area, and direct savings of reduced potential flood damages to residences, businesses, and public infrastructure. Direct benefits may also include increases in land value because of the increase in market value of property protected from flood damages. Avoided costs coming about as a result of not having to use resources in some other way are generally counted as direct benefits. Indirect or external benefits accrue to parties not directly associated with entities responsible for developing water-resources projects; for example, creating a lake for water-supply purposes may also provide recreation benefits, habitat for wildlife, and flood-control benefits, and new businesses or residences may be established in direct response to a newly completed project. However, if a new activity is a transfer of an existing activity from another location, it should not be counted as a benefit from a national point of view except for any value added to the new location. Intangible benefits are those that cannot be quantified and are not included directly in the benefit-cost analysis. These benefits must be incorporated in some other way if they are to be part of the overall study.

A concern in using benefits and costs to quantitatively assess the economics of a project is that not all of them can be assigned monetary values. For example, aesthetic values vary widely among people and cannot be assigned a true monetary value. The value of fish and wildlife, aquatic ecosystems, and vegetation cannot be assigned true monetary values, and estimates are frequently made using surrogate measures such as the commercial value of fish or trees or the value of fishing and hunting experiences.

Spending on water-resources projects often consists of multiple-year capital programs, with operating and maintenance costs spread out over time, and benefits that may be even further spread out over time. To provide a common basis for cost and benefit comparisons, financial analysts commonly use the concept of *present value* which adjusts future amounts of costs or savings to equivalent values today. To make this conversion, a planner needs to know the earnings that are possible for investing available funds. The interest rate used is called the *discount rate*, and is defined as the highest rate of return that could be earned by investing available funds with the same level of risk. The discount rate should reflect either the best alternative use of the funds available or the cost of capital, which for most public agencies is equivalent to the interest rate on long-term debt. For U.S. federal government projects, the Office of Management and Budget (OMB) provides guidelines for choosing the discount rates, based on interest rates for treasury notes and bonds, with maturities that can be matched to project lives. From a water-supply utility perspective, the discount rate would likely be based on long-term tax-exempt bonds of the utility (AWWA, 2001). An additional consideration in choosing a discount rate is whether it should be the nominal or real discount rate. The nominal discount rate reflects current market conditions, taking into account current inflationary impacts of money. Market rates are nominal rates. The *real discount rate* is the nominal rate less expected inflation, which erodes the purchasing power of money over time. Since the real discount rate is an inflationary adjusted value, it is sometimes referred to as the *constant-dollar rate* (AWWA, 2001). The choice of real or nominal discount rates depends on whether costs and benefits are evaluated based on a real (constant-dollar) or nominal basis. Many federal water-resources project planners, the U.S. Army Corps of Engineers, and the U.S. Bureau of Reclamation have been criticized for using low discount rates. The actual figure is set by legislation enacted by the U.S. Congress; therefore all U.S. agencies are consistent in their selection of a discount rate.

The economic life of a project ends when the incremental benefit from continued use no longer exceeds the cost of continued operation. The period of economic (benefit-cost) analysis should not exceed the economic life, and the same period of analysis must be adopted for all alternatives, even if their economic lives differ.

7.4.2   Compound-Interest Factors

In order to use one of the discounting techniques, alternative discounting factors must be considered to convert cash flows to a single number. Variables used in these analyses include the present value (P), future value (F), uniform annual value (A), interest rate (i), and number of payment periods (typically years or months) denoted by n.

7.4.2.1   Single-payment factors

The *single-payment compound amount factor* gives the amount that will have accumulated after n years per unit initial investment at a return rate of i percent per year. Common notation for this factor is ($F/P,i,n$), which indicates a future value, F, of a present amount, P, invested at i percent for n years or payment periods. Based on this definition,

$$F = P(1 + i)^n \tag{7.1}$$

and

$$\boxed{\left(\frac{F}{P}, i, n\right) = (1 + i)^n = \frac{F}{P}} \tag{7.2}$$

The *single-payment present-worth factor* is the inverse of the single-payment compound amount factor (Equation 7.2), and gives the amount that must be invested initially at i percent in order to have unit (e.g., one dollar) return at the end of n years or other payment periods. The notation for this factor is $(P/F,i,n)$, which indicates a present value, P, which must be invested at i percent for n years or time periods in order to yield a future value, F. Based on this definition, Equation 7.1 gives

$$\left(\frac{P}{F}, i, n\right) = \frac{1}{(1 + i)^n} = \frac{P}{F} \tag{7.3}$$

This factor is important when finding the present value of future costs or benefits that appear as discrete values.

7.4.2.2   Uniform-series factors

In many cases, costs and benefits occur in uniform amounts year after year, and a better alternative is to use *uniform annual series factors* to show the present or future values of equal annual costs and/or benefits.

The *sinking-fund factor* gives the uniform amount, A, that must be invested at i percent at the end of each of n years to accumulate a unit (e.g., one dollar) return. Based on this definition, the sinking-fund factor $(A/F, i, n)$ is given by

$$\left(\frac{A}{F}, i, n\right) = \frac{i}{(1 + i)^n - 1} = \frac{A}{F} \tag{7.4}$$

The *uniform-series compound-amount factor*, which is the reciprocal of the sinking-fund factor, gives the amount that will accumulate from unit (e.g., one dollar) investments at the end of each of n years at i percent. Based on this definition, the uniform-series compound-amount factor $(F/A, i, n)$ is given by

$$\left(\frac{F}{A}, i, n\right) = \frac{(1 + i)^n - 1}{i} = \frac{F}{A} \tag{7.5}$$

The *capital-recovery factor* is the annual value that, after n years, will yield the equivalent of a unit (e.g., one dollar) initial investment at i percent. Based on this definition, the capital-recovery factor $(A/P, i, n)$ is given by

$$\left(\frac{A}{P}, i, n\right) = \frac{i(1 + i)^n}{(1 + i)^n - 1} = \frac{A}{P} \tag{7.6}$$

The *uniform-series present-worth factor*, which is the reciprocal of the capital-recovery factor, gives the present value (or the amount that must be invested initially) at i percent to yield a unit (e.g. one dollar) return at the end of each of n years. Based on this definition, the uniform-series present-worth factor $(P/A, i, n)$ is given by

$$\left(\frac{P}{A}, i, n\right) = \frac{(1 + i)^n - 1}{i(1 + i)^n} = \frac{P}{A} \tag{7.7}$$

7.4.2.3 Arithmetic-gradient factors

The *arithmetic-gradient present-worth factor* gives the amount that must be invested initially at i percent in order to obtain an incremental unit (e.g., one dollar) return at the end of the second year, an incremental two-unit (e.g., two dollars) return at the end of the third year, and continuing to an incremental $n - 1$-unit (e.g., $n - 1$ dollars) at the end of the nth year. Based on this definition, the arithmetic-gradient present-worth factor $(P/G, i, n)$ is given by

$$\left(\frac{P}{G}, i, n\right) = \frac{(1 + i)^n - in - 1}{i^2(1 + i)^n} = \frac{P}{G} \tag{7.8}$$

where G is the increment in investment return from one year to the next. To determine the present worth of a uniformly decreasing series, subtract a uniformly increasing series from a uniform annual series. The *arithmetic-gradient uniform-series factor* $(A/G, i, n)$ can be derived by multiplying the A/P and P/G factors and is given by

$$\left(\frac{A}{G}, i, n\right) = \frac{1}{i} - \frac{n}{(1 + i)^n - 1} = \frac{A}{G} \tag{7.9}$$

The *arithmetic-gradient future-worth factor* $(F/G, i, n)$ can be derived by multiplying the P/G and F/P factors and is given by

$$\left(\frac{F}{G}, i, n\right) = \frac{1}{i}\left[\frac{(1 + i)^n - 1}{i} - n\right] = \frac{A}{G} \tag{7.10}$$

In cases where the initial return at the end of the first year is A_0 and the annual return increases by an increment G in subsequent years, the present worth of these is calculated by adding the present worth of the annual series A_0 over n years to the present worth of the uniform-gradient series with annual increments of G.

7.4.2.4 Geometric-gradient factors

In cases where investment returns or expenditures change from period to period by a constant percentage, this defines a geometric-gradient series. To characterize a geometric-gradient series, it is convenient to define the parameter, g, as the constant rate of change, in decimal form, by which amounts increase or decrease from one time period to the next. Defining P as the present worth of the entire cash-flow series, and A_0 as the initial cost or revenue, then the *geometric-gradient present-worth factor* $(P/A_0, g, i, n)$ is given by (Blank and Tarquin, 2002)

$$\left(\frac{P}{A_0}, g, i, n\right) = \frac{P}{A_0} = \begin{cases} \dfrac{1 - \left(\frac{1+g}{1+i}\right)^n}{i - g}, & g \neq i \\[3mm] \dfrac{n}{1+i}, & g = i \end{cases} \tag{7.11}$$

Corresponding factors for equivalent A and F values can be derived; however, it is easier to determine P and then multiply by A/P or F/P factors to determine the required information.

EXAMPLE 7.1

A project to develop the water-supply infrastructure in western Montana is being considered, and four alternatives have been proposed. All alternatives have a 20-year design life, and projected economic conditions indicate a 6% interest rate should be used in comparing alternatives. The first alternative will lead to a lump-sum return of $100,000 at the end of the first 10 years, and a lump-sum return of $200,000 at the end of the second 10 years. The second alternative yields annual returns of $15,000 for all 20 years of the project. The third alternative yields a return of $6000 at the end of the first year, and the return increases by $1000 per year in subsequent years. The fourth alternative yields a return of $6000 at the end of the first year, and the returns are projected to increase by 8% per year in subsequent years. Determine the equivalent present worth for each alternative. If all projects have approximately the same cost, which of the four alternatives provides the greatest return on investment?

Solution

Alternative 1:

This alternative produces a lump-sum return of $100,000 at the end of year 10, and $200,000 at the end of year 20. The present worth of each of these lump-sum returns can be determined using the single-payment present-worth factor (Equation 7.3), which can be put in the form

$$P = \left[\frac{1}{(1 + i)^n} \right] F$$

For the lump-sum return of $100,000 at the end of year 10: $F = \$100,000$, $i = 0.06$, $n = 10$, and

$$P = \left[\frac{1}{(1 + 0.06)^{10}} \right] (\$100,000) = \$55,839$$

For the lump sum return of $200,000 at the end of year 20: $F = \$200,000$, $i = 0.06$, $n = 20$, and

$$P = \left[\frac{1}{(1 + 0.06)^{20}} \right] (\$200,000) = \$62,361$$

Therefore, the total present worth of the lump-sum returns at the end of years 10 and 20 is $55,839 + $62,361 = $118,200.

Alternative 2:

This alternative produces annual returns of $15,000 per year for all 20 years of the project. The present worth of these returns can be determined using the uniform-series present-worth factor (Equation 7.7), which can be put in the form

$$P = \left[\frac{(1 + i)^n - 1}{i(1 + i)^n} \right] A$$

In this case, $A = \$15,000$, $i = 0.06$, $n = 20$, and

$$P = \left[\frac{(1 + 0.06)^{20} - 1}{0.06(1 + 0.06)^{20}} \right] (\$15,000) = \$172,049$$

Therefore the present worth of the uniform annual returns is $172,049.

Alternative 3:
This alternative produces a return of $6000 at the end of the first year, with returns increasing by $1000 per year thereafter. The present worth of the annual incremental returns can be determined using the arithmetic-gradient present-worth factor (Equation 7.8), which can be put in the form

$$P = \left[\frac{(1 + i)^n - in - 1}{i^2(1 + i)^n} \right] G$$

In this case, $G = \$1000$, $i = 0.06$, and $n = 20$, and

$$P = \left[\frac{(1 + 0.06)^n - (0.06)(20) - 1}{(0.06)^2(1 + 0.06)^n} \right] (\$1000) = \$87{,}230$$

The incremental return of $1000 per year (starting in year 2) is superimposed on a uniform annual return of $6000 per year beginning at the end of year 1. The present worth of the uniform annual series is given by the uniform-series present-worth factor (Equation 7.7), which can be put in the form

$$P = \left[\frac{(1 + i)^n - 1}{i(1 + i)^n} \right] A$$

In this case, $A = \$6000$, $i = 0.06$, $n = 20$, and

$$P = \left[\frac{(1 + 0.06)^n - 1}{0.06(1 + 0.06)^{20}} \right] (\$6000) = \$68{,}820$$

Therefore, the total present worth of the uniform-gradient annual returns is $87,230 + $68,820 = $156,050.

Alternative 4:
This alternative produces a return of $6000 at the end of the first year and increases by 8% each year for the duration of the 20-year design life. The present worth of these returns can be determined using the geometric-gradient present-worth factor (Equation 7.11), which can be put in the form

$$P = \left[\frac{1 - \left(\frac{1+g}{1+i} \right)^n}{i - g} \right] A_1$$

In this case, $A_1 = \$6000$, $g = 0.08$, $i = 0.06$, $n = 20$, and

$$P = \left[\frac{1 - \left(\frac{1+0.08}{1+0.06} \right)^{20}}{0.06 - 0.08} \right] (\$6000) = \$135{,}993$$

Therefore the present worth of the geometric-gradient annual returns is $135,993.

Comparison of Alternatives:
The present worths of the four alternative proposals are as follows:

Alternative	Present Worth
1	$118,200
2	$172,049
3	$156,050
4	$135,993

Based on these results, and the fact that all proposed alternatives have approximately the same cost, Alternative 2 produces the greatest return on investment.

7.4.3   Evaluating Alternatives

Four different approaches can be used to deal with comparisons of time and value in considering the economic merits of alternative plans. These approaches are: present-worth analysis, annual-worth analysis, rate-of-return analysis, and benefit-cost analysis. They are different ways of analyzing the same information. Each emphasizes a particular economic aspect, and each has its own advantages and disadvantages. However, each approach should identify the same alternative as the best choice from an economic perspective.

Present-worth analysis.   A present-worth analysis uses the net present value (NPV) of costs minus benefits as the basis for comparing alternatives, and the alternative with the highest NPV is ranked highest, as it contributes the greatest amount to net benefits. In using present-worth analysis, annual net benefits (benefits minus costs) are calculated for each year of the project, the present value of each is determined, and the net sum of all present values is calculated. Suggested rules in performing present-worth analyses are as follows (James and Lee, 1971):

- Use the same time period and discount rates for all alternatives.
- Calculate the present worth of each alternative. Choose all alternatives having a positive present worth. Reject the rest.
- In a set of mutually exclusive alternatives choose the one having the greatest present worth.
- If the alternatives in the set of mutually exclusive alternatives have benefits which cannot be quantified but are approximately equal, choose the alternative having the least cost.

Individual future amounts are discounted using the present-worth factor $(P/F, i, n)$; series of equal amounts over n years (periods) are discounted using the uniform-series present-worth factor $(P/A, i, n)$; and for a nonuniform series of payments, present worth can be determined by calculating all individual present worths, or by using an arithmetic-gradient present-worth factor, if appropriate. By using net present worth, comparisons can be made of costs and benefits on an equivalent basis throughout the life of the project.

EXAMPLE 7.2

Three alternative proposals for developing the water resources of a region are being evaluated for funding. The design life of all alternatives is 30 years and market conditions indicate an interest rate of 5%. The first alternative requires an initial investment of $100,000, produces annual revenues beginning with $25,000 in year 1 and increasing by $1000 per year, operating costs begin with $3000 in the first year and increase by $2000 per year, and the salvage value of the capital investment is $10,000 at the end of 30 years. The second alternative requires an initial investment of $50,000, produces annual revenues beginning with $10,000 in year 1 and increasing by 7% per year, operating costs begin with $5500 in the first year and increase by 8% per year, and the salvage value of the capital investment is $5000 at the end of 30 years. The third alternative requires an initial investment of $65,000, produces annual revenues of $10,000 per year and operating costs of $6000 per year, and the salvage value of the capital investment is $12,000 at the end of 30 years. Compare these alternatives using the present-worth method and identify any that are not economically feasible.

Solution   For each alternative, the present worth of each cost and revenue item is calculated individually, and the net present value (NPV) is calculated as the sum of the present values of the cost and revenue items as follows:

Alternative 1:

Item	Formula	Parameters	Present worth
Initial investment, I	$P = I$	$I = -\$100,000$	$-\$100,000$
Base revenue, A_r	$P = \left[\frac{(1+i)^n - 1}{i(1+i)^n} \right] A_r$	$i = 5\%, n = 30, A_r = \$25,000$	$\$384,250$
Revenue gradient, G_r	$P = \left[\frac{(1+i)^n - in - 1}{i^2(1+i)^n} \right] G_r$	$i = 5\%, n = 30, G_r = \$1000$	$\$168,623$
Base operating cost, A_c	$P = \left[\frac{(1+i)^n - 1}{i(1+i)^n} \right] A_c$	$i = 5\%, n = 30, A_c = -\$3000$	$-\$46,110$
Operating cost gradient, G_c	$P = \left[\frac{(1+i)^n - in - 1}{i^2(1+i)^n} \right] G_c$	$i = 5\%, n = 30, G_c = -\$2000$	$-\$337,240$
Salvage value, F	$P = \left[\frac{1}{(1+i)^n} \right] F$	$i = 5\%, n = 30, F = \$10,000$	$\$2314$
Net present value			$\$71,837$

Alternative 2:

Item	Formula	Parameters	Present worth
Initial investment, I	$P = I$	$I = -\$50,000$	$-\$50,000$
Revenue, A_r, g	$P = \left[\frac{1 - \left(\frac{1+g}{1+i}\right)^n}{i - g} \right] A_r$	$i = 5\%, n = 30, A_r = \$10,000, g = 7\%$	$\$380,652$
Cost, A_c, g	$P = \left[\frac{1 - \left(\frac{1+g}{1+i}\right)^n}{i - g} \right] A_c$	$i = 5\%, n = 30, A_c = -\$5500, g = 8\%$	$-\$243,540$
Salvage value, F	$P = \left[\frac{1}{(1+i)^n} \right] F$	$i = 5\%, n = 30, F = \$5000$	$\$1155$
Net present value			$\$88,267$

Alternative 3:

Item	Formula	Parameters	Present worth
Initial investment, I	$P = I$	$I = -\$65,000$	$-\$65,000$
Revenue, A_r	$P = \left[\dfrac{(1+i)^n - 1}{i(1+i)^n}\right] A_r$	$i = 5\%, n = 30, A_r = \$10,000$	$\$153,724$
Cost, A_c	$P = \left[\dfrac{(1+i)^n - 1}{i(1+i)^n}\right] A_c$	$i = 5\%, n = 30, A_r = -\$6000$	$-\$92,235$
Salvage value, F	$P = \left[\dfrac{1}{(1+i)^n}\right] F$	$i = 5\%, n = 30, F = \$12,000$	$\$2772$
Net present value			$-\$739$

Based on these results, Alternative 2 has the greatest net present value ($88,267), and Alternative 3 is not economically feasible.

Annual-worth analysis.   This method is similar to the present-worth method, but it converts present worths to equivalent uniform annual values. Annual net benefits are first calculated for each year of the project, the present value of each is determined, and the net sum of all present values in obtained. Once this value is obtained, the appropriate capital recovery factor (A/P, i, n) is applied to determine equivalent annual figures. The appropriate decision rule in this method is to select the alternative having the greatest annual benefit. Although the annual-cost method is much like the present-worth method, it is often preferred by people more accustomed to thinking of annual costs rather than present worth.

EXAMPLE 7.3

Two proposed alternatives have present worths of $72,000 and $88,000, respectively. If the design life of these alternatives is 30 years and the interest rate is 5%, express these alternatives in terms of annual worths.

Solution   The present worth, P, is related to annual worth, A, by the capital recovery factor (Equation 7.6), which can be put in the form

$$A = \left[\frac{i(1 + i)^n}{(1 + i)^n - 1}\right] P$$

In this case, $i = 5\% = 0.05$, and $n = 30$, which yields

$$A = \left[\frac{0.05(1 + 0.05)^{30}}{(1 + 0.05)^{30} - 1}\right] P = 0.0651P$$

For $P = \$72,000$, $A = 0.0651(\$72,000) = \$4684$; and for $P = \$88,000$, $A = 0.0651(\$88,000) = \$5725$. Hence, the annual worths of the proposed alternatives are $4684 and $5725 respectively.

Rate-of-return analysis.   This approach identifies the rate of return at which the net present worth of all benefits and costs over the project life is equal to zero. The *rate of* *return* is also called the *internal rate of return, return on investment*, and the *profitability index*. Rate-of-return analysis provides for a direct comparison between the earning power, or return from the proposed investment, and alternative forms of investment. Suggested rules in rate-of-return analysis are as follows (James and Lee, 1971):

- Compare all alternatives over the same period of analysis.
- Calculate the rate of return for each alternative. Choose all alternatives having a rate of return exceeding the minimum acceptable value. Reject the rest.
- Rank the alternatives in the set of mutually exclusive alternatives in order of increasing cost. Calculate the rate of return on the incremental cost and incremental benefits of the next alternative above the least costly alternative. Choose the more costly alternative if the incremental rate of return exceeds the minimum acceptable discount rate. Otherwise choose the less costly alternative. Continue the analysis by considering the alternatives in order of increased costliness, the alternative on the less costly side of each increment being the most costly project chosen thus far.

The incremental procedure described above must be used in place of choosing the mutually exclusive alternative having the highest rate of return in order to have the same decisions as provided by a present-worth analysis. If the rate of return of a proposed investment exceeds the minimum attractive rate of return, the project is considered to be economically feasible. Typically, a target rate of return is determined based on some alternative investment opportunity. Rate-of-return analysis is used less than other methods because it requires prior calculation of net present worth, and it must be used with caution in comparing alternatives. Rate-of-return analysis provides a means of screening projects for economic feasibility but should not be used to rank projects for implementation.

EXAMPLE 7.4

Three alternative proposals to develop the water resources of a region are being reviewed. All proposals require an initial capital investment, generate revenues that increase at a constant rate, have operating costs that increase at a constant rate, and have a residual value at the end of the 20-year design life. These costs and revenues are summarized in the following table

Alternative	Initial cost	First-year revenue	Revenue growth rate	First-year operating cost	Operating-cost growth rate	Residual value
1	$100,000	$15,000	5%	$5000	6%	$5000
2	$50,000	$10,000	4%	$3000	5%	$1000
3	$75,000	$12,000	3%	$7500	5%	$1500

If the minimum attractive rate of return on the investment is 8%, identify the alternatives that are economically justified. What is the preferred alternative?

Solution   All proposed alternatives are described by an initial cost, I, first-year revenue, R_0, revenue growth rate, g_r, first-year cost, C_0, cost growth rate, g_c, residual value, F, and design life, n, equal to 20 years. The return rate, i^*, for each alternative satisfies the relation

$$-I + \left(\frac{P}{R_0}, g_r, i^*, n\right) R_0 - \left(\frac{P}{C_0}, g_c, i^*, n\right) C_0 + \left(\frac{P}{F}, i^*, n\right) F = 0 \qquad (7.12)$$

where $(P/A_0, g, i, n)$ is the geometric-gradient present-worth factor given by Equation 7.11, and $(P/F, i, n)$ is the single-payment present-worth factor given by Equation 7.3. Combining Equations 7.3, 7.11, and 7.12 gives

$$-I + \left[\frac{1 - \left(\frac{1+g_r}{1+i^*}\right)^n}{i^* - g_r}\right] R_0 - \left[\frac{1 - \left(\frac{1+g_c}{1+i^*}\right)^n}{i^* - g_c}\right] C_o + \left[\frac{1}{(1 + i^*)^n}\right] F = 0 \quad (7.13)$$

Substituting given values of I, R_0, g_r, C_0, g_c, and F for each of the proposed alternatives yields the following results:

Alternative	I	R_0	g_r	C_0	g_c	F	i^*
1	$100,000	$15,000	5%	$5000	6%	$5000	12%
2	$50,000	$10,000	4%	$3000	5%	$1000	16%
3	$75,000	$12,000	3%	$7500	5%	$1500	0.1%

Since the minimum attractive rate of return is 8%, only alternatives 1 and 2 are economically feasible and deserve further consideration.

To compare alternatives, they must first be ranked by initial cost, starting with the smaller cost. Alternative 2 has the lower initial cost of the two viable alternatives and will be ranked number one, and Alternative 1 has the next-lowest initial cost and will be ranked number two. The next step is to calculate the incremental cash flow between the number-two ranked alternative (Alternative 1) and the number-one ranked alternative (Alternative 2), and these increments are given in the following table:

Year	Alternative 2	Alternative 1	Increment
0	$-\$50,000$	$-\$100,000$	$-\$50,000$
$i = 1, 20$	$R_{02}(1 + g_{r2})^{i-1}$	$R_{01}(1 + g_{r1})^{i-1}$	$R_{01}(1 + g_{r1})^{i-1} - R_{02}(1 + g_{r2})^{i-1}$
$i = 1, 20$	$-C_{02}(1 + g_{c2})^{i-1}$	$-C_{01}(1 + g_{c1})^{i-1}$	$-C_{01}(1 + g_{c1})^{i-1} + C_{02}(1 + g_{c2})^{i-1}$
20	$1000	$5000	$4000

where R_{01} and R_{02} are the initial revenues for Alternatives 1 and 2, respectively, and g_{r1} and g_{r2} are the corresponding growth rates; C_{01} and C_{02} are the initial operating costs for Alternatives 1 and 2, respectively, and g_{c1} and g_{c2} are the corresponding growth rates.

The next step is to calculate the rate of return, i^*, of the incremental costs, which requires that

$$-50,000 + \left(\frac{P}{R_{01}}, g_{r1}, i^*, n\right) R_{01} - \left(\frac{P}{R_{02}}, g_{r2}, i^*, n\right) R_{02}$$

$$- \left(\frac{P}{C_{01}}, g_{c1}, i^*, n\right) C_{01} + \left(\frac{P}{C_{02}}, g_{c2}, i^*, n\right) C_{02}$$

$$+ \left(\frac{P}{F}, i^*, n\right)(4000) = 0$$

which can be put in the form

$$-50,000 + \left[\frac{1 - \left(\frac{1+g_{r1}}{1+i^*}\right)^n}{i^* - g_{r1}}\right] R_{01} - \left[\frac{1 - \left(\frac{1+g_{r2}}{1+i^*}\right)^n}{i^* - g_{r2}}\right] R_{02} - \left[\frac{1 - \left(\frac{1+g_{c1}}{1+i^*}\right)^n}{i^* - g_{c1}}\right] C_{01}$$

$$+ \left[\frac{1 - \left(\frac{1+g_{c2}}{1+i^*}\right)^n}{i^* - g_{c2}}\right] C_{02} + \left[\frac{1}{(1 + i^*)^n}\right] (4000) = 0$$

Substituting $g_{r1} = 0.05$, $R_{01} = \$15,000$, $g_{r2} = 0.04$, $R_{02} = \$10,000$, $g_{c1} = 0.06$, $C_{01} = \$5000$, $g_{c2} = 0.05$, and $C_{02} = \$3000$, and $n = 20$ gives

$$-50,000 + \left[\frac{1 - \left(\frac{1+0.05}{1+i^*}\right)^{20}}{i^* - 0.05}\right] (15,000) - \left[\frac{1 - \left(\frac{1+0.04}{1+i^*}\right)^{20}}{i^* - 0.04}\right] (10,000)$$

$$- \left[\frac{1 - \left(\frac{1+0.06}{1+i^*}\right)^{20}}{i^* - 0.06}\right] (5000) + \left[\frac{1 - \left(\frac{1+0.05}{1+i^*}\right)^{20}}{i^* - 0.05}\right] (3000)$$

$$+ \left[\frac{1}{(1 + i^*)^{20}}\right] (4000) = 0$$

which yields

$$i^* = 0.078$$

Since the rate of return of the incremental costs (7.8%) is less than the minimum attractive rate of return (8%), the additional investment returns generated by selecting Alternative 1 over Alternative 2 are not justified. Therefore, Alternative 2 is preferred.

If more alternatives are to be considered, then the incremental costs and associated rate of return relative to the currently preferred alternative (Alternative 2) must be considered.

Benefit-cost analysis. This approach is based on the premise that the ratio of benefits to costs must exceed 1.0 for a project to be considered economically feasible. Public-investment analysis is commonly done using the benefit-cost ratio method, particularly by U.S. government agencies. All benefits and costs are identified individually, with appropriate monetary values assigned and time periods determined, then all benefits and costs are brought to a comparable time (using present-worth analysis), and the benefit-cost ratio computed at this time. Benefits are viewed as all the positive returns from a project, regardless of who benefits, while costs are measured as the outlays made by the project sponsors as well as losses suffered by groups directly affected by the project. The benefit-cost ratio can have either of the following forms:

$$\frac{B - D}{C} \quad \text{or} \quad \frac{B}{C + D} \tag{7.14}$$

where B is the benefit (\$), C is the cost (\$), and D measures the disbenefits, hardships, and losses caused by the project (\$). The alternative benefit-cost ratios in Equation 7.14 give slightly different results for the same set of numbers. Benefit-cost ratios can be based on either present values (AWWA, 2001), annual values (Prakash, 2004), or future values (Blank and Tarquin, 2002); the present-value approach is more common. It is very important to note that the benefit-cost ratio provides information about the economic feasibility of the project and the efficiency of allocating monetary resources, but benefit-cost ratios should not be used to develop a ranking of feasible projects; for that purpose the net-present-value approach is preferable.

EXAMPLE 7.5

Two proposals are being considered for funding a public-works project with a design life of 20 years. The first proposal requires an initial capital cost of \$400,000 and will produce annual net benefits of \$35,000 for the life of the project. The second proposal will require an initial capital cost of \$600,000 and produce annual net benefits of \$60,000, but funding it will require that spending on other public-works projects be reduced by \$45,000 per year for the first five years of the project. Using an interest rate of 4%, assess the economic feasibility of each of the proposed projects.

Solution   From the given data: the design life, n, is 20 years, and the interest rate, i, to be used in the analysis is 4%. Considering the first proposal, the present value of the cost, C, is \$400,000, the annual benefit, A, is \$35,000, and the present value of the benefits, B is given by

$$B = \left(\frac{P}{A}, i, n\right)A = \left[\frac{(1 + i)^n - 1}{i(1 + i)^n}\right]A = \left[\frac{(1 + 0.04)^{20} - 1}{0.04(1 + 0.04)^{20}}\right](\$35,000) = \$475,661$$

Therefore the benefit-cost ratio, B/C, is given by

$$\frac{B}{C} = \frac{\$475,661}{\$400,000} = 1.19$$

Since the benefit-cost ratio (1.19) is greater than one, the proposed project is economically feasible.

Considering the second proposal, the present value of the cost, C, is \$600,000, the annual benefit, A, is \$60,000, and the present value of the benefits, B, is given by

$$B = \left(\frac{P}{A}, i, n\right)A = \left[\frac{(1 + i)^n - 1}{i(1 + i)^n}\right]A = \left[\frac{(1 + 0.04)^{20} - 1}{0.04(1 + 0.04)^{20}}\right](\$60,000) = \$815,419$$

The disbenefit to the second proposal is that annual spending on other projects, A_D, is reduced by \$45,000 for the first five years of the project, so the present value of the disbenefit, D is given by

$$D = \left(\frac{P}{A}, i, n\right)A_D = \left[\frac{(1 + i)^n - 1}{i(1 + i)^n}\right]$$

$$A_D = \left[\frac{(1 + 0.04)^5 - 1}{0.04(1 + 0.04)^5}\right](\$45,000) = \$200,332$$

The benefit-cost ratio of the second proposal, taking the disbenefit into account, is given by

$$\frac{B - D}{C} = \frac{\$815,419 - \$200,332}{\$600,000} = 1.03 \quad \text{or}$$

$$\frac{B}{C + D} = \frac{\$815,419}{\$600,000 + \$200,332} = 1.02$$

Since the benefit-cost ratio (1.02) is greater than one, this second proposal is economically feasible.

Both proposed projects are economically feasible, and present-worth analysis must be done to determine their ranking.

7.5 Water-Supply Projects

A dependable supply of water is essential to human activity, and planning for adequate water supplies is one of the most important activities of water-resources engineers. Such planning requires knowledge of present consumption, future needs, and available supplies. In some cases, water supply can expand, whereas in other cases reuse and conservation strategies may be required. Forecasting the supply side of water use requires information on the sources available to meet projected demand, the amount that can be provided from these sources, and the associated costs and environmental effects.

There are numerous examples of inadequate planning for water supplies. A particularly extreme example is utilization of the Ogallala aquifer, which extends from north Texas to South Dakota, and was once thought to be inexhaustible. As a result of overexploitation, the average thickness of the Ogallala aquifer has dropped from 17.5 m (58 ft) in 1930 to 2.5 m (8 ft) in 2002, resulting primarily from increased withdrawals for the Texas-Gulf and Rio Grande water-resources regions. A second extreme example is the case of the Colorado River, which flows through several states, is used extensively for irrigation, and has enabled the rapid growth of desert cities such as Las Vegas, Phoenix, and San Diego. Demands have so drained the river that it no longer consistently reaches its mouth in the Gulf of California. Overuse of the Colorado River has long been a source of contention between the United States and Mexico.

In the United States, the three main categories of water use are: agricultural, industrial, and public-supply use.

Agricultural Use. Agriculture is the largest user of water in the United States, with irrigation accounting for 97% of total agriculture use and the remainder going to rural domestic use and livestock production. Most irrigation water is used in the western United States (west of the Mississippi River), where the nine western water-resources regions (excluding Alaska and Hawaii) account for 89% of the total irrigation water use in the United States. Ground-water withdrawals for irrigation account for approximately 40% of all water used for irrigation. The high withdrawal and consumption needs of agriculture also mean that the potential for water conservation and reuse is very high. Such practices include the development of more productive and/or salt-tolerant crop species; lining of irrigation canals or the use of pipelines to prevent seepage loss; "trickle" or "drip" irrigation practices to reduce the amount of water necessary for continued crop yield; and the reuse of municipal and/or agricultural wastewater for direct irrigation or for recharge of ground water.

Industrial Use. Industrial use is for purposes such as processing, washing, and cooling in manufacturing plants. Some of the major water-using industries are steel, chemical and allied products, paper and allied products, and petroleum refining. In the United States, self-supplied industrial uses of water, excluding thermoelectric power plants, account for almost half of the total water withdrawn. A lesser portion of industrial water supplies is obtained from public water systems.

Public-Supply Use. Public water supply is water withdrawn by public and private suppliers and delivered to various users for residential, commercial, industrial, and thermoelectric power uses. In the United States, 1995 data indicate that public-supply users were 56% residential, 17% commercial, 12% industrial, 0.3% thermoelectric power, and 15% unaccounted water or public use and losses. In a typical public water-supply system, residential water use is 50% to 65% of the system water demand, with residential demand tending to make up a larger portion of system water use in utilities serving smaller urban areas. In larger urban areas, a greater proportion of total water is devoted to commercial, industrial, and public-sector use; consequently, per-capita water usage rates generally increase with service-area population (Billings and Jones, 1996). Although the public water-use sector is vital to our well being, since it furnishes much of our drinking water, the total amount of water used by the public-use sector is small when compared with water-using sectors such as irrigation and thermoelectric cooling. In 2000, the United States Geological Survey (USGS) reported that this sector represented about 12.5% of the nation's fresh-water withdrawals.

Analyzing and evaluating public water-supply projects is complex, involving technical issues about the yield of potential new sources as well as regulations and permits that require extensive environmental analysis.

7.5.1 Public-Supply Use

Public water supplies are typically grouped into four categories: residential, commercial, industrial, and public use. In some cases, residential, commercial, and institutional use by facilities such as schools and hospitals are collectively called *domestic use* (AWWA, 2003a). Brief descriptions of the main categories of public-supply water use are as follows:

Residential Use. Water used by residential households includes both interior uses (e.g., toilets, showers, clothes washers) and exterior uses (e.g., lawn watering, car washing). The amount of water used by households varies with factors such as the age and number of occupants in a household, income level, and geographic location. Outdoor water use is strongly influenced by annual weather patterns, with hot, dry climates having a significantly higher quantity of use than cold wet climates.

Commercial Use. Water use for commercial facilities can vary widely depending on the type of establishment (e.g., office, restaurant). In addition to water used for drinking and sanitary purposes by employees and customers, many commercial users have special water needs, such as for cooling, humidifying, washing, ice making, vegetable and produce watering, ornamental fountains, and landscape irrigation.

Industrial Use. Water use by industries varies widely. It depends on the type of industry, water cost, wastewater disposal practices, types of processes and equipment, and local water conservation and reuse practices. In general, industries that use large quantities of water are located in communities where water quality is good and the cost is reasonable. Historically, major industrial water users such as steel, petroleum products, pulp and paper, and power industries have provided their own water supplies. Smaller industries and industries with low water usage generally purchase water from public systems.

Public Use. Municipalities and other public entities provide public services that require varying amounts of water. Some typical public uses are: public parks, municipal buildings, fire fighting, and public works uses such as street cleaning, sewer flushing, and water-system flushing.

Unaccounted-for system leakage and losses are typically 10%–12% of production for new distribution systems (less than 25 years old) and 15%–30% for older systems (AWWA, 2003a).

The billing systems in most large water utilities permit at least some disaggregation of total billings, by residential, commercial, industrial, and municipal or public facilities. However, while most water utilities have meters installed on all water services, some systems do not have customer meters and charge customers a flat rate regardless of the quantity of water they use. Forecast accuracy is generally improved by disaggregation as long as water-use or trend patterns are different among the segments.

Planning for water utilities relies heavily on water-demand forecasting, which establishes how much water the planning area will need in the future. Water-use forecasts in most cases are long-term, cover more than 10 years into the future, and typically forecast average daily use. These long-term forecasts are generally required for identifying needed raw-water supplies and system capacity. Medium-term forecasts for 1 to 10 years are used for planning improvements to the water distribution and treatment system and for setting water rates.

Most agencies charged with planning for water supplies use simple forecasting techniques. However, a wide range of techniques are available, several of which are reviewed in the following sections. The forecasting method selected will depend largely on the data available, the complexity of the utilities service area, and the degree of forecast reliability required. When data are limited, a simple model should be used.

Extrapolation Models. This method considers only past water-use records and extrapolates into the future. A continuation of past trends is generally assumed, and extrapolation in time can take a variety of functional forms such as linear, exponential, and logarithmic forms. In some cases, complex autoregressive integrated moving average (ARIMA) time-series models are developed to describe the behavior of water demand over time and project future behavior. Water use and time are the only variables considered using this technique. Extrapolation

FIGURE 7.2: Water-demand curve

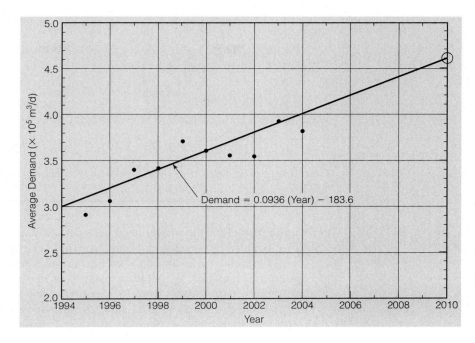

Demand = 0.0936 (Year) − 183.6

models can provide reasonably accurate forecasts as long as the future is similar to the past; however, since future trends are unlikely to repeat past trends, time-extrapolation techniques are not highly reliable, especially for long-term projections.

EXAMPLE 7.6

The average water demand for a city in central Texas has been growing steadily from 1995 to 2004, and water-demand data for these years are shown in the following table:

Year	Average demand $(\times 10^5 \text{ m}^3/\text{d})$
1995	2.91
1996	3.05
1997	3.39
1998	3.41
1999	3.70
2000	3.62
2001	3.55
2002	3.54
2003	3.93
2004	3.82

Estimate the average water demand in the year 2010.

Solution   The water-demand data are plotted in Figure 7.2, and a linear trend is indicated.

Fitting a straight line to the data yields a correlation coefficient, r, of 0.89 ($r^2 = 0.79$) and the linear-regression equation

$$\text{Demand} = 0.0936(\text{Year}) - 183.6$$

which has a standard error of 0.16 ($\times 10^5$ m³/d). Taking Year = 2010 yields

$$\text{Demand} = 0.0936(2010) - 183.6 = 4.54 \ (\times 10^5 \text{ m}^3/\text{d})$$

Therefore, assuming that the linear growth trend continues, the average water demand in 2010 is estimated to be 4.54×10^5 m³/d.

Per-Capita Models. This widely used method estimates future water demand as the product of the service-area population and per-capita water use. The per-capita water use may be assumed to be constant, or it may be projected to increase (or decrease) over time. Population projections may be obtained for the service area through original work, from local sources such as local planning departments, or from higher-level sources such as state-agency projections. The per-capita approach is usually applied to municipal water use, and many studies have shown population to be a reliable indicator of water use. This method may be refined by using separate per-capita coefficients for different use categories such as residential, commercial, industrial, and public. The coefficients can also be disaggregated by geographic area and by season. Commercial-use forecasts may be done in terms of water use per employee or per square foot; industrial-use forecasts may be done in terms of water use per employee and projections of employment in the industrial sector; and agricultural-use forecasts may be done by projecting land area in specific crops and multiplying by irrigation requirements per unit area for each crop type. The per-capita models produce satisfactory results as long as the customer mix does not change substantially. The more critical element in forecast accuracy is usually the population projection, not the per-capita water-use factor (Billings and Jones, 1996). The obvious exception is when a very heavy water user enters or leaves a utilities service area. In general, the per-capita model is reasonably reliable for short-range forecasts but becomes increasingly questionable for long-term projections. Per-capita models are mostly used by smaller utilities that do not have the resources or need to develop more sophisticated models. When there are changes in water demand factors, such as scheduled rate increases or implementation of conservation programs, adjustments should be made to per-capita water-usage factors. Water conservation is usually included in forecast adjustments by subtracting an estimated reduction in water use from per-capita demand.

The distribution of per-capita rates among 392 water-supply systems serving approximately 95 million people in the United States is shown in Table 7.2 (AWWA, 1986). The mean per-capita use in this sample was 662 L/day with a standard deviation of 273 L/day. Generally, high per-capita rates are found in

TABLE 7.2: Average Per-Capita Rates of Water Use

Range (liters/person/day)	Number of systems	Percent of systems
189–375	30	7.7
378–564	132	33.7
568–753	133	33.9
757–943	51	13.0
946–1132	19	4.8
≥ 1136	27	6.9
Total	392	100.0

Source: AWWA (1986).

water-supply systems servicing large industrial and commercial sectors. There-fore, more meaningful comparisons between water systems would require the disaggregation of total use into homogeneous sectors of water users.

EXAMPLE 7.7

The planning department of a city estimates that the current population is 98,000, and the population is expected to increase by approximately 10,000 in the next five years. The city water department estimates that the per-capita demand has been approximately constant at 700 L/day for the past five years, and the customer mix is not expected to change significantly over the next five years. Estimate the current water demand, and the average water demand in five years. The city water supply originates from ground water, and the current pump units have a capacity of 2000 L/s. Considering that water-supply pumps are typically designed to accommodate the maximum daily demand, and that the ratio of maximum daily to average daily demand reported by the water department is 2.3, assess the adequacy of the existing pump capacity for the next five years.

Solution   For a population of 98,000 people and a per-capita rate of 700 L/d, the average water demand is given by

$$\text{current demand} = \text{current population} \times \text{per-capita rate}$$

$$= 98{,}000 \text{ persons} \times 700 \text{ L/(d·person)}$$

$$= 6.86 \times 10^7 \text{ L/d} = 6.86 \times 10^4 \text{ m}^3\text{/d}$$

In five years, the population is estimated to be $98{,}000 + 10{,}000 = 108{,}000$ people and, assuming the per-capita demand does not change significantly, the average water demand in five years is given by

$$\text{demand in five years} = \text{population in five years} \times \text{per-capita rate}$$

$$= 108{,}000 \text{ persons} \times 700 \text{ L/(d·person)}$$

$$= 7.56 \times 10^7 \text{ L/d} = 7.66 \times 10^4 \text{ m}^3\text{/d}$$

Therefore, over the next five years the (annual) average water demand is expected to increase from 6.86×10^4 m³/d to 7.66×10^4 m³/d.

Based on these results, assuming a peaking factor of 2.3, and designing the water-supply pumps to deliver the maximum-daily flow,

$$\text{minimum pumping capacity now} = 2.3 \times 6.86 \times 10^4 \text{ m}^3/\text{d}$$

$$= 15.8 \times 10^4 \text{ m}^3/\text{d} = 1830 \text{ L/s}$$

$$\text{minimum pumping capacity in five years} = 2.3 \times 7.66 \times 10^4 \text{ m}^3/\text{d}$$

$$= 17.6 \times 10^4 \text{ m}^3/\text{d} = 2040 \text{ L/s}$$

The existing pump capacity of 2000 L/s is adequate to meet the current capacity requirements (1830 L/s), but the required pump capacity in five years (2040 L/s) is predicted to exceed the installed pump capacity (2000 L/s). Additional pump capacity should be planned for installation within the next five years.

Causal/Structural Models. *Causal/structural models* define water use as a function of two or more variables associated with water use. These models are sometimes referred to as *multiple-regression* models, since multiple-regression techniques are used to estimate the relevant coefficients. To forecast water use, future values of the independent variables must be determined by other means. Independent variables used in causal/structural models to predict household water use can include the marginal price of water, fixed customer charges, household income, household size, lot size, the presence or absence of a swimming pool, and various weather parameters. The availability of municipal sewer systems has been found to increase water use, due to less concern about overloading an onsite wastewater treatment and disposal system (AWWA, 2003a). In many cases, logarithmic transformation of the raw data is appropriate, and a regression relationship for annual water use of a residential customer (household) could be expressed as (Billings and Jones, 1996)

$$\ln Q = a_0 + a_1 \ln X_1 + a_2 \ln X_2 + a_3 \ln X_3 + a_4 X_4 + a_5 X_5 + \varepsilon \quad (7.15)$$

where Q is the annual water use per household (m³), a_o to a_7 are regression constants, X_1 is the marginal price of water ($/m³), X_2 is the annual household income ($), X_3 is the number of people in the household (persons), X_4 indicates the presence of a swimming pool (1 if present, 0 otherwise), X_5 indicates the presence of indoor water-conserving fixtures (1 if present, 0 otherwise), and ε is the residual error of the estimate. Average monthly demands are commonly derived from average annual demand using a *seasonal index*, which is defined as the average fraction of total annual water use to be expected during a given calendar month (Dziegielewski and Opitz, 2002). Multiple-regression models for water-demand forecasts rarely include more than four or five variables, and these models are also used to estimate seasonal and peak water demands (AWWA, 2001). Multiple-regression methods where the price of water is included as an explanatory variable are sometimes called *demand models*, since there is a relationship between price, P, and demand, Q, which is generally quantified by the *price elasticity* of demand, e_d, given by

$$e_d = -\frac{\frac{dQ}{Q}}{\frac{dP}{P}} \tag{7.16}$$

Previous studies have shown values of e_d on the order of -0.4 for residential water demand, which indicates a 40% reduction in the demand for water if the price of water were doubled (Howe and Linaweaver, 1967). Measured price elasticity is normally less in the winter months than in summer months for residential accounts (AWWA, 2001). Substantial price increases in water may reflect changes in water-supply costs as well as deliberate pricing policy. However, because price elasticities are low and inflation typically erodes the impact of price increases, and because water is relatively cheap, rate effects are frequently underplayed in water-resource planning discussions (Billings and Jones, 1996).

EXAMPLE 7.8

An analysis of data collected from residential customers yields the following empirical expression for the annual water use in each household:

$$\ln Q = 5.68 + 0.099 \ln M + 0.0084 \ln I + 0.78 \ln N$$

where Q is the annual water use in m^3/(year·household); M is the marginal water rate in $/m^3$, I is the household income in $, and N is the number of persons in the household. An econometric model projects that in year 2050 water rates will be $0.80/m^3$ (in current dollars), and a typical household in the service area will have an annual income of $90,000 (in current dollars) and consist of 2.7 persons. If the land-use plan for 2050 indicates that there will be 4000 households in the service area, estimate the average daily water demand for the service area in the year 2050. What will be the price elasticity of demand for a typical household in the year 2050? Based on this price elasticity, would a pricing adjustment be able to correct for a 5% shortfall in available water supplies?

Solution   From the given data: $M = $0.80/m^3$, $I = $90,000$, $N = 2.7$, and the water-demand regression equation for the service area gives

$$\ln Q = 5.68 + 0.099 \ln M + 0.0084 \ln I + 0.78 \ln N$$
$$= 5.68 + 0.099 \ln (\$0.80/m^3) + 0.0084 \ln (\$90,000) + 0.78 \ln (2.7) = 6.53$$

which yields $Q = e^{6.53} = 685$ m^3/(year · household). Since the service area contains 4000 households, the average water demand is given by

$$\text{average demand} = 685 \frac{m^3}{\text{year} \cdot \text{household}} \times 4000 \text{ households}$$
$$= 2.74 \times 10^6 \text{ m}^3/\text{year} = 7500 \text{ m}^3/\text{d}$$

Therefore, the average water demand in 2050 for this service area is predicted to be 7500 m^3/d.

The price elasticity of demand, e_d, is given by Equation 7.16. In the year 2050, $I = \$90,000$, $N = 2.7$, and the annual water demand, Q, in a typical household is related to the marginal rate, M, by

$$\ln Q = 5.68 + 0.099 \ln M + 0.0084 \ln I + 0.78 \ln N$$

$$= 5.68 + 0.099 \ln M + 0.0084 \ln (\$90,000) + 0.78 \ln (2.7)$$

$$= 6.55 + 0.099 \ln M$$

Differentiating with respect to the marginal water rate, M, yields

$$\frac{1}{Q}\frac{dQ}{dM} = 0.099 \frac{1}{M}$$

which can be put in the form

$$-\frac{\frac{dQ}{Q}}{\frac{dM}{M}} = -0.099$$

Comparing this expression with the definition of the price elasticity given by Equation 7.16 indicates that the price elasticity is equal to 0.099. Therefore, doubling the price of water will cause a 9.9% reduction in water demand, which more than adequately covers a 5% shortfall in supply.

Per-capita models and causal/structural models depend explicitly on population and economic forecasts. There are a variety of approaches to population forecasts, and the most common ones are:

- Extrapolation of the current average growth rate, perhaps using some adjustment factor throughout the forecast period.
- Structural models using causal variables in which population growth is the dependent variable and independent variables include social and economic factors.
- Land-use models, which concentrate on current and projected uses of residential, commercial, industrial, and public lands within the ultimate boundaries of the water utility. Growth rates are established in consultation with city planning departments for each land-use segment, usually by census tract, allowing for public land, infill development, and growth into undeveloped areas.
- Economic-base models, which link population growth to growth in economic-base industries and employment
- Cohort-component technique based on assigning different fertility, migration, and death rates to various age groups or cohorts

---

EXAMPLE 7.9

The residential service area of a water utility contains four census tracts and, based on past trends and planned zoning regulations, a city planning department estimates the following number of households in each census tract in the year 2030:

Tract	Households
1	2040
2	3010
3	1560
4	1840

If the number of persons per household is expected to be 2.5, what is the population in the residential service area in 2030?

Solution   The total number of households is 2040 + 3010 + 1560 + 1840 = 8450. Since there are 2.5 persons per household, the expected population is 2.5(8450) = 21,125 persons.

Although this example seems very simple, the key point to note here is that planning departments tend to focus on predicting the growth in the number of households within census tracts. The projected numbers of households are conveniently derived from minimum lot sizes and projected land uses in residentially zoned areas, and census tracts are convenient because this is the unit in which demographic data are reported by every 10 years by the Census Bureau. The population forecast is derived from the projected number of households and is the key input to per-capita models of water demand.

It is important to keep in mind that total population is not necessarily the whole story for purposes of forecasting water use. Other demographic features such as household size and age structure can significantly influence residential water use (Billings and Jones, 1996). Errors in demographic forecasts, on which water-demand forecasts are necessarily based, translate directly into sufficient or excess capacity, either of which imposes substantial political and economic costs on the utility and its managers. Sensitivity or risk analysis is often helpful in evaluating the potential range of supply capability. Forecasts generally gain credibility by offering a confidence interval rather than a point forecast. Economic models that project industry growth and personal income are usually most effective.

In a typical water-demand forecast application, system demand is disaggregated into residential, commercial, industrial, and public water components. Water uses for each of these customer classes might be forecast using a different approach or technique. Residential demand, for example, could be forecast with a structural regression model relating household water usage to weather, household size, and annual income, and then the per household usage combined with projections of the number of households. Commercial water-demand forecasts might be based on water use per customer connection, developed from company records on past usage and extrapolations of business growth. There could be a single large industrial user in the community that provides the utility with a forecast of its water needs. Public water use could be forecast based on a per-capita approach. Per-capita models are typically used by small utilities to estimate total water consumption, and this model does not attempt to disaggregate the customer components.

In purely statistical terms, a minimum of 20 years of annual data and 4 to 5 years of monthly data are necessary to support the development of forecasting models.

More reliable results can be obtained from these models with 30 to 40 years of annual data and 10 to 15 years of monthly data (Billings and Jones, 1996). In general, the more distant the planning horizon, the more questionable the forecast. It is therefore important that great care be exercised in selecting a forecasting technique, and the limitations of the forecasting method should be fully understood. It is recommended that an array of alternative scenarios and assumptions be used to guide decision-making processes.

Larger water utilities routinely build sophisticated forecasting models, not only because they have larger staffs and budgets, but because the capital at risk with planning decisions is greater.

7.5.2 Conservation

Water conservation reduces the demand for water through improvements in efficiency and reductions in water waste. About 30% to 50% of the water used in the United States is wasted unnecessarily (Chang, 2002), and a carefully planned and implemented long-term water-conservation program can typically reduce consumption by 10%–20% over 10 to 20 years (Maddaus et al., 1996). In addition to reduced demand for water, additional benefits of water-conservation programs are a reduction of wastewater flows, reduced costs for customers, potential capital and operations savings, and protection of the environment. An essential part of planning a conservation program is to take inventory of the types of customers, how many customers of each type, and review their associated water-use patterns. The inventory can range in detail from a full-scale water audit to querying metered water consumption. Each community will have a different mix of customer types: residential, commercial, industrial, and institutional. To achieve the desired conservation levels from a program, the selection of conservation measures should be based on the larger-use sectors and categories, while taking into account the ability to implement the measures and acceptability by customers. Conservation of public water supplies can be accomplished in a number of ways, including:

- repair of antiquated and deteriorated water-supply systems
- increased water prices
- increased metering of water use or updating of existing meters
- plumbing codes specifying low-water-use fixtures or appliances
- distribution of water-saving devices
- short-term restrictions on water use at certain times of the day or week
- targeted decreases in water pressure during high-irrigation periods
- promotion of low-water-use landscaping (Xeriscape)
- monitoring customer compliance during droughts
- public-education campaign
- promotion of water conservation by businesses and industries

Most analysts suggest that water conservation or efficiency programs, policies, practices, or measures should be evaluated using a benefit-cost analysis to determine if the program is worth its cost. Costs for programs can vary depending on the implementation method, size of the program, and the extent of the evaluation planned (e.g., surveys, billing data reviews, water audits). Loss of revenue from water sales, costly delays in developing additional source capacity, and additional costs in dealing with drought conditions are all associated with developing and implementing conservation programs. To facilitate the process of choosing a conservation program, measures with expected positive net benefits can be ranked according to social acceptability based on local values, practices, and experience. Typically there is broad consensus about the value of pursuing leak control in the water system, providing incentives to customers to install and use low-water-use appliances and fixtures, and installing in-house recycling equipment for industrial cooling systems (Billings and Jones, 1976). Pricing policies are often politically controversial, but can be effective. Many pricing policies are designed to reduce the peak level of demand, since peak demands often contribute to escalating system capital and operating costs. In the case where the availability of water is limited due to drought, water conservation is a viable target. Marginal pricing of water demand is found to be one of the best alternatives in urban water-conservation efforts (Ejeta and Mays, 2002).

The basic method of including water-conservation and efficiency savings in demand forecasts is to deduct the estimated savings associated with each measure or program from system water demand. In a regression-estimated structural/causal model, the conservation measures are represented by independent variables and their effects directly incorporated into the results.

The potential for water conservation is enormous, since only a small percent of delivered water is actually used directly by people. In Los Angeles, for example, only 11% of urban water is used by households for personal purposes (Johnson, 1988).

7.5.3 Supply Sources

Once demand forecasts are developed, the next step is to assess the need for added supplies. This requires estimating the yields available from current supply sources. In some cases, yields will decrease over time as a result of a number of factors, including falling ground-water levels, increased upstream diversions, and sedimentation of reservoirs. The difference between the water demand and supply at a future time is called the *supply deficit*. Feasible new sources that can be exploited usually include new or expanded surface supplies, new or expanded ground-water supplies, reclaimed water, and desalinated water. Conserved water can also reasonably be counted as an equivalent supply source. In developing new sources of water supply, the following issues will need to be addressed (AWWA, 2001):

1. Water quality of new sources.
2. Protection of current and new sources.
3. Regulations.
4. Water rights and policies.
5. Yield of new sources and impacts on other sources.
6. Environmental impact assessments.
7. Economic feasibility and financing considerations.
8. Long-term viability (design life).

Under requirements of the Safe Drinking Water Act (SDWA), state regulatory agencies are required to develop Source Water Assessment Plans (SWAPs). This requires working with water-supply utilities to delineate watershed boundaries and identify sources of regulated and high-risk unregulated drinking-water contaminants.

Watershed Protection Plans (WPPs) are developed to protect sources of drinking water (both surface water and ground water).

7.5.3.1 Surface-water sources

Common surface-water sources include direct river withdrawals, on-stream reservoirs, and pumped-storage reservoirs. Where large rivers are located within a reasonable distance to the service area, direct withdrawals for water supply may be practical. Primary issues to be considered in investigating river withdrawals are diversion allocations, river-water quality, contamination potential (e.g., highway crossings, barge traffic, industrial discharge potential), watershed-protection measures, existing water rights, permits, the need for backup or independent short-term supply, and transmission costs. On-stream reservoirs use reservoir storage capacity and natural runoff to meet water-supply needs. During periods of low flow, water-supply requirements are met or supplemented by reservoir drawdown, and during periods of high flow water-supply needs are met by streamflow, with storage recharge provided by excess flow. If significant pumped diversions are provided to a reservoir as a reserve for meeting demands during droughts or other periodic service interruptions, this type of project is called a pumped-storage reservoir project. These facilities are also known in some regions as *skimmer reservoirs* or *side-stream reservoirs* (AWWA, 2001).

A key concept of a watershed area or surface reservoir system is *safe yield*, which is the maximum quantity of water that can be withdrawn continuously over a long period of time, including very dry periods. For example, without storage, the safe yield of a stream is simply the amount of water that can be withdrawn during a period of lowest flow. In many states, the safe yield of a direct stream withdrawal for public water supply is taken as the "1Q30," defined as the lowest one-day flow with a return period of 30 years.

7.5.3.2 Ground-water sources

The identification of potential ground-water sources for a public water supply broadly involves the following three steps (AWWA, 2003b): (1) identifying regions that have low pollution potential, high recharge capability, and a favorable location to the utility and its customers; (2) performing field investigations to confirm site-specific characteristics; and (3) dealing with land-use and wellhead-protection issues.

The primary options for ground-water sources of supply include: wells, infiltration galleries, and aquifer storage and recovery (ASR) wells. Ground-water wells permit water to be extracted using pumps, and the design and operation of wells have been discussed extensively in Chapter 6. *Infiltration galleries* are subsurface drains (or horizontal wells) that intercept interflow in permeable materials or infiltrating surface water and discharge into a sump whose bottom is below the invert of the gallery screen and casing. *Aquifer storage and recovery* (ASR) wells are special wells that can inject water into the ground and extract water from the ground. Water is injected into empty storage space of an underlying aquifer and later extracted to meet water demands. Infiltration galleries and ASR wells are described is more detail below.

Infiltration galleries are typically constructed when subsurface conditions do not permit ground-water development using vertical production wells. Such conditions occur in thin aquifers or where a thin fresh-water layer is underlain by saline water. A typical example is a river valley where the alluvial deposits lie above bedrock where the water is plentiful, and the hydraulic conductivities are high, but the aquifer transmissivities are inadequate for well development because of the thinness of the aquifer. In such cases a subsurface drain or horizontal well is placed in permeable alluvial material and water is collected in a sump connected to a pump. Significant amounts of water can often be extracted through infiltration galleries because of high hydraulic conductivity of alluvial material and their proximity to a recharge source, such as a stream. The yield from infiltration galleries beneath a water body is normally twice that from galleries adjacent to the water body; however, constructing infiltration galleries is usually more difficult under a water body (AWWA, 2001).

Aquifer storage and recovery (ASR) wells are combination wells that can inject water into and extract water from an aquifer. ASR wells inject water into an aquifer during periods of excess supply, and later extract the water during periods of short supply. This process is illustrated in Figure 7.3. ASR is used in both potable and nonpotable water systems. In the case of potable systems, unused water-treatment capacity can be used to treat water during low-demand periods of the year, and this excess water injected into the aquifer. If demand increases beyond capacity, the stored water is recovered from the aquifer, disinfected, and blended with treated water. In nonpotable ASR systems, surface water, runoff, or water pumped from a surficial aquifer is stored for later treatment and use. The pump used for extraction is also periodically used to redevelop the well, thus maintaining its capacity. The concept of aquifer storage and recovery has been utilized or investigated at more than 60 sites in the United States.

In developing ground-water resources it is of primary importance to determine the maximum *safe yield* of a ground-water basin and develop management strategies to protect the aquifer from overexploitation. Although there is no precise definition of the safe yield of an aquifer, it is commonly defined as the annual amount of water that can be withdrawn from the aquifer without producing any undesirable effects, including lowering the water table below acceptable limits, salinity intrusion, undue induced recharge from nearby surface water bodies, wetlands impacts, and degradation of ground-water quality (Lohman, 1979; Dingman, 2002). A variety of methods have been proposed for estimating the safe yield of an aquifer, and the differences between them are typically associated with the type and amount of hydrologic data that are available and can be used in the analysis. Safe-yield estimates are frequently based on the conservation of mass equation given by

$$\boxed{\bar{I} - \bar{O} = \frac{\Delta S}{\Delta t}}$$

(7.17)

where $\bar{I}$ and $\bar{O}$ are the average inflow and outflow from the ground-water basin over the time interval Δt, and ΔS is the change in storage during this same time interval. Inflow, $\bar{I}$, includes direct recharge from rainfall, and indirect recharge from rivers, and lakes; outflow, $\bar{O}$, includes evapotranspiration, seepage to surface waters such as rivers and lakes, and water pumped from the aquifer; and changes in storage, ΔS, are taken as either change in piezometric head multiplied by the storage coefficient (for confined aquifers) or the change in water-table elevation multiplied by the specific yield (for unconfined aquifers). Any of the following three methods can be used to estimate the safe yield, depending on the available data (McCuen, 2005):

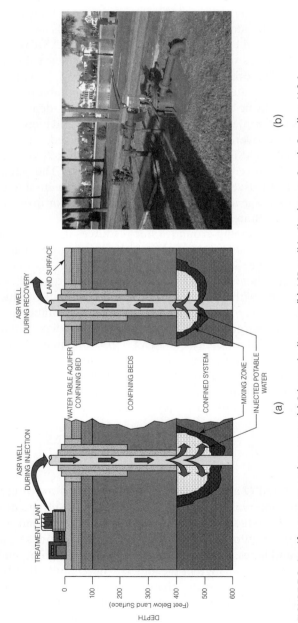

FIGURE 7.3: Aquifer storage and recovery: (a) Schematic diagram; (b) ASR well in Charleston, South Carolina, U.S.A.

Source: Bloetscher et al. (2005).

Zero-Fluctuation Method. For the zero-fluctuation method, the data required to estimate the safe yield include the total draft over a period of record when the water-table elevation at the beginning and end of the period is the same. The longer the period of record, the more accurate the estimate of the safe yield. In practice, the data will probably consist of annual records of water-table elevations and annual draft.

Average-Draft Method. The average-draft method uses the same type of data as the zero-fluctuation method; however, instead of finding just the total draft for a period during which the water-table elevation is the same at the beginning and end of the period, annual values of the change in water-table elevation and annual draft are required. To compute the safe yield, the annual draft is plotted as the abscissa, the change in the annual water-table elevation is plotted as the ordinate, and a line that best represents the data is fitted by regression. The safe yield is the value of the annual draft corresponding to zero elevation change.

Simplified Water-Balance Method. The simplified water-balance method uses conservation of mass (Equation 7.17) to compute the safe yield. In many cases, annual precipitation and streamflow are the only data available to represent the inflow and outflow terms in Equation 7.17. Observation wells can be used to estimate the annual changes in the water-table elevation, and the water-balance equation can be reduced to

$$P - (Q + ET + Q_p) = S_y \Delta e \qquad (7.18)$$

where P is the annual precipitation, Q is the annual surface runoff, ET is the annual evapotranspiration, Q_p is the annual draft, S_y is the specific yield, and Δe is the annual change in water-table elevation. If values of the annual draft are not available, $P - (Q + ET)$ can be plotted against the change in water-table elevation to estimate the safe yield, with the value of Δe used as the ordinate. A line can be fitted to the data, and the safe yield can be estimated from the value of $P - (Q + ET)$ corresponding to the value of Δe equal to zero.

It is important to emphasize that the widely accepted definition of safe yield is "the rate at which ground water can be withdrawn without producing undesirable effects." Based on this definition, it should be clear to water-resources planners that undesirable effects will generally occur as a result of ground-water development, even though the net change in aquifer storage is zero—for example, undesirable redistribution of ground-water/surface-water interactions. The social and technical acceptability of these undesirable effects must be considered, and these considerations are central to determining the "safe yield." Estimates of safe yield must be revisited periodically, since changes in land use or urbanization can cause changes in recharge and thus affect the safe yield.

EXAMPLE 7.10

A water-management agency is charged with managing the resources of a 5120-km² region where ground water is the only source of drinking water. Permits specifying allowable withdrawal rates are required for all water pumped from the aquifer, and

USGS monitoring stations provide data on water-table elevations, surface runoff (in rivers and streams), evapotranspiration, and precipitation. In the period from 1994 to 2003, these data indicate the following data for each year:

Year	Annual draft $(\times 10^8 \ m^3)$	Water-table elevation* (m)	Surface runoff + evapotranspiration $(\times 10^8 \ m^3)$	Rainfall (cm)
1993	–	3.342	–	–
1994	2.76	3.462	62.1	13.10
1995	2.99	3.530	60.8	12.40
1996	3.20	3.540	56.6	11.92
1997	2.91	3.295	52.3	10.14
1998	3.27	3.198	62.3	12.78
1999	2.95	3.336	71.4	14.62
2000	3.01	3.255	59.8	12.31
2001	3.15	3.053	60.8	11.88
2002	2.80	3.150	65.3	13.72
2003	3.36	2.998	52.9	10.47

*Water-table elevation at the end of the year.

Estimate the safe yield of the aquifer using the zero-fluctuation, average-draft, and simplified water-balance methods.

Solution

Zero-Fluctuation Method:
This method estimates the safe yield from the total draft during an interval in which water table at the beginning and end of the interval is the same. In cases where several such "zero-fluctuation" intervals exist, the longer the interval the better. In this case, the starting water-table elevation at the end of 1993 is 3.342 m and the water table returns to this elevation approximately 3.81 years later (in 1997). Considering the water-table elevation at the end of each year, the maximum time intervals for "zero fluctuation" are given in the following table:

Year	Start elevation (m)	Zero-fluctuation interval (years)
1993	3.342	3.81
1994	3.462	2.32
1995	3.530	1.04
1996	3.540	–
1997	3.295	2.51
1998	3.198	2.28
1999	3.336	–
2000	3.255	–
2001	3.053	1.64
2002	3.150	–
2003	2.998	–

Based on these results, the maximum time interval between "zero fluctuations" is 3.81 years, between a water-table elevation of 3.342 m at the end of 1993 and the same water-table elevation within the year 1997 (81% into 1997). From the given data, the total draft during this period is $(2.76 + 2.99 + 3.20 + 0.81 \times 2.91) \times 10^8$ m^3 = 1.13$\times 10^9$ m^3, and hence the safe yield is given by

$$\text{safe yield} = \frac{1.13 \times 10^9 \text{ m}^3}{3.81 \text{ years}} = 2.97 \times 10^8 \text{ m}^3/\text{year} = 8.13 \times 10^5 \text{ m}^3/\text{d}$$

Therefore, according to the zero-fluctuation method, the average withdrawal rate from the aquifer should not be permitted to exceed 8.13×10^5 m^3/d.

Average-Draft Method:

The average-draft method requires plotting the annual change in water-table elevation versus the annual draft and extrapolating the best-fit line to where the annual change in water-table elevation is equal to zero; the annual draft at this point is equal to the safe yield. From the given data, the annual change in water-table elevation versus draft is given in the following table, and these data are plotted in Figure 7.4.

Year	Change in elevation (m)	Draft ($\times 10^8$ m^3)
1994	0.120	2.76
1995	0.068	2.99
1996	0.010	3.20
1997	-0.245	2.91
1998	-0.097	3.27
1999	0.138	2.95
2000	-0.081	3.01
2001	-0.202	3.15
2002	0.097	2.80
2003	-0.152	3.36

The equation of the best-fit line is given by

$$\text{change in elevation} = -0.364 \, (\text{draft}) + 1.07$$

which indicates an annual average draft of 2.94×10^8 m^3/year ($=8.05 \times 10^5$ m^3/d) when the annual change in water-table elevation is equal to zero. Therefore, according to the average-draft method, the average withdrawal rate from the aquifer should not be permitted to exceed 8.05×10^5 m^3/d.

Simplified Water-Balance Method:

The simplified water-balance method requires plotting the annual change in water-table elevation versus $P - (Q + ET)$, where P is the precipitation, Q is the surface runoff, and ET is the evapotranspiration. The best-fit line is extrapolated to where the annual change in water-table elevation is equal to zero, and the value of $P - (Q + ET)$ at this point is equal to the safe yield. The volume of precipitation is equal to the depth of precipitation multiplied by the plan area of the aquifer, which is given as

FIGURE 7.4: Estimation of safe yield using average-draft method

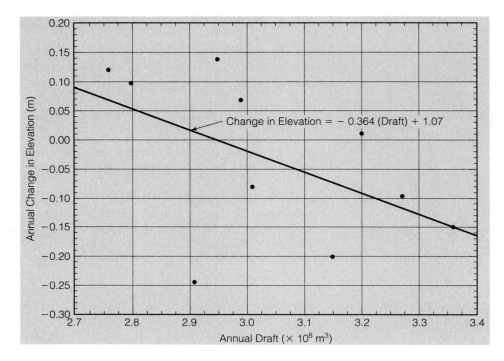

5120 km². From the given data, the annual change in water-table elevation versus $P - (Q + ET)$ is given in the following table, and these data are plotted in Figure 7.5.

Year	Change in elevation (m)	P ($\times 10^8$ m³)	$Q + ET$ ($\times 10^8$ m³)	$P - (Q + ET)$ ($\times 10^8$ m³)
1994	0.120	67.1	62.1	5.0
1995	0.068	63.5	60.8	2.7
1996	0.010	61.0	56.6	4.4
1997	−0.245	51.9	52.3	−0.4
1998	−0.097	65.4	62.3	3.1
1999	0.138	74.9	71.4	3.5
2000	−0.081	63.0	59.8	3.2
2001	−0.202	60.8	60.8	0.0
2002	0.097	70.2	65.3	4.9
2003	−0.152	53.6	52.9	0.7

The equation of the best-fit line is given by

$$\text{change in elevation} = 0.0615(P - Q - ET) - 0.201$$

which indicates an annual draft of 3.27×10^8 m³/year ($= 8.96 \times 10^5$ m³/d) when the annual change in water-table elevation is equal to zero. Therefore, according to the simplified water-balance method, the average withdrawal rate from the aquifer should not be permitted to exceed 8.96×10^5 m³/d.

FIGURE 7.5: Estimation of safe yield using simplified water-balance method

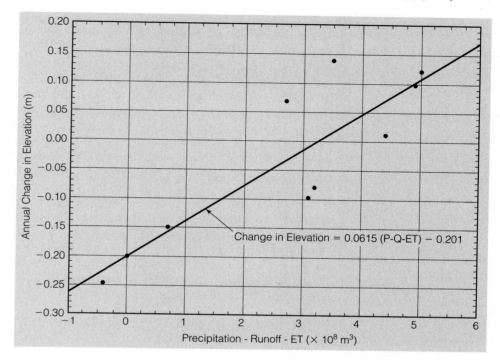

The zero-fluctuation, average-draft, and simplified water-balance methods indicate safe yields of 8.13×10^5 m^3/d, 8.05×10^5 m^3/d, and 8.96×10^5 m^3/d, respectively. Based on these results, a conservative estimate of the safe yield of the aquifer is 8.05×10^5 m^3/d.

A significant consequence of ground-water development can be downward movement of the land surface, called subsidence. Subsidence can occur when the ground water, which exerts a buoyancy effect on the surrounding soil/rock matrix, is removed, thereby increasing the stress on the soil/rock matrix. This additional stress may cause the soil/rock matrix to collapse, resulting in an altered surface topography. Development of ground water needs to include consideration of possible land-surface subsidence. In some areas, clays such as montmorillonite might exist beneath the surface, causing significant problems for water suppliers. The compressibility of clay is 1–2 orders of magnitude greater than that of sand, and 2–3 orders of magnitude greater than that of gravel. When water levels decline, most of the compression occurs in clay units, causing land subsidence.

7.5.3.3 Source-water protection

In accordance with the Safe Drinking Water Act (SDWA), all public water suppliers (surface and ground-water systems) in the United States must be covered under a source-water protection program (SWPP). The basic assumption for implementing the SWPP is that multiple-barrier protection of public water supplies will provide for high-quality water supplies and protect public health. Multiple barriers include source-water protection, treatment, distribution-system maintenance, and monitoring.

Source-water assessments are the centerpiece of the current SDWA focus on prevention. They identify the potential threats to the source of a community's drinking water. States can use these assessments to issue water suppliers monitoring waivers for many regulated chemicals. All states are required to develop source-water assessments for all public water supplies that will include: (1) delineating the ground-water area or surface watershed contributing water to the water-supply intake; (2) inventory the contaminant sources in the delineated water-supply area; and (3) determine the susceptibility of the water system to contamination. For ground-water systems, delineations should be conducted using one or more of the following methods: arbitrary radii, calculated fixed radii, simplified variable shapes, analytical methods, hydrogeologic mapping, or numerical flow and transport models (AWWA, 2001). Sources of contaminants that are regulated under the Safe Drinking Water Act (SDWA) should be identified. This includes contaminants for which there are primary drinking-water standards, contaminants regulated under the Surface Water Treatment Rule, and the microorganism *Cryptosporidium*. Susceptibility assessments are the least understood (and most important) aspects of source-water protection. Assessment methodologies include simple analytical techniques such as hydrogeologic and hydrologic mapping to identify the relative vulnerability of ground-water and surface-water supplies, as well as complex contaminant transport models linked to risk-assessment matrices.

7.5.3.4   Reclaimed water

Analysis of reclaimed water as a possible water-supply source begins with the raw wastewater from which it is derived and ends with its ultimate reuse. According to the American Water Works Association (AWWA, 2001), "recycled water" is preferred as a general label over "reclaimed water," since water must first be reclaimed from wastewater in order to be recycled for reuse. Water reclamation often provides both water supply and pollution control.

The most widely available and least variable source of wastewater for reuse is municipal wastewater. The level of treatment required to produce recycled water depends on both the quality of raw wastewater influent to the treatment plant and the quality of plant effluent required to satisfy the most constraining reuse. In the United States, individual states promulgate specific treatment requirements for recycled water, and Table 7.3 lists a variety of recycled water uses and the minimum treatment levels required.

Definitions of treatment levels vary in common usage and even overlap. The following levels are guidelines, and raw municipal wastewater is used as a standard of comparison (AWWA, 2001):

Pretreatment. Pretreatment is the level of treatment required to produce wastewater of an acceptable quality to be discharged into the public sanitary sewer system. Pretreatment requirements may be stated in terms of water quality or required pretreatment processes. Examples include total suspended solids (TSS) removal from brewery waste, and application of grease traps to restaurant wastewater. Pretreatment is often a requirement imposed on a wastewater source through a sewerage agency's industrial pretreatment program.

Primary Treatment. Primary treatment is the level of treatment required to produce effluent free of most settleable and floatable solids present in raw municipal wastewater. A substantial reduction in oxygen demand is usually accomplished

TABLE 7.3: Recycled Water Use and Treatment Level

Use	Treatment Level
Fodder, fiber, seed irrigation	Primary
Orchard, vineyard irrigation	Primary
Pasture irrigation for milking animals	Disinfected* secondary
Controlled impoundment	Disinfected* secondary
Landscape irrigation with restricted public access	Disinfected* secondary
Food crop surface irrigation	Disinfected† secondary
Uncontrolled impoundment	Disinfected† tertiary
Landscape irrigation with unrestricted public access	Disinfected† tertiary
Food crop spray irrigation	Disinfected† tertiary
In-building toilet and urinal flushing	Disinfected† tertiary
Ground-water recharge for nonpotable supply	Disinfected† tertiary
Cooling-tower supply	Disinfected tertiary plus advanced‡
Ground-water recharge for potable supply	Disinfected tertiary plus advanced§
Live-stream discharge (environmental enhancement)	Disinfected tertiary plus advanced§§

Source: AWWA (2001).*A lower level of disinfection may be allowed. †A higher level of disinfection may be required.
‡A demineralization process may be required.
§Demineralization and activated carbon processes may be required.
§§Nutrient removal and dechlorination processes may be required.

at the same time. Primary treatment is usually provided by the gravity sedimentation process, and its performance often assessed by the extent of TSS and BOD removal.

Secondary Treatment. Secondary treatment is the level of treatment required to produce an oxidized effluent in which organic matter has been stabilized, is nonputrescible, and contains dissolved oxygen. Secondary treatment is often characterized by and effluent with TSS and BOD concentrations of 30 mg/L or less. This level of treatment is usually provided by an aerobic biological process, such as activated sludge, performed on good-quality primary effluent.

Advanced Secondary (Tertiary) Treatment. Advanced secondary (tertiary) treatment is the level of treatment required to produce an effluent capable of being used to irrigate crops, golf courses, home lawns, and other areas where the public may come into contact with the treated wastewater. Tertiary effluent is often characterized by turbidities almost as low as those of a good-quality potable water supply. This level of treatment is usually provided by filtration of good-quality secondary effluent and high levels of disinfection. Total suspended solids (TSS) of this effluent are generally below 5 mg/L.

Advanced Wastewater Treatment (AWT). Advanced wastewater treatment involves the reduction or removal of other wastewater constituents such as nutrients. The goal is generally to have total nitrogen below 3 mg/L, total ammonia below 2 mg/L, and total phosphorus below 1 mg/L. Modifying conventional activated-sludge treatment to completely nitrify process water is considered advanced treatment for nitrogen reduction or removal. Chemical augmentation can be used for advanced treatment for phosphorus reduction or solids treatment optimization. Similarly, chemical conditioning and reverse osmosis applied to good-quality filtered effluent will achieve specific ion or total dissolved-solids reduction.

Health risks from water reuse are generally associated with exposure to particular chemicals or microbial agents present in recycled water. Exposure can occur directly, such as through skin contact or aerosol ingestion, or indirectly, such as from eating food from an outdoor table surface sprayed with recycled water or food crops irrigated with recycled water. Reuse regulations have been developed that minimize direct and indirect exposure, and guard against better-known and unstudied agents. Recycled water, appropriately treated, can be used as a source to offset some current or future potable-water demands. For example, golf courses can be irrigated with recycled water, removing them from the potable supply and freeing up water to support growth. In California, municipal wastewaters are reused for fiber-crop and seed-crop irrigation, landscape irrigation, orchard and vineyard irrigation, processed-food crop irrigation, ground-water recharge, and recreational impoundments.

The three ways in which water reuse is currently considered are indirect reuse, direct reuse, and provision of dual water systems.

Indirect Reuse. Indirect potable reuse refers to the use of water from streams that have upstream discharges of wastewater. This type of reuse has been practiced without clear recognition of the concept for a very long time. Other examples of indirect reuse include the water discharged from individual septic systems that percolates into aquifers and at some later time is withdrawn from wells. Sewage and industrial wastes discharged to rivers are diluted and somewhat treated by natural processes, and the water is later withdrawn by downstream communities for water-supply use. Every day millions of people worldwide consume water that is partially reuse water.

Direct Reuse. A large potential source of additional water for water-short areas of the world is the direct reuse of wastewater. At the present time, the principal direct reuse of wastewater is for agricultural irrigation (AWWA, 2003a). Certain areas of the United States have successfully been applying reclaimed water to golf courses, orchards, nonedible crops, and flushing toilets for decades.

The use of wastewater for potable purposes is not considered acceptable at the present time, and the American Water Works Association opposes the direct use of reclaimed water as a potable-water supply source until more research can be completed (AWWA, 2003a). Cost-effective treatment technology and monitoring techniques are not typically available to ensure that all processed wastewater is completely safe for human consumption.

Dual Water Systems. One method of achieving water reuse is to deliver potable water and nonpotable water to customers through separate distribution systems. This is generally referred to as a dual water system. It is a possible remedy where potable water is in very limited supply but a lower-quality, nonpotable water supply is more plentiful. This concept is not new and has been used occasionally by water systems having unusual source-water conditions. There are a number of areas in the United States where dual systems are in use—for example: Tucson and Phoenix, Arizona; Irvine, California; Colorado Springs and Denver, Colorado; and St. Petersburg, Florida, which probably has the most extensive system in the United States. It is likely that more dual systems will be constructed in the future, but health considerations must be paramount.

7.5.3.5 Desalination

Historically, desalination processes were used to treat seawater. However, they are also useful for treating other water sources, and desalination is now becoming a more significant option for providing potable water. Raw water sources with high concentrations of total dissolved solids (TDS) typically include seawater, brackish ground water, connate water, and ground water contaminated by saltwater intrusion into coastal aquifers. Raw water with high TDS is not limited to coastal areas, as TDS generally increases in aquifers as the depth increases. Membrane-separation processes (such as electrodialysis reversal and reverse osmosis) and thermal processes (such as multistage flash distillation) are the two major desalination techniques that are used, and each requires varying degrees of energy to convert high-TDS water into potable water for municipal use. Wastes from desalination plants are highly regulated. In some areas they can be disposed of by surface-water discharge, ocean discharge, deep-well injection, oilfield injection, or any other legal disposal method. Federal regulations that apply to such discharges include Clean Water Act requirements for NPDES permits for discharges into navigable waters, Safe Drinking Water Act requirements for underground injection control (UIC) permits (concentrate injection is prohibited if it causes a potable ground-water source to exceed maximum contaminant levels or affect public health), and Resource Conservation and Recovery Act requirements, which may be relevant if the waste is determined to be hazardous.

7.6 Floodplain Management

A flood occurs when surface runoff exceeds the capacity of the drainage infrastructure, resulting in the inundation of areas served by these drainage systems. The elevation at which the flow overtops the embankments of a river is called the *flood stage*, and the *floodplain* is the (normally dry) land adjacent to rivers that is inundated during a specified flood event. In analyzing floods, engineers and hydrologists commonly utilize the so-called *annual flood series*, which is a series of the maximum instantaneous flowrates for each year of record. Typically, annual peak flows with return periods from 1 to 10 years represent bankfull conditions, with larger flows causing inundation of the floodplain (McCuen, 1989; Wurbs and James, 2002).

Floodplains are usually relatively flat lands where roads and buildings are easily constructed. These areas are typically attractive, yet hazardous, sites for development. To appreciate the concern about floodplains, it should be noted that approximately 7%–10% of the land in the United States is located in floodplains; and riverine floods are the most lethal and costly natural hazard in the United States, causing an average of 140 fatalities and \$5 billion damage each year (Schildgen, 1999; O'Connor and Costa, 2004). Floods are the number-one killers among natural disasters, especially in developing countries. A flood along the Yellow River (Huang Ho) in China in 1931 inundated 1.1×10^5 km^2, caused 1 million fatalities, and left 80 million people homeless (Chang, 2002). This is the worst recorded natural disaster. The largest floodplain areas

FIGURE 7.6: Cross section of floodplain

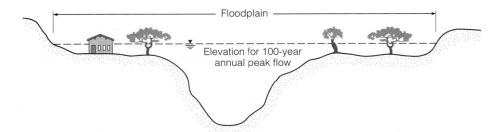

in the United States are in the south, and the most populated floodplains are along the north Atlantic coast, the Great Lakes region, and in California.

Regulatory mechanisms used to control development in floodplains include flood-insurance requirements, open-space reservations, and building restrictions. In the United States, *regulatory floods* for floodplain management are specified by the National Flood Insurance Program (NFIP), managed by the Federal Emergency Management Agency (FEMA). The purpose of the NFIP is to minimize future flood loss and to allow floodplain occupants to be mostly responsible for flood-damage costs instead of the taxpayer. Under the NFIP, part of the insurance is paid by the federal government, and this subsidy is used to entice local communities to adopt floodplain zoning practices. The 100-year flood (1% annual exceedance probability) has been adopted by FEMA as the base flood for delineating floodplains, and the area covered by the base flood is identified on a Flood Insurance Rate Map (FIRM). Communities participating in the NFIP are required to regulate development in the floodplain, and a flood-insurance study is required to delineate the 10-, 50-, 100-, and 500-year recurrence flood flows in flood-prone areas. The cross section of a typical floodplain is shown in Figure 7.6. For specified flows, floodplains are delineated using open-channel backwater computations by computer codes such as HEC-RAS and WSP2, described in Chapter 3. Floods larger than and smaller than the regulatory (100-year) flood are identified on the FIRM for informational purposes. Updates to FIRMs are frequently necessary, since encroachment onto floodplains reduces the capacity of the watercourse and increases the spatial extent of the floodplain. Participants in the NFIP must require new buildings to be elevated or floodproofed up to or above the 100-year flood level. FEMA has a service center that can be accessed on the Internet to determine whether any location in the United States has been mapped and to order copies of available maps.* As of 2002, the National Flood Insurance Program had grown to cover 20,000 participating communities with over 5 million properties insured, producing over $600 billion in coverage. Though successful, it is estimated that only 25% of the 12 million households in the U.S. floodplains are protected by flood insurance.

In flood-insurance studies, the *floodway* is defined as the channel of a watercourse and the adjacent land areas that must be reserved in order to discharge the base flood (i.e., 100-year flood) without cumulatively increasing the water-surface elevation more than a designated height. The designated maximum rise is usually taken to be 0.31 m (1 ft) above the 100-year flood elevation under prefloodway conditions. The area between the floodway and the boundary of the 100-year flood is called the *floodway fringe*. The floodway fringe is the portion of the floodplain that could be totally

---

*http://web1.msc.fema.gov/webapp/commerce/command/ExecMacro/MSC/macros/welcome.d2w/report.

obstructed without increasing the 100-year flood elevation by more than 0.31 m (1 ft) at any point (Prakash, 2004). Development is often allowed in the floodway fringe for political and economic reasons, so the floodway fringe area may not be available to convey floodwater. Development within the floodway is allowed only if compensated by relocating the floodway or mitigating the water-surface increase due to development. Normally, the width of the floodway is determined using equal loss of channel conveyance on opposite sides of the watercourse. If equal loss of conveyance is not practical, unequal conveyances may be used subject to acceptance by the community, state and federal agencies, and the insuring agency. Communities often require that if the floodway fringe is developed, the channel must be improved so that there is no change in the 100-year return-period water-surface elevation.

Issues relating to floodplain management do not call for building flood-control structures at the one extreme or prohibiting the use of the floodplain at the other, but rather a total program of floodplain management involving the appropriate mix of structural and nonstructural measures for any particular situation. The Federal Emergency Management Agency (FEMA) has categorized three alternative strategies for flood-loss reduction: (1) modification of the susceptibility to flood damage; (2) modification of flooding; and (3) modification of the impact of flooding. Each method is described in the following sections, but the first is emphasized because of its planning and management focus.

7.6.1 Modification of the Susceptibility to Flood Damage

This strategy avoids unwise use of the floodplain and thus avoids the dangers associated with flooding. Measures that modify the damage susceptibility of floodplains are usually called *nonstructural measures*, and include land-use regulations, development and redevelopment policies, disaster or emergency preparedness and response plans, flood forecasting and warning, floodproofing, and evacuation. In many communities, urban growth pressures encroach on flood-prone areas, and developments occur in the floodplain because of recreational and aesthetic benefits obtained from being on or near open water. The use of flood-prone lands and the construction of dwellings on such lands should be given special consideration by urban communities in order to reduce potential flood damages. Lowlands subject to flooding and unsuitable for high-density development should be used for open-space purposes. To discourage development in flood-hazard areas, local governments should adopt policies that resist the extension of utilities and streets and the construction of schools and other facilities in these hazard areas. Infrastructure development on lands that are not flood-prone should be given priority, and land-management tools, such as zoning, should be used to supplement infrastructure policies.

Land-use controls/flood insurance.   Regulation of land in the face of the hazard of inundation is based on well-founded legal principles. The original act authorizing the federal role in flood insurance granted federal subsidies in flood insurance to flood-prone communities that instituted certain kinds of floodplain-management regulations. No community is required to participate, nor are property owners in participating communities required to purchase the insurance when it is available. The only incentive provided is that the subsidized insurance is available only to prospective purchasers in communities whose floodplain management programs meet the requirements of the act.

Flood insurance involves categorization of flood-prone areas in accordance with their flooding potential. For purposes of flood insurance, areas within the 100-year floodplain boundary are termed *special flood hazard areas*; areas between the 100-year and 500-year flood boundaries are termed *areas of moderate flood hazard*; and the remaining areas outside the 500-year floodplain boundary are termed *areas of minimal flood hazard*. In flood insurance studies, flood-hazard areas are divided into the following flood-risk zones (FEMA, 1993):

- Zone A: Areas that correspond to 100-year floodplains which are determined by approximate hydraulic analyses. Base flood elevations and depths for this zone are not shown on flood insurance rate maps.

- Zone AE: Areas that correspond to 100-year floodplains which are determined by detailed hydraulic analyses. Base flood elevations for this zone are shown on flood insurance rate maps.

- Zone AH: Areas that correspond to 100-year shallow flooding with a constant water surface elevation (usually areas of ponding) where average depths are between 0.31 and 0.91 meters. The flood elevations derived from detailed hydraulic analyses are shown at selected intervals within this zone.

- Zone AO: Areas that correspond to 100-year shallow flooding (usually sheet flow on sloping terrain) where average depths are between 0.31 and 0.91 meters. Average depths derived from detailed hydraulic analyses are shown within this zone. Alluvial fan flood hazards are also shown as Zone AO.

- Zone A99: Areas that correspond to areas of the 100-year floodplain which will be protected by a federal flood-protection system and for which no base flood elevations or depths are shown on flood insurance rate maps.

- Zone V: Areas that correspond to the 100-year coastal floodplains determined by approximate hydraulic analyses that have additional hazards associated with storm waves for which no base flood elevations are shown on flood insurance rate maps.

- Zone VE: Areas that correspond to the 100-year coastal floodplains determined by detailed hydraulic analyses that have additional hazards associated with storm waves for which base flood elevations are shown on flood insurance rate maps.

- Zone X: Areas that may be outside the 100-year floodplain or within the 100-year floodplain where average flood depths are less than 0.31 m, areas of 100-year flooding where the contributing drainage area is less the 2.59 km^2, or areas protected from the 100-year flood by levees. No base flood elevations or depths are shown on flood insurance rate maps.

- Zone D: Unstudied areas where flood hazards are undetermined but possible.

Over the long term, the land-use regulations required for participation in the flood insurance program should reduce floodplain development. To participate in the NFIP a community must agree to require building permits in the identified flood hazard areas of the community and to review building permit applications in that area to determine whether the proposed building sites will be reasonably safe from flooding. Building permits are sometimes required only in identified flood-prone areas of the community. It is generally required that structures within the flood-hazard areas be designed and anchored to prevent floatation, collapse, or lateral movement of the structure. For residential structures within the area of flood hazard, communities must require new construction and substantial improvements to existing structures to have the lowest floor elevated to or above the level of the base flood. Further regulations state that building in a coastal hazard area is not permitted unless the site is landward of the mean high-tide level and the lowest floor is elevated to the level of the base flood plus an allowance for wave action on adequately anchored piles. The space below the lowest floor must be kept open and free of obstruction. The low structural members of the floor system of a new building in such an area, or any part of the outside wall, should be above the base flood elevation.

Stormwater regulations and ordinances are enacted by local agencies to regulate developments within their jurisdictions so that there are no adverse impacts on flooding conditions in the area. Examples of local flood-control regulations include requirements to provide dry- or wet-detention basins such that peak flows released to a receiving watercourse during a prescribed storm do not exceed predevelopment levels.

Zoning.   Zoning and subdivision regulation are major regulatory devices used in land-use control. Zoning programs are sometimes subject to significant political pressure and may represent a very limited and static approach.

Eminent domain.   Eminent domain is the traditional device used by governments to take property for public purposes. The power of eminent domain allows for the taking of private property for public purposes with just compensation provided to the owner. For purposes of modifying susceptibility to flood damage, isolated units may be relocated from within the floodplain or flood-hazard zone to safer areas with appropriate compensation for resettlement.

Easements.   An easement can be used as an alternative to the outright purchase of properties located in the floodplain. Using the easement process, several rights normally associated with land ownership can be purchased to accomplish land-use objectives while leaving title with the owner. The easement approach offers a form of equity that is not possible through zoning or other forms of regulation by government.

Tax policies.   Tax policies can be used to encourage property owners to use flood-prone lands in a manner consistent with a proposed land-use plan. This may include tax assessment that encourages low-density or no development versus the usual market-value assessment approach which considers the potential for development and the sale price of similar properties.

Transferable development rights.   Ownership of property commonly includes the right to develop and sell the land. In communities that adopt transferable development rights (TDR), development is restricted in some parts of the jurisdiction but compensation is made in the form of development rights, which may be sold to property owners in other parts of the community who wish to exceed the usual zoning limits on density. The basic purpose of TDR is to provide a mechanism for equitable treatment of those landowners whose rights may be restricted as a result of inequities embodied in zoning.

7.6.2 Modification of Flooding

Constructed flood-control facilities are collectively referred to as *structural measures* of flood control, and structural measures taken to modify floods include construction of dams and reservoirs, levees, and floodwalls; channel alterations; onsite detention measures; bridge and culvert modifications; and tidal barriers. A levee is an earth dike that is usually constructed of material excavated near the levee, while a floodwall is usually constructed of concrete and requires less right-of-way than a levee. Levees and floodwalls generally run parallel to watercourses and provide 100% protection until they are overtopped, then they provide no flood protection. In most instances, flood modifications acting alone, without other nonstructural strategies, leave a residual flood-loss potential and may encourage inappropriate uses of the land being protected through an false sense of security.

7.6.3 Modification of the Impact of Flooding

This approach is designed to assist the individual and the community in the preparatory, survival, and recovery phases of floods. Tools include information dissemination, arrangements for spreading the costs of the loss over a period of time, and purposeful transfer of some of the individual's loss to the community through flood emergency and postflood recovery measures.

7.7 Drought Management

Droughts are a normal part of any climate, and government agencies must have plans that reduce the impacts of droughts on society. Drought plans generally include mechanisms to collect and analyze drought-related information, establish criteria for declaring drought emergencies and triggering various response actions to minimize economic stress, environmental losses, and social hardships resulting from droughts. In developing drought plans, it is generally recognized that social, economic, and environmental values often clash as competition for scarce water resources intensifies, and therefore it is essential that all stakeholders be involved in developing and implementing drought plans.

A drought is loosely defined as a prolonged and abnormal moisture deficiency (Palmer, 1965). Droughts are associated with sustained periods of significantly lower rainfall, soil moisture, surface-water storage, streamflow, and ground-water levels. Droughts are typically grouped into three categories: meteorological drought, agricultural drought, and hydrologic drought (Singh, 1992). Droughts begin with a deficit in precipitation that is usually extreme and prolonged relative to the usual climatic conditions; this is called a *meteorological drought*. These conditions commonly produce extended periods of unusually low soil moisture, which adversely affect agriculture and natural plant growth. This is called an *agricultural drought*. As the precipitation deficit continues, stream discharge, lake, wetland, reservoir, and water-table elevations decline to unusually low levels, which is called a *hydrological drought*. Other types of drought, such as *ground-water drought* (Peters et al., 2005), are included in the hydrological drought classification. Because of the sequencing of meteorological, agricultural, and hydrological droughts, a short-term meteorological drought with a duration of 3 to 6 months may cause a minimal agricultural drought and no hydrological drought at all. When meteorological drought conditions abate, drought impacts tend to diminish rapidly in the agricultural sector because of its reliance on soil water, but other impacts may linger much longer because of dependence on stored surface or subsurface supplies. Ground-water users are often the last to be affected by drought and the last to experience a return to normal water levels. The longest recorded period of no rain in the United States is 767 days (3 October 1912 to 8 November 1914) at Bagdad, California; and the longest recorded period of no rainfall in the world is 14 years (October 1903 to January 1918), held by Arica, Chile (Wurbs and James, 2002).

Droughts are frequently measured by their magnitude and duration. The magnitude of a drought is usually measured by the cumulative deficit in some quantity over a defined averaging period, and the duration of a drought is equal to the continuous time interval over which the cumulative deficit exceeds a specified threshold value. The beginning and end of a drought are usually defined by a threshold limit; the drought begins when the cumulative deficit exceeds the threshold limit, and ends when the cumulative deficit falls below the threshold limit. The intensity of a drought is commonly defined as the cumulative deficit during the drought divided by the duration of the drought (Salas, 1993).

Droughts are described statistically by a three-dimensional probability distribution that assigns a return period to droughts with a given magnitude (or intensity) and duration. Methodologies for describing the probability distributions of droughts can be found in Loáiciga (2005), Shiau and Shen (2001), and Dingman (2002).

Regions most subject to droughts are those with the greatest variability in annual rainfall (Ponce, 1989), and studies have shown that regions where the coefficient of variation of annual rainfall exceeds 0.35 are more likely to have frequent droughts (Chow, 1964). High annual-rainfall variability is typical of arid and semiarid regions. Analysis of past records in the United States indicates that the sequence of dry years is not random, and a prolonged drought in the western United States occurs about every 22 years (Chang, 2002).

Droughts are typically characterized by *drought indices*, and a variety of such indices are used in practice. For example, the U.S. Department of Agriculture uses the Palmer Drought Severity Index (Palmer, 1965) to determine when to grant emergency drought assistance, states in the western United States frequently supplement the Palmer Drought Severity Index with the Surface Water Supply Index (Shafer and Dezman, 1982), and National Drought Mitigation Center uses the Standardized Precipitation Index (McKee et al., 1993) to monitor moisture-supply conditions. The most widely used drought index is the Palmer Drought Severity Index (PDSI), described in detail below.

Palmer Drought Severity Index (PDSI)

The Palmer Drought Severity Index (PDSI), originally proposed by Palmer (1965), is the most widely used index to characterize meteorological droughts. The PDSI is based on monthly hydrologic accounting using the following relationship:

$$P = ET + R + Q - L \qquad (7.19)$$

where P is the precipitation (assumed to be rainfall), ET is the evapotranspiration, R is the recharge to the soil moisture, Q is the runoff, and L is the moisture loss from the soil. Palmer (1965) divided the soil into two layers: upper and lower. The upper layer corresponds to the surface soil and is assumed to contain 2.54 cm (1 in.) of available moisture at field capacity. This is the layer into which rain infiltrates and from

TABLE 7.4: Hydrologic Accounting in Central Iowa

1	2	3	4	5	6	7	8	9	10	11	12	13	14
Month	P (cm)	ET_p (cm)	ΔS_s (cm)	ΔS_u (cm)	S_s (cm)	S_u (cm)	S (cm)	R (cm)	L (cm)	ET (cm)	Q (cm)	PR (cm)	PL (cm)
January	2.29	0.03	0	0	2.54	22.86	25.40	0	0	0.03	2.26	0	0.03
February	0.53	0	0	0	2.54	22.86	25.40	0	0	0	0.53	0	0
March	8.18	0.79	0	0	2.54	22.86	25.40	0	0	0.79	7.39	0	0.79
April	2.95	4.14	−1.19	0	1.35	22.86	24.21	0	1.19	4.14	0	0	3.99
May	14.22	9.04	1.19	0	2.54	22.86	25.40	1.19	0	9.04	3.99	1.19	8.28
June	2.62	16.18	−2.54	−9.93	0	12.93	12.93	0	12.47	15.09	0	0	14.81
July	9.07	15.80	0	−3.43	0	9.50	9.50	0	3.43	12.50	0	12.47	8.05
August	4.67	12.29	0	−2.82	0	6.68	6.68	0	2.82	7.49	0	15.90	4.60
September	9.07	10.62	0	−0.41	0	6.27	6.27	0	0.41	9.47	0	18.72	2.79
October	4.95	3.86	1.09	0	1.09	6.27	7.37	1.09	0	3.86	0	19.13	0.97
November	0.66	0.86	−0.20	0	0.89	6.27	7.16	0	0.20	0.86	0	18.03	0.86
December	2.26	0	1.65	0.61	2.54	6.88	9.42	2.26	0	0	0	18.24	0

which evaporation takes place. It is assumed that evapotranspiration takes place at the potential rate from the upper soil layer until all of the available moisture has been removed, and only then can moisture be removed from the lower soil layer. It is assumed there is no further recharge to the underlying (lower) soil until the surface layer has been brought to field capacity. Moisture loss from the lower soil layer depends on the initial moisture content, the potential evapotranspiration, ET_p, and the available water capacity, w_{ac}, of the soil. The following equations are used to account for soil moisture:

$$L_s = \min\left[S'_s, (ET_p - P)\right] \tag{7.20}$$

$$L_u = (ET_p - P - L_s)\frac{S'_u}{w_{ac}}, \qquad \text{provided that} \quad L_u \le S'_u \tag{7.21}$$

where L_s is the moisture loss from the surface layer, S'_s is the available moisture in the surface layer at the start of the month, L_u is the moisture loss from the underlying soil, and S'_u is the available moisture stored in the underlying soil at the start of the month. An accurate accounting for soil moisture is important, since it is assumed that no runoff occurs until both soil layers reach field capacity.

Application of Equation 7.19 to central Iowa for a particularly dry year is illustrated in Table 7.4. Column 2 gives the measured rainfall (P) for each month, column 3 gives the potential evapotranspiration (ET_p) estimated from measured climatic parameters, column 4 gives the incremental moisture storage in the surface soil layer (ΔS_s), column 5 gives the incremental moisture storage in the underlying soil layer (ΔS_u), column 6 gives the total soil moisture stored in the surface soil layer (S_s), column 7 gives the total soil moisture stored in the underlying soil layer (S_u), column 8 gives the total moisture storage in the soil (S), column 9 gives the soil-moisture recharge (R), column 10 gives the soil-moisture loss (L), column 11 gives the evapotranspiration (ET), column 12 gives the runoff (Q), column 13 gives the potential recharge (PR), and column 14 gives the potential loss (PL). An available water capacity, w_{sc}, of 25.40 cm (10 in.) is characteristic of the root zone in central Iowa (Palmer, 1965), with 2.54 cm (1 in.) assigned to the surface layer and 22.86 cm (9 in.) to the lower layer. The year prior to that shown in Table 7.4 was relatively wet in central Iowa, and both soil layers are assumed to have been at field capacity at the end of the previous year. The hydrologic accounting for January through June is as follows:

January. The rainfall is 2.29 cm (column 2) and the potential ET is 0.03 cm (column 3). Since the soil is at field capacity, the actual ET is equal to the potential ET of 0.03 cm (column 11), and the runoff is 2.29 cm − 0.03 cm = 2.26 cm (column 12).

February. The rainfall is 0.53 cm (column 2) and the potential ET is 0 cm (column 3). Since the soil is at field capacity, the actual ET is equal to the potential ET of 0 cm (column 11), and the runoff is 0.53 cm − 0 cm = 0.53 cm (column 12).

March. The rainfall is 8.18 cm (column 2) and the potential ET is 0.79 cm (column 3). Since the soil is at field capacity, the actual ET is equal to the potential ET of 0.79 cm (column 11), and the runoff is 8.18 cm − 0.79 cm = 7.39 cm (column 12).

April. The rainfall is 2.95 cm (column 2) and the potential ET is 4.14 cm (column 3). Since the potential ET exceeds the rainfall amount, then ET occurs from the soil moisture. The ET demand from the soil moisture is equal to 4.14 cm − 2.95 cm = 1.19 cm, and since 2.54 cm of moisture is available in the upper layer, then 1.19 cm is extracted from the surface soil layer to meet the ET demand (column 4), and 0 cm is extracted from the lower soil (column 5), leaving 2.54 cm − 1.19 cm = 1.35 cm stored in the surface layer (column 6). Since 22.86 cm of moisture remains stored in the lower soil layer (column 7), the total soil storage is 1.35 cm + 22.86 cm = 24.21 cm (column 8). Since the rainfall and soil moisture meet the potential ET demand, the recharge is zero (column 9), the total moisture loss is 1.19 cm (column 10), the actual ET is equal to the potential ET of 4.14 cm (column 11), and the runoff is zero (column 12).

May. The rainfall is 14.22 cm (column 2) and the potential ET is 9.04 cm (column 3). Since the rainfall exceeds the potential ET, the excess rainfall fills the upper-layer soil-moisture deficit of 1.19 cm (column 4) and the moisture stored in the upper soil layer returns to 2.54 cm (column 6), and the total moisture recharge is 1.19 cm (column 9). The remainder of the rainfall produces a runoff of 14.22 cm − 9.04 cm − 1.19 cm = 3.99 cm (column 12).

June. The rainfall is 2.62 cm (column 2) and the potential ET is 16.18 cm (column 3). Since the potential ET exceeds the rainfall amount, ET occurs from the soil moisture. The ET demand from the soil moisture is equal to 16.18 cm − 2.62 cm = 13.56 cm, and since 2.54 cm of moisture is available in the upper layer, then all 2.54 cm is extracted from the surface soil layer to meet the ET demand (column 4), and using Equation 7.21 (with L_s = 2.54 cm, S'_u = 22.86 cm, and w_{ac} = 25.4 cm), 9.93 cm is extracted from the lower soil (column 5), leaving 2.54 cm − 2.54 cm = 0 cm stored in the surface layer (column 6), and 22.86 cm − 9.93 cm = 12.93 cm stored in the lower layer (column 7). The total soil storage is therefore 0 cm + 12.93 cm = 12.93 cm (column 8), and the total moisture loss is 2.54 cm + 9.93 cm = 12.47 cm (column 10). The total ET is therefore equal to 2.62 cm + 2.54 cm + 9.93 cm = 15.09 cm (column 11), and since the ET exceeds precipitation, the runoff is equal to zero (column 12).

The computations for July through December follow the same procedure as for January through June.

For each of the water-balance variables in Table 7.4 there is a potential value that characterizes the maximum value of the variable. For example, the potential evapotranspiration, ET_p (in column 3), characterizes the maximum value of the actual evapotranspiration, ET (in column 11). The *potential recharge*, PR (in column 13), is defined as the amount of moisture required to bring the soil to field capacity and is defined by

$$PR = w_{ac} - S' \tag{7.22}$$

where S' is the amount of moisture stored in both soil layers at the beginning of the month. The *potential loss*, PL (in column 14), is defined as the amount of moisture that could be lost from the soil provided that the precipitation during the period were zero, and is defined by

$$PL = PL_s + PL_u \tag{7.23}$$

where

$$PL_s = \min[ET_p, S'_s] \tag{7.24}$$

$$PL_u = (ET_p - PL_s)\frac{S'_u}{w_{ac}} \tag{7.25}$$

The *potential runoff*, PQ, is defined as the maximum runoff expected, assuming the precipitation to be equal to the soil-moisture storage capacity, w_{ac}, hence

$$PQ = w_{ac} - PR = S' \tag{7.26}$$

Potential values for all hydrologic variables provide reference values relative to which actual values of the hydrologic variables can be measured. These potential values are estimated for each month in Table 7.4 (columns 13 and 14). Averaged values of the potential and actual hydrologic variables for central Iowa for the 1931–1957 period are given in Table 7.5 (Palmer, 1965). On the basis of these results, the ratios of actual

TABLE 7.5: Mean Hydrologic Variables in Central Iowa

Month	$\overline{P}$ (cm)	$\overline{ET}$ (cm)	$\overline{ET_p}$ (cm)	$\overline{R}$ (cm)	$\overline{PR}$ (cm)	$\overline{Q}$ (cm)	$\overline{S'}$ (cm)	$\overline{L}$ (cm)	$\overline{PL}$ (cm)
January	2.95	0	0	1.47	6.38	1.47	19.02	0	0
February	2.64	0.05	0.05	1.24	4.90	1.37	20.50	0	0.05
March	5.18	0.66	0.66	1.14	3.66	3.38	21.74	0	0.66
April	6.55	4.34	4.37	0.71	2.51	2.11	22.89	0.58	3.99
May	10.39	8.86	9.12	0.66	2.41	2.29	22.99	1.45	7.87
June	12.85	12.42	13.18	0.23	3.18	1.98	22.23	1.80	10.77
July	8.74	14.12	15.57	0	4.75	0	20.65	5.38	11.86
August	9.53	11.18	13.36	1.04	10.06	0.05	15.34	2.72	7.95
September	8.20	7.82	8.94	1.88	11.76	0	13.64	1.50	4.60
October	5.31	4.52	5.00	1.14	11.35	0.33	14.05	0.69	2.87
November	4.75	0.76	0.76	3.45	10.90	0.58	14.50	0.05	0.53
December	2.92	0	0	1.85	7.52	1.07	17.88	0	0

to potential values of the hydrologic variables are defined as follows:

$$\text{coefficient of evapotranspiration, } \alpha = \frac{\overline{ET}}{\overline{ET_p}} \tag{7.27}$$

$$\text{coefficient of recharge, } \beta = \frac{\overline{R}}{\overline{PR}} \tag{7.28}$$

$$\text{coefficient of runoff, } \gamma = \frac{\overline{Q}}{\overline{PQ}} = \frac{\overline{Q}}{\overline{S'}} \tag{7.29}$$

$$\text{coefficient of loss, } \delta = \frac{\overline{L}}{\overline{PL}} \tag{7.30}$$

where the overbar indicates the monthly average. These coefficients give a measure of the expected value of the hydrologic variables in each month relative to their potential values for the month. A key point to note here is that the potential value of a hydrologic variable for any month may vary based on unusual weather or soil-moisture conditions occurring during that month and, assuming a constant coefficient, the expected value of the hydrologic variable will vary in proportion. The expected value of the hydrologic variable is commonly referred to as "climatically appropriate for existing conditions" (CAFEC), and the values of the coefficients given by Equations 7.27 to 7.30 for central Iowa are shown in Table 7.6. Using these coefficients and the actual values of the hydrologic variables for any month, the CAFEC quantities (denoted by a circumflex) for evapotranspiration, recharge, runoff, loss, and precipitation are given by

$$\widehat{ET} = \alpha ET_p \tag{7.31}$$

$$\widehat{R} = \beta PR \tag{7.32}$$

$$\widehat{Q} = \gamma PQ \tag{7.33}$$

$$\widehat{L} = \delta PL \tag{7.34}$$

$$\widehat{P} = \widehat{ET} + \widehat{R} + \widehat{Q} - \widehat{L} \tag{7.35}$$

TABLE 7.6: Climate Coefficients and Constants in Central Iowa

Month	α	β	γ	δ	D (cm)	K
January	1.00	0.2315	0.0776	0	1.85	1.55
February	1.00	0.2530	0.0670	0.1538	1.68	1.61
March	1.00	0.3129	0.1554	0	2.29	1.41
April	0.9968	0.2804	0.0919	0.1495	3.53	1.14
May	0.9727	0.2790	0.0996	0.1835	4.55	0.97
June	0.9425	0.0709	0.0897	0.1677	4.83	0.94
July	0.9081	0	0	0.4535	2.84	1.28
August	0.8357	0.1027	0.0028	0.3418	5.05	0.93
September	0.8738	0.1603	0	0.3246	4.22	1.04
October	0.9035	0.1012	0.0244	0.2420	2.84	1.28
November	1.00	0.3171	0.0408	0.1150	3.38	1.16
December	1.00	0.2453	0.0603	0	1.88	1.52

The difference between the actual precipitation, P, and the CAFEC precipitation, $\widehat{P}$, for each month provides a meaningful measure of the departure of the actual precipitation from the expected precipitation, and this departure, d, is given by

$$d = P - \widehat{P} \tag{7.36}$$

A given moisture departure, d, does not have the same significance at all locations. Palmer (1965) suggested that moisture departures be weighed according the ratio of moisture demand to supply at a particular location. Defining such a weighting factor for month i, K_i, as the *climatic characteristic*, then a *moisture anomaly index* for month i, z_i, is defined as

$$z_i = K_i d_i \tag{7.37}$$

where d_i is the precipitation departure for month i. The moisture anomaly index defined by Equation 7.37 expresses on a monthly basis the weighted departure of the weather of the month from the average moisture climate of the month. Based on comparing drought impacts at a variety of locations within the United States, Palmer (1965) recommended the following equation for estimating K_i:

$$K_i = \frac{44.88}{\sum_{i=1}^{12} \overline{D}_i K_i'} K_i' \tag{7.38}$$

where

$$K_i' = 1.5 \log_{10} \left[\frac{\frac{\overline{ET_{pi}} + \overline{R}_i + \overline{Q}_i}{\overline{P}_i + \overline{L}_i} + 2.80}{\overline{D}_i} \right] + 1.11 \tag{7.39}$$

where the i subscripts denote the month, and D_i is the mean of the absolute values of d_i (in cm) for the period of record. Values of K_i for central Iowa using Equation 7.39 are given in Table 7.6. The severity of a drought at any time is determined by the sum of the moisture anomaly indices for previous months, and Palmer (1965) defined the *drought severity index* for month i, X_i, based on the cumulative moisture anomaly index, z, by

$$\boxed{X_i = 0.897 X_{i-1} + \frac{z_i}{3}} \tag{7.40}$$

where the series is started by taking

$$X_1 = \frac{z_1}{3} \tag{7.41}$$

The Palmer drought severity classification for month i is related to the drought severity index, X_i, according to Table 7.7.

The importance of the Palmer Drought Severity Index (PDSI) is evidenced by the many U.S. government agencies and states that rely on it to trigger drought relief programs. The distribution of the PDSI in the United States is updated weekly and can be found at:

`http://lwf.ncdc.noaa.gov/oa/climate/research/dm/pdi.html`

TABLE 7.7: Drought Classifications Based on Palmer Drought Severity Index

Index, X	Drought Classification	Possible Impacts
≥ 4.00	Extremely wet	—
3.00 to 3.99	Very wet	—
2.00 to 2.99	Moderately wet	—
1.00 to 1.99	Slightly wet	—
0.50 to 0.99	Incipient wet spell	—
0.49 to -0.49	Near normal	—
-0.50 to -0.99	Incipient dry spell	—
-1.00 to -1.99	Mild drought	**Going into drought:** Short-term dryness slowing planting, growth of crops or pastures; fire risk above average. **Coming out of drought:** Some lingering water deficits, pastures or crops not fully recovered.
-2.00 to -2.99	Moderate drought	Some damage to crops, pastures; fire risk high; streams, reservoirs, or wells low, some water shortages developing or imminent, voluntary water-use restrictions requested.
-3.00 to -3.99	Severe drought	Crop or pasture losses likely; fire risk high; water shortages common; water restrictions imposed.
≤ -4.00	Extreme drought	Major crop/pasture losses; extreme fire danger; widespread water shortages or restrictions.

From a scientific viewpoint, it should be kept in mind that the PDSI is less well suited for mountainous land, since it does not include snowfall, snow cover, and frozen ground.

EXAMPLE 7.11

In western Kansas, analysis of historical hydrologic data indicates that the climatic coefficients and constants for the month of July are as follows:

Parameter	Value*
Coefficient of evapotranspiration, α	0.5660
Coefficient of recharge, β	0.0071
Coefficient of runoff, γ	0
Coefficient of loss, δ	0.5151
Climatic characteristic, K_{July}	1.38

*For July.

During July 2004, rainfall was 4.67 cm and potential evapotranspiration was 17.55 cm. Soils in western Kansas are predominantly of Colby series with an available water capacity of 15.24 cm, of which 2.54 cm is in the surface layer and 12.70 cm in the lower layers. At the end of June, moisture accounting indicated that the surface soil had

0.50 cm of stored moisture, and the underlying soil was at the field capacity. If the Palmer Drought Severity Index at the end of June was -0.90, characterize the drought conditions at the end of July.

Solution   For the month of July, $P = 4.67$ cm, $ET_p = 17.55$ cm, $S_s' = 0.50$ cm, $S_u' = 12.70$ cm, and $w_{ac} = 15.24$ cm. These data indicate that $ET_p \gg P$, all of the July rainfall evaporates, moisture is extracted from the soil layers, and the runoff is zero. The potential recharge, PR, is given by Equation 7.22 as

$$PR = w_{ac} - S'$$

$$= 15.24 \text{ cm} - (0.50 \text{ cm} + 12.70 \text{ cm}) = 2.04 \text{ cm}$$

the potential runoff, PQ, is given by Equation 7.26 as

$$PQ = w_{ac} - PR = S'$$

$$= (0.50 \text{ cm} + 12.70 \text{ cm}) = 13.20 \text{ cm}$$

and the potential loss, PL, is given by Equations 7.23 to 7.25 as

$$PL = \min[ET_p, S_s'] + (ET_p - PL_s)\frac{S_u'}{w_{ac}}$$

$$= \min[17.55 \text{ cm}, 0.50 \text{ cm}] + (17.55 \text{ cm} - 0.50 \text{ cm})\frac{12.70 \text{ cm}}{15.24 \text{ cm}} = 14.71 \text{ cm}$$

The calculated potential values ($PR = 2.04$ cm, $PQ = 13.20$ cm, and $PL = 14.71$ cm) are combined with the given coefficients of evapotranspiration ($\alpha = 0.5660$), recharge ($\beta = 0.0071$), runoff ($\gamma = 0$), and loss ($\delta = 0.5151$) to calculate the CAFEC precipitation for July, $\widehat{P}_{July}$, using Equations 7.31 to 7.35, which yield

$$\widehat{P}_{July} = \alpha ET_p + \beta PR + \gamma PQ - \delta PL$$

$$= 0.5660(17.55 \text{ cm}) + 0.0071(2.04 \text{ cm}) + 0(13.20 \text{ cm}) - 0.5151(14.71 \text{ cm})$$

$$= 2.37 \text{ cm}$$

Hence, the departure, d_{July}, of the July rainfall from expected conditions is given by Equation 7.36 as

$$d_{July} = P - \widehat{P}_{July}$$

$$= 4.67 \text{ cm} - 2.37 \text{ cm} = 2.30 \text{ cm}$$

and the moisture anomaly index, z_{July}, is given by Equation 7.37 as

$$z_{July} = K_{July}d_{July}$$

$$= 1.38(2.30 \text{ cm}) = 3.17 \text{ cm}$$

Since the drought severity index at the end of June, X_{June}, is -0.90, the index at the end of July, X_{July}, is given by Equation 7.40 as

$$X_{July} = 0.897X_{June} + \frac{z_{July}}{3}$$

$$= 0.897(-0.90) + \frac{3.17 \text{ cm}}{3} = 0.25$$

A drought index of 0.25 indicates "near normal" conditions according to the Palmer drought classifications given in Table 7.7. Therefore, conditions have changed from an "incipient dry spell" ($X = -0.90$) at the end of June to an "near normal" at the end of July.

7.8 Irrigation

Irrigation is the artificial application of water for crop production. Irrigation water is usually obtained from either surface water or ground water. Melted snow can be captured in reservoirs or diverted directly from a river for irrigation.

Ground water is often used for irrigation if the depth to the water table is not excessive (≤ 100 m) and aquifer transmissivity is adequate to provide sufficient quantities of water. If the depth of the water table is too high, pumping costs may be excessive. Ground water is used almost exclusively as the source of irrigation water east of the 100th meridian in the United States, while surface water is the major irrigation source in the western part of the country.

Surface-water irrigators use a variety of methods to deliver water to crops, and these methods generally include diverting water from a river or reservoir into a *delivery canal*. Irrigation delivery canals vary greatly in size, ranging from widths and depths on the order of 1 m to extensive systems like the All-American Canal in the Imperial Valley of southern California, which is 61 m wide and 6 m deep at some locations (Cech, 2002). Water is diverted from a delivery canal to a farm through a *headgate*, which typically consists of a metal structure with a vertically sliding gate that can be raised or lowered. An open headgate allows irrigation water to flow from a delivery canal into a *farm ditch*, which is usually 0.6 to 0.9 m wide and carries water to individual farm fields.

Methods used in irrigation can be categorized as gravity systems, sprinkler systems, or drip-irrigation systems.

Gravity Irrigation. The two primary gravity methods that are used to apply irrigation water to farm fields are *furrow irrigation* and *wild-flood irrigation*. Furrow irrigation is a gravity method that utilizes crop rows to convey irrigation water across a field. Furrow widths are usually between 61 and 107 cm to allow adequate room for a tractor tire to fit between a crop row without crushing plants. Typical row crops include corn, soy and pinto beans, sugar beets, and a wide variety of vegetables. Wild-flood irrigation is used on non-row crops such as alfalfa, oats, and wheat, where water is released from a farm ditch into channels or between low ridges that are carved at regular intervals across the field. Common gravity-irrigated field lengths are 200 to 400 m long due to infiltration of applied irrigation water as it moves from one side of the field to the other (Cech, 2002).

Sprinkler Irrigation. Sprinkler irrigation opened the way to irrigate lands that were unsuitable for gravity irrigation. Sandy soils, land with excessive slope, and regions where surface-water supplies are inadequate (but ground water is plentiful) can be irrigated using sprinklers where gravity irrigation is not an option. Sprinkler systems can be stationary, such as in a fruit orchard, or may involve long towers that move across fields on wheels. Nonstationary sprinkler systems include lateral and pivot systems. *Lateral sprinkler systems* are sprinkler

FIGURE 7.7: Lateral sprinkler system

Source: David A. Chin.

machines that move suspended irrigation pipes across a field (see Figure 7.7), while in pivot-irrigation systems the irrigation system rotates around a fixed pivot point.

Center-pivot circles can be seen from the air and are common in Nebraska, eastern Colorado, Arizona, Kansas, Oklahoma, the Texas panhandle, northwest Mississippi, and numerous countries in the Middle East.

Drip Irrigation. Drip irrigation is a relatively new technology that is gaining wider acceptance as the cost of irrigation water increases. Drip irrigation is used primarily on fruit and vegetable crops because the economic return is much higher for these than for crops such as corn or beans. Rigid plastic pipes are either laid on the surface or buried below ground in each crop row at the root line. Small holes allow water to drip from the pipe directly to each plant root system. The amount of water applied is regulated by the hole size in the distribution tubes and the water pressure in the system. Although installation of drip systems is very expensive, evaporation rates are greatly reduced when compared to gravity and sprinkler irrigation. Efficiencies can reach 90 percent for drip systems.

Water loss due to inefficient irrigation is an important consideration in water conservation. In many cases, as much as 50% of the irrigation water is not used by crops due to evaporation loss or seepage. Improving the efficiency of irrigation practices can increase crop yields, reduce weed growth, and cut irrigation water by 50% to 60% (Chang, 2002). Techniques to improve efficiency include: better scheduling of water application, use of closed conduits in water-conveyance systems, use of trickle or drip-irrigation systems, reducing irrigation water runoff by using contour cultivation and terracing, and covering the soil with mulch.

7.9 Dams and Reservoirs

Reservoir impoundments are created by building dams, and these impoundments (reservoirs) typically serve multiple purposes, including water supply, flood control,

FIGURE 7.8: Principal parts of a dam
Source: Cech (2002).

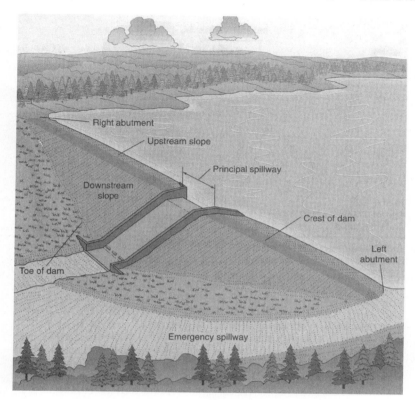

hydroelectric power generation, navigation, and water-based recreation activities. In many cases, reservoirs are needed to augment natural systems during periods of low flow, and to regulate the distribution surface-water flows and volumes during periods of high flow. Dams and reservoirs have enhanced the health and economic prosperity of people around the world for centuries; however, dam construction can also have negative effects such as reduced streamflows, degraded water quality, and impacts on migrating fish.

The principal parts of a dam are illustrated in Figure 7.8. The *face* is the exposed surface of the structure, which contains materials such as rockfill, concrete, or earth. There are both an upstream and a downstream face. *Abutments* are the sides of the dam structure that tie into canyon walls or contact the far sides of a river valley. Left and right abutments are identified by facing downstream from the reservoir. The top of the dam is called the *crest*, while the point of intersection between the downstream face and the natural ground surface is called the *toe*. The crest usually has a road or walkway along the top that follows the entire length of the dam, where the length is measured from abutment to abutment. A parapet wall is often built along the dam crest for ornamental, safety, or wave-control purposes. All dams have a spillway to allow excess water to flow past the dam in a safe manner. Dams on rivers used for navigation often include *locks*, which are rectangular box-like structures with gates at both ends that allow vessels to move upstream and downstream through the dam. The highest lock in the United States is the John Day lock on the Columbia River with a lift of 34.5 meters.

FIGURE 7.9: Basic dam designs
Source: Cech (2002).

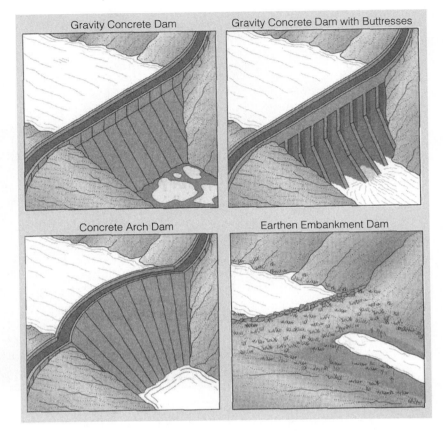

7.9.1 Types of Dams

The four primary types of dams are: gravity concrete, buttress, concrete arch, and earthen embankment dams, as illustrated in Figure 7.9.

Gravity Concrete Dam. A *gravity concrete dam* is a solid concrete structure that uses its mass (weight) to hold back water. It requires massive amounts of concrete, especially at its base, to provide the weight necessary to withstand the hydrostatic force exerted by the water impounded behind the dam. An example of a gravity concrete dam is the Grand Coulee Dam on the Columbia River. The Grand Coulee Dam has a height of 168 m, a crest length of 1730 m, a spillway discharge capacity of 28,300 m³/s, and a hydroelectric power generation capacity of 6.48 GW. It is operated by the U.S. Bureau of Reclamation and impounds the Franklin D. Roosevelt reservoir (Lake Roosevelt), with a storage capacity of 6.40×10^9 m³, which extends 240 km upstream to the Canadian border. The Grand Coulee Dam is the largest concrete structure and hydroelectric facility in the United States, and is designated as a National Historic Civil Engineering Landmark by the American Society of Civil Engineers.

Buttress Dam. A gravity concrete dam may have triangular supports called *buttresses* on the downstream side to strengthen it and to distribute water pressure to the foundation. Such dams are commonly placed in a separate category from concrete gravity dams and are called *buttress dams*. Typical configurations for buttress dams include the flat slab and multiple arch. The face of a flat-slab buttress dam is a series of flat reinforced concrete slabs, while the face of a multiple-arch dam consists of a series of arches that permit wider spacing of the buttresses. Buttress dams usually require only one-third to one-half as much concrete as gravity dams of similar height. Consequently, they may be used on foundations that are too weak to support a gravity dam.

Concrete Arch Dam. A *concrete arch dam* has a curvature design that arches across a canyon and has abutments embedded into solid rock walls. Hydrostatic forces push against the curve of the concrete arch and compress the material inside the structure. This pressure actually makes the dam more solid and dissipates water pressure into the canyon walls. Arch dams require less concrete but must have solid rock walls as anchors for the abutments. Arch dams have thinner sections than comparable gravity dams, and are feasible only in canyons with walls capable of withstanding the thrust produced by the arch action. An example of a concrete arch dam is the Hoover Dam on the Colorado River. The Hoover Dam has a height of 221 m, a crest length of 379 m, a spillway discharge capacity of 11,300 m^3/s, and a hydroelectric power generation capacity of 1.34 GW. It is operated by the U.S. Bureau of Reclamation and impounds Lake Mead, which is the largest reservoir in the United States with a storage capacity of 3.67×10^{10} m^3 and a surface area of 658 km^2. The Hoover Dam is the highest concrete dam in the Western Hemisphere and is designated as a National Historic Civil Engineering Landmark by the American Society of Civil Engineers.

Earthen Embankment Dam. More than 50% of the total volume of an *earthen embankment dam* or *earthfill dam* consists of compacted earth material. Large earthen dams have an impervious core of clay, or other material of low permeability, that prevents reservoir water from rapidly seeping through or beneath the foundation of the structure. Most large earthen dams also have drains installed along the downstream toe of the dam. These small parallel conduits transport any seepage water inside the earth dam to locations downstream and away from the toe of the dam. This prevents water from building up inside the structure and eventually eroding the material within the dam. The base of an earthen dam is very broad when compared to the crest of the dam. A 2 : 1 (H : V) downstream slope and a 3 : 1 (H : V) upstream slope are fairly common. The upstream face of the dam is often covered with riprap to prevent erosion by wave action. Earthen embankment dams are usually the most economical to build in small watersheds or across broad valleys. Almost 80% of all dams in the world are earthen dams (Cech, 2002). An example of an earthen embankment dam is the Somerville Dam in Texas on Yequa Creek, a tributary of the Brazos River. The Somerville Dam has a height of 24 m, a crest length of 6160 m, and impounds the Somerville Reservoir having a storage capacity of 6.26×10^8 m^3.

In addition to the above classifications, dams may also be categorized as overflow or nonoverflow dams. Overflow dams are concrete structures designed for water to flow over their crests, while nonoverflow dams have spillways to prevent overtopping. Earthen embankment dams are damaged by the erosive action of overflowing water

FIGURE 7.10: Reservoir storage zones
Source: Loucks (1976).

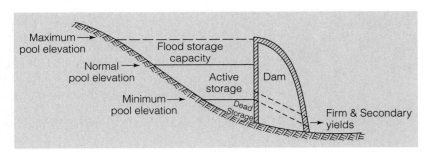

and are generally designed as nonoverflow dams. The geology, topography, and streamflow at a site generally dictate the type of dam that is appropriate for a particular location.

7.9.2   Reservoir Storage

The three major components of reservoir storage are shown in Figure 7.10. These components are: (1) *dead storage* for sediment collection; (2) *active storage* or *conservation storage* for firm and secondary yields; and (3) *flood storage* for reduction of downstream flood damages. *Firm yield*, which is also called *safe yield*, can be defined as the maximum amount of water that will be consistently available from a reservoir based on historical streamflow records, whereas *secondary yield* is any amount greater than firm yield.

Dead/Inactive Storage. Dead or inactive storage is equal to the capacity at the bottom of the reservoir that is reserved for sediment accumulation during the anticipated life of the reservoir. The top of the inactive pool elevation may be fixed by the invert of the lowest outlet, or by conditions of operating efficiency for hydroelectric turbines. Typically, the dead-storage capacity is lost to sedimentation gradually over many decades. However, in extreme cases, small reservoirs have been almost completely filled with sediment during a single major flood event.

Active Storage. Active storage is the storage between the top of the dead storage pool and the normal reservoir water-surface elevation. The active storage capacity is sometimes referred to as the *active conservation pool* or *conservation storage*. This storage is available for various uses such as water supply, irrigation, navigation, recreation, and hydropower generation. The reservoir water surface is maintained at or near the designated top of the conservation pool level as streamflows and water demands allow. Drawdowns are made as required to meet the various needs for water.

Flood Storage. Flood storage is the storage between the top of the conservation pool and the maximum permissible water level in the reservoir. This storage is reserved for flood control and is usually kept unused or empty most of the time to be used for temporary storage of floodwater during storm events. The flood-storage volume includes the *flood-control pool* and *surcharge capacity*. Under normal operating conditions surface runoff is handled within the flood-control pool. There is an additional surcharge capacity, which is the storage available between the normal flood-control pool and the maximum permissible water level in the reservoir. The surcharge capacity is not used except during abnormal conditions, and is reserved for extreme flood events that are larger than the design basis flood. The crest of the emergency spillway is typically located at the top of the flood-control pool, with normal flood releases being made through other outlet structures. The top of the surcharge storage is established during project design from the perspective of dam safety.

Freeboard. The *freeboard* is the difference in elevation between the dam crest and the maximum permissible water level in the reservoir. This allowance provides for wave action and an additional factor of safety against overtopping.

For many reservoir projects, a full range of outflow rates are discharged through a single spillway. However some reservoirs have more than one spillway, with a *service spillway* conveying smaller, frequently occurring releases, and an *emergency spillway* that is used only rarely during extreme floods. The crest elevation of the service spillway is typically located at the top of the conservation pool, and the crest of the emergency spillway located at the top of the flood-control pool. The detailed design of spillways is covered in Section 3.4.3. *Outlet works* are used for releases from storage both below and above the spillway crest; however, discharge capacities for outlet works are typically much smaller than for spillways. The components of an outlet-works facility include an intake structure in the reservoir, one or more conduits or sluices through the dam, gates located either in the intake structure or conduits, and a stilling basin or other energy-dissipation structure at the downstream end.

The storage zones illustrated in Figure 7.10 present a simplified view of reservoir capacity, since sedimentation storage capacity must generally be provided in all storage zones. Typically, sediment reserve storage capacity is provided to accommodate sediment deposition expected to occur over a specified design life which, for large projects, is typically on the order of 50–100 years. Reservoir sedimentation amounts are predicted as the sediment yield entering the reservoir multiplied by a *trap efficiency*. The reservoir trap efficiency is a measure of the proportion of the inflowing sediment that is deposited, where

$$\text{trap efficiency (\%)} = \frac{\text{sediment amount deposited}}{\text{sediment amount entering}} \times 100 \tag{7.42}$$

Analyses of sediment measurements for a number of reservoirs resulted in Figure 7.11, which may be used to estimate the trap efficiency as a function of the ratio of the reservoir storage capacity to the average annual inflow for normal-ponded reservoirs (USBR, 1987).

The reservoir trap-efficiency curve shown in Figure 7.11 was originally developed by Brune (1953) and has been widely used and revised (Heinemann, 1981). The relation shown in Figure 7.11 gives only a general estimate of the trap efficiency, because there is little consideration of reservoir flow dynamics (Ward and Trimble, 2004).

EXAMPLE 7.12

A reservoir covers an area of 400 km² and has an average depth of 24.8 m. The inflow to the reservoir is from a river with an average flowrate of 3000 m³/s and a suspended-sediment concentration of 150 mg/L. Estimate the rate at which sediment is accumulating in the reservoir, the rate at which the depth of the reservoir is decreasing due to sediment accumulation, and the average suspended sediment concentration in the water released from the reservoir. Assume that the accumulated sediment has a bulk density of 1600 kg/m^3.

Solution   From the given data, the average flowrate into the reservoir is 3000 m^3/s = 9.46 × 10^{10} m^3/year, and the average suspended-sediment concentration is 150 mg/L = 0.150 kg/m^3, therefore the average sediment load entering the reservoir is given by

$$\text{sediment load} = \text{inflow rate} \times \text{suspended-sediment concentration}$$

$$= 9.46 \times 10^{10}\, \frac{\text{m}^3}{\text{year}} \times 0.150\, \frac{\text{kg}}{\text{m}^3} = 1.42 \times 10^{10}\ \text{kg/year}$$

The area of the reservoir is 400 km^2 = 4 × 10^8 m^2, and the average depth of the reservoir is 24.8 m, therefore the reservoir storage capacity is given by

$$\text{storage capacity} = \text{area of reservoir} \times \text{average depth}$$

$$= 4 \times 10^8\ \text{m}^2 \times 28.4\ \text{m} = 9.92 \times 10^9\ \text{m}^3$$

and

$$\frac{\text{storage capacity}}{\text{annual inflow}} = \frac{9.92 \times 10^9\ \text{m}^3}{9.46 \times 10^{10}\ \text{m}^3} = 0.10$$

Based on this ratio of storage capacity to annual inflow (=0.10), the percent of sediment trapped in the reservoir is estimated from Figure 7.11 as 87%. Since the average sediment load delivered by the river to the reservoir is 1.42 × 10^{10} kg/year, the rate at which sediment is accumulating in the reservoir is 0.87 × 1.42 × 10^{10} kg/year = 1.24 × 10^{10} kg/year.

Since the bulk density of the sediment accumulating at the bottom of the reservoir is 1600 kg/m^3, and the area of the reservoir is 400 km^2 = 4 × 10^8 m^2, the

FIGURE 7.11: Reservoir trap efficiency

Source: USBR (1987).

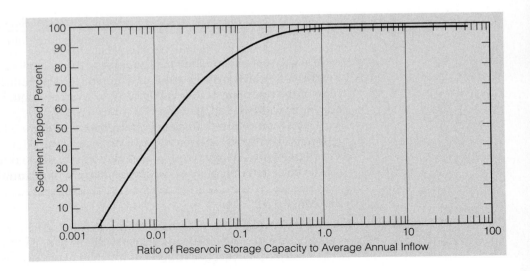

rate at which sediment volume is accumulating is given by

$$\text{sediment volume accumulation rate} = \frac{\text{sediment trap rate}}{\text{sediment bulk density}}$$

$$= \frac{1.24 \times 10^{10} \text{ kg/year}}{1600 \text{ kg/m}^3} = 7.75 \times 10^6 \text{ m}^3/\text{year}$$

Since the plan area of the reservoir is $400 \text{ km}^2 = 4 \times 10^8 \text{ m}^2$, the rate of sediment accumulation on the bottom of the reservoir is given by

$$\text{rate of sediment accumulation} = \frac{\text{sediment volume accumulation rate}}{\text{reservoir area}}$$

$$= \frac{7.75 \times 10^6 \text{ m}^3/\text{year}}{4 \times 10^8 \text{ m}^2} = 0.019 \text{ m/year} = 1.9 \text{ cm/year}$$

At this rate, it will take approximately 1300 years for the reservoir capacity to decrease by 10% due to sediment accumulation.

Since 87% of the incoming sediment is trapped by the reservoir, and the sediment load delivered by the river to the reservoir is 1.42×10^{10} kg/year, the sediment load released from the reservoir is $(1 - 0.87)(1.42 \times 10^{10})$ kg/year $= 5.36 \times 10^9$ kg/year. Assuming that the average flowrate of water released from the reservoir is equal to the average flowrate entering the reservoir ($=9.46 \times 10^{10}$ m^3/year), then

$$\text{sediment concentration downstream} = \frac{\text{sediment load release from reservoir}}{\text{flowrate from reservoir}}$$

$$= \frac{5.36 \times 10^9 \text{ kg/year}}{9.46 \times 10^{10} \text{ m}^3/\text{year}} = 0.057 \text{ kg/m}^3 = 57 \text{ mg/L}$$

Therefore, the reservoir reduces the suspended-sediment concentration from 150 mg/L to 57 mg/L, a reduction of 62%. It is interesting to note that the reservoir trap efficiency ($=87\%$) is not equal to the reduction in suspended-solids concentration ($=62\%$).

A relatively simple method for determining reservoir storage requirements is based on a mass diagram, which shows the cumulative reservoir inflows and demands plotted against time. Water demand includes water withdrawn directly from the reservoir, required downstream releases, and evaporation of water stored in the reservoir. When the cumulative differences between inflow and demand are plotted on a mass diagram, the maximum vertical distance between the highest point of the cumulative difference curve and the lowest subsequent point represents the required capacity. This method, sometimes referred to as a *mass-curve analysis* or *Rippl analysis* (Rippl, 1883), assumes that future inflows to a reservoir will be a duplicate of the historical record. Commonly, storage calculations are based on inflows during a critical low-flow period such as the most severe drought of record. Once the critical period is chosen, the required storage is determined using the mass-curve analysis (Viessman and Hammer, 2005). Reservoirs in humid regions tend to refill annually and function principally to smooth out intraannual (seasonal) fluctuations in flows. In more arid regions, additional carry-over storage is required to smooth out interannual variations, and reservoirs in these regions fill only rarely.

EXAMPLE 7.13

A reservoir is to be sized to stabilize the flows in a river and provide a dependable source of irrigation water. During a critical 24-month drought, the average monthly inflows to the reservoir and the average water demand for downstream flow, irrigation, and evaporation are given in the following table:

Month*	Average inflow (m^3/s)	Average demand (m^3/s)	Month	Average inflow (m^3/s)	Average demand (m^3/s)	Month	Average inflow (m^3/s)	Average demand (m^3/s)
1	1450	1600	9	2150	1900	17	1950	2000
2	1730	1700	10	2200	1800	18	1870	2000
3	1910	1800	11	1970	1700	19	1720	2000
4	1600	1900	12	1810	1600	20	1830	2000
5	1770	2000	13	1540	1600	21	1920	1900
6	1860	2000	14	1420	1700	22	1850	1800
7	1480	2000	15	1640	1800	23	1840	1700
8	1530	2000	16	1890	1900	24	1680	1600

*Month "1" is January.

Use these data to estimate the required storage volume of the reservoir.

Solution From the given data, the cumulative inflow, demand, and difference between inflow and demand are tabulated below, and a plot of cumulative inflow minus demand versus time is shown.

Month	Cumulative inflow $(\times 10^{10} \ m^3)$	Cumulative demand $(\times 10^{10} \ m^3)$	Cumulative inflow − demand $(\times 10^{10} \ m^3)$
1	0.39	0.43	−0.04
2	0.81	0.84	−0.03
3	1.32	1.32	0.00
4	1.73	1.81	−0.08
5	2.21	2.35	−0.14
6	2.69	2.87	−0.18
7	3.09	3.40	−0.32
8	3.50	3.94	−0.44
9	4.05	4.43	−0.38
10	4.64	4.91	−0.27
11	5.15	5.36	−0.20
12	5.64	5.78	−0.15
13	6.05	6.21	−0.16
14	6.39	6.62	−0.23
15	6.83	7.11	−0.27
16	7.32	7.60	−0.28
17	7.84	8.13	−0.29
18	8.33	8.65	−0.32
19	8.79	9.19	−0.40

Month	Cumulative inflow ($\times 10^{10}$ m^3)	Cumulative demand ($\times 10^{10}$ m^3)	Cumulative inflow–demand ($\times 10^{10}$ m^3)
20	9.28	9.72	−0.44
21	9.78	10.22	−0.44
22	10.27	10.70	−0.42
23	10.75	11.14	−0.39
24	11.20	11.57	−0.37

The plotted data in Figure 7.12 indicate that the reservoir storage must be sufficient to accommodate the cumulative deficit between demand and reservoir inflow that occurs between Month 3 and Month 21, and this deficit is equal to 0.44×10^{10} m^3. Assuming that the reservoir is full at the beginning of this critical interval, then an active reservoir storage of 0.44×10^{10} m^3 will be sufficient. If the reservoir is assumed to be partially full at the beginning of the critical interval between Months 3 and 21, then the unfilled volume must be added to the deficit volume (0.44×10^{10} m^3) to determine the required storage volume of the reservoir.

7.9.3 Planning Guidelines

Recommended steps in the planning and investigation of dam and reservoir projects are as follows (Prakash, 2004):

1. **Identification of project objectives including approximate magnitudes.** As an example, project objectives could include water supply for a specified

FIGURE 7.12: Cumulative difference between inflow and outflow volumes

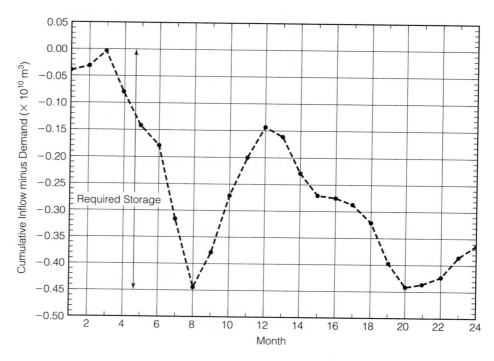

community or agricultural area, hydropower generation to meet a specified demand, and flood control for a specified community.

2. **Selection of a dam and reservoir site.** This is conducted by a multidisciplinary team, usually including a water-resources engineer, geologist, geotechnical engineer, and community leader. Additional members may be added, depending on specific circumstances. Generally, more than one site is identified on topographic maps, and the preliminary choice is narrowed down after field visits. Items to be observed include:

   - Suitability of foundations for an earth, earth and rockfill, or gravity dam.
   - Existence of a relatively narrow valley to avoid unduly large embankments and inundated areas.
   - Suitability of the reservoir bottom to hold the anticipated volume of water without undue seepage losses.
   - Availability of construction materials within reasonable hauling distances (e.g., embankment fill materials, rockfill, concrete aggregate).
   - Proximity to the service area (e.g., agricultural area and community to be served).
   - Availability of suitable sites for spillways.

3. **Preliminary sizing and determination of dam type.** Preliminary estimation of the dam height, length, and reservoir capacity may be made using available contoured topographic maps of the area. This estimation may be used to reduce the number of alternative sites identified in the previous steps. A suitable type of dam (e.g., earth, earth and rockfill, or gravity) for each promising site is determined by a multidisciplinary team using preliminary geological, geotechnical, and economic analyses.

4. **Preliminary surveys.** Preliminary field surveys are conducted for selected promising sites. These include preliminary geological surveys to assess the rock and soil conditions for the dam, reservoir, spillway, and borrow areas, and topographic surveys to estimate reservoir capacity with different dam heights. The topographic surveys include cross sections across the valley covering the potential reservoir area. These cross sections are used to prepare elevation-area and elevation-capacity tables or curves for the reservoir.

5. **Hydrologic investigations.** These include demarcation of the selected sites and estimation of drainage areas upstream of each location from available topographic maps of the respective watersheds. Available data on streamflows for the stream intended to serve as the source of water, and rainfall data from precipitation gages, are assembled. A hydrologic monitoring plan is prepared that includes installation of stream and precipitation gages at suitable locations. Available streamflow and precipitation data for monitoring stations in adjacent watersheds are collected, along with information such as the hydraulic characteristics of the watershed, location and datum of existing stream gages, and location and altitude of existing precipitation gages.

6. **Hydrologic analyses.** These include flood-routing computations for several combinations of dam heights and spillway widths.

A probabilistic analysis of reservoir capacity is done by calculating the required reservoir capacity for each year of streamflow record, calculating the probability distribution of reservoir capacities, and then selecting the reservoir capacity with the required design return period. Dams for water supply should be designed to pass at least a 100-year flood event (AWWA, 2003a). However, if failure of the dam could cause loss of life to people downstream, it should be designed for a larger flood. In many states, such dams must be designed to pass the "probable maximum flood," which can be up to five times the size of the 100-year flood. Nearly all states have dam-safety regulations that specify the standards that dams must meet.

Offstream reservoirs can be constructed next to major rivers to add storage to a water system. Water can be pumped from the river to the reservoir during high-flow periods, then withdrawn for water supply during low-flow periods. This type of impoundment is commonly referred to as a *pumped-storage reservoir* and can involve constructing a dam across the stream of a small tributary, thereby avoiding many of the adverse environmental impacts of building a dam on a major river.

After a dam has been built and the reservoir filled, the exposed water-surface area is increased significantly over that of the natural stream. The resultant effect is a greatly increased opportunity for evaporation. The opportunity for the generation of runoff from the flooded land is also eliminated, but this loss is countered by gains made through the catchment of direct precipitation. These water-surface effects tend to result in net gains in well-watered regions, but in arid lands, losses are typical, since evaporation greatly exceeds precipitation (Viessman and Hammer, 2005).

7.10 Hydropower

Hydroelectric power is an important source of energy, with hydropower responsible for about 15% of the electric energy generated in the United States and over 70% of electric energy generated in Brazil and Norway. There are two types of hydropower plants: (1) run-of-river plants that use direct streamflow, and where energy output is directly related to the instantaneous flow of the river; and (2) storage plants that use a reservoir to store water, and where energy is produced using controlled water releases. Run-of-river plants use the sustained flow of a stream or river to turn the turbines for electricity generation. This type usually has limited storage capacity and provides a continuous output of electricity. Storage plants use a reservoir of sufficient size to increase the amount of water available for power generation and to carry over water through dry periods. A *pumped-storage* plant uses power generated during low periods of demand to pump water back up the system to a headwater pond for use during peak demand periods. This recycling of water represents an economic efficiency between peak and off-peak demand requirements.

Turbines are the central components of all hydropower facilities. Their role is to extract energy from flowing water and convert it to mechanical energy to drive electric generators. There are two basic types of hydraulic turbines: impulse turbines and reaction turbines.

Impulse Turbines. In an impulse turbine, a free jet of water impinges on a revolving element, called a *runner*, which is exposed to atmospheric pressure. The runner in an impulse turbine is sometimes called an impulse wheel or *Pelton wheel*, in honor of Lester A. Pelton (1829–1908), who contributed much to its development in the early gold-mining days in California. A typical impulse turbine is shown in

FIGURE 7.13: Impulse turbine.

Source: Topomatika (2005).

Figure 7.13, where the runner has a series of split buckets located around its periphery. When the jet strikes the dividing ridge of the bucket, it is split into two parts that discharge at both sides of the bucket. Only one jet is used on small turbines, but two or more jets impinging at different points around the runner are often used on large units. The jet flows are usually controlled by a needle nozzle, and some jet velocities exceed 150 m/s. The generator rotor is usually mounted on a horizontal shaft between two bearings with the runner installed on the projecting end of the shaft; this is called a *single-overhung* installation. In some cases, runners are installed on both sides of the generator to equalize bearing loads, and this is called a *double-overhung* installation. The diameters of runners range up to about 5 m.

The schematic layout of an impulse-turbine installation is shown in Figure 7.14. The intake to the turbine is from the *forebay* of the upstream reservoir, and the water is delivered to the turbine through a large-diameter pipe called a *penstock*. The energy grade line (EGL) and the hydraulic grade line (HGL) are both shown in Figure 7.14. The head loss in the penstock between the forebay and the nozzle entrance (point B) is h_L, and the pressure head and velocity head at

FIGURE 7.14: Energy losses in impulse-turbine system.

Source: Finnemore and Franzini (2002).

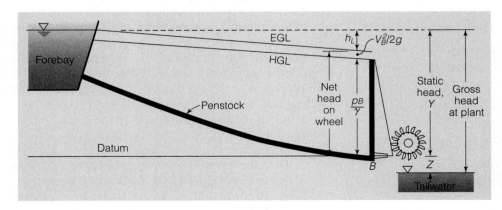

the nozzle entrance are p_B/γ and $V_B^2/2g$, respectively. The difference between the water elevation in the forebay and the elevation of the nozzle is called the *static head*, Y, and the difference between the nozzle elevation and the tailwater elevation is denoted by Z. The nozzle is considered to be an integral part of the impulse turbine, hence the *net head* or *effective head*, h, on the turbine is equal to the static head minus the pipe friction losses and is given by

$$h = \frac{p_B}{\gamma} + \frac{V_B^2}{2g} = Y - h_L \tag{7.43}$$

The head at the entrance to the nozzle is expended in four ways. Some energy is lost in fluid friction in the nozzle, known as *nozzle loss*; some energy is expended in fluid friction over the buckets of the turbine runner; kinetic energy is carried away in the water discharged from the buckets; and the rest is available to the buckets. Hence, the head directly available to the runner, h', is given by

$$h' = h - \left(\frac{1}{C_v^2} - 1\right)\left[1 - \left(\frac{A_j}{A_B}\right)^2\right]\frac{V_j^2}{2g} + k\frac{v_2^2}{2g} + \frac{V_2^2}{2g} \tag{7.44}$$

where C_v is the coefficient of velocity of the nozzle, A_j is the cross-sectional flow area of the jet, A_B is the cross-sectional flow area of the pipe upstream of the nozzle, V_j is the jet velocity, v_2 is the velocity of water relative to the bucket at the exit from the bucket, and V_2 is the absolute velocity of the water leaving the bucket. The coefficient of velocity is defined as the ratio of the actual velocity exiting the nozzle to the velocity that would exist in the absence of frictional effects. Typical values of the bucket friction loss coefficient, k, are in the range of 0.2 to 0.6. It is important to note that, for a given pipeline, there is a unique jet diameter that will deliver maximum power to a jet. This is apparent by noting that the power of the jet, P_{jet}, issuing from the nozzle is given by

$$P_{jet} = \gamma Q \frac{V_j^2}{2g} \tag{7.45}$$

As the size of the nozzle opening is increased, the flow rate Q increases while the jet velocity, V_j, decreases, hence there is some intermediate size of nozzle opening that will provide maximum power to the jet.

The *hydraulic efficiency*, η', of the impulse turbine is the ratio of the power delivered to the turbine buckets to the power in the flow at the entrance to the nozzle. Thus, for impulse turbines,

$$\eta' = \frac{\gamma Q h'}{\gamma Q h} = \frac{h'}{h} \tag{7.46}$$

The overall efficiency, η, of an impulse turbine is less than the hydraulic efficiency, η', because of that part of the energy delivered to the buckets that is lost in the mechanical friction in the bearings and the air resistance associated with the spinning runner. The efficiency of an impulse turbine, η, is given by

$$\eta = \frac{\text{output (shaft) power}}{\text{input power}} = \frac{T\omega}{\gamma Q h} \tag{7.47}$$

where T is the torque delivered to the shaft by the turbine, and ω is the angular speed of the runner.

Reaction Turbines. A reaction turbine is one in which flow takes place in a closed chamber under pressure, and the flow through a reaction turbine may be radially inward, axial, or mixed. There are two types of radial turbines in general use, the *Francis turbine* and the *axial-flow turbine*, also called a *propeller turbine*.

The Francis turbine is named after the American engineer James B. Francis (1815–1892), who designed, built, and tested the first efficient inward-flow turbine in 1849. All inward-flow turbines are called Francis turbines. In the usual Francis turbine, water enters a *scroll case* and moves into the runner through a series of guide vanes with contracting passages that convert pressure head to velocity head. These vanes, called *wicket gates*, are adjustable so that the quantity and direction of flow can be controlled. A Francis turbine runner located in the center of a scroll case is shown in Figure 7.15. When installed, the scroll case/runner combination shown in Figure 7.15 will be turned on its side with the runner facing downward. Flow through the Francis runner is at first inward in the radial direction, gradually changing to axial (vertically downward); such turbines are therefore called mixed-flow turbines. The scroll case is designed to decrease the cross-sectional area in proportion to the decreasing flow rate passing a given section of the casing. Constant rotative speed of the runner under varying load is achieved by a governor that actuates a mechanism that regulates the gate openings. A relief valve or surge tank is generally necessary to prevent serious water-hammer pressures.

The propeller turbine is an axial-flow machine with its runner confined in a closed conduit. The usual runner has four to eight blades mounted on a hub, with very little clearance between the blades and the conduit wall. A *Kaplan turbine* is a propeller turbine with movable blades whose pitch can be adjusted to suit existing operating conditions. A typical Kaplan turbine runner is shown in Figure 7.16. Other types of propeller turbine include the *Deriaz turbine*, an adjustable-blade, diagonal-flow turbine where the flow is directed inward as it passes through the blades, and the *tube turbine*, an inclined-axis type that is particularly well adapted to low-head installations, since the water passages can be formed directly in the concrete structure of the low dam.

FIGURE 7.15: Francis turbine, Mavel FSH 450 installed at SHP Fujiyoshida, Japan

Source: Mavel, a.s. (2005).

FIGURE 7.16: Kaplan axial flow turbine

Source: U.S. Army Corps of Engineers (2005g).

FIGURE 7.17: Energy losses in impulse-turbine system

Source: Finnemore and Franzini (2002).

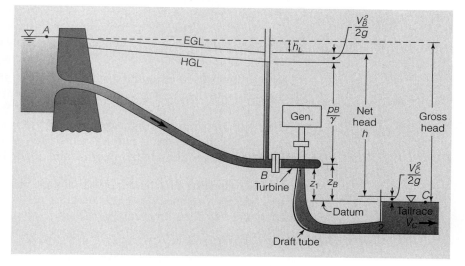

To operate properly, reaction turbines must have a submerged discharge. After passing through the runner, the water enters a *draft tube*, which directs the water to the discharge location in the downstream channel called the *tailrace*. A schematic diagram of a reaction-turbine system is shown in Figure 7.17. Applying the energy equation to the draft tube, where z_1 is the height of the entrance of the draft tube above the surface of the water in the tailrace, the absolute pressure at the draft tube entrance (section 1 in Figure 7.17) is given by

$$\frac{(p_1)_{\text{abs}}}{\gamma} = \frac{p_{\text{atm}}}{\gamma} - z_1 - \frac{V_1^2}{2g} + h_D + \frac{V_2^2}{2g} \qquad (7.48)$$

where p_{atm} is the atmospheric pressure, h_D is the head loss in the draft tube, $V_1^2/2g$ is the kinetic energy at the entrance to the draft tube, and $V_2^2/2g$ is the kinetic energy at the exit from the draft tube (section 2). To prevent cavitation, the vertical distance z_1 from the tailwater to the draft-tube inlet should be limited such that at no point within the draft tube or turbine will the absolute pressure drop to the vapor pressure of the water. For a reaction turbine, the *net head*, h, is the difference between the energy level just upstream of the turbine and that of the tailrace. In terms of the variables shown in Figure 7.17,

$$h = \left(z_B + \frac{p_B}{\gamma} + \frac{V_B^2}{2g} \right) - \frac{V_C^2}{2g} \tag{7.49}$$

where z_B is defined as the *draft head*, and V_C is the velocity in the tailrace. In most cases, $V_C^2/2g$ is relatively small and may be neglected. The *effective head*, h', that is available to act on the runner of a reaction turbine is

$$h' = h - k'\frac{(V_1 - V_2)^2}{2g} - \left[\frac{V_2^2}{2g} - \frac{V_C^2}{2g} \right] \tag{7.50}$$

where the latter two terms are the head loss in the draft tube and the loss from the submerged part of the tube, respectively. The head h'' that is actually extracted from the water by the runner is smaller than h' by an amount equal to the *shock loss* at entry to the runner and the hydraulic friction losses in the scroll case, guide vanes, and runner. The efficiency of a reaction turbine is given by Equation 7.47, which also describes the efficiency of impulse turbines.

Pump turbines used at pumped-storage facilities are very similar in design and construction to the Francis turbine. When water enters the rotor at the periphery and flows inward, the machine acts as a turbine. When water enters the center (or eye) and flows outward, the machine performs as a pump. The pump-turbine is connected to a motor generator, which acts as either a motor or generator, depending on whether the pump or turbine mode is being used.

Similarity laws derived in Chapter 2 for pumps also apply to reaction turbines. Therefore, the performance of a homologous series of reaction turbines can be described by the following functional relation:

$$\frac{gh}{\omega^2 D^2} = f\left(\frac{Q}{\omega D^3} \right) \tag{7.51}$$

where h is the net head on the turbine, ω is the angular speed of the runner, D is the characteristic size of the runner, and Q is the volumetric flowrate through the turbine. In a similar fashion to pumps, the operating point of greatest efficiency is defined by the dimensionless specific speed, n_s, where

$$n_s = \frac{\omega Q^{\frac{1}{2}}}{(gh)^{\frac{3}{4}}} \tag{7.52}$$

and any consistent set of units can be used. In the United States, it is common to define the specific speed of a turbine in U.S. Customary units, in which case the specific speed, N_s, is expressed in the form

$$N_s = \frac{n_e\sqrt{\text{bhp}}}{h^{5/4}} \tag{7.53}$$

where n_e is the speed of the runner in revolutions per minute, bhp is the shaft or brake horsepower ($=T\omega$) in horsepower, and h is the net head in feet, where all performance variables are at the point of maximum efficiency. For $N_s < 10$, impulse turbines are typically most efficient; for $10 \le N_s < 100$, Francis turbines are most efficient, and for $100 \le N_s \le 200$, propeller turbines are most efficient. These efficiencies indicate that high-head installations work best with impulse turbines and low-head installations work best with propeller turbines. Impulse turbines are commonly used for heads greater than 150 to 300 m, while the upper limit for using a Francis turbine is on the order of 450 m because of possible cavitation and the difficulty of building casings to withstand such high pressures (Finnemore and Franzini, 2002).

Turbines are operated at constant speed. In the United States, 60 Hz electric current is most common, and under such conditions the rotative speed n of a turbine is related to the number of pairs of poles, N, by the relation

$$n = \frac{3600}{N} \tag{7.54}$$

where n is in revolutions per minute.

The capacity of a hydroelectric generating plant is defined as the maximum rate at which the plant can produce electricity. The *installed capacity* or *hydropower generation potential* for a given site is estimated by the relation

$$\boxed{P = \gamma Q h \eta} \tag{7.55}$$

where P is the hydropower generation potential, γ is the specific weight of water ($=9.79$ kN/m^3), Q is the volumetric flowrate (L^3/T), h is the available head (L), and η is the turbine efficiency, which is usually in the range of 0.80 to 0.90. Design values of Q and h are estimated from flow-duration and head-discharge tables or curves for the water-supply source. To maximize use of available water, a value of Q which is exceeded 10% to 30% of the time may be selected to estimate the installed capacity of the facility. Several values of Q and the corresponding h may be examined to estimate the values that result in optimum installed capacity from water use and economic considerations. For preliminary planning, optimum installed capacity may be that beyond which relatively large increases in Q are required to obtain relatively small increases in P. It is considered good practice to have at least two turbines at an installation so that the plant can continue operation while one of the turbines is shut down for repairs or inspection.

To estimate the annual hydropower generation potential of a facility, a reservoir and power-plant operation study has to be conducted using codes such as HEC-3 (U.S. Army Corps of Engineers, 1981) and HEC-5 (U.S. Army Corps of Engineers, 1981) with sequences of available flows and corresponding heads for relatively long periods of time—for example, 10 to 50 years or more. For preliminary estimates, flow-duration and head-discharge tables may be used. For these estimates, the overall water-to-wire efficiency should be used, which varies from about 0.70 to 0.86 and includes efficiencies of the turbine, generator, transformers, and other equipment. In addition, adjustments to the efficiency may be made for tailwater fluctuations and unscheduled down time.

EXAMPLE 7.14

A hydroelectric project is to be developed along a river where the 90-percentile flowrate is 2240 m^3/s, and a Rippl analysis indicates that storage to a height of 30 m above the downstream stage is required to meet all the demands. Estimate the installed capacity that would be appropriate for these conditions.

Solution   Assuming that the capacity of the turbines will be sufficient to pass a flow, Q, of 2240 m^3/s at a head of 30 m, and taking the turbine efficiency, η, as 0.85, and $\gamma =$ 9.79 kN/m^3, yields

$$P = \gamma Q H \eta = (9.79)(2240)(30)(0.85) = 5.59 \times 10^5 \text{ kW} = 559 \text{ MW}$$

Therefore, to fully utilize the available head and anticipated flowrates, an installed capacity of 559 MW is required.

The extraction of hydropower from surface runoff is potentially feasible in cases where water supplies are in excess of demand requirements, and in cases where the total quantity of water available over a period of time is equal to or in excess of the demand requirements, although the demand requirements may at times exceed the availability of water. In both of these cases and where the sites are within reasonable transmission distance to a power market, there is a potential for hydropower development. Detailed guidelines for designing the components of hydroelectric power plants can be found in several reputable publications, including: U.S. Army Corps of Engineers (1985), ASCE (1989), and Zipparro and Hansen (1993). The design and construction of a hydroelectric plant is very complex and can easily take more than a decade from the drawing board to the actual generation of electricity. Depending on the installed capacity, hydroelectric power plants are classified as follows (Prakash, 2004):

- Conventional: Installed capacity greater than 15 MW
- Small-scale: Installed capacity between 1 and 15 MW
- Mini: Installed capacity between 100 KW and 1 MW
- Micro: Installed capacity less than 100 KW

The largest hydroelectric project in the world is the Three Gorges Dam in China, designed to produce 18 GW of electrical energy.

7.11   Navigation

The goal of navigation planning is to improve a waterway or harbor in order to maximize the net benefits to a community. There are three different methods to provide navigable waterways: (1) river regulation; (2) lock-and-dam; and (3) artificial canalization. In river regulation, water depths are maintained by the release of water from upstream reservoirs, and large reservoir capacities are usually needed. In the lock-and-dam approach, the depth of water is increased behind a series of dams through a succession of backwater curves. At each dam, a shiplock is provided to negotiate the difference in water levels upstream and downstream of the dam. In the artificial canalization approach, a humanmade channel is combined with shiplocks.

Summary

Water-resources planning and management addresses issues related to developing the water resources of a region. Key steps in the water-resources planning process include development of goals and objectives, identification of alternative solutions, and evaluation and recommendation of actions. In developing a water-resources plan, a variety of legal and regulatory issues must be addressed. These include water rights, and a myriad of regulations associated with federal environmental laws such as the Clean Water Act, Fish and Wildlife Coordination Act, and the Safe Drinking Water Act. Regulations associated with environmental laws typically require that construction permits be obtained and completed projects be inspected. In cases where federal funding is involved, the environmental impacts of water-resources projects must be documented in environmental impact reports. After alternatives have been identified, economic feasibility is a major factor in identifying the preferred alternative. Economic feasibility analysis usually consists of some form of benefit-cost analysis in which the lowest net present value of benefits and costs is used as a basis to identify the preferred alternative.

Common objectives associated with water-resources projects include: water supply, flood control, drought management, irrigation, hydropower, and navigation. In planning water-supply projects, key components include forecasting water demand, utilization of conservation practices, identification of new water supplies, development of regulatory source-water protection plans, and the possible utilization of recycled water. In planning flood-control projects, key components include identification of flood-prone areas, implementation of land-use controls to modify the susceptibility to flood damage, and structural controls to minimize the amount of flooding. Drought-management plans relate the severity of droughts to actions needed to mitigate drought effects, and such plans typically use the Palmer drought severity index to classify droughts and infer possible impacts. In irrigation projects, a key consideration is the type of irrigation system that is most appropriate and economically feasible for local conditions. In hydropower projects, a key component is the determination of hydropower generation potential, which is derived from local flow-duration and head-discharge relations. In navigation projects, a key component is to identify the best method to provide a navigable waterway; considerations may include river regulation, lock-and-dam, and artificial canalization. Dams and reservoirs are commonly at the heart of multiobjective projects, with reservoir storage being a key component of water-supply, flood-control, drought-management, irrigation, hydropower, and navigation projects.

Problems

7.1. Derive the expression for the sinking-fund factor given by Equation 7.4.

7.2. Derive the expression for the capital-recovery factor given by Equation 7.6.

7.3. Derive the expression for the arithmetic-gradient present-worth factor given by Equation 7.8.

7.4. Derive the expression for the geometric-gradient present-worth factor given by Equation 7.11.

7.5. Two alternatives for developing the water resources of a region are being considered. The design life of both alternatives is 30 years, and economic conditions indicate a 4% interest rate should be used in comparing alternatives. The first alternative will yield annual returns of $50,000 for all 30 years of the project, while the second alternative yields a return of $17,000 at the end of the first year, and the returns are projected to increase by 6% per year in subsequent years. Determine the equivalent present worth for each alternative.

7.6. Two alternative proposals are being evaluated. The design life of both alternatives is 20 years, and an interest rate of 4% can be assumed. The first alternative requires an initial investment of $200,000 and produces annual revenues beginning with $50,000 in year 1 and increasing by $2000 per year, operating costs begin with $6000 in the first year and increase by $4000 per year; and the salvage value of the capital investment is $20,000 at the end of 20 years. The second alternative requires an initial investment of $100,000 and produces annual revenues beginning with $20,000 in year 1 and increasing by 6% per year; operating costs begin with $11,000 in the first year and increase by 6% per year; and the salvage value of the capital investment is $10,000 at the end of 20 years. Compare these alternatives using the present-worth method.

7.7. Two alternative proposals to develop the water resources of a region are being reviewed. Both require an initial capital investment, generate revenues that increase at a constant rate, have operating costs that increase at a constant rate, and have a design life of 15 years. These projects have no residual value at the end of their design life, and the alternatives are summarized in the following table

Alternative	Initial cost	First-year revenue	Revenue growth rate	First-year operating cost	Operating-cost growth rate
1	$130,000	$16,000	4.5%	$6000	5%
2	$75,000	$12,000	3.5%	$3200	3.5%

If the minimum attractive rate of return on the investment is 6%, what is the preferred alternative?

7.8. A proposal for a project with a 25-year design life will require an initial capital cost of $800,000 and produce annual net benefits of $80,000. Funding this project will require that spending on other projects be reduced by $55,000 per year for the first eight years of the project. Using an interest rate of 5%, determine the benefit-cost ratio of the project.

7.9. The average water demand for a city during the 7-year interval from 1998 to 2004 is shown in the following table:

Year	Average demand $(\times 10^5 \text{ m}^3/\text{d})$
1998	1.51
1999	1.78
2000	1.82
2001	1.85
2002	2.05
2003	2.02
2004	2.21

Use an extrapolation method to estimate the average water demand in the year 2008.

7.10. The current population of a city is 50,000, and the population is expected to increase by approximately 7000 in the next four years. The average annual per-capita demand is approximately constant at 530 L/day, and pumps at the water-supply source have a capacity of 1500 L/s. If the ratio of maximum daily to average daily demand is 2.5, assess the adequacy of the existing pump capacity for the next four years.

7.11. Data collected from residential customers yield the following empirical expression for the annual water use in each household:

$$\ln Q = 3.21 + 0.0.243 \ln M + 0.0172 \ln I + 1.54 \ln N$$

where Q is the annual water use in m^3/(year·household); M is the marginal water rate in \$/m^3, I is the household income in \$, and N is the number of persons in the household. In 2040, water rates will be \$1.00/m^3, a typical household in the service area will have an annual income of \$65,000 and consist of 3.4 persons, and there will be 5000 households in the service area. Estimate the average daily water demand and the price elasticity for the service area in the year 2040.

7.12. A city planning department estimates the following number of households per census tract in a water-distribution service area:

Tract	Households
1	550
2	1260
3	740
4	1110

If the number of persons per household is expected to be 2.1, what is the expected population in the service area?

7.13. The safe yield of an aquifer with an areal extent of 10,200-km^2 is to be determined. In the period from 1994 to 2003, these are the data for each year:

Year	Annual draft ($\times 10^8$ m^3)	Water-table elevation* (m)	Surface runoff + evapotranspiration ($\times 10^8$ m^3)	Rainfall (cm)
1993	–	5.442	–	–
1994	2.90	5.577	50.0	6.50
1995	2.84	5.630	63.8	6.20
1996	3.36	5.655	53.8	5.96
1997	2.76	5.395	54.9	5.07
1998	3.43	5.313	59.2	6.39
1999	2.80	5.436	75.0	7.31
2000	3.16	5.370	56.8	6.16
2001	2.99	5.153	63.8	5.94
2002	2.94	5.265	62.0	6.86
2003	3.19	5.098	55.5	5.24

*Water-table elevation at the end of the year.

Estimate the safe yield of the aquifer using the zero-fluctuation, average-draft, and simplified water-balance methods.

7.14. In western Kansas, analysis of historical hydrologic data indicates that the climatic coefficients and constants for the month of April are as follows:

Parameter	Value*
Coefficient of evapotranspiration, α	0.9218
Coefficient of recharge, β	0.1499
Coefficient of runoff, γ	0.0359
Coefficient of loss, δ	0.2688
Climatic characteristic, K_{April}	1.54

*For April.

During April 2004, rainfall was 5.05 cm and potential evapotranspiration was 4.47 cm. Soils in western Kansas typically have an available water capacity of 15.24 cm, of which 2.54 cm is in the surface layer and 12.70 cm in the lower layers. At the end of March, moisture accounting indicated that the surface soil was dry, and the underlying soil had a moisture content of 5.94 cm. If the Palmer Drought Severity Index at the end of March was -2.1, characterize the drought conditions at the end of April.

7.15. A reservoir covers an area of 850 km², and has an average depth of 18.7 m. The inflow to the reservoir is from a river with an average flowrate of 2500 m³/s and a suspended-sediment concentration of 250 mg/L. Estimate the rate at which the depth of the reservoir is decreasing due to sediment accumulation, and the time it will take for the reservoir storage to decrease by 10%. Assume that the accumulated sediment has a bulk density of 1600 kg/m³.

7.16. During a critical 24-month period, the average monthly inflows to a reservoir and the average water demand are given in the following table:

Month*	Average inflow (m³/s)	Average demand (m³/s)	Month	Average inflow (m³/s)	Average demand (m³/s)	Month	Average inflow (m³/s)	Average demand (m³/s)
1	850	1000	9	1550	1300	17	1350	1400
2	1130	1100	10	1600	1200	18	1270	1400
3	1310	1200	11	1370	1100	19	1120	1400
4	1000	1300	12	1210	1000	20	1230	1400
5	1170	1400	13	940	1000	21	1320	1300
6	1260	1400	14	820	1100	22	1250	1200
7	880	1400	15	1040	1200	23	1240	1100
8	930	1400	16	1290	1300	24	1080	1000

*Month "1" is January.

Use these data to estimate the required storage volume of the reservoir.

7.17. A hydroelectric project is to be developed where the 90-percentile flowrate is 1630 m³/s, and storage to a height of 15 m above the downstream stage is required to meet all the demands. Estimate the installed capacity that would be appropriate for these conditions.

Units and Conversion Factors

A.1 Units

The Système International d'Unités (International System of Units or SI system) was adopted by the 11th General Conference on Weights and Measures (CGPM) in 1960 and is now used by almost the entire world. In the SI system, all quantities are expressed in terms of seven base (fundamental) units. These base units and their standard abbreviations are as follows (Dean, 1985):

Meter (m): Distance light travels in a vacuum during $1/299\,792\,458$ of a second.*

Kilogram (kg): Mass of a cylinder of platinum-iridium alloy kept in Paris.

Second (s): Duration of $9\,192\,631\,770$ cycles of the radiation corresponding to the transition between two hyperfine levels of the ground state of the cesium-133 atom.

Ampere (A): Magnitude of the current that, when flowing through each of two long parallel wires of negligible cross section separated by one meter in a vacuum, results in a force between the two wires of 2×10^{-7} newtons for each meter of length.

Kelvin (K): Defined in the thermodynamic scale by assigning 273.16 K to the triple point of water (freezing point, $273.16\,\text{K} = 0°\text{C}$).

Candela (cd): Luminous intensity of $1/600\,000$ of a square meter of a radiating cavity at the temperature of freezing platinum (2042 K).

Mole (mol): Amount of substance which contains as many specified entities (molecules, atoms, ions, electrons, photons, etc.) as there are there are atoms in exactly 0.012 kg of carbon-12.

In addition to the seven base units of the SI system, there are two supplementary SI units: the radian and the steradian. The radian (rad) is defined as the angle at the center of a circle subtended by an arc equal in length to the radius, and the steradian (sr) is defined as the solid angle with its vertex at the center of a sphere that is subtended by an area of the spherical surface equal to the radius squared. The SI units should not be confused with the now obsolete *metric units*, which were developed in Napoleonic France approximately 200 years ago. The primary difference between the metric and SI units is that the former uses centimeters and grams to measure length and mass, while these quantities are measured in meters and kilograms in SI units. The United States is gradually moving toward the use of SI units, but there is still widespread use

---

*"Meter" is the accepted spelling in the United States, the rest of the world uses the spelling "metre."

TABLE A.1: SI Derived Units

Unit name	Quantity	Symbol	In terms of base units
becquerel	Activity of a radionuclide	Bq	s^{-1}
coulomb	Quantity of electricity, electric charge	C	$A \cdot s$
degree Celsius	Celsius temperature	°C	K
farad	Capacitance	F	C/V
gray	Absorbed dose of ionizing radiation	Gy	J/kg
henry	Inductance	H	Wb/A
hertz	Frequency	Hz	s^{-1}
joule	Energy, work, quantity of heat	J	$N \cdot m$
lumen	Luminous flux	lm	$cd \cdot sr$
lux	Illuminance	lx	lm/m^2
newton	Force	N	$kg \cdot m/s^2$
ohm	Electric resistance	Ω	V/A
pascal	Pressure, stress	Pa	N/m^2
siemens*	Conductance	S	A/V
sievert	Dose equivalent of ionizing radiation	Sv	J/kg
telsa	Magnetic flux density	T	Wb/m^2
volt	Electric potential, potential difference	V	W/A
watt	Power, radiant flux	W	J/s
weber	Magnetic flux	Wb	$V \cdot s$

*The siemens was previously called the mho.

of the "English" system of units, which are also referred to as "U.S. Customary" or "British Gravitational" units.

In addition to the seven SI base units, there are several derived units that are used for convenience rather than necessity. They are listed in Table A.1 (Wandmacher and Johnson, 1995; Szirtes, 1998).

A.2 Conversion Factors

Conversion factors for various units are given in Table A.2. In most cases, application of unit conversion factors results in converted numbers that have more significant digits than the original numbers. In these cases, the converted number should be rounded such that the rounding error is consistent with that of the original number (Taylor, 1995).

EXAMPLE A.1

The height of a water-control structure is reported as 19.3 feet. Convert this dimension to meters.

Solution   The conversion factor is given in Table A.2 as 1 foot = 0.3048 meter, hence,

$$19.3 \text{ ft} = 19.3 \times 0.3048 = 5.88264 \text{ m}$$

Since 19.3 ft could have resulted in rounding any number between 19.25 ft and 19.35 ft, the maximum possible rounding error is $\pm 0.05/19.3 = \pm 0.26\%$. Similarly, rounding 5.88264 m to 5.88 m gives a maximum rounding error of $\pm 0.005/5.88 = \pm 0.085\%$, and rounding to 5.9 m gives an error of $\pm 0.05/5.9 = \pm 0.85\%$. Hence, accuracy is lost by

TABLE A.2: Multiplicative Factors for Unit Conversion

Quantity	Convert From	Convert To	Multiply By
Area	ac	ha	0.404687
	mi^2	km^2	2.59000
Energy	Btu	J	1,054.350 264
	cal	J	4.184[†]
Energy/area	ly[‡]	kJ/m^2	41.84[†]
Flowrate	cfs	m^3/s	0.028 316 85
	gpm[‡]	L/s	0.06309
	mgd[‡]	m^3/s	0.04381
		m^3/d	3,785.412
Force	lbf	N	4.448 221 615 260 5[†]
Length	ft	m	0.3048[†]
	in.	m	0.0254[†]
	mi (U.S. statute)	km	1.609 344[†]
	mi (U.S. nautical)	km	1.852 000[†]
	yd	m	0.9144[†]
Mass	g	kg	0.001[†]
	lbm	kg	0.453 592 37[†]
Permeability	darcy	m^2	0.987×10^{-12}
Power	hp	W	745.699 87
Pressure	atm	kPa	101.325[†]
	bar	kPa	100.000[†]
	mm Hg (at $0°C$)	kPa	0.133322
	psi	kPa	6.894757
	torr	kPa	0.133322
Speed	knot	m/s	0.514 444 444
Viscosity (dynamic)	cp	Pa·s	0.001[†]
Viscosity (kinematic)	cs	m^2/s	10^{-6}[†]
Volume	gal	L	3.785 411 784[†]

[†] Exact conversion
[‡] gpm ≡ gallons per minute, ly ≡ langley, mgd ≡ million gallons per day.

taking 19.3 ft as 5.9 m, while 5.88 m is more accurate than indicated by 19.3 ft. It is usually prudent not to discard accuracy, so take 19.3 ft = 5.88 m.

A good rule of thumb is that the converted number should have the same number of significant digits as the original number, assuming that the conversion factor is always more accurate than the original number.

APPENDIX B

Fluid Properties

B.1 Water

Temperature (°C)	Density (kg/m^3)	Dynamic viscosity (mPa·s)	Heat of vaporization (MJ/kg)	Saturation vapor pressure (kPa)	Surface tension (mN/m)	Bulk modulus (10^6 kPa)
0	999.8	1.781	2.499	0.611	75.6	2.02
5	1000.0	1.518	2.487	0.872	74.9	2.06
10	999.7	1.307	2.476	1.227	74.2	2.10
15	999.1	1.139	2.464	1.704	73.5	2.14
20	998.2	1.002	2.452	2.337	72.8	2.18
25	997.0	0.890	2.440	3.167	72.0	2.22
30	995.7	0.798	2.428	4.243	71.2	2.25
40	992.2	0.653	2.405	7.378	69.6	2.28
50	988.0	0.547	2.381	12.340	67.9	2.29
60	983.2	0.466	2.356	19.926	66.2	2.28
70	977.8	0.404	2.332	31.169	64.4	2.25
80	971.8	0.354	2.307	47.367	62.6	2.20
90	965.3	0.315	2.282	70.113	60.8	2.14
100	958.4	0.282	2.256	101.325	58.9	2.07

Density, viscosity, surface tension, and bulk modulus are from Finnemore and Franzini (2002), and heat of vaporization and saturation vapor pressure are from Viessman and Lewis (1996).

The properties given in Table B.1 are for pure water, which seldom exists in nature. The general dependence of water density on temperature has been found to be approximately parabolic, with a maximum at 4°C. However, the temperature corresponding to the maximum density of water changes with increasing salinity, decreasing to about 0°C for highly saline systems. To a first-order approximation, density is linearly dependent on salinity over much of the normal range of interest.

B.2 Organic Compounds Found in Water

The properties of several organic compounds commonly found in contaminated waters are given in Table B.2. Substances with boiling points at or below 100°C are classified as *volatile organic compounds* (VOCs). The chemical properties given in Table B.2 are those most frequently used in the quantitative analysis of the fate of organic contaminants in the environment.

Compound	Formula	Molecular weight	Boiling point (°C)	Density @ 20°C (kg/m³)	Dynamic viscosity @ 20°C (mPa·s)	Solubility in water @ 20°C (mg/L)	Sorption coefficient, log K_{OC} (log mL/g)	Saturation vapor pressure @ 20°C (kPa)	Henry's Law constant @ 20°C (Pa·m³/mol)
Acetone (dimethyl ketone)	CH3COCH3	58.08	56.1[j]–56.2[b]	790[b]	—	440,600[a]	−0.59[b]–0.34[h]	3.07[g]–35.97[h]	3.3[a]–10[l]
Benzene	C6H6	78.11	80.1[b]	870[h]–880[b]	—	1,710[b]–1,796[b]	1.39[b]–2.95[b]	1.20[g]–12.67[b]	458[b]–557[p]
Bis(2-ethylhexyl)phthalate (DEHP)	C24H38O4	390.57	230[f]–387[f]	985[b]	—	0.041[b]–0.285[h]	3.77[h]–5.15[b]	1.09×10^{-9b}–26.7×10^{-9b}	0.0011[a]
Chlorobenzene	C6H5Cl	112.56	132[l]	1106[b]	0.8[m]	466[h]–500[m]	1.92[b]–2.63[d]	1.17[l]–1.2[b]	263[p]–288[b]
Chloroethane	CH3CH2Cl	64.5	62[l]	897[b]–920[j]	—	5,710[b]–5740[d]	0.51[b]–1.57[h]	133.3[i]–133.7[h]	1,030[n]
Chloroform (Trichloromethane)	CHCl3	119.4	62[l]	1,480[c]–1,490[h]	0.56[m]	8000[j]–8200[b]	1.46[b]–1.94[b]	20.2[h]–25.9[m]	278[l]–486[p]
1,1-Dichloroethane	C2H4Cl2	99.0	57.3[l]	1,174[l]–1,180[c]	0.5[m]	5,100[m]–5500[b]	1.48[b]–1.80[d]	23.9[l]–29.5[m]	435[l]–550[m]
1,2-Dichloroethane	C2H4Cl2	99.0	83.5[l]	1,235[b]–1,253[b]	0.84[m]	8,000[d]–8800[c]	1.15[h]–1.57[d]	8.13[l]–10.9[m]	92[p]–152[m]
Ethylbenzene	C8H10	106.2	—	867[b]–870[c]	—	150[l]–152[c]	1.98[h]–3.13[d]	0.911[h]–1.28[g]	559[n]–882[p]
Gasoline[e]	—	—	—	680[g]–750[f]	0.29[s]–0.31[f]	—	—	55.2[s]	64.3[o]
Methyl tertiary-butyl ether (MTBE)	(CH3)COCH3	88.2	—	740[k]	—	48,000[h]	0.55[k]–1.05[h]	32.7[h]	229[b]
Methylene chloride (Dichloromethane)	CH2Cl2	84.9	—	1,327[b]–1,336[b]	—	13,000[c]–20,000[h]	0.94[b]	48.2[h]	56.0[o]
Naphthalene	C10H8	128.2	—	1,030[h]–1,162[b]	—	31.7[h]–38[b]	2.62[b]–5.00[b]	0.0113[g]–0.037[h]	56.0
Phenol	C6H6O	94.1	—	1,058[b]–1,070[b]	0.9[m]	93,000[l]	1.15[h]–3.49[b]	0.027[p]–0.045[l]	0.030
Tetrachloroethene (PCE)	CCl2CCl2	165.8	121.4[l]	1,623[b]–1,630[m]	—	149[b]–200[d]	2.25[b]–2.99[b]	1.87[q]–2.53[g]	1327[l]–1763[m]
Toluene (Methylbenzene)	C6H5CH3	92.1	—	867[b]–870[c]	—	500[l]–535[d]	1.57[b]–3.05[b]	2.93[b]–3.75[h]	529[n]–578[p]
1,1,1-Trichloroethane	CCl3CH3	133.4	—	1,339[c]–1,350[l]	0.84[m]	480[c]–4400[d]	2.02[b]–3.40[b]	13.3[l]–16.6[m]	1337[l]–1824[p]
Trichloroethene (TCE)	C2HCl3	131.5	86.7[l]	1,460[c]–1,460[m]	0.57[m]	1,000[c]–1,100[m]	1.66[b]–2.83[b]	7.70[h]–10.0[m]	722[l]–1013[p]
Vinyl chloride (Chloroethene)	CH2CHCl	62.5	—	910[b]–912[b]	—	90[c]–267[h]	0.39[b]–1.76[h]	355[h]–400[g]	2,200[l]
o-Xylene (1,2 Dimethylbenzene)	C6H4(CH3)2	106.2	—	880[b]	—	152[b]–175[c]	1.92[b]–2.92[b]	0.667[p]–0.912[h]	409[p]–537[p]
p-Xylene (1,4 Dimethylbenzene)	C6H4(CH3)2	106.2	—	861[b]–881[b]	—	160[b]	2.31[b]–2.72[b]	—	555[n]

[a] Montgomery (2000) at 25°C.
[b] Montgomery (2000).
[c] Hemond and Fechner (1993).
[d] Schnoor (1996).
[e] Typical properties at 15.6°C. Properties of petroleum products vary.
[f] Munson et al. (1990).
[g] USEPA (1996).
[h] Charbeneau (2000).
[i] Fetter (1999).
[j] Sage and Sage (2000).
[k] U.S. Environmental Protection Agency (1998).
[l] Pankow and Cherry (1996).
[m] Pankow and Cherry (1996) at 25°C.
[n] Rathbun (1998).
[o] Rathbun (1998) at 25°C.
[p] Jackson et al. (1985).
[q] Nyer et al. (1991).
[r] Montgomery (2000) at 5 mm Hg.
[s] Finnemore and Franzini (2002) at 20°C.
[t] Finnemore and Franzini (2002) at 20°C.
[u] Davis and Masten (2004).

B.3   Air at Standard Atmospheric Pressure

Temperature (^{o}C)	Density (kg/m^3)	Dynamic viscosity (mPa·s)	Specific heat ratio	Speed of sound (m/s)
-40	1.514	0.0157	1.401	306.2
-20	1.395	0.0163	1.401	319.1
0	1.292	0.0171	1.401	331.4
5	1.269	0.0173	1.401	334.4
10	1.247	0.0176	1.401	337.4
15	1.225	0.0180	1.401	340.4
20	1.204	0.0182	1.401	343.3
25	1.184	0.0185	1.401	346.3
30	1.165	0.0186	1.400	349.1
40	1.127	0.0187	1.400	354.7
50	1.109	0.0195	1.400	360.3
60	1.060	0.0197	1.399	365.7
70	1.029	0.0203	1.399	371.2
80	0.9996	0.0207	1.399	376.6
90	0.9721	0.0214	1.398	381.7
100	0.9461	0.0217	1.397	386.9
200	0.7461	0.0253	1.390	434.5
300	0.6159	0.0298	1.379	476.3
400	0.5243	0.0332	1.368	514.1
500	0.4565	0.0364	1.357	548.8
1000	0.2772	0.0504	1.321	694.8

Properties of air derived from Blevins (1984).

APPENDIX C

Statistical Tables

C.1 Areas Under Standard Normal Curve

TABLE C.1: Areas Under Standard Normal Curve

z	0.00	0.01	0.02	0.03	0.04	0.05	0.06	0.07	0.08	0.09
−4.0	.0000	.0000	.0000	.0000	.0000	.0000	.0000	.0000	.0000	.0000
−3.9	.0000	.0000	.0000	.0000	.0000	.0000	.0000	.0000	.0000	.0000
−3.8	.0001	.0001	.0001	.0001	.0001	.0001	.0001	.0001	.0001	.0001
−3.7	.0001	.0001	.0001	.0001	.0001	.0001	.0001	.0001	.0001	.0001
−3.6	.0002	.0002	.0001	.0001	.0001	.0001	.0001	.0001	.0001	.0001
−3.5	.0002	.0002	.0002	.0002	.0002	.0002	.0002	.0002	.0002	.0002
−3.4	.0003	.0003	.0003	.0003	.0003	.0003	.0003	.0003	.0003	.0002
−3.3	.0005	.0005	.0005	.0004	.0004	.0004	.0004	.0004	.0004	.0003
−3.2	.0007	.0007	.0006	.0006	.0006	.0006	.0006	.0005	.0005	.0005
−3.1	.0010	.0009	.0009	.0009	.0008	.0008	.0008	.0008	.0007	.0007
−3.0	.0013	.0013	.0013	.0012	.0012	.0011	.0011	.0011	.0010	.0010
−2.9	.0019	.0018	.0018	.0017	.0016	.0016	.0015	.0015	.0014	.0014
−2.8	.0026	.0025	.0024	.0023	.0023	.0022	.0021	.0021	.0020	.0019
−2.7	.0035	.0034	.0033	.0032	.0031	.0030	.0029	.0028	.0027	.0026
−2.6	.0047	.0045	.0044	.0043	.0041	.0040	.0039	.0038	.0037	.0036
−2.5	.0062	.0060	.0059	.0057	.0055	.0054	.0052	.0051	.0049	.0048
−2.4	.0082	.0080	.0078	.0075	.0073	.0071	.0069	.0068	.0066	.0064
−2.3	.0107	.0104	.0102	.0099	.0096	.0094	.0091	.0089	.0087	.0084
−2.2	.0139	.0136	.0132	.0129	.0125	.0122	.0119	.0116	.0113	.0110
−2.1	.0179	.0174	.0170	.0166	.0162	.0158	.0154	.0150	.0146	.0143
−2.0	.0228	.0222	.0217	.0212	.0207	.0202	.0197	.0192	.0188	.0183
−1.9	.0287	.0281	.0274	.0268	.0262	.0256	.0250	.0244	.0239	.0233
−1.8	.0359	.0351	.0344	.0336	.0329	.0322	.0314	.0307	.0301	.0294
−1.7	.0446	.0436	.0427	.0418	.0409	.0401	.0392	.0384	.0375	.0367
−1.6	.0548	.0537	.0526	.0516	.0505	.0495	.0485	.0475	.0465	.0455
−1.5	.0668	.0655	.0643	.0630	.0618	.0606	.0594	.0582	.0571	.0559
−1.4	.0808	.0793	.0778	.0764	.0749	.0735	.0721	.0708	.0694	.0681
−1.3	.0968	.0951	.0934	.0918	.0901	.0885	.0869	.0853	.0838	.0823
−1.2	.1151	.1131	.1112	.1093	.1075	.1056	.1038	.1020	.1003	.0985
−1.1	.1357	.1335	.1314	.1292	.1271	.1251	.1230	.1210	.1190	.1170
−1.0	.1587	.1562	.1539	.1515	.1492	.1469	.1446	.1423	.1401	.1379
−.9	.1841	.1814	.1788	.1762	.1736	.1711	.1685	.1660	.1635	.1611
−.8	.2119	.2090	.2061	.2033	.2005	.1977	.1949	.1922	.1894	.1867
−.7	.2420	.2389	.2358	.2327	.2297	.2266	.2236	.2206	.2177	.2148
−.6	.2743	.2709	.2676	.2643	.2611	.2578	.2546	.2514	.2483	.2451
−.5	.3085	.3050	.3015	.2981	.2946	.2912	.2877	.2843	.2810	.2776

TABLE C.1: Areas Under Standard Normal Curve (Cont'd)

z	0.00	0.01	0.02	0.03	0.04	0.05	0.06	0.07	0.08	0.09
−.4	.3446	.3409	.3372	.3336	.3300	.3264	.3228	.3192	.3156	.3121
−.3	.3821	.3783	.3745	.3707	.3669	.3632	.3594	.3557	.3520	.3483
−.2	.4207	.4168	.4129	.4090	.4052	.4013	.3974	.3936	.3897	.3859
−.1	.4602	.4562	.4522	.4483	.4443	.4404	.4364	.4325	.4286	.4247
.0	.5000	.5040	.5080	.5120	.5160	.5199	.5239	.5279	.5319	.5359
.1	.5398	.5438	.5478	.5517	.5557	.5596	.5636	.5675	.5714	.5753
.2	.5793	.5832	.5871	.5910	.5948	.5987	.6026	.6064	.6103	.6141
.3	.6179	.6217	.6255	.6293	.6331	.6368	.6406	.6443	.6480	.6517
.4	.6554	.6591	.6628	.6664	.6700	.6736	.6772	.6808	.6844	.6879
.5	.6915	.6950	.6985	.7019	.7054	.7088	.7123	.7157	.7190	.7224
.6	.7257	.7291	.7324	.7357	.7389	.7422	.7454	.7486	.7517	.7549
.7	.7580	.7611	.7642	.7673	.7704	.7734	.7764	.7794	.7823	.7852
.8	.7881	.7910	.7939	.7967	.7995	.8023	.8051	.8078	.8106	.8133
.9	.8159	.8186	.8212	.8238	.8264	.8289	.8315	.8340	.8365	.8389
1.0	.8413	.8438	.8461	.8485	.8508	.8531	.8554	.8577	.8599	.8621
1.1	.8643	.8665	.8686	.8708	.8729	.8749	.8770	.8790	.8810	.8830
1.2	.8849	.8869	.8888	.8907	.8925	.8944	.8962	.8980	.8997	.9015
1.3	.9032	.9049	.9066	.9082	.9099	.9115	.9131	.9147	.9162	.9177
1.4	.9192	.9207	.9222	.9236	.9251	.9265	.9279	.9292	.9306	.9319
1.5	.9332	.9345	.9357	.9370	.9382	.9394	.9406	.9418	.9429	.9441
1.6	.9452	.9463	.9474	.9484	.9495	.9505	.9515	.9525	.9535	.9545
1.7	.9554	.9564	.9573	.9582	.9591	.9599	.9608	.9616	.9625	.9633
1.8	.9641	.9649	.9656	.9664	.9671	.9678	.9686	.9693	.9699	.9706
1.9	.9713	.9719	.9726	.9732	.9738	.9744	.9750	.9756	.9761	.9767
2.0	.9772	.9778	.9783	.9788	.9793	.9798	.9803	.9808	.9812	.9817
2.1	.9821	.9826	.9830	.9834	.9838	.9842	.9846	.9850	.9854	.9857
2.2	.9861	.9864	.9868	.9871	.9875	.9878	.9881	.9884	.9887	.9890
2.3	.9893	.9896	.9898	.9901	.9904	.9906	.9909	.9911	.9913	.9916
2.4	.9918	.9920	.9922	.9925	.9927	.9929	.9931	.9932	.9934	.9936
2.5	.9938	.9940	.9941	.9943	.9945	.9946	.9948	.9949	.9951	.9952
2.6	.9953	.9955	.9956	.9957	.9959	.9960	.9961	.9962	.9963	.9964
2.7	.9965	.9966	.9967	.9968	.9969	.9970	.9971	.9972	.9973	.9974
2.8	.9974	.9975	.9976	.9977	.9977	.9978	.9979	.9979	.9980	.9981
2.9	.9981	.9982	.9982	.9983	.9984	.9984	.9985	.9985	.9986	.9986
3.0	.9987	.9987	.9987	.9988	.9988	.9989	.9989	.9989	.9990	.9990
3.1	.9990	.9991	.9991	.9991	.9992	.9992	.9992	.9992	.9993	.9993
3.2	.9993	.9993	.9994	.9994	.9994	.9994	.9994	.9995	.9995	.9995
3.3	.9995	.9995	.9995	.9996	.9996	.9996	.9996	.9996	.9996	.9997
3.4	.9997	.9997	.9997	.9997	.9997	.9997	.9997	.9997	.9997	.9998
3.5	.9998	.9998	.9998	.9998	.9998	.9998	.9998	.9998	.9998	.9998
3.6	.9998	.9998	.9999	.9999	.9999	.9999	.9999	.9999	.9999	.9999
3.7	.9999	.9999	.9999	.9999	.9999	.9999	.9999	.9999	.9999	.9999
3.8	.9999	.9999	.9999	.9999	.9999	.9999	.9999	.9999	.9999	.9999
3.9	1.0000	1.0000	1.0000	1.0000	1.0000	1.0000	1.0000	1.0000	1.0000	1.0000
4.0	1.0000	1.0000	1.0000	1.0000	1.0000	1.0000	1.0000	1.0000	1.0000	1.0000

C.2 Frequency Factors for Pearson Type III Distribution

TABLE C.2: Pearson Type III Frequency Factors

Skew	Return period, T						
	2	5	10	25	50	100	200
3.0	−0.396	0.420	1.180	2.278	3.152	4.051	4.970
2.9	−0.390	0.440	1.195	2.277	3.134	4.013	4.909
2.8	−0.384	0.460	1.210	2.275	3.114	3.973	4.847
2.7	−0.376	0.479	1.224	2.272	3.093	3.932	4.783
2.6	−0.368	0.499	1.238	2.267	3.071	3.889	4.718
2.5	−0.360	0.518	1.250	2.262	3.048	3.845	4.652
2.4	−0.351	0.537	1.262	2.256	3.023	3.800	4.584
2.3	−0.341	0.555	1.274	2.248	2.997	3.753	4.515
2.2	−0.330	0.574	1.284	2.240	2.970	3.705	4.444
2.1	−0.319	0.592	1.294	2.230	2.942	3.656	4.372
2.0	−0.307	0.609	1.302	2.219	2.912	3.605	4.398
1.9	−0.294	0.627	1.310	2.207	2.881	3.553	4.223
1.8	−0.282	0.643	1.318	2.193	2.848	3.499	4.147
1.7	−0.268	0.660	1.324	2.179	2.815	3.444	4.069
1.6	−0.254	0.675	1.329	2.163	2.780	3.388	3.990
1.5	−0.240	0.690	1.333	2.146	2.743	3.330	3.910
1.4	−0.225	0.705	1.337	2.128	2.706	3.271	3.828
1.3	−0.210	0.719	1.339	2.108	2.666	3.211	3.745
1.2	−0.195	0.732	1.340	2.087	2.626	3.149	3.661
1.1	−0.180	0.745	1.341	2.066	2.585	3.087	3.575
1.0	−0.164	0.758	1.340	2.043	2.542	3.022	3.489
0.9	−0.148	0.769	1.339	2.018	2.498	2.957	3.401
0.8	−0.132	0.780	1.336	1.993	2.453	2.891	3.312
0.7	−0.116	0.790	1.333	1.967	2.407	2.824	3.223
0.6	−0.099	0.800	1.328	1.939	2.359	2.755	3.132
0.5	−0.083	0.808	1.323	1.910	2.311	2.686	3.041
0.4	−0.066	0.816	1.317	1.880	2.261	2.615	2.949
0.3	−0.050	0.824	1.309	1.849	2.211	2.544	2.856
0.2	−0.033	0.830	1.301	1.818	2.159	2.472	2.763
0.1	−0.017	0.836	1.292	1.785	2.107	2.400	2.670
0.0	−0.000	0.842	1.282	1.751	2.054	2.326	2.576
−0.1	0.017	0.846	1.270	1.716	2.000	2.252	2.482
−0.2	0.033	0.850	1.258	1.680	1.945	2.178	2.388
−0.3	0.050	0.853	1.245	1.643	1.890	2.104	2.294
−0.4	0.066	0.855	1.231	1.606	1.834	2.029	2.201
−0.5	0.083	0.856	1.216	1.567	1.777	1.955	2.108
−0.6	0.099	0.857	1.200	1.528	1.720	1.880	2.016
−0.7	0.116	0.857	1.183	1.488	1.663	1.806	1.926
−0.8	0.132	0.856	1.166	1.448	1.606	1.733	1.837
−0.9	0.148	0.854	1.147	1.407	1.549	1.660	1.749
−1.0	0.164	0.852	1.128	1.366	1.492	1.588	1.664
−1.1	0.180	0.848	1.107	1.324	1.435	1.518	1.581

TABLE C.2: Pearson Type III Frequency Factors (Cont'd)

Skew	Return period, T						
	2	5	10	25	50	100	200
−1.2	0.195	0.844	1.086	1.282	1.379	1.449	1.501
−1.3	0.210	0.838	1.064	1.240	1.324	1.383	1.424
−1.4	0.225	0.832	1.041	1.198	1.270	1.318	1.351
−1.5	0.240	0.825	1.018	1.157	1.217	1.256	1.282
−1.6	0.254	0.817	0.994	1.116	1.166	1.197	1.216
−1.7	0.268	0.808	0.970	1.075	1.116	1.140	1.155
−1.8	0.282	0.799	0.945	1.035	1.069	1.087	1.097
−1.9	0.294	0.788	0.920	0.996	1.023	1.037	1.044
−2.0	0.307	0.777	0.895	0.959	0.980	0.990	0.995
−2.1	0.319	0.765	0.869	0.923	0.939	0.946	0.949
−2.2	0.330	0.752	0.844	0.888	0.900	0.905	0.907
−2.3	0.341	0.739	0.819	0.855	0.864	0.867	0.869
−2.4	0.351	0.725	0.795	0.823	0.830	0.832	0.833
−2.5	0.360	0.711	0.771	0.793	0.798	0.799	0.800
−2.6	0.368	0.696	0.747	0.764	0.768	0.769	0.769
−2.7	0.376	0.681	0.724	0.738	0.740	0.740	0.741
−2.8	0.384	0.666	0.702	0.712	0.714	0.714	0.714
−2.9	0.390	0.651	0.681	0.683	0.689	0.690	0.690
−3.0	0.396	0.636	0.660	0.666	0.666	0.667	0.667

C.3   Critical Values of the Chi-Square Distribution

TABLE C.3: Critical Values of the Chi-Square Distribution

ν	α							
	0.995	0.990	0.975	0.950	0.050	0.025	0.010	0.005
1	.000	.000	.001	.004	3.841	5.024	6.635	7.880
2	.010	.020	.051	.103	5.991	7.378	9.210	10.597
3	.072	.115	.216	.352	7.815	9.348	11.345	12.838
4	.207	.297	.484	.711	9.488	11.143	13.277	14.861
5	.412	.554	.831	1.145	11.071	12.833	15.086	16.750
6	.676	.872	1.237	1.635	12.592	14.449	16.812	18.548
7	.989	1.239	1.690	2.167	14.067	16.013	18.476	20.279
8	1.344	1.646	2.180	2.733	15.507	17.535	20.090	21.956
9	1.735	2.088	2.700	3.325	16.919	19.023	21.666	23.590
10	2.156	2.558	3.247	3.940	18.307	20.483	23.210	25.189
11	2.603	3.053	3.816	4.575	19.675	21.920	24.725	26.757
12	3.074	3.571	4.404	5.226	21.026	23.337	26.217	28.300
13	3.565	4.107	5.009	5.892	22.362	24.736	27.688	29.819
14	4.075	4.660	5.629	6.571	23.685	26.119	29.141	31.319
15	4.601	5.229	6.262	7.261	24.996	27.488	30.578	32.801
16	5.142	5.812	6.908	7.962	26.296	28.845	32.000	34.267
17	5.697	6.408	7.564	8.672	27.587	30.191	33.409	35.718
18	6.265	7.015	8.231	9.390	28.869	31.526	34.805	37.156
19	6.844	7.633	8.907	10.117	30.144	32.852	36.191	38.582

TABLE C.3: Critical Values of the Chi-Square Distribution (Cont'd)

ν	0.995	0.990	0.975	0.950	0.050	0.025	0.010	0.005
20	7.434	8.260	9.591	10.851	31.410	34.170	37.566	39.997
21	8.034	8.897	10.283	11.591	32.671	35.479	38.932	41.401
22	8.643	9.542	10.982	12.338	33.924	36.781	40.289	42.796
23	9.260	10.196	11.689	13.091	35.172	38.076	41.638	44.181
24	9.886	10.856	12.401	13.848	36.415	39.364	42.980	45.559
25	10.520	11.524	13.120	14.611	37.652	40.646	44.314	46.928
26	11.160	12.198	13.844	15.379	38.885	41.923	45.642	48.290
27	11.808	12.879	14.573	16.151	40.113	43.195	46.963	49.645
28	12.461	13.565	15.308	16.928	41.337	44.461	48.278	50.993
29	13.121	14.256	16.047	17.708	42.557	45.722	49.588	52.336
30	13.787	14.953	16.791	18.493	43.773	46.979	50.892	53.672
40	20.707	22.164	24.433	26.509	55.758	59.342	63.691	66.766
50	27.991	29.707	32.357	34.764	67.505	71.420	76.154	79.490
60	35.534	37.485	40.482	43.188	79.082	83.298	88.379	91.952
70	43.275	45.442	48.758	51.739	90.531	95.023	100.425	104.215
80	51.172	53.540	57.153	60.391	101.879	106.629	112.329	116.321
90	59.196	61.754	65.647	69.126	113.145	118.136	124.116	128.299
100	67.328	70.065	74.222	77.929	124.342	129.561	135.807	140.170

C.4 Critical Values for the Kolmogorov–Smirnov Test Statistic

Sample size (n)	Significance level				
	0.20	0.15	0.10	0.05	0.01
1	0.900	0.925	0.950	0.975	0.995
2	0.684	0.726	0.776	0.842	0.929
3	0.585	0.597	0.642	0.708	0.829
4	0.494	0.525	0.564	0.624	0.734
5	0.446	0.474	0.510	0.563	0.669
6	0.410	0.436	0.470	0.521	0.618
7	0.381	0.405	0.438	0.486	0.577
8	0.358	0.381	0.411	0.457	0.543
9	0.339	0.360	0.388	0.432	0.514
10	0.322	0.342	0.368	0.409	0.486
11	0.307	0.326	0.352	0.391	0.468
12	0.295	0.313	0.338	0.375	0.450
13	0.284	0.302	0.325	0.361	0.433
14	0.274	0.292	0.314	0.349	0.418
15	0.266	0.283	0.304	0.338	0.404
16	0.258	0.274	0.295	0.328	0.391
17	0.250	0.266	0.286	0.318	0.380
18	0.244	0.259	0.278	0.309	0.370
19	0.237	0.252	0.272	0.301	0.361
20	0.231	0.246	0.264	0.294	0.352

Sample size (n)	Significance level				
	0.20	0.15	0.10	0.05	0.01
25	0.210	0.220	0.240	0.264	0.320
30	0.190	0.200	0.220	0.242	0.290
35	0.180	0.190	0.210	0.230	0.270
40				0.210	0.250
50				0.190	0.230
60				0.170	0.210
70				0.160	0.190
80				0.150	0.180
90				0.140	
100				0.140	
Asymptotic formula:	$\dfrac{1.07}{\sqrt{n}}$	$\dfrac{1.14}{\sqrt{n}}$	$\dfrac{1.22}{\sqrt{n}}$	$\dfrac{1.36}{\sqrt{n}}$	$\dfrac{1.63}{\sqrt{n}}$

APPENDIX D

Special Functions

D.1 Error Function

The error function, $\text{erf}(z)$, is defined by the relation

$$\text{erf}(z) = \frac{2}{\sqrt{\pi}} \int_0^z e^{-x^2} dx \qquad (D.1)$$

The error function is closely related to the cumulative distribution function of a normal probability distribution, and is defined for $-\infty \leq z \leq \infty$. The error function is antisymmetric, such that

$$\text{erf}(-z) = -\text{erf}(z) \qquad (D.2)$$

The constant before the integral sign in Equation D.1 is simply a normalizing constant such that $\text{erf}(z)$ approaches 1 as z approaches infinity. For small values of z, it is convenient to use the series expansion for e^{-x^2} to obtain (Carslaw and Jaeger, 1959)

$$\text{erf}(z) = \frac{2}{\sqrt{\pi}} \int_0^z \left[\sum_{n=0}^{\infty} \frac{(-1)^n x^{2n}}{n!} \right] dx \qquad (D.3)$$

Since the series is uniformly convergent, it can be integrated term by term to yield

$$\text{erf}(z) = \frac{2}{\sqrt{\pi}} \left[\sum_{n=0}^{\infty} \frac{(-1)^n z^{2n+1}}{(2n+1)n!} \right] \qquad (D.4)$$

which can also be written in the form (Hermance, 1999)

$$\text{erf}(z) = \frac{2}{\sqrt{\pi}} \left[z - \frac{z^3}{3} + \frac{z^5}{2!5} - \frac{z^7}{3!7} + \dots \right] \qquad (D.5)$$

This relationship is particularly useful in estimating $\text{erf}(z)$ for small values of z. A closely related function to the error function is the *complementary error function*, $\text{erfc}(z)$, which is defined by

$$\text{erfc}(z) = 1 - \text{erf}(z) \qquad (D.6)$$

Values of the error function, $\text{erf}(z)$, are listed in Table D.1.

TABLE D.1: Error Function

z	erf (z)
0.0	0.00000
0.1	0.11246
0.2	0.22270
0.3	0.32863
0.4	0.42839
0.5	0.52050
0.6	0.60386
0.7	0.67780
0.8	0.74210
0.9	0.79691
1.0	0.84270
1.1	0.88021
1.2	0.91031
1.3	0.93401
1.4	0.95229
1.5	0.96611
1.6	0.97635
1.7	0.98379
1.8	0.98909
1.9	0.99279
2.0	0.99532
2.1	0.99702
2.2	0.99814
2.3	0.99886
2.4	0.99931
2.5	0.99959
2.6	0.99976
2.7	0.99987
2.8	0.99992
2.9	0.99996
3.0	0.99998
∞	1.00000

D.2   Bessel Functions

D.2.1   Definition

A second-order linear homogeneous differential equation of the form

$$x^2\frac{d^2y}{dx^2} + x\frac{dy}{dx} + (x^2 - n^2)y = 0, \qquad n \geq 0 \tag{D.7}$$

is called *Bessel's equation*. The general solutions of Bessel's equation are

$$y = AJ_n(x) + BJ_{-n}(x), \qquad n \neq 1, 2, \ldots \tag{D.8}$$

$$y = AJ_n(x) + BY_n(x), \qquad \text{all } n \tag{D.9}$$

where $J_n(x)$ is called the *Bessel function of the first kind of order n*, and $Y_n(x)$ is called the *Bessel function of the second kind of order n*.

If Bessel's equation (Equation D.7) is slightly modified and written in the form

$$x^2 \frac{d^2 y}{dx^2} + x \frac{dy}{dx} - (x^2 + n^2)y = 0, \qquad n \geq 0 \tag{D.10}$$

then this equation is called the *modified Bessel's equation*. The general solutions of the modified Bessel's equation are

$$y = A I_n(x) + B I_{-n}(x), \qquad n \neq 1, 2, \ldots \tag{D.11}$$

$$y = A I_n(x) + B K_n(x), \qquad \text{all } n \tag{D.12}$$

where $I_n(x)$ is called the *modified Bessel function of the first kind of order n*, and $K_n(x)$ is called the *modified Bessel function of the second kind of order n*.

D.2.2 Evaluation of Bessel Functions

The solution of Bessel's equations can be found in most calculus texts—for example, Hildebrand (1976). The Bessel functions cannot generally be expressed in closed form, and are usually presented as infinite series.

D.2.2.1 Bessel function of the first kind of order n

This function is given by

$$J_n(x) = \frac{x^n}{2^n \Gamma(n + 1)} \left[1 - \frac{x^2}{2(2n + 2)} + \frac{x^4}{2 \cdot 4(2n + 2)(2n + 4)} - \cdots \right] \tag{D.13}$$

$$= \sum_{k=0}^{\infty} \frac{(-1)^k (x/2)^{n+2k}}{k! \Gamma(n + k + 1)} \tag{D.14}$$

The Bessel function $J_{-n}(x)$ can be derived from Equation D.14 by simply replacing n by $-n$ in the formula. A convenient relationship to note is

$$J_{-n}(x) = (-1)^n J_n(x), \qquad n = 0, 1, 2, \ldots \tag{D.15}$$

D.2.2.2 Bessel function of the second kind of order n

This function is given by

$$Y_n(x) = \begin{cases} \frac{J_n(x) \cos n\pi - J_{-n}(x)}{\sin n\pi}, & n \neq 0, 1, 2, \ldots \\ \lim_{p \to n} \frac{J_p(x) \cos p\pi - J_{-p}(x)}{\sin p\pi}, & n = 0, 1, 2, \ldots \end{cases} \tag{D.16}$$

D.2.2.3 Modified Bessel function of the first kind of order n

This function is given by

$$I_n(x) = \frac{x^n}{2^n \Gamma(n + 1)} \left[1 + \frac{x^2}{2(2n + 2)} + \frac{x^4}{2 \cdot 4(2n + 2)(2n + 4)} + \cdots \right] \tag{D.17}$$

$$= \sum_{k=0}^{\infty} \frac{(x/2)^{n+2k}}{k! \Gamma(n + k + 1)} \tag{D.18}$$

The Bessel function $I_{-n}(x)$ can be derived from Equation D.18 by simply replacing n by $-n$ in the formula. A convenient relationship to note is

$$I_{-n}(x) = I_n(x), \qquad n = 0, 1, 2, \ldots \tag{D.19}$$

D.2.2.4   Modified Bessel function of the second kind of order n

This function is given by

$$K_n(x) = \begin{cases} \dfrac{\pi}{2 \sin n\pi}[I_{-n}(x) - I_n(x)], & n \neq 0, 1, 2, \ldots \\ \lim_{p \to n} \dfrac{\pi}{2 \sin p\pi}[I_{-p}(x) - I_p(x)], & n = 0, 1, 2, \ldots \end{cases} \tag{D.20}$$

D.2.2.5   Tabulated values of useful Bessel functions

TABLE D.2: Useful Bessel Functions

x	$I_0(x)$	$K_0(x)$	$I_1(x)$	$K_1(x)$
.001	1.0000	7.0237	.0005	999.9962
.002	1.0000	6.3305	.0010	499.9932
.003	1.0000	5.9251	.0015	333.3237
.004	1.0000	5.6374	.0020	249.9877
.005	1.0000	5.4143	.0025	199.9852
.006	1.0000	5.2320	.0030	166.6495
.007	1.0000	5.0779	.0035	142.8376
.008	1.0000	4.9443	.0040	124.9782
.009	1.0000	4.8266	.0045	111.0871
.010	1.0000	4.7212	.0050	99.9739
.020	1.0001	4.0285	.0100	49.9547
.030	1.0002	3.6235	.0150	33.2715
.040	1.0004	3.3365	.0200	24.9233
.050	1.0006	3.1142	.0250	19.9097
.060	1.0009	2.9329	.0300	16.5637
.070	1.0012	2.7798	.0350	14.1710
.080	1.0016	2.6475	.0400	12.3742
.090	1.0020	2.5310	.0450	10.9749
.100	1.0025	2.4271	.0501	9.8538
.110	1.0030	2.3333	.0551	8.9353
.120	1.0036	2.2479	.0601	8.1688
.130	1.0042	2.1695	.0651	7.5192
.140	1.0049	2.0972	.0702	6.9615
.150	1.0056	2.0300	.0752	6.4775
.160	1.0064	1.9674	.0803	6.0533
.170	1.0072	1.9088	.0853	5.6784
.180	1.0081	1.8537	.0904	5.3447
.190	1.0090	1.8018	.0954	5.0456
.200	1.0100	1.7527	.1005	4.7760
.210	1.0111	1.7062	.1056	4.5317
.220	1.0121	1.6620	.1107	4.3092

TABLE D.2: Useful Bessel Functions (Cont'd)

x	$I_0(x)$	$K_0(x)$	$I_1(x)$	$K_1(x)$
.230	1.0133	1.6199	.1158	4.1058
.240	1.0145	1.5798	.1209	3.9191
.250	1.0157	1.5415	.1260	3.7470
.260	1.0170	1.5048	.1311	3.5880
.270	1.0183	1.4697	.1362	3.4405
.280	1.0197	1.4360	.1414	3.3033
.290	1.0211	1.4036	.1465	3.1755
.300	1.0226	1.3725	.1517	3.0560
.310	1.0242	1.3425	.1569	2.9441
.320	1.0258	1.3136	.1621	2.8390
.330	1.0274	1.2857	.1673	2.7402
.340	1.0291	1.2587	.1725	2.6470
.350	1.0309	1.2327	.1777	2.5591
.360	1.0327	1.2075	.1829	2.4760
.370	1.0345	1.1832	.1882	2.3973
.380	1.0364	1.1596	.1935	2.3227
.390	1.0384	1.1367	.1987	2.2518
.400	1.0404	1.1145	.2040	2.1844
.410	1.0425	1.0930	.2093	2.1202
.420	1.0446	1.0721	.2147	2.0590
.430	1.0468	1.0518	.2200	2.0006
.440	1.0490	1.0321	.2254	1.9449
.450	1.0513	1.0129	.2307	1.8915
.460	1.0536	.9943	.2361	1.8405
.470	1.0560	.9761	.2415	1.7916
.480	1.0584	.9584	.2470	1.7447
.490	1.0609	.9412	.2524	1.6997
.500	1.0635	.9244	.2579	1.6564
.510	1.0661	.9081	.2634	1.6149
.520	1.0688	.8921	.2689	1.5749
.530	1.0715	.8766	.2744	1.5364
.540	1.0742	.8614	.2800	1.4994
.550	1.0771	.8466	.2855	1.4637
.560	1.0800	.8321	.2911	1.4292
.570	1.0829	.8180	.2967	1.3960
.580	1.0859	.8042	.3024	1.3638
.590	1.0889	.7907	.3080	1.3328
.600	1.0920	.7775	.3137	1.3028
.610	1.0952	.7646	.3194	1.2738
.620	1.0984	.7520	.3251	1.2458
.630	1.1017	.7397	.3309	1.2186
.640	1.1051	.7277	.3367	1.1923
.650	1.1084	.7159	.3425	1.1668
.660	1.1119	.7043	.3483	1.1420

TABLE D.2: Useful Bessel Functions (Cont'd)

x	$I_0(x)$	$K_0(x)$	$I_1(x)$	$K_1(x)$
.670	1.1154	.6930	.3542	1.1181
.680	1.1190	.6820	.3600	1.0948
.690	1.1226	.6711	.3659	1.0722
.700	1.1263	.6605	.3719	1.0503
.710	1.1301	.6501	.3778	1.0290
.720	1.1339	.6399	.3838	1.0083
.730	1.1377	.6300	.3899	.9882
.740	1.1417	.6202	.3959	.9686
.750	1.1456	.6106	.4020	.9496
.760	1.1497	.6012	.4081	.9311
.770	1.1538	.5920	.4142	.9130
.780	1.1580	.5829	.4204	.8955
.790	1.1622	.5740	.4266	.8784
.800	1.1665	.5653	.4329	.8618
.810	1.1709	.5568	.4391	.8456
.820	1.1753	.5484	.4454	.8298
.830	1.1798	.5402	.4518	.8144
.840	1.1843	.5321	.4581	.7993
.850	1.1889	.5242	.4646	.7847
.860	1.1936	.5165	.4710	.7704
.870	1.1984	.5088	.4775	.7564
.880	1.2032	.5013	.4840	.7428
.890	1.2080	.4940	.4905	.7295
.900	1.2130	.4867	.4971	.7165
.910	1.2180	.4796	.5038	.7039
.920	1.2231	.4727	.5104	.6915
.930	1.2282	.4658	.5171	.6794
.940	1.2334	.4591	.5239	.6675
.950	1.2387	.4524	.5306	.6560
.960	1.2440	.4459	.5375	.6447
.970	1.2494	.4396	.5443	.6336
.980	1.2549	.4333	.5512	.6228
.990	1.2604	.4271	.5582	.6122
1.000	1.2661	.4210	.5652	.6019
1.100	1.3262	.3656	.6375	.5098
1.200	1.3937	.3185	.7147	.4346
1.300	1.4693	.2782	.7973	.3725
1.400	1.5534	.2437	.8861	.3208
1.500	1.6467	.2138	.9817	.2774
1.600	1.7500	.1880	1.0848	.2406
1.700	1.8640	.1655	1.1963	.2094
1.800	1.9896	.1459	1.3172	.1826
1.900	2.1277	.1288	1.4482	.1597
2.000	2.2796	.1139	1.5906	.1399

TABLE D.2: Useful Bessel Functions (Cont'd)

x	$I_0(x)$	$K_0(x)$	$I_1(x)$	$K_1(x)$
2.100	2.4463	.1008	1.7455	.1227
2.200	2.6291	.0893	1.9141	.1079
2.300	2.8296	.0791	2.0978	.0950
2.400	3.0493	.0702	2.2981	.0837
2.500	3.2898	.0623	2.5167	.0739
2.600	3.5533	.0554	2.7554	.0653
2.700	3.8417	.0493	3.0161	.0577
2.800	4.1573	.0438	3.3011	.0511
2.900	4.5027	.0390	3.6126	.0453
3.000	4.8808	.0347	3.9534	.0402
3.100	5.2945	.0310	4.3262	.0356
3.200	5.7472	.0276	4.7343	.0316
3.300	6.2426	.0246	5.1810	.0281
3.400	6.7848	.0220	5.6701	.0250
3.500	7.3782	.0196	6.2058	.0222
3.600	8.0277	.0175	6.7927	.0198
3.700	8.7386	.0156	7.4357	.0176
3.800	9.5169	.0140	8.1404	.0157
3.900	10.3690	.0125	8.9128	.0140
4.000	11.3019	.0112	9.7595	.0125
5.000	27.2399	.0037	24.3356	.0040
6.000	67.2344	.0012	61.3419	.0013
7.000	168.5939	.0004	156.0391	.0005
8.000	427.5641	.0001	399.8731	.0002
9.000	1093.5880	.0001	1030.9150	.0001
10.000	2815.7170	.0000	2670.9880	.0000

D.3 Gamma Function

TABLE D.3: Gamma Function

z	$\Gamma(z)$	z	$\Gamma(z)$	z	$\Gamma(z)$
1.00	1.00000	1.34	0.89222	1.68	0.90500
1.01	0.99433	1.35	0.89115	1.69	0.90678
1.02	0.98884	1.36	0.89018	1.70	0.90864
1.03	0.98355	1.37	0.88931	1.71	0.91057
1.04	0.97844	1.38	0.88854	1.72	0.91258
1.05	0.97350	1.39	0.88785	1.73	0.91467
1.06	0.96874	1.40	0.88726	1.74	0.91683
1.07	0.96415	1.41	0.88676	1.75	0.91906
1.08	0.95973	1.42	0.88636	1.76	0.92137
1.09	0.95546	1.43	0.88604	1.77	0.92376
1.10	0.95135	1.44	0.88581	1.78	0.92623
1.11	0.94740	1.45	0.88566	1.79	0.92877

TABLE D.3: Gamma Function (Cont'd)

z	$\Gamma(z)$	z	$\Gamma(z)$	z	$\Gamma(z)$
1.12	0.94359	1.46	0.88560	1.80	0.93138
1.13	0.93993	1.47	0.88563	1.81	0.93408
1.14	0.93642	1.48	0.88575	1.82	0.93685
1.15	0.93304	1.49	0.88595	1.83	0.93969
1.16	0.92980	1.50	0.88623	1.84	0.94261
1.17	0.92670	1.51	0.88659	1.85	0.94561
1.18	0.92373	1.52	0.88704	1.86	0.94869
1.19	0.92089	1.53	0.88757	1.87	0.95184
1.20	0.91817	1.54	0.88818	1.88	0.95507
1.21	0.91558	1.55	0.88887	1.89	0.95838
1.22	0.91311	1.56	0.88964	1.90	0.96177
1.23	0.91075	1.57	0.89049	1.91	0.96523
1.24	0.90852	1.58	0.89142	1.92	0.96877
1.25	0.90640	1.59	0.89243	1.93	0.97240
1.26	0.90440	1.60	0.89352	1.94	0.97610
1.27	0.90250	1.61	0.89468	1.95	0.97988
1.28	0.90072	1.62	0.89592	1.96	0.98374
1.29	0.89904	1.63	0.89724	1.97	0.98768
1.30	0.89747	1.64	0.89864	1.98	0.99171
1.31	0.89600	1.65	0.90012	1.99	0.99581
1.32	0.89464	1.66	0.90167	2.00	1.00000
1.33	0.89338	1.67	0.90330		

APPENDIX E

Pipe Specifications

E.1 PVC Pipe

Pipe dimensions of interest to engineers are usually the diameter and wall thickness. The pipe diameter is typically specified by the *nominal pipe size* and the wall thickness by the *schedule*. Nominal pipe sizes, typically given in either inches or millimeters, represent rounded approximations to the inside diameter of the pipe. The schedule of a pipe is a number that approximates the value of the expression $1000P/S$, where P is the service pressure and S is the allowable stress. Higher schedule numbers correspond to thicker pipes, and schedule numbers in common use are: 5, 5S, 10, 10S, 20, 20S, 30, 40, 40S, 60, 80, 80S, 100, 120, 140, and 160. The schedule numbers followed by the letter "S" are primarily intended for use with stainless steel pipe (ASME B36.19M).

		Schedule 5		Schedule 10		Schedule 40		Schedule 80	
Nominal pipe size (mm)	Outside diameter (mm)	Wall thickness (mm)	Inside diameter (mm)	Wall thickness (mm)	Inside diameter (mm)	Wall thickness (mm)	Inside diameter (mm)	Wall thickness (mm)	Inside diameter (mm)
50	60	1.7	57	2.8	55	3.9	53	5.5	49
80	90	2.1	85	3.0	83	5.5	78	7.6	74
100	114	2.1	110	3.0	108	6.0	102	8.6	97
125	141	2.8	136	3.4	134	6.6	128	9.5	122
150	168	2.8	163	3.4	161	7.1	154	11.0	146

Source: Fetter (1999).

		Schedule 5		Schedule 10		Schedule 40		Schedule 80	
Nominal pipe size (in.)	Outside diameter (in.)	Wall thickness (in.)	Inside diameter (in.)	Wall thickness (in.)	Inside diameter (in.)	Wall thickness (in.)	Inside diameter (in.)	Wall thickness (in.)	Inside diameter (in.)
2	2.375	0.065	2.245	0.109	2.157	0.154	2.067	0.218	1.939
3	3.500	0.083	3.334	0.120	3.260	0.216	3.068	0.300	2.900
4	4.500	0.083	4.334	0.120	4.260	0.237	4.026	0.337	3.826
5	5.563	0.109	5.345	0.134	5.295	0.258	5.047	0.375	4.813
6	6.625	0.109	6.407	0.134	6.357	0.280	6.065	0.432	5.761

Source: Fetter (1999).

E.2 Ductile Iron Pipe

Ductile iron pipe is manufactured in diameters from 100 to 1200 mm (4 to 48 in.), and for diameters from 100 to 500 mm (4 to 20 in.) standard commercial sizes are available in 50-mm (2 in.) increments, while for diameters from 600 to 1200 mm (24 to 48 in.) the size increments are 150 mm (6 in.). The standard lengths of ductile iron pipe are 5.5 m (18 ft) and 6.1 m (20 ft).

E.3 Concrete Pipe

TABLE E.1: Commercially Available Sizes of Concrete Pipe

Nonreinforced pipe		Reinforced pipe	
Diameter (mm)	Diameter (in.)	Diameter (mm)	Diameter (in.)
100	4	—	—
150	6	—	—
205	8	—	—
255	10	—	—
305	12	305	12
380	15	380	15
455	18	455	18
535	21	535	21
610	24	610	24
685	27	685	27
760	30	760	30
840	33	840	33
915	36	915	36
—	—	1065	42
—	—	1220	48
—	—	1370	54
—	—	1525	60
—	—	1675	66
—	—	1830	72
—	—	1980	78
—	—	2135	84
—	—	2285	90
—	—	2440	96
—	—	2590	102
—	—	2745	108

E.4 Physical Properties of Common Pipe Materials

Material	Young's modulus, E (GPa)	Poisson's ratio
Concrete	14–30	0.10–0.15
Concrete (reinforced)	30–60	—
Ductile iron	165–172	0.28–0.30
PVC	2.4–3.5	0.45–0.46
Steel	200–207	0.30

Bibliography

[1] M. Abramowitz and I.A. Stegun. *Handbook of Mathematical Functions*. Dover, New York, 1965.

[2] M. Abramowitz and I.A. Stegun. *Handbook of Mathematical Functions*. Dover, New York, 1972.

[3] W. Abtew. Evapotranspiration measurements and modeling for three wetland systems in South Florida. *Journal of the American Water Resources Association*, 32(3):465–473, 1996.

[4] W. Abtew, J. Obeysekera, M. Irizarry-Ortiz, D. Lyons, and A. Reardon. Evapotranspiration estimation for South Florida. EMA 407, South Florida Water Management District, West Palm Beach, Florida, January 2003.

[5] W. Abtew and J. Obeysekera. Lysimeter study of evapotranspiration of cattails and comparison of three estimation methods. *Transactions of the American Society of Agricultural Engineers*, 38(1):121–129, 1995.

[6] M.M. Abu-Seida and A.A. Quraishi. A flow equation for submerged rectangular weirs. *Proceedings of the Institution of Civil Engineers*, 61(2):685–696, 1976.

[7] P. Ackers, W.R. White, J.A. Perkins, and A.J.M. Harrison. *Weirs and Flumes for Flow Measurement*. Wiley, Chichester, 1978.

[8] N. Ahmed and D.K. Sunada. Nonlinear flow in porous media. *Journal of the Hydraulics Division, ASCE*, 95(HY6):1847–1857, 1969.

[9] T.P. Ahrens. Well design criteria. *Water Well Journal, National Water Well Association*, September and November 1957.

[10] A.J. Aisenbrey, Jr., R.B. Hayes, H.J. Warren, D.L. Winsett, and R.B. Young. *Design of Small Canal Structures*. United States Department of the Interior, Bureau of Reclamation, Denver, CO, 1974.

[11] A.O. Akan. *Urban Stormwater Hydrology*. Technomic Publishing Company, Inc., Lancaster, PA, 1993.

[12] R. Aldridge and R.J. Jackson. Interception of rainfall by hard beech. *New Zealand Journal of Science*, 16:185–198, 1973.

[13] A. Alexandrou. *Principles of Fluid Mechanics*. Prentice Hall, Upper Saddle River, NJ, 2001.

[14] Y. Alila. Regional rainfall depth-duration-frequency equations for Canada. *Water Resources Research*, 36(7):1767–1778, July 2000.

[15] R.G. Allen. A Penman for all seasons. *Journal of Irrigation and Drainage Engineering, ASCE*, 112(4):348–368, 1986.

[16] R.G. Allen. Self-calibrating method for estimating solar radiation from air temperature. *Journal of Irrigation and Drainage Engineering, ASCE*, 2(2):56–67, April 1997.

[17] R.G. Allen, A.J. Clemmens, C.M. Burt, K. Solomon, and T. O'Halloran. Prediction accuracy for projectwide evapotranspiration using crop coefficients and reference evapotranspiration. *Journal of Irrigation and Drainage Engineering, ASCE*, 131(1):24–36, February 2005d.

[18] R.G. Allen, M.E. Jensen, J.L. Wright, and R.D. Burman. Operational estimates of reference evapotranspiration. *Journal of Agronomy*, 81:650–662, 1989.

[19] R.G. Allen, L.S. Pereira, D. Raes, and M. Smith. Crop evapotranspiration. Guidelines for computer crop water requirements. Irrigation and Drainage Paper 56, Food and Agricultural Organization of the United Nations (FAO), Rome, Italy, 1998.

[20] R.G. Allen, L.S. Pereira, M. Smith, D. Raes, and J.L. Wright. FAO-56 dual crop coefficient method for estimating evaporation from soil and application extensions. *Journal of Irrigation and Drainage Engineering, ASCE*, 131(1):2–13, February 2005b.

[21] R.G. Allen and W.O. Pruitt. Rational use of the FAO Blaney–Criddle formula. *Journal of Irrigation and Drainage Engineering, ASCE*, 112(IR2):139–155, 1986.

[22] R.G. Allen, W.O. Pruitt, D. Raes, M. Smith, and L.S. Pereira. Estimating evaporation from bare soil and the crop coefficient for the initial period using common soils information. *Journal of Irrigation and Drainage Engineering, ASCE*, 131(1):14–23, February 2005c.

[23] R.G. Allen, M. Smith, L.S. Pereira, and A. Perrier. An update for the calculation of reference evapotranspiration. *ICID Bulletin*, 43(2):35–92, 1994b.

[24] R.G. Allen, M. Smith, A. Perrier, and L.S. Pereira. An update for the definition of reference evapotranspiration. *ICID Bulletin*, 43(2):1–34, 1994a.

[25] R.J. Allen and A.T. DeGaetano. Areal reduction factors for two Eastern United States regions with high rain-gauge density. *Journal of Hydraulic Engineering*, 10(4):327–335, July 2005.

[26] R.J. Allen and W.O. Pruitt. FAO-24 reference evapotranspiration factors. *Journal of Irrigation and Drainage Engineering, ASCE*, 117(5):758–773, 1991.

[27] M. Amein. Stream flow routing on computer by characteristics. *Water Resources Research*, 2(1):123–130, 1966.

[28] M. Amein and C.S. Fang. Stream flow routing (with applications to North Carolina rivers). Report 17, Water Resources Research Institute of the University of North Carolina, Raleigh, NC, 1969.

[29] American Concrete Pipe Association. *Concrete Pipe Handbook*. American Concrete Pipe Association, Vienna, VA 1981.

[30] American Concrete Pipe Association. *Concrete Pipe Design Manual*. American Concrete Pipe Association, Vienna, VA 1985.

[31] American Iron and Steel Institute. *Modern Sewer Design*. AISI, Washington, DC, 3d ed., 1995.

[32] American Society for Testing and Materials. Standard practice for design and installation of ground water monitoring wells in aquifers. Technical Report D 5092-1990, ASTM Committee D-18 on Soil and Rock, 1990.

[33] American Society for Testing and Materials. Standard guide for selection of aquifer test method in determining of hydraulic properties by well techniques. Technical Report D 4043-91, ASTM Committee D-18 on Soil and Rock, 1991a.

[34] American Society for Testing and Materials. Standard test method (field procedure) for withdrawal and injection well tests for determining hydraulic properties of aquifer systems. Technical Report D 4050-91, ASTM Committee D-18 on Soil and Rock, 1991b.

[35] American Society for Testing and Materials. Standard test method (analytical procedure) for determining transmissivity and storage coefficient of nonleaky confined aquifers by overdamped well response to instantaneous change in head (slug test). Technical Report D 4106-91, ASTM Committee D-18 on Soil and Rock, 1991c.

[36] American Society for Testing and Materials. Standard method for open-channel flow measurement of water with thin-plate weirs. Technical Report D5242, ASTM, West Conshohocken, PA, 1993.

[37] American Society of Civil Engineers. Report of ASCE Task Force on Friction Factors in Open Channels. *Proceedings of the American Society of Civil Engineers*, 89(HY2):97, March 1963.

[38] American Society of Civil Engineers. *Gravity Sanitary Sewer Design and Construction*. American Society of Civil Engineers, New York, 1982. Manual of Practice No. 60.

[39] American Society of Civil Engineers. *Civil Engineering Guidelines for Planning and Designing Hydroelectric Developments*. ASCE, New York, 1989.

[40] American Society of Civil Engineers. *Design and Construction of Urban Stormwater Management Systems*. American Society of Civil Engineers, New York, 1992.

[41] American Society of Civil Engineers. *Hydrology Handbook*. American Society of Civil Engineers, New York, 2d ed., 1996. Manual of Practice No. 28.

[42] American Society of Civil Engineers. *Technical Engineering and Design Guides as Adapted from the US Army Corps of Engineers, No. 19: Flood-Runoff Analysis*. ASCE Press, New York, 1996.

[43] American Society of Civil Engineers. *Urban Runoff Quality Management*. American Society of Civil Engineers, New York, 1998. Manual of Practice No. 23.

[44] American Society of Heating, Refrigerating and Air Conditioning Engineers. *ASHRAE Handbook, 1981. Fundamentals*. ASHRAE, New York, 1981.

[45] American Water Works Association. *1984 Water Utility Operating Data*. AWWA, Denver, Colorado, 1986.

[46] American Water Works Association. *Distribution System Requirements for Fire Protection, Manual of Water Supply Practices M31*. AWWA, Denver, CO, 1992.

[47] American Water Works Association. *Water Resources Planning, Manual of Water Supply Practices M50*. AWWA, Denver, CO, 1st ed., 2001.

[48] American Water Works Association. *PVC Pipe–Design and Installation, Manual of Water Supply Practices M23*. AWWA, Denver, CO, 2d ed., 2002b.

[49] American Water Works Association. *Water Sources. Principles and Practices of Water Supply Operations*. AWWA, Denver, CO, 3d ed., 2003a.

[50] American Water Works Association. *Groundwater, Manual of Water Supply Practices M21*. AWWA, Denver, CO, 3d ed., 2003b.

[51] American Water Works Association. *Water Transmission and Distribution*. AWWA, Denver, CO, 3d ed., 2003c.

[52] American Water Works Association. *Ductile-Iron Pipe and Fittings, Manual of Water Supply Practices M41*. AWWA, CO, 2d ed., 2003d.

[53] American Water Works Association. *Sizing Water Service Lines and Meters, Manual of Water Supply Practices M22*. AWWA, Denver, CO, 2d ed., 2004.

[54] American Water Works Association. *American National Standard for Polyethylene Encasement for Ductile-Iron Pipe Systems*. ANSI/AWWA C105/A21.5. AWWA, Denver, CO, latest edition.

[55] American Water Works Association. *AWWA Standard for Cement-Mortar Protective Lining and Coating for Steel Water Pipe—4 in. (100 mm) and Larger—Shop Applied*. ANSI/AWWA C205. AWWA, Denver, CO, latest edition.

[56] American Water Works Association. *AWWA Standard for Concrete Pressure Pipe, Bar-Wrapped, Steel Cylinder Type.* ANSI/AWWA C303. AWWA, Denver, CO, latest edition.

[57] American Water Works Association. *AWWA Standard for Design of Prestressed Concrete Cylinder Pipe.* ANSI/AWWA C304. AWWA, Denver, CO, latest edition.

[58] American Water Works Association. *AWWA Standard for Fiberglass Pressure Pipe.* ANSI/AWWA C950. AWWA, Denver, CO, latest edition.

[59] American Water Works Association. *AWWA Standard for Liquid-Epoxy Coating Systems for the Interior and Exterior of Steel Water Pipelines.* ANSI/AWWA C210. AWWA, Denver, CO, latest edition.

[60] American Water Works Association. *AWWA Standard for Polyethylene (PE) Pressure Pipe and Fittings, 4 in. (100 mm) through 63 in. (1575 mm), for Water Distribution and Transmission.* ANSI/AWWA C906. AWWA, Denver, CO, latest edition.

[61] American Water Works Association. *AWWA Standard for Prestressed Concrete Pressure Pipe, Steel Cylinder Type.* ANSI/AWWA C301. AWWA, Denver, CO, latest edition.

[62] American Water Works Association. *AWWA Standard for Reinforced Concrete Pressure Pipe, Noncylinder Type.* ANSI/AWWA C302. AWWA, Denver, CO, latest edition.

[63] American Water Works Association. *AWWA Standard for Reinforced Concrete Pressure Pipe, Steel-Cylinder Type.* ANSI/AWWA C300. AWWA, Denver, CO, latest edition.

[64] American Water Works Association. *Manual M11, Steel Pipe–A Guide for Design and Installation.* AWWA, Denver, CO, latest edition.

[65] American Water Works Association. *Manual M9, Concrete Pressure Pipe.* AWWA, Denver, CO, latest edition.

[66] W.F. Ames. *Numerical Methods for Partial Differential Equations.* Academic Press, New York, 2d ed., 1977.

[67] A.G. Anderson, A. Painta, and J.T. Davenport. Tentative design procedure for riprap lined channels. Technical Report, Highway Research Board, National Academy of Sciences, Washington, DC, 1970. NCHRP Report No. 108.

[68] M.P. Anderson and W.W. Woessner. *Applied Groundwater Modeling, Simulation of Flow and Transport.* Academic Press, New York, 1992.

[69] G. Aronica, G. Freni, and E. Oliveri. Uncertainty analysis of the influence of rainfall time resolution in the modelling of urban drainage systems. *Hydrological Processes,* 19:1055–1071, 2005b.

[70] G.T. Aronica and L.G. Lanza. Drainage efficiency in urban areas: A case study. *Hydrological Processes,* 19:1105–1119, 2005a.

[71] ASCE. Elevated water tank rises to new levels. *Civil Engineering,* 70(1):19, January 2000.

[72] ASCE Task Committee on Ground Water Protection. *Ground Water, Protection Alternatives and Strategies in the U.S.A.* ASCE, New York, 1997.

[73] ASME B36.19M. *Stainless Steel Pipe.* American Society of Mechanical Engineers, New York, 1985.

[74] Atlanta Regional Commission/Georgia Department of Natural Resources-Environmental Protection Division, Atlanta, Georgia. *Georgia Stormwater Management Manual,* 2003. www.georgiastormwater.com.

[75] M. Aubertin, R.P. Chapuis, and M. Mbonimpa. Discussion of 'Goodbye, Hazen; Hello, Kozeny–Carman' by W. David Carrier III. *Journal of Geotechnical and Geoenvironmental Engineering,* pp. 1056–1057, 2005.

[76] G. Aussenac and C. Boulangeat. Interception des precipitation et evapotranspiration reelle dans des peuplements de feuillu (*fagus silvatica* L.) et de resineux (*pseudotsuga menziesii* (Mirb.) Franco). *Annales des Sciences Forestieres*, 37(2):91–107, 1980.

[77] H.E. Babbitt and E.R. Baumann. *Sewerage and Sewage Treatment*. John Wiley & Sons, New York, 1958.

[78] W. Badon-Ghyben. Nota in Verband met de Voorgenomen Putboring Nabij Amsterdam (Notes on the probable results of well drilling near Amsterdam). Technical Report 1888/9, Tijdschrift van het Koninklijk Instituut van Ingenieurs, The Hague, 1888.

[79] B.A. Bakhmeteff. *Hydraulics of Open Channel Flow*. McGraw-Hill, Inc., New York, 1932.

[80] R.C. Balling and S.W. Brazel. Recent changes in Phoenix, Arizona, summertime diurnal precipitation patterns. *Theoretical and Applied Climatology*, 38:50–54, 1987.

[81] D.W. Barr. Discussion of "Goodbye, Hazen; Hello, Kozeny–Carman" by W. David Carrier III. *Journal of Geotechnical and Geoenvironmental Engineering*, p. 1057, 2005.

[82] Barre de Saint-Venant. Theory of unsteady water flow, with application to river floods and to propagation of tides in river channels. *French Academy of Science*, 73:148–154, 237–240, 1871.

[83] W.G.M. Bastiaanssen, E.J.M. Noordman, H. Pelgrum, G. Davids, B.P. Thoreson, and R.G. Allen. SEBAL model with remotely sensed data to improve water-resources management under actual field conditions. *Journal of Irrigation and Drainage Engineering, ASCE*, 131(1):85–93, February 2005.

[84] V. Batu. *Aquifer Hydraulics*. John Wiley & Sons, New York, 1998.

[85] J. Bear. *Dynamics of Fluids in Porous Media*. Dover Publications, Inc., New York, 1972.

[86] J. Bear. *Hydraulics of Groundwater*. McGraw-Hill Book Company, New York, 1979.

[87] J. Bear and G. Dagan. The transition zone between fresh and salt waters in a coastal aquifer, Progress Report 1: The steady interface between two immiscible fluids in a two-dimensional field of flow. Technical report, Hydraulic Lab, Technion, Haifa, Israel, 1962.

[88] J.S. Beard. Rainfall interception by grass. *South African Forestry Journal*, 42:12–25, 1962.

[89] P.B. Bedient and W.C. Huber. *Hydrology and Floodplain Analysis*. Prentice Hall, Upper Saddle River, NJ, 1992.

[90] P.B. Bedient and W.C. Huber. *Hydrology and Floodplain Analysis*. Prentice Hall, Upper Saddle River, NJ, 3d ed., 2002.

[91] P.B. Bedient, H.S. Rifai, and C.J. Newell. *Ground Water Contamination*. Prentice Hall, Upper Saddle River, NJ, 2d ed., 1999.

[92] T.V. Belanger and M.T. Montgomery. Seepage meter errors. *Limnology and Oceanography*, 37:1787–1795, 1992.

[93] J.R. Benjamin and C. Allin Cornell. *Probability, Statistics, and Decision for Civil Engineers*. McGraw-Hill, Inc., New York, 1970.

[94] R.P. Betson. What is watershed runoff? *Journal of Geophysical Research*, 69(8):1541–1552, 1964.

[95] W.R. Bidlake, W.M. Woodham, and M.A. Lopez. Evapotranspiration from areas of native vegetation in west-central Florida. Water-Supply Paper 2430, U.S. Geological Survey, 1996.

[96] R.B. Billings and C.V. Jones. *Forecasting Urban Water Demand.* American Water Works Association, Denver, CO, 1996.

[97] D.M. Bjerklie, D. Moller, L.C. Smith, and S.L. Dingman. Estimating discharge in rivers using remotely sensed hydraulic information. *Journal of Hydrology,* 309:191–209, 2005.

[98] J.H. Black and K.L. Kipp. Observation well response time and its effect upon aquifer test results. *Journal of Hydrology,* 34:297–306, 1977.

[99] P.E. Black. *Watershed Hydrology.* Ann Arbor Press, Inc., Chelsea, MI, 2d ed., 1996.

[100] L. Blank and A. Tarquin. *Engineering Economy.* McGraw-Hill, Inc., New York, 2002.

[101] H. Blasius. Das Ähnlichkeitsgesetz bei Reibungsvorgängen in Flüssigkeiten. *Forschungs-Arbeit des Ingenieur-Wesens,* 131, 1913.

[102] R.D. Blevins. *Applied Fluid Dynamics Handbook.* Van Nostrand Reinhold, New York, 1984.

[103] F. Bloetscher, A. Muniz, and G.M. Witt. *Groundwater Injection.* McGraw-Hill, Inc., New York, 2005.

[104] F.E. Blow. Quantity and hydrologic characteristics of litter upon upland oak forests in eastern Tennessee. *Journal of Forestry,* 53:190–195, 1955.

[105] B.B. Bobée and R. Robitaille. Correction of bias in estimation of the coefficient of skewness. *Water Resources Research,* 11(6):851–854, December 1975.

[106] G.L. Bodhaine. Measurement of peak discharge at culverts by indirect methods. Techniques of water resources investigations. Chapter A3, Book 3, U.S. Geological Survey, Washington, DC, 1976.

[107] G.H. Bolt and P.H. Groenvelt. Coupling phenomena as a possible cause for non-darcian behavior of water in soil. *Bulletin of the International Association of Scientific Hydrology,* 14:17–26, 1969.

[108] F.A. Bombardelli and M.H. García. Hydraulic design of large-diameter pipes. *Journal of Hydraulic Engineering,* 129(11):839–846, November 2003.

[109] J. Boonstra. Well hydraulics and aquifer tests. In J.W. Delleur, ed., *The Handbook of Groundwater Engineering,* pp. 8.1–8.34. CRC Press, Inc., Boca Raton, FL, 1998a.

[110] L.E. Borgman. Risk criteria. *Journal of Waterways and Harbors Division, ASCE,* 89(WW3):1–35, August 1963.

[111] M.G. Bos. *Discharge Measurement Structures.* Wageningen, The Netherlands, 3d rev'd ed., 1988. ILRI Publication 20.

[112] Boulder County. *Drainage Criteria Manual.* Boulder County, CO, 1984.

[113] D. Bousmar and Y. Zech. Momentum transfer for practical flow computation in compound channels. *Journal of Water Resources Planning and Management,* 125(7):696–706, July 1999.

[114] H. Bouwer. *Groundwater Hydrology.* McGraw-Hill Book Company, New York, 1978.

[115] H. Bouwer. Elements of soil science and groundwater hydrology. In G. Bitton and C.P. Gerba, eds., *Groundwater Pollution Microbiology,* pp. 9–38. Krieger Publishing Company, Malabar, FL, 1984.

[116] H. Bouwer. The Bouwer and Rice slug test: An update. *Ground Water,* 27:304–309, 1989.

[117] H. Bouwer. Discussion of Bouwer and Rice slug test review articles. *Ground Water,* 34:171, 1996.

[118] H. Bouwer, J.T. Back, and J.M. Oliver. Predicting infiltration and ground-water mounds for artificial recharge. *Journal of Hydrologic Engineering,* 4(4):350–357, 1999.

[119] H. Bouwer and R.C. Rice. A slug test for determining hydraulic conductivity of unconfined aquifers with completely or partially penetrating wells. *Water Resources Research*, 12:423–428, 1976.

[120] N.C. Brady. *The Nature and Properties of Soils*. Macmillan, New York, 8th ed., 1974.

[121] E.F. Brater, H.W. King, J.E. Lindell, and C.Y. Wei. *Handbook of Hydraulics*. McGraw-Hill, Inc., New York, 7th ed., 1996.

[122] W.C. Bridges. Techniques for estimating magnitude and frequency of floods on natural-flow streams in Florida. Water-Resources Investigations Report 82-4012, United States Geological Survey, Reston, VA, 1982.

[123] J. Bromley, M. Robinson, and J.A. Barker. Scale-dependency of hydraulic conductivity: An example from Thorne Moor, a raised mire in South Yorkshire, UK. *Hydrological Processes*, 18:973–985, 2004.

[124] K.N. Brooks, P.F. Folliott, H.M. Gregersen, and J.L. Thames. *Hydrology and the Management of Watersheds*. Iowa State University Press, Ames, IA, 1991.

[125] D.S. Brookshire and D. Whittington. Water resources issues in developing countries. *Water Resources Research*, 29(7):1883–1888, July 1993.

[126] G.O. Brown. Henry Darcy and the making of a law. *Water Resources Research*, 38(7), doi:10.1029/2001WR000727, 2002.

[127] G.M. Brune. Trap efficiency of reservoirs. *Transactions of the American Geophysical Union*, 34(3):407–418, 1953.

[128] D. Brunt. Notes on radiation in the atmosphere. *Quarterly Journal of Royal Meteorological Society*, 58:389–418, 1932.

[129] W. Brutsaert. *Evaporation into the Atmosphere: Theory, History, and Applications*. R. Deidel, Dordrecht, The Netherlands, 1982.

[130] P.H. Burgi. Bicycle-Safe Grate Inlets Study: Volume 2—Hydraulic Characteristics of Three Selected Grate Inlets on Continuous Grades. Technical Report FHWA-RD-78-4, Federal Highway Administration, U.S. Department of Transportation, Washington, DC, 1978.

[131] S.J. Burian and J.M. Shepherd. Effect of urbanization on the diurnal rainfall pattern in Houston. *Hydrological Processes*, 19:1089–1103, 2005.

[132] C.M. Burt, A.J. Mutziger, R.G. Allen, and T.A. Howell. Evaporation research: Review and interpretation. *Journal of Irrigation and Drainage Engineering, ASCE*, 131(1):37–58, February 2005.

[133] F.L. Burton. Wastewater-collection systems. In L.W. Mays, ed., *Water Resources Handbook*, pp. 19.1–19.53. McGraw-Hill, Inc., New York, 1996.

[134] K. Bury. *Statistical Distributions in Engineering*. Cambridge University Press, New York, 1999.

[135] J.A. Businger. Some remarks on Penman's equations for the evapotranspiration. *Netherlands Journal of Agricultural Science*, 4:77, 1956.

[136] D. Butler and J.W. Davies. *Urban Drainage*. E & FN Spon, London, 2000.

[137] J.J. Butler Jr., C.D. McElwee, and W. Liu. Improving the quality of parameter estimates obtained from slug tests. *Ground Water*, 34(3):480–490, 1996.

[138] T.R. Camp. Design of sewers to facilitate flow. *Sewer Works Journal*, 18:3, 1946.

[139] G.S. Campbell. *An Introduction to Environmental Physics*. Springer, New York, 1977.

[140] M.D. Campbell and J.H. Lehr. *Water Well Technology*. McGraw-Hill, Inc., New York, 1972.

[141] P.C. Carman. Fluid flow through a granular bed. *Transaction of the Institutions of Chemical Engineers*, 15:150–156, 1937.

[142] P.C. Carman. *Flow of Gases Through Porous Media*. Butterworths, London, 1956.

[143] F.G. Carollo, V. Ferro, and D. Termini. Flow resistance law in channels with flexible submerged vegetation. *Journal of Hydraulic Engineering*, 131(7):554–564, July 2005.

[144] D. Carroll. Rainwater as a chemical agent and geologic processes—A review. Water-Supply Paper 1535-G, United States Geological Survey, 1962.

[145] H.W. Carslaw and J.C. Jaeger. *Conduction of Heat in Solids*. Oxford University Press, New York, 2d ed., 1959.

[146] T.V. Cech. *Principles of Water Resources*. John Wiley & Sons, New York, 2002.

[147] H.R. Cedergren, K.H. O'Brien, and J.A. Arman. Guidelines for the design of subsurface drainage systems for highway structural sections. Technical Report FHWA-RD-72-30, Federal Highway Administration, June 1972.

[148] A. Chadwick and J. Morfett. *Hydraulics in Civil and Environmental Engineering*. E & FN Spon, London, 2d ed., 1993.

[149] N. Chahinian, R. Moussa, P. Andrieux, and M. Voltz. Comparison of infiltration models to simulate flood events at the field scale. *Journal of Hydrology*, 306:191–214, 2005.

[150] R.A. Chandler and D.B. McWhorter. Upconing of the salt-water–fresh-water interface beneath a pumping well. *Ground Water*, 13:354–359, 1975.

[151] M. Chang. *Forest Hydrology: An Introduction to Water and Forests*. CRC Press, Inc., Boca Raton, FL, 2002.

[152] R.J. Charbeneau. *Groundwater Hydraulics and Pollutant Transport*. Prentice-Hall, Upper Saddle River, NJ, 2000.

[153] D.V. Chase. Operation of water distribution systems. In L.W. Mays, ed., *Water Distribution Systems Handbook*, pp. 15.1–15.16. McGraw-Hill Book Company, New York, 2000.

[154] M.H. Chaudhry. *Open-Channel Flow*. Prentice-Hall, Upper Saddle River, NJ, 1993.

[155] C. Chen. Rainfall intensity-duration-frequency formulas. *Journal of Hydraulic Engineering*, 109(12):1603–1621, 1983.

[156] C.-N. Chen and T.S.W. Wong. Critical rainfall duration for maximum discharge from overland plane. *Journal of Hydraulic Engineering*, 119(9):1040–1045, September 1993.

[157] A.H.-D. Cheng. *Multilayered Aquifer Systems: Fundamentals and Applications*. Marcel Dekker, Inc., New York, 2000.

[158] D.A. Chin. Leakage of clogged channels that partially penetrate surficial aquifers. *Journal of Hydraulic Engineering*, 117(4):467–488, April 1991.

[159] D.A. Chin. Analysis and prediction of South-Florida rainfall. Technical Report CEN-93-2, University of Miami, Coral Gables, FL, 1993.

[160] D.A. Chin. A risk management strategy for wellhead protection. Technical Report CEN93-3, University of Miami, Department of Civil Engineering, Coral Gables, FL, 1993.

[161] D.A. Chin and P.V.K. Chittaluru. Risk management in wellhead protection. *Journal of Water Resources Planning and Management*, 120(3):294–315, 1994.

[162] D.A. Chin. *Water-Quality Engineering in Natural Systems*. John Wiley & Sons, New York, 2006.

[163] C.-L. Chiu and N.-C. Tung. Maximum velocity and regularities in open-channel flow. *Journal of Hydraulic Engineering*, 128(4):390–398, April 2002.

[164] V.T. Chow. Frequency analysis of hydrologic data with special application to rainfall intensities. Bulletin 414, University of Illinois Engineering Experiment Station, 1953.

[165] V.T. Chow. The log-probability law and its emerging applications. *Proceedings of the American Society of Civil Engineers*, 80:536–1 to 536–25, 1954.

[166] V.T. Chow. *Open-Channel Hydraulics*. McGraw-Hill, Inc., New York, 1959.

[167] V.T. Chow. *Handbook of Applied Hydrology*. McGraw-Hill, Inc., New York, 1964.

[168] V.T. Chow, D.R. Maidment, and L.W. Mays. *Applied Hydrology*. McGraw-Hill, Inc., New York, 1988.

[169] J.E. Christiansen. Pan evaporation and evapotranspiration from climatic data. *Journal of Irrigation and Drainage Engineering, ASCE*, 94:243–265, 1968.

[170] J.E. Christiansen and G.H. Hargreaves. Irrigation requirements from evaporation. *Transactions of the International Commission on Irrigation & Drainage*, III:23.569–23.596, 1969.

[171] C.O. Clark. Storage and the unit hydrograph. *Transactions of the American Society of Civil Engineers*, 110:1419–1446, 1945.

[172] R.M. Clark. Water supply. In R.A. Corbitt, ed., *Standard Handbook of Environmental Engineering*, pp. 5.1–5.225. McGraw-Hill, Inc., New York, 1990.

[173] C.F. Colebrook. Turbulent flow in pipes, with particular reference to the transition region between the smooth and rough pipe laws. *Journal, Institution of Civil Engineers (London)*, 11:133–156, February 1939.

[174] H.H. Cooper, J.D. Bredehoeft, and I.S. Papadopulos. Response of a finite-diameter well to an instantaneous charge of water. *Water Resources Research*, 3(1):263–269, 1967.

[175] H.H. Cooper and C.E. Jacob. A generalized graphical method for evaluating formation constants and summarizing well-field history. *Transactions of the American Geophysical Union*, 27:526–534, 1946.

[176] R.A. Corbitt. Wastewater disposal. In R.A. Corbitt, ed., *Standard Handbook of Environmental Engineering*, pp. 6.1–6.274. McGraw-Hill, Inc., New York, 1990.

[177] I. Cordery. Synthetic unit graphs for small catchments in Eastern New South Wales. *Civil Engineering Transactions, Institution of Engineers (Australia)*, 10:47–58, 1968.

[178] G.K. Cotton. Hydraulic design of flood control channels. In L.W. Mays, ed., *Stormwater Collection Systems Design Handbook*, pp. 16.1–16.42. McGraw-Hill, Inc., New York, 2001.

[179] N.H. Crawford and R.K. Linsley. Digital simulation in hydrology, Stanford Watershed Model IV. Technical Report 39, Department of Civil Engineering, Stanford University, Stanford, CA, 1966.

[180] H. Cross. Analysis of flow in networks of conduits or conductors. Bulletin 286, University of Illinois Engineering Experiment Station, 1936.

[181] Z. Şen. *Applied Hydrogeology for Scientists and Engineers*. Lewis Publishers, Boca Raton, FL, 1995.

[182] J.A. Cunge. On the subject of a flood propagation computation method (Muskingum method). *Journal of Hydraulic Research, IAHR*, 7(2):205–230, 1969.

[183] C. Cunnane. Methods and merits of regional flood frequency analysis. *Journal of Hydrology*, 100:269–290, 1988.

[184] G. Dagan. *Flow and Transport in Porous Formations*. Springer-Verlag, New York, 1989.

[185] M. Dagg. Evaporation Pans in East Africa, 1968. *Proceedings Fourth Specialist Meeting on Applied Meteorology in East Africa*.

[186] J.W. Daily and D.R.F. Harleman. *Fluid Dynamics*. Addison-Wesley Publishing Company, Inc., Reading, MA, 1966.

[187] H. Darcy. *Les Fontaines Publiques de la Ville de Dijon*. Victor Dalmont, Paris, 1856.

[188] T. Davie. *Fundamentals of Hydrology*. Routledge, London, 2002.

[189] G. de Marsily. *Quantitative Hydrogeology, Groundwater Hydrology for Engineers*. Academic Press, Inc., San Diego, CA, 1986.

[190] T.N. Debo and A.J. Reese. *Municipal Storm Water Management*. Lewis Publishers, Inc., Boca Raton, FL, 1995.

[191] M. Dechesne, S. Barraud, and J.-P. Bardin. Experimental assessment of stormwater infiltration basin evolution. *Journal of Environmental Engineering*, 131(7):1090–1098, July 2005.

[192] G.J. DeGlee. *Over Grondwaterstromingen bij Wateronttrekking Door Middel van Putten* (in Dutch). J. Waltman, Delft, 1930.

[193] J. Delleur. The evolution of urban hydrology: Past, present, and future. *Journal of Hydraulic Engineering*, 129(8):563–573, August 2003.

[194] J.L. Devore. *Probability and Statistics for Engineering and the Sciences*. Brooks/Cole, Pacific Grove, CA, 2000.

[195] S.L. Dingman. *Fluvial Hydrology*. W.H. Freeman and Company, New York, 1984.

[196] S.L. Dingman. *Physical Hydrology*. Prentice Hall, Upper Saddle River, NJ, 2d ed., 2002.

[197] R.D. Dodson. *Storm Water Pollution Control*. McGraw-Hill, Inc., New York, 2d ed., 1998.

[198] R.D. Dodson. Floodplain hydraulics. In L.W. Mays, ed., *Hydraulic Design Handbook*, pp. 19.1–19.50. McGraw-Hill Book Company, New York, 1999.

[199] R.D. Dodson and C.T. Maske. Regulation of stormwater collection systems in the United States. In L.W. Mays, ed., *Stormwater Collection Systems Design Handbook*, pp. 2.1–2.25. McGraw-Hill, Inc., New York, 2001.

[200] P.A. Domenico and M.D. Mifflin. Water from low-permeability sediments and land subsidence. *Water Resources Research*, 1:563–576, 1965.

[201] P.A. Domenico and F.W. Schwartz. *Physical and Chemical Hydrogeology*. John Wiley & Sons, New York, 2d ed., 1998.

[202] J. Doorenbos and W.O. Pruitt. Guidelines for the prediction of crop water requirements. Irrigation and Drainage Paper 24, UN Food and Agriculture Organization, Rome, Italy, 1975.

[203] J. Doorenbos and W.O. Pruitt. Guidelines for the prediction of crop water requirements. Irrigation and Drainage Paper 24, UN Food and Agriculture Organization, Rome, Italy, 2d ed., 1977.

[204] J.F. Douglas, J.M. Gasiorek, and J.A. Swaffield. *Fluid Mechanics*. Prentice Hall, Upper Saddle River, New Jersey, 4th ed., 2001.

[205] F.G. Driscoll. *Ground Water and Wells*. Johnson Filtration System, Inc., St. Paul, MN, 2d ed., 1986.

[206] N.E. Driver. Summary of nationwide analysis of storm-runoff quality and quantity in urban watersheds. In Y. Iwasa and T. Sueishi, eds., *Proceedings of the Fifth International Conference on Urban Storm Drainage, Osaka, Japan*, vol. 1, pp. 333–338, 1990.

[207] N.E. Driver and D.J. Lystrom. Estimation of urban storm runoff loads, urban runoff quality–Impact and quality enhancement technology. In B. Urbonas and L.A. Roesner, eds., *Proceedings of an Engineering Foundation Conference, New Hampshire*, pp. 222–232, Reston, VA, 1986. American Society of Civil Engineers.

[208] N.E. Driver and G.D. Tasker. Techniques for estimation of storm-runoff loads, volumes, and selected constituent concentrations in urban watersheds in the United States. Open File Report 88-191, U.S. Geological Survey, Denver, CO, 1988.

[209] N.E. Driver and G.D. Tasker. Techniques for estimation of storm-runoff loads, volumes, and selected constituent concentrations in urban watersheds in the United States. Water-Supply Paper 2363, U.S. Geological Survey, Washington, DC, 1990.

[210] M. Duchene, E.A. McBean, and N.R. Thomson. Modeling of infiltration from trenches for storm-water control. *Journal of Water Resources Planning and Management*, 120(3):276–293, May/June 1994.

[211] J.A. Duffie and W.A. Beckman. *Solar Engineering of Thermal Processes*. John Wiley and Sons, Inc., New York, 1980.

[212] T. Dunne and L.B. Leopold. *Water in Environmental Planning*. W.H. Freeman and Company, San Francisco, 1978.

[213] J. Dupuit. *Études Théoriques et Pratiques sur le Mouvement des Eaux dans les Canaux DéCauverts et á Travers les Terrains Perméables*. Dunod, Paris, 2d ed., 1863.

[214] H.B. Dwight. *Tables of Integrals and Other Mathematical Data*. Macmillan Publishing Company, New York, 4th ed., 1961.

[215] B. Dziegielewski and E. Opitz. Water Demand Analysis. In L.W. Mays, ed., *Urban Water Supply Handbook*, pp. 5.1–5.55. McGraw-Hill Book Company, New York, 2002.

[216] B. Dziegielewski, E.M. Opitz, and D. Maidment. Water demand analysis. In L.W. Mays, ed., *Water Resources Handbook*, pp. 23.1–23.62. McGraw-Hill, Inc., New York, 1996.

[217] A.A. Dzurik. *Water Resources Planning*. Rowman and Littlefield Publishers, Inc., Lanham, MD, 3d ed., 2003.

[218] P.S. Eagleson. Dynamics of flood frequency. *Water Resources Research*, 8:878–898, 1972.

[219] S.M. Easa. Geometric design. In W.F. Chen, ed., *The Civil Engineering Handbook*, pp. 2287–2319. CRC Press, Boca Raton, FL, 1995.

[220] G. Echávez. Increase in losses coefficient with age for small diameter pipes. *Journal of Hydraulic Engineering*, 123(2):157–159, February 1997.

[221] J.W. Eheart and J.R. Lund. Water-use management: Permit and water-transfer systems. In L.W. Mays, ed., *Urban Water Supply Handbook*, pp. 4.1–4.41. McGraw-Hill Book Company, New York, 2002.

[222] H.A. Einstein. Der hydraulische oder profil-radius. *Schweizerische Bauzeitung*, 103(8):89–91, February 1934.

[223] H.A. Einstein. The bed load function for sediment transport in open channel flows. Technical Bulletin 1026, U.S. Department of Agriculture, Washington, DC, 1950.

[224] H.A. Einstein and R.B. Banks. Fluid resistance of composite roughness. *Transactions of the American Geophysical Union*, 31(4):603–610, 1951.

[225] M.Z. Ejeta and L.W. Mays. Water pricing and drought management. In L.W. Mays, ed., *Urban Water Supply Handbook*, pp. 6.1–6.43. McGraw-Hill Book Company, New York, 2002.

[226] N. El-Jabi and S. Sarraf. Effect of maximum rainfall position on rainfall-runoff relationship. *Journal of Hydraulic Engineering*, 117(5):681–685, May 1991.

[227] W.J. Elliot. Precipitation. In A.D. Ward and W.J. Elliot, eds., *Environmental Hydrology*, pp. 19–49. Lewis Publishers, Inc., Boca Raton, FL, 1995.

[228] E.T. Engman. Roughness coefficients for routing surface runoff. *Journal of Irrigation and Drainage Engineering, ASCE*, 112(1):39–53, 1986.

[229] Environmental and Water Resources Institute, Standardization of Reference Evapotranspiration Task Committee. The ASCE Standardized Reference Evapotranspiration Equation, Draft Report July 9, 2002, American Society of Civil Engineers, 2002.

[230] Environmental Protection Agency. Results of the nationwide urban runoff program, final report. Technical Report Accession No. PB84-185552, NTIS, Washington, DC, December 1983.

[231] D.R. Erickson. A study of littoral groundwater seepage at Williams Lake, Minnesota using seepage meters and wells. Master's thesis, University of Minnesota, Minneapolis, 1981.

[232] W.H. Espey, Jr., and D.G. Altman. Nomographs for ten-minute unit hydrographs for small urban watersheds. Addendum 3 of Urban Runoff Control Planning EPA-600/9-78-035, U.S. Environmental Protection Agency, Washington, DC, 1978.

[233] W.H. Espey, Jr., D.G. Altman, and C.B. Graves. Nomographs for ten-minute unit hydrographs for small urban watersheds. ASCE Urban Water Resources Research Program, Technical Memorandum 32 (NTIS PB-282158), ASCE, New York, 1977.

[234] G.M. Fair and L.P. Hatch. Fundamental factors governing the streamline flow of water through sand. *Journal of the American Water Works Association*, 25:1551–1565, 1933.

[235] C.-Y. Fan, R. Field, W.C. Pisano, J. Barsanti, J.J. Joyce, and H. Sorenson. Sewer and tank flushing for sediment, corrosion, and pollution control. *Journal of Water Resources Planning and Management*, 127(3):194–201, May/June 2001.

[236] J.A. Fay. *Introduction to Fluid Mechanics*. MIT Press, Cambridge, MA, 1994.

[237] Federal Emergency Management Agency. Flood insurance study guidelines and specifications for study contractors. Report 37, FEMA, Washington, DC, 1993.

[238] Water Pollution Control Federation. *Gravity Sanitary Sewer Design and Construction, WPCF Manual of Practice No. FD-5*. Water Pollution Control Federation, Washington, DC, 1982.

[239] M.C. Feller. Water balances in *eucalyptus regnans, e.obliqua*, and *pinus radiata* forests in Victoria. *Australian Forestry*, 44(3):153–161, 1981.

[240] V. Ferro. Friction factor for gravel-bed channel with high boulder concentration. *Journal of Hydraulic Engineering*, 125(7):771–778, July 1999.

[241] C.W. Fetter. *Contaminant Hydrogeology*. Prentice Hall, Upper Saddle River, NJ, 2d ed., 1999.

[242] C.W. Fetter. *Applied Hydrogeology*. Prentice Hall, Upper Saddle River, NJ, 4th ed., 2001.

[243] L. Feyen, P.J. Ribeiro Jr., F. DeSmedt, and P.J. Diggle. Stochastic delineation of capture zones: Classical versus Bayesian approach. *Journal of Hydrology*, 281:313–324, 2003.

[244] F.R. Fiedler. Simple, practical method for determining station weights using Thiessen polygons and isohyetal maps. *Journal of Hydrologic Engineering*, 8(4):219–221, 2003.

[245] R. Field, J.P. Heaney, and R. Pitt. *Innovative Urban Wet-Weather Flow Management Systems*. Technomic Publishing Company, Inc., Lancaster, PA, 2000.

[246] E.J. Finnemore and J.B. Franzini. *Fluid Mechanics with Engineering Applications.* McGraw-Hill, Inc., New York, NY, 10th ed., 2002.

[247] K. Fisher. The hydraulic roughness of vegetated channels. Technical Report SR 305, Hydraulics Research, Ltd., Wallingford, England, 1992.

[248] R.A. Fisher and L.H.C. Tippett. Limiting forms of the frequency distribution of the largest or smallest member of a sample. *Proceedings of the Cambridge Philosophical Society*, 24, part II:180–191, 1928.

[249] C.R. Fitts. *Groundwater Science.* Academic Press, New York, 2002.

[250] Florida Department of Transportation. Drainage Manual. Technical report, FDOT, Tallahassee, FL, 2000. English Units.

[251] Florida Geological Survey. *Florida's Ground Water Quality Monitoring Program, Hydrogeological Framework.* Tallahassee, FL, 1991. Special Publication No. 32.

[252] P. Forchheimer. *Hydraulik.* Teubner Verlagsgesellschaft, Stuttgart, Germany, 1930.

[253] S. Fortier and F.C. Scobey. Permissible canal velocities. *Transactions of the American Society of Civil Engineers*, 89:940–984, 1926.

[254] H.A. Foster. Theoretical frequency curves and their application to engineering problems. *Transactions of the American Society of Civil Engineers*, 87:142–173, 1924.

[255] R.W. Fox and A.T. McDonald. *Introduction to Fluid Mechanics.* John Wiley & Sons, New York, 4th ed., 1992.

[256] M.A. Franklin and G.T. Losey. Magnitude and frequency of floods from urban streams in Leon County, Florida. Water-Resources Investigations Report 84-4004, United States Geological Survey, Reston, VA, 1984.

[257] J.B. Franzini and E.J. Finnemore. *Fluid Mechanics with Engineering Applications.* McGraw-Hill, Inc., New York, 9th ed., 1997.

[258] R.H. Frederick, V.A. Myers, and E.P. Auciello. Five- to 60-minute precipitation frequency for the eastern and central United States. NOAA Technical Memorandum NWS HYDRO-35, National Oceanic and Atmospheric Administration, National Weather Service, Silver Spring, MD, June 1977.

[259] R.A. Freeze and J.A. Cherry. *Groundwater.* Prentice-Hall, Englewood Cliffs, NJ, 1979.

[260] R.H. French. *Open-Channel Hydraulics.* McGraw-Hill, Inc., New York, 1985.

[261] R.H. French. Hydraulics of Open Channel Flow. In L.W. Mays, ed., *Stormwater Collection Systems Design Handbook*, pp. 3.1–3.35. McGraw-Hill, Inc., New York, 2001.

[262] M. Frére and G.F. Popov. Agrometeorological crop monitoring and forecasting. FAO Plant Production and Protection Paper 17, FAO, Rome, Italy, 1979, 38–43.

[263] M.C. Friedman. Verification and control of low pressure transients in distribution systems. In *Proceedings of the 18th Annual ASDWA Conference, Association of State Drinking Water Officials*, Boston, MA, 2003.

[264] G.E. Galloway. 2004 Julian Hinds Water Resources Development Award Lecture. *Journal of Water Resources Planning and Management*, pp. 251–252, July/August 2005.

[265] E. Ganguillet and W.R. Kutter. Versuch zur Aufstellung einer neuen allegemeinen Formel für die gleichförmige Bewegung des Wassers in Canälen und Flüssen (An investigation to establish a new general formula for uniform flow of water in canals and rivers). *Zeitschrift des Oesterreichischen Ingenieur- und Architekten Vereines*, 21(1,2-3):6–25,46–59, 1869. (Published as a book in Bern, Switzerland, 1877; translated into English by Rudolph Hering and John C. Trautwine, Jr., as *A General Formula for the Uniform Flow of Water in Rivers and Other Channels*, John Wiley & Sons, Inc., New York, 1st ed., 1888; 2d ed., 1891 and 1901.)

[266] W.F. Geiger and H.R. Dorsch. Quantity-quality simulation (QQS): A detailed continuous planning model for urban runoff control, in *Model Description, Testing and Application*, Volume I. Technical Report EPA-600/2-80-011, U.S. Environmental Protection Agency, Cincinnati, OH, 1980.

[267] P.M. Gerhart, R.J. Gross, and J.I. Hochstein. *Fundamentals of Fluid Mechanics*. Addison-Wesley Publishing Company, Inc., Reading, MA, 2d ed., 1992.

[268] P.H. Gleick. An introduction to global fresh water issues. In P.H. Gleick, ed., *Water in Crisis*, pp. 3–12. Oxford University Press, New York, 1993.

[269] GLUMB. *Recommended Standards for Water Works, 1987*. Great Lakes Upper Mississippi River Board of State Public Health and Environmental Managers, Health Research, Inc., 1987.

[270] L. Gordon. *Mississippi River Discharge*. RD Instruments, San Diego, CA, 1992.

[271] S.M. Gorelick, R.A. Freeze, D. Donohue, and J.F. Keely. *Groundwater Contamination*. Lewis Publishers, Inc., Boca Raton, FL, 1993.

[272] R.A. Granger. *Fluid Mechanics*. Holt, Rinehart and Winston, New York, 1985.

[273] D.M. Gray, ed. *Handbook on the Principles of Hydrology*. National Research Council, Port Washington, Canada, 1973. Water Information Center, Inc.

[274] Great Lakes-Upper Mississippi River Board of Sanitary Engineers. *Recommended Standards for Sewage Works*. Health Education Service, Albany, NY, 1978.

[275] J.C. Green. Modelling flow resistance in vegetated streams: Review and development of new theory. *Hydrological Processes*, 19:1245–1259, 2005.

[276] W.H. Green and G.A. Ampt. Studies of soil physics, Part I: The flow of air and water through soils. *Journal of Agricultural Science*, 4(1):1–24, 1911.

[277] J.E. Gribbin. *Hydraulics and Hydrology for Stormwater Management*. Delmar Publishers, Albany, NY, 1997.

[278] I.I. Gringorten. A plotting rule for extreme probability paper. *Journal of Geophysical Research*, 68(3):813–814, 1963.

[279] T.L. Grizzard, C.W. Randall, B.L. Weand, and K.L. Ellis. Effectiveness of extended detention ponds. In *Urban Runoff Quality*. American Society of Civil Engineers, 1986.

[280] E.J. Gumbel. Statistical theory of extreme values and some practical applications. Applied Mathematics Series 33, U.S. National Bureau of Standards, Washington, DC, 1954.

[281] E.J. Gumbel. *Statistics of Extremes*. Columbia University Press, New York, 1958.

[282] J.C.Y. Guo. Street storm water conveyance capacity. *Journal of Irrigation and Drainage Engineering, ASCE*, 126(2):119–123, March/April 2000.

[283] J.C.Y. Guo. Design of infiltration basins for stormwater. In L.W. Mays, ed., *Stormwater Collection Systems Design Handbook*, pp. 9.1–9.35. McGraw-Hill, Inc., New York, 2001a.

[284] R.S. Gupta. *Hydrology and Hydraulic Systems*. Waveland Press, Inc., Prospect Heights, IL, 2d ed., 2001.

[285] Y. Gyasi-Agyei. Stochastic disaggregation of daily rainfall into one-hour time scale. *Journal of Hydrology*, 309:178–190, 2005.

[286] C.T. Haan. *Statistical Methods in Hydrology*. The Iowa State University Press, Ames, IA, 1977.

[287] Haestad Methods, Inc. *1997 Practical Guide to Hydraulics and Hydrology*. Haestad Press, Waterbury, CT, 1997.

[288] Haestad Methods, Inc. *Computer Applications in Hydraulic Engineering.* Haestad Press, Waterbury, CT, 1997.

[289] Haestad Methods, Inc. *Computer Applications in Hydraulic Engineering.* Haestad Press, Waterbury, CT, 5th ed., 2002.

[290] W.H. Hager. *Energy Dissipators and Hydraulic Jump.* Kluwer Academic Publishers, Dordrecht, Netherlands, 1991.

[291] W.H. Hager. *Wastewater Hydraulics.* Springer-Verlag, Berlin, 1999.

[292] A. Haldar and S. Mahadevan. *Probability, Reliability and Statistical Methods in Engineering Design.* John Wiley & Sons, New York, 2000.

[293] R.L. Hall. Interception loss as a function of rainfall and forest types: Stochastic modelling for tropical canopies revisited. *Journal of Hydrology*, 280:1–12, 2003.

[294] M.S. Hantush. Analysis of data from pumping tests in leaky aquifers. *Transactions of the American Geophysical Union*, 37(6):702–714, 1956.

[295] M.S. Hantush. Modification of the theory of leaky aquifers. *Journal of Geophysical Research*, 65:3713–3725, 1960.

[296] M.S. Hantush. Hydraulics of wells. In V.T. Chow, ed., *Advances in Hydroscience*, vol. 1, pp. 281–432. Academic Press, New York, 1964.

[297] M.S. Hantush. Growth and decay of groundwater-mounds in response to uniform percolation. *Water Resources Research*, 3(1):227–234, 1967.

[298] M.S. Hantush and C.E. Jacob. Plane potential flow of ground-water with linear leakage. *Transactions of the American Geophysical Union*, 35:917–936, 1954.

[299] M.S. Hantush and C.E. Jacob. Non-steady radial flow in an infinite leaky aquifer. *Transactions of the American Geophysical Union*, 36(1):95–100, 1955.

[300] A.W. Harbaugh and M.G. McDonald. User's documentation for MODFLOW-96, an update to the U.S. Geological Survey modular finite-difference ground-water flow model. Open File Report 96-485, United States Geological Survey, 1996.

[301] G.H. Hargreaves and Z.A. Samani. Estimating potential evapotranspiration. *Journal of Irrigation and Drainage Engineering, ASCE*, 108(3):225–230, 1982.

[302] G.L. Hargreaves, G.H. Hargreaves, and J.P. Riley. Agricultural benefits for Senegal River Basin. *Journal of Irrigation and Drainage Engineering, ASCE*, 111(2):113–124, 1985.

[303] G.L. Hargreaves and Z.A. Samani. Reference-crop evapotranspiration from temperature. *Applied Engineering in Agriculture*, 1(2):96–99, 1985.

[304] D.R.F. Harleman, P.F. Melhorn, and R.R. Rumer. Dispersion-permeability correlation in porous media. *Journal of the Hydraulics Division, ASCE*, 89:67–85, 1963.

[305] W.G. Harmon. Forecasting sewage at Toledo under dry weather conditions. *Engineering News Record*, 80, 1918.

[306] B.W. Harrington. Design and construction of infiltration trenches. In L.A. Roesner, B. Urbonas, and M.B. Sonnen, eds., *Design of Urban Runoff Quality Controls*, pp. 290–304. ASCE, New York, 1989.

[307] L.P. Harrison. Fundamental concepts and definitions relating to humidity. In A. Wexler, ed., *Humidity and Moisture.* Reinhold Publishing Company, New York, 1963.

[308] J.P. Hartigan. Basis for design of wet detention basin BMP's. In L.A. Roesner, B. Urbonas, and M.B. Sonnen, eds., *Design of Urban Runoff Quality Controls*, pp. 122–143. ASCE, New York, 1989.

[309] G.A. Hathaway. Design of drainage facilities. *Transactions of the American Society of Civil Engineers*, 110, 1945.

[310] R.H. Hawkins, A.T. Hjelmfelt, and A.W. Zevenbergen. Runoff probability, storm depth, and curve numbers. *Journal of the Irrigation and Drainage Division, ASCE*, 111(4):330–340, December 1985.

[311] A. Hazen. Discussion of "Dams on sand foundations, by A.C. Koenig." *Transactions of the American Society of Civil Engineers*, 73:199, 1911.

[312] J.P. Heany, W.C. Huber, and S.J. Nix. Storm water management model: Level I–Preliminary screening procedures. Report EPA-600/2-76-275, U.S. Environmental Protection Agency, Cincinnati, Ohio, 1976.

[313] J.P. Heany, W.C. Huber, H. Sheikh, M.A. Medina, J.R. Doyle, W.A. Peltz, and J.E. Darling. Nationwide evaluation of combined sewer overflows and urban stormwater discharges, Volume 2, Cost assessment and impacts. Report EPA-600/2-77-064, U.S. Environmental Protection Agency, Washington, DC, 1977.

[314] H.G. Heinemann. A new sediment trap efficiency curve for small reservoirs. *Water Resources Bulletin*, 17:825–830, 1981.

[315] J.D. Helvey. Rainfall interception by hardwood forest litter in the Southern Appalachians. Research Paper SE-8, U.S. Department of Agriculture, Forest Service, Southeastern Forest Experiment Station, Asheville, NC, 1964. 9 pp.

[316] J.D. Helvey. Interception by eastern white pine. *Water Resources Research*, 3(3):723–729, 1967.

[317] J.D. Helvey. Biological effects in the hydrological cycle. In E.J. Monke, ed., *Proceedings of the Third International Seminar for Hydrology Professors*, pp. 103–113, West Lafayette, IN, 1971. Purdue University.

[318] J.D. Helvey and J.J. Patric. Canopy and litter interception of rainfall by hardwoods of eastern United States. *Water Resources Research*, 1:193–206, 1965.

[319] J.C. Hemain. Statistically based modelling of urban runoff quality: State of the art. In H.C. Torno, J. Marsalek, and M. Desbordes, eds., *Urban Runoff Pollution*, Volume 10 of *NATO ASI Ser., Ser. G*, pp. 277–304. Springer-Verlag, Berlin Heidelberg, 1986.

[320] H.F. Hemond and E.J. Fechner. *Chemical Fate and Transport in the Environment*. Academic Press, New York, 1993.

[321] F.M. Henderson. *Open Channel Flow*. Macmillan Publishing Co., Inc., New York, 1966.

[322] H.R. Henry. Discussion of "Diffusion of submerged jets," by M.L. Albertson, Y.B. Dai, R.A. Jensen, and H. Rouse. *Transactions of the American Society of Civil Engineers*, 115:687–694, 1960.

[323] J.A. Henry, K.M. Portier, and J. Coyne. *The Climate and Weather of Florida*. Pineapple Press, Inc., Sarasota, FL, 1994.

[324] J.F. Hermance. *A Mathematical Primer on Groundwater Flow*. Prentice Hall, Upper Saddle River, NJ, 1999.

[325] D.M. Hershfield. Rainfall frequency atlas of the United States for durations from 30 minutes to 24 hours and return periods from 1 to 100 years. Technical Paper 40, U.S. Department of Commerce, Weather Bureau, Washington, DC, May 1961.

[326] D.M. Hershfield. Method for estimating probable maximum precipitation. *Journal of the American Water Works Association*, 57:965–972, 1965.

[327] A. Herzberg. Die Wasserversorgung einiger Nordseebaden (The water supply on parts of the North Sea coast in Germany). *Z. Gasbeleucht. Wasserversorg.*, 44:815–819, 824–844, 1901.

[328] B.L. Herzog. Slug tests for determining hydraulic conductivity of natural geologic deposits. In D.E. Daniel and S.J. Trautwein, eds., *Hydraulic Conductivity and Waste Contaminant Transport in Soil*, pp. 95–110. ASTM, Philadelphia, PA, 1994.

[329] J.D. Hewlett and A.R. Hibbert. Factors affecting the response of small watersheds to precipitation in humid areas. In W.E. Sopper and H.W. Lull, eds., *Forest Hydrology*, pp. 275–290. Pergamon, New York, 1967.

[330] W.I. Hicks. A method of computing urban runoff. *Transactions of the American Society of Civil Engineers*, 109:1217, 1944.

[331] F. Hildebrand. *Advanced Calculus for Applications*. Prentice-Hall, Inc., Englewood Cliffs, NJ, 2d ed., 1976.

[332] A.J. Hill and V.S. Neary. Factors affecting estimates of average watershed slope. *Journal of Hydraulic Engineering*, 10(2):133–140, March 2005.

[333] A.T. Hjelmfelt, Jr. Negative outflows from Muskingum flood routing. *Journal of Hydraulic Engineering*, 111(6), June 1985.

[334] A.T. Hjelmfelt, Jr. Investigation of curve number procedure. *Journal of Hydraulic Engineering*, 117(6):725–737, 1991.

[335] R.A. Horton. Separate roughness coefficients for channel bottom and sides. *Engineering News Record*, 111(22):652–653, November 1933a.

[336] R.E. Horton. Rainfall interception. *Monthly Weather Review*, 47:603–623, 1919.

[337] R.E. Horton. The role of infiltration in the hydrologic cycle. *Transactions of the American Geophysical Union*, 14:446–460, 1933b.

[338] R.E. Horton. Analysis of runoff-plat experiments with varying infiltration-capacity. *Transactions of the American Geophysical Union*, 20:693–711, 1939.

[339] R.E. Horton. An approach towards physical interpretation of infiltration capacity. *Soil Science Society of America Proceedings*, 5:399–417, 1940.

[340] R.E. Horton. Erosional development of streams and their drainage basins–Hydrophysical approach to quantitative morphology. *Geological Society of America Bulletin*, 56:275–370, 1945.

[341] J.R.M. Hosking and J.R. Wallis. *Regional Frequency Analysis*. Cambridge University Press, New York, 1997.

[342] C.H. Howe and F.P. Linaweaver. The impact of price on residential water demand and its relation to system design and price structure. *Water Resources Research*, 13(1), February 1967.

[343] T.A. Howell, S.R. Evett, A.D. Schneider, D.A. Dusek, and K.S. Copeland. Irrigated fescue grass ET compared with calculated reference grass ET. In *Proceedings, Fourth National Irrigation Symposium*, pp. 228–242, St. Joseph, MI, 2000. American Society of Agricultural Engineers.

[344] W.C. Huber and R.E. Dickinson. Storm water management model, Version 4, User's manual, with addendums. Technical Report EPA-600/3-88/001a, U.S. Environmental Protection Agency, Athens, Georgia, 1988.

[345] M.D. Hudlow and R.A. Clark. Hydrological synthesis by digital computer. *Journal of the Hydraulics Division, ASCE*, 95(HY3):839–860, 1969.

[346] F.A. Huff. Characteristics and contributing causes of an abnormal frequency of flood-producing rainstorms at Chicago. *Water Resources Bulletin*, 31:703–714, 1995.

[347] F.A. Huff and J.L. Vogel. Urban, topographic and diurnal effects on rainfall in the St. Louis region. *Journal of Applied Meteorology*, 17:565–577, 1978.

[348] L. Huisman. *Groundwater Recovery*. Macmillan, London, 1972.

[349] B. Hunt. Dispersion model for mountain streams. *Journal of Hydraulic Engineering*, 125(2):99–105, February 1999.

[350] M.J. Hvorslev. Time lag and soil permeability in ground-water observation. Bulletin 36, Waterways Experiment Station, U.S. Army Corps of Engineers, Vicksburg, MS, 1951.

[351] Hydrologic Engineering Center. Probable Maximum Storm (Eastern U.S.) HMR 52, User's Manual. Technical report, U.S. Army Corps of Engineers, Davis, CA, March 1984.

[352] Hydrologic Engineering Center. HEC-FFA Flood Frequency Analysis, User's Manual. Report CPD-13, U.S. Army Corps of Engineers, Davis, CA, 1992.

[353] Hydrologic Engineering Center. HEC-1 Flood Hydrograph Package User's Manual. Technical report, U.S. Army Corps of Engineers, Davis, CA, June 1998.

[354] Hydrologic Engineering Center. HMS Hydrologic Modeling System Technical Reference Manual. Technical report, U.S. Army Corps of Engineers, Davis, CA, March 2000.

[355] Hydrologic Engineering Center. HEC-HMS Hydrologic Modeling System User's Manual. Technical report, U.S. Army Corps of Engineers, Davis, CA, January 2001.

[356] Insurance Services Office. Fire Suppression Rating Schedule. New York, 1980.

[357] Interagency Advisory Committee on Water Data. Guidelines for determining flood flow frequency. Bulletin 17B of the Hydrology Committee, U.S. Geological Survey, Reston, VA, 1982.

[358] S. Irmak, R.G. Allen, and E.B. Whitty. Daily grass and alfalfa-reference evapotranspiration estimates and alfalfa-to-grass evapotranspiration ratios in Florida. *Journal of Irrigation and Drainage Engineering, ASCE*, 129(5):360–370, 2003c.

[359] S. Irmak, A. Irmak, R.G. Allen, and J.W. Jones. Solar and net radiation-based equations to estimate reference evapotranspiration in humid climates. *Journal of Irrigation and Drainage Engineering, ASCE*, 129(5):336–347, September/October 2003b.

[360] S. Irmak, A. Irmak, J.W. Jones, T.A. Howell, J.M. Jacobs, R.G. Allen, and G. Hoogenboom. Predicting daily net radiation using minimum climatological data. *Journal of Irrigation and Drainage Engineering, ASCE*, 129(4):256–269, July/August 2003a.

[361] S.A. Isiorho and J.H. Meyer. The effects of bag type and meter size on seepage meter measurements. *Ground Water*, 37:411–413, 1999.

[362] O.W. Israelsen and R.C. Reeve. Canal lining experiments in the Delta Area. Utah Technical Bulletin 313, Utah Agricultural Experiment Station, Logan, UT, 1944.

[363] D. Itenfisu, R.L. Elliott, R.G. Allen, and I.A. Walter. Comparison of reference evapotranspiration calculations as part of the ASCE standardization effort. *Journal of Irrigation and Drainage Engineering, ASCE*, 129(6):440–448, November/December 2003.

[364] C.F. Izzard. The surface-profile of overland flow. *Transactions of the American Geophysical Union*, 25:959–969, 1944.

[365] C.F. Izzard. Hydraulics of runoff from developed surfaces. *Proceedings of the Highway Research Board*, 26:129–146, 1946.

[366] C.R. Jackson, S.J. Burges, X. Liang, K.M. Leytham, K.R. Whiting, D.M. Hartley, C.W. Crawford, B.N. Johnson, and R.R. Horner. Development and application of simplified continuous hydrologic modeling for drainage design and analysis. In M.S. Wigmosta and S.J. Burges, eds., *Land Use and Watersheds*, pp. 39–58. American Geophysical Union, Washington, DC, 2001.

[367] I.J. Jackson. Problems of throughfall and interception assessment under tropical forest. *Journal of Hydrology*, 12:234–254, 1971.

[368] Jackson et al. Contaminant hydrogeology of toxic organic chemicals at a disposal site, Gloucester, Ontario: 1. Chemical concepts and site assessment. IWD Scientific Series 141, Environment Canada, 1985. 114 pp.

[369] C.E. Jacob. Drawdown test to determine effective radius of artesian well. *Transactions of the American Society of Civil Engineers*, 112:1047–1064, 1947. Paper 2321.

[370] A.K. Jain. Accurate explicit equation for friction factor. *ASCE, Journal of the Hydraulics Division*, 102(HY5):674–677, May 1976.

[371] A.K. Jain, D.M. Mohan, and P. Khanna. Modified Hazen–Williams formula. *Journal of the Environmental Engineering Division, ASCE*, 104(EE1):137–146, February 1978.

[372] S.C. Jain. *Open-Channel Flow*. John Wiley & Sons, New York, 2001.

[373] S.V. Jakovlev, J.A.A. Karelm, A.I. Zukov, and S. Kolobanov. *Kanalizacja*. Stroizdat, Moscow, 1975.

[374] L.D. James and R.R. Lee. *Economics of Water Resources Planning*. McGraw-Hill, Inc., New York, 1971.

[375] W.S. Janna. *Introduction to Fluid Mechanics*. PWS-KENT Publishing Company, Boston, MA, 3d ed., 1993.

[376] A.F. Jenkinson. The frequency distribution of the annual maximum (or minimum) value of meteorological elements. *Quarterly Journal of Royal Meteorological Society*, 81:158–171, 1955.

[377] M.E. Jennings, W.O. Thomas, and H.C. Riggs. Nationwide summary of U.S. Geological Survey regional regression equations for estimating magnitude and frequency of floods for ungaged sites. Water-Resources Investigations Report 94-4002, United States Geological Survey, Reston, VA, 1994.

[378] S.W. Jens. Design of urban highway drainage—The state of the art. Report FHWA-TS-79-225, U.S. Department of Transportation, Federal Highway Administration, Washington, DC, 1979.

[379] M.E. Jensen. Empirical methods of estimating or predicting evapotranspiration using radiation. In *Proceedings of Conference on Evapotranspiration*, pp. 57–61, 64, Chicago, IL, December 1966. American Society of Agricultural Engineers.

[380] M.E. Jensen and H.R. Haise. Estimating evapotranspiration from solar radiation. *Journal of Irrigation and Drainage Engineering, ASCE*, 89:15–41, 1963.

[381] M.E. Jensen, R.D. Burman, and R.G. Allen, eds., *Evapotranspiration and Irrigation Water Requirements*. American Society of Civil Engineers, New York, 1990. Manual of Practice No. 70.

[382] M.E. Jensen, J.L. Wright, and B.J. Pratt. Estimating soil moisture depletion from climate, crop, and soil data. *Transactions of the American Society of Agricultural Engineers*, 14:954–959, 1971.

[383] T.K. Jewell and D.D. Adrian. Development of improved stormwater quality models. *Journal of the Environmental Engineering Division, ASCE*, 107:957–974, 1981.

[384] T.K. Jewell and D.D. Adrian. Statistical analysis to derive improved stormwater quality models. *Journal of the Water Pollution Control Federation*, 54:489–499, 1982.

[385] R.C. Johanson, J.C. Imhoff, J.L. Kittle, and S. Donigan. Hydrological Simulation Program-Fortran (HSPF): User manual for release 8.0. Technical Report EPA/600/3-84/066, U.S. Environmental Protection Agency, Athens, GA, 1984.

[386] D. Johnson. Down the drain. *Los Angeles Times*, V1, February 2, 1988.

[387] D. Johnson, M. Smith, V. Koren, and B. Finnerty. Comparing mean areal precipitation estimates from NEXRAD and rain gage networks. *Journal of Hydrologic Engineering*, 4(2):117–124, April 1999.

[388] F.L. Johnson and F.M. Chang. Drainage of highway pavements. Hydraulic Engineering Circular 12, Federal Highway Administration, McLean, VA, 1984.

[389] J.L. Johnson. Local wellhead protection in Florida. Technical report, City of Tallahassee, Water Quality Division, Aquifer Protection Section, Tallahassee, FL, 1997.

[390] P.A. Johnson. Uncertainty in hydraulic parameters. *Journal of Hydraulic Engineering*, 22(2):112–114, February 1996.

[391] J.E. Jones, T.A. Earles, E.A. Fassman, E.E. Herricks, B. Urbonas, and J.K. Clary. Urban storm-water regulations–Are impervious area limits a good idea? *Journal of Environmental Engineering*, February 2005.

[392] C. Jothityangkoon and M. Sivapalan. Towards estimation of extreme floods: Examination of the roles of runoff process changes and floodplain flows. *Journal of Hydrology*, 281:206–229, 2003.

[393] A.G. Journel. *Fundamentals of Geostatistics in Five Lessons*, Volume 8 of *Short Course in Geology*. American Geophysical Union, Washington, DC, 1989.

[394] Z.J. Kabala. Well response in a leaky aquifer and computational interpretation of pumping tests. In W.S. Hsieh, S.T. Su, and F. Wen, eds., *Hydraulic Engineering, Proceedings of ASCE National Conference on Hydraulic Engineering and International Symposium on Engineering Hydrology*, p. 21, New York, 1993. ASCE.

[395] A.I. Kashef. *Groundwater Engineering*. McGraw-Hill Book Company, Inc., New York, 1986.

[396] M. Kay. *Practical Hydraulics*. E & FN Spon, London, 1998.

[397] M.W. Kemblowki and G.E. Urroz. Subsurface flow and transport. In L.W. Mays, ed., *Hydraulic Design Handbook*, pp. 4.1–4.26. McGraw-Hill Book Company, New York, 1999.

[398] W.S. Kerby. Time of concentration for overland flow. *Civil Engineering*, 29(3):174, March 1959.

[399] G.H. Keulegan. Laws of turbulent flow in open channels. *Journal of Research of N.B.S.*, 21:707–741, 1938.

[400] D.F. Kibler. Desk-top runoff methods. In D.F. Kibler, ed., *Urban Stormwater Hydrology*, pp. 87–135. American Geophysical Union, Water Resources Monograph Series, Washington, DC, 1982.

[401] G.J. Kirmeyer, L. Kirby, B.M. Murphy, P.F. Noran, K. Martel, T.W. Lund, J.L. Anderson, and R. Medhurst. *Maintaining Water Quality in Finished Water Storage Facilities*. AWWA Research Foundation and American Water Works Association, Denver, CO, 1999.

[402] P.Z. Kirpich. Time of concentration of small agricultural watersheds. *Civil Engineering*, 10(6):362, June 1940.

[403] G.W. Kite. *Frequency and Risk Analysis in Hydrology*. Water Resources Publications, Fort Collins, CO, 1977.

[404] R.T. Knapp, J.W. Daily, and F.G. Hammitt. *Cavitation*. McGraw-Hill, Inc., New York, 1970.

[405] S.F. Korom, K.F. Bekker, and O.J. Helweg. Influence of pump intake location on well efficiency. *Journal of Hydrologic Engineering*, 8(4):197–203, July/August 2003.

[406] N.T. Kottegoda and R. Rosso. *Statistics, Probability, and Reliability for Civil and Environmental Engineers*. McGraw-Hill, Inc., New York, 1997.

[407] D. Koutsoyiannis. A probabilistic view of Hershfield's method for estimating probable maximum precipitation. *Water Resources Research*, 35(4):1313–1322, April 1999.

[408] N. Kouwen, R.M. Li, and D.B. Simons. Velocity measurements in a channel lined with flexible plastic roughness elements. Technical report, Department of Civil Engineering, Colorado State University, Fort Collins, CO, 1980. Technical Report No. CER79-80-RML-DBS-11.

[409] J. Kozeny. Über Kapillare Leitung des Wassers im Boden. *Sitzungsberichte der Akadamie der Wissenschaften in Wein*, 136(2a):271–306, 1927 (in German).

[410] J. Kozeny. Theorie und Berechnung der Brunnen. *Wasserkraft und Wassenwirtschaft*, 28:88–92, 101–105, 113–116, 1933 (in German).

[411] M. Krishnamurthy and B.A. Christensen. Equivalent roughness for shallow channels. *Journal of the Hydraulics Division, ASCE*, 98(12):2257–2263, 1972.

[412] W.C. Krumbein and G.D. Monk. Permeability as a function of the size parameters of unconsolidated sand. *Technical Publications of the American Institute of Mining and Metallurgical Engineers*, 150(11):1492, 1942.

[413] W.C. Krumbein and G.D. Monk. Permeability as a function of the size parameters of unconsolidated sand. *Technical Publications of the American Institute of Mining and Metallurgical Engineers*, 151:153–163, 1943.

[414] E. Kuichling. The relation between the rainfall and the discharge of sewers in populous districts. *Transactions of the American Society of Civil Engineers*, 20:1–56, 1889.

[415] A.L. Lagvankar and J.P. Velon. Wastewater and stormwater piping systems. In M.L. Nayyar, ed., *Piping Handbook*, pp. C.619–C.665. McGraw-Hill Book Company, New York, 7th ed., 2000.

[416] E.W. Lane. Design of stable channels. *Transactions of the American Society of Civil Engineers*, 120:1234–1279, 1955.

[417] K. Lansey and W. El-Shorbagy. Design of pumps and pump facilities. In L.W. Mays, ed., *Stormwater Collection Systems Design Handbook*, pp. 12.1–12.41. McGraw-Hill, Inc., New York, 2001.

[418] K. Lansey and L.M. Mays. Hydraulics of water distribution systems. In L.W. Mays, ed., *Hydraulic Design Handbook*, pp. 9.1–9.38. McGraw-Hill Book Company, New York, 1999.

[419] B.E. Larock, R.W. Jeppson, and G.Z. Watters. *Hydraulics of Pipeline Systems*. CRC Press, Inc., Boca Raton, FL, 2000.

[420] G. Leclerc and J.C. Schaake. Derivation of hydrologic frequency curves. Report 142, R.M. Parsons Laboratory of Hydrodynamics and Water Resources, Massachusetts Institute of Technology, Cambridge, MA, 1972.

[421] D.R. Lee. A device for measuring seepage flux in lakes and estuaries. *Limnology and Oceanography*, 22:140–147, 1977.

[422] J.G. Lee and J.P. Heany. Estimation of urban imperviousness and its impacts on storm water systems. *Journal of Water Resources Planning and Management*, 129(5):419–426, September/October 2003.

[423] T.-C. Lee. *Applied Mathematics in Hydrogeology*. Lewis Publishers, Inc., Boca Raton, FL, 1999.

[424] J. Lehr, S. Hurlburt, B. Gallagher, and J. Voytek. *Design and Construction of Water Wells, A Guide for Engineers*. Van Nostrand Reinhold, New York, 1988.

[425] D.H. Lennox. Analysis of step-drawdown test. *Journal of the Hydraulics Division, ASCE*, 92(HY6):25–48, 1966.

[426] B. Levy and R. McCuen. Assessment of storm duration for hydrologic design. *Journal of Hydrologic Engineering*, 4(3):209–213, July 1999.

[427] L. Leyton, E.R.C. Reynolds, and F.B. Thompson. Rainfall interception in forest and moorland. In W.E. Sopper and H.W. Lull, eds., *International Symposium on Forest Hydrology; 1965 August 29–September 10; Pennsylvania State University, University Park, PA*, pp. 163–178. Oxford: Pergamon Press, 1967.

[428] J.A. Liggett. *Fluid Mechanics*. McGraw-Hill, Inc., New York, 1994.

[429] J.T. Limerinos. Determination of the Manning coefficient from measured bed roughness in natural channels. Water-Supply Paper 1898-B, U.S. Geological Survey, Washington, DC, 1970.

[430] C.H. Lin, J.F. Yean, and C.T. Tsai. Influence of sluice gate contraction coefficient on distinguishing condition. *Journal of Irrigation and Drainage Engineering, ASCE*, 128(4):249–252, July/August 2002.

[431] S.S.T. Lin and W.A. Perkins. Review of pre-development runoff analysis methods, Volume I. Report DRE 270, South Florida Water Management District, West Palm Beach, FL, 1989.

[432] R.K. Linsley. Application to the synthetic unit graph in the Western Mountain States. *Transactions of the American Geophysical Union*, 24:581–587, 1943.

[433] R.K. Linsley, J.B. Franzini, D.L. Freyberg, and G. Tchobanoglous. *Water-Resources Engineering*. McGraw-Hill, Inc., New York, 4th ed., 1992.

[434] R.K. Linsley, M.A. Kohler, and J.L.H. Paulhus. *Hydrology for Engineers*. McGraw-Hill, Inc., New York, 3d ed., 1982.

[435] C.P. Liou. Limitations and proper use of the Hazen–Williams equation. *Journal of Hydraulic Engineering*, 124(9):951–954, September 1998.

[436] D.E. Lloyd-Davies. The elimination of storm water from sewerage systems. *Minutes of the Proceedings of the Institution of Civil Engineers*, 164:41–67, 1906.

[437] LMNO Engineering Research Software, Ltd. Focus on open channel flow measurement: V-notch weirs. *Newsletter*, 1, 1999.

[438] H.A. Loáiciga. On the probability of droughts: The compound renewal model. *Water Resources Research*, 41, W01009, doi:10.1029/2004WR003075, 2005.

[439] G.V. Loganathan, D.F. Kibler, and T.J. Grizzard. Urban stormwater management. In L.W. Mays, ed., *Water Resources Handbook*, pp. 26.1–26.35. McGraw-Hill, Inc., New York, 1996.

[440] S.W. Lohman. Ground-water hydraulics. Professional Paper 708, U.S. Geological Survey, 1979.

[441] M.A. Lopez and W.M. Woodham. Magnitude and frequency of flooding on small urban watersheds in the Tampa Bay area, west-central Florida. Water-Resources Investigations Report 82-42, United States Geological Survey, Reston, VA, 1982.

[442] G.K. Lotter. Considerations of hydraulic design of channels with different roughness of walls. *Trans. All Union Scientific Research, Institute of Hydraulic Engineering*, 9:238–241, 1933.

[443] D.P. Loucks. Surface-water quality management models. In A.K. Biswas, ed., *Systems Approach to Water Management*. McGraw-Hill Book Company, New York, 1976.

[444] T.L. Lovell and E. Atkinson. From slide rules to GIS: A 35-year evolution in hydrologic and hydraulic engineering. In J.R. Rogers, G.O. Brown, and J.D. Garbrecht, eds., *Water Resources and Environmental History*, pp. 5–10. American Society of Civil Engineers, Reston, VA, 2004.

[445] W. Maddaus, G. Gleason, and J. Darmody. Integrating conservation into water supply planning: How can water suppliers achieve an appropriate balance between capacity expansion and conservation? *Journal of the American Water Works Association*, November 1996.

[446] R. Manning. Flow of water in open channels and pipes. *Transactions of the Institute of Civil Engineers (Ireland)*, 20, 1890.

[447] A. Mantoglou. Pumping management of coastal aquifers using analytic models of saltwater intrusion. *Water Resources Research*, 39(12), 1335, doi:10.1029/2002WR001891, 2003.

[448] J. Marsalek. Calibration of the tipping bucket raingage. *Journal of Hydrology*, 53:343–354, 1981.

[449] C.S. Martin. Hydraulic transient design for pipeline systems. In L.W. Mays, ed., *Water Distribution Systems Handbook*. McGraw-Hill Book Company, New York, 2000.

[450] D.L. Martin and J. Gilley. Irrigation water requirements. In *National Engineering Handbook*, chapter 2, part 623. U.S. Department of Agriculture, Natural Resources Conservation Service, 1993.

[451] J. Martinez, J. Reca, M.T. Morillas, and J.G. López. Design and calibration of a compound sharp-crested weir. *Journal of Hydraulic Engineering*, 131(2):112–116, February 2005.

[452] E.S. Martins and J.R. Stedinger. Generalized maximum-likelihood generalized extreme-value quantile estimators for hydrologic data. *Water Resources Research*, 36(3):737–744, March 2000.

[453] Mavel, a.s. *Mavel Turbine*. http://www.waterpowermagazine.com/graphic.asp?sc=2024017&seq=2, 2005.

[454] L.W. Mays. Water resources: An introduction. In L.W. Mays, ed., *Water Resources Handbook*, pp. 1.3–1.35. McGraw-Hill, Inc., New York, 1996.

[455] L.W. Mays. Introduction. In L.W. Mays, ed., *Water Distribution Systems Handbook*, pp. 1.1–1.30. McGraw-Hill Book Company, New York, 2000.

[456] L.W. Mays. *Water Resources Engineering*. John Wiley & Sons, New York, 2001.

[457] G.T. McCarthy. The unit-hydrograph and flood routing. Unpublished article at *U.S. Army Corps of Engineers North Atlantic Division Conference*, 1938.

[458] R.H. McCuen. *Hydrologic Analysis and Design*. Prentice Hall, Englewood Cliffs, NJ, 1989.

[459] R.H. McCuen. *Hydrologic Analysis and Design*. Prentice Hall, Upper Saddle River, NJ, 2d ed., 1998.

[460] R.H. McCuen. Do water resources engineers need factors of safety for design? *Journal of Water Resources Planning and Management*, 127(1):4–5, 2001.

[461] R.H. McCuen. *Modeling Hydrologic Change: Statistical Methods*. Lewis Publishers, Inc., Boca Raton, FL, 2002a.

[462] R.H. McCuen. *Hydrologic Analysis and Design*. Prentice Hall, Upper Saddle River, NJ, 3d ed., 2005.

[463] R.H. McCuen and R.E. Beighley. Seasonal flow frequency analysis. *Journal of Hydrology*, 279:43–56, 2003.

[464] R.H. McCuen, P.A. Johnson, and R.M. Ragan. Hydrology. Technical Report HDS-2, U.S. Federal Highway Administration, Washington, DC, September 1996.

[465] R.H. McCuen and O. Okunola. Extension of TR-55 for microwatersheds. *Journal of Hydrologic Engineering*, 7(4):319–325, July/August 2002.

[466] R.H. McCuen, S.L. Wong, and W.J. Rawls. Estimating urban time of concentration. *Journal of Hydraulic Engineering*, 110(7):887–904, 1984.

[467] M.G. McDonald and A.W. Harbaugh. A modular three-dimensional finite-difference ground-water flow model. Techniques of Water Resources Investigations of the United States Geological Survey, Book 6, Chapter A1, U.S. Geological Survey, Reston, VA, 1988.

[468] B.M. McEnroe. Characteristics of intense storms in Kansas. In *World Water Issues in Evolution*, pp. 933–940. ASCE, New York, 1986.

[469] T.J. McGhee. *Water Supply and Sewerage*. McGraw-Hill, Inc., New York, 6th ed., 1991.

[470] T.B. McKee, N.J. Doesken, and J. Kleist. The relationship of drought frequency and duration to time scales. In *Proceedings of Eighth Conference on Applied Climatology*, pp. 179–184, Anaheim, CA, January 17–22, 1993.

[471] R.G. Mein and C.L. Larson. Modeling infiltration during a steady rain. *Water Resources Research*, 9(2):384–394, 1973.

[472] R.A. Meriam. A note on the interception loss equation. *Journal of Geophysical Research*, 65:3850–3851, 1960.

[473] B. Merz and A.H. Thieken. Separating natural and epistemic uncertainty in flood frequency analysis. *Journal of Hydrology*, 309:114–132, 2005.

[474] Metcalf & Eddy, Inc. *Wastewater Engineering: Collection and Pumping of Wastewater*. McGraw-Hill, Inc., New York, 1981.

[475] P.E. Meyer-Peter and R. Muller. Formulas for bed load transport. In *Proceedings of the Third International Association for Hydraulic Research, Stockholm, Sweden*, pp. 39–64, 1948.

[476] Miami-Dade Public Works Department. *Public Works Manual, Part 2: Design and Construction*, 2001.

[477] C. Michel, V. Andréassian, and C. Perrin. Soil Conservation Service curve number method: How to mend a wrong soil moisture accounting procedure. *Water Resources Research*, 41, W02011, doi:10.1029/2004WR003191, 2005.

[478] A.C. Miller, S.N. Kerr, and D.J. Spaeder. Calibration of Snyder coefficients for Pennsylvania. *Water Resources Bulletin*, 19(4):625–630, 1983.

[479] D.W. Miller, J.J. Geraghty, and R.S. Collins. *Water Atlas of the United States*. Water Information Center, Inc., Port Washington, NY, 1962.

[480] J.F. Miller. Probable maximum precipitation and rainfall frequency data for Alaska. Technical report, U.S. Department of Commerce, Weather Bureau, Washington, DC, 1963.

[481] J.F. Miller, R.H. Frederick, and R.J. Tracey. Precipitation frequency atlas of the western United States. NOAA Atlas 2, U.S. National Weather Service, Silver Spring, MD, 1973. 11 volumes.

[482] N. Miller and R. Cronshey. Runoff curve numbers—the next step. In B.C. Yen, ed., *Channel Flow and Catchment Runoff*, pp. 910–916. Department of Civil Engineering, University of Virginia, Charlottesville, VA, 1989.

[483] S.K. Mishra, J.V. Tyagi, and V.P. Singh. Comparison of infiltration models. *Hydrological Processes*, 17:2629–2652, August 2003.

[484] A. Molini, L.G. Lanza, and P. La Barbera. The impact of tipping-bucket raingage measurement errors on design rainfall for urban-scale applications. *Hydrological Processes*, 19:1073–1088, 2005.

[485] J.L. Monteith. Evaporation and the environment. *Symposium of the Society of Experimental Biologists*, 19:205–234, 1965.

[486] J.L. Monteith. *Principles of Environmental Physics*. Edward Arnold, London, 1973.

[487] J.L. Monteith. Evaporation and surface temperature. *Quarterly Journal of Royal Meteorological Society*, 107:1–27, 1981.

[488] S. Montes. *Hydraulics of Open Channel Flow*. ASCE Press, Reston, VA, 1998.

[489] J.H. Montgomery. *Groundwater Chemicals Desk Reference*. Lewis Publishers, Inc., Boca Raton, FL, 3d ed., 2000.

[490] L.F. Moody. Friction factors for pipe flow. *Transactions of the ASME*, 66(8), 1944.

[491] L.F. Moody. Some pipe characteristics of engineering interest. *Houille Blanche*, May–June 1950.

[492] L.F. Moody and T. Zowski. Hydraulic machinery. In C.V. Davis and K.E. Sorensen, eds., *Handbook of Applied Hydraulics*. McGraw-Hill, Inc., New York, 3d ed., 1989.

[493] H.J. Morel-Seytoux and J.P. Verdin. Extension of Soil Conservation Service rainfall runoff methodology for ungaged watersheds. Report FHWA/RD-81/060, Federal Highway Administration, Washington, DC, 1981.

[494] R. Morgan and D.W. Hulinghorst. Unit hydrographs for gauged and ungauged watersheds. Technical report, U.S. Engineers Office, Binghamton, New York, July 1939.

[495] D.A. Morris and A.I. Johnson. Summary of hydrologic and physical properties of rock and soil materials, as analyzed by the Hydrologic Laboratory of the U.S. Geologic Survey, 1948–1960. Water-Supply Paper 1839-D, U.S. Geological Survey, 1967.

[496] A.K. Motayed and M. Krishnamurthy. Composite roughness of natural channels. *Journal of the Hydraulics Division, ASCE*, 106(6):1111–1116, 1980.

[497] R.L. Mott. *Applied Fluid Mechanics*. Merrill, New York, 1994.

[498] K.N. Moutsopoulos and V.A. Tsihrintzis. Approximate analytical solutions of the Forcheimer equation. *Journal of Hydrology*, 309:93–103, 2005.

[499] L. Muhlhofer. Rauhigkeitsuntersuchungen in einem stollen mit betonierter sohle und unverkleideten wanden. *Wasserkraft und Wasserwirtschaft*, 28(8):85–88, 1933.

[500] T.J. Mulvaney. On the use of self-registering rain and flood gauges in making observations on the relations of rainfall and flood discharges in a given catchment. *Transactions of the Institute of Civil Engineers (Ireland)*, 4(2):18–31, 1850.

[501] B.R. Munson, D.F. Young, and T.H. Okiishi. *Fundamentals of Fluid Mechanics*. John Wiley & Sons, New York, 2d ed., 1994.

[502] L.C. Murdoch and S.E. Kelly. Factors affecting the performance of conventional seepage meters. *Water Resources Research*, 39(6), 1163, doi:10.1029/2002WR001347, 2003.

[503] T.E. Murphy. Spillway crest design. Misc. Paper H-73-5, Waterways Experiment Station, U.S. Army Corps of Engineers, Vicksburg, MS, 1973.

[504] M. Muskat. Potential distribution in large cylindrical discs (homogeneous sands) with partially penetrating electrodes (partially penetrating wells). *Physics*, 2(5):329–384, 1932.

[505] M. Muskat. *The Flow of Homogeneous Fluids through Porous Media*. McGraw-Hill, Inc., New York, 1937.

[506] W.R.C. Myers. Influence of geometry on discharge capacity of open channels. *Journal of Hydraulic Engineering*, 117(5):676–680, May 1991.

[507] L. Nandagiri and G.M. Kovoor. Sensitivity of the Food and Agriculture Organization Penman–Monteith evapotranspiration estimates to alternative procedures for estimation of parameters. *Journal of Irrigation and Drainage Engineering, ASCE*, 131(3):238–248, June 2005.

[508] S. Naoum and I.K. Tsanis. Temporal and spatial variation of annual rainfall on the island of Crete, Greece. *Hydrological Processes*, 17:1899–1922, 2003.

[509] National Research Council. *Safety of Dams, Flood and Earthquake Criteria*. National Academy Press, Washington, DC, 1985.

[510] National Research Council. *Opportunities in the Hydrologic Sciences*. National Academy Press, Washington, DC, 1991.

[511] National Weather Service (NWS). *NWSRFS User's Manual Documentation, Release Number 64*, June 2002 ⟨http://www.nws.noaa.gov/oh/hrl/nwsrfs/users_manual/htm/xrfsdochtm.htm⟩.

[512] L.C. Neale and R.E. Price. Flow characteristics of PVC sewer pipe. *Journal of the Sanitary Engineering Division, ACSE*, 90(SA3):109, 1964.

[513] S.P. Neuman. Theory of flow in unconfined aquifers considering delayed response of the water table. *Water Resources Research*, 8(4):1031–1045, August 1972.

[514] S.P. Neuman. Supplementary comments on "Theory of flow in unconfined aquifers considering delayed response of the water table." *Water Resources Research*, 9:1102–1103, 1973.

[515] S.P. Neuman. Analysis of pumping test data from anisotropic unconfined aquifers considering delayed gravity response. *Water Resources Research*, 11:329–342, 1975.

[516] S.P. Neuman and P.A. Witherspoon. Theory of flow in aquicludes adjacent to slightly leaky aquifers. *Water Resources Research*, 4(1):103–112, 1968.

[517] S.P. Neuman and P.A. Witherspoon. Theory of flow in a confined two aquifers system. *Water Resources Research*, 5(2):803–816, 1969a.

[518] S.P. Neuman and P.A. Witherspoon. Applicability of current theories of flow in leaky aquifers. *Water Resources Research*, 5(4):817–829, 1969b.

[519] S.P. Neuman and P.A. Witherspoon. Field determination of the hydraulic properties of leaky multiple aquifer systems. *Water Resources Research*, 8(5):1284–1298, 1972.

[520] B.D. Newport. Salt water intrusion in the United States. Technical Report 600/8-77-011, U.S. Environmental Protection Agency, Washington, DC, 1977.

[521] V. Nguyen and G.F. Pinder. Direct calculation of aquifer parameters in slug test analysis. In J. Rosenshein and G.D. Bennett, eds., *Groundwater Hydraulics*, Water Resources Monograph 9, pp. 222–239. American Geophysical Union, 1984.

[522] J.W. Nicklow. Design of stormwater inlets. In L.W. Mays, ed., *Stormwater Collection Systems Design Handbook*, pp. 5.1–5.42. McGraw-Hill, Inc., New York, 2001.

[523] J. Niemczynowicz. The dynamic calibration of tipping-bucket raingauges. *Nordic Hydrology*, 17:203–214, 1986.

[524] J. Nikuradse. Gesetzmässigkeiten der turbulenten strömung in glatten röhren. *VDI-Forschungshaft*, 356, 1932.

[525] J. Nikuradse. Strömungsgesetze in rauhen Röhren. *VDI-Forschungshaft*, 362, 1933.

[526] N.S. Norman, E.J. Nelson, and A.K. Zundel. Improved process for floodplain delineation from digital terrain models. *Journal of Water Resources Planning and Management*, 129(5):427–436, September/October 2003.

[527] M.A. Nouh and R.D. Townsend. Shear stress distribution in stable channel bends. *Journal of the Hydraulics Division, ASCE*, 105(HY10):1233–1245, October 1979.

[528] P. Novak. Improvement of water quality in rivers by aeration at hydraulic structures. In M. Hino, ed., *Water Quality and its Control*, pp. 147–168. A.A. Balkema Publishers, Brookfield, VT, 1994.

[529] V. Novotny, K.R. Imhoff, M. Olthof, and P.A. Krenkel. *Karl Imhoff's Handbook of Urban Drainage and Wastewater Disposal*. John Wiley & Sons, New York, 1989.

[530] E. Nyer, G. Boettcher, and B. Morello. Using the properties of organic compounds to help design a treatment system. *Ground Water Monitoring Review*, 11(4):115–120, 1991.

[531] J.E. O'Connor and J.E. Costa. Spatial distribution of the largest rainfall-runoff floods from basins between 2.6 and 26,000 km^2 in the United States and Puerto Rico. *Water Resources Research*, 40, W01107, doi:10.1029/2003WR002247, 2004.

[532] American Society of Civil Engineers. *Quality of Ground Water, Guidelines for Selection and Application of Frequently Used Models*. American Society of Civil Engineers, New York, 1996. Manual of Practice No. 85.

[533] U.S. Bureau of Reclamation. *Ground Water Manual*. U.S. Department of the Interior, 2d ed., 1995.

[534] I.C. Olmsted. Stomatal resistance and water stress in melaleuca. Final Report, Contract with USDA, Forest Service, August 1978.

[535] R.M. Olson and S.J. Wright. *Essentials of Engineering Fluid Mechanics*. Harper & Row, Publishers, Inc., New York, 5th ed., 1990.

[536] P.S. Osbourne. Suggested operating procedures for aquifer pumping tests. EPA Ground Water Issue EPA/540/S-93/503, U.S. Environmental Protection Agency, February 1993.

[537] D.E. Overton and M.E. Meadows. *Storm Water Modeling*. Academic Press, New York, 1976.

[538] J.N. Paine and A.O. Akan. Design of detention systems. In L.W. Mays, ed., *Stormwater Collection Systems Design Handbook*, pp. 7.1–7.66. McGraw-Hill, Inc., New York, 2001.

[539] W.C. Palmer. Meteorological drought. Technical Report Research Paper No. 45, U.S. Department of Commerce, Weather Bureau, 1965.

[540] J.F. Pankow and J.A. Cherry. *Dense Chlorinated Solvents and Other DNAPLs in Groundwater*. Waterloo Press, Portland, OR, 1996.

[541] I.S. Papadopulos and H.H. Cooper, Jr. Drawdown in a well of large diameter. *Water Resources Research*, 3(1):241–244, 1967.

[542] R.G. Patil, J.S.R. Murthy, and L.K. Ghosh. Uniform and critical flow computations. *Journal of Irrigation and Drainage Engineering, ASCE*, 131(4):375–378, August 2005.

[543] N.N. Pavlovskii. K Voporosu o Raschetnoi dlia Ravnomernogo Dvizheniia v Vodotokahk s Neodnorodnymi Stenkami (On a design formula for uniform flow in channels with nonhomogeneous walls). *Izvestiia Vsesoiuznogo Nauchno-Issledovatel'skogo Instituta Gidrotekhniki (Transactions, All-Union Scientific Research Institute of Hydraulic Engineering)*, 3:157–164, 1931.

[544] K. Pearson. *Tables for Statisticians and Biometricians*. Columbia University Press, New York, 3d ed. 1930.

[545] R. Pecher. The runoff coefficient and its dependence on rain duration. *Berichte aus dem Institut fur Wasserwirtschaft und Gesundheitsingenieurwesen*, No. 2, 1969. TU Munich.

[546] H. Penman. Natural evaporation from open water, bare soil, and grass. *Proceedings of the Royal Society of London, Series A*, A193:120–146, 1948.

[547] H.L. Penman. Evaporation: An introductory survey. *Netherlands Journal of Agricultural Sciences, Series A*, 4:9–29, 1956.

[548] H.L. Penman. Vegetation and hydrology. Technical Communication 53, Commonwealth Bureau of Soils, Harpenden, England, 1963. 125 pp.

[549] L.S. Pereira, A. Perrier, R.G. Allen, and I. Alves. Evapotranspiration: Concepts and future trends. *Journal of Irrigation and Drainage Engineering, ASCE*, 125(2):45–51, 1999.

[550] E. Peters, H.A.J. van Lanen, and P.J.J.F. Bier. Drought in groundwater—Drought distribution and performance indicators. *Journal of Hydrology*, 306:302–317, 2005.

[551] J.R. Philip. An infiltration equation with physical significance. *Soil Science*, 77:153–157, 1954.

[552] J.R. Philip. Variable-head ponded infiltration under constant or variable rainfall. *Water Resources Research*, 29(7):2155–2165, July 1993.

[553] N.N. Pillai. Effect of shape on uniform flow through smooth rectangular open channels. *Journal of Hydraulic Engineering*, 123(7):656–658, July 1997.

[554] R.D. Pomeroy and J.D. Parkhurst. The forecasting of sulfide buildup rates in sewers. *Progress in Water Technology*, 9, 1977.

[555] V.M. Ponce. *Engineering Hydrology, Principles and Practices*. Prentice Hall, Englewood Cliffs, NJ, 1989.

[556] V.M. Ponce, A.K. Lohani, and P.T. Huston. Surface albedo and water resources: Hydroclimatological impact of human activities. *Journal of Hydrologic Engineering*, 2(4):197–203, October 1997.

[557] V.M. Ponce, R.P. Pandey, and Sezar Ercan. Characterization of drought across climatic spectrum. *Journal of Hydrologic Engineering*, 5(2):222–224, April 2000.

[558] M.C. Potter and D.C. Wiggert. *Mechanics of Fluids*. Prentice Hall, Englewood Cliffs, NJ, 1991.

[559] M.C. Potter and D.C. Wiggert. *Mechanics of Fluids*. Brooks/Cole, Pacific Grove, CA, 3d ed., 2001.

[560] A. Prakash. *Water Resources Engineering*. ASCE Press, New York, 2004.

[561] C.H.B. Priestly and R.J. Taylor. On the assessment of surface heat flux and evaporation using large-scale parameters. *Monthly Weather Review*, 100:81–92, 1972.

[562] L.G. Puls. Construction of flood flow routing curves, 1928. U.S. 70th Congress, First Session, House Document 185, pp. 46–52, and U.S. 71st Congress, Second Session, House Document 328, pp. 190–191.

[563] V.L. Quisenberry and R.E. Phillips. Percolation of surface-applied water in the field. *Soil Science Society of America Journal*, 40:484–489, 1976.

[564] M. Radojkovic and C. Maksimovic. On standardization of computational models for overland flow. In B.C. Yen, ed., *Proceedings of the Fourth International Conference on Urban Storm Drainage*, pp. 100–105, Lausanne, Switzerland, 1987. International Association for Hydraulic Research.

[565] C.D. Rail. *Groundwater Contamination*. Technomic Publishing Company, Inc., Lancaster, PA, 1989.

[566] N. Rajaratnam and K. Subramanya. Flow equation for the sluice gate. *Journal of Irrigation and Drainage Engineering, ASCE*, 93(3):167–186, 1967.

[567] C.E. Ramser. Runoff from small agricultural areas. *Journal of Agricultural Research*, 34(9), 1927.

[568] R.E. Rathbun. Transport, behavior, and fate of volatile organic compounds in streams. Professional Paper 1589, U.S. Geological Survey, Washington, DC, 1998.

[569] A.J. Raudkivi and R.A. Callander. *Analysis of Groundwater Flow*. John Wiley & Sons, Inc., New York, 1976.

[570] W. Rawls, P. Yates, and L. Asmussen. Calibration of selected infiltration equations for the Georgia Coastal Plain. Technical Report ARS-S-113, U.S. Department of Agriculture, Agricultural Research Service, Washington, DC, 1976.

[571] W.J. Rawls and D.L. Brakensiek. Estimating soil water retention from soil properties. *Journal of Irrigation and Drainage Engineering, ASCE*, 108(2):166–171, 1982.

[572] W.J. Rawls, D.L. Brakensiek, and N. Miller. Green–Ampt infiltration parameters from soils data. *Journal of Hydraulic Engineering*, 109(1):1316–1320, 1983.

[573] W.J. Rawls, D.L. Brakensiek, and K.E. Saxton. Estimation of soil water properties. *Transactions of the American Society of Agricultural Engineers*, 25(5):1316–1320, 1328, 1982.

[574] A.J. Reese and S.T. Maynord. Design of spillway crest. *Journal of Hydraulic Engineering*, 113(4):476–490, 1987.

[575] O. Reynolds. An experimental investigation of the circumstances which determine whether the motion of water shall be direct or sinuous and of the law of resistance in parallel channels. *Philosophical Transactions, Royal Society of London*, 174:935–982, 1883.

[576] W. Rippl. The capacity of storage reservoirs for water supply. *Proceedings of the Institution of Civil Engineers, London*, 71:270, 1883.

[577] J.A. Roberson and C.T. Crowe. *Engineering Fluid Mechanics*. John Wiley & Sons, New York, 6th ed., 1997.

[578] F. Rodriguez, H. Andrieu, and J.-D. Creutin. Surface runoff in urban catchments: Morphological identification of unit hydrographs from urban databanks. *Journal of Hydrology*, 283:146–168, 2003.

[579] M.I. Rorabaugh. Graphical and theoretical analysis of step-drawdown test of artesian well. *Proceedings of the American Society of Civil Engineers*, 79(362):1–23 pp., 1953.

[580] Roscoe Moss Company. *Handbook of Groundwater Development*. John Wiley & Sons, New York, 1990.

[581] J.R. Rossing. Identification of critical runoff generating areas using a variable source area model. PhD thesis, Cornell University, Ithaca, NY, 1996.

[582] L.A. Rossman. Computer models/epanet. In L.W. Mays, ed., *Water Distribution Systems Handbook*, pp. 12.1–12.23. McGraw-Hill Book Company, New York, 2000.

[583] W.M. Grayman, L.A. Rossman, and E.E. Geldreich. Water quality. In L.W. Mays, ed., *Water Distribution Systems Handbook*, pp. 9.1–9.24. McGraw-Hill Book Company, New York, 2000.

[584] R.L. Rossmiller. The rational formula revisited. In *Proceedings of the International Symposium on Urban Storm Runoff, July 28-31*, Lexington, Kentucky, 1980. University of Kentucky.

[585]   J. Rothacher. Net precipitation under a Douglas-fir forest. *Forest Science*, 9(4):423–429, 1963.

[586]   H. Rouse. *Elementary Fluid Mechanics*. John Wiley & Sons, New York, 1946.

[587]   H. Rouse and S. Ince. *History of Hydraulics*. Iowa Institute of Hydraulic Research, University of Iowa, Iowa City, 1957.

[588]   L.K. Rowe. Rainfall interception by a beech-podocarp-hardwood forest near Reefton, North Westland, New Zealand. *Journal of Hydrology (N.Z.)*, 18(2):63–72, 1979.

[589]   H. Rubin and J. Atkinson. *Environmental Fluid Mechanics*. Marcel Dekker, Inc., New York, 2001.

[590]   M.L. Sage and G.W. Sage. Vapor pressure. In R.S. Boethling and D. Mackay, eds., *Handbook of Property Estimation Methods for Chemicals*, pp. 53–65. Lewis Publishers, Inc., Boca Raton, FL, 2000.

[591]   J.D. Salas. Analysis and modeling of hydrologic time series. In D.R. Maidment, ed., *Handbook of Hydrology*, pp. 19.1–19.72. McGraw-Hill Book Company, New York, 1993.

[592]   G.D. Salvucci and D. Entekhabi. Explicit expressions for Green–Ampt (delta function diffusivity) infiltration rate and cumulative storage. *Water Resources Research*, 30(9):2661–2663, September 1994.

[593]   Z. Samani. Estimating solar radiation and evapotranspiration using minimum climatological data. *Journal of Irrigation and Drainage Engineering, ASCE*, 126(4):265–267, 2000.

[594]   X. Sánchez-Vila, P.M. Meier, and J. Carrera. Pumping tests in heterogeneous aquifers: An analytical study of what can be obtained from their interpretation using Jacob's method. *Water Resources Research*, 35(4):943–952, April 1999.

[595]   R.L. Sanks. *Pump Station Design*. Butterworth Publishers, London, 2d ed., 1998.

[596]   R. Schildgen. Unnatural disasters. *Sierra*, 84:48–57, 1999.

[597]   H. Schlichting. *Boundary Layer Theory*. McGraw-Hill, Inc., New York, 7th ed., 1979.

[598]   S. Schmorak and A. Mercado. Upconing of fresh water–sea water interface below pumping wells, field study. *Water Resources Research*, 5:1290–1311, 1969.

[599]   L.E. Schneider and R.H. McCuen. Statistical guidelines for curve number generation. *Journal of Irrigation and Drainage Engineering, ASCE*, 131(3):282–290, June 2005.

[600]   J.L. Schnoor. *Environmental Modeling: Fate and Transport of Pollutants in Water, Air, and Soil*. John Wiley & Sons, New York, 1996.

[601]   T.R. Schueler. *Controlling Urban Runoff: A Practical Manual for Planning and Designing Urban BMPs*. Metropolitan Washington Council of Governments, Washington, DC, July 1987.

[602]   T.R. Schueler, P.A. Kumble, and M.A. Heraty. *Current Assessment of Urban Best Management Practices: Techniques for Reducing Non-point Source Pollution in the Coastal Zones*. Metropolitan Washington Council of Governments, Washington, DC, 1991. Office of Wetlands, Oceans, and Watersheds, U.S. Environmental Protection Agency.

[603]   C.K. Sehgal. Design guidelines for spillway gates. *Journal of Hydraulic Engineering*, 122(3):155–165, March 1996.

[604]   Soil Conservation Service. *SCS National Engineering Handbook, Section 4: Hydrology*. U.S. Department of Agriculture, Washington, DC, 1993.

[605] B.A. Shafer and L.E. Dezman. Development of a Surface Water Supply Index (SWSI) to assess the severity of drought conditions in snowpack runoff areas. In *Proceedings of the Western Snow Conference*, pp. 164–175, Colorado State University, Fort Collins, CO, 1982.

[606] R.M. Shane and W.R. Lynn. Mathematical model for flood risk evaluation. *Journal of the Hydraulics Division, ASCE*, 90(HY6):1–20, November 1964.

[607] L.K. Sherman. Stream-flow from rainfall by the unit graph method. *Engineering News Record*, 108:501–505, 1932.

[608] J.-T. Shiau and H.W. Shen. Recurrence analysis of hydrologic droughts of differing severity. *Journal of Water Resources Planning and Management*, 127(1):30–40, January/February 2001.

[609] V. Sichard. *Das Fassungsvermögen von Bohrbrunnen und Eine Bedeutung für die Grundwasserersenkung inbesondere für grossere Absentiefen*. PhD thesis, Technische Hochschule, Berlin, 1927.

[610] J.R. Simanton and H.B. Osborn. Reciprocal-distance estimate of point rainfall. *Journal of the Hydraulic Engineering Division, ASCE*, 106(HY7), July 1980.

[611] D.B. Simons and E.V. Richardson. Resistance to flow in alluvial channels. Professional Paper 422-J, U.S. Geological Survey, Washington, DC, 1966.

[612] V.P. Singh. *Hydrologic Systems, Rainfall-Runoff Modeling, Volume I*. Prentice Hall, Englewood Cliffs, NJ, 1989.

[613] V.P. Singh. *Elementary Hydrology*. Prentice Hall, Englewood Cliffs, NJ, 1992.

[614] V.P. Singh and W.G. Strupczewski. On the status of flood frequency analysis. *Hydrological Processes*, 16:3737–3740, 2002b.

[615] M.K. Smith. Throughfall, stemflow, and interception in pine and eucalypt forests. *Australian Forestry*, 36:190–197, 1974.

[616] M.R. Smith, R.G. Allen, J.L. Monteith, L.S. Pereira, and A. Segeren. Report on the expert consultation on procedures for revision of FAO guidelines for predicting crop water requirements. Technical report, Land and Water Development Division, Food and Agricultural Organization of the United Nations, Rome, 1991.

[617] R.E. Smith. Discussion of "Runoff curve number: Has it reached maturity?" by V.M. Ponce and R.H. Hawkins. *Journal of Hydrologic Engineering*, 2(3):145–147, 1997.

[618] F.F Snyder. Synthetic unit hydrographs. *Transactions of the American Geophysical Union*, 19:447, 1938.

[619] R.L. Snyder, M. Orang, S. Matyac, and M.E. Grismer. Simplified estimation of reference evapotranspiration from pan evaporation data in California. *Journal of Irrigation and Drainage Engineering, ASCE*, 131(3):249–253, June 2005.

[620] Soil Conservation Service. Computer program for project formulation hydrology (draft). Technical Release 20, Soil Conservation Service, U.S. Department of Agriculture, Washington, DC, 1983.

[621] Soil Conservation Service. Urban hydrology for small watersheds. Technical Release 55, Soil Conservation Service, U.S. Department of Agriculture, Washington, DC, 1986.

[622] W.B. Solley. Estimated Use of Water in the United States in 1995. Circular 1200, U.S. Geological Survey, Washington, DC, 1998.

[623] South Carolina Land Resources Conservation Commission. South Carolina stormwater management and sediment reduction regulations. Technical report, Division of Engineering, Columbia, SC, 1992.

[624] South Florida Water Management District. *Management and Storage of Surface Waters, Permit Information Manual, Volume IV*. West Palm Beach, FL, May 1994.

[625] South Florida Water Management District. *Best Management Practices for South Florida Urban Stormwater Management Systems*. West Palm Beach, FL, April 2002.

[626] M.S. Sperry and J.J. Pierce. A model for estimating the hydraulic conductivity of granular material based on grain shape, grain size, and porosity. *Ground Water*, 33(6):892–898, November–December 1995.

[627] P. Stahre and B. Urbonas. Swedish approach to infiltration and percolation design. In L.A. Roesner, B. Urbonas, and M.B. Sonnen, eds., *Design of Urban Runoff Quality Controls*, pp. 307–323. ASCE, New York, 1989.

[628] R.W. Stallman. Aquifer-test design, observation and data analysis. Techniques of Water Resources Investigations, Book 3, Chapter B1, U.S. Geological Survey, 1971.

[629] G. Stanhill. The control of field irrigation practice from measurements of evaporation. *Israel Journal of Agricultural Research*, 12:51–62, 1962.

[630] T.E. Stanton and J.R. Pannell. Similarity of motion in relation to surface friction of fluids. *Philosophical Transactions, Royal Society of London*, 214A:199–224, 1914.

[631] A.J. Stepanoff. *Centrifugal and Axial Flow Pumps*. John Wiley & Sons, New York, 2d ed., 1957.

[632] W.J. Stone. *Hydrogeology in Practice*. Prentice Hall, Upper Saddle River, NJ, 1999.

[633] R.L. Street, G.Z. Watters, and J.K. Vennard. *Elementary Fluid Mechanics*. John Wiley & Sons, New York, 7th ed., 1996.

[634] V.L. Streeter and E.B. Wylie. *Fluid Mechanics*. McGraw-Hill, Inc., New York, 8th ed., 1985.

[635] V.L. Streeter, E.B. Wylie, and K.W. Bedford. *Fluid Mechanics*. McGraw-Hill, Inc., New York, 9th ed., 1998.

[636] A. Strickler. Contributions to the question of a velocity formula and roughness data for streams, channels and closed pipelines. Technical report, W.M. Keck Laboratory, California Institute of Technology, Pasadena, California, 1923. Translation from the original German by T. Roesgen and W.R. Brownlie (1981).

[637] W.G. Strupczewski, V.P. Singh, and S. Weglarczyk. Asymptotic bias of estimation methods caused by the assumption of false probability distribution. *Journal of Hydrology*, 258:122–148, 2002.

[638] J.M. Stubchaer. The Santa Barbara urban hydrograph method. *Proceedings, National Symposium of Hydrology and Sediment Control*, University of Kentucky, Lexington, Kentucky, 1980.

[639] T. Sturm. *Open Channel Hydraulics*. McGraw-Hill, Inc., New York, 2001.

[640] D.M. Sumner and J.M. Jacobs. Utility of Penman–Monteith, Priestly–Taylor, reference evapotranspiration, and pan evaporation methods to estimate pasture evapotranspiration. *Journal of Hydrology*, 308:81–104, 2005.

[641] P.K. Swamee. Generalized rectangular weir equations. *Journal of Hydraulic Engineering*, 114(8):945–949, 1988.

[642] P.K. Swamee and A.K. Jain. Explicit equations for pipe-flow problems. *ASCE, Journal of the Hydraulics Division*, 102(HY5):657–664, May 1976.

[643] W.T. Swank, N.B. Goebel, and J.D. Helvey. Interception loss in loblolly pine stands in the piedmont of South Carolina. *Journal of Soil and Water Conservation*, 27(4):160–164, 1972.

[644] R.M. Sykes. Water and wastewater planning. In W.F. Chen, ed., *The Civil Engineering Handbook*, pp. 169–223. CRC Press, Inc., Boca Raton, FL, 1995.

[645] T. Szirtes. *Applied Dimensional Analysis and Modeling*. McGraw-Hill, Inc., New York, 1998.

[646] L.M. Tallaksen. A review of baseflow recession analysis. *Journal of Hydrology*, 165, 1995.

[647] G.D. Tasker and J.R. Stedinger. Regional skew with weighted LS regression. *Journal of Water Resources Planning and Management*, 112(2):225–237, April 1986.

[648] M. Tasumi, R.G. Allen, R. Trezza, and J.L. Wright. Satellite-based energy balance to assess within-population variance of crop coefficient curves. *Journal of Irrigation and Drainage Engineering, ASCE*, 131(1):94–109, February 2005.

[649] B.N. Taylor. Guide for the use of the international system of units (SI). Technical report, National Institute of Standards and Technology, 1995. NIST Special Publication 811, 1995 Edition.

[650] D.W. Taylor. *Fundamentals of Soil Mechanics*. Wiley, New York, 1948.

[651] B. Temesgen, S. Eching, B. Davidoff, and K. Frame. Comparison of some reference evapotranspiration equations for California. *Journal of Irrigation and Drainage Engineering, ASCE*, 131(1):73–84, February 2005.

[652] C.V. Theis. The relation between lowering of the piezometric surface and the rate and duration of discharge of a well using ground water storage. In *Transactions of the American Geophysical Union, 16th Annual Meeting, Part 2*, pp. 519–524, 1935.

[653] H.J. Thiébaux. *Statistical Data Analysis for Ocean and Atmospheric Sciences*. Academic Press, Inc., San Diego, CA, 1994.

[654] G. Thiem. *Hydrologische*. Gebhardt, Leipzig, Germany, 1906.

[655] A.H. Thiessen. Precipitation for large areas. *Monthly Weather Review*, 39:1082–1084, July 1911.

[656] A.L. Tholin and C.J. Keifer. The hydrology of urban runoff. *Transactions of the American Society of Civil Engineers*, 125:1308, 1960.

[657] A.S. Thom and H.R. Oliver. On Penman's equation for estimating regional evapotranspiration. *Quarterly Journal of Royal Meteorological Society*, 103:345–357, 1977.

[658] G.W. Thomas and R.E. Phillips. Consequences of water movement in macropores. *J. Environ. Qual.*, 8(2):149–152, 1979.

[659] L. Thomas. What Causes Wells to Lose Capacity? *Opflow*, 28(1):8–15, January 2002. American Water Works Association.

[660] C.W. Thornthwaite. An approach toward a rational classification of climate. *Geographical Review*, 38:55, 1948.

[661] C.W. Thornthwaite and J.R. Mather. The water balance. *Climatology*, 8(1), 1955. Laboratory of Climatology, Seabrook, NJ.

[662] J.A. Tindall and J.R. Kunkel. *Unsaturated Zone Hydrology for Scientists and Engineers*. Prentice Hall, Upper Saddle River, NJ, 1999.

[663] D.K. Todd. *Groundwater Hydrology*. John Wiley & Sons, New York, 2d ed., 1980.

[664] D.K. Todd and L.W. Mays. *Groundwater Hydrology*. John Wiley & Sons, New York, 3d ed., 2005.

[665] M. Todorovic. Single-layer evapotranspiration model with variable canopy resistance. *Journal of Irrigation and Drainage Engineering, ASCE*, 125(5):235–245, 1999.

[666] Topomatika. 3D digitizing of the rotor of a Pelton turbine, 2005. http://www. topomatika.hr/Applications/turbine.htm.

[667] S. Trajkovic. Temperature-based approaches for estimating reference evapotranspiration. *Journal of Irrigation and Drainage Engineering, ASCE,* 131(4):316–323, 2005.

[668] P.A. Troch, F.P. De Troch, and W. Brutsaert. Effective water table depth to describe initial conditions prior to storm rainfall in humid regions. *Water Resources Research,* 29(2):427–434, 1993.

[669] I.K. Tuncock and L.W. Mays. Hydraulic design of culverts and highway structures. In L.W. Mays, ed., *Hydraulic Design Handbook,* pp. 15.1–15.71. McGraw-Hill Book Company, New York, 1999.

[670] L. Turc. Evaluation des besoins en eau d'irrigation, evapotranspiration potentielle, formule climatique simplifice et mise a jour (in French). (English title: Estimation of irrigation water requirements, potential evapotranspiration: A simple climatic formula evolved up to date). *Annales Agronomiques,* 12:13–49, 1961.

[671] G.J.G. Upton and A.R. Rahimi. On-line detection of errors in tipping-bucket raingages. *Journal of Hydrology,* 278:197–212, 2003.

[672] Urban Drainage and Flood Control District. *Urban Storm Drainage Criteria Manual.* Denver Regional Council of Governments, Denver, CO, 1984.

[673] B. Urbonas and P. Stahre. *Stormwater: Best Management Practices and Detention for Water Quality, Drainage, and CSO Management.* Prentice Hall, Englewood Cliffs, NJ, 1993.

[674] B.R. Urbonas and L.A. Roesner. Hydrologic design for urban drainage and flood control. In D.R. Maidment, ed., *Handbook of Hydrology,* pp. 28.1–28.52. McGraw-Hill, Inc., New York, 1993.

[675] U.S. Army Corps of Engineers. Standard project flood determination. Report EM 1110-2-1411, U.S. Army Corps of Engineers, Washington, DC, March 1965.

[676] U.S. Army Corps of Engineers. *Feasibility Studies for Small Scale Hydropower Additions, Volume 3.* Hydrologic Engineering Center, Davis, CA, 1979.

[677] U.S. Army Corps of Engineers. *Reservoir System Analysis for Conservation, HEC-3.* Hydrologic Engineering Center, Davis, CA, 1981.

[678] U.S. Army Corps of Engineers. *Simulation of Flood Control and Conservation Systems, HEC-5.* Hydrologic Engineering Center, Davis, CA, 1982.

[679] U.S. Army Corps of Engineers. Hydropower engineering manual. Report EM-1110-2-1701, Engineering Design, U.S. Army Corps of Engineers, Washington, DC, 1985.

[680] U.S. Army Corps of Engineers. *Accuracy of Computed Water Surface Profiles.* Hydrologic Engineering Center, Davis, CA, 1986.

[681] U.S. Army Corps of Engineers. Digest of water resources policies and authorities. Report EP 1165-2-1, U.S. Army Corps of Engineers, Washington, DC, 1989.

[682] U.S. Army Corps of Engineers. *Interior Flood Hydrology Package, HEC-IFH.* Hydrologic Engineering Center, Davis, CA, 1992.

[683] U.S. Army Corps of Engineers. Hydrologic frequency analysis. Report EM 1110-2-1415, U.S. Army Corps of Engineers, Washington, DC, 1993.

[684] U.S. Army Corps of Engineers. Hydraulic design of flood control channels. Report EM 1110-2-1601, U.S. Army Corps of Engineers, Washington, DC, 1994.

[685] U.S. Army Corps of Engineers. *Hydraulic Design of Flood Control Channels.* American Society of Civil Engineers, New York, 1995.

[686] U.S. Army Corps of Engineers. HEC-RAS Hydraulic Reference Manual, Version 2.2. Technical Reference Manual, Hydrologic Engineering Center, Davis, CA, 1998.

[687] U.S. Army Corps of Engineers. Hydrologic Modeling System HEC–HMS. Technical Reference Manual CPD-74B, Hydrologic Engineering Center, Davis, CA, March 2000.

[688] U.S. Army Corps of Engineers. HEC-RAS River Analysis System, Version 3.1. Hydraulic Reference Manual CPD-69, Hydrologic Engineering Center, Davis, CA, November 2002.

[689] U.S. Army Corps of Engineers. Propeller turbine runner, 2005g. http://www.hq. usace.army.mil/cepa/pubs/mar00/story11.htm.

[690] U.S. Bureau of Reclamation. Chapter 6.10: Flood routing. Technical Report Water Studies, Part 6: Flood Hydrology, Volume IV, USBR, Washington, DC, 1949.

[691] U.S. Bureau of Reclamation. Research studies on stilling basins, energy dissipators, and associated appurtenances. Technical Report Hydraulics Laboratory Report, Hyd-399, USBR, Washington, DC, 1955.

[692] U.S. Bureau of Reclamation. *Design of Small Dams*. U.S. Department of the Interior, Washington, DC, 1977.

[693] U.S. Bureau of Reclamation. *Ground Water Manual*. U.S. Department of the Interior, Washington, DC, 1977.

[694] U.S. Bureau of Reclamation. *Design of Small Canal Structures*. U.S. Department of the Interior, Denver, CO, 1978.

[695] U.S. Bureau of Reclamation. *Design of Small Dams*. U.S. Government Printing Office, Washington, DC, 3d ed., 1987.

[696] U.S. Bureau of Reclamation. *Water Measurement Manual*. U.S. Government Printing Office, Washington, DC, 3d ed., 1997.

[697] U.S. Department of Agriculture. Irrigation water requirements. Technical Release 21 (rev.), Soil Conservation Service, 1970. 92 pp.

[698] U.S. Department of Agriculture. Design of open channels. Technical Release No. 25, Soil Conservation Service, Washington, DC, October 1977.

[699] U.S. Department of Commerce. Generalized estimates of probable maximum precipitation and rainfall frequency data for Puerto Rico and Virgin Islands. Technical Paper 42, Weather Bureau, Washington, DC, 1961.

[700] U.S. Department of Commerce. Rainfall-frequency atlas of the Hawaiian Islands. Technical Paper 43, Weather Bureau, Washington, DC, 1962.

[701] U.S. Department of the Army. Hydraulic design of spillways. Technical Report Engineering Manual 1110-2-1603, Office of the Chief of Engineers, USDOA, Washington, DC, 1986.

[702] U.S. Department of Transportation, Federal Highway Administration. HEC-12. U.S. Department of Transportation, McLean, VA, 1984a.

[703] U.S. Department of Transportation, Federal Highway Administration. Hydraulic design of highway culverts. Hydraulic Design Series 5 (HDS-5) FHWA-IP-85-15, FHWA, Washington, DC, 1985. Reprinted 1998.

[704] U.S. Department of Transportation, Federal Highway Administration. HEC-12, Design of roadside channels with flexible linings. Technical Report FHWA-IP-87-7, FHWA, Washington, DC, April 1988.

[705] U.S. Department of Transportation, Federal Highway Administration. HEC-19, Hydrology. Technical Report FHWA-IP-95, FHWA, Washington, DC, 1995.

[706] U.S. Department of Transportation, Federal Highway Administration. HEC-22, Urban drainage design manual. Technical Report FHWA-SA-96-078, FHWA, Washington, DC, 1996.

[707] U.S. Department of Transportation, Federal Highway Administration. Hydraulic Design Series 2 (HDS-2). Hydrology. Technical report, U.S. Federal Highway Administration, Washington, DC, September 1996.

[708] U.S. Department of Transportation, Federal Highway Administration. User's manual for WSPRO. Technical Report FHWA-SA-98-080, Office of Technology Applications, Washington, DC, June 1998.

[709] U.S. Environmental Protection Agency. Guidelines for delineation of wellhead protection areas. Technical Report EPA 440/6-87-010, Office of Ground-Water Protection, Washington, DC, June 1987.

[710] U.S. Environmental Protection Agency. Guide to ground-water supply contingency planning for local and state governments. Technical Assistance Document EPA 440/6-90-003, Office of Water, Washington, DC, 1990.

[711] U.S. Environmental Protection Agency. Superfund chemical data matrix (scdm), 1996. Downloaded from USEPA web site www.epa.gov, cited in Johnson (1999).

[712] U.S. Environmental Protection Agency. Oxygenates in water: Critical information and research needs. Technical Report EPA/600/R-98/048, USEPA, Washington, DC, 1998. Office of Research and Development.

[713] U.S. Environmental Protection Agency. Wastewater technology fact sheet: Sewers, force main. Technical Report EPA 832-F-00-071, USEPA, Washington, DC, September 2000e. Office of Water.

[714] U.S. Geological Survey. A modular three-dimensional finite-difference ground-water flow model: MODFLOW. Technical report, USGS, Reston, VA, 2000.

[715] U.S. National Weather Service. Probable maximum precipitation estimates–United States east of the 105th meridian. Report 51, U.S. Department of Commerce, National Oceanic and Atmospheric Administration, Washington, DC, 1978. (Reprinted 1986).

[716] U.S. National Weather Service. Application of probable maximum precipitation estimates–United States east of the 105th meridian. Report 52, U.S. Department of Commerce, National Oceanic and Atmospheric Administration, Washington, DC, 1982.

[717] U.S. Weather Bureau. Rainfall intensity-frequency regime—Part 1, The Ohio Valley. Technical Paper 29, U.S. Department of Commerce, Washington, DC, 1957.

[718] U.S. Weather Bureau. Two-to-ten-day precipitation for a return period of 2 to 100 years in the contiguous United States. Technical Paper 49, U.S. Department of Commerce, Washington, DC, 1964.

[719] U.S.S.R. National Committee for the International Hydrological Decade. World water balance and water resources of the earth. In *Studies and Reports in Hydrology*, volume 25. UNESCO, Paris, 1978. English translation.

[720] J.D. Valiantzas. Modified Hazen–Williams and Darcy–Weisbach equations for friction and local head losses along irrigation laterals. *Journal of Irrigation and Drainage Engineering, ASCE*, 131(4):342–350, August 2005.

[721] C.H.M. van Bavel. Potential evaporation: The combination concept and its experimental verification. *Water Resources Research*, 2(3):455–467, 1966.

[722] G. Van der Kamp. Determining aquifer transmissivity by means of well response tests: The underdamped case. *Water Resources Research*, 12(1):71–77, 1976.

[723] J.A. Van Mullem. Runoff and peak discharges using Green–Ampt infiltration model. *Journal of Hydraulic Engineering*, 117(3):354–370, March 1991.

[724] V.A. Vanoni. Velocity distribution in open channels. *Civil Engineering*, 11:356–357, 1941.

[725] J. Vaze and H.S. Chiew. Comparative evaluation of urban storm water quality models. *Water Resources Research*, 39(10), 1280, doi:10.1029/2002WR001788, 2003.

[726] J.P. Velon and T.J. Johnson. Water distribution and treatment. In V.J. Zipparro and H. Hasen, eds., *Davis' Handbook of Applied Hydraulics*, pp. 27.1–27.50. McGraw-Hill, Inc., New York, 4th ed., 1993.

[727] P. Venkataraman and P.R.M. Rao. Darcian, transitional, and turbulent flow through porous media. *Journal of Hydraulic Engineering*, 124(8):840–846, August 1998.

[728] W. Viessman, Jr., and M.J. Hammer. *Water Supply and Pollution Control*. Prentice Hall, Upper Saddle River, NJ, 7th ed., 2005.

[729] W. Viessman, Jr., J. Knapp, and G.L. Lewis. *Introduction to Hydrology*. Harper & Row, Publishers, Inc., New York, 2d ed., 1977.

[730] W. Viessman, Jr., and G.L. Lewis. *Introduction to Hydrology*. HarperCollins College Publishers, New York, 4th ed., 1996.

[731] W. Viessman, Jr., and G.L. Lewis. *Introduction to Hydrology*. Prentice Hall, Upper Saddle River, NJ, 5th ed., 2003.

[732] W. Viessman, Jr., and C. Welty. *Water Management Technology and Institutions*. Harper & Row, Publishers, Inc., New York, 1985.

[733] J.R. Villemonte. Submerged weir discharge studies. *Engineering News Record*, December 25, 1947.

[734] F.N. Visher and J.F. Mink. Ground-water resources in southern Oahu, Hawaii. Water-Supply Paper 1778, United States Geological Survey, 1964.

[735] G.K. Voigt. Distribution of rainfall under forest stands. *Forest Science*, 6:2–10, 1960.

[736] F. Von Hofe and O.J. Helweg. Modeling well hydrodynamics. *Journal of Hydraulic Engineering*, 124(12):197–203, July/August 1998.

[737] J.R. Wallis, N.C. Matalas, and J.R. Slack. Just a moment. *Water Resources Research*, 10(2):211–219, 1974.

[738] T.M. Walski. Water distribution. In L.W. Mays, ed., *Water Resources Handbook*, pp. 18.1–18.45. McGraw-Hill, Inc., New York, 1996.

[739] T.M. Walski. Hydraulic design of water distribution storage tanks. In L.W. Mays, ed., *Water Distribution Systems Handbook*, pp. 10.1–10.20. McGraw-Hill Book Company, New York, 2000.

[740] T.M. Walski, J. Gessler, and J.A.W. Sjostrom. *Water Distribution Systems: Simulation and Sizing*. Lewis Publishers, Inc., Chelsea, MI, 1990.

[741] I.A. Walter, R.G. Allen, R. Elliott, M.E. Jensen, D. Itenfisu, B. Mecham, T.A. Howell, R. Snyder, P. Brown, S. Eching, T. Spofford, M. Hattendorf, R.H. Cuenca, J.L.

Wright, and D. Martin. ASCE's standardized reference evapotranspiration equation. In *Proceedings, Fourth National Irrigation Symposium*, pp. 209–215, St. Joseph, MI, 2000. American Society of Agricultural Engineers.

[742] M.T. Walter, V.K. Metha, A.M. Marrone, J. Boll, P. Gérard-Marchant, T.S. Steenhuis, and M.F. Walter. Simple estimation of prevalence of Hortonian flow in New York City watersheds. *Journal of Hydrologic Engineering*, 8(4):214–218, July/August 2003.

[743] W.C. Walton. Selected analytical methods for well and aquifer evaluation. Bulletin 49, Illinois State Water Survey, Urbana, IL, 1962.

[744] W.C. Walton. *Principles of Groundwater Engineering*. CRC Press, Inc., Boca Raton, FL, 1991.

[745] C. Wandmacher and A.I. Johnson. *Metric Units in Engineering, Going SI*. ASCE Press, New York, 1995.

[746] N.H.C. Wang and R.J. Houghtalen. *Fundamentals of Hydraulic Engineering Systems*. Prentice Hall, Upper Saddle River, NJ, 3d ed., 1996.

[747] M. Wanielista, R. Kersten, and R. Eaglin. *Hydrology: Water Quantity and Quality Control*. John Wiley & Sons, New York, 2d ed., 1997.

[748] M.P. Wanielista. *Stormwater Management: Quantity and Quality*. Ann Arbor Science, Ann Arbor, MI, 1978.

[749] M.P. Wanielista and Y.A. Yousef. *Stormwater Management*. John Wiley & Sons, New York, 1993.

[750] A.D. Ward and J. Dorsey. Infiltration and soil water processes. In A.D. Ward and W.J. Elliot, eds., *Environmental Hydrology*, pp. 51–90. Lewis Publishers, Inc., Boca Raton, FL, 1995.

[751] A.D. Ward and S.W. Trimble. *Environmental Hydrology*. CRC Press, Inc., Boca Raton, FL, 2d ed., 2004.

[752] R.C. Ward and M. Robinson. *Principles of Hydrology*. McGraw-Hill, Inc., New York, 4th ed., 1970.

[753] D.L. Warner and J.H. Lehr. *Subsurface Wastewater Injection*. Premier Press, Berkeley, CA, 1981.

[754] M.J. Waterloo, L.A. Bruijnzeel, H.F. Vugts, and T.T. Rawaqa. Evaporation from *Pinus caribaea* plantations on former grassland soils under maritime tropical conditions. *Water Resources Research*, 35(7):2133–2144, July 1999.

[755] T.C. Wei and J.L. McGuinness. Reciprocal distance squared method, a computer technique for estimating area precipitation. Technical Report ARS-NC-8, U.S. Agricultural Research Service, North Central Region, Coshocton, OH, 1973.

[756] W. Weibull. A statistical theory of the strength of materials. *Ingenioers Vetenskapsakad Handl.* Stockholm, 151:15, 1939.

[757] H.G. Wenzel. The effect of raindrop impact and surface roughness on sheet flow. WRC Research Report 34, Water Resources Center, University of Illinois, Urbana, IL, 1970.

[758] H.G. Wenzel, Jr. Rainfall for urban design. In D.F. Kibler, ed., *Urban Stormwater Hydrology*, pp. 35–67. American Geophysical Union, Water Resources Monograph Series, Washington, DC, 1982.

[759] J.A. Westphal. Hydrology for drainage system design and analysis. In L.W. Mays, ed., *Stormwater Collection Systems Design Handbook*, pp. 4.1–4.44. McGraw-Hill, Inc., New York, 2001.

[760] P.J. Whalen and M.G. Cullum. An assessment of urban land use/stormwater runoff quality relationships and treatment efficiencies of selected stormwater management systems. Technical Publication 88-9, South Florida Water Management District, West Palm Beach, FL, July 1988.

[761] W. Whipple and C.W. Randall. *Detention and Flow Retardation Devices*. Prentice-Hall, Inc., Englewood Cliffs, NJ, 1983. Stormwater management in urbanizing areas.

[762] F.M. White. *Fluid Mechanics*. McGraw-Hill, Inc., New York, 3d ed., 1994.

[763] E.R. Wilcox. A comparative test of the flow of water in 8-inch concrete and vitrified clay sewer pipe. Experiment Station Series Bulletin 27, University of Washington, 1924.

[764] G.V. Wilkerson and J.L. McGahan. Depth-averaged velocity distribution in straight trapezoidal channels. *Journal of Hydraulic Engineering*, 131(6):509–512, June 2005.

[765] J.O. Wilkes. *Fluid Mechanics for Chemical Engineers*. Prentice Hall, Upper Saddle River, NJ, 1999.

[766] Willeke. Discussion of "Runoff curve number: Has it reached maturity?" by V.M. Ponce and R.H. Hawkins. *Journal of Hydrologic Engineering*, 2(3):147, 1997.

[767] G.S. Williams and A.H. Hazen. *Hydraulic Tables*. John Wiley & Sons, New York, 1920.

[768] J. Williamson. The laws of motion in rough pipes. *La Houille Blanche*, 6(5):738, September–October 1951.

[769] M.D. Winsberg. *Florida Weather*. University of Central Florida Press, Orlando, FL, 1990.

[770] E. Wohl. *Mountain Rivers*. American Geophysical Union, Washington, DC, 2000.

[771] T.S.W. Wong. Assessment of time of concentration formulas for overland flow. *Journal of Irrigation and Drainage Engineering, ASCE*, 131(4):383–387, August 2005.

[772] T.S.W. Wong and C.-N. Chen. Time of concentration formula for sheet flow of varying flow regime. *Journal of Hydrologic Engineering*, 2(3):136–139, July 1997.

[773] D.J. Wood. Waterhammer analysis—Essential and easy (and efficient). *Journal of Environmental Engineering*, 131(8):1123–1131, August 2005d.

[774] E.F. Wood. Global scale hydrology: Advances in land surface modeling, U.S. National Report to International Union of Geodesy and Geophysics 1987–1990. Technical report, American Geophysical Union, Washington, DC, 1991.

[775] S.L. Woodall. Rainfall interception losses from melaleuca forest in Florida. Research Note SE-323, United States Department of Agriculture, Southeastern Forest Experiment Station, Forest Resources Laboratory, Lehigh Acres, FL 33936, February 1984.

[776] D.A. Woolhiser, R.E. Smith, and J.-V. Giraldez. Effects of spatial variability of saturated hydraulic conductivity on Hortonian overland flow. *Water Resources Research*, 32(3):671–678, March 1996.

[777] World Meteorological Organization. Guide to hydrological practices, Volume II, Analysis, forecasting and other applications. Technical Report No. 168, 4th ed., WMO, Geneva, Switzerland, 1983.

[778] World Meteorological Organization. Manual for estimation of probable maximum precipitation. Operational Hydrology Report No. 1, 2d ed., Publ. 332, WMO, Geneva, Switzerland, 1986.

[779] J.L. Wright. Crop coefficients for estimates of daily crop evapotranspiration. In *Irrigation Scheduling for Water and Energy Conservation in the 80s*. ASAE, 1981.

[780] J.L. Wright. New evapotranspiration crop coefficients. *Journal of Irrigation and Drainage Engineering, ASCE*, 108(IR2):57–74, 1982.

[781] J.L. Wright, R.G. Allen, and T.A. Howell. Conversion between evapotranspiration references and methods. In *Proceedings, Fourth National Irrigation Symposium*, pp. 251–259, St. Joseph, MI, 2000. American Society of Agricultural Engineers.

[782] J.L. Wright and M.E. Jensen. Peak water requirements of crops in southern Idaho. *Journal of Irrigation and Drainage Engineering, ASCE*, 96(IR1):193–201, 1972.

[783] S. Wu and N. Rajaratnam. Transition from hydraulic jump to open channel flow. *Journal of Hydraulic Engineering*, 122(9):526–528, September 1996.

[784] R.A. Wurbs and W.P. James. *Water Resources Engineering*. Prentice Hall, Upper Saddle River, NJ, 2002.

[785] C.T. Yang. *Sediment Transport: Theory and Practice*. McGraw-Hill, Inc., New York, 1996.

[786] D.L. Yarnell and S.M. Woodward. The flow of water in drain tile. Bulletin 854, Department of Agriculture, Washington, DC, 1924.

[787] M.V. Yates. Septic tank siting to minimize the contamination of ground water by microorganisms. Technical Report EPA 440/6-87-007, U.S. Environmental Protection Agency, Office of Ground-Water Protection, Washington, DC, 1987.

[788] B.C. Yen. Hydraulic resistance in open channels. In B.C. Yen, ed., *Channel Flow Resistance: Centennial of Manning's Formula*, pp. 1–135. Water Resource Publications, Highlands Ranch, CO, 1991.

[789] B.C. Yen. Open channel flow resistance. *Journal of Hydraulic Engineering*, 128(1):20–39, January 2002.

[790] B.C. Yen and A.O. Akan. Hydraulic design of urban drainage systems. In L.W. Mays, ed., *Hydraulic Design Handbook*, pp. 14.1–14.114. McGraw-Hill Book Company, New York, 1999.

[791] B.C. Yen and V.T. Chow. Design hyetographs for small drainage structures. *Journal of Hydraulic Engineering*, 106(HY6), 1980.

[792] V. Yevjevich. *Probability and Statistics in Hydrology*. Water Resources Publications, Littleton, CO, 1972.

[793] R.E. Yoder, L.O. Odhiambo, and W.C. Wright. Effects of vapor-pressure deficit and net-irradiance calculation methods on accuracy of standardized Penman–Monteith equation in a humid climate. *Journal of Irrigation and Drainage Engineering, ASCE*, 131(3):228–237, June 2005.

[794] D.H. Yoo and V.P. Singh. Two methods for the computation of commercial pipe friction factors. *Journal of Hydraulic Engineering*, 131(8):694–704, 2005.

[795] G. K. Young, Jr., and S.M. Stein. Hydraulic design of drainage for highways. In L.W. Mays, ed., *Hydraulic Design Handbook*, pp. 13.1–13.44. McGraw-Hill Book Company, New York, 1999.

[796] Y.A. Yousef and M.P. Wanielista. Efficiency optimization of wet-detention ponds for urban stormwater management. Florida Department of Environmental Regulation, Tallahassee, FL, 1989.

[797] S.L. Yu, R.J. Kaighn Jr., and S. Liao. Testing of best management practices for controlling highway runoff. Phase II final report. Technical Report FHWA/VA-94-R21, Virginia Transportation Research Council, Richmond, VA, 1994.

[798] S.L. Yu, J.-T. Kuo, E.A. Fassman, and H. Pan. Field test of grassed-swale performance in removing runoff pollution. *Journal of Water Resources Planning and Management*, 127(3):168–171, May/June 2001.

[799] L. Zhang, K. Hickel, W.R. Dawes, F.H.S. Chiew, A.W. Western, and P.R. Briggs. A rational function approach for estimating mean annual evapotranspiration. *Water Resources Research*, 40, W02502, doi:10.1029/2003WR002710, 2004.

[800] V.J. Zipparro and H. Hasen. *Davis' Handbook of Applied Hydraulics*. McGraw-Hill, Inc., New York, 4th ed., 1993.

Index